Springer Monographs in Mathematics

Patrice Tauvel
Rupert W. T. Yu

Lie Algebras
and Algebraic Groups

With 44 Figures

 Springer

Patrice Tauvel
Rupert W. T. Yu

Département de Mathématiques
Université de Poitiers
Boulevard Marie et Pierre Curie,
Téléport 2 – BP 30179
86962 Futuroscope Chasseneuil cedex, France
e-mail: tauvel@math.univ-poitiers.fr
 yuyu@math.univ-poitiers.fr

Library of Congress Control Number: 2005922400

Mathematics Subject Classification (2000): 17-01, 17-02, 17Bxx, 20Gxx

ISSN 1439-7382
ISBN-10 3-540-24170-1 Springer Berlin Heidelberg New York
ISBN-13 978-3-540-24170-6 Springer Berlin Heidelberg New York

Springer is a part of Springer Science+Business Media
springeronline.com
© Springer-Verlag Berlin Heidelberg 2005
Printed in Germany

Typesetting: by the authors and TechBooks using a Springer LaTeX macro package
Cover design: Erich Kirchner, Heidelberg

Printed on acid-free paper   41/sz - 5 4 3 2 1 0

Preface

The theory of groups and Lie algebras is interesting for many reasons. In the mathematical viewpoint, it employs at the same time algebra, analysis and geometry. On the other hand, it intervenes in other areas of science, in particular in different branches of physics and chemistry. It is an active domain of current research.

One of the difficulties that graduate students or mathematicians interested in the theory come across, is the fact that the theory has very much advanced, and consequently, they need to read a vast amount of books and articles before they could tackle interesting problems.

One of the goals we wish to achieve with this book is to assemble in a single volume the basis of the algebraic aspects of the theory of groups and Lie algebras. More precisely, we have presented the foundation of the study of finite-dimensional Lie algebras over an algebraically closed field of characteristic zero.

Here, the geometrical aspect is fundamental, and consequently, we need to use the notion of algebraic groups. One of the main differences between this book and many other books on the subject is that we give complete proofs for the relationships between algebraic groups and Lie algebras, instead of admitting them.

We have also given the proofs of certain results on commutative algebra and algebraic geometry that we needed so as to make this book as self-contained as possible. We believe that in this way, the book can be useful for both graduate students and mathematicians working in this area.

Let us give a brief description of the material treated in this book.

As we have stated earlier, our goal is to study Lie algebras over an algebraically closed field of characteristic zero. This allows us to avoid, in considering questions concerning algebraic geometry, the notion of separability, which simplifies considerably our presentation. In fact, under certain conditions of separability, the correspondence between Lie algebras and algebraic groups described in chapter 24 has a very nice generalization when the algebraically closed base field has prime characteristic.

Chapters 1 to 9 treat basic results on topology, commutative algebra and sheaves of functions that are required in the rest of the book.

In chapter 10, we recall some standard results on Jordan decompositions and the theory of abstract groups and group actions. Here, the base field is assumed to be algebraically closed in order to obtain a Jordan decomposition.

Chapters 11 to 17 give an introduction to the theory of algebraic geometry which we shall encounter continually in the chapters which follow. We have selected only the notions that we require in this book. The reader should by no means consider these chapters as a thorough introduction to the theory of algebraic geometry.

Chapters 18 and 36 are dedicated to root systems which are fundamental to the study of semisimple Lie algebras.

We introduce Lie algebras in chapter 19. In this chapter, we prove important results on the structure of Lie algebras such as Engel's theorem, Lie's theorem and Cartan's criterion on solvability.

In chapter 20, we define the notions of semisimple and reductive Lie algebras. In addition to characterizing these Lie algebras, we discover in this chapter how the structure of semisimple Lie algebras can be related to root systems.

The general theory of algebraic groups is studied in chapters 21 to 28. The relations between Lie algebras and algebraic groups, which are fundamental to us, are established in chapters 23 and 24. Chapter 29 presents applications of these relations to tackle the systematic study of Lie algebras. The reader will observe that the geometrical aspects have an important part in this study. In particular, the orbits of points under the action of an algebraic group plays a central role.

Chapter 30 gives a short introduction of the theory of representations of semisimple Lie algebras which we need in order to prove Chevalley's theorem on invariants in chapter 31.

We define in chapter 32 S-triples which are essential to the study of semisimple Lie algebras. Another fundamental notion, treated in chapters 33 to 35, is the notion of nilpotent orbits in semisimple Lie algebras.

We introduce symmetric Lie algebras in chapter 37, and semisimple symmetric Lie algebras in chapter 38. In these chapters, we give generalizations of certain results of chapters 32 to 35.

In addition to presenting the essential classical results of the theory, some of the results we have included in the final chapters are recent, and some are yet to be published.

At the end of each chapter, the reader may find a list of relevant references, and in some cases, remarks concerning the contents of the chapter.

There are many approaches to reading this book. We need not read this book linearly. A reader familiar with the theory of commutative algebra may skip chapters 2 to 8, and consider these chapters for references only. Let us also point out that chapters 18, 19 and 20 constitute a short introduction to the theory of finite-dimensional Lie algebras and the structure of semisimple Lie algebras.

We wish to thank our colleagues A. Bouaziz and H. Sabourin of the university of Poitiers with whom we had many useful discussion during the preparation of this book.

Poitiers,   *Patrice Tauvel*
January 2005   *Rupert W.T. Yu*

Contents

1

Results on topological spaces

In this chapter, we treat some basic notions of topology such as irreducible and constructible sets, dimension of a topological space, Noetherian space, which are fundamental in algebraic geometry.

1.1 Irreducible sets and spaces

1.1.1 Definition. *A topological space X is said to be* irreducible *if any finite intersection of non-empty open subsets is non-empty.*

1.1.2 It follows from the definition that an irreducible topological space is not empty.

1.1.3 Proposition. *Let X be a non-empty topological space. Then the following conditions are equivalent:*
   (i) *X is irreducible.*
   (ii) *X is not the finite union of distinct proper closed subsets.*
   (iii) *X is not the union of two proper closed subsets.*
   (iv) *Any non-empty open subset of X is dense in X.*
   (v) *Any open subset of X is connected.*

Proof. The implications (i) $\Rightarrow$ (ii) $\Rightarrow$ (iii) are clear and (iii) $\Rightarrow$ (iv) follows from the fact that a subset in X is dense if and only if it meets all non-empty open subsets. Now if U is a non-connected non-empty open subset, then $U = U_1 \cup U_2$ where U_1, U_2 are non-empty open subsets and $U_1 \cap U_2 = \emptyset$. Thus (iv) $\Rightarrow$ (v). The same argument gives (v) $\Rightarrow$ (i). $\square$

Remark. If X is irreducible, then it is connected. The converse is not true.

1.1.4 In the rest of this chapter, X is a topological space.
   A subset of X is called *irreducible* if it is non-empty and irreducible as a topological space. From the above definitions, the following result is clear.

Proposition. *Let A be a non-empty subset of X. Then the following conditions are equivalent:*

(i) *A is irreducible.*

(ii) *Let $F_1, \ldots, F_n$ be closed subsets of X such that A is contained in the union of the F_i's, then there exists $j \in \{1, \ldots, n\}$ such that $A \subset F_j$.*

(iii) *Let U, V be open subsets of X such that $U \cap A$ and $V \cap A$ are non-empty, then $U \cap V \cap A \neq \emptyset$.*

1.1.5 Proposition. *Let A, B be subsets of X.*

(i) *A is irreducible if and only if its closure $\overline{A}$ is irreducible.*

(ii) *If A is irreducible and $A \subset B \subset \overline{A}$, then B is irreducible.*

Proof. For any open subset U, we have $U \cap A \neq \emptyset$ if and only if $U \cap \overline{A} \neq \emptyset$. So (i) and (ii) follow. □

1.1.6 Proposition. (i) *If X is irreducible, then any non-empty open subset of X is also irreducible.*

(ii) *Let $(U_i)_{i \in I}$ be a covering of X by open subsets such that $U_i \cap U_j \neq \emptyset$ for all $i, j \in I$. If all the U_i's are irreducible, then X is irreducible.*

Proof. (i) Let U, V be non-empty open subsets of X such that $V \subset U$. If X is irreducible, then V is dense in U. Thus U is irreducible.

(ii) Let V be a non-empty open subset of X. There exists $k \in I$ such that $V \cap U_k \neq \emptyset$. Since $U_i \cap U_k \neq \emptyset$ for all $i \in I$ and $V \cap U_k$ is dense in U_k, $V \cap U_i \cap U_k \neq \emptyset$. Hence $V \cap U_i \neq \emptyset$ for all i. It follows that $V \cap U_i$ is dense in U_i for all $i \in I$, so V is dense in X. □

1.1.7 Proposition. *Let Y be a topological space and $f : X \rightarrow Y$ a continuous map.*

(i) *If $A \subset X$ is irreducible, then $f(A)$ is irreducible in Y.*

(ii) *Suppose that Y is irreducible, f is an open map and that $f^{-1}(y)$ is irreducible for all $y \in Y$. Then X is irreducible.*

Proof. (i) Let U, V be open subsets of Y such that $U \cap f(A)$ and $V \cap f(A)$ are non-empty. Then $f^{-1}(U)$ and $f^{-1}(V)$ are open subsets whose intersection with A is non-empty. It follows that $f^{-1}(U \cap V) = f^{-1}(U) \cap f^{-1}(V)$ meets A. Therefore $U \cap V$ meets $f(A)$ and assertion (i) follows.

(ii) Let U, V be non-empty open subsets of X. Since f is open and Y is irreducible, $f(U)$ meets $f(V)$ at some point y. Further, $f^{-1}(y)$ is irreducible, therefore the open subsets $U \cap f^{-1}(y)$ and $V \cap f^{-1}(y)$ of $f^{-1}(y)$ have non-empty intersection. Hence $U \cap V \neq \emptyset$. □

1.1.8 Remark. A map $f : X \rightarrow Y$ is called *dominant* if $f(X)$ is dense in Y. It follows from 1.1.5 and 1.1.7 that if X is irreducible and f is continuous and dominant, then Y is irreducible.

1.1.9 Definition. *A maximal irreducible subset of X will be called an* irreducible component *of X.*

1.1.10 Proposition. (i) *All irreducible components of X are closed.*

(ii) *Any irreducible subset of X is contained in an irreducible component and X is the union of its irreducible components.*

Proof. (i) This is clear from 1.1.5.

(ii) Let $(A_i)_{i \in I}$ be a family of irreducible subsets of X totally ordered by inclusion and $A = \bigcup_{i \in I} A_i$. Let U, V be open subsets which meet A. Since the family is totally ordered, there exist $k \in I$ such that $U \cap A_k \neq \emptyset$ and $V \cap A_k \neq \emptyset$. Thus $U \cap V \cap A_k \neq \emptyset$ because A_k is irreducible. Now the first assertion follows from Zorn's lemma. The second assertion follows from the fact that any singleton in X is irreducible. $\square$

1.1.11 Corollary. *Any connected component of X is a union of irreducible components of X.*

Proof. Any irreducible subset of X is connected, so it is contained in a connected component of X. $\square$

Remark. Two distinct irreducible components of X can have a non-empty intersection.

1.1.12 Proposition. *Let $U \subset X$ be an open subset. The map $Z \mapsto Z \cap U$ is a bijection between the set of irreducible closed subsets of X meeting U and the set of irreducible subsets of U, closed in U. The reciprocal map is $V \mapsto \overline{V}$. In particular, $C \mapsto C \cap U$ induces a bijection between the set of irreducible components of X meeting U and the set of irreducible components of U.*

Proof. Let Z be an irreducible closed subset of X meeting U. Since $U \cap Z$ is open in Z, it is irreducible by 1.1.6 and dense in Z. Therefore $Z = \overline{Z \cap U}$.

Conversely, if V is an irreducible closed subset of U, then $\overline{V}$ is irreducible by 1.1.5 and $V = U \cap \overline{V}$. $\square$

1.1.13 Recall that $A \subset X$ is *nowhere dense* in X if the interior of $\overline{A}$ is empty.

Let us suppose that X has only a finite number of distinct irreducible components $X_1, \ldots, X_n$. For $1 \leqslant i \leqslant n$, denote by $U_i = X \setminus (\bigcup_{j \neq i} X_j)$ the set of elements of X which are not in $\bigcup_{j \neq i} X_j$.

Lemma. *The set U_i is open in X and dense in X_i. The disjoint union of the U_i's is dense in X. For $i \neq j$, $X_i \cap X_j$ is nowhere dense in X.*

Proof. It is obvious that U_i is open in X and is contained in X_i. If U_i is empty, then $X_i \subset \bigcup_{j \neq i} X_j$. Since X_i is irreducible, there exists $k \neq i$ such that $X_i = X_k$ which is impossible by our hypothesis. Thus U_i is dense in X_i since it is a non-empty open subset. The second assertion is now obvious.

Let $i \neq j$ and U an open subset contained in $X_i \cap X_j$. So U is contained in the union of the X_k's, $k \neq i$. Hence $U \cap U_i = \emptyset$. In the same manner, we have $U \cap U_j = \emptyset$. Thus $U = \emptyset$ since $U_i \cup U_j$ is dense in $X_i \cup X_j$. $\square$

1.1.14 Lemma. *Using the hypotheses of 1.1.13, suppose that X is a subset of a topological space Y. Then $\overline{X_1}, \ldots, \overline{X_n}$ are the distinct irreducible components of $\overline{X}$.*

Proof. By 1.1.5, the $\overline{X_i}$'s are irreducible. Clearly $\overline{X} = \overline{X_1} \cup \cdots \cup \overline{X_n}$. If A is an irreducible subset of $\overline{X}$, then it is contained in one of the $\overline{X_i}$'s. Finally, if $\overline{X_i} \subset \overline{X_j}$, then $X_i \subset X \cap \overline{X_j} = X_j$ since X_j is closed in X. This contradicts our hypothesis and so the lemma follows. $\square$

1.2 Dimension

1.2.1 Definition. *Let E be a set. A* chain *of subsets of length n of E is a strictly increasing sequence of subsets $E_0 \subset E_1 \subset \cdots \subset E_n$.*

1.2.2 Definition. *The* dimension *of a non-empty topological space X is the supremum of the lengths of chains consisting of irreducible closed subsets. We shall denote it by $\dim X$ which is in $\mathbb{N} \cup \{+\infty\}$.*

1.2.3 Proposition. *Let us assume that X is non-empty.*
(i) Let Y be a non-empty subset of X, then $\dim Y \leqslant \dim X$.
(ii) Let $X_1, \ldots, X_r$ be closed subsets of X such that $X = \bigcup_{i=1}^r X_i$, then $\dim X = \max\{\dim X_i; 1 \leqslant i \leqslant r\}$.
(iii) Let $X_1, \ldots, X_r$ be open subsets of X such that $X = \bigcup_{i=1}^r X_i$, then $\dim X = \max\{\dim X_i; 1 \leqslant i \leqslant r\}$.
(iv) Suppose that X is irreducible and $\dim X$ is finite. If Y is a closed subset such that $\dim Y = \dim X$, then $X = Y$.

Proof. (i) Let $Y_0 \subset Y_1 \subset \cdots \subset Y_n$ be a chain of irreducible closed subsets in Y. The closure $\overline{Y_i}$ of Y_i in X is irreducible by 1.1.5 and since $Y_i = \overline{Y_i} \cap Y$, $\overline{Y_i}$ is strictly contained in $\overline{Y_{i+1}}$. Thus $\dim Y \leqslant \dim X$.

(ii) Let $Z_0 \subset Z_1 \subset \cdots \subset Z_n$ be a chain of irreducible closed subsets in X. Since Z_n is irreducible and $Z_n = (Z_n \cap X_1) \cup \cdots \cup (Z_n \cap X_r)$, there exists k such that $Z_n \subset X_k$. It follows that $\dim X_k \geqslant n$. But $\dim X_k \leqslant \dim X$ by (i), so (ii) follows.

(iii) Let $Z_0 \subset Z_1 \subset \cdots \subset Z_n$ be a chain of irreducible closed subsets in X. There exists k such that $X_k \cap Z_0 \neq \emptyset$. It follows that $X_k \cap Z_i$ is a non-empty open irreducible subset in Z_k for all $1 \leqslant i \leqslant r$. By 1.1.3 (iv), $\overline{X_k \cap Z_i} = Z_i$ and $X_k \cap Z_0 \subset \cdots \subset X_k \cap Z_n$ is a chain of irreducible closed subsets in X_k. Hence $\dim X \leqslant \dim X_k$ and (iii) follows from (i).

(iv) Let $\dim Y = n$ and $Y_0 \subset Y_1 \subset \cdots \subset Y_n$ be a chain of irreducible closed subsets in Y. If $Y \neq X$, then $Y_0 \subset Y_1 \subset \cdots \subset Y_n \subset X$ is a chain of irreducible closed subsets in X. So $\dim X > \dim Y$ which contradicts our hypothesis. $\square$

1.2.4 Definition. *Let X be non-empty and finite-dimensional and Y a non-empty subset of X. We define the* codimension *of Y in X, denoted by $\operatorname{codim}_X Y$, to be $\dim X - \dim Y$.*

1.3 Noetherian spaces

1.3.1 Proposition. *The following conditions are equivalent:*
(i) *Any decreasing sequence of closed subsets of X is stationary.*
(ii) *Any ascending sequence of open subsets of X is stationary.*
(iii) *Any non-empty family of closed subsets of X has a minimal element.*
(iv) *Any non-empty family of open subsets of X has a maximal element.*
If X satisfies the above conditions, then we shall say that X is Noetherian.

Proof. This is straightforward.   □

Remark. A Noetherian topological space can be infinite-dimensional.

1.3.2 Proposition. (i) *Any subset of a Noetherian space is Noetherian.*
(ii) *Let $(X_i)_{1 \leqslant i \leqslant p}$ be a finite covering of X such that the X_i's are Noetherian. Then X is also Noetherian.*

Proof. (i) Let Y be a subset of X and $(F_n)_n$ be a decreasing sequence of closed subsets of Y. Then $(\overline{F_n})_n$ is a decreasing sequence of closed subsets of X. Therefore there exists p such that $\overline{F_p} = \overline{F_n}$ for all $n \geqslant p$. This implies that $F_n = Y \cap \overline{F_n} = Y \cap \overline{F_p} = F_p$, and (i) follows.

(ii) Let $(F_n)_n$ be a decreasing sequence of closed subsets of X. Then for $1 \leqslant i \leqslant p$, $(X_i \cap F_n)_n$ is a decreasing sequence of closed subsets of X_i. Hence there exists n_i such that $X_i \cap F_{n_i} = X_i \cap F_n$ for $n \geqslant n_i$. Set $q = \max(n_1, \ldots, n_p)$, then $F_n = F_q$ for $n \geqslant q$.   □

1.3.3 Proposition. *The following conditions are equivalent:*
(i) *X is Noetherian.*
(ii) *Any open subset of X is quasi-compact.*

Proof. (i) $\Rightarrow$ (ii) By 1.3.2, it suffices to show that X is quasi-compact. Let $(U_i)_i$ be an open covering of X. The set of all finite unions of the U_i's has a maximal element U since X is Noetherian. It is clear that $U = X$.

(ii) $\Rightarrow$ (i) Let $(U_n)_n$ be an ascending sequence of open subsets of X. The union U of the U_n's is open and therefore it is quasi-compact. Since $(U_n)_n$ is an open covering of U, it is clear that the sequence $(U_n)_n$ is stationary.   □

1.3.4 The following result is called the *principle of transfinite induction*.

Lemma. *Let E be a well-ordered set, that is an ordered set such that any non-empty subset has a minimal element. Let F be a subset of E such that: if $a \in E$ satisfies $\{x \in E; x < a\} \subset F$, then $a \in F$. Then $F = E$.*

Proof. Suppose that $F \neq E$. Let b be a minimal element in $E \setminus F$. Then $\{x \in E; x < b\} \subset F$. Thus $b \in F$ which contradicts our hypothesis on b.   □

1.3.5 Proposition. *Let X be Noetherian. Then the set of irreducible components of X is finite.*

Proof. Let E be the set of all closed subsets of X ordered by inclusion, and F be the set of all finite unions of irreducible closed subsets of X. Now let $Y \in E$ be such that any proper closed subset of Y is in F. If Y is irreducible, then $Y \in F$. If Y is not irreducible, then Y is the union of two distinct proper closed subsets Y_1, Y_2. But $Y_1, Y_2 \in F$, so $Y \in F$. Thus we can apply 1.3.4, and we have $E = F$.

Thus X is the finite union of irreducible closed subsets $X_1, \ldots, X_n$. Further, we can assume that they are non-comparable. Let Z be an irreducible component of X. Then $Z = (Z \cap X_1) \cup \cdots \cup (Z \cap X_n)$ which implies that there exists k such that $Z = Z \cap X_k$. Thus $Z = X_k$. This shows that $\{X_1, \ldots, X_n\}$ is the set of irreducible components of X. □

1.3.6 Proposition. *Let X be non-empty Noetherian, and $X_1, \ldots, X_n$ its irreducible components. For $1 \leqslant i \leqslant n$, set:*

$$Y_i = X_i \setminus \bigcup_{j \neq i} (X_i \cap X_j) = X \setminus \bigcup_{j \neq i} X_j.$$

(i) *An open subset U is dense in X if and only if $U \cap X_i \neq \emptyset$ for $1 \leqslant i \leqslant n$.*

(ii) *For $1 \leqslant i \leqslant n$, Y_i is open in X. Further, $Y_0 = Y_1 \cup \cdots \cup Y_n$ is open dense in X and the irreducible (connected) components of Y_0 are $Y_1, \ldots, Y_n$.*

Proof. Clearly, Y_i is open in X and it is non-empty. So it is dense in X_i. Thus Y_0 is open dense in X. Further, the Y_i's are pairwise disjoint. So (ii) follows from 1.1.11.

Now if U is open dense in X, then $U \cap X_i \supset U \cap Y_i \neq \emptyset$.

Conversely, if an open subset V meets each X_i, then $V \cap X_i$ is dense in X_i and $X_i \subset \overline{V}$. Hence V is dense. □

1.4 Constructible sets

1.4.1 Proposition. *Let Y be a subset of X. The following conditions are equivalent:*

(i) *There exists a family $(U_i)_{i \in I}$ of open subsets of X, whose union contains Y, such that $Y \cap U_i$ is closed in U_i for all $i \in I$.*

(ii) *Any point of Y admits an open neighbourhood U in X such that $U \cap Y$ is closed in U.*

(iii) *Y is open in $\overline{Y}$.*

(iv) *Y is the intersection of an open subset and a closed subset of X.*

If Y satisfies these conditions, then it is called a locally closed *subset of X.*

Proof. Only (ii) $\Rightarrow$ (iii) needs attention. Let $x \in Y$, and U be an open neighbourhood of x in X such that $U \cap Y$ is closed in U. Then $U \cap Y = U \cap \overline{Y}$, which shows that in $\overline{Y}$, x is in the interior of Y. □

1.4.2 Let $\mathfrak{F}$ be a set of subsets of X such that

(i) Any open subset of X is in $\mathfrak{F}$.

(ii) If $Z \in \mathfrak{F}$, then $X \setminus Z \in \mathfrak{F}$.

(iii) Any finite union of elements of $\mathfrak{F}$ is in $\mathfrak{F}$.

For example, the set of subsets of X satisfies these conditions.

The intersection $\mathfrak{C}$ of all such sets satisfies these conditions. An element of $\mathfrak{C}$ is called a *constructible* subset of X.

1.4.3 Proposition. *The set $\mathfrak{C}$ is the set of finite unions of locally closed subsets of X.*

Proof. Denote by $\mathfrak{C}_0$ the set of finite unions of locally closed subsets of X. It is clear that $\mathfrak{C}_0 \subset \mathfrak{C}$.

Let $C = (U_1 \cap F_1) \cup \cdots \cup (U_n \cap F_n) \in \mathfrak{C}_0$ where the U_i's (resp. F_i's) are open (resp. closed) subsets of X. Then it is clear that $X \setminus C$ is the union of subsets of the form $E_1 \cap \cdots \cap E_n$ where for each j, there exists i such that E_j is either $X \setminus U_i$ or $X \setminus F_i$. So the proposition follows. $\square$

1.4.4 Proposition. *Let Y be a subset of X.*

(i) *If C be a constructible subset of X, then $C \cap Y$ is a constructible subset of Y.*

(ii) *Suppose that Y is closed in X. If C is a constructible subset of Y, then C is a constructible subset of X.*

Proof. Part (i) is obvious and (ii) follows from the fact that if F is closed in Y and U is open in Y, then F is closed in X and U is locally closed in X. $\square$

1.4.5 Lemma. *Let C be a constructible subset of X whose closure is irreducible. Then C contains an open subset of $\overline{C}$.*

Proof. Let $C = (U_1 \cap F_1) \cup \cdots \cup (U_n \cap F_n) \in \mathfrak{C}_0$ where the U_i's (resp. F_i's) are open (resp. closed) subsets of X. We can assume further that $U_i \cap F_i \neq \emptyset$ for $1 \leqslant i \leqslant n$. Since $\overline{C}$ is irreducible and contained in $\bigcup_{i=1}^{n} F_i$, there exists j such that $\overline{C} \subset F_j$. It follows that $U_j \cap F_j = U_j \cap \overline{C}$ and $U_j \cap F_j$ is a non-empty subset of C, open in $\overline{C}$. $\square$

1.4.6 Proposition. *Let X be Noetherian. Then any constructible subset Y contains an open and dense subset of $\overline{Y}$.*

Proof. Let $Y = Y_1 \cup \cdots \cup Y_n$ be the decomposition of Y into irreducible components. Since irreducible components are closed, they are constructible in Y. By 1.1.14, the $\overline{Y_i}$'s are the irreducible components of $\overline{Y}$. For $1 \leqslant i \leqslant n$, set:

$$T_i = \overline{Y_i} \setminus \bigcup_{j \neq i} (\overline{Y_i} \cap \overline{Y_j}).$$

We saw in 1.3.6 that the T_i's are non-empty and open in $\overline{Y}$. By 1.4.5, there exists a subset U_i in Y_i which is open and dense in $\overline{Y_i}$. Let $V_i = U_i \cap T_i$. Then

V_i is open and dense in $\overline{Y_i}$. But the V_j's are pairwise disjoint, so V_i is in fact open in $\overline{Y}$. Now it follows from 1.3.6 (i) that $V = V_1 \cup \cdots \cup V_n$ is a subset of Y which is open and dense in $\overline{Y}$. □

1.5 Gluing topological spaces

1.5.1 Let $(X_i)_{i \in I}$ be a family of sets and X be the sum of the X_i's. For $i \in I$, we denote by $\theta_i : X_i \to X$ the canonical injection and we shall identify X_i with its image via θ_i.

Suppose that for all $i,j \in I$, there are subsets $X_{ij} \subset X_i$, $X_{ji} \subset X_j$ and a map $h_{ji} : X_{ij} \to X_{ji}$ such that

a) For all $i \in I$, $X_{ii} = X_i$ and $h_{ii} = \mathrm{id}_{X_i}$ is the identity map.

b) For all $i,j,k \in I$ and $x \in X_{ij} \cap X_{ik}$, we have $h_{ji}(x) \in X_{jk}$ and:

$$h_{ki}(x) = h_{kj}(h_{ji}(x)).$$

By applying b) to the triples (i,j,i) and (j,i,j), we remark that h_{ij} is bijective with reciprocal map h_{ji}.

Let us define a binary relation $\mathcal{R}$ on X by $x\mathcal{R}y$ if and only if there exist $i,j \in I$ such that

$$x \in X_{ij} \ , \ y \in X_{ji} \ , \ y = h_{ji}(x).$$

The preceding remark implies that $\mathcal{R}$ is reflexive and symmetric. Now suppose that $x,y,z \in X$, $x\mathcal{R}y$ and $y\mathcal{R}z$. There exist $i,j,k \in I$ such that

$$x \in X_{ij} \ , \ y \in X_{ji} \cap X_{jk} \ , \ z \in X_{kj} \ , \ y = h_{ji}(x) \ , \ z = h_{kj}(y).$$

By b), we have $x = h_{ij}(y) \in X_{ik}$ and:

$$z = h_{kj}(y) = h_{kj}(h_{ji}(x)) = h_{ki}(x).$$

Thus $\mathcal{R}$ is transitive and it defines an equivalence relation on X.

We shall say that the set of equivalence classes $X/\mathcal{R}$ is obtained by *gluing the X_i's along the X_{ij}'s via the bijections h_{ji}*. Let us denote by $\varphi : X \to X/\mathcal{R}$ the canonical surjection.

Note that we have the following:

(i) Each equivalence class has at most one element in each of the X_i's.

(ii) For $i,j \in I$, X_{ij} is the set of $x \in X_i$ such that there exists a $y \in X_j$ with $x\mathcal{R}y$.

We deduce from these remarks that for $i \in I$, $\varphi \circ \theta_i$ induces a bijection from X_i to $\varphi \circ \theta_i(X_i)$.

1.5.2 Conserving the hypotheses of 1.5.1, let us suppose further that the following properties are satisfied:

c) Each X_i is a topological space with topology $\mathcal{T}_i$.

d) For $i,j \in I$, X_{ij} is open in X_i and h_{ji} is a homeomorphism from X_{ij} to X_{ji}.

We can equip X with the topology $\mathcal{S}$, the sum of the topologies $\mathcal{T}_i$'s (this is the finest topology on X for which the injections θ_i are continuous). Thus a subset $A \subset X$ is open (resp. closed) if and only if each $\theta_i^{-1}(A)$ is open (resp. closed) in X_i. In particular, each $\theta_i(X_i)$ is open and closed in X.

Let $\mathcal{T}$ be the topology on $X/\mathcal{R}$, quotient of the topology $\mathcal{S}$. This is the finest topology for which the maps $\varphi \circ \theta_i$ are continuous. Therefore a subset B in $X/\mathcal{R}$ is open (resp. closed) if and only if $B = \varphi(C)$ where C is an open (resp. closed) subset of X, saturated with respect to $\mathcal{R}$.

The topological space $X/\mathcal{R}$ is said to be obtained by *gluing the X_i's along the X_{ij}'s via the maps h_{ji}*.

1.5.3 Proposition. *Let us suppose that the hypotheses* a), b), c), d) *of 1.5.1 and 1.5.2 are satisfied. Then each $\varphi(X_i)$ is open in $X/\mathcal{R}$. Further, the restriction of φ to X_i is a homeomorphism from X_i onto $\varphi(X_i)$.*

Proof. Set
$$Y_i = \varphi(X_i) \ , \ Z_i = \bigcup_{j \in I} \theta_j(X_{ij}).$$

Then Z_i is a subset of X saturated with respect to $\mathcal{R}$, and we have $Y_i = \varphi(Z_i)$. Since $Z_i \cap X_j = X_{ij}$ and Z_i is open in X by our hypotheses, Y_i is open in $X/\mathcal{R}$.

Let A_i be an open subset of X_i. Then each $A_i \cap X_{ij}$ is open in X_{ij}. Consequently, each $h_{ji}(A_i \cap X_{ij})$ is open in X_{ji}, and therefore open in X_j. It follows that $\bigcup_{j \in I} \theta_j \circ h_{ji}(A_i \cap X_{ij})$ is open in X. So $\varphi(A_i)$ is open in $X/\mathcal{R}$, and hence open in Y_i. Now the proposition follows. $\square$

1.5.4 Let us conserve the hypotheses and notations of 1.5.1 and 1.5.2. To summarize, we have constructed a topological space $Y = X/\mathcal{R}$, and for each $i \in I$, a homeomorphism ψ_i from X_i onto an open subset $\psi_i(X_i)$ in Y such that $Y = \bigcup_{i \in I} \psi_i(X_i)$, and such that

$$\psi_i(X_{ij}) = \psi_j(X_{ji}) = \psi_i(X_i) \cap \psi_j(X_j) \ , \ h_{ji}(x) = \psi_j^{-1}(\psi_i(x))$$

for all $i, j \in I$ and $x \in X_{ij}$.

References

- [8], [26].

2

Rings and modules

Unless otherwise stated, all algebras over a commutative ring considered in this book are associative.

We recall here basic concepts of the theory of commutative rings which will allow us to fix vocabulary and notations for the following chapters. Notions such as local rings, field of fractions and module of differentials will often appear in algebraic geometry.

All rings considered in this chapter are commutative. Let A be a ring.

2.1 Ideals

2.1.1 Let us fix some notations.

• If B is a ring, we denote by $\mathrm{Hom}(A, B)$ the set of ring homomorphisms from A to B.

• An element $x \in A$ is called a *zero divisor* if there exists $y \in A \setminus \{0\}$ such that $xy = 0$. A ring A is called an *integral domain* if $A \neq \{0\}$ and if A has no zero divisors other than 0.

• An element $x \in A$ is called *nilpotent* if there exists $n \in \mathbb{N}^*$ such that $x^n = 0$. The set of nilpotent elements of A is an ideal called the *nilradical* of A. We say that A is *reduced* if its nilradical is $\{0\}$.

2.1.2 Proposition. (Chinese Remainder Theorem) *Let* $\mathfrak{a}_1, \ldots, \mathfrak{a}_n$ *be ideals of A such that* $\mathfrak{a}_i + \mathfrak{a}_j = A$ *for all* $i \neq j$. *Let* $x_1, \ldots, x_n \in A$. *Then there exists* $x \in A$ *such that* $x - x_i \in \mathfrak{a}_i$ *for* $1 \leqslant i \leqslant n$.

Proof. First, suppose that $n = 2$. Then there exist $a_1 \in \mathfrak{a}_1$, $a_2 \in \mathfrak{a}_2$ such that $a_1 + a_2 = 1$. Hence $x = x_2 a_1 + x_1 a_2$ satisfies the conclusion of the proposition.

Suppose now that $n \geqslant 3$. For $i \geqslant 2$, there exist $a_i \in \mathfrak{a}_1$ and $b_i \in \mathfrak{a}_i$ such that $a_i + b_i = 1$. The product $\prod_{i \geqslant 2}(a_i + b_i)$ is equal to 1 and it is contained in $\mathfrak{a}_1 + \mathfrak{a}_2 \cdots \mathfrak{a}_n$. Thus $\mathfrak{a}_1 + \mathfrak{a}_2 \cdots \mathfrak{a}_n = A$ and from the previous paragraph, there exists $y_1 \in A$ such that $y_1 - 1 \in \mathfrak{a}_1$ and $y_1 \in \mathfrak{a}_2 \cdots \mathfrak{a}_n$. So $y_1 - 1 \in \mathfrak{a}_1$ and $y_1 \in \mathfrak{a}_i$ for all $i \geqslant 2$.

Proceeding in the same manner for an index $j \geqslant 2$, we see that there exists $y_j \in A$ such that $y_j - 1 \in \mathfrak{a}_j$ and $y_j \in \mathfrak{a}_i$ for all $i \neq j$.

We verify easily that $x = x_1 y_1 + \cdots + x_n y_n$ satisfies the conclusion of the proposition. □

2.1.3 Corollary. *Let* $\mathfrak{a}_1, \ldots, \mathfrak{a}_n$ *be ideals of* A *such that* $\mathfrak{a}_i + \mathfrak{a}_j = A$ *for all* $i \neq j$. *Set* $\mathfrak{b} = \mathfrak{a}_1 \cap \cdots \cap \mathfrak{a}_n$.

(i) *We have* $\mathfrak{b} = \mathfrak{a}_1 \cdots \mathfrak{a}_n$.

(ii) *The homomorphism* $f : A \to (A/\mathfrak{a}_1) \times \cdots \times (A/\mathfrak{a}_n)$ *induced by the canonical surjections on each factor is surjective and its kernel is* $\mathfrak{b}$. *In particular* f *induces an isomorphism between* $A/\mathfrak{b}$ *and* $(A/\mathfrak{a}_1) \times \cdots \times (A/\mathfrak{a}_n)$.

Proof. (i) From the proof of 2.1.2, $\mathfrak{a}_1 + \mathfrak{a}_2 \cdots \mathfrak{a}_n = A$. Therefore by induction, it suffices to consider the case $n = 2$. Clearly, $\mathfrak{a}_1 \mathfrak{a}_2 \subset \mathfrak{a}_1 \cap \mathfrak{a}_2$. Let $a_1 \in \mathfrak{a}_1$, $a_2 \in \mathfrak{a}_2$ be such that $a_1 + a_2 = 1$. If $x \in \mathfrak{a}_1 \cap \mathfrak{a}_2$, then $x = xa_1 + xa_2 \in \mathfrak{a}_1 \mathfrak{a}_2$.

(ii) It is clear that $\mathfrak{b}$ is the kernel of f. The surjectivity of f follows from 2.1.2 and the last assertion is now obvious. □

2.2 Prime and maximal ideals

2.2.1 An ideal $\mathfrak{m}$ of A is called *maximal* if it is maximal in the set of proper ideals of A. This is equivalent to the condition that $A/\mathfrak{m}$ is a field. Any ideal of A is contained in a maximal ideal and we shall denote by $\mathrm{Spm}(A)$ the set of maximal ideals of A.

2.2.2 An ideal $\mathfrak{p}$ of A is called *prime* if it satisfies one of the following equivalent conditions:

(i) $\mathfrak{p} \neq A$ and if $x, y \in A \setminus \mathfrak{p}$, then $xy \in A \setminus \mathfrak{p}$.

(ii) The ring $A/\mathfrak{p}$ is an integral domain.

A maximal ideal is a prime ideal. Let us denote by $\mathrm{Spec}(A)$ the set of prime ideals of A.

2.2.3 A *chain* of prime ideals of *length* n of A is a sequence

$$\mathfrak{p}_0 \subset \mathfrak{p}_1 \subset \cdots \subset \mathfrak{p}_n$$

of pairwise distinct elements of $\mathrm{Spec}(A)$.

The *(Krull) dimension* of A, denoted by $\dim A$, is the supremum of the lengths of chains of prime ideals. It is an element of $\mathbb{N} \cup \{+\infty\}$. Thus $\dim A = 0$ if and only if prime ideals of A are maximal; this is the case for a commutative field.

2.2.4 Proposition. *Let* $B = A_1 \times \cdots \times A_n$ *be a product of rings. Any prime (resp. maximal) ideal of* B *is of the form* $A_1 \times \cdots \times A_{i-1} \times \mathfrak{a}_i \times A_{i+1} \times \cdots \times A_n$, *where* $1 \leqslant i \leqslant n$ *and* $\mathfrak{a}_i$ *is a prime (resp. maximal) ideal of* A_i.

Proof. If C and D are rings, then $C \times D$ is not an integral domain since $(0, 1)(1, 0) = (0, 0)$. The result now follows easily. □

2.2.5 Proposition. (i) *Let* $\mathfrak{a}_1, \ldots, \mathfrak{a}_n$ *be ideals of* A *and* $\mathfrak{p}$ *a prime ideal of* A *such that* $\mathfrak{a}_1 \cdots \mathfrak{a}_n \subset \mathfrak{p}$. *Then* $\mathfrak{p}$ *contains one of the* $\mathfrak{a}_i$'s.

(ii) *Let* $\mathfrak{a}$ *be an ideal of* A *and* $\mathfrak{p}_1, \ldots, \mathfrak{p}_r$ *be prime ideals of* A *such that* $\mathfrak{a} \subset \mathfrak{p}_1 \cup \cdots \cup \mathfrak{p}_r$. *Then* $\mathfrak{a}$ *is contained in one of the* $\mathfrak{p}_i$'s.

Proof. (i) Suppose that $\mathfrak{a}_i \not\subset \mathfrak{p}$ for $1 \leqslant i \leqslant n$. There exists $a_i \in \mathfrak{a}_i \setminus \mathfrak{p}$. But then $a = a_1 \cdots a_n \in \mathfrak{a}_1 \cdots \mathfrak{a}_n \setminus \mathfrak{p}$, which contradicts our hypotheses.

(ii) We can assume that $\mathfrak{p}_i \not\subset \mathfrak{p}_j$ for $i \neq j$. Let us proceed by induction on r. The result is obvious when $r = 1$. If $\mathfrak{a} \subset \mathfrak{p}_1 \cup \cdots \cup \mathfrak{p}_{r-1}$, then we obtain our result by induction. So let us suppose that there exists $a \in \mathfrak{a} \cap \mathfrak{p}_r$ such that $a \notin \mathfrak{p}_1 \cup \cdots \cup \mathfrak{p}_{r-1}$.

Suppose that there exists $y \in \mathfrak{a}\mathfrak{p}_1 \cdots \mathfrak{p}_{r-1} \setminus \mathfrak{p}_r$. Then $a + y \in \mathfrak{a}$ and $a + y \notin \mathfrak{p}_1 \cup \cdots \cup \mathfrak{p}_{r-1}$. It follows that $a + y \in \mathfrak{p}_r$ and so $y \in \mathfrak{p}_r$ which gives us a contradiction.

So we have $\mathfrak{a}\mathfrak{p}_1 \cdots \mathfrak{p}_{r-1} \subset \mathfrak{p}_r$. By our assumption, for $i < r$, we have $\mathfrak{p}_i \not\subset \mathfrak{p}_r$. So part (i) implies that $\mathfrak{a} \subset \mathfrak{p}_r$. $\square$

2.2.6 Corollary. *Let* $\mathfrak{m} \in \operatorname{Spm}(A)$ *and* $n \in \mathbb{N}^*$. *Then the only prime ideal containing* $\mathfrak{m}^n$ *is* $\mathfrak{m}$.

Proof. If $\mathfrak{p} \in \operatorname{Spec}(A)$ contains $\mathfrak{m}^n$, then by 2.2.5 (i), $\mathfrak{m} \subset \mathfrak{p}$. Hence $\mathfrak{m} = \mathfrak{p}$. $\square$

2.2.7 A prime ideal in A is said to be *minimal* if it is minimal by inclusion in $\operatorname{Spec}(A)$.

Proposition. *If* $A \neq \{0\}$, *then the set of minimal prime ideals of* A *is non-empty.*

Proof. Since $A \neq \{0\}$, $\operatorname{Spec}(A)$ is non-empty. We shall endow $\operatorname{Spec}(A)$ with the partial order given by reverse inclusion.

Let $(\mathfrak{p}_i)_{i \in I}$ be a totally ordered family of prime ideals. Set $\mathfrak{r} = \bigcap_{i \in I} \mathfrak{p}_i$. Let $a, b \in A$ be such that $ab \in \mathfrak{r}$ and $a \notin \mathfrak{r}$. So there exists $j \in I$ such that $a \notin \mathfrak{p}_j$. Now for all $k \in I$ such that $\mathfrak{p}_k \subset \mathfrak{p}_j$, $a \notin \mathfrak{p}_k$ and since $\mathfrak{p}_k$ is prime, $b \in \mathfrak{p}_k$. Consequently $b \in \mathfrak{r}$ and the result follows from Zorn's lemma. $\square$

2.3 Rings of fractions and localization

2.3.1 Definition. *A subset* S *of* A *is said to be* multiplicative *if* $1 \in S$ *and it is closed under multiplication.*

2.3.2 Lemma. *Let* S *be a multiplicative subset of* A *not containing* 0 *and* $\mathfrak{a}$ *an ideal of* A *such that* $\mathfrak{a} \cap S = \emptyset$. *Then there exists* $\mathfrak{p} \in \operatorname{Spec}(A)$ *such that* $\mathfrak{a} \subset \mathfrak{p}$ *and* $\mathfrak{p} \cap S = \emptyset$.

Proof. Let $\mathcal{A}$ be the set, partially ordered by inclusion, of ideals $\mathfrak{b}$ of A such that $\mathfrak{a} \subset \mathfrak{b}$ and $\mathfrak{b} \cap S = \emptyset$. Since $\mathfrak{a} \in \mathcal{A}$, $\mathcal{A}$ is non-empty and it is clearly inductive. By Zorn's lemma, $\mathcal{A}$ has a maximal element $\mathfrak{p}$.

Suppose that there exist $a, b \in A \setminus \mathfrak{p}$ such that $ab \in \mathfrak{p}$. Then the ideals $Aa + \mathfrak{p}$ and $Ab + \mathfrak{p}$ contain strictly $\mathfrak{p}$. By the maximality of $\mathfrak{p}$, there exist

$s \in S \cap (Aa + \mathfrak{p})$ and $t \in S \cap (Ab + \mathfrak{p})$. It follows that $st \in S \cap \mathfrak{p}$ which is a contradiction. We have therefore shown that $\mathfrak{p}$ is a prime ideal. $\square$

2.3.3 Let S be a multiplicative subset of A. We define the binary relation $\mathcal{R}$ on $S \times A$ as follows: $(s, a)\mathcal{R}(t, b)$ if and only if there exists $u \in S$ such that $u(sb - ta) = 0$. This defines an equivalence relation on $S \times A$.

We shall endow $S \times A$ with the following operations:

$$(s, a) + (t, b) = (st, sb + ta) , \quad (s, a).(t, b) = (st, ab).$$

These operations are compatible with $\mathcal{R}$ and we see clearly that they induce a ring structure on $(S \times A)/\mathcal{R}$. We call the ring $(S \times A)/\mathcal{R}$, denoted by $S^{-1}A$, the *ring of fractions of A by S*. We shall denote the class of (s, a) by a/s or $\dfrac{a}{s}$.

The map $i : A \to S^{-1}A$ defined by $a \mapsto a/1$ is a ring homomorphism. We shall call i the *canonical* homomorphism.

We have the following results:

• The canonical homomorphism i is injective if and only if S contains no zero divisor.

• We have $S^{-1}A = \{0\}$ if and only if $0 \in S$. This is the case if S contains a nilpotent element.

• Let $s \in S$. The element $i(s) = s/1$ is invertible in $S^{-1}A$. Its inverse is $1/s$.

• If $\mathfrak{p} \in \mathrm{Spec}(A)$, then $S = A \setminus \mathfrak{p}$ is a multiplicative subset of A. The ring of fractions $S^{-1}A$, denoted by $A_{\mathfrak{p}}$, is called the *localization* of A at $\mathfrak{p}$. In particular, $A_{\mathfrak{p}} \neq \{0\}$.

• Suppose that $A \neq \{0\}$ and let S be the set of non zero divisors of A. Then S is a multiplicative subset of A. The ring $S^{-1}A$, denoted by $\mathrm{Fract}(A)$, is called the *full ring of fractions* of A. Further, if A is an integral domain, then $\mathrm{Fract}(A)$ is a field which is called the *quotient field* or *field of fractions* of A.

2.3.4 Theorem. *Let A, B be rings, S a multiplicative subset of A, $i : A \to S^{-1}A$ the canonical homomorphism and $f \in \mathrm{Hom}(A, B)$. Suppose further that any element of $f(S)$ is invertible in B. Then there exists a unique homomorphism $g : S^{-1}A \to B$ such that $g \circ i = f$.*

Proof. Let $p : S \times A \to S^{-1}A$ be the map sending (s, a) to a/s and $h : S \times A \to B$ be the map defined by $(s, a) \mapsto f(a)f(s)^{-1}$. We verify easily that $h(s, a) = h(t, b)$ if $a/s = b/t$. It follows therefore that there is a unique map $g : S^{-1}A \to B$ such that $g \circ p = h$. A direct computation show that g is a ring homomorphism and $g \circ i = f$. The uniqueness of g is immediate. $\square$

2.3.5 Corollary. *Let S, T be multiplicative subsets of A such that $S \subset T$, $i_S : A \to S^{-1}A$ and $i_T : A \to T^{-1}A$ the corresponding canonical homomorphisms. There exists a unique homomorphism $f : S^{-1}A \to T^{-1}A$ such that $f \circ i_S = i_T$. Further, if T contains no zero divisors, then f is injective.*

Remark. If A is an integral domain and S a multiplicative subset of A which does not contain 0, then $S^{-1}A$ embeds canonically into the quotient field $\mathrm{Fract}(A)$.

2.3.6 Corollary. *Suppose that A is a subring of a ring B and $f : A \to B$ the canonical injection. Let S be a multiplicative subset of A (hence of B), and $i : A \to S^{-1}A$, $j : B \to S^{-1}B$ the canonical homomorphisms. There exists a unique homomorphism $g : S^{-1}A \to S^{-1}B$ such that $g \circ i = j \circ f$. The homomorphism g is injective and we can therefore identify canonically $S^{-1}A$ as a subring of $S^{-1}B$.*

Proof. The existence and uniqueness of g follows from 2.3.4. Let $(s, a) \in S \times A$ be such that $g(a/s) = 0$. Then

$$0 = g((a/1).(1/s)) = g \circ i(a)[g \circ i(s)]^{-1}$$
$$\Rightarrow g \circ i(a) = 0 \Rightarrow f \circ j(a) = 0 \Rightarrow j(a) = 0.$$

It follows that there exists $t \in S$ such that $ta = 0$. Hence $(s, a)\mathcal{R}(t, 0)$ and $a/s = 0$. $\square$

2.3.7 Let S be a multiplicative subset of A, $i : A \to S^{-1}A$ be the canonical homomorphism and $\mathcal{J}(A)$ (resp. $\mathcal{J}(S^{-1}A)$) the set of ideals of A (resp. $S^{-1}A$). Denote by $\mathcal{J}'(A)$ the set of ideals $\mathfrak{a}$ of A which verifies the condition:

(1) $\qquad\qquad\qquad s \in S \ , \ a \in A \ , \ sa \in \mathfrak{a} \Rightarrow a \in \mathfrak{a}.$

For $\mathfrak{a} \in \mathcal{J}(A)$, let $S^{-1}\mathfrak{a}$ be the set of elements a/s of $S^{-1}A$ such that $a \in \mathfrak{a}$, $s \in S$. We verify easily that $S^{-1}\mathfrak{a} \in \mathcal{J}(S^{-1}A)$ and thus we have a map $\psi : \mathcal{J}(A) \to \mathcal{J}(S^{-1}A)$ sending $\mathfrak{a}$ to $S^{-1}\mathfrak{a}$.

The proof of the following result is straightforward.

Proposition. (i) *If $\mathfrak{a} \in \mathcal{J}(A)$, then $S^{-1}\mathfrak{a}$ is the ideal of $S^{-1}A$ generated by $i(\mathfrak{a})$. Any ideal of $S^{-1}A$ is of the form $S^{-1}\mathfrak{b}$ for some $\mathfrak{b} \in \mathcal{J}(A)$.*
(ii) *If $\mathfrak{n}$ is the nilradical of A, then $S^{-1}\mathfrak{n}$ is the nilradical of $S^{-1}A$.*
(iii) *For $\mathfrak{a}, \mathfrak{b} \in \mathcal{J}(A)$, we have:*

$$S^{-1}(\mathfrak{a} + \mathfrak{b}) = S^{-1}\mathfrak{a} + S^{-1}\mathfrak{b},$$

$$S^{-1}(\mathfrak{ab}) = (S^{-1}\mathfrak{a})(S^{-1}\mathfrak{b}), S^{-1}(\mathfrak{a} \cap \mathfrak{b}) = S^{-1}\mathfrak{a} \cap S^{-1}\mathfrak{b}.$$

(iv) *The map ψ induces a bijection between $\mathcal{J}'(A)$ and $\mathcal{J}(S^{-1}A)$, whose reciprocal map is given by $\mathfrak{b} \mapsto i^{-1}(\mathfrak{b})$.*

2.3.8 Corollary. *The map ψ induces a bijection between the set of prime ideals of A not meeting S and the set of prime ideals of $S^{-1}A$.*

Proof. A prime ideal $\mathfrak{p}$ of A which does not meet S satisfies the condition (1) of 2.3.7. So it suffices to show that $S^{-1}\mathfrak{p} \in \mathrm{Spec}\, S^{-1}A$.

Let $a, b \in A$, $s, t \in S$ be such that $(a/s).(b/t) \in S^{-1}\mathfrak{p}$. There exists $u \in S$ such that $uab \in \mathfrak{p}$. Since $u \notin \mathfrak{p}$, $ab \in \mathfrak{p}$. So either $a \in \mathfrak{p}$ or $b \in \mathfrak{p}$. Thus either $a/s \in S^{-1}\mathfrak{p}$ or $b/t \in S^{-1}\mathfrak{p}$. $\square$

2.3.9 Let A be a subring of a ring B and $\mathfrak{a}$ be an ideal of A. Then an ideal $\mathfrak{b}$ of B is said to *lie above* $\mathfrak{a}$ if $\mathfrak{b} \cap A = \mathfrak{a}$.

Let S be a multiplicative subset of A. Let us conserve the notations i, j, f, g of 2.3.6 and let $\mathcal{J}(S^{-1}A)$, $\mathcal{J}(S^{-1}B)$, $\mathcal{J}'(A)$, $\mathcal{J}'(B)$ be as in 2.3.7.

Proposition. *Let $\mathfrak{p} \in \operatorname{Spec} A$ be such that $S \cap \mathfrak{p} = \emptyset$.*

(i) *The map $\mathfrak{b} \mapsto S^{-1}\mathfrak{b}$ induces a surjection from the set $\mathcal{E}$ of ideals of B lying above $\mathfrak{p}$ and the set $\mathcal{F}$ of ideals of $S^{-1}B$ lying above $S^{-1}\mathfrak{p}$, and the map $\mathfrak{b} \mapsto j^{-1}(\mathfrak{b})$ is a bijection between $\mathcal{F}$ and $\mathcal{E} \cap \mathcal{J}'(B)$.*

(ii) *The map $\mathfrak{q} \mapsto S^{-1}\mathfrak{q}$ induces a bijection between the set of prime ideals in B lying above $\mathfrak{p}$ and the set of prime ideals of $S^{-1}B$ lying above $S^{-1}\mathfrak{p}$.*

Proof. By 2.3.8, we have $S^{-1}\mathfrak{p} \in \operatorname{Spec} S^{-1}A$ and $i^{-1}(S^{-1}\mathfrak{p}) = \mathfrak{p}$.

Let $\mathfrak{b}' \in \mathcal{J}(S^{-1}B)$ be such that $\mathfrak{b}' \cap S^{-1}A = S^{-1}\mathfrak{p}$. Then we have:

$$\mathfrak{p} = i^{-1}(S^{-1}\mathfrak{p}) = i^{-1}(\mathfrak{b}' \cap S^{-1}A) = A \cap j^{-1}(\mathfrak{b}').$$

By 2.3.7, $S^{-1}(j^{-1}(\mathfrak{b}')) = \mathfrak{b}'$ and therefore the image of $\mathcal{E}$ via the map $\mathfrak{b} \mapsto S^{-1}\mathfrak{b}$ contains $\mathcal{F}$.

Let $\mathfrak{b} \in \mathcal{E}$, $a \in A$ and $s \in S$. The following conditions are equivalent:

(i) $g(a/s) \in S^{-1}\mathfrak{b}$.

(ii) $f(a)/f(s) \in S^{-1}\mathfrak{b}$.

(iii) There exists $t \in S$ such that $f(t)f(a) \in \mathfrak{b}$.

(iv) There exists $t \in S$ such that $ta \in \mathfrak{p}$.

(v) $a/s \in S^{-1}\mathfrak{p}$.

We deduce that $S^{-1}\mathfrak{b} \cap S^{-1}A = S^{-1}\mathfrak{p}$. Thus the image of $\mathcal{E}$ via the map $\mathfrak{b} \mapsto S^{-1}\mathfrak{b}$ is $\mathcal{F}$. The rest of the proposition follows from 2.3.7 and 2.3.8. $\square$

2.3.10 Proposition. *Let S be a multiplicative subset of A and $\mathfrak{a} \in \mathcal{J}(A)$ be such that $\mathfrak{a} \cap S = \emptyset$. Denote by $\pi : A \to A/\mathfrak{a}$ the canonical projection. Then $\pi(S)$ is a multiplicative subset of $A/\mathfrak{a}$, and the rings $S^{-1}A/S^{-1}\mathfrak{a}$ and $\pi(S)^{-1}(A/\mathfrak{a})$ are isomorphic.*

Proof. It is clear that $T = \pi(S)$ is a multiplicative subset of $B = A/\mathfrak{a}$, and that $0 \notin T$ since $\mathfrak{a} \cap S = \emptyset$. Let $i_S : A \to S^{-1}A$ and $i_T : B \to T^{-1}B$ be the canonical homomorphisms. If $s \in S$, then $i_T \circ \pi(s)$ is invertible in $T^{-1}B$. By 2.3.4, there exists a unique homomorphism $f : S^{-1}A \to T^{-1}B$ such that $f \circ i_S = i_T \circ \pi$. For $a \in A$, $s \in S$, we have:

$$\pi(a)/\pi(s) = (\pi(a)/\pi(1)).(\pi(1)/\pi(s)) = (i_T \circ \pi(a))(i_T \circ \pi(s))^{-1}$$
$$= (f \circ i_S(a))(f \circ i_S(s))^{-1} = f((a/1).(1/s)) = f(a/s).$$

Thus f is surjective.

Now suppose that $f(a/s) = 0$. The preceding computation implies that $\pi(a)/\pi(s) = 0$, and so there exists $t \in S$ such that $\pi(ta) = \pi(a)\pi(t) = 0$. Thus $ta \in \mathfrak{a}$ and it follows that $a/s = (ta)/(ts) \in S^{-1}\mathfrak{a}$. $\square$

2.3.11 Definition. *Let* $\mathfrak{p}$ *be a prime ideal of* A. *We define the* height of $\mathfrak{p}$, *denoted by* ht $\mathfrak{p}$, *to be the dimension of the ring* $A_{\mathfrak{p}}$.

2.3.12 Proposition. *The height of a prime ideal* $\mathfrak{p}$ *is the supremum of chains* $\mathfrak{p}_0 \subset \mathfrak{p}_1 \subset \cdots \subset \mathfrak{p}_r$ *of prime ideals of* A *such that* $\mathfrak{p}_r = \mathfrak{p}$.

Proof. This is immediate from 2.3.7 (iii) and 2.3.8. □

2.4 Localizations of modules

2.4.1 In this section, S will denote a multiplicative subset of A.

Let E be an A-module. We define the binary relation $\mathcal{S}$ on $E \times S$ by: for $x, y \in E$, $s, t \in S$, we have $(x, s)\mathcal{S}(y, t)$ if and only if there exists $u \in S$ such that $u(sy - tx) = 0$.

It is easy to verify that $\mathcal{S}$ defines an equivalence relation on $E \times S$. Let us denote by $S^{-1}E$ the set of equivalence classes and by x/s the class of (x, s).

The proof of the following is similar to the one for rings of fractions.

Theorem. *Let* E *be an* A-module.
(i) *Define a structure of* $S^{-1}A$-module *on* $S^{-1}E$ *by*

$$x/s + y/t = (tx + sy)/st \ , \ (a/s)(y/t) = (ay)/(st)$$

where $x, y \in E$, $s, t \in S$ *and* $a \in A$. *The map*

$$j : E \to S^{-1}E \ , \ x \mapsto x/1$$

is a homomorphism of A-modules, *called the* canonical homomorphism *of* E *to* $S^{-1}E$.

(ii) *Let* F *be a* $S^{-1}A$-module *and* $f : E \to F$ *be a homomorphism of* A-modules. *Then there exists a unique homomorphism of* $S^{-1}A$-modules $g :$ $S^{-1}E \to F$ *such that* $f = g \circ j$.

2.4.2 Proposition. *Let* $f : E \to F$ *be a homomorphism of* A-modules. *There exists a unique homomorphism of* $S^{-1}A$-modules $g : S^{-1}E \to S^{-1}F$ *such that* $g(x/1) = f(x)/1$ *for all* $x \in E$.

Proof. Let $j_E : E \to S^{-1}E$ and $j_F : F \to S^{-1}F$ be the canonical homomorphisms. By 2.4.1, there exists a $S^{-1}A$-module homomorphism $g : S^{-1}E \to$ $S^{-1}F$ such that $j_F \circ f = g \circ j_E$. So $g(x/1) = f(x)/1$ for all $x \in E$. The uniqueness is clear. □

2.4.3 Proposition. *Let* F *be a submodule of* E. *The homomorphism* $S^{-1}F \to S^{-1}E$ *induced by the canonical injection* $F \to E$ *in 2.4.2 is injective, and the* $S^{-1}A$-modules $S^{-1}(E/F)$ *and* $S^{-1}E/S^{-1}F$ *are isomorphic.*

Proof. Let x/s be in the kernel of the homomorphism $S^{-1}F \to S^{-1}E$. There exists $t \in S$ such that $tx = 0$ which in turn implies that $x/s = 0$ in $S^{-1}F$.

The map $f : E/F \to S^{-1}E/S^{-1}F$ defined by sending $x+F$ to $x/1+S^{-1}F$ is well-defined and we verify easily that it is a homomorphism of A-modules. By 2.4.1, there exists a unique homomorphism of $S^{-1}A$-modules

$$g : S^{-1}(E/F) \to S^{-1}E/S^{-1}F \; , \; (x+F)/s \mapsto x/s + S^{-1}F.$$

By construction, g is surjective. Let $x \in E$, $s \in S$ be such that $x/s \in S^{-1}F$. Then there exists $t \in S$ such that $tx \in F$. So we have $t(x+F) = 0$ in E/F and hence $(x+F)/s = 0$ in $S^{-1}(E/F)$. Thus g is injective. $\square$

2.4.4 Proposition. *Let E be an A-module, $j_E : E \to S^{-1}E$ the canonical homomorphism and $\mathfrak{a}$ an ideal of A. We have*

$$(S^{-1}\mathfrak{a})(S^{-1}E) = \mathfrak{a}(S^{-1}E) = (S^{-1}\mathfrak{a})j_E(E).$$

Proof. Let $a \in \mathfrak{a}$, $s,t \in S$ and $x \in E$. We have:

$$(a/s).(x/t) = (a/1).(x/st) = (a/st)(x/1).$$

So the proposition follows. $\square$

2.4.5 Remarks. 1) Proposition 2.4.3 justifies the notation $S^{-1}\mathfrak{a}$ of 2.3.7.

2) Let M be an A-module, then we see easily that $S^{-1}M$ and $S^{-1}A \otimes_A M$ are isomorphic $S^{-1}A$-modules where we consider $S^{-1}A$ as an A-algebra via the canonical map $A \to S^{-1}A$.

2.5 Radical of an ideal

2.5.1 Definition. *Let $\mathfrak{a}$ be an ideal of A. We define the* radical *of $\mathfrak{a}$, denoted by $\sqrt{\mathfrak{a}}$, to be the set of elements $a \in A$ for which there exists $n \in \mathbb{N}^*$ such that $a^n \in \mathfrak{a}$. If $\mathfrak{a} = \sqrt{\mathfrak{a}}$, then we say that $\mathfrak{a}$ is a* radical *ideal.*

2.5.2 Let $\mathfrak{a}$ be an ideal of A, then by using the formula of binomial expansion, we obtain easily that its radical $\sqrt{\mathfrak{a}}$ is an ideal of A. Note that the nilradical of A is the radical of the ideal $\{0\}$.

2.5.3 Proposition. *Let $\mathfrak{a}$, $\mathfrak{b}$ be ideals of A.*

(i) *We have $\sqrt{\mathfrak{a}} = A$ if and only if $\mathfrak{a} = A$. Further,*

$$\sqrt{\sqrt{\mathfrak{a}}} = \sqrt{\mathfrak{a}}, \sqrt{\mathfrak{a}+\mathfrak{b}} = \sqrt{\sqrt{\mathfrak{a}}+\sqrt{\mathfrak{b}}}, \sqrt{\mathfrak{a}\mathfrak{b}} = \sqrt{\sqrt{\mathfrak{a}}\sqrt{\mathfrak{b}}} = \sqrt{\mathfrak{a}} \cap \sqrt{\mathfrak{b}}.$$

(ii) *If $\sqrt{\mathfrak{a}}$ is finitely generated, then $(\sqrt{\mathfrak{a}})^n \subset \mathfrak{a}$ for some $n \in \mathbb{N}^*$.*

(iii) *If $\mathfrak{p}$ is a prime ideal of A and $n \in \mathbb{N}^*$, then $\sqrt{\mathfrak{p}^n} = \mathfrak{p}$.*

Proof. (i) Let us prove the last equality. From $\mathfrak{a}\mathfrak{b} \subset \mathfrak{a} \cap \mathfrak{b} \subset \mathfrak{a}$, we deduced that $\sqrt{\mathfrak{a}\mathfrak{b}} \subset \sqrt{\mathfrak{a} \cap \mathfrak{b}} \subset \sqrt{\mathfrak{a}} \cap \sqrt{\mathfrak{b}}$. Let $a \in \sqrt{\mathfrak{a}} \cap \sqrt{\mathfrak{b}}$, then there exist $n, p \in \mathbb{N}^*$ such that $a^n \in \mathfrak{a}$ and $a^p \in \mathfrak{b}$. Hence $a^{n+p} \in \mathfrak{a}\mathfrak{b}$ and so $a \in \sqrt{\mathfrak{a}\mathfrak{b}}$ as required.

(ii) Let us suppose that $\sqrt{\mathfrak{a}} = Aa_1 + \cdots + Aa_p$. There exists $m \in \mathbb{N}^*$ such that $a_i^m \in \mathfrak{a}$ for all $1 \leqslant i \leqslant p$. It follows that $(\sqrt{\mathfrak{a}})^n \subset \mathfrak{a}$ if $n > mp$.

(iii) We have $\mathfrak{p} \subset \sqrt{\mathfrak{p}}$ and $\sqrt{\mathfrak{p}^n} = \sqrt{\mathfrak{p}}$ by (i). Now if $a \in \sqrt{\mathfrak{p}}$, then there exists $n \in \mathbb{N}^*$ such that $a^n \in \mathfrak{p}$. Since $\mathfrak{p}$ is prime, $a \in \mathfrak{p}$. $\square$

2.5.4 Proposition. *Let $\mathfrak{a}$ be an ideal of A.*

(i) *The nilradical of $A/\mathfrak{a}$ is $\sqrt{\mathfrak{a}}/\mathfrak{a}$.*

(ii) *The ring $A/\sqrt{\mathfrak{a}}$ is reduced.*

(iii) *If $A/\mathfrak{a}$ is reduced, then $\mathfrak{a}$ contains the nilradical of A.*

Proof. (i) It is clear that the nilradical of $A/\mathfrak{a}$ is contained in $\sqrt{\mathfrak{a}}/\mathfrak{a}$. Let $a \in A$ and $n \in \mathbb{N}^*$ be such that $a^n \in \mathfrak{a}$. Then $(a + \mathfrak{a})^n \subset \mathfrak{a}$ and hence $a + \mathfrak{a} \in \sqrt{\mathfrak{a}}/\mathfrak{a}$.

(ii) This is immediate from (i) and 2.5.3.

(iii) If $A/\mathfrak{a}$ is reduced, then the kernel of the canonical surjection $A \to A/\mathfrak{a}$ contains the nilradical of A. □

2.5.5 Proposition. *Let $\mathfrak{a}$ be an ideal of A. Then $\sqrt{\mathfrak{a}}$ is the intersection of prime ideals of A containing $\mathfrak{a}$. In particular, the nilradical of A is the intersection of prime ideals of A.*

Proof. By 2.5.4, we can replace A by $A/\mathfrak{a}$ and assume that $\mathfrak{a} = \{0\}$. Denote by $\mathfrak{q}$ the intersection of prime ideals of A and by $\mathfrak{n}$ the nilradical of A.

Clearly, $\mathfrak{n} \subset \mathfrak{q}$. Now suppose that there exists an element $a \in \mathfrak{q}$ which is not nilpotent. Then $S = \{a^n; n \in \mathbb{N}\}$ is a multiplicative subset of A which does not contain 0. It follows from 2.3.2 that there exists $\mathfrak{p} \in \operatorname{Spec} A$ such that $\mathfrak{p} \cap S = \emptyset$. So $a \notin \mathfrak{p}$ which contradicts the fact that $a \in \mathfrak{q}$. □

2.6 Local rings

2.6.1 Definition. *The intersection of maximal ideals of A, denoted by $\operatorname{rad} A$, is an ideal called the* Jacobson radical *of A.*

2.6.2 Proposition. *Let $a \in A$. Then $a \in \operatorname{rad} A$ if and only if $1 - ab$ is invertible for all $b \in A$.*

Proof. If $a \in \operatorname{rad} A$, then $ab \in \operatorname{rad} A$. Therefore $1 - ab$ is not contained in any maximal ideal. So $1 - ab$ is invertible.

If $a \notin \operatorname{rad} A$, then there exists a maximal ideal $\mathfrak{m}$ which does not contain a. So $\mathfrak{m} + Aa = A$ and there exists $b \in A$, $c \in \mathfrak{m}$ such that $c + ab = 1$. It follows that $1 - ab = c$ is not invertible. □

2.6.3 Proposition. *The following conditions are equivalent:*

(i) *The ring A has a unique maximal ideal.*

(ii) *The set $A \setminus \operatorname{rad} A$ is the set of invertible elements of A.*

(iii) *There exists a proper ideal $\mathfrak{a}$ of A such that any element of $A \setminus \mathfrak{a}$ is invertible.*

Proof. (i) $\Rightarrow$ (ii) If $\mathfrak{m}$ is the unique maximal ideal of A, then $\mathfrak{m} = \operatorname{rad} A$. So (ii) follows.

(ii) $\Rightarrow$ (iii) Take $\mathfrak{a} = \operatorname{rad} A$.

(iii) $\Rightarrow$ (i) Suppose that $\mathfrak{a}$ satisfies the hypotheses of (iii). If $a \in \mathfrak{a}$, then $1 - ab$ is invertible for all $b \in A$. So by 2.6.2, $\mathfrak{a} \subset \operatorname{rad} A$. Now let $\mathfrak{b}$ be an ideal of A containing $\mathfrak{a}$. If $\mathfrak{b} \neq \mathfrak{a}$, then there exists $b \in \mathfrak{b} \setminus \mathfrak{a}$, which is invertible. This implies that $\mathfrak{b} = A$. So $\mathfrak{a} = \operatorname{rad} A$ is the unique maximal ideal of A. □

2.6.4 Definition. *A ring A satisfying the conditions of 2.6.3 is called a* local ring. *If A is a local ring with maximal ideal $\mathfrak{m}_A$, we call $A/\mathfrak{m}_A$ the* residual field *of A. A homomorphism $u : A \to B$ of local rings is said to be* local *if $u(\mathfrak{m}_A) \subset \mathfrak{m}_B$.*

2.6.5 Remark. Let A be a local ring with maximal ideal $\mathfrak{m}$. Then the rings A and $A_\mathfrak{m}$ are canonically isomorphic by 2.3.3 and 2.6.3.

2.6.6 Proposition. *Let $\mathfrak{p} \in \operatorname{Spec} A$. Then $A_\mathfrak{p}$ is a local ring with maximal ideal $\mathfrak{p}A_\mathfrak{p}$, and the residual field is canonically isomorphic to the field of fractions of $A/\mathfrak{p}$.*

Proof. Since $A_\mathfrak{p} \setminus \mathfrak{p}A_\mathfrak{p}$ consists of invertible elements, the first assertion follows by 2.6.3. Let $f : A \to A/\mathfrak{p}$ be the canonical projection. Then $f(A \setminus \mathfrak{p})$ is the set $A/\mathfrak{p} \setminus \{0\}$. So the other assertions follow from 2.3.4 and 2.3.10. $\square$

2.6.7 Proposition. (Nakayama's Lemma) *Let $\mathfrak{a}$ be an ideal of A and E be a finitely generated A-module such that $\mathfrak{a}E = E$. Then there exists $f \in 1 + \mathfrak{a}$ such that $fE = 0$.*

Proof. Denote by $\mathrm{M}_r(A)$ the ring of r-by-r matrices with coefficients in A, and by I_r the identity matrix in $\mathrm{M}_r(A)$. Let $\{e_1, \ldots, e_r\}$ be a system of generators of E. Our hypothesis says that we can write for $1 \leqslant i \leqslant r$:

$$e_i = a_{i1}e_1 + \cdots + a_{ir}e_r$$

where $a_{ij} \in \mathfrak{a}$. Set $U = [a_{ij}] \in \mathrm{M}_r(A)$ and $\Lambda = I_r - U$. Then

$$\Lambda \begin{pmatrix} e_1 \\ \vdots \\ e_r \end{pmatrix} = \begin{pmatrix} 0 \\ \vdots \\ 0 \end{pmatrix}.$$

If $\widetilde{\Lambda}$ is the transpose of the matrix of cofactors of Λ, then $\widetilde{\Lambda}\Lambda = (\det \Lambda)I_r$. This implies that $(\det \Lambda)e_j = 0$ for $1 \leqslant j \leqslant r$. Since $\det \Lambda$ is of the form $1 + f$ with $f \in \mathfrak{a}$, our result follows. $\square$

2.6.8 Corollary. *Let $\mathfrak{a} \subset \operatorname{rad} A$ be an ideal of A, E a finitely generated A-module and F a submodule of E.*
  (i) *If $\mathfrak{a}E = E$, then $E = \{0\}$.*
  (ii) *If $E = \mathfrak{a}E + F$, then $E = F$.*

Proof. (i) By 2.6.7, there exists $f \in \mathfrak{a}$ such that $(1 + f)E = \{0\}$. But $\mathfrak{a}$ is contained in $\operatorname{rad} A$, so $1 + f$ is invertible and hence $E = \{0\}$.
  (ii) Since $E/F = \mathfrak{a}(E/F)$, the result follows from (i). $\square$

2.6.9 Corollary. *Let A be a local ring with maximal ideal $\mathfrak{m}$, E a finitely generated A-module. Let $e_1, \ldots, e_n \in E$ be such that their images in the $A/\mathfrak{m}$-module $E/\mathfrak{m}E$ form a system of generators. Then $e_1, \ldots, e_n$ generate the A-module E.*

Proof. Let F be the submodule of E generated by $e_1, \ldots, e_n$. We have $E = \mathfrak{m}E + F$. Since $\mathfrak{m} = \operatorname{rad} A$ by 2.6.3, the result follows from 2.6.8.   □

2.6.10 Let $\mathfrak{p} \in \operatorname{Spec} A$, $S = A \setminus \mathfrak{p}$ and E an A-module. We shall denote by $E_\mathfrak{p}$ the $A_\mathfrak{p}$-module $S^{-1}E$.

Proposition. *Let $\mathfrak{m}$ be a maximal ideal of A and E an A-module. Suppose that there exists an ideal $\mathfrak{a}$ of A satisfying the following conditions:*
   (i) *The only maximal ideal containing $\mathfrak{a}$ is $\mathfrak{m}$.*
   (ii) *We have $\mathfrak{a}E = \{0\}$.*
Then the canonical homomorphism $E \to E_\mathfrak{m}$ is bijective.

Proof. Condition (i) implies that $A/\mathfrak{a}$ is local with maximal ideal $\mathfrak{m}/\mathfrak{a}$. Condition (ii) allows us to consider E as an $A/\mathfrak{a}$-module. If $s \in A \setminus \mathfrak{m}$, then $s + \mathfrak{a}$ is invertible in $A/\mathfrak{a}$. Thus the map $\mu_s : E \to E$ defined by $x \mapsto sx$ is bijective.

If $x, y \in E$ satisfy $x/1 = y/1$, then there exists $s \in A \setminus \mathfrak{m}$ such that $s(x-y) = 0 = \mu_s(x-y)$. So $x = y$. Now let $y \in E$ and $s \in A \setminus \mathfrak{m}$. There exists $x \in E$ such that $\mu_s(x) = sx = y$ and so $x/1 = y/s$.   □

2.6.11 Corollary. *Let $\mathfrak{m}$ be a maximal ideal of A, E an A-module and $k \in \mathbb{N}$ be such that $\mathfrak{m}^k E = \{0\}$. Then the canonical homomorphism $E \to E_\mathfrak{m}$ is bijective.*

Proof. This is clear if $k = 0$. For $k > 0$, it follows from 2.2.6 and 2.6.10.   □

2.6.12 Proposition. *Let $\mathfrak{m}$ be a maximal ideal of A, E an A-module and $k \in \mathbb{N}$. The canonical homomorphism $E \to E_\mathfrak{m}/\mathfrak{m}^k E_\mathfrak{m}$ is surjective. Its kernel is $\mathfrak{m}^k E$ and it induces an isomorphism between $E/\mathfrak{m}^k E$ and $E_\mathfrak{m}/\mathfrak{m}^k E_\mathfrak{m}$.*

Proof. By 2.6.10, the canonical homomorphism $E/\mathfrak{m}^k E \to (E/\mathfrak{m}^k E)_\mathfrak{m}$ is bijective. Further, we can identify canonically $(E/\mathfrak{m}^k E)_\mathfrak{m}$ and $E_\mathfrak{m}/(\mathfrak{m}^k E)_\mathfrak{m}$ by 2.4.3, and $(\mathfrak{m}^k E)_\mathfrak{m} = \mathfrak{m}^k E_\mathfrak{m}$ by 2.4.4. Thus the map $E/\mathfrak{m}^k E \to E_\mathfrak{m}/\mathfrak{m}^k E_\mathfrak{m}$ given by $x + \mathfrak{m}^k E \mapsto x/1 + \mathfrak{m}^k E_\mathfrak{m}$ is an isomorphism.   □

2.7 Noetherian rings and modules

2.7.1 Proposition. *For any A-module E, the following conditions are equivalent:*
   (i) *Any submodule of E is finitely generated.*
   (ii) *Any ascending sequence of submodules of E is stationary.*
   (iii) *Any non-empty family of submodules of E has a maximal element.*

Proof. This is straightforward.   □

2.7.2 Definition. *An A-module is called* Noetherian *if it satisfies the conditions of* 2.7.1. *The ring A is said to be* Noetherian *if A itself is a Noetherian A-module.*

2.7.3 Proposition. *Let E be an A-module and F a submodule of E. The following conditions are equivalent:*

(i) *E is Noetherian.*

(ii) *F and E/F are Noetherian.*

Proof. Again this is straightforward.   □

2.7.4 Corollary. (i) *Any finite direct sum of Noetherian modules is Noetherian.*

(ii) *Let F, G be submodules of an A-module E such that $E = F + G$. If F and G are Noetherian, then E is Noetherian.*

Proof. This is a direct consequence of 2.7.3.   □

2.7.5 Proposition. *Let us suppose that A is Noetherian. For any A-module E, the following conditions are equivalent:*

(i) *E is Noetherian.*

(ii) *E is finitely generated.*

Proof. The implication (i) $\Rightarrow$ (ii) is clear. Suppose that (ii) is verified, and let $x \in E$. The submodule Ax is isomorphic to a quotient of A, so it is Noetherian by 2.7.3. Now by applying 2.7.4 (ii), any finitely generated submodule of E is Noetherian.   □

2.7.6 Proposition. *Let $\mathfrak{a}$ be an ideal of a Noetherian ring A and S a multiplicative subset of A.*

(i) *The nilradical of A is a nilpotent ideal.*

(ii) *The rings $A/\mathfrak{a}$ and $S^{-1}A$ are Noetherian.*

Proof. Assertion (i) follows from 2.5.3. We deduce from 2.7.3 that $A/\mathfrak{a}$ is Noetherian because the $A/\mathfrak{a}$-submodules of $A/\mathfrak{a}$ are the A-submodules of $A/\mathfrak{a}$. Finally $S^{-1}A$ is Noetherian by 2.3.7 (iii) and (iv).   □

2.7.7 Theorem. *Let $\mathfrak{a}$ be a radical ideal of a Noetherian ring A. There exist $\mathfrak{p}_1, \ldots, \mathfrak{p}_n \in \operatorname{Spec} A$ such that:*

$$\mathfrak{a} = \mathfrak{p}_1 \cap \cdots \cap \mathfrak{p}_n.$$

If we assume that $\mathfrak{p}_i \not\subset \mathfrak{p}_j$ for $i \neq j$, then the preceding decomposition is unique up to a permutation.

Proof. Let $\mathcal{F}$ be the set of proper radical ideals of A which are not finite intersections of prime ideals. Let us suppose that $\mathcal{F}$ is not empty. By 2.7.1, there is a maximal element in $\mathcal{F}$, say $\mathfrak{r}$.

Since $\mathfrak{r}$ is not prime, there exist $a, b \in A \setminus \mathfrak{r}$ such that $ab \in \mathfrak{r}$, and the ideals $\mathfrak{a} = \mathfrak{r} + Aa$, $\mathfrak{b} = \mathfrak{r} + Ab$ contain strictly $\mathfrak{r}$. We claim that $\mathfrak{a}$ and $\mathfrak{b}$ are proper. If

$\mathfrak{a} = A$, then $1 = xa + r$ for some $x \in A$ and $r \in \mathfrak{r}$. But this would imply that $b = xab + rb \in \mathfrak{r}$, which is impossible.

It follows that there exist prime ideals $\mathfrak{p}_1, \ldots, \mathfrak{p}_n, \mathfrak{q}_1, \ldots, \mathfrak{q}_m$ such that $\sqrt{\mathfrak{a}} = \mathfrak{p}_1 \cap \cdots \cap \mathfrak{p}_n$ and $\sqrt{\mathfrak{b}} = \mathfrak{q}_1 \cap \cdots \cap \mathfrak{q}_n$.

Set $\mathfrak{s} = \sqrt{\mathfrak{a}} \cap \sqrt{\mathfrak{b}}$. We have $\mathfrak{r} \subset \mathfrak{s}$ and if $u \in \mathfrak{s}$, there exists $n \in \mathbb{N}^*$ such that $u^n \in \mathfrak{a} \cap \mathfrak{b}$. This implies in turn that $u^{2n} \in \mathfrak{ab} \subset \mathfrak{r}$, and so $u \in \mathfrak{r}$. Hence $\mathfrak{r} = \mathfrak{s}$ which is absurd since $\mathfrak{s} \notin \mathcal{F}$.

Now suppose that $\mathfrak{p}_1, \ldots, \mathfrak{p}_n, \mathfrak{q} \in \operatorname{Spec} A$ be such that $\mathfrak{p}_1 \cap \cdots \cap \mathfrak{p}_n \subset \mathfrak{q}$. Then $\mathfrak{p}_1 \cdots \mathfrak{p}_n \subset \mathfrak{q}$, and by 2.2.5 (i), there exists i such that $\mathfrak{p}_i \subset \mathfrak{q}$. Now the uniqueness of the decomposition follows easily. $\square$

2.7.8 Corollary. *Let A be a Noetherian ring which is not reduced to $\{0\}$.*
  (i) *The set of minimal prime ideals is finite.*
  (ii) *Suppose that A is an integral domain. Then any proper non-zero ideal $\mathfrak{a}$ of A contains a non-zero finite product of prime ideals.*

Proof. (i) Let $\mathfrak{n}$ be the nilradical of A. By 2.7.7, there exist $\mathfrak{p}_1, \ldots, \mathfrak{p}_r \in \operatorname{Spec} A$ such that $\mathfrak{n} = \mathfrak{p}_1 \cap \cdots \cap \mathfrak{p}_r$. If $\mathfrak{q}$ is a minimal prime ideal, then $\mathfrak{n} \subset \mathfrak{q}$ by 2.5.5. So by 2.2.5 (i), there exists i such that $\mathfrak{q} = \mathfrak{p}_i$. Thus any minimal prime ideal is one of the $\mathfrak{p}_i$'s.

(ii) We have $\sqrt{\mathfrak{a}} = \mathfrak{q}_1 \cap \cdots \cap \mathfrak{q}_s$ where $\mathfrak{q}_1, \ldots, \mathfrak{q}_s \in \operatorname{Spec} A$ are non-zero. Since A is Noetherian, by 2.5.3 (ii), for n large, $(\sqrt{\mathfrak{a}})^n \subset \mathfrak{a}$. So $(\mathfrak{q}_1 \cdots \mathfrak{q}_s)^n \subset \mathfrak{a}$. The result now follows since A is an integral domain. $\square$

2.7.9 Theorem. *Let A be a Noetherian ring and $n \in \mathbb{N}^*$. The polynomial ring over A in n variables is a Noetherian ring.*

Proof. By induction on n, it suffices to prove that $A[X]$ is Noetherian. Further, we can suppose that A is not reduced to $\{0\}$.

Suppose that there exists an ideal $\mathfrak{a}$ of A which is not finitely generated. Then it is non-zero and let $P_1 \in \mathfrak{a} \setminus \{0\}$ be such that its degree is minimal among the non-zero elements of $\mathfrak{a}$. Since $\mathfrak{a} \neq A[X]P_1$, there exists $P_2 \in \mathfrak{a} \setminus A[X]P_1$ such that its degree is minimal among the elements of $\mathfrak{a} \setminus A[X]P_1$. Repeating this process, we obtain a sequence of elements $(P_n)_{n \geqslant 1}$ of polynomials in $\mathfrak{a}$ such that, for any $n \in \mathbb{N}^*$, we have:

$$P_{n+1} \in \mathfrak{a} \setminus (A[X]P_1 + \cdots + A[X]P_n),$$
$$\deg P_{n+1} \leqslant \deg Q \text{ for all } Q \in \mathfrak{a} \setminus (A[X]P_1 + \cdots + A[X]P_n).$$

In particular, $\deg P_{n+1} \geqslant \deg P_n$ for all $n \in \mathbb{N}^*$.

Let a_n be the leading coefficient of P_n and set $\mathfrak{b}_n = Aa_1 + \cdots + Aa_n$. Then $(\mathfrak{b}_n)_{n \geqslant 1}$ is an ascending sequence of ideals of A. Since A is Noetherian, there exists $n \in \mathbb{N}^*$ such that $a_{n+1} \in \mathfrak{b}_n$. So we have $a_{n+1} = \lambda_1 a_1 + \cdots + \lambda_n a_n$ for some $\lambda_1, \ldots, \lambda_n \in A$. Let

$$Q = P_{n+1} - \sum_{i=1}^{n} \lambda_i X^{\deg P_{n+1} - \deg P_i} P_i.$$

We have $\deg Q < \deg P_{n+1}$ and $Q \in \mathfrak{a} \setminus (A[X]P_1 + \cdots + A[X]P_n)$. Thus we have obtained a contradiction. $\square$

2.8 Derivations

2.8.1 In the rest of this chapter, k will denote a commutative ring and A will denote a commutative k-algebra.

Definition. *Let M be an A-module. A* derivation *of A into M is a map $\delta : A \to M$ such that for all $a, b \in A$:*

$$\delta(a + b) = \delta(a) + \delta(b) , \quad \delta(ab) = a\delta(b) + b\delta(a).$$

Denote by $\mathrm{Der}(A, M)$ the set of such derivations. If $A = M$, we shall write $\mathrm{Der}\, A$ for $\mathrm{Der}(A, A)$, and an element of $\mathrm{Der}\, A$ is called a derivation *of A.*

2.8.2 Proposition. *Let $\delta \in \mathrm{Der}(A, M)$, $x \in A$ and $n \in \mathbb{N}^*$. Then:*
(i) $\delta(1) = 0$.
(ii) $\delta(x^n) = nx^{n-1}\delta(x)$.
(iii) *If x is invertible, then $\delta(x^{-1}) = -x^{-2}\delta(x)$.*

Proof. Since $\delta(1) = \delta(1.1) = \delta(1) + \delta(1)$, part (i) follows. Part (ii) is a simple induction on n and part (iii) follows directly from (i). $\square$

2.8.3 Observe that if $\delta \in \mathrm{Der}(A, M)$, then the set B of elements $a \in A$ such that $\delta(x) = 0$, is a subring of A and δ is B-linear. Conversely, if B is a subring of A such that δ is B-linear, then $\delta(b) = 0$ for all $b \in B$.

2.8.4 Definition. *A* derivation *of A over k into M is a derivation δ of A into M such that δ is k-linear. The set of such derivations will be denoted by $\mathrm{Der}_k(A, M)$. If $A = M$, then we write $\mathrm{Der}_k A$ for $\mathrm{Der}_k(A, A)$ and an element of $\mathrm{Der}_k A$ is called a* derivation *of A over k.*

2.8.5 Remarks. 1) The partial derivations of the ring $k[T_1, \ldots, T_n]$ are k-derivations of $k[T_1, \ldots, T_n]$.

2) Let $\delta \in \mathrm{Der}_k(A, M)$ and $a \in A$. Then the map $A \to M$ defined by $b \mapsto a\delta(b)$ is a k-derivation of A with values in M. This induces a natural structure of A-module on $\mathrm{Der}_k(A, M)$.

3) If B is a commutative A-algebra and $\delta \in \mathrm{Der}(A, B)$, then for $b \in B$, the map $A \to B$ given by $a \mapsto b\delta(a)$ is a derivation of A into B. So $\mathrm{Der}(A, B)$ admits a natural structure of B-module.

4) Let B be a commutative k-algebra, $u : A \to B$ be a homomorphism of k-algebras and N a B-module. We can consider N as an A-module via u. If $\delta \in \mathrm{Der}_k(B, N)$, then we verify easily that $\delta \circ u \in \mathrm{Der}_k(A, N)$. The map $C(u) : \mathrm{Der}_k(B, N) \to \mathrm{Der}_k(A, N)$, $\delta \mapsto \delta \circ u$ is a homomorphism of A-modules. We claim that $\ker C(u) = \mathrm{Der}_A(B, N)$, that is, the following sequence of A-modules is exact:

(2) $0 \to \mathrm{Der}_A(B, N) \to \mathrm{Der}_k(B, N) \to \mathrm{Der}_k(A, N).$

Clearly $\mathrm{Der}_A(B, N) \subset \ker C(u)$ since $\delta(u(a)) = u(a)\delta(1) = 0$. Conversely, if $\delta \in \ker C(u)$, then $\delta(u(a)b) = u(a)\delta(b)$. So $\ker C(u) = \mathrm{Der}_A(B, N)$.

2.8.6 Proposition. *Let A be an integral domain, F the quotient field of A and N a F-vector space (considered also as an A-module). Then any derivation of A over k into N can be extended in a unique way to a derivation of F over k into N.*

Proof. Let $\delta \in \mathrm{Der}_k(A, N)$ and $a, b \in A$, $c, d \in A \setminus \{0\}$ be such that $ad = bc$. Then $a\delta(d) + d\delta(a) = b\delta(c) + c\delta(b)$. It follows that:

$$
\begin{aligned}
d^2 b\delta(a) - d^2 a\delta(b) &= bd^2\delta(a) - bcd\delta(b) \\
&= b^2 d\delta(c) + bcd\delta(b) - abd\delta(d) - bcd\delta(b) \\
&= b^2 d\delta(c) - b^2 c\delta(d).
\end{aligned}
$$

Define $\Delta : F \to N$ as follows, if $\alpha = a/b$, then:

$$
\Delta(\alpha) = \delta(a)/b - a\delta(b)/b^2.
$$

The preceding computation implies that Δ is well-defined and we verify easily that $\Delta \in \mathrm{Der}_k(F, N)$.

Now if $\Delta' \in \mathrm{Der}_k(F, N)$ extends δ to F, then by applying 2.8.2 (iii):

$$
\Delta'(a/b) = a\Delta'(1/b) + \Delta'(a)/b = \Delta'(a)/b - a\Delta'(b)/b^2 = \delta(a)/b - a\delta(b)/b^2.
$$

So $\Delta' = \Delta$ and we have uniqueness. $\square$

2.9 Module of differentials

2.9.1 Recall that A is a commutative k-algebra. If M, N are A-modules, then we shall write $\mathrm{Hom}_A(M, N)$ the set of homomorphisms of A-modules from M to N.

By a *universal derivation for A over k*, we mean a pair (Ω, d) verifying the following properties:

(i) Ω is an A-module and $d \in \mathrm{Der}_k(A, \Omega)$.

(ii) For any A-module M and any $\delta \in \mathrm{Der}_k(A, M)$, there exists a unique $f \in \mathrm{Hom}_A(\Omega, M)$ such that $f \circ d = \delta$.

Lemma. *Let (Ω, d) and (Γ, D) be universal derivations for A over k.*

(i) *There exists a unique $\lambda \in \mathrm{Hom}_A(\Omega, \Gamma)$ such that $D = \lambda \circ d$. Further, λ is an isomorphism.*

(ii) *Let M be an A-module. If $f \in \mathrm{Hom}_A(\Omega, M)$, then $f \circ d \in \mathrm{Der}_k(A, M)$ and the map*

$$
\psi : \mathrm{Hom}_A(\Omega, M) \to \mathrm{Der}_k(A, M) , \quad f \mapsto f \circ d
$$

is an isomorphism of A-modules.

Proof. (i) By definition, there exist $\lambda \in \mathrm{Hom}_A(\Omega, \Gamma)$, $\mu \in \mathrm{Hom}_A(\Gamma, \Omega)$ such that $D = \lambda \circ d$ and $d = \mu \circ D$. So $d = \mu \circ \lambda \circ d$ and $D = \lambda \circ \mu \circ D$. By the uniqueness property, $\mu \circ \lambda = \mathrm{id}_\Omega$ and $\lambda \circ \mu = \mathrm{id}_\Gamma$.

(ii) That $f \circ d \in \mathrm{Der}_k(A, M)$ and ψ is a surjective homomorphism of A-modules are straightforward verifications. The injectivity of ψ follows from the uniqueness property. $\square$

2.9.2 The tensor product $A \otimes_k A$ has a natural structure of (A, A)-bimodule given by

$$a.(x \otimes y).b = (ax) \otimes (by)$$

for all $a, b, x, y \in A$. Since A is commutative, we can identify the above structure of (A, A)-bimodule with the structure of $A \otimes_k A$-module induced by the ring structure on $A \otimes_k A$, namely:

$$a.(x \otimes y).b = (a \otimes b)(x \otimes y)$$

Thus $A \otimes_k A$ is a left (resp. right) A-module via the ring homomorphism $A \to A \otimes_k A$, $a \mapsto a \otimes 1$ (resp. $a \mapsto 1 \otimes a$).

Denote by $m : A \otimes_k A \to A$ the k-linear map $m(x \otimes y) = xy$. The kernel $\mathcal{J}$ of m is an ideal of $A \otimes_k A$ and the ring $A \otimes_k A/\mathcal{J}$ is isomorphic to A.

Let $\delta_A : A \to A \otimes_k A$ be the map given by $x \mapsto x \otimes 1 - 1 \otimes x$. Clearly $\delta_A(x) \in \mathcal{J}$ for any $x \in A$.

Lemma. *As a left (or right) A-module, $\mathcal{J}$ is generated by the image of δ_A.*

Proof. Let $x_1, \ldots, x_r, y_1, \ldots, y_r \in A$. Then $\sum_{i=1}^r x_i \otimes y_i \in \mathcal{J}$ implies that $\sum_{i=1}^r x_i y_i = 0$. Thus the lemma follows since

$$\sum_{i=1}^r x_i \otimes y_i = -\sum_{i=1}^r x_i \delta_A(y_i) = \sum_{i=1}^r \delta_A(x_i) y_i.$$

$\square$

2.9.3 Set

$$\Omega_k(A) = \mathcal{J}/\mathcal{J}^2.$$

It is clear that $\Omega_k(A)$ is an $A \otimes_k A$-module. For $a, b, x \in A$, we have:

$$ab\delta_A(x) - b\delta_A(x)a = b\delta_A(a)\delta_A(x) \in \mathcal{J}^2.$$

In view of 2.9.2, the structures of left and right A-module on $\Omega_k(A)$ induced by the one on $A \otimes_k A$ are the same. We can therefore talk about the A-module $\Omega_k(A)$, that we call the *module of (Kähler) differentials of A over k*.

If $x \in A$, denote by $d_{A/k}(x)$ (or dx if there is no confusion), the image of $\delta_A(x)$ in $\Omega_k(A)$. We say that dx is the *differential* of x. By 2.9.2, the set of dx, $x \in A$, is a system of generators of the A-module $\Omega_k(A)$.

Theorem. (i) *We have $d_{A/k} \in \mathrm{Der}_k(A, \Omega_k(A))$.*
(ii) *The pair $(\Omega_k(A), d_{A/k})$ is a universal derivation for A over k.*

Proof. (i) Let $a, b \in A$. We have:

$$\delta_A(ab) = a\delta_A(b) + \delta_A(a)b = a\delta_A(b) + b\delta_A(a) + \alpha$$

with $\alpha \in \mathcal{J}^2$. So $d_{A/k} \in \mathrm{Der}_k(A, \Omega_k(A))$.

(ii) Let M be an A-module and $\delta \in \mathrm{Der}_k(A, M)$. Define

$$\varphi : A \otimes_k A \to M \ , \ a \otimes b \mapsto b\delta(a).$$

If $a, x \in A$, then

$$\varphi(\delta_A(x)) = \delta(x) - x\delta(1) = \delta(x),$$
$$\varphi(a\delta_A(x)) = \delta(ax) - x\delta(a) = a\delta(x).$$

So $\varphi|_{\mathcal{J}} \in \mathrm{Hom}_A(\mathcal{J}, M)$. On the other hand:

$$\varphi(\delta_A(a)\delta_A(x)) = \delta(ax) - a\delta(x) - x\delta(a) = 0.$$

We deduce that there exists $f \in \mathrm{Hom}_A(\Omega_k(A), M)$ such that $\delta = f \circ d_{A/k}$. Uniqueness follows from 2.9.2. $\square$

2.9.4 Example. Let $A = k[(T_i)_{i \in I}]$ be the ring of polynomials over k in the variables T_i, $i \in I$. Denote $d_{A/k}$ by d. We shall show that $\Omega_k(A)$ is a free A-module with basis $(dT_i)_{i \in I}$.

Let $P \in A$. Since $d \in \mathrm{Der}_k(A, \Omega_k(A))$, we have:

$$dP = \sum_{i \in I} \frac{\partial P}{\partial T_i} dT_i.$$

It follows that $(dT_i)_{i \in I}$ is a set of generators of $\Omega_k(A)$.

Now let M be a free A-module with basis $(e_i)_{i \in I}$. Let $\delta \in \mathrm{Der}_k(A, M)$ be defined as follows: for $P \in A$:

$$\delta P = \sum_{i \in I} \frac{\partial P}{\partial T_i} e_i.$$

Let $f \in \mathrm{Hom}_A(\Omega_k(A), M)$ be such that $\delta = f \circ d$. Then $e_i = \delta(T_i) = f(dT_i)$ for all $i \in I$. So the family $(dT_i)_{i \in I}$ is linearly independent over A.

2.9.5 Let $u : A \to B$ be a homomorphism of commutative k-algebras.

1) By 2.9.1 and 2.9.3, for any A-module M, there exists an isomorphism of A-modules:

$$\varphi_{k,A} : \mathrm{Hom}_A(\Omega_k(A), M) \to \mathrm{Der}_k(A, M) \ , \ f \mapsto f \circ d_{A/k}.$$

2) We have $d_{B/k} \circ u \in \mathrm{Der}_k(A, \Omega_k(B))$. So there exists a unique A-module homomorphism $\Omega(u) : \Omega_k(A) \to \Omega_k(B)$ such that $\Omega(u) \circ d_{A/k} = d_{B/k} \circ u$.

3) Let $i_u : \Omega_k(A) \to \Omega_k(A) \otimes_A B$ be the canonical homomorphism $\alpha \mapsto \alpha \otimes 1$. Since $\Omega_k(B)$ is a B-module and $\Omega_k(A) \otimes_A B$ is a B-module via the right action, there exists a B-linear map

$$\Omega_0(u) : \Omega_k(A) \otimes_A B \to \Omega_k(B)$$

such that $\Omega(u) = \Omega_0(u) \circ i_u$.

4) Note that $d_{B/A}$ is a derivation over A, so it is a derivation over k. It follows that there exists a unique B-linear map Ω_u such that $d_{B/A} = \Omega_u \circ d_{B/k}$.

Let N be a B-module. From the preceding discussion, we have the following commutative diagram of B-modules:

(3)
$$
\begin{array}{ccc}
\mathrm{Hom}_B(\Omega_B(A), N) & \xrightarrow{\mathrm{Hom}(\Omega_u, \mathrm{id}_N)} & \mathrm{Hom}_B(\Omega_k(B), N) \\
{\scriptstyle \varphi_{A,B}} \downarrow & & \downarrow {\scriptstyle \varphi_{k,B}} \\
\mathrm{Der}_A(B, N) & \xrightarrow[j_u]{} & \mathrm{Der}_k(B, N)
\end{array}
$$

where j_u is the canonical injection and $\mathrm{Hom}(\Omega_u, \mathrm{id}_N)$ is the map $h \mapsto h \circ \Omega_u$.

Proposition. *The following sequence of B-linear maps is exact:*

(4)
$$ \Omega_k(A) \otimes_A B \xrightarrow{\Omega_0(u)} \Omega_k(B) \xrightarrow{\Omega_u} \Omega_A(B) \longrightarrow 0. $$

Proof. It suffices to prove that, for any B-module N, the following sequence is exact:

$$ 0 \longrightarrow \mathrm{Hom}_B(\Omega_A(B), N) \xrightarrow{h_u} \mathrm{Hom}_B(\Omega_k(B), N) \xrightarrow{\ell_u} \mathrm{Hom}_B(\Omega_k(A) \otimes_A B, N) $$

where $h_u = \mathrm{Hom}(\Omega_u, \mathrm{id}_N)$ and $\ell_u = \mathrm{Hom}(\Omega_0(u), \mathrm{id}_N)$.

We have the following commutative diagram:

(5)
$$
\begin{array}{ccc}
\mathrm{Hom}_B(\Omega_k(B), N) & \xrightarrow{\mathrm{Hom}(\Omega_0(u), \mathrm{id}_N)} & \mathrm{Hom}_B(\Omega_k(A) \otimes_A B, N) \\
{\scriptstyle \varphi_{k,B}} \downarrow & & \downarrow {\scriptstyle \varphi_{k,A} \circ s_u} \\
\mathrm{Der}_k(B, N) & \xrightarrow[C(u)]{} & \mathrm{Der}_k(A, N)
\end{array}
$$

where $C(u)$ is as defined in 2.8.5, and s_u is the canonical isomorphism:

$$ s_u = \mathrm{Hom}(i_u, \mathrm{id}_N) : \mathrm{Hom}_B(\Omega_k(A) \otimes_A B, N) \to \mathrm{Hom}_A(\Omega_k(A), N). $$

The vertical arrows of diagram (5) are isomorphisms, and therefore, in view of the diagrams (3) and (5), the result follows from the fact that the sequence (2) in 2.8.5 is exact. $\square$

2.9.6 Proposition. *Suppose that A is an integral domain. Let B be its field of fractions and $u : A \to B$ the canonical injection. Then $\Omega_0(u)$ is an isomorphism from $\Omega_k(A) \otimes_A B$ to $\Omega_k(B)$.*

Proof. Since the vertical arrows of diagram (5) are isomorphisms, it suffices to prove that the map $C(u) : \mathrm{Der}_k(B, N) \to \mathrm{Der}_k(A, N)$ is bijective for any B-vector space N. So the proposition follows from 2.8.6. $\square$

2.9.7 Corollary. *Let K be a commutative field and $E = K((T_i)_{i \in I})$ the field of fractions of the ring of polynomials in the variables $(T_i)_{i \in I}$. Then the family $(dT_i)_{i \in I}$ is a basis for the E-vector space $\Omega_K(E)$.*

Proof. This is an immediate consequence of 2.9.4 and 2.9.6. $\square$

2.9.8 Let $k_n[\mathbf{T}] = k[T_1, \ldots, T_n]$, $P_1, \ldots, P_m \in k_n[\mathbf{T}]$, $\mathfrak{a}$ the ideal generated by $P_1, \ldots, P_n$ and $A = k_n[\mathbf{T}]/\mathfrak{a}$. For $1 \leqslant i \leqslant n$, let t_i be the image of T_i in A. Set $t = (t_1, \ldots, t_n)$. Recall that the A-module $\Omega_k(A)$ is generated by $d_{A/k}(t_i)$, $1 \leqslant i \leqslant n$.

Let $(e_1, \ldots, e_n)$ be the canonical basis of A^n and $u : A^n \to \Omega_k(A)$ the surjective homomorphism given by $u(e_i) = d_{A/k}(t_i)$ for $1 \leqslant i \leqslant n$. Denote by M the submodule of A^n generated by

$$\frac{\partial P_j}{\partial T_1}(t)e_1 + \cdots + \frac{\partial P_j}{\partial T_n}(t)e_n$$

where $1 \leqslant j \leqslant m$. The following lemma extends the result of 2.9.4.

Lemma. *The kernel of u is M. In other terms, the A-modules $\Omega_k(A)$ and A^n/M are isomorphic.*

Proof. Let $\alpha : k_n[\mathbf{T}] \to A$ and $\beta : A^n \to A^n/M$ be the canonical surjections. Since $d_{A/k} \in \mathrm{Der}_k(A, \Omega_k(A))$, if $Q \in k_n[\mathbf{T}]$, we have:

$$d_{A/k}(Q(t)) = \frac{\partial Q}{\partial T_1}(t)d_{A/k}(t_1) + \cdots + \frac{\partial Q}{\partial T_n}(t)d_{A/k}(t_n).$$

In particular, by taking Q to be one of the P_i's, we obtain that $M \subset \ker u$. So there exists a unique $\mu \in \mathrm{Hom}_A(A^n/M, \Omega_k(A))$ such that $\mu \circ \beta = u$.

Let $\theta : k_n[\mathbf{T}] \to A^n$ be the k-linear map defined by:

$$\theta(Q) = \frac{\partial Q}{\partial T_1}(t)e_1 + \cdots + \frac{\partial Q}{\partial T_n}(t)e_n.$$

We have $\mathfrak{a} \subset \ker \beta \circ \theta$, and hence there exists a unique $\varphi \in \mathrm{Hom}_A(A, A^n/M)$ such that $\varphi \circ \alpha = \beta \circ \theta$. It is obvious that $\varphi \in \mathrm{Der}_k(A, A^n/M)$. It follows from 2.9.1 and 2.9.5 that there exists a unique $\psi \in \mathrm{Hom}_A(\Omega_k(A), A^n/M)$ such that $\varphi = \psi \circ d_{A/k}$. We obtain that $\psi(d_{A/k}(t_i)) = \psi \circ d_{A/k} \circ \alpha(T_i) = \beta \circ \theta(T_i) = \beta(e_i)$ for $1 \leqslant i \leqslant n$. Since $\mu \circ \beta(e_i) = d_{A/k}(t_i)$ and the $\beta(e_i)$'s generate the A-module A^n/M, the lemma follows. $\square$

References

- [8], [10], [11], [52], [56].

3

Integral extensions

In this chapter, we study the notion of integral extensions. In particular, we prove the Going Up Theorem concerning the extension of prime ideals.

All rings considered in this chapter are assumed to be commutative. Let A be a subring of a ring B. For $x_1, \ldots, x_n \in B$, we denote by $A[x_1, \ldots, x_n]$ the subring of B generated by A and $x_1, \ldots, x_n$.

3.1 Integral dependence

3.1.1 For $n \in \mathbb{N}$, denote by $\mathrm{M}_n(A)$ the A-algebra of n-by-n matrices with coefficients in A. If $M \in \mathrm{M}_n(A)$, then we write $\chi_M(X)$ its characteristic polynomial. So $\chi_M(X) = \det(X I_n - M)$ where I_n is the identity matrix.

3.1.2 Definition. *An element $x \in B$ is said to be* integral *over A if there exist $n \in \mathbb{N}^*$ and $a_0, a_1, \ldots, a_{n-1} \in A$ such that:*

$$x^n + a_{n-1}x^{n-1} + \cdots + a_1 x + a_0 = 0.$$

Such a relation is called an integral equation *or* equation of integral dependence *for x over A.*

3.1.3 Let C be a commutative A-algebra, $f : A \to C$ the canonical A-algebra homomorphism. If an element $c \in C$ is integral over $f(A)$, then we shall say that c is integral over A.

3.1.4 Theorem. *Let A be a subring of B and $x \in B$. Then the following conditions are equivalent:*

(i) *x is integral over A.*
(ii) *$A[x]$ is a finitely generated A-module.*
(iii) *There exists a subring C of B, containing $A[x]$ such that C is a finitely generated A-module.*

Proof. (i) $\Rightarrow$ (ii) Let $x^n + a_{n-1}x^{n-1} + \cdots + a_1 x + a_0 = 0$ be an integral equation for x over A. Then it is clear that $A[x]$ is generated by $1, x, \ldots, x^{n-1}$.

(ii) $\Rightarrow$ (iii) This is obvious.

(iii) $\Rightarrow$ (i) Let $\{y_1, \ldots, y_n\}$ be a system of generators of the A-module C. For $1 \leqslant i \leqslant n$, there exist $a_{ij} \in A$, $1 \leqslant j \leqslant n$, such that:

$$xy_i = a_{i1}y_1 + \cdots + a_{in}y_n.$$

Set $M = [a_{ij}] \in M_n(A)$ and for $k \in \mathbb{N}^*$, let $M^k = [a_{ijk}]$. Then we have:

$$x^k y_i = a_{i1k}y_1 + \cdots + a_{ink}y_n.$$

Now $\chi_M(M) = 0$ and so $\chi_M(x)y_i = 0$ for $1 \leqslant i \leqslant n$. Thus $\chi_M(x)u = 0$ for any $u \in C$. By setting $u = 1$, we obtain an integral equation for x over A. $\square$

3.1.5 Corollary. (i) *Let $x_1, \ldots, x_n \in B$. Suppose that x_1 is integral over A and that for $2 \leqslant i \leqslant n$, x_i is integral over $A[x_1, \ldots x_{i-1}]$. Then $A[x_1, \ldots, x_n]$ is a finitely generated A-module.*

(ii) *The set of elements of B integral over A is a subring of B which contains A.*

Proof. (i) Let us prove this by induction on n. For $n = 1$, the result follows from 3.1.4. Suppose that $n \geqslant 2$, then by induction hypothesis, $D = A[x_1, \ldots, x_{n-1}]$ is a finitely generated A-module. Let $\{y_1, \ldots, y_r\}$ be a system of generators of the A-module D. Now x_n is integral over D, so $D[x_n]$ is a finitely generated D-module. Let $\{z_1, \ldots, z_s\}$ be a system of generators of the D-module $D[x_n]$. Then clearly, $\{y_i z_j; 1 \leqslant i \leqslant r, 1 \leqslant j \leqslant s\}$ is a system of generators of the A-module $A[x_1, \ldots, x_n]$.

(ii) Let $x, y \in B$ be integral over A. Since $xy, x + y \in A[x, y]$, the result follows from (i) and 3.1.4. $\square$

3.1.6 Definition. *Let A be a subring of B. The set C of elements of B integral over A is called the* integral closure *of A in B. By 3.1.5, C is a subring of B which contains A. If $C = B$, we shall say that B is* integral *over A. If $C = A$, we shall say that A is* integrally closed *in B.*

3.1.7 Proposition. *Let A and D be subrings of a ring B.*

(i) *Suppose that $A \subset D$. If D is integral over A, and B is integral over D, then B is integral over A.*

(ii) *If C is the integral closure of A in B, then C is integrally closed in B.*

Proof. Let $x \in B$ and $x^n + d_{n-1}x^{n-1} + \cdots + d_1 x + d_0 = 0$ be an integral equation of x over D. By 3.1.5, the A-module $A[d_0, \ldots, d_{n-1}, x]$ is finitely generated. So x is integral over A. This proves (i), and part (ii) follows. $\square$

3.1.8 Proposition. *Let $\mathfrak{b}$ be an ideal of B, $\mathfrak{a} = \mathfrak{b} \cap A$ and S a multiplicative subset of A. Let us suppose that B is integral over A. Then:*

(i) *$B/\mathfrak{b}$ is integral over $A/\mathfrak{a}$.*

(ii) *$S^{-1}B$ is integral over $S^{-1}A$.*

Proof. (i) Let us denote $x + \mathfrak{b}$ by $\overline{x}$, and identify $A/\mathfrak{a}$ as a subring of $B/\mathfrak{b}$. If $x^n + a_{n-1}x^{n-1} + \cdots + a_1 x + a_0 = 0$ is an integral equation of x over A, then $\overline{x}^n + \overline{a_{n-1}}\overline{x}^{n-1} + \cdots + \overline{a_1 x} + \overline{a_0} = 0$ is an integral equation of $\overline{x}$ over $A/\mathfrak{a}$.

(ii) By 2.3.6, we identify $S^{-1}A$ as a subring of $S^{-1}B$. Let $x/s \in S^{-1}B$. If $x^n + a_{n-1}x^{n-1} + \cdots + a_1 x + a_0 = 0$ is an integral equation of x over A, then $(x/s)^n + (a_{n-1}/s)(x/s)^{n-1} + \cdots + (a_1/s^{n-1})(x/s) + a_0/s^n = 0$ is an integral equation of x/s over $S^{-1}A$. $\square$

3.1.9 Proposition. *Let S be a multiplicative subset of A and C the integral closure of A in B. Then the integral closure of $S^{-1}A$ in $S^{-1}B$ is $S^{-1}C$. In particular, if A is integrally closed in B, then $S^{-1}A$ is integrally closed in $S^{-1}B$.*

Proof. By 3.1.8, the integral closure of $S^{-1}A$ in $S^{-1}B$ contains $S^{-1}C$. Now, if $x/s \in S^{-1}B$ is integral over $S^{-1}A$, then there exist $a_0, \ldots, a_{n-1} \in A$, $t_0, \ldots, t_{n-1} \in S$ such that

$$x^n/s^n + (a_{n-1}/t_{n-1})(x^{n-1}/s^{n-1}) + \cdots + (a_1/t_1)(x/s) + a_0/t_0 = 0.$$

Set $b = t_0 \cdots t_{n-1}x$ and $u = (st_0 \cdots t_{n-1})^n$, then

$$(b^n + s_{n-1}a_{n-1}b^{n-1} + \cdots + s_1 a_1 b + s_0 a_0)/u = 0$$

where $s_0, \ldots, s_{n-1} \in S$. This implies that there exists $v \in S$ such that:

$$v(b^n + s_{n-1}a_{n-1}b^{n-1} + \cdots + s_1 a_1 b + s_0 a_0) = 0.$$

From this, we deduce that $vb = vt_0 \cdots t_{n-1}x \in C$ and so $x/s \in S^{-1}C$. $\square$

3.2 Integrally closed domains

3.2.1 Lemma. *Let E be an A-module. Then the following conditions are equivalent:*

(i) *The module E is $\{0\}$.*

(ii) *For any prime ideal $\mathfrak{p}$ of A, $E_\mathfrak{p} = \{0\}$.*

(iii) *For any maximal ideal $\mathfrak{m}$ of A, $E_\mathfrak{m} = \{0\}$.*

Proof. The implications (i) $\Rightarrow$ (ii) $\Rightarrow$ (iii) are obvious. Suppose now that (iii) is satisfied. If there exists $x \in E \setminus \{0\}$, then $\mathfrak{a} = \{a \in A; ax = 0\}$ is a proper ideal of A. Let $\mathfrak{m}$ be a maximal ideal containing $\mathfrak{a}$. Since $E_\mathfrak{m} = \{0\}$, we have $x/1 = 0$, or equivalently, there exists $s \in A \setminus \mathfrak{m}$ such that $sx = 0$ which implies that $s \in \mathfrak{a} \subset \mathfrak{m}$. We have therefore a contradiction. $\square$

3.2.2 Lemma. *Let A be an integral domain. Then*

$$A = \bigcap_{\mathfrak{m} \in \mathrm{Spm}\, A} A_\mathfrak{m}.$$

Proof. For any $\mathfrak{m} \in \operatorname{Spm} A$, $A_\mathfrak{m}$ is a subring of its quotient field K by 2.3.5. The intersection B of the $A_\mathfrak{m}$'s is therefore a subring of K containing A.

Let $\mathfrak{m} \in \operatorname{Spm} A$, $S = A \setminus \mathfrak{m}$. We have $A_\mathfrak{m} \subset S^{-1}B \subset A_\mathfrak{m}$. So $A_\mathfrak{m} = S^{-1}B$. It follows from 2.4.3 that $S^{-1}(B/A) = \{0\}$. So by 3.2.1, $B/A = \{0\}$. $\square$

3.2.3 Definition. (i) *Let A be an integral domain and K its quotient field. By the* integral closure of A, *we mean the integral closure of A in K.*

(ii) *We say that A is* integrally closed domain *if A is an integral domain, and it is equal to its integral closure.*

3.2.4 Lemma. *Let B be an integral domain and $(A_i)_{i \in I}$ a family of subrings which are integrally closed domains. Then $A = \bigcap_{i \in I} A_i$ is an integrally closed domain.*

Proof. Let K, H, K_i be the field of fractions of A, B, A_i respectively. We have $K \subset K_i \subset H$. If $x \in K$ is integral over A, then it is integral over A_i for all $i \in I$. Thus $x \in A_i$ for all $i \in I$. $\square$

3.2.5 Proposition. *Let A be an integral domain. The following conditions are equivalent:*

(i) *The ring A is an integrally closed domain.*

(ii) *For any $\mathfrak{p} \in \operatorname{Spec} A$, $A_\mathfrak{p}$ is an integrally closed domain.*

(iii) *For any $\mathfrak{m} \in \operatorname{Spm} A$, $A_\mathfrak{m}$ is an integrally closed domain.*

Proof. We have (i) $\Rightarrow$ (ii) by 3.1.9. The implication (ii) $\Rightarrow$ (iii) is obvious. Finally, (iii) $\Rightarrow$ (i) follows from 3.2.2 and 3.2.4. $\square$

3.2.6 Lemma. *Let A be an integrally closed domain, K its quotient field and $P, Q \in K[X]$ monic polynomials such that $PQ \in A[X]$. Then $P, Q \in A[X]$.*

Proof. Let L be an algebraic closure of K. Then

$$P(X) = (X - \alpha_1) \cdots (X - \alpha_m) , \; Q(X) = (X - \beta_1) \cdots (X - \beta_n),$$

with $\alpha_1, \ldots, \alpha_m, \beta_1, \ldots, \beta_n \in L$. The α_i's and the β_j's are integral over A because they are roots of $PQ \in A[X]$. Since the coefficients of P and Q are polynomials in the α_i's and the β_i's with coefficients in A, it follows from 3.1.5 that they are integral over A. Hence $P, Q \in A[X]$. $\square$

3.2.7 Lemma. *If K is a commutative field, then $K[X]$ is an integrally closed domain.*

Proof. Let $P, Q \in K[X] \setminus \{0\}$ be relatively prime and let us suppose that P/Q is integral over $K[X]$. Let $(P/Q)^n + F_{n-1}(P/Q)^{n-1} + \cdots + F_0 = 0$ be an integral equation of P/Q over $K[X]$. Suppose further that n is minimal. So $F_0 \neq 0$, and we have:

$$P^n + F_{n-1}P^{n-1}Q + \cdots + F_0 Q^n = 0.$$

It follows that Q divides P^n and so Q is invertible in $K[X]$ because P, Q are relatively prime. Thus $Q \in K \setminus \{0\}$ which implies that $P/Q \in K[X]$. $\square$

3.2.8 Theorem. *Let A be an integrally closed domain, then $A[X]$ is also an integrally closed domain.*

Proof. Denote by K and L, the quotient field of A and $A[X]$ respectively. Let $\alpha \in L$ be integral over $A[X]$. Then it is integral over $K[X]$. So by 3.2.7, $\alpha \in K[X]$. Let $Q = T^m + p_{m-1}T^{m-1} + \cdots + p_0 \in A[X][T]$ be such that $Q(\alpha)$ is an integral equation of α over $A[X]$. Let $s \in \mathbb{N}$ be such that $\deg \alpha < s$ and set $\beta = \alpha - X^s$. Then $-\beta$ is monic and β is a root of $Q(T + X^s) = T^m + q_{m-1}T^{m-1} + \cdots + q_0$ where $q_0, \ldots, q_{m-1} \in A[X]$. Now

$$q_0 = Q(X^s) = X^{sm} + p_{m-1}X^{s(m-1)} + \cdots + p_0,$$

and therefore q_0 is monic for s large. But β is a root of $Q(T + X^s)$, so:

$$q_0 = -\beta(\beta^{m-1} + q_{m-1}\beta^{m-2} + \cdots + q_1) = -\beta\gamma,$$

where $\gamma \in A[X]$. Thus γ is monic in $K[X]$ for s large. Since $q_0 \in A[X]$, it follows from 3.2.6 that $-\beta \in A[X]$. Hence $\alpha \in A[X]$. $\square$

3.3 Extensions of prime ideals

3.3.1 Proposition. *Let A be a subring of an integral domain B such that B is integral over A. Then A is a field if and only if B is a field.*

Proof. Suppose that A is a field. Let $b \in B \setminus \{0\}$ and $b^n + a_{n-1}b^{n-1} + \cdots + a_0 = 0$ an integral equation of b over A. Since B is an integral domain, we can suppose that $a_0 \neq 0$. Then in $\mathrm{Fract}(B)$:

$$b^{-1} = -(a_0^{-1}b^{n-1} + a_0^{-1}a_{n-1}b^{n-2} + \cdots a_0^{-1}a_1) \in B.$$

Now let us suppose that B is a field. Let $a \in A \setminus \{0\}$, so $a^{-1} \in B$. Let $a^{-n} + a_{n-1}a^{-(n-1)} + \cdots + a_0 = 0$ be an integral equation of a^{-1} over A. Then $a^{-1} = -(a_{n-1} + a_{n-2}a + \cdots + a_0 a^{n-1}) \in A$. $\square$

3.3.2 Corollary. *Let A be a subring of a ring B such that B is integral over A and $\mathfrak{q}, \mathfrak{q}' \in \mathrm{Spec}\, B$.*
   (i) *We have $\mathfrak{q} \in \mathrm{Spm}\, B$ if and only if $\mathfrak{p} = \mathfrak{q} \cap A \in \mathrm{Spm}\, A$.*
   (ii) *Suppose that $\mathfrak{q} \subset \mathfrak{q}'$. Then $A \cap \mathfrak{q} = A \cap \mathfrak{q}'$ if and only if $\mathfrak{q} = \mathfrak{q}'$.*

Proof. (i) By 3.1.8, $B/\mathfrak{q}$ is integral over $A/\mathfrak{p}$. So (i) follows from 3.3.1.
   (ii) Suppose that $\mathfrak{p} = A \cap \mathfrak{q} = A \cap \mathfrak{q}'$. Set $S = A \setminus \mathfrak{p}$. Since $S \cap \mathfrak{q} = S \cap \mathfrak{q} = \emptyset$, by 2.3.8, $\mathfrak{r} = S^{-1}\mathfrak{q}$ and $\mathfrak{r}' = S^{-1}\mathfrak{q}'$ are prime ideals of $S^{-1}B$ such that $\mathfrak{r} \subset \mathfrak{r}'$.
   Since $\mathfrak{p} \subset \mathfrak{q} \subset \mathfrak{q}'$, we have $S^{-1}\mathfrak{p} \subset \mathfrak{r} \subset \mathfrak{r}'$. Hence $S^{-1}\mathfrak{p} \subset \mathfrak{r} \cap A_\mathfrak{p} \subset \mathfrak{r}' \cap A_\mathfrak{p}$. But $A_\mathfrak{p}$ is a local ring with maximal ideal $S^{-1}\mathfrak{p}$, so $S^{-1}\mathfrak{p} = \mathfrak{r} \cap A_\mathfrak{p} = \mathfrak{r}' \cap A_\mathfrak{p}$.
   It follows from (i) that $\mathfrak{r}, \mathfrak{r}' \in \mathrm{Spm}\, S^{-1}B$. Since $\mathfrak{r} \subset \mathfrak{r}'$, we have equality and we conclude by 2.3.7 that $\mathfrak{q} = \mathfrak{q}'$. $\square$

3.3.3 Theorem. (Going Up Theorem) *Let A be a subring of a ring B such that B is integral over A.*

(i) *If $\mathfrak{p} \in \operatorname{Spec} A$, then there exists $\mathfrak{q} \in \operatorname{Spec} B$ lying above $\mathfrak{p}$.*

(ii) *Let $\mathfrak{p}_1, \mathfrak{p}_2 \in \operatorname{Spec} A$ and $\mathfrak{q}_1 \in \operatorname{Spec} B$ lying above $\mathfrak{p}_1$. Suppose that $\mathfrak{p}_1 \subset \mathfrak{p}_2$. Then there exists $\mathfrak{q}_2 \in \operatorname{Spec} B$ lying above $\mathfrak{p}_2$ such that $\mathfrak{q}_1 \subset \mathfrak{q}_2$.*

Proof. (i) Let $S = A \setminus \mathfrak{p}$, $i : A \to A_{\mathfrak{p}}$, $j : B \to S^{-1}B$ the canonical homomorphisms, $\lambda : A \to B$ and $\mu : A_{\mathfrak{p}} \to S^{-1}B$ the canonical injections. By 2.3.6, we have $j \circ \lambda = \mu \circ i$.

By 3.1.8, $S^{-1}B$ is integral over $A_{\mathfrak{p}}$. Let $\mathfrak{n} \in \operatorname{Spm} S^{-1}B$, then by 3.3.2, $\mathfrak{n} \cap A_{\mathfrak{p}} \in \operatorname{Spm} A_{\mathfrak{p}}$. Since $A_{\mathfrak{p}}$ is a local ring with maximal ideal $S^{-1}\mathfrak{p}$, we have $S^{-1}\mathfrak{p} = \mathfrak{n} \cap A_{\mathfrak{p}}$. So $\mathfrak{q} = j^{-1}(\mathfrak{n}) \in \operatorname{Spec} B$ and $\mathfrak{q} \cap A = i^{-1}(S^{-1}\mathfrak{p}) = \mathfrak{p}$.

(ii) By 3.1.8, $B' = B/\mathfrak{q}_1$ is integral over $A' = A/\mathfrak{p}_1$. Denote by $p : A \to A'$ and $q : B \to B'$ the canonical surjections, $\lambda : A \to B$ and $\mu : A' \to B'$ the canonical injections. We have $q \circ \lambda = \mu \circ p$.

Now $p(\mathfrak{p}_2) \in \operatorname{Spec} A'$. It follows from (i) that there exists $\mathfrak{q}'_2 \in \operatorname{Spec} B'$ such that $A' \cap \mathfrak{q}'_2 = p(\mathfrak{p}_2)$. So $\mathfrak{q}_2 = q^{-1}(\mathfrak{q}'_2) \in \operatorname{Spec} B$ satisfies $\mathfrak{q}_1 \subset \mathfrak{q}_2$ and $A \cap \mathfrak{q}_2 = p^{-1}(A' \cap \mathfrak{q}'_2) = \mathfrak{p}_2$. $\square$

3.3.4 Theorem. *Let A be a subring of a ring B such that B is integral over A. Then the following conditions are equivalent:*

(i) *The dimension of A is finite.*

(ii) *The dimension of B is finite.*

Further, if these conditions are satisfied, then $\dim A = \dim B$.

Proof. By 3.3.3, any chain of length n of prime ideals of A gives a chain of length n of prime ideals of B. So $\dim A \leqslant \dim B$. Conversely, by using 3.3.2 (ii), we obtain that a chain of length m of prime ideals of B gives a chain of length m of prime ideals of A, so $\dim B \leqslant \dim A$. $\square$

3.3.5 Let A be a subring of a ring B. We say that B is a *finite A-algebra* if B is a finitely generated A-module. By 3.1.4, a finite A-algebra is integral over A. Conversely, by 3.1.5, a finitely generated A-algebra B, integral over A is a finite A-algebra.

In general, let C be an A-algebra and $f : A \to C$ the ring homomorphism inducing the A-algebra structure on C. We say that C is a *finite A-algebra* if C is a finite $f(A)$-algebra.

3.3.6 Lemma. *Let K be a subring of A. Suppose that K is a field and A is integral over K.*

(i) *We have $\operatorname{Spm} A = \operatorname{Spec} A$.*

(ii) *If A is a finite K-algebra, then $\operatorname{Spec} A$ is a finite set.*

Proof. (i) By 3.3.4, $\dim A = 0$. So $\operatorname{Spec} A = \operatorname{Spm} A$.

(ii) Since $\operatorname{Spec} A = \operatorname{Spm} A$, any prime ideal is a minimal prime ideal. If A is a finite K-algebra, then A is Noetherian by 2.7.5. So the result follows from 2.7.8 (i). $\square$

3.3.7 Proposition. *Let A be a subring of a ring B and $\mathfrak{p} \in \operatorname{Spec} A$. If B is a finite A-algebra, then the set of prime ideals lying above $\mathfrak{p}$ is finite.*

Proof. Let $S = A \setminus \mathfrak{p}$. In view of 2.3.9, we can replace A, B, $\mathfrak{p}$ by $S^{-1}A$, $S^{-1}B$ (which is a finite $S^{-1}A$-algebra) and $S^{-1}\mathfrak{p}$. So A is now a local ring with maximal ideal $\mathfrak{p}$. Next, replace A and B by $A/\mathfrak{p}A$ and $B/\mathfrak{p}B$ so that we can apply 3.3.6 since A is now a field and B a finite A-algebra. □

References

- [10], [11], [52], [56].

4

Factorial rings

All rings considered in this chapter are commutative. For a ring A, we shall denote by $U(A)$ the group of invertible elements of A.

4.1 Generalities

4.1.1 Let A be a ring. If $a \in A$, then we shall denote the ideal Aa by (a). An ideal of this form is called *principal*. Let us denote by $\mathcal{P}_A$ the set of principal ideals of A, and $\mathcal{P}_A^*$ the set of non-zero principal ideals of A. These sets are partially ordered by inclusion.

4.1.2 Let us suppose that A is an integral domain. Let $a, b \in A$. Then we say that a *divides* b or b is a *multiple* of a if there exists $c \in A$ such that $ac = b$. We shall write $a \mid b$ (resp. $a \nmid b$) if a divides (resp. does not divide) b. It is clear that
$$a \mid b \Leftrightarrow (b) \subset (a).$$
Note that 0 divides b if and only if $b = 0$. Further, since A is an integral domain, c is unique when $a \neq 0$. We shall call c the *quotient* of b by a.

4.1.3 Proposition. *Let A be an integral domain. The following conditions are equivalent for $a, b \in A$:*
  (i) *There exists $u \in U(A)$ such that $b = au$.*
  (ii) *$a \mid b$ and $b \mid a$.*
  (iii) *$(a) = (b)$.*
When these conditions are verified, we say that a and b are equivalent, *and we shall write $a \sim b$.*

Proof. This is straightforward. $\square$

4.1.4 Definition. *An element a in an integral domain A is called* irreducible *if the following conditions are satisfied:*
  (i) *$a \notin U(A)$.*
  (ii) *If $b, c \in A$ are such that $a = bc$, then $b \in U(A)$ or $c \in U(A)$.*

4.1.5 Proposition. *Let a be a non-zero element in an integral domain A. If (a) is a prime ideal, then a is an irreducible element of A.*

Proof. Since (a) is a prime ideal, $(a) \neq A$, so $a \notin U(A)$.

Now let $b, c \in A$ be such that $a = bc$. Then $b \in (a)$ or $c \in (a)$, *i.e.* $a \mid b$ or $a \mid c$. By 4.1.3, this implies that either $c \in U(A)$ or $b \in U(A)$. $\square$

4.1.6 Definition. *Let $\mathbf{a} = (a_i)_{i \in I}$ be a non-empty family of elements of an integral domain A. Denote by $\mathfrak{U}$ the family $(Aa_i)_{i \in I}$ in $\mathcal{P}_A$.*

(i) *If $\mathfrak{U}$ has a least upper bound $\mathfrak{d}$ in $\mathcal{P}_A$, then a generator δ of $\mathfrak{d}$ is called a* greatest common divisor, *abbreviated to* gcd *of $\mathbf{a}$. We shall write:*

$$(1) \qquad\qquad \delta = \gcd((a_i)_{i \in I}) = \curlywedge_{i \in I}\, a_i.$$

(ii) *If $\mathfrak{U}$ has a greatest lower bound $\mathfrak{d}$ in $\mathcal{P}_A$, then a generator μ of $\mathfrak{m}$ is called a* least common multiple, *abbreviated to* lcm *of $\mathbf{a}$. We shall write:*

$$(2) \qquad\qquad \mu = \operatorname{lcm}((a_i)_{i \in I}) = \curlyvee_{i \in I}\, a_i.$$

4.1.7 Remarks. 1) The notations in (1) and (2) are not rigorous since δ (resp. μ) is not unique. However, the ideals $\mathfrak{d}$ and $\mathfrak{m}$ are unique. So by 4.1.3, a generator of $\mathfrak{d}$ (resp. $\mathfrak{m}$) is unique up to multiplication by an element in $U(A)$.

2) By convention, when I is empty, we set $\mathfrak{d} = \{0\}$, $\delta = 0$, $\mathfrak{m} = A$ and $\mu = 1$.

4.1.8 Proposition. *Let $\mathbf{a} = (a_i)_{i \in I}$, δ, μ be elements of an integral domain A.*

(i) *δ is the gcd of the family $\mathbf{a}$ if and only if δ divides each a_i, and if any common divisor of the a_i's is a divisor of δ.*

(ii) *μ is the lcm of the family $\mathbf{a}$ if and only if μ is a multiple of each a_i, and if any common multiple of the a_i's is a multiple of δ.*

Proof. We can assume that I is non-empty.

If $\delta = \gcd(\mathbf{a})$, then $(a_i) \subset (\delta)$, thus δ divides a_i for all $i \in I$. Now if r is a common divisor of the a_i's, then $(a_i) \subset (r)$. It follows from the definition of δ that $(\delta) \subset (r)$, which implies that r divides δ.

Conversely, if δ divides each a_i, and if any common divisor r of the a_i's is a divisor of δ, then $(a_i) \subset (\delta)$ for all $i \in I$ and $(\delta) \subset (r)$. It follows that $\delta = \gcd(\mathbf{a})$.

This proves (i). The proof of part (ii) is analogue. $\square$

4.1.9 Definition. *Let $\mathbf{a} = (a_i)_{i \in I}$ be elements of an integral domain A.*

(i) *We say that the a_i's are* relatively prime *if $\gcd(\mathbf{a}) = 1$.*

(ii) *We say that the a_i's are* pairwise relatively prime *if $\gcd(a_i, a_j) = 1$ for all $i, j \in I$, $i \neq j$.*

4.1.10 Proposition. *Let a, b be non-zero elements in an integral domain A. If a is irreducible, then a, b are relatively prime if and only if a does not divide b.*

Proof. If a divides b, then a is a greatest common divisor of a and b. So a, b are not relatively prime since $a \notin U(A)$.

Conversely, if a does not divide b, then any common divisor d of a and b must be invertible since it divides a. It follows that a and b are relatively prime. $\square$

4.2 Unique factorization

4.2.1 Definition. *A commutative ring A is called* factorial *(or a* unique factorization domain*) if it is an integral domain and if it verifies the following conditions:*

(i) *For any $a \in A \setminus \{0\}$, there exist $m \in \mathbb{N}$, $p_1, \ldots, p_m$ irreducible elements in A, and $u \in U(A)$ such that:*

$$(3) \qquad\qquad a = up_1 \cdots p_m.$$

(ii) *Let $m, n \in \mathbb{N}$, $p_1, \ldots, p_m, q_1, \ldots, q_n$ irreducible elements in A and $u, v \in U(A)$ be such that*

$$up_1 \cdots p_m = vq_1 \cdots q_n.$$

Then $m = n$ and there exists a permutation σ of the set $\{1, \ldots, m\}$ such that $q_i \sim p_{\sigma(i)}$ for all i. We shall express this condition by saying that the decomposition (3) is essentially unique.

4.2.2 Definition. *A subset $\mathcal{I}$ of an integral domain A is called a* complete system of irreducible elements *of A if*

(i) *Any element of $\mathcal{I}$ is irreducible.*

(ii) *If $q \in A$ is irreducible, then there exists a unique $p \in \mathcal{I}$ such that $p \sim q$.*

4.2.3 By 4.2.2, we have another definition of a factorial ring:

Definition. *Let A be an integral domain and $\mathcal{I} \subset A$ a complete system of irreducible elements of A. We say that A is* factorial *if for any $a \in A$ non-zero, we can associate a unique pair $(u_a, \nu(a))$ where $u_a \in U(A)$ and $\nu(a) : \mathcal{I} \to \mathbb{N}$, $p \mapsto \nu_p(a)$, which is zero almost everywhere such that:*

$$(4) \qquad\qquad a = u_a \prod_{p \in \mathcal{I}} p^{\nu_p(a)}.$$

We call (4) the decomposition of a into irreducible elements (relative to $\mathcal{I}$), and we call $\nu_p(a)$, the p-adic valuation of a.

4.2.4 Definition. *Let A be an integral domain.*

(i) *We say that A satisfies* Euclid's Lemma *if for $a, b, p \in A$ with p irreducible, we have*

$$p \mid ab \Rightarrow p \mid a \text{ or } p \mid b.$$

(ii) *We say that A satisfies* Gauss' Lemma *if for $a, b, c \in A$, we have:*

$$a \mid bc \text{ and } a \frown b = 1 \Rightarrow a \mid c.$$

(iii) *Let $\mathcal{I}$ be a complete system of irreducible elements of A. We say that A satisfies the condition $(\mathcal{D})$ for $\mathcal{I}$ if for any $a \in A$ non-zero, there exists a pair $(u_a, \nu(a))$ where $u_a \in U(A)$ and $\nu(a) : \mathcal{I} \to \mathbb{N}$, $p \mapsto \nu_p(a)$, such that a decomposes as in (4).*

Remark. It is clear that if A satisfies condition $(\mathcal{D})$ for $\mathcal{I}$, then it satisfies condition $(\mathcal{D})$ for any other complete system of irreducible elements of A. We shall therefore state condition $(\mathcal{D})$ without specifying $\mathcal{I}$.

4.2.5 Theorem. *Let A be an integral domain satisfying condition $(\mathcal{D})$. Then the following conditions are equivalent:*
  (i) *A is factorial.*
  (ii) *A satisfies Euclid's Lemma.*
  (iii) *An element $p \in A \setminus \{0\}$ is irreducible if and only if the ideal (p) is prime.*
  (iv) *A satisfies Gauss' Lemma.*

Proof. Let us fix a complete system $\mathcal{I}$ of irreducible elements of A.

(i) $\Rightarrow$ (iv) Let $a, b, c \in A$ be such that $a \mid bc$ and $a \frown b = 1$. If a, b or c is 0, then clearly, $a \mid c$. So let us assume that a, b, c are non-zero. Let $d \in A$ be such that $ad = bc$. Since A is factorial, we have $\nu_p(a) + \nu_p(d) = \nu_p(b) + \nu_p(c)$ for all $p \in \mathcal{I}$. But $a \frown b = 1$, so either $\nu_p(a)$ or $\nu_p(b)$ is zero. It follows in both cases that $\nu_p(a) \leqslant \nu_p(c)$, and so $a \mid c$.

(iv) $\Rightarrow$ (ii) This is clear by 4.1.9.

(ii) $\Leftrightarrow$ (iii) This is straightforward from Euclid's Lemma and 4.1.7.

(ii) $\Rightarrow$ (i) Let $a \in A \setminus \{0\}$ and

$$a = u \prod_{p \in \mathcal{I}} p^{\nu_p(a)} = v \prod_{p \in \mathcal{I}} p^{\omega_p(a)}$$

be two decompositions of a into irreducible elements. Write $a = p^{\nu_p(a)}q = p^{\omega_p(a)}r$ where $q \frown p = 1 = r \frown p$. Then by Euclid's Lemma, we must have $\nu_p(a) = \omega_p(a)$. Now by induction, we have that $\nu_p(a) = \omega_p(a)$ for all $p \in \mathcal{I}$ and $u = v$. $\square$

4.2.6 Proposition. *Let A be a factorial ring and $\mathcal{I}$ a complete system of irreducible elements of A. Let a, b be non-zero elements of A. Then a divides b if and only if $\nu_p(a) \leqslant \nu_p(b)$ for all $p \in \mathcal{I}$.*

Proof. This is immediate by 4.2.5. $\square$

4.2.7 Theorem. *Let A be a factorial ring. Any family $\mathbf{a} = (a_i)_{i \in I}$ of elements of A has a greatest common divisor and a least common multiple.*

Proof. We can assume that $I \neq \emptyset$ and $a_i \neq 0$ for all $i \in I$. Let $\mathcal{I}$ be a complete system of irreducible elements of A.

Set $\delta = \prod_{p \in \mathcal{I}} p^{\nu_p}$ where for $p \in \mathcal{I}$, $\nu_p = \inf\{\nu_p(a_i), i \in I\}$. Set $\mu = 0$ if there exists $q \in \mathcal{I}$ such that $\nu_q(a_i)$ is not bounded, otherwise, set $\mu = \prod_{p \in \mathcal{I}} p^{\omega_p}$ where $\omega_p = \sup\{\nu_p(a_i), i \in I\}$.

By 4.1.7 and 4.2.6, it is clear that $\delta = \gcd(\mathbf{a})$ and $\mu = \operatorname{lcm}(\mathbf{a})$. $\square$

4.2.8 Proposition. *Let a, b be non-zero elements of a factorial ring A. Then*

$$(a \frown b)(a \smile b) \sim ab.$$

Proof. This is an immediate consequence of the proof of 4.2.7. $\square$

4.2.9 Theorem. *A factorial ring A is an integrally closed domain.*

Proof. Let $x \in \operatorname{Fract}(A) \setminus \{0\}$ be integral over A, and $a_0, \cdots, a_{n-1} \in A$ be such that $x^n + a_{n-1}x^{n-1} + \cdots + a_0 = 0$. Write $x = p/q$ where $p, q \in A$ and $p \frown q = 1$. Then $p^n + a_{n-1}qp^{n-1} + \cdots + a_0q^n = 0$, which implies that $p \mid q^n$. So $q \in U(A)$ and $x \in A$. $\square$

4.3 Principal ideal domains and Euclidean domains

4.3.1 Definition. *A commutative ring is called a* principal ideal domain *if it is an integral domain and if all ideals of A are principal.*

4.3.2 Remarks. 1) A principal ideal domain is Noetherian.

2) Examples of principal ideal domains are $\mathbb{Z}$ and $K[X]$, where K is a commutative field.

4.3.3 Proposition. *Let A be a principal ideal domain. Suppose that A is not a field and $\mathfrak{a}$ an ideal of A. Then the following conditions are equivalent:*

(i) *$\mathfrak{a}$ is maximal.*

(ii) *$\mathfrak{a}$ is prime and non-trivial.*

(iii) *There exists an irreducible element $p \in A$ such that $\mathfrak{a} = (p)$.*

Proof. This is straightforward from the definitions and 4.1.5. $\square$

4.3.4 Corollary. (i) *Let p be an irreducible element of a principal ideal domain A, then $A/(p)$ is a field.*

(ii) *The ring $A[X]$ is a principal ideal domain if and only if A is a field.*

Proof. Part (i) is clear from 4.3.3 and the "if" part of (ii) is also clear. Now if $A[X]$ is a principal ideal domain, then clearly A must be an integral domain. Hence $U(A[X]) = U(A)$. It follows that X is an irreducible element of $A[X]$. By part (i), $A[X]/(X) \simeq A$ is a field. $\square$

4.3.5 Lemma. *A Noetherian ring A satisfies condition $(\mathcal{D})$.*

Proof. Let E be the set of elements $a \in A \setminus (U(A) \cup \{0\})$ which are not products of irreducible elements. Suppose that $E \neq \emptyset$, and let $a_1 \in E$. In particular, a_1 is not irreducible. So $a_1 = a_2 a_2'$ where a_2, a_2' are both non-equivalent to a_1. Further, since $a_1 \in E$, one of them must be in E, say a_2. Now, by repeating the argument with a_2, we obtain a strictly increasing sequence $(a_n)_n$ of ideals of A which contradicts the hypothesis that A is Noetherian. $\square$

4.3.6 Theorem. *A principal ideal domain is factorial.*

Proof. Let A be a principal ideal domain. By 4.3.2 and 4.3.5, A satisfies condition $(\mathcal{D})$. Now the result follows from 4.3.3 and 4.2.5. $\square$

4.3.7 Let A be a principal ideal domain and $\mathcal{P}$ the set of ideals of A. If $\mathbf{a} = (a_i)_{i \in I}$ is a family of elements in A, and $\delta = \gcd(\mathbf{a})$, $\mu = \operatorname{lcm}(\mathbf{a})$, then:

$$(\delta) = \sum_{i \in I} (a_i) \, , \ (\mu) = \bigcap_{i \in I} (a_i).$$

We obtain in particular the following well-known result:

Theorem. (Bezout's Theorem) *Let a, b be elements of a principal ideal domain A and $\delta = \gcd(a, b)$. Then there exist $x, y \in A$ such that $\delta = ax + by$.*

4.3.8 Definition. *A commutative integral domain A is called an* Euclidean domain *if there exists a map $\nu : A \setminus \{0\} \to \mathbb{N}$ such that:*
  (i) *For all $x, y \in A \setminus \{0\}$ such that $x \mid y$, we have $\nu(x) \leqslant \nu(y)$.*
  (ii) *If $a, b \in A \setminus \{0\}$, there exist $q, r \in A$ such that $a = bq + r$ with $r = 0$ or $\nu(r) < \nu(b)$.*

4.3.9 The principal ideal domains $\mathbb{Z}$ and $K[X]$ are examples of Euclidean domains. In fact, we obtain easily (as for $\mathbb{Z}$) that:

Theorem. *An Euclidean domain is a principal ideal domain.*

4.3.10 Theorem. *Let A be an integral domain, S a multiplicative set of A and $B = S^{-1}A$.*
  (i) *If A is factorial, then so is B.*
  (ii) *If A is a principal ideal domain, then so is B.*
  (iii) *If A is an Euclidean domain, then so is B.*

Proof. Clearly, the statements are trivial if B is a field. So let us assume that B is not a field.

(i) Let $\mathcal{I}$ be a complete system of irreducible elements of A, and $\mathcal{I}_S$ the set of elements of $\mathcal{I}$ which do not divide any element of S. Since B is not a field, $\mathcal{I}_S \neq \emptyset$.

Let $q \in \mathcal{I}_S$. Let $a, b \in A$, $s, t \in S$ be such that $q = (a/s)(b/t)$ in B. Thus $qst = ab$ and the factoriality of A implies that q divides a or b. Let us suppose that $q \mid a$ and $a = qc$ where $c \in A$. Then $(c/s)(b/t) = 1$ and so $b/t \in U(B)$. Thus q is irreducible in B.

Further, it is easy to see that non-zero elements of B are products of elements of $\mathcal{I}_S$ and elements of $U(B)$. Thus $\mathcal{I}_S$ is a complete system of irreducibles in B.

Finally, to show that B is factorial, we shall show that (ii) of 4.2.5 is satisfied. Let $q \in \mathcal{I}_S$, $b, c \in A$, $s, t \in S$ be such that $q/1$ divides $(b/s)(c/t)$.

We have $(qd)/u = (bc)/(st)$ for some $d \in A$, $u \in S$. Thus $qdst = bcu$. Further, if $qr = u$ for some $r \in A$, then $(q/1)(r/u) = 1$ which is absurd since $q/1$ is irreducible in B. It follows that q divides bc.

By Euclid's Lemma, we have $q \mid b$ or $q \mid c$. It follows that $q/1$ divides b/s or c/t. Hence B is factorial by 4.2.5.

(ii) Let $\mathfrak{b}$ be an ideal of B and $A \cap \mathfrak{b} = \mathfrak{a}$. There exists $a \in A$ such that $\mathfrak{a} = Aa$. So by 2.3.7, $\mathfrak{b} = Ba$.

(iii) Let $\nu : A \setminus \{0\} \to \mathbb{N}$ be the map which makes A an Euclidean domain. By the proof of part (i), B is factorial and $\mathcal{I}_S$ is a complete system of irreducible elements of A. It follows that for any $x \in B \setminus \{0\}$, there exist unique $u_x \in U(B)$ and a_x, a product of irreducibles in $\mathcal{I}_S$, such that $x = a_x u_x$.

Define $\nu_S : B \setminus \{0\} \to \mathbb{N}$ by $\nu_S(x) = \nu(a_x)$. This is a well-defined map. On verifies easily that $\nu_S(x) = 1$ if $x \in U(B)$ and $\nu_S(x) \leqslant \nu_S(y)$ if $x \mid y$.

Finally, if $x, y \in B \setminus \{0\}$ and $x = u_x a_x$, $y = u_y a_y$ as above. Then there exist $q, r \in A$ such that $a_x = qa_y + r$ with $r = 0$ or $\nu(r) < \nu(a_y)$. It follows that $x = (qu_x/u_y)y + u_x r$ with $u_x r = 0$ or $\nu_S(u_x r) = \nu_S(r) \leqslant \nu(r) < \nu(a_y) = \nu_S(y)$. $\square$

4.4 Polynomials and factorial rings

4.4.1 Let A be a factorial ring. The *content* of a polynomial $P \in A[X] \setminus \{0\}$ is defined to be any greatest common multiple of the coefficients of P, and we shall denote it by $c(P)$. Of course, $c(P)$ is unique up to a multiple of an element of $U(A)$. A polynomial P is called *primitive* if $c(P) \in U(A)$. Clearly, $P = c(P)P_1$ where $P_1 \in A[X]$ is primitive.

4.4.2 Lemma. *Let A be a factorial ring and $P, Q \in A[X] \setminus \{0\}$. Then* $c(PQ) \sim c(P)c(Q)$.

Proof. We can assume that P, Q are primitive. If $c(PQ) \notin U(A)$, then there exists an irreducible element p of A which divides $c(PQ)$. Let $B = A/(p)$ and consider the algebra morphism $\psi : A[X] \to B[X]$ induced from the canonical surjection $A \to B$. Then $\psi(P)\psi(Q) = \psi(PQ) = 0$. Since B is an integral domain by 4.2.5, we have that either $\psi(P) = 0$ or $\psi(Q) = 0$. This implies that p divides $c(P)$ or $c(Q)$ which is absurd. $\square$

4.4.3 Proposition. *Let K be the field of fractions of a factorial ring A and $P \in A[X]$.*

(i) *If $P \in A$, then P is irreducible in $A[X]$ if and only if P is irreducible in A.*

(ii) *If $P \notin A$, then P is irreducible in $A[X]$ if and only if P is primitive in $A[X]$ and irreducible in $K[X]$.*

Proof. (i) If $P = QR$ with $Q, R \in A[X]$, then $\deg Q = \deg R = 0$. So Q, R belong to A. Since $U(A) = U(A[X])$, the result is clear.

(ii) If P is irreducible in $A[X]$, then clearly, it is primitive. Now if $P = QR$ where $Q, R \in K[X]$, then there exist primitive polynomials $Q', R' \in A[X]$ and elements $a, b, c, d \in A$ such that

$$Q = ab^{-1}Q' \; , \;\; R = cd^{-1}Q' \; , \;\; a \frown b = 1 = c \frown d.$$

Hence $bdP = acQ'R'$ and so $(bd) = (ac)$ by 4.4.2. It follows that $P = uQ'R'$ where $u \in U(A)$. Since P is irreducible in $A[X]$, either Q' or R' is in $U(A)$. This implies that P is irreducible in $K[X]$.

Conversely, suppose that P is primitive in $A[X]$ and irreducible in $K[X]$. If $P = QR$ with $Q, R \in A[X] \setminus \{0\}$, then the irreducibility of P in $K[X]$ implies that either Q or R is in K. Let us suppose that $Q \in K$, then $Q \in A \setminus \{0\}$. Now $Q \mid c(P)$. Thus $Q \in U(A)$ and we conclude that P is irreducible in $A[X]$.
$\square$

4.4.4 Corollary. *Let the notations be as in 4.4.3. If $P \notin A$ and P is irreducible in $A[X]$, then P is irreducible in $K[X]$.*

4.4.5 Theorem. *Let A be a factorial ring, and $\mathbf{X} = (X_i)_{i \in I}$ be a non-empty family of indeterminates over A. The ring $A[\mathbf{X}]$ is factorial.*

Proof. If $P \in A[\mathbf{X}]$, then there exists a finite subset J of I such that $P \in A[(X_j)_{j \in J}]$. Thus it suffices to prove the theorem for I finite. Further, by induction, we are reduced to prove that $A[X]$ is factorial.

Let $P \in A[X]$ and $n = \deg P \geqslant 0$. We shall first prove by induction on n that P is a product of irreducible elements of $A[X]$. Since A is factorial, the result is clear for $n = 0$ by part (i) of 4.4.3.

Let us suppose that $n \geqslant 1$ and P is not irreducible. By 4.4.1, we can assume further that $c(P) = 1$. By part (ii) of 4.4.3, $P = QR$ where $1 \leqslant \deg Q, \deg R < n$. Applying our induction hypothesis on Q and R, we obtain that P is a product of irreducible elements.

We shall show that $A[X]$ satisfies Euclid's Lemma, which, by 4.2.5, would imply that $A[X]$ is factorial.

Let $P, Q, R \in A[X] \setminus \{0\}$ be such that P is irreducible and $P \mid QR$. If $P \in A$, then $P \mid c(QR)$. Hence P divides $c(Q)$ or $c(R)$ since A is factorial. So P divides Q or R as required.

If $P \notin A$, then P is irreducible in $K[X]$ by 4.4.4 and $c(P) = 1$. Now, $K[X]$ is factorial, so P divides Q or R in $K[X]$. Let us suppose that $PF = Q$ where $F \in K[X]$. There exist $a, b \in A$ with $a \frown b = 1$ such that $a^{-1}bF = G \in A[X]$ is primitive. It follows that $aPG = bQ$ and so $a = bc(Q)$. Thus $b \in U(A)$ and $P \mid Q$ in $A[X]$. $\square$

4.4.6 Remarks. 1) Let K be a (commutative) field. The ring $K[X, Y]$ is a factorial ring which is not a principal ideal domain. In fact, Bezout's identity is not verified here, for $X \frown Y = 1$ and $(X) + (Y) \neq K[X, Y]$.

2) If I is an infinite set, then $A[\mathbf{X}]$ is a factorial ring which is not Noetherian.

4.4.7 Theorem. (Eisenstein's Criterion) *Let A be a factorial ring with field of fractions K. Let*

$$P = a_n X^n + a_{n-1} X^{n-1} + \cdots + a_0 \in A[X]$$

where a_n and a_0 are non-zero. Suppose that there exists an irreducible element p in A such that

  (i) $p \mid a_i$, *for* $0 \leqslant i \leqslant n-1$,
  (ii) a_n *is not divisible by* p,
  (iii) a_0 *is not divisible by* p^2.
  Then P is irreducible in $K[X]$. Further, if $c(P) = 1$, then P is irreducible in $A[X]$.

Proof. Let us write $P = c(P)(a'_n X^n + a'_{n-1} X^{n-1} + \cdots + a'_0)$. By (ii), we have $p \frown c(P) = 1$, so the polynomial $a'_n X^n + a'_{n-1} X^{n-1} + \cdots + a'_0$ satisfies conditions (i),(ii) and (iii). So by 4.4.3, we are reduced to the case where P is primitive.

If P is not irreducible in $K[X]$, then by 4.4.3, we have $P = QR$ with $Q = b_q X^q + b_{q-1} X^{q-1} + \cdots + b_0, R = c_r X^r + c_{r-1} X^{r-1} + \cdots + c_0 \in A[X]$, where b_q and c_r are non-zero, $q, r \in \mathbb{N}^*$ and $q + r = n$.

Since $a_0 = b_0 c_0$, conditions (i) and (iii) imply that p divides only one of b_0, c_0. Let us suppose that p divides b_0 and that c_0 is not divisible by p. Let $s = \min\{i \text{ such that } p \nmid b_i\} > 0$. This is well-defined since $c(P) = 1$, so $c(Q) \in U(A)$ by 4.4.2.

We have $s \leqslant q \leqslant n-1$. But $a_s = b_0 c_s + b_1 c_{s-1} + \cdots + b_s c_0$ where we set $c_j = 0$ if $j > r$. So condition (i) implies that $p \mid b_s$ which is absurd.   □

4.4.8 Let $\mathfrak{a}$ be an ideal of A, $B = A/\mathfrak{a}$ and $\varphi : A \to B$ the canonical surjection. Denote also by φ the induced surjection $A[X] \to B[X]$. The image of $P \in A[X]$ under φ is called the *reduction of P modulo* $\mathfrak{a}$.

4.4.9 Theorem. *Let K be the field of fractions of a factorial ring A, $\mathfrak{p}$ a prime ideal of A, $B = A/\mathfrak{p}$, $L = \mathrm{Fract}(B)$, $\varphi : A[X] \to B[X]$ the reduction map as above.*

Let $P \in A[X]$ be such that $\deg P = \deg \varphi(P)$. If $\varphi(P)$ is irreducible in $B[X]$ or $L[X]$, then P is irreducible in $K[X]$.

Proof. Suppose that $P = QR$ where $Q, R \in A[X]$ with:

$$Q = b_q X^q + b_{q-1} X^{q-1} + \cdots + b_0 , \ R = c_r X^r + c_{r-1} X^{r-1} + \cdots + c_0$$

where $q + r = \deg P$. We have $b_q c_r \notin \mathfrak{p}$ since $\deg P = \deg \varphi(P)$. Further $\mathfrak{p}$ is prime, so $b_q, c_r \notin \mathfrak{p}$ which implies that $\deg Q = \deg \varphi(Q)$ and $\deg R = \deg \varphi(R)$. Now $\varphi(P) = \varphi(Q)\varphi(R)$ is irreducible in $B[X]$ or $L[X]$. We have therefore that either $\varphi(Q)$ or $\varphi(R)$ has degree 0. Thus either $\deg Q$ or $\deg R$ is zero. By 4.4.3, P is irreducible in $K[X]$.   □

4.5 Symmetric polynomials

4.5.1 In this section, A will be a ring and K a commutative field. We shall fix an integer $n > 0$ and write $A[\mathbf{X}]$ for $A[X_1, \ldots, X_n]$. For $q \in \mathbb{N}$, we denote by $\mathcal{P}_q$ the A-submodule of $A[\mathbf{X}]$ of homogeneous polynomials of degree q. We shall also denote by $\mathfrak{S}_n$ the symmetric group of $\{1, \ldots, n\}$.

4.5.2 Let $\sigma \in \mathfrak{S}_n$. For $P \in A[\mathbf{X}]$, we define the polynomial P^σ as follows:

$$P^\sigma(X_1, \ldots, X_n) = P(X_{\sigma(1)}, \ldots, X_{\sigma(n)}).$$

Clearly, P^σ is homogeneous of degree q if and only if P is homogeneous of degree q. Further, the map

$$\theta_\sigma : A[\mathbf{X}] \to A[\mathbf{X}] \ , \ P \mapsto P^\sigma$$

is an automorphism of the algebra $A[\mathbf{X}]$. We have $\theta_\sigma \circ \theta_\tau = \theta_{\sigma\tau}$ for $\sigma, \tau \in \mathfrak{S}_n$.

Definition. *A polynomial* $P \in A[\mathbf{X}]$ *is called* symmetric *if* $P^\sigma = P$ *for all* $\sigma \in \mathfrak{S}$. *We denote by* $A[\mathbf{X}]^{\mathrm{sym}}$ *the subalgebra of* $A[\mathbf{X}]$ *consisting of symmetric polynomials.*

4.5.3 For $1 \leqslant p \leqslant n$, set

$$E_p = \sum_{1 \leqslant i_1 < \cdots < i_p \leqslant n} X_{i_1} \cdots X_{i_p}.$$

This is a symmetric homogeneous polynomial of degree p. We call this polynomial *the elementary symmetric polynomial of degree p in* $X_1, \ldots, X_n$. We set $E_0 = 1$ and $E_p = 0$ if $p > n$.

We see easily that, in $A[X_1, \cdots, X_n, U, V]$, we have

$$(5) \qquad \prod_{i=1}^{n} (U + VX_i) = \sum_{k=0}^{n} U^{n-k} V^k E_k.$$

Thus, in particular, in $A[X_1, \ldots, X_n, T]$:

$$(6) \qquad \prod_{i=1}^{n} (1 + TX_i) = \sum_{k=0}^{n} E_k T^k.$$

$$(7) \qquad \prod_{i=1}^{n} (T - X_i) = \sum_{k=0}^{n} (-1)^{n-k} E_{n-k} T^k.$$

4.5.4 Lemma. *Let* $P \in A[\mathbf{X}]^{\mathrm{sym}}$ *be such that* $P(X_1, \ldots, X_{n-1}, 0) = 0$. *Then P is divisible by* E_n.

Proof. We proceed by induction on n. For $n = 1$, the result is clear. So let $n \geqslant 2$.

Since $P(X_1, \ldots, X_{n-1}, 0) = 0$, we have $P = P_1 X_n + \cdots + P_k X_n^k$ where $P_1, \ldots, P_k \in A[X_1, \ldots, X_{n-1}]$. Further since P is symmetric, the P_j's are symmetric in $X_1, \ldots, X_{n-1}$.

We have $P(X_1, \ldots, X_{n-2}, 0, X_n) = 0$, since $P^\tau = P$ where τ denotes the transposition $(n - 1, n)$. It follows that $P_j(X_1, \ldots, X_{n-2}, 0) = 0$ for all j. Thus the P_j's are divisible by $X_1 \cdots X_{n-1}$. Hence $E_n = X_1 \cdots X_n$ divides P as required. $\square$

4.5.5 Let $\nu = (\nu_1, \ldots, \nu_n) \in \mathbb{N}^n$ and $\lambda \in A \setminus \{0\}$. We define the *weight* $\pi(\lambda X_1^{\nu_1} \cdots X_n^{\nu_n})$ of the monomial $\lambda X_1^{\nu_1} \cdots X_n^{\nu_n}$ to be:

$$\pi(\lambda X_1^{\nu_1} \cdots X_n^{\nu_n}) = \nu_1 + 2\nu_2 + \cdots + n\nu_n.$$

The weight $\pi(P)$ of a non-zero polynomial P is the maximum of the weights associated to the non-zero terms in P. We set $\pi(0) = +\infty$.

4.5.6 Lemma. *Let P be a symmetric polynomial in $A[\mathbf{X}]$ of degree d. There exists $Q \in A[Y_1, \ldots, Y_n]$ such that $\pi(Q) \leqslant d$ and $P(X_1, \ldots, X_n) = Q(E_1, \ldots, E_n)$.*

Proof. We can suppose that $n \geqslant 2$ and $d \geqslant 1$. Let $e_1, \ldots, e_{n-1}$ be elementary symmetric polynomials in $X_1, \ldots, X_{n-1}$. Set $B = A[X_1, \ldots, X_{n-1}]$ and for $1 \leqslant j \leqslant n - 1$, we have:

$$e_j = E_j(X_1, \ldots, X_{n-1}, 0).$$

We shall proceed by induction on n and d.

Since $P(X_1, \ldots, X_{n-1}, 0) \in B^{\mathrm{sym}}$, we have by our induction hypothesis that there exists $Q_1 \in B$ such that:

$$\pi(Q_1) \leqslant d \text{ and } P(X_1, \ldots, X_{n-1}, 0) = Q_1(e_1, \ldots, e_{n-1}).$$

Clearly, $\deg Q_1(E_1, \ldots, E_{n-1}) \leqslant d$ and the polynomial

$$P_1(X_1, \ldots, X_n) = P(X_1, \ldots, X_n) - Q_1(E_1, \ldots, E_{n-1})$$

has degree $\leqslant d$, and $P_1(X_1, \ldots, X_{n-1}, 0) = 0$. It follows from 4.5.4 that E_n divides P_1. Let $P_2 \in A[\mathbf{X}]$ be such that $P_1 = E_n P_2$. Then $\deg P_2 < d$ and $P_2 \in A[\mathbf{X}]^{\mathrm{sym}}$. By induction hypothesis, there exists $Q_2 \in A[\mathbf{X}]$ such that:

$$\pi(Q_2) \leqslant n - d \text{ and } P_2(X_1, \ldots, X_n) = Q_2(E_1, \ldots, E_n)$$

and hence

$$P(X_1, \ldots, X_n) = Q_1(E_1, \ldots, E_{n-1}) + E_n Q_2(E_1, \ldots, E_n).$$

Finally, $Q_1(Y_1, \ldots, Y_{n-1}) + Y_n Q_2(Y_1, \ldots, Y_n)$ has weight $\leqslant d$. $\square$

4.5.7 Theorem. (i) *The algebra $A[\mathbf{X}]^{\mathrm{sym}}$ is generated by the elementary symmetric polynomials in $X_1, \ldots, X_n$.*

(ii) *The map $P \mapsto P(E_1, \ldots, E_n)$ defines an isomorphism from the polynomial ring $A[T_1, \ldots, T_n]$ to $A[\mathbf{X}]^{\mathrm{sym}}$.*

Proof. Part (i) is a corollary of 4.5.5. The map in part (ii) defines clearly a morphism, and part (i) implies that it is surjective. Let us prove the injectivity of the map by induction on n. The case $n = 1$ is trivial.

Let $n \geqslant 2$ and suppose that the map is not injective. Choose $P \in A[T_1, \ldots, T_n] \setminus A$ of minimal degree such that $P(E_1, \ldots, E_n) = 0$. Write $P = P_0 + P_1 T_n + \cdots + P_k T_n^k$ with $P_j \in A[T_1, \ldots, T_{n-1}]$.

Since $P(e_1, \ldots, e_{n-1}, 0) = P_0(e_1, \ldots, e_{n-1}) = 0$ where the e_i's are the elementary symmetric polynomials in $X_1, \ldots, X_{n-1}$, we have $P_0 = 0$ by induction. Thus $P = T_n Q$ for some $Q \in A[T_1, \ldots, T_n]$. But $\deg(Q) < \deg(P)$ and $Q(E_1, \ldots, E_n) = 0$ which contradicts the minimality of $\deg(P)$. $\square$

4.6 Resultant and discriminant

4.6.1 Let A be a ring and $P, Q \in A[X] \setminus A$ of degree m and n respectively. Let us write:

$$P = a_0 + a_1 X + \cdots + a_m X^m \ , \ Q = b_0 + b_1 X + \cdots + b_n X^n.$$

We define the *resultant* of P and Q, denoted by $\mathrm{res}(P, Q)$, to be the determinant of the following $m + n$ by $m + n$ matrix:

$$\begin{vmatrix}
a_0 & a_1 & \cdots & a_{n-1} & a_n & \cdots & a_{m-1} & a_m & 0 & \cdots & 0 \\
0 & a_0 & \cdots & a_{n-2} & a_{n-1} & \cdots & a_{m-2} & a_{m-1} & a_m & \cdots & 0 \\
0 & 0 & \cdots & a_{n-3} & a_{n-2} & \cdots & a_{m-3} & a_{m-2} & a_{m-1} & \cdots & 0 \\
\hdotsfor{11} \\
0 & 0 & \cdots & a_0 & a_1 & \cdots & \cdots & \cdots & \cdots & \cdots & a_m \\
b_0 & b_1 & \cdots & b_{n-1} & b_n & \cdots & 0 & 0 & \cdots & \cdots & 0 \\
0 & b_0 & \cdots & b_{n-2} & b_{n-1} & \cdots & 0 & 0 & \cdots & \cdots & 0 \\
0 & 0 & \cdots & b_{n-3} & b_{n-2} & \cdots & \cdots & \cdots & \cdots & \cdots & 0 \\
\hdotsfor{11} \\
0 & 0 & \cdots & 0 & 0 & \cdots & b_0 & b_1 & b_2 & \cdots & b_n
\end{vmatrix}$$

In the first n rows, we have the coefficients of P, while in the last m rows, we have the coefficients of Q.

Let us suppose that A is an integral domain. For $\lambda, \mu \in A \setminus \{0\}$, we have:

$$\mathrm{res}(\lambda P, \mu Q) = \lambda^n \mu^m \, \mathrm{res}(P, Q) \ , \ \mathrm{res}(Q, P) = (-1)^{mn} \, \mathrm{res}(P, Q).$$

Observe also that $\mathrm{res}(P, Q)$ is a homogeneous polynomial in the a_i's and the b_j's of degree n (resp. m) in the a_i's (resp. in the b_j's).

4.6.2 Lemma. *If A is factorial, then the following conditions are equivalent:*

(i) *P and Q has a non-constant common factor.*

(ii) *There exist $F, G \in A[X] \setminus \{0\}$ such that*

$$PG = QF \ , \ \deg(F) < \deg(P) \ , \ \deg(G) < \deg(Q).$$

Proof. Let $R \in A[X]$ be a non constant polynomial dividing P and Q. Then there exist $F, G \in A[X]$ such that $P = FR$ and $Q = GR$. Clearly, F and G satisfies (ii). So we have (i) $\Rightarrow$ (ii).

Conversely, let F, G be as in (ii). Since $\deg G < \deg Q$, there exists an irreducible divisor R of Q and $n \in \mathbb{N}^*$ such that R^n divides Q, but R^n does not divide G. It follows that R divides P. $\square$

4.6.3 Theorem. *Let A be factorial, then the following conditions are equivalent:*

(i) *P and Q has a non-constant common factor.*

(ii) $\operatorname{res}(P, Q) = 0$.

Proof. By 4.6.2, (i) is equivalent to the existence of non zero polynomials $F = \alpha_0 + \alpha_1 X + \cdots + \alpha_{m-1} X^{m-1}$ and $G = \beta_0 + \beta_1 X + \cdots + \beta_{n-1} X^{n-1}$ such that $PG = QF$. This means that the following homogeneous system in $x_0, \ldots, x_{m-1}, y_0, \ldots, y_{n-1}$

$$(8) \qquad \begin{cases} a_0 y_0 & = b_0 x_0 \\ a_1 y_0 + a_0 y_1 & = b_1 x_0 + b_0 x_1 \\ \quad \ddots & \quad \vdots \quad \vdots \\ a_m y_{n-1} & = b_n x_{m-1} \end{cases}$$

has a non trivial solution in A. This is equivalent to the existence of a non trivial solution in $\operatorname{Fract}(A)$. Now the determinant of (8) is $\pm \operatorname{res}(P, Q)$, so we have proved our result. $\square$

4.6.4 Corollary. *Let A be a factorial subring of a factorial ring B, and $P, Q \in A[X]$. If P, Q has a non constant common factor in $B[X]$, then P, Q has a non constant common factor in $A[X]$.*

4.6.5 Corollary. *Let A be factorial. There exist $F, G \in A[X]$ such that:*

$$\deg(F) < \deg(Q) \ , \ \deg(G) < \deg(P) \ , \ \operatorname{res}(P, Q) = PF + QG.$$

Proof. If $\operatorname{res}(P, Q) = 0$, then the assertion follows from 4.6.2 and 4.6.3. So let us suppose that $\operatorname{res}(P, Q) \neq 0$. Set

$$\mathbf{X} = {}^t(1, X, \ldots, X^{m+n-1}) \in M_{m+n,1}(A[X]),$$
$$\mathbf{Y} = {}^t(P, XP, \ldots, X^{n-1}P, Q, XQ, \ldots, X^{m-1}Q) \in M_{n+m,1}(A[X]).$$

Let $M \in M_{m+n}(A)$ be the matrix in 4.6.1 defining $\operatorname{res}(P, Q)$. We have $\mathbf{Y} = M\mathbf{X}$. Let M' be the transpose of the matrix of cofactors of M. Then,

(9) $$M'\mathbf{Y} = M'M\mathbf{X} = \text{res}(P,Q)\mathbf{X}.$$

Let $(c_0,\ldots,c_{n-1},d_0,\ldots,d_{m-1})$ be the first row of M' and set :

$$F = c_0 + c_1 X + \cdots + c_{n-1}X^{n-1} \; , \; G = d_0 + d_1 X + \cdots + d_{m-1}X^{m-1}.$$

It follows from (9) that $PF + QG = \text{res}(P,Q)$. □

4.6.6 Theorem. *Let* $\alpha_1,\ldots,\alpha_m,\beta_1,\ldots,\beta_n$ *be elements of a factorial ring* A, $a_0,b_0 \in A \setminus \{0\}$ *and* :

$$P = a_0(X - \alpha_1)\cdots(X - \alpha_n) \; , \; Q = b_0(X - \beta_1)\cdots(X - \beta_n).$$

We have :

$$\text{res}(P,Q) = a_0^n b_0^m \prod_{i=1}^{m}\prod_{j=1}^{n}(\alpha_i - \beta_j) = a_0^n \prod_{i=1}^{m} Q(\alpha_i) = (-1)^{mn} b_0^m \prod_{j=1}^{n} P(\beta_j).$$

Proof. Let $T_1,\ldots,T_m,U_1,\ldots,U_n,V_0,W_0,X$ be indeterminates over A. Consider the following polynomials in $B = A[T_1,\ldots,T_m,U_1,\ldots,U_n]$:

$$f = V_0(X - T_1)\cdots(X - T_m) = V_0 X^n + \cdots + V_{m-1}X + V_m,$$
$$g = W_0(X - U_1)\cdots(X - U_n) = W_0 X^n + \cdots + W_{n-1}X + W_n.$$

With the notations of 4.5.3, we have :

$$V_i = (-1)^i V_0 E_i(T_1,\ldots,T_m) \; , \; W_j = (-1)^j W_0 E_j(U_1,\ldots,U_n).$$

By 4.5.6, we can also view f and g as polynomials over A in the indeterminates $V_0,V_1,\ldots,V_m,W_0,W_1,\ldots,W_n,X$. Set :

$$S = V_0^n W_0^m \prod_{i=1}^{m}\prod_{j=1}^{n}(T_i - U_j).$$

Clearly, we have :

$$S = V_0^n \prod_{i=1}^{m} g(T_i) = (-1)^{mn} W_0^m \prod_{j=1}^{n} f(U_j).$$

Let $R = \text{res}(f,g)$. Since V_0 divides the V_i's, and W_0 divides the W_j's, it follows from the definition of R that $R = V_0^n W_0^m H$, where $H \in B$. On the other hand, by 4.6.3, it is clear that R is divisible by $T_i - U_j$. Hence, S divides R. Since R and S are both homogeneous of degree m (resp. n) in V_i (resp. W_j), we deduce that there exists $a \in A$ such that $R = aS$. Finally, since the coefficient of R and S in the monomial $V_0^n W_0^m$ is equal to 1, we have $R = S$. Our theorem follows immediately. □

4.6.7 Definition. *Let $P \in K[X]$ be a polynomial of degree $n \geqslant 2$ whose dominant coefficient is a_n. We define the* discriminant *of P, denoted by $\operatorname{dis}(P)$, to be the scalar:*

$$\operatorname{dis}(P) = (-1)^{n(n-1)/2} a_n^{-1} \operatorname{res}(P, P').$$

4.6.8 Proposition. *Let $P \in K[X]$ be a polynomial of degree $n \geqslant 2$ whose dominant coefficient is a_n. Suppose further that P is split over K and let $\alpha_1, \ldots, \alpha_n$ be the roots (not necessarily distinct) of P. Then*

$$\operatorname{dis}(P) = a_n^{2n-2} \prod_{i<j} (\alpha_i - \alpha_j)^2.$$

Proof. This is a direct consequence of 4.6.6 and the definition of the discriminant. □

4.6.9 Proposition. *Let $P \in K[X]$ be a polynomial of degree $n \geqslant 2$.*
  (i) *P and P' are coprime if and only if $\operatorname{dis}(P) \neq 0$.*
  (ii) *Suppose that the characteristic of K is zero and that P is split over K. Then all the roots of P are simple if and only if $\operatorname{dis}(P) \neq 0$.*

Proof. These are direct consequences of 4.6.3 and 4.6.8. □

References

- [11], [52], [56].

5

Field extensions

We study certain properties of field extensions in this chapter. In particular, we prove the well-known Going Down Theorem.

All fields considered here are assumed to be commutative. Let K be a field.

5.1 Extensions

5.1.1 An *extension* of K is a pair (L, i) where L is a field and $i : K \to L$ is a ring homomorphism. It follows that i is injective and when there is no confusion, we shall identify K with its image $i(K)$. Thus L is an extension of K if K is a subfield of L, and we shall write $K \subset L$.

We define the *degree* of an extension $K \subset L$ to be the dimension of L (finite or not) as a vector space over K. We shall denote it by $[L : K]$. An extension of finite degree is called *finite*.

5.1.2 Let $K \subset L$ be a field extension and S a subset of L. Denote by $K[S]$ (resp. $K(S)$) the subring (resp. subfield) of L generated by K and S. Clearly, $K(S)$ is the quotient field of $K[S]$. If $S = \{x_1, \ldots, x_n\}$, we shall write $K[x_1, \ldots, x_n]$ and $K(x_1, \ldots, x_n)$ for $K[S]$ and $K(S)$. An extension $K \subset L$ is *finitely generated* if there is a finite subset S in L such that $L = K(S)$. If there exists $x \in L$ such that $L = K(x)$, then the extension is called *cyclic*.

5.1.3 The following propositions are straightforward.

Proposition. *Let $K \subset L$ be a field extension, S, T subsets of L and $\mathcal{F}$ the set of finite subsets of S.*
(i) *We have $K(S \cup T) = K(S)(T) = K(T)(S)$.*
(ii) *$K(S)$ is the union of $K(U)$ for $U \in \mathcal{F}$.*

5.1.4 Proposition. *Let $K \subset L \subset M$ be field extensions.*
(i) *If $(a_i)_{i \in I}$ is a K-basis for L and $(b_j)_{j \in J}$ is a L-basis for M, then $(a_i b_j)_{i \in I, j \in J}$ is a K-basis for M.*
(ii) *The extension $K \subset M$ is finite if and only if the extensions $K \subset L$ and $L \subset M$ are finite. Further, if $K \subset M$ is finite, then*

$$[M : K] = [M : L][L : K].$$

5.2 Algebraic and transcendental elements

5.2.1 Let A be a commutative ring containing K as a subring and $S = \{x_1, \ldots, x_n\}$ a finite subset of A. We say that S is *algebraically independent over K* if $P(x_1, \ldots, x_n) \neq 0$ for all $P \in K[X_1, \ldots, X_n] \setminus \{0\}$.

A subset T of A is said to be *algebraically independent over K* if any finite subset of T is algebraically independent over K.

5.2.2 Let $K \subset L$ be a field extension and $x \in L$. We define a K-algebra homomorphism :

$$\varphi_x : K[X] \to L \ , \ P \mapsto P(x).$$

If $\ker \varphi_x = 0$, then x is algebraically independent over K, and we say that x is *transcendental* over K. In particular, φ_x induces an isomorphism between $K[X]$ and $K[x]$ which extends to an isomorphism between $K(X)$ and $K(x)$. Thus the extension $K \subset K(x)$ is infinite.

If $\ker \varphi_x \neq 0$, then x is algebraically dependent over K, and we say that x is *algebraic* over K. In this case, $\ker \varphi_x$ is a prime ideal and by 4.1.4, there exists a unique irreducible unitary polynomial P_x such that $\ker \varphi_x = (P_x)$. We shall call P_x, the *minimal polynomial* of x over K. The homomorphism φ_x then induces an isomorphism between $K[X]/(P_x)$ and $K[x] = K(x)$. Thus $\deg P_x = [K(x) : K]$ and $\{1, x, \ldots, x^{\deg P_x - 1}\}$ is a K-basis of $K(x)$.

5.2.3 Proposition. *Let $K \subset L$ be a field extension, S a subset of L algebraically independent over K and $x \in L \setminus S$. The following conditions are equivalent :*

(i) *The element x is algebraic over $K(S)$.*

(ii) *The set $S \cup \{x\}$ is algebraically dependent over K.*

Proof. This is straightforward. $\square$

5.3 Algebraic extensions

5.3.1 Definition. *An extension $K \subset L$ is* algebraic *if any element of L is algebraic over K. An extension which is not algebraic will be called* transcendental.

5.3.2 Clearly, by 5.1.4 and 5.2.2, we have :

Proposition. (i) *Let $K \subset L$ be a field extension. An element $x \in L$ is algebraic over K if and only if $[K(x) : K]$ is finite.*

(ii) *A finite extension is algebraic.*

5.3.3 Proposition. *Let $K \subset L$ be a field extension and S a subset of L. Suppose that all the elements of S are algebraic over K. Then :*

(i) *The extension $K \subset K(S)$ is algebraic.*

(ii) *If S is finite, then the extension $K \subset K(S)$ is finite.*

Proof. By 5.1.3 and 5.3.2, it suffices to prove (ii). Part (ii) is clear if S is empty. Let us therefore suppose that S is not empty and proceed by induction on the cardinality of S. Let $x \in S$ and set $T = S \setminus \{s\}$. We have $K(S) = K(T)(x)$ and $[K(S) : K] = [K(T)(x) : K(T)][K(T) : K]$ by 5.1.4. By induction, $[K(T) : K]$ is finite, and x, being algebraic over K, is also algebraic over $K(T)$. Hence $[K(T)(x) : K(T)]$ is finite and we are done. $\square$

5.3.4 Proposition. *Let $K \subset L \subset M$ be field extensions. The following conditions are equivalent :*
   (i) *The extension $K \subset M$ is algebraic.*
   (ii) *The extensions $K \subset L$ and $L \subset M$ are algebraic.*

Proof. The implication (i) $\Rightarrow$ (ii) is trivial.

Conversely, suppose that (ii) is satisfied. Let $x \in M$. There exist elements $a_0, \ldots, a_{n-1} \in L$ such that $x^n + a_{n-1}x^{n-1} + \cdots + a_0 = 0$. We have by 5.3.3 that $K \subset K(a_0, \ldots, a_{n-1})$ is finite. On the other hand, x is algebraic over $K(a_0, \ldots, a_{n-1})$, so $K(a_0, \ldots, a_{n-1}) \subset K(a_0, \ldots, a_{n-1})(x)$ is also finite. Hence $K \subset K(x)$ is finite and part (i) follows. $\square$

5.3.5 Let $K \subset L$ be a field extension. By 5.3.3 and 5.3.4, the set of elements of L algebraic over K is a subfield M containing K. We shall call M, the *algebraic closure of K in L*. If $M = K$, then we shall say that K is *algebraically closed in L*.

5.3.6 Lemma. *Let $K \subset L \subset M$ be field extensions and S a subset of M.*
   (i) *If $K \subset L$ is algebraic, then $K(S) \subset L(S)$ is also algebraic.*
   (ii) *If S is a finite algebraically independent subset over L, then $[L : K] \leqslant [L(S) : K(S)]$.*

Proof. (i) Let $x \in L(S)$, then there exist $x_1, \ldots, x_n \in L$ such that $x \in K(S)(x_1, \ldots, x_n)$. Since the x_i's are algebraic over K, x is algebraic over $K(S)$ by 5.3.3.

(ii) Let $S = \{x_1, \ldots, x_r\}$ and B a K-basis of L. We shall show that B is linearly independent over $K(S)$. Let $b_1, \ldots, b_n$ be pairwise distinct elements of B and $f_1, \ldots, f_n \in K(S)$ be such that $f_1 b_1 + \cdots + f_n b_n = 0$. By multiplying by a suitable element of $K(S)$, we can assume that $f_i \in K[S]$ for all i. We can therefore write for each i:

$$f_i = \sum_{\alpha \in \mathbb{N}^r} \lambda_{\alpha,i} x^\alpha, \quad \sum_{\alpha \in \mathbb{N}^r} \left(\sum_{i=1}^n \lambda_{\alpha,i} b_i \right) x^\alpha = 0$$

where the $\lambda_{\alpha,i}$'s are in K. By the algebraic independence of S over L, we have that

$$\sum_{i=1}^n \lambda_{\alpha,i} b_i = 0$$

for all $\alpha \in \mathbb{N}^r$. Finally, B is a K-basis, so all the $\lambda_{\alpha,i}$'s are 0, and the result follows. $\square$

5.4 Transcendence basis

5.4.1 Definition. *Let $K \subset L$ be a field extension. A subset S of L is called a* transcendence basis *of L over K if :*

(i) *S is algebraically independent over K.*

(ii) *The extension $K(S) \subset L$ is algebraic.*

Example. Let $L = K(X_1, \ldots, X_n)$ be the quotient field of the polynomial ring $K[X_1, \ldots, X_n]$. Then $\{X_1, \ldots, X_n\}$ is a transcendence basis of L over K.

5.4.2 Definition. *Let $K \subset L$ be a field extension. A subset B of L is called a* purely transcendence basis *of L over K if B is a transcendence basis of L over K and $L = K(B)$. If L has a purely transcendence basis, then the extension $K \subset L$ is called a* purely transcendental extension.

5.4.3 Theorem. *Let $K \subset L$ be a field extension and S, T subsets of L verifying the following conditions :*

(i) *L is algebraic over $K(T)$.*

(ii) *S is algebraically independent over K.*

Then there exists a transcendence basis B of L over K such that $S \subset B \subset S \cup T$.

Proof. By Zorn's lemma, there is a maximal (by inclusion) element B among the set of subsets in $S \cup T$ algebraically independent over K which contain S. Clearly, by 5.2.3, this implies that $K(B) \subset K(S \cup T)$ is algebraic. On the other hand, by 5.3.3, $K(T) \subset K(S \cup T) \subset L$ are algebraic extensions. Hence $K(B) \subset L$ is algebraic by 5.3.3 and the result now follows. $\square$

5.4.4 Corollary. *Let $K \subset L$ be a field extension. There exists a transcendence basis of L over K.*

Proof. This follows from 5.4.3 by taking $S = \emptyset$ and $T = L$. $\square$

5.4.5 Proposition. *Let $K \subset L$ be a field extension and S a subset of L.*

(i) *If S is a transcendence basis of L over K, then it is maximal among subsets of L algebraically independent over K.*

(ii) *If L is algebraic over $K(S)$, then any maximal subset of S algebraically independent over K is a transcendence basis of L over K.*

Proof. These are easy consequences of 5.4.3. $\square$

5.4.6 Theorem. *Let $K \subset L$ be a field extension and B, B' two transcendence bases of L over K. Then B and B' are equipotent.*

Proof. It suffices to prove that card $B \geqslant$ card B'.

First, suppose that B is finite of cardinal n. If $n = 0$, then $K \subset L$ is algebraic and we have $B' = \emptyset$. So let $n \geqslant 1$ and proceed by induction on n.

Let $x \in B'$, then applying 5.4.3 with $S = \{x\}$ and $T = B$, there exists a proper subset C of B such that $C \cup \{x\}$ is a transcendence basis of L over K. Let $K_1 = K(x)$ and $C' = B' \setminus \{x\}$.

Clearly, C, C' are algebraically independent subsets of L over K_1. It follows that they are transcendence bases of L over K_1. By induction, $\operatorname{card} C' = \operatorname{card} C$ and hence $\operatorname{card} B' \leqslant \operatorname{card} B$ as required.

Let us now consider the case where B is infinite. Since any x in B is algebraic over $K(B')$, there exists a finite subset S_x of B' such that x is algebraic over $K(S_x)$. Denote by S the union of the S_x for $x \in B$. We have $S \subset B'$ and since B is infinite, $\operatorname{card} S \leqslant \operatorname{card} B$. But $K(B) \subset K(S) \subset L$ are algebraic, it follows from 5.4.5 that $S = B'$ and hence $\operatorname{card} B' \leqslant \operatorname{card} B$ as required. $\square$

5.4.7 Definition. *Let $K \subset L$ be a field extension. We define the* transcendence degree *of L over K, denoted by $\operatorname{tr deg}_K L$, to be the cardinality of any transcendence basis of L over K.*

Remark. Let $L = K(x_1, \ldots, x_n)$ be a finitely generated extension over K. Then by 5.4.3, $\operatorname{tr deg}_K L \leqslant n$.

5.4.8 From the preceding discussion, we obtain easily the following proposition :

Proposition. *Let $K \subset L$ be an extension such that $\operatorname{tr deg}_K L = n$.*

(i) *Let $S \subset L$ be such that $K(S) \subset L$ is algebraic. Then $\operatorname{card} S \geqslant n$, and if $\operatorname{card} S = n$, then S is a transcendence basis of L over K.*

(ii) *Any subset of L algebraically independent over K has at most n elements. If it has n elements, then it is a transcendence basis of L over K.*

(iii) *Suppose that $L = K(x_1, \ldots, x_m)$. Then $m \geqslant n$, and if $m = n$, then $K \subset L$ is a purely transcendental extension, and $x_1, \ldots, x_m$ is a purely transcendence basis of L over K.*

5.4.9 Theorem. *Let $K \subset L \subset M$ be field extensions, S a transcendence basis of L over K and T a transcendence basis of M over L. Then $S \cap T = \emptyset$ and $S \cup T$ is a transcendence basis of M over K.*

Proof. Since T is algebraically independent over L, it is also algebraically independent over $K(S)$. Hence $S \cap T = \emptyset$ and $S \cup T$ is algebraically independent over K.

Now, $K(S) \subset L$ is algebraic, so $L(T)$ is algebraic over $K(S \cup T)$. This implies in turn that M is algebraic over $K(S \cup T)$ since M is algebraic over $L(T)$. This finishes the proof. $\square$

5.4.10 Corollary. *Let $K \subset L \subset M$ be field extensions. Then :*

$$\operatorname{tr deg}_K M = \operatorname{tr deg}_K L + \operatorname{tr deg}_L M.$$

5.4.11 Lemma. *Let $K \subset M$ be a finitely generated field extension and L the algebraic closure of K in M. Then $K \subset L$ is a finite extension.*

Proof. The extension $L \subset M$, being also finitely generated, has a finite transcendence basis B. Since $K \subset L$ is algebraic, $K(B) \subset L(B)$ is algebraic by 5.3.6. Now, $L(B) \subset M$ is algebraic, so 5.3.4 implies that $K(B) \subset M$ is also algebraic. Being also finitely generated, the extension $K(B) \subset M$ is finite by 5.3.3. Finally, by 5.3.6, $[L : K] \leqslant [L(B) : K(B)] \leqslant [M : K(B)]$, and so the result follows. $\square$

5.4.12 Proposition. *Let $K \subset L \subset M$ be field extensions. If $K \subset M$ is finitely generated, then so is $K \subset L$.*

Proof. By 5.4.3, any transcendence basis B of L over K is contained in a transcendence basis of M over K. So the hypothesis implies that B is finite.

By 5.4.11, $[L : K(B)]$ is finite since $K(B) \subset L$ is algebraic. It follows that if S is a $K(B)$-basis of L, then $S \cup B$ is a finite system of generators for the extension $K \subset L$. $\square$

5.5 Norm and trace

5.5.1 Let $K \subset L$ be a finite extension of degree n. For $x \in L$, the map $u_x : L \to L$, $y \mapsto xy$ is a homomorphism of vector spaces over K. Set :

$$\text{Norm}_{L/K}(x) = \det u_x , \quad \text{Tr}_{L/K}(x) = \text{tr}\, u_x.$$

We call $\text{Norm}_{L/K}(x)$ (resp. $\text{Tr}_{L/K}(x)$) the *norm* (resp. *trace*) of x in L relative to K. We have clearly that for $\lambda \in K$ and $x, y \in L$:

$$\text{Tr}_{L/K}(\lambda x + y) = \lambda \, \text{Tr}_{L/K}(x) + \text{Tr}_{L/K}(y) , \quad \text{Tr}_{L/K}(\lambda) = n\lambda,$$
$$\text{Norm}_{L/K}(\lambda xy) = \lambda^n \, \text{Norm}_{L/K}(x) \, \text{Norm}_{L/K}(y) , \quad \text{Norm}_{L/K}(\lambda) = \lambda^n.$$

These equalities implies that $\text{Norm}_{L/K}(x) = 0$ if and only if $x = 0$, and if $x \neq 0$, then $[\text{Norm}_{L/K}(x)]^{-1} = \text{Norm}_{L/K}(x^{-1})$.

5.5.2 Proposition. *Let $K \subset L$ be a finite extension of degree n, $x \in L$ and $X^r + a_{r-1}X^{r-1} + \cdots + a_0$ the minimal polynomial of x over K.*

(i) *$s = n/r$ is an integer.*

(ii) *We have $\text{Tr}_{L/K}(x) = -s a_{r-1}$ and $\text{Norm}_{L/K}(x) = [(-1)^r a_0]^s$.*

Proof. Part (i) is clear since $n/r = [L : K(x)]$.

Now let $(b_1, \ldots, b_s)$ be a $K(x)$-basis of L, then $(x^i b_j)_{0 \leqslant i < n, 1 \leqslant j \leqslant s}$ is a K-basis of L. Set E_j to be the K-span of $(x^i b_j)_{0 \leqslant i < n}$. Then $E = E_1 \oplus \cdots \oplus E_s$ as a decomposition of u_x-stable subspaces over K. Now, the matrix of $u_x|_{E_j}$ in the basis $(x^i b_j)_{0 \leqslant i < n}$ is :

$$\begin{pmatrix} 0 & 0 & \cdots & 0 & -a_0 \\ 1 & 0 & \cdots & 0 & -a_1 \\ 0 & 1 & \cdots & 0 & -a_2 \\ \vdots & \vdots & & \vdots & \vdots \\ 0 & 0 & \cdots & 1 & -a_{r-1} \end{pmatrix},$$

so part (ii) follows.   □

5.5.3 Let $K \subset L$ be a finite extension of degree n. The map

$$\varphi : L \times L \to K \ , \ (x,y) \mapsto \mathrm{Tr}_{L/K}(xy)$$

is a symmetric K-bilinear form on L. If $B = (a_1, \ldots, a_n)$ is a K-basis of L, the determinant of $(\varphi(a_i, a_j))_{i,j}$ is called the *discriminant* with respect to the basis B of L. The discriminant is non-zero exactly when φ is non-degenerated. Indeed, this is the case when the characteristic of K is zero since for any $x \neq 0$, $\varphi(x, x^{-1}) = n$.

5.5.4 Proposition. *Let A be an integral domain, K its quotient field, $K \subset L$ a field extension, $x \in L$ algebraic over K and*

$$P(X) = X^r + a_{r-1}X^{r-1} + \cdots + a_0$$

the minimal polynomial of x over K. Then the following conditions are equivalent :

(i) *x is integral over A.*

(ii) *$a_0, \ldots, a_{r-1}$ are integral over A.*

In particular, if x is integral over A and the extension $K \subset L$ is finite, then $\mathrm{Norm}_{L/K}(x)$ and $\mathrm{Tr}_{L/K}(x)$ are integral over A.

Proof. (ii) $\Rightarrow$ (i) This is clear by 3.1.5 and 3.1.7.

(i) $\Rightarrow$ (ii) Let Ω be an algebraically closed field containing L and $x_1, \ldots, x_r$ the roots of P in Ω. Since x is integral over A, there exists a polynomial $Q(X) = X^n + b_{n-1}X^{n-1} + \cdots + b_0 \in A[X]$ such that $Q(x) = 0$. Then P divides Q in $K[X]$ and $x_1, \ldots, x_r$ are roots of Q. Thus $x_1, \ldots, x_r$ are integral over A. Since $\pm a_i$ is the i-th elementary symmetric polynomial in $x_1, \ldots, x_r$, the a_i's are integral over A by 3.1.5. The last statement follows therefore from 5.5.2.   □

5.5.5 Theorem. *Let A be an integral domain, K its quotient field, $K \subset L$ a field extension, B the integral closure of A in L and F the algebraic closure of K in L.*

(i) *The ring B is an integrally closed domain.*

(ii) *We have $F = \mathrm{Fract}(B)$.*

Proof. Clearly, $B \subset F$ and so $\mathrm{Fract}(B) \subset F$.

Conversely let $x \in F$. There exists $P(X) = X^r + a_{r-1}X^{r-1} + \cdots + a_0 \in K[X]$ such that $P(x) = 0$. Let us write $a_i = b_i c_i^{-1}$ where b_i, $c_i \in A$ for all i. Set $d = c_0 \cdots c_{r-1} \in A$, we have that $a_i d \in A$ for all i. It follows that dx is a root of

$$X^r + a_{r-1}dX^{r-1} + a_{r-2}d^2X^{r-2} + \cdots + a_0d^r \in A[X].$$

This implies that $dx \in B$. Thus $F \subset \mathrm{Fract}(B)$, which completes the proof of part (ii).

By 3.1.7, B is integrally closed in L, hence it is integrally closed in F. So (i) follows from (ii).   □

5.5.6 Theorem. *Let A be an integrally closed domain, $K = \mathrm{Fract}(A)$, $K \subset L$ a finite extension of degree n and B the integral closure of A in L. If the characteristic of K is zero, then there exists a K-basis $(v_1, \ldots, v_n)$ of L such that B is a submodule of the free A-module $Av_1 \oplus \cdots \oplus Av_n$.*

Proof. Let $(u_1, \ldots, u_n)$ be a K-basis of L. By 5.5.5, we can suppose that $u_1, \ldots, u_n \in B$.

Since the characteristic of K is zero, the bilinear form φ (5.5.3) is non-degenerated. Thus there exists a basis $(v_1, \ldots, v_n)$ of L over K verifying $\varphi(u_i, v_j) = \delta_{ij}$ for all $1 \leqslant i, j \leqslant n$. Clearly the sum $Av_1 \oplus \cdots \oplus Av_n$ is direct.

Let $z = a_1 v_1 + \cdots + a_n v_n \in B$ where $a_1, \ldots, a_n \in K$. Since $a_i = \mathrm{Tr}_{L/K}(zu_i)$, we have $a_i \in A$ by 5.5.4. Hence $B \subset Av_1 \oplus \cdots \oplus Av_n$ as required. $\square$

5.5.7 Proposition. *Let notations be as in 5.5.6. If A is Noetherian, then so is B.*

Proof. By 5.5.6, B is a Noetherian A-module. So it is also a Noetherian B-module. $\square$

5.6 Theorem of the primitive element

5.6.1 Let E, F be field extensions over K, a K-homomorphism from E to F is a K-algebra homomorphism, *i.e.* a ring homomorphism leaving invariant the elements of K. Clearly, such a homomorphism is injective. A bijective K-homomorphism will be called a K-isomorphism and a bijective K-homomorphism of E onto itself is called a K-automorphism.

5.6.2 Proposition. *Let Ω be an algebraically closed field containing K, $x, y \in \Omega$ algebraic over K. The following conditions are equivalent:*
   (i) *x and y have the same minimal polynomial over K.*
   (ii) *There exists a K-isomorphism $\sigma : K(x) \to K(y)$ such that $\sigma(x) = y$.*
   (iii) *There exists a K-homomorphism $\sigma : K(x) \to \Omega$ such that $\sigma(x) = y$.*
If these conditions are verified, we shall say that x and y are conjugate *over K.*

Proof. We have (i) $\Rightarrow$ (ii) by 5.2.2, and (ii) $\Rightarrow$ (iii) is obvious. Finally if P is the minimal polynomial of x over K. Then $P(\sigma(x)) = 0$. Since P is irreducible over K, we have (iii) $\Rightarrow$ (i). $\square$

5.6.3 Lemma. *Let Ω be an algebraically closed field containing a field K of characteristic zero, and $P \in K[T] \setminus \{0\}$ be an irreducible polynomial of degree n. Then P has exactly n distinct roots in Ω.*

Proof. We can suppose that $n \geqslant 1$ and that P is unitary. If P has a multiple root $\alpha \in \Omega$, then $P'(\alpha) = 0$. Since P is irreducible, it is the minimal polynomial of α over K. It follows that P' divides P, and therefore $P' = 0$. This is absurd since $P \notin K$ and the characteristic of K is zero. $\square$

5.6.4 Let $\sigma : K \to L$ be a ring homomorphism. This induces a ring homomorphism $K[X] \to L[X]$ given by :

$$K[X] \ni P = a_n X^n + \cdots + a_0 \mapsto P^\sigma = \sigma(a_n)X^n + \cdots + \sigma(a_0) \in L[X].$$

Lemma. *Let $K \subset L$ be a field extension, $x \in L$ algebraic over K, Ω an algebraically closed field and σ a homomorphism from K to Ω. There exists a homomorphism $\tau : K(x) \to \Omega$ extending σ.*

Proof. Let P be the minimal polynomial of x over K. Then $P^\sigma \in \Omega[X]$ has non zero degree. It has a root, say y, in Ω.

Now if $Q(x) = 0$, then P divides Q. Thus P^σ divides Q^σ and therefore $Q^\sigma(y) = 0$. It follows that the map $K(x) = K[x] \to \Omega$ defined by sending $\sum_{i=0}^n a_i x^i$ to $\sum_{i=0}^n \sigma(a_i)y^i$ is well-defined. Clearly, this is a homomorphism extending σ. $\square$

5.6.5 Theorem. *Let $K \subset L$ be a finite extension of degree n and σ a homomorphism from K to an algebraically closed field Ω. Then :*

  (i) *There exist k extensions of σ to L where $1 \leqslant k \leqslant n$.*
  (ii) *If the characteristic of K is zero, then $k = n$.*

Proof. We have $L = K(x_1, \ldots, x_p) = K[x_1, \ldots, x_p]$ where the x_i's are algebraic over K. We proceed by induction on p. If $p = 1$, then the theorem follows from 5.6.3 and the proof of 5.6.4.

Suppose that $p > 1$. Let $F = K[x_1, \ldots, x_{p-1}]$, then by induction, σ admits r extensions $\sigma_1, \ldots, \sigma_r$ to F where $1 \leqslant r \leqslant [F : K]$, and $r = [F : K]$ if the characteristic of K is zero. Again by induction, each σ_i admits s_i extensions to L where $1 \leqslant s_i \leqslant [L : F]$, and $s_i = [L : F]$ if the characteristic of K is zero. The theorem follows immediately. $\square$

5.6.6 Corollary. *Let $K \subset \Omega$ be a field extension where Ω is an algebraically closed field, and $K \subset L$ a finite extension of degree n. The cardinality k of the set of K-homomorphism from L to Ω verifies $1 \leqslant k \leqslant n$. Further, if the characteristic of K is zero, then $k = n$.*

5.6.7 Theorem. *(Theorem of the Primitive Element) Let $K \subset L$ be a finite extension of characteristic zero fields. Then there exists $\xi \in L$ such that $L = K(\xi)$.*

Proof. Clearly, it suffices to prove the theorem in the case where $L = K(x, y)$. Let $n = [L : K]$ and Ω an algebraically closed field containing K. Let $\{\sigma_1, \ldots, \sigma_n\}$ be the set of K-homomorphisms from L to Ω (5.6.6). For $1 \leqslant i \neq j \leqslant n$, set:

$$A_{ij} = \{a \in K \; ; \; \sigma_i(x) + a\sigma_i(y) = \sigma_j(x) + a\sigma_j(y)\}.$$

Clearly, A_{ij} contains at most one element. Since the characteristic of K is zero, it is infinite. So there exists $a \in K$ not contained in any of the A_{ij}'s. It follows that $\sigma_1, \ldots, \sigma_n$ induce distinct K-homomorphisms from $K(x + ay)$ to Ω. By 5.6.6, we have $[K(x + ay) : K] \geqslant n$. Hence $L = K(x + ay)$. $\square$

Remark. Let $K \subset L$ be a finite extension. An element $\xi \in L$ is called *primitive* if $L = K(\xi)$.

5.6.8 Corollary. *Let $K \subset L$ be a finite extension of characteristic zero fields. Then the set of fields E verifying $K \subset E \subset L$ is finite.*

Proof. By 5.6.7, there exists $x \in L$ such that $L = K(x)$. Let E be a subfield of L containing K. The minimal polynomial Q of x over E is a unitary divisor in $L[X]$ of the minimal polynomial P of x over K. It suffices therefore to show that E is completely determined by Q because the set of unitary divisors of P in $L[X]$ is finite.

Let E_0 be the field generated by K and the coefficients of Q. Since $E_0 \subset E$, Q is also irreducible in $E_0[X]$. So Q is the minimal polynomial of x over E_0. But $E_0(x) = E(x) = L$, so $\deg Q = [L : E_0] = [L : E]$. Thus $E = E_0$ and E is completely determined by Q. $\square$

5.6.9 Theorem. *Let Ω, Ω' be two algebraic closures of K. There exists a K-isomorphism from Ω to Ω'. More precisely, if $K \subset L \subset \Omega$ are field extensions and $\theta : L \to \Omega'$ is a K-homomorphism, then there exists a K-isomorphism from Ω to Ω' extending θ.*

Proof. By Zorn's Lemma, there exists a field E verifying $L \subset E \subset \Omega$, maximal such that there exists a K-homomorphism $\sigma : E \to \Omega'$ extending θ.

Let $x \in \Omega$. Since x is algebraic over K, it is algebraic over E, so by 5.6.5, σ extends to a K-homomorphism $\psi : E(x) \to \Omega'$. It follows from the maximality of E that $x \in E$. Hence $E = \Omega$. Finally $\sigma(\Omega)$, being isomorphic to Ω, is an algebraically closed subfield of Ω' containing K. We deduce that $\sigma(\Omega) = \Omega'$. $\square$

5.7 Going Down Theorem

5.7.1 Let $K \subset E \subset F$ be field extensions. Denote by $\mathrm{Hom}_K(E, F)$ the set of K-homomorphisms from E to F, and $\mathrm{Gal}(E/K)$ the group of K-automorphisms of E.

5.7.2 Lemma. *Let $K \subset E \subset \Omega$ be field extensions. Suppose that Ω is algebraically closed and E is algebraic over K. If $\sigma \in \mathrm{Hom}_K(E, \Omega)$ verifies $\sigma(E) \subset E$, then $\sigma(E) = E$.*

Proof. Let $x \in E$, $P \in K[X]$ its minimal polynomial over K, and $x_1, \ldots, x_n$ the (distinct) roots of P contained in E. Since σ is injective and $\sigma(x_i)$ is again a root of P contained in E for all i, it follows that σ is a permutation on the set $\{x_1, \ldots, x_n\}$. Hence $x \in \sigma(E)$. $\square$

5.7.3 Definition. *A field extension $K \subset E$ is called* normal *if it is algebraic and if any irreducible polynomial in $K[X]$ admitting a root in E is split over E.*

5.7.4 Proposition. *Let $K \subset E \subset \Omega$ be field extensions. Suppose that Ω is algebraically closed and E is algebraic over K. The following conditions are equivalent:*

(i) *The extension $K \subset E$ is normal.*

(ii) *For all $x \in E$, the conjugates of x in Ω over K are in E.*

(iii) *For all $\sigma \in \mathrm{Hom}_K(E, \Omega)$, we have $\sigma(E) \subset E$.*

(iv) *For all $\sigma \in \mathrm{Hom}_K(E, \Omega)$, we have $\sigma(E) = E$.*

Proof. The equivalence (i) $\Leftrightarrow$ (ii) is clear. Further we have (iii) $\Leftrightarrow$ (iv) by 5.7.2.

Now let $x \in E$, since $\sigma(x)$ and x share the same minimal polynomial, we have (ii) $\Rightarrow$ (iii). Conversely, if $y \in \Omega$ is conjugate to x, then by 5.6.2, there exists a K-homomorphism $\tau : K(x) \to K(y)$ verifying $\tau(x) = y$. We have seen in 5.6.9 that τ can be extended to a K-homomorphism $\sigma : E \to \Omega$. So $y = \tau(x) = \sigma(x) \in E$ and (iii) $\Rightarrow$ (ii). $\square$

5.7.5 Proposition. *Let $K \subset E \subset \Omega$ be field extensions. Suppose that Ω is algebraically closed and $K \subset E$ is a finite extension. Let F be the subfield of Ω generated by the $\sigma(E)$ where σ runs over the set $\mathrm{Hom}_K(E, \Omega)$. Then $K \subset F$ is a finite normal extension. Further, if $K \subset L$ is a normal extension verifying $E \subset L \subset \Omega$, then $F \subset L$.*

Proof. By 5.6.6, the set $\mathrm{Hom}_K(E, \Omega)$ is finite. Since $K \subset \sigma(E)$ is a finite extension for any σ, $K \subset F$ is a finitely generated algebraic extension. It follows from 5.4.11 that $K \subset F$ is a finite extension. Since $\tau(\sigma(E)) \subset F$ for all $\tau \in \mathrm{Hom}_K(F, \Omega)$, $\sigma \in \mathrm{Hom}_K(E, \Omega)$, normality follows from 5.7.4.

The last assertion is a direct consequence of 5.7.4 since any element σ of $\mathrm{Hom}_K(E, \Omega)$ can be extended to an element $\tau \in \mathrm{Hom}_K(L, \Omega)$ (5.6.9), so $\sigma(E) = \tau(E) \subset \tau(L) \subset L$. $\square$

5.7.6 Let G be a subgroup of the group of automorphisms of a field E. We shall denote the subfield of G-fixed points by E^G.

Lemma. *Let K be a field of characteristic zero, $K \subset E$ a finite normal extension and $G = \mathrm{Gal}(E/K)$. Then $K = E^G$.*

Proof. Clearly, $K \subset E^G$. Now suppose that there exists $x \in E^G \setminus K$. Then the minimal polynomial of x over K is at least of degree two. Since the characteristic of K is zero and $K \subset E$ is normal, there exists $y \in E$, $x \neq y$, sharing same minimal polynomial. By 5.6.2, 5.6.9 and 5.7.4, there exists $\sigma \in G$ such that $\sigma(x) = y$ which is absurd since $x \in E^G$. $\square$

5.7.7 Proposition. *Let A be an integrally closed domain of characteristic zero, $K = \mathrm{Fract}\, A$, $K \subset E$ a finite normal extension and C the integral closure of A in E. If $\mathfrak{q}, \mathfrak{q}' \in \mathrm{Spec}\, C$ verifies $A \cap \mathfrak{q} = A \cap \mathfrak{q}'$, then there exists $\sigma \in \mathrm{Gal}(E/K)$ such that $\sigma(\mathfrak{q}) = \mathfrak{q}'$.*

Proof. Let $\mathrm{Gal}(E/K) = \{\sigma_1, \ldots, \sigma_n\}$ and $\mathfrak{q}_i = \sigma_i(\mathfrak{q})$. It is clear that $\sigma_i(C) = C$, so $\mathfrak{q}_i \in \mathrm{Spec}\, C$. Suppose that $\mathfrak{q}' \neq \mathfrak{q}_i$ for all $1 \leqslant i \leqslant n$. Then by 3.3.2 and 2.2.5, $\mathfrak{q}' \not\subset \mathfrak{q}_1 \cup \cdots \cup \mathfrak{q}_n$.

Let $x \in \mathfrak{q}' \setminus \mathfrak{q}_1 \cup \cdots \cup \mathfrak{q}_n$ and $y = \sigma_1(x) \cdots \sigma_n(x)$. We have $y \in E^G = K$ (5.7.5), and since $y \in C$ and A is an integrally closed domain, $y \in A$. It follows that $y \in A \cap \mathfrak{q}'$ and by the choice of x, we have $y \notin A \cap \mathfrak{q}$ which contradicts the hypothesis. □

5.7.8 Theorem. (Going Down Theorem) *Let A be an integrally closed domain of characteristic zero, $K = \mathrm{Fract}\, A$, and B an integral extension of A whose field of fractions E is a finite extension of K. Let $\mathfrak{p}, \mathfrak{p}' \in \mathrm{Spec}\, A$ be such that $\mathfrak{p}' \subset \mathfrak{p}$. If $\mathfrak{q} \in \mathrm{Spec}\, B$ verifies $\mathfrak{p} = A \cap \mathfrak{q}$, then there exists $\mathfrak{q}' \in \mathrm{Spec}\, B$ such that $\mathfrak{p}' = A \cap \mathfrak{q}'$ and $\mathfrak{q}' \subset \mathfrak{q}$. In particular, $\mathrm{ht}\, \mathfrak{p} = \mathrm{ht}\, \mathfrak{q}$.*

Proof. Let F be a finite normal extension of K containing E (5.7.5) and denote by C the integral closure of A (so of B also) in F. By 3.3.3, there exist $\mathfrak{r}, \mathfrak{r}' \in \mathrm{Spec}\, C$ verifying $\mathfrak{r}' \subset \mathfrak{r}$ and $A \cap \mathfrak{r} = \mathfrak{p}$, $A \cap \mathfrak{r}' = \mathfrak{p}'$. Let $\mathfrak{s} \in \mathrm{Spec}\, C$ be such that $B \cap \mathfrak{s} = \mathfrak{q}$. Then $A \cap \mathfrak{s} = \mathfrak{p} = A \cap \mathfrak{r}$, so by 5.7.7, there exists $\sigma \in \mathrm{Gal}(F/K)$ such that $\sigma(\mathfrak{r}) = \mathfrak{s}$. Since $A \cap \sigma(\mathfrak{r}') = \mathfrak{p}'$, we have that $\mathfrak{q}' = B \cap \sigma(\mathfrak{r}')$ verifies $\mathfrak{q}' \subset \mathfrak{q}$ and $\mathfrak{p}' = A \cap \mathfrak{q}'$. □

5.7.9 Let A be an integrally closed domain of characteristic zero, $K = \mathrm{Fract}\, A$, $K \subset L$ a finite extension of degree n, B a subring of L integral over A such that $L = \mathrm{Fract}\, B$.

By 5.6.7, there exists $\xi = \alpha/\beta \in L$, where $\alpha, \beta \in B$, such that $L = K(\xi)$. Suppose that $\beta \notin A$ and let $\gamma_0, \ldots, \gamma_{m-1} \in A$ be such that $\gamma_0 \neq 0$ and $\beta^m + \gamma_{m-1}\beta^{m-1} + \cdots + \gamma_0 = 0$. It follows that $\gamma_0 \xi = \alpha(\beta^{m-1} + \gamma_{m-1}\beta^{m-2} + \cdots + \gamma_1)$. Since $K(\gamma_0\xi) = K(\xi)$, there exists $b \in B$ such that $L = K(b)$.

Let $P(T) = T^n + a_{n-1}T^{n-1} + \cdots + a_0$ be the minimal polynomial of b over K. By 5.5.4, we have $a_0, \ldots, a_{n-1} \in A$. Let Ω be an algebraic closure of L and $b_1, \ldots, b_n$ the conjugates of b in Ω over K. Recall from 5.6.3 that the b_i's are the n distinct roots of P in Ω. Set

$$\Delta = \prod_{i<j}(b_i - b_j) \neq 0$$

whose square is the discriminant d of P. Recall from 4.6.7 that $d \in A$.

Lemma. *Let us suppose that d is invertible in A. Then:*
  (i) *B is a free A-module and $\{1, b, \ldots, b^{n-1}\}$ is a basis of B over A.*
  (ii) *B is the integral closure of A in L.*

Proof. The set $\{1, b, \ldots, b^{n-1}\}$ is a basis of L over K. Let $x = a_0 + a_1 b + \cdots + a_{n-1}b^{n-1} \in L$, $a_0, \ldots, a_{n-1} \in K$, be integral over A. Then the conjugates of x in Ω over K are: for $1 \leqslant i \leqslant n$

$$x_i = a_0 + a_1 b_i + \cdots + a_{n-1}b_i^{n-1}.$$

This forms a linear system in the a_i's whose determinant is Δ. It follows that $a_i = \Delta_i/\Delta$ where $\Delta_i \in L$ is integral over A since the x_i's are integral over A. On the other hand, $a_i = \Delta_i\Delta/d$, and d is invertible in A, so the a_i's are integral over A. Since A is an integrally closed domain, $a_i \in A$ for all i. Hence $x \in B$ and the assertions follow. $\square$

5.7.10 Let us conserve the notations and hypotheses of 5.7.9. We suppose further that d is invertible in A. Let $\mathfrak{m}$ be a maximal ideal of A, $k = A/\mathfrak{m}$ and for $x \in A$, denote by $\overline{x}$ its class in k. Let $S(T) = T^n + \overline{a_{n-1}}T^{n-1} + \cdots + \overline{a_0} \in k[T]$. The discriminant of $\overline{d}$ of S is non-zero because d is invertible in A.

Let us suppose that S splits over k and let $\gamma_1, \ldots, \gamma_n$ be the distinct roots of S. For $1 \leqslant i \leqslant n$, set

$$Q_i(T) = \prod_{j \neq i} \frac{T - \gamma_j}{\gamma_i - \gamma_j} \in k[T].$$

Clearly, $Q_i(\gamma_i) = 1$ and $Q_i(\gamma_j) = 0$ if $i \neq j$. If $R \in k[T]$ verifies $\deg R \leqslant n-1$, then

(1) $$R(T) = R(\gamma_1)Q_1(T) + \cdots + R(\gamma_n)Q_n(T).$$

Let θ denote the class of b in $B/\mathfrak{m}B = B \otimes_A k$ and $e_i = Q_i(\theta)$. Then since $\{1, \theta, \ldots, \theta^{n-1}\}$ is a basis of $B/\mathfrak{m}B$ over k by 5.7.9, (1) implies that $\{e_1, \ldots, e_n\}$ is also a basis of $B/\mathfrak{m}B$ over k.

If $i \neq j$, then Q_iQ_j is divisible by S, so $e_ie_j = 0$. On the other hand, $Q_i^2 = SM + R$ where $R \in k[T]$ and $\deg R \leqslant n - 1$. By (1), we obtain that $R = Q_i$ and therefore, $e_ie_i = e_i$.

Thus the ring $B/\mathfrak{m}B$ is the direct sum of fields $ke_i = B/\mathfrak{n}_i$ where $\mathfrak{n}_i$ is the inverse image of $\sum_{j \neq i} ke_j$. It follows that $\mathfrak{n}_1, \ldots, \mathfrak{n}_n$ are the n maximal ideals of B lying above $\mathfrak{m}$.

5.8 Fields and derivations

5.8.1 We shall use the notations of 2.8 and 2.9. Further let K be a field of characteristic zero.

Lemma. *Let $K \subset L$ be a finite extension. Then any K-derivation of L in a L-vector space N is zero. In particular, $\Omega_K(L) = \{0\}$.*

Proof. By 5.6.7, there exists $x \in L$ algebraic over K such that $L = K(x)$. Let $P(T)$ be the minimal polynomial of x over K. Let $\delta \in \mathrm{Der}_K(L, N)$. If $Q \in K[T]$, then $\delta(Q(x)) = Q'(x)\delta(x)$. In particular, $0 = \delta(P(x)) = P'(x)\delta(x)$. By 5.6.3, $P'(x) \neq 0$. So $\delta(x) = 0$ and the lemma follows since any element of L can be written as $Q(x)$ for some $Q \in K[T]$. $\square$

5.8.2 Lemma. *Let $K \subset E \subset F$ be field extensions such that $[F : E]$ is finite, and N a F-vector space. Any element of $\mathrm{Der}_K(E, N)$ extends uniquely to an element of $\mathrm{Der}_K(F, N)$.*

Proof. By 5.6.7, there exists $x \in F$ such that $F = E(x)$. Let P be the minimal polynomial of x over E and $\delta \in \operatorname{Der}_K(E, N)$.

For $Q(T) = \sum_n a_n T^n \in E[T]$, we define $Q^\delta(x) = \sum_n x^n \delta(a_n) \in N$. Suppose that $\Delta \in \operatorname{Der}_K(F, N)$ extends δ, then for $Q \in E[T]$, we have:

$$(2) \qquad \Delta(Q(x)) = Q^\delta(x) + Q'(x)\Delta(x).$$

In particular, since $P'(x) \neq 0$, $u = \Delta(x)$ is the unique solution in N of

$$(3) \qquad 0 = P^\delta(x) + P'(x)u.$$

So uniqueness follows.

Now, let $u = -P'(x)^{-1}P^\delta(x) \in N$ be the unique solution of (3). We claim that the map $\Delta : F \to N$ given by (2) and $\Delta(x) = u$ is a well-defined map extending δ. This is just an easy consequence of the following equality:

$$(QR)^\delta(x) + (QR)'(x)u = R(x)(Q^\delta(x) + Q'(x)u) + Q(x)(R^\delta(x) + R'(x)u)$$

for all $Q, R \in E[T]$. $\square$

5.8.3 Let us conserve the notations of 5.8.2. By 2.8.5 and 5.8.2, we obtain an exact sequence of E-vector spaces:

$$0 \to \operatorname{Der}_E(F, N) \to \operatorname{Der}_K(F, N) \to \operatorname{Der}_K(E, N) \to 0.$$

By 2.9.5, this induces an exact sequence of F-vector spaces:

$$0 \to \Omega_K(E) \otimes_E F \xrightarrow{\alpha} \Omega_K(F) \xrightarrow{\beta} \Omega_E(F) \to 0$$

where $\alpha(d_{E/K}(x) \otimes 1) = d_{F/K}(x)$ and $\beta(d_{F/K}(y)) = d_{F/E}(y)$ for all $x \in E$, $y \in F$. It follows therefore from 5.8.1 that the F-vector spaces $\Omega_K(E) \otimes_E F$ and $\Omega_K(F)$ are isomorphic.

Now if $K \subset E$ is purely transcendental and $\operatorname{tr\,deg}_K(E) = n$, then 2.9.7 implies that the dimension of the E-vector space $\Omega_K(E)$ is n. So, we have obtained:

Theorem. *Let $K \subset F$ be finitely generated. Then the dimension of the F-vector space $\Omega_K(F)$ is equal to the transcendence degree of F over K.*

5.8.4 Proposition. *Let $u : A \to B$ be a homomorphism of finitely generated commutative K-algebras which are integral domains. Denote by $\Omega(u) : \Omega_K(A) \to \Omega_K(B)$ the homomorphism of A-modules defined in 2.9.5. Suppose that $\Omega_K(A)$ (resp. $\Omega_K(B)$) is a free A-module (resp. free B-module). Then the following conditions are equivalent:*
  (i) u is injective.
  (ii) $\Omega(u)$ is injective.

Proof. (ii) $\Rightarrow$ (i) In the notations of 2.9.5, $\Omega(u)$ is the unique A-module homomorphism such that the following diagram commutes:

$$
\begin{array}{ccc}
A & \xrightarrow{\;u\;} & B \\
{\scriptstyle d_{A/K}}\downarrow & & \downarrow{\scriptstyle d_{B/K}} \\
\Omega_K(A) & \xrightarrow{\;\Omega(u)\;} & \Omega_K(B)
\end{array}
$$

Let $a \in \ker u$, then $d_{A/K}(a) = 0$. For any $x \in A$, we have $ax \in \ker u$, and therefore $0 = d_{A/K}(ax) = a\,d_{A/K}(x)$. But the elements $d_{A/K}(x)$, $x \in A$, generate the A-module $\Omega_K(A)$. So $a = 0$ because $\Omega_K(A)$ is a free A-module.

(i) $\Rightarrow$ (ii) Let E (resp. F) be the field of fractions of A (resp. B). Since u is injective, we may assume that $A \subset B$ and $E \subset F$. By 5.8.3, we have

$$
\dim_F \Omega_K(F) = \operatorname{tr} \deg_K F \;,\quad \dim_F \Omega_E(F) = \operatorname{tr} \deg_E F \;,
$$
$$
\dim_F \Omega_K(E) \otimes_E F = \dim_E \Omega_K(E) = \operatorname{tr} \deg_K E.
$$

It follows from a dimension count that the following sequence (2.9.5), induced by the canonical injection $v : E \to F$, is exact:

$$
0 \to \Omega_K(E) \otimes_E F \to \Omega_K(F) \to \Omega_E(F) \to 0.
$$

We deduce therefore from 2.9.5 that $\Omega(v) : \Omega_K(E) \to \Omega_K(F)$ is injective.

Let $\iota_A : A \to E$ and $\iota_B : B \to F$ denote the canonical injections. By 2.9.5, 2.9.6, the uniqueness of $\Omega(\iota_A)$ and the hypothesis that $\Omega_K(A)$ is a free A-module, we obtain that $\Omega(\iota_A)$ is the composition :

$$
\Omega_K(A) \hookrightarrow \Omega_K(A) \otimes_A E \to \Omega_K(E).
$$

In particular, $\Omega(\iota_A)$ is injective.

Now we have the following commutative diagram:

$$
\begin{array}{ccc}
A & \xrightarrow{\;u\;} & B \\
{\scriptstyle \iota_A}\downarrow & & \downarrow{\scriptstyle \iota_B} \\
E & \xrightarrow{\;v\;} & F
\end{array}
$$

which induces, from the uniqueness of $\Omega(\iota_B \circ u) = \Omega(v \circ \iota_A)$, the following commutative diagram:

$$
\begin{array}{ccc}
\Omega_K(A) & \xrightarrow{\;\Omega(u)\;} & \Omega_K(B) \\
{\scriptstyle \Omega(\iota_A)}\downarrow & & \downarrow{\scriptstyle \Omega(\iota_B)} \\
\Omega_K(E) & \xrightarrow{\;\Omega(v)\;} & \Omega_K(F)
\end{array}
$$

Since $\Omega(v)$ and $\Omega(\iota_A)$ are injective, we deduce that $\Omega(u)$ is also injective. $\quad\square$

5.9 Conductor

5.9.1 In this section, A will be a subring of a ring B. The annihilator of the A-module B/A is an ideal of A called the *conductor* of B in A. It is easy to verify that this is the largest ideal of B contained in A.

Proposition. *Let S be a multiplicative subset of A, $\mathfrak{f}$ the conductor of B in A and $\mathfrak{f}_1$ the conductor of $S^{-1}B$ in $S^{-1}A$. Then $S^{-1}\mathfrak{f} \subset \mathfrak{f}_1$. Furthermore, we have $S^{-1}\mathfrak{f} = \mathfrak{f}_1$ if B is a finite A-algebra.*

Proof. Since $\mathfrak{f}B \subset A$, we have $(S^{-1}\mathfrak{f})(S^{-1}B) \subset S^{-1}A$. So $S^{-1}\mathfrak{f} \subset \mathfrak{f}_1$.

Now let us suppose that $\{x_1, \ldots, x_n\}$ is a system of generators of the A-module B. Let $a/s \in \mathfrak{f}_1$ where $a \in A$, $s \in S$. Then for each i, $(a/s)(x_i/1) \in S^{-1}A$, *i.e.* there exists $t_i \in S$ such that $t_i a x_i \in A$. Set $t = t_1 \cdots t_n$, then clearly $t a x_i \in A$. Hence $t a \in \mathfrak{f}$ and $a/s \in S^{-1}\mathfrak{f}$. $\square$

5.9.2 Corollary. *Let $\mathfrak{f}$ be the conductor of B in A, $\mathfrak{p} \in \operatorname{Spec} A$ be such that $\mathfrak{f} \not\subset \mathfrak{p}$, and $S = A \setminus \mathfrak{p}$. If B is a finite A-algebra, then $A_{\mathfrak{p}} = S^{-1}B$.*

Proof. By 5.9.1, $S^{-1}\mathfrak{f}$ is the conductor of $S^{-1}B$ in $A_{\mathfrak{p}}$. Since $\mathfrak{f} \not\subset \mathfrak{p}$, there exists $s \in S \cap \mathfrak{f}$. It follows that $1 = s/s \in S^{-1}\mathfrak{f}$. Hence $S^{-1}\mathfrak{f} = A_{\mathfrak{p}}$ and $A_{\mathfrak{p}} = S^{-1}B$ by the definition of the conductor. $\square$

5.9.3 Proposition. *Let A be an integral domain, B be the integral closure of A, $\mathfrak{f}$ the conductor of B in A and S a multiplicative subset of A not containing 0.*

(i) A sufficient condition for $S^{-1}A$ to be an integrally closed domain is that $S \cap \mathfrak{f} \neq \emptyset$. If B is a finite A-algebra, then this is also a necessary condition.

(ii) Suppose that B is a finite A-algebra. Let $\mathfrak{p} \in \operatorname{Spec} A$, then $A_{\mathfrak{p}}$ is not an integrally closed domain if and only if $\mathfrak{f} \subset \mathfrak{p}$.

Proof. By 3.1.9, $S^{-1}B$ is the integral closure of $S^{-1}A$. Since $S^{-1}\mathfrak{f} = S^{-1}A$ if and only if $S \cap \mathfrak{f} \neq \emptyset$, part (i) follows from 5.9.1 and 5.9.2. Finally, part (ii) is just a special case of (i). $\square$

5.9.4 Proposition. *Let B be an integral domain, A integrally closed in B and $x \in B$ such that B is integral over $A[x]$. Suppose that there exists a unitary polynomial $P \in A[T]$ such that $P(x)$ is contained in the conductor of B in $A[x]$. Then $B = A[x]$.*

Proof. Let $b \in B$. Then $bP(x) = Q(x) \in A[x]$ for some $Q \in A[T]$. Since P is unitary, we have $Q = FP + R$ with $F, R \in A[T]$ and $\deg R < \deg P$. Let $y = b - F(x)$, thus $yP(x) = R(x)$. If $y = 0$, then clearly $b = F(x) \in A[x]$. If $y \neq 0$, then $P(x) = R(x)/y$ in $\operatorname{Fract}(B)$. Since $\deg R < \deg P$, we deduce that x is integral over $A[1/y]$. On the other hand, y is integral over $A[x]$, and hence over $A[x, 1/y]$. It follows from 3.1.7 that $A[x, y, 1/y]$ is integral over $A[1/y]$. So y is integral over $A[1/y]$, and there exist $a_0, \ldots, a_{m-1} \in A[1/y]$ such that

$$y^m + a_{m-1}y^{m-1} + \cdots + a_0 = 0.$$

Clearly, this implies that y is integral over A. Since A is integrally closed in B, we have $y \in A$, and therefore $b = y + F(x) \in A[x]$. $\square$

5.9.5 Lemma. *Let B be an integral domain and $x \in B$ an element transcendental over $\mathrm{Fract}\, A$ such that B is a finite $A[x]$-algebra. For any maximal ideal $\mathfrak{n}$ of B, there exists $\mathfrak{q} \in \mathrm{Spec}\, B$ such that*

$$\mathfrak{q} \subset \mathfrak{n}\, , \ \mathfrak{q} \neq \mathfrak{n}\, , \ \mathfrak{n} \cap A = \mathfrak{q} \cap A.$$

Proof. Let A' (resp. B') be the integral closure of A in $\mathrm{Fract}\, A$ (resp. B in $\mathrm{Fract}\, B$). It is clear that B' is integral over $A'[x]$ and x is transcendental over $\mathrm{Fract}\, A' = \mathrm{Fract}\, A$.

Let $\mathfrak{n}'$ be a prime ideal of B' lying above $\mathfrak{n}$. By 3.3.2, $\mathfrak{n}' \in \mathrm{Spm}\, B'$. Set $\mathfrak{m}' = \mathfrak{n}' \cap A'$ and $\mathfrak{r}' = \mathfrak{n}' \cap A'[x]$. Note that $\mathfrak{r}' \in \mathrm{Spm}\, A'[x]$ by 3.3.2, and so $\mathfrak{m}'A'[x] \subset \mathfrak{r}'$, $\mathfrak{m}'A'[x] \neq \mathfrak{r}'$ since $A'[x]/\mathfrak{m}'A'[x] \simeq (A'/\mathfrak{m}')[T]$ is not a field. Since $A'[x]$ is an integrally closed domain by 3.2.8 and $\mathfrak{m}'A'[x] \in \mathrm{Spec}\, A'[x]$, we obtain by 5.7.8 that there exists $\mathfrak{q}' \in \mathrm{Spec}\, B'$ verifying $\mathfrak{q}' \subset \mathfrak{n}'$, $\mathfrak{q}' \neq \mathfrak{n}'$ and $\mathfrak{q}' \cap A'[x] = \mathfrak{m}'A'[x]$. Hence $\mathfrak{q}' \cap A' = \mathfrak{m}'$. Now $\mathfrak{q} = \mathfrak{q}' \cap B$ is not a maximal ideal of B by 3.3.2, so we have $\mathfrak{q} \neq \mathfrak{n}$ and $\mathfrak{q} \cap A = \mathfrak{n} \cap A$. $\square$

5.9.6 Lemma. *Let B be an integral domain, A integrally closed in B and $x \in B$ such that B is a finite $A[x]$-algebra. Suppose that there exists a maximal ideal $\mathfrak{n}$ of B which is minimal in the set of $\mathfrak{q} \in \mathrm{Spec}\, B$ such that $\mathfrak{q} \cap A = \mathfrak{n} \cap A$.*

(i) For all $\mathfrak{q} \in \mathrm{Spec}\, B$ verifying $\mathfrak{q} \subset \mathfrak{n}$, the class of x modulo $\mathfrak{q}$ is algebraic over $\mathrm{Fract}(A/(\mathfrak{q} \cap A))$.

(ii) The conductor $\mathfrak{f}$ of B in $A[x]$ is not contained in $\mathfrak{n}$.

Proof. (i) Denote by x_1 the class of x modulo $\mathfrak{q}$. The ring $B/\mathfrak{q}$ is a finite $(A/(\mathfrak{q} \cap A))[x_1]$-algebra and $\mathfrak{n}/\mathfrak{q} \in \mathrm{Spm}(B/\mathfrak{q})$ is minimal in the set of prime ideals in $B/\mathfrak{q}$ lying above $(\mathfrak{n} \cap A)/(\mathfrak{q} \cap A)$. So x_1 is algebraic over $A/(\mathfrak{q} \cap A)$ by 5.9.5.

(ii) Suppose that $\mathfrak{f} \subset \mathfrak{n}$. Let $\mathfrak{q} \in \mathrm{Spec}\, B$ be minimal verifying $\mathfrak{f} \subset \mathfrak{q} \subset \mathfrak{n}$, and denote $\mathfrak{p} = \mathfrak{q} \cap A$.

Recall from 2.6.6 that since $A_{\mathfrak{p}}$ is a local ring with maximal ideal $\mathfrak{p}A_{\mathfrak{p}}$, its residue field K is isomorphic to $\mathrm{Fract}(A/\mathfrak{p})$. Further the ideal $\mathfrak{q}_{\mathfrak{p}}$ in $B_{\mathfrak{p}}$ is prime and verifies $\mathfrak{q}_{\mathfrak{p}} \cap A_{\mathfrak{p}} = \mathfrak{p}A_{\mathfrak{p}}$ (2.3.7).

By part (i), the class of x modulo $\mathfrak{q}$ is algebraic over K. Since $B_{\mathfrak{p}}$ is integral over $A_{\mathfrak{p}}$ by 3.1.8, all elements of $B_{\mathfrak{q}}/\mathfrak{q}_{\mathfrak{p}}$ are algebraic over K. It follows therefore that $B_{\mathfrak{q}}/\mathfrak{q}_{\mathfrak{p}}$ is a field, and so $\mathfrak{q}_{\mathfrak{p}} \in \mathrm{Spm}\, B_{\mathfrak{q}}$. Further $A_{\mathfrak{p}}$ is integrally closed in $B_{\mathfrak{p}}$ by 3.1.9, $\mathfrak{f}_{\mathfrak{p}}$ is the conductor of $B_{\mathfrak{p}}$ in $A_{\mathfrak{p}}$ by 5.9.1, and $\mathfrak{q}_{\mathfrak{p}}$ is minimal among the prime ideals of $B_{\mathfrak{p}}$ containing $\mathfrak{f}_{\mathfrak{p}}$. So by replacing A, B by $A_{\mathfrak{p}}$, $B_{\mathfrak{p}}$, we can assume that $\mathfrak{q} = \mathfrak{n}$.

Under this assumption, $\mathfrak{n}/\mathfrak{f}$ is a minimal prime ideal of $B/\mathfrak{f}$ and $(\mathfrak{n}/\mathfrak{f})_{\mathfrak{n}/\mathfrak{f}}$ is the minimal prime ideal of $(B/\mathfrak{f})_{\mathfrak{n}/\mathfrak{f}}$. So $(\mathfrak{n}/\mathfrak{f})_{\mathfrak{n}/\mathfrak{f}}$ is the nilradical of $(B/\mathfrak{f})_{\mathfrak{n}/\mathfrak{f}}$.

Since the class of x modulo $\mathfrak{n}$ is algebraic over K, there exist $y \in A \setminus \mathfrak{p} \subset B \setminus \mathfrak{n}$ and $P \in K[T]$ unitary such that $yP(x) \in \mathfrak{n}$. By the preceding paragraph, the image in $(B/\mathfrak{f})_{\mathfrak{n}/\mathfrak{f}}$ of the class of $yP(x)$ in $B/\mathfrak{f}$ is nilpotent. So there exist $z \in B \setminus \mathfrak{n}$ and $k \in \mathbb{N}^*$ such that $zy^k P(x)^k \in \mathfrak{f}$, hence $P(x)^k$ belongs to the conductor of $A[x][zy^k B]$ in $A[x]$. Now by applying 5.9.4 to A, $A[x]$ and $A[x][zy^k B]$, we obtain that $zy^k B \subset A[x]$. But this implies that $zy^k \in \mathfrak{f} \subset \mathfrak{n}$ which is absurd since $y, z \notin \mathfrak{n}$ and $\mathfrak{n}$ is prime. $\square$

5.9.7 Lemma. *Let B be an integral domain and A a local ring with maximal ideal $\mathfrak{m}$. Let us suppose that:*

(i) The ring A is integrally closed in B and there exists $\mathfrak{n} \in \operatorname{Spm} B$ lying above $\mathfrak{m}$.

(ii) There exists $x \in B$ such that $B = A[x]$ and the class of x modulo $\mathfrak{n}$ is algebraic over $K = A/\mathfrak{m}$.

Then $B = A$.

Proof. By (ii), there exists $P \in A[T] \setminus A$ unitary such that $P(x) \in \mathfrak{m}A[x]$. Set $y = 1 + P(x)$ and θ, the image of y in $C = A[y]/\mathfrak{m}A[y]$.

If θ is not invertible in C, then θ is contained in a maximal ideal of C. But $A[x]/\mathfrak{m}A[x]$ is integral over C, so the image $\overline{y}$ of y in $A[x]/\mathfrak{m}A[x]$ is also contained in a maximal ideal by 3.3.2 and 3.3.3. This is absurd since $\overline{y} = \overline{1}$. Thus we have shown that θ is invertible in C.

Next, let $X^m + \alpha_{m-1}X^{m-1} + \cdots + \alpha_0$ be the minimal polynomial of θ over K. Since θ is invertible in C, $\alpha_0 \neq 0$. For $0 \leqslant i \leqslant m-1$, let $a_i \in A$ be a representative of α_i, then there exist $n \in \mathbb{N}$ and $u_0, \ldots, u_n \in \mathfrak{m}$ such that:

$$y^m + a_{m-1}y^{m-1} + \cdots + a_0 = u_n y^n + u_{n-1}y^{n-1} + \cdots + u_0.$$

Thus there exists $z \in A[y]$ such that $yz = a_0 - u_0 \notin \mathfrak{m}$. Since A is a local ring, this implies that yz is invertible in A. Hence y is invertible in $A[y] \subset B$, i.e. there exist $b_1, \ldots, b_r \in A$ such that $b_r y^r + \cdots + b_1 y = 1$. It follows that $1/y$ is integral over A and (i) implies that $1/y \in A$.

Now if $1/y \in \mathfrak{m}$, then $1 \in y\mathfrak{m} \subset \mathfrak{n}$ which is absurd. So $1/y \in A \setminus \mathfrak{m}$ is invertible in A. Thus $y \in A$, which implies that $P(x) \in A$. So the hypotheses on P implies that x is integral over A. Hence $x \in A$ and $B = A$. $\square$

5.9.8 Proposition. *Let B be an integral domain, A integrally closed in B and $x \in B$ such that B is a finite $A[x]$-algebra. Let $\mathfrak{m} \in \operatorname{Spm}(A)$ and $\mathfrak{n}$ be a maximal ideal of B which is minimal among the prime ideals of B lying above $\mathfrak{m}$. Then $A_{\mathfrak{m}} = B_{\mathfrak{m}} = B_{\mathfrak{n}}$.*

Proof. By 3.1.9, $A_{\mathfrak{m}}$ is integrally closed in $B_{\mathfrak{m}}$ and $B_{\mathfrak{m}}$ is a finite $A_{\mathfrak{m}}[x]$-algebra. Further, $\mathfrak{n}_{\mathfrak{m}}$ is a maximal ideal of $B_{\mathfrak{m}}$ which is minimal among the prime ideals of $B_{\mathfrak{m}}$ lying above $\mathfrak{m}A_{\mathfrak{m}}$ (2.3.7 and 3.3.2). We are therefore reduced to the case where A is a local ring with maximal ideal $\mathfrak{m}$ and $B = B_{\mathfrak{m}}$.

By 5.9.6, $\mathfrak{n}$ does not contain the conductor of B in $A[x]$. It follows from 5.9.2 that $B = B_{\mathfrak{r}} = A[x]_{\mathfrak{r}}$ where $\mathfrak{r} = \mathfrak{n} \cap A[x]$. By 3.3.2, $\mathfrak{r} \in \mathrm{Spm}\, A[x]$, so $\mathfrak{n} = \mathfrak{r}A[x]_{\mathfrak{r}}$. Our hypothesis on $\mathfrak{n}$ implies via 3.3.3 that $\mathfrak{r}$ is minimal among the prime ideals of $A[x]$ lying above $\mathfrak{m}$. Now, 5.9.6 and 5.9.7 imply that $A = A[x]$, and so $\mathfrak{r} = \mathfrak{m}$. Hence $A_{\mathfrak{m}} = B = B_{\mathfrak{m}} = B_{\mathfrak{n}}$ as required. $\square$

References

- [9],[19], [22], [26], [52], [56].

6

Finitely generated algebras

We prove three fundamental results in this chapter : Noether's Normalization Theorem, Krull's Principal Ideal Theorem, and Hilbert's Nullstellensatz. We introduce also the Zariski topology on the set of maximal ideals of a ring.

All fields and rings considered are commutative. Let us denote by K a field. If A is a ring, then we shall denote by $\mathrm{Spec}_m(A)$ the set of minimal prime ideals of A.

6.1 Dimension

6.1.1 Let A be a finitely generated K-algebra and $\{x_1, \ldots, x_n\}$ a system of generators of A. Since A is isomorphic to a quotient of the polynomial ring $K[T_1, \ldots, T_n]$, A is Noetherian by 2.7.9 and 2.7.3.

Suppose that A is an integral domain and let F be its quotient field. Then by 5.4.3, we can extract from $\{x_1, \ldots, x_n\}$ a transcendence basis of F over K, and hence $\mathrm{tr\,deg}_K F \leqslant n$. In particular, there exists a transcendence basis of F over K contained in A.

6.1.2 Proposition. *Let A be a finitely generated K-algebra and $\mathfrak{p}, \mathfrak{q}$ be distinct prime ideals of A such that $\mathfrak{p} \subset \mathfrak{q}$. Then*

$$\mathrm{tr\,deg}_K \mathrm{Fract}(A/\mathfrak{q}) < \mathrm{tr\,deg}_K \mathrm{Fract}(A/\mathfrak{p}).$$

Proof. Replace A by $A/\mathfrak{p}$, we are reduced to the case where A is an integral domain, $\mathfrak{p} = \{0\}$ and $\mathfrak{q} \neq \{0\}$. Let $\pi : A \to A/\mathfrak{q}$ be the canonical surjection.

Let $y_1, \ldots, y_r \in A/\mathfrak{q}$ be a transcendence basis of $\mathrm{Fract}(A/\mathfrak{q})$ over K. For each i, fix $x_i \in A$ such that $\pi(x_i) = y_i$. Clearly, the x_i's are algebraically independent over K. It follows from 6.1.1 that:

$$r = \mathrm{tr\,deg}_K \mathrm{Fract}(A/\mathfrak{q}) \leqslant \mathrm{tr\,deg}_K \mathrm{Fract}\, A.$$

If equality holds, then π induces an isomorphism from $K[x_1, \ldots, x_r]$ onto $K[y_1, \ldots, y_r]$. Further for any $a \in \mathfrak{q}$, a is algebraic over $K(x_1, \ldots, x_r)$. So there exist $P_0, \ldots, P_n \in K[X_1, \ldots, X_r]$ such that $P_0(x_1, \ldots, x_r) \neq 0$ and

$$P_0(x_1, \ldots, x_r) + P_1(x_1, \ldots, x_r)a + \cdots + P_n(x_1, \ldots, x_r)a^n = 0.$$

Applying π to this equality, we obtain that $P_0(y_1, \ldots, y_r) = 0$ which contradicts the algebraic independence of the y_i's.   □

6.1.3 Corollary. *Let A be a finitely generated K-algebra. Then:*

$$\dim A \leqslant \max_{\mathfrak{p} \in \mathrm{Spec}(A)} \mathrm{tr}\deg_K \mathrm{Fract}(A/\mathfrak{p}) = \max_{\mathfrak{p} \in \mathrm{Spec}_m(A)} \mathrm{tr}\deg_K \mathrm{Fract}(A/\mathfrak{p}).$$

Proof. This is immediate from 6.1.2.   □

6.1.4 Corollary. *We have*

$$\dim K[X_1, \ldots, X_n] = \mathrm{tr}\deg_K K(X_1, \ldots, X_n) = n.$$

Proof. Let $A = K[X_1, \ldots, X_n]$. Since $X_1, \ldots, X_n$ form a transcendence basis of $K(X_1, \ldots, X_n)$, we have $\dim A \leqslant \mathrm{tr}\deg_K K(X_1, \ldots, X_n) = n$.

On the other hand, let $\mathfrak{p}_i$ be the ideal of A generated by $X_1, \ldots, X_i$. Clearly $A/\mathfrak{p}_i$ is isomorphic to $K[X_{i+1}, \ldots, X_n]$. So we have a chain of prime ideals $\{0\} \subset \mathfrak{p}_1 \subset \cdots \subset \mathfrak{p}_n$ in A. So $\dim A \geqslant n$.   □

6.1.5 Corollary. *Let $\mathfrak{p}_i$, $1 \leqslant i \leqslant n$ be as in the proof of 6.1.4. Then the height of $\mathfrak{p}_i$ is equal to i.*

Proof. The chain $\{0\} \subset \mathfrak{p}_1 \subset \cdots \subset \mathfrak{p}_i$ implies that $\mathrm{ht}\,\mathfrak{p}_i \geqslant i$. Now the result follows from 6.1.2 and 6.1.4.   □

6.2 Noether's Normalization Theorem

6.2.1 Theorem. *Let $n \in \mathbb{N}^*$, $A = K[X_1, \ldots, X_n]$ and $\mathfrak{a}$ a proper ideal of A. Then there exist $x_1, \ldots, x_n \in A$ algebraically independent over K verifying the following conditions:*

(i) The ring A is integral over $B = K[x_1, \ldots, x_n]$.

(ii) If $\mathfrak{a}$ is non trivial, then there exists $s \in \{1, \ldots, n\}$ such that the ideal $B \cap \mathfrak{a}$ of B is generated by $x_1, \ldots, x_s$.

Proof. If $\mathfrak{a} = \{0\}$, then it suffices to take $x_i = X_i$ for all i. So let us suppose that $\mathfrak{a} \neq \{0\}$.

We shall proceed by induction on n. For $n = 1$, there exists $x \in A$ non-constant such that $\mathfrak{a} = Ax$. Let $x_1 = x$, then (i) and (ii) are satisfied.

Let us suppose that $n > 1$ and $x_1 \in \mathfrak{a} \setminus \{0\}$. Then $x_1 \notin K$ and we can write

$$x_1 = P(X_1, \ldots, X_n) = \sum_{\lambda = (i_1, \ldots, i_n) \in \mathbb{N}^n} a_\lambda X_1^{i_1} \cdots X_n^{i_n}$$

where $a_\lambda \in K$. Let $k \in \mathbb{N}^*$ be such that $k > i_1 + \cdots + i_n$ for all $\lambda = (i_1, \ldots, i_n)$ such that $a_\lambda \neq 0$. For $\lambda = (i_1, \ldots, i_n)$, write $l(\lambda) = i_1 + i_2 k + \cdots + i_n k^{n-1}$. Then by the choice of k and the uniqueness of the development of an integer in base k, we have that for $\lambda = (i_1, \ldots, i_n)$, $\lambda' = (j_1, \ldots, j_n)$ such that $a_\lambda \neq 0$, $a_{\lambda'} \neq 0$, $l(\lambda) = l(\lambda')$ implies that $\lambda = \lambda'$. So there exists a unique $\mu \in \mathbb{N}^n$ verifying $a_\mu \neq 0$ such that $l(\mu)$ is maximal.

Set $t_i = X_i - X_1^{k^{i-1}}$ for $2 \leqslant i \leqslant n$. Then

$$
\begin{aligned}
0 &= P(X_1, t_2 + X_1^k, \ldots, t_n + X_1^{k^{n-1}}) - x_1 \\
&= a_\mu X_1^{l(\mu)} + \sum_{j < l(\mu)} Q_j(x_1, t_2, \ldots, t_n) X_1^j
\end{aligned}
$$

where the Q_j's are polynomials. It follows that X_1 is integral over the algebra $C = K[x_1, t_2, \ldots, t_n]$. Since $X_i = t_i + X_1^{k^{i-1}}$ for $2 \leqslant i \leqslant n$, X_i is integral over C by 3.1.7. Thus A is integral over C.

Further, by 5.3.3, the extension $\operatorname{Fract}(C) \subset \operatorname{Fract}(A)$ is algebraic and $\operatorname{tr} \deg_K \operatorname{Fract}(A) = n$. Therefore by 5.4.3, $x_1, t_2, \ldots, t_n$ are algebraically independent over K. In particular, C is isomorphic to A, so it is factorial and an integrally closed domain.

Let $y \in A$ be such that $yx_1 \in C$, then $y \in \operatorname{Fract}(C)$ and since C is an integrally closed domain and y is integral over C, $y \in C$. Hence $C \cap Ax_1 = Cx_1$.

Suppose first that $\mathfrak{a} \cap K[t_2, \ldots, t_n] = \{0\}$. Let $S = K[t_2, \ldots, t_n] \setminus \{0\}$. The ideal generated by x_1 in $S^{-1}C = K(t_2, \ldots, t_n)[x_1]$ is maximal. So $S^{-1}\mathfrak{a} = S^{-1}Cx_1$. It follows that if $y \in \mathfrak{a} \cap C$, then there exists $s \in S$ such that $sy \in Cx_1$. Since $x_1, t_2, \ldots, t_n$ are algebraically independent, $Cx_1 \in \operatorname{Spec}(C)$. Hence $y \in Cx_1$, and $\mathfrak{a} \cap C = Cx_1$. So we obtain the result by setting $x_2 = t_2, \ldots, x_n = t_n$.

Suppose now that $\mathfrak{a} \cap K[t_2, \ldots, t_n] \neq \{0\}$. Let $D = K[t_2, \ldots, t_n]$. By induction, there exist $x_2, \ldots, x_n \in D$ algebraically independent over K, such that D is integral over $E = K[x_2, \ldots, x_n]$ and $\mathfrak{a} \cap E = Ex_2 + \cdots Ex_k$ for some k with $2 \leqslant k \leqslant n$.

By 3.1.7, A is integral over $B = K[x_1, x_2, \ldots, x_n]$, and so $x_1, x_2, \ldots, x_n$ are algebraically independent over K. Since $x_1 \in \mathfrak{a}$, it follows that $\mathfrak{a} \cap B = Bx_1 + \mathfrak{a} \cap E = Bx_1 + Bx_2 + \cdots + Bx_k$. $\square$

6.2.2 Theorem. (Noether's Normalization Theorem) *Let A be a finitely generated K-algebra and $\mathfrak{a}_1 \subset \mathfrak{a}_2 \subset \cdots \subset \mathfrak{a}_p$ a chain of ideals of A such that $\mathfrak{a}_p \neq A$. There exist $x_1, \ldots, x_n \in A$ verifying the following conditions:*

(i) $x_1, \ldots, x_n$ are algebraically independent over K, A is integral over $B = K[x_1, \ldots, x_n]$ and $\dim A = n$.

(ii) There exists a chain of integers $0 \leqslant h(1) \leqslant \cdots \leqslant h(p)$ such that for $1 \leqslant k \leqslant p$, $B \cap \mathfrak{a}_k$ is $\{0\}$ if $h(k) = 0$ and it is the ideal of B generated by $x_1, \ldots, x_{h(k)}$ if $h(k) > 0$.

Proof. First of all, observe that if $x_1, \ldots, x_n \in A$ are algebraically independent over K and A is integral over $K[x_1, \ldots, x_n]$, then by 6.1.4 and 3.3.4, we have $\dim A = n$.

If A is integral over K, then the conditions are void, *i.e.* $n = 0$ and $B = K$. So let us suppose that A is not integral over K.

Let us first establish the theorem for $A = K[X_1, \ldots, X_n]$, the ring of polynomials over K.

If $p = 1$, then the theorem is just a consequence of 6.2.1. So let us suppose that $p \geqslant 2$. By induction on p, there exist $t_1, \ldots, t_n \in A$ verifying conditions (i) and (ii) with respect to the chain $\mathfrak{a}_1 \subset \cdots \subset \mathfrak{a}_{p-1}$. Set $r = h(p-1)$, $B = K[t_1, \ldots, t_n]$ and $C = K[t_{r+1}, \ldots, t_n]$.

Applying 6.2.1 to C and the ideal $C \cap \mathfrak{a}_p$ of C, we obtain elements $x_{r+1}, \ldots, x_n \in C$ algebraically independent over K such that C is integral over $D = K[x_{r+1}, \ldots, x_n]$, and there is an integer $h(p)$ such that $D \cap \mathfrak{a}_p$ is generated by $x_{r+1}, \ldots, x_{h(p)}$.

Let $x_i = t_i$ for $1 \leqslant i \leqslant r$ and $E = K[x_1, \ldots, x_n]$. Then by construction, $\mathrm{Fract}(B)$ is algebraic over $\mathrm{Fract}(E)$. Thus $x_1, \ldots, x_n$ are algebraically independent over K. Further, by 3.1.7, A is integral over E. So we have (i).

For $1 \leqslant i \leqslant p$, $E \cap \mathfrak{a}_i$ contains obviously the ideal $Ex_1 + \cdots + Ex_{h(i)}$ of E. If $i \leqslant p-1$, then $E \cap \mathfrak{a}_i = E \cap (Bx_1 + \cdots + Bx_{h(i)})$. But by 5.7.8 and 6.1.5, $E \cap \mathfrak{a}_i$ and $Ex_1 + \cdots + Ex_{h(i)}$ are prime ideals of E of height $h(i)$. So they must be equal.

Let $z \in E \cap \mathfrak{a}_p$. We can write

$$z = \sum_{\lambda = (i_1, \ldots, i_r)} a_\lambda x_1^{i_1} \cdots x_r^{i_r}$$

where $a_\lambda \in D$. Clearly $a_\lambda x_1^{i_1} \cdots x_r^{i_r} \in \mathfrak{a}_{p-1} \subset \mathfrak{a}_p$ if $\lambda \neq \lambda_0 = (0, \ldots, 0)$. It follows that $a_{\lambda_0} \in D \cap \mathfrak{a}_p$. This implies that $E \cap \mathfrak{a}_p \subset Ex_1 + \cdots + Ex_{h(p)}$, and since clearly $x_1, \ldots, x_{h(p)} \in E \cap \mathfrak{a}_p$, we have condition (ii).

Now let us suppose that A is a finitely generated K-algebra. Then there exists a surjective ring homomorphism $\pi : A' = K[X_1, \ldots, X_m] \to A$ for some $m \in \mathbb{N}$. Let $\mathfrak{a}'_0 = \ker \pi$ and $\mathfrak{a}'_i = \pi^{-1}(\mathfrak{a}_i)$ for $1 \leqslant i \leqslant p$.

Since A' is a polynomial ring, there exist $x'_1, \ldots, x'_m \in A'$ verifying conditions (i) and (ii) with respect to the chain $\mathfrak{a}'_0 \subset \cdots \subset \mathfrak{a}'_p$. By 3.1.8, A is integral over the K-algebra generated by $x_j = \pi(x'_j)$, $j > h(0)$. Thus $s > h(0)$ since A is not integral over K. To prove the theorem, it suffices therefore to show that $x_{h(0)+1}, \ldots, x_m$ are algebraically independent over K.

Let $t = h(0) + 1$. If $P(x_t, \ldots, x_m) = 0$ for some non zero polynomial over K, then

$$P(x'_t, \ldots, x'_m) \in \mathfrak{a}'_0 \cap K[x'_1, \ldots, x'_m] = Bx'_1 + \cdots + Bx'_{h(0)}$$

where $B = K[x'_1, \ldots, x'_m]$. But this contradicts the fact that $x'_1, \ldots, x'_m$ are algebraically independent. Hence $x_t, \ldots, x_m$ are algebraically independent over K and the theorem follows. $\square$

6.2.3 Corollary. *Let A be a finitely generated K-algebra. We have*

$$\dim A = \max_{\mathfrak{p} \in \mathrm{Spec}(A)} \mathrm{tr}\deg_K \mathrm{Fract}(A/\mathfrak{p}) = \max_{\mathfrak{p} \in \mathrm{Spec}_m(A)} \mathrm{tr}\deg_K \mathrm{Fract}(A/\mathfrak{p}).$$

In particular, if A is an integral domain, then $\dim A = \mathrm{tr}\deg_K \mathrm{Fract}\, A$.

Proof. By 6.2.2, A is integral over a polynomial ring B contained in A, and $\dim A = \dim B$. By 3.1.8, $A/\mathfrak{p}$ is integral over $B/(B \cap \mathfrak{p})$ for any $\mathfrak{p} \in \operatorname{Spec}(A)$. So $\operatorname{tr} \deg_K \operatorname{Fract}(A/\mathfrak{p}) = \operatorname{tr} \deg_K \operatorname{Fract}(B/(B \cap \mathfrak{p})) \leqslant \operatorname{tr} \deg_K \operatorname{Fract} B$. Thus the result follows from 6.1.3 and 6.1.4.

Finally, if A is an integral domain, then $\{0\}$ is the unique minimal prime ideal of A. $\square$

6.2.4 Corollary. *Let A be a finitely generated K-algebra such that $\dim A = 0$. If A is an integral domain, then A is a field and the extension $K \subset A$ is finite.*

Proof. If A is an integral domain and $\dim A = 0$, then $\{0\}$ is a maximal ideal. So A is a field. By 6.2.3, the extension $K \subset A$ is algebraic. So it is finite since A is a finitely generated K-algebra. $\square$

6.2.5 Corollary. *Let A be a finitely generated K-algebra. Then we have $\dim A[X] = 1 + \dim A$.*

Proof. By 6.2.2, A is integral over a polynomial ring B contained in A. So $A[X]$ is integral over $B[X]$. Hence the result follows from 6.2.2. $\square$

6.2.6 Proposition. *Let A be a finitely generated K-algebra which is an integral domain.*

(i) *For any $\mathfrak{p} \in \operatorname{Spec}(A)$, we have:*

$$\operatorname{ht} \mathfrak{p} + \dim(A/\mathfrak{p}) = \dim A.$$

(ii) *Let $\mathfrak{p}, \mathfrak{q} \in \operatorname{Spec}(A)$ be such that $\mathfrak{p} \subset \mathfrak{q}$. Then:*

$$\operatorname{ht} \mathfrak{q} = \operatorname{ht} \mathfrak{p} + \operatorname{ht}(\mathfrak{q}/\mathfrak{p}).$$

Proof. Part (i) is clear for $\mathfrak{p} = \{0\}$. So let us suppose that $\mathfrak{p} \neq \{0\}$. By 6.2.2, there exist $x_1, \ldots, x_n \in A$ such that $B = K[x_1, \ldots, x_n]$ is isomorphic to a polynomial ring, A is integral over B and $B \cap \mathfrak{p} = Bx_1 + \cdots + Bx_r$ where $1 \leqslant r \leqslant n$.

By 5.7.8 and 6.1.5, $\operatorname{ht} \mathfrak{p} = \operatorname{ht}(B \cap \mathfrak{p}) = r$. On the other hand, $B/(B \cap \mathfrak{p})$ is isomorphic to $K[x_{r+1}, \ldots, x_n]$, so its dimension is $n - r$. Since A is integral over B, $A/\mathfrak{p}$ is integral over $B/(B \cap \mathfrak{p})$ (3.1.8). This proves part (i).

Finally, since

$$\dim A = \operatorname{ht} \mathfrak{p} + \dim(A/\mathfrak{p}) = \operatorname{ht} \mathfrak{q} + \dim(A/\mathfrak{q}),$$
$$\dim(A/\mathfrak{p}) = \operatorname{ht}(\mathfrak{q}/\mathfrak{p}) + \dim(A/\mathfrak{q}),$$

part (ii) follows from part (i). $\square$

6.2.7 Definition. *A chain $\mathfrak{p}_0 \subset \cdots \subset \mathfrak{p}_n$ of prime ideals of length n of a ring A is called* maximal *if it satisfies the following conditions:*

(i) $\mathfrak{p}_0 \in \operatorname{Spec}_m(A)$ *and* $\mathfrak{p}_n \in \operatorname{Spm}(A)$.

(ii) *For any integer $i \in \{0, 1, \ldots, n-1\}$, the set of $\mathfrak{q} \in \operatorname{Spec} A$ verifying $\mathfrak{p}_i \subset \mathfrak{q} \subset \mathfrak{p}_{i+1}$ is $\{\mathfrak{p}_i, \mathfrak{p}_{i+1}\}$.*

6.2.8 Proposition. *Let A be a finitely generated K-algebra which is an integral domain. Then all maximal chains of prime ideals of A has length $\dim A$.*

Proof. Let $\{0\} = \mathfrak{p}_0 \subset \cdots \subset \mathfrak{p}_n$ be a maximal chain of prime ideals of A. For $1 \leqslant i \leqslant n$, the height of $\mathfrak{p}_i/\mathfrak{p}_{i-1} \in \operatorname{Spec}(A/\mathfrak{p}_{i-1})$ is 1. Hence by 6.2.6, $\dim(A/\mathfrak{p}_{i-1}) - \dim(A/\mathfrak{p}_i) = 1$. On the other hand, $\dim(A/\mathfrak{p}_n) = 0$ since, $A/\mathfrak{p}_n$ is a field. It follows that $\dim A = \dim(A/\mathfrak{p}_0) = n$. $\square$

6.2.9 Proposition. *Let B be a finitely generated K-algebra which is an integral domain, A a finitely generated K-subalgebra of B. There exist a non-zero element $t \in A$ and a subalgebra $C = A[y_1, \ldots, y_n]$ of B verifying the following conditions:*

(i) *The elements $y_1, \ldots, y_n$ are algebraically independent over K.*
(ii) *The subalgebra $B[1/t]$ of $\operatorname{Fract}(B)$ is integral over $C[1/t]$.*

Proof. Let $L = \operatorname{Fract}(A)$ and $z_1, \ldots, z_r \in B$ be such that $B = A[z_1, \ldots, z_r]$. By applying 6.2.2 to the L-algebra $L[z_1, \ldots, z_r]$, we can find $x_1, \ldots, x_n \in B$ algebraically independent over L such that $L[z_1, \ldots, z_r]$ is integral over $L[x_1, \ldots, x_n]$. Let

$$z_i^{m_i} + \sum_{j=0}^{m_i-1} P_{ij}(x_1, \ldots, x_n) z_i^j = 0$$

be an integral equation for z_i over $L[x_1, \ldots, x_n]$, where $P_{ij} \in L[T_1, \ldots, T_n]$.

Let $k = \max_{1 \leqslant j \leqslant m_i - 1} \{\deg P_{ij}\}$ and $t \in A$ non-zero such that $tP_{ij} \in A[T_1, \ldots, T_n]$ for all j. Then there exist $Q_{ij} \in A[T_1, \ldots, T_n]$ such that $t^{k+1} P_{ij}(x_1, \ldots, x_n) = Q_{ij}(tx_1, \ldots, tx_n)$ and so

$$t^{k+1} z_i^{m_i} + \sum_{j=0}^{m_i-1} Q_{ij}(tx_1, \ldots, tx_n) z_i^j = 0.$$

Set $y_i = tx_i$. Clearly, the y_i's are algebraically independent over L, hence over K. The above equality implies that z_i is integral over $A[y_1, \ldots, y_n][1/t]$. Hence $B[1/t]$ is integral over $C[1/t]$. $\square$

6.2.10 Proposition. *Let A be a finitely generated K-algebra which is an integral domain, $F = \operatorname{Fract}(A)$, $F \subset L$ a finite field extension and B the integral closure of A in L. Suppose that the characteristic of K is zero. Then B is a finitely generated A-module (so B is a finitely generated K-algebra which is an integral domain).*

Proof. By 6.2.2, A contains a polynomial ring C over which A is integral. Clearly, B is also the integral closure of C in L. Since $\operatorname{Fract}(C) \subset F$ is a finite extension, we can therefore assume that A is a polynomial ring. The result then follows from 4.2.9 and 5.5.6. $\square$

6.3 Krull's Principal Ideal Theorem

6.3.1 Lemma. *Let $C \subset A$ be integral domains, $F = \mathrm{Fract}(C)$ and $K = \mathrm{Fract}(A)$. Suppose that C is an integrally closed domain, A is integral over C and $F \subset K$ is a finite extension. Let $x \in A$ be such that $\sqrt{Ax} \in \mathrm{Spec}\, A$. Then $y = \mathrm{Norm}_{K/F}(x) \in C$ and $C \cap \sqrt{Ax} = \sqrt{Cy}$.*

Proof. Let $P = X^m + c_{m-1}X^{m-1} + \cdots + c_0$ be the minimal polynomial of x over F. By 5.5.4, the c_i's are integral over C. So $P \in C[X]$ and since $P(x) = 0$, $c_0 \in Ax$. It follows therefore from 5.5.2 that $y \in C \cap \sqrt{Ax}$. Hence $\sqrt{Cy} \subset C \cap \sqrt{Ax}$.

Conversely, if $u \in C \cap \sqrt{Ax}$, then there exist $n \in \mathbb{N}^*$ and $a \in A$ such that $u^n = ax$. Hence $u \in \sqrt{Cy}$ because $u^{n[K:F]} = \mathrm{Norm}_{K/F}(u^n) = \mathrm{Norm}_{K/F}(a)\,\mathrm{Norm}_{K/F}(x) \in Cy$. $\square$

6.3.2 The following theorem is also called *Krull's Hauptidealsatz.*

Theorem. (Krull's Principal Ideal Theorem) *Let A be a finitely generated K-algebra, $x \in A$ non invertible and $\mathfrak{p} \in \mathrm{Spec}\, A$ minimal such that $x \in \mathfrak{p}$. Then $\mathrm{ht}\,\mathfrak{p} \leqslant 1$ and equality holds if x is not a zero divisor.*

Proof. Note that such a $\mathfrak{p}$ exists by applying 2.2.7 to A/Ax.

Let us suppose that $\mathrm{ht}\,\mathfrak{p} = r \geqslant 1$. Let $\mathfrak{q}_0 \subset \cdots \subset \mathfrak{q}_r = \mathfrak{p}$ be a maximal chain of prime ideals. It suffices therefore to show that $\mathrm{ht}\,\mathfrak{p}/\mathfrak{q}_0 = 1$ in $A/\mathfrak{q}_0$. We are therefore reduced to the case where A is an integral domain and $x \neq 0$.

By 2.7.8, the set of prime ideals containing x has a finite number of minimal elements, say $\mathfrak{p}, \mathfrak{p}_2, \ldots, \mathfrak{p}_s$. Let $u \in (\mathfrak{p}_2 \cap \cdots \cap \mathfrak{p}_s) \setminus \mathfrak{p}$, $S = \{u^n; u \in \mathbb{N}\}$ and $B = S^{-1}A$. Since B is a finitely generated K-algebra and $\mathrm{Fract}\, A = \mathrm{Fract}\, B$, we have $\dim A = \dim B$ and $\mathrm{ht}\,\mathfrak{p} = \mathrm{ht}(B\mathfrak{p})$ by 6.2.3 and 2.3.9. Further, $B\mathfrak{p}$ is now the unique prime ideal of B among those containing x. So we can suppose that $\mathfrak{p}$ is the unique minimal element of the set of prime ideals containing x. By 2.7.7, $\sqrt{Ax} = \mathfrak{p}$.

By 6.2.2, A contains a polynomial ring C over which A is integral. Let $F = \mathrm{Fract}(C)$, $K = \mathrm{Fract}(A)$ and $y = \mathrm{Norm}_{K/F}(x)$. Then 6.3.1 implies that $C \cap \mathfrak{p} = \sqrt{Cy}$. Now C is factorial, so there exists $z \in C$ irreducible such that $\sqrt{Cy} = Cz$. So $\mathrm{ht}\,\mathfrak{p} = \mathrm{ht}\,\sqrt{Cy} = 1$ by 5.7.8. $\square$

6.3.3 Corollary. *Let A be a finitely generated K-algebra and $\mathfrak{a}$ a proper ideal of A generated by $x_1, \ldots, x_n$. If $\mathfrak{p}$ is minimal among the prime ideals of A containing $\mathfrak{a}$, then $\mathrm{ht}\,\mathfrak{p} \leqslant n$.*

Proof. Let us proceed by induction on n. We have treated the case $n = 1$ in 6.3.2. So let us suppose that $n \geqslant 2$. By the same argument as in the proof of 6.3.2, we are reduced to the case where A is an integral domain.

Now the set $\mathcal{E}$ of $\mathfrak{q} \in \mathrm{Spec}\, A$ such that $Ax_1 + \cdots + Ax_{n-1} \subset \mathfrak{q} \subset \mathfrak{p}$ is not empty. Let $\mathfrak{q}$ be minimal in $\mathcal{E}$. By our induction hypothesis, $\mathrm{ht}\,\mathfrak{q} \leqslant n - 1$. Further, $\mathfrak{p}/\mathfrak{q}$ is minimal among the prime ideals of $A/\mathfrak{q}$ containing the class of x_n modulo $\mathfrak{q}$. So the result follows from 6.2.6 and 6.3.2. $\square$

6.3.4 Corollary. *Any maximal ideal of $A = K[X_1, \ldots, X_n]$, the polynomial ring in n indeterminates, can be generated by n elements.*

Proof. Let $\mathfrak{m} \in \mathrm{Spm}(A)$. Then by 6.1.4 et 6.2.6, $\mathrm{ht}\,\mathfrak{m} = n$. Hence 6.3.3 implies that any system of generators of $\mathfrak{m}$ has at least n elements.

If $n = 1$, the result is obvious since $K[X_1]$ is a principal ideal domain. Let us suppose that $n \geqslant 2$ and proceed by induction on n.

If $\mathfrak{m} \cap K[X_1] = \{0\}$, then let $S = K[X_1] \setminus \{0\}$ and $S^{-1}\mathfrak{m}$ is a maximal ideal of $K(X_1)[X_2, \ldots, X_n]$. Therefore $\mathrm{ht}\,S^{-1}\mathfrak{m} = n - 1$. This is not possible since $\mathrm{ht}\,S^{-1}\mathfrak{m} = \mathrm{ht}\,\mathfrak{m} = n$.

Thus $\mathfrak{m} \cap K[X_1]$ is a (non-zero) maximal ideal $K[X_1]x$ where $x \in K[X_1]$ is irreducible. Clearly, x is also irreducible in A and consequently, Ax is a prime ideal of A.

Let $y_2, \ldots, y_n$ be the images of $X_2, \ldots, X_n$ in A/Ax. We have $A/Ax = L[y_2, \ldots, y_n]$ where $L = K[X_1]/(K[X_1]x)$ and by 6.2.6, $\dim A/Ax = n - 1$. It follows from 6.2.3 that the y_i's are algebraically independent over L.

By the induction hypothesis, the maximal ideal $\mathfrak{m}/Ax$ of A/Ax is generated by $n - 1$ elements. Consequently, if we choose representatives of these $n - 1$ elements in A, then these $n - 1$ representatives together with x form a system of generators for $\mathfrak{m}$. $\square$

6.4 Maximal ideals

6.4.1 The following result is a direct consequence of 6.2.4.

Proposition. *Let A be a finitely generated K-algebra and $\mathfrak{m} \in \mathrm{Spm}(A)$. The extension $K \subset A/\mathfrak{m}$ is finite. Moreover, if K is algebraically closed, then the natural map $K \to A \to A/\mathfrak{m}$ is an isomorphism of K onto $A/\mathfrak{m}$.*

6.4.2 Theorem. (Hilbert's Nullstellensatz) *Let $\mathfrak{a}$ be a proper ideal of a finitely generated K-algebra A. Then $\sqrt{\mathfrak{a}}$ is the intersection of maximal ideals of A containing $\mathfrak{a}$.*

Proof. By replacing A by $A/\mathfrak{a}$, it suffices to prove that the nilradical $\mathfrak{n}$ of A is the intersection $\mathfrak{r}$ of maximal ideals of A. Clearly, $\mathfrak{n} \subset \mathfrak{r}$ (see also 2.5.5).

Let $a \in \mathfrak{r}$. If $a \notin \mathfrak{n}$, then $S = \{a^n; n \in \mathbb{N}\}$ does not contain 0. The K-algebra $B = S^{-1}A$ is again finitely generated (just add $1/a$ to a system of generators of A).

Let $i : A \to B$ be the canonical map (see 2.3.3), $\mathfrak{m} \in \mathrm{Spm}(B)$ and $i^{-1}(\mathfrak{m}) = \mathfrak{p} \in \mathrm{Spec}(A)$ (see 2.3.8). By 6.4.1, $K \subset B/\mathfrak{m}$ is finite. So we deduce, via the canonical injections $K \to A/\mathfrak{p} \to B/\mathfrak{m}$, that $\dim A/\mathfrak{p} = 0$. So $\mathfrak{p} \in \mathrm{Spm}(A)$ by 6.2.4. But $a \notin \mathfrak{m}$, so $a \notin \mathfrak{p}$ which is absurd since $a \in \mathfrak{r}$. $\square$

6.4.3 Proposition. *Let $A = K[X_1, \ldots, X_n]$ be the polynomial ring in n indeterminates, and $\mathfrak{a}$ an ideal of A. Suppose that K is an algebraically closed field. The following conditions are equivalent:*

(i) *There exist $a_1, \ldots, a_n \in K$ such that $\mathfrak{a} = \sum_{i=1}^{n} A(X_i - a_i)$.*

(ii) *We have $\mathfrak{a} \in \mathrm{Spm}(A)$.*

Proof. (i) $\Rightarrow$ (ii) This is clear since $A/\mathfrak{a}$ is isomorphic to K.

(ii) $\Rightarrow$ (i) Let $\mathfrak{a} \in \mathrm{Spm}\, A$. Then 6.2.6 implies that $K \subset A/\mathfrak{a}$ is finite and since K is algebraically closed, we have $K = A/\mathfrak{a}$. Let a_i be the image of X_i in $A/\mathfrak{a} = K$. Then $\mathfrak{m} = A(X_1 - a_1) + \cdots + A(X_n - a_n) \subset \mathfrak{a}$. But $\mathfrak{m} \in \mathrm{Spm}(A)$, so $\mathfrak{m} = \mathfrak{a}$. $\square$

6.4.4 Corollary. *Let A, B be finitely generated K-algebras. Suppose that K is algebraically closed.*

(i) *Let $\mathfrak{m} \in \mathrm{Spm}(A)$. Then $A = K.1 \oplus \mathfrak{m}$.*

(ii) *Let $\rho : A \to B$ be a K-algebra homomorphism. If $\mathfrak{m} \in \mathrm{Spm}(B)$, then $\rho^{-1}(\mathfrak{m}) \in \mathrm{Spm}(A)$.*

Proof. Part (i) is clear since $1 \notin \mathfrak{m}$ and $A/\mathfrak{m} = K$ by 6.4.1.

(ii) Since $\rho(1) = 1$ and $B/\mathfrak{m} = K$ by part (i), $\rho^{-1}(\mathfrak{m}) \neq A$ and the induced map $A/\rho^{-1}(\mathfrak{m}) \to B/\mathfrak{m}$ is an isomorphism. So $\rho^{-1}(\mathfrak{m}) \in \mathrm{Spm}(A)$. $\square$

6.4.5 For any K-algebras A, B, let us denote by $\mathrm{Hom}_{\mathrm{alg}}(A, B)$ the set of K-algebra homomorphisms from A to B.

Given a finitely generated K-algebra A, it is clear from 6.4.4 that there is a canonical bijection:

$$\mathrm{Hom}_{\mathrm{alg}}(A, K) \to \mathrm{Spm}(A) \ , \ \chi \mapsto \ker \chi$$

whose inverse is given by the projection $A = K.1 \oplus \mathfrak{m} \to K = A/\mathfrak{m}$ for $\mathfrak{m} \in \mathrm{Spm}\, A$.

Now if B is another finitely generated K-algebra and $\varphi \in \mathrm{Hom}_{\mathrm{alg}}(A, B)$, then the map

$$\varphi_* : \mathrm{Hom}_{\mathrm{alg}}(B, K) \to \mathrm{Hom}_{\mathrm{alg}}(A, K) \ , \ \chi \mapsto \chi \circ \varphi$$

induces a map $\mathrm{Spm}(\varphi) : \mathrm{Spm}(B) \to \mathrm{Spm}(A)$ via the canonical bijections above. More precisely, if $\mathfrak{n} \in \mathrm{Spm}(B)$, then $\mathrm{Spm}(\varphi)(\mathfrak{n}) = \varphi^{-1}(\mathfrak{n})$.

6.4.6 We can view A as an algebra of functions on $\mathrm{Spm}(A)$ as follows: let $f \in A$ and $\mathfrak{m} \in \mathrm{Spm}(A)$, then we define $f(\mathfrak{m})$ to be the class of f in $A/\mathfrak{m} = K$.

Note that if $\varphi \in \mathrm{Hom}_{\mathrm{alg}}(A, B)$, then for $\mathfrak{n} \in \mathrm{Spm}(B)$ and $f \in A$, we have:

$$(f \circ \mathrm{Spm}(\varphi))(\mathfrak{n}) = f(\varphi^{-1}(\mathfrak{n})) = \varphi(f)(\mathfrak{n})$$

since the homomorphism $A/\varphi^{-1}(\mathfrak{n}) \to B/\mathfrak{n}$ induced by φ is just the identity map from K to itself.

6.5 Zariski topology

6.5.1 Let A be a finitely generated K-algebra. For any subset $\mathcal{E}$ of A, we shall denote by $\mathcal{V}(\mathcal{E})$ the set of maximal ideals of A which contain $\mathcal{E}$. When $\mathcal{E} = \{f_1, \ldots, f_n\}$, we shall also write $\mathcal{V}(f_1, \ldots, f_n)$ for $\mathcal{V}(\mathcal{E})$.

6.5.2 Remark. Suppose that K is algebraically closed. Then considering A as functions as in 6.4.6, $\mathcal{V}(\mathcal{E})$ can be identified as the set of common zeros of the elements in $\mathcal{E}$.

Note also that if $\mathfrak{a}$ is an ideal of A generated by $\mathcal{E}$, then $\mathcal{V}(\mathfrak{a}) = \mathcal{V}(\mathcal{E})$.

6.5.3 Proposition. *Let A be a finitely generated K-algebra, $\mathfrak{a}, \mathfrak{b}$ and $(\mathfrak{a}_i)_{i \in I}$ ideals of A. We have:*

(i) $\mathcal{V}(\mathfrak{a}) = \mathcal{V}(\sqrt{\mathfrak{a}})$ *and if $\mathfrak{a} \subset \mathfrak{b}$, then $\mathcal{V}(\mathfrak{b}) \subset \mathcal{V}(\mathfrak{a})$.*

(ii) $\mathcal{V}(\mathfrak{a}) = \emptyset$ *if and only if $\mathfrak{a} = A$.*

(iii) $\mathcal{V}(\mathfrak{a}) = \mathrm{Spm}(A)$ *if and only if $\mathfrak{a} \subset \sqrt{0}$.*

(iv) $\mathcal{V}(\sum_{i \in I} \mathfrak{a}_i) = \bigcap_{i \in I} \mathcal{V}(\mathfrak{a}_i)$ *and* $\mathcal{V}(\mathfrak{a}\mathfrak{b}) = \mathcal{V}(\mathfrak{a} \cap \mathfrak{b}) = \mathcal{V}(\mathfrak{a}) \cup \mathcal{V}(\mathfrak{b})$.

Proof. The first point of (i) and part (iii) are consequences of 6.4.2, while the second point of (i), part (ii) and the first point of (iv) are obvious. If $\mathfrak{m} \in \mathrm{Spm}(A)$, then $\mathfrak{m} \supset \mathfrak{a}\mathfrak{b}$ if and only if $\mathfrak{m}$ contains either $\mathfrak{a}$ or $\mathfrak{b}$. So $\mathcal{V}(\mathfrak{a}\mathfrak{b}) = \mathcal{V}(\mathfrak{a}) \cup \mathcal{V}(\mathfrak{b})$. The other equalities follow from part (i) and 2.5.3. $\square$

6.5.4 In view of 6.5.3, we define a topology, called *Zariski topology* on $\mathrm{Spm}(A)$, by letting closed subsets to be the $\mathcal{V}(\mathfrak{a})$'s where $\mathfrak{a}$ is an ideal of A. Observe that by part (i) of 6.5.3, we can restrict ourselves to radical ideals. By convention, when we consider $\mathrm{Spm}(A)$ as a topological space, we assume that the underlying topology is the Zariski topology.

Let $\mathfrak{a}$ be an ideal of A and $B = A/\mathfrak{a}$. Then the map given by $\mathfrak{m} \mapsto \mathfrak{m}/\mathfrak{a}$ is a bijection between the set $\mathcal{V}(\mathfrak{a})$ and $\mathrm{Spm}\,B$. Clearly, via this identification, the Zariski topology of $\mathrm{Spm}(B)$ is the one induced by the Zariski topology of $\mathrm{Spm}(A)$.

Suppose that K is algebraically closed and $\varphi : A \to B$ is the canonical surjection. Then the map $\mathrm{Spm}(\varphi)$ of 6.4.5 is the canonical injection $\mathcal{V}(\mathfrak{a}) \to \mathrm{Spm}\,A$.

6.5.5 Let A be a finitely generated K-algebra. For $M \subset \mathrm{Spm}(A)$, we denote by $\mathcal{I}(M)$ the set of elements $a \in A$ such that $a \in \mathfrak{m}$ for all $\mathfrak{m} \in M$. Clearly,

$$\mathcal{I}(M) = \bigcap_{\mathfrak{m} \in M} \mathfrak{m}.$$

Proposition. *Let $\mathfrak{a}, \mathfrak{b}$ be ideals of A and $M \subset \mathrm{Spm}\,A$.*

(i) *We have $\mathcal{I}(\mathcal{V}(\mathfrak{a})) = \sqrt{\mathfrak{a}}$.*

(ii) *We have $\mathcal{V}(\mathfrak{a}) \subset \mathcal{V}(\mathfrak{b})$ (resp. $\mathcal{V}(\mathfrak{a}) = \mathcal{V}(\mathfrak{b})$) if and only if $\sqrt{\mathfrak{b}} \subset \sqrt{\mathfrak{a}}$ (resp. $\sqrt{\mathfrak{b}} = \sqrt{\mathfrak{a}}$).*

(iii) *The Zariski closure of M in $\mathrm{Spm}(A)$ is $\mathcal{V}(\mathcal{I}(M))$.*

Proof. Part (i) is proved in 6.4.2 and part (ii) is a direct consequence of 6.5.3 (i) and the fact that the map $M \mapsto \mathcal{I}(M)$ reverses inclusions.

To prove part (iii), let $\mathfrak{a}$ be a radical ideal of A such that $M \subset \mathcal{V}(\mathfrak{a})$. Then $\mathfrak{a} = \mathcal{I}(\mathcal{V}(\mathfrak{a})) \subset \mathcal{I}(M)$ by (i). It follows that $\mathcal{V}(\mathcal{I}(M)) \subset \mathcal{V}(\mathfrak{a})$. On the other hand, it is clear that $M \subset \mathcal{V}(\mathcal{I}(M))$. Hence we have proved (iii).   □

6.5.6 Proposition. *Let A be a finitely generated K-algebra.*

(i) $\mathrm{Spm}(A)$ *is a Noetherian topological space whose points are closed.*

(ii) *Let $\mathfrak{a}$ be a radical ideal of A. Then $\mathfrak{a} \in \mathrm{Spec}(A)$ if and only if $\mathcal{V}(\mathfrak{a})$ is irreducible.*

Proof. (i) Since $\mathcal{V}(\mathfrak{m}) = \{\mathfrak{m}\}$ for all $\mathfrak{m} \in \mathrm{Spm}(A)$, points are closed.

Let $(F_p)_p$ be a decreasing sequence of closed subsets of $\mathrm{Spm}(A)$ where $F_p = \mathcal{V}(\mathfrak{a}_p)$, $\mathfrak{a}_p$ a radical ideal of A. Then by 6.5.5, $(\mathfrak{a}_p)_p$ is an increasing sequence of radical ideals of A. Since A is Noetherian (6.1.1), this sequence is stationary. It follows that $(F_p)_p$ is also stationary.

(ii) Suppose that $\mathcal{V}(\mathfrak{a})$ is irreducible and let $f, g \in A$ be such that $fg \in \mathfrak{a}$. We have $\mathcal{V}(\mathfrak{a}) \subset \mathcal{V}(fg) = \mathcal{V}(f) \cup \mathcal{V}(g)$. Hence $\mathcal{V}(\mathfrak{a})$ is contained in either $\mathcal{V}(f)$ or $\mathcal{V}(g)$. Now 6.5.5 implies that $f \in \sqrt{\mathfrak{a}} = \mathfrak{a}$ or $g \in \mathfrak{a}$. So $\mathfrak{a} \in \mathrm{Spec}(A)$.

Conversely, suppose that $\mathfrak{a} \in \mathrm{Spec}(A)$. Let $\mathfrak{b}, \mathfrak{c}$ be radical ideals of A such that $\mathcal{V}(\mathfrak{a}) \subset \mathcal{V}(\mathfrak{b}) \cup \mathcal{V}(\mathfrak{c})$. So $\mathfrak{bc} \subset \sqrt{\mathfrak{a}} = \mathfrak{a}$ by 6.5.3 and 6.5.5. Since $\mathfrak{a}$ is prime, $\mathfrak{a}$ contains either $\mathfrak{b}$ or $\mathfrak{c}$. Hence $\mathcal{V}(\mathfrak{a})$ is contained in either $\mathcal{V}(\mathfrak{b})$ or $\mathcal{V}(\mathfrak{c})$.   □

6.5.7 Corollary. *Let $\mathfrak{a}$ be an ideal of A and $\mathfrak{p}_1, \ldots, \mathfrak{p}_n$ be the minimal elements of the set of prime ideals of A containing $\mathfrak{a}$. Then the set of irreducible components of $\mathcal{V}(\mathfrak{a})$ is the finite set $\{\mathcal{V}(\mathfrak{p}_i) \; ; \; 1 \leqslant i \leqslant n\}$.*

In particular, the irreducible components of $\mathrm{Spm}(A)$ are the closed sets associated to minimal prime ideals of A. So $\mathrm{Spm}(A)$ is irreducible if and only if $\sqrt{0} \in \mathrm{Spec}(A)$. Furthermore, if A is reduced, then $\mathrm{Spm}(A)$ is irreducible if and only if A is an integral domain.

Proof. By 2.7.7 and 2.7.8, we have $\sqrt{\mathfrak{a}} = \mathfrak{p}_1 \cap \cdots \cap \mathfrak{p}_n$, and so 6.5.3 implies that $\mathcal{V}(\mathfrak{a}) = \mathcal{V}(\mathfrak{p}_1) \cup \cdots \cup \mathcal{V}(\mathfrak{p}_n)$. The result follows from 6.5.6 and 6.5.5.   □

6.5.8 Proposition. *Suppose that K is algebraically closed. Let A, B be finitely generated K-algebras and $\varphi \in \mathrm{Hom}_{\mathrm{alg}}(B, A)$.*

(i) *The map $\mathrm{Spm}(\varphi) : \mathrm{Spm}(A) \to \mathrm{Spm}(B)$ defined in 6.4.5 is continuous.*

(ii) *If $\mathfrak{b}$ is an ideal of B, then $\mathrm{Spm}(\varphi)^{-1}(\mathcal{V}(\mathfrak{b})) = \mathcal{V}(A\varphi(\mathfrak{b}))$.*

(iii) *For any ideal $\mathfrak{a}$ of A, the Zariski closure of $\mathrm{Spm}(\varphi)(\mathcal{V}(\mathfrak{a}))$ is the set $\mathcal{V}(\varphi^{-1}(\mathfrak{a}))$.*

Proof. It suffices to prove (ii) and (iii). Let $\mathfrak{m} \in \mathrm{Spm}(A)$, then

$$\mathfrak{m} \in \mathcal{V}(A\varphi(\mathfrak{b})) \Leftrightarrow A\varphi(\mathfrak{b}) \subset \mathfrak{m} \Leftrightarrow \varphi(\mathfrak{b}) \subset \mathfrak{m} \Leftrightarrow \mathfrak{b} \subset \varphi^{-1}(\mathfrak{m}) = \mathrm{Spm}(\varphi)(\mathfrak{m})$$
$$\Leftrightarrow \mathrm{Spm}(\varphi)(\mathfrak{m}) \in \mathcal{V}(\mathfrak{b}).$$

This proves (ii).

Let $\mathfrak{a}$ be an ideal of A. By 6.5.5, the Zariski closure of $\mathrm{Spm}(\varphi)(\mathcal{V}(\mathfrak{a}))$ is $\mathcal{V}(\mathcal{I}(\varphi(\mathcal{V}(\mathfrak{a}))))$, the set of maximal ideals of B containing the intersection of the $\varphi^{-1}(\mathfrak{m})$'s where $\mathfrak{m} \in \mathcal{V}(\mathfrak{a})$. Now

$$\bigcap_{\mathfrak{m}\in\mathcal{V}(\mathfrak{a})} \varphi^{-1}(\mathfrak{m}) = \varphi^{-1}\Big(\bigcap_{\mathfrak{m}\in\mathcal{V}(\mathfrak{a})} \mathfrak{m}\Big) = \varphi^{-1}(\sqrt{\mathfrak{a}}) = \sqrt{\varphi^{-1}(\mathfrak{a})}.$$

Hence (iii) follows from 6.5.3. □

6.5.9 Corollary. *Let us keep the notations and hypotheses of 6.5.8. Suppose further that A and B are reduced. Then φ is injective if and only if $\mathrm{Spm}(\varphi)$ is dominant.*

Proof. By 6.5.8 (iii), $\overline{\mathrm{Spm}(\varphi)(\mathrm{Spm}(A))} = \mathcal{V}(\varphi^{-1}(\{0\}))$. Note that $\{0\}$ is radical in the reduced ring A. So the equality $\varphi^{-1}(\sqrt{\mathfrak{a}}) = \sqrt{\varphi^{-1}(\mathfrak{a})}$ implies that $\varphi^{-1}(\{0\})$ is radical in B. Since B is also reduced, it follows from 6.5.3 and 6.5.5 that $\mathcal{V}(\varphi^{-1}(\{0\})) = \mathrm{Spm}(B)$ if and only if $\varphi^{-1}(\{0\}) = \{0\}$. □

References

- [9], [26], [52], [56].

7

Gradings and filtrations

In this chapter, we study the notions of gradings and filtrations on rings and modules. These notions are useful in many areas of mathematics. We shall use them in particular in Chapters 13, 30 and 31.

7.1 Graded rings and graded modules

7.1.1 Definition. *Let G be an abelian group. A* grading *on G is a family $(G_n)_{n \in \mathbb{Z}}$ of subgroups of G such that G is the direct sum of the G_n's. A group equipped with a grading is called a* graded group.

7.1.2 Let G be a graded group. An element $x \in G$ is called *homogeneous of degree n* if $x \in G_n$. If x is non-zero and homogeneous, then x is contained in a unique G_n and we shall denote by $n = \deg(x)$ the *degree* of x.

Any $y \in G$ can be written uniquely in the form $\sum_n y_n$ where $y_n \in G_n$. We call y_n the *homogeneous component of degree n* of y.

When there is no confusion, we shall denote the grading of a graded group G by $(G_n)_{n \in \mathbb{Z}}$.

7.1.3 Definition. *Let A be a ring. A grading $(A_n)_{n \in \mathbb{Z}}$ on A is said to be* compatible *with the ring structure if $A_n A_m \subset A_{n+m}$ for all $m, n \in \mathbb{Z}$. A ring equipped with a compatible grading is called a* graded ring.

7.1.4 Definition. *Let A be a graded ring and M a left A-module. A grading $(M_n)_{n \in \mathbb{Z}}$ on M is said to be* compatible *with the A-module structure if $A_m M_n \subset M_{m+n}$ for all $m, n \in \mathbb{Z}$. A left A-module equipped with a compatible grading is called a* graded left A-module.

7.1.5 Proposition. *Let A be a graded ring. Then A_0 is a subring of A.*

Proof. Clearly, $A_0 A_0 \subset A_0$. Now let $(e_n)_{n \in \mathbb{Z}}$ be the set of homogeneous components of 1. For any $x \in A_p$, we have

$$x = x.1 = \sum_{n \in \mathbb{Z}} x e_n = 1.x = \sum_{n \in \mathbb{Z}} e_n x.$$

By comparing the homogeneous components of degree p, we obtain that $x = xe_0 = e_0x$. Since any element of A is the sum of its homogeneous components, the equality $x = xe_0 = e_0x$ is valid for all $x \in A$. Hence $1 = e_0 \in A_0$. $\square$

7.1.6 It follows from 7.1.5 that if M is a graded left A-module, then each M_n is a left A_0-module. We define graded right A-modules similarly. All modules considered in the rest of this chapter are left modules.

7.1.7 Definition. (i) *Let A and B be graded rings. A ring homomorphism $f : A \to B$ is called* graded *if $f(A_n) \subset B_n$ for all $n \in \mathbb{Z}$.*

(ii) *Let A be a graded ring and M, N graded A-modules. A homomorphism $u : M \to N$ of A-modules is called* graded of degree $p \in \mathbb{Z}$ *if $u(M_n) \subset N_{n+p}$ for all $n \in \mathbb{Z}$.*

A homomorphism $u : M \to N$ of A-modules is called graded *if it is graded of degree p for some $p \in \mathbb{Z}$. Note that if u is non-zero, then p is uniquely determined by u.*

7.2 Graded submodules

7.2.1 Proposition. *Let M be a graded A-module and N a submodule of M. The following conditions are equivalent:*

(i) *N is a graded submodule, i.e. $N = \bigoplus_{n \in \mathbb{Z}}(N \cap M_n)$.*

(ii) *The homogeneous components of any element of N are in N.*

(iii) *N is generated by homogeneous elements.*

Proof. This is straightforward. $\square$

7.2.2 Corollary. *Let N be a graded submodule of a graded A-module M.*

(i) *If $(x_i)_{i \in I}$ is a system of generators of N, then the homogeneous components of the x_i's form a system of homogeneous generators of N.*

(ii) *If N is finitely generated, then it has a finite system of homogeneous generators.*

7.2.3 Let N be a graded submodule of a graded A-module M, then one verifies easily that $M/N = \bigoplus_{n \in \mathbb{Z}}(N + M_n)/N$ is a graded *(quotient)* module.

7.2.4 Let A be a graded ring. A graded submodule of A is called a *graded (left) ideal*. If $\mathfrak{a}$ is a bilateral ideal of A, then the preceding paragraph says that the quotient $A/\mathfrak{a}$ is again a graded ring.

7.2.5 In the rest of this section, A is a commutative graded ring. We call a graded A-module $(M_n)_{n \in \mathbb{Z}}$ *positively graded* if $M_n = \{0\}$ for all $n < 0$.

7.2.6 Proposition. *Let $\mathfrak{a}$ be a graded ideal of A. Then $\mathfrak{a}$ is prime if and only if given any homogeneous elements $x, y \in A$ such that $xy \in \mathfrak{a}$, we have $x \in \mathfrak{a}$ or $y \in \mathfrak{a}$.*

Proof. The condition is obviously necessary. Let us prove that it is sufficient. Let $a, b \in A$ be such that $ab \in \mathfrak{a}$. Write $a = a_{i_1} + \cdots + a_{i_r}$, $b = b_{j_1} + \cdots + b_{j_s}$ where $i_1 < \cdots < i_r$, $j_1 < \cdots < j_s$ and the a_{i_k}'s and b_{j_l}'s are homogeneous of degree i_k and j_l.

Now we can assume that none of the b_{j_l}'s belong to $\mathfrak{a}$. Since $ab \in \mathfrak{a}$ and $\mathfrak{a}$ is graded, the homogeneous component of degree $i_r j_s$ of ab, which is $a_{i_r} b_{j_s}$, belongs to $\mathfrak{a}$. It follows that $a_{i_r} \in \mathfrak{a}$, and so $(a - a_{i_r})b \in \mathfrak{a}$. The result follows by induction. $\square$

7.2.7 Proposition. *Let $\mathfrak{a}$ be a graded ideal of A. Then $\sqrt{\mathfrak{a}}$ is graded.*

Proof. Let $x \in \sqrt{\mathfrak{a}}$. Write $x = x_{i_1} + \cdots + x_{i_r}$ where $i_1 < \cdots < i_r$ and x_{i_k} is homogeneous of degree i_k. There exists $n \in \mathbb{N}^*$ such that $x^n \in \mathfrak{a}$. Since $x_{i_r}^n$ is the homogeneous component of degree $n i_r$ of x^n and $\mathfrak{a}$ is graded, $x_{i_r}^n \in \mathfrak{a}$. So $x_{i_r} \in \sqrt{\mathfrak{a}}$ and $x - x_{i_r} \in \sqrt{\mathfrak{a}}$. By induction, $x_{i_k} \in \sqrt{\mathfrak{a}}$ for $k = 1, \ldots, r$. $\square$

7.2.8 Theorem. *Let A be positively graded. The following are equivalent:*
(i) *A is Noetherian.*
(ii) *A_0 is Noetherian and A is a finitely generated A_0-algebra.*

Proof. (i) $\Rightarrow$ (ii) Set $A_+ = \bigoplus_{n \in \mathbb{N}^*} A_n$. This is a graded ideal of A and $A_0 \simeq A/A_+$. So A_0 is Noetherian. Further, A_+ is finitely generated, so by 7.2.2, it is generated by a finite number of homogeneous elements $x_{i_1}, \ldots, x_{i_r}$ where $x_{i_k} \in A_{i_k} \subset A_+$. Denote by B the A_0-subalgebra of A generated by the x_{i_k}'s. Clearly $A_0 \subset B$. Now assume that $A_k \subset B$ for $k = 0, \ldots, n$. Let $y \in A_{n+1}$. Then we can find $a_1 \in A_{n+1-i_1}, \ldots, a_r \in A_{n+1-i_r}$ such that $y = a_1 x_{i_1} + \cdots + a_r x_{i_r}$. By our assumption, $a_k \in B$ for all $k = 1, \ldots, r$. Hence $A_{n+1} \subset B$, and by induction, we have $A = B$.

(ii) $\Rightarrow$ (i) Since A is a finitely generated A_0-algebra, A is isomorphic to a quotient of a polynomial algebra $A_0[X_1, \ldots, X_s]$. The result follows by 2.7.3 and 2.7.9. $\square$

7.2.9 Let K be a commutative field and $A = K[T_1, \ldots, T_n]$ together with its natural grading. The following result is a graded version of 2.2.5.

Proposition. *Let $\mathfrak{p}_1, \ldots, \mathfrak{p}_r$ be graded prime ideals of A. If $\mathfrak{a}$ is a graded ideal of A verifying $\mathfrak{a} \not\subset \mathfrak{p}_i$ for $1 \leqslant i \leqslant r$, then there exists a homogeneous element $a \in \mathfrak{a}$ such that $a \notin \mathfrak{p}_1 \cup \cdots \cup \mathfrak{p}_r$.*

Proof. If $\mathfrak{a} = A$, then $a = 1$ works. So let us assume that $\mathfrak{a} \neq A$. We can further assume that $\mathfrak{p}_i \not\subset \mathfrak{p}_j$ for $i \neq j$. We shall proceed by induction on r. The case $r = 1$ is obvious. So let $r \geqslant 2$.

For $1 \leqslant i \leqslant r$, by induction, there exists a homogeneous element $a_i \in \mathfrak{a}$ such that $a_i \notin \bigcup_{j \neq i} \mathfrak{p}_j = \mathcal{E}_i$.

If there exists i such that $\mathfrak{a} \cap \mathfrak{p}_i \subset \mathcal{E}_i$, then $a_i \notin \mathfrak{p}_1 \cup \cdots \cup \mathfrak{p}_r$.

If $\mathfrak{a} \cap \mathfrak{p}_i \not\subset \mathcal{E}_i$ for $1 \leqslant i \leqslant r$, then we may take $a_i \in \mathfrak{a} \cap \mathfrak{p}_i$, and by replacing the a_i's by some suitable powers, we can assume that $\deg(a_1) =$

$\deg(a_2 \cdots a_r)$. Then $a = a_1 + a_2 \cdots a_r$ is a homogeneous element of $\mathfrak{a}$ and clearly $a \notin \mathfrak{p}_1 \cup \cdots \cup \mathfrak{p}_r$. $\quad \square$

7.3 Applications

7.3.1 In this section, A is a commutative ring.

7.3.2 Theorem. (Artin-Rees Theorem) *Let A be Noetherian, $\mathfrak{a}$ an ideal of A, M a finitely generated A-module et N a submodule of M. There exists $q \in \mathbb{N}^*$ such that, for all $n \geqslant q$, we have*

$$N \cap (\mathfrak{a}^n M) = \mathfrak{a}^{n-q}(N \cap \mathfrak{a}^q(M)).$$

Proof. Set

$$B = \bigoplus_{n \geqslant 0} \mathfrak{a}^n \, , \, S = \bigoplus_{n \geqslant 0} \mathfrak{a}^n M \, , \, T = \bigoplus_{n \geqslant 0} (N \cap \mathfrak{a}^n M).$$

One verifies easily that B is a (positively) graded ring, and S is a (positively) graded B-module. Now B is generated by $\mathfrak{a}$ as an A-algebra. Since A is Noetherian, B is a finitely generated A-algebra. It follows from 7.2.8 that B is Noetherian. Hence S is finitely generated, and T is a finitely generated graded submodule of S. For $q \in \mathbb{N}$, set

$$F_q = \bigoplus_{i=0}^{q} (N \cap \mathfrak{a}^i M), \, G_q = F_q \oplus \left(\bigoplus_{i \geqslant 1} \mathfrak{a}^i (N \cap \mathfrak{a}^q M) \right).$$

Since $\mathfrak{a}^i(N \cap \mathfrak{a}^q M) \subset N \cap \mathfrak{a}^{q+i} M$, $(G_q)_q$ is an increasing sequence of submodules of T. Further, the union of the G_q's is T. So there exists q such that $T = G_q$. Therefore, for $n \geqslant q$, we have $N \cap \mathfrak{a}^n M = \mathfrak{a}^{n-q}(N \cap \mathfrak{a}^q M)$. $\quad \square$

7.3.3 Lemma. *Let $\mathfrak{a}$ be an ideal of A and N a finitely generated A-module. Then $N = \mathfrak{a}N$ if and only if there exists $a \in \mathfrak{a}$ such that $(1 - a)N = 0$.*

Proof. Let $x_1, \ldots, x_s$ be a system of generators of N. If $N = \mathfrak{a}N$, then for $1 \leqslant i \leqslant s$, there exist $a_{ij} \in \mathfrak{a}$ such that

$$x_i = a_{i1}x_1 + \cdots + a_{is}x_s.$$

The determinant of $[\delta_{ij} - a_{ij}]_{1 \leqslant i,j \leqslant s}$ is of the form $1 - a$ for some $a \in \mathfrak{a}$. But for $1 \leqslant i \leqslant s$, $(1 - a)x_i = 0$. Hence $(1 - a)N = \{0\}$. The converse is obvious. $\square$

7.3.4 Theorem. *Let A be Noetherian, $\mathfrak{a}$ an ideal of A and M a finitely generated A-module. Then $x \in \bigcap_{n \geqslant 0} \mathfrak{a}^n M$ if and only if there exists $a \in \mathfrak{a}$ such that $ax = x$. In particular, we have $\bigcap_{n \geqslant 0} \mathfrak{a}^n M = \{0\}$ if and only if for any $a \in \mathfrak{a} \setminus \{0\}$, $x = 0$ is the unique solution in M of $ax = x$.*

Proof. This is a consequence of 7.3.2 and 7.3.3 with $N = \bigcap_{n \geqslant 0} \mathfrak{a}^n M$. $\quad \square$

7.3.5 Corollary. *Let A be a Noetherian integral domain and $\mathfrak{a}$ an ideal of A distinct from A. Then $\bigcap_{n \geqslant 0} \mathfrak{a}^n = \{0\}$.*

7.3.6 Corollary. *Let A be Noetherian and $\mathfrak{b}$ an ideal of A. Then $\mathfrak{b}$ is contained in the Jacobson radical $\operatorname{rad} A$ of A if and only if*

$$\bigcap_{n \geqslant 0} (\mathfrak{a} + \mathfrak{b}^n) = \mathfrak{a}$$

for any ideal $\mathfrak{a}$ of A. Further, if $\mathfrak{b} \subset \operatorname{rad} A$, then

$$\bigcap_{n \geqslant 0} (N + \mathfrak{b}^n M) = N$$

for any submodule N of a finitely generated A-module M.

Proof. Let $\mathfrak{b} \subset \operatorname{rad} A$, M a finitely generated A-module and N a submodule of M. One checks easily that $\mathfrak{b}^n(M/N) = (N + \mathfrak{b}^n M)/N$ for all $n \in \mathbb{N}$. By 7.3.4, if $x \in \bigcap_{n \geqslant 0} \mathfrak{b}^n(M/N)$, then there exists $b \in \mathfrak{b}$ such that $(1 - b)x = 0$. But $1 - b$ is invertible since $\mathfrak{b} \subset \operatorname{rad} A$. So $x = 0$ and N is the intersection of the $(N + \mathfrak{b}^n M)$'s, for $n \geqslant 0$. In particular, $\mathfrak{a}$ is the intersection of the $\mathfrak{a} + \mathfrak{b}^n$, for $n \geqslant 0$.

Conversely, if $\mathfrak{b} \not\subset \operatorname{rad} A$, then $\mathfrak{b} \not\subset \mathfrak{m}$ for some $\mathfrak{m} \in \operatorname{Spm}(A)$. Hence $\mathfrak{b}^n \not\subset \mathfrak{m}$, and $\mathfrak{b}^n + \mathfrak{m} = A$ for all $n \in \mathbb{N}$. Thus $A = \bigcap_{n \geqslant 0}(\mathfrak{m} + \mathfrak{b}^n) \neq \mathfrak{m}$. $\square$

7.3.7 Corollary. *Let A be Noetherian and $\mathfrak{a}$ an ideal of A such that $\mathfrak{a} \subset \operatorname{rad} A$. Then, for any finitely generated A-module M:*

$$\bigcap_{n \geqslant 0} \mathfrak{a}^n M = \{0\}.$$

7.4 Filtrations

7.4.1 Definition. *A filtration on a group G is an increasing sequence $(G_n)_{n \in \mathbb{Z}}$ of subgroups of G such that $G = \bigcup_{n \in \mathbb{Z}} G_n$.*

7.4.2 Definition. *A filtration on a ring A is a filtration $(A_n)_{n \in \mathbb{Z}}$ of the additive group A which is compatible with the ring structure, i.e.:*
(1) $A_n A_m \subset A_{m+n}$ *for all $m, n \in \mathbb{Z}$.*
(2) $1 \in A_0$.

7.4.3 Definition. *Let $(A_n)_{n \in \mathbb{Z}}$ be a filtration on a ring A, M an A-module. A filtration $(M_n)_{n \in \mathbb{Z}}$ is said to be compatible (with the filtration on A) if $A_m M_n \subset M_{m+n}$ for all $m, n \in \mathbb{Z}$.*

7.4.4 Let $(G_n)_{n \in \mathbb{Z}}$ be a filtration on G, and H a subgroup of G. Then $(H \cap G_n)_{n \in \mathbb{Z}}$ is called the *induced filtration* on H. Similarly, if H is normal, then $(HG_n/H)_{n \in \mathbb{Z}}$ is a filtration on G/H.

Let $(A_n)_{n \in \mathbb{Z}}$ be a filtration of a ring A. Then the above procedure provides filtrations on any subring of A and any quotient of A by a bilateral ideal $\mathfrak{a}$.

It is easy to check that the same procedure works for any submodule N of an A-module M with a compatible filtration.

7.4.5 Let $(A_n)_{n\in\mathbb{Z}}$ be a filtration on the ring A and M, N two A-modules with compatible filtrations. A homomorphism $u : M \to N$ is said to be *compatible* if $u(M_n) \subset N_n$ for all $n \in \mathbb{Z}$.

7.5 Grading associated to a filtration

7.5.1 Let $(G_n)_{n\in\mathbb{Z}}$ be a filtration on an additive group G. For $n \in \mathbb{Z}$, let $\mathrm{gr}_n(G) = G_n/G_{n-1}$ and

$$\mathrm{gr}(G) = \bigoplus_{n\in\mathbb{Z}} \mathrm{gr}_n(G)$$

is a graded commutative group. We call $\mathrm{gr}(G)$ the *graded group associated to the filtration*.

7.5.2 Let $(A_n)_{n\in\mathbb{Z}}$ be a filtration on a ring A, and $(M_n)_{n\in\mathbb{Z}}$ a compatible filtration on an A-module M. Let $p, q \in \mathbb{Z}$, we define:

(1) $\mathrm{gr}_p(A) \times \mathrm{gr}_q(M) \to \mathrm{gr}_{p+q}(M)$, $(a + A_{p-1}, m + M_{q-1}) \mapsto am + M_{p+q-1}$.

One verifies easily that this is a well-defined $\mathbb{Z}$-bilinear map. This induces a $\mathbb{Z}$-bilinear map:

(2) $\mathrm{gr}(A) \times \mathrm{gr}(M) \to \mathrm{gr}(M)$.

In particular, when $M = A$, we obtain a ring structure on $\mathrm{gr}(A)$ via this map. We call $\mathrm{gr}(A)$ the *graded ring associated to A*.

It follows that $\mathrm{gr}(M)$ is a graded $\mathrm{gr}(A)$-module called the *graded module associated to M*.

7.5.3 Let $(A_n)_{n\in\mathbb{Z}}$ be a filtration on a ring A, and M, N be A-modules with compatible filtrations. If $u : M \to N$ is a compatible homomorphism, then it is easy to see that u induces a homomorphism $\mathrm{gr}(u)$ of the associated graded modules given by $\mathrm{gr}(u)(m + M_{p-1}) = u(m) + N_{p-1}$ for $m + M_{p-1} \in \mathrm{gr}_p(M)$. Further, if $v : N \to P$ is another compatible homomorphism of A-modules with compatible filtrations, then $\mathrm{gr}(v \circ u) = \mathrm{gr}(v) \circ \mathrm{gr}(u)$.

In the rest of this section, let $(A_n)_{n\in\mathbb{Z}}$ be a filtration on a ring A and let M be an A-module with a compatible filtration $(M_n)_{n\in\mathbb{Z}}$.

Proposition. *Let N be a submodule of M. Then we have the following exact sequence of graded $\mathrm{gr}(A)$-modules:*

$$0 \longrightarrow \mathrm{gr}(N) \xrightarrow{\mathrm{gr}(j)} \mathrm{gr}(M) \xrightarrow{\mathrm{gr}(p)} \mathrm{gr}(M/N) \longrightarrow 0$$

where $j : N \to M$ is the canonical inclusion, $p : M \to M/N$ is the canonical surjection, and we endow N with the induced filtration.

Proof. The homomorphisms j and p are clearly compatible and the exactness of the sequence is a straightforward verification. $\square$

7.5.4 Let $N = \bigcap_{n \in \mathbb{Z}} M_n$. If $N = \{0\}$, then we say that the filtration (on M) is *separated*.

Let $x \in M$, we define the *order of x*, denoted by $\nu(x)$ or $\nu_M(x)$, to be:

$$\nu(x) = +\infty \text{ if } x \in N \, , \; \nu(x) = p \text{ if } x \in M_p, x \notin M_{p-1}.$$

7.5.5 Proposition. *Assume that* $\mathrm{gr}(A)$ *has no zero divisors other than* 0, *then* $\nu(ab) = \nu(a) + \nu(b)$ *for all* $a, b \in A$.

Proof. Since $\bigcap_{n \in \mathbb{Z}} A_n$ is a bilateral ideal of A, the equality is clear if a or b belongs to $\bigcap_{n \in \mathbb{Z}} A_n$. Now if $\nu(a) = r$ and $\nu(b) = s$ where $r, s \in \mathbb{Z}$, then clearly $ab \in A_{r+s}$. Since $\mathrm{gr}(A)$ has no zero divisors other than 0, $ab \notin A_{r+s-1}$. So $\nu(ab) = \nu(a) + \nu(b)$. $\square$

7.5.6 Corollary. *If* $\mathrm{gr}(A)$ *has no zero divisors other than* 0, *then neither has* $A / \bigcap_{n \in \mathbb{Z}} A_n$.

Proof. By 7.5.4, we have $\nu(ab) \neq +\infty$ for any $a, b \in A \setminus \bigcap_{n \in \mathbb{Z}} A_n$. Hence $ab \notin \bigcap_{n \in \mathbb{Z}} A_n$. $\square$

7.5.7 Proposition. *Let* $u : M \to N$ *be a compatible homomorphism of A-modules with compatible filtrations. If the filtration on M is separated and* $\mathrm{gr}(u)$ *is injective, then* u *is also injective.*

Proof. Note that, in general, we have $u(\bigcap_{n \in \mathbb{Z}} M_n) \subset \bigcap_{n \in \mathbb{Z}} N_n$. Let us assume that the filtration on M is separated and $\mathrm{gr}(u)$ is injective. The injectivity of $\mathrm{gr}(u)$ implies that $\nu_M(x) = \nu_N(u(x))$ for any $x \in M$. So if $x \in \ker u$, then $\nu_M(x) = +\infty$. Since the filtration on M is separated, u is injective. $\square$

7.5.8 If there exists $n \in \mathbb{Z}$ such that $M_n = \{0\}$, then $M_p = \{0\}$ for all $p \leqslant n$, and we say that the filtration on M is *discrete*. Note that a discrete filtration is separated.

Proposition. *Let* $u : M \to N$ *be a compatible homomorphism of A-modules with compatible filtrations. If the filtration on N is discrete and* $\mathrm{gr}(u)$ *is surjective, then* u *is also surjective.*

Proof. Let us assume that the filtration on N is discrete and $\mathrm{gr}(u)$ is surjective. There exists $q \in \mathbb{Z}$ such that $N_q = \{0\}$. This implies that $u(M_n) = N_n$ for $n \leqslant q$. Now let us assume that $u(M_n) = N_n$ for $n \leqslant p$. Since $\mathrm{gr}(u)$ is surjective, $N_{p+1} = u(M_{p+1}) + N_p = u(M_{p+1}) + u(M_p) = u(M_{p+1})$. It follows by induction that $u(M_n) = N_n$ for all $n \in \mathbb{Z}$. Hence the surjectivity of u. $\square$

7.5.9 In the rest of this section, all the filtrations considered are discrete.

Let $x \in M$ and $n = \nu(x)$. We shall denote the image of x in $\mathrm{gr}_n(M)$ by $\sigma_M(x)$ or simply $\sigma(x)$, and we shall call $\sigma(x)$ the *principal symbol* of x. This induces a map $\sigma_M : M \to \mathrm{gr}(M)$.

7.5.10 Proposition. *Let $(x_i)_{i \in I}$ be a family of elements of M.*

(i) *If $(\sigma(x_i))_{i \in I}$ is a system of generators of the $\mathrm{gr}(A)$-module $\mathrm{gr}(M)$, then $(x_i)_{i \in I}$ generates the A-module M.*

(ii) *If $\mathrm{gr}(M)$ is a free $\mathrm{gr}(A)$-module and $(\sigma(x_i))_{i \in I}$ is a $\mathrm{gr}(A)$-basis of $\mathrm{gr}(M)$, then M is a free A-module and $(x_i)_{i \in I}$ is a A-basis of M.*

Proof. (i) Let $q \in \mathbb{Z}$ be such that $A_q = \{0\}$ and $M_q = \{0\}$. Let $x \in M_n$. We shall prove by induction on n that x is contained in the A-submodule generated by $(x_i)_{i \in I}$. This is true for $n \leqslant q$. So we may assume that $n > q$ and $\nu(x) = n$.

There exist non-zero elements $\alpha_1, \dots, \alpha_r \in \mathrm{gr}(A)$ and indices $i_1, \dots, i_r \in I$ such that

$$\sigma(x) = \alpha_1 \sigma(x_1) + \cdots + \alpha_r \sigma(x_r).$$

Set $n_i = \nu(x_i)$, then we can find $a_1, \dots, a_r \in A$ such that $\sigma(a_j) = \alpha_j$ and $\nu(a_j) = n - n_j$ for all $1 \leqslant j \leqslant r$. It follows that:

$$x - \sum_{j=1}^r a_j x_j \in M_{n-1}.$$

By our induction hypothesis, x is contained in the A-submodule generated by $(x_i)_{i \in I}$.

(ii) Let N be a free A-module with basis $(y_i)_{i \in I}$. Define a (discrete) filtration on N as follows: for $n \in \mathbb{Z}$,

$$N_n = \sum_{j + n_i \leqslant n} A_j y_i$$

where $n_i = \nu(x_i)$ for $i \in I$. One verifies easily that this is a compatible filtration and the map $u : N \to M$ defined by sending y_i to x_i is a compatible homomorphism of A-modules. The associated map $\mathrm{gr}(u)$ is clearly bijective and it follows from 7.5.7 and 7.5.8 that u is an isomorphism. Hence M is free and $(x_i)_{i \in I}$ is a basis of M. $\square$

7.5.11 Corollary. *If $\mathrm{gr}(M)$ is a finitely generated (resp. Noetherian) $\mathrm{gr}(A)$-module, then M is a finitely generated (resp. Noetherian) A-module.*

Proof. These are direct consequences of 7.2.2, 7.5.3 and 7.5.10. $\square$

7.5.12 Corollary. *If $\mathrm{gr}(A)$ is a left Noetherian ring, then so is A.*

References

- [10], [52], [56].

8

Inductive limits

8.1 Generalities

8.1.1 Recall that a *preorder* on a set I is a binary relation which is reflexive and transitive. Let $\leqslant$ be a preorder on I, then it is *filtered* or *directed* if for all $i, j \in I$, there exists $k \in I$ such that $i \leqslant k$ and $j \leqslant k$. A subset $J \subset I$ is *cofinal* if for all $i \in I$, there exists $j \in J$ such that $i \leqslant j$.

A set with a filtered preorder is called a *filtered set* or *directed set*.

8.1.2 Remarks. 1) A preorder on a set I is filtered if and only if any non-empty finite subset of I has an upper bound.

2) Let X be a topological space and $x \in X$. Then the set of neighbourhoods of x is a filtered set via the preorder $U \leqslant V$ if and only if $V \subset U$.

8.1.3 In the rest of this chapter, I is a filtered set with preorder $\leqslant$.

8.1.4 Definition. *An* inductive system *of sets indexed by I is a family of sets $(E_i)_{i \in I}$ and maps $(f_{ji})_{i,j \in I, i \leqslant j}$ verifying: for $i, j, k \in I$ such that $i \leqslant j \leqslant k$,*

(i) *f_{ji} is a map from E_i to E_j.*
(ii) *f_{ii} is the identity map on E_i.*
(iii) *$f_{kj} \circ f_{ji} = f_{ki}$.*

When there is no confusion, we shall denote an inductive system of sets indexed by I simply by (E_i, f_{ji}).

8.1.5 Let (E_i, f_{ji}) be an inductive system of sets indexed by I, and F be the sum of the E_i's. If $x \in F$, we define $\theta(x)$ to be the unique index $i \in I$ such that $x \in E_i$.

Define a binary relation $\mathfrak{R}$ on F as follows: for $x, y \in F$, let $i = \theta(x)$ and $j = \theta(y)$, then

$$x \mathfrak{R} y \Leftrightarrow \text{there exists } k \in I \text{ such that } k \geqslant i, k \geqslant j \text{ and } f_{ki}(x) = f_{kj}(y).$$

One verifies easily that $\mathfrak{R}$ is an equivalence relation and the quotient $\varinjlim(E_i, f_{ji}) = F/\mathfrak{R}$ is called the *inductive limit* of the inductive system. To simply notations, we shall write $\varinjlim E_i$ for $\varinjlim(E_i, f_{ji})$ when there is no confusion.

Note that $\varinjlim(E_i, f_{ji})$ is non-empty if at least one of the E_i's is non-empty.

For $i \in I$, we write f_i for the restriction to E_i of the canonical surjection $F \to \varinjlim E_i$. In particular, one verifies easily that for $i, j \in I$,

$$(1) \qquad\qquad f_i = f_j \circ f_{ji} \text{ whenever } i \leqslant j.$$

8.1.6 Examples. 1) Let G be a set and consider the inductive system (E_i, f_{ji}) where $E_i = G$ for all $i \in I$ and if $i \leqslant j$, then f_{ji} is the identity map on G. The maps f_i are identical for all $i \in I$, and therefore there is a canonical bijection from G to $\varinjlim E_i$.

2) Let A, B be sets and $(V_i)_{i \in I}$ a family of subsets of A such that $V_j \subset V_i$ if $i \leqslant j$. Let E_i be the set of maps from V_i to B and define, for $i, j \in I$, $i \leqslant j$, $f_{ji} : E_i \to E_j$ by $f_{ji}(u) = u|_{V_j}$. Then (E_i, f_{ji}) is an inductive system. We call $\varinjlim E_i$ the set of *germs of functions* from V_i to B. This generalizes the case where $(V_i)_{i \in I}$ is the set of neighbourhoods of a point of a topological space A.

8.1.7 Proposition. *Let (E_i, f_{ji}) be an inductive system of sets, $E = \varinjlim E_i$ and for $i \in I$, $f_i : E_i \to E$ the canonical map.*

(i) *Let $x_1, \ldots, x_n \in E$. There exist $i \in I$ and $y_1, \ldots, y_n \in E_i$ such that $x_k = f_i(y_k)$ for $1 \leqslant k \leqslant n$.*

(ii) *Let $i \in I$ and $y_1, \ldots, y_n \in E_i$ be such that $f_i(y_1) = \cdots = f_i(y_n)$. There exists $j \in I$ such that $i \leqslant j$ and $f_{ji}(y_1) = \cdots = f_{ji}(y_n)$.*

Proof. (i) For $1 \leqslant k \leqslant n$, there exist $i_k \in I$ and $z_k \in E_{i_k}$ such that $x_k = f_{i_k}(z_k)$. Since I is a directed set, there exists $i \in I$ such that $i \geqslant i_k$ for $1 \leqslant k \leqslant n$. Set $y_k = f_{i,i_k}(z_k)$. Then (1) implies that $x_k = f_{i_k}(z_k) = f_i \circ f_{i,i_k}(z_k) = f_i(y_k)$.

(ii) For $1 \leqslant h, k \leqslant n$, there exists $\alpha_{h,k} \in I$ such that $\alpha_{h,k} \geqslant i$ and $f_{\alpha_{h,k},i}(y_h) = f_{\alpha_{h,k},i}(y_k)$. Let $j \in I$ be such that $j \geqslant \alpha_{h,k}$ for all $1 \leqslant h, k \leqslant n$. Then $f_{ji}(y_h) = f_{ji}(y_k)$ since $f_{ji} = f_{j,\alpha_{h,k}} \circ f_{\alpha_{h,k},i}$. $\square$

8.2 Inductive systems of maps

8.2.1 Let (E_i, f_{ji}) be an inductive system of sets indexed by I. Set $E = \varinjlim E_i$ and for $i \in I$, $f_i : E_i \to E$ the canonical map. Let G be a set.

Proposition. *Let $(u_i)_{i \in I}$ be a family of maps verifying $u_i : E_i \to G$, and*

$$(2) \qquad\qquad u_j \circ f_{ji} = u_i \text{ if } i \leqslant j.$$

(i) *There exists a unique map $u : E \to G$ such that*

$$(3) \qquad\qquad u_i = u \circ f_i \text{ for all } i \in I.$$

(ii) *The map u is surjective if and only if G is the union of the $u_i(E_i)$'s.*

(iii) *The map u is injective if and only if for any $i \in I$ and $x, y \in E_i$, $u_i(x) = u_i(y)$ implies that $f_{ji}(x) = f_{ji}(y)$ for some $j \geqslant i$.*

Proof. (i) Let F be the disjoint union of the E_i's and $f : F \to E$ the canonical surjection. There is a unique map $v : F \to G$ such that $v|_{E_i} = u_i$ for $i \in I$. Now (2) says that v is compatible with the equivalence relation $\mathfrak{R}$. So there is a unique map $u : E \to G$ verifying $v = u \circ f$.

Part (ii) is obvious and part (iii) is just a consequence of 8.1.7. $\square$

8.2.2 Remarks. 1) When u is bijective, we shall also call G the inductive limit of (E_i, f_{ji}).

2) If all the f_{ji}'s are injective, then so are the f_i's. In this case, E_i can be identified with $f_i(E_i)$, and E can be viewed as the union of the E_i's.

Conversely, let $(F_i)_{i \in I}$ be a family of subsets of a set F such that $F = \bigcup_{i \in I} F_i$ and $F_i \subset F_j$ if $i \leqslant j$. For $i \leqslant j$, denote by $f_{ji} : F_i \to F_j$ the canonical injection. Then F is the inductive limit of the inductive system (F_i, f_{ji}), and for $i \in I$, the canonical map $f_i : F_i \to F$ is just the canonical injection.

8.2.3 Proposition. *Let (E_i, f_{ji}) and (F_i, g_{ji}) be inductive systems of sets indexed by I. Let $(u_i : E_i \to F_i)_{i \in I}$ be a family of maps such that for all $i, j \in I$ verifying $i \leqslant j$, the following diagram is commutative:*

$$
\begin{array}{ccc}
E_i & \xrightarrow{\ u_i\ } & F_i \\
{\scriptstyle f_{ji}}\Big\downarrow & & \Big\downarrow{\scriptstyle g_{ji}} \\
E_j & \xrightarrow[\ u_j\ ]{} & F_j
\end{array}
$$

Let $E = \varinjlim E_i$, $F = \varinjlim F_i$ and for $i \in I$, let f_i and g_i be the corresponding canonical maps.

There exists a unique map $u : E \to F$ such that for any $i \in I$, the following diagram is commutative:

$$
\begin{array}{ccc}
E_i & \xrightarrow{\ u_i\ } & F_i \\
{\scriptstyle f_i}\Big\downarrow & & \Big\downarrow{\scriptstyle g_i} \\
E & \xrightarrow[\ u\ ]{} & F
\end{array}
$$

Further, u is injective (resp. surjective) if all the u_i's are injective (resp. surjective).

Proof. The family $(g_i \circ u_i)_{i \in I}$ satisfies the hypothesis of 8.2.1, so there exists a unique map $u : E \to F$ such that $u \circ f_i = g_i \circ u_i$.

If the u_i's are surjective, then

$$F = \bigcup_{i \in I} g_i \circ u_i(E_i) = \bigcup_{i \in I} u \circ f_i(E_i) = u\Big(\bigcup_{i \in I} f_i(E_i)\Big) = u(E).$$

If the u_i's are injective, then for any $i \in I$, $x, y \in E_i$ such that $u(f_i(x)) = u(f_i(y))$, we have $g_i(u_i(x)) = g_i(u_i(y))$. Now by 8.1.7, there exists $j \geqslant i$ such that $g_{ji}(u_i(x)) = g_{ji}(u_i(y))$. Thus $u_j(f_{ji}(x)) = u_j(f_{ji}(y))$, and since u_j is injective, $f_{ji}(x) = f_{ji}(y)$. So u is injective (8.2.1). $\square$

8.2.4 A family $(u_i : E_i \to F_i)_{i \in I}$ of maps verifying the hypothesis of 8.2.3 shall be called an *inductive system of maps from* (E_i, f_{ji}) *to* (F_i, g_{ji}). The map u is called the *inductive limit* of the family $(u_i)_{i \in I}$ and we shall denote it by $\varinjlim u_i$ when there is no confusion.

8.2.5 Corollary. *Let* $(u_i : E_i \to F_i)_{i \in I}$ *be an inductive system of maps from* (E_i, f_{ji}) *to* (F_i, g_{ji}), *and* $(v_i : F_i \to G_i)_{i \in I}$ *an inductive system of maps from* (F_i, g_{ji}) *to* (G_i, h_{ji}). *Then* $(v_i \circ u_i)_{i \in I}$ *is an inductive system of maps from* (E_i, f_{ji}) *to* (G_i, h_{ji}) *and*

$$(4) \qquad \varinjlim(v_i \circ u_i) = (\varinjlim v_i) \circ (\varinjlim u_i).$$

Proof. This is a straightforward verification. □

8.2.6 Let J be a directed subset of I, *i.e.* J is filtered with respect to the preorder induced from I. Then the subfamily of sets $(E_i)_{i \in J}$ and the subfamily of maps $(f_{ji})_{i \leqslant j, i, j \in J}$ form an inductive system of sets indexed by J, denoted by $(E_i, f_{ji})_J$. We call $(E_i, f_{ji})_J$ the *inductive subsystem relative to* J.

Now let $E = \varinjlim E_i$ and $E_J = \varinjlim(E_i, f_{ji})_J$. For $i \in I$, denote by f_i the canonical map. Then $(f_i)_{i \in J}$ is an inductive system of maps from $(E_i, f_{ji})_J$ to (E_i, f_{ji}), and $g = \varinjlim(f_i)_{i \in J} : E_J \to E$.

Let $K \subset J$ be another directed subset of I, and $g' = \varinjlim(f_i)_{i \in K} : E_K \to E_J$, $g'' = \varinjlim(f_i)_{i \in K} : E_K \to E$, then by 8.2.5, we have

$$(5) \qquad g'' = g \circ g'.$$

8.2.7 Proposition. *Let* J *be a cofinal subset of* I. *The map* $g = \varinjlim(f_i)_J : E_J \to E$ *is bijective.*

Proof. Let $i \in J$ and $x, y \in E_i$ be such that $f_i(x) = f_j(y)$. By 8.1.7, there exists $j \in J$ verifying $j \geqslant i$ and $f_{ji}(x) = f_{ji}(y)$. By 8.2.1(iii), g is injective.

For $i \in J$, let $h_i : E_i \to E_J$ be the canonical map. If $x \in E$, then there exist $i \in I$ and $y \in E_i$ such that $x = f_i(y)$. If $j \in J$ is such that $j \geqslant i$, then $x = f_j \circ f_{ji}(y) = g(h_j \circ f_{ji}(y))$. It follows that g is surjective. □

8.2.8 Remark. The previous statement says that in order to determine the inductive limit, it suffices to determine the inductive limit of the inductive subsystem relative to a cofinal subset of I.

8.3 Inductive systems of magmas, groups and rings

8.3.1 Definition. *Let* $(E_i)_{i \in I}$ *be a family of magmas (resp. semigroups, monoids, groups, rings). An inductive system* (E_i, f_{ji}) *indexed by* I *is called an* inductive system of magmas *(resp. semigroups, monoids, groups, rings) if the* f_{ji}'s *are homomorphisms of magmas (resp. semigroups, monoids, groups, rings).*

8.3.2 Let us denote $E = \varinjlim E_i$ and $f_i : E_i \to E$ the canonical maps.

Proposition. *There exists a unique structure of magma (resp. semigroups, monoids, groups, rings) on E such that for $i \in I$, f_i is a homomorphism of magmas (resp. semigroups, monoids, groups, rings). Further, if the E_i's are commutative, then so is E.*

Proof. Let $x, y \in E$, $i \in I$ and $x_i, y_i \in E_i$ be such that $x = f_i(x_i)$, $y = f_i(y_i)$.

Uniqueness follows from the fact that if the f_i's are homomorphisms of magmas, $f_i(x_i y_i)$ is the composition of x and y for any law of composition on E.

Let us prove its existence. Suppose that $a_j, b_j \in E_j$ verify $f_j(a_j) = x$ and $f_j(b_j) = y$. By definition, there exists $k \in I$, $k \geqslant i$, $k \geqslant j$ such that $f_{kj}(a_j) = f_{ki}(x_i)$ and $f_{kj}(b_j) = f_{ki}(y_i)$. Since the f_{ji}'s are homomorphisms, it follows that:

$$f_i(x_i y_i) = f_k(f_{ki}(x_i y_i)) = f_k(f_{kj}(a_j b_j)) = f_j(a_j b_j).$$

Hence $xy = f_i(x_i y_i)$ defines a structure of magma on E and it is clear that with respect to this law of composition, the f_i's are homomorphisms of magma.

The proofs for semigroups, monoids, groups, rings and commutativity are similar. $\square$

8.3.3 Proposition. *Let (E_i, f_{ji}) be an inductive system of magmas (resp. semigroups, monoids, groups, rings), $E = \varinjlim E_i$ and $f_i : E_i \to E$ the canonical maps. Let F be a magma (resp. semigroup, monoid, group, ring) and $u_i : E_i \to F$ be homomorphisms verifying $u_i = u_j \circ f_{ji}$ whenever $i \leqslant j$. There exists a unique homomorphism $u : E \to F$ such that $u_i = u \circ f_i$.*

Proof. By 8.2.1, there exists a unique map $u : E \to F$ such that $u_i = u \circ f_i$. Therefore, we only have to check that u is a homomorphism. Let $x, y \in E$, $i \in I$ and $x_i, y_i \in E_i$ be such that $x = f_i(x_i)$, $y = f_i(y_i)$. Then

$$u(xy) = u(f_i(x_i y_i)) = u_i(x_i y_i) = u_i(x_i)u_i(y_i) = u(x)u(y).$$

So the result follows. $\square$

8.3.4 Remark. As in 8.2.3, if (E_i, f_{ji}) and (F_i, g_{ji}) are inductive systems of magmas (resp. semigroups, monoids, groups, rings), and $(u_i : E_i \to F_i)_{i \in I}$ is an inductive system of maps which are homomorphisms, then $\varinjlim u_i : \varinjlim E_i \to \varinjlim F_i$ is a homomorphism.

8.3.5 Proposition. *Let (A_i, f_{ji}) be an inductive system of rings and $A = \varinjlim A_i$.*

(i) *If the A_i's are non-zero, then A is non-zero.*
(ii) *If the A_i's are integral domains, then A is an integral domain.*
(iii) *If the A_i's are fields, then A is a field.*

Proof. (i) Let 0_i, 1_i (resp. 0, 1) be the zero and the identity in A_i (resp. A). There exists $i \in I$ such that $f_i(0_i) = 0$ and $f_i(1_i) = 1$. If $0 = 1$, then there exists $j \geqslant i$ such that $f_{ji}(0_i) = f_{ji}(1_i)$. This implies that $0_i = 1_i$ which contradicts the fact that A_i is non-zero.

(ii) Let $x, y \in A$ be such that $xy = 0$. There exist $i \in I$, $x_i, y_i \in A_i$ such that $f_i(x_i) = x$, $f_i(y_i) = y$. It follows that $f_i(x_i y_i) = 0_i$ and there exists $j \geqslant i$ such that $f_{ji}(x_i y_i) = f_{ji}(x_i) f_{ji}(y_i) = 0_j$. Since A_j is an integral domain, $f_{ji}(x_i) = 0_j$ or $f_{ji}(y_i) = 0_j$. But this implies that $f_j(f_{ji}(x_i)) = x = 0$ or $f_j(f_{ji}(y_i)) = y = 0$. So we are done.

(iii) Let $x \in A \setminus \{0\}$, $i \in I$ and $x_i \in A_i$ with $x = f_i(x_i)$. Since the A_i's are fields, $x_i \neq 0_i$ and $f_i(x_i^{-1})$ is the inverse of x. So A is a field. $\square$

8.4 An example

8.4.1 Let A be a commutative ring, $\mathcal{S}$ be the set of multiplicative subsets S of A such that $1 \in S$, $0 \notin S$. For $S \in \mathcal{S}$, let $i_S : A \to S^{-1}A$ denote the canonical homomorphism and $U(S^{-1}A)$ the set of invertible elements in $S^{-1}A$.

Let $S, T, V \in \mathcal{S}$ verify $S \subset T \subset V$. Note that $i_T(S) \subset U(T^{-1}A)$. By 2.3.5, there exists a unique homomorphism $f_{TS} : S^{-1}A \to T^{-1}A$ such that $f_{TS} \circ i_S = i_T$. Since $(f_{VT} \circ f_{TS}) \circ i_S = i_V = f_{VS} \circ i_S$, it follows that: $f_{VT} \circ f_{TS} = f_{VS}$.

8.4.2 Let us fix $\mathfrak{p} \in \mathrm{Spec}(A)$ and let $\mathcal{S}_1$ be the set of $S \in \mathcal{S}$ such that $S \cap \mathfrak{p} = \emptyset$.

Let $S, T \in \mathcal{S}_1$ and $ST = \{st; s \in S, t \in T\}$. Since $\mathfrak{p}$ is prime, $S, T \subset ST \in \mathcal{S}_1$. We conclude that $\mathcal{S}_1$ is a directed set, and $(S^{-1}A, f_{TS})_{S \in \mathcal{S}_1}$ is an inductive system of commutative rings. Let $B = \varinjlim S^{-1}A$ and $f_S : S^{-1}A \to B$ the canonical maps. We claim that B is isomorphic to $A_{\mathfrak{p}}$.

Let $P = A \setminus \mathfrak{p}$, then $P \in \mathcal{S}$ and $A_{\mathfrak{p}} = P^{-1}A$. From 8.4.1, we obtain that the family $(f_{PS} : S^{-1}A \to A_{\mathfrak{p}})_{S \in \mathcal{S}_1}$ satisfies the hypothesis of 8.2.1. It follows that there is a unique ring homomorphism $u : B \to A_{\mathfrak{p}}$ such that $f_{PS} = u \circ f_S$ for all $S \in \mathcal{S}_1$.

Now P is the biggest element in $\mathcal{S}_1$ and f_{PP} is just the identity map on $A_{\mathfrak{p}}$. So u is surjective.

Finally u is injective by 8.2.1(iii). Hence B is isomorphic to $A_{\mathfrak{p}}$.

8.4.3 Let $\mathcal{S}_2$ be the set of $S \in \mathcal{S}$ not containing any zero divisors of A. Then a similar argument shows that $\mathcal{S}_2$ is a directed set and the limit of the inductive system $(S^{-1}A, f_{TS})_{S \in \mathcal{S}_2}$ is isomorphic to the full ring of fractions of A.

8.5 Inductive systems of algebras

8.5.1 Let A be a ring and (E_i, f_{ji}) be an inductive system of abelian groups with $E = \varinjlim E_i$, and f_i the canonical maps.

Let us suppose further that for $i \in I$, E_i is an A-module and whenever $i \leqslant j$, f_{ji} is a homomorphism of A-modules.

Then a simple verification shows that E admits a structure of A-module given by: for any $a \in A$, $x \in E$, $i \in I$, $x_i \in E_i$ verifying $f_i(x_i) = x$, we set $ax = f_i(ax_i)$.

Note that if (F_i, g_{ji}) is another such inductive system of abelian groups, and $(u_i : E_i \to F_i)_{i \in I}$ is an inductive system of homomorphisms of A-modules, then we verify readily that $\varinjlim u_i : \varinjlim E_i \to \varinjlim F_i$ is a homomorphism of A-modules.

8.5.2 Let us conserve the notations and hypotheses of 8.5.1. Suppose further that A is a commutative ring, the E_i's are A-algebras and the f_{ji}'s are A-algebra homomorphisms.

Then the ring structure and the A-module structure on E make E into an A-algebra and the f_i's are A-algebra homomorphisms. Again, E is associative (resp. commutative) if the E_i's are associative (resp. commutative), and if (F_i, g_{ji}) is another such inductive system and $(u_i : E_i \to F_i)_{i \in I}$ is an inductive system of maps such that u_i is a homomorphism of A-algebras, then $\varinjlim u_i : \varinjlim E_i \to \varinjlim F_i$ is a homomorphism of A-algebras.

References

- [8], [9], [34], [52].

9

Sheaves of functions

The notion of a sheaf is essential in many areas in mathematics, and most particularly in algebraic geometry. Indeed, the theory of sheaves plays an important part in the foundation of modern algebraic geometry. We describe in this chapter some basic facts on sheaves of functions.

9.1 Sheaves

9.1.1 In this chapter, X will be a topological space and E a set. We shall denote the set of open subsets of X by Ω^X or Ω, and by Ω_x^X or Ω_x the set of open neighbourhoods in X of an element $x \in X$.

Let us fix the partial order $\leqslant$ on Ω and Ω_x given by: $U \leqslant V$ if and only if $V \subset U$.

9.1.2 Definition. *A presheaf of functions on X with values in E is a map $\mathcal{F}$ which associates to each $U \in \Omega$, a set $\mathcal{F}(U)$ of maps from U to E, such that whenever $U, V \in \Omega$ verify $V \subset U$, $f|_V \in \mathcal{F}(V)$ for all $f \in \mathcal{F}(U)$.*

This last condition says that the restriction map, denoted by $r_{VU}^{\mathcal{F}}$ or r_{VU}, from $\mathcal{F}(U)$ to $\mathcal{F}(V)$ is well-defined.

In the sequel, by a presheaf on X, we shall mean a presheaf of functions on X with values in E.

9.1.3 Remarks. 1) A presheaf on X is an inductive system of sets indexed by Ω.

2) Let $\mathcal{F}$ be a presheaf on X. If all the $\mathcal{F}(U)$'s are groups (resp. rings, etc...) and all the r_{VU}'s are homomorphisms of groups (resp. rings, etc...), we shall say that $\mathcal{F}$ is a presheaf of groups (resp. rings, etc...) on X.

3) Let $\mathcal{F}$ be a presheaf on X. We shall also use the notation $\Gamma(U, \mathcal{F})$ for $\mathcal{F}(U)$. The elements of $\Gamma(U, \mathcal{F})$ are called *sections of $\mathcal{F}$ over U*, and those of $\Gamma(X, \mathcal{F})$ are called *global sections*.

9.1.4 Let $\mathcal{F}$ be a presheaf on X and $x \in X$. Then $(\mathcal{F}(U), r_{VU})_{U \in \Omega_x}$ is an inductive system of sets indexed by Ω_x. Let us denote the inductive limit of this system by $\mathcal{F}_x$, and we call $\mathcal{F}_x$ the *fibre* of $\mathcal{F}$ at x. The elements of $\mathcal{F}_x$ are called the *germs of sections* of $\mathcal{F}$ at x.

Further, if $\mathcal{F}$ is a presheaf of groups (resp. ring, etc...), then by 8.3.2, $\mathcal{F}_x$ is a group (resp. ring, etc...).

For $U \in \Omega_x$, let $r_{U,x} : U \to \mathcal{F}_x$ be the canonical map. We call $r_{U,x}(f)$ the germ of f at x.

Now let $U, V \in \Omega_x$, $s \in \mathcal{F}_x$, $f \in \mathcal{F}(U)$ and $g \in \mathcal{F}(V)$. Then $s = r_{U,x}(f) = r_{V,x}(g)$ if and only if there exists $W \in \Omega_x$ such that $W \subset U \cap V$ and $r_{WU}(f) = r_{WV}(g)$. It follows that $f(x)$ is the same for any f having s as germ at x. We shall call this the *value* of the germ s at x, and we shall denote it by $s(x)$.

9.1.5 Definition. *A presheaf $\mathcal{F}$ on X is a sheaf if it satisfies the following condition: if $(U_i)_{i \in I}$ is a family of open subsets of X, U their union and $f : U \to E$ a map such that $f|_{U_i} \in \mathcal{F}(U_i)$, then $f \in \mathcal{F}(U)$.*

9.1.6 Examples. 1) Let G be an abelian group and $\mathcal{F}$ the presheaf of abelian groups on X defined by $\mathcal{F}(U) = \mathrm{Hom}(U, G)$ for all $U \in \Omega$, here $\mathrm{Hom}(U, G)$ denotes the set of maps from U to G. One verifies easily that $\mathcal{F}$ is a sheaf of abelian groups on X.

2) For $U \in \Omega$, let $\mathcal{F}(U) = \mathcal{C}(U, \mathbb{R})$, the ring of continuous real-valued functions on X. Then $\mathcal{F}$ is a sheaf of rings on X.

3) Let $X = \mathbb{C}$ and for $U \in \Omega$, let $\mathcal{F}(U)$ be the ring of holomorphic functions on U. Then we obtain the sheaf of holomorphic functions on $\mathbb{C}$.

4) Let $X = \mathbb{R}$ and for $U \in \Omega$, let $\mathcal{F}(U)$ be the ring of bounded continuous real-valued functions on X. It is a presheaf, but not a sheaf. In fact, the identity function is bounded on $U_n =]-n, n[$, but it is not bounded on $\mathbb{R}$.

9.2 Morphisms

9.2.1 Let A, B be sets (resp. groups, rings, etc...), we denote by $\mathrm{Hom}(A, B)$ the set of maps (resp. homomorphisms of groups, homomorphisms of rings, etc...) from A to B.

9.2.2 Definition. *Let $\mathcal{F}$ and $\mathcal{G}$ be presheaves on X. A morphism $\varphi : \mathcal{F} \to \mathcal{G}$ is a family of maps $(\varphi(U) : \mathcal{F}(U) \to \mathcal{G}(U))_{U \in \Omega}$ such that whenever $U, V \in \Omega$ verify $V \subset U$, the following diagram is commutative:*

$$
(1) \qquad
\begin{array}{ccc}
\mathcal{F}(U) & \xrightarrow{\varphi(U)} & \mathcal{G}(U) \\
{\scriptstyle r^{\mathcal{F}}_{VU}} \downarrow & & \downarrow {\scriptstyle r^{\mathcal{G}}_{VU}} \\
\mathcal{F}(V) & \xrightarrow[\varphi(V)]{} & \mathcal{G}(V)
\end{array}
$$

A morphism of sheaves is a morphism of the underlying presheaves. The morphism φ is an isomorphism if $\varphi(U)$ is an isomorphism for all $U \in \Omega$.

9.2.3 Remarks. 1) We shall denote the set of morphisms of (pre)sheaves from $\mathcal{F}$ to $\mathcal{G}$ by $\mathrm{Hom}(\mathcal{F}, \mathcal{G})$. It is clear that the composition of two morphisms is again a morphism.

Let us denote by $\mathrm{id}_{\mathcal{F}}$ the morphism where $\mathrm{id}_{\mathcal{F}}(U) = \mathrm{id}_{\mathcal{F}(U)}$ for all $U \in \Omega$. It follows that given $\varphi \in \mathrm{Hom}(\mathcal{F}, \mathcal{G})$, φ is an isomorphism if and only if there exists $\psi \in \mathrm{Hom}(\mathcal{G}, \mathcal{F})$ such that $\varphi \circ \psi = \mathrm{id}_{\mathcal{G}}$ and $\psi \circ \varphi = \mathrm{id}_{\mathcal{F}}$.

2) Let $x \in X$ and $\varphi \in \mathrm{Hom}(\mathcal{F}, \mathcal{G})$. Recall from 8.2.3 that there is a unique homomorphism $\varphi_x : \mathcal{F}_x \to \mathcal{G}_x$ such that for all $U \in \Omega_x$, the following diagram is commutative:

(2)

$$
\begin{array}{ccc}
\mathcal{F}(U) & \xrightarrow{\ \varphi(U)\ } & \mathcal{G}(U) \\
{\scriptstyle r^{\mathcal{F}}_{U,x}} \big\downarrow & & \big\downarrow {\scriptstyle r^{\mathcal{G}}_{U,x}} \\
\mathcal{F}_x & \xrightarrow[\varphi_x]{} & \mathcal{G}_x
\end{array}
$$

We call φ_x the morphism induced by φ.

9.2.4 Theorem. *Let $\varphi : \mathcal{F} \to \mathcal{G}$ be a morphism of presheaves on X. The following conditions are equivalent:*
 (i) *For all $U \in \Omega$, $\varphi(U)$ is injective.*
 (ii) *For all $x \in X$, φ_x is injective.*
When these conditions are satisfied, we say that φ is injective.

Proof. The implication (i) $\Rightarrow$ (ii) was proved in 8.2.3. Let us prove (ii) $\Rightarrow$ (i). Let $U \in \Omega$ and $f, g \in \mathcal{F}(U)$ be such that $\varphi(U)(f) = \varphi(U)(g)$. Then by (2) and the injectivity of φ_x, we have that $r^{\mathcal{F}}_{U,x}(f) = r^{\mathcal{F}}_{U,x}(g)$. So there is an open neighbourhood $V_x \in \Omega_x$ of x such that $V_x \subset U$ and $f|_{V_x} = g|_{V_x}$. Since the V_x's, $x \in U$ form a covering of U, we deduce that $f = g$. $\square$

9.2.5 Theorem. *Let $\varphi : \mathcal{F} \to \mathcal{G}$ be a morphism of sheaves on X. The following conditions are equivalent:*
 (i) *For all $U \in \Omega$ and $g \in \mathcal{G}(U)$, there exists an open covering $(U_i)_{i \in I}$ of U and elements $f_i \in \mathcal{F}(U_i)$ such that $\varphi(U_i)(f_i) = g|_{U_i}$ for all $i \in I$.*
 (ii) *For all $x \in X$, φ_x is surjective.*
When these conditions are satisfied, we say that φ is surjective.

Proof. (i) $\Rightarrow$ (ii) Let $x \in X$, $\theta \in \mathcal{G}_x$, $U \in \Omega$ and $g \in \mathcal{G}(U)$ be such that $\theta = r^{\mathcal{G}}_{U,x}(g)$. By our hypotheses, there exist an open covering $(U_i)_{i \in I}$ of U and elements $f_i \in \mathcal{F}(U_i)$ such that $\varphi(U_i)(f_i) = g|_{U_i}$ for all $i \in I$. Fix $k \in I$ such that $x \in U_k$, then

$$
r^{\mathcal{G}}_{U,x}(g) = r^{\mathcal{G}}_{U_k,x} \circ r^{\mathcal{G}}_{U_k,U}(g) = r^{\mathcal{G}}_{U_k,x} \circ \varphi(U_k)(f_k) = \varphi_x(r^{\mathcal{G}}_{U_k,x}(f_k)).
$$

Since $r^{\mathcal{G}}_{U_k,x}(f_k) \in \mathcal{F}_x$, φ_x is surjective.

(ii) $\Rightarrow$ (i) Let $U \in \Omega$ and $g \in \mathcal{G}(U)$. For $x \in U$, there exists $\rho \in \mathcal{F}_x$ such that $\varphi_x(\rho) = r^{\mathcal{G}}_{U,x}(g)$. Let $U_x \in \Omega_x$ and $f^x \in \mathcal{F}(U_x)$ be such that $U_x \subset U$ and $\rho = r^{\mathcal{F}}_{U_x,x}(f^x)$. Then:

$$
r^{\mathcal{G}}_{U,x}(g) = \varphi_x \circ r^{\mathcal{F}}_{U_x,x}(f^x) = r^{\mathcal{G}}_{U_x,x} \circ \varphi(U_x)(f^x).
$$

This implies that there exists an open neighbourhood $V_x \in \Omega_x$ such that $V_x \subset U_x$ and $g|_{V_x} = \varphi(U_x)(f^x)|_{V_x}$. Hence

$$g|_{V_x} = r^{\mathcal{G}}_{V_x,U_x} \circ \varphi(U_x)(f^x) = \varphi(V_x) \circ r^{\mathcal{F}}_{V_x,U_x}(f^x).$$

It follows that the family $(V_x, r^{\mathcal{F}}_{V_x,U_x}(f^x))_{x \in U}$ satisfies (i).   □

9.2.6 Remark. Let $\varphi : \mathcal{F} \to \mathcal{G}$ be a morphism of sheaves on X. If $\varphi(U)$ is surjective for all $U \in \Omega$, then by 8.2.3, φ_x is surjective for all $x \in X$. However, the converse is not true.

For example, let $X = \mathbb{C} \setminus \{0\}$ and for $U \in \Omega$, let $\mathcal{F}(U)$ be the set of holomorphic functions on U. Consider the morphism of sheaves $\varphi : \mathcal{F} \to \mathcal{F}$ given by $g \mapsto g'$. Then $\varphi(X)$ is not surjective since there is no holomorphic function on X such that $h'(z) = 1/z$ for all $z \in X$.

On the other hand, let $U \in \Omega_x$ and $f \in \mathcal{F}(U)$. For any point $z \in U$, fix an open ball B_z of X with centre z, contained in U. Since B_z is simply connected, there exists $g_z \in \mathcal{F}(B_z)$ such that $g'_z = f|_{B_z}$. Hence condition (i) of 9.2.5 is satisfied, and φ_x is surjective.

9.2.7 Theorem. *Let $\varphi : \mathcal{F} \to \mathcal{G}$ be a morphism of sheaves on X. Then φ is an isomorphism if and only if φ_x is bijective for all $x \in X$.*

Proof. By 9.2.4 and 9.2.6, we only have to prove that if φ_x is bijective for all $x \in X$, then $\varphi(U)$ is surjective for all $U \in \Omega$.

Let $U \in \Omega$ and $g \in \mathcal{G}(U)$. By 9.2.5, for any $x \in U$, there exists $U_x \in \Omega_x$ and $f^x \in \mathcal{F}(U_x)$ such that $\varphi(U_x)(f^x) = g|_{U_x}$. It follows that for $x, y \in U$, we have:

$$\varphi(U_x \cap U_y)(f^x|_{U_x \cap U_y}) = g|_{U_x \cap U_y} = \varphi(U_x \cap U_y)(f^y|_{U_x \cap U_y}).$$

The injectivity of φ implies that $f^x|_{U_x \cap U_y} = f^y|_{U_x \cap U_y}$. Since $\mathcal{F}$ is a sheaf, there exists $f \in \mathcal{F}(U)$ such that $f|_{U_x} = f^x$ for all $x \in U$. It follows that $\varphi(U)(f) = g$ since $\varphi(U)(f)|_{U_x} = g|_{U_x}$ for all $x \in U$.   □

9.2.8 Lemma. *Let $\mathcal{F}$ be a sheaf on X, $U \in \Omega$ and $f, g \in \mathcal{F}(U)$ Then $f = g$ if and only if $r_{U,x}(f) = r_{U,x}(g)$ for all $x \in U$.*

Proof. It is clear that $f = g$ implies that $r_{U,x}(f) = r_{U,x}(g)$ for all $x \in U$. Conversely, if $r_{U,x}(f) = r_{U,x}(g)$ for all $x \in U$, then for any $x \in U$, there exists $V_x \in \Omega_x$ such that $V_x \subset U$ and $f|_{V_x} = g|_{V_x}$. So $f = g$.   □

9.3 Sheaf associated to a presheaf

9.3.1 Let $\mathcal{F}$ be a presheaf on X and $U \in \Omega$. Denote by $\mathcal{F}^+(U)$ the set of maps $f : U \to E$ verifying: for any $x \in U$, there exist $V \in \Omega_x$ and $g \in \mathcal{F}(V)$ such that $V \subset U$ and $f|_V = g$.

We see immediately that $\mathcal{F}^+$ is a sheaf on X that we shall call the *sheaf associated to the presheaf* $\mathcal{F}$. If $\mathcal{F}$ is a presheaf of groups (resp. rings, etc...), then the above definition implies that $\mathcal{F}^+$ is a sheaf of groups (resp. rings, etc...).

There is a canonical injective morphism of presheaves $\eta^{\mathcal{F}}$ or $\eta : \mathcal{F} \to \mathcal{F}^+$ given by $f \mapsto f$.

If $\mathcal{F}$ is a sheaf, then $\mathcal{F} = \mathcal{F}^+$ since given $U \in \Omega$, $f \in \mathcal{F}^+$, $x \in U$, there exist $V_x \in \Omega_x$, $V_x \subset U$ and $g^x \in \mathcal{F}(V_x)$ such that $f|_{V_x} = g^x$. Since the V_x's is a covering of U, $f \in \mathcal{F}(U)$.

9.3.2 Theorem. *Let $\mathcal{F}$ be a presheaf on X.*

(i) *For all $x \in X$, the map $\eta_x : \mathcal{F}_x \to \mathcal{F}_x^+$ is bijective.*

(ii) *If $\varphi : \mathcal{F} \to \mathcal{G}$ is a morphism of presheaves, then there is a unique morphism of sheaves $\varphi^+ : \mathcal{F}^+ \to \mathcal{G}^+$ such that $\varphi^+ \circ \eta^{\mathcal{F}} = \eta^{\mathcal{G}} \circ \varphi$.*

(iii) *If $\varphi : \mathcal{F} \to \mathcal{G}$ is a morphism of presheaves and $\mathcal{G}$ is a sheaf, then there is a unique morphism of sheaves $\psi : \mathcal{F}^+ \to \mathcal{G}$ such that $\psi \circ \eta^{\mathcal{F}} = \varphi$.*

Proof. (i) Let $x \in X$ and $f_x \in \mathcal{F}_x^+$. There exist $U \in \Omega_x$, $f \in \mathcal{F}^+(U)$ such that $r_{U,x}^{\mathcal{F}^+}(f) = f_x$. By definition, there exist $V \in \Omega_x$, $V \subset U$ and $g \in \mathcal{F}(V)$ such that $f|_V = g$. So $\eta_x(r_{V,x}^{\mathcal{F}}(g)) = f_x$. Hence η_x is surjective. The injectivity of η_x is a consequence of 9.2.4 and 9.3.1.

(ii) Uniqueness follows from (i) and 9.2.8.

Now let $U \in \Omega$ and $f \in \mathcal{F}^+(U)$. There exist an open covering $(U_i)_{i \in I}$ of U and elements $f_i \in \mathcal{F}(U_i)$ such that $f|_{U_i} = f_i$. Set $g_i = \varphi(U_i)(f_i)$. Since $f_i|_{U_i \cap U_j} = f_j|_{U_i \cap U_j}$, we have $g_i|_{U_i \cap U_j} = g_j|_{U_i \cap U_j}$. It follows that there exists $g \in \mathcal{G}^+(U)$ such that $g|_{U_i} = g_i$. It is easy to see that g does not depend on the choice of the covering $(U_i)_{i \in I}$ and of the elements f_i. Now set $\varphi^+(U)(f) = g$. Then it is clear that φ^+ is a morphism of sheaves from $\mathcal{F}^+$ to $\mathcal{G}^+$ verifying $\varphi^+ \circ \eta^{\mathcal{F}} = \eta^{\mathcal{G}} \circ \varphi$.

(iii) This is a direct consequence of (ii) since $\mathcal{G}^+ = \mathcal{G}$. $\square$

9.3.3 Remarks. 1) Assume that $\mathcal{F}$ is not a sheaf. Then η_x is bijective for all $x \in X$ (9.3.2), but η is not an isomorphism. Thus 9.2.7 is false for morphisms of presheaves.

2) Let $\mathcal{F}$ be the presheaf of 9.1.6 (4), then $\mathcal{F}^+$ is the sheaf of 9.1.6 (2) with $X = \mathbb{R}$.

3) Let $\mathcal{F}, \mathcal{G}$ be presheaves on X, $\varphi \in \mathrm{Hom}(\mathcal{F}, \mathcal{G})$. If φ_x is bijective for all $x \in X$, then so is φ_x^+ (9.3.2). Consequently, φ^+ is an isomorphism (9.2.7).

9.3.4 Let $\mathcal{F}$ be a presheaf and $\mathcal{G}$ a sheaf on X. The results in 9.3.2 induce a map:

$$\xi : \mathrm{Hom}(\mathcal{F}, \mathcal{G}) \to \mathrm{Hom}(\mathcal{F}^+, \mathcal{G}), \quad \varphi \mapsto \varphi^+.$$

Proposition. *The map ξ is bijective.*

Proof. Let $\varphi, \psi \in \mathrm{Hom}(\mathcal{F}, \mathcal{G})$ be such that $\xi(\varphi) = \xi(\psi)$. Then for $U \in \Omega$ and $f \in \mathcal{F}(U)$, we have $\varphi^+(U) \circ \eta^{\mathcal{F}}(U)(f) = \psi^+(U) \circ \eta^{\mathcal{F}}(U)(f)$. Hence $\varphi(U)(f) = \psi(U)(f)$ and ξ is injective.

Now if $\mu \in \mathrm{Hom}(\mathcal{F}^+, \mathcal{G})$. Let $\nu(U)(f) = \mu(U) \circ \eta^{\mathcal{F}}(U)(f)$ for $U \in \Omega$ and $f \in \mathcal{F}(U)$. Then $\nu \in \mathrm{Hom}(\mathcal{F}, \mathcal{G})$ and $\nu^+ = \mu$. So ξ is surjective. $\square$

9.3.5 Proposition. *Let $\mathcal{F}$ be a presheaf, $\mathcal{H}$ a sheaf on X and $\rho : \mathcal{F} \to \mathcal{H}$ a morphism of presheaves. Suppose that for any morphism of presheaves $\varphi : \mathcal{F} \to \mathcal{G}$, there exists a unique morphism of sheaves $\psi : \mathcal{H} \to \mathcal{G}$ such that $\varphi = \psi \circ \rho$. Then $\mathcal{F}^+$ and $\mathcal{H}$ are isomorphic.*

Proof. There exist $\psi \in \text{Hom}(\mathcal{H}, \mathcal{F}^+)$ and $\mu \in \text{Hom}(\mathcal{F}^+, \mathcal{H})$ such that $\eta^{\mathcal{F}} = \psi \circ \rho$ and $\rho = \mu \circ \eta^{\mathcal{F}}$. It follows that $\eta^{\mathcal{F}} = \psi \circ \mu \circ \eta^{\mathcal{F}}$ and $\rho = \mu \circ \psi \circ \rho$. By the uniqueness property of our hypothesis, we obtain that $\psi \circ \mu = \text{id}_{\mathcal{F}^+}$ and $\mu \circ \psi = \text{id}_{\mathcal{H}}$. $\square$

9.3.6 Proposition. *Let $\varphi : \mathcal{G} \to \mathcal{F}$ be a morphism of presheaves on X. Suppose that $\mathcal{F}$ is a sheaf and that X has a base of open subsets $\mathfrak{B}$ such that $\varphi(U)$ is bijective for all $U \in \mathfrak{B}$. Then $\mathcal{F}$ and $\mathcal{G}^+$ are isomorphic.*

Proof. For $x \in X$, let $\mathcal{W}_x = \Omega_x \cap \mathfrak{B}$. Then $\mathcal{W}_x$ is a cofinal subset of Ω_x, it follows from 8.2.7 that:

$$\varinjlim_{U \in \mathcal{W}_x} \mathcal{F}(U) = \mathcal{F}_x \ , \quad \varinjlim_{U \in \mathcal{W}_x} \mathcal{G}(U) = \mathcal{G}_x.$$

Our hypothesis and 8.2.3 imply that φ_x is bijective. Hence φ_x^+ is bijective by 9.3.2 and the result follows from 9.2.7. $\square$

9.3.7 Let $Y \in \Omega^X$ and $\mathcal{F}$ a presheaf on X. The *restriction* of $\mathcal{F}$ to Y, denoted by $\mathcal{F}|_Y$ is a presheaf on Y endowed with the induced topology. More precisely, $\mathcal{F}|_Y(U) = \mathcal{F}(U)$ for $U \in \Omega^Y$. If $\mathcal{F}$ is a sheaf, then $\mathcal{F}|_Y$ is also a sheaf. Further, if $Z \in \Omega^Y$, then $\mathcal{F}|_Z = (\mathcal{F}|_Y)|_Z$.

Let $\varphi : \mathcal{F} \to \mathcal{G}$ be a morphism of presheaves on X. Then we can define the *restriction* $\varphi|_Y : \mathcal{F}|_Y \to \mathcal{G}|_Y$ in an obvious way. Clearly, $\varphi|_Y$ is a morphism of presheaves and it is an isomorphism if φ is an isomorphism.

9.3.8 In general, let Y be any subset of X endowed with the induced topology. Let $\mathcal{F}$ be a sheaf on X. We define, for $V \in \Omega^Y$, $\mathcal{H}(V)$ to be the set of maps $f : V \to E$ such that there exist $U \in \Omega^X$ and $g \in \mathcal{F}(U)$ verifying $V = Y \cap U$ and $f = g|_V$. It is easy to verify that $\mathcal{H}$ is a presheaf on Y. However, it is not a sheaf in general.

The sheaf $\mathcal{H}^+$ associated to $\mathcal{H}$ is then defined as follows: for $V \in \Omega^Y$, $\mathcal{H}^+(V)$ is the set of maps $f : V \to E$ such that for all $x \in V$, there exist $U_x \in \Omega^X$ and $g \in \mathcal{F}(U_x)$ verifying $f|_{V \cap U_x} = g|_{V \cap U_x}$.

We shall call $\mathcal{H}^+$ the *restriction* of $\mathcal{F}$ to Y, and we shall denote it by $\mathcal{F}|_Y$. Again, it is easy to see that if Y is open, then this coincides with the definition in 9.3.7, and that $\mathcal{F}|_Z = (\mathcal{F}|_Y)|_Z$ if $Z \subset Y \subset X$.

9.4 Gluing

9.4.1 Proposition. *Let $\mathcal{F}$ be a presheaf, $\mathcal{G}$ a sheaf on X, $(X_i)_{i \in I}$ an open covering of X and for $i \in I$, $\varphi_i : \mathcal{F}|_{X_i} \to \mathcal{G}|_{X_i}$ a morphism of presheaves such that $\varphi_i|_{X_i \cap X_j} = \varphi_j|_{X_i \cap X_j}$ for all $i, j \in I$. There exists a unique morphism $\varphi : \mathcal{F} \to \mathcal{G}$ such that $\varphi|_{X_i} = \varphi_i$ for all $i \in I$. We say that φ is obtained by gluing the φ_i's.*

Proof. Let $U \in \Omega$, $f \in \mathcal{F}(U)$ and for $i, j \in I$, let $U_i = U \cap X_i$, $f_i = f|_{U_i}$ and $f_{ij} = f|_{U_i \cap U_j}$. If $\varphi \in \operatorname{Hom}(\mathcal{F}, \mathcal{G})$ is such that $\varphi|_{X_i} = \varphi_i$ for all i, then

$$r^{\mathcal{G}}_{U_i, U} \circ \varphi(U)(f) = \varphi_i(U_i)(f_i).$$

Thus φ is unique if it exists.

Let us prove the existence of φ. By our hypotheses, we have:

$$r^{\mathcal{G}}_{U_i \cap U_j, U_i} \circ \varphi_i(U_i)(f_i) = \varphi_i(U_i \cap U_j)(f_{ij}) = \varphi_j(U_i \cap U_j)(f_{ij})$$
$$= r^{\mathcal{G}}_{U_i \cap U_j, U_j} \circ \varphi_j(U_j)(f_j).$$

Since $\mathcal{G}$ is a sheaf, there exists $g \in \mathcal{G}(U)$ such that $r^{\mathcal{G}}_{U_i, U}(g) = \varphi_i(U_i)(f_i)$ for all i. If we define $\varphi(U)(f) = g$, then we verify easily that $\varphi : \mathcal{F} \to \mathcal{G}$ is a morphism such that $\varphi|_{X_i} = \varphi_i$ for all i. $\square$

9.4.2 Proposition. *Let $(X_i)_{i \in I}$ be an open covering of X and for $i \in I$, $\mathcal{F}_i$ a presheaf on X_i. Suppose that the following conditions are satisfied:*

(i) *There is an isomorphism of sheaves $\varepsilon_{ij} : \mathcal{F}_i|_{X_i \cap X_j} \to \mathcal{F}_j|_{X_i \cap X_j}$ for all $i, j \in I$.*

(ii) *For all $i, j, k \in I$, we have*

$$\varepsilon_{ii} = \operatorname{id}_{\mathcal{F}_i}, \quad \varepsilon_{jk}|_{X_i \cap X_j \cap X_k} \circ \varepsilon_{ij}|_{X_i \cap X_j \cap X_k} = \varepsilon_{ik}|_{X_i \cap X_j \cap X_k}.$$

Then there exist a sheaf $\mathcal{F}$ on X and isomorphisms $\varepsilon_i : \mathcal{F}|_{X_i} \to \mathcal{F}_i$ such that

$$\varepsilon_j|_{X_i \cap X_j} = \varepsilon_{ij} \circ \varepsilon_i|_{X_i \cap X_j}.$$

Further, the ε_i's and $\mathcal{F}$ are unique in the following sense: if $\mathcal{F}'$ and $(\varepsilon'_i)_{i \in I}$ verify the same conclusion, then there is a unique isomorphism $\varepsilon : \mathcal{F} \to \mathcal{F}'$ such that $\varepsilon'_i \circ \varepsilon|_{X_i} = \varepsilon_i$ for all $i \in I$. We say that $\mathcal{F}$ is obtained by gluing the sheaves $\mathcal{F}_i$.

Proof. Let $U \in \Omega$, and for $i, j \in I$, let $U_i = U \cap X_i$, $U_{ij} = U \cap X_i \cap X_j$. Set $\mathcal{F}(U)$ to be the set of families $(f_i) \in \prod_{i \in I} \mathcal{F}_i(U_i)$ verifying $\varepsilon_{ij}(U_{ij})(f_i|_{U_{ij}}) = f_j|_{U_{ij}}$ for all $i, j \in I$.

For $V \in \Omega$ and $V \subset U$, define $r^{\mathcal{F}}_{VU} : \mathcal{F} \to \mathcal{G}$ by $r^{\mathcal{F}}_{VU}((f_i)) = (r^{\mathcal{F}_i}_{V_i, U_i}(f_i))$ where $V_i = U_i \cap V$. Then we verify easily that $\mathcal{F}$ is a sheaf on X. Finally, define the morphisms $\varepsilon_i : \mathcal{F}|_{X_i} \to \mathcal{F}_i$ by $\varepsilon_i(U)((f_j)) = f_i$. Then a straightforward verification proves the proposition. $\square$

9.5 Ringed space

9.5.1 Lemma. *Let $\mathfrak{B}$ be a base of open subsets of X. For $U \in \mathfrak{B}$, let $\mathcal{L}(U)$ be a set of maps from U to E. Further, suppose that the following conditions are verified:*

(i) *If $U, V \in \mathfrak{B}$ are such that $V \subset U$, then $f|_V \in \mathcal{L}(V)$ for all $f \in \mathcal{L}(U)$.*

(ii) *If $(U_i)_{i \in I}$ is a family of open subsets in $\mathfrak{B}$ whose union U is in $\mathfrak{B}$, and $f : U \to E$ is a map verifying $f|_{U_i} \in \mathcal{L}(U_i)$ for all i, then $f \in \mathcal{L}(U)$.*

Then there exists a unique sheaf $\mathcal{F}$ on X such that $\mathcal{F}(U) = \mathcal{L}(U)$ for all $U \in \mathfrak{B}$.

Proof. Any $U \in \Omega$ is the union of $V \in \mathfrak{B}$ such that $V \subset U$. Let $\mathcal{M}(U)$ be the set of maps $f : U \to E$ verifying $f|_V \in \mathcal{L}(V)$ for all $V \in \mathfrak{B}$ contained in U. It is then clear that if $\mathcal{F}$ exists, then $\mathcal{F}(U) = \mathcal{M}(U)$ for all $U \in \Omega$. So we only need to show that $\mathcal{M}$ is a sheaf.

Clearly, $\mathcal{M}$ is a presheaf. Let $(U_i)_{i \in I}$ be a family of open subsets of X whose union is U and the elements $f_i \in \mathcal{M}(U_i)$ be such that $f_i|_{U_i \cap U_j} = f_j|_{U_i \cap U_j}$ for all $i, j \in I$. Then there exists a unique map $f : U \to E$ such that $f|_{U_i} = f_i$ for all $i \in I$.

Now if $V \in \mathfrak{B}$ is contained in U, then for any $W \in \mathfrak{B}$ contained in $V \cap U_i$, $f|_W \in \mathcal{L}(W)$. Thus $f|_{V \cap U_i} \in \mathcal{M}(V \cap U_i)$. Since V is the union of the set of $W \in \mathfrak{B}$ such that $W \subset V \cap U_i$ for some $i \in I$, we obtain by (ii) that $f|_V \in \mathcal{L}(V)$. We have therefore proved that $\mathcal{M}$ is a sheaf. $\square$

9.5.2 Definition. (i) *A Cartan space relative to E is a pair $(X, \mathcal{F})$ where X is a topological space and $\mathcal{F}$ a sheaf of functions on X with values in E.*

(ii) *A morphism of Cartan spaces from $(X, \mathcal{F})$ to $(Y, \mathcal{G})$ is a continuous map $u : X \to Y$ such that for all $V \in \Omega^Y$ and $g \in \mathcal{G}(V)$, the map defined by $x \mapsto g \circ u(x)$ belongs to $\mathcal{F}(u^{-1}(V))$.*

9.5.3 Proposition. *Let $(X, \mathcal{F})$ and $(Y, \mathcal{G})$ be Cartan spaces, $\mathfrak{B}$ and $\mathfrak{C}$ bases of open subsets for X and Y respectively. The following conditions are equivalent for a map $u : X \to Y$:*

(i) *u is a morphism of Cartan spaces.*

(ii) *For $x \in X$, $V \in \mathfrak{C} \cap \Omega_{u(x)}$, there exists an open subset $U \in \mathfrak{B} \cap \Omega_x$ such that $U \subset u^{-1}(V)$ and for all $g \in \mathcal{G}(V)$, the map $(g \circ u)|_U \in \mathcal{F}(U)$.*

Proof. Obviously, (i) $\Rightarrow$ (ii). Conversely, suppose that (ii) is verified. Then u is continuous. Let $W \in \Omega^Y$, $g \in \mathcal{G}(W)$ and $x \in u^{-1}(W)$. There exist an open neighbourhood $U_x \in \mathfrak{B}$ of x and an open neighbourhood $V_{u(x)} \in \mathfrak{C}$ of $u(x)$ such that:

$$U_x \subset u^{-1}(W) \, , \ u(U_x) \subset V_{u(x)} \subset W \ \text{ and } \ g \circ u|_{U_x} \in \mathcal{F}(U_x).$$

Since the union of the U_x's, with $x \in u^{-1}(W)$, is $u^{-1}(W)$, we have that $g \circ u \in \mathcal{F}(u^{-1}(W))$. $\square$

9.5.4 Let $(X, \mathcal{F})$ be a Cartan space and $Y \subset X$, then we have seen in 9.3.8 that $(Y, \mathcal{F}|_Y)$ is again a Cartan space and the canonical injection $j : Y \to X$ is a morphism of Cartan spaces.

If $u : (X, \mathcal{F}) \to (Z, \mathcal{G})$ is a morphism of Cartan spaces, then $u \circ j$ is again a morphism of Cartan spaces. This is just the *restriction* of u to $(Y, \mathcal{F}|_Y)$.

Now let $u : (Z, \mathcal{G}) \to (X, \mathcal{F})$ be a morphism of Cartan spaces such that $u(Z) \subset Y$. Let us denote by $\nu : Z \to Y$ the induced map of topological spaces. We claim that ν is a morphism of Cartan spaces.

Let $V \in \Omega^Y$, $f \in \mathcal{F}|_Y(V)$ and $U \in \Omega^X$ be such that $V = U \cap Y$. Since there is an open covering $(U_i)_{i \in I}$ of V in X and $g_i \in \mathcal{F}(U_i)$ such that $f|_{V \cap U_i} = g_i|_{V \cap U_i}$, we can suppose (by replacing U_i by $U \cap U_i$ if necessary) that the unions of the $U_i \cap Y$ is equal to V.

Now $g_i \circ u \in \mathcal{G}(u^{-1}(U_i)) = \mathcal{G}(u^{-1}(U_i \cap Y))$. On the other hand, if x is an element of $u^{-1}(U_i) = u^{-1}(U_i \cap Y)$, then $g_i \circ u(x) = f \circ u(x)$. Since $\mathcal{G}$ is a sheaf, we obtain that $f \circ u \in \mathcal{G}(u^{-1}(V))$. So we have proved our claim.

9.5.5 Let $(X, \mathcal{F})$ be a Cartan space and $(Y_i)_{i \in I}$ an open covering of X. Suppose that for each $i \in I$, we have a morphism of Cartan spaces $u_i : (Y_i, \mathcal{F}|_{Y_i}) \to (Z, \mathcal{G})$ and the u_i's verify $u_i|_{Y_i \cap Y_j} = u_j|_{Y_i \cap Y_j}$ for all $i, j \in I$. then there exists a unique map $u : X \to Z$ such that $u|_{Y_i} = u_i$ for all $i \in I$.

Now if $\mathfrak{B}_i$ is a base of open subsets of Y_i, then their union is a base of open subsets of X. It follows from the definitions and 9.5.3 that u is a morphism of Cartan spaces.

9.5.6 As a consequence of the preceding remarks, we have:

Proposition. *Let $(X, \mathcal{F})$, $(Y, \mathcal{G})$ be Cartan spaces, $u : X \to Y$ a map, $(V_i)_{i \in I}$ an open covering of Y and for each $i \in I$, $(U_{ij})_{j \in J}$ is a family of open subsets of X whose union is $u^{-1}(V_i)$. For $(i, j) \in I \times J$, let $u_{ij} : U_{ij} \to V_i$ be the map induced by u. Then u is a morphism of Cartan spaces if and only if for all $(i, j) \in I \times J$, the map u_{ij} is a morphism of Cartan spaces from $(U_{ij}, \mathcal{F}|_{U_{ij}})$ to $(V_i, \mathcal{G}|_{V_i})$.*

9.5.7 Let us suppose that E is a commutative ring. A Cartan space $(X, \mathcal{F})$ is a *ringed space* relative to E if for all $U \in \Omega$, $\mathcal{F}(U)$ is a subring of the ring of E-valued functions on U. In particular, $\mathcal{F}$ is a sheaf of commutative rings on X.

Let $(X, \mathcal{F})$, $(Y, \mathcal{G})$ be ringed spaces and u a morphism from $(X, \mathcal{F})$ to $(Y, \mathcal{G})$ as Cartan spaces. Then for any $V \in \Omega^Y$, the map $\mathcal{G}(V) \to \mathcal{F}(u^{-1}(V))$, $g \mapsto g \circ u$, is a ring homomorphism.

9.5.8 Remark. Let us suppose that E is a commutative ring and $\mathfrak{B}$ a base of open subsets of X. Then a Cartan space $(X, \mathcal{F})$ is a ringed space if and only if $\mathcal{F}(U)$ is a subring of the ring of E-valued functions on U for all $U \in \Omega^X$ (9.5.1).

References and comments

- [5], [26], [34], [37].

For our purposes, we have restricted ourselves to sheaf of functions on a topological space. For a more thorough treatment of the theory of sheaves, the reader may refer to [15], [34] and [37].

10

Jordan decomposition and some basic results on groups

The patient reader will soon be rewarded. After this chapter, we shall begin to study algebraic geometry, Lie algebras and algebraic groups. A large part of this chapter covers basic material on the theory of abstract groups.

In this chapter, k will denote an algebraically closed (commutative) field.

10.1 Jordan decomposition

10.1.1 Let V be a finite-dimensional k-vector space. We shall denote by $\text{End}(V)$ (resp. $GL(V)$) the set of endomorphisms (resp. invertible endomorphisms) of V. An element $x \in \text{End}(V)$ is said to be:

- *semisimple* if V has a basis consisting of eigenvectors for x.
- *nilpotent* if $x^n = 0$ for some $n \in \mathbb{N}^*$.
- *unipotent* if $x - \text{id}_V$ is nilpotent.

10.1.2 Theorem. *Let $x \in \text{End}(V)$. There exists a unique pair (x_s, x_n) of endomorphisms of V such that:*

(i) *x_s is semisimple, x_n is nilpotent and $x = x_s + x_n$.*

(ii) *$x_s \circ x_n = x_n \circ x_s$.*

Furthermore, there exist $P, Q \in k[T]$ such that $P(0) = Q(0) = 0$ and $P(x) = x_s$, $Q(x) = x_n$. We say that $x = x_s + x_n$ is the additive Jordan decomposition *of x and x_s (resp. x_n) is the* semisimple *(resp. nilpotent) component of x.*

Proof. Let $\chi_x(T) = (T - \lambda_1)^{m_1} \cdots (T - \lambda_r)^{m_r}$ be the characteristic polynomial of x where $\lambda_1, \ldots, \lambda_r$ are pairwise distinct eigenvalues of x. Let $\mathfrak{a}_i$ be the ideal of $k[T]$ generated by $(T - \lambda_i)^{m_i}$ and $V_i = \ker(x - \lambda_i \text{id}_V)^{m_i}$. The V_i's are x-stable and $V = V_1 \oplus \cdots \oplus V_r$.

By 2.1.2, there exists $P \in k[T]$ such that $P(T) - \lambda_i \in \mathfrak{a}_i$ for all $1 \leqslant i \leqslant r$ and $P(T) \in \mathfrak{a}_0 = Tk[T]$ (this last condition is redundant if $0 \in \{\lambda_1, \ldots, \lambda_r\}$). Set $x_s = P(x)$. Then $x_s(v_i) = \lambda_i v_i$ for $v_i \in V_i$. So x_s is semisimple. Now let $x_n = x - x_s = Q(x)$ where $Q(T) = T - P(T)$. Then x_n is nilpotent and the pair (x_s, x_n) verifies conditions (i) and (ii).

If (y_s, y_n) is another such pair, then y_s, y_n commute with x, and hence they commute with x_s, x_n. It follows that $x_s - y_s = y_n - x_n$ is both semisimple and nilpotent. Thus $x_s = y_s$ and $x_n = y_n$.   □

10.1.3 Corollary. (i) *Let* $x, y \in \mathrm{End}(V)$ *be such that* $x \circ y = y \circ x$. *Then* y *commutes with* x_s *and* x_n. *Further,* $(x+y)_s = x_s + y_s$ *and* $(x+y)_n = x_n + y_n$.

(ii) *Let* $x \in \mathrm{End}(V)$ *and* W *a* x-*stable subspace of* V. *Then* W *is* x_s-*stable,* x_n-*stable and* $x|_W = x_s|_W + x_n|_W$ *is the Jordan decomposition of* $x|_W$.

(iii) *Let* x, W *be as in* (ii) *and* z, z_s, z_n *be endomorphisms of* V/W *induced by* x, x_s, x_n. *Then* $z = z_s + z_n$ *is the Jordan decomposition of* z.

10.1.4 Theorem. *Let* $x \in \mathrm{GL}(V)$. *There exists a unique pair* (x_s, x_u) *of elements of* $\mathrm{GL}(V)$ *such that:*

(i) x_s *is semisimple,* x_u *is unipotent and* $x = x_s \circ x_u$.

(ii) $x_s \circ x_u = x_u \circ x_s$.

We say that $x = x_s \circ x_u$ *is the* multiplicative Jordan decomposition *of* x *and* x_u *is the* unipotent component *of* x.

Proof. Let $x = x_s + x_n$ be the additive Jordan decomposition of x. Since $x \in \mathrm{GL}(V)$ and the eigenvalues of x and x_s are identical, $x_s \in \mathrm{GL}(V)$. Hence $x = x_s \circ (\mathrm{id}_V + x_s^{-1} \circ x_n)$. Let $x_u = \mathrm{id}_V + x_s^{-1} \circ x_n$, then clearly (x_s, x_u) satisfies conditions (i) and (ii). Finally, the proof of uniqueness is as in 10.1.2.   □

10.1.5 We have also analogues of the results of 10.1.3 for the multiplicative Jordan decomposition.

10.1.6 Let E be a k-vector space (not necessarily finite-dimensional). If $x \in \mathrm{End}(E)$ and $v \in E$, we set $E_x(v)$ to be the subspace generated by $x^n(v)$, $n \in \mathbb{N}$. This is clearly a x-stable subspace of E.

Proposition. *Let* $x \in \mathrm{End}(E)$. *The following conditions are equivalent:*

(i) *The space* E *is the union of a family of finite-dimensional* x-*stable subspaces.*

(ii) *For any* $v \in E$, $E_x(v)$ *is finite-dimensional.*

If these conditions are satisfied, we say that x *is* locally finite.

Proof. Clearly E is the union of the $E_x(v)$, $v \in E$, so (ii) ⇒ (i). Now given (i), any element v is contained in some finite-dimensional x-stable subspace V. Hence $E_x(v) \subset V$, and $\dim E_x(v)$ is finite.   □

10.1.7 Remark. Let x be a diagonalisable endomorphism of E and $(v_i)_{i \in I}$ a basis of eigenvectors. Let $v \in E$, then there exists a finite subset $J \subset I$ such that v belongs to V_J, the span of the v_j's, $j \in J$. Since V_J is clearly x-stable, it follows that x is locally finite.

10.1.8 Proposition. *Let* $x \in \mathrm{End}(E)$. *The following conditions are equivalent:*

(i) x *is diagonalisable.*

(ii) x *is locally finite, and the restriction of x to any finite-dimensional x-stable subspace is semisimple.*

If these conditions are satisfied, we say that x is semisimple.

Proof. (i) $\Rightarrow$ (ii) This is clear by 10.1.7 since any finite-dimensional x-stable subspace is contained in some V_J where J is a finite subset of I.

(ii) $\Rightarrow$ (i) Since x is locally finite, V is the union of a family $(V_\alpha)_{\alpha \in \mathcal{A}}$ of finite-dimensional x-stable subspaces. The restriction of x to V_α is semisimple and therefore there exists a basis $\mathcal{B}_\alpha$ of eigenvectors for V_α. The union $\mathcal{B}$ of the $\mathcal{B}_\alpha$'s, $\alpha \in \mathcal{A}$, is a family of eigenvectors spanning V. We can therefore extract a basis, from $\mathcal{B}$, of eigenvectors for x. $\square$

10.1.9 Proposition. *Let $x \in \mathrm{End}(E)$ (resp. $x \in \mathrm{GL}(E)$). The following conditions are equivalent:*

(i) *E is the union of a family $(V_\alpha)_{\alpha \in \mathcal{A}}$ of finite-dimensional x-stable subspaces such that $x|_{V_\alpha}$ is nilpotent (resp. unipotent) for all $\alpha \in \mathcal{A}$.*

(ii) *x is locally finite, and the restriction of x to any finite-dimensional x-stable subspace is nilpotent (resp. unipotent).*

If these conditions are satisfied, we say that x is locally nilpotent *(resp.* locally unipotent*).*

Proof. Clearly, (ii) $\Rightarrow$ (i). Conversely, suppose that (i) is verified. Let V be a finite-dimensional x-stable subspace and $(v_1, \dots, v_n)$ be a basis of V. Then $x|_{E_x(v_i)}$ is nilpotent (resp. unipotent) for $1 \leqslant i \leqslant n$. It follows clearly that $x|_V$ is nilpotent (resp. unipotent). $\square$

10.1.10 Theorem. *Let x be a locally finite endomorphism of E. There exists a unique pair (x_s, x_n) of elements of $\mathrm{End}(E)$ such that:*

(i) *x_s is semisimple, x_n is locally nilpotent and $x = x_s + x_n$.*

(ii) *$x_s \circ x_n = x_n \circ x_s$.*

Any x-stable subspace is x_s-stable and x_n-stable. Moreover, if V is a finite-dimensional x-stable subspace of E, then $x|_V = x_s|_V + x_n|_V$ is the Jordan decomposition (10.1.2). We shall call $x = x_s + x_n$ the additive Jordan decomposition *of x.*

Proof. First of all, the restriction of x to any finite-dimensional x-stable subspace has a Jordan decomposition. Further, given two finite-dimensional x-stable subspaces V and W of E, by 10.1.3 (ii) and the uniqueness of the Jordan decomposition, we have

$$(x|_V)_s|_{V \cap W} = (x|_W)_s|_{V \cap W} \ , \ (x|_V)_n|_{V \cap W} = (x|_W)_n|_{V \cap W}.$$

Since x is locally finite, it follows that there exists a unique endomorphism x_s (resp. x_n) of E such that $x_s|_V = (x|_V)_s$ (resp. $x_n|_V = (x|_V)_n$) for all finite-dimensional x-stable subspace V of E. It is clear by construction that the pair (x_s, x_n) satisfies conditions (i) and (ii), and that any x-stable subspace of E is x_s-stable and x_n-stable. $\square$

10.1.11 We prove in a similar way the following:

Theorem. *Let x be a locally finite invertible endomorphism of E. There exists a unique pair (x_s, x_u) of elements of* $\mathrm{GL}(E)$ *such that:*

(i) x_s *is semisimple,* x_u *is locally unipotent and* $x = x_s \circ x_u$.

(ii) $x_s \circ x_u = x_u \circ x_s$.

Any x-stable subspace is x_s-stable and x_u-stable. Moreover, if V is a finite-dimensional x-stable subspace of E, then $x|_V = x_s|_V \circ x_u|_V$ is the Jordan decomposition (10.1.4). *We shall call $x = x_s \circ x_u$ the* multiplicative Jordan decomposition *of x.*

10.1.12 Lemma. *Let E, F be vector spaces, $x \in \mathrm{End}(E)$, $y \in \mathrm{End}(F)$ and $z : E \to F$ a surjective linear map. Suppose further that x is locally finite and $z \circ x = y \circ z$. Then y is locally finite. Furthermore:*

(i) *We have $z \circ x_s = y_s \circ z$ and $z \circ x_n = y_n \circ z$.*

(ii) *If x, y are automorphisms, then $z \circ x_s = y_s \circ z$ and $z \circ x_u = y_u \circ z$.*

Proof. Since $z \circ R(x) = R(y) \circ z$ for any $R \in k[T]$, it is clear from the surjectivity of z that y is locally finite and we can assume that E and F are finite-dimensional. Let $P, Q \in k[T]$ be such that $x_s = P(x)$ and $x_n = Q(x)$.

Let $(e_i)_{i \in I}$ be a basis of E consisting of eigenvectors of x_s. Then $(z(e_i))_{i \in I}$ is a family of eigenvectors of $P(y)$ and it follows from the surjectivity of z that we can extract from this family a basis of F. Hence $P(y)$ is semisimple.

Finally $Q(y)$ is clearly nilpotent and $y \circ z = (P(y) + Q(y)) \circ z$, again the surjectivity of z implies that $y = P(y) + Q(y)$. It is now clear that $y_s = P(y)$ and $y_n = Q(y)$. The proof of part (ii) is similar. $\quad\square$

10.1.13 Let E, F be vector spaces, $x \in \mathrm{End}(E)$, $y \in \mathrm{End}(F)$. We denote by $x \otimes y$ the endomorphism of $E \otimes_k F$ defined as follows: for $v \in E$ and $w \in F$:

$$x \otimes y(v \otimes w) = x(v) \otimes y(w).$$

Lemma. *Let x, y be locally finite automorphisms of E and F respectively. Then $x \otimes y$ is a locally finite automorphism of $E \otimes_k F$ and $(x \otimes y)_s = x_s \otimes y_s$, $(x \otimes y)_u = x_u \otimes y_u$.*

Proof. It is clear that $x \otimes y$ is locally finite and $x \otimes y = (x_s \otimes y_s) \circ (x_u \otimes y_u) = (x_u \otimes y_u) \circ (x_s \otimes y_s)$.

Let $(e_i)_{i \in I}$ (resp. $(f_j)_{j \in J}$) be a basis of E (resp. F) consisting of eigenvectors of x_s (resp. y_s). Then clearly $(e_i \otimes f_j)_{(i,j) \in I \times J}$ is a basis of $E \otimes_k F$ consisting of eigenvectors of $x_s \otimes y_s$. So $x_s \otimes y_s$ is semisimple.

Finally, we have $x_u \otimes y_u - \mathrm{id}_{E \otimes_k F} = (x_u - \mathrm{id}_E) \otimes y_u + \mathrm{id}_E \otimes (y_u - \mathrm{id}_F)$. Since $(x_u - \mathrm{id}_E) \otimes y_u$ commutes with $\mathrm{id}_E \otimes (y_u - \mathrm{id}_F)$, it follows that $x_u \otimes y_u$ is locally unipotent. $\quad\square$

10.1.14 Proposition. *Let x be a locally finite endomorphism of a k-algebra $\mathfrak{A}$.*

    (i) *If x is a k-algebra automorphism, then so are x_s and x_u.*

    (ii) *If x is a derivation of $\mathfrak{A}$, then so are x_s and x_n.*

Proof. Let $\mathfrak{B}$ be the image of the linear map $q : \mathfrak{A} \otimes_k \mathfrak{A} \to \mathfrak{A}$ given by the multiplication of the k-algebra $\mathfrak{A}$.

(i) If x is a k-algebra automorphism, then $q \circ (x \otimes x) = x \circ q$. By 10.1.13, $x \otimes x$ is locally finite and since q is surjective onto $\mathfrak{B}$, we have:

$$q \circ (x_s \otimes x_s) = q \circ (x \otimes x)_s = x_s \circ q \; , \; q \circ (x_u \otimes x_u) = q \circ (x \otimes x)_u = x_u \circ q$$

by 10.1.12 and 10.1.13. Finally, if $\mathfrak{A}$ has an identity element e, then $x(e) = e$ implies that the line spanned by e is x-stable, hence it is also x_s- and x_u-stable. Now, x_u is unipotent, so $x_u(e) = e$ and it follows that $x_s(e) = e$. Hence, x_s and x_u are k-algebra automorphisms of $\mathfrak{A}$.

(ii) If x is a derivation, then $q \circ (\iota \otimes x + x \otimes \iota) = x \circ q$ where ι denotes the identity map on $\mathfrak{A}$. Now we see easily that $\iota \otimes x + x \otimes \iota$ is locally finite and that $(\iota \otimes x + x \otimes \iota)_s = \iota \otimes x_s + x_s \otimes \iota$ and $(\iota \otimes x + x \otimes \iota)_n = \iota \otimes x_n + x_n \otimes \iota$. It follows from 10.1.12 that:

$$q \circ (\iota \otimes x_s + x_s \otimes \iota) = x_s \circ q \; , \; q \circ (\iota \otimes x_n + x_n \otimes \iota) = x_n \circ q.$$

Consequently, x_s and x_n are derivations of $\mathfrak{A}$. $\quad\square$

10.2 Generalities on groups

10.2.1 In general, we shall present group laws multiplicatively.

Let G be a group. Denote by $\mu_G : G \times G \to G$, $(\alpha, \beta) \mapsto \alpha\beta$ the group multiplication and $\iota_G : G \to G$, $\alpha \mapsto \alpha^{-1}$, the inverse. When there is no confusion, we shall simply write μ, ι for μ_G, ι_G.

Let U, V be subsets of G, then we set $UV = \{\alpha\beta \; ; \; \alpha \in U, \, \beta \in V\}$. and $U^{-1} = \{\alpha^{-1}; \, \alpha \in U\}$.

Given a subgroup H of G, recall that $N_G(H) = \{\alpha \in G; \, \alpha H \alpha^{-1} = H\}$ is a subgroup of G called the *normalizer* of H in G. If $K \subset N_G(H)$, then we say that K normalizes H. If $N_G(H) = G$, then H is a *normal* subgroup of G.

Similarly $C_G(H) = \{\alpha \in G; \, \alpha\beta = \beta\alpha \text{ for all } \beta \in H\}$ is a subgroup of G called the *centralizer* of H in G. If $K \subset C_G(H)$, then we say that K centralizes H.

10.2.2 Let us denote the group of automorphisms of G by $\mathrm{Aut}(G)$. A subgroup H of G is called *characteristic* if $\sigma(H) = H$ for all $\sigma \in \mathrm{Aut}(G)$.

Given $\alpha \in G$, the map $i_\alpha : G \to G$ given by $\beta \mapsto \alpha\beta\alpha^{-1}$, is an automorphism of G called an *inner automorphism* of G. The set $\mathrm{Int}(G)$ of inner automorphisms of G is a subgroup of $\mathrm{Aut}(G)$. Note that a subgroup of G is normal if and only if it is stable under all inner automorphisms of G.

10.2.3 Let H be a subgroup of G. We define the quotient G/H (resp. $H\backslash G$) to be the set of equivalence classes, called *left* (resp. *right*) *cosets*, of the equivalence relation $\mathcal{R}_l$ (resp. $\mathcal{R}_r$) defined by $\alpha\mathcal{R}_l\beta \Leftrightarrow \alpha^{-1}\beta \in H$ (resp. $\alpha\mathcal{R}_r\beta \Leftrightarrow \beta\alpha^{-1} \in H$). Note that a left (resp. right) coset is of the form αH (resp. $H\alpha$) for some $\alpha \in G$. It follows that if H is normal, then $G/H = H\backslash G$ and we have an induced group structure on G/H such that the projection $G \to G/H$ is a group homomorphism.

In general, we have a bijection between G/H and $H\backslash G$ given by $\alpha H \mapsto H\alpha^{-1}$. We define the *index* of H in G to be the cardinal, denoted by $(G : H)$, of the set G/H (or $H\backslash G$).

Proposition. *Let $H \supset K$ be subgroups of G.*
(i) *We have* $\mathrm{card}(G) = (G : H)\,\mathrm{card}(H)$.
(ii) *We have* $(G : K) = (G : H)(H : K)$.

10.2.4 Let H, K be groups and $\sigma : K \to \mathrm{Aut}(H)$ a group homomorphism. We define a law on $H \times K$ by:

$$(\alpha, \beta)(\alpha', \beta') = (\alpha\sigma(\beta)(\alpha'), \beta\beta')$$

for $\alpha, \alpha' \in H$, $\beta, \beta' \in K$.

We check easily that there is a group law on $H \times K$, denoted by $H \rtimes_\sigma K$ or simply $H \rtimes K$, that we call the *semi-direct product* of K by H with respect to σ. If e_H, e_K denote the identity elements of H and K respectively, then (e_H, e_K) is the identity element of $H \rtimes K$ and $(\alpha, \beta)^{-1} = (\sigma(\beta^{-1})(\alpha^{-1}), \beta^{-1})$.

Proposition. *The maps* $i_H : H \to H \rtimes_\sigma K$, $\alpha \mapsto (\alpha, e_K)$, $i_K : K \to H \rtimes_\sigma K$, $\beta \mapsto (e_H, \beta)$ *and* $p_K : H \rtimes_\sigma K \to K$, $(\alpha, \beta) \mapsto \beta$ *are group homomorphisms. Further we have the following exact sequence of group homomorphisms*

$$1 \longrightarrow H \xrightarrow{i_H} H \rtimes_\sigma K \xrightarrow{p_K} K \longrightarrow 1.$$

10.2.5 Suppose that H, K are subgroups of G such that K normalizes H and $K \cap H = \{e_G\}$. Let $\sigma : K \to \mathrm{Aut}(H)$, be the group homomorphism given by $\sigma(\beta)(\alpha) = \beta\alpha\beta^{-1}$. Then we have:

Proposition. *The map* $(\alpha, \beta) \mapsto \alpha\beta$ *is an isomorphism of* $H \rtimes_\sigma K$ *onto the subgroup HK of G.*

10.3 Commutators

10.3.1 Let $\alpha, \beta \in G$. We define the *commutator* of α and β to be $(\alpha, \beta) = \alpha\beta\alpha^{-1}\beta^{-1}$. Clearly, $(\alpha, \beta)^{-1} = (\beta, \alpha)$ and $\gamma(\alpha, \beta)\gamma^{-1} = (\gamma\alpha\gamma^{-1}, \gamma\beta\gamma^{-1})$ for $\gamma \in G$.

If H, K are subgroups of G, we shall denote by (H, K) the subgroup generated by the commutators (α, β) where $\alpha \in H$, $\beta \in K$. The group (H, K) is normal (resp. characteristic) if H and K are normal (resp. characteristic). In particular, the subgroup (G, G) is a characteristic subgroup of G called the *derived subgroup* or the *group of commutators*. We shall denote (G, G) also by $\mathcal{D}(G)$. It is easy to see that:

Proposition. *A subgroup H of G contains $\mathcal{D}(G)$ if and only if G/H is abelian and H is normal.*

10.3.2 Proposition. *If the index of the centre Z of G is finite, then $\mathcal{D}(G)$ is finite.*

Proof. Let $n = (G : Z)$ and $a_1, \ldots, a_n$ be representatives of G/Z. Since $(a_p, a_q) = (a_p z, a_q z')$ for any $z, z' \in Z$, the set C of commutators is finite and $\operatorname{card}(C) \leqslant n^2$. Let $c_1, \ldots, c_r$ be the (distinct) elements of C.

Since $(G : Z) = n$, $(\alpha, \beta)^n \in Z$ for all $\alpha, \beta \in G$. This implies that $(\alpha, \beta)^n = \beta^{-1}(\alpha, \beta)^n \beta = (\beta^{-1}\alpha\beta, \beta)^n$, and hence:

$$(\alpha, \beta)^{n+1} = (\beta^{-1}\alpha\beta, \beta)^n (\alpha, \beta) = (\beta^{-1}\alpha\beta, \beta)^{n-1}(\beta^{-1}\alpha\beta, \beta)(\alpha, \beta)$$
$$= (\beta^{-1}\alpha\beta, \beta)^{n-1}(\beta^{-1}\alpha\beta, \beta^2).$$

Thus $(\alpha, \beta)^{n+1}$ can be written as a product of n commutators.

On the other hand, $c_i c_j = c_j (c_j^{-1} c_i c_j)$ and $c_j^{-1} c_i c_j = c_k$ for some k.

Now let $x = c_{i_1} \cdots c_{i_s} \in \mathcal{D}(G)$, $1 \leqslant i_1, \ldots, i_s \leqslant r$, be such that s is minimal for this property. Then the preceding paragraph says that $x = c_1^{m_1} \cdots c_r^{m_r}$ where $m_1 + \cdots + m_r = s$. Since c_i^{n+1} is a product of n commutators, it follows from the minimality of s that $0 \leqslant m_1, \ldots, m_r \leqslant n$. Thus $\operatorname{card}(\mathcal{D}(G)) \leqslant (n+1)^r \leqslant (n+1)^{n^2}$. $\square$

10.3.3 Proposition. *Let H, K be subgroups of G such that H normalizes K. If the set $S = \{(\alpha, \beta); \alpha \in H, \beta \in K\}$ is finite, then so is (H, K).*

Proof. We can assume that $G = HK$ and so K is normal in G. For $\beta, \beta_1 \in H$ and $\gamma, \gamma_1 \in K$, we have

$$(\gamma_1\beta_1\beta\beta_1^{-1}\gamma_1^{-1}, \gamma) = (\gamma_1, \beta_1\beta\beta_1^{-1})(\beta_1\beta\beta_1^{-1}, \gamma\gamma_1).$$

Thus the set T of commutators $(\alpha\beta\alpha^{-1}, \gamma)$, where $\alpha \in G$, $\beta \in H$ and $\gamma \in K$, is finite and is contained in (H, K). Moreover, T is stable under inner automorphisms. Consequently, (H, K) is normal in G since $S \subset T$.

Since an inner automorphism defines a permutation of the set T, this induces a homomorphism from G to the group of permutations of T whose kernel L is exactly $C_G((H, K))$. This implies that L is normal and $(G : L)$ is finite. Now $L \cap (H, K)$ is central of finite index in (H, K), so by 10.3.2, $\mathcal{D}((H, K))$ is finite. Since $\mathcal{D}((H, K))$ is normal in G, we can replace G by $G/\mathcal{D}((H, K))$ and assume that (H, K) is abelian.

The set $U = \{(\alpha, \beta); \alpha \in H, \beta \in (H, K)\} \subset S$ is finite and generates the subgroup $(H, (H, K))$. Given $\alpha \in H$, $\beta \in (H, K)$, we have $\alpha\beta\alpha^{-1} \in (H, K)$ and

$$(\alpha, \beta)^2 = (\alpha, \beta)(\alpha\beta\alpha^{-1})\beta^{-1} = (\alpha\beta\alpha^{-1})(\alpha, \beta)\beta^{-1} = (\alpha, \beta^2) \in U.$$

Since $(H, (H, K))$ is abelian and generated by U, it is finite. Further, $(H, (H, K))$ is normal in G, therefore we can replace G by $G/(H, (H, K))$ and assume that H centralizes (H, K). Given $\alpha \in H$, $\beta \in K$, it follows that

$$(\alpha, \beta)^2 = (\alpha, \beta)\alpha^{-1}(\alpha, \beta)\alpha = \alpha^{-1}(\alpha^2, \beta)\alpha = (\alpha^2, \beta) \in S.$$

Since S generates (H, K) and (H, K) is abelian, we obtain that (H, K) is finite. $\square$

10.4 Solvable groups

10.4.1 Definition. *The* derived series *of G is the sequence $(\mathcal{D}^n(G))_{n \geqslant 0}$ of subgroups of G defined by $\mathcal{D}^0(G) = G$ and $\mathcal{D}^n(G) = (\mathcal{D}^{n-1}(G), \mathcal{D}^{n-1}(G))$ for $n \geqslant 1$.*
A group G is solvable *if there exists an integer n such that $\mathcal{D}^n(G) = \{e_G\}$.*

10.4.2 Proposition. (i) *Let $f : G \to H$ be a homomorphism of groups. We have $f(\mathcal{D}^n(G)) \subset \mathcal{D}^n(H)$ for all $n \in \mathbb{N}$. Furthermore, if f is surjective, then $f(\mathcal{D}^n(G)) = \mathcal{D}^n(H)$.*
(ii) *For any $n \in \mathbb{N}$, $\mathcal{D}^n(G)$ is a characteristic subgroup of G.*

Proof. Part (ii) is clear and part (i) is a simple induction. $\square$

10.4.3 Definition. *A sequence $G = G_0 \supset G_1 \supset \cdots \supset G_m$ of subgroups of G is called a* normal chain *if G_{i+1} is normal in G_i for $0 \leqslant i \leqslant m - 1$.*

10.4.4 Proposition. *The following conditions are equivalent for a group G:*
   (i) *G is solvable.*
   (ii) *There is a sequence $G = G_0 \supset G_1 \supset \cdots \supset G_m = \{e_G\}$ of normal subgroups of G such that G_i/G_{i+1} is abelian for $0 \leqslant i \leqslant m - 1$.*
   (iii) *There is a normal chain $G = G_0 \supset G_1 \supset \cdots \supset G_m = \{e_G\}$ of subgroups of G such that G_i/G_{i+1} is abelian for $0 \leqslant i \leqslant m - 1$.*

Proof. For (i) $\Rightarrow$ (ii), take $G_i = \mathcal{D}^i(G)$, while (ii) $\Rightarrow$ (iii) is trivial. Finally, given (iii), $\mathcal{D}(G_i) \subset G_{i+1}$. By induction, we obtain $\mathcal{D}^i(G) \subset G_i$ for all i. $\square$

10.4.5 Proposition. *Let H be a normal subgroup of G. Then G is solvable if and only if H and G/H are solvable.*

Proof. Clearly if G is solvable, H and G/H are solvable. Conversely suppose that H and G/H are solvable. Since the canonical surjection $\pi : G \to G/H$ is a homomorphism of groups, by 10.4.2 (i) there exists $p \in \mathbb{N}$ such that $\pi(\mathcal{D}^p(G)) = \{e_{G/H}\}$, i.e. $\mathcal{D}^p(G) \subset H$. Now there exists $q \in \mathbb{N}$ such that $\mathcal{D}^{p+q}(G) \subset \mathcal{D}^q(H) = \{e_G\}$. Thus G is solvable. $\square$

10.4.6 Corollary. *Let H, K be normal subgroups of G. If H, K are solvable, then HK is also solvable.*

Proof. Since HK/H is isomorphic to $K/(H \cap K)$, it is solvable. Now apply 10.4.5 to HK and H. $\square$

10.5 Nilpotent groups

10.5.1 Definition. *The* central descending series *of G is the sequence $(\mathcal{C}^n(G))_{n \geqslant 1}$ of subgroups of G defined by $\mathcal{C}^1(G) = G$ and for $n \geqslant 1$, $\mathcal{C}^{n+1}(G) = (G, \mathcal{C}^n(G))$.*

A group G is nilpotent *if there exists an integer n such that $\mathcal{C}^n(G) = \{e_G\}$.*

10.5.2 Remarks. 1) An abelian group is clearly nilpotent.

2) We verify easily that $(\mathcal{C}^p(G), \mathcal{C}^q(G)) \subset \mathcal{C}^{p+q}(G)$, $p, q \in \mathbb{N}^*$. Thus $\mathcal{D}^n(G) \subset \mathcal{C}^{2^n}(G)$, $n \in \mathbb{N}^*$. So a nilpotent group is solvable.

3) Let G be nilpotent and n be minimal such that $\mathcal{C}^{n+1}(G) = \{e_G\}$. Then $\mathcal{C}^n(G)$ is contained in the centre of G. In particular, the centre of G is non-trivial if G is non-trivial.

4) Let $f : G \to H$ be a homomorphism of groups. As in 10.4.2, we have by induction that $f(\mathcal{C}^n(G)) \subset \mathcal{C}^n(H)$ for all $n \in \mathbb{N}^*$, and equality holds if f is surjective. It follows that $\mathcal{C}^n(G)$ is a characteristic subgroup of G for $n \in \mathbb{N}^*$. Further, the quotient of a nilpotent group is nilpotent.

10.5.3 Proposition. *Let H be a subgroup of G contained in the centre of G. Then G is nilpotent if and only if G/H is nilpotent.*

Proof. We have already remarked (10.5.2) that the quotient of a nilpotent group is nilpotent. Conversely, if G/H is nilpotent, then by 10.5.2 there exists $n \in \mathbb{N}^*$ such that $\mathcal{C}^n(G) \subset H \subset Z(G)$. Thus $\mathcal{C}^{n+1}(G) = \{e_G\}$. $\square$

10.5.4 Proposition. *A group G is nilpotent if and only if there exists a sequence $G = G_1 \supset G_2 \supset \cdots \supset G_{n+1} = \{e_G\}$ of subgroups of G such that $(G, G_k) \subset G_{k+1}$ for $1 \leqslant k \leqslant n$.*

Proof. If G is nilpotent, then take $G_k = \mathcal{C}^k(G)$. Conversely, we verify easily by induction that $\mathcal{C}^k(G) \subset G_k$ for all k. $\square$

10.5.5 Proposition. *Let G be a nilpotent group, H a subgroup and K a normal subgroup of G.*

(i) *There exists a normal chain $G = H_1 \supset H_2 \supset \cdots \supset H_{n+1} = H$ of subgroups of G such that H_k/H_{k+1} is abelian for $1 \leqslant k \leqslant n$.*

(ii) *There exists a sequence $K = K_1 \supset K_2 \supset \cdots \supset K_{m+1} = \{e_G\}$ of subgroups of G such that $(G, K_i) \subset K_{i+1}$ for $1 \leqslant i \leqslant m$.*

Proof. Set $G_k = \mathcal{C}^k(G)$ and n minimal such that $G_{n+1} = \{e_G\}$.

(i) Let $H_k = G_k H = H G_k$ since G_k is normal in G. If $\alpha \in H$, $\beta \in G_k$, $\gamma \in G_{k+1}$, then

$$\alpha H_{k+1}\alpha^{-1} = (\alpha H \alpha^{-1})(\alpha G_{k+1}\alpha^{-1}) \subset H_{k+1}$$
$$\beta\alpha\gamma\beta^{-1} = (\beta, \alpha\gamma)\alpha\gamma \in (G_k, G)HG_{k+1} \subset G_{k+1}HG_{k+1} = H_{k+1}.$$

Thus H_{k+1} is normal in H_k. Finally, G_k/G_{k+1} is abelian and the canonical homomorphism $G_k/G_{k+1} \to H_k/H_{k+1}$ is surjective. So H_k/H_{k+1} is abelian.

(ii) It suffices to take $K_i = K \cap G_i$. $\square$

10.5.6 Corollary. *Let G be a nilpotent group.*

(i) *Let $H \neq G$ be a subgroup of G, then $N_G(H) \neq H$ and there is a normal subgroup L containing H, $L \neq G$ and such that G/L is abelian.*

(ii) *Let $K \neq \{e_G\}$ be a normal subgroup of G. Then $K \cap Z(G) \neq \{e_G\}$.*

Proof. (i) Let (H_k) be a sequence of subgroups verifying condition (i) of 10.5.5. If k is maximal such that $H_k \neq H$, then $H_k \subset N_G(H)$ and $H \neq N_G(H)$. If i is minimal such that $H_i \neq G$, then $L = H_i$ is a normal subgroup of G containing H and G/L is abelian as required.

(ii) Let (K_i) be a sequence of subgroups verifying 10.5.5 (ii). If j is maximal such that $K_j \neq \{e_G\}$, then $K \cap Z(G) \supset K_j \cap Z(G) = K_j \neq \{e_G\}$. $\square$

10.6 Group actions

10.6.1 We shall denote the group of permutations of a set E by $\mathfrak{S}(E)$.

If we have a group homomorphism θ from G to $\mathfrak{S}(E)$, then we say that G *acts* on E. For simplicity, we shall often write $\alpha.x$ for $\theta(\alpha)(x)$ if $\alpha \in G$ and $x \in E$. The set $G.x = \{\alpha.x; \ \alpha \in G\}$ is called the G-*orbit* of x, and $G_x = \{\alpha \in G; \ \alpha.x = x\}$ the *stabilizer* (or *isotropy group*) of x. We verify easily that G_x is a subgroup of G.

10.6.2 1) Let G acts on E, $x, y \in E$ and $\alpha \in G$ be such that $\alpha.x = y$. Then $G_y = \alpha G_x \alpha^{-1}$.

2) We shall denote by E^G the set of fixed points of E. Thus $x \in E^G$ if and only if $G_x = G$ if and only if $G.x = \{x\}$.

3) Let $k \in \mathbb{N}^*$. An action of G on E is called k-*transitive* if given $x_1, \ldots, x_k, y_1, \ldots, y_k$ where $x_i \neq x_j$ and $y_i \neq y_j$ if $i \neq j$, there exists $\alpha \in G$ such that $\alpha.x_i = y_i$, $1 \leqslant i \leqslant k$. When $k = 1$, we say that G acts *transitively* on E and that E is a G-*homogeneous space*.

4) Let E', E'' be subsets of E. We define the *transporter* $\text{Tran}_G(E', E'')$ of E' to E'' to be the set of $\alpha \in G$ such that $\alpha(E') \subset E''$.

5) Let G acts on E and F. A map $f : E \to F$ is called G-*equivariant* if $f(\alpha.x) = \alpha.f(x)$ for all $x \in E$ and $\alpha \in G$.

10.6.3 Theorem. *Let G acts on E and $x \in E$. There is a bijection between $G.x$ and G/G_x. Thus $\text{card}(G.x) = (G : G_x)$.*

Proof. The map $f : G \to G.x$ defined by $\alpha \mapsto \alpha.x$ is surjective and we have $f^{-1}(\alpha.x) = \alpha G_x$. $\square$

10.6.4 Examples. 1) Left multiplication is an action of G on G called the *left regular* action. The *right regular* action is given by $\alpha.\beta = \beta\alpha^{-1}$.

2) The homomorphism $G \to \mathrm{Aut}(G)$ induced by inner automorphisms is an action called *conjugation*. The stabilizer of α is just the centralizer $C_G(\alpha)$. Two elements α, β are conjugate in G if they are in the same G-orbit.

3) Let H be a subgroup of G. Then G acts naturally on G/H via left multiplication: $\alpha.(\beta H) = \alpha\beta H$.

10.6.5 Let V be a k-vector space, $\varphi : G \to \mathrm{GL}(V)$ a group homomorphism. Then G acts on V and V is called a G-module and (V, φ) (or φ) a *representation* of G in V.

Let (V, φ) be a representation of G and V^* be the dual of V. Then there is a natural action of G on V^* as follows: let $f \in V^*$, $v \in V$ and $\alpha \in G$, then

$$(\alpha.f)(v) = f(\alpha^{-1}.v).$$

We call this the *contragredient* or *dual* representation of φ.

If (W, ψ) is another representation of G, then $\mathrm{Hom}_k(V, W)$ also has a natural G-module structure as follows: let $f \in \mathrm{Hom}_k(V, W)$, $v \in V$, $\alpha \in G$, then

$$(\alpha.f)(v) = \psi(\alpha)f(\varphi(\alpha^{-1})(v)).$$

A homomorphism $f \in \mathrm{Hom}_k(V, W)$ is called a G-*module homomorphism* (or G-*homomorphism*) if $\psi(\alpha)f(v) = f(\varphi(\alpha)(v))$ for all $\alpha \in G$. The set of homomorphisms of G-modules will be denoted by $\mathrm{Hom}_G(V, W)$. We shall also write $\mathrm{End}_G(V)$ for $\mathrm{Hom}_G(V, V)$. Clearly, $\mathrm{Hom}_G(V, W) = \mathrm{Hom}_k(V, W)^G$.

We can also put a representation on the tensor product $V \otimes_k W$ as follows: let $v \in V$, $w \in W$, $\alpha \in G$, then

$$\alpha.(v \otimes w) = (\varphi(\alpha)(v)) \otimes (\psi(\alpha)(w)).$$

We can extend this to the n-th tensor product $\mathrm{T}^n(V)$, and consequently to $\mathrm{S}^n(V)$, the n-th symmetric product and $\bigwedge^n V$, the n-th alternating product.

10.7 Generalities on representations

10.7.1 Definition. *A G-module E is simple if $E \neq \{0\}$ and the only submodules of E are $\{0\}$ and E.*

10.7.2 Proposition. *(Schur's Lemma) (i) If E, F are non-isomorphic simple G-modules, then $\mathrm{Hom}_G(E, F) = \{0\}$.*

(ii) If E is a simple G-module, then $\mathrm{End}_G(E)$ is a field. Furthermore, if E is finite-dimensional, then $\mathrm{End}_G(E) = k\,\mathrm{id}_E$.

Proof. Since the image and the kernel of a homomorphism of G-modules are G-modules, a homomorphism of simple G-modules is either zero or an isomorphism. So part (i) follows and $\mathrm{End}_G(E)$ is a field which contains clearly $k\,\mathrm{id}_E$.

Now, if E is finite-dimensional, then any $u \in \text{End}_G(E)$ has an eigenvalue λ (k being algebraically closed). This implies that $\ker(u - \lambda \, \text{id}_E)$ is a non-trivial submodule of E. Hence $u = \lambda \, \text{id}_E$. $\quad \square$

10.7.3 Lemma. *Let E be the sum of a family $(E_i)_{i \in I}$ of simple G-modules, and F a submodule of E. There exists $J \subset I$ such that $E = F \oplus \sum_{i \in J} E_i$.*

Proof. Let $\mathcal{F}$ be the set of subsets K of I such that the sum $F + \sum_{i \in K} E_i$ is direct. Together with the partial order given by inclusion, $\mathcal{F}$ is an inductive system. Let J be a maximal element in $\mathcal{F}$. Let $V = F \oplus \sum_{i \in J} E_i$. For any $k \in I$, the sum $E_k + V$ is not direct, so $E_k \cap V \neq \{0\}$ and therefore $E_k \subset V$ since E_k is simple. Hence $E = V$. $\quad \square$

10.7.4 Proposition. *The following conditions are equivalent for a G-module E:*

(i) *E is a sum of simple submodules.*

(ii) *E is a direct sum of simple submodules.*

(iii) *Any submodule F of E is a direct summand, i.e. $E = F \oplus F'$ for some submodule F'.*

If these conditions are satisfied, we say that E is semisimple *or* completely reducible.

Proof. We may assume that $E \neq \{0\}$.

(i) $\Rightarrow$ (ii) Suppose that E is the sum of a family $(E_i)_{i \in I}$ of simple submodules. Let $\mathcal{F}$ be the set (partially ordered by inclusion) of subsets J of I such that the sum $\sum_{j \in J} E_j$ is direct. Then $\mathcal{F}$ is an inductive system If K is a maximal element of $\mathcal{F}$, then clearly we have $E = \sum_{k \in K} E_k$.

(ii) $\Rightarrow$ (iii) This is obvious by 10.7.3.

(iii) $\Rightarrow$ (i) Let F be a non trivial submodule of E, $x \in F \setminus \{0\}$ and V_x the submodule of F generated by x.

Let $\mathcal{F}$ be the set (partially ordered by inclusion) of submodules of V_x which are distinct from V_x. Then it is easy to check that $\mathcal{F}$ is an inductive system. If M is a maximal element of $\mathcal{F}$, then V_x/M is clearly a simple module.

Let M' be a submodule of E such that $E = M \oplus M'$. Then V_x is the direct sum of M and $M' \cap V_x$ because $M \subset V_x$. Since the submodule $M' \cap V_x$ is isomorphic to V_x/M, it is simple. We have therefore proved that F contains a simple submodule.

Now let S be the sum of all simple submodules of E and T a submodule such that $E = S \oplus T$. So T has no simple submodule. If $T \neq \{0\}$, then the preceding paragraph says that T contains a simple submodule. So $S \cap T \neq \{0\}$ which is absurd. Thus $E = S$. $\quad \square$

10.7.5 Proposition. *Let E be the sum of a family $(E_i)_{i \in I}$ of simple G-modules.*

(i) *Any submodule (resp. quotient) F of E is semisimple. Furthermore, there exists $J \subset I$ such that F is isomorphic to $\bigoplus_{i \in J} E_i$.*

(ii) *Any simple submodule of E is isomorphic to some E_i, $i \in I$.*

Proof. By 10.7.4, there exists a submodule F' of E such that $E = F \oplus F'$. Then by 10.7.3, there exists $J \subset I$ such that $E = F' \oplus (\bigoplus_{i \in J} E_i)$. Since F is isomorphic to E/F', we have part (i). Part (ii) a just a special case of (i). □

10.7.6 Proposition. *Any representation of a finite group is semisimple.*

Proof. Let (E, ρ) be a G-module, F a submodule of E and $p \in \text{End}(E)$ a projector with respect to F and $n = \text{card}(G)$. Set:

$$q = \frac{1}{n} \sum_{\alpha \in G} \rho(\alpha)^{-1} \circ p \circ \rho(\alpha).$$

We have $q(E) \subset F$ and $q(x) = x$ for $x \in F$. Now for $\beta \in G$,

$$\rho(\beta) \circ q = \frac{1}{n} \sum_{\alpha \in G} \rho(\alpha\beta^{-1})^{-1} \circ p \circ \rho(\alpha\beta^{-1}) \circ \rho(\beta) = q \circ \rho(\beta).$$

Thus q is a G-module homomorphism and $E = F \oplus \ker q$. □

10.7.7 Let V be a k-vector space of dimension n.

Lemma. *Let W be a subspace of V of dimension d, $y \in \bigwedge^d W \setminus \{0\}$ and a subspace $U \subset \bigwedge^d V$ such that $\bigwedge^d V = ky \oplus U$. Then $V = W \oplus W'$ where $W' = \{x \in V ; x \wedge z \in U \text{ for all } z \in \bigwedge^{d-1} W\}$.*

Proof. Let $\{u_1, \ldots, u_d\}$ be a basis of $\bigwedge^{d-1} W$ and $\varphi \in (\bigwedge^d V)^*$ be such that $\varphi(y) = 1$ and $\varphi|_U = 0$. Then $W' = \{x \in V ; \varphi(x \wedge u_i) = 0 \text{ for } 1 \leqslant i \leqslant d\}$. It follows that $\dim W' \geqslant n - d$. So it suffices to prove that $W \cap W' = \{0\}$.

Let $x \in W \setminus \{0\}$ and $\{x, x_2, \ldots, x_d\}$ be a basis of W. There exists $\lambda \in k \setminus \{0\}$ such that $x \wedge (x_2 \wedge \cdots \wedge x_d) = \lambda y \notin U$. Since $x_2 \wedge \cdots \wedge x_d \in \bigwedge^{d-1} W$, we deduce that $x \notin W'$. □

10.7.8 Lemma. *Let G be a subgroup of $\text{GL}(V)$ such that V is a semisimple G-module. Let $\chi : G \to k^*$ be a homomorphism of groups, $U = \{x \in V ; \alpha(x) = \chi(\alpha)x \text{ for all } \alpha \in G\}$, and W the subspace generated by $\alpha(x) - \chi(\alpha)(x)$, where $\alpha \in G$ and $x \in V$.*

(i) The subspaces U and W are G-submodules. Furthermore, U (resp. W) is the unique G-submodule such that $V = U \oplus W$.

(ii) Let H be a subgroup of $\text{GL}(V)$ containing G as a normal subgroup. If χ extends to a homomorphism of H into k^*, then U and W are H-stable.

Proof. (i) The subspace U is clearly G-stable. On the other hand,

$$\beta(\alpha(x) - \chi(\alpha)(x)) = (\beta\alpha\beta^{-1})(\beta(x)) - \chi(\beta\alpha\beta^{-1})(\beta(x)) \in W,$$

so W is also G-stable. Now V is semisimple, there exist submodules U', W' such that $V = U \oplus U' = W \oplus W'$.

If $x \in W'$, then $\alpha(x) - \chi(\alpha)x \in W \cap W' = \{0\}$. Hence $W' \subset U$. On the other hand, if $x \in V$, there exist $y \in U$, $z \in U'$ such that $x = y + z$. For $\alpha \in G$, $\alpha(x) - \chi(\alpha)(x) = \alpha(z) - \chi(\alpha)(z) \in U'$. Hence $W \subset U'$. Since $U \cap W = \{0\}$, part (i) follows.

(ii) The proof is analogue to the one for (i). $\square$

10.7.9 Let V be a semisimple G-module and W the subspace of V generated by $\alpha.x - x$ where $\alpha \in G$, $x \in V$. By 10.7.8, W is a direct summand of V and V^G is the unique submodule such that $V = V^G \oplus W$. The projection of V onto V^G corresponding to this direct sum decomposition, denoted by p_V, is called the *Reynolds operator* associated to V.

10.8 Examples

10.8.1 For $n, p \in \mathbb{N}^*$, we denote by $\mathrm{M}_{n,p}(k)$ the space of n by p matrices with coefficients in k, and we write $\mathrm{M}_n(k) = \mathrm{M}_{n,n}(k)$. Let I_n denote the identity matrix and $(E_{ij})_{1 \leqslant i,j \leqslant n}$ the canonical basis of $\mathrm{M}_n(k)$. Denote by $\mathrm{GL}_n(k)$, $\mathrm{SL}_n(k)$, $\mathrm{D}_n(k)$, $\mathrm{T}_n(k)$ and $\mathrm{U}_n(k)$ respectively the group of invertible matrices, the group of invertible matrices of determinant 1, the group of invertible diagonal matrices, the group of invertible upper triangular matrices and the group of upper triangular unipotent matrices.

We shall denote by $\mathrm{diag}(\lambda_1, \ldots, \lambda_n)$ the matrix $\lambda_1 E_{11} + \cdots + \lambda_n E_{nn}$.

10.8.2 It is clear that any matrix $A \in \mathrm{GL}_n(k)$ can be written uniquely as a product $\mathrm{diag}(1, \ldots, 1, \det A)B$ where $B \in \mathrm{SL}_n(k)$. It follows that $\mathrm{GL}_n(k)$ is isomorphic to the semidirect product $\mathrm{SL}_k(n) \rtimes H$ where $H = \{\mathrm{diag}(1, \ldots, 1, \lambda); \lambda \in k^*\} \simeq k^*$.

Further, we can verify easily that $\mathrm{T}_n(k) = \mathrm{U}_n(k) \rtimes \mathrm{D}_n(k)$.

10.8.3 Let $S_1 = \{1\}$, and for $n \geqslant 2$, let

$$S_n = \{B_{ij}(\lambda) = I_n + \lambda E_{ij}; \ 1 \leqslant i \neq j \leqslant n \ , \ \lambda \in k\} \subset \mathrm{SL}_n(k).$$

A straightforward computation gives: for $i \neq j$,

$$B_{ij}(\lambda)^{-1} = B_{ij}(-\lambda) \ , \ B_{ij}(\lambda)B_{ij}(\mu) = B_{ij}(\lambda + \mu).$$

10.8.4 Proposition. *Let $\lambda \in k^*$ and $A \in \mathrm{M}_n(k)$. If A commutes with $B_{ij}(\lambda)$ for all $i \neq j$, then $A \in kI_n$.*

Proof. If A commutes with the $B_{ij}(\lambda)$'s, then $AE_{ij} = E_{ij}A$ for all $i \neq j$. Thus,

$$a_{1i}E_{1j} + \cdots + a_{ni}E_{nj} = a_{j1}E_{i1} + \cdots + a_{jn}E_{in}.$$

So $a_{ii} = a_{jj}$ and $a_{ki} = 0 = a_{jl}$ for $k \neq i$ and $l \neq j$. $\square$

10.8.5 Corollary. (i) *The centre of* $\mathrm{GL}_n(k)$ *is* $k^* I_n$.
(ii) *The centre of* $\mathrm{SL}_n(k)$ *is* $k^* I_n \cap \mathrm{SL}_n(k) = \{\mu I_n; \mu^n = 1\}$.

10.8.6 Theorem. (i) *The group* $\mathrm{SL}_n(k)$ *is generated by* S_n.
(ii) *The group* $\mathrm{U}_n(k)$ *is generated by* $S_n \cap \mathrm{U}_n(k)$.

Proof. We shall prove (i) by induction on n. The result is obvious for $n = 1$.
Let us suppose that $n \geqslant 2$, and let $A = [a_{ij}] \in \mathrm{SL}_n(k)$.

If $a_{1j} = 0$ for all $j \geqslant 2$, then $a_{11} \neq 0$. The matrix $A' = AB_{12}(1) = [a'_{ij}]$
verifies $a'_{12} \neq 0$. So we can assume that there exists $j \geqslant 2$ such that $a_{1j} \neq 0$.
Now $A'' = AB_{1p}(\dfrac{1 - a_{11}}{a_{1p}}) = (a''_{ij})$ verifies $a''_{11} = 1$. Hence

$$B_{n1}(-a''_{n1}) \cdots B_{21}(-a''_{21}) A'' B_{12}(-a''_{12}) \cdots B_{1n}(-a''_{1n}) = \begin{pmatrix} 1 & 0 \\ 0 & C \end{pmatrix}$$

where $C \in \mathrm{SL}_{n-1}(k)$. By induction, A is a product of elements of S_n (recall
that $B_{ij}(\lambda)^{-1} = B_{ij}(-\lambda)$).

(ii) The proof of part (i) applies since given $A = [a_{ij}] \in \mathrm{U}_n(k)$, $a_{11} = 1$
and $a_{ij} = 0$ if $i > j$. □

10.8.7 Since k is algebraically closed, $k \setminus \{0, 1, -1\}$ is non-empty. We have
by straightforward computations the following result.

Lemma. *Let* $\lambda \in k$.
(i) *If* $n \geqslant 3$ *and* $i, j, k \in \{1, \ldots, n\}$ *are distinct, then*

$$B_{ij}(\lambda) = (B_{ik}(\lambda), B_{kj}(1)).$$

(ii) *Let* $n = 2$, $\mu \in k \setminus \{0, -1, 1\}$ *and* $\nu = \lambda(\mu^2 - 1)^{-1}$. *Then,*

$$B_{12}(\lambda) = (\mathrm{diag}(\mu, \mu^{-1}), B_{12}(\nu)) \ , \ B_{21}(\lambda) = (B_{21}(\nu)^{-1}, \mathrm{diag}(\mu^{-1}, \mu)).$$

10.8.8 Theorem. *The derived subgroup of* $\mathrm{GL}_n(k)$ *is* $\mathrm{SL}_n(k)$ *and* $\mathrm{SL}_n(k)$
is equal to its derived subgroup.

Proof. By 10.8.7, any element of S_n is a commutator of elements of $\mathrm{SL}_n(k)$.
Since S_n generates $\mathrm{SL}_n(k)$ by 10.8.6, and $\mathcal{D}(\mathrm{SL}_n(k)) \subset \mathcal{D}(\mathrm{GL}_n(k)) \subset \mathrm{SL}_n(k)$,
the result follows. □

10.8.9 Since $(\mathrm{diag}(\mu_1, \ldots, \mu_n), B_{ij}(\lambda)) = B_{ij}(\lambda - \dfrac{\lambda \mu_i}{\mu_j})$ for $i \neq j$, we obtain
via 10.8.6 (i):

Proposition. *The derived subgroup of* $\mathrm{T}_n(k)$ *is* $\mathrm{U}_n(k)$.

10.8.10 Let $\mathrm{T}_n^+(k)$ denote the subspace of strictly upper triangular ma-
trices. Then $\mathrm{U}_n(k) = I_n + \mathrm{T}_n^+(k)$. For $r \geqslant 1$, set $\mathrm{U}_{n,r}(k) = I_n + (\mathrm{T}_n^+(k))^r$.

Clearly, $\mathrm{U}_{n,n}(k) = \{I_n\}$. Further, it is easy to verify that $\mathrm{U}_{n,r}(k)$ are normal subgroups of $\mathrm{U}_n(k)$ and for $r, s \in \mathbb{N}^*$,

$$(\mathrm{U}_{n,r}(k), \mathrm{U}_{n,s}(k)) \subset \mathrm{U}_{n,r+s}(k).$$

Hence, by 10.4.6, 10.5.4 and 10.8.2, we have that:

Theorem. *The group $\mathrm{U}_n(k)$ (resp. $\mathrm{T}_n(k)$) is nilpotent (resp. solvable).*

10.8.11 Lemma. (Burnside's Theorem) *Let E be a finite-dimensional k-vector space and A a subalgebra of $\mathrm{End}(E)$. If the only A-stable subspaces of E are $\{0\}$ and E, then $A = \mathrm{End}(E)$.*

Proof. Let $n = \dim E$. The result being obvious for $n \leqslant 1$. We can assume that $n \geqslant 2$. Note that by our hypothesis, $Ax = E$ for all $x \in E \setminus \{0\}$.

Let $u \in A \setminus \{0\}$ be of minimal rank r. We claim that $r = 1$. If $r > 1$, there exist $x, y \in E$ such that $u(x)$ and $u(y)$ are linearly independent. Since $Au(x) = E$, there exists $v \in A$ such that $v \circ u(x) = y$. Since $u \circ v \circ u(x)$ and $u(x)$ are linearly independent, $u \circ v \circ u$ and u are linearly independent.

Let $F = u(E)$ which is $u \circ v$-stable. Since k is algebraically closed, $u \circ v$ has an eigenvalue λ and an eigenvector $z = u(t)$ for some $t \in E \setminus \ker u$. So $t \in \ker(u \circ v \circ u - \lambda u)$ and $\ker u$ is strictly contained in $\ker(u \circ v \circ u - \lambda u)$. We deduce that $u \circ v \circ u - \lambda u$ is an element of A of rank strictly less than r. But $u \circ v \circ u - \lambda u \neq 0$ because $u \circ v \circ u$ and u are linearly independent. This contradicts the minimality of r. Thus $r = 1$ and we have proved our claim.

Let $u_0 \in A$ and $u \in \mathrm{End}(E)$ be both of rank 1. There exist $y_0, y \in E \setminus \{0\}$ and $\varphi_0, \varphi \in E^*$ such that $u_0(x) = \varphi_0(x)y_0$ and $u(x) = \varphi(x)y$ for $x \in E$. Let v be the endomorphism of E defined by $v(x) = \varphi(x)y_0$.

The subspace $L = \{x \in E; \; \varphi_0(w(x)) = 0 \text{ for all } w \in A\}$ is A-stable and $L \subset \ker \varphi_0 \neq E$. It follows from our hypothesis that $L = \{0\}$ and $\{\varphi_0 \circ w; \; w \in A\} = E^*$. This implies that there exists $w \in A$ such that $\varphi = \varphi_0 \circ w$, and hence $v = u_0 \circ w \in A$. Since $Ay_0 = E$, there exists $s \in A$ such that $y = s(y_0)$. We obtain that $u = s \circ v \in A$ since $v \in A$.

We have therefore proved that the algebra A contains all endomorphisms of E of rank 1. It is now clear that $A = \mathrm{End}(E)$. $\square$

10.8.12 Definition. *Let E be a finite-dimensional k-vector space and G a subgroup of $\mathrm{GL}(E)$. We say that G is unipotent if all the elements of G are unipotent.*

10.8.13 Theorem. *Let E be a finite-dimensional k-vector space and G a unipotent subgroup of $\mathrm{GL}(E)$. There exists $x \in E \setminus \{0\}$ such that $\alpha(x) = x$ for all $\alpha \in G$.*

Proof. Let us proceed by induction on $n = \dim E$. The case $n = 1$ is trivial. So let us suppose that $n \geqslant 2$. Furthermore, if F is a G-stable subspace of E of dimension $< n$. We can apply the induction hypothesis on F. We can therefore assume that the only G-stable subspaces of E are $\{0\}$ and E.

By Burnside's Theorem (10.8.11), the subalgebra generated by G is $\mathrm{End}(E)$. Now, given $u, v \in G$,

$$\mathrm{tr}((\mathrm{id}_E - u) \circ v) = \mathrm{tr}(v) - \mathrm{tr}(u \circ v) = n - n = 0.$$

It follows that $\mathrm{tr}((\mathrm{id}_E - u) \circ w) = 0$ for all $w \in \mathrm{End}(E)$. Hence $u = \mathrm{id}_E$ and so $G = \{\mathrm{id}_E\}$, which implies that $n = 1$. Contradiction. $\square$

10.8.14 Let E be a k-vector space of dimension n and for $u \in \mathrm{End}(E)$, we shall denote the matrix of u with respect to a basis $\mathcal{E}$ of E by $\mathrm{Mat}(u, \mathcal{E})$.

Corollary. *Let G be a unipotent subgroup of $\mathrm{GL}(E)$.*
 (i) *There exists a basis $\mathcal{E}$ of E such that $\mathrm{Mat}(u, \mathcal{E}) \in \mathrm{U}_n(k)$ for all $u \in G$.*
 (ii) *The group G is nilpotent.*

Proof. By applying 10.8.13, part (i) is immediate by induction on n. Now 10.8.10 implies (ii). $\square$

References

- [5], [8], [10], [23], [40], [52], [56], [78].

11

Algebraic sets

In this chapter, we define algebraic sets, regular functions on them and morphisms between them. These notions will be generalized in the next chapters.

From now on, $\Bbbk$ denotes a commutative algebraically closed field of *characteristic zero*. Let $n \in \mathbb{N}^*$.

11.1 Affine algebraic sets

11.1.1 We call $\Bbbk^n$ the *affine space* of dimension n (*affine line* if $n = 1$ and *affine plane* if $n = 2$). Since $\Bbbk$ is algebraically closed, $\Bbbk$ is infinite and therefore an element $P \in \Bbbk[\mathbf{T}] = \Bbbk[T_1, \ldots, T_n]$ can be identified as a $\Bbbk$-valued polynomial function on $\Bbbk^n$ sending $t = (t_1, \ldots t_n)$ to $P(t) = P(t_1, \ldots, t_n)$.

11.1.2 Definition. *Let $\mathcal{A} \subset \Bbbk[\mathbf{T}]$. Denote by $\mathcal{V}(\mathcal{A})$ the set of points $t \in \Bbbk^n$ such that $P(t) = 0$ for all $P \in \mathcal{A}$.*

A subset X of $\Bbbk^n$ is called an affine algebraic set *if there exists $\mathcal{A} \subset \Bbbk[\mathbf{T}]$ such that $X = \mathcal{V}(\mathcal{A})$.*

11.1.3 Let $\mathcal{A} \subset \Bbbk[\mathbf{T}]$. We verify easily that if $\mathfrak{a}$ is the ideal of $\Bbbk[\mathbf{T}]$ generated by $\mathcal{A}$, then $\mathcal{V}(\mathcal{A}) = \mathcal{V}(\mathfrak{a})$. Since $\Bbbk[\mathbf{T}]$ is Noetherian, $\mathfrak{a}$ is finitely generated. Therefore, there exist $P_1, \ldots, P_n \in \Bbbk[\mathbf{T}]$ such that $\mathcal{V}(\mathcal{A}) = \mathcal{V}(\{P_1, \ldots, P_n\})$ (denoted simply by $\mathcal{V}(P_1, \ldots, P_n)$). A subset of the form $\mathcal{V}(P)$, where $P \in \Bbbk[\mathbf{T}]$, is called a *hypersurface* of $\Bbbk^n$.

11.1.4 Proposition. *Let $\mathfrak{a}, \mathfrak{b}, (\mathfrak{a}_i)_{i \in I}$ be ideals of $\Bbbk[\mathbf{T}]$. Then:*
  (i) $\mathcal{V}(0) = \Bbbk^n$, $\mathcal{V}(\Bbbk[\mathbf{T}]) = \emptyset$, *and if $\mathfrak{a} \subset \mathfrak{b}$, $\mathcal{V}(\mathfrak{b}) \subset \mathcal{V}(\mathfrak{a})$.*
  (ii) $\mathcal{V}(\sum_{i \in I} \mathfrak{a}_i) = \bigcap_{i \in I} \mathcal{V}(\mathfrak{a}_i)$ *and* $\mathcal{V}(\mathfrak{a}\mathfrak{b}) = \mathcal{V}(\mathfrak{a} \cap \mathfrak{b}) = \mathcal{V}(\mathfrak{a}) \cup \mathcal{V}(\mathfrak{b})$.

Proof. This is analogue to the proof of 6.5.3. $\quad\square$

11.1.5 Let $\mathfrak{m} \in \mathrm{Spm}\,\Bbbk[\mathbf{T}]$. By 6.4.3, there exists $a = (a_1, \ldots, a_n) \in \Bbbk^n$ such that $\mathfrak{m} = (T_1 - a_1)\Bbbk[\mathbf{T}] + \cdots + (T_n - a_n)\Bbbk[\mathbf{T}]$. We deduce that $\mathcal{V}(\mathfrak{m}) = \{a\}$. It follows from 11.1.4 that:

Proposition. *We have $\mathcal{V}(\mathfrak{a}) = \emptyset$ if and only if $\mathfrak{a} = \Bbbk[\mathbf{T}]$.*

11.1.6 For a subset $M \subset \Bbbk^n$, we denote by $\mathcal{I}(M)$ the set of polynomials $P \in \Bbbk[\mathbf{T}]$ such that $P(a) = 0$ for all $a \in M$. It is clear that $\mathcal{I}(M)$ is a radical ideal of $\Bbbk[\mathbf{T}]$. We shall call $\mathcal{I}(M)$, the *ideal* of M.

Proposition. (i) *We have $\mathcal{I}(\emptyset) = \Bbbk[\mathbf{T}]$, and if $M \subset N \subset \Bbbk^n$, then $\mathcal{I}(M) \supset \mathcal{I}(N)$.*
(ii) *If M is an affine algebraic set, then $\mathcal{V}(\mathcal{I}(M)) = M$.*
(iii) *The map sending an affine algebraic set to its ideal is injective.*
(iv) *If $\mathfrak{a}$ is an ideal of $\Bbbk[\mathbf{T}]$, then $\mathcal{I}(\mathcal{V}(\mathfrak{a})) = \sqrt{\mathfrak{a}}$.*
(v) *We have $\mathcal{I}(\Bbbk^n) = \{0\}$.*

Proof. Part (i) is obvious. Clearly $M \subset \mathcal{V}(\mathcal{I}(M))$. Conversely, if $M = \mathcal{V}(\mathfrak{a})$ for some ideal $\mathfrak{a} \subset \Bbbk[\mathbf{T}]$, then $\mathfrak{a} \subset \mathcal{I}(M)$. Hence $\mathcal{V}(\mathcal{I}(M)) \subset \mathcal{V}(\mathfrak{a}) = M$ (11.1.4). Now (iii) follows immediately from (ii).

Since $\mathfrak{a} \subset \mathcal{I}(\mathcal{V}(\mathfrak{a}))$, part (iv) is clear for $\mathfrak{a} = \Bbbk[\mathbf{T}]$. Furthermore, for any $\mathfrak{m} \in \mathrm{Spm}\,\Bbbk[\mathbf{T}]$, $\mathcal{V}(\mathfrak{m}) \neq \emptyset$ (11.1.5). It follows that $\mathfrak{m} \subset \mathcal{V}(\mathfrak{m}) \neq \Bbbk[\mathbf{T}]$. Hence $\mathcal{I}(\mathcal{V}(\mathfrak{m})) = \mathfrak{m}$.

Now suppose that $\mathfrak{a} \neq \Bbbk[\mathbf{T}]$. Since $\mathcal{I}(\mathcal{V}(\mathfrak{a}))$ is radical, $\sqrt{\mathfrak{a}} \subset \mathcal{I}(\mathcal{V}(\mathfrak{a}))$. Conversely, if $\mathfrak{m}$ is a maximal ideal containing $\mathfrak{a}$, then $\mathcal{I}(\mathcal{V}(\mathfrak{a})) \subset \mathcal{I}(\mathcal{V}(\mathfrak{m})) = \mathfrak{m}$ by 11.1.4 and part (i). Since $\sqrt{\mathfrak{a}}$ is the intersection of maximal ideals containing $\mathfrak{a}$ by 6.4.2, it follows that $\mathcal{I}(\mathcal{V}(\mathfrak{a})) \subset \sqrt{\mathfrak{a}}$.

Finally, part (v) follows from (iv) and 11.1.4. $\square$

11.1.7 Corollary. *The map $\mathfrak{a} \mapsto \mathcal{V}(\mathfrak{a})$ induces a bijection between the set of radical ideals of $\Bbbk[\mathbf{T}]$ and the set of affine algebraic subsets of $\Bbbk^n$.*

11.2 Zariski topology

11.2.1 By 11.1.4, we have a topology, called the *Zariski topology*, on $\Bbbk^n$ where the closed subsets are affine algebraic subsets of $\Bbbk^n$ (or the $\mathcal{V}(\mathfrak{a})$'s by 11.1.7).

In general, this topology is not Hausdorff. Indeed, closed subsets $\mathcal{V}(\mathfrak{a})$ of $\Bbbk$ are finite, thus two non-empty open subsets of $\Bbbk$ have non-empty intersection.

Let $\mathfrak{a}$ be a radical ideal of $\Bbbk[\mathbf{T}]$ and $V = \mathcal{V}(\mathfrak{a})$ an affine algebraic subset of $\Bbbk^n$. The topology induced by the Zariski topology of $\Bbbk^n$ shall also be called the *Zariski topology* on V. In view of 11.1.4, the closed subsets of V are the $\mathcal{V}(\mathfrak{b})$'s where $\mathfrak{b}$ is an ideal of $\Bbbk[\mathbf{T}]$ containing $\mathfrak{a}$.

11.2.2 Proposition. *Let M be a subset of $\Bbbk^n$. The closure of M, with respect to the Zariski topology, is $\mathcal{V}(\mathcal{I}(M))$.*

Proof. Clearly, $M \subset \mathcal{V}(\mathcal{I}(M))$. Now if $\mathfrak{a}$ is a radical ideal of $\Bbbk[\mathbf{T}]$ such that $M \subset \mathcal{V}(\mathfrak{a})$, then $\mathfrak{a} = \mathcal{I}(\mathcal{V}(\mathfrak{a})) \subset \mathcal{I}(M)$ by 11.1.6. Thus $\mathcal{V}(\mathcal{I}(M)) \subset \mathcal{V}(\mathfrak{a})$. $\square$

11.2.3 Proposition. (i) *Endowed with the Zariski topology, $\Bbbk^n$ is a Noetherian space whose points are closed.*

(ii) *Let* $\mathfrak{a}$ *be a radical ideal of* $\Bbbk[\mathbf{T}]$. *Then* $\mathcal{V}(\mathfrak{a})$ *is irreducible if and only if* $\mathfrak{a}$ *is prime.*

Proof. The proof is analogous to the one for 6.5.6. □

11.2.4 Corollary. *For any ideal* $\mathfrak{a}$ *of* $\Bbbk[\mathbf{T}]$, *the set of irreducible components of* $\mathcal{V}(\mathfrak{a})$ *is finite and it is the set of* $\mathcal{V}(\mathfrak{p}_i)$'s *where the* $\mathfrak{p}_i$'s, $1 \leqslant i \leqslant r$, *are the minimal prime ideals of* $\Bbbk[\mathbf{T}]$ *containing* $\mathfrak{a}$.

Proof. These are direct consequences of 1.3.5 and 11.2.3. □

11.3 Regular functions

11.3.1 Definition. *Let* V *be an affine algebraic subset of* $\Bbbk^n$. *A function from* V *to* $\Bbbk$ *is a* regular function *if it is the restriction to* V *of a polynomial function on* $\Bbbk^n$.

11.3.2 For any affine algebraic subset $V \subset \Bbbk^n$, we denote by $\mathbf{A}(V)$ the algebra of regular functions on V. Note that $\mathbf{A}(\Bbbk^n) = \Bbbk[\mathbf{T}]$.

Let $\mathfrak{a} = \mathcal{I}(V)$. Then $V = \mathcal{V}(\mathfrak{a})$ and $\mathfrak{a} = \sqrt{\mathfrak{a}}$ (11.1.6). Thus the restrictions to V of $P, Q \in \Bbbk[\mathbf{T}]$ are equal if and only if $P - Q \in \mathfrak{a}$. It follows that the algebras $\mathbf{A}(V)$ and $\Bbbk[\mathbf{T}]/\mathfrak{a}$ are isomorphic. In particular, $\mathbf{A}(V)$ is finitely generated and reduced.

11.3.3 Proposition. *Let* $V \subset \Bbbk^n$ *be an affine algebraic subset.*

(i) *The algebra* $\mathbf{A}(V)$ *separates the points of* V, *i.e. if* a, b *are distinct points of* V, *then there exists* $f \in \mathbf{A}(V)$ *such that* $f(a) \neq f(b)$.

(ii) *Elements of* $\mathbf{A}(V)$ *are continuous maps with respect to the Zariski topology.*

Proof. (i) Let $a = (a_1, \ldots, a_n)$, $b = (b_1, \ldots, b_n)$ be distinct points in V. Without loss of generality, we can assume that $a_1 \neq b_1$. The restriction to V of the coordinate function $\Bbbk^n \to \Bbbk$, $(x_1, \ldots, x_n) \mapsto x_1$ clearly separates a and b.

(ii) For $\lambda \in \Bbbk$ and $\varphi \in \mathbf{A}(V)$, we have $\varphi^{-1}(\lambda) = \mathcal{V}(\mathcal{I}(V)) \cap \mathcal{V}(P - \lambda)$ where P is an element of $\Bbbk[\mathbf{T}]$ whose restriction to V is φ. Thus $\varphi^{-1}(\lambda)$ is closed. Since a closed subset of $\Bbbk$ is either $\Bbbk$ or a finite set, φ is continuous. □

11.3.4 Let $V \subset \Bbbk^n$ be an affine algebraic subset. Recall from 11.3.2 that $\mathbf{A}(V)$ is a reduced and finitely generated $\Bbbk$-algebra.

By 6.4.5, the map

$$\mathrm{Hom}_{\mathrm{alg}}(\mathbf{A}(V), \Bbbk) \to \mathrm{Spm}(\mathbf{A}(V)) \,, \quad \chi \mapsto \ker \chi$$

is a bijection.

There is a canonical bijection from $\mathrm{Spm}(\mathbf{A}(V))$ to V. Namely, given $\mathfrak{m} \in \mathrm{Spm}(\mathbf{A}(V))$, there exists $\mathfrak{m}' \in \mathrm{Spm}(\Bbbk[\mathbf{T}])$ such that $\mathfrak{m}' \supset \mathcal{I}(V)$ and $\mathfrak{m}'/\mathcal{I}(V) = \mathfrak{m}$. By 6.4.3, there exist $a_1, \ldots, a_n \in \Bbbk$ such that

$$\mathfrak{m}' = (T_1 - a_1)\Bbbk[\mathbf{T}] + \cdots + (T_n - a_n)\Bbbk[\mathbf{T}].$$

Now $\mathfrak{m}'$ depends only on $\mathfrak{m}$ and since $a = (a_1, \ldots, a_n) \in \mathcal{V}(\mathfrak{m}')$, $a \in \mathcal{V}(\mathcal{I}(V)) = V$. So we have a well-defined map from $\mathrm{Spm}(\mathbf{A}(V))$ to V.

Conversely, given $a = (a_1, \ldots, a_n) \in V$, $\mathcal{I}(V) \subset \mathcal{I}(\{a\}) = \mathfrak{m}'$ where $\mathfrak{m}'$ denotes the maximal ideal of $\Bbbk[\mathbf{T}]$ generated by the $T_i - a_i$, $1 \leqslant i \leqslant n$. Then a corresponds to the maximal ideal $\mathfrak{m} = \mathfrak{m}'/\mathcal{I}(V)$ of $\mathbf{A}(V)$.

Let us consider the composition of the above bijective maps:

$$V \xrightarrow{\ \theta\ } \mathrm{Spm}(\mathbf{A}(V)) \xrightarrow{\ \psi\ } \mathrm{Hom}_{\mathrm{alg}}(\mathbf{A}(V), \Bbbk).$$

We verify easily that for $a \in V$, $\psi \circ \theta(a)$ is the homomorphism defined by the evaluation at the point a, namely $f \mapsto f(a)$, that we shall denote by χ_a.

11.4 Morphisms

11.4.1 Definition. *Let $V \subset \Bbbk^n$ and $W \subset \Bbbk^m$ be affine algebraic subsets. A map $u : V \to W$ is called a* morphism *(of affine algebraic sets) if for any $g \in \mathbf{A}(W)$, $g \circ u \in \mathbf{A}(V)$.*

11.4.2 We shall denote by $\mathrm{Mor}(V, W)$ the set of morphisms from V to W. Clearly, the composition of morphisms is again a morphism. An *isomorphism* of V to W is a bijective morphism whose inverse is a morphism.

The function $\Bbbk \to \Bbbk$, $t \mapsto t$, is regular on $\Bbbk$. It follows that $\mathbf{A}(V)$ is the set of morphisms from V to $\Bbbk$.

For $u \in \mathrm{Mor}(V, W)$, we shall denote by $\mathbf{A}(u) : \mathbf{A}(W) \to \mathbf{A}(V)$ the map $g \mapsto g \circ u$. We call $\mathbf{A}(u)$ the *comorphism* of u. If $v \in \mathrm{Mor}(W, Z)$, then we have $\mathbf{A}(v \circ u) = \mathbf{A}(u) \circ \mathbf{A}(v)$.

11.4.3 Proposition. *Let $V \subset \Bbbk^n$ and $W \subset \Bbbk^m$ be affine algebraic subsets. For any map $u : V \to W$, denote by $u_i : V \to \Bbbk$, $1 \leqslant i \leqslant m$, the coordinate maps of u. The following conditions are equivalent:*
  (i) $u \in \mathrm{Mor}(V, W)$.
  (ii) $u_1, \ldots, u_m \in \mathbf{A}(V)$.

Proof. (i) $\Rightarrow$ (ii) Let $p_i : \Bbbk^m \to \Bbbk$ denote the i-th canonical projection. Then $u_i = p_i \circ u = \mathbf{A}(u)(p_i)$ and since $p_i|_W \in \mathbf{A}(W)$, we have (ii).

(ii) $\Rightarrow$ (i) Let $P_1, \ldots, P_m \in \Bbbk[\mathbf{T}]$ be such that the restriction of P_i to V is u_i. Let $g \in \mathbf{A}(W)$ and $Q \in \Bbbk[X_1, \ldots, X_m]$ be such that its restriction to W is g. For any $t \in V$, we have $g \circ u(t) = Q(P_1(t), \ldots, P_m(t))$. Thus $g \circ u$ is the restriction of a polynomial function on $\Bbbk^n$. Hence $g \circ u \in \mathbf{A}(V)$. $\square$

11.4.4 Theorem. *Let $V \subset \Bbbk^n$, $W \subset \Bbbk^m$ be affine algebraic subsets. The map $u \mapsto \mathbf{A}(u)$ is a bijection between $\mathrm{Mor}(V, W)$ and $\mathrm{Hom}_{\mathrm{alg}}(\mathbf{A}(W), \mathbf{A}(V))$.*

Proof. For $\varphi \in \mathrm{Hom}_{\mathrm{alg}}(\mathbf{A}(W), \mathbf{A}(V))$, we define

$$h_\varphi : \mathrm{Hom}_{\mathrm{alg}}(\mathbf{A}(V), \Bbbk) \to \mathrm{Hom}_{\mathrm{alg}}(\mathbf{A}(W), \Bbbk) , \ \chi \mapsto \chi \circ \varphi.$$

There exists a unique map $m(\varphi) : V \to W$ such that the following diagram is commutative:

$$
\begin{array}{ccc}
V & \xrightarrow{\;m(\varphi)\;} & W \\[2mm]
\downarrow & & \downarrow \\[2mm]
\mathrm{Hom}_{\mathrm{alg}}(\mathbf{A}(V), \Bbbk) & \xrightarrow[h_\varphi]{} & \mathrm{Hom}_{\mathrm{alg}}(\mathbf{A}(W), \Bbbk)
\end{array}
$$

where the vertical maps are the isomorphisms $x \mapsto \chi_x$ and $y \mapsto \chi_y$ of 11.3.4. For $x \in V$, we have therefore:

$$
(1) \qquad\qquad \chi_{m(\varphi)(x)} = \chi_x \circ \varphi.
$$

It follows that for $x \in V$ and $g \in \mathbf{A}(W)$,

$$
g(m(\varphi)(x)) = \chi_{m(\varphi)(x)}(g) = \chi_x \circ \varphi(g) = \chi_x(\varphi(g)) = \varphi(g)(x).
$$

We deduce that:

$$
g \circ m(\varphi) = \varphi(g).
$$

In particular, $g \circ m(\varphi) \in \mathbf{A}(V)$ which shows that $m(\varphi) \in \mathrm{Mor}(V, W)$. Furthermore, we have shown that $\mathbf{A}(m(\varphi)) = \varphi$.

Now given $u \in \mathrm{Mor}(V, W)$, it is clear that $\mathbf{A}(u) \in \mathrm{Hom}_{\mathrm{alg}}(\mathbf{A}(W), \mathbf{A}(V))$. Taking $\varphi = \mathbf{A}(u)$ in the previous argument, we have:

$$
\chi_{m(\mathbf{A}(u))(x)}(g) = (\mathbf{A}(u)(g))(x) = g(u(x)) = \chi_{u(x)}(g).
$$

Hence $m(\mathbf{A}(u)) = u$.

We have therefore proved that the maps $\varphi \mapsto m(\varphi)$ and $u \mapsto \mathbf{A}(u)$ are mutually inverse. $\square$

11.4.5 Corollary. *Let $u \in \mathrm{Mor}(V, W)$. Then u is an isomorphism if and only if the comorphism $\mathbf{A}(u)$ of u is an isomorphism of $\Bbbk$-algebras.*

Proof. If u is an isomorphism and $v \in \mathrm{Mor}(W, V)$ is such that $u \circ v = \mathrm{id}_W$, $v \circ u = \mathrm{id}_V$, then $\mathbf{A}(u) \circ \mathbf{A}(v) = \mathrm{id}_{\mathbf{A}(V)}$ and $\mathbf{A}(v) \circ \mathbf{A}(u) = \mathrm{id}_{\mathbf{A}(W)}$. This implies that $\mathbf{A}(u)$ is an isomorphism of $\Bbbk$-algebras.

Conversely, if $\varphi \in \mathrm{Hom}_{\mathrm{alg}}(\mathbf{A}(V), \mathbf{A}(W))$ verifies $\mathbf{A}(u) \circ \varphi = \mathrm{id}_{\mathbf{A}(V)}$ and $\varphi \circ \mathbf{A}(u) = \mathrm{id}_{\mathbf{A}(W)}$, then $m(\varphi) \circ u = \mathrm{id}_V$ and $u \circ m(\varphi) = \mathrm{id}_W$. Hence u is an isomorphism. $\square$

11.5 Examples of morphisms

11.5.1 It follows from 11.4.4 that $\mathrm{Mor}(\Bbbk^n, \Bbbk^m)$ is in bijection with the set of $\Bbbk$-algebra homomorphisms

$$\varphi : \mathbf{A}(\Bbbk^m) \to \mathbf{A}(\Bbbk^n).$$

Let $\mathbf{A}(\Bbbk^n) = \Bbbk[T_1, \ldots, T_n]$ and $\mathbf{A}(\Bbbk^m) = \Bbbk[X_1, \ldots, X_m]$. Then φ is uniquely determined by m polynomials $P_1, \ldots, P_m \in \Bbbk[T_1, \ldots, T_n]$ verifying:

$$\varphi(X_i) = P_i(T_1, \ldots, T_n) \text{ for } 1 \leqslant i \leqslant m.$$

By formula (1) of 11.4.4, the morphism $u = m(\varphi)$ (hence $\mathbf{A}(u) = \varphi$) is defined by:

$$\chi_{u(x)} = \chi_x \circ \varphi \text{ for } x = (x_1, \ldots, x_n) \in \Bbbk^n.$$

Let $u(x) = (y_1, \ldots, y_m)$. Then for $1 \leqslant i \leqslant m$,

$$y_i = \chi_{u(x)}(X_i) = \chi_x(\varphi(X_i)) = P_i(x_1, \ldots, x_n).$$

Thus

$$(2) \qquad u(x) = (P_1(x_1, \ldots, x_n), \ldots, P_m(x_1, \ldots, x_n)).$$

11.5.2 Let us suppose that $n \leqslant m$ in 11.5.1, then the map

$$u : \Bbbk^n \to \Bbbk^m \ , \ (x_1, \ldots, x_n) \mapsto (x_1, \ldots, x_n, 0, \ldots, 0)$$

is a morphism. Its comorphism $\mathbf{A}(u)$ is given by:

$$\mathbf{A}(u)(X_i) = T_i \text{ if } 1 \leqslant i \leqslant n \ , \ \mathbf{A}(u)(X_i) = 0 \text{ if } n + 1 \leqslant i \leqslant m.$$

11.5.3 Let us suppose that $m \leqslant n$ in 11.5.1, then the map

$$v : \Bbbk^n \to \Bbbk^m \ , \ (x_1, \ldots, x_n) \mapsto (x_1, \ldots, x_m)$$

is a morphism. For $1 \leqslant i \leqslant m$, we have:

$$\mathbf{A}(v)(X_i) = T_i.$$

11.5.4 Let $V \subset \Bbbk^n$ be an affine algebraic subset and $\mathfrak{a} = \mathcal{I}(V) \subset \mathbf{A}(\Bbbk^n)$. By the definition of $\mathbf{A}(V)$, the canonical injection $j : V \to \Bbbk^n$ is a morphism and $\mathbf{A}(j)$ is the canonical surjection $\mathbf{A}(\Bbbk^n) \to \mathbf{A}(V) = \mathbf{A}(\Bbbk^n)/\mathfrak{a}$.

11.5.5 Let $u \in \mathrm{Mor}(\Bbbk^n, \Bbbk^m)$ be given by formula (2) of 11.5.1. Denote by $v : \Bbbk^n \to \Bbbk^n \times \Bbbk^m = \Bbbk^{n+m}$ the map given by:

$$x = (x_1, \ldots, x_n) \mapsto (x, u(x)) = (x_1, \ldots, x_n, P_1(x), \ldots, P_m(x)).$$

The image G of v is called the *graph* of u.

Note that $y = (y_1, \ldots, y_{n+m}) \in G$ if and only if y is a zero of the polynomials

$$T_{n+i} - P_i(T_1, \ldots, T_n)$$

for $1 \leqslant i \leqslant m$. It follows that G is an affine algebraic set. Further, if we consider v as a map from $\Bbbk^n$ to G, then $v \in \mathrm{Mor}(\Bbbk^n, G)$ and v is bijective.

Now let $p : \Bbbk^{n+m} \to \Bbbk^n$ be the canonical projection (11.5.3) and let $j : G \to \Bbbk^{n+m}$ be the canonical injection. Then by 11.5.3 and 11.5.4, p and j are morphisms (hence $p \circ j$ is a morphism). But

$$v \circ (p \circ j) = \mathrm{id}_G \ , \ \ (p \circ j) \circ v = \mathrm{id}_{\Bbbk^n} \ .$$

Consequently, v is an isomorphism between $\Bbbk^n$ and G.

11.5.6 Again, let $u, P_1, \ldots, P_m$ be as in 11.5.1. The ideal $\mathcal{I}(u(\Bbbk^n))$ is the set of polynomials $Q \in \Bbbk[X_1, \ldots, X_m]$ such that for all $x \in \Bbbk^n$,

$$Q(P_1(x), \ldots, P_m(x)) = 0.$$

Since $\Bbbk$ is infinite, $\mathcal{I}(u(\Bbbk^n))$ is the set of polynomials $Q \in \Bbbk[X_1, \ldots, X_m]$ such that

$$Q(P_1(T_1, \ldots, T_n), \ldots, P_m(T_1, \ldots, T_n)) = 0.$$

The set $W = \mathcal{V}(\mathcal{I}(u(\Bbbk^n)))$ is the smallest affine algebraic subset containing $u(\Bbbk^n)$. Hence W is the Zariski closure of $u(\Bbbk^n)$.

Denote by $v : \Bbbk^n \to W$ the morphism induced by u, and $j : W \to \Bbbk^m$ the canonical injection. We have $u = j \circ v$, so $\mathbf{A}(u) = \mathbf{A}(v) \circ \mathbf{A}(j)$ (by 11.5.4, $\mathbf{A}(j)$ is the canonical surjection $\mathbf{A}(\Bbbk^n) \to \mathbf{A}(W)$). On the other hand, $\mathbf{A}(u)(X_i) = P_i$ for $1 \leqslant i \leqslant m$ (11.5.1). Let $g \in \mathbf{A}(W)$ be such that $\mathbf{A}(v)(g) = 0$. For all $x \in \Bbbk^n$, we have $g(v(x)) = 0$. But by construction, $u(\Bbbk^n)$ is dense in W, hence $g = 0$. We deduced therefore that $\mathbf{A}(W)$ is isomorphic to the subalgebra of $\mathbf{A}(\Bbbk^n)$ generated by $P_1, \ldots, P_m$.

Note that in general, $u(\Bbbk^n) \neq W$ (*i.e.* the image of a morphism is not necessarily an affine algebraic set). For example, let

$$u : \Bbbk^2 \to \Bbbk^2 \ , \ \ (t_1, t_2) \mapsto (t_1 t_2, t_2).$$

We see easily that the only polynomial $Q \in \Bbbk[X_1, X_2]$ such that $Q(T_1 T_2, T_2) = 0$ is zero. Thus the Zariski closure of $u(\Bbbk^2)$ is $\Bbbk^2$. But clearly, $(1,0) \notin u(\Bbbk^2)$ and therefore $u(\Bbbk^2)$ is not an affine algebraic set.

11.5.7 Let us consider the following morphism

$$u : \Bbbk \to \Bbbk^2 \ , \ \ t \mapsto (t^2, t^3).$$

The image C of u is contained in $\mathcal{V}(T_1^3 - T_2^2)$.

Conversely, given $(x_1, x_2) \in \mathcal{V}(T_1^3 - T_2^2)$, we have $(x_1, x_2) = u(0)$ if $x_1 = 0$ and $(x_1, x_2) = u(x_2/x_1)$ if $x_1 \neq 0$. It follows that $C = \mathcal{V}(T_1^3 - T_2^2)$ and $\mathbf{A}(C)$ is isomorphic to the $\Bbbk$-algebra $\Bbbk[T^2, T^3]$ (11.5.6). Note however that this algebra, which is an integral domain, is not isomorphic to $\Bbbk[T]$, since $T \notin \Bbbk[T^2, T^3]$ and T is integral over $\Bbbk[T^2, T^3]$. This is an example of a bijective morphism which is not an isomorphism.

11.6 Abstract algebraic sets

11.6.1 Proposition. *Any finitely generated reduced $\Bbbk$-algebra A is isomorphic to the algebra of regular functions over some affine algebraic set.*

Proof. Let $s_1, \ldots, s_n$ be generators for A. The surjection $\theta : \Bbbk[T_1, \ldots, T_n] \to A$, $T_i \mapsto s_i$, induces an isomorphism between A and $\Bbbk[T_1, \ldots, T_n]/\ker \theta$. Since A is reduced, $\ker \theta$ is a radical ideal. Hence $\mathcal{I}(\mathcal{V}(\ker \theta)) = \ker \theta$ and A is isomorphic to $\mathbf{A}(\mathcal{V}(\ker \theta))$. $\square$

11.6.2 Remark. Let A be a finitely generated reduced $\Bbbk$-algebra and V, W be affine algebraic sets such that A is isomorphic to $\mathbf{A}(V)$ and $\mathbf{A}(W)$. Then by 11.4.5, V and W are isomorphic.

11.6.3 Let $V \subset \Bbbk^n$ be an affine algebraic set, $\mathfrak{a} = \mathcal{I}(V) \subset \mathbf{A}(\Bbbk^n)$. To avoid confusion between the notations of 6.5.1 and 11.1.3, for $\mathfrak{b}$ an ideal of $\mathbf{A}(V)$, we shall denote in this section by $\mathcal{V}'(\mathfrak{b})$, the set of maximal ideals of $\mathbf{A}(V)$ containing $\mathfrak{b}$.

By 11.3.4, there is a canonical bijection:

$$\gamma_V : V \to \mathrm{Hom}_{\mathrm{alg}}(\mathbf{A}(V), \Bbbk) \, , \, a \mapsto \chi_a.$$

On the other hand, we have another canonical bijection (6.4.5):

$$\delta_V : \mathrm{Hom}_{\mathrm{alg}}(\mathbf{A}(V), \Bbbk) \to \mathrm{Spm}(\mathbf{A}(V)) \, , \, \chi \mapsto \ker \chi.$$

Thus $\varepsilon_V = \delta_V \circ \gamma_V$ is a bijection between V and $\mathrm{Spm}(\mathbf{A}(V))$.

It is clear that if $\mathfrak{b}$ is an ideal of $\mathbf{A}(\Bbbk^n)$ containing $\mathfrak{a}$, then ε_V induces a bijection between $\mathcal{V}(\mathfrak{b})$ and $\mathcal{V}'(\mathfrak{b}/\mathfrak{a})$. It follows that ε_V is a homeomorphism with respect to the Zariski topology (6.5.4 and 11.2.1).

Let $W \subset \Bbbk^m$ be another affine algebraic set. Then by 6.4.5 and 11.4.4, we have for $\varphi \in \mathrm{Hom}_{\mathrm{alg}}(\mathbf{A}(W), \mathbf{A}(V))$, the following commutative diagram:

$$
\begin{array}{ccc}
V & \xrightarrow{\ m(\varphi)\ } & W \\
\gamma_V \downarrow & & \downarrow \gamma_W \\
\mathrm{Hom}_{\mathrm{alg}}(\mathbf{A}(V), \Bbbk) & \xrightarrow{\ h_\varphi\ } & \mathrm{Hom}_{\mathrm{alg}}(\mathbf{A}(W), \Bbbk) \\
\delta_V \downarrow & & \downarrow \delta_W \\
\mathrm{Spm}(\mathbf{A}(V)) & \xrightarrow{\ \mathrm{Spm}(\varphi)\ } & \mathrm{Spm}(\mathbf{A}(W))
\end{array}
$$

11.6.4 The preceding discussion allows us to consider the category of affine algebraic sets as a subcategory of a bigger (but equivalent) category, namely (*abstract*) *algebraic sets* which are not necessarily subsets of $\Bbbk^n$.

The objects of this new category are the triples (X, A, β_X) where:

(i) X is a set.

(ii) A is a finitely generated reduced $\Bbbk$-algebra.

(iii) β_X is a bijection from $\mathrm{Spm}(A)$ to X.

Morphisms of (X, A, β_X) to (Y, B, β_Y) are pairs (u, φ) where $u : X \to Y$ and $\varphi \in \mathrm{Hom}_{\mathrm{alg}}(B, A)$ are such that the following diagram is commutative:

$$
\begin{array}{ccc}
\mathrm{Spm}(A) & \xrightarrow{\ \mathrm{Spm}(\varphi)\ } & \mathrm{Spm}(B) \\
\beta_X \downarrow & & \downarrow \beta_Y \\
X & \xrightarrow{\quad u \quad} & Y
\end{array}
$$

The composition of morphisms (v, ψ) and (u, φ) is $(v \circ u, \varphi \circ \psi)$. By abuse of notations, we shall write X and u for (X, A, β_X) and (u, φ); we shall also call φ the *comorphism* of u.

11.6.5 To identify the category of affine algebraic sets as a subcategory of the category of algebraic sets, it suffices to identify an affine algebraic set V with the triple $(V, \mathbf{A}(V), \varepsilon_V^{-1})$ (11.6.3), and a morphism $u \in \mathrm{Mor}(V, W)$ with the pair $(u, \mathbf{A}(u))$. By 11.4.4, 11.6.1 and 11.6.3, this category is equivalent to the category of affine algebraic sets.

Similarly, we can identify a finitely generated reduced $\Bbbk$-algebra A with the triple $(\mathrm{Spm}(A), A, \mathrm{id}_{\mathrm{Spm}(A)})$, and a homomorphism φ with the pair $(\mathrm{Spm}(\varphi), \varphi)$. Again this induces an equivalence between the category of finitely generated reduced $\Bbbk$-algebras and the category of algebraic sets.

We define regular functions on an algebraic set (X, A, β_X) via transport of structure. If $x \in X$, $\beta_X^{-1}(x)$ is a maximal ideal of A. For $f \in A$, we set $f(x) = f(\beta_X^{-1}(x))$ as defined in 6.4.6. Then regular functions on X are morphisms from X to $\Bbbk$, and we can identify them as elements of A. For any morphism $u : X \to Y$, we define $\mathbf{A}(u) : B \to A$ as in 11.4.2, by $g \mapsto g \circ u$.

11.6.6 Let (X, A, β_X) be an algebraic set. We define the *Zariski topology* on X to be the topology induced, via the bijection β_X, by the Zariski topology on $(\mathrm{Spm}(A), A, \mathrm{id}_{\mathrm{Spm}(A)})$ (6.5.4). In the case where $A = \mathbf{A}(V)$ for some affine algebraic set V, the Zariski topology of $\mathrm{Spm}(A)$ is the one obtained from the Zariski topology on V via the canonical bijection from $\mathrm{Spm}(\mathbf{A}(V))$ to V (11.6.3).

11.6.7 Let A be a finitely generated reduced $\Bbbk$-algebra, $\mathfrak{a}$ a radical ideal of A and $\beta_{\mathfrak{a}}$ the bijection $\mathrm{Spm}(A/\mathfrak{a}) \to \mathcal{V}(\mathfrak{a})$ which is the inverse of the canonical bijection $\mathfrak{m} \mapsto \mathfrak{m}/\mathfrak{a}$. It is clear that $(\mathcal{V}(\mathfrak{a}), A/\mathfrak{a}, \beta_{\mathfrak{a}})$ is an algebraic set, that we shall denote by $\mathcal{V}(\mathfrak{a})$ or $\mathrm{Spm}(A/\mathfrak{a})$. The Zariski topology on this set is induced by the one on $\mathrm{Spm}(A)$. The canonical injection $\mathcal{V}(\mathfrak{a}) \to \mathrm{Spm}(A)$ is a morphism, whose comorphism is $A \to A/\mathfrak{a}$. Thus regular functions on $\mathcal{V}(\mathfrak{a})$ can be viewed as restrictions of regular functions on $\mathrm{Spm}(A)$.

11.6.8 Proposition. *Let A, B be finitely generated reduced $\Bbbk$-algebras and $\varphi \in \mathrm{Hom}_{\mathrm{alg}}(B, A)$. The following conditions are equivalent:*

(i) *There exist a closed subset $V \subset \mathrm{Spm}(B)$ and an isomorphism $v : \mathrm{Spm}(A) \to V$ such that $\mathrm{Spm}(\varphi) = j \circ v$ where $j : V \to \mathrm{Spm}(B)$ is the canonical injection.*

(ii) *φ is surjective.*

Proof. (i) $\Rightarrow$ (ii) Let $\mathfrak{b}$ be a radical ideal of B such that $V = \mathcal{V}(\mathfrak{b})$, $\psi : B \to B/\mathfrak{b}$ the canonical surjection and $\theta \in \mathrm{Hom}_{\mathrm{alg}}(B/\mathfrak{b}, A)$ such that $\varphi = \theta \circ \psi$. It follows that $j = \mathrm{Spm}(\psi)$, $v = \mathrm{Spm}(\theta)$ and θ is an isomorphism by 11.4.5 and our hypotheses. Hence φ is surjective.

(ii) $\Rightarrow$ (i) Let $\mathfrak{b} = \ker \varphi$, $\theta : B/\mathfrak{b} \to A$ the induced isomorphism and $\psi : B \to B/\mathfrak{b}$ the canonical surjection. The ideal $\mathfrak{b}$ is radical and by 11.4.5, $\mathrm{Spm}(\theta) : \mathrm{Spm}(A) \to \mathrm{Spm}(B/\mathfrak{b})$ is an isomorphism. $\square$

11.7 Principal open subsets

11.7.1 Let A be a finitely generated reduced $\Bbbk$-algebra and $f \in A$. We set

$$D(f) = \mathrm{Spm}(A) \setminus \mathcal{V}(Af).$$

Thus $D(f)$ is the set of maximal ideals $\mathfrak{m}$ which do not contain f (or equivalently $f(\mathfrak{m}) \neq 0$). This is an open subset of $\mathrm{Spm}(A)$ and an open subset of this form is called a *principal open subset* of $\mathrm{Spm}(A)$. Note that:

$$D(0) = \emptyset \;,\; D(A) = \mathrm{Spm}(A) \;,\; D(fg) = D(f) \cap D(g).$$

If $\varphi \in \mathrm{Hom}_{\mathrm{alg}}(B, A)$ and $g \in B$, then by 6.5.8, we have:

$$\mathrm{Spm}(\varphi)^{-1}(D(g)) = D(\varphi(g)).$$

11.7.2 Proposition. *Principal open subsets of $\mathrm{Spm}(A)$ form a base for the Zariski topology on $\mathrm{Spm}(A)$.*

Proof. Let V be a closed subset of $\mathrm{Spm}(A)$. If $x \notin V$, then there exists $f \in \mathcal{I}(V)$ such that $f(x) \neq 0$. Thus $x \in D(f)$ and $D(f) \subset \mathrm{Spm}(A) \setminus V$. $\square$

11.7.3 Let A be a finitely generated reduced $\Bbbk$-algebra and $f \in A$. Denote by A_f the set of maps from $D(f)$ to $\Bbbk$ of the form

$$(3) \qquad\qquad x \mapsto \frac{g(x)}{f(x)^n}$$

where $g \in A$ and $n \in \mathbb{N}$. It is clear that A_f is a finitely generated $\Bbbk$-algebra.

Proposition. (i) *An element g of A_f is identically zero on $D(f)$ if and only if $fg = 0$.*

(ii) *The algebra A_f is reduced.*

(iii) *Let $S = \{f^n ; n \in \mathbb{N}\}$. The algebras A_f and $S^{-1}A$ are isomorphic.*

Proof. Part (i) is obvious. Now if $(g(x)/f(x)^n)^m = 0$ for all $x \in D(f)$. Then $fg = 0$ and $g(x)/f(x)^n = 0$ for all $x \in D(f)$. Hence A_f is reduced.

Let $\varepsilon : A \to A_f$ be the restriction map $g \mapsto g|_{D(f)}$. If $g \in S$, then $\varepsilon(g)$ is invertible. By 2.3.4, there exists a unique homomorphism $\theta : S^{-1}A \to A_f$ such that $\theta \circ i = \varepsilon$ where $i : A \to S^{-1}A$ is the canonical homomorphism. It is clear that θ is surjective. On the other hand, by part (i), θ is injective. Thus θ is an isomorphism. $\square$

11.7.4 We shall prove that a principal open subset $D(f)$ of $\mathrm{Spm}(A)$ can be considered canonically as an algebraic set whose topology is the one induced by the topology of $\mathrm{Spm}(A)$.

Theorem. *There exists a canonical bijection $\beta : \mathrm{Spm}(A_f) \to D(f)$ which is a homeomorphism when $D(f)$ is endowed with the topology induced from the Zariski topology of $\mathrm{Spm}(A)$.*

Proof. Let $\varepsilon : A \to A_f$, $g \mapsto g|_{D(f)}$, and $\gamma = \mathrm{Spm}(\varepsilon) : \mathrm{Spm}(A_f) \to \mathrm{Spm}(A)$. Set $F = \varepsilon(f)$.

If $\mathfrak{m} \in \mathrm{Spm}(A_f)$, then $F \notin \mathfrak{m}$ and $f \notin \varepsilon^{-1}(\mathfrak{m})$. This implies that the image of γ is contained in $D(f)$.

For $\mathfrak{m} \in D(f)$, set $I(\mathfrak{m}) = \{\varepsilon(g)/F^n; g \in \mathfrak{m}, n \in \mathbb{N}\}$. The set $I(\mathfrak{m})$ is an ideal of A_f.

If $I(\mathfrak{m}) = A_f$, then there exist $g \in \mathfrak{m}$ and $n \in \mathbb{N}$ such that $\varepsilon(g)/F^n = 1$. So $g/f^n = 1$ or equivalently $(g - f^n)/f^n = 0$. So $(g - f^n)f = 0$ by 11.7.3. Thus $gf = f^{n+1}$, hence $f^{n+1} \in \mathfrak{m}$ and $f \in \mathfrak{m}$. Contradiction.

So $I(\mathfrak{m}) \neq A_f$. There exists $\mathfrak{n} \in \mathrm{Spm}(A_f)$ such that $\mathfrak{n} \supset I(\mathfrak{m})$. It follows that $\mathfrak{m} \subset \varepsilon^{-1}(\mathfrak{n})$, so $\mathfrak{m} = \varepsilon^{-1}(\mathfrak{n})$ and $I(\mathfrak{m}) = \mathfrak{n}$. Thus the image of γ is $D(f)$.

Since $I(\gamma(\mathfrak{n})) = \mathfrak{n}$ for $\mathfrak{n} \in \mathrm{Spm}(A_f)$, γ is injective and therefore it induces a bijection β from $\mathrm{Spm}(A_f)$ onto $D(f)$.

As γ is continuous (6.5.8), to obtain the result, it suffices to prove that β maps open sets to open sets. By 11.7.2, we are reduced to prove that $\gamma(D(\varepsilon(g)/F^n)) = \gamma(D(\varepsilon(g)))$ is open for all $g \in A$.

Let $\mathfrak{n} \in \mathrm{Spm}(A_f)$ be such that $\varepsilon(g) \notin \mathfrak{n}$. Then $g \notin \varepsilon^{-1}(\mathfrak{n})$ and $\gamma(D(\varepsilon(g)))$ is contained in $D(f) \cap D(g)$.

Conversely, let $\mathfrak{m} \in \mathrm{Spm}(A)$ be such that $f \notin \mathfrak{m}$ and $g \notin \mathfrak{m}$. There exists $\mathfrak{n} \in \mathrm{Spm}(A_f)$ such that $\beta(\mathfrak{n}) = \mathfrak{m}$. Thus $\mathfrak{m} = \varepsilon^{-1}(\mathfrak{n})$. Now there exist $h \in A$, $\theta \in \mathfrak{m}$ such that $1 = hg + \theta$. So $1 = \varepsilon(h)\varepsilon(g) + \varepsilon(\theta)$. Since $\varepsilon(\theta) \in \mathfrak{n}$, $\varepsilon(g) \notin \mathfrak{n}$. Thus $\gamma(D(\varepsilon(g))) = D(f) \cap D(g)$, and we have obtained our result. $\square$

11.7.5 Remarks. 1) If A is an integral domain and $f \neq 0$, then A_f can be identified with subring $A[1/f]$ of $\mathrm{Fract}(A)$ (since $fg = 0$ implies that $g = 0$, the homomorphism $A[1/f] \to A_f$, $g/f^n \mapsto \varepsilon(g)/\varepsilon(f)^n$ is injective).

2) Let $j : D(f) \to \mathrm{Spm}(A)$ be the canonical injection. The diagram

$$\begin{array}{ccc}
\mathrm{Spm}(A_f) & \xrightarrow{\ \gamma\ } & \mathrm{Spm}(A) \\
\beta\downarrow & & \downarrow \mathrm{id}_{\mathrm{Spm}(A)} \\
D(f) & \xrightarrow{\ j\ } & \mathrm{Spm}(A)
\end{array}$$

is commutative. Consequently, j is a morphism whose comorphism is ε.

3) Note that regular functions on $D(f)$ are not necessarily restrictions of regular functions on $\mathrm{Spm}(A)$. For example, let $A = \Bbbk[T]$, then $D(T) = \Bbbk \setminus \{0\}$ and $x \mapsto 1/x$ is a regular function on $D(f)$ which is not the restriction of a polynomial.

4) The map $\delta : \mathrm{M}_n(\Bbbk) \to \Bbbk$, $M \mapsto \det M$ is regular and $D(\delta) = \mathrm{GL}_n(\Bbbk)$. Thus $\mathrm{GL}_n(\Bbbk)$ can be considered as an algebraic set.

Left multiplication or right multiplication by $A \in \mathrm{GL}_n(\Bbbk)$, and the map sending M to M^{-1} are clearly isomorphisms of the algebraic set $\mathrm{GL}_n(\Bbbk)$.

11.8 Products of algebraic sets

11.8.1 Let $V \subset \Bbbk^m$ and $W \subset \Bbbk^n$ be affine algebraic sets. Let us identify $\mathbf{A}(\Bbbk^m)$ (resp. $\mathbf{A}(\Bbbk^n)$) as the subring $\Bbbk[T_1, \ldots, T_m]$ (resp. $\Bbbk[T_{m+1}, \ldots, T_{n+m}]$) of $\Bbbk[T_1, \ldots, T_{n+m}]$.

Let $\mathfrak{a} = \mathcal{I}(V)$, $\mathfrak{b} = \mathcal{I}(W)$ and $\mathfrak{c}$ the ideal in $\Bbbk[T_1, \ldots, T_{n+m}]$ generated by $\mathfrak{a} \cup \mathfrak{b}$. The affine algebraic set $\mathcal{V}(\mathfrak{c})$ is defined by the equations:

$$P(x_1, \ldots, x_m) = 0 \ , \ Q(x_{m+1}, \ldots, x_{n+m}) = 0$$

for all $P \in \mathfrak{a}$ and $Q \in \mathfrak{b}$. Thus

$$\mathcal{V}(\mathfrak{c}) = V \times W \ , \ \mathbf{A}(V \times W) = \Bbbk[T_1, \ldots, T_{n+m}]/\sqrt{\mathfrak{c}}.$$

Since we have the canonical isomorphisms

$$\mathbf{A}(V) \otimes_\Bbbk \mathbf{A}(W) \simeq (\Bbbk[T_1, \ldots, T_m]/\mathfrak{a}) \otimes_\Bbbk (\Bbbk[T_{m+1}, \ldots, T_{m+n}]/\mathfrak{b}),$$
$$\Bbbk[T_1, \ldots, T_{n+m}] \simeq \Bbbk[T_1, \ldots, T_m] \otimes_\Bbbk \Bbbk[T_{m+1}, \ldots, T_{m+n}],$$

the ideal $\mathfrak{c}$ can be written as:

$$\mathfrak{c} = \mathfrak{a} \otimes_\Bbbk \Bbbk[T_{m+1}, \ldots, T_{m+n}] + \Bbbk[T_1, \ldots, T_m] \otimes_\Bbbk \mathfrak{b}.$$

Hence $\mathbf{A}(V) \otimes_\Bbbk \mathbf{A}(W)$ can be identified canonically with $\Bbbk[T_1, \ldots, T_{m+n}]/\mathfrak{c}$.

It follows that there is a canonical surjective homomorphism:

$$\varphi : \mathbf{A}(V) \otimes_\Bbbk \mathbf{A}(W) \to \mathbf{A}(V \times W)$$

where for $f \in \mathbf{A}(V)$, $g \in \mathbf{A}(W)$, $(x, y) \in V \times W$,

$$\varphi(f \otimes g)(x, y) = f(x)g(y).$$

Proposition. *The homomorphism φ is an isomorphism.*

Proof. It suffices to prove that φ is injective. Let $h = f_1 \otimes g_1 + \cdots + f_r \otimes g_r$ be a non-zero element in $\ker \varphi$. We may assume that h is written with r minimal. It follows that the f_i's are linearly independent. Now, given any $y \in W$,

$$f_1(x)g_1(y) + \cdots + f_r(x)g_r(y) = 0$$

for all $x \in V$. Since the f_i's are linearly independent, $g_i(y) = 0$ for all $y \in W$, $i = 1,\dots,r$. It follows that the g_i's are zero, which implies that $h = 0$. Contradiction. $\square$

11.8.2 Corollary. *Let A, B be finitely generated reduced $\Bbbk$-algebras, then $A \otimes_{\Bbbk} B$ is a finitely generated reduced $\Bbbk$-algebra.*

Proof. It is clear by 11.6.1 and 11.8.1. $\square$

11.8.3 From now on, we shall identify $\mathbf{A}(V \times W)$ and $\mathbf{A}(V) \otimes_{\Bbbk} \mathbf{A}(W)$ via φ. Denote by $\mathrm{pr}_1 : V \times W \to V$ and $\mathrm{pr}_2 : V \times W \to W$ the canonical projections. These are morphisms of affine algebraic sets.

For $(x,y) \in V \times W$, $f \in \mathbf{A}(V)$, we have that

$$f \circ \mathrm{pr}_1(x,y) = f(x).$$

It follows that the comorphisms of pr_1 and pr_2 are given by:

$$\mathbf{A}(\mathrm{pr}_1) : \mathbf{A}(V) \to \mathbf{A}(V) \otimes_{\Bbbk} \mathbf{A}(W) \,,\; f \mapsto f \otimes 1,$$
$$\mathbf{A}(\mathrm{pr}_2) : \mathbf{A}(W) \to \mathbf{A}(V) \otimes_{\Bbbk} \mathbf{A}(W) \,,\; g \mapsto 1 \otimes g.$$

Proposition. *Let U, V, W be affine algebraic sets and $v \in \mathrm{Mor}(U,V)$, $w \in \mathrm{Mor}(U,W)$. The map*

$$u : U \to V \times W \,,\; z \mapsto (v(z), w(z))$$

is the unique morphism such that $v = \mathrm{pr}_1 \circ u$ and $w = \mathrm{pr}_2 \circ u$.

Proof. Clearly u is the unique map having the required properties. Now let $f \in \mathbf{A}(V)$, $g \in \mathbf{A}(W)$ and $z \in U$. We have:

$$(f \otimes g)(u(z)) = (f \otimes g)(v(z), w(z)) = (f \circ v(z))(g \circ w(z)).$$

Since $f \circ v, g \circ w \in \mathbf{A}(U)$, u is a morphism. $\square$

11.8.4 Proposition. *Let V, W, V', W' be affine algebraic sets. For any $v \in \mathrm{Mor}(V,V')$ and $w \in \mathrm{Mor}(W,W')$, the map*

$$u : V \times W \to V' \times W' \,,\; (x,y) \mapsto (v(x), w(y))$$

is a morphism.

Proof. Note that $u(x,y) = (v \circ \mathrm{pr}_1(x,y), w \circ \mathrm{pr}_2(x,y))$. So the result follows from 11.8.3. $\square$

11.8.5 Proposition. *Let V, W be affine algebraic sets. The canonical projections $\mathrm{pr}_1, \mathrm{pr}_2$ of $V \times W$ onto V and W are open maps.*

Proof. It suffices to show that the image of a principal open subset is open. Let $h = f_1 \otimes g_1 + \cdots + f_n \otimes g_n \in \mathbf{A}(V \times W)$, and $x_0 \in U = \mathrm{pr}_1(D(h))$. Then there exists $y_0 \in W$ such that $(x_0, y_0) \in D(h)$. Set $u = g_1(y_0)f_1 + \cdots + g_n(y_0)f_n \in \mathbf{A}(V)$. Then $x_0 \in D(u)$ and $D(u) \times \{y_0\} \subset D(h)$. Hence $D(u) \subset U$ and U is an open subset of V. The proof for pr_2 is analogue. $\square$

11.8.6 Proposition. *Let V, W be affine algebraic sets. Then $V \times W$ is irreducible if and only if V and W are irreducible.*

Proof. Since the canonical projections are morphisms, V and W are irreducible if $V \times W$ is irreducible.

Conversely, suppose that V and W are irreducible. The projections $\mathrm{pr}_1, \mathrm{pr}_2$ of $V \times W$ onto V and W are continuous open maps (6.5.8 and 11.8.5). For $x \in V$, $\mathrm{pr}_1^{-1}(x) = \{x\} \times W$ is clearly irreducible. It follows therefore from 1.1.7 that $V \times W$ is irreducible. $\square$

11.8.7 Corollary. *Let A, B be finitely generated $\Bbbk$-algebras. If A and B are integral domains, then $A \otimes_{\Bbbk} B$ is also an integral domain.*

11.8.8 Proposition. *Let V, W be affine algebraic sets, V_1 (resp. W_1) a closed subset of V (resp. W). Then $V_1 \times W_1$ is a closed subset of $V \times W$.*

Proof. Let $\mathfrak{a} \subset \mathbf{A}(V)$, $\mathfrak{b} \subset \mathbf{A}(W)$ be ideals such that $\mathbf{A}(V_1) = \mathbf{A}(V)/\mathfrak{a}$ and $\mathbf{A}(W_1) = \mathbf{A}(W)/\mathfrak{b}$. Then

$$\mathbf{A}(V_1 \times W_1) = \mathbf{A}(V_1) \otimes_{\Bbbk} \mathbf{A}(W_1) = \mathbf{A}(V \times W)/\mathfrak{c}$$

where $\mathfrak{c} = \mathfrak{a} \otimes_{\Bbbk} \mathbf{A}(W) + \mathbf{A}(V) \otimes_{\Bbbk} \mathfrak{b}$. Hence $V_1 \times W_1 = \mathcal{V}(\mathfrak{c})$. $\square$

11.8.9 Remarks. 1) The Zariski topology on $V \times W$ is in general strictly finer than the product of the Zariski topologies on V and W.

The open subsets $D(f \otimes g) = D(f) \times D(g)$, $f \in \mathbf{A}(V)$ $g \in \mathbf{A}(W)$, form a base with respect to the product topology. However, there exists Zariski-open subsets of $V \times W$ which does not contain any non-empty subsets of this form. For example, if $V = W = \Bbbk$, the non-empty subsets $D(f \otimes g)$ are the complements of finite unions of subsets of the form $\{x\} \times W$ and $V \times \{y\}$. Let $U = \Bbbk^2 \setminus \mathcal{V}(T_1 - T_2)$. Since $\Bbbk$ is infinite, none of the preceding subsets contain U.

2) With the notations of example 4 of 11.7.5, the multiplication

$$\mathrm{GL}_n(\Bbbk) \times \mathrm{GL}_n(\Bbbk) \to \mathrm{GL}_n(\Bbbk) \ , \ (M, N) \mapsto MN$$

is a morphism of algebraic sets.

References and comments

- [5], [19], [22], [26], [37], [40], [78].

All the results in this chapter are classic and can be found in the books cited above.

Prevarieties and varieties

In this chapter, we define an algebraic variety via the theory of sheaves. Local behaviour of an algebraic variety is considered in the last section.

12.1 Structure sheaf

12.1.1 Let A be a finitely generated reduced $\Bbbk$-algebra and $X = \mathrm{Spm}(A)$ be endowed with the Zariski topology. For $f \in A$, we shall conserve the notations $A_f, \mathcal{V}(f)$ and $D(f)$ of chapter 11.

Theorem. *There exists a unique sheaf of functions $\mathcal{O}_X$ on X, with values in $\Bbbk$ such that for any $f \in A$, we have:*

$$\Gamma(D(f), \mathcal{O}_X) = A_f.$$

We call $\mathcal{O}_X$ the structure sheaf *of the algebraic set X.*

Proof. By 11.7.2, the $D(f)$'s, $f \in A$, form a base $\mathfrak{B}$ for the topology on X. It suffices therefore to verify that $\mathfrak{B}$ satisfies the hypotheses of 9.5.1.

Let $f, g \in A$ be such that $D(f) \subset D(g)$. Then $\mathcal{V}(g) \subset \mathcal{V}(f)$, hence by 11.1.6:

$$\sqrt{Af} = \mathcal{I}(\mathcal{V}(f)) \subset \mathcal{I}(\mathcal{V}(g)) = \sqrt{Ag}.$$

Thus there exist $n \in \mathbb{N}^*$ and $h \in A$ such that $f^n = gh$. If $u/g^i \in A_g$, then its restriction to $D(f)$ is $(uh^i)/(g^ih^i) = ug^i/f^{ni}$, which belongs to A_f.

It follows therefore that if $D(f) = D(g)$, then $A_f = A_g$.

Now let f and $(f_i)_{i \in I}$ be non zero elements of A such that $D(f_i)$ is a covering of $D(f)$. Then $\mathcal{V}(f)$ is the intersection of the $\mathcal{V}(f_i)$'s, so $\mathcal{V}(f) = \mathcal{V}(\mathfrak{a})$, where $\mathfrak{a}$ is the ideal generated by the f_i's. Since A is Noetherian, we may assume that $I = \{1, \ldots, r\}$.

Let $g : D(f) \to \Bbbk$ be a function such that $g|_{D(f_i)} \in A_{f_i}$ for $1 \leqslant i \leqslant r$. There exist $a_1, \ldots, a_r \in A$ and $n \in \mathbb{N}^*$ such that $g(x) = a_i(x)/f_i(x)^n$ for all $x \in D(f_i)$.

If $x \in D(f_i) \cap D(f_j) = D(f_i f_j)$, then $a_i(x) f_j(x)^n = a_j(x) f_i(x)^n$. By 11.7.3, it follows that $f_i f_j (a_i f_j^n - a_j f_i^n) = 0$.

We have $\mathcal{V}(f_1^n, \ldots, f_r^n) = \mathcal{V}(f_1, \ldots, f_r) \subset \mathcal{V}(f)$. So 11.1.6 implies that there exist $m \in \mathbb{N}^*$ and $b_1, \ldots, b_r \in A$ such that $f^m = b_1 f_1^{n+1} + \cdots + b_r f_r^{n+1}$. Hence

$$a_i f_i f^m = \sum_{j=1}^{r} a_i b_j f_i f_j^{n+1} = f_i^{n+1} \sum_{j=1}^{r} a_j b_j f_j.$$

Set

$$a = \sum_{j=1}^{r} a_j b_j f_j.$$

For $1 \leqslant i \leqslant r$, we have therefore $a_i f_i f^m = a f_i^{n+1}$ which implies that if $x \in D(f_i)$,

$$\frac{a(x)}{f(x)^m} = \frac{a_i(x)}{f_i(x)^n}.$$

Thus for $x \in D(f)$,

$$g(x) = \frac{a(x)}{f(x)^m},$$

and $g \in A_f$. $\square$

12.1.2 By 9.5.8 and 12.1.1, $(X, \mathcal{O}_X)$ is a ringed space. Note that $\mathcal{O}_X$ determines A since $\Gamma(X, \mathcal{O}_X) = A$.

12.1.3 Proposition. *Let $X = \mathrm{Spm}(A)$, $Y = \mathrm{Spm}(B)$ where A, B are finitely generated reduced $\Bbbk$-algebras. The set of morphisms of the algebraic set X to the algebraic set Y and the set of morphisms of the ringed space $(X, \mathcal{O}_X)$ to the ringed space $(Y, \mathcal{O}_Y)$ are identical.*

Proof. Let $u : (X, \mathcal{O}_X) \to (Y, \mathcal{O}_Y)$ be a morphism of ringed spaces. For any $g \in B = \Gamma(Y, \mathcal{O}_Y)$, $g \circ u \in A = \Gamma(X, \mathcal{O}_X)$. Hence u is a morphism of algebraic sets.

Conversely, let $u : X \to Y$ be a morphism of algebraic sets and $\mathbf{A}(u)$ its comorphism. For $g \in B$, we have $u^{-1}(D(g)) = D(\mathbf{A}(u)(g))$ (11.7.1). On the other hand, for $f = h/g^n \in \Gamma(D(g), \mathcal{O}_Y)$ where $h \in B$, $n \in \mathbb{N}^*$, we have for $x \in D(\mathbf{A}(u)(g))$,

$$(f \circ u)(x) = \frac{h(u(x))}{g(u(x))^n} = \frac{(\mathbf{A}(u)(h))(x)}{((\mathbf{A}(u)(g))(x))^n}.$$

Thus $f \mapsto f \circ u$ is a homomorphism (of $\Bbbk$-algebras) from $\Gamma(D(g), \mathcal{O}_Y) = B_g$ to $\Gamma(u^{-1}(D(g)), \mathcal{O}_X) = A_{\mathbf{A}(u)(g)}$. Hence by 9.5.3, u is a morphism of ringed spaces. $\square$

12.1.4 We can define by transport of structure, the structure sheaf of an (abstract) algebraic set (11.6.4). Proposition 12.1.3 is again valid in this setting.

Definition. *An* affine algebraic variety *over* $\Bbbk$, *or simply an* affine variety *is a ringed space* $(X, \mathcal{O}_X)$ *where* X *is an algebraic set. A morphism of affine varieties is a morphism of the underlying ringed spaces.*

12.2 Algebraic prevarieties

12.2.1 Definition. (i) *An* (algebraic) prevariety *over* $\Bbbk$ *is a ringed space* $(X, \mathcal{O}_X)$ *relative to* $\Bbbk$ *verifying the following conditions:*

a) X *is a Noetherian topological space.*

b) *There is an open covering* $(U_i)_{i \in I}$ *of* X *such that each induced ringed space* $(U_i, \mathcal{O}_X|_{U_i})$ *is an affine algebraic variety over* $\Bbbk$.

(ii) *An open subset* U *of an algebraic prevariety* $(X, \mathcal{O}_X)$ *is called an* affine open subset *if the induced ringed space* $(U, \mathcal{O}_X|_U)$ *is an affine variety.*

Remarks. 1) An affine variety is a prevariety.

2) When there is no confusion, we shall simply say that X is a prevariety, instead of $(X, \mathcal{O}_X)$.

3) A prevariety is quasicompact (1.3.3). Thus there exists a finite covering of X by open affine subsets.

12.2.2 We define the category of algebraic prevarieties over $\Bbbk$ by letting morphisms of prevarieties to be morphisms of ringed spaces. By 9.5.3, we have the following result:

Proposition. *Let* X, Y *be prevarieties and* $u : X \to Y$ *a map. Then* u *is a morphism of prevarieties if and only if for all open affine subset* V *of* Y *and* $x \in u^{-1}(V)$, *there exists an open affine subset* U *of* X *containing* x, *and contained in* $u^{-1}(V)$, *such that the restriction of* u *to* U *is a morphism of algebraic sets from* U *to* V.

12.2.3 It follows from the previous discussion that:

Proposition. *Let us conserve the notations of* 12.2.2. *The map* u *is an* isomorphism *of prevarieties if and only if* u *is bijective and for all* $x \in X$, *there exists an affine open subset* U *of* X *containing* x, *such that* u *induces an isomorphism from* U *to an affine open subset of* Y.

12.2.4 Remark. When u satisfies the condition of 12.2.3 without being bijective, we say that u is a *local isomorphism* from X to Y.

12.2.5 Proposition. *Let* X *be a prevariety.*

(i) *The topological space* X *admits a base consisting of affine open subsets.*

(ii) *All points of* X *are closed.*

(iii) *If* U *is open in* X, *then the ringed space* $(U, \mathcal{O}_X|_U)$ *is a prevariety.*

Proof. Let $(U_i)_{i \in I}$ be a covering of X by affine open subsets.

(i) Each U_i has a base consisting of affine open subsets (11.7.2 and 11.7.4). So X also has a base consisting of affine open subsets.

(ii) A subset F of X is closed if and only if $F \cap U_i$ is closed for all $i \in I$. Thus the assertion follows from 11.2.3.

(iii) By 1.3.2, U is Noetherian. If $\mathfrak{B}$ is a base for X consisting of affine open subsets, then those open subsets in $\mathfrak{B}$ contained in U form a base for the topology of U. The result now follows.   □

12.2.6 Proposition. *Let $(X, \mathcal{O}_X)$ be a prevariety and Y a locally closed subset of X. Then the ringed space $(Y, \mathcal{O}_X|_Y)$ is a prevariety.*

Proof. By 1.3.2, Y is Noetherian. We may further assume by 9.5.4 and 12.2.5 (iii) that Y is closed in X. Let $(U_i)_{i \in I}$ be a covering of X by affine open subsets. To obtain the result, it suffices to prove that for all $i \in I$, the ringed space $(Y \cap U_i, \mathcal{O}_X|_{Y \cap U_i})$ is an affine variety.

It follows that we may take $X = \mathrm{Spm}(A)$ to be an affine variety and $Y = \mathcal{V}(\mathfrak{a})$ where $\mathfrak{a}$ is a radical ideal of A. Let us denote $\varphi : A \to A/\mathfrak{a}$ the canonical surjection and $j = \mathrm{Spm}(\varphi) : Y \to X$ the canonical injection (11.5.4).

Let V be an open subset of Y, and $\mathcal{G}(V)$ be the set of maps $h : V \to \Bbbk$ such that there exist an open subset $U \subset X$ and $g \in \mathcal{O}_X(U)$ verifying $V = Y \cap U$ and $g|_V = h$. The sheaf associated to the presheaf $\mathcal{G}$ is $\mathcal{O}_X|_Y$ (9.3.8).

Let $f \in A$. We have:

$$D(\varphi(f)) = j^{-1}(D(f)) = Y \cap D(f).$$

It follows from 12.1.1 that $\mathcal{O}_Y(D(\varphi(f))) = (A/\mathfrak{a})_{\varphi(f)}$ where $\mathcal{O}_Y$ denotes the structure sheaf of Y. Thus if $h \in \mathcal{O}_Y(D(\varphi(f)))$, then there exist $g \in A$ and $n \in \mathbb{N}$ such that $h(x) = g(x)/f(x)^n$, $x \in Y \cap D(f)$. So the map $y \mapsto g(y)/f(y)^n$ on $D(f)$ belongs to $\mathcal{O}_X(D(f))$, consequently we obtained that $\mathcal{O}_Y(D(\varphi(f))) \subset \mathcal{G}(D(\varphi(f)))$.

Conversely, given $h \in \mathcal{G}(D(\varphi(f)))$. There exist an open subset $U \subset X$, and $g \in \mathcal{O}_X(U)$ such that $U \cap Y = D(\varphi(f)) = Y \cap D(f)$ and $g|_{Y \cap D(f)} = h$. Cover U by principal open subsets $U_i = D(g_i)$, $g_i \in A$. Then the $D(\varphi(g_i))$'s cover $D(\varphi(f))$ and since $g|_{U_i} \in \mathcal{O}_X(U_i) = A_{g_i}$, $\varphi(g)|_{D(\varphi(g_i))}$ belongs to $(A/\mathfrak{a})_{\varphi(g_i)} = \mathcal{O}_Y(D(\varphi(g_i)))$. As $\mathcal{O}_Y$ is a sheaf, we deduce that $\varphi(g)|_{D(\varphi(f))} \in \mathcal{O}_Y(D(\varphi(f)))$. Hence $h \in \mathcal{O}_Y(\varphi(f))$.

The $D(\varphi(f))$'s form a base for the topology on Y. It follows therefore (using for example 9.3.6) that $\mathcal{O}_Y = \mathcal{O}_X|_Y$.   □

12.2.7 Let $(X, \mathcal{O}_X)$ be a prevariety and Y a locally closed subset of X. We call the prevariety $(Y, \mathcal{O}_X|_Y)$ a *sub-prevariety* of $(X, \mathcal{O}_X)$.

From now on, when we considered a locally closed subset of X as a prevariety, it will be as a sub-prevariety of X.

12.2.8 Let $u : (X, \mathcal{O}_X) \to (Y, \mathcal{O}_Y)$ be a morphism of prevarieties, Z a locally closed subset of Y containing $u(X)$. Then by 9.5.4, u induces a morphism of prevarieties from $(X, \mathcal{O}_X)$ to $(Z, \mathcal{O}_Y|_Z)$.

12.2.9 Let X, Y be prevarieties, Z the sum of the topological spaces X and Y. We identify X and Y as two disjoint open subsets of Z. By 9.5.1, we can define a structure of prevariety on Z by setting $\mathcal{O}_Z(U) = \mathcal{O}_X(U)$ and

$\mathcal{O}_Z(V) = \mathcal{O}_Y(V)$ for open subsets $U \subset X$ and $V \subset Y$. This prevariety is called the *sum* of X and Y, and we shall denote it by $X \sqcup Y$.

If X and Y are affine varieties, then so is $X \sqcup Y$, and we have:

$$\mathbf{A}(X \sqcup Y) = \mathbf{A}(X) \times \mathbf{A}(Y) \ , \ X = \mathcal{V}(\{0\} \times \mathbf{A}(Y)) \ , \ Y = \mathcal{V}(\mathbf{A}(X) \times \{0\}).$$

12.3 Morphisms of prevarieties

12.3.1 Let $(X, \mathcal{O}_X)$ be a prevariety. An element in $\Gamma(X, \mathcal{O}_X)$ is called a *regular function* on X.

Let f be a regular function on X. If U is an affine open subset of X, then the restriction of f to U is a regular function on U. By 12.2.2, we deduce that f is a morphism from X to the affine line $\Bbbk$.

12.3.2 Let X, Y be prevarieties and $y_0 \in Y$. Then the map $u : X \to Y$ sending any $x \in X$ to y_0 is a morphism (this is clear by 11.4.3 since we may assume that X and Y are affine varieties by 12.2.2).

12.3.3 Let $u : (X, \mathcal{O}_X) \to (Y, \mathcal{O}_Y)$ be a morphism of prevarieties. Then we can associated to u the map

$$\Gamma(u) : \Gamma(Y, \mathcal{O}_Y) \to \Gamma(X, \mathcal{O}_X) \ , \ g \mapsto g \circ u$$

which is a homomorphism of $\Bbbk$-algebras. But contrary to 11.4.4, $\Gamma(u)$ does not in general determine u. However,

Proposition. *Let X be a prevariety and Y an affine variety. The map $u \mapsto \Gamma(u)$ is a bijection between the set of morphisms of X to Y and the set of $\Bbbk$-algebra homomorphisms from $\Gamma(Y, \mathcal{O}_Y)$ to $\Gamma(X, \mathcal{O}_X)$.*

Proof. Let $(U_i)_{i \in I}$ be an covering of X by affine open subsets.

1) Given $u : X \to Y$ a morphism of prevarieties and $i \in I$. The map

$$\Gamma(Y, \mathcal{O}_Y) \to \Gamma(U_i, \mathcal{O}_X) \ , \ g \mapsto g \circ u|_{U_i}$$

is a homomorphism of $\Bbbk$-algebras. Clearly this is $\mathbf{A}(u_i)$ where $u_i : U_i \to Y$ is the morphism induced by u.

If $v : X \to Y$ is a morphism verifying $g \circ u = g \circ v$ for all $g \in \Gamma(Y, \mathcal{O}_Y)$, then $u_i = v_i$ for all $i \in I$ (11.4.4). Hence $u = v$ since the U_i's cover X.

2) Let $\varphi \in \mathrm{Hom}_{\mathrm{alg}}(\Gamma(Y, \mathcal{O}_Y), \Gamma(X, \mathcal{O}_X))$ and φ_i the composition

$$\Gamma(Y, \mathcal{O}_Y) \ \xrightarrow{\ \varphi\ } \ \Gamma(X, \mathcal{O}_X) \ \longrightarrow \ \Gamma(U_i, \mathcal{O}_X)$$

where the second map is the restriction homomorphism. By 11.4.4, we have $\varphi_i = \mathbf{A}(u_i)$, where $u_i \in \mathrm{Mor}(U_i, Y)$. To obtain $\varphi = \mathbf{A}(u)$, it suffices to show that $u_i|_{U_i \cap U_j} = u_j|_{U_i \cap U_j}$.

Since affine open subsets form a base for the topology on X, it suffices to show that $u_i|_V = u_j|_V$ for any affine open subset $V \subset U_i \cap U_j$. But these two morphisms have the same comorphism

$$\Gamma(Y, \mathcal{O}_Y) \xrightarrow{\;\;\varphi\;\;} \Gamma(X, \mathcal{O}_X) \longrightarrow \Gamma(V, \mathcal{O}_X)$$

where the second map is the restriction homomorphism. So they are identical by 11.4.4. □

12.3.4 Corollary. *Let* $(X, \mathcal{O}_X)$ *be a prevariety,* U *an open subset of* X *and* $A = \Gamma(U, \mathcal{O}_X)$. *The following conditions are equivalent:*

(i) U *is an affine open subset.*

(ii) *The algebra* A *is finitely generated and the morphism* $u : U \to \operatorname{Spm}(A)$, *associated to the identity map of* A *(see 12.3.3), is an isomorphism.*

Proof. Clearly A is reduced. We have (i) $\Rightarrow$ (ii) by 11.4.5. Conversely, given (ii), $\operatorname{Spm}(A)$ is an affine variety. So by 12.3.3, the morphism u is unique such that $\Gamma(u) = \operatorname{id}_A$. Since u is an isomorphism, (i) follows. □

12.3.5 Let us give an example of an open subset which is not affine.

Let $X = \Bbbk^2 = \operatorname{Spm}(\Bbbk[T_1, T_2])$ and $U = X \setminus \{(0,0)\}$. We claim that U is not affine. Set $U_1 = D(T_1)$ and $U_2 = D(T_2)$, then $U = U_1 \cup U_2$.

Let $f \in \Gamma(U, \mathcal{O}_X)$. Then $g = f|_{U_1} \in \Gamma(U_1, \mathcal{O}_X)$, so by 12.1.1, it is of the form

$$g(x_1, x_2) = \frac{P(x_1, x_2)}{x_1^m}$$

where $P \in \Bbbk[T_1, T_2]$ and $m \in \mathbb{N}$. We may assume that T_1 does not divide P. If $(x_1, x_2) \in U_1$, then

$$x_1^m f(x_1, x_2) - P(x_1, x_2) = 0.$$

Now $h = T_1^m f - P$ defines a regular function on U. Thus $h^{-1}(0) = F$ is a closed subset of U containing U_1. But U_1 is dense in X (so in U), we deduce that h is identically zero on U. In particular, $m = 0$ for otherwise $P(0, x_2) = 0$ for all $x_2 \neq 0$ which contradicts the assumption that T_1 does not divide P.

We have therefore proved that f is the restriction to U of a polynomial, which is unique since $\Bbbk$ is infinite. It follows that $\Gamma(U, \mathcal{O}_X) = \Gamma(X, \mathcal{O}_X)$.

Now let $\varphi : \Gamma(X, \mathcal{O}_X) \to \Gamma(U, \mathcal{O}_X)$ be the restriction homomorphism which is the identity map from the preceding discussion. Let $u : U \to X$ be the morphism such that $\Gamma(u) = \varphi$ (12.3.3). If $x \in U$ and $g \in \Gamma(X, \mathcal{O}_X)$, we have $g(x) = g(u(x))$. Thus $x = u(x)$ (11.3.3) and u is the canonical injection. So u is not bijective and 12.3.4 says that U is not affine.

12.4 Products of prevarieties

12.4.1 Let X, Y be prevarieties over $\Bbbk$, $U \subset X$ and $V \subset Y$ affine open subsets. We have seen in 11.8 that $U \times V$ has a canonical structure of affine variety for which the elements of $\mathbf{A}(U \times V)$ are functions of the form

$$(x, y) \mapsto f_1(x)g_1(y) + \cdots + f_n(x)g_n(y)$$

where $n \in \mathbb{N}^*$, $f_1, \ldots, f_n \in \mathbf{A}(U)$ and $g_1, \ldots, g_n \in \mathbf{A}(V)$.

For $h \in \mathbf{A}(U \times V)$, we set

$$D_{U,V}(h) = \{ z \in U \times V \ ; \ h(z) \neq 0 \} \subset X \times Y.$$

The $D_{U,V}(h)$'s, $h \in \mathbf{A}(U \times V)$, form a base for the Zariski topology in $U \times V$ (11.7.2). Denote by $\mathfrak{C}$ the set of $D_{U,V}(h)$'s for U an open affine subset of X, V an open affine subset of Y and $h \in \mathbf{A}(U \times V)$.

Let $D_{U,V}(h)$, $D_{U',V'}(h') \in \mathfrak{C}$ with

$$h(x,y) = \alpha_1(x)\beta_1(y) + \cdots + \alpha_m(x)\beta_m(y)$$
$$h'(x',y') = \gamma_1(x')\delta_1(y') + \cdots + \gamma_n(x')\delta_n(y')$$

for $(x,y) \in U \times V$ and $(x',y') \in U' \times V'$ where $\alpha_1, \ldots, \alpha_m \in \mathbf{A}(U)$, $\beta_1, \ldots, \beta_m \in \mathbf{A}(V)$, $\gamma_1, \ldots, \gamma_n \in \mathbf{A}(U')$ and $\delta_1, \ldots, \delta_n \in \mathbf{A}(V')$.

Let $(x_0, y_0) \in D_{U,V}(h) \cap D_{U',V'}(h')$. There exist affine open subsets $U'' \subset U \cap U'$ and $V'' \subset V \cap V'$ such that $(x_0, y_0) \in U'' \times V''$. Denote by a_i, c_i (resp. b_i, d_i) the restrictions to U'' (resp. V'') of α_i, γ_i (resp. β_i, δ_i). Then $a_i, c_i \in \mathbf{A}(U'')$ and $b_i, d_i \in \mathbf{A}(V'')$. Thus the maps defined, for $(x,y) \in U'' \times V''$, by

$$s(x,y) = a_1(x)b_1(y) + \cdots + a_m(x)b_m(y)$$
$$t(x,y) = c_1(x)d_1(y) + \cdots + c_n(x)d_n(y)$$

belong to $\mathbf{A}(U'' \times V'')$. On the other hand:

$$(x_0, y_0) \in D_{U'',V''}(st) \ , \ D_{U'',V''}(st) \subset D_{U,V}(h) \cap D_{U',V'}(h').$$

It follows that $\mathfrak{C}$ is a base for a topology $\mathcal{T}$ on $X \times Y$. Let us endow $X \times Y$ with this topology. By construction, if $U \subset X$ and $V \subset Y$ are affine open subsets, then $\mathcal{T}$ induces on $U \times V$ the Zariski topology on $U \times V$.

Affine open subsets in the product of affine open subsets $U \times V$ form a base for the Zariski topology on $U \times V$. Consequently, the set $\mathfrak{B}$ of affine open subsets of a product $U \times V$ of affine open subsets is a base for the topology $\mathcal{T}$ on $X \times Y$.

Since X (resp. Y) is Noetherian, it is the union of a finite number of affine open subsets $U_1, \ldots, U_m$ (resp. $V_1, \ldots, V_n$). So $X \times Y$ is the finite union of the Noetherian subsets $U_i \times V_j$. Hence $X \times Y$ is Noetherian (1.3.2).

For $W \in \mathfrak{B}$, set $\mathcal{F}(W) = \mathbf{A}(W)$. Suppose that $\mathcal{F}$ verifies the conditions of 9.5.1. Then there exists a sheaf of functions $\mathcal{O}_{X \times Y}$ on $X \times Y$ such that $\Gamma(W, \mathcal{O}_{X \times Y}) = \mathcal{F}(W)$ for all $W \in \mathfrak{B}$.

The ringed space $(X \times Y, \mathcal{O}_{X \times Y})$ is therefore an algebraic prevariety over $\Bbbk$ that we shall call the *product* of the prevarieties X and Y. From now on, when we consider $X \times Y$ as a prevariety, it will be with respect to this structure.

12.4.2 Proposition. *Let X, Y be prevarieties, $X_1 \subset X$ and $Y_1 \subset Y$. If X_1 and Y_1 are open (resp. closed, locally closed) in X and Y, then $X_1 \times Y_1$ is open (resp. closed, locally closed) in $X \times Y$. Furthermore, the subvariety $X_1 \times Y_1$ of $X \times Y$ is the product of the sub-prevarieties X_1 and Y_1 of X and Y.*

Proof. If X_1 and Y_1 are open, then the result is clear by the definition of the topology on $X \times Y$. Next, a locally closed subset is open in its closure, so we are reduced to the case where X_1 and Y_1 are closed.

We have to show that if U and V are affine open subsets of X and Y, then

$$(U \times V) \cap (X_1 \times Y_1) = (U \cap X_1) \times (V \cap Y_1)$$

is closed in $U \times V$, and as an affine variety, it is the product of $U \cap X_1$ and of $V \cap Y_1$. But this is clear by 11.8.8. $\square$

12.4.3 Proposition. *Let X, X', Y, Y', Z be prevarieties.*
   (i) *The canonical projections* $\mathrm{pr}_1 : X \times Y \to X$ *and* $\mathrm{pr}_2 : X \times Y \to Y$ *are morphisms.*
   (ii) *If $u : Z \to X$ and $v : Z \to Y$ are morphisms, then the map*

$$w : Z \to X \times Y \ , \ z \mapsto (u(z), v(z))$$

is also a morphism.
   (iii) *If $u : X' \to X$ and $v : Y' \to Y$ are morphisms, then*

$$t : X' \times Y' \to X \times Y \ , \ (x', y') \mapsto (u(x'), v(y'))$$

is also a morphism.
   (iv) *If $x_0 \in X$ (resp. $y_0 \in Y$), then the map $y \mapsto (x_0, y)$ (resp. $x \mapsto (x, y_0)$) is an isomorphism from Y (resp. X) to the closed sub-prevariety $\{x_0\} \times Y$ (resp. $X \times \{y_0\}$) of $X \times Y$.*

Proof. (i) Any point $(x, y) \in X \times Y$ has an open neighbourhood of the form $U \times V$ where U, V are affine open subsets of X and Y. The restrictions of pr_1 and pr_2 are the canonical projections of this product. So the result is a consequence of 11.8.3.
   (ii) Let $z \in Z$. There exist an affine open neighbourhood U (resp. V) of $u(z)$ (resp. $v(z)$) in X (resp. Y), and an affine open neighbourhood W of z in Z such that $u(W) \subset U$ and $v(W) \subset V$. So (ii) follows from 11.8.3.
   (iii) The argument is analogue to the one used in 11.8.4.
   (iv) The map $Y \to \{x_0\} \times Y$, $y \mapsto (x_0, y)$, is clearly bijective. Taking affine open neighbourhoods, we are reduced to the case where $X = \mathrm{Spm}(A)$ and $Y = \mathrm{Spm}(B)$ are affine varieties. Let $\mathfrak{m} \in \mathrm{Spm}(A)$ be such that $x = \mathcal{V}(\mathfrak{m})$. Since $A/\mathfrak{m}$ is isomorphic to $\Bbbk$, the canonical homomorphism $B \to (A/\mathfrak{m}) \otimes_{\Bbbk} B$ is an isomorphism. Hence we have (iv). $\square$

12.4.4 Remark. Let X, Y, Z be prevarieties. By 12.4.3, the bijections

$$X \times Y \to Y \times X \ , \ (x, y) \mapsto (y, x),$$
$$(X \times Y) \times Z \to X \times (Y \times Z) \ , \ ((x, y), z) \mapsto (x, (y, z))$$

are isomorphisms of prevarieties. We deduce in particular the definition of the product $X_1 \times \cdots \times X_n$ for a finite number of prevarieties.

12.4.5 Proposition. *Let X and Y be prevarieties.*

(i) *If X and Y are non-empty, then the topology on X (resp. on Y) is the quotient topology of the one on $X \times Y$ with respect to the relation $\mathrm{pr}_1(z) = \mathrm{pr}_1(z')$ (resp. $\mathrm{pr}_2(z) = \mathrm{pr}_2(z')$).*

(ii) *If X and Y are irreducible, then $X \times Y$ is also irreducible.*

Proof. (i) Let $F \subset X$ be such that $F \times Y$ is closed in $X \times Y$. If $y \in Y$, we have $(F \times Y) \cap (X \times \{y\}) = F \times \{y\}$. So $F \times \{y\}$ is closed in $X \times \{y\}$. By 12.4.3 (iv), F is closed in X. So part (i) follows.

(ii) Let us write $X = U_1 \cup \cdots \cup U_m$, $Y = V_1 \cup \cdots \cup V_n$ where the U_i's and the V_j's are non-empty affine open subsets. Then

$$X \times Y = \bigcup_{i,j} U_i \times V_j.$$

Since X, Y are irreducible, $(U_i \times V_j) \cap (U_k \times V_l) \neq \emptyset$ for $1 \leqslant i, k \leqslant m$ and $1 \leqslant j, l \leqslant n$. By 1.1.6, it suffices to prove that $U_i \times V_j$ is irreducible. So we are reduced to the case where X and Y are affine varieties, and the result follows from 11.8.6. $\square$

12.5 Algebraic varieties

12.5.1 The *diagonal* of a set X, denoted by Δ_X, is the subset of $X \times X$ defined by

$$\Delta_X = \{(x,x) \; ; \; x \in X\}.$$

Proposition. (i) *Let X be an affine variety, then Δ_X is closed in $X \times X$.*

(ii) *Let X be a prevariety, then Δ_X is locally closed in $X \times X$.*

Proof. (i) Since X is affine, it is closed in an affine space $\Bbbk^n$. So $X \times X$ is closed in $\Bbbk^n \times \Bbbk^n$ (11.8.8). Now,

$$\Delta_X = (X \times X) \cap \Delta_{\Bbbk^n}.$$

So it suffices to show that $\Delta_{\Bbbk^n}$ is closed in $\Bbbk^n \times \Bbbk^n$. But this is clear since it is the affine algebraic subset defined by the polynomials $x_i - x_{n+i} = 0$ for $1 \leqslant i \leqslant n$.

(ii) Let $z \in \Delta_X$, there is an affine open subset $U \subset X$ such that $U \times U$ is an open neighbourhood of z in $X \times X$. Now (ii) follows from (i) since $\Delta_X \cap (U \times U) = \Delta_U$. $\square$

12.5.2 Corollary. *Let X, Y be prevarieties (resp. affine varieties) and $u : X \to Y$ a morphism. The graph G_u of u is locally closed (resp. closed) in $X \times Y$. Furthermore, the graph morphism of u*

$$X \to X \times Y \; , \; x \mapsto (x, u(x))$$

is an isomorphism onto the sub-prevariety G_u of $X \times Y$.

Proof. Set $v : X \times Y \to Y \times Y$, $(x,y) \mapsto (u(x),y)$. By 12.4.3 (iii), v is a morphism and $G_u = v^{-1}(\Delta_Y)$. So the first part follows from 12.5.1 and the second by observing that $\mathrm{pr}_1 |_{G_u}$ is the inverse of the graph morphism. □

12.5.3 Definition. *An algebraic prevariety X over $\Bbbk$ is called an* (algebraic) *variety over $\Bbbk$ if Δ_X is closed in $X \times X$.*

12.5.4 Proposition. (i) *Any sub-prevariety of a variety is a variety.*
(ii) *The product of two varieties is a variety.*
(iii) *Let X be a prevariety, Y a variety and $u : X \to Y$ a morphism. Then the graph of u is closed in $X \times Y$.*
(iv) *Let X be a variety, $n \in \mathbb{N}^*$ and*

$$\Delta_X^n = \{(x,\dots,x) \in X^n \; ; \; x \in X \}.$$

Then Δ_X^n is closed in X^n.

Proof. Part (i) follows from the fact that if Y is a sub-prevariety of a variety X, then $\Delta_Y = (Y \times Y) \cap \Delta_X$.

Let X,Y be varieties, then we can identify $\Delta_{X \times Y} \subset (X \times Y) \times (X \times Y)$ with $\Delta_X \times \Delta_Y \subset (X \times X) \times (Y \times Y)$ (12.4.4). So part (ii) follows from 12.4.2.

Part (iii) is proved as in 12.5.2, and part (iv) follows from (ii) and (iii) since Δ_X^n is the graph of the morphism $X \to X^{n-1}$, $x \mapsto (x,\dots,x)$. □

12.5.5 Remarks. 1) An affine variety is a variety.

2) A sub-prevariety of a variety X shall be called a *subvariety* of X.

3) A variety is *quasi-affine* if it is isomorphic to a subvariety of an affine variety. We have seen in 12.3.5 that a quasi-affine variety is not necessary affine.

12.5.6 Theorem. *Let X be a prevariety. The following conditions are equivalent:*

(i) *X is a variety.*

(ii) *For all morphisms $u,v : Y \to X$ of prevarieties, the set of $y \in Y$ verifying $u(y) = v(y)$ is closed in Y.*

(iii) *For all affine open subsets U,V of X, $U \cap V$ is an open affine subset and the algebra $\mathbf{A}(U \cap V)$ is generated by the functions $x \mapsto f(x)g(x)$, for $x \in U \cap V$, where $f \in \mathbf{A}(U)$ and $g \in \mathbf{A}(V)$.*

Proof. (i) ⇒ (ii) Let Y, u,v as in (ii), set:

$$w : Y \to X \times X \; , \; y \mapsto (u(y),v(y)).$$

Then $\{y \in Y \; ; \; u(y) = v(y)\} = w^{-1}(\Delta_X)$, and (ii) follows since X is a variety and w is a morphism (12.4.3).

(ii) ⇒ (i) This is clear by taking $Y = X \times X$, $u = \mathrm{pr}_1$ and $v = \mathrm{pr}_2$.

(iii) ⇔ (i) The subsets $U \times V$, where U,V are affine open subsets of X, form an open covering of $X \times X$. So Δ_X is closed if and only if $\Delta_X \cap (U \times V)$ is closed in $U \times V$ for all affine open subsets U, V of X. Now the map $u : U \cap V \to U \times V$, $x \mapsto (x, x)$, induces an isomorphism of the sub-prevariety $U \cap V$ onto the sub-prevariety $\Delta_X \cap (U \cap V)$ of the affine variety $U \times V$ (12.5.2).

If we have (i), then the open subset $U \cap V$ is affine since $(U \times V) \cap \Delta_X$ is closed in $U \cap V$.

Conversely, suppose that the open subset $U \cap V$ is affine. Denote by j the canonical injection from $(U \times V) \cap \Delta_X$ into $U \times V$. Then

$$U \cap V \xrightarrow{\ w\ } (U \times V) \cap \Delta_X \xrightarrow{\ j\ } U \times V,$$

where w denotes the isomorphism induced by u.

By 11.6.8, $(U \times V) \cap \Delta_X$ is a closed subset of $U \times V$ if and only if the comorphism $\mathbf{A}(U \times V) \to \mathbf{A}(U \cap V)$ of u is surjective. Hence we have obtained the equivalence of (i) and (iii). $\square$

12.5.7 Remark. Let $(U_i)_{i \in I}$ be a covering of a prevariety X by affine open subsets. If for all $i, j \in I$, the open subsets U_i and U_j verify condition (iii) of 12.5.6, then the preceding proof says that X is a variety.

12.5.8 Corollary. *Let X be a variety, Y a prevariety, Z a subset of Y and u, v be morphisms from Y to X such that $u(y) = v(y)$ for all $y \in Z$. Then $u(y) = v(y)$ for all $y \in \overline{Z}$.*

Proof. This is clear by 12.5.6 (ii). $\square$

12.5.9 Proposition. *Let X be a prevariety such that any two points in X are contained in some affine open subset. Then X is a variety.*

Proof. Let x, y be two distinct points of X and U an affine open subset of X containing x and y. The set $(U \times U) \cap \Delta_X$ is closed in $U \times U$ and its complement Ω in $U \times U$ is open in $U \times U$, so in $X \times X$ as well. Since $(x, y) \in \Omega$ and $\Omega \cap \Delta_X = \emptyset$, Δ_X is closed in $X \times X$. $\square$

12.5.10 Proposition. *Let $u : X \to Y$ be a morphism of prevarieties. Suppose that the following conditions are satisfied:*

(i) *Y is a variety.*

(ii) *There exists an open covering $(V_i)_{i \in I}$ of Y such that each of the open subsets $U_i = u^{-1}(V_i)$ is a variety.*

Then X is a variety.

Proof. Since $(U_i)_{i \in I}$ is an open covering of X, it suffices to show that the set $(U_i \times U_j) \cap \Delta_X$ is closed in $U_i \times U_j$.

Let $v : X \times X \to Y \times Y$, $(x, x') \mapsto (u(x), u(x'))$. Then

$$(U_i \times U_j) \cap \Delta_X = [v^{-1}((V_i \times V_j) \cap \Delta_Y)] \cap [(U_i \times U_i) \cap \Delta_X].$$

By condition (i), $(V_i \times V_j) \cap \Delta_Y$ is closed in $V_i \times V_j$. So $v^{-1}((V_i \times V_j) \cap \Delta_Y)$ is closed in $U_i \cap U_j$. Now condition (ii) implies that $(U_i \times U_i) \cap \Delta_X$ is closed in $U_i \times U_i$. Hence $(U_i \times U_j) \cap \Delta_X$ is closed in $U_i \times U_j$ as required. $\square$

12.5.11 Proposition. *Let X, Y be irreducible varieties. A morphism $u :$ $X \to Y$ which is a local isomorphism is injective. Thus it is an isomorphism from X onto an open subvariety of Y.*

Proof. Let $x, x' \in X$ be distinct points such that $u(x) = u(x') = y$.

There exists, by our hypothesis, an open subset U (resp. U') of X containing x (resp. x') such that $u|_U$ (resp. $u|_{U'}$) is an isomorphism from U (resp. U') onto an open subset V (resp. V') of Y. Let g, g' be the inverse isomorphisms. Then g, g' induce morphisms h, h' from the open subset $V \cap V'$ of Y to X. Since X is irreducible, $U \cap U'$ is non-empty, and so $u(U \cap U')$ is a non-empty open subset of Y contained in $V \cap V'$. Hence $u(U \cap U')$ is dense in $V \cap V'$ since Y is irreducible. We deduce that h and h' are identical on $V \cap V'$ which contradicts the fact that $h(y) \neq h'(y)$. $\square$

12.6 Gluing

12.6.1 Let $X_1, \ldots, X_n$ be prevarieties. Assume that for $1 \leqslant i, j \leqslant n$, there exist open subsets X_{ij} of X_i, X_{ji} of X_j, and an isomorphism $h_{ji} : X_{ij} \to X_{ji}$ of prevarieties, such that

a) $X_{ii} = X_i$ and $h_{ii} = \mathrm{id}_{X_i}$ for $1 \leqslant i \leqslant n$.

b) h_{ij} and h_{ji} are mutually inverse isomorphisms for $1 \leqslant i, j \leqslant n$.

c) For $1 \leqslant i, j, k \leqslant n$, $h_{ji}|_{X_{ij} \cap X_{ik}}$ induces an isomorphism of $X_{ij} \cap X_{ik}$ onto $X_{ji} \cap X_{jk}$, and $h_{ki} = h_{kj} \circ h_{ji}$ are identical on $X_{ij} \cap X_{ik}$.

12.6.2 With the above assumption, there exist by 1.5.4 a topological space X, and for $1 \leqslant i \leqslant n$, a homeomorphism ψ_i from X_i to an open subset $\psi_i(X_i)$ of X such that:

1) $X = \psi_1(X_1) \cup \cdots \cup \psi_n(X_n)$.

2) $\psi_i(X_{ij}) = \psi_j(X_{ji}) = \psi_i(X_i) \cap \psi_j(X_j)$ for $1 \leqslant i, j \leqslant n$.

3) $h_{ji}(x) = \psi_j^{-1}(\psi_i(x))$ for $1 \leqslant i, j \leqslant n$ and $x \in X_{ij}$.

We shall endow X with a structure of prevariety.

The sheaf $\mathcal{O}_{X_i}$ on X_i transports via ψ_i into a sheaf $\mathcal{F}_i$ on $\psi_i(X_i)$. Thus if U is an open subset of $\psi_i(X_i)$, $\mathcal{F}_i(U)$ is the set of functions $z \mapsto g(\psi_i^{-1}(z))$, with $g \in \Gamma(\psi_i^{-1}(U), \mathcal{O}_{X_i})$.

We claim that there exists a unique sheaf $\mathcal{O}_X$ on X such that $\mathcal{O}_X(U) = \mathcal{F}_i(U)$ if $1 \leqslant i \leqslant n$ and U an open subset of $\psi_i(X_i)$.

For $1 \leqslant i \leqslant n$, let $\mathfrak{B}_i$ be a base of open subsets of $\psi_i(X_i)$. The union $\mathfrak{B}$ of the $\mathfrak{B}_i$'s is a base of open subsets of X. In view of 9.5.1, it suffices to prove that $\mathcal{F}_i$ and $\mathcal{F}_j$ are identical when restricted to $\psi_i(X_i) \cap \psi_j(X_j)$.

Let W be an open subset of $\psi_i(X_i) \cap \psi_j(X_j)$. For $x \in W$, we have $\psi_j^{-1}(x) = h_{ji}(\psi_i^{-1}(x))$. But the elements of $\mathcal{F}_j(W)$ are the functions $g \circ \psi_j^{-1}$, with $g \in \Gamma(\psi_j^{-1}(W), \mathcal{O}_{X_j})$. So, by the hypothesis, the functions $g \circ h_{ji}$ are in $\Gamma(\psi_i^{-1}(W), \mathcal{O}_{X_i})$ (since the h_{ji}'s are isomorphisms of prevarieties). So we have proved our claim (in fact, we have glued the $\mathcal{F}_i$'s in the sense of 9.4.2).

Since X is a finite union of Noetherian open subsets, it is a Noetherian topological space. Further, if we take $\mathfrak{B}_i$ a base of affine open subsets of

$\psi_i(X_i)$, then $\mathfrak{B}$ is a base of affine open subsets of X. This implies that X is a prevariety. With respect to this structure of prevariety on X, it is clear that each $\psi_i(X_i)$ is an open sub-prevariety of X and each ψ_i is an isomorphism of prevarieties.

12.6.3 With the above notations, we say that the prevariety X is obtained by *gluing the prevarieties X_i along the X_i using the isomorphisms h_{ji}.* In general, we identify via ψ_i, the prevarieties X_i and $\psi_i(X_i)$.

12.6.4 Using the preceding construction, we give below examples of prevarieties which are not varieties.

Let $X_1 = X_2 = \Bbbk$ and denote by O_1, O_2 the point 0 in X_1 and X_2 respectively. Let $U_1 = X_1 \setminus \{O_1\}$, $U_2 = X_2 \setminus \{O_2\}$ and glue X_1 and X_2 along U_1 and U_2 via the identity map $U_1 \to U_2$. The prevariety X thus obtained is the union $X_1 \cup \{O_2\} = X_2 \cup \{O_1\}$ and X_1, X_2 are affine open subsets of X.

Now the open subset $X_1 \cap X_2$ is just U_1 (or U_2), so it is affine. Moreover, $X \setminus (X_1 \cap X_2)$ consists of the two distinct points O_1 and O_2.

The intersection $(X_1 \times X_2) \cap \Delta_X$ can be identified with $\Delta_\Bbbk \setminus \{(0,0)\}$ where we identify $X_1 \times X_2$ with $\Bbbk^2$. It follows that Δ_X is not closed in $X \times X$, and so X is not a variety.

Note that if $x, y \in X \setminus \{O_1\}$ (resp. $X \setminus \{O_2\}$), then x, y are contained in the affine open subset X_1 (resp. X_2). Since X is not a variety, it follows from 12.5.9 that no affine open subset of X contains simultaneously O_1 and O_2.

The preceding discussion says that any affine open subset of X is contained in X_1 or X_2. Hence the intersection of two affine open subsets of X is again an affine open subset. Compare this with condition (iii) of 12.5.6.

12.6.5 Let $X_1 = X_2 = \Bbbk^2$, O_1, O_2 the point $(0,0)$ of X_1 and X_2 respectively, and X the prevariety obtained by gluing X_1 and X_2 along $X_1 \setminus \{O_1\}$ and $X_2 \setminus \{O_2\}$ via the identity map $X_1 \setminus \{O_1\} \to X_2 \setminus \{O_2\}$.

Again we may identify X_1 and X_2 as affine open subsets of X, and identify $X_1 \cap X_2$ with $\Bbbk^2 \setminus \{(0,0)\}$, which is not affine (12.3.5). So by 12.5.6, X is not a variety.

12.7 Rational functions

12.7.1 In this section, X shall be an *irreducible* algebraic variety.

Lemma. *Let U, V be non-empty open subsets of X, $u : U \to Y$ and $v : V \to Y$ morphisms to a variety Y. If there exists an open subset W in $U \cap V$ such that $u|_W = v|_W$, then $u|_{U \cap V} = v|_{U \cap V}$.*

Proof. Since X is irreducible, W is dense in X and so in $U \cap V$. The result follows therefore from 12.5.8. □

12.7.2 Let U be a non-empty open subset of X and $u : U \to Y$ a morphism to a variety Y. By 12.7.1, there exists a largest open subset U_0 containing U

such that u extends to a morphism $u_0 : U_0 \to Y$. Namely, U_0 is the union of all the open subsets containing U where u extends to a morphism to Y.

When $U_0 = U$, we say that u is a *rational map from X to Y* and that U is the *domain of definition* of the rational map u. So a morphism of X to Y is a rational map whose domain of definition is X, and by 12.7.1, any morphism from a non-empty open subset of X to Y extends uniquely to a rational map from X to Y.

A rational map from X to $\Bbbk$ is called a *rational function* on X. Such a function is therefore the maximal extension of a regular function on a non-empty open subset of X.

There is another way to define rational maps:

Let E be the set of pairs (U, u), where U is a non-empty open subset of X and u a morphism from U to Y. Define the following relation on E

$$(U, u) \sim (V, v) \Leftrightarrow u|_{U \cap V} = v|_{U \cap V}.$$

This is an equivalence relation, and a rational map is simply an equivalence class.

There is a third way to define rational maps. Order E by setting

$$(U, u) \leqslant (V, v) \Leftrightarrow V \subset U \ \text{ and } \ u|_V = v.$$

Then we have an inductive system, and the set of rational maps is the set of inductive limits of this system.

12.7.3 Example. Let $f, g \in \Gamma(X, \mathcal{O}_X)$ with $g \neq 0$. The set U of $x \in X$ such that $g(x) \neq 0$ is a non-empty open subset. The map $x \mapsto f(x)/g(x)$ is regular on U since it is regular on all affine open subsets contained in U (11.7.5 and 12.1.1). It follows that the preceding map extends uniquely to a rational function on X.

In general, we do not obtain all the rational functions in this way.

12.7.4 From now on, we shall denote by $\mathbf{R}(X)$ the set of rational functions on X. We claim that $\mathbf{R}(X)$ has a canonical structure of commutative field.

Firstly, the interpretation of rational functions in terms of inductive limit and the results of 8.3 show that $\mathbf{R}(X)$ has a canonical structure of commutative ring.

Let $f \in \mathbf{R}(X) \setminus \{0\}$. The set U of points in X where f is non-zero is a non-empty open subset, and the function $x \mapsto 1/f(x)$ is regular on U (12.7.3). Denote by $1/f$ the unique rational function such that its restriction to U is $x \mapsto 1/f(x)$. Then it is obvious that $1/f$ is the inverse of f in the ring $\mathbf{R}(X)$. Thus we have proved our claim.

12.7.5 Remark. Let $f \in \mathbf{R}(X)$ and $x \in X$. If f is not defined at x, it may happen that $g = 1/f$ is defined at x. Then $g(x) = 0$ for otherwise $f = 1/g$ would be defined in a neighbourhood of x. We shall say that x is a *pole* of f.

For example, the rational function x_2/x_1 on $\Bbbk^2$ is not defined at the points $(0, x_2)$. We see that such a point $(0, x_2)$ is a pole if and only if $x_2 \neq 0$.

12.7.6 Theorem. *Let X be an irreducible variety. For all affine open subset U of X, the ring $\Gamma(U, \mathcal{O}_X)$ is an integral domain and the canonical map $\Gamma(U, \mathcal{O}_X) \rightarrow \mathbf{R}(X)$, sending a regular function on U to the unique rational function extending it, is injective and extends to an isomorphism from the quotient field of $\Gamma(U, \mathcal{O}_X)$ onto $\mathbf{R}(X)$.*

Proof. Since X is irreducible, so is U. Hence $\Gamma(U, \mathcal{O}_X)$ is an integral domain (11.2.3 and 11.3.2). The injectivity of the map $\Gamma(U, \mathcal{O}_X) \rightarrow \mathbf{R}(X)$ is a consequence of 12.7.1. To finish our proof, we only have to show that the extension of this map to the quotient field of $\Gamma(U, \mathcal{O}_X)$ is surjective.

Let $h \in \mathbf{R}(X)$. There exists an affine open subset $D(g) \subset U$, where $g \in \Gamma(U, \mathcal{O}_X)$, such that $h|_{D(g)} \in \Gamma(D(g), \mathcal{O}_X)$. So $h|_{D(g)}$ is of the form $x \mapsto f(x)/g(x)^m$, with $f \in \Gamma(U, \mathcal{O}_X)$. If $\widetilde{f}, \widetilde{g} \in \mathbf{R}(X)$ are the images of f and g, then $h = \widetilde{f}/\widetilde{g}^m$. Thus we have obtained our result. □

12.7.7 In view of 12.7.6, we can identify $\mathbf{R}(X)$ with the quotient field of $\Gamma(U, \mathcal{O}_X)$, for an affine open subset U of X. In particular, if $X = \operatorname{Spm}(A)$ is an irreducible affine variety, then $\mathbf{R}(X)$ can be identified with the quotient field of A.

12.7.8 Remark. It follows from 12.7.6 that when X is an irreducible variety, $\mathbf{R}(X)$ is a finitely generated extension of $\Bbbk$.

Conversely, let K be a finitely generated extension of $\Bbbk$ and $y_1, \ldots, y_n$ a system of generators of K, then K is the quotient field of the finitely generated $\Bbbk$-algebra $A = \Bbbk[y_1, \ldots, y_n]$ which is also an integral domain. It follows that there exists an irreducible affine variety $X \subset \Bbbk^n$ such that $\mathbf{A}(X)$ is isomorphic to A. Thus K is isomorphic to $\mathbf{R}(X)$.

An irreducible variety Y such that $\mathbf{R}(Y)$ is isomorphic to K is called a *model* of K. The preceding paragraph says that there is an affine model of K.

12.7.9 Proposition. *Let $X = \operatorname{Spm}(A)$ be an irreducible variety such that A is a factorial ring, and $h \in \mathbf{R}(X)$. There exist $f, g \in A$ verifying the following conditions:*
  (i) *The domain of definition of h is $D(g)$.*
  (ii) *For all $x \in D(g)$, we have $h(x) = f(x)/g(x)$.*

Proof. By 12.7.6, there exist $f, g \in A$, $g \neq 0$, relatively prime, such that $h = f/g$ (and f, g are unique up to a multiple of an invertible element). Denote by U the domain of definition of h, we have $D(g) \subset U$.

If h is defined on a point $x \notin D(g)$, then $h = f_1/g_1$ where $f_1, g_1 \in A$ with $g_1(x) \neq 0$. So there exists $u \in A$ such that $f_1 = uf$ and $g_1 = ug$. This is absurd since $g(x) = 0$. □

12.7.10 Remark. We can not drop the hypothesis that A is factorial. In general, the domain of definition of h can contain strictly $D(g)$ for any $g \in A$ such that $h = f/g$ with $f \in A$.

12.7.11 Let X, Y be irreducible varieties, $u : X \to Y$ a dominant morphism and $V \subset Y$ an affine open subset. There exists an affine open subset U of X such that $U \subset u^{-1}(V)$. Since u is dominant, $u(U)$ is dense in V. By 6.5.9, the comorphism

$$\varphi : \Gamma(V, \mathcal{O}_Y) \to \Gamma(U, \mathcal{O}_X)$$

of the restriction $U \to V$ of u is injective. Hence by 12.7.6, φ can be extended to an injective homomorphism of the fields $\theta : \mathbf{R}(Y) \to \mathbf{R}(X)$.

Let $U' \subset X$ and $V' \subset Y$ be affine open subsets such that $U' \subset u^{-1}(V')$. The comorphism $\varphi' : \Gamma(V', \mathcal{O}_Y) \to \Gamma(U', \mathcal{O}_X)$ extends to the same homomorphism θ. In fact, replacing V' (resp. U') by an affine open subset in $V \cap V'$ (resp. $U \cap U'$) if necessary, we may assume that $V' \subset V$ and $U' \subset U$. It is then clear that φ' extends φ.

When there is no confusion, we shall say that the field homomorphism $\theta : \mathbf{R}(Y) \to \mathbf{R}(X)$ is the *comorphism* associated to u.

12.7.12 Definition. *A dominant morphism* $u : X \to Y$ *of irreducible varieties is called* birational *if the comorphism* $\theta : \mathbf{R}(Y) \to \mathbf{R}(X)$ *associated to u is an isomorphism.*

12.7.13 Remarks. 1) By 11.5.6, the morphism

$$u : \Bbbk^2 \to \Bbbk^2 , \ (x_1, x_2) \mapsto (x_1 x_2, x_2)$$

is dominant. It is birational since $\Bbbk(T_1, T_2) = \Bbbk(T_1 T_2, T_2)$. However, it is neither surjective (11.5.6) nor injective (since $u(x_1, 0) = (0, 0)$ for all $x_1 \in \Bbbk$).

2) Recall from 11.5.7 the morphism

$$u : \Bbbk \to \Bbbk^2 , \ t \mapsto (t^2, t^3),$$

whose image C is the algebraic set defined by $x_1^3 - x_2^2 = 0$.

Recall that the morphism $v : \Bbbk \to C$ is bijective, bicontinuous but is not an isomorphism. However, it is a birational morphism since $\Bbbk(T) = \Bbbk(T^2, T^3)$.

3) So a birational morphism can be neither injective nor surjective. If it is bijective and bicontinuous, it does not need to be an isomorphism. Note that a bijective and bicontinuous morphism is not necessarily birational.

12.8 Local rings of a variety

12.8.1 Let $(X, \mathcal{O}_X)$ be a variety and $x \in X$. We denote by $\mathcal{O}_{X,x}$, or $\mathcal{O}_x$, the fibre at x of the sheaf $\mathcal{O}_X$ (9.1.4). If U is an open subset of X containing x and $f \in \Gamma(U, \mathcal{O}_X)$, we denote by f_x the image of f in $\mathcal{O}_{X,x}$. We defined in 9.1.4 the value of a germ at x. By 8.3, $\mathcal{O}_{X,x}$ has a natural structure of commutative ring.

12.8.2 Remarks. 1) If U is an open subset of X containing x, then $\mathcal{O}_{X,x} = \mathcal{O}_{U,x}$ (8.2.7).

2) Suppose that X is irreducible, then any regular function on an open neighbourhood of x extends uniquely to a rational function on X defined at x. Moreover, two such functions having the same germ at x extend to the same rational function. Thus $\mathcal{O}_{X,x}$ can be identified as a subring of $\mathbf{R}(X)$ of rational functions defined at x. In particular, since X is irreducible, the ring $\mathcal{O}_{X,x}$ is an integral domain, and if U is an open subset of X, then

$$\Gamma(U, \mathcal{O}_X) = \bigcap_{x \in U} \mathcal{O}_{X,x}.$$

12.8.3 Proposition. *Let $(X, \mathcal{O}_X)$ be a variety and $x \in X$.*

(i) *The ring $\mathcal{O}_{X,x}$ is local and its maximal ideal $\mathfrak{m}_{X,x}$ consists of germs of regular functions which are zero at x. The fields $\mathcal{O}_{X,x}/\mathfrak{m}_{X,x}$ and $\Bbbk$ are isomorphic.*

(ii) *Let U be an affine open subset of X containing x, $A = \Gamma(U, \mathcal{O}_X)$ and $\mathfrak{n}$ the maximal ideal of A corresponding to x. The rings $A_\mathfrak{n}$ and $\mathcal{O}_{X,x}$ are isomorphic. In particular, there is a bijection between the set of prime ideal of $\mathcal{O}_{X,x}$ and the set of closed irreducible subsets of U containing x.*

(iii) *The ring $\mathcal{O}_{X,x}$ is Noetherian.*

Proof. (i) Let $\rho : \mathcal{O}_{X,x} \to \Bbbk$ be the map sending a germ to its value at x. This is a surjective ring homomorphism. Thus $\mathfrak{m}_{X,x} = \ker \rho$ is a maximal ideal and the fields $\mathcal{O}_{X,x}/\mathfrak{m}_{X,x}$ and $\Bbbk$ are isomorphic.

If $s \in \mathcal{O}_{X,x} \setminus \mathfrak{m}_{X,x}$ and (U, f) a representative of s (9.1.4) where U is an affine open subset of X containing x, then $x \in D(f) = \{y \in U; f(y) \neq 0\}$. Since f is invertible in $\Gamma(D(f), \mathcal{O}_U) = \Gamma(D(f), \mathcal{O}_X)$, s is invertible in $\mathcal{O}_{X,x}$. So $\mathcal{O}_{X,x}$ is local.

(ii) Let $\varphi : A \to \mathcal{O}_{X,x}$ be the homomorphism sending $f \in A$ to its germ at x. If $f(x) \neq 0$, that is $f \notin \mathfrak{n}$, then $\varphi(f)$ is invertible by (i). By 2.3.4, φ extends uniquely to a homomorphism $\psi : A_\mathfrak{n} \to \mathcal{O}_{X,x}$.

Let $s \in \mathcal{O}_{X,x}$. There exist a principal open subset $D(f)$ of U containing x and a function $g \in \Gamma(D(f), \mathcal{O}_X)$ such that $(D(f), g)$ is a representative of s. Since $x \in D(f)$, we have $f \in A \setminus \mathfrak{n}$. Moreover, if $y \in D(f)$, then $g(y) = h(y)/f(y)^n$, for some $n \in \mathbb{N}$, $h \in A$. Thus $\psi(h/f^n) = s$, and ψ is surjective.

Next, if $g \in \ker \psi$, there exist $h \in A$, $f \in A \setminus \mathfrak{n}$ such that $g = h/f$. Then $\varphi(h) = 0$. Thus there exists $t \in A \setminus \mathfrak{n}$ verifying $h|_{D(t)} = 0$, and hence $ht = 0$. So $g = 0$.

We have therefore proved that ψ is an isomorphism, and the second part follows from 2.3.8 and 11.2.3.

(iii) This is clear by (ii), 1.3.2 and 2.7.6. $\square$

12.8.4 Remark. In general, $\mathcal{O}_{X,x}$ is not a finitely generated $\Bbbk$-algebra. Take $X = \Bbbk$ and $x = 0$, then $\mathcal{O}_{X,x}$ is isomorphic to the subring of $\Bbbk(T)$ consisting of rational functions of the form P/Q with $Q(x) \neq 0$. Since a non-zero polynomial in one variable has a finite number of zeros, it is clear that $\mathcal{O}_{X,x}$ is not finitely generated as a $\Bbbk$-algebra.

12.8.5 Proposition. *Let $(X, \mathcal{O}_X)$ be a variety and $x \in X$ Then the following conditions are equivalent:*

(i) *x belongs to a unique irreducible component of X.*

(ii) *The ring $\mathcal{O}_{X,x}$ is an integral domain.*

Proof. Let $Y_1, \ldots, Y_n$ be the irreducible components of X.

(i) $\Rightarrow$ (ii) Let $U = Y_2 \cup \cdots \cup Y_n$ and $V = X \setminus U$. By 1.1.13, $V \subset Y_1$ is open in X, so it is irreducible. So if $x \in V$, then $\mathcal{O}_{X,x} = \mathcal{O}_{V,x}$, and $\mathcal{O}_{V,x}$ is an integral domain (12.8.2).

(ii) $\Rightarrow$ (i) Replacing X by an affine open subset containing x, we may assume that $X = \mathrm{Spm}(A)$ is affine. Then $Y_i = \mathcal{V}(\mathfrak{p}_i)$ where $\mathfrak{p}_1, \ldots, \mathfrak{p}_n$ are the minimal prime ideals of A. Suppose that $x \in Y_1 \cap Y_2$.

Again, there is a non-empty affine open subset $U_1 \subset Y_1$ (resp. $U_2 \subset Y_2$) such that $U_1 \cap Y_i = \emptyset$ if $i \geqslant 2$ (resp. $U_2 \cap Y_i = \emptyset$ if $i \neq 2$). So we can find $f_1, f_2 \in A \setminus \{0\}$ such that $D(f_1) \subset U_1$ and $D(f_2) \subset U_2$. Thus $f_1 f_2 = 0$. Now Y_1 and Y_2 are irreducible, so U_1 and U_2 are dense in Y_1 and Y_2, which implies that the germs of f_1, f_2 at x are non-zero. Hence $\mathcal{O}_{X,x}$ is not an integral domain. $\square$

12.8.6 Let $u : X \to Y$ be a morphism of varieties and $x \in X$. If V is an open neighbourhood of $u(x)$ and $g \in \Gamma(V, \mathcal{O}_Y)$, then $g \circ u$ is a regular function in the open subset $u^{-1}(V)$. Moreover, if g_1, g_2 are equal in a neighbourhood of $u(x)$, then $g_1 \circ u$ and $g_2 \circ u$ are equal in a neighbourhood of x. We obtain therefore a map

$$u_x : \mathcal{O}_{Y,u(x)} \to \mathcal{O}_{X,x}$$

which is clearly a $\Bbbk$-algebra homomorphism. We can also verify easily that u_x is a local homomorphism.

Let U be an affine open subset of X containing x. Any element of $\mathcal{O}_{X,x}$ is of the form f_x / h_x (notations of 12.8.1), where $f, h \in \mathbf{A}(U)$ and $h(x) \neq 0$. It follows that u_x is completely determined by its values on the germs of functions which belong to $\mathbf{A}(U)$.

12.8.7 Lemma. *Let X, Y be varieties, $x \in X$, $y \in Y$ and $U = \mathrm{Spm}(A)$ (resp. $V = \mathrm{Spm}(B)$) an affine open subset of X (resp. Y) containing x (resp. y). Suppose further that there is a local homomorphism $\varphi : \mathcal{O}_{Y,y} \to \mathcal{O}_{X,x}$.*

(i) *There exist an affine open neighbourhood $U_0 \subset U$ of x, an affine open neighbourhood $V_0 \subset V$ of y and a morphism $u_0 : U_0 \to V_0$ such that $u_0(x) = y$ and φ is the local homomorphism $(u_0)_x$.*

(ii) *Let $U_1 \subset U$ (resp. $V_1 \subset V$) be an affine open neighbourhood of x (resp. y), $u_1 : U_1 \to V_1$ a morphism such that $u_1(x) = y$ and φ is the local homomorphism $(u_1)_x$. Then u_0 and u_1 coincide in a neighbourhood of x.*

Proof. (i) For $b \in B$, denote by $i_y(b)$ the canonical image of b in $\mathcal{O}_{Y,y}$.

Let $\{b_1, \ldots, b_m\}$ be a system of generators of B and $\alpha : \Bbbk[T_1, \ldots, T_m] \to B$ the homomorphism defined by $\alpha(T_i) = b_i$ for $1 \leqslant i \leqslant m$. Let $\beta = \varphi \circ i_y \circ \alpha$,

$\mathfrak{a} = \ker \beta$, $\{Q_1, \ldots, Q_r\}$ a system of generators of $\mathfrak{a}$, T_0 a new indeterminate, and $P_1, \ldots, P_r$ homogeneous polynomials of degree h, such that for $1 \leqslant j \leqslant r$,

$$Q_j(T_1/T_0, \ldots, T_m/T_0) = P_j(T_0, T_1, \ldots, T_m)/T_0^h.$$

Each $\varphi(i_y(b_j))$ is the germ of a regular function at x which can be written as a_j/s where $a_j, s \in A$ and $s(x) \neq 0$. Since the Q_j's are in $\mathfrak{a}$, the regular functions $Q_j(a_1/s, \ldots, a_m/s) = P_j(s, a_1, \ldots, a_m)/s^h$ are identically zero in a neighbourhood of x. We deduce therefore that there exists $t \in A$ verifying $t(x) \neq 0$ and $P_j(ts, ta_1, \ldots, ta_m) = 0$ in A for $1 \leqslant j \leqslant r$.

Let $\rho : \Bbbk[T_1, \ldots, T_m] \rightarrow A_{ts}$ be the homomorphism such that $\rho(T_i) = (ta_i)/(ts)$, $1 \leqslant i \leqslant m$. We have $\rho(\mathfrak{a}) = \{0\}$, so $\rho(\ker \alpha) = \{0\}$. Thus there exists a homomorphism $\gamma_0 : B \rightarrow A_{ts}$ such that $\rho = \gamma_0 \circ \alpha$; so $\gamma_0(b_i) = (ta_i)/(ts)$ for $1 \leqslant i \leqslant m$. Finally, let $u_0 = \mathrm{Spm}(\gamma_0) : U_0 = D(st) \rightarrow V$ be the associated morphism.

If $i_x : A_{ts} = \Gamma(D(ts), \mathcal{O}_X) \rightarrow \mathcal{O}_{X,x}$ is the canonical homomorphism, then $i_x \circ \gamma_0 = \varphi \circ i_y$. It follows that, using the notations of 12.8.3,

$$\gamma_0^{-1}(\mathcal{I}(\{x\})) = \gamma_0^{-1}(i_x^{-1}(\mathfrak{m}_{X,x})) = i_y^{-1}(\mathfrak{m}_{Y,y}) = \mathcal{I}(\{y\}).$$

Hence φ is the local homomorphism $(u_0)_x$.

(ii) Replace U_0, U_1 (resp. V_0, V_1) by an affine open subset contained in $U_0 \cap U_1$ (resp. $V_0 \cap V_1$), we may assume that $U = U_0 = U_1$ (resp. $V = V_0 = V_1$). Let $\gamma_0 = \mathbf{A}(u_0)$ and $\gamma_1 = \mathbf{A}(u_1)$. For $1 \leqslant i \leqslant m$, $\gamma_0(b_i)$ and $\gamma_1(b_i)$ are regular functions in a neighbourhood of x having the same germ at x. So there exists $s \in A$ verifying $s(x) \neq 0$ such that the restriction of $\gamma_0(b_i)$ and $\gamma_1(b_i)$ to $D(s)$ are equal. We deduce therefore that $u_0|_{D(s)} = u_1|_{D(s)}$. $\square$

12.8.8 Proposition. *Let X, Y be varieties, $x \in X$ and $y \in Y$. Suppose further that there is a $\Bbbk$-algebra isomorphism $\varphi : \mathcal{O}_{Y,y} \rightarrow \mathcal{O}_{X,x}$.*

(i) There exist an open neighbourhood $U_0 \subset X$ of x, an open neighbourhood $V_0 \subset Y$ of y, and an isomorphism $u : U_0 \rightarrow V_0$ such that $u(x) = y$.

(ii) Let $U_1 \subset X$ (resp. $V_1 \subset Y$) be an open neighbourhood of x (resp. y) and $u_1 : U_1 \rightarrow V_1$ an isomorphism. Suppose that the isomorphisms u_x and $(u_1)_x$ coincide. Then there exists an open neighbourhood $U \subset U_0 \cap U_1$ of x such that $u|_U = u_1|_U$.

Proof. By 12.8.7, it suffices to prove (i). Let ψ be the inverse isomorphism of φ. Again by 12.8.7, we can find affine open neighbourhoods U_0 of x, V_0 of y and a morphism v from V_0 to an open subset of X such that φ and ψ are the local homomorphisms u_x and v_y induced by u and v. Replacing U_0 by an affine open subset, we may assume that $u(U_0) \subset V_0$. Then $v \circ u$ is a morphism from U_0 to an open neighbourhood of x such that $(v \circ u)_x$ is the identity map of $\mathcal{O}_{X,x}$. By 12.8.7, $(v \circ u)(z) = z$ for all z in an open neighbourhood of x.

The same arguments show that there exist affine open neighbourhoods U_1 of x, V_1 of y such that $u|_{U_1}$ is a morphism from U_1 to V_1, $v|_{V_1}$ is a morphism from V_1 to U_1 and $(u \circ v)(z) = z$ for all $z \in V_1$. Thus our result is proved. $\square$

Remark. We see from 12.8.8 that the local ring $\mathcal{O}_{X,x}$ determines a neighbourhood of x in X up to an isomorphism.

References

- [5], [19], [22], [26], [37], [40], [78].

13

Projective varieties

We attach to a projective space a natural structure of an irreducible algebraic variety. More generally, we define the notion of projective varieties, and discuss properties of these varieties. In particular, we introduce complete varieties. Certain properties of complete varieties are analogue to those of compact topological spaces.

13.1 Projective spaces

13.1.1 Let n be a non-zero positive integer. We define an equivalence relation $\mathcal{R}$ on $\Bbbk^{n+1} \setminus \{0\}$ by $x\mathcal{R}y$ if and only if $y = \lambda x$ for some $\lambda \in \Bbbk \setminus \{0\}$. The quotient space $(\Bbbk^{n+1} \setminus \{0\})/\mathcal{R}$ is called the *projective space of dimension* n, and we shall denote it by $\mathbb{P}_n(\Bbbk)$ or $\mathbb{P}_n$.

By definition, $\mathbb{P}_n$ is the set of 1-dimensional vector subspaces of $\Bbbk^{n+1}$.

Let $z \in \mathbb{P}_n$, its *homogeneous coordinates* are the coordinates $x_0, x_1, \ldots, x_n$ of a point x in $\Bbbk^{n+1}$ whose class is z. Homogeneous coordinates are unique up to multiplication by a non-zero scalar. We shall denote by

$$\pi : \Bbbk^{n+1} \setminus \{0\} \to \mathbb{P}_n$$

the canonical surjection.

13.1.2 We shall call a subset C in $\Bbbk^{n+1} \setminus \{0\}$ a *cone* if $\lambda x \in C$ for all $x \in C$ and for all $\lambda \in \Bbbk \setminus \{0\}$, or equivalently,

$$C = \pi^{-1}(\pi(C)).$$

The union of $\{0\}$ and a cone C in $\Bbbk^{n+1} \setminus \{0\}$ shall be called the *affine cone over* $\pi(C)$ in $\Bbbk^{n+1}$.

Lemma. *Let C be a non-empty closed cone in $\Bbbk^{n+1} \setminus \{0\}$. Then $C \cup \{0\}$ is the closure $\overline{C}$ of C in $\Bbbk^{n+1}$.*

Proof. Let D be a 1-dimensional subspace of $\Bbbk^{n+1}$. Then $0 \in \overline{D \setminus \{0\}}$. Since C is non-empty, $C \cup \{0\} \subset \overline{C}$. Now C is closed, so $C = \overline{C} \cap (\Bbbk^{n+1} \setminus \{0\})$. Hence $\overline{C} = C \cup \{0\}$. $\square$

13.1.3 By 13.1.2, a closed cone in $\Bbbk^{n+1}\setminus\{0\}$ is the intersection of $\Bbbk^{n+1}\setminus\{0\}$ and a closed affine cone in $\Bbbk^{n+1}$.

13.1.4 Let C be a closed affine cone in $\Bbbk^{n+1}$ and $\mathfrak{a} = \mathcal{I}(C)$ the associated radical ideal in $\Bbbk[T_0,\dots,T_n]$. Let $P \in \mathfrak{a}$. Since C is a cone, the condition $P(x_0,\dots,x_n) = 0$ implies that $P(\lambda x_0,\dots,\lambda x_n) = 0$ for all $\lambda \in \Bbbk$. It follows that all the homogeneous components of P are in $\mathfrak{a}$, and so $\mathfrak{a}$ is a graded ideal.

Conversely, given a graded ideal $\mathfrak{a} \neq \Bbbk[T_0,\dots,T_n]$, the set $\mathcal{V}(\mathfrak{a})$ is an affine cone in $\Bbbk^{n+1}$. The only maximal graded ideal in $\Bbbk[T_0,\dots,T_n]$ is the ideal $\mathfrak{m}_0$ generated by $T_0,\dots,T_n$. It follows that for any graded ideal $\mathfrak{a} \neq \Bbbk[T_0,\dots,T_n]$,

$$\mathcal{V}(\mathfrak{a}) = \{0\} \Leftrightarrow \sqrt{\mathfrak{a}} = \mathfrak{m}_0.$$

By 7.2.7, we see that the map $\mathfrak{a} \mapsto \mathcal{V}(\mathfrak{a})$ induces a bijection between the set of graded radical ideals of $\Bbbk[T_0,\dots,T_n]$ which do not contain $\mathfrak{m}_0$ and the set of closed affine cones of $\Bbbk^{n+1}$ which are not reduced to $\{0\}$.

13.2 Projective spaces and varieties

13.2.1 For $0 \leqslant i \leqslant n$, let us denote by H_i^0 (resp. H_i^1) the hyperplane (resp. affine hyperplane) of $\Bbbk^{n+1}$ defined by $x_i = 0$ (resp. $x_i = 1$).

Let U_i be the set of 1-dimensional subspaces of $\Bbbk^{n+1}$ which are not contained in H_i^0. Thus U_i is the set of points z in $\mathbb{P}_n$ whose homogeneous coordinates $(x_0,\dots,x_n)$ verify $x_i \neq 0$. Now any 1-dimensional subspace in U_i intersects H_i^1 at a unique point and this induces a bijection φ_i between U_i and H_i^1. Namely, φ_i sends the point z of $U_i \subset \mathbb{P}_n$ with homogeneous coordinates $(x_0,\dots,x_n)$ to the point

$$\left(\frac{x_0}{x_i},\dots,\frac{x_{i-1}}{x_i},1,\frac{x_{i+1}}{x_i},\dots,\frac{x_n}{x_i}\right).$$

It follows clearly that:

$$\mathbb{P}_n = U_0 \cup U_1 \cup \cdots \cup U_n.$$

13.2.2 For $i,j \in \{0,1,\dots,n\}$, $i \neq j$, $U_i \cap U_j$ is the set of points whose homogeneous coordinates $(x_0,\dots,x_n)$ verify $x_i \neq 0$ and $x_j \neq 0$. Such a point corresponds to the points:

$$\left(\frac{x_0}{x_i},\dots,\frac{x_{i-1}}{x_i},1,\frac{x_{i+1}}{x_i},\dots,\frac{x_n}{x_i}\right) \text{ and } \left(\frac{x_0}{x_j},\dots,\frac{x_{j-1}}{x_j},1,\frac{x_{j+1}}{x_j},\dots,\frac{x_n}{x_j}\right)$$

in the hyperplanes H_i^1 and H_j^1 respectively.

Set $H_{ij}^1 = H_i^1 \setminus H_j^0$ and $H_{ji}^1 = H_j^1 \setminus H_i^0$. We define the bijection $p_{ji} : H_{ij}^1 \to H_{ji}^1$ by:

$$(1) \quad (x_0,\dots,x_{i-1},1,x_{i+1},\dots,x_n) \mapsto \left(\frac{x_0}{x_j},\dots,\frac{x_{i-1}}{x_j},\frac{1}{x_j},\frac{x_{i+1}}{x_j},\dots,\frac{x_n}{x_j}\right).$$

Let $z \in U_i \cap U_j \cap U_l$, then $\varphi_i(z)$ is contained in $H^1_{ij} \cap H^1_{il}$. We verify readily that $\varphi_j(z) = p_{ji}(\varphi_i(z))$ and $\varphi_l(z) = p_{li}(\varphi_i(z))$. Hence, for any $x \in H^1_{ij} \cap H^1_{il}$:

$$(2) \qquad\qquad p_{li}(x) = p_{lj}(p_{ji}(x)).$$

Let $H^1_{ii} = H^1_i$ and $p_{ii} = \mathrm{id}_{H^1_i}$. Then we have shown that $\mathbb{P}_n$ can be considered as the set obtained by gluing the sets H^1_i, for $0 \leqslant i \leqslant n$, along the H^1_{ij}'s via the bijections p_{ji} (1.5.1).

13.2.3 Using the preceding consideration, we can define a structure of algebraic prevariety on $\mathbb{P}_n$.

The set H^1_i is an affine variety, and H^1_{ij} is an open subvariety of H^1_i ($H^1_{ij} = D(\mathrm{pr}_j) \cap H^1_i$ where $\mathrm{pr}_j : \Bbbk^{n+1} \to \Bbbk$ denotes the canonical projection $(x_0, \ldots, x_n) \mapsto x_j$).

Since regular functions on H^1_i are restrictions of polynomial functions on $\Bbbk^{n+1}$, it follows from 11.7.5, 12.1.1 that for $i \neq j$, the algebra of regular functions on H^1_{ij} is generated by the functions:

$$(x_0, \ldots, x_n) \mapsto \frac{x_l}{x_j},$$

for $0 \leqslant l \leqslant n$. We conclude that p_{ji} is an isomorphism between the affine varieties H^1_{ij} and H^1_{ji}.

Now equation (2) above is condition c) of 12.6.1. Hence we can endow $\mathbb{P}_n$ with a structure of algebraic prevariety over $\Bbbk$. From now on, when we consider the projective space as a prevariety, we shall mean the above structure.

The main point of the preceding discussion is that the maps φ_i of 13.2.1 allow us to identify U_i with H^1_i, and that we use these maps to define a structure of prevariety on $\mathbb{P}_n$ where the U_i's are affine open subsets.

13.2.4 For $0 \leqslant i \leqslant n$, the map $\psi_i : H^1_i \to \Bbbk^n$

$$(x_0, \ldots, x_{i-1}, 1, x_{i+1}, \ldots, x_n) \mapsto (x_0, \ldots, x_{i-1}, x_{i+1}, \ldots, x_n)$$

is an isomorphism of varieties.

Let $z \in U_i$, with homogeneous coordinates $(x_0, \ldots, x_n)$. The map

$$\psi_i \circ \varphi_i : U_i \to \Bbbk^n \ , \ z \mapsto \left(\frac{x_0}{x_i}, \ldots, \frac{x_{i-1}}{x_i}, \frac{x_{i+1}}{x_i}, \ldots, \frac{x_n}{x_i} \right)$$

is an isomorphism of U_i onto $\Bbbk^n$. Thus regular functions on U_i are of the form:

$$z \mapsto P \left(\frac{x_0}{x_i}, \ldots, \frac{x_{i-1}}{x_i}, \frac{x_{i+1}}{x_i}, \ldots, \frac{x_n}{x_i} \right)$$

where $P \in \Bbbk[T_0, \ldots, T_{i-1}, T_{i+1}, \ldots, T_n]$. Or equivalently, they are of the form:

$$z \mapsto \frac{R(x_0, \ldots, x_n)}{x_i^m}$$

where $m \in \mathbb{N}$ and $R \in \mathbb{k}[T_0, \ldots, T_n]$ is homogeneous of degree m.

The algebra $\mathbf{A}(U_i)$ is therefore generated by the $\dfrac{x_l}{x_i}$, where $0 \leqslant l \leqslant n$. The intersection $U_i \cap U_j$, $i \neq j$, corresponds via the isomorphism $\psi_i \circ \varphi_i$ to the affine open subset $D(T_j)$ of $\mathbb{k}^n$. From this, we deduce that $\mathbf{A}(U_i \cap U_j)$ is generated by the functions

$$\frac{x_l}{x_i} \text{ for } 0 \leqslant l \leqslant n \text{ and } \frac{x_i}{x_j} = \frac{1}{(x_j/x_i)}$$

on $U_i \cap U_j$.

13.2.5 For $0 \leqslant i \leqslant n$, denote by $\theta_i : D(T_i) \to H_i^1 \times (\mathbb{k} \setminus \{0\})$, the map

$$(x_0, \ldots, x_n) \mapsto \left(\left(\frac{x_0}{x_i}, \ldots, \frac{x_{i-1}}{x_i}, 1, \frac{x_{i+1}}{x_i}, \ldots, \frac{x_n}{x_i} \right), x_i \right).$$

By 12.1.1 and 12.4.3 (ii), θ_i is a bijective morphism. Its inverse

$$((y_0, \ldots, y_{i-1}, 1, y_{i+1}, \ldots, y_n), y_i) \mapsto (y_0 y_i, \ldots, y_{i-1} y_i, y_i, y_{i+1} y_i, \ldots, y_n y_i)$$

is clearly a morphism by the definition of regular functions on $D(T_i)$ and $H_i^1 \times (\mathbb{k} \setminus \{0\})$. Hence θ_i is an isomorphism.

13.2.6 Proposition. *The projective space $\mathbb{P}_n$ is an irreducible variety.*

Proof. That $\mathbb{P}_n$ is a variety results from 13.2.4 and 12.5.7. Now U_i is irreducible and $U_i \cap U_j \neq \emptyset$ for all i, j. Hence $\mathbb{P}_n$ is irreducible (1.1.6). $\square$

13.2.7 Proposition. *The map π is a morphism of varieties.*

Proof. Since $\mathbb{k}^{n+1} \setminus \{0\} = D(T_0) \cup D(T_1) \cup \cdots \cup D(T_n)$, it suffices to prove that the restriction of π to $D(T_i)$, $0 \leqslant i \leqslant n$, is a morphism (9.5.5). Now the image of $D(T_i)$ via π is the affine open subset U_i. Hence by 9.5.4 and 12.2.8, it suffices to show that the map $\rho_i : D(T_i) \to U_i$ induced by the restriction of π, is a morphism. But we have seen in 13.2.4 that regular functions on U_i are of the form:

$$z \mapsto \frac{R(x_0, \ldots, x_n)}{x_i^m}$$

where $z \in U_i$ with homogeneous coordinates $(x_0, \ldots, x_n)$, $R \in \mathbb{k}[T_0, \ldots, T_n]$ homogeneous of degree m and $m \in \mathbb{N}$. So it is clear that if $f \in \mathbf{A}(U_i)$, then $f \circ \rho_i \in \mathbf{A}(D(T_i))$. The result now follows. $\square$

13.3 Cones and projective varieties

13.3.1 Definition. (i) *Any algebraic variety isomorphic to a closed sub-variety of the projective space $\mathbb{P}_n$ is called a* projective variety.

(ii) *An algebraic variety isomorphic to a (not necessarily closed) subvariety of $\mathbb{P}_n$ is called a* quasi-projective variety.

Remark. An affine or quasi-affine variety is a quasi-projective variety.

13.3.2 Proposition. *Two points in a projective variety X are contained in an affine open subset.*

Proof. A closed subvariety of an affine variety is affine, so we may assume that $X = \mathbb{P}_n$, $n \geqslant 1$.

Let $x, y \in \mathbb{P}_n$. There exists a hyperplane in $\mathbb{k}^{n+1}$ which does not contain the subspaces $\{0\} \cup \pi^{-1}(x)$ and $\{0\} \cup \pi^{-1}(y)$. By applying a linear transformation, we may further assume that this hyperplane is H_i^0. Hence x, y belong to U_i. $\square$

13.3.3 Proposition. *The map $C \mapsto \pi(C)$ is a bijection between the set of closed cones in $\mathbb{k}^{n+1} \setminus \{0\}$ and the set of closed subsets of $\mathbb{P}_n$. Furthermore, for $0 \leqslant i \leqslant n$, $C \cap \pi^{-1}(U_i)$ is isomorphic to the product $(\pi(C) \cap U_i) \times (\mathbb{k} \setminus \{0\})$.*

Proof. If F is a closed subset of $\mathbb{P}_n$, then 13.2.7 says that $\pi^{-1}(F)$ is a closed cone in $\mathbb{k}^{n+1} \setminus \{0\}$.

Conversely, given a closed cone C in $\mathbb{k}^{n+1} \setminus \{0\}$, we have by identifying U_i and H_i^1 (13.2.1):

$$\theta_i(C \cup D(T_i)) = (\pi(C) \cap U_i) \times (\mathbb{k} \setminus \{0\})$$

where θ_i, T_i are as in 13.2.5. Since $C \cap D(T_i)$ is closed in $D(T_i) = \pi^{-1}(U_i)$ and θ_i is an isomorphism of $D(T_i)$ onto $U_i \times (\mathbb{k} \setminus \{0\})$ (13.2.5), $\pi(C) \cap U_i$ is closed in U_i by 12.4.5 (i). Hence $\pi(C)$ is closed in $\mathbb{P}_n$. The last statement follows from 9.5.4 and 12.2.8. $\square$

13.3.4 Corollary. (i) *The topology on $\mathbb{P}_n$ is the quotient topology of the topology on $\mathbb{k}^{n+1} \setminus \{0\}$ with respect to the equivalence relation $\pi(x) = \pi(x')$.*

(ii) *The map π is open.*

Proof. (i) By 13.3.3, closed subsets of $\mathbb{P}_n$ are images of closed subsets saturated by the equivalence relation $\pi(x) = \pi(x')$. Hence we have our result.

(ii) Let U be an open subset of $\mathbb{k}^{n+1} \setminus \{0\}$ and

$$V = \bigcup_{\lambda \in \mathbb{k} \setminus \{0\}} \lambda U = \pi^{-1}(\pi(U)).$$

Since $x \mapsto \lambda x$ is an automorphism of the variety $\mathbb{k}^{n+1} \setminus \{0\}$, it follows that V is open in $\mathbb{k}^{n+1} \setminus \{0\}$. So (i) implies that π is open. $\square$

13.3.5 It follows from 13.1.4 and 13.3.3 that any closed subvariety of $\mathbb{P}_n$ is defined by a system of equations:

$$(3) \qquad\qquad P_i(x_0, \ldots, x_n) = 0$$

where $(x_0, \ldots, x_n)$ are homogeneous coordinates of its points and the P_i's are homogeneous polynomials in $\Bbbk[T_0, \ldots, T_n]$. A closed subvariety defined by a single non-constant homogeneous polynomial (resp. of degree 1) shall be called a *hypersurface* (resp. *hyperplane*) of $\mathbb{P}_n$. Thus a hyperplane of $\mathbb{P}_n$ is the image $\pi(H \setminus \{0\})$ where H is a hyperplane in $\Bbbk^{n+1}$. For example, the complement H_i of U_i is the hyperplane $\pi(H_i^0 \setminus \{0\})$.

13.3.6 Let C be a closed cone in $\Bbbk^{n+1} \setminus \{0\}$ and $X = \pi(C)$. By 13.1.2 and 13.1.4, $C = \mathcal{V}(\mathfrak{a}) \setminus \{0\}$ for some graded radical ideal $\mathfrak{a}$ of $\Bbbk[T_0, \ldots, T_n]$. We shall say that X is *defined by the graded ideal* $\mathfrak{a}$.

The intersection of X with U_i, identified as $\Bbbk^n$ as in 13.2.4, is a subvariety $\mathcal{V}(\mathfrak{a}_i)$ of $\Bbbk^n$ where $\mathfrak{a}_i$ is the ideal of $\Bbbk[T_0, \ldots, T_{i-1}, T_{i+1}, \ldots, T_n]$ generated by the polynomials $P(T_0, \ldots, T_{i-1}, 1, T_{i+1}, \ldots, T_n)$, $P \in \mathfrak{a}$.

Let $Q \in \Bbbk[T_0, \ldots, T_{i-1}, T_{i+1}, \ldots, T_n]$ be a polynomial of degree r. Then

$$P(T_0, \ldots, T_n) = T_i^r Q\left(\frac{T_0}{T_i}, \ldots, \frac{T_{i-1}}{T_i}, \frac{T_{i+1}}{T_i}, \ldots, \frac{T_n}{T_i}\right)$$

is a homogeneous polynomial in $\Bbbk[T_0, \ldots, T_n]$. So we have *homogenized* Q.

Now let $Z = \mathcal{V}(\mathfrak{b})$ be a closed subvariety of U_i, where $\mathfrak{b}$ is an ideal of $\Bbbk[T_0, \ldots, T_{i-1}, T_{i+1}, \ldots, T_n]$. The closure $\overline{Z}$ of Z in $\mathbb{P}_n$ is equal to $\pi(C)$ for some closed cone $C = \mathcal{V}(\mathfrak{a}) \setminus \{0\}$ where $\mathfrak{a}$ is a graded ideal of $\Bbbk[T_0, \ldots, T_n]$. It is easy to see that we can take $\mathfrak{a}$ to be the graded ideal obtained by homogenizing the elements of $\mathfrak{b}$. We can also verify easily that $Z = \overline{Z} \cap U_i$.

Conversely, let X be a closed irreducible subvariety of $\mathbb{P}_n$ such that X is not contained in the complement H_i of U_i. Then $X \cap U_i$ is a non-empty open subset of X. So $X = \overline{X \cap U_i}$.

13.3.7 Let $A = \Bbbk[T_0, \ldots, T_n]$ and $Q \in A$ a homogeneous polynomial of degree $d > 0$. Denote by $D_+(Q)$ the set of points of $\mathbb{P}_n$ whose homogeneous coordinates $(x_0, \ldots, x_n)$ verify $Q(x_0, \ldots, x_n) \neq 0$ (this is well-defined since Q is homogeneous). Thus $D_+(Q)$, being the complement of a hypersurface, is open in $\mathbb{P}_n$.

Lemma. *Any non-empty open subset U of $\mathbb{P}_n$ is of the form*

$$U = D_+(Q_1) \cup \cdots \cup D_+(Q_r)$$

where $r \in \mathbb{N}^*$ and $Q_1, \ldots, Q_r \in A$ are homogeneous and non-constant.

Proof. The closed subset $V = \mathbb{P}_n \setminus U$ is defined by a graded ideal $\mathfrak{a}$ of A (13.3.6). By 7.2.2, there exist homogeneous elements $Q_1, \ldots, Q_r \in A$ such that $\mathfrak{a} = AQ_1 + \cdots + AQ_r$. Clearly, we have $U = D_+(Q_1) \cup \cdots \cup D_+(Q_r)$. $\square$

13.3.8 Proposition. *Let u be a map from $\mathbb{P}_n$ to a prevariety X. The following conditions are equivalent:*

(i) *u is a morphism.*

(ii) *$u \circ \pi$ is a morphism from $\mathbb{k}^{n+1} \setminus \{0\}$ to X.*

Proof. (i) $\Rightarrow$ (ii) This is clear by 13.2.7.

(ii) $\Rightarrow$ (i) Suppose that $u \circ \pi$ is a morphism. To show that u is a morphism, it suffices to prove that $u|_{U_i}$ is a morphism for $0 \leqslant i \leqslant n$.

For $Q \in \mathbb{k}[T_0, \ldots, T_n]$, we set:

$$D'(Q) = \{x \in \mathbb{k}^{n+1} \setminus \{0\} \ ; \ Q(x) \neq 0\}.$$

In view of 11.7.2, the $D'(Q)$'s form a base for the topology on $\mathbb{k}^{n+1} \setminus \{0\}$.

If $x \in \pi^{-1}(U_i)$, then our hypothesis says that there exist an affine open subset V of X containing $u(\pi(x))$ and $Q \in \mathbb{k}[T_0, \ldots, T_n]$ verifying:

1) $x \in D'(Q)$ and $u(\pi(D'(Q))) \subset V$.

2) For all regular function f on V, the function $y \mapsto f(u(\pi(y)))$ is regular on $D'(T_i Q) = D'(Q) \cap \pi^{-1}(U_i)$.

Let $f \in \Gamma(V, \mathcal{O}_X)$ and g the function $y \mapsto f(u(\pi(y)))$ defined on $D'(T_i Q)$. Then there exist $r \in \mathbb{N}$ and a polynomial $R \in \mathbb{k}[T_0, \ldots, T_n]$ such that for $y = (y_0, \ldots, y_n) \in D'(T_i Q)$,

$$(4) \qquad g(y) = \frac{R(y_0, \ldots, y_n)}{y_i^r Q^r(y_0, \ldots, y_n)}.$$

Let $\lambda \in \mathbb{k} \setminus \{0\}$. There exists $S_\lambda \in \mathbb{k}[T_0, \ldots, T_n]$ such that

$$S_\lambda(y_0, \ldots, y_n) = Q\left(\frac{y_0}{\lambda}, \ldots, \frac{y_n}{\lambda}\right)$$

for all $(y_0, \ldots, y_n) \in \mathbb{k}^{n+1} \setminus \{0\}$. Set $W_\lambda = \lambda D'(T_i Q)$.

A point $y = (y_0, \ldots, y_n)$ belongs to W_λ if and only if $y_i \neq 0$ and $S_\lambda(y) \neq 0$. Hence, $W_\lambda = D(T_i S_\lambda)$. Let

$$h_\lambda : W_\lambda \to \mathbb{k} \ , \ y \mapsto g\left(\frac{y}{\lambda}\right).$$

There exists $R_\lambda \in \mathbb{k}[T_0, \ldots, T_n]$ such that if $y \in W_\lambda$, then:

$$h_\lambda(y) = \frac{R_\lambda(y_0, \ldots, y_n)}{y_i^r S_\lambda^r(y_0, \ldots, y_n)}.$$

We deduce that h_λ is a regular function on W_λ.

If $x \in W_\lambda \cap W_\mu$, then $x = \lambda y = \mu z$ for some $y, z \in D'(T_i Q)$. Thus $\pi(z) = \pi(y)$, and from the definition of g, we obtain that $h_\lambda(x) = h_\mu(x)$. It follows that g can be extended to a regular function ℓ on the open subset

$$U = \bigcup_{\lambda \in \mathbb{k} \setminus \{0\}} W_\lambda = \pi^{-1}(\pi(D'(Q)) \cap U_i),$$

such that $\ell(\lambda y) = \ell(y)$ for all $y \in U$ and $\lambda \in \mathbb{k} \setminus \{0\}$.

By 13.3.4 (ii), $\pi(D'(Q))$ is open in $\mathbb{P}_n$, so $\pi(D'(Q)) \cap U_i$ is also open in $\mathbb{P}_n$. By 13.3.7, there exist homogeneous polynomials $Q_1, \ldots, Q_r \in \mathbb{k}[T_0, \ldots, T_n]$ such that

$$\pi(D'(Q)) \cap U_i = D_+(Q_1) \cup \cdots \cup D_+(Q_r).$$

Replacing Q by one of the Q_i's, we can reduced to the case where Q is homogeneous. Then the function g on $D'(T_i Q)$ is invariant by the automorphisms $y \mapsto \lambda y$. This implies that the polynomial R in (4) is homogeneous. From the definition of regular functions on U_i (13.2.4), we conclude that the restriction of u to $\pi(D'(Q))$ is a morphism. Hence we have our result. $\quad\square$

13.3.9 Corollary. *A subvariety X of $\mathbb{P}_n$ is irreducible if and only if $\pi^{-1}(X)$ is an irreducible subvariety of $\mathbb{k}^{n+1} \setminus \{0\}$.*

Proof. Since $\pi(\pi^{-1}(X)) = X$, X is irreducible if $\pi^{-1}(X)$ is irreducible. Conversely, suppose that X is irreducible. If $x \in \mathbb{P}_n$, then $\pi^{-1}(x)$, being isomorphic to $\mathbb{k} \setminus \{0\}$, is irreducible. It follows from 1.1.7, 13.2.7 and 13.3.4 (ii) that $\pi^{-1}(X)$ is irreducible. $\quad\square$

13.3.10 Let $m, n \in \mathbb{N}$, $E = (\mathbb{k}^{n+1} \setminus \{0\}) \times (\mathbb{k}^{m+1} \setminus \{0\})$ and

$$\pi : \mathbb{k}^{n+1} \setminus \{0\} \to \mathbb{P}_n \; , \; \pi' : \mathbb{k}^{m+1} \setminus \{0\} \to \mathbb{P}_m$$

the canonical morphisms. Define

$$p : E \to \mathbb{P}_n \times \mathbb{P}_m \; , \; (x, x') \mapsto (\pi(x), \pi'(x')).$$

The map p is a morphism by 12.4.3. We show as in the previous paragraphs that:
- The map p is open.
- A subset F in $\mathbb{P}_n \times \mathbb{P}_m$ is irreducible if and only if $p^{-1}(F)$ is irreducible in E.
- Closed subsets of $\mathbb{P}_n \times \mathbb{P}_m$ are the subsets F such that $p^{-1}(F)$ is closed in E and invariant by the maps $(x, x') \mapsto (\lambda x, \lambda' x')$ where $\lambda, \lambda' \in \mathbb{k} \setminus \{0\}$.

13.3.11 Let $u : \mathbb{P}_n \to \mathbb{P}_m$ be a morphism of varieties. By 13.3.8, $v = u \circ \pi$ is a morphism of $\mathbb{k}^{n+1} \setminus \{0\}$ to $\mathbb{P}_m$. Denote by $z_0, \ldots, z_m$ the homogeneous coordinates of points in $\mathbb{P}_m$, V_j the affine open subset of $\mathbb{P}_m$ defined by $z_j \neq 0$.

For $0 \leqslant i \leqslant n$ and $x \in \pi^{-1}(U_i)$, the point $v(x)$ belongs to some V_j. The open subset $\pi^{-1}(U_i) \cap v^{-1}(V_j)$ is then a neighbourhood of x, invariant by the automorphisms $y \mapsto \lambda y$, for $\lambda \in \mathbb{k} \setminus \{0\}$. It follows from 13.3.4 (i) and 13.3.7 that it contains a neighbourhood W_x of x in $\mathbb{k}^{n+1} \setminus \{0\}$ defined by $Q(y_0, \ldots, y_n) \neq 0$, where Q is a homogeneous polynomial.

Let us identify canonically V_j with the set of points $(x_0, \ldots, x_m) \in \mathbb{k}^{m+1} \setminus \{0\}$ such that $x_j = 1$ (13.2.1). The coordinates of $v(y)$, for $y \in W_x$, are regular functions, invariant by the automorphisms $y \mapsto \lambda y$. Replacing Q by a suitable power, we may write

$$v(y) = \left(\frac{P_0(y)}{Q(y)}, \dots, \frac{P_{j-1}(y)}{Q(y)}, 1, \frac{P_{j+1}(y)}{Q(y)}, \dots, \frac{P_m(y)}{Q(y)} \right)$$

where the P_l's and Q are homogeneous polynomials of the same degree. Set $P_j = Q$, then if $y \in W_x$,

(5) $$u(\pi(y)) = \pi'(w(y))$$

with

(6) $$w(y) = (P_0(y), \dots, P_m(y)).$$

We may further assume that $P_0, \dots, P_m$ are relatively prime.

Proceed in the same manner with another point x'. Since $\Bbbk^{n+1} \setminus \{0\}$ is irreducible, $W_x \cap W_{x'}$ is non-empty and open. It follows that, multiply the P_l's by the non-zero scalar if necessary, we may suppose that (5) is verified in $W_{x'}$ with the same polynomials P_l. We deduce therefore that (5) is verified for all $y \in \Bbbk^{n+1} \setminus \{0\}$. Further, by construction, $P_0, \dots, P_n$ have no common zero in $\Bbbk^{n+1} \setminus \{0\}$.

Conversely, let $P_0, \dots, P_n$ be homogeneous polynomials of the same degree having no common zero in $\Bbbk^{n+1} \setminus \{0\}$. By 13.3.8, formulas (5) and (6) define a morphism from $\mathbb{P}_n$ to $\mathbb{P}_m$.

13.3.12 Proposition. *The only regular functions on $\mathbb{P}_n$, that is the elements of $\Gamma(\mathbb{P}_n, \mathcal{O}_{\mathbb{P}_n})$, are the constant functions.*

Proof. Such a function is a morphism from $\mathbb{P}_n$ to $\Bbbk \subset \mathbb{P}_1$. If $z_0, \dots, z_n$ are homogeneous coordinates of $z \in \mathbb{P}_n$, then 13.3.11 says that there exist homogeneous polynomials P, Q of the same degree such that

$$f(z) = \frac{P(z_0, \dots, z_n)}{Q(z_0, \dots, z_n)}$$

for any $z \in \mathbb{P}_n$ and $Q(z_0, \dots, z_n) \neq 0$ if $(z_0, \dots, z_n) \neq (0, \dots, 0)$. Since Q is homogeneous in at least two indeterminates, it is constant. Hence P is also constant. $\square$

13.3.13 Let us conserve the notations of 13.3.11, and identify $\Bbbk^n$ with U_0 and $\Bbbk^m$ with V_0. Let $u : \Bbbk^n \to \Bbbk^m$ be a morphism. Under what condition can we extend u to a morphism from $\mathbb{P}_n$ to $\mathbb{P}_m$? Note that if such an extension exists, it is unique by 12.5.6 and 13.2.6. Now we have from 11.5.1 that

$$u(x_1, \dots, x_n) = (Q_1(x_1, \dots, x_n), \dots Q_m(x_1, \dots, x_n))$$

where $Q_1, \dots, Q_m$ are polynomials. Let $r = \max\{\deg Q_i \; ; \; 1 \leqslant i \leqslant m \}$. Then the extension, if it exists, is necessary the morphism:

$$(x_0, x_1, \ldots, x_n) \mapsto \left(x_0^r, x_0^r Q_1 \left(\frac{x_1}{x_0}, \ldots, \frac{x_n}{x_0} \right), \ldots, x_0^r Q_m \left(\frac{x_1}{x_0}, \ldots, \frac{x_n}{x_0} \right) \right).$$

We deduce that the condition for the extension of u is that the degree r homogeneous components of the Q_i's are not simultaneously zero on $\Bbbk^n \setminus \{0\}$.

13.4 Complete varieties

13.4.1 Definition. *An algebraic variety X is said to be* complete *if for all algebraic variety Y, the projection* $\mathrm{pr}_2 : X \times Y \to Y$ *is a closed morphism.*

13.4.2 The subset $\{(x, y) \in \Bbbk \times \Bbbk \ ; \ xy = 1\} \subset \Bbbk \times \Bbbk$ is closed, but its image under pr_2 is $\Bbbk \setminus \{0\}$, which is not closed in $\Bbbk$. So the affine line $\Bbbk$ is not complete.

13.4.3 Proposition. (i) *A closed subvariety of a complete variety is complete.*

(ii) *The product of two complete varieties is complete.*

(iii) *Let X be a complete variety and u a morphism from X to a variety Y. Then $u(X)$ is a closed and complete subvariety in Y.*

Proof. (i) Let X be a complete variety, Y a closed subvariety of X and Z a variety. We know from 12.4.5 and 12.5.4 that $Y \times Z$ is a closed subvariety of $X \times Z$. Thus closed subsets of $Y \times Z$ are closed in $X \times Z$. It follows that the restriction to $Y \times Z$ of $\mathrm{pr}_2 : X \times Z \to Z$ is closed.

(ii) Let X, Y be complete varieties and Z a variety. Then $X \times Y$ is a variety (12.5.4). If F is a closed subset of $(X \times Y) \times Z = X \times (Y \times Z)$, its projection F' on $Y \times Z$ is closed, and the projection F'' of F' on Z is also closed. Since F'' is the projection of F on Z. We are done.

(iii) Let G_u be the graph of u. Since Y is a variety, G_u is closed in $X \times Y$ (12.5.4). It follows that $u(X)$, the projection of G_u on Y is closed in Y.

Let Z be a variety, F a closed subset in $u(X) \times Z$ and F' the projection of F on Z. Consider the following projections

$$\mathrm{pr}_{13} : X \times Y \times Z \to X \times Z \ , \ \mathrm{pr}_{23} : X \times Y \times Z \to Y \times Z$$

and set $F'' = \mathrm{pr}_{23}^{-1}(F) \cap (G_u \times Z)$. Since $u(X)$ is closed in Y and F is closed in $u(X) \times Z$, F is closed in $Y \times Z$. This implies that F'' is closed in $G_u \times Z$. But the restriction of pr_{13} to $G_u \times Z$ is an isomorphism to $X \times Z$ (12.5.2). Hence $\mathrm{pr}_{13}(F'')$ is closed in $X \times Z$. On the other hand, F' is the projection of $\mathrm{pr}_{13}(F'')$ on Z. So F' is closed in Z because X is complete. $\square$

13.4.4 Corollary. *A complete affine variety is a finite set.*

Proof. It suffices to prove that if X is a non-empty, irreducible and complete affine variety, then X is a singleton, or equivalently, $\mathbf{A}(X) = \Bbbk$.

Let $f \in \mathbf{A}(X)$, then by 13.4.3, $f(X)$ is a closed irreducible and complete subvariety of $\Bbbk$. Since $\Bbbk$ is not complete and proper closed subsets of $\Bbbk$ are finite subsets, $f(X)$ is a singleton, and the result follows. $\square$

13.4.5 Theorem. *A projective variety is complete.*

Proof. By 13.4.3 (i), it suffices to prove that $\mathbb{P}_n$ is complete. Let Y be a variety, $p : \mathbb{P}_n \times Y \to Y$ the canonical projection and Z a closed subset of $\mathbb{P}_n \times Y$. We need to show that $p(Z)$ is closed.

a) It suffices to show that for all affine open subset V of Y, $V \cap p(Z)$ is closed in V, that is $p(Z \cap p^{-1}(V))$ is closed in V. Now $Z \cap p^{-1}(V)$ is closed in $\mathbb{P}_n \times V$, so we are reduced to the case where $Y = \mathrm{Spm}(A)$ is affine. Set $B = A[T_0, \dots, T_n]$.

b) Using the notations of 13.2.1, $\mathbb{P}_n \times Y$ is the union of the affine open subsets $W_i = U_i \times Y$, $0 \leqslant i \leqslant n$. The algebra $A_i = \mathbf{A}(W_i)$ is given by:

$$A_i = \Bbbk \left[\frac{T_0}{T_i}, \dots, \frac{T_n}{T_i} \right] \otimes_\Bbbk A = A \left[\frac{T_0}{T_i}, \dots, \frac{T_n}{T_i} \right].$$

Denote by $\mathfrak{a}$ the graded ideal of B generated by the homogeneous polynomials Q such that, for $0 \leqslant i \leqslant n$,

$$Q \left(\frac{T_0}{T_i}, \dots, \frac{T_n}{T_i} \right) \in \mathcal{I}(Z \cap W_i).$$

Let B_r (resp. $\mathfrak{a}_r$) be the A-module of homogeneous polynomials of degree r in B (resp. $\mathfrak{a}$).

c) For $i \in \{0, \dots, n\}$ and $g \in \mathcal{I}(Z \cap W_i)$, we shall prove that there exist integers k, r such that $T_i^k g \in \mathfrak{a}_r$.

Since g is a polynomial in T_l / T_i, $0 \leqslant l \leqslant n$, for a sufficiently large m, $T_i^m g$ is a homogeneous polynomial of degree m in $T_0, \dots, T_n$. On the other hand, for any j, $(T_i^m / T_j^m)g$ is zero on $Z \cap W_i \cap W_j$, and $(T_i^{m+1} / T_j^{m+1})g$ is zero on $(Z \cap W_j) \setminus W_i$. Since this holds for any j, we have $T_i^{m+1} g \in \mathfrak{a}_{m+1}$.

d) Let $y_0 \in Y \setminus p(Z)$ and $\mathfrak{m} = \mathcal{I}(\{y_0\}) \in \mathrm{Spm}\, A$. To finish our proof, we need to show that there exists $f \in A$ verifying $y_0 \in D(f)$ and $D(f) \cap p(Z) = \emptyset$.

Now $Z \cap W_i$ and $U_i \times \{y_0\}$ are disjoint closed subsets of the affine open set W_i, and $\mathcal{I}(U_i \times \{y_0\}) = \mathfrak{m}A_i$, it follows from 11.1.4 and 11.1.6 that:

$$A_i = \mathfrak{m}A_i + \mathcal{I}(Z \cap W_i).$$

Thus there exist $g_i \in \mathcal{I}(Z \cap W_i)$, $\alpha_{ij} \in \mathfrak{m}$ and $g_{ij} \in A_i$ such that for all $0 \leqslant i \leqslant n$:

$$1 = g_i + \sum_j \alpha_{ij} g_{ij}.$$

Paragraph c) says that by multiplying the above equality by T_i^m with m suitably large, we obtain

$$T_i^m = h_i + \sum_j \alpha_{ij} h_{ij}$$

where $h_i \in \mathfrak{a}_m$ and $h_{ij} \in B_m$. This implies that for N suitably large, any monomial of degree N in $T_0, \dots, T_n$ belongs to $\mathfrak{a}_N + \mathfrak{m}B_N$, that is, $B_N = \mathfrak{a}_N + \mathfrak{m}B_N$.

Now apply 2.6.7 to the finitely generated A-module $B_N/\mathfrak{a}_N$, there exists $f \in A \setminus \mathfrak{m}$ such that $fB_N \subset \mathfrak{a}_N$. Consequently, $f(y_0) \neq 0$ and $f.T_i^N \in \mathfrak{a}_N$ for all i. Hence f is zero on $p(Z)$ as required. $\square$

13.4.6 Remarks. 1) Let X, Y be complete varieties and $u : X \to Y$ a morphism. Contrary to the case of compact spaces, u can be a bicontinuous bijection without being an isomorphism. For example, let $u : \mathbb{P}_1 \to \mathbb{P}_2$ be the map defined by:

$$(x, y) \mapsto (x^3, xy^2, y^3).$$

This is a morphism by 13.3.11. Its image Y is a closed subvariety of $\mathbb{P}_2$ (13.4.3 (iii) and 13.4.5). Denote by $v : \mathbb{P}_1 \to Y$ the induced morphism. We check easily that v is bijective. Again by using 13.4.3 and 13.4.5, we see that v is bicontinuous. But the image of the affine open subset U_0 of $\mathbb{P}_1$ is the algebraic set $\{(x, y) \in \mathbb{k}^2 \; ; \; x^3 - y^2 = 0\}$. So by 11.5.7, v is not an isomorphism.

2) There are complete varieties which are not projective varieties.

13.5 Products

13.5.1 Let $z \in \mathbb{P}_n$ (resp. $z' \in \mathbb{P}_m$) with homogeneous coordinates $(x_0, \dots, x_n)$ (resp. $(x_0', \dots, x_m')$). Denote by $s(z, z') \in \mathbb{P}_{nm+n+m}$ the point with homogeneous coordinates the $(n+1)(m+1)$ scalars $y_{(i,j)}$ where $y_{(i,j)} = x_i x_j'$ for $0 \leqslant i \leqslant n$, $0 \leqslant j \leqslant m$. Thus we have a map

$$s : \mathbb{P}_n \times \mathbb{P}_m \to \mathbb{P}_{nm+n+m}.$$

We shall show that s is a morphism, called the *Segre embedding*. More precisely, we have the following result:

Proposition. *The map s is a morphism. Its image S is a closed subvariety of $\mathbb{P}_{nm+n+m}$ and s induces an isomorphism of $\mathbb{P}_n \times \mathbb{P}_m$ onto S.*

Proof. Let U_i (resp. U_j') be the affine open subset of $\mathbb{P}_n$ (resp. $\mathbb{P}_m$) defined by $x_i \neq 0$ (resp. $x_j' \neq 0$) and let W_{ij} be the affine open subset of $\mathbb{P}_{nm+n+m}$ defined by $y_{(i,j)} \neq 0$. The restriction of s to $U_i \times U_j'$ (identified as $\mathbb{k}^n \times \mathbb{k}^m$) is clearly a morphism to W_{ij} (identified as $\mathbb{k}^{nm+n+m}$). Hence u is a morphism.

Now $\mathbb{P}_n \times \mathbb{P}_m$ is complete (13.4.3 and 13.4.5), so S is a closed subvariety of $\mathbb{P}_{nm+n+m}$ (13.4.3).

From the definition of s, it is straightforward to check that:
- $s^{-1}(W_{ij}) = U_i \times U_j'$.
- The restriction of s to each $U_i \times U_j'$ is injective. So u is injective.
- Let $t : W_{ij} \to U_i \times U_j'$ be the map defined by sending a point with homogeneous coordinates $y_{(l,h)}$ to the point (z, z') where z (resp. z') has homogeneous coordinates the $y_{l,j}$'s for $0 \leqslant l \leqslant n$ (resp. the $y_{i,h}$'s for $0 \leqslant h \leqslant m$). Then t is a morphism such that $t(s(z, z')) = (z, z')$ if $(z, z') \in U_i \times U_j'$. Hence we have our result. $\square$

13.5.2 Proposition. *The product of two projective varieties is projective.*

Proof. This is clear from 12.4.2 and 13.5.1. $\square$

13.5.3 Let $d \in \mathbb{N}^*$ and I the set of $\alpha = (\alpha(0), \dots, \alpha(n)) \in \mathbb{N}^{n+1}$ such that

$$|\alpha| = \alpha(0) + \cdots + \alpha(n) = d.$$

The cardinality of I is equal to $\binom{n+d}{d}$.

Let J be a set containing I and N its cardinality. By 13.3.13, we can define a morphism $v : \mathbb{P}_n \to \mathbb{P}_{N-1}$ by sending $z \in \mathbb{P}_n$ with homogeneous coordinates $(x_0, \dots, x_n)$ to the point of $\mathbb{P}_{N-1}$ with homogeneous coordinates given by the following system:

$$\begin{cases} x^\alpha = x_0^{\alpha(0)} \cdots x_n^{\alpha(n)} & \text{if } \alpha \in I, \\ P_h(x_0, \dots, x_n) & \text{if } h \in J \setminus I, \end{cases}$$

where the P_h's are homogeneous polynomials of degree d. Since $\mathbb{P}_n$ is complete, the image V of v is a closed subvariety of $\mathbb{P}_{N-1}$.

Proposition. *The morphism v induces an isomorphism of $\mathbb{P}_n$ onto the closed subvariety V of $\mathbb{P}_{N-1}$.*

Proof. For $0 \leqslant i \leqslant n$, denote by $\beta_i = (\beta(0), \dots, \beta(n)) \in I$ the elements such that $\beta(i) = d$ and $\beta(j) = 0$ if $j \neq i$. Let W_i be the affine open subset of $\mathbb{P}_{N-1}$ consisting of points whose homogeneous coordinate y_{β_i} is non-zero. If U_i is the affine open subset of $\mathbb{P}_n$ defined by $x_i \neq 0$, then

$$z \in U_i \Rightarrow v(z) \in W_i \ , \ z \notin U_i \Rightarrow v(z) \notin W_i.$$

We deduce that $v^{-1}(W_i) = U_i$. Now let $w : W_i \to \mathbb{P}_n$ be the map sending a point with homogeneous coordinates y_h, $h \in J$, to the point of $\mathbb{P}_n$ with homogeneous coordinates y_{γ_j}, $0 \leqslant j \leqslant n$, where $\gamma_i = \beta_i$ and if $i \neq j$, $\gamma_j \in I$ is such that:

$$\gamma_j(i) = d - 1 \ , \ \gamma_j(j) = 1 \ , \ \gamma_j(h) = 0 \ \text{ if } \ h \notin \{i, j\}.$$

The map w is a morphism from W_i to U_i and we verify easily that $w(v(z)) = z$ if $z \in U_i$. Hence the result follows. $\square$

13.5.4 Remarks. 1) In the case where $J = I$, the morphism v is called the *Veronese embedding of degree d*.

2) Let X be a hypersurface of $\mathbb{P}_n$ defined by the homogeneous polynomial P of degree $d > 0$. Using the notations of 13.5.3, we have

$$P = \sum_{\alpha \in L} a_\alpha T^\alpha$$

where $L \subset I$ and $a_\alpha \in \Bbbk \setminus \{0\}$ for $\alpha \in L$. If v is the Veronese embedding of degree d, then $v(X)$ is contained in the hyperplane of $\mathbb{P}_{N-1}$ defined by

$$\sum_{\alpha \in L} a_\alpha Y_\alpha = 0.$$

This is one of the reason why the Veronese embedding is useful: it allows us to reduce a problem concerning a hypersurface to a problem concerning a hyperplane.

13.5.5 The following result completes the one in 13.3.7.

Proposition. *Let $Q \in \Bbbk[T_0, \ldots, T_n]$ be a homogeneous polynomial of degree $d > 0$. The subset $D_+(Q)$ of $\mathbb{P}_n$ is an affine open subset.*

Proof. Let $N = \begin{pmatrix} n + d \\ d \end{pmatrix}$, $v : \mathbb{P}_n \to \mathbb{P}_N$ be defined as in 13.5.3 by

$$v(z) = ((x^\alpha)_{\alpha \in I}, Q(x_0, \ldots, x_n)),$$

and $V = v(\mathbb{P}_n)$ which is closed by 13.5.3.

Let D_{N+1} be the affine open subsets of $\mathbb{P}_N$ consisting of points such that the last homogeneous coordinate is non-zero. Since $V \cap D_{N+1}$ is closed in D_{N+1}, it is an affine open subset of V. But $V \cap D_{N+1} = v(D_+(Q))$. Hence the result follows since v is an isomorphism of $\mathbb{P}_n$ onto V. □

13.6 Grassmannian variety

13.6.1 Let E_r denote the $\Bbbk$-vector space $\bigwedge^r(\Bbbk^n)$. If $(e_1, \ldots, e_n)$ is the canonical basis of $\Bbbk^n$, then the canonical basis of E_r is, by definition, the $N = \begin{pmatrix} n \\ r \end{pmatrix}$ vectors

$$e_H = e_{i_1} \wedge e_{i_2} \wedge \cdots \wedge e_{i_r}$$

for any subset H of $\{1, \ldots, n\}$ consisting of r elements $i_1 < \cdots < i_r$. We have $E_0 = \Bbbk$, $E_1 = \Bbbk^n$ and $E_r = \{0\}$ if $r > n$.

13.6.2 Recall that if V is a subspace of $\Bbbk$, then $\bigwedge^r V$ can be identified canonically as a subspace of E_r.

Let $\mathcal{S}_{n,r}$ be the set of r-dimensional subspaces of $\Bbbk^n$. If $V \in \mathcal{S}_{n,r}$, then $\dim \bigwedge^r V = 1$, and so $\bigwedge^r V$ is a line in E_r, or a point in the projective space $\mathbb{P}(E_r) = \mathbb{P}_{N-1}$. Thus we have a map $\psi : \mathcal{S}_{n,r} \to \mathbb{P}_{N-1}$. Denote by $\mathbb{G}_{n,r}$ the image of ψ.

Note that ψ is injective. Indeed, if $V, W \in \mathcal{S}_{n,r}$ and $(v_1, \ldots, v_n)$ a basis of $\Bbbk^n$ such that $(v_1, \ldots, v_r)$ (resp. $(v_s, \ldots, v_{s+r-1})$) is a basis of V (resp. W). Then $\psi(V) = \psi(W)$ implies that $v_1 \wedge \cdots \wedge v_r$ and $v_s \wedge \cdots \wedge v_{s+r-1}$ are proportional. Hence $s = 1$ (13.6.1) and $V = W$.

13.6.3 Proposition. *The set $\mathbb{G}_{n,r}$ is closed in $\mathbb{P}_{N-1}$.*

Proof. Using the notations of 13.6.1, let W_H be the affine open subsets of $\mathbb{P}_{N-1}$ defined by $e_H \neq 0$. Then the N subsets W_H cover E_r. It suffices therefore to prove that $\mathbb{G}_{n,r} \cap W_H$ is closed in W_H for all H.

Take for example $H = \{1, \ldots, r\}$, and write $W = W_H$. Let F_0 (resp. F_0') be the subspace generated by $e_1, \ldots, e_r$ (resp. $e_{r+1}, \ldots, e_n$) and p the projection on F_0 with respect to the direct sum $\Bbbk^n = F_0 \oplus F_0'$. If $(v_1, \ldots, v_r)$ is a basis of $V \in \mathcal{S}_{n,r}$, then writing $v_i = p(v_i) + w_i$, $w_i \in F_0'$, we have

$$\psi(V) \in W \Leftrightarrow p(v_1) \wedge \cdots \wedge p(v_r) \neq 0,$$

which is equivalent to the condition that p induces an isomorphism of the vector spaces V and F_0. So if $\psi(V) \in W$, then V has a unique basis $(v_1, \ldots, v_r)$ of the form

$$(7) \qquad v_i = e_i + \sum_{j=r+1}^{n} a_{ij} e_j = e_i + u_i$$

for $1 \leqslant i \leqslant r$. Thus

$$v_1 \wedge \cdots \wedge v_r = e_1 \wedge \cdots \wedge e_r + \sum_{1 \leqslant j \leqslant r} (e_1 \wedge \cdots \wedge u_i \wedge \cdots \wedge e_r) + w,$$

where w is a linear combination of vectors e_H such that H contains at least 2 elements in $\{r+1, \ldots, n\}$. Moreover, it is clear that the coefficient of such a vector e_H is a polynomial P_H in the a_{ij}'s, and this polynomial does not depend on V. Finally, we have

$$e_1 \wedge \cdots \wedge u_i \wedge \cdots \wedge e_r = \sum_{j>r} \varepsilon a_{ij} (e_1 \wedge \cdots \wedge e_{i-1} \wedge e_{i+1} \wedge \cdots \wedge e_r \wedge e_j)$$

where $\varepsilon \in \{-1, 1\}$. Hence the a_{ij}'s determine completely the decomposition of $v_1 \wedge \cdots \wedge v_r$ in the basis $(e_H)_H$.

Conversely, given $r(n - r)$ scalars a_{ij} and define the v_i as in (7), the subspace V generated by the vectors v_i verifies $V \in \mathcal{S}_{n,r}$ and $\psi(V) \in W$.

Identifying W with $\Bbbk^{N-1}$, the preceding arguments show that $W \cap \mathbb{G}_{n,r}$ is isomorphic to the graph of the morphism

$$\Bbbk^{r(n-r)} \longrightarrow \Bbbk^{N-r(n-r)-1} \ , \ (a_{ij}) \mapsto (P_H(a_{ij})).$$

Hence $W \cap \mathbb{G}_{n,r}$ is closed in W. $\square$

13.6.4 Definition. *The closed subvariety $\mathbb{G}_{n,r}$ of $\mathbb{P}_{N-1}$ is called the Grassmannian variety of r-dimensional subspaces of $\Bbbk^n$.*

13.6.5 Remark. Let $U_H = W_H \cap \mathbb{G}_{n,r}$. The proof of 13.6.3 shows that U_H is an affine open subset of $\mathbb{G}_{n,r}$, which is isomorphic to $\Bbbk^{r(n-r)}$.

13.6.6 Proposition. *The Grassmannian variety $\mathbb{G}_{n,r}$ is irreducible.*

Proof. Being isomorphic to $\Bbbk^{r(n-r)}$, the affine open subsets U_H are irreducible. Since the U_H's cover $\mathbb{G}_{n,r}$, it suffices to show that $U_H \cap U_J \neq \emptyset$ for $H \neq J$. Let

$$H \cap J = \{\alpha_1, \ldots, \alpha_s\},$$
$$H \setminus (H \cap J) = \{\beta_1, \ldots, \beta_{r-s}\}, \quad J \setminus (H \cap J) = \{\gamma_1, \ldots, \gamma_{r-s}\},$$
$$V = \Bbbk e_{\alpha_1} \oplus \cdots \oplus \Bbbk e_{\alpha_s} \oplus \Bbbk(e_{\beta_1} + e_{\gamma_1}) \oplus \cdots \oplus \Bbbk(e_{\beta_{r-s}} + e_{\gamma_{r-s}}).$$

Then it is clear from the proof of 13.6.3 that $\psi(V) \in U_H \cap U_J$. $\quad\square$

13.6.7 A *flag* in $\Bbbk^n$ is a chain of subspaces of $\Bbbk^n$

$$V_1 \subset V_2 \subset \cdots \subset V_n = \Bbbk^n$$

such that $\dim V_i = i$ for $1 \leqslant i \leqslant n$. Denote by $\mathfrak{F}_n$ the set of flags of $\Bbbk^n$. Then $\mathfrak{F}_n$ can be identified naturally with a subset of $\mathbb{G} = \mathbb{G}_{n,1} \times \mathbb{G}_{n,2} \times \cdots \times \mathbb{G}_{n,n}$.

By 13.5.2, $\mathbb{G}$ has a canonical structure of projective variety. Using methods analogue to those in 13.6.3, we can prove that $\mathfrak{F}_n$ is closed in $\mathbb{G}$. Thus $\mathfrak{F}_n$ has a structure of projective variety. We shall call $\mathfrak{F}_n$, the *flag variety* of $\Bbbk^n$.

References and comments

- [5], [19], [22], [26], [37], [40], [78].

Being projective, Grassmannian varieties are complete varieties. This fact will prove to be very useful later on.

14

Dimension

The first section of this chapter gives an algebraic interpretation of the dimension of a variety as a topological space. We then use this to establish certain results concerning the dimension of algebraic varieties.

All the algebraic varieties considered in this chapter are defined over $\Bbbk$.

14.1 Dimension of varieties

14.1.1 Recall that the dimension of a topological space and the dimension of an algebra were defined in chapter 1 and chapter 2 respectively.

14.1.2 Proposition. *Let X be an affine variety.*
(i) *We have:*
$$\dim X = \dim \mathbf{A}(X).$$

(ii) *If X is irreducible, then:*
$$\dim X = \operatorname{tr\,deg}_{\Bbbk} \operatorname{Fract}(\mathbf{A}(X)).$$

Proof. (i) The map $\mathfrak{p} \mapsto \mathcal{V}(\mathfrak{p})$ from the set of prime ideals to the set of irreducible subsets of X is an inclusion-reversing bijection (11.1.4, 11.1.7 and 11.2.3). It follows from the definition of $\dim X$ (1.2.2) and the definition of $\dim \mathbf{A}(X)$ (2.2.3) that $\dim X = \dim \mathbf{A}(X)$.

(ii) Recall from 11.2.3 that if X is an affine variety, then the algebra $\mathbf{A}(X)$ of regular functions on X is an integral domain if and only if X is irreducible. The result follows therefore from part (i) and 6.2.3. $\quad\square$

14.1.3 Theorem. *Let X be an irreducible variety and U a non-empty open subset of X. Then $\dim X = \dim U$ is finite and:*
$$\dim X = \operatorname{tr\,deg}_{\Bbbk} \mathbf{R}(X).$$

Proof. Let us first suppose that X is affine. Let U, V be non-empty affine open subsets of X, then $U \cap V \neq \emptyset$ because X is irreducible. So there exists $f \in \mathbf{A}(X)$ such that $D(f) \neq \emptyset$ and $D(f) \subset U \cap V$ (11.7.2).

It follows from 1.2.3 that:

$$\dim D(f) \leqslant \dim U \leqslant \dim X \ , \ \dim D(f) \leqslant \dim V \leqslant \dim X.$$

Hence by remark 1 of 11.7.5, $\mathrm{Fract}(\mathbf{A}(X)) = \mathrm{Fract}(\mathbf{A}(D(f)))$, and 14.1.2 (ii) implies that
$$\dim D(f) = \dim U = \dim V = \dim X.$$

Now any non-empty open subset W of X contains an affine open subset U of X. Using the preceding arguments and 1.2.3, we have $\dim W = \dim X$.

Let us now suppose that X is not affine. The preceding arguments show that all the affine open subsets of X have the same dimension. Since X is the union of a finite number of affine open subsets, $\dim X = \dim U$ for any affine open subset U of X (1.2.3 (iii)). Finally any non-empty open subset W of X contains a non-empty affine open subset. Hence $\dim W = \dim X$ (1.2.3 (i)).

The equality $\dim X = \mathrm{tr} \deg_{\Bbbk} \mathbf{R}(X)$ follows immediately from 12.7.6 and 14.1.2. $\square$

14.1.4 Corollary. *Let X be a non-empty algebraic variety. Then its dimension is finite and if $X_1, \ldots, X_n$ are the irreducible components of X, we have:*
$$\dim X = \max\{\dim X_i \ ; \ 1 \leqslant i \leqslant n \}.$$

Proof. This follows from 1.2.3 (i), 1.3.5 and 14.1.3. $\square$

14.1.5 Examples. 1) A variety of dimension zero is a finite set (since an irreducible variety of dimension zero is clearly a singleton).

2) We have $\dim \Bbbk^n = n$ (6.1.4 and 14.1.2).

3) The projective space $\mathbb{P}_n$ is irreducible (13.2.6) and contains an open subset isomorphic to $\Bbbk^n$ (13.2.4). Hence $\dim \mathbb{P}_n = n$.

4) The Grassmannian variety $\mathbb{G}_{n,r}$ is irreducible (13.6.6) and contains an open subset isomorphic to $\Bbbk^{r(n-r)}$ (13.6.5). Hence $\dim \mathbb{G}_{n,r} = r(n-r)$.

14.1.6 Proposition. *Let Y be a non-empty subvariety of a variety X.*

(i) *We have $\dim Y \leqslant \dim X$.*

(ii) *If X is irreducible and Y is closed, then $\dim Y = \dim X$ if and only if $Y = X$.*

(iii) *We have $\dim Y = \dim \overline{Y}$.*

(iv) *If X is irreducible, then $\dim Y = \dim X$ if and only if Y is open.*

Proof. Parts (i) and (ii) have been proved in 1.2.3.

(iii) Let $Y_1, \ldots, Y_n$ be the irreducible components of Y. Then by 1.1.14, $\overline{Y_1}, \ldots, \overline{Y_n}$ are the irreducible components of $\overline{Y}$. Since Y_i is closed in Y (1.1.5), it is locally closed in X. So Y_i is open in $\overline{Y_i}$ (1.4.1), and $\dim Y_i = \dim \overline{Y_i}$ (14.1.3). The result follows therefore from 14.1.4.

(iv) Let us write $Y = F \cap U$ where U (resp. F) is an open (resp. closed) subset of X. By 1.2.3, $\dim Y \leqslant \dim F \leqslant \dim X$. It is now clear by part (ii) and 14.1.3 that $\dim X = \dim Y$ if and only if $Y = U$ is open. $\square$

14.1.7 Proposition. *Let X, Y be irreducible varieties. Then:*

$$\dim(X \times Y) = \dim X + \dim Y.$$

Proof. By 12.4.5, $X \times Y$ is irreducible. Further, if $U \subset X$ and $V \subset Y$ are affine open subsets, then $U \times V$ is an affine open subset of $X \times Y$. We may therefore assume that $X = \mathrm{Spm}(A)$ and $Y = \mathrm{Spm}(B)$ are affine. Let $m = \dim X$ and $n = \dim Y$.

By 6.2.2 and 14.1.2, A (resp. B) contains a subalgebra A_0 (resp. B_0) isomorphic to $\Bbbk[T_1, \ldots, T_m]$ (resp. $\Bbbk[T_{m+1}, \ldots, T_{m+n}]$), and A (resp. B) is generated over A_0 (resp. B_0) by a finite number of elements $f_1, \ldots, f_r$ (resp. $g_1, \ldots, g_s$) algebraic over E_0 (resp. F_0), the quotient field of A_0 (resp. B_0).

Since $\mathbf{A}(X \times Y) = A \otimes_{\Bbbk} B$, it contains the subalgebra $A_0 \otimes B_0 \simeq \Bbbk[T_1, \ldots, T_{n+m}]$ whose quotient field L_0 has transcendence degree $n + m$ over $\Bbbk$. Now, the quotient field L of $A \otimes_{\Bbbk} B$ is an extension of L_0 generated by $f_1, \ldots, f_r, g_1, \ldots, g_s$. So the degree of transcendence of L over $\Bbbk$ is $n + m$. $\square$

14.1.8 Let us use the notations $\pi : \Bbbk^{n+1} \setminus \{0\} \to \mathbb{P}_n$ and U_i, $0 \leqslant i \leqslant n$, of 13.1.1 and 13.2.1.

Corollary. *If X is a closed and irreducible subvariety of $\mathbb{P}_n$, then*

$$\dim \pi^{-1}(X) = 1 + \dim X.$$

Proof. If $X \cap U_i \neq \emptyset$, then by 13.3.3, $\pi^{-1}(X \cap U_i)$ is isomorphic to the product $(X \cap U_i) \times (\Bbbk \setminus \{0\})$. Thus the result follows from 14.1.7. $\square$

14.1.9 Remarks. 1) Let Y be a closed and irreducible subvariety of an irreducible variety X. The *codimension* of Y in X, denoted by $\mathrm{codim}_X(Y)$ or simply $\mathrm{codim}(Y)$ when there is no confusion, is defined to be $\dim X - \dim Y$. Hence $\mathrm{codim}_X(Y) = 0$ if and only if $Y = X$ (14.1.6).

2) Let $X_1, \ldots, X_n$ be the irreducible components of a variety X. Then by 14.1.6 and 1.1.14, we have $\mathrm{codim}_{X_i}(X_i \cap X_j) \geqslant 1$ whenever $i \neq j$.

14.2 Dimension and the number of equations

14.2.1 Definition. *A non-empty variety is called of pure dimension or equidimensional if all of its irreducible components has the same dimension. We shall call an irreducible variety of dimension 1 (resp. 2) an algebraic curve (resp. algebraic surface) over $\Bbbk$.*

14.2.2 Let X be a variety. As in the affine case, if $f \in \Gamma(X, \mathcal{O}_X)$, we denote by $\mathcal{V}(f)$ the closed subvariety of X consisting of elements $x \in X$ such that $f(x) = 0$. If $f_1, \ldots, f_r \in \Gamma(X, \mathcal{O}_X)$, we set:

$$\mathcal{V}(f_1, \ldots, f_r) = \mathcal{V}(f_1) \cap \cdots \cap \mathcal{V}(f_r).$$

Lemma. *We have $\mathcal{V}(f) = \emptyset$ if and only if f is invertible in $\Gamma(X, \mathcal{O}_X)$.*

Proof. If f is invertible, then clearly $\mathcal{V}(f) = \emptyset$. Conversely, if $\mathcal{V}(f) = \emptyset$, then for any affine open subset $U \subset X$, $f|_U \in \Gamma(U, \mathcal{O}_X(U))$ verifies $\mathcal{V}(f|_U) = \emptyset$. Thus $f|_U$ is invertible with inverse, say g_U. If V is another affine open subset, then $g_U|_{U \cap V} = g_V|_{U \cap V}$ since it is the inverse of $f|_{U \cap V}$. So there exists an element $g \in \Gamma(X, \mathcal{O}_X)$ such that $g|_U = g_U$ for any affine open subset U. Hence $fg = 1$. $\square$

14.2.3 The following theorem, also called *Krull's Principal Ideal Theorem (or Hauptidealsatz)*, is a geometrical translation of 6.3.2.

Theorem. *Let X be an irreducible variety and f be a non-zero and non-invertible element of $\Gamma(X, \mathcal{O}_X)$. Then $\mathcal{V}(f)$ is a non-empty subvariety of X of pure dimension $\dim X - 1$.*

Proof. By 14.2.2, $\mathcal{V}(f) \neq \emptyset$. Let $Y_1, \ldots, Y_r$ be the irreducible components of $\mathcal{V}(f)$. There exists $y \in Y_1$ such that $y \notin Y_2 \cup \cdots \cup Y_r$ (1.1.13), and since $Y_2 \cup \cdots \cup Y_r$ is closed, there exists an affine open subset $U \subset X$ such that $y \in U$ and $U \cap \mathcal{V}(f) = U \cap Y_1$ is irreducible. By 14.1.3, $\dim U \cap Y_1 = \dim Y_1$ and $\dim U = \dim X$. So we may assume that $X = \mathrm{Spm}(A)$ is affine with A an integral domain, $f \in A$ is non-zero and non-invertible, and $\sqrt{Af}$ is a prime ideal. The result follows immediately from 6.2.6 and 6.3.2. $\square$

14.2.4 Remark. If X is not irreducible, then $\mathcal{V}(f)$ is not necessary of pure dimension (even if X is of pure dimension). For example, let $X = \mathcal{V}(T_1 T_2) \subset \Bbbk^2$ and f the restriction to X of the map $(t_1, t_2) \mapsto t_1(t_1 + t_2 + 1)$, then we check easily that $\mathcal{V}(f)$ is the disjoint union of a line and a point.

14.2.5 Corollary. *Let X be a closed irreducible subvariety of $\mathbb{P}_n$ such that $\dim X \geqslant 1$. Given any homogeneous polynomial $P \in \Bbbk[T_0, \ldots, T_n] \setminus \Bbbk$, the hypersurface $\mathcal{V}(P)$ defined by $P(x_0, \ldots, x_n) = 0$ has a non-empty intersection with X. If $\mathcal{V}(P)$ does not contain X, then $X \cap \mathcal{V}(P)$ is a subvariety of pure dimension $\dim X - 1$.*

Proof. Let $C = \pi^{-1}(X)$ where π is as in 13.1.1, and $\overline{C}$ the closure of C in $\Bbbk^{n+1}$. Then $\dim \overline{C} \geqslant 2$ (14.1.8) and $0 \in \overline{C}$ is a zero of P, so the restriction of P to $\overline{C}$ is non-invertible. Now let C' be the zero set of P in $\overline{C}$; it is an affine cone and a variety of pure dimension with $\dim C' \geqslant \dim X \geqslant 1$ (14.2.3). It follows that $C' \cap (\Bbbk^{n+1} \setminus \{0\}) \neq \emptyset$, which implies that $\pi(C') = \mathcal{V}(P) \cap X \neq \emptyset$. The last statement is a direct consequence of 13.3.3. $\square$

14.2.6 Proposition. *Let X be an irreducible variety and $f_1, \ldots, f_r$ be elements of $\Gamma(X, \mathcal{O}_X)$. Then for any irreducible component Y of $\mathcal{V}(f_1, \ldots, f_r)$, we have $\operatorname{codim}_X(Y) \leqslant r$.*

Proof. Induction on r. The case $r = 1$ is just 14.2.3. So let $r > 1$. There exists an irreducible component Y' of $\mathcal{V}(f_1, \ldots, f_{r-1})$ such that $Y \subset Y'$. So:

$$Y \subset Y' \cap \mathcal{V}(f_r) \subset \mathcal{V}(f_1, \ldots, f_r).$$

It follows that Y is an irreducible component of $Y' \cap \mathcal{V}(f_r)$. By induction hypothesis, $\operatorname{codim}_X(Y') \leqslant r - 1$. Now if $f_r|_{Y'} = 0$, then $Y = Y'$. Otherwise, $\dim Y = \dim Y' - 1$ by 14.2.3. In both cases, we have $\operatorname{codim}_X(Y) \leqslant r$. $\square$

14.2.7 Remark. In 14.2.6, irreducible components of $\mathcal{V}(f_1, \ldots, f_r)$ do not necessary have the same dimension, and it is possible that they all have codimension strictly less than r.

14.2.8 Proposition. *Let X be a closed and irreducible subvariety of $\mathbb{P}_n$ and $P_1, \ldots, P_r \in \Bbbk[T_0, \ldots, T_n]$ be non-constant homogeneous polynomials. We denote by $\mathcal{V}(P_1, \ldots, P_r)$ the set of points of X whose homogeneous coordinates are zeros of $P_1, \ldots, P_r$.*
   (i) *Any irreducible component Y of $\mathcal{V}(P_1, \ldots, P_r)$ verifies $\operatorname{codim}_X Y \leqslant r$.*
   (ii) *If $\dim X \geqslant r$, then $\mathcal{V}(P_1, \ldots, P_r)$ is non-empty.*

Proof. Part (i) follows from 14.2.6 applied to $\pi^{-1}(X)$, while (ii) follows from 14.2.5 and a simple induction. $\square$

14.2.9 Corollary. *If $m < n$, then the only morphisms from $\mathbb{P}_n$ to $\mathbb{P}_m$ are the constant maps.*

Proof. By 14.2.8 (ii), if $P_1, \ldots, P_r \in \Bbbk[T_0, \ldots, T_n]$ are non-constant homogeneous polynomials, with $r < n$, then $\mathcal{V}(P_1, \ldots, P_r) \neq \emptyset$. So the result is a consequence of the description of morphisms given in 13.3.11. $\square$

14.3 System of parameters

14.3.1 Proposition. *Let X be an irreducible variety and Y be a closed irreducible subvariety of codimension 1 in X.*
   (i) *There exist an affine open subset $U \subset X$ and $f \in \Gamma(U, \mathcal{O}_X) \setminus \{0\}$ such that $Y \cap U \neq \emptyset$ and $f|_{Y \cap U} = 0$.*
   (ii) *For any open subset $V \subset X$ and any $f \in \Gamma(V, \mathcal{O}_X) \setminus \{0\}$ verifying $Y \cap V \neq \emptyset$ and $f|_{Y \cap V} = 0$, $Y \cap V$ is an irreducible component of $\mathcal{V}(f)$.*

Proof. (i) Being irreducible, Y is non-empty, so there exists an affine open subset $U = \operatorname{Spm}(A)$ of X such that $Y \cap U \neq \emptyset$. Let $\mathfrak{a}$ be the radical ideal of A such that $Y \cap U = \mathcal{V}(\mathfrak{a})$. Then $\mathfrak{a} \neq \{0\}$ by 14.1.6 (iv). Thus any $f \in \mathfrak{a} \setminus \{0\}$ works.

(ii) The irreducibility of Y implies that $Y \cap V$ is irreducible for any open subset V such that $Y \cap V \neq \emptyset$. So the open subset $Y \cap V$ of Y is contained in an irreducible component W of $\mathcal{V}(f)$. By 14.1.6, we have:

$$\dim V = \dim X \, , \ \dim(Y \cap V) = \dim Y = \dim X - 1.$$

So we obtain by 14.2.3 that:

$$\dim V > \dim W \geqslant \dim(Y \cap V) = \dim V - 1.$$

This implies that $\dim W = \dim(Y \cap V)$, and since W and $Y \cap V$ are closed irreducible subsets of V, we have $W = Y \cap V$ (14.1.6). $\square$

14.3.2 Corollary. *Let X be an irreducible variety, $\mathfrak{F}$ the set of closed irreducible subvarieties of X which are distinct from X, and $\mathfrak{F}^*$ the set of maximal elements of $\mathfrak{F}$. If $Y \in \mathfrak{F}$, then $Y \in \mathfrak{F}^*$ if and only if $\mathrm{codim}_X(Y) = 1$.*

Proof. If $\mathrm{codim}_X(Y) = 1$, then $Y \in \mathfrak{F}^*$ by 14.1.6. Conversely, let $Y \in \mathfrak{F}^*$. For any affine open subset $U = \mathrm{Spm}(A)$ of X, $Y \cap U$ is a closed irreducible subset of U, distinct from U. Thus $Y \cap U = \mathcal{V}(\mathfrak{p})$ for some $\mathfrak{p} \in \mathrm{Spec}\, A$. It follows that $Y \cap U \subset \mathcal{V}(f)$ for any $f \in \mathfrak{p} \setminus \{0\}$, and so $Y \cap U$ is contained in an irreducible component Z of $\mathcal{V}(f)$, and $\dim Z = \dim X - 1$.

By 1.1.12, $Y = \overline{Y \cap U}$ and so $Y \subset \overline{Z}$. Since Z is closed in U, it is locally closed in X. So Z is open in $\overline{Z}$ and $\dim Z = \dim \overline{Z}$ (14.1.6). So the result follows from the maximality of Y. $\square$

14.3.3 Using the notations of 14.3.1, we can ask if for any $x \in Y$, there is an open subset U of X containing x and $f \in \Gamma(X, \mathcal{O}_X)$ such that the variety $Y \cap U$ is exactly $\mathcal{V}(f)$. In general, the answer is no. However:

Proposition. *Let X be an irreducible affine variety such that $\mathbf{A}(X)$ is factorial. If Y is a closed subvariety of X of pure dimension $\dim X - 1$, then there exists $f \in \mathbf{A}(X)$ such that $Y = \mathcal{V}(f)$.*

Proof. Let $Y_1, \ldots, Y_r$ be the irreducible components of Y and $\mathfrak{p}_i = \mathcal{I}(Y_i) \in \mathrm{Spec}(\mathbf{A}(X))$, $1 \leqslant i \leqslant r$. By 11.2.4, the $\mathfrak{p}_i$'s are minimal among the prime ideals of $\mathbf{A}(X)$. Hence $\mathfrak{p}_i = \mathbf{A}(X)f_i$, where f_i is an irreducible element of $\mathbf{A}(X)$. Finally, we obtain that $Y = \mathcal{V}(f)$ where $f = f_1 \cdots f_r$. $\square$

14.3.4 Corollary. *Any closed subvariety Y of $\mathbb{P}_n$ of pure dimension $n-1$ is defined by a single equation $P(x_0, \ldots, x_n) = 0$ where $P \in \Bbbk[T_0, \ldots, T_n]$ is a non-constant homogeneous polynomial. In other words, Y is a hypersurface.*

Proof. Let $Y_1, \ldots, Y_r$ be the irreducible components of Y and $C_i = \pi^{-1}(Y_i)$, $C = \pi^{-1}(Y)$. By 14.1.8, 13.3.3 and 13.3.9, the C_i's are closed irreducible cones of dimension n in $\Bbbk^{n+1} \setminus \{0\}$. The closure $\overline{C_i}$ of C_i in $\Bbbk^{n+1}$ is again irreducible of dimension n. We see easily that the $\overline{C_i}$'s are the irreducible components of $\overline{C}$. Thus $\overline{C}$ is of pure dimension n in $\Bbbk^{n+1}$. Since it is a closed affine cone, there exists a non-constant homogeneous polynomial $P \in \Bbbk[T_0, \ldots, T_n]$ such that $\overline{C} = \mathcal{V}(P)$ (14.3.3). Hence our result. $\square$

14.3.5 Proposition. *Let $X = \mathrm{Spm}(A)$ be an irreducible affine variety and Y a closed irreducible subvariety of codimension $r \geqslant 1$ in X. For any integer s between 1 and r, there exist $f_1, \ldots, f_s \in A$ such that:*
  (i) $Y \subset \mathcal{V}(f_1, \ldots, f_s)$.
  (ii) *The irreducible components of $\mathcal{V}(f_1, \ldots, f_s)$ are of codimension s.*
  Consequently, there exist $f_1, \ldots, f_r \in A$ such that Y is an irreducible component of $\mathcal{V}(f_1, \ldots, f_r)$. We say that the f_i's form a system of parameters *of Y.*

Proof. Let us proceed by induction on s. Since $Y \neq X$, $\mathcal{I}(Y) \neq \{0\}$. If $s = 1$, any $f_1 \in \mathcal{I}(Y) \setminus \{0\}$ works by 14.2.3. Let us suppose that $s > 1$.

The induction hypothesis says that there exist $f_1, \ldots, f_{s-1} \in A$ such that $Z = \mathcal{V}(f_1, \ldots, f_{s-1})$ is a subvariety of pure dimension containing Y, and its irreducible components $Z_1, \ldots, Z_n$ have codimension $s - 1$ in X. Since $s - 1 < r$, $Z_i \not\subset Y$, so $\mathcal{I}(Y) \not\subset \mathcal{I}(Z_i)$. Hence $\mathcal{I}(Y) \not\subset \mathcal{I}(Z_1) \cup \cdots \cup \mathcal{I}(Z_n)$ (2.2.5). Let $f_s \in \mathcal{I}(Y) \setminus (\mathcal{I}(Z_1) \cup \cdots \cup \mathcal{I}(Z_n))$. We have $Y \subset \mathcal{V}(f_1, \ldots, f_s)$.

If W is an irreducible component of $\mathcal{V}(f_1, \ldots, f_s)$, then by 14.2.6, we have $\mathrm{codim}_X(W) \leqslant s$. But $W \subset \mathcal{V}(f_1, \ldots, f_{s-1})$, so $W \subset Z_i$ for some i and $W \neq Z_i$ since $f_s \notin \mathcal{I}(Z_i)$. We conclude by 14.1.6 that $\mathrm{codim}_X(W) \geqslant s$. Hence $\mathrm{codim}_X(W) = s$. $\square$

14.3.6 Corollary. *Let A be a finitely generated $\Bbbk$-algebra which is an integral domain, $\mathfrak{p} \in \mathrm{Spec}(A)$, $n = \dim A$, and $r = \dim A/\mathfrak{p}$. Then there exists a chain*

$$\{0\} = \mathfrak{p}_0 \subset \mathfrak{p}_1 \subset \cdots \subset \mathfrak{p}_{n-r} = \mathfrak{p} \subset \mathfrak{p}_{n-r+1} \subset \cdots \subset \mathfrak{p}_n$$

of prime ideals of A, containing $\mathfrak{p}$, of length n.

Proof. The existence of the $\mathfrak{p}_k$'s, $k \geqslant n - r$, is just the definition of $\dim A/\mathfrak{p}$. Consider A as the algebra of regular functions on an irreducible affine variety X, $\mathfrak{p}$ corresponds to a closed irreducible subvariety of X of dimension r. So the result follows by applying 14.3.5. $\square$

14.3.7 Corollary. *Let X be a closed irreducible subvariety of $\mathbb{P}_n$. For any closed irreducible subvariety Y of codimension r in X, there exist r non-constant homogeneous polynomials $P_1, \ldots, P_r \in \Bbbk[T_0, \ldots, T_n]$ such that Y is an irreducible component of $X \cap \mathcal{V}(P_1, \ldots, P_r)$.*

Proof. We prove, as in 14.3.5, that there exist non-constant homogeneous polynomials $P_1, \ldots, P_s$, $1 \leqslant s \leqslant r$, such that $Y \subset \mathcal{V}(P_1, \ldots, P_s)$ and $X \cap \mathcal{V}(P_1, \ldots, P_s)$ is a subvariety of pure dimension. If $Z_1, \ldots, Z_m$ are the irreducible components of $X \cap \mathcal{V}(P_1, \ldots, P_s)$, then the ideals $\mathcal{I}(\pi^{-1}(Z_j))$ and $\mathcal{I}(\pi^{-1}(Y))$ are graded. We can finish the proof as in 14.3.5 by using 7.2.9. $\square$

14.4 Counterexamples

14.4.1 Let us return to the point mentioned in 12.7.10.

Denote by $s : \mathbb{P}_1 \times \mathbb{P}_1 \to \mathbb{P}_3$ the Segre embedding (13.5.1). Thus the points $z, z' \in \mathbb{P}_1$ with homogeneous coordinates (x_0, x_1) and (x'_0, x'_1) are sent to the point $s(z, z')$ of $\mathbb{P}_3$ with homogeneous coordinates:

$$(x_0 x'_0, x_0 x'_1, x_1 x'_0, x_1 x'_1).$$

It is easy to check that the image C of s is the set of points of $\mathbb{P}_3$ whose homogeneous coordinates y_0, y_1, y_2, y_3 verify $y_0 y_3 - y_1 y_2 = 0$. By 13.5.1, C is irreducible.

Let $\pi : \mathbb{k}^4 \setminus \{0\} \to \mathbb{P}_3$ be the canonical surjection (13.1.1), and X the affine cone in $\mathbb{k}^4$ defined by $x_1 x_4 - x_2 x_3 = 0$. Then $\pi(X \setminus \{0\}) = C$. It follows from 13.1.2 and 13.3.9 that X is an irreducible subvariety of $\mathbb{k}^4$.

On the open subset $D(x_2) \cap D(x_4) = D(x_2 x_4)$ of X, the functions x_1/x_2 and x_3/x_4 are identical. They are therefore the restriction of a rational function h on X whose domain of definition U contains $D(x_2)$ and $D(x_4)$.

Suppose that $h \in \mathbf{A}(X)$. Then $x_2 h$ is regular and is zero on the points $(1, 0, \gamma, 0)$ of X. But $x_2 h$ and x_1 are identical on $D(x_2 x_4)$, so the irreducibility of X implies that $x_2 h$ and x_1 are identical on X. Contradiction.

Now suppose that there exists $g \in \mathbf{A}(X)$ non-invertible such that $U = D(g)$. The complement $\mathcal{V}(g)$ of U in X is contained in the plane $P \subset X$ of $\mathbb{k}^4$ defined by the equations $x_2 = 0$ and $x_4 = 0$. Since P is irreducible, we obtain by 14.1.6 and 14.2.3 that $P = \mathcal{V}(g)$. So the restriction of g to the plane $P' \subset X$ defined by $x_1 = 0$ and $x_3 = 0$ is regular and $(0, 0, 0, 0)$ is the only zero. This is absurd in view of 14.2.3.

The preceding argument says that U is contained strictly in $D(g)$ for all functions $f, g \in \mathbf{A}(X)$ such that $h = f/g$. As we have seen in 12.7.10, this proves that $\mathbf{A}(X)$ is not factorial.

14.4.2 Let us conserve the notations of 14.4.1 and set $x = (0, 0, 0, 0)$. If U is an open subset of X and $f \in \Gamma(U, \mathcal{O}_X)$, let $\mathcal{V}_U(f) = \{y \in U \; ; \; f(y) = 0\}$.

Let us suppose that there exist an open subset U of X containing x, and $f \in \Gamma(U, \mathcal{O}_X)$ such that $\mathcal{V}_U(f) = P \cap U$.

Let $W = P' \cap U$. Then W contains x and is an open subset of P'. So $\dim W = 2$. Let $g = f|_W$. Then $g \in \Gamma(W, \mathcal{O}_X|_{P'})$ and $\mathcal{V}_W(g) = \mathcal{V}_U(f) \cap P'$. But $\mathcal{V}_U(f) \cap P' = \{x\}$, so its dimension is zero. This contradicts 14.2.3.

Thus we have proved that there does not exist any open subset U of X containing x and any $f \in \Gamma(U, \mathcal{O}_X)$ verifying $\mathcal{V}_U(f) = P \cap U$ (see 14.3.3).

References

- [5], [19], [22], [26], [37], [40], [78].

15

Morphisms and dimension

We study properties of morphisms of algebraic varieties in this chapter. The notions of an affine morphism and a finite morphism are introduced. We also consider the relation between dimension and morphism. In particular, the dimension of a fibre of a morphism is considered.

15.1 Criterion of affineness

15.1.1 Let X be a variety and U an open subset of X, we shall denote often $\Gamma(U, \mathcal{O}_X)$ by $\mathcal{O}_X(U)$. Furthermore, if U is affine, then we shall also write $\mathbf{A}(U)$ for $\mathcal{O}_X(U)$. If $f \in \mathcal{O}_X(U)$, we set:

$$D_U(f) = \{x \in U \ ; \ f(x) \neq 0\}.$$

15.1.2 Lemma. *Let X be a variety and $f \in \mathcal{O}_X(X)$. Then we have $D_X(f) = X$ if and only if f is invertible in $\mathcal{O}_X(X)$.*

Proof. This is just a reformulation of 14.2.2. $\square$

15.1.3 Lemma. *Let X be a variety, $A = \mathcal{O}_X(X)$ and $f \in A$.*
(i) The restriction to $D_X(f)$ of f is invertible in $\mathcal{O}_X(D_X(f))$.
(ii) The identity map $A \rightarrow A$ induces an isomorphism of A_f onto $\mathcal{O}_X(D_X(f))$.

Proof. Let $Y = D_X(f)$ and $\mathcal{O}_Y = \mathcal{O}_X|_Y$ (9.3.7 and 9.3.8). By 12.2.6 and 12.5.4, $(Y, \mathcal{O}_Y)$ is a variety. If $g = f|_Y$, then $D_Y(g) = Y$ and so g is invertible in $\mathcal{O}_X(Y)$ by 15.1.2.

It follows from 2.3.4 that there is a unique $\varphi \in \mathrm{Hom}_{\mathrm{alg}}(A_f, \mathcal{O}_X(Y))$ such that $\varphi(h/f^n) = (h|_Y)/(f^n|_Y)$. We shall prove that φ is an isomorphism. Let $(U_i)_{i \in I}$ be a covering of X by affine open subsets. Since $D_X(f) \cap U_i = D_{U_i}(f|_{U_i})$, we deduce that $\mathcal{O}_X(Y \cap U_i) = \mathbf{A}(U_i)_{f|_{U_i}}$ (12.1.1).

Let $h/f^n \in \ker \varphi$. Then $(h|_Y)/(f^n|_Y) = 0$, so $(h|_{Y \cap U_i})/(f^n|_{Y \cap U_i}) = 0$. Thus $(f|_{U_i})(h|_{U_i}) = 0$ (11.7.3), which implies that $fh = 0$. Hence $h/f^n = 0$, and φ is injective.

Let $g \in \mathcal{O}_X(Y)$. Then $g|_{Y \cap U_i} \in \mathcal{O}_X(Y \cap U_i) = \mathbf{A}(U_i)_{f|_{U_i}}$. So there exist $n_i \in \mathbb{N}$ and $h_i \in \mathbf{A}(U_i)$ such that $(f^{n_i}|_{Y \cap U_i})(g|_{Y \cap U_i}) = h_i|_{Y \cap U_i}$. Let m be maximal among the n_i's and $\ell_i = (f^{m-n_i}|_{U_i}).h_i \in \mathbf{A}(U_i)$. We have $(\ell_i - \ell_j)|_{Y \cap U_i \cap U_j} = 0$. Since $U_i \cap U_j$ is an affine open subset (12.5.6), $(f|_{U_i \cap U_j}).((\ell_i - \ell_j)|_{U_i \cap U_j}) = 0$ (11.7.3), and so $(f\ell_i)|_{U_i \cap U_j} = (f\ell_j)|_{U_i \cap U_j}$. We deduce therefore that there exists $\ell \in A$ such that $\ell|_{U_i} = (f|_{U_i})\ell_i$ for all i. It follows that $(f^{m+1}|_Y)g = \ell|_Y$, and so φ is surjective. $\square$

15.1.4 Let X, Y be varieties with Y affine. Recall that given a morphism $v : X \to Y$, we have the associated $\Bbbk$-algebra homomorphism:

$$\Gamma(u) : \mathcal{O}_Y(Y) \to \mathcal{O}_X(X) , \; g \mapsto g \circ v.$$

By 12.3.3, the map $v \mapsto \Gamma(v)$ is a bijection between $\mathrm{Mor}(X, Y)$ and $\mathrm{Hom}_{\mathrm{alg}}(\mathcal{O}_Y(Y), \mathcal{O}_X(X))$.

Suppose further that $A = \mathcal{O}_X(X)$ is a finitely generated $\Bbbk$-algebra, then the identity map $A \to A = \mathcal{O}_X(X)$ induces a morphism $u : X \to \mathrm{Spm}(A)$.

Proposition. *The following conditions are equivalent:*
(i) *X is an affine variety.*
(ii) *The $\Bbbk$-algebra A is finitely generated and u is an isomorphism.*
(iii) *The $\Bbbk$-algebra A is finitely generated and u is a homeomorphism.*

Proof. We have seen in 12.3.4 that (i) $\Rightarrow$ (ii), and (ii) $\Rightarrow$ (iii) is clear. Suppose that we have (iii). Set $Y = \mathrm{Spm}(A)$. Since the $D_Y(f)$'s, $f \in A$, form a base of open subsets of Y, the subsets $D_X(f) = u^{-1}(D_Y(f))$ form a base of open subsets of X. Moreover, if we denote $u_f : D_X(f) \to D_Y(f)$ the morphism induced by u, it follows from 15.1.3 and the bijectivity of u that

$$\Gamma(u_f) : \mathcal{O}_Y(D_Y(f)) = A_f \to \mathcal{O}_X(D_X(f))$$

is an isomorphism. Hence u is an isomorphism (12.2.3) and X is affine. $\square$

15.1.5 Theorem. *Let X be a variety, $A = \mathcal{O}_X(X)$ and $f_1, \ldots, f_r \in A$ be such that $X = D_X(f_1) \cup \cdots \cup D_X(f_r)$ and each $D_X(f_i)$, $1 \leqslant i \leqslant r$, is an affine variety. The following conditions are equivalent:*
(i) *X is an affine variety.*
(ii) *$A = Af_1 + \cdots + Af_r$.*

Proof. (i) $\Rightarrow$ (ii) If X is affine, the identity map id_A induces an isomorphism $u : X \to Y = \mathrm{Spm}(A)$ (15.1.4). Since $u^{-1}(D_Y(f)) = D_X(f)$, we have

$$\mathrm{Spm}(A) = D_Y(f_1) \cup \cdots \cup D_Y(f_r) , \; \mathcal{V}(Af_1 + \cdots + Af_r) = \emptyset.$$

Hence $A = Af_1 + \cdots + Af_r$ as required (6.5.3).

(ii) $\Rightarrow$ (i) Suppose that $A = Af_1 + \cdots + Af_r$. First, we shall show that A is a finitely generated $\Bbbk$-algebra.

By our hypotheses and 15.1.3, A_{f_i}, being isomorphic to $\mathcal{O}_X(D_X(f_i))$, is finitely generated. Let $\{s_{i,1}/f_i^{n_i}, \ldots, s_{i,r_i}/f_i^{n_i}\}$ be a system of generators of A_{f_i} where $s_{i,k} \in A$. Now there exist $\ell_1, \ldots, \ell_r \in A$ such that:

$$(1) \qquad\qquad 1 = \ell_1 f_1 + \cdots + \ell_r f_r.$$

Denote by B the subalgebra of A generated by the f_i's, the ℓ_i's and the $s_{i,k}$'s.

Let $g \in A$, then $g/1 \in A_{f_i}$. So for each i, there exists $m_i \in \mathbb{N}$ such that $f_i^{m_i} g = h_i \in B$. Suppose that $m_i \geqslant 1$. Then by taking a suitable power of (1), there exist $t_1, \cdots, t_r \in B$ such that:

$$1 = t_1 f_1^{m_1} + \cdots + t_r f_r^{m_r}.$$

Hence $g = t_1 h_1 + \cdots + t_r h_r \in B$, and $B = A$ is finitely generated.

Finally, the identity map id_A induces an isomorphism $\varphi_i : A_{f_i} \rightarrow \mathcal{O}_X(D_X(f_i))$ by 15.1.3, which in turn induces an isomorphism (since $D_X(f_i)$ is affine) $v_i : D_X(f_i) \rightarrow \mathrm{Spm}(A_{f_i})$. Moreover, $v_i(x) = u(x)$ for $x \in D_X(f_i)$ and $u^{-1}(D_Y(f_i)) = D_X(f_i)$. As v_i is a homeomorphism, so is u. Hence X is affine (15.1.4). $\square$

15.2 Affine morphisms

15.2.1 Definition. *Let X, Y be varieties. A morphism $u : X \rightarrow Y$ is called* affine *if there is a covering $(V_i)_{i \in I}$ of Y by affine open subsets such that $u^{-1}(V_i)$ is an affine open subset of X for all $i \in I$.*

15.2.2 Lemma. *Let X, Y be affine varieties and $u : X \rightarrow Y$ a morphism. Then for all affine open subset V in Y, $u^{-1}(V)$ is an affine open subset. In particular, u is affine.*

Proof. Since Y is affine, there exist $g_1, \ldots, g_n \in \mathbf{A}(Y)$ such that $V = D_Y(g_1) \cup \cdots \cup D_Y(g_n)$. Let $f_i = g_i \circ u \in \mathbf{A}(X)$ and $v : u^{-1}(V) \rightarrow V$ the morphism induced by u. We have $u^{-1}(D_Y(g_i)) = D_X(f_i)$, $D_Y(g_i) = D_V(g_i|_V)$ and so $u^{-1}(D_Y(g_i)) = D_{u^{-1}(V)}(h_i)$ where $h_i = \Gamma(v)(g_i|_V)$ (here, the notations are as in 15.1.4).

Since V is affine, there exist, by 15.1.5, $t_1, \ldots, t_n \in \mathcal{O}_Y(V)$ such that

$$1 = t_1(g_1|_V) + \cdots + t_n(g_n|_V).$$

Hence

$$1 = \Gamma(v)(t_1)h_1 + \cdots + \Gamma(v)(t_n)h_n.$$

Since X is affine, so is $D_X(f_i) = D_{u^{-1}(V)}(h_i)$. It follows from 15.1.5 that $u^{-1}(V)$ is affine. $\square$

15.2.3 Proposition. *The following conditions are equivalent for a morphism $u : X \to Y$ of varieties:*

(i) *u is affine.*

(ii) *For any affine open subset V of Y, $u^{-1}(V)$ is an affine open subset.*

Proof. Clearly, (ii) implies (i). Conversely, let u be affine and $(V_i)_{i \in I}$ a covering of Y by affine open subsets such that $u^{-1}(V_i)$ is affine for all $i \in I$. Given any affine open subset V of Y and $v : u^{-1}(V) \to V$ the morphism induced by u, we have

$$V \cap V_i = \bigcup_{j=1}^{n_i} D_V(g_{ij}),$$

where $g_{ij} \in \mathcal{O}_Y(V)$.

We have $u^{-1}(D_V(g_{ij})) = D_{u^{-1}(V)}(h_{ij})$ where $h_{ij} = \Gamma(v)(g_{ij})$. Applying 15.2.2 to the morphism $u^{-1}(V_i) \to V_i$, we see that $u^{-1}(D_v(g_{ij}))$ is affine.

Since V is affine, there exist, by 15.1.5, $t_{ij} \in \mathcal{O}_Y(V)$ verifying

$$1 = \sum_{i,j} t_{ij} g_{ij}.$$

Hence

$$1 = \sum_{i,j} \Gamma(v)(t_{ij}) h_{ij}.$$

Thus $u^{-1}(V)$ is affine (15.1.5). $\square$

15.3 Finite morphisms

15.3.1 Definition. *A morphism $u : X \to Y$ of varieties is called* finite *if there is a covering $(V_i)_{i \in I}$ of Y by affine open subsets such that*

(i) *$u^{-1}(V_i)$ is an affine open subset of X for all $i \in I$.*

(ii) *For all $i \in I$, $\mathbf{A}(u^{-1}(V_i))$ is, via the comorphism $\mathbf{A}(V_i) \to \mathbf{A}(u^{-1}(V_i))$ of the morphism $u^{-1}(V_i) \to V_i$ induced by u, a finite $\mathbf{A}(V_i)$-algebra.*

Remarks. 1) In 15.3.1, we do not exclude the possibility that $u^{-1}(V_i)$ may be empty.

2) A finite morphism is affine. So the inverse image of an affine open subset under a finite morphism is affine (15.2.3).

15.3.2 Proposition. *Let $X = \mathrm{Spm}(A)$, $Y = \mathrm{Spm}(B)$ be affine varieties, $\varphi \in \mathrm{Hom}_{\mathrm{alg}}(B, A)$ and $u = \mathrm{Spm}(\varphi) : X \to Y$ the morphism associated to φ.*

(i) *If A is (via φ) a finite B-algebra, then u is finite, and for any open subset V of Y, the morphism $u^{-1}(V) \to V$, induced by u, is also finite.*

(ii) *Conversely, if u is finite, then A is (via φ) a finite B-algebra.*

Proof. (i) The first part is clear. As for the second, it suffices to prove the result in the case where $V = D(g)$, $g \in B$. Then $u^{-1}(V) = D(h)$ where $h = \varphi(g)$.

Let $f_1, \ldots, f_m \in A$ be a system of generators of A viewed as a B-module. If $f \in A$, then there exist $t_1, \ldots, t_m \in B$ such that:

$$f = \varphi(t_1)f_1 + \cdots + \varphi(t_m)f_m.$$

Since regular functions on $D(h)$ are of the form $x \mapsto f(x)/h^n(x)$, they can be written as:

$$x \mapsto [t_1(u(x))/(g(u(x)))^n]f_1(x) + \cdots + [t_m(u(x))/(g(u(x)))^n]f_m(x).$$

Thus the restriction to $D(h)$ of the f_i's generates $\mathbf{A}(D(h))$ as a $\mathbf{A}(D(g))$-module.

(ii) By the second part of (i), we can write $Y = D(g_1) \cup \cdots \cup D(g_m)$ with $g_i \in B$, and the ring A_{h_i} is a finitely generated B_{g_i}-module where $h_i = \varphi(g_i)$. Denote by $(f_{ij}/h_i^{n_i})_j$ a finite system of generators of A_{h_i}. Multiplying the f_{ij} by some power of h_i, we may assume that all the n_i's are equal to the some integer n.

Let $f \in A$, and f_i its restriction to $D(h_i)$. We have

$$f_i = \sum_j \varphi(t_{ij})f_{ij}/h_i^n \Rightarrow h_i^n f = \sum_j \varphi(t_{ij})f_{ij}.$$

Since the set of $D(g_i) = D(g_i^n)$ covers Y, there exist $s_1, \ldots, s_m \in B$ such that $1 = s_1 g_1^n + \cdots + s_m g_m^n$. But

$$\varphi(s_i g_i^n)f = \sum_j \varphi(s_i t_{ij})f_{ij},$$

so we have

$$f = \sum_{i,j} \varphi(s_i t_{ij})f_{ij}.$$

Hence A is a finite B-algebra. $\square$

15.3.3 Corollary. *Let $u : X \to Y$ be a morphism of varieties. The following conditions are equivalent:*

(i) *u is finite.*

(ii) *For all affine open subset $V \subset Y$, $u^{-1}(V)$ is an affine open subset of X and $\mathbf{A}(u^{-1}(V))$ is a finite $\mathbf{A}(V)$-algebra.*

Proof. This is clear by 15.2.3 and 15.3.2. $\square$

15.3.4 Proposition. *Let $u : X \to Y$ be a finite morphism of varieties.*

(i) *The map u is closed.*

(ii) *For all $y \in Y$, the set $u^{-1}(y)$ is finite.*

Proof. (i) Let $(V_i)_{i \in I}$ be a covering of Y by affine open subsets.

If F is a closed subset of X, we have $u(F) \cap V_i = u(F \cap u^{-1}(V_i))$. It suffices therefore to show that $u(F) \cap V_i$ is closed in V_i for all i. So we are reduced to the case where $X = \mathrm{Spm}(A)$ and $Y = \mathrm{Spm}(B)$ are affine and $u = \mathrm{Spm}(\varphi)$ for some $\varphi \in \mathrm{Hom}_{\mathrm{alg}}(B, A)$.

Let $Z = \mathcal{V}(\mathfrak{a})$ be a closed subset of X and $\mathfrak{b} = \varphi^{-1}(\mathfrak{a})$. The points of $u(Z)$ are $\varphi^{-1}(\mathfrak{m})$, with $\mathfrak{m}$ a maximal ideal of A containing $\mathfrak{a}$. Identifying $B/\mathfrak{b}$ with a subring of $A/\mathfrak{a}$ via φ, $A/\mathfrak{a}$ is a finitely generated $B/\mathfrak{b}$-module. Now by 3.3.2 and 3.3.3, if $\mathfrak{n}' \in \mathrm{Spm}(B/\mathfrak{b})$, then $\mathfrak{n}' = \mathfrak{m}' \cap (B/\mathfrak{b})$ for some $\mathfrak{m}' \in \mathrm{Spm}(A/\mathfrak{a})$. Let $\mathfrak{m} \in \mathrm{Spm}(A)$ and $\mathfrak{n} \in \mathrm{Spm}(B)$ be such that $\mathfrak{m}' = \mathfrak{m}/\mathfrak{a}$ and $\mathfrak{n}' = \mathfrak{n}/\mathfrak{b}$. Then $\mathfrak{n} = \varphi^{-1}(\mathfrak{m})$. It follows that $u(Z) = \mathcal{V}(\mathfrak{b})$, and $u(Z)$ is closed in X.

(ii) Again, we may assume that $X = \mathrm{Spm}(A)$ and $Y = \mathrm{Spm}(B)$ are affine and we only need to consider $y \in u(X)$. By (i), $u(X) = \mathcal{V}(\mathfrak{b})$ where $\mathfrak{b} = \ker \varphi$. Identify $B/\mathfrak{b}$ as a subring of A as above. According to 3.3.7, there is a finite number of maximal ideal of A lying above a maximal ideal of $B/\mathfrak{b}$. So the result follows.   □

15.3.5 Proposition. (i) *The composition of finite morphisms is again finite.*

(ii) *Let Y be a closed subvariety of X. The canonical injection $Y \to X$ is finite.*

(iii) *Let $u : X \to Y$ be a finite morphism, Z be a closed subvariety of Y containing $u(X)$, $j : Z \to Y$ the canonical injection and $v : X \to Z$ the morphism such that $u = j \circ v$. Then v is finite.*

(iv) *Let $u_i : X_i \to Y_i$, $i = 1, 2$, be finite morphisms. Then the morphism $v : X_1 \times X_2 \to Y_1 \times Y_2$ defined by $(x_1, x_2) \mapsto (u_1(x_1), u_2(x_2))$, is finite.*

Proof. (i) This is clear from 15.3.3 and 3.1.7.

(ii) Since the intersection of an affine open subset U of X with Y is an affine open subset of Y which is closed in U, we are reduced to the case where $X = \mathrm{Spm}(A)$ is affine and $Y = \mathcal{V}(\mathfrak{a})$ with $\mathfrak{a}$ a radical ideal of A. The result now follows from the fact that the A-module $A/\mathfrak{a}$ is generated by 1.

(iii) Let $z \in Z$ and V_z an affine open neighbourhood of z in Y. Then $u^{-1}(V_z) = v^{-1}(Z \cap V_z)$ is an affine open subset and the morphism $u^{-1}(V_z) \to V_z$ induced by u is finite (15.3.3). Thus we may assume that $X = \mathrm{Spm}(A)$, $Y = \mathrm{Spm}(B)$ are affine and $Z = \mathrm{Spm}(B/\mathfrak{b})$ where $\mathfrak{b}$ is an ideal of B. Since u is finite, A is a finitely generated B-module via $\mathbf{A}(u)$ which is just the composition $B \to B/\mathfrak{b} \to A$. Hence A is also a finitely generated $B/\mathfrak{b}$-module.

(iv) The affine open subsets $V \times W$, with V (resp. W) an affine open subset of Y_1 (resp. Y_2), cover $Y_1 \times Y_2$. So we may assume that $X_i = \mathrm{Spm}(A_i)$ and $Y_i = \mathrm{Spm}(B_i)$, $i = 1, 2$, are affine. If A_i is a finitely generated B_i-module, then it is clear that $A_1 \otimes_{\Bbbk} A_2$ is a finitely generated $B_1 \otimes_{\Bbbk} B_2$-module.   □

15.3.6 Remark. Let Y be a subvariety of X. In general, the canonical injection $Y \to X$ is not finite. For example, the injection $\Bbbk \setminus \{0\} \to \Bbbk$ is not finite since its image is not closed (15.3.4).

15.3.7 Example. Let us consider the morphism $u : \Bbbk \to \Bbbk^2$, $t \mapsto (t^2, t^3)$, in 11.5.7. Recall that its image is $C = \mathcal{V}(T_1^3 - T_2^2)$, and $\mathbf{A}(C) = \Bbbk[T^2, T^3]$. Let $v : \Bbbk \to C$ be the morphism induced by u. Since the $\mathbf{A}(C)$-algebra $\Bbbk[T]$

is finite, v is finite. The canonical injection $j : C \to \Bbbk^2$ is finite by 15.3.5 (ii), and so by 15.3.5 (i), $u = j \circ v$ is finite.

15.3.8 Corollary. *Let* $u : X \to Y$ *be a finite dominant morphism of irreducible varieties.*

(i) *u is surjective and $\mathbf{R}(X)$ is a finite algebraic field extension of $\mathbf{R}(Y)$. In particular,* $\dim X = \dim Y$.

(ii) *If $C \subset X$ is nowhere dense in X, then $u(C)$ is nowhere dense in Y.*

Proof. (i) Since $u(X)$ is dense in Y and closed (15.3.4), u is surjective.

Let V be a non-empty affine open subset of Y. Then $\mathbf{A}(u^{-1}(V))$ is a finitely generated $\mathbf{A}(V)$-module. Let $x_1, \ldots, x_n$ be a system of generators, then each x_i is integral over $\mathbf{A}(V)$. This implies that the x_i's are algebraic over $\mathbf{R}(Y) = \mathrm{Fract}(\mathbf{A}(V))$ (12.7.6). Since $\mathbf{R}(X) = \mathrm{Fract}(\mathbf{A}(u^{-1}(V)))$, the result follows from 14.1.2.

(ii) Since $u(\overline{C}) = \overline{u(C)}$, we may assume that C is closed. It suffices then to prove that if Z is an irreducible component of C (so Z is closed in X by 1.1.5), then $u(Z)$ is nowhere dense in Y. Since $u(Z)$ is an irreducible closed subset of Y (1.1.7 and 15.3.4), it follows from 15.3.5 that the restriction $Z \to u(Z)$ of u is a surjective finite morphism. According to (i) and 14.1.6,

$$\dim u(Z) = \dim Z < \dim X = \dim Y.$$

Using 14.1.6 again, we deduce that $u(Z)$ is nowhere dense in Y. $\square$

15.3.9 Corollary. *Let* $u : X \to Y$ *be a finite morphism of varieties. If Y is complete, then X is complete.*

Proof. Let Z be a variety, $p : X \times Z \to Z$ and $q : Y \times Z \to Z$ the canonical projections. The morphism $v : X \times Z \to Y \times Z$ defined by $(x, z) \mapsto (u(x), z)$, satisfies $p = q \circ v$. By 15.3.5 (iv), v is finite. So v is a closed map (15.3.4), and since q is closed by hypothesis, so is p. Thus X is a complete variety. $\square$

15.4 Factorization and applications

15.4.1 Proposition. *let X be an irreducible affine variety of dimension n. There exists a surjective finite morphism $u : X \to \Bbbk^n$.*

Proof. Let $X = \mathrm{Spm}(A)$, where A is a finitely generated $\Bbbk$-algebra which is an integral domain. According to 6.2.2 and 14.1.2, there is a subalgebra B of A, isomorphic to $\Bbbk[T_1, \ldots, T_n]$, such that A is integral over B. So A is a finite B-algebra (3.3.5), and the morphism $u : X \to \mathrm{Spm}(B) = \Bbbk^n$ which corresponds to the canonical injection $B \to A$, is then a finite morphism (15.3.2). It is dominant by 6.5.9, and so it is surjective (15.3.4). $\square$

15.4.2 Proposition. *Let* $u : X \to Y$ *be a dominant morphism of irreducible varieties.*

(i) *There exist* $m \in \mathbb{N}$, $V \subset Y$ *a non-empty affine open subset and a covering* $(U_i)_{1 \leqslant i \leqslant r}$ *of* $u^{-1}(V)$ *by affine open subsets such that the restriction* $u_i : U_i \to V$ *factors through*

$$U_i \xrightarrow{v_i} V \times \mathbb{k}^m \xrightarrow{\mathrm{pr}} V,$$

where pr *is the canonical surjection and* v_i *a surjective finite morphism.*

(ii) *The set* $u(X)$ *contains a non-empty open subset of* Y.

Proof. We only need to prove (i). Given $W = \mathrm{Spm}(B)$ a non-empty affine open subset of Y, $u^{-1}(W)$ is a non-empty open subset of X since u is dominant. So $u^{-1}(W)$ is the finite union of non-empty affine open subsets $U'_1, \ldots, U'_r$. Again, since u is dominant, the $u(U'_i)$'s are dense in Y.

Let $U'_i = \mathrm{Spm}(A_i)$ where A_i is a finitely generated $\mathbb{k}$-algebra which is an integral domain. The morphism $B \to A_i$ induced by u is injective (6.5.9). There exist, by 6.2.9, $s_i \in B \setminus \{0\}$ and a subalgebra C_i of A_i containing B such that $C_i \simeq B[T_1, \ldots, T_{m_i}] = B \otimes_{\mathbb{k}} \mathbb{k}[T_1, \ldots, T_{m_i}]$ and $A_i[1/s_i]$ is a finitely generated $C_i[1/s_i]$-module.

The quotient field of $A_i[1/s_i]$ is $L_i = \mathrm{Fract}(A_i)$ and the quotient field of C_i is isomorphic to $K(T_1, \ldots, T_{m_i}) = L'_i$ where $K = \mathrm{Fract}(B)$. It follows that L_i and L'_i have the same transcendence degree over $\mathbb{k}$ (3.3.4 and 6.2.3). This degree is equal to $\dim U'_i = \dim X$ (14.1.2). We deduce therefore that all the m_i's are equal to some integer m.

Let $s = s_1 \cdots s_r$. Then $C_i[1/s] = B[1/s] \otimes_{\mathbb{k}} \mathbb{k}[T_1, \ldots, T_m]$. Take

$$V = \mathrm{Spm}(B[1/s]) = D_Y(s) , \; U_i = \mathrm{Spm}(A_i[1/s]) , \; 1 \leqslant i \leqslant r.$$

Then we are done since the morphism v_i is the one associated to the injection $C_i[1/s] \to A_i[1/s]$. $\square$

15.4.3 Proposition. *Let* $u : X \to Y$ *be a morphism of varieties and* C *a constructible subset of* X. *Then* $u(C)$ *is constructible.*

Proof. Let $X_1, \ldots, X_r$ be the irreducible components of X. Since $u(C) = u(C \cap X_1) \cup \cdots \cup u(C \cap X_r)$ and $C \cap X_i$ is constructible in X_i, we may assume that X is irreducible. We shall proceed by induction on the dimension n of X. The case $n = 0$ is obvious.

Since C is Noetherian, it is the finite union of its irreducible components. They are constructible in X since they are closed in C. Thus we may further assume that C is irreducible. So $\overline{C}$ is an irreducible closed subvariety of X.

Suppose that $\overline{C} \neq X$. Then $\dim \overline{C} < X$ and C is constructible in $\overline{C}$. By the induction hypothesis, $u(C)$ is constructible.

Suppose that $\overline{C} = X$. It suffices to show that $u(C)$ is constructible in $u(X)$. So we may assume that Y is irreducible and u is dominant.

Since C is constructible and dense in X, C contains a non-empty open subset U of X (1.4.5), and the restriction of u to U is still dominant. By 15.4.2, $u(U) \subset u(C)$ contains a non-empty open subset V of Y. So we have:

$$u(C) = V \cup u(C \cap u^{-1}(Y \setminus V)).$$

The set $u^{-1}(Y \setminus V)$ is a closed subvariety of X distinct from X since u is dominant and $Y \setminus V$ is nowhere dense in Y. Moreover, $C \cap u^{-1}(Y \setminus V)$ is constructible in $u^{-1}(Y \setminus V)$. By the induction hypothesis, $u(C \cap u^{-1}(Y \setminus V))$ is constructible in Y, and we have finished our proof. $\square$

15.5 Dimension of fibres of a morphism

15.5.1 Let $u : X \to Y$ be a morphism of varieties and X', Y' subvarieties of X and Y such that $u(X') \subset Y'$. We say that X' *dominates* Y' if the morphism $X' \to Y'$ induced by u is dominant.

15.5.2 Lemma. *Let $u : X \to Y$ be a surjective finite morphism of irreducible varieties and W an irreducible closed subvariety of Y. There is at least one irreducible component of $u^{-1}(W)$ which dominates W, and any such irreducible component has dimension $\dim W$.*

Proof. Let $Z_1, \ldots, Z_r$ be the irreducible components of $u^{-1}(W)$. Then $u(Z_i)$ is closed in Y (15.3.4) and since u is surjective, $W = u(Z_1) \cup \cdots \cup u(Z_r)$. The irreducibility of W implies that $W = u(Z_i)$ for at least one i.

Now if $W = u(Z_i)$, then 15.3.5 implies that the morphism $Z \to W$ induced by u is finite, and so $\dim W = \dim Z_i$ (15.3.8). $\square$

15.5.3 Theorem. *Let $u : X \to Y$ be a dominant morphism of irreducible varieties.*

(i) *We have $\dim X \geqslant \dim Y$.*

(ii) *If W is an irreducible closed subvariety of Y and Z an irreducible component of $u^{-1}(W)$ which dominates W. Then*

$$(2) \qquad\qquad \dim Z \geqslant \dim W + \dim X - \dim Y.$$

(iii) *There exists a non-empty open subset V of Y such that for all irreducible closed subvariety W of Y verifying $W \cap V \neq \emptyset$, there is at least one irreducible component of $u^{-1}(W)$ which dominates W, and for such an irreducible component Z, we have*

$$(3) \qquad\qquad \dim Z = \dim W + \dim X - \dim Y.$$

Proof. Since u is dominant, the comorphism (of fields) $\mathbf{R}(Y) \to \mathbf{R}(X)$ is injective (12.7.11). Therefore, $\dim Y = \operatorname{tr} \deg_k \mathbf{R}(Y) \leqslant \operatorname{tr} \deg_k \mathbf{R}(X) = \dim X$.

By 15.4.2 (i), there exist an integer $m \in \mathbb{N}$, a non-empty open subset $V \subset Y$ and a covering $(U_i)_{1 \leqslant i \leqslant r}$ of $u^{-1}(V)$ by affine open subsets such that $u|_{U_i} = \operatorname{pr} \circ v_i$ where $v_i : U_i \to V \times \mathbb{k}^m$ is a surjective finite morphism and $\operatorname{pr} : V \times \mathbb{k}^m \to V$ is the canonical surjection. Since v_i is finite and surjective, we have $m = \dim X - \dim Y$ (15.3.8).

Let W be an irreducible subvariety of Y verifying $W \cap V \neq \emptyset$. Then $(W \cap V) \times \mathbb{k}^m$ is an irreducible closed subvariety of dimension $m + \dim W$

in $V \times \Bbbk^m$. Applying 15.5.2 to v_j, we obtain that there is at least one irreducible component of $v_i^{-1}((V \cap W) \times \Bbbk^m)$ which dominates W, and such an irreducible component has dimension $m + \dim W$. Since $u^{-1}(V \cap W)$ is the union of $v_i^{-1}((V \cap W) \times \Bbbk^m)$, $1 \leqslant i \leqslant r$, we have proved that there exists an irreducible component of $u^{-1}(W)$ which dominates W. If Z is such an irreducible component, then $Z \cap U_i$ is dense in Z for some i and $Z \cap U_i$ dominates $(V \cap W) \times \Bbbk^m$. Hence $\dim Z = m + \dim W$. So we have part (iii).

To prove (ii), replace Y by an affine open subset V which meets W, and replace X by $u^{-1}(V)$, we may assume that Y is affine. Let $s = \mathrm{codim}_Y(W)$. According to 14.3.5, there exist $f_1, \ldots, f_s \in \mathbf{A}(Y)$ such that W is an irreducible component of $\mathcal{V}(f_1, \ldots, f_s)$. Let $g_i = f_i \circ u \in \mathbf{A}(X)$. If Z is an irreducible component of $u^{-1}(W)$ dominating W, then $Z \subset T = \mathcal{V}(g_1, \ldots, g_s)$. We claim that Z is an irreducible component of T.

Let Z' be an irreducible component of T containing Z, then $W = \overline{u(Z)} \subset \overline{u(Z')} \subset \mathcal{V}(f_1, \ldots, f_s)$. Since W is an irreducible component of the latter, we have $u(Z') \subset W$ and so $Z \subset Z' \subset u^{-1}(W)$. Since Z is an irreducible component of $u^{-1}(W)$, we have $Z = Z'$ as claimed. From our claim, we deduce that $\mathrm{codim}_X(Z) \leqslant s = \mathrm{codim}_Y(W)$ (14.2.6). Hence we have (2). $\square$

15.5.4 Corollary. *Let $u : X \to Y$ be a dominant morphism of irreducible varieties.*

(i) *If $y \in u(X)$, each irreducible component of $u^{-1}(y)$ has dimension at least $\dim X - \dim Y$.*

(ii) *There exists a non-empty open subset of $V \subset u(X)$ such that for any $y \in V$, each irreducible component of $u^{-1}(y)$ has dimension $\dim X - \dim Y$.*

15.5.5 Corollary. *Let $u : X \to Y$ be a morphism of varieties.*

(i) *If $\dim u^{-1}(y) \leqslant r$ for all $y \in Y$, then $\dim X \leqslant r + \dim Y$.*

(ii) *If u is dominant and $\dim u^{-1}(y) = r$ for all $y \in u(X)$, then $\dim X = r + \dim Y$.*

Proof. (i) Let $Y_1, \ldots, Y_n$ be the irreducible components of Y and for $1 \leqslant i \leqslant n$, $(X_{ij})_j$ be the irreducible components of $u^{-1}(Y_i)$. Let $z \in u(X_{ij})$. Applying 15.5.4 to the restriction $X_{ij} \to \overline{u(X_{ij})}$, we have

$$\dim X_{ij} \leqslant \dim \overline{u(X_{ij})} + \dim u^{-1}(z) \leqslant r + \dim Y.$$

Since X is the union of the X_{ij}'s, the result follows.

(ii) Let us conserve the notations above and let Y_i be an irreducible component of Y of dimension $\dim Y$. Since $Y_i \setminus (\bigcup_{j \neq i} Y_j)$ is open and non-empty in Y (1.3.6), the restriction $u^{-1}(Y_i) \to Y_i$ is dominant. Removing certain X_{ij}'s if necessary, we may assume that all the X_{ij}'s dominate Y_i. If $\dim X_{ij} < r + \dim Y$ for all j, then applying 15.5.4 (ii) to the restriction $X_{ij} \to Y_i$, we can find points y such that $\dim u^{-1}(y) < r$ which contradicts our hypothesis. Hence we have $\dim X = r + \dim Y$. $\square$

15.5.6 Corollary. *Let $u : X \to Y$ be a dominant birational morphism of irreducible varieties. There exists a non-empty open subset V of Y such that the restriction $u^{-1}(V) \to V$ of u is an isomorphism.*

Proof. By our hypothesis, $\dim X = \dim Y$. Replace Y by an affine open subset W and X by $u^{-1}(W)$, we are reduced to the case where $Y = \mathrm{Spm}(B)$ is affine.

Let $U = \mathrm{Spm}(A)$ be a non-empty affine open subset of X (so $u(U)$ is dense in Y) and $W = \overline{u(X \setminus U)}$. If T is an irreducible component of $X \setminus U$, then by 14.1.6, $\dim T < \dim X = \dim Y$. Thus $\dim Z < \dim Y$ for any irreducible component Z of W (15.5.3). It follows from 14.1.6 that there exists $g \in B$ such that $W \cap D_Y(g) = \emptyset$, and so $u^{-1}(D_Y(g)) \subset U$. If $f = g \circ u \in A$, then $u^{-1}(D_Y(g)) = D_U(f)$. Replacing Y by $D_Y(g)$ and X by $D_U(f)$, we may suppose that $X = \mathrm{Spm}(A)$ and $Y = \mathrm{Spm}(B)$ are affine where A, B are integral domains, and the comorphism $\varphi : B \to A$ extends to an isomorphism of fields $\mathrm{Fract}(B) \to \mathrm{Fract}(A)$. It follows that if $a_1, \ldots, a_m \in A$ are such that $A = \Bbbk[a_1, \ldots, a_m]$, then there exist $b_1, \ldots, b_m \in B$ and $s \in B \setminus \{0\}$ such that $a_i = b_i/s$ for $1 \leqslant i \leqslant m$. So $A[1/s]$ and $B[1/s]$ are isomorphic. Thus $V = D_Y(s)$ works. $\square$

15.5.7 Let E be a topological space and $f : E \to \mathbb{N}$ a map. We say that f is *upper semi-continuous* if $\{x \in E; \ f(x) \geqslant n\}$ is closed in E for all $n \in \mathbb{N}$.

Theorem. *Let $u : X \to Y$ be a morphism of irreducible varieties. For $x \in X$, set $e(x)$ to be the maximum dimension of any irreducible component of $\dim u^{-1}(u(x))$ containing x. Then the map $e : X \to \mathbb{N}$ is upper semi-continuous.*

Proof. Replacing Y by $\overline{u(X)}$, we may assume that u is dominant. Set $r = \dim X - \dim Y \geqslant 0$ (15.5.3), and for $n \in \mathbb{N}$,

$$S_n(u) = \{x \in X \ ; \ e(x) \geqslant n\}.$$

We proceed by induction on $p = \dim Y$. The case $p = 0$ is obvious since $S_n(u)$ is the union of irreducible components of X of dimension at least n.

If $n \leqslant r$, then $S_n(u) = X$ (15.5.4 (i)). So suppose that $n > r$. By 15.5.4, there is a non-empty open subset V of Y such that $S_n(u) \subset X \setminus u^{-1}(V)$. Let $W_1, \ldots, W_s$ be the irreducible components of $Y \setminus V$. We have $\dim W_j < \dim Y$ (14.1.6). For $1 \leqslant j \leqslant s$, let $(Z_{ij})_j$ be the irreducible components of $u^{-1}(W_i)$. Denote by $v_{ij} : Z_{ij} \to W_i$ the restriction of u.

By the induction hypothesis, the $S_n(v_{ij})$'s are closed in Z_{ij}, and so in X. But if $x \in X \setminus u^{-1}(V)$, any irreducible components of $u^{-1}(u(x))$ containing x is contained in one of the Z_{ij}'s, so it is an irreducible component of $v_{ij}^{-1}(v_{ij}(x))$. Thus $S_n(u)$ is the union of the $S_n(v_{ij})$, and so $S_n(u)$ is closed in X. $\square$

15.5.8 Corollary. *Let X, Y be varieties. The canonical projections $\mathrm{pr}_1 : X \times Y \to X$, $\mathrm{pr}_2 : X \times Y \to Y$ are open.*

Proof. Let $X_1, \ldots, X_m$ (resp. $Y_1, \ldots, Y_n$) be the irreducible components of X (resp. Y). If U is open in $X \times Y$, then $\mathrm{pr}_1(U)$ is open if $\mathrm{pr}_1((X_i \times Y) \cap U) = X_i \cap \mathrm{pr}_1(U)$ is open in X_i for all i. Moreover,

$$\mathrm{pr}_1(U \cap (X_i \times Y)) = \mathrm{pr}_1(U \cap (X_i \times Y_1)) \cup \cdots \cup \mathrm{pr}_1(U \cap (X_i \times Y_n)).$$

So we may assume that X, Y are irreducible.

Let U be an open subset of $X \times Y$ and $F = X \setminus \mathrm{pr}_1(U)$. If $x \in F$ and $z \in \mathrm{pr}_1^{-1}(x)$, then $\mathrm{pr}_1^{-1}(x) = \{x\} \times Y$ is the unique irreducible component of $\mathrm{pr}_1^{-1}(\mathrm{pr}_1(z)) \cap ((X \times Y) \setminus U)$ containing z. Its dimension is $\dim Y$.

Now if $x \notin F$, $U \cap \mathrm{pr}_1^{-1}(x)$ is not empty, so the irreducible components of $((X \times Y) \setminus U) \cap \mathrm{pr}_1^{-1}(x)$ are all of dimension strictly less than $\dim Y$ (14.1.6).

Hence $F \times Y = \mathrm{pr}_1^{-1}(F)$ is the set of points z of the variety $(X \times Y) \setminus U$ such that $e(z) \geq \dim Y$ (with respect to the restriction of pr_1). By 15.5.7, $F \times Y$ is closed, and so F is closed (12.4.5). □

15.5.9 Examples. 1) Let $X \subset \Bbbk^3$ be the cone defined by the equation $x_1 x_3 - (x_1^2 + x_2^2) = 0$ and $u : X \to \Bbbk^2 = Y$ the projection $(x_1, x_2, x_3) \mapsto (x_1, x_2)$. The morphism u is dominant and birational since

$$\Bbbk(T_1, T_2, (T_1^2 + T_2^2)T_1^{-1}) = \Bbbk(T_1, T_2).$$

Clearly $V = D(T_1) \subset Y$ satisfies the conclusion of 15.5.6. Denote by W the line $x_1 = 0$, which is just the complement of V in $\Bbbk^2$.

The set $u^{-1}(W)$ is the line defined by $x_1 = 0$ and $x_2 = 0$ in $\Bbbk^3$. Thus $u(u^{-1}(W)) = \{(0,0)\}$, and no irreducible components of $u^{-1}(W)$ dominates W. Moreover $\dim u^{-1}(W) = 1 > \dim X - \dim Y = 0$. Since $u^{-1}(W) = u^{-1}(0,0)$, we see that, in 15.5.4, we can have $\dim u^{-1}(y) > \dim X - \dim Y$.

2) Let us consider the following morphism

$$v : X = \Bbbk^2 \to \Bbbk^3 \ , \ (t_1, t_2) \mapsto (t_1^2 - 1, t_1(t_1^2 - 1), t_2).$$

The image Y of v is the cylinder $x_2^2 - x_1^2 - x_1^3 = 0$. Let $u : X \to Y$ be the morphism induced by v. It is a finite morphism.

The image W of the morphism $w : \Bbbk \to Y$, $t \mapsto (t^2 - 1, t(t^2 - 1), t)$, is irreducible of dimension 1. We verify easily that the diagonal Δ of $\Bbbk$ and the point $(1, -1)$ are the irreducible components of $u^{-1}(W)$. Clearly the irreducible components consisting of the point $(1, -1)$ does not dominate W since $\dim W = 1$.

Note that $u(\Bbbk^2 \setminus \Delta) = (Y \setminus W) \cup \{(0, 0, 1)\}$. Thus the image of the open set $\Bbbk^2 \setminus \Delta$ is not open. So a surjective finite morphism is not necessarily open.

15.6 An example

15.6.1 For $p, q \in \mathbb{N}^*$, let $\mathrm{M}_{p,q}$ be the set of p by q matrices with coefficients in $\Bbbk$ and we write M_p for $\mathrm{M}_{p,p}$. Endow $\mathrm{M}_{p,q}$ with its natural structure of irreducible affine algebraic variety of dimension pq. Then the map

$$\mathrm{M}_{r,p} \times \mathrm{M}_{p,q} \times \mathrm{M}_{q,s} \to \mathrm{M}_{r,s} \ , \ (A, B, C) \mapsto ABC$$

is a morphism.

15.6.2 The group $\mathrm{GL}_n(\Bbbk)$ is a principal open subset of M_n (11.7.5), so it is an irreducible affine variety of dimension n^2. The map $\mathrm{GL}_n(\Bbbk) \to \mathrm{GL}_n(\Bbbk)$, $A \mapsto A^{-1}$, is an automorphism of the variety $\mathrm{GL}_n(\Bbbk)$.

15.6.3 Let $r \in \{0, 1, \ldots, \min(p, q)\}$. Denote by

$$C_r = \{A \in \mathrm{M}_{p,q} \ ; \ \mathrm{rk}(A) \leqslant r\} \ , \ C_r' = \{A \in \mathrm{M}_{p,q} \ ; \ \mathrm{rk}(A) = r\}.$$

In other words, $A \in C_r$ if and only if all the $(r+1)$-minors of A are zero. It follows that C_r is closed in $\mathrm{M}_{p,q}$.

Similarly, if $A \in C_r$, then $A \in C_r'$ if and only if at least one of the r-minors of A is non-zero. Thus C_r' is open in C_r.

Let $I_r \in \mathrm{M}_r$ be the identity matrix and

$$J = \begin{pmatrix} I_r & 0 \\ 0 & 0 \end{pmatrix} \in \mathrm{M}_{p,q} \ .$$

It is well-known that the image of the morphism $w : \mathrm{M}_p \times \mathrm{M}_q \to \mathrm{M}_{p,q}$, $(P, Q) \mapsto PJQ$, is C_r. So C_r is irreducible in $\mathrm{M}_{p,q}$. Let $u : \mathrm{M}_p \times \mathrm{M}_q \to C_r$ be the morphism induced by w. We have $u(\mathrm{GL}_p(\Bbbk) \times \mathrm{GL}_q(\Bbbk)) = C_r'$. Let $v : \mathrm{GL}_p(\Bbbk) \times \mathrm{GL}_q(\Bbbk) \to C_r'$ be the morphism induced by u.

Let $A \in C_r'$ and $(U, V) \in \mathrm{GL}_p(\Bbbk) \times \mathrm{GL}_q(\Bbbk)$ be such that $A = UJV$. If $(P, Q) \in \mathrm{GL}_p(\Bbbk) \times \mathrm{GL}_q(\Bbbk)$, then

$$(P, Q) \in v^{-1}(A) \Leftrightarrow (U^{-1}P, QV^{-1}) \in v^{-1}(J).$$

We deduce that all the fibres of v are isomorphic.

Writing

$$P = \begin{pmatrix} A & B \\ C & D \end{pmatrix} \ , \ Q = \begin{pmatrix} X & Y \\ Z & T \end{pmatrix}$$

where $A, X \in \mathrm{M}_r$, we have

$$PJQ = \begin{pmatrix} AX & AY \\ CX & CY \end{pmatrix} .$$

Hence

$$(P, Q) \in v^{-1}(J) \Leftrightarrow A \in \mathrm{GL}_r(\Bbbk) \ , \ X = A^{-1} \ , \ C = 0 \ , \ Y = 0.$$

It is now clear that the map

$$\mathrm{GL}_r(\Bbbk) \times \mathrm{GL}_{p-r}(\Bbbk) \times \mathrm{GL}_{q-r}(\Bbbk) \times \mathrm{M}_{r,p-r} \times \mathrm{M}_{q-r,r} \to v^{-1}(J)$$

$$(A, B, C, D, E) \mapsto \left(\begin{pmatrix} A & D \\ 0 & B \end{pmatrix}, \begin{pmatrix} A^{-1} & 0 \\ E & C \end{pmatrix} \right)$$

is an isomorphism.

We deduce from the above discussion that for any $A \in C'_r$, $v^{-1}(A)$ is irreducible and:

$$\dim v^{-1}(A) = r^2 + (p-r)^2 + (q-r)^2 + r(p-r) + r(q-r).$$

Since C'_r is open in the irreducible variety C_r, we have

$$\dim C_r = \dim C'_r = r(p+q-r).$$

References and comments

- [5], [19], [22], [26], [37], [40], [78].

Many results of this chapter are due to Chevalley, for example: 15.4.2, 15.4.3, 15.5.7.

16

Tangent spaces

The reader may have come across the notion of a tangent space in differential geometry. We define here the analogous algebraic object, the Zariski tangent space, for an algebraic variety. We shall see later that the Zariski tangent space appears in relating algebraic groups and Lie algebras.

16.1 A first approach

16.1.1 Given $P \in \Bbbk[T_1, \ldots, T_n] = \Bbbk[\mathbf{T}]$ and $x \in \Bbbk^n$, set:

$$D_x(P) = \frac{\partial P}{\partial T_1}(x)T_1 + \cdots + \frac{\partial P}{\partial T_n}(x)T_n.$$

The map $P \mapsto D_x(P)$ is $\Bbbk$-linear, and if $P, Q \in \Bbbk[\mathbf{T}]$, then

(1) $$D_x(PQ) = P(x)D_x(Q) + Q(x)D_x(P).$$

16.1.2 Let X be a closed subvariety of $\Bbbk^n$, $\mathfrak{a} = \mathcal{I}(X)$ the associated radical ideal in $\Bbbk[\mathbf{T}]$ and $x = (x_1, \ldots, x_n) \in X$. Denote by $\mathfrak{n}_x = \mathcal{I}(\{x\})$ the maximal ideal of $A = \mathbf{A}(X) = \Bbbk[\mathbf{T}]/\mathfrak{a}$ associated to x.

16.1.3 Let $\mathfrak{a}'_x$ be the ideal of $\Bbbk[\mathbf{T}]$ generated by the $D_x(P)$, $P \in \mathfrak{a}$. Then by (1), if $(P_i)_{i \in I}$ is a system of generators of $\mathfrak{a}$, then $(D_x(P_i))_{i \in I}$ is a system of generators of $\mathfrak{a}'_x$.

We call *the tangent space of X at x*, denoted by $\mathrm{Tan}_x(X)$, the vector subspace $\mathcal{V}(\mathfrak{a}'_x)$ of $\Bbbk^n$. If r is the rank of the set of linear forms $D_x(P)$, $P \in \mathfrak{a}$, then the dimension of $\mathrm{Tan}_x(X)$ is $n - r$.

Let $\{Q_1, \ldots, Q_r\}$ be a basis of the $\Bbbk$-vector space spanned by the $D_x(P)$, $P \in \mathfrak{a}$. There exist indices $i_1, \ldots, i_{n-r}$ such that $\{Q_1, \ldots, Q_r, T_{i_1}, \ldots, T_{i_{n-r}}\}$ is a basis of the space of homogeneous polynomials of degree 1. Clearly we can identify $\Bbbk[\mathbf{T}]/\mathfrak{a}'_x$ with $\Bbbk[T_{i_1}, \ldots, T_{i_{n-r}}]$. It follows that $\mathfrak{a}'_x$ is prime and the algebra of regular functions on $\mathrm{Tan}_x(X)$ is $\Bbbk[\mathbf{T}]/\mathfrak{a}'_x$.

16.1.4 Now let $L \subset \mathbb{k}^n$ be a line containing x and $v = (v_1, \ldots, v_n) \in \mathbb{k}^n$ be such that L is the set of points of the form $x + tv$, $t \in \mathbb{k}$. Let $(P_i)_{1 \leqslant i \leqslant s}$ be a system of generators of $\mathfrak{a}$. Then the points in $L \cap X$ are obtained by solving the following system of equations in t:

$$(2) \qquad\qquad P_1(x + tv) = \cdots = P_s(x + tv) = 0.$$

For $1 \leqslant i \leqslant n$, we have

$$P_i(x + tv) = t \sum_{j=1}^{n} \frac{\partial P_i}{\partial T_j}(x)v_j + t^2 Q_i(t)$$

where $Q_i \in \mathbb{k}[T]$. Thus, $t = 0$ is a "multiple root" of (2) if and only if,

$$\frac{\partial P_i}{\partial T_1}(x)v_1 + \cdots + \frac{\partial P_i}{\partial T_n}(x)v_n = 0$$

for $1 \leqslant i \leqslant s$. Hence $t = 0$ is a multiple root of (2) if and only if $v \in \mathrm{Tan}_x(X)$.

16.1.5 Let $P, Q \in \mathbb{k}[\mathbf{T}]$ be such that $P - Q \in \mathfrak{a}$. By the definition of $\mathrm{Tan}_x(X)$, the restriction of $D_x(P) - D_x(Q)$ to $\mathrm{Tan}_x(X)$ is zero. So we can define a linear map $\overline{D}_x : A \rightarrow (\mathrm{Tan}_x(X))^*$, the dual space of $\mathrm{Tan}_x(X)$, by setting $\overline{D}_x(f) = D_x(F)|_{\mathrm{Tan}_x(X)}$ where $f \in A$ and $F \in \mathbb{k}[\mathbf{T}]$ is a representative of f. The map $\overline{D}_x$ is obviously surjective.

Furthermore, since $A = \mathbb{k} \oplus \mathfrak{n}_x$ and $\overline{D}_x(\lambda) = 0$ for any $\lambda \in \mathbb{k}$, $\overline{D}_x$ induces a linear surjection

$$d_x : \mathfrak{n}_x \rightarrow (\mathrm{Tan}_x(X))^*.$$

Proposition. *The kernel of d_x is $\mathfrak{n}_x^2$. Thus d_x induces an isomorphism $\mathfrak{n}_x/\mathfrak{n}_x^2 \rightarrow (\mathrm{Tan}_x(X))^*$ of $\mathbb{k}$-vector spaces.*

Proof. Let $\mathfrak{N}_x = \mathbb{k}[\mathbf{T}](T_1 - x_1) + \cdots + \mathbb{k}[\mathbf{T}](T_n - x_n)$ be the maximal ideal of $\mathbb{k}[\mathbf{T}]$ associated to x. Then $\mathfrak{a} \subset \mathfrak{N}_x$ and $\mathfrak{n}_x = \mathfrak{N}_x/\mathfrak{a}$. It is clear from (1) that $\mathfrak{n}_x^2 \subset \ker d_x$.

Conversely, let $f \in \ker d_x$ and $F \in \mathfrak{N}_x$ be a representative of f. Since $\mathcal{I}(\mathrm{Tan}_x(X)) = \mathfrak{a}_x'$ (16.1.3), there exist $F_1, \ldots, F_s \in \mathfrak{a}$ and $\lambda_1, \ldots, \lambda_s \in \mathbb{k}$ such that

$$D_x(F) = \lambda_1 D_x(F_1) + \cdots + \lambda_s D_x(F_s).$$

Set $G = F - \lambda_1 F_1 - \cdots - \lambda_s F_s$, then $D_x(G) = 0$. So

$$\frac{\partial G}{\partial T_1}(x) = \cdots = \frac{\partial G}{\partial T_n}(x) = 0.$$

Since $G(x) = 0$, it follows from Taylor's formula that G is contained in the ideal generated by $(T_i - x_i)(T_j - x_j)$, $1 \leqslant i, j \leqslant n$, which is $\mathfrak{N}_x^2$. Finally, G is also a representative of f because $F_1, \ldots, F_s \in \mathfrak{a}$. Hence $f \in \mathfrak{n}_x^2$. $\qquad\square$

16.1.6 It follows from the preceding discussion that we can identify $\text{Tan}_x(X)$ with $(\mathfrak{n}_x/\mathfrak{n}_x^2)^*$. We now give another interpretation of $\text{Tan}_x(X)$.

Let $\varphi : A \to \mathcal{O}_{X,x}$ be the map sending $f \in A$ to its germ at x, and $\mathfrak{m}_{X,x}$ the maximal ideal of $\mathcal{O}_{X,x}$. We have seen in 12.8.3 that φ induces a local isomorphism $\psi : A_{\mathfrak{n}_x} \to \mathcal{O}_{X,x}$. By 2.6.12, the vector spaces $\mathfrak{n}_x/\mathfrak{n}_x^2$ and $\mathfrak{n}_x A_{\mathfrak{n}_x}/\mathfrak{n}_x^2 A_{\mathfrak{n}}$ are isomorphic. So we have proved:

Proposition. *The vector spaces* $\text{Tan}_x(X)$ *and* $(\mathfrak{m}_{X,x}/\mathfrak{m}_{X,x}^2)^*$ *are isomorphic.*

16.2 Zariski tangent space

16.2.1 Using the discussion in 16.1, we can define the tangent space at a point of an arbitrary variety as follows:

Definition. *Let X be a variety and $x \in X$. We define the* Zariski tangent space *at x, denoted by $\text{T}_x(X)$, the vector space* $(\mathfrak{m}_{X,x}/\mathfrak{m}_{X,x}^2)^*$.

16.2.2 Let us denote by $\rho : \mathfrak{m}_{X,x} \to \mathfrak{m}_{X,x}/\mathfrak{m}_{X,x}^2$ the canonical surjection. If $s \in \mathcal{O}_{X,x}$, we denote by $s(x)$ its value at x. Recall that $\mathcal{O}_{X,x} = \Bbbk \oplus \mathfrak{m}_{X,x}$.

Let us consider $\Bbbk$ as an $\mathcal{O}_{X,x}$-module via the homomorphism defined by the evaluation at x, that is $s.\lambda = s(x)\lambda$ for all $s \in \mathcal{O}_{X,x}$, $\lambda \in \Bbbk$. A $\Bbbk$-derivation of $\mathcal{O}_{X,x}$ in $\Bbbk$ (2.8.1) is then a $\Bbbk$-linear map $\delta : \mathcal{O}_{X,x} \to \Bbbk$ verifying

$$\delta(st) = s(x)\delta(t) + t(x)\delta(x)$$

where $s, t \in \mathcal{O}_{X,x}$. Such a derivation shall be called a *point derivation* of $\mathcal{O}_{X,x}$, and we denote by $\text{Der}_{\Bbbk}^x(\mathcal{O}_{X,x}, \Bbbk)$ the set of these derivations.

In the same way, if U is an affine open subset containing x, we can consider $\Bbbk$ as an $\mathbf{A}(U)$-module by setting $f.\lambda = f(x)\lambda$, for $f \in \mathbf{A}(U)$ and $\lambda \in \Bbbk$. We shall also call a $\Bbbk$-derivation of $\mathbf{A}(U)$ to $\Bbbk$, a point derivation of $\mathbf{A}(U)$, that is, a $\Bbbk$-linear map $\Delta : \mathbf{A}(U) \to \Bbbk$ verifying for $f, g \in \mathbf{A}(U)$,

$$\Delta(fg) = f(x)\Delta(g) + g(x)\Delta(f).$$

We shall denote by $\text{Der}_{\Bbbk}^x(\mathbf{A}(U), \Bbbk)$ the set of these derivations.

16.2.3 Let $\lambda \in \text{T}_x(X)$ and define a map $L_\lambda : \mathcal{O}_{X,x} \to \Bbbk$ by

$$L_\lambda|_{\Bbbk} = 0 \ , \ L_\lambda|_{\mathfrak{m}_{X,x}} = \lambda \circ \rho.$$

If $s \in \mathcal{O}_{X,x}$, then $s - s(x) \in \mathfrak{m}_{X,x}$. We deduce easily that if $s, t \in \mathcal{O}_{X,x}$, then

(3) $$L_\lambda(st) = s(x)L_\lambda(t) + t(x)L_\lambda(s).$$

Thus $L_\lambda \in \text{Der}_{\Bbbk}^x(\mathcal{O}_{X,x}, \Bbbk)$.

Conversely, let $\delta \in \text{Der}_{\Bbbk}^x(\mathcal{O}_{X,x}, \Bbbk)$. If s is the germ at x of the constant function 1, then

$$\delta(s) = \delta(ss) = s(x)\delta(s) + s(x)\delta(s) = 2\delta(s).$$

Hence $\delta(s) = 0$.

Further, if $s, t \in \mathcal{O}_{X,x}$ verify $s(x) = t(x) = 0$, then $\delta(st) = 0$. Thus $\delta|_{\mathfrak{m}^2_{X,x}} = 0$. It follows that there is a linear form λ on $\mathfrak{m}_{X,x}/\mathfrak{m}^2_{X,x}$ such that $\delta|_{\mathfrak{m}_{X,x}} = \lambda \circ \rho = L_\lambda$. Hence $\delta = L_\lambda$.

We have therefore identified the tangent space $T_x(X)$ of X at x with the vector space $\mathrm{Der}^x_{\Bbbk}(\mathcal{O}_{X,x}, \Bbbk)$. An element of $\mathrm{Der}^x_{\Bbbk}(\mathcal{O}_{X,x}, \Bbbk)$ shall be called a *tangent vector* of X at x.

16.2.4 Let U be an affine open subset of X containing x and let $\delta \in \mathrm{Der}^x_{\Bbbk}(\mathcal{O}_{X,x}, \Bbbk)$. For $f \in \mathbf{A}(U)$, set $\Delta^U_\delta(f) = \delta(f_x)$, where f_x denotes the germ of f at x. We verify easily that $\Delta^U_\delta \in \mathrm{Der}^x_{\Bbbk}(\mathbf{A}(U), \Bbbk)$.

Conversely, let $\Delta \in \mathrm{Der}^x_{\Bbbk}(\mathbf{A}(U), \Bbbk)$. If $g \in \mathbf{A}(U)$ is such that its germ at x is 0, then by 11.7.3, there exists $f \in \mathbf{A}(U)$ verifying $f(x) \neq 0$ and $fg = 0$. So

$$0 = \Delta(fg) = f(x)\Delta(g) + g(x)\Delta(f) = f(x)\Delta(g).$$

Hence $\Delta(g) = 0$.

Now let $s \in \mathcal{O}_{X,x}$, there exists $f \in \mathbf{A}(U)$ such that $f(x) = 0$ and s is the germ of some function g/f on $D(f)$. Let us assume that s is also the germ of a function g_1/f_1 with $f_1(x) \neq 0$. Then $gf_1 - g_1 f$ is identically zero in a neighbourhood of x. From the preceding paragraph, we have $\Delta(gf_1 - g_1 f) = 0$, which implies that

$$g(x)\Delta(f_1) - f(x)\Delta(g_1) = g_1(x)\Delta(f) - f_1(x)\Delta(g).$$

A simple computation gives:

$$f_1^2(x)[f(x)\Delta(g) - g(x)\Delta(f)] = f^2(x)[f_1(x)\Delta(g_1) - g_1(x)\Delta(f_1)].$$

From this, we deduce that if $h = g/f$ is defined on $D(f)$ where $f, g \in \mathbf{A}(U)$ and $f(x) \neq 0$, we can define a linear form δ on $\mathcal{O}_{X,x}$ by setting:

$$\delta(h_x) = f^{-2}(x)[f(x)\Delta(g) - g(x)\Delta(f)]$$

where h_x is the germ of h at x. We now verify easily that $\delta \in \mathrm{Der}^x_{\Bbbk}(\mathcal{O}_{X,x}, \Bbbk)$ and $\Delta^U_\delta = \Delta$.

We have therefore shown that the tangent space $T_x(X)$ can be identified with $\mathrm{Der}^x_{\Bbbk}(\mathbf{A}(U), \Bbbk)$. In particular, if U is an affine open subset of X containing x, then we can identify $T_x(X)$ with $T_x(U)$.

16.2.5 We shall give yet another interpretation of $T_x(X)$. Let U be an affine open subset of X containing x. Consider $\Bbbk$ as an $\mathbf{A}(U)$-module as in 16.2.2, and denote it by $\Bbbk_x$.

Let $\Omega_U = \Omega_{\Bbbk}(\mathbf{A}(U))$ be the module of differentials of $\mathbf{A}(U)$ over $\Bbbk$ (see 2.9.3). Recall from 2.9.5 that the map $f \mapsto f \circ d_{\mathbf{A}(U)/\Bbbk}$ is a $\Bbbk$-linear isomorphism from $\mathrm{Hom}_{\mathbf{A}(U)}(\Omega_U, \Bbbk_x)$ to $\mathrm{Der}_{\Bbbk}(\mathbf{A}(U), \Bbbk_x) = \mathrm{Der}^x_{\Bbbk}(\mathbf{A}(U), \Bbbk)$.

Let $\mathfrak{n}_x = \mathcal{I}(\{x\})$ be the maximal ideal of $\mathbf{A}(U)$ associated to x, and set

$$\Omega_U^x = \Omega_U/\mathbf{n}_x\Omega_U.$$

Since $\mathbf{A}(U) = \Bbbk \oplus \mathbf{n}_x$ as vector spaces, we can identify the $\mathbf{A}(U)$-module $\Bbbk_x$ with $\mathbf{A}(U)/\mathbf{n}_x$ via the canonical surjection $\mathbf{A}(U) \to \mathbf{A}(U)/\mathbf{n}_x$. So the vector spaces Ω_U^x and $\Bbbk_x \otimes_{\mathbf{A}(U)} \Omega_U$ are isomorphic. On the other hand, it is easy to verify that the map $\Omega_U \to \Bbbk_x \otimes_{\mathbf{A}(U)} \Omega_U$, $\alpha \mapsto 1 \otimes \alpha$, induces a $\Bbbk$-linear bijection $\Psi : \mathrm{Hom}_\Bbbk(\Bbbk_x \otimes_{\mathbf{A}(U)} \Omega_U, \Bbbk) \to \mathrm{Hom}_{\mathbf{A}(U)}(\Omega_U, \Bbbk_x)$, defined by $(\Psi(u))(\alpha) = u(1 \otimes \alpha)$ for all $u \in \mathrm{Hom}_\Bbbk(\Bbbk_x \otimes_{\mathbf{A}(U)} \Omega_U, \Bbbk)$ and $\alpha \in \Omega_U$.

It follows from the preceding paragraphs that there is a $\Bbbk$-linear isomorphism from $\mathrm{Der}_\Bbbk^x(\mathbf{A}(U), \Bbbk)$ to $\mathrm{Hom}_\Bbbk(\Omega_U^x, \Bbbk)$. So by 16.2.4, we may also identify the tangent space $\mathrm{T}_x(X)$ with the dual of Ω_U^x.

16.3 Differential of a morphism

16.3.1 Let X, Y be varieties, $u \in \mathrm{Mor}(X, Y)$ and $x \in X$. We saw in 12.8.6 that u induces a local morphism $u_x : \mathcal{O}_{Y,u(x)} \to \mathcal{O}_{X,x}$. Since $u_x(\mathfrak{m}_{Y,u(x)})$ is contained in $\mathfrak{m}_{X,x}$, we have $u_x(\mathfrak{m}_{Y,u(x)}^2) \subset \mathfrak{m}_{X,x}^2$. It follows that u_x induces a map $u^x : \mathfrak{m}_{Y,u(x)}/\mathfrak{m}_{Y,u(x)}^2 \to \mathfrak{m}_{X,x}/\mathfrak{m}_{X,x}^2$.

We shall call the dual map

$$\mathrm{T}_x(u) : \mathrm{T}_x(X) \to \mathrm{T}_{u(x)}(Y)$$

of u^x, the *differential* of the morphism u at x. It is also sometimes denoted by du_x. To avoid confusions with the elements of $\Omega_\Bbbk(\mathbf{A}(U))$, we shall not use the notation du_x in this chapter.

A straightforward verification shows that if $v : Y \to Z$ is a morphism, then:

$$\mathrm{T}_x(v \circ u) = \mathrm{T}_{u(x)}(v) \circ \mathrm{T}_x(u).$$

16.3.2 Let us conserve the notations of 16.3.1. Let $y = u(x)$, and

$$\theta_x : \mathrm{T}_x(X) \to \mathrm{Der}_\Bbbk^x(\mathcal{O}_{X,x}, \Bbbk) \ , \ \theta_y : \mathrm{T}_y(Y) \to \mathrm{Der}_\Bbbk^y(\mathcal{O}_{Y,y}, \Bbbk)$$

be the isomorphisms defined in 16.2.3. Denote by $\mathrm{T}'_x(u) : \mathrm{Der}_\Bbbk^x(\mathcal{O}_{X,x}, \Bbbk) \to \mathrm{Der}_\Bbbk^y(\mathcal{O}_{Y,y}, \Bbbk)$ the unique linear map such that $\mathrm{T}'_x(u) \circ \theta_x = \theta_y \circ \mathrm{T}_x(u)$. If $\lambda \in \mathrm{T}_x(X)$ and $\mu = \mathrm{T}_x(u)(\lambda)$, then

$$\theta_x(\lambda)|_\Bbbk = 0 \ , \ \theta_x(\lambda)|_{\mathfrak{m}_{X,x}} = \lambda \circ \rho_x \ , \ \theta_y(\mu)|_\Bbbk = 0 \ , \ \theta_y(\mu)|_{\mathfrak{m}_{Y,y}} = \mu \circ \rho_y$$

where $\rho_x : \mathfrak{m}_{X,x} \to \mathfrak{m}_{X,x}/\mathfrak{m}_{X,x}^2$ and $\rho_y : \mathfrak{m}_{Y,y} \to \mathfrak{m}_{Y,y}/\mathfrak{m}_{Y,y}^2$ denote the canonical surjections. Since $\mu = \lambda \circ u^x$, we have

$$\theta_y(\mu)|_{\mathfrak{m}_{Y,y}} = \lambda \circ u^x \circ \rho_y = \lambda \circ \rho_x \circ (u_x|_{\mathfrak{m}_{Y,y}}).$$

We deduce therefore that $\mathrm{T}'_x(u)$ is the map:

$$\mathrm{Der}_\Bbbk^x(\mathcal{O}_{X,x}, \Bbbk) \to \mathrm{Der}_\Bbbk^y(\mathcal{O}_{Y,y}, \Bbbk) \ , \ \delta \mapsto \delta \circ u_x.$$

16.3.3 Let V be an affine open neighbourhood of $y = u(x)$ in Y and U an affine open neighbourhood of x in X such that $u(U) \subset V$. Denote by

$$\alpha_x : \mathrm{Der}_{\Bbbk}^x(\mathcal{O}_{X,x}, \Bbbk) \to \mathrm{Der}_{\Bbbk}^x(\mathbf{A}(U), \Bbbk) \; , \;\; \alpha_y : \mathrm{Der}_{\Bbbk}^y(\mathcal{O}_{Y,y}, \Bbbk) \to \mathrm{Der}_{\Bbbk}^y(\mathbf{A}(V), \Bbbk)$$

the isomorphisms in 16.2.4. Let $\mathrm{T}_x''(u) : \mathrm{Der}_{\Bbbk}^x(\mathbf{A}(U), \Bbbk) \to \mathrm{Der}_{\Bbbk}^y(\mathbf{A}(V), \Bbbk)$ be the unique linear map such that $\alpha_y \circ \mathrm{T}_x'(u) = \mathrm{T}_x''(u) \circ \alpha_x$.

It follows from 16.2.4 and 16.3.2 that if $\delta \in \mathrm{Der}_{\Bbbk}^x(\mathcal{O}_{X,x}, \Bbbk)$ and $g \in \mathbf{A}(V)$,

$$(\alpha_y \circ \mathrm{T}_x'(u)(\delta))(g) = (\alpha_y \circ \delta \circ u_x)(g) = \delta \circ u_x(g_y) = \delta((g \circ u)_x) = \alpha_x(\delta)(g \circ u).$$

where g_y is the germ of g at y. It follows that if $\mathbf{A}(u)$ is the comorphism of u, then

$$\mathrm{T}_x''(u) : \mathrm{Der}_{\Bbbk}^x(\mathbf{A}(U), \Bbbk) \to \mathrm{Der}_{\Bbbk}^y(\mathbf{A}(V), \Bbbk) \; , \;\; \Delta \mapsto \Delta \circ \mathbf{A}(u).$$

Remark. When there is no confusion, we shall also denote $\mathrm{T}_x'(u)$ and $\mathrm{T}_x''(u)$ by $\mathrm{T}_x(u)$.

16.3.4 Let us take $X = \Bbbk^m$ and $Y = \Bbbk^n$. Then

$$\mathbf{A}(X) = \Bbbk[T_1, \dots, T_m], \mathbf{A}(Y) = \Bbbk[S_1, \dots, S_n], \mathrm{Tan}_x(X) = \Bbbk^m, \mathrm{Tan}_y(Y) = \Bbbk^n.$$

Let $P_1, \dots, P_n \in \mathbf{A}(X)$ be such that $u(x) = (P_1(x), \dots, P_n(x))$ for all $x \in X$. We have the following bijective linear maps:

$$\beta_x : \mathrm{Der}_{\Bbbk}^x(\mathbf{A}(X), \Bbbk) \to \mathrm{Tan}_x(X) = \Bbbk^m \; , \;\; \Delta \mapsto (\Delta(T_1), \dots, \Delta(T_m)) \, ,$$
$$\beta_y : \mathrm{Der}_{\Bbbk}^y(\mathbf{A}(Y), \Bbbk) \to \mathrm{Tan}_y(Y) = \Bbbk^n \; , \;\; \Delta \mapsto (\Delta(S_1), \dots, \Delta(S_n)).$$

Let $F_x(u) : \Bbbk^m \to \Bbbk^n$ be the linear map such that $F_x(u) \circ \beta_x = \beta_y \circ \mathrm{T}_x(u)$. If $Q \in \mathbf{A}(Y)$ and $\Delta \in \mathrm{Der}_{\Bbbk}^x(\mathbf{A}(X), \Bbbk)$, then 16.3.3 implies that:

$$
\begin{aligned}
(\mathrm{T}_x(u)(\Delta)) (Q) = \Delta(Q \circ u) &= \sum_{i=1}^m \frac{\partial(Q \circ u)}{\partial T_i}(x)\Delta(T_i) \\
&= \sum_{i=1}^m \left(\sum_{j=1}^n \frac{\partial Q}{\partial S_j}(u(x)) \frac{\partial P_j}{\partial T_i}(x) \right) \Delta(T_i) \\
&= \sum_{j=1}^n \frac{\partial Q}{\partial S_j}(y) \left(\sum_{i=1}^m \frac{\partial P_j}{\partial T_i}(x)\Delta(T_i) \right).
\end{aligned}
$$

Hence $F_x(u)$ is the linear map

$$\Bbbk^m \to \Bbbk^n \; , \;\; (v_1, \dots, v_m) \mapsto \left(\sum_{i=1}^m \frac{\partial P_1}{\partial T_i}(x)v_i, \dots \sum_{i=1}^m \frac{\partial P_n}{\partial T_i}(x)v_i \right).$$

Let $M_x(u) = [a_{ij}] \in \mathrm{M}_{n,m}(\Bbbk)$ be the matrix of $F_x(u)$ with respect to the canonical bases of $\Bbbk^m$ and $\Bbbk^n$. Then for $1 \leqslant i \leqslant m$, $1 \leqslant j \leqslant n$, we have:

$$a_{ij} = \frac{\partial P_j}{\partial T_i}(x).$$

16.3.5 Now assume that X is a closed subvariety of $Y = \mathbb{k}^n$. Let $j : X \to Y$ be the canonical injection, $\mathfrak{a} = \mathcal{I}(X) \subset \mathbf{A}(Y) = \mathbb{k}[T_1, \ldots, T_n]$ and, for $1 \leqslant i \leqslant n$, let t_i be the image of T_i in $\mathbf{A}(X) = \mathbf{A}(Y)/\mathfrak{a}$.

Let $v = (v_1, \ldots, v_n) \in Y$. For $P \in \mathfrak{a}$ and $v \in \mathrm{Tan}_x(X)$, we saw in 16.1 that

$$\frac{\partial P}{\partial T_1}(x)v_1 + \cdots + \frac{\partial P}{\partial T_n}(x)v_n = 0.$$

We can then define an element $\Delta_v \in \mathrm{Der}_{\mathbb{k}}^x(\mathbf{A}(X), \mathbb{k})$ by setting for $f \in \mathbf{A}(X)$ and P any representative of f in $\mathbf{A}(Y)$,

$$\Delta_v(f) = \frac{\partial P}{\partial T_1}(x)v_1 + \cdots + \frac{\partial P}{\partial T_n}(x)v_n.$$

Conversely, given $\Delta \in \mathrm{Der}_{\mathbb{k}}^x(\mathbf{A}(X), \mathbb{k})$. We have, for $P \in \mathbf{A}(Y)$,

$$\Delta(P(t_1, \ldots, t_n)) = \frac{\partial P}{\partial T_1}(x)\Delta(t_1) + \cdots + \frac{\partial P}{\partial T_n}(x)\Delta(t_n).$$

In particular, if $P \in \mathfrak{a}$, then $\Delta(P(t_1, \ldots, t_n)) = 0$. It follows that we have a bijective linear map

$$\gamma_x : \mathrm{Der}_{\mathbb{k}}^x(\mathbf{A}(X), \mathbb{k}) \to \mathrm{Tan}_x(X) \ , \ \Delta \mapsto (\Delta(t_1), \ldots, \Delta(t_n)).$$

Let $G_x(j) : \mathrm{Tan}_x(X) \to \mathrm{Tan}_y(Y)$ be the unique linear map such that $G_x(j) \circ \gamma_x = \beta_y \circ \mathrm{T}_x(j)$ where β_y is as in 16.3.4. It follows immediately from the preceding discussion that $G_x(j)$ is the canonical injection from $\mathrm{Tan}_x(X)$ to $\mathbb{k}^n = \mathrm{Tan}_y(Y)$.

16.3.6 Now let X and Y be closed subvarieties of $\mathbb{k}^m$ and $\mathbb{k}^n$ respectively and denote by $j_1 : X \to \mathbb{k}^m$ and $j_2 : Y \to \mathbb{k}^n$ the canonical injections.

There exist $P_1, \ldots, P_n \in \mathbb{k}[T_1, \ldots, T_m]$ (not necessarily unique in general) such that $u(z) = (P_1(z), \ldots, P_n(z))$ for all $z \in X$. Let $v : \mathbb{k}^m \to \mathbb{k}^n$ be the morphism defined by $z \mapsto (P_1(z), \ldots, P_n(z))$. Then we have $j_2 \circ u = v \circ j_1$. From this, we obtain the following commutative diagram:

$$
\begin{array}{ccc}
\mathrm{Tan}_x(X) & \xrightarrow{H_x(u)} & \mathrm{Tan}_y(Y) \\
\gamma_x \uparrow & & \uparrow \gamma_y \\
\mathrm{Der}_{\mathbb{k}}^x(\mathbf{A}(X), \mathbb{k}) & \xrightarrow{\mathrm{T}_x(u)} & \mathrm{Der}_{\mathbb{k}}^y(\mathbf{A}(Y), \mathbb{k}) \\
\mathrm{T}_x(j_1) \downarrow & & \downarrow \mathrm{T}_y(j_2) \\
\mathrm{Der}_{\mathbb{k}}^x(\mathbb{k}[T_1, \ldots, T_m], \mathbb{k}) & \xrightarrow{\mathrm{T}_x(v)} & \mathrm{Der}_{\mathbb{k}}^y(\mathbb{k}[S_1, \ldots, S_n], \mathbb{k})
\end{array}
$$

where $H_x(u)$ is the unique linear map such that $H_x(u) \circ \gamma_x = \gamma_y \circ \mathrm{T}_x(u)$. Consequently, $H_x(u)$ is the linear map

$$\mathrm{Tan}_x(X) \to \mathrm{Tan}_y(Y) \; , \; (v_1,\ldots,v_n) \mapsto \left(\sum_{i=1}^{n} \frac{\partial P_1}{\partial T_i}(x)v_i, \ldots, \sum_{i=1}^{n} \frac{\partial P_n}{\partial T_i}(x)v_i \right).$$

In particular, if $\ell : \Bbbk^n \to \Bbbk^m$ is the linear map whose matrix $[a_{ij}]$ with respect to the canonical bases of $\Bbbk^m$ and $\Bbbk^n$ is defined by $a_{ij} = \dfrac{\partial P_i}{\partial T_j}(x)$, then $\ell(\mathrm{Tan}_x(X)) \subset \mathrm{Tan}_y(Y)$ and $\ell(v) = H_x(u)(v)$ for all $v \in \mathrm{Tan}_x(X)$.

16.3.7 Proposition. *Let X be a subvariety of a variety Y, $j : X \to Y$ the canonical injection and $x \in X$. The map $\mathrm{T}_x(j) : \mathrm{T}_x(X) \to \mathrm{T}_x(Y)$ is injective.*

Proof. The set X is locally closed in Y (12.2.7), so it is the intersection of a closed subset F and an open subset U of Y. Let $j_1 : X \to U$, $j_2 : U \to Y$ be the canonical injections. Since $j = j_2 \circ j_1$, we have $\mathrm{T}_x(j) = \mathrm{T}_x(j_2) \circ \mathrm{T}_x(j_1)$. By 16.2.4 and 16.3.3, $\mathrm{T}_x(j_2)$ is an isomorphism. So we are reduced to the case where X is closed in Y. We may assume that $Y = \mathrm{Spm}(A)$ is affine and $X = \mathrm{Spm}(A/\mathfrak{a})$ where $\mathfrak{a}$ is a radical ideal of A. By 16.3.3, if $\varphi : A \to A/\mathfrak{a}$ is the canonical surjection, then $\mathrm{T}_x(j)$ is the map $\mathrm{Der}_{\Bbbk}^x(A/\mathfrak{a}, \Bbbk) \to \mathrm{Der}_{\Bbbk}^x(A, \Bbbk)$, $\Delta \mapsto \Delta \circ \varphi$. It is now clear that $\mathrm{T}_x(j)$ is injective. $\square$

16.3.8 Let U, V be affine varieties and $(x, y) \in U \times V$. Recall from 11.8.1 that the map associating $f \otimes g \in \mathbf{A}(U) \otimes_{\Bbbk} \mathbf{A}(V)$ to the function $(u, v) \mapsto f(u)g(v)$ on $U \times V$ is an isomorphism from $\mathbf{A}(U) \otimes_{\Bbbk} \mathbf{A}(V)$ to $\mathbf{A}(U \times V)$.

For $D \in \mathrm{Der}_{\Bbbk}^x(\mathbf{A}(U), \Bbbk)$ and $D' \in \mathrm{Der}_{\Bbbk}^y(\mathbf{A}(V), \Bbbk)$, we define a map $\theta_{D,D'} : \mathbf{A}(U \times V) \to \Bbbk$ by setting, for $f \in \mathbf{A}(U)$, $g \in \mathbf{A}(V)$,

$$\theta_{D,D'}(f \otimes g) = D(f)g(y) + f(x)D'(g).$$

We verify easily that $\theta_{D,D'} \in \mathrm{Der}_{\Bbbk}^{(x,y)}(\mathbf{A}(U \times V), \Bbbk)$. So we have a linear map $\theta : \mathrm{Der}_{\Bbbk}^x(\mathbf{A}(U), \Bbbk) \times \mathrm{Der}_{\Bbbk}^y(\mathbf{A}(V), \Bbbk) \to \mathrm{Der}_{\Bbbk}^{(x,y)}(\mathbf{A}(U \times V), \Bbbk)$, $(D, D') \mapsto \theta_{D,D'}$. Since $\theta_{D,D'}(f \otimes 1) = D(f)$ and $\theta_{D,D'}(1 \otimes g) = D'(g)$, the map θ is injective.

Now let $\Delta \in \mathrm{Der}_{\Bbbk}^{(x,y)}(\mathbf{A}(U \times V), \Bbbk)$. For $f \in \mathbf{A}(U)$ and $g \in \mathbf{A}(V)$, we set:

$$D(f) = \Delta(f \otimes 1) \; , \; D'(g) = \Delta(1 \otimes g).$$

Then $D \in \mathrm{Der}_{\Bbbk}^x(\mathbf{A}(U), \Bbbk)$, $D' \in \mathrm{Der}_{\Bbbk}^y(\mathbf{A}(V), \Bbbk)$ and $\Delta = \theta_{D,D'}$. We have therefore proved that θ is an isomorphism.

Proposition. *Let X, Y be varieties and $(x, y) \in X \times Y$. Denote by*

$$j : X \times \{y\} \to X \times Y \; , \; j' : \{x\} \times Y \to X \times Y$$

the canonical injections. Then the map

$$\mathrm{T}_{(x,y)}(X \times \{y\}) \times \mathrm{T}_{(x,y)}(\{x\} \times Y) \to \mathrm{T}_{(x,y)}(X \times Y)$$
$$(D, D') \mapsto \mathrm{T}_{(x,y)}(j)(D) + \mathrm{T}_{(x,y)}(j')(D')$$

is a bijection. In particular, $\mathrm{T}_x(X) \times \mathrm{T}_y(Y)$ and $\mathrm{T}_{(x,y)}(X \times Y)$ are isomorphic as $\Bbbk$-vector spaces.

Proof. By 16.2.4, we may assume that X and Y are affine. Let us identify $X \times \{y\}$ with X and $\{x\} \times Y$ with Y. The comorphism of j (resp. j') is the map sending $f \otimes g$ to the function $u \mapsto f(u)g(y)$ on X (resp. $v \mapsto f(x)g(v)$ on Y). It follows from 16.3.3 that $\mathrm{T}_{(x,y)}(j)(D) = \theta_{D,0}$ and $\mathrm{T}_{(x,y)}(j')(D') = \theta_{0,D'}$. So the result follows. $\square$

16.4 Some lemmas

16.4.1 In this section, k is a commutative integral domain and K its quotient field. For $\lambda \in k \setminus \{0\}$, set $S_\lambda = \{\lambda^n; n \in \mathbb{N}\}$ and $k_\lambda = S_\lambda^{-1} k$. Since k is an integral domain, we may identify $S_\lambda^{-1} k$ with the subring of K consisting of elements of the form $\lambda^{-n} a$ with $n \in \mathbb{N}$ and $a \in k$. For any k-module M, denote by M_λ the k_λ-module $k_\lambda \otimes_k M$, which we may identify with $S_\lambda^{-1} M$ (2.4.1).

Let $m, p \in \mathbb{N}^*$ and $r \in \{1, \ldots, \min(m, p)\}$. We define the following matrix:

$$J_r = \begin{pmatrix} I_r & 0 \\ 0 & 0 \end{pmatrix} \in \mathrm{M}_{m,p}(k).$$

16.4.2 Let $\varLambda = [\alpha_{ij}] \in \mathrm{M}_{m,p}(k)$ and $(e_1, \ldots, e_p)$ be the canonical basis of k^p. Denote by $\mathfrak{M}_k(\varLambda)$ or $\mathfrak{M}(\varLambda)$, the quotient of k^p by the submodule generated by the elements

$$\varepsilon_i = \alpha_{i1} e_1 + \cdots + \alpha_{ip} e_p , \ 1 \leqslant i \leqslant m.$$

Lemma. (i) *Let* $P \in \mathrm{GL}_m(k)$, $Q \in \mathrm{GL}_p(k)$. *Then* $\mathfrak{M}(P\varLambda) = \mathfrak{M}(\varLambda)$, *and the k-modules* $\mathfrak{M}(\varLambda Q)$ *and* $\mathfrak{M}(\varLambda)$ *are isomorphic.*

(ii) *There exist* $\lambda \in k \setminus \{0\}$, $P \in \mathrm{GL}_m(k_\lambda)$ *and* $Q \in \mathrm{GL}_p(k_\lambda)$ *such that* $\varLambda = P J_r Q$, *where r is the rank of the matrix $\varLambda$ considered as an element of* $\mathrm{M}_{m,p}(K)$.

Proof. (i) Let $P = [p_{ij}]$, $P\varLambda = [\alpha'_{ij}]$ and for $1 \leqslant i \leqslant m$, set

$$\varepsilon'_i = \alpha'_{i1} e_1 + \cdots + \alpha'_{ip} e_p.$$

Denote by L (resp. L') the submodule of k^p generated by the ε_i (resp. ε'_i). Since $\varepsilon'_i = p_{i1} \varepsilon_1 + \cdots + p_{im} \varepsilon_m$ for $1 \leqslant i \leqslant m$, $L' \subset L$. Now the same argument applied to P^{-1} implies that $L \subset L'$. Therefore $L = L'$ and $\mathfrak{M}(P\varLambda) = \mathfrak{M}(\varLambda)$.

Next let $Q = [q_{ij}]$, $\varLambda Q = [\alpha''_{ij}]$ and for $1 \leqslant i \leqslant m$ and $1 \leqslant j \leqslant p$, set:

$$\varepsilon''_i = \alpha''_{i1} e_1 + \cdots + \alpha''_{ip} e_p , \ e''_j = q_{j1} e_1 + \cdots + q_{jp} e_p.$$

So $(e''_1, \ldots, e''_p)$ is a basis of k^p and for $1 \leqslant i \leqslant m$,

$$\varepsilon''_i = \alpha_{i1} e''_1 + \cdots + \alpha_{ip} e''_p.$$

Let u be the automorphism of k^p verifying $u(e_j) = e''_j$ for $1 \leqslant j \leqslant p$. Then u induces an isomorphism from $\mathfrak{M}(\varLambda)$ to $\mathfrak{M}(\varLambda Q)$.

(ii) It is well-known that there exist $P \in \mathrm{GL}_m(K)$ and $Q \in \mathrm{GL}_p(K)$ such that $\varLambda = P J_r Q$. To obtain the result, it suffices to take λ such that the coefficients of P, Q, P^{-1}, Q^{-1} belong to k_λ. $\square$

16.4.3 Lemma. *Let us conserve the notations of* 16.4.2.

(i) *There exists* $\lambda \in k \setminus \{0\}$ *such that* $\mathfrak{M}(\Lambda)_\lambda$ *is a free* k_λ*-module of rank* $p - r$.

(ii) *We can choose* λ *such that* $p - r$ *images of the elements* $1 \otimes e_i$ *of* $k_\lambda \otimes_k k^p$ *form a basis of the* k_λ*-module* $\mathfrak{M}(\Lambda)_\lambda$.

Proof. Part (i) is clear by 16.4.2. As for part (ii), for $1 \leqslant i \leqslant p$, denote by ν_i the canonical image of $1 \otimes e_i$ in $\mathfrak{M}(\Lambda)_\lambda$. Let $(f_1, \ldots, f_{p-r})$ be a basis of the k_λ-module $\mathfrak{M}(\Lambda)_\lambda$. Since the ν_i's generate $\mathfrak{M}(\Lambda)_\lambda$, we may assume that there exists $M = [\beta_{ij}] \in \mathrm{M}_{p-r}(k_\lambda)$ verifying $\det M \neq 0$ and

$$\nu_j = \beta_{1j} f_1 + \cdots + \beta_{p-r,j} f_{p-r}$$

for $1 \leqslant j \leqslant p - r$. Modify λ if necessary, we may assume that the coefficients of the matrix $M^{-1} \in \mathrm{GL}_{p-r}(K)$ are in k_λ. So part (ii) follows. $\square$

16.4.4 Let X be an irreducible affine variety and $x \in X$. The ring $k = \mathbf{A}(X)$ is an integral domain and recall from 12.7.6 that we can identify its quotient field K with the field of rational functions $\mathbf{R}(X)$ on X.

Let $\Lambda = [\alpha_{ij}] \in \mathrm{M}_{m,p}(k)$ be as in 16.4.3, r its rank considered as an element of $\mathrm{M}_{m,p}(K)$ and let $\Lambda(x) \in \mathrm{M}_{m,p}(\Bbbk)$ be the matrix $[\alpha_{ij}(x)]$. Consider $\Bbbk$ as a k-module, denoted by $\Bbbk_x$, as in 16.2.2.

Set $\mathfrak{M}_k(\Lambda)^x = \mathfrak{M}_k(\Lambda)/\mathfrak{n}_x \mathfrak{M}_k(\Lambda)$. Then $\mathfrak{M}_k(\Lambda)^x = \Bbbk_x \otimes_k \mathfrak{M}_k(\Lambda)$, and it is clear that we may identify $\mathfrak{M}_k(\Lambda)^x$ with $\mathfrak{M}_\Bbbk(\Lambda(x))$ as $\Bbbk$-vector spaces.

Lemma. *Let* $\mathcal{E} = \{x \in X;\ \dim_\Bbbk \mathfrak{M}_k(\Lambda)^x = p - r\}$.

(i) $\mathcal{E}$ *is a non-empty open subset of* X.

(ii) *We have* $\dim_K K \otimes_k \mathfrak{M}_k(\Lambda) = p - r$.

(iii) *If* $x \in \mathcal{E}$, *then there exists* $f \in k = \mathbf{A}(X)$ *verifying* $f(x) \neq 0$, *and* $\mathfrak{M}_k(\Lambda)_f$ *is a free* k_f*-module of rank* $p - r$.

Proof. (i) Let $\mathcal{L}_s$ (resp. $\mathcal{L}_s(x)$) be the set of square submatrices of size s in Λ (resp. $\Lambda(x)$). Then $\det \Theta = 0$ if $\Theta \in \mathcal{L}_s$ with $s > r$, and there exists $\Theta \in \mathcal{L}_r$ such that $\det \Theta \neq 0$.

Similarly, since $\det \Theta \in k$, we have $\det A = 0$ if $A \in \mathcal{L}_s(x)$ with $s > r$, and the set $\mathcal{F}$ of $x \in X$ for which there exists $A \in \mathcal{L}_r(x)$ verifying $\det A \neq 0$, is a non-empty open subset of X.

Since we can identify $\mathfrak{M}_k(\Lambda)^x$ with $\mathfrak{M}_\Bbbk(\Lambda(x))$, we have $\mathcal{E} = \mathcal{F}$. So we have proved part (i).

(ii) This is clear because the rank of Λ in $\mathrm{M}_{m,p}(K)$ is r.

(iii) Let $x \in \mathcal{E}$. There exists $A \in \mathcal{L}_r(x)$ such that $\det A \neq 0$. Reindexing if necessary, we may assume that $A = [\alpha_{ij}(x)]_{1 \leqslant i,j \leqslant r}$ and denote by Θ the matrix $[\alpha_{ij}]_{1 \leqslant i,j \leqslant r}$. We have $\Theta \in \mathrm{GL}_r(k_f)$ where $f = \det \Theta$.

Let $(e_1, \ldots, e_p)$ be the canonical basis of k^p and for $1 \leqslant j \leqslant p$, ν_j the image of $1 \otimes e_j$ in $\mathfrak{M}_k(\Lambda)_f$. Then for $1 \leqslant i \leqslant m$, $\alpha_{i1} \nu_1 + \cdots + \alpha_{ip} \nu_p = 0$, or equivalently

$$\alpha_{i1}\nu_1 + \cdots + \alpha_{ir}\nu_r = -(\alpha_{i,r+1}\nu_{r+1} + \cdots + \alpha_{ip}\nu_p).$$

It follows from our choice of A that $\nu_1, \ldots, \nu_r$ are k_f-linear combinations of $\nu_{r+1}, \ldots, \nu_p$. So by (ii), the elements $1 \otimes \nu_{r+1}, \ldots, 1 \otimes \nu_p$ form a basis of the K-vector space $K \otimes_{k_f} \mathfrak{M}_k(\Lambda)_f$. Consequently $\nu_{r+1}, \ldots, \nu_p$ are linearly independent over k_f. Thus $\mathfrak{M}_k(\Lambda)_f$ is a free k_f-module of rank $p - r$. $\quad\square$

16.5 Smooth points

16.5.1 Proposition. *Let X be a variety and $x \in X$. The dimension of $\mathrm{T}_x(X)$ is greater than or equal to the dimension of any irreducible component of X containing x.*

Proof. By 16.2.4 and 16.3.7, we may assume that X is irreducible and affine. Let $X = \mathrm{Spm}(A)$ and $\mathfrak{n}_x = \mathcal{I}(\{x\})$. Since the k-vector spaces $\mathfrak{n}_x/\mathfrak{n}_x^2$ and $\mathfrak{m}_{X,x}/\mathfrak{m}_{X,x}^2$ are isomorphic, it suffices to show that $\dim_k \mathfrak{n}_x/\mathfrak{n}_x^2 \geqslant \dim X$.

Let $f_1, \ldots, f_r \in \mathfrak{n}_x$ be such that their images in $\mathfrak{n}_x/\mathfrak{n}_x^2$ form a basis of the k-vector space $\mathfrak{n}_x/\mathfrak{n}_x^2$. By 2.6.9, we have $\mathfrak{n}_x = Af_1 + \cdots + Af_r$. Consider the morphism

$$u : X \to k^r \, , \, y \mapsto (f_1(y), \ldots, f_r(y)).$$

If $u(y) = 0$, then $\mathfrak{n}_x \subset \mathfrak{n}_y$ and so $\mathfrak{n}_x = \mathfrak{n}_y$, which in turn implies that $x = y$.

For $y \in X$, denote by $e(y)$ the maximum dimension of an irreducible component of $u^{-1}(u(y))$ containing y. By 15.5.7, the set of $y \in X$ verifying $e(y) \geqslant 1$ is closed in X, so the set U of $y \in X$ such that $e(y) = 0$ is open in X. From the previous paragraph, we have $x \in U$. Let $v : U \to k^r$ denote the restriction of u to U. Then $\dim v^{-1}(z) = 0$ for all $z \in v(U)$. Thus $\dim X = \dim U \leqslant \dim k^r = r$ (15.5.4 (i)) which proves our result. $\quad\square$

16.5.2 Definition. *Let X be an irreducible variety of dimension n. A point $x \in X$ is called* smooth *(or* simple*) if $\dim \mathrm{T}_x(X) = n$. The variety X is called* smooth *(or* non-singular*) if all its points are smooth.*

16.5.3 Remarks. 1) When X is not irreducible, let $\dim_x(X)$ denote the maximum dimension of irreducible components of X containing x. Then x is a *smooth point* of X if $\mathrm{T}_x(X) = \dim_x X$. Similarly, the variety X is *smooth* if all its points are smooth.

2) We saw in 16.1 that the affine space k^n is smooth. Similarly the projective space $\mathbb{P}_n$ is also smooth, because any point of $\mathbb{P}_n$ is contained in an open affine neighbourhood isomorphic to k^n. By using the same argument, we see that the Grassmannian varieties $\mathbb{G}_{n,r}$ are smooth.

3) If x is a smooth point in an irreducible variety X, then we can show that the ring $\mathcal{O}_{X,x}$ is factorial, and hence an integrally closed domain.

16.5.4 Proposition. *Let $X = \mathrm{Spm}(A)$ be an irreducible affine variety, $x \in X$ a smooth point, $\mathfrak{n}_x$ the maximal ideal of A associated to x, and $f_1, \ldots, f_n \in \mathfrak{n}_x$ be such that their images in $\mathfrak{n}_x/\mathfrak{n}_x^2$ form a basis of the k-vector space $\mathfrak{n}_x/\mathfrak{n}_x^2$. Then $f_1, \ldots, f_n$ are algebraically independent over k, and* *so they form a transcendence basis of the field of rational functions* $\mathbf{R}(X)$ *over* $\Bbbk$.

Proof. Consider the following morphism:

$$u : X \to \Bbbk^n \ , \ y \mapsto (f_1(y), \dots, f_n(y)).$$

Suppose that there exists $P \in \Bbbk[T_1, \dots, T_n] \setminus \{0\}$ such that $P(f_1, \dots, f_n) = 0$, then $\overline{u(X)} \subset \mathcal{V}(P)$ and so $\dim \overline{u(X)} \leqslant n - 1$ (14.2.3). Now proceed as in the proof of 16.5.1, we deduce that $\dim X \leqslant n - 1$ which is absurd. So the f_i's are algebraically independent over $\Bbbk$. The last part follows from 12.7.6 and 5.4.8. $\square$

16.5.5 Let X be an irreducible variety, $x \in X$ and U an affine open subset of X containing x. Recall that $\Omega_U = \Omega_\Bbbk(\mathbf{A}(U))$ is the module of differentials of $\mathbf{A}(U)$ over $\Bbbk$.

Theorem. *Let X be an irreducible variety of dimension n.*
   (i) *Let x be a smooth point of X. There is an affine open neighbourhood U of x such that Ω_U is a free $\mathbf{A}(U)$-module of rank n.*
   (ii) *The set of smooth points in X is a non-empty open subset of X.*

Proof. Let U be an affine open subset of X. By 12.7.6, we may identify the quotient field of $\mathbf{A}(U)$ with $\mathbf{R}(X)$. Further, the $\mathbf{R}(X)$-vector spaces $\mathbf{R}(X) \otimes_{\mathbf{A}(U)} \Omega_U$ and $\Omega_\Bbbk(\mathbf{R}(X))$ are isomorphic (2.9.6). Since the characteristic of $\Bbbk$ is zero, it follows from 5.8.3 and 14.1.2 that

$$\dim_{\mathbf{R}(X)} \Omega_\Bbbk(\mathbf{R}(X)) = \operatorname{tr} \deg_\Bbbk \mathbf{R}(X) = n.$$

So we may assume that $X = \operatorname{Spm}(A)$ is affine with $A = \Bbbk[T_1, \dots, T_p]/\mathfrak{a}$ where $\mathfrak{a}$ is a radical ideal of $\Bbbk[T_1, \dots, T_p]$. Let $\{P_1, \dots, P_m\}$ be a system of generators for $\mathfrak{a}$, $\Lambda = \left[\dfrac{\partial P_i}{\partial T_j}\right]_{ij} \in \mathrm{M}_{m,p}(\mathbf{A}(X))$ and r the rank of Λ considered as an element of $\mathrm{M}_{m,p}(\mathbf{R}(X))$. Using the notations of 16.4.2, the $\mathbf{A}(X)$-modules Ω_X and $\mathfrak{M}_{\mathbf{A}(X)}(\Lambda)$ are isomorphic (2.9.8). In view of 16.4.4, we obtain that $n = p - r$.

By 16.2.5 and 16.4.4, the $\Bbbk$-vector spaces $\mathrm{T}_x(X)$ and $\mathfrak{M}_{\mathbf{A}(X)}(\Lambda)^x$ have the same dimension. So part (ii) follows by 16.4.4.

Finally, let $x \in X$ be a smooth point. By 16.4.4, there exists $f \in \mathbf{A}(X)$ such that $f(x) \neq 0$ and $\mathfrak{M}_{\mathbf{A}(X)}(\Lambda)_f$ is a free $\mathbf{A}(X)_f$-module of rank n. If $U = D(f)$, then U is an affine open subset of X containing f such that $\mathbf{A}(U) = \mathbf{A}(X)_f$. So we have part (i). $\square$

16.5.6 Remark. If X is equidimensional and if X is the disjoint union of its irreducible components, then 16.5.5 implies that the set of smooth points points of X is a dense open subset of X.

16.5.7 Let $u : X \to Y$ be a morphism of irreducible affine varieties, $x \in X$ and $y = u(x)$. Recall that any $\mathbf{A}(X)$-module can be considered as an $\mathbf{A}(Y)$-module via the comorphism $\mathbf{A}(u)$ of u.

Denote by $\Bbbk_x$ (resp. $\Bbbk_y$) the $\mathbf{A}(X)$-module (resp. $\mathbf{A}(Y)$-module) such that $f.\lambda = f(x)\lambda$ (resp. $g.\lambda = g(y)\lambda$) for all $f \in \mathbf{A}(X)$, $g \in \mathbf{A}(Y)$ and $\lambda \in \Bbbk$. Observe that $\Bbbk_x$ and $\Bbbk_y$ are isomorphic as $\mathbf{A}(Y)$-modules.

Denote $d_{\mathbf{A}(X)/\Bbbk}$ and $d_{\mathbf{A}(Y)/\Bbbk}$ by d_X and d_Y. We saw in 2.9.5 that there exists a unique homomorphism $\theta(u) : \Omega_Y \to \Omega_X$ of $\mathbf{A}(Y)$-modules such that $\theta(u) \circ d_Y = d_X \circ \mathbf{A}(u)$.

Recall from 16.3.3 that $\mathbf{T}_x(u)$ can be identified with the map

$$\mathrm{Der}_{\Bbbk}(\mathbf{A}(X), \Bbbk_x) \to \mathrm{Der}_{\Bbbk}(\mathbf{A}(Y), \Bbbk_y) \ , \ \Delta \mapsto \Delta \circ \mathbf{A}(u).$$

Now, using the following canonical isomorphisms (2.9.5):

$$\mathrm{Hom}_{\mathbf{A}(X)}(\Omega_X, \Bbbk_x) \to \mathrm{Der}_{\Bbbk}(\mathbf{A}(X), \Bbbk_x), \mathrm{Hom}_{\mathbf{A}(Y)}(\Omega_Y, \Bbbk_y) \to \mathrm{Der}_{\Bbbk}(\mathbf{A}(Y), \Bbbk_y),$$

we can identify $\mathbf{T}_x(u)$ with the map

$$\mathrm{Hom}_{\mathbf{A}(X)}(\Omega_X, \Bbbk_x) \to \mathrm{Hom}_{\mathbf{A}(Y)}(\Omega_Y, \Bbbk_y) \ , \ \varphi \mapsto \varphi \circ \theta(u).$$

Finally, with the identifications in 16.2.5, $\mathbf{T}_x(u)$ can be identified with the map from $\mathrm{Hom}_{\Bbbk}(\Omega_X^x, \Bbbk) \to \mathrm{Hom}_{\Bbbk}(\Omega_Y^y, \Bbbk)$, $\varphi \mapsto \psi$, where

$$\psi(\lambda_1 d_Y(g_1) + \cdots + \lambda_s d_Y(g_s)) = \varphi(\lambda_1 d_X(g_1 \circ u) + \cdots + \lambda_s d_X(g_s \circ u))$$

for $\lambda_1, \ldots, \lambda_s \in \Bbbk$ and $g_1, \ldots, g_s \in \mathbf{A}(Y)$.

Theorem. *Let $u : X \to Y$ be a morphism of irreducible varieties.*

(i) Suppose that there exists a smooth point $x \in X$ such that $y = u(x)$ is a smooth point in Y and $\mathbf{T}_x(u) : \mathbf{T}_x(X) \to \mathbf{T}_y(Y)$ is surjective. Then u is dominant.

(ii) Suppose that u is dominant. There exists a non-empty open subset U of X such that for all $x \in U$, $u(x)$ is a smooth point in Y and $\mathbf{T}_x(u)$ is a surjection from $\mathbf{T}_x(X)$ onto $\mathbf{T}_{u(x)}(Y)$.

Proof. In view of 16.5.5, we are reduced to the case where X and Y are affine, smooth and irreducible, and Ω_X (resp. Ω_Y) is a free $\mathbf{A}(X)$-module (resp. $\mathbf{A}(Y)$-module) of rank $m = \dim X$ (resp. $n = \dim Y$).

We have the following homomorphism of free $\mathbf{A}(X)$-modules:

$$\varphi(u) : \mathbf{A}(X) \otimes_{\mathbf{A}(Y)} \Omega_Y \to \Omega_X \ , \ 1 \otimes \alpha \mapsto \theta(u)(\alpha).$$

Let us fix bases for these free modules and let $\Lambda = [a_{ij}] \in \mathrm{M}_{m,n}(\mathbf{A}(X))$ be the matrix of $\varphi(u)$ with respect to these bases. For $x \in X$, set $\Lambda(x) = [a_{ij}(x)] \in \mathrm{M}_{m,n}(\Bbbk)$.

With the notations of 16.2.5, let

$$\varphi(u)^x : \Omega_Y^y = \Bbbk_x \otimes_{\mathbf{A}(X)} (\mathbf{A}(X) \otimes_{\mathbf{A}(Y)} \Omega_Y) \to \Bbbk_x \otimes_{\mathbf{A}(X)} \Omega_X = \Omega_X^x$$

be the homomorphism induced by $\varphi(u)$ by scalar extension. In suitable bases, $\Lambda(x)$ is the matrix of $\varphi(u)^x$. The preceding description shows that $\varphi(u)^x$ is the dual map of $T_x(u)$. Thus $\varphi(u)^x$ and $T_x(u)$ have the same rank.

(i) Suppose that $T_x(u)$ is surjective. Then $\varphi(u)^x$ is injective, so the rank of $\Lambda(x)$ is n. As in the proof of 16.4.4, we deduce that the rank of Λ (over $\mathbf{R}(X)$) is at least n. But the rank of Λ is at most n, so we have equality.

It follows that $\varphi(u)$ is injective, which implies in turn that $\theta(u)$ is injective. Finally, since Ω_X (resp. Ω_Y) is a free $\mathbf{A}(X)$-module (resp. $\mathbf{A}(Y)$-module), $\mathbf{A}(u)$ is injective (5.8.4). Hence u is dominant (6.5.9).

(ii) If u is dominant, then $\mathbf{A}(u)$ is injective. By 5.8.4, $\varphi(u)$ is injective. Hence the rank of Λ (over $\mathbf{R}(X)$) is n. There exists therefore a non-empty open subset U of X such that for $x \in U$, the rank of $\Lambda(x)$ is n. It follows from the discussion above that $T_x(u)$ is surjective for all $x \in U$. $\square$

References

- [5], [19], [22], [26], [37], [40], [78].

Normal varieties

17.1 Normal varieties

17.1.1 Let us begin by a remark which will be useful later on. Let X be a prevariety and U, V affine open subsets of X such that $U \cap V \neq \emptyset$. Any $x \in U \cap V$ has a fundamental system of neighbourhoods consisting of subsets $D(s)$ such that $D(s) \subset U \cap V$ and $s \in \mathbf{A}(U)$. For each of these subsets $D(s)$, there exists $t \in \mathbf{A}(V)$ such that $D(t)$ is a neighbourhood of x in $D(s)$. Denote by $u : \mathbf{A}(V) \to \mathbf{A}(U)_s = \mathbf{A}(D(s))$ the comorphism of the canonical injection $D(s) \to V$. We have $D(t) = D(u(t))$, and $u(t)$ is of the form r/s^n for some $r \in \mathbf{A}(U)$. Thus $D(u(t)) = D(sr)$ where $sr \in \mathbf{A}(U)$.

Hence open subsets contained in $U \cap V$ which are principal in both U and V, form a base for the topology on $U \cap V$.

17.1.2 Definition. *A point x in a variety X is said to be* normal *if the local ring $\mathcal{O}_{X,x}$ of X at x is an integrally closed domain. The variety X is* normal *if all its points are normal.*

17.1.3 Remarks. 1) If X is irreducible, then it is normal if and only if it admits an open covering consisting of normal subvarieties.

2) If x is a normal point of X, then it is contained in a unique irreducible component of X (12.8.5). We deduce that if X is normal, then irreducible components of X are pairwise disjoint. So we may reduce the study of normal varieties to irreducible normal varieties.

17.1.4 Proposition. *Let $X = \mathrm{Spm}(A)$ be an irreducible affine variety. Then X is normal if and only if A is an integrally closed domain.*

Proof. Suppose that X is normal. By 12.8.2, we have

$$A = \bigcap_{x \in X} \mathcal{O}_{X,x}.$$

It follows from 3.2.4 that A is an integrally closed domain.

Suppose that A is an integrally closed domain. For $\mathfrak{p} \in \mathrm{Spec}(A)$, the ring $A_{\mathfrak{p}}$ is an integrally closed domain (3.2.5). So X is normal since $\mathcal{O}_{X,x}$ is isomorphic to $A_{\mathfrak{n}_x}$, where $\mathfrak{n}_x = \mathcal{I}(\{x\})$ is the maximal ideal of A associated to x (12.8.3).   □

17.1.5 Corollary. *The set of normal points of an irreducible variety X contains a non-empty open subset of X.*

Proof. We may assume that $X = \mathrm{Spm}(A)$ is affine. By 6.2.2, there exist $t_1, \ldots, t_n \in A$ algebraically independent over $\Bbbk$ such that A is integral over $C = \Bbbk[t_1, \ldots, t_n]$. Let B be the integral closure of A. Then B is integral over C (3.1.7), and by 5.5.6, B is a finitely generated C-module, hence B is a finitely generated A-module. Let $Y = \mathrm{Spm}(B)$ and $u : Y \to X$ be the morphism induced by the injection $A \to B$. The morphism u is dominant (6.5.9) and birational since $\mathrm{Fract}(A) = \mathrm{Fract}(B)$. It follows from 15.5.6 that there exists a non-empty open subset U of X such that the restriction $u^{-1}(U) \to U$ of u is an isomorphism. By 17.1.4, Y is normal, and the result follows.   □

17.1.6 Example. The affine space $\Bbbk^n$ is normal (4.2.9 and 17.1.4). Since $\mathbb{P}_n$ has an open covering by subsets isomorphic to $\Bbbk^n$, it is normal. In the same way, the Grassmannian varieties $\mathbb{G}_{n,r}$ are also normal.

17.1.7 Proposition. *Let X, Y be irreducible varieties and $u : X \to Y$ a surjective finite morphism. Assume further that Y is normal. Then for all $x \in X$ and for all irreducible subvariety W of Y containing $u(x)$, there exists an irreducible component of $u^{-1}(W)$ containing x which dominates W.*

Proof. Since u is finite, there is an affine open subset V containing $u(x)$ such that $U = u^{-1}(V)$ is an affine open neighbourhood of x. We may therefore assume that $X = \mathrm{Spm}(A)$, $Y = \mathrm{Spm}(B)$ are affine where $B \subset A$ are integral domains, B is an integrally closed domain and A is a finitely generated B-module.

Let $\mathfrak{p} = \mathcal{I}(W)$, $\mathfrak{m} = \mathcal{I}(\{x\})$, $\mathfrak{n} = \mathcal{I}(\{u(x)\})$. We have $\mathfrak{n} = B \cap \mathfrak{m}$ and $\mathfrak{p} \subset \mathfrak{n}$. By 5.7.8, there exists $\mathfrak{q} \in \mathrm{Spec}(A)$ such that $\mathfrak{q} \subset \mathfrak{m}$ and $\mathfrak{p} = B \cap \mathfrak{q}$. So $x \in \mathcal{V}(\mathfrak{q})$ and $u(\mathcal{V}(\mathfrak{q})) = W$. Hence there is an irreducible component of $u^{-1}(W)$ containing $\mathcal{V}(\mathfrak{q})$ and it dominates W.   □

17.1.8 Corollary. *Let $u : X \to Y$ be a dominant morphism of irreducible varieties. There exists a non-empty open subset V of Y such that:*

(i) *The variety V is normal.*

(ii) *For all $y \in V$, $x \in u^{-1}(y)$ and irreducible subvariety W of Y containing y, there exists an irreducible component of $u^{-1}(W)$ containing x which dominates W.*

Proof. By 15.4.2, there exist a non-empty affine open subset V of Y, $m \in \mathbb{N}$, and a covering $(U_i)_{i \in I}$ of $u^{-1}(V)$ by affine open subsets such that the restriction $v_i : U_i \to V$ of u factors through $U_i \xrightarrow{v_i} V \times \Bbbk^m \xrightarrow{\text{pr}} V$, where pr denotes the canonical projection. Note that v_i is a surjective finite morphism.

We may assume that V is normal (17.1.5), and x belongs to one of the U_i's. We are reduced to the case where X is one of the U_i's, so X is affine.

Since $\mathbf{A}(V \times \Bbbk^m) = \mathbf{A}(V) \otimes_{\Bbbk} \Bbbk[T_1, \ldots, T_m] = \mathbf{A}(V)[T_1, \ldots, T_m]$, it follows from 3.2.8 that $\mathbf{A}(V \times \Bbbk^m)$ is an integrally closed domain. So $V \times \Bbbk^m$ is normal (17.1.4). Now v_i is a surjective finite morphism, so the result follows from 17.1.7. $\square$

17.2 Normalization

17.2.1 Theorem. *Let X be an irreducible variety and K a finite field extension of $\mathbf{R}(X)$, the field of rational functions on X. There exist a normal irreducible variety Y such that $\mathbf{R}(Y)$ is isomorphic to K, and a surjective finite morphism $u : Y \to X$ such that there is a covering $(U_i)_{i \in I}$ of X by affine open subsets verifying $u^{-1}(U_i)$ is affine and $\mathbf{A}(u^{-1}(U_i))$ is the integral closure of $\mathbf{A}(U_i)$ in $\mathbf{R}(Y)$. Furthermore, if Y_1 is an irreducible variety and $u_1 : Y_1 \to X$ is a surjective finite morphism having the same properties, then there is a unique isomorphism $v : Y \to Y_1$ such that $u = u_1 \circ v$.*

Proof. Let us prove the existence of Y.

First, suppose that $X = \mathrm{Spm}(A)$ is affine. Let B be the integral closure of A in K, and $Y = \mathrm{Spm}(B)$. Then B is a finitely generated A-module (6.2.10). Take $u : Y \to X$ to be the morphism induced by the canonical injection $A \to B$. The result follows since K is the quotient field of B (5.5.5).

Now let us treat the general case. Let $(U_i)_{i \in I}$ be a covering of X by non-empty affine open subsets. For $i \in I$, let $V_i = \mathrm{Spm}(B_i)$ where B_i is the integral closure of $A_i = \mathbf{A}(U_i)$ in K. Denote by $\theta_i : A_i \to B_i$ the canonical injection and $u_i = \mathrm{Spm}(\theta_i)$.

Let $i, j \in I$. By 12.5.6, $U_i \cap U_j$ is a non-empty affine open subset, and the algebra $A_{ij} = \mathbf{A}(U_i \cap U_j)$ is generated by the functions $x \mapsto f(x)g(x)$ where $x \in U_i \cap U_j$, $f \in A_i$ and $g \in A_j$. Let B_{ij} be the integral closure of A_{ij} in K, $\theta_{ij} : A_{ij} \to B_{ij}$ the canonical injection and $u_{ij} = \mathrm{Spm}(\theta_{ij})$.

Let $U \subset U_i \cap U_j$ be a principal open subset of both U_i and U_j (17.1.1). We have $U = D(t) = D(s_i) = D(s_j)$ with $t \in A_{ij}$, $s_i \in A_i$ and $s_j \in A_j$.

Observe that if $s \in A_i \setminus \{0\}$, then the ring of regular functions on $u_i^{-1}(D(s)) = D(\theta_i(s))$ is equal to $B_i[1/\theta_i(s)]$, the integral closure in K of the ring $(A_i)_s = A_i[1/s]$ of regular functions on $D(s)$ (3.1.9). The same applies for A_{ij}, B_{ij}, θ_{ij}.

It follows from the above observation that the rings $\mathbf{A}(u_i^{-1}(U))$, $\mathbf{A}(u_j^{-1}(U))$ and $\mathbf{A}(u_{ij}^{-1}(U))$ are all equal to $B_{ij}[1/\theta_{ij}(t)]$, which is the integral closure in K of $A_{ij}[1/t]$. Since the set of open subsets in $U_i \cap U_j$ which are principal in both U_i and U_j form a base of the topology on $U_i \cap U_j$ (17.1.1), it follows from 9.5.1 and 12.1.1 that the open subsets $u_i^{-1}(U_i \cap U_j)$ and $u_j^{-1}(U_i \cap U_j)$ can be identified canonically.

Let Y be the prevariety defined by gluing the V_i's along the intersections $V_i \cap V_j = \mathrm{Spm}(B_{ij})$ via these canonical identifications. Then the map $u : Y \to X$, which is u_i on V_i, is a morphism.

Since X is a variety and each $V_i = u^{-1}(U_i)$ is a variety, Y is also a variety (12.5.10). Furthermore, the V_i's are irreducible and the $V_i \cap V_j$'s are non-empty, so Y is irreducible (1.1.6).

Let us now prove that Y is unique up to an isomorphism. Let Y_1 be another such variety. Let $(U'_j)_{j \in J}$ be a covering of X by affine open subsets. The open subsets contained in $U_i \cap U'_j$ which are principal in both U_i and U'_j form a base of the topology of $U_i \cap U'_j$ (17.1.1). Using the same arguments, we have our result. □

17.2.2 The normal variety Y in 17.2.1 is called the *normalization* of X in K. When $K = \mathbf{R}(X)$, we shall simply say that Y is the *normalization* of X; in this case, the morphism u is birational.

17.2.3 Corollary. *Let X be an irreducible complete variety. If Y is the normalization of X in a finite extension of $\mathbf{R}(X)$, then Y is complete.*

Proof. Since the morphism $Y \to X$ is finite (17.2.1), the result is a consequence of 15.3.9. □

17.2.4 Proposition. *Let X, Y be irreducible varieties, $u : X \to Y$ a surjective finite morphism and $n = [\mathbf{R}(X) : \mathbf{R}(Y)]$. Suppose that Y is normal.*

(i) *For all $y \in Y$, the cardinality of $u^{-1}(y)$ is at most n.*

(ii) *There exists a non-empty open subset V of Y such that $u^{-1}(V)$ is isomorphic to the normalization of V in $\mathbf{R}(X)$, and the cardinality of $u^{-1}(y)$ is n for all $y \in V$.*

Proof. (i) Since u is finite, $u^{-1}(y)$ is a finite set for all $y \in Y$ (15.3.4). Further, if V is an affine open subset of Y, $u^{-1}(V)$ is an affine open subset of X, and $\mathbf{A}(u^{-1}(V))$ is a finite $\mathbf{A}(V)$-algebra (15.3.3). So we are reduced to the case where $X = \mathrm{Spm}(A)$ and $Y = \mathrm{Spm}(B)$ are affine.

Let $y \in Y$ and $u^{-1}(y) = \{x_1, \ldots, x_m\}$. There exists $a \in \mathbf{A}(X)$ such that the $a(x_i)'s$, $1 \leqslant i \leqslant m$, are pairwise distinct. Let

$$P(T) = T^p + \alpha_{p-1}T^{p-1} + \cdots + \alpha_0$$

the minimal polynomial of a over $\mathbf{R}(Y)$. We have $\deg(P) \leqslant n$, and by 5.5.4, $\alpha_{p-1}, \ldots, \alpha_0$ are integral over B. Hence $\alpha_{p-1}, \ldots, \alpha_0 \in B$. Let

$$Q(T) = T^p + \alpha_{p-1}(y)T^{p-1} + \cdots + \alpha_0(y).$$

Then $a(x_1), \ldots, a(x_m)$ are roots of Q. Hence $m \leqslant n$.

(ii) Again, we may assume that $X = \mathrm{Spm}(A)$ and $Y = \mathrm{Spm}(B)$ are affine. By 5.6.7, there exists $a \in \mathbf{R}(X)$ such that $\mathbf{R}(X) = \mathbf{R}(Y)(a)$. We may assume further that $a \in A$ (proceed as in 5.7.9). The degree of the minimal polynomial P of a over $\mathbf{R}(Y)$ is n, and as above,

$$P(T) = T^p + \alpha_{p-1}T^{p-1} + \cdots + \alpha_0$$

where $\alpha_{p-1}, \ldots, \alpha_0 \in B$.

Let $d \in B$ be the discriminant of P. Then $d \neq 0$ since the characteristic of $\Bbbk$ is zero. Replace Y by $D(d)$ and X by $u^{-1}(D(d))$, we may assume that d is invertible in B. In view of 5.7.9, A is the integral closure of B in $\mathbf{R}(X)$ and $\{1, a, \ldots, a^{n-1}\}$ is a basis of the B-module A. Further, by 5.7.10, the cardinality of $u^{-1}(y)$ is n for all $y \in Y$.

By 17.1.5, there exists a non-empty open subset U of X which is a normal variety. The set $u(X \setminus U)$ is closed and nowhere dense in Y (15.3.4 and 15.3.8). So there exists an affine open subset V of Y such that $u^{-1}(V)$ is an affine open subset contained in U. Thus the affine variety $u^{-1}(V)$ is normal; its field of rational functions is $\mathbf{R}(X)$. Finally, $\mathbf{A}(u^{-1}(V))$ is the integral closure of $\mathbf{A}(V)$ in $\mathbf{R}(X)$. Hence we have our result. $\square$

17.2.5 Example. Let $X = \{(x_1, x_2) \in \Bbbk^2; x_1^3 - x_2^2 = 0\}$. We saw in 11.5.7 that X is the image of the morphism $u : \Bbbk \to \Bbbk^2$, $t \mapsto (t^2, t^3)$. Thus X is irreducible. However, $\mathbf{A}(X) = \Bbbk[T^2, T^3]$ is not an integrally closed domain (11.5.7). So X is not normal.

The set $Y = X \setminus \{(0,0)\}$ is the principal open subset $D_X(T^2)$ of X, and $\mathbf{A}(Y) = \Bbbk[T, T^{-1}]$. It follows from 3.2.5 that Y is normal. Thus $(0,0)$ is the only point in X which is not normal.

The morphism $v : \Bbbk \to X$, $t \mapsto u(t)$, is bijective, bicontinuous, birational (12.7.13) and finite (15.3.7). The integral closure of $\Bbbk[T^2, T^3]$ in $\Bbbk(T)$ is $\Bbbk[T]$. Thus $\Bbbk$ is the normalization of X.

17.3 Products of normal varieties

17.3.1 Lemma. *Let $X = \mathrm{Spm}(A)$ be an irreducible affine variety and E a finite-dimensional $\Bbbk$-vector subspace of A. Denote by F the subspace of A generated by the elements of A which, in $\mathbf{R}(X)$, can be represented in the form uv^{-1}, with $u \in E$ and $v \in E \setminus \{0\}$. Then F is finite-dimensional.*

Proof. Let us fix a finite normal extension L of $\mathrm{Fract}(A)$ (5.7.5).

There exist $t_1, \ldots, t_n \in A$ algebraically independent over $\Bbbk$ such that A is integral over $B = \Bbbk[t_1, \ldots, t_n]$ (6.2.2). The integral closure B' of B in L is also the integral closure of A in L (3.1.7). By 5.5.6, B' is a finitely generated B-module, so it is a finitely generated $\Bbbk$-algebra. Replace X by $\mathrm{Spm}(B')$, we may assume that A is the integral closure of B in L. Denote by $K = \mathrm{Fract}(B)$. The extension $K \subset L$ is finite and normal. Let d be its degree, which is also the order of the Galois group $\mathrm{Gal}(L/K)$ (5.6.6 and 5.7.4). Let $\mathrm{Gal}(L/K) = \{\sigma_1, \ldots, \sigma_d\}$ with $\sigma_1 = \mathrm{id}_L$. If $u \in L$, then $\sigma_1(u) \cdots \sigma_d(u) \in K$. Replace E by the sum of the $\sigma_i(E)$'s, we may assume that E is $\mathrm{Gal}(L/K)$-stable.

Let E_1 be the $\Bbbk$-vector space spanned by the products of k elements of E, with $k \leqslant d$. Clearly E_1 is finite-dimensional, and the same goes for $E_1 \cap K$. Since any element of E_1 is integral over the integrally closed domain B, we deduce that $E_1 \cap K \subset B$. Denote by B_r of elements in $B[t_1, \ldots, t_n]$ of degree at most r in the t_i's. Then there exists $r \in \mathbb{N}$ such that $E_1 \cap K \subset B_r$.

Now let $w \in A$, $u, v \in E$ be such that $w = uv^{-1}$. Then:

$$w = u\sigma_2(v) \cdots \sigma_d(v)/(\sigma_1(v) \cdots \sigma_d(v)).$$

We deduce that $w = u'(v')^{-1}$ where $u' \in A$ and $v' \in B_r \setminus \{0\}$.

There exists a basis $(z_1, \ldots, z_d)$ of L over K such that $A \subset Bz_1 \oplus \cdots \oplus Bz_d$ (5.5.6). It follows that there exists $s \in \mathbb{N}$ such that $E_1 \subset B_s z_1 \oplus \cdots \oplus B_s z_d$. Let us write

$$u' = p_1 z_1 + \cdots + p_d z_d \ , \ w = q_1 z_1 + \cdots + q_d z_d$$

where $p_1, q_1, \ldots, p_d, q_d \in B$. Then $q_i v' = p_i$ for $1 \leqslant i \leqslant d$, and we have $p_i \in B_s$, for $1 \leqslant i \leqslant d$, because $u' = u\sigma_2(v) \cdots \sigma_d(v) \in E_1$. Thus $q_i \in B_s$, and hence F is finite-dimensional. $\square$

17.3.2 Proposition. *Let X, Y be irreducible normal varieties. Then the variety $X \times Y$ is normal.*

Proof. We may assume that $X = \mathrm{Spm}(A)$ and $Y = \mathrm{Spm}(B)$ are affine. Let $f \in \mathbf{R}(X \times Y)$ be integral over $A \otimes_{\Bbbk} B$ and $\mathcal{D}$ its domain of definition. Then f can be represented in the form g/h where

$$g = u_1 \otimes v_1 + \cdots + u_n \otimes v_n \ , \ h = u_1 \otimes w_1 + \cdots + u_n \otimes w_n$$

with $u_1, \ldots, u_n \in A$ linearly independent and $v_1, w_1, \ldots, v_n, w_n \in B$.

For $y \in Y$, define $g_y, h_y \in A$ by:

$$g_y = v_1(y)u_1 + \cdots + v_n(y)u_n \ , \ h_y = w_1(y)u_1 + \cdots + w_n(y)u_n.$$

The set V of elements $y \in Y$ such that $w_i(y)$ is non-zero for some i is a non-empty open subset of Y. If $y \in V$, then h_y is a non-zero element of A. So there exists $x \in X$ such that $(x, y) \in D_{X \times Y}(h) \subset \mathcal{D}$ and $f_y = g_y/h_y \in \mathbf{R}(X)$ is a regular function on $D_X(h_y)$. Denote by $\mathcal{F}$ the set of maps from V to $\Bbbk$, $\mathcal{G}$ the subset of $\mathcal{F}$ consisting of maps u such that there exist a non-empty open subset $W \subset V$ and $v \in B$ verifying $u|_W = v|_W$. Since Y is irreducible, it is easy to see that $\mathcal{G}$ is a subspace of $\mathcal{F}$.

Let $y \in V$. Since f is integral over $A \otimes_{\Bbbk} B$, we verify easily that f_y is integral over A. Since X is normal, there exists a function $t_y \in A$ which represents f_y. On the other hand, f_y is the quotient of two elements of the vector subspace E of A generated by the u_i's. By 17.3.1, there exist $s_1, \ldots, s_m \in A$ linearly independent, such that $t_y = \rho_1(y)s_1 + \cdots + \rho_m(y)s_m$ where $\rho_1, \ldots, \rho_m \in \mathcal{F}$.

In particular, for all $x \in X$ such that $(x, y) \in \mathcal{D}$, $t_y(x) = f(x, y)$.

A similar argument shows that there is a non-empty open subset U of X verifying the following conditions: for all $x \in U$, there exist $y \in Y$ such that $(x, y) \in \mathcal{D}$, and $t^x \in B$ such that $t^x(y) = f(x, y)$ for all $(x, y) \in \mathcal{D}$. For $x \in U$, denote by V_x the set of $y \in V$ such that $(x, y) \in \mathcal{D}$. It is a non-empty open subset of V.

Let $x \in U$. For $y \in V_x$, we have

$$\rho_1(y)s_1(x) + \cdots + \rho_m(y)s_m(x) = t_y(x) = f(x, y) = t^x(y).$$

So the restrictions of t^x and $s_1(x)\rho_1 + \cdots + s_m(x)\rho_m$ to V_x are identical. It follows that $s_1(x)\rho_1 + \cdots + s_m(x)\rho_m \in \mathcal{G}$. Hence, if λ is a linear form on $\mathcal{F}$ which is zero on $\mathcal{G}$, then $s_1(x)\lambda(\rho_1) + \cdots + s_m(x)\lambda(\rho_m) = 0$. This being true for all x in the dense open subset U of X, we have:

$$\lambda(\rho_1)s_1 + \cdots + \lambda(\rho_m)s_m = 0.$$

But the s_i's are linearly independent, so $\lambda(\rho_i) = 0$ for $1 \leqslant i \leqslant m$. Hence $\rho_i \in \mathcal{G}$, $1 \leqslant i \leqslant m$. Thus there exist $t_1, \ldots, t_m \in B$ such that f can be represented in the form $s_1 \otimes t_1 + \cdots + s_m \otimes t_m \in A \otimes_{\Bbbk} B$. So our result follows. $\square$

17.4 Properties of normal varieties

17.4.1 Lemma. *Let $A \subset B$ be finitely generated $\Bbbk$-algebras, $X = \mathrm{Spm}(A)$, $Y = \mathrm{Spm}(B)$, $u : Y \to X$ the morphism associated to the canonical injection $A \to B$, $\mathfrak{n} \in \mathrm{Spm}(B)$, $\mathfrak{m} = \mathfrak{n} \cap A$, and $y \in Y$ the point associated to $\mathfrak{n}$. The following conditions are equivalent:*

(i) *$\mathfrak{n}$ is minimal among the prime ideals $\mathfrak{q} \in \mathrm{Spec}(B)$ such that $\mathfrak{q} \cap A = \mathfrak{m}$.*
(ii) *The point y is isolated in the fibre $u^{-1}(u(y))$ of the morphism u.*

Proof. The algebra $A/\mathfrak{m}$ is a subalgebra of $B/\mathfrak{n}$, which is isomorphic to $\Bbbk$ (6.4.1). So $\mathfrak{m} \in \mathrm{Spm}(A)$. Thus any proper ideal $\mathfrak{b}$ of B containing $\mathfrak{m}$ verifies $\mathfrak{b} \cap A = \mathfrak{m}$. Let $\sqrt{B\mathfrak{m}} = \mathfrak{q}_1 \cap \cdots \cap \mathfrak{q}_s$ where $\mathfrak{q}_1, \ldots, \mathfrak{q}_s \in \mathrm{Spec}(B)$ with $\mathfrak{q}_i \not\subset \mathfrak{q}_j$ whenever $i \neq j$. Then $\mathfrak{q}_1, \ldots, \mathfrak{q}_s$ are exactly the minimal prime ideal lying above $\mathfrak{m}$. Also, $Z = u^{-1}(u(y)) = \mathcal{V}(\mathfrak{q}_1) \cup \cdots \cup \mathcal{V}(\mathfrak{q}_s)$ is the decomposition of Z into irreducible components.

If part (i) is true, then we may take $\mathfrak{n} = \mathfrak{q}_1$, and Z is the disjoint union of $\{y\}$ and $\mathcal{V}(\mathfrak{q}_2) \cup \cdots \cup \mathcal{V}(\mathfrak{q}_s)$. Hence $\{y\}$ is open in Z, which proves (ii).

Conversely, if y is isolated in Z, then it is an irreducible component of Z. In view of the above discussion, we have (i). $\square$

17.4.2 Proposition. *Let $A \subset C \subset B$ be finitely generated $\Bbbk$-algebras verifying: they are integral domains, B is integral over C, A is integrally closed in B and $\mathrm{Fract}(B)$ is a finite extension of $\mathrm{Fract}(A)$. Let $\mathfrak{m} \in \mathrm{Spm}(A)$. Suppose that there exists $\mathfrak{n} \in \mathrm{Spm}(B)$ minimal in the set of $\mathfrak{q} \in \mathrm{Spec}(B)$ verifying $\mathfrak{q} \cap A = \mathfrak{m}$. Then the canonical homomorphism $A_{\mathfrak{m}} \to B_{\mathfrak{n}}$ is bijective.*

Proof. From our hypotheses, C is a finitely generated A-algebra, so $C = A[u_1, \ldots, u_m]$, where $u_1, \ldots, u_m \in C$. The case $m = 1$ is exactly 5.9.8. Let us proceed by induction on m.

Suppose that $m > 1$. Let B' be the integral closure of $A[u_1, \ldots, u_{m-1}]$ in B. It is finitely generated (3.3.5) and integrally closed in B. We have $C \subset B'[u_m] \subset B$, so B is integral over $B'[u_m]$ and $\mathrm{Fract}(B)$ is a finite extension of $\mathrm{Fract}(B')$. The ideal $\mathfrak{n}' = \mathfrak{n} \cap B' \in \mathrm{Spm}(B')$ and if $\mathfrak{q} \in \mathrm{Spec}(B)$ verifies $\mathfrak{q} \subset \mathfrak{n}$ and $\mathfrak{q} \cap B' = \mathfrak{n}'$, then $\mathfrak{q} \cap A = \mathfrak{n}' \cap A = \mathfrak{n} \cap A = \mathfrak{m}$, so it follows from our hypothesis on $\mathfrak{n}$ that $\mathfrak{q} = \mathfrak{n}$ and $\mathfrak{n}$ is minimal among the prime ideals of B lying above $\mathfrak{n}'$. We may therefore apply 5.9.8 to $B' \subset B'[u_m] \subset B$, and we obtain that the homomorphism $B'_{\mathfrak{n}'} \to B_{\mathfrak{n}}$ is bijective. Denote by

$$u : \mathrm{Spm}(B) \to \mathrm{Spm}(A) \ , \ v : \mathrm{Spm}(B') \to \mathrm{Spm}(A) \ , \ w : \mathrm{Spm}(B) \to \mathrm{Spm}(B')$$

the morphisms induced by the canonical injections. Note that $u = v \circ w$.

By 17.4.1, the point y corresponding to $\mathfrak{n}$ is isolated in $u^{-1}(u(y))$. It follows that y is also isolated in $w^{-1}(w(y))$. Since the homomorphism $B'_{\mathfrak{n}'} \to B_{\mathfrak{n}}$ is bijective, it follows from 12.8.8 that w induces an isomorphism from a neighbourhood of y in $\mathrm{Spm}(B)$ onto a neighbourhood of $w(y)$ in $\mathrm{Spm}(B')$. Hence $w(y)$ is isolated in $v^{-1}(v(w(y)))$. Applying 17.4.1, we deduce that $\mathfrak{n}'$ is minimal among the prime ideals of B' lying above $\mathfrak{m}$. By induction on $A \subset A[u_1, \ldots, u_{m-1}] \subset B'$, the homomorphism $A_{\mathfrak{m}} \to B'_{\mathfrak{n}'}$ is bijective. So we are done. $\square$

17.4.3 Let $u : X \to Y$ be a morphism of irreducible varieties. Suppose that there exist $x \in X$ and an open neighbourhood U of x in X such that u induces an isomorphism from U onto an open subset V of Y. We may assume that U is affine. We see in particular that u is dominant and birational. Further $U \cap u^{-1}(u(x)) = \{x\}$, and so x is isolated in $u^{-1}(u(x))$. In this situation, we have the following more precise statement.

Theorem. (Zariski's Main Theorem) *Let $u : X \to Y$ be a dominant morphism of irreducible varieties. Suppose that there exists $x \in X$ such that x is isolated in the $u^{-1}(u(x))$.*

(i) *We have $\dim X = \dim Y$, so $\mathbf{R}(X)$ is a finite extension of $\mathbf{R}(Y)$.*

(ii) *There is an open neighbourhood V of $u(x)$ in Y and an open neighbourhood U of x in $u^{-1}(V)$ such that the restriction of u factorizes through:*

$$U \xrightarrow{\ j\ } V' \xrightarrow{\ v\ } V$$

where v is finite and surjective, and j is an isomorphism from U onto an open subset of V'.

(iii) *If X is normal, we may suppose that $V = Y$ and V' is the normalization of Y in $\mathbf{R}(X)$.*

Proof. Since $\{x\}$ is an irreducible component of $u^{-1}(u(x))$, we have $\dim X = \dim Y$ by 15.5.3 and 15.5.4. In particular, $\mathbf{R}(X)$ is a finite extension of $\mathbf{R}(Y)$.

To prove the other parts, we may assume that $X = \mathrm{Spm}(B)$, $Y = \mathrm{Spm}(A)$ are affine, and $A \subset B$. Let A' (resp. A'') be the integral closure of A in B (resp. $\mathrm{Fract}(B)$). By 6.2.10, A'' is a finitely generated A-algebra and since A is Noetherian, A' is a finitely generated A-module. Thus A' is a finitely generated $\Bbbk$-algebra which is also an integral domain. Hence $Y' = \mathrm{Spm}(A')$ is irreducible and the morphism $v : Y' \to Y$, associated to the canonical injection $A \to A'$, is finite (15.3.2). It follows from the injections $A \to A' \to B$ that u factorizes through:

$$X \xrightarrow{\;w\;} Y' \xrightarrow{\;v\;} Y.$$

Since $w^{-1}(w(x)) \subset u^{-1}(u(x))$, the point x is isolated in $w^{-1}(w(x))$. So by 17.4.1, $\mathfrak{n} = \mathcal{I}(\{x\})$ is minimal among the prime ideals $\mathfrak{q}$ of B verifying $\mathfrak{q} \cap A' = \mathfrak{n} \cap A' = \mathfrak{n}' = \mathcal{I}(\{w(x)\})$. Applying 17.4.2 to $A' \subset B \subset B$, we obtain that the canonical homomorphism $A'_{\mathfrak{n}'} \to B_{\mathfrak{n}}$ is bijective. So w induces an isomorphism from a neighbourhood of x in X onto a neighbourhood of $w(x)$ of Y' (12.8.8). So we have proved (ii).

Let Z be the normalization of Y in $\mathbf{R}(X)$. There exist a surjective finite morphism $s : Z \to Y$ and a covering $(V_i)_{i \in I}$ of Y by affine open subsets such that $s^{-1}(V_i)$ is an affine open subset and $\mathbf{A}(s^{-1}(V_i))$ is the integral closure of $\mathbf{A}(V_i)$ in $\mathbf{R}(Z) \simeq \mathbf{R}(X)$.

Suppose that X is normal. Replacing Y by a V_i containing $u(x)$ and U by an affine open neighbourhood U' of x in X such that $U' \subset u^{-1}(V_i)$, we are reduced to the case where $Y = \mathrm{Spm}(A)$, $X = \mathrm{Spm}(B)$ with B is an integrally closed domain. Then A' (as defined above) is isomorphic to $\mathbf{A}(Z)$. So (iii) follows. $\square$

17.4.4 Corollary. *Let $u : X \to Y$ be a dominant morphism of irreducible varieties. The following conditions are equivalent:*

(i) *X is normal and for all $x \in X$, x is isolated in $u^{-1}(u(x))$.*

(ii) *The extension $\mathbf{R}(Y) \subset \mathbf{R}(X)$ is finite and u factorizes through:*

$$X \xrightarrow{\;j\;} Y' \xrightarrow{\;v\;} Y$$

where Y' is the normalization of Y in $\mathbf{R}(X)$, v is a surjective finite morphism as in 17.2.1, and j is an isomorphism from X onto an open subset of Y'.

If these conditions are satisfied, there exists an open subset V of Y such that V is normal, and for all $y \in V$, $\mathrm{card}\, u^{-1}(y) = [\mathbf{R}(X) : \mathbf{R}(Y)]$.

Proof. The fact that (ii) implies (i) is clear. Furthermore, the last part follows from 17.1.5 and 17.2.4.

Let us suppose that (i) is true. Let Y' be the normalization of Y in $\mathbf{R}(X)$, $v : Y' \to Y$ be as in 17.2.1 and $x \in X$. By 17.4.3, there exists an isomorphism j_U from an affine open neighbourhood U of x in X onto an open subset of Y' such that $u|_U = v \circ j_U$.

Let U' be another affine open neighbourhood of x and $j_{U'}$ an isomorphism from U' to an open subset of Y' such that $u|_{U'} = v \circ j_{U'}$. Then there exists an affine open neighbourhood $U'' \subset U \cap U'$ of x such that $u(U'')$ is contained in an affine open subset V of Y. Let $A = \mathbf{A}(V)$, $B = \mathbf{A}(U'')$ with B an integrally closed domain, and A' the integral closure of A in $\mathbf{R}(X)$. The restrictions $j_U|_{U''}$ and $j_{U'}|_{U''}$ correspond both to the injection $A' \to B$. So they are identical. We deduce that there exists a morphism $j : X \to Y'$ such that $j|_U = j_U$ for all affine open subset U with the above properties. Thus j is a local isomorphism, and (ii) follows by 12.5.11. $\square$

17.4.5 Corollary. *Let $u : X \to Y$ be a birational morphism of irreducible varieties. Suppose that Y is normal and for all $x \in X$, x is isolated in $u^{-1}(u(x))$. Then*

(i) *u is an isomorphism from X onto an open subset of Y.*

(ii) *If X is complete, then u is an isomorphism from X to Y.*

Proof. Let $v : X' \to X$ be the normalization of X (17.2.1). The morphism $u \circ v$ has the same properties as u. Since Y is normal and $\mathbf{R}(X') = \mathbf{R}(X) = \mathbf{R}(Y)$, it follows from 17.4.4 that $u \circ v$ induces an isomorphism w from X' onto an open subset of Y. So $w^{-1} \circ u$ is the reciprocal isomorphism of v. Hence $X' = X$ and (i) follows.

If X is complete, then $u(X)$ is closed in Y (13.4.3), so $u(X) = Y$. $\square$

17.4.6 Corollary. *Let $u : X \to Y$ be a bijective birational morphism of irreducible varieties. If Y is normal, then u is an isomorphism.*

17.4.7 Corollary. *Let $u : X \to Y$ be a morphism of irreducible normal varieties. Suppose that there exists $n \in \mathbb{N}^*$ such that $\operatorname{card} u^{-1}(y) = n$ for all $y \in Y$. Then u is a finite morphism and u factorizes through:*

$$X \xrightarrow{\ j\ } Y' \xrightarrow{\ v\ } Y$$

where $v : Y' \to Y$ is the normalization of Y in $\mathbf{R}(X)$ and j is an isomorphism from X to Y'.

Proof. Note that our hypothesis implies that $u(X) = Y$ and that for any $x \in X$, x is isolated in $u^{-1}(u(x))$ (since it is a finite set). So by 17.4.4, u factorizes through $X \xrightarrow{j} Y' \xrightarrow{v} Y$, where v is a surjective finite morphism and j is an isomorphism from X onto an open subset of Y'. So to obtain the result, we only need to show that $j(X) = Y'$.

Suppose that there exists $y' \in Y' \setminus j(X)$. Then $v^{-1}(v(y'))$ is a finite set (15.3.4). Since $j^{-1}(v^{-1}(v(y'))) = u^{-1}(v(y'))$, $\operatorname{card} v^{-1}(v(y')) \geqslant n$. If $\operatorname{card} v^{-1}(v(y')) = n$, then $y' \in j(X)$. So $\operatorname{card} v^{-1}(v(y')) \geqslant n + 1$.

On the other hand, $Y' \setminus j(X)$ is closed and nowhere dense in Y'. By 15.3.4 and 15.3.8, $v(Y' \setminus j(X))$ is also closed and nowhere dense in Y. Let $V = Y \setminus v(Y' \setminus j(X))$. Then V is open. But $v^{-1}(V) \subset j(X)$ and $\operatorname{card} v^{-1}(y) = n$ if $y \in V$. It follows from 17.2.4 that $\operatorname{card} v^{-1}(v(y')) \leqslant n$. Contradiction. $\square$

17.4.8 Corollary. *Let* $u : X \to Y$ *be a bijective morphism of irreducible varieties. If* Y *is normal, then* u *is an isomorphism.*

Proof. Let $v : X' \to X$ be the normalization of X (17.2.1), and $w = u \circ v$. Since v is finite and u is bijective, $w^{-1}(y)$ is finite for all $y \in Y$ (15.3.4). By 17.4.4, $\mathbf{R}(X) = \mathbf{R}(X')$ is a finite extension of $\mathbf{R}(Y)$. Let V be a non-empty open subset of Y such that $u^{-1}(V)$ is normal (17.1.5), and $v_1 : v^{-1}(u^{-1}(V)) \to u^{-1}(V)$ the morphism induced by v. Then v_1 is finite and surjective (15.3.3), and in view of 17.2.4, card $v_1^{-1}(z) = [\mathbf{R}(X') : \mathbf{R}(X)] = 1$ for all $z \in u^{-1}(V)$. Thus card $w^{-1}(y) = 1$ for all $y \in V$. The morphism $u \circ v_1 : v^{-1}(u^{-1}(V)) \to V$ is surjective and finite (17.4.7), so by applying 17.2.4, we obtain that $\mathbf{R}(X) = \mathbf{R}(Y)$. Hence u is birational, and the result follows from 17.4.6. □

17.4.9 Corollary. *Let* $u : X \to Y$ *be a dominant morphism of irreducible varieties. Suppose that* Y *is normal. If* $x \in X$ *is isolated in* $u^{-1}(u(x))$, *then for all neighbourhood* U *of* x *in* X, $u(U)$ *is a neighbourhood of* $u(x)$ *in* Y. *In particular, if* x *is isolated in* $u^{-1}(u(x))$ *for all* $x \in X$, *then* u *is open.*

Proof. Let $v : X' \to X$ be the normalization of X (17.2.1), $w = u \circ v$ and $x \in X$ be isolated in $u^{-1}(u(x))$. Then any $x' \in X'$ verifying $v(x') = x$ is isolated in $w^{-1}(w(x'))$. If U is a neighbourhood of x in X, then $v^{-1}(U)$ is a neighbourhood of x' in X'. So it suffices to prove that $w(v^{-1}(U))$ is a neighbourhood of $u(x)$ in Y. Therefore, we may assume that X is normal.

By 17.4.3, there exists an open neighbourhood V of $u(x)$ in Y and an open neighbourhood U of x in X such that the restriction $u|_U : U \to V$ factorizes through $U \xrightarrow{j} V' \xrightarrow{v} V$, where V' is the normalization of V in $\mathbf{R}(X)$ and j is an isomorphism from U onto an open subset of V'. So we can further reduce to the case where $X = Y'$ is the normalization of Y in a finite extension of $\mathbf{R}(X)$.

Let W be an open neighbourhood of x in X. Then $u(W) = Y \setminus T$ where $T = \{y \in Y; u^{-1}(y) \subset X \setminus W\}$. To obtain the result, it suffices to prove that $u(x) \notin \overline{T}$.

Suppose that $u(x) \in \overline{T}$. Then $u(x)$ is contained in the closure of an irreducible component T_0 of T. By 17.1.7, there exists an irreducible component Z of $u^{-1}(\overline{T_0})$ containing x verifying $\overline{u(Z)} = \overline{T_0}$. If $S \subset Z$ is an open subset containing x, then $u(S)$ contains a non-empty open subset of $\overline{T_0}$ (15.4.2), hence $u(S) \cap T_0 \neq \emptyset$. It follows that $S \cap u^{-1}(T_0) \neq \emptyset$, so $x \in \overline{u^{-1}(T_0)} \subset X \setminus W$ which contradicts the fact that W is a neighbourhood of x in X. □

17.4.10 Theorem. *Let* $u : X \to Y$ *be a dominant morphism of irreducible varieties and* $r = \dim X - \dim Y$. *Suppose that* Y *is normal. If* $x \in X$ *is such that all the irreducible components of* $u^{-1}(u(x))$ *containing* x *has dimension* r, *then the image via* u *of a neighbourhood of* x *in* X *is a neighbourhood of* $u(x)$ *in* Y.

Proof. Let $U \subset X$ (resp. $V \subset Y$) be an affine open neighbourhood of x (resp. $u(x)$) such that $U \subset u^{-1}(V)$. Since the intersection of U and an irreducible component of $u^{-1}(u(x))$ containing x is dense in the irreducible component, the restriction $U \to V$ satisfies the same hypotheses. So we may assume that $X = \mathrm{Spm}(B)$ and $Y = \mathrm{Spm}(A)$ are affine, where $A \subset B$ are integral domains and A is an integrally closed domain. Also, we only need to prove that $u(X)$ is a neighbourhood of $u(x)$ in Y. Replacing X by a smaller affine open subsets, we may assume further that all the irreducible components of $Z = u^{-1}(u(x))$ have dimension r.

Denote by $\mathfrak{m} = \mathcal{I}(\{u(x)\}) \in \mathrm{Spm}(A)$. We have $Z = \mathcal{V}(B\mathfrak{m})$, and if $\mathfrak{a} = \sqrt{B\mathfrak{m}}$, then $\mathbf{A}(Z) = B/\mathfrak{a}$. Since all the irreducible components of Z have the same dimension r, it follows from 6.2.2 and 6.2.3 that there exist elements $s_1, \ldots, s_r \in B/\mathfrak{a}$ algebraically independent over $\Bbbk$ such that $B/\mathfrak{a}$ is a finitely generated C-module, where $C = \Bbbk[s_1, \ldots, s_r]$.

For $1 \leqslant i \leqslant r$, fix $t_i \in B$ such that its class in $B/\mathfrak{a}$ is s_i. Denote by $\varphi : A[T_1, \ldots, T_r] \to B$ the A-module homomorphism such that $\varphi(T_i) = t_i$. Since the canonical injection $A \to B$ factorizes through $A \to A[T_1, \ldots, T_r] \xrightarrow{\varphi} B$, u factorizes through:

$$X \xrightarrow{v} Y \times \Bbbk^r \xrightarrow{\mathrm{pr}} Y.$$

If $W = \mathrm{pr}^{-1}(u(x)) = \{u(x)\} \times \Bbbk^r$, then $Z = v^{-1}(W)$. Since $W = \mathrm{Spm}(C)$, the restriction $Z \to W$ of v is finite and surjective. It follows that any point y of $u^{-1}(u(x))$ is isolated in $u^{-1}(u(y))$.

Let $S = \{x' \in X; x' \text{ is isolated in } v^{-1}(v(x'))\}$. The set S contains Z and is open in X (15.5.7). Since $Y \times \Bbbk^r$ is normal (17.1.6 and 17.3.2), the restriction of v to S is open (17.4.9). But pr is also open (15.5.8), we deduce therefore that $u(S)$ is open in Y, and hence $u(X)$ is a neighbourhood of $u(x)$ in Y.   □

17.4.11 Corollary. *Let $u : X \to Y$ be a dominant morphism of irreducible varieties. Suppose that Y is normal and for all $y \in u(X)$, all the irreducible components of $u^{-1}(y)$ have dimension $\dim X - \dim Y$. Then u is open.*

17.4.12 Proposition. *Let $u : X \to Y$ be a dominant morphism of irreducible varieties. If f is a regular function which is constant on the fibres $u^{-1}(y)$, $y \in Y$, then f, considered as an element of $\mathbf{R}(X)$, belongs to $\mathbf{R}(Y)$.*

Proof. Let $v : X \to Y \times \Bbbk$ be the morphism $x \mapsto (u(x), f(x))$ and $p : Y \times \Bbbk \to Y$ the canonical projection. Our hypothesis implies that the restriction of p to $v(X)$ is injective. Let $W \subset v(X)$ be such that W is open and dense in $\overline{v(X)}$ (15.4.2). Denote by w the restriction p to W, then w is injective and $p(W) = Y$. Replacing W by a smaller open subset if necessary, we may assume that W is normal (17.1.5). The set $u(W)$ contains a non-empty open subset V of Y (15.4.2) that we may also assume to be normal. Replacing W by $w^{-1}(W)$, we may further assume that $w : W \to V$ is bijective. By 17.4.8, w induces an isomorphism from W onto V. The restriction q to W of the canonical projection $Y \times \Bbbk \to \Bbbk$ is a regular function on W, and $q \circ (w|_W)^{-1}$ is regular on V. For $x \in u^{-1}(W)$, we have $q \circ (w|_W)^{-1}(u(x)) = f(x)$. So we are done.   □

17.4.13 Proposition. *Let x be a normal point of an irreducible variety X and f be a rational function on X whose domain of definition is $\mathcal{D}$. Suppose that $x \notin \mathcal{D}$, and denote by $\mathcal{E}$ the domain of definition of $f' = 1/f$. There exists a subvariety Y of X containing x verifying $Y \cap \mathcal{E} \neq \emptyset$ and such that $f'(y) = 0$ for all $y \in Y \cap \mathcal{E}$.*

Proof. Recall from 12.8.3 that $A = \mathcal{O}_{X,x}$ is Noetherian. By our hypothesis, $f \notin A$. Let $\mathfrak{a} = \{h \in A; fh \in A\}$. We observe that $\mathfrak{a}$ is an ideal of A contained in the maximal ideal $\mathfrak{m}_{X,x}$ of A.

Let $\mathfrak{p}_1, \ldots, \mathfrak{p}_r$ be the (pairwise distinct) prime ideals minimal among the prime ideals containing $\mathfrak{a}$. By 2.5.3 and 2.7.7, for some $n \in \mathbb{N}^*$, we have $\mathfrak{p}_1^n \mathfrak{p}_2^n \cdots \mathfrak{p}_r^n \subset \mathfrak{a}$. It follows that $\mathfrak{p}_i A_{\mathfrak{p}_1} = A_{\mathfrak{p}_1}$ for $i \geqslant 2$ since $\mathfrak{p}_1 A_{\mathfrak{p}_1}$ is the unique maximal ideal of the local ring $A_{\mathfrak{p}_1}$. Let $k \in \mathbb{N}$ be minimal such that $\mathfrak{p}_1^k f \subset A_{\mathfrak{p}_1}$. Our hypothesis implies that $k > 0$. Let $g \in \mathfrak{p}_1^{k-1} f \setminus A_{\mathfrak{p}_1}$, so $\mathfrak{p}_1 g \subset A_{\mathfrak{p}_1}$.

Since $A_{\mathfrak{p}_1}$ is an integrally closed domain (3.2.5), g is not integral over $A_{\mathfrak{p}_1}$. If $\mathfrak{p}_1 g \subset \mathfrak{p}_1 A_{\mathfrak{p}_1}$, then multiplication by g induces a homomorphism of the finitely generated $A_{\mathfrak{p}_1}$-module $\mathfrak{p}_1 A_{\mathfrak{p}_1}$. Proceed as in the proof of 3.1.4, we would obtain that g in integral over $A_{\mathfrak{p}_1}$. Contradiction. So we deduce that $\mathfrak{p}_1 A_{\mathfrak{p}_1} g \not\subset \mathfrak{p}_1 A_{\mathfrak{p}_1}$. Thus $\mathfrak{p}_1 g$ contains an invertible element of $A_{\mathfrak{p}_1}$, hence $1/g \in \mathfrak{p}_1 A_{\mathfrak{p}_1}$ and $(1/g) A_{\mathfrak{p}_1} = \mathfrak{p}_1 A_{\mathfrak{p}_1}$.

Let $h = f/g^k$, then $h \in f \mathfrak{p}_1^k A_{\mathfrak{p}_1} \subset A_{\mathfrak{p}_1}$. If $h \in \mathfrak{p}_1 A_{\mathfrak{p}_1}$, then $f/g^{k-1} \in A_{\mathfrak{p}_1}$ which contradicts the minimality of k. So h is invertible in $A_{\mathfrak{p}_1}$ and $f' = 1/f = h^{-1}(1/g^k) \in \mathfrak{p}_1 A_{\mathfrak{p}_1}$.

Let $\ell_1, \ldots, \ell_s \in A$ be generators of the ideal $\mathfrak{p}_1$ and U an affine open neighbourhood of x contained in the domain of definition of the ℓ_i's. Let $Y = \{y \in U; \ell_i(y) = 0 \text{ for all } 1 \leqslant i \leqslant s\}$. Since $\mathfrak{p}_1 \subset \mathfrak{m}_{X,x}$, $x \in Y$. But $f' \in \mathfrak{p}_1 A_{\mathfrak{p}_1}$, so $f'(z) = 0$ for all $z \in Y \cap \mathcal{E}$. $\square$

17.4.14 Proposition. *Let $u : X \to Y$ be a finite morphism of irreducible varieties and $x \in X$ a simple point such that $u^{-1}(u(x)) = \{x\}$. Suppose that the map $T_x(u) : T_x(X) \to T_{u(x)}(Y)$ is injective. Then there exists an open neighbourhood V of $y = u(x)$ in Y such that the restriction of u to $u^{-1}(V)$ is an isomorphism onto a closed subvariety of V.*

Proof. We may assume that $X = \mathrm{Spm}(B)$ and $Y = \mathrm{Spm}(A)$ are affine where B is a finite A-algebra. Let φ be the comorphism of u and $\mathfrak{m}, \mathfrak{n}$ the maximal ideals of A and B corresponding to y and x respectively. By our hypothesis, the dual map, $\mathfrak{m}/\mathfrak{m}^2 \to \mathfrak{n}/\mathfrak{n}^2$, of $T_x(u)$ is surjective, and so a system of generators of $\mathfrak{m}$ gives a system of generators of $\mathfrak{n}/\mathfrak{n}^2$. Thus $\mathfrak{n} = B\varphi(\mathfrak{m}) + \mathfrak{n}^2$. It follows that $\mathfrak{n}^2 \subset B\varphi(\mathfrak{m}) + \mathfrak{n}^3$, so $\mathfrak{n} = B\varphi(\mathfrak{m}) + \mathfrak{n}^3$. By induction, we obtain that $\mathfrak{n} = B\varphi(\mathfrak{m}) + \mathfrak{n}^k$ for all $k \in \mathbb{N}^*$.

By 17.4.1, $\mathfrak{n}$ is the unique prime ideal of B containing $B\varphi(\mathfrak{m})$. Thus $\mathfrak{n}/B\varphi(\mathfrak{m})$ is the nilradical of $B/B\varphi(\mathfrak{m})$. Since $B/B\varphi(\mathfrak{m})$ is Noetherian, it follows from 2.5.2 and 2.5.3 that $\mathfrak{n}^k \subset B\varphi(\mathfrak{m})$ for some $k \in \mathbb{N}^*$. We deduce therefore that $\mathfrak{n} = B\varphi(\mathfrak{m})$, and so $B\varphi(\mathfrak{m}) + \varphi(A) = B$ because $B = \mathfrak{n} \oplus \Bbbk$. The

A-module $B/\varphi(A)$ is finitely generated and verifies $\varphi(\mathfrak{m})(B/\varphi(A)) = B/\varphi(A)$. So by 2.6.7, there exists $f \in A \setminus \mathfrak{m}$ such that $\varphi(f)B = \varphi(A)$. Hence the homomorphism $\varphi_f : A_f \to B_{\varphi(f)}$ is surjective. By 11.6.8, u is an isomorphism from the open subset $D(\varphi(f))$ of X onto a closed subvariety of $D(f)$ in Y. $\square$

References and comments

- [5], [19], [22], [26], [37], [40], [78].

The result stated and proved in 17.4.11 is very useful for showing that a morphism is an open map.

18

Root systems

In this chapter, we are concerned with the theory of root systems. This simple geometric notion will play an amazingly important role in the structure theory of semisimple Lie algebras.

Throughout this chapter, k is a field of characteristic zero and V is a k-vector space of dimension l. If $u \in \text{End}(V)$, we shall denote by $\text{rk}(u)$ the rank of u.

18.1 Reflections

18.1.1 Definition. *We say that $s \in \text{End}(V)$ is a reflection if $s^2 = \text{id}_V$ and $\text{rk}(\text{id}_V - s) = 1$.*

18.1.2 Let s be a reflection of V and denote by V_s^+ (resp. V_s^-) the eigenspace of s with eigenvalue 1 (resp. -1). Then

$$V = V_s^+ \oplus V_s^- \ , \ \dim V_s^- = 1 \ , \ \det s = -1.$$

18.1.3 Lemma. *The following conditions are equivalent for $s \in \text{End}(V)$:*
(i) *s is a reflection.*
(ii) *There exist $a \in V$ and $a^* \in V^*$ such that $a^*(a) = 2$ and $s(x) = x - a^*(x)a$ for all $x \in V$.*

Proof. The implication (ii) $\Rightarrow$ (i) is obvious. Let us prove that (i) $\Rightarrow$ (ii). Since $\text{rk}(\text{id}_V - s) = 1$, there exist $a \in V$ and $a^* \in V^*$ such that $x - s(x) = a^*(x)a$ for all $x \in V$. Finally, $s^2 = \text{id}_V$ implies that $a^*(a) = 2$. $\square$

18.1.4 If $a \in V$ and $a^* \in V$ are such that $a^*(a) = 2$, then we shall denote by s_{a,a^*} the reflection of V defined by:

$$s_{a,a^*}(x) = x - a^*(x)a.$$

We can also define a reflection $s_{a^*,a}$ of V^* by:

$$s_{a^*,a}(x^*) = x^* - x^*(a)a^*.$$

We have $s_{a^*,a} = {}^t(s_{a,a^*}) = {}^t(s_{a,a^*})^{-1}$.

18.1.5 Proposition. *Let s be a reflection of V, $u \in \mathrm{End}(V)$ and $W \subset V$ a subspace.*

(i) *We have $u \circ s = s \circ u$ if and only if V_s^+ and V_s^- are u-stable.*

(ii) *We have $s(W) \subset W$ if and only if $V_s^- \subset W$ or $W \subset V_s^+$.*

Proof. Part (i) is clear. Let us prove part (ii). If $W \subset V_s^+$, then $s(W) \subset W$. Now if $V_s^- \subset W$, then $x - s(x) \in V_s^- \subset W$ for all $x \in W$. Hence $s(W) \subset W$.

Conversely, suppose that $s(W) \subset W$ and $W \not\subset V_s^+$. Then there exists $x \in W$ such that $y = x - s(x) \neq 0$. It follows that $y \in V_s^- \cap W$, hence $V_s^- \subset W$ since $\dim V_s^- = 1$. $\square$

18.1.6 Proposition. *Let R be a finite subset of V which generates V, s, t reflections of V and $\alpha \in R \setminus \{0\}$ verifying $s(\alpha) = t(\alpha) = -\alpha$ and $s(R) = t(R) = R$. Then $s = t$.*

Proof. Let $G \subset \mathrm{GL}(V)$ be the stabilizer of R. Since R generates V, G can be identified with a subgroup of the symmetric group on R, so G is finite.

By 18.1.4, we can write $s = s_{\alpha,\alpha^*}$ and $t = s_{\alpha,\beta^*}$. A simple induction gives:

$$(s \circ t)^n(x) = x + n\left(\beta^*(x) - \alpha^*(x)\right)\alpha.$$

Taking n to be the order of $s \circ t$ in G, we see that $\alpha^* = \beta^*$. So $s = t$. $\square$

18.1.7 Let B be a bilinear form on V. An endomorphism u of V is an *orthogonal transformation* (with respect to B) if $B(u(x), u(y)) = B(x, y)$ for all $x, y \in V$.

Suppose that B is symmetric or antisymmetric. Then $U, W \subset V$ are *orthogonal* if $B(U, W) = \{0\}$. A vector $x \in V$ is *isotropic* if $B(x, x) = 0$. A subspace W of V is *non-degenerate* if $W \cap W^\perp = \{0\}$ where $W^\perp$ is the set of $x \in V$ such that $B(x, y) = 0$ for all $y \in W$.

Proposition. *Let B be a non-degenerate symmetric bilinear form on V.*

(i) *A reflection s of V is an orthogonal transformation if and only if V_s^+ and V_s^- are orthogonal. In this case, V_s^+ and V_s^- are non-degenerate.*

(ii) *Let H be a non-degenerate hyperplane of V. Then there exists a unique reflection s of V such that s is an orthogonal transformation and $s|_H = \mathrm{id}_H$. If $\alpha \in H^\perp \setminus \{0\}$, then $H^\perp = k\alpha$ is non-degenerate and:*

$$s(x) = x - 2\frac{B(x, \alpha)}{B(\alpha, \alpha)}\alpha.$$

We say that s is the reflection orthogonal to the hyperplane H.

Proof. (i) Let $x \in V_s^+$, $y \in V_s^-$. If s is an orthogonal transformation, then

$$B(x, y) = B(s(x), s(y)) = -B(x, y).$$

Hence V_s^+ and V_s^- are orthogonal. These eigenspaces are non-degenerate since B is non-degenerate and $V = V_s^+ \oplus V_s^-$. The converse is immediate.

(ii) By our assumptions, $k\alpha$ is non-degenerate and $V = H \oplus k\alpha$. Clearly, the reflection s described above verifies $V_s^+ = H$ and $V_s^- = k\alpha$. So part (i) says that s is an orthogonal transformation.

Now let t be another such orthogonal transformation, then $V_t^+ = H$ and $V_t^- = k\alpha$ by (i). Let $\alpha^* \in V^*$ be such that $\alpha^*(\alpha) = 2$ and $t = s_{\alpha,\alpha^*}$. Then $H = \ker \alpha^*$ and there exists $\lambda \in k$ such that $\alpha^*(x) = \lambda B(x, \alpha)$ for all $x \in V$. It follows that $2 = \alpha^*(\alpha) = \lambda B(\alpha, \alpha)$. So $s = t$. $\square$

18.1.8 Proposition. *Let G be a subgroup of $\mathrm{GL}(V)$. Suppose that V is a simple G-module and that G contains a reflection s.*

(i) *The centralizer of G in $\mathrm{End}(V)$ is $k\,\mathrm{id}_V$.*

(ii) *Let B be a G-invariant non-zero bilinear form on V. Then B is non-degenerate, symmetric or antisymmetric. Any bilinear form G-invariant on V is proportional to B.*

Proof. (i) Let $D = (s - \mathrm{id}_V)(V)$. Then $\dim D = 1$. By 18.1.5, if $u \in \mathrm{End}(V)$ commutes with s, then $u(D) \subset D$. It follows that there exists $\lambda \in k$ such that $(u - \lambda\,\mathrm{id}_V)(D) = 0$. Hence $\ker(u - \lambda\,\mathrm{id}_V) = V$ because V is a simple G-module.

(ii) Let $M = \{x \in V; B(x, V) = \{0\}\}$ and $N = \{x \in V; B(V, x) = \{0\}\}$. Then M, N are G-submodules of V since B is G-invariant. Hence $M = N = \{0\}$ and B is non-degenerate.

Let b be a G-invariant bilinear form on V. Since B is non-degenerate, there exists $u \in \mathrm{End}(V)$ such that $b(x, y) = B(u(x), y)$ for all $x, y \in V$. If $\alpha \in G$ and $x, y \in V$, then:

$$B(u \circ \alpha(x), y) = b(\alpha(x), y) = b(x, \alpha^{-1}(y)) = B(u(x), \alpha^{-1}(y)) = B(\alpha \circ u(x), y).$$

So u commutes with G. By (i), $u = \lambda\,\mathrm{id}_V$ for some $\lambda \in k$. Take, in particular, the bilinear form $b(x, y) = B(y, x)$, then

$$B(y, x) = b(x, y) = \lambda B(x, y) = \lambda b(y, x) = \lambda^2 B(y, x).$$

Hence $\lambda^2 = 1$ and B is symmetric or antisymmetric. $\square$

18.2 Root systems

18.2.1 Definition. *A subset $R \subset V$ is called a* root system *in V if the following conditions are satisfied:*

(R1) *R is a finite subset of V which spans V, and $0 \notin R$.*

(R2) *For all $\alpha \in R$, there exists $\alpha^\vee \in V^*$ such that $\alpha^\vee(\alpha) = 2$ and $s_{\alpha,\alpha^\vee}(R) = R$.*

(R3) *For all $\alpha \in R$, $\alpha^\vee(R) \subset \mathbb{Z}$.*

Further, if for all $\alpha \in R$, the only elements of R proportional to α are α and $-\alpha$, we say that R is reduced.

18.2.2 From now on, let R be a root system in V.

The elements of R are called the *roots* of R, the dimension of V is the *rank* of R that we shall denote by rk R.

Let $\alpha \in R$. By 18.1.6, there exists a unique $\alpha^\vee$ satisfying (R2). It follows that $(-\alpha)^\vee = -\alpha^\vee$. So we may write s_α for $s_{\alpha,\alpha^\vee}$ and $R^\vee = \{\alpha^\vee ; \alpha \in R\}$.

By (R1), the set of $u \in \mathrm{GL}(V)$ verifying $u(R) \subset R$ is a finite group, denoted by $A(R)$. The subgroup $W(R)$ of $A(R)$ generated by the s_α's is called the *Weyl group* of R.

Let $\alpha \in R$. We have $-\alpha = s_\alpha(\alpha) \in R$ by (R2), and so $-R = R$. Hence $-\mathrm{id}_V \in A(R)$. However, $-\mathrm{id}_V$ is not necessarily contained in $W(R)$.

For $\alpha, \beta \in R$, set:
$$a_{\alpha,\beta} = \beta^\vee(\alpha).$$

From the definition of a root system, we have:

(1)
$$\begin{cases} a_{\alpha,\beta} \in \mathbb{Z}\,,\ a_{\alpha,\alpha} = 2, \\ a_{-\alpha,\beta} = a_{\alpha,-\beta} = -a_{\alpha,\beta}, \\ s_\beta(\alpha) = \alpha - a_{\alpha,\beta}\beta. \end{cases}$$

18.2.3 Let $V = V_1 \oplus \cdots \oplus V_n$. For $1 \leqslant i \leqslant n$, let R_i be a root system in V_i. Identify V^* with the direct sum of the V_i^*'s. If $\alpha \in R_i$, $\alpha^\vee$ can be identified with an element of V^* which is zero on the V_j for $j \neq i$. Thus the union R of the R_i's is a root system in V that we shall call the *direct sum* of the R_i's. We have $R^\vee = R_1^\vee \cup \cdots \cup R_n^\vee$. If $\alpha \in R_i$ and $j \neq i$, then $V_j \subset \ker \alpha^\vee$ and $k\alpha \subset V_i$. We deduce that $s_\alpha|_{V_j} = \mathrm{id}_{V_j}$ and $s_\alpha(V_i) \subset V_i$ by 18.1.5. The group $W(R)$ is the product of the groups $W(R_i)$.

Definition. *A root system is* irreducible *if it is not the direct sum of two non-empty root systems.*

18.2.4 Proposition. *A root system R is the direct sum of a finite family $(R_i)_{i \in I}$ of irreducible root systems, which is unique up to a permutation of indices. We call the R_i's, the* irreducible components *of R.*

Proof. If R is non-empty and non-irreducible, then it is the direct sum of two non-empty root systems. By induction on the cardinality of R, we obtain that R is the direct sum of a finite family $(R_i)_{i \in I}$ of irreducible root systems.

To prove the uniqueness, it suffices to show that if R is the direct sum of R' and R'', then $R_i \subset R'$ or $R_i \subset R''$. Let $R_i' = R_i \cap R'$, $R_i'' = R_i \cap R''$ and V', V'', V_i', V_i'' the vector subspaces spanned by R', R'', R_i', R_i''.

If $\alpha \in R_i'$, then $s_\alpha(R') \subset R'$, $s_\alpha(R_i') \subset R_i'$ (18.2.3), so R_i' is a root system in V_i'. Similarly, R_i'' is a root system in V_i''. It follows easily that R_i is the direct sum of R_i' and R_i''. But R_i is irreducible, so $R_i' = \emptyset$ or $R_i'' = \emptyset$. So we are done. $\square$

18.2.5 Lemma. *Let U be a subspace of V and W the subspace of V spanned by $R \cap U$. Then $R \cap U$ is a root system in W.*

Proof. If $\alpha \in R \cap U$, then $\alpha_W^\vee = \alpha^\vee|_W$. For $\beta, \gamma \in R \cap U$, we have $\beta_W^\vee(\beta) = 2$ and $s_{\beta,\beta_W^\vee}(\gamma) = \gamma - \beta^\vee(\gamma)\beta \in U$. So $s_{\beta,\beta_W^\vee}(\gamma) \in R \cap U$ by (R2). The other properties for a root system are clear. $\square$

18.2.6 Proposition. *Let $V = V_1 \oplus \cdots \oplus V_n$ and $R_i = R \cap V_i$ for $1 \leqslant i \leqslant n$. The following conditions are equivalent:*

(i) *The subspaces $V_1, \ldots, V_n$ are $W(R)$-stable.*

(ii) *We have $R \subset V_1 \cup \cdots \cup V_n$.*

(iii) *For $1 \leqslant i \leqslant n$, R_i is a root system in V_i and R is the direct sum of the R_i's.*

Proof. (i) $\Rightarrow$ (ii) Let $\alpha \in R$. If $s_\alpha(V_i) \subset V_i$, then $k\alpha \subset V_i$ or $V_i \subset \ker \alpha^\vee$ (18.1.5). Since V is the direct sum of the V_i's, $V_j \not\subset \ker \alpha^\vee$ for some j. So $\alpha \in V_j$.

(ii) $\Rightarrow$ (iii) This is clear by 18.2.5.

(iii) $\Rightarrow$ (i) This follows from 18.2.3. $\square$

18.2.7 Corollary. *The root system R is irreducible if and only if V is a simple $W(R)$-module.*

Proof. This follows from 10.7.6 and 18.2.6. $\square$

18.2.8 Proposition. *If R is a root system in V, then $R^\vee$ is a root system in V^*. Furthermore, $\alpha^{\vee\vee} = \alpha$ for all $\alpha \in R$.*

Proof. Let us first show that $R^\vee$ spans V^*. We may assume by 18.2.3 and 18.2.4 that R is irreducible. The result is clear if $\operatorname{rk} R = 1$. Suppose that $\operatorname{rk} R \geqslant 2$. If there exists $x \in V \setminus \{0\}$ such that $\alpha^\vee(x) = 0$ for all $\alpha \in R$, then the subspace kx is $W(R)$-stable. But this is absurd by 18.2.7. So $R^\vee$ spans V^*.

Now, let $\alpha, \beta \in R$, $\gamma = s_\alpha(\beta)$ and $\theta = \beta^\vee - \beta^\vee(\alpha)\alpha^\vee$. Then

$$\theta(\gamma) = 2 , \quad s_{\gamma,\theta} = s_\alpha \circ s_\beta \circ s_\alpha.$$

Hence $s_{\gamma,\theta}(R) = R$, and so $\theta = \gamma^\vee = s_{\alpha^\vee,\alpha}(\beta^\vee)$ by 18.1.6. Thus $s_{\alpha^\vee,\alpha}(R^\vee) = R^\vee$. It follows that condition (R2) is satisfied and $\alpha^{\vee\vee} = \alpha$. Finally, we see readily that (R3) is also satisfied. $\square$

18.2.9 We call $R^\vee$ the *dual root system* of R. The map $R \to R^\vee$, $\alpha \mapsto \alpha^\vee$, is a bijection, that we shall call *canonical*. We observe that $a_{\alpha,\beta} = a_{\beta^\vee,\alpha^\vee}$ for $\alpha, \beta \in R$. Since $s_{\alpha^\vee,\alpha} = {}^t(s_{\alpha,\alpha^\vee})^{-1}$, the map

$$W(R) \to W(R^\vee) , \quad w \mapsto {}^t w^{-1}$$

is an isomorphism of groups. So we may identify $W(R)$ and $W(R^\vee)$ via this isomorphism and consider that $W(R)$ acts on V and V^*.

Let $\alpha \in R$ and $g \in A(R)$. Set $\beta = g(\alpha)$, $\theta = {}^t g^{-1}(\alpha^\vee)$. Then

$$\theta(\beta) = 2 \; , \; s_{\beta,\theta} = g \circ s_\alpha \circ g^{-1}.$$

Hence $s_{\beta,\theta}(R) = R$ and $\theta = \beta^\vee$. Hence ${}^t g(R^\vee) = R^\vee$ and the map $g \mapsto {}^t g^{-1}$ is an isomorphism from $A(R)$ onto $A(R^\vee)$. So we may consider that $A(R)$ acts on V^*. Note that the previous discussion shows that $W(R)$ is a normal subgroup of $A(R)$.

18.3 Root systems and bilinear forms

18.3.1 Proposition. *For $x, y \in V$, the map $V \times V \to k$ defined by:*

$$(x,y) \mapsto (x|y) = \sum_{\alpha \in R} \alpha^\vee(x)\alpha^\vee(y)$$

is a non-degenerate $A(R)$-invariant symmetric bilinear form on V.

Proof. If $g \in A(R)$, we saw in 18.2.9 that ${}^t g(R^\vee) = R^\vee$. Thus

$$(g(x)|g(y)) = \sum_{\alpha \in R} {}^t g(\alpha^\vee)(x) {}^t g(\alpha^\vee)(y) = (x|y).$$

So the bilinear form $(.|.)$ is $A(R)$-invariant. Clearly it is symmetric.

For $\alpha, \beta \in R$, we have $\alpha^\vee(\beta) \in \mathbb{Z}$ and $\beta^\vee(\beta) = 2$. So $(\beta|\beta) \in \mathbb{N}$ and $(\beta|\beta) \geqslant 4$. So this bilinear form is non-zero.

Now let $R_1, \ldots, R_n$ be the irreducible components of R and V_i the subspace spanned by R_i. Then by 18.2.3, for $i \neq j$, $\alpha^\vee(x)\alpha^\vee(y) = 0$ for all $x \in V_i$, $y \in V_j$ and $\alpha \in R$. Thus V_i and V_j are orthogonal with respect to $(.|.)$. We are therefore reduced to the case where R is irreducible and the proposition follows from 18.1.8 and 18.2.7. $\square$

18.3.2 Proposition. *Let $R_1, \ldots, R_n$ be the irreducible components of R, V_i the subspace spanned by R_i and B a $W(R)$-invariant symmetric bilinear form on V.*

(i) *The restriction of B to V_i is proportional to the one defined in 18.3.1.*
(ii) *The V_i's are pairwise orthogonal with respect to B.*

Proof. (i) By 18.2.7, for $1 \leqslant i \leqslant n$, V_i is a simple $W(R_i)$-module. So (i) follows from 18.1.8 and 18.3.1.

(ii) Let U_i be the subspace of V_i spanned by $w(x) - x$, $w \in W(R_i)$ and $x \in V_i$. If $w, w' \in W(R_i)$ and $x \in V_i$, then:

$$w'(w(x)) - w'(x) = [w'(w(x)) - w(x)] + [w(x) - x] - [w'(x) - x].$$

So U_i is a $W(R_i)$-module. Further, $s_\alpha(\alpha) - \alpha = -2\alpha \neq 0$ if $\alpha \in R_i$. So $U_i \neq \{0\}$. We deduce from 18.2.7 that $U_i = V_i$.

Now let $x \in V_i$, $y \in V_j$ with $i \neq j$. We saw in 18.2.3 that $w|_{V_j} = \mathrm{id}_{V_j}$ for all $w \in W(R_i)$. It follows that:

$$B(w(x), y) = B(w(x), w(y)) = B(x, y).$$

So $w(x) - x$ is orthogonal to y. Thus V_i and V_j are orthogonal with respect to B. $\square$

18.3.3 Let $(.|.)$ be a non-degenerate $W(R)$-invariant bilinear form on V. For $\alpha, \beta \in R$, it follows from 18.1.7 and 18.3.2 that

$$a_{\beta,\alpha} = 2\frac{(\alpha|\beta)}{(\alpha|\alpha)}.$$

Hence:

$$(\alpha|\beta) \neq 0 \Rightarrow \frac{a_{\beta,\alpha}}{a_{\alpha,\beta}} = \frac{(\beta|\beta)}{(\alpha|\alpha)} \; , \; a_{\alpha,\beta} = 0 \Leftrightarrow a_{\beta,\alpha} = 0 \Leftrightarrow (\alpha|\beta) = 0.$$

If $x \in V$, then:

$$s_\beta \circ s_\alpha(x) - s_\alpha \circ s_\beta(x) = a_{\alpha,\beta}\alpha^\vee(x)\beta - a_{\beta,\alpha}\beta^\vee(x)\alpha.$$

Consequently, if α and β are not proportional, then

$$a_{\alpha,\beta} = 0 \Leftrightarrow a_{\beta,\alpha} = 0 \Leftrightarrow s_\alpha \circ s_\beta = s_\beta \circ s_\alpha.$$

18.4 Passage to the field of real numbers

18.4.1 Let R be a root system in V. Denote by $V_\mathbb{Q}$ (resp. $V_\mathbb{Q}^*$) the $\mathbb{Q}$-vector subspace spanned in V (resp. V^*) by R (resp. $R^\vee$).

Let K be an extension of k. We can identify canonically V as a subset of $K \otimes_k V$, and V^* as a subset of $K \otimes_k V^* = (K \otimes_k V)^*$. Then R is a root system in $K \otimes_k V$ and the $\alpha^\vee$'s are the same.

18.4.2 Proposition. (i) *R is a root system in $V_\mathbb{Q}$.*

(ii) *The vector space V (resp. V^*) identifies canonically with $k \otimes_\mathbb{Q} V_\mathbb{Q}$ (resp. $k \otimes_\mathbb{Q} V_\mathbb{Q}^*$). The canonical bilinear form on $V \times V^*$ allows us to identify $V_\mathbb{Q}^*$ with $(V_\mathbb{Q})^*$ and $V_\mathbb{Q}$ with $(V_\mathbb{Q}^*)^*$.*

Proof. (i) By (R3), $\alpha^\vee(V_\mathbb{Q}) \subset \mathbb{Q}$, so $\alpha^\vee$ defines a linear form $\alpha_\mathbb{Q}^\vee$ on $V_\mathbb{Q}$. It is now easy to verify that R is a root system in $V_\mathbb{Q}$, and $\alpha_\mathbb{Q}^\vee$ is the linear form associated to α in $(V_\mathbb{Q})^*$.

(ii) Let $i : k \otimes_\mathbb{Q} V_\mathbb{Q} \to V$ be the canonical homomorphism and ${}^t i : V^* \to k \otimes_\mathbb{Q} (V_\mathbb{Q})^*$ its dual. The map i is surjective since R spans V. Further, if $\alpha \in R$, then ${}^t i(\alpha^\vee) = 1 \otimes \alpha_\mathbb{Q}^\vee$. By (i) and 18.2.8, the $\alpha_\mathbb{Q}^\vee$'s span $(V_\mathbb{Q})^*$, and so ${}^t i$ is surjective. Hence i and ${}^t i$ are isomorphisms. They allow us to identify $V_\mathbb{Q}^*$ with $(V_\mathbb{Q})^*$ and $V_\mathbb{Q}$ with $(V_\mathbb{Q}^*)^*$. $\square$

18.4.3 By 18.4.1 and 18.4.2, we can consider R as a root system in $V_\mathbb{Q}$ and in $V_\mathbb{R} = \mathbb{R} \otimes_\mathbb{Q} V_\mathbb{Q}$, and the corresponding Weyl groups are canonically isomorphic.

Let us use the notations of 18.3.1. If $x, y \in V_\mathbb{Q}$, we have $(x|y) \in \mathbb{Q}$. Denote again by $(.|.)$ the canonical extension of this bilinear form to $V_\mathbb{R}$. Then for $z \in V_\mathbb{R}$:

$$(z|z) = \sum_{\alpha \in R} [\alpha^\vee(z)]^2 \geqslant 0.$$

By 18.3.1, equality holds if and only if $z = 0$. Thus $(.|.)$ is a $W(R)$-invariant inner product on $V_{\mathbb{R}}$.

Conversely, let us fix a $W(R)$-invariant inner product on $V_{\mathbb{R}}$. We can consider the length of a root and the angle between two roots. By 18.3.2, this angle does not depend on the choice of the inner product. Similarly, if two roots belong to the same irreducible component, then the ratio of their lengths does not depend on the choice of the inner product.

18.5 Relations between two roots

18.5.1 We shall assume in the rest of this chapter that $k = \mathbb{R}$. Let us fix a root system R in V and a $W(R)$-invariant inner product $(.|.)$ on V. If $\alpha, \beta \in R$, denote by $\theta_{\alpha\beta} \in [0, \pi[$ the angle between α and β, $\omega_{\alpha\beta}$ the order of $s_\alpha \circ s_\beta$ and $\|\alpha\|$ the length of α.

18.5.2 If $\alpha, \beta \in R$, then 18.3.3 implies that:

$$a_{\alpha,\beta} a_{\beta,\alpha} = 4\cos^2 \theta_{\alpha,\beta} \leqslant 4.$$

Furthermore, $a_{\alpha,\beta} a_{\beta,\alpha} \in \mathbb{Z}$. Assuming $\|\alpha\| \leqslant \|\beta\|$, we see easily that the only possibilities are listed in the following table:

$a_{\alpha,\beta}$	$a_{\beta,\alpha}$	$\theta_{\alpha\beta}$	$\|\beta\|^2/\|\alpha\|^2$	
0	0	$\dfrac{\pi}{2}$	undetermined	$\omega_{\alpha\beta} = 2$
1	1	$\dfrac{\pi}{3}$	1	$\omega_{\alpha\beta} = 3$
-1	-1	$\dfrac{2\pi}{3}$	1	$\omega_{\alpha\beta} = 3$
1	2	$\dfrac{\pi}{4}$	2	$\omega_{\alpha\beta} = 4$
-1	-2	$\dfrac{3\pi}{4}$	2	$\omega_{\alpha\beta} = 4$
1	3	$\dfrac{\pi}{6}$	3	$\omega_{\alpha\beta} = 6$
-1	-3	$\dfrac{5\pi}{6}$	3	$\omega_{\alpha\beta} = 6$
2	2	0	1	$\beta = \alpha$
-2	-2	π	1	$\beta = -\alpha$
1	4	0	4	$\beta = 2\alpha$
-1	-4	π	4	$\beta = -2\alpha$

18.5.3 We deduce from the table above the following proposition.

Proposition. *Let α and β be roots.*

(i) *If $\beta = \lambda\alpha$ with $\lambda \in k$, then $\lambda \in \left\{ \pm 1, \pm 2, \pm\frac{1}{2} \right\}$.*

(ii) *If $\beta \notin k\alpha$ and $\|\alpha\| \leqslant \|\beta\|$, then $a_{\alpha,\beta} \in \{-1, 0, 1\}$.*

(iii) *If $a_{\alpha,\beta} > 0$, that is $(\alpha|\beta) > 0$, then $\alpha - \beta \in R$ unless $\alpha = \beta$.*

(iv) *If $a_{\alpha,\beta} < 0$, that is $(\alpha|\beta) < 0$, then $\alpha + \beta \in R$ unless $\alpha = -\beta$.*

(v) *If $\alpha - \beta$ and $\alpha + \beta$ do not belong to $R \cup \{0\}$, then $(\alpha|\beta) = 0$.*

Remarks. 1) A root in R is *reduced* if $\frac{1}{2}\alpha \notin R$.

2) It may happen that $\alpha + \beta \in R$ and $(\alpha|\beta) = 0$. Also, if $\alpha - \beta$ and $\alpha + \beta$ do not belong to $R \cup \{0\}$, we say that α and β are *strongly orthogonal*.

18.5.4 Proposition. *Let α, β be roots which are not proportional. Set $I_{\alpha,\beta} = \{j \in \mathbb{Z}; \beta + j\alpha \in R\}$ and $S_{\alpha,\beta} = \{\beta + j\alpha; j \in I_{\alpha,\beta}\}$.*

(i) *The set $I_{\alpha,\beta}$ is an interval $[-p, q]$ of $\mathbb{Z}$ containing 0.*

(ii) *The set $S_{\alpha,\beta}$ is stable under s_α and*

$$s_\alpha(\beta + q\alpha) = \beta - p\alpha \; , \; s_\alpha(\beta - p\alpha) = \beta + q\alpha.$$

We say that $S_{\alpha,\beta}$ is the α-string through β, and that $\beta - p\alpha$ is the source, $\beta + q\alpha$ is the target and $p + q$ the length.

(iii) *We have $a_{\beta,\alpha} = p - q$.*

Proof. Let $q = \max I_{\alpha,\beta}$, $-p = \min I_{\alpha,\beta}$. Since $\beta \in R$, $0 \in I_{\alpha,\beta}$. Suppose that $I_{\alpha,\beta} \neq [-p, q]$. Then there exist $-p \leqslant r, s \leqslant q$ such that

$$r, s \in I_{\alpha,\beta} \; , \; r < s - 1 \; , \; r + i \notin I_{\alpha,\beta} \text{ if } 1 \leqslant i \leqslant s - r - 1.$$

It follows from 18.5.3 that $(\alpha|\beta + s\alpha) \leqslant 0$ and $(\alpha|\beta + r\alpha) \geqslant 0$. But this is absurd since:

$$(\alpha|\beta + s\alpha) - (\alpha|\beta + r\alpha) = (s - r)\|\alpha\|^2 > 0.$$

So $I_{\alpha,\beta} = [-p, q]$.

Next, we have $s_\alpha(\beta + j\alpha) = \beta - (a_{\beta,\alpha} + j)\alpha$. Since $s_\alpha(R) = R$, $s_\alpha(S_{\alpha,\beta}) = S_{\alpha,\beta}$. So (ii) follows from the fact that $j \mapsto -a_{\beta,\alpha} - j$ is an order-reversing bijection from $I_{\alpha,\beta}$ to itself.

Finally, since $s_\alpha(\beta + q\alpha) = \beta - p\alpha = \beta - (a_{\beta,\alpha} + q)\alpha$, part (iii) follows from (ii). $\square$

18.5.5 Corollary. *Let α, β be roots that are not proportional, S the α-string through β and γ its source. The length of S is $-a_{\gamma,\alpha}$ and it is equal to $0, 1, 2$ or 3.*

Proof. Note that $S = S_{\alpha,\gamma}$, so the first part follows from 18.5.4 (iii) and the second follows from the table in 18.5.2. $\square$

18.5.6 Proposition. *Let* α, β *be roots which are not proportional and* p, q *be the integers such that* $I_{\alpha,\beta} = [-p, q]$. *If* $\alpha + \beta \in R$, *then*

$$\frac{\|\beta + \alpha\|^2}{\|\beta\|^2} = \frac{p+1}{q}.$$

Proof. Let $\gamma = \beta - p\alpha$ be the source of the α-string through β. Since $\alpha + \beta \in R$, the length $l = p + q$ of $S_{\alpha,\gamma} = S_{\alpha,\beta}$ is in $\{1, 2, 3\}$ (18.5.5). The results follows since by using 18.5.2 and 18.5.4, the only possibilities are listed in the following table:

l		p	q	$a_{\beta,\alpha}$	
1	$\beta = \gamma$	0	1	-1	$\|\beta + \alpha\|^2 = \|\beta\|^2$
2	$\beta = \gamma$	0	2	-2	$2\|\beta + \alpha\|^2 = \|\beta\|^2$
2	$\beta = \gamma + \alpha$	1	1	0	$\|\beta + \alpha\|^2 = 2\|\beta\|^2$
3	$\beta = \gamma$	0	3	-3	$3\|\beta + \alpha\|^2 = \|\beta\|^2$
3	$\beta = \gamma + \alpha$	1	2	-1	$\|\beta + \alpha\|^2 = \|\beta\|^2$
3	$\beta = \gamma + 2\alpha$	2	1	1	$\|\beta + \alpha\|^2 = 3\|\beta\|^2$

Let us take for example the case $p = 0$ and $q = 2$, then by 18.5.2, $a_{\alpha,\beta} = -1$. Now

$$\|\beta + \alpha\|^2 = \|\beta\|^2 + \|\alpha\|^2 + 2(\beta|\alpha).$$

Replacing $2(\beta|\alpha)$ by $\|\beta\|^2 a_{\alpha,\beta}$, we obtain $\|\beta + \alpha\|^2 = \|\alpha\|^2$. So we have $2\|\beta + \alpha\|^2 = \|\beta\|^2$ by replacing $2(\beta|\alpha)$ by $\|\alpha\|^2 a_{\beta,\alpha}$ in the above equality. $\square$

18.5.7 Proposition. *Assume that* R *is irreducible. Let* $\alpha, \beta \in R$, *then there exists* $w \in W(R)$ *such that* $w(\beta) = \alpha$ *if and only if* $\|\alpha\| = \|\beta\|$.

Proof. If $\|\alpha\| = \|\beta\|$, then since R is irreducible, there exists $w \in W(R)$ such that α and $w(\beta)$ are not orthogonal. So we may assume that $(\alpha|\beta) \neq 0$. Replacing β by $s_\beta(\beta)$ if necessary, we may further assume that $(\alpha|\beta) > 0$. By 18.5.2, either $\alpha = \beta$ or $a_{\alpha,\beta} = a_{\beta,\alpha} = 1$. In the latter,

$$s_\alpha \circ s_\beta \circ s_\alpha(\beta) = s_\alpha \circ s_\beta(\beta - \alpha) = s_\alpha(-\beta - \alpha + \beta) = \alpha.$$

So we are done. The converse is clear since the inner product is $W(R)$-invariant. $\square$

18.5.8 Proposition. *Assume that* R *is irreducible and reduced.*
(i) *For* $\alpha, \beta \in R$, *we have:*

$$\frac{\|\beta\|^2}{\|\alpha\|^2} \in \left\{1, 2, 3, \frac{1}{2}, \frac{1}{3}\right\}.$$

(ii) *The set* $\{\|\alpha\|; \alpha \in R\}$ *contains at most two elements.*

Proof. (i) In view of 18.2.7, the subspace spanned by $w(\beta)$, $w \in W(R)$, is V. Thus there exists $w \in W(R)$ such that $(\alpha|w(\beta)) \neq 0$. Since $\|w(\beta)\| = \|\beta\|$, the result follows from 18.5.2.

(ii) This is a simple consequence of part (i). $\square$

18.5.9 Assume that R is irreducible and reduced. If $\{\|\alpha\|; \alpha \in R\} = \{\lambda, \mu\}$ where $\lambda < \mu$, then a root of length λ (resp. μ) is called a *short* (resp. *long*) root. If all the roots have the same length, then, by convention, they are called long.

18.6 Examples of root systems

18.6.1 In view of the results of the previous sections, we give some examples of root systems of small rank over the real numbers. The reader may verify easily that the examples below are indeed roots systems.

- Rank 1.

$$A_1$$

$$-\alpha \longleftarrow \!\!\!\!\!\!\!\!-\!\!\!\!\circ\!\!\!\!-\!\!\!\!\!\!\!\!\longrightarrow \alpha$$

Clearly, this is the only irreducible reduced root system of rank 1.

- Rank 2.

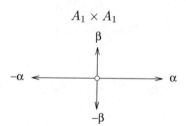

$$A_1 \times A_1$$

The roots α and β are orthogonal, and this root system is not irreducible.

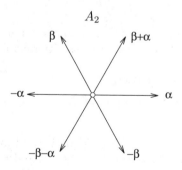

$$A_2$$

Here, all the roots have the same length, and the angle between two adjacent roots is $\pi/3$. This is an irreducible reduced root system of rank 2.

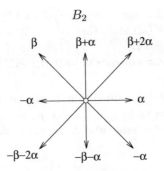

This is again an irreducible reduced root system of rank 2. We have $\|\beta\| = \sqrt{2}\|\alpha\|$, and the angle between two adjacent roots is $\pi/4$.

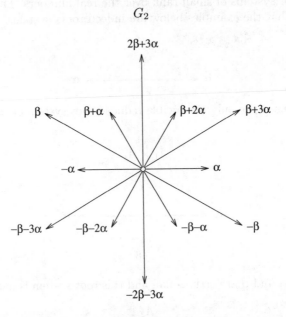

In this root system, $\|\beta\| = \sqrt{3}\|\alpha\|$, and the angle between two adjacent roots is $\pi/6$. This rank 2 root system is also irreducible and reduced.

18.7 Base of a root system

18.7.1 In this rest of this chapter, all root systems considered are assumed to be reduced.

18.7.2 We define a *lexicographic order* on V to be an order obtained in the following way: choose a $\mathbb{R}$-basis $(e_1, \ldots, e_l)$ of V, and write $\lambda_1 e_1 + \cdots + \lambda_l e_l \prec \mu_1 e_1 + \cdots + \mu_l e_l$ if for some integer n, we have $\lambda_1 = \mu_1$, ..., $\lambda_n = \mu_n$, and $\lambda_{n+1} < \mu_{n+1}$. Clearly, this is a total order on V which is compatible with addition.

18.7.3 Let $\prec$ be a lexicographic order on V. Set:

$$R_+ = \{\alpha \in R; \alpha \succ 0\} \ , \ R_- = \{\alpha \in R; \alpha \prec 0\}.$$

An element of R_+ (resp. R_-) is called a *positive* (resp. *negative*) root.

Denote by B the set of elements of R_+ which is not the sum of two positive roots. The set B is called a *base* of R.

18.7.4 Theorem. *Let R be a root system, $\prec$ a lexicographic order on V, R_+ the set of positive roots and B the base of R with respect to this order. Then:*

(i) *Any positive root is the sum of roots in B.*

(ii) *If $\alpha, \beta \in B$ are distinct, then $a_{\alpha,\beta} \leqslant 0$ and $\alpha - \beta \notin R$.*

(iii) *If $B = \{\beta_1, \ldots, \beta_q\}$, then $(\beta_1, \ldots, \beta_q)$ is a basis of V.*

(iv) *If $\alpha \in R_+ \setminus B$, then there exists $\beta \in B$ such that $a_{\alpha,\beta} > 0$ and $\alpha - \beta \in R_+$.*

(v) *If $\alpha \in R_+$, $\beta \in B$ and $\alpha \neq \beta$, then $s_\beta(\alpha) \in R_+$. In particular, s_β permutes the elements of $R_+ \setminus \{\beta\}$.*

Proof. (i) Let $\alpha_1 \prec \cdots \prec \alpha_n$ be the elements in R_+. It is clear that $\alpha_1 \in B$. Suppose that $\alpha_1, \ldots, \alpha_{i-1}$ can all be expressed as a sum of elements of B. If $\alpha_i \notin B$. then $\alpha_i = \alpha_r + \alpha_s$ where $\alpha_r, \alpha_s \in R$ and $r, s < i$. We deduce therefore that α_i is also a sum of elements of B. So by induction, we have (i).

(ii) If $\alpha - \beta \in R$, then $\beta - \alpha \in R$. So either $\alpha - \beta$ or $\beta - \alpha$ is positive. This is impossible since $\alpha, \beta \in B$ and $\alpha = (\alpha - \beta) + \beta$, $\beta = (\beta - \alpha) + \alpha$. Thus $\alpha - \beta \notin R$, and $a_{\alpha,\beta} \leqslant 0$ by 18.5.3 (iii).

(iii) Let $\lambda_1, \ldots, \lambda_q \in \mathbb{R}_+$ be such that, reindexing if necessary,

$$(2) \qquad v = \lambda_1 \beta_1 + \cdots + \lambda_p \beta_p = \lambda_{p+1} \beta_{p+1} + \cdots + \lambda_q \beta_q.$$

Then by part (ii),

$$(v|v) = (\lambda_1 \beta_1 + \cdots + \lambda_p \beta_p | \lambda_{p+1} \beta_{p+1} + \cdots + \lambda_q \beta_q) \leqslant 0.$$

Hence $v = 0$. Since $\beta_i \succ 0$, it follows that $\lambda_i = 0$ for all i. So the elements of B are linearly independent, and part (i) implies that B is a basis of V.

(iv) By (i), there exist $\beta_1, \ldots, \beta_p \in B$ pairwise distinct, and $n_1, \ldots, n_p \in \mathbb{N}^*$ such that $\alpha = n_1 \beta_1 + \cdots + n_p \beta_p$. Since $(\alpha|\alpha) > 0$, it follows that $(\beta_i|\alpha) > 0$ for some i. Then 18.5.3 says that $\alpha - \beta_i$ and $\beta_i - \alpha$ are roots. Since $\beta_i \in B$, this forces $\alpha - \beta_i$ to be positive.

(v) Let us write $\alpha = n_1 \beta_1 + \cdots + n_p \beta_p$ as above. Since $\alpha \neq \beta$, there exists r such that $\beta_r \neq \beta$. It follows that the coefficient of β_r in $s_\beta(\alpha) = \alpha - a_{\beta,\alpha}\beta$ is n_r. But by (i) and (iii), the coefficients of a root in the basis B are either all positive or all negative. So $s_\beta(\alpha) \in R_+$. $\square$

18.7.5 Remarks. 1) An element of B is called a *simple* root of R with respect to B.

2) In general, let V be a k-vector space and consider R as a root system in $V_{\mathbb{R}}$. The same proof (18.7.4) shows that B is a basis of V, $V_{\mathbb{Q}}$ and $V_{\mathbb{R}}$.

18.7.6 Corollary. *Set:*

$$\rho = \frac{1}{2} \sum_{\alpha \in R_+} \alpha.$$

Then $s_\beta(\rho) = \rho - \beta$ and $\beta^\vee(\rho) = 1$ for all $\beta \in B$.

Proof. This is a simple consequence of 18.7.4 (v). $\square$

18.7.7 Proposition. *Let $\prec$ be a lexicographic order on V and $\alpha_1, \ldots, \alpha_n$ positive roots such that $\alpha = \alpha_1 + \cdots + \alpha_n \in R_+$. Then there is a permutation σ of $\{1, 2, \ldots, n\}$ such that $\alpha_{\sigma(1)} + \cdots + \alpha_{\sigma(i)} \in R_+$ for $1 \leqslant i \leqslant n$.*

Proof. Since $(\alpha|\alpha) > 0$, there exists j such that $(\alpha|\alpha_j) > 0$. If $\alpha = \alpha_j$, then $n = 1$ and the result is obvious. Otherwise by 18.5.3, $\alpha - \alpha_j \in R_+$, and the result follows by induction on n. $\square$

18.7.8 Proposition. *Let $\prec$ be a lexicographic order on V, B be a base of R and W_B the subgroup of $W(R)$ generated by s_β, $\beta \in B$.*
   (i) *We have $R = \{w(\beta); w \in W_B, \beta \in B\}$.*
   (ii) *We have $W_B = W(R)$.*

Proof. Let $S = \{w(\beta); w \in W_B, \beta \in B\} \subset R$.
   (i) Since $s_\beta(\beta) = -\beta$, it suffices to show that $R_+ \subset S$. Let $\alpha_1 \prec \cdots \prec \alpha_n$ be the elements of R_+. Since $\alpha_1 \in B$ and $B \subset S$, $\alpha_1 \in S$. Let us suppose that $\alpha_1, \ldots, \alpha_{i-1} \in S$. If $\alpha_i \in B$, then $\alpha_i \in S$. Otherwise, by 18.7.4 (iv), there exists $\beta \in B$ such that $a_{\alpha,\beta} > 0$. Thus $\delta = s_\beta(\alpha) = \alpha - a_{\alpha,\beta}\beta \prec \alpha$. Further $\delta \in R_+$ (18.7.4 (v)). So $\delta \in S$ and hence $\alpha = s_\beta(\delta) \in S$. Now (i) follows by induction.
   (ii) It suffices to prove that $s_\alpha \in W_B$ for all $\alpha \in R$. By part (i), there exist $w \in W_B$ and $\beta \in B$ such that $\alpha = w(\beta)$. Hence $s_\alpha = w \circ s_\beta \circ w^{-1} \in W_B$. $\square$

18.7.9 Proposition. (i) *Let B be a base of R, $B' \subset B$, V' the subspace spanned by B' and $R' = R \cap V'$. Then B' is a base for the root system R'.*
   (ii) *Let W be a subspace of V, $R' = R \cap W$ and B' a base of R'. Then there exists a base of R containing B'.*

Proof. (i) We may assume that $B' \neq \emptyset$. By 18.2.5, R' is a root system in V'. Let $B = (\beta_1, \ldots, \beta_l)$, $B' = (\beta_1, \ldots, \beta_p)$ and $R'_+ = V' \cap R_+$. Then it is clear that with respect to the lexicographic order associated to B (resp. B') on V (resp. V'), R_+ (resp. R'_+) is the set of positive roots. Hence B' is the base of R' with respect to this order.
   (ii) Let $B' = (\beta_1, \ldots, \beta_p)$ and complete to a basis $(\beta_1, \ldots, \beta_l)$ of V. The lexicographic order on V associated to this basis defines a base B of R which clearly contains B'. $\square$

18.8 Weyl chambers

18.8.1 We shall endow V with the usual topology defined for example by the $W(R)$-invariant inner product. Let $\mathcal{E} = (e_1, \ldots, e_l)$ be a basis of V and $\lambda_1, \ldots, \lambda_l \in V^*$ such that for any $x \in V$,

$$x = \lambda_1(x)e_1 + \cdots + \lambda_l(x)e_l.$$

The set $\{x \in V; \lambda_i(x) > 0, 1 \leqslant i \leqslant l\}$ will be called the *open simplicial cone* associated to the basis $\mathcal{E}$.

18.8.2 Let $B = \{\beta_1, \ldots, \beta_l\}$ be a base of R and $s_i = s_{\beta_i}$ for $1 \leqslant i \leqslant l$. For $w \in W(R)$, by 18.7.8 (ii), there exist $1 \leqslant i_1, \ldots, i_n \leqslant l$ such that $w = s_{i_1} \circ \cdots \circ s_{i_n}$. Such an expression is called a *reduced decomposition* if n is minimal, and we say that n is the *length* of w that we shall denote by $\ell(w)$. By convention, $\ell(\mathrm{id}_V) = 0$.

18.8.3 Lemma. *Let B be a base of R.*

(i) *Let $\beta_1, \ldots, \beta_p \in B$ not necessary distinct. If $s_1 \circ \cdots \circ s_{p-1}(\beta_p) \in R_-$, $s_i = s_{\beta_i}$, then $s_1 \circ \cdots \circ s_p = s_1 \circ \cdots \circ s_{q-1} \circ s_{q+1} \circ \cdots \circ s_{p-1}$ for some $q \in \{1, \ldots, p-1\}$.*

(ii) *Let $w \in W(R)$ and $\beta_1, \ldots, \beta_p \in B$ such that $w = s_1 \circ \cdots \circ s_p$, $s_i = s_{\beta_i}$. If $p = \ell(w)$, then $w(\beta_p) \in R_-$.*

(iii) *Let $w \in W(R)$ and $n(w)$ the number of positive roots α such that $w(\alpha) \in R_-$. Then $n(w) = \ell(w)$.*

Proof. (i) Let $\alpha_{p-1} = \beta_p$ and $\alpha_i = s_{i+1} \circ \cdots \circ s_{p-1}(\beta_p)$, $0 \leqslant i \leqslant p-2$. We have $\alpha_0 \in R_-$ and $\alpha_{p-1} \in R_+$. Let q be the smallest integer such that $\alpha_q \in R_+$. Then $s_q(\alpha_q) = \alpha_{q-1} \in R_-$. So $\alpha_q = \beta_q$ by 18.7.4 (v). Since $s_{w(\alpha)} = w \circ s_\alpha \circ w^{-1}$ for $w \in W(R)$ and $\alpha \in R$, we have:

$$s_q = (s_{q+1} \circ \cdots \circ s_{p-1}) \circ s_p \circ (s_{p-1} \circ \cdots \circ s_{q+1}).$$

So part (i) follows.

(ii) We have $w(\beta_p) = -s_1 \circ \cdots \circ s_{p-1}(\beta_p)$. If $s_1 \circ \cdots \circ s_{p-1}(\beta_p) \in R_-$, then $p > \ell(w)$ by part (i). So we have proved (ii).

(iii) Proceed by induction on $\ell(w)$. The case $\ell(w) = 0$ is obvious. Let $w = s_1 \circ \cdots \circ s_p$ be a reduced decomposition of w where the notations are as above. By (ii), $w(\beta_p) \in R_-$. So it follows from 18.7.4 (v) that $n(w \circ s_p) = n(w) - 1$. But it is clear that $\ell(w \circ s_p) = \ell(w) - 1$, so we have by induction that $n(w) = \ell(w)$. $\square$

18.8.4 For $\alpha \in R$, define $P_\alpha = \ker \alpha^\vee = \{x \in V; (x|\alpha) = 0\}$. Let P be the union of P_α, $\alpha \in R$. It is a closed subset of V, and the connected components of $P' = V \setminus P$ are open subsets of V, called the *Weyl chambers* (or chambers) of R. By 18.3.2, they do not depend on the choice of the $W(R)$-invariant inner product on V. Since the inner product is $W(R)$-invariant, $W(R)$ acts on P and P', and $W(R)$ permutes the chambers of R.

18.8.5 Lemma. *The group $W(R)$ acts transitively on the set of Weyl chambers of R.*

Proof. Let C_1, C_2 be chambers in R, $x_1 \in C_1$ and $x_2 \in C_2$. Let $w \in W(R)$ be such that
$$\|x_1 - w(x_2)\| = \inf\{\|x_1 - w'(x_2)\|; w' \in W(R)\}.$$
Set $I = \{tx_1 + (1-t)w(x_2); t \in [0,1]\}$.

Suppose that $I \cap P_\alpha \neq \emptyset$ for some $\alpha \in R$, then since $x_1, w(x_2) \notin P_\alpha$, there exists $t_0 \in]0,1[$ such that $(t_0 x_1 + (1-t_0)w(x_2)|\alpha) = 0$. Thus

$$\|x_1 - s_\alpha \circ w(x_2)\|^2 = \|x_1 - w(x_2)\|^2 - \frac{4(1-t_0)}{t_0\|\alpha\|^2}(w(x_2)|\alpha)^2$$
$$< \|x_1 - w(x_2)\|^2.$$

But this contradicts the assumption on w, and since I is connected, I is contained in a chamber. Hence $x_1, w(x_2) \in C_1$. Since w permutes the chambers and $C_1 \cap w(C_2) \neq \emptyset$, we have $C_1 = w(C_2)$. $\square$

18.8.6 Proposition. *Let $B = \{\beta_1, \ldots, \beta_l\}$ be a base of R, $(\beta_1', \ldots, \beta_l')$ the basis of V dual to B with respect to the inner product $(.|.)$, and*

$$C(B) = \{x \in V; (x|\beta_i) > 0, 1 \leqslant i \leqslant l\} = \left\{ \sum_{i=1}^{l} x_i \beta_i'; x_1, \cdots, x_l > 0 \right\}.$$

Then $C(B)$ is a Weyl chamber of R.

Proof. It is clear from 18.7.4 that $C(B) \subset P'$. Since $C(B)$ is convex, it is connected. So $C(B)$ is contained in a chamber, say C, of R.

Suppose that there exists $x \in C \setminus C(B)$. Then $(x|\beta) < 0$ for some $\beta \in B$. Set $C_1 = \{x \in C; (x|\beta) > 0\}$ and $C_2 = \{x \in C; (x|\beta) < 0\}$. Then C is the disjoint union of the open subsets C_1 and C_2. Now $x \in C_2$ and $C(B) \subset C_1$. This implies that C is not connected. Contradiction. $\square$

18.8.7 Theorem. *Let R be a root system in V.*

(i) *Any chamber of R is of the form $C(B)$ for some base B of R. In particular, any chamber of R is an open simplicial cone.*

(ii) *The map $B \mapsto C(B)$ induces a bijection between the set of bases of R and the set of chambers of R.*

(iii) *The group $W(R)$ acts simply transitively on the set of bases of R and on the set of chambers of R.*

Proof. (i) If $w \in W(R)$ and B a base of R, then clearly $w(B)$ is another base of R. So (i) follows from 18.8.5 and 18.8.6.

(ii) The map $B \mapsto C(B)$ is surjective by (i). Since $C(B)$ is uniquely determined by $\mathbb{R}_+ \beta_i'$, $1 \leqslant i \leqslant l$, (notations of 18.8.6), the map is injective.

(iii) By 18.8.5 and (ii), $W(R)$ acts transitively on the set of bases of R. Let B be a base and $w \in W(R)$ such that $w(B) = B$. Then 18.8.3 (ii) forces w to be id_V. Hence the action is simply transitive on the set of the bases and, by (ii), on the set of chambers as well. $\square$

18.8.8 Remarks. 1) For a fixed base B of R, the chamber $C(B)$ is called the *fundamental Weyl chamber*.

2) If C is a Weyl chamber, denote by $B = B(C)$ the corresponding base of R, $R_+ = R_+(C)$ and $R_- = R_-(C)$. For $\beta \in B$, the hyperplane P_β is called a *wall* of the chamber C.

3) Suppose that R is the direct sum of root systems $R_1, \ldots, R_n$. Let C_i be a chamber of R_i and B_i the corresponding base of R_i, then $C = C_1 \times \cdots \times C_n$ is a chamber of R corresponding to the base $B = B_1 \cup \cdots \cup B_n$. By 18.8.7 (iii), all the chambers and bases of R are obtained in this way.

18.8.9 Proposition. *Let $B = \{\beta_1, \ldots, \beta_l\}$ be a base of R and $C = C(B)$.*
  (i) *For all $x \in V$, there exists $w \in W(R)$ such that $w(x) \in \overline{C}$.*
  (ii) *If $x, y \in \overline{C}$ and $w \in W(R)$ verify $w(x) = y$, then $x = y$.*
  (iii) *Any $W(R)$-orbit of V intersects $\overline{C}$ at a unique point.*

Proof. Let us define the following partial order on V:

$$x \preccurlyeq y \;\Leftrightarrow\; y - x \in \mathbb{R}_+\beta_1 + \cdots + \mathbb{R}_+\beta_l.$$

(i) Let w be maximal with respect to $\preccurlyeq$ in the set $\{w'(x); w' \in W(R)\}$. If $w(x) \notin \overline{C}$, then there exists $\beta \in B$ such that $\beta^\vee(w(x)) = (\beta|w(x)) < 0$. But $s_\beta \circ w(x) - w(x) = -\beta^\vee(w(x))\beta$, so $w(x) \prec s_\beta \circ w(x)$ which contradicts the maximality of w. So $w(x) \in \overline{C}$.

(ii) We shall show by induction on $\ell(w)$ that w is the product of reflections s_β, $\beta \in B$ such that $s_\beta(x) = x$. The case $\ell(w) = 0$ is trivial. So let us suppose that $\ell(w) > 0$. Then by 18.7.8 (ii) and 18.8.3 (ii), there exists $\beta \in B$ such that $w(\beta) \in R_-$ and $\ell(w \circ s_\beta) = \ell(w) - 1$. Since $x, y \in \overline{C}$,

$$0 \leqslant (\beta|x) = (\beta|w^{-1}(y)) = (w(\beta)|y) \leqslant 0.$$

Hence $(\beta|x) = 0$. Thus $s_\beta(x) = x$ and $w \circ s_\beta(x) = y$. The result follows by induction. Thus $x = y$.

(iii) This is clear from (i) and (ii). $\square$

18.8.10 Remark. The partial order introduced in the proof of 18.8.9 is called the partial order *associated* to B (or to C).

18.8.11 Remark. The notions of bases and Weyl chambers can easily be extended to a non-reduced root system R. In particular, R has a base, and $W(R)$ acts simply transitively on the set of bases and the set of Weyl chambers.

18.9 Highest root

18.9.1 Lemma. *Assume that R is irreducible. Let B be a base of R. Then B is not the (disjoint) union of two orthogonal subsets.*

Proof. Suppose that B is the disjoint union of (non-empty) subsets B_1 and B_2 such that $(\beta_1|\beta_2) = 0$ for $\beta_1 \in B_1$, $\beta_2 \in B_2$. By 18.7.8 (ii), the subspace of V spanned by B_1 is proper and $W(R)$-stable. This contradicts 18.2.7. $\square$

18.9.2 Theorem. *Assume that R is irreducible. Let $B = \{\beta_1, \ldots, \beta_l\}$ be a base of R, $C = C(B)$ the corresponding chamber.*

(i) *There exists a root $\widetilde{\alpha} = n_1\beta_1 + \cdots + n_l\beta_l$ such that for any root $\alpha = p_1\beta_1 + \cdots + p_l\beta_l$, we have $n_i \geqslant p_i$, $1 \leqslant i \leqslant l$.*

(ii) *We have $\widetilde{\alpha} \in \overline{C}$.*

(iii) *We have $\|\widetilde{\alpha}\| \geqslant \|\alpha\|$ for all $\alpha \in R$. In particular, $\widetilde{\alpha}$ is a long root.*

(iv) *If $\alpha \in R_+ \setminus \{\widetilde{\alpha}\}$, then $a_{\alpha,\widetilde{\alpha}} \in \{0, 1\}$.*

Proof. Let $\alpha = n_1\beta_1 + \cdots + n_l\beta_l$ be a maximal element of R with respect to the partial order $\preccurlyeq$ associated to B. Clearly, $\alpha \in R_+$. Let

$$I = \{1, \ldots, l\}\,, \ I_1 = \{i \in I; n_i > 0\}\,, \ I_2 = \{i \in I; n_i = 0\} = I \setminus I_1.$$

We claim that $I_2 = \emptyset$, because otherwise, for each $i \in I$, $(\alpha|\beta_i) \leqslant 0$ (18.7.4), and so by 18.9.1, $(\beta_i|\beta_j) \neq 0$ for some $i \in I_2$ and $j \in I_1$. It follows that $(\alpha|\beta_i) < 0$ and hence $\alpha \prec \alpha + \beta_i \in R_+$ (18.5.3), which contradicts the maximality of α. So we have our claim.

Using 18.5.3 again, we deduce that $(\alpha|\beta_i) \geqslant 0$ for all $i \in I$, and so $(\alpha|\beta_i) > 0$ for some $i \in I$. Now, if α' is another maximal element of R with respect to $\preccurlyeq$, then $(\alpha|\alpha') > 0$. So if $\alpha \neq \alpha'$, then $\alpha - \alpha' \in R$ (18.5.3), and therefore $\alpha \preccurlyeq \alpha'$ or $\alpha' \preccurlyeq \alpha$ which again contradicts the maximality of α and α'. Thus we have proved (i), and part (ii) follows from 18.5.3.

Let $\alpha \in R$. To prove (iii), we may assume that $\alpha \neq \widetilde{\alpha}$ and $\alpha \in \overline{C}$ (18.8.9). Since $0 \prec \widetilde{\alpha} - \alpha$, $(x|\widetilde{\alpha} - \alpha) \geqslant 0$ for all $x \in \overline{C}$. In particular, $(\widetilde{\alpha}|\widetilde{\alpha} - \alpha) \geqslant 0$ and $(\alpha|\widetilde{\alpha} - \alpha) \geqslant 0$. Hence $(\widetilde{\alpha}|\widetilde{\alpha}) \geqslant (\alpha|\widetilde{\alpha}) \geqslant (\alpha|\alpha)$.

Finally, if $\alpha \in R_+ \setminus \{\widetilde{\alpha}\}$, then $\|\widetilde{\alpha}\| \geqslant \|\alpha\|$ and $(\widetilde{\alpha}|\alpha) \geqslant 0$ since $\widetilde{\alpha} \in \overline{C}$. So (iv) is a consequence of 18.5.2. $\square$

18.9.3 Assume that R is irreducible. The root $\widetilde{\alpha}$ in 18.9.2 is called the *highest root* of R with respect to the base B (or chamber C).

18.10 Closed subsets of roots

18.10.1 Definition. *Let P be a subset of R.*

(i) *We say that P is* symmetric *if $P = -P$.*

(ii) *We say that P is* closed *if given $\alpha, \beta \in P$ verifying $\alpha + \beta \in R$, then $\alpha + \beta \in P$.*

(iii) *We say that P is* parabolic *if P is closed and $R = P \cup (-P)$.*

18.10.2 Proposition. *Let $P \subset R$ be a closed subset such that $P \cap (-P) = \emptyset$. There exists a chamber C of R such that $P \subset R_+(C)$.*

Proof. 1) Let $n \in \mathbb{N}^*$, $\alpha_1, \ldots, \alpha_n \in P$ and $\alpha = \alpha_1 + \cdots + \alpha_n$. We claim that $\alpha \neq 0$. This is obvious if $n = 1$. Let us proceed by induction on n.

If $n \geqslant 2$ and $\alpha = 0$, then $(\alpha_1 | \alpha_2 + \cdots + \alpha_n) < 0$. So there exists some j such that $(\alpha_1 | \alpha_j) < 0$. Since P is closed, it follows from 18.5.3 that $\alpha_1 + \alpha_j \in P$. Thus $\alpha = 0$ is the sum of $n - 1$ elements of P, which contradicts the induction hypothesis. So we have proved our claim.

2) We claim there exists $\alpha \in P$ such that $(\alpha | \beta) \geqslant 0$ for all $\beta \in P$. For otherwise, there exist $\alpha_1, \alpha_2 \in P$ such that $(\alpha_1 | \alpha_2) < 0$, then $\alpha_1 + \alpha_2 \in P$, and there exists $\alpha_3 \in P$ such that $(\alpha_1 + \alpha_2 | \alpha_3) < 0$, and so $\alpha_1 + \alpha_2 + \alpha_3 \in P$. Repeating this process, we obtain a sequence $(\alpha_i)_{i \geqslant 1}$ of elements of P such that $\alpha_1 + \cdots + \alpha_i \in P$ for all $i \in \mathbb{N}^*$. It follows that there exists $i < j$ such that $\alpha_1 + \cdots + \alpha_i = \alpha_1 + \cdots + \alpha_j$. Hence $\alpha_{i+1} + \cdots + \alpha_j = 0$ which contradicts 1).

3) Finally, let us prove that there is a basis $\mathcal{E}$ of V such that the elements of P are positive with respect to the lexicographic order associated to $\mathcal{E}$. The case $l = 1$ is trivial. Let us proceed by induction on l and assume that $l \geqslant 2$.

From 2), there exists $v_1 \in V \setminus \{0\}$, such that $(v_1 | \alpha) \geqslant 0$ for all $\alpha \in P$. Let H be the hyperplane orthogonal to v_1 and V_1 the subspace of V spanned by $R \cap H$. It follows from 18.2.5 that $P \cap H$ is a closed subset of $R \cap H$.

By our induction hypothesis, there exists a basis $(v_2, \ldots, v_l)$ of H such that the elements of $P \cap H$ are positive with respect to the associated lexicographic order. It is then clear that the elements of P are positive with respect to the lexicographic order associated to the basis $(v_1, \ldots, v_l)$ of V. □

18.10.3 Corollary. *The following conditions are equivalent for a subset P of R.*

(i) *There exists a chamber C of R such that $P = R_+(C)$.*

(ii) *P is parabolic and $P \cap (-P) = \emptyset$.*

If these conditions are verified, the chamber C such that $P = R_+(C)$ is unique.

Proof. The equivalence is clear from 18.10.2. If $P = R_+(C)$, then $C^\vee$ is the set of $\varphi \in V^*$ verifying $\varphi(\alpha) > 0$ for all $\alpha \in P$. Hence the uniqueness of C. □

18.10.4 Lemma. *Let C be a chamber of R, $B = B(C)$, $R_+ = R_+(C)$, P a closed subset of R containing R_+, $\Sigma = B \cap (-P)$ and Q the set of roots which is the sum of elements of $-\Sigma$. Then $P = R_+ \cup Q$.*

Proof. Since P is closed, $Q \subset P$ (18.7.7). So $R_+ \cup Q \subset P$.

Now let $\alpha \in P \setminus R_+$. Then α is the sum of p elements of $-B$. We shall show by induction on p that $\alpha \in Q$. The case $p = 1$ is obvious. So let us assume that $p \geqslant 2$.

By 18.7.7, there exist $\beta \in -B$, $\gamma \in R_-$ such that $\alpha = \beta + \gamma$, and γ is the sum of $p-1$ elements of $-B$. Since $R_+ \subset P$, $-\gamma, -\beta \in P$. So $\beta = (-\gamma) + \alpha \in P$ and $\gamma = (-\beta) + \alpha \in P$, because P is closed. Thus $\beta \in -\Sigma$, and $\gamma \in Q$ by our induction hypothesis. Again since P is closed, $\alpha = \beta + \gamma \in Q$. □

18.10.5 Proposition. *The following conditions are equivalent for a subset P of R.*

(i) *P is parabolic.*

(ii) *P is closed and there exists a chamber C of R such that $R_+(C) \subset P$.*

(iii) *There exists a chamber C of R and a subset Σ of $B(C)$ such that P is union of $R_+(C)$ and the set of roots which can be expressed as a sum of elements of $-\Sigma$.*

Proof. (i) $\Rightarrow$ (ii) Let C be a chamber such that the cardinal of $P \cap R_+(C)$ is maximal. Set $B = B(C)$, $R_+ = R_+(C)$.

Suppose that there exists $\beta \in B \setminus P$. So $-\beta \in P$ and $-\beta = s_\beta(\beta) \in s_\beta(R_+) = R'_+$ where $C' = s_\beta(C)$ and $R'_+ = R_+(C')$. Hence $-\beta \in P \cap R'_+$.

If $\alpha \in R_+ \setminus \{\beta\}$, then $s_\beta(\alpha) \in R_+$ (18.7.4 (v)). So $\alpha = s_\beta \circ s_\beta(\alpha) \in s_\beta(R_+) = R'_+$. Hence $P \cap R_+ \subset P \cap R'_+$.

We deduce that $\{-\beta\} \cup (P \cap R_+) \subset P \cap R'_+$ which contradicts our choice of C. Hence $B \subset P$ and $R_+ \subset P$.

(ii) $\Rightarrow$ (iii) This is clear by 18.10.4.

(iii) $\Rightarrow$ (i) We only need to show that P is closed. Let $\Sigma = \{\beta_1, \ldots, \beta_p\}$ and $B = B(C) = \Sigma \cup \{\beta_{p+1}, \ldots, \beta_l\}$.

Let $\alpha, \beta \in P$ be such that $\alpha + \beta \in R$. Since $R_+(C) \subset P$, we may assume that $\alpha + \beta = -n_1\beta_1 - \cdots - n_l\beta_l \in R_-(C)$, where $n_1, \ldots, n_l \in \mathbb{N}$. Since $P = R_+(C) \cup Q$, $n_i = 0$ for $p + 1 \leqslant i \leqslant l$. Hence $\alpha + \beta \in Q \subset P$. $\square$

18.10.6 Proposition. *Let P be a subset of R, Γ (resp. V') the subgroup (resp. subspace) of V generated (resp. spanned) by P. Then the following conditions are equivalent:*

(i) $\Gamma \cap R = P$.

(ii) *P is closed and symmetric.*

(iii) *P is closed and is a root system in V'.*

If these conditions are verified, then the map $\alpha \mapsto \alpha^\vee|_{V'}$ is a bijection from P onto $P^\vee$.

Proof. (i) $\Rightarrow$ (ii) This is clear.

(ii) $\Rightarrow$ (iii) With the notations of 18.2.1, P verifies (R1) in V'. Let $\alpha, \beta \in P$. If $\beta = \pm\alpha$, then $s_\alpha(\beta) = -\beta \in P$. If $\beta \neq \pm\alpha$, then for $0 \leqslant j \leqslant a_{\beta,\alpha}$, $\beta - j\alpha \in R$. Since P is symmetric and closed, $s_\alpha(\beta) = \beta - a_{\beta,\alpha}\alpha \in P$. Thus $s_\alpha(P) \subset P$, and (R2) is verified. Finally, it is clear that (R3) is verified. So P is a root system V' and last part of the proposition follows also.

(iii) $\Rightarrow$ (i) Clearly, $P \subset \Gamma \cap R$. Conversely, if $\alpha \in \Gamma \cap R$, then $\alpha = \alpha_1 + \cdots + \alpha_n$ where $\alpha_i \in P$, since $P = -P$. From $0 < (\alpha|\alpha) = (\alpha|\alpha_1 + \cdots + \alpha_n)$, we deduce that $(\alpha|\alpha_i) > 0$ for some i. If $\alpha = \alpha_i$, then $\alpha \in P$. Otherwise, $\alpha - \alpha_i \in R$ (18.5.3) and $\alpha - \alpha_i \in \Gamma$. By induction on n, $\alpha - \alpha_i \in P$ and since P is closed, $\alpha = (\alpha - \alpha_i) + \alpha_i \in P$. $\square$

18.11 Weights

18.11.1 Let us use the notations of 18.5.1 and 18.8.1. We define an isomorphism $\varphi : V \to V^*$, $\varphi(x)(y) = (x|y)$ for all $x, y \in V$. Denoting $(\varphi(x)|\varphi(y)) = (x|y)$, we obtain a $W(R^\vee)$-invariant inner product on V^*. For $\alpha \in R$, we have:

$$\varphi(\alpha) = \frac{2\alpha^\vee}{(\alpha^\vee|\alpha^\vee)}.$$

Thus $\varphi(P_\alpha) = P_{\alpha^\vee}$ (18.8.4). From this, we deduce that the image of a chamber of R via φ is a chamber of $R^\vee$.

18.11.2 Let C be a chamber of R, $B = B(C) = \{\beta_1, \ldots, \beta_l\}$, $C^\vee = \varphi(C)$ and $B' = B(C^\vee) = \{\gamma_1, \ldots, \gamma_l\}$, $S(C) = \{x \in V; (x|y) \geqslant 0 \text{ for all } y \in C\}$ and $S(C^\vee) = \{\lambda \in V^*; (\lambda|\mu) \geqslant 0 \text{ for all } \mu \in C^\vee\}$. By 18.8.6, we have:

$$S(C) = \mathbb{R}_+\beta_1 + \cdots + \mathbb{R}_+\beta_l \ , \ \ S(C^\vee) = \mathbb{R}_+\gamma_1 + \cdots + \mathbb{R}_+\gamma_l.$$

It is then clear that $S(C^\vee) = \mathbb{R}_+\beta_1^\vee + \cdots + \mathbb{R}_+\beta_l^\vee$, and we deduce that $B' = B^\vee = \{\beta_1^\vee, \ldots, \beta_l^\vee\}$. In particular, $B^\vee$ is a base of $R^\vee$.

18.11.3 Let $\alpha = n_1\beta_1 + \cdots + n_l\beta_l \in R$. Then:

$$\frac{2\alpha^\vee}{(\alpha^\vee|\alpha^\vee)} = \varphi(\alpha) = \sum_{i=1}^l n_i \frac{2\beta_i^\vee}{(\beta_i^\vee|\beta_i^\vee)}.$$

Hence

$$\alpha^\vee = \sum_{i=1}^l n_i\|\beta_i\|^2\|\alpha\|^{-2}\beta_i^\vee.$$

By 18.11.2, $n_i\|\beta_i\|^2\|\alpha\|^{-2} \in \mathbb{Z}$ for $1 \leqslant i \leqslant l$.

18.11.4 Denote by $Q(R)$ the sublattice of V spanned by R. By 18.7.4, it is a free abelian subgroup of rank l in V. Any base of R is a basis of $Q(R)$. In the same way, $Q(R^\vee)$ is a sublattice of rank l in V^*. The lattice $Q(R)$ is called the *root lattice* of R and its elements are called *radical weights*.

18.11.5 Proposition. *Let $P(R)$ be the set of $x \in V$ such that $\alpha^\vee(x) \in \mathbb{Z}$ for all $\alpha \in R$. We call $P(R)$ the* weight lattice *of R and its elements are called* weights.

(i) $P(R)$ *is a sublattice of rank l in V.*

(ii) *We have $Q(R) \subset P(R) \subset V_\mathbb{Q}$.*

(iii) *If $B^\vee$ is a base of $R^\vee$, the dual basis of $B^\vee$ in V is a basis of $P(R)$.*

Proof. If $x \in V$, the following conditions are equivalent:

(i) $a^*(x) \in \mathbb{Z}$ for all $a^* \in Q(R^\vee)$.

(ii) $\beta^\vee(x) \in \mathbb{Z}$ for all $\beta^\vee \in B^\vee$.

(iii) The coordinates of x in the dual basis of $B^\vee$ are integers.

Consequently, we have parts (i) and (iii). By (R3), we have $Q(R) \subset P(R)$.

Let $B = \{\beta_1, \ldots, \beta_l\}$ be a base of R, $B^\vee = \{\beta_1^\vee, \ldots, \beta_l^\vee\}$ as in 18.11.2 and $\{\alpha_1, \ldots, \alpha_l\}$ the dual basis of $B^\vee$. For $1 \leqslant p, q \leqslant l$, write

$$\alpha_p = \sum_{i=1}^{l} \lambda_{pj}\beta_j, \quad \text{and} \quad \delta_{pq} = \sum_{j=1}^{l} \lambda_{pj}\beta_q^\vee(\beta_j).$$

This is a Cramer's system with coefficients in $\mathbb{Q}$, so $P(R) \subset V_{\mathbb{Q}}$. $\square$

18.11.6 Let C be a chamber in R, $B = B(C)$, $B^\vee = \{\beta_1^\vee, \ldots, \beta_l^\vee\}$ the base of $R^\vee$ as in 18.11.2. The dual basis $\{\varpi_1, \ldots, \varpi_l\}$ of $B^\vee$ is a basis of $P(R)$, and they are called the *fundamental weights* of R with respect to the base B (or chamber C). We shall conserve this notation for fundamental weights.

If $x \in V$, then $x \in C$ if and only if $\beta^\vee(x) > 0$ for all $\beta \in B$. Hence

$$C = \mathbb{R}_+^* \varpi_1 + \cdots + \mathbb{R}_+^* \varpi_l \,, \quad \overline{C} = \mathbb{R}_+ \varpi_1 + \cdots + \mathbb{R}_+ \varpi_l.$$

18.11.7 Proposition. *Let $B = (\beta_1, \ldots, \beta_l)$ be a base of R, $\varpi_1, \ldots, \varpi_l$ the corresponding fundamental weights and $C = C(B)$. For $x \in V$, the following conditions are equivalent:*

(i) $\beta^\vee(x) \in \mathbb{N}$ *for all* $\beta \in B$.

(ii) $x = n_1\varpi_1 + \cdots + n_l\varpi_l$ *where* $n_1, \ldots, n_l \in \mathbb{N}$.

(iii) $x \in \overline{C} \cap P(R)$.

(iv) *For all* $w \in W(R)$, $x - w(x) \in Q(R)$ *and* $x - w(x) \succcurlyeq 0$ *in the partial order associated to B.*

If $x \in V$ verifies these conditions, then x is called a dominant weight *of R with respect to B.*

Proof. The implications (i) $\Rightarrow$ (ii) $\Rightarrow$ (iii) are obvious.

(iii) $\Rightarrow$ (iv) Let $w \in W(R)$ be such that $w(x)$ is maximal in $X = \{w'(x); w' \in W(R)\}$ with respect to the partial order associated to B. We saw in the proof of 18.8.9 (i) that $w(x) \in \overline{C}$. So 18.8.9 (iii) implies that $w(x) = x$. Thus x is the unique maximal element of X. Since $x \in P(R)$, (iv) follows from 18.7.8 (ii).

(iv) $\Rightarrow$ (i) This is clear since $x - s_\beta(x) = \beta^\vee(x)\beta$ for $\beta \in B$. $\square$

18.11.8 Proposition. *Let B be a base of R, $C = C(B)$ and $\preccurlyeq$ the partial order associated to B. If $x \in V$, the following conditions are equivalent:*

(i) $x \in \overline{C}$.

(ii) $s_\beta(x) \preccurlyeq x$ *for all* $\beta \in B$.

(iii) $w(x) \preccurlyeq x$ *for all* $w \in W(R)$.

Proof. Since $x - s_\beta(x) = \beta^\vee(x)\beta$, we have (i) $\Leftrightarrow$ (ii). Finally, (iii) $\Rightarrow$ (ii) is obvious and (i) $\Rightarrow$ (iii) is similar to the implication (iii) $\Rightarrow$ (iv) of 18.11.7. $\square$

18.11.9 Corollary. *With the hypotheses of 18.11.8, we have $x \in C$ if and only if $w(x) \prec x$ for all $w \in W(R) \setminus \{id_V\}$.*

Proof. The "if" part is clear by 18.11.8 and the definition of C. Conversely, if $x \in C$, then $w(x) \preccurlyeq x$ for all $w \in W(R)$. If $w(x) = x$ and $w \neq \mathrm{id}_V$, then it follows from the proof of 18.8.9 (ii) that w is the product of reflections s_β, $\beta \in B$, verifying $s_\beta(x) = x$. Thus $\beta^\vee(x) = 0$ which is absurd since $x \in C$. $\square$

18.11.10 Proposition. *With the notations of 18.11.6, let ρ be the half sum of the positive roots in R. Then:*
  (i) $\rho = \varpi_1 + \cdots + \varpi_l \in C$.
  (ii) *We have $\|\beta\|^2 = 2(\rho|\beta)$ for all $\beta \in B$.*

Proof. This is clear since $\beta^\vee(\rho) = 1 = 2(\rho|\beta)/(\beta|\beta)$ for $\beta \in B$ (18.7.6). $\square$

18.11.11 With the notations of 18.2.2 and 18.11.6, if $1 \leqslant i \leqslant l$, then:

$$\beta_i = \sum_{j=1}^{l} \alpha_{ij} \varpi_j.$$

Then $A = [\alpha_{ij}] \in \mathrm{GL}_l(\Bbbk)$. On the other hand,

$$\delta_{ij} = \beta_i^\vee(\varpi_j) = 2\frac{(\beta_i|\varpi_j)}{(\beta_i|\beta_i)}.$$

Thus $\alpha_{ij} = \beta_j^\vee(\beta_i) = a_{\beta_i,\beta_j}$, and the matrix $[a_{\beta_i,\beta_j}]$ is invertible.

18.12 Graphs

18.12.1 Definition. *A graph is a pair (S, A) where S is a finite set and A is a subset of $\mathfrak{P}(S)$ (the set of subsets of S) whose elements are subsets of cardinal 2.*

A normed graph is a pair (Γ, f) with the following properties:
  (a) *$\Gamma = (S, A)$ is a graph.*
  (b) *Let E be the set of pairs (i, j) such that $\{i, j\} \in A$. Then f is a map from E to $\mathbb{R}$ such that $f(i, j)f(j, i) = 1$ for all $(i, j) \in E$.*

18.12.2 Let $\Gamma = (S, A)$ be a graph. An element of S is called a *vertex* and an element of A is called an *edge*. Two vertices x, y are *linked* if $\{x, y\}$ is an edge. A vertex is called a *terminal* vertex (resp. a *ramification point*) if it is linked to at most one vertex (resp. at least three vertices).

In general, we represent a graph by a picture consisting of points corresponding to the elements of S and two points are linked by a line if and only if the corresponding vertices are linked.

18.12.3 Let $\Gamma = (S, A)$ and $\Gamma' = (S', A')$ be graphs.

An isomorphism of Γ onto Γ' is a bijection from S to S' which transports A to A'.

If $S' \subset S$ and $A' \subset A$, then we say that Γ' is a *subgraph* of Γ. Furthermore, if $A' = A \cap \mathfrak{P}(S')$, then we say that Γ' is a *full* subgraph of Γ.

18.12.4 Let $\Gamma = (S, A)$ be a graph.

If $a, b \in S$, then a *path* from a to b is a sequence $C = (x_0, \ldots, x_n)$ of elements of S verifying:

(i) $x_0 = a$ and $x_n = b$.

(ii) For $0 \leqslant i \leqslant n - 1$, $\{x_i, x_{i+1}\} \in A$.

The integer n is the *length* of the path C. The path C is *injective* if the x_i's are pairwise distinct. Thus a path from a to b of minimal length is injective.

We define an equivalence relation $\sim$ on S as follows: $a \sim b$ if and only if there is a path from a to b. The equivalence classes are called the *connected components*. A graph is *connected* if it has at most one connected component. Thus in a connected graph, there is always a path between two vertices.

Let $\Gamma_1, \ldots, \Gamma_n$ be the connected components of Γ, considered as full subgraphs of Γ. Then each Γ_i is connected and for $i \neq j$, there is no path between a vertex of Γ_i and a vertex of Γ_j.

A *cycle* in Γ is a path $C = (x_0, \ldots, x_n)$ of length $n \geqslant 2$ verifying $x_0 = x_n$ and $C' = (x_0, \ldots, x_{n-1})$ is an injective path. A graph is a *forest* if Γ has no cycles. A connected forest is called a *tree*.

For $n \in \mathbb{N}^*$, we denote by A_n the graph with vertices $\{1, 2, \ldots, n\}$ and edges are the subsets $\{i, j\}$ such that $i - j = \pm 1$:

A graph which is isomorphic to A_{n+1} is called a *chain* of length $n \geqslant 0$.

18.12.5 Proposition. *Let Γ be a non-empty graph. Then Γ is a chain if and only if Γ is a tree without ramification points.*

Proof. It is clear that if Γ is chain, then it is a tree without ramification points. Conversely, suppose that Γ is a tree without ramification points. Let $C = (x_0, \ldots, x_n)$ be an injective path of maximal length in Γ. Suppose that a is a vertex which is distinct from the x_i's. Since Γ is connected, there is a path C' from a to one of the x_i's. Replacing a by a vertex in C', we may assume that a and x_j are linked.

If $j = 0$ (resp. n), then $(a, x_0, \ldots, x_n)$ (resp. $(x_0, \ldots, x_n, a)$) is an injective path of length $n + 1$. But this contradicts our choice of C.

If $0 < j < n$, then x_j is a ramification point which is absurd.

Hence $\{x_0, \ldots, x_n\}$ is the set of vertices of Γ and since Γ is a tree without ramification points, Γ is necessarily a chain. $\square$

18.13 Dynkin diagrams

18.13.1 Definition. *Let R, R' be root systems in V and V'. We say that R and R' are isomorphic if there exists a bijective linear map $F : V \to V'$ such that $F(R) = R'$ and for all $\alpha, \beta \in R$, we have:*

$$\frac{(F(\alpha)|F(\beta))}{(F(\beta)|F(\beta))} = \frac{(\alpha|\beta)}{(\beta|\beta)}.$$

18.13.2 Observe that the notion of isomorphism in 18.13.1 does not depend on the choice of invariant inner products (18.3.2).

Suppose that R and R' are isomorphic and $F : V \to V'$ the corresponding bijective linear map. Then the map $w \mapsto F \circ w \circ F^{-1}$ is an isomorphism of the groups $W(R)$ and $W(R')$.

In particular, let $R_1, \ldots, R_m$ be the irreducible components of R. Then given $\lambda_1, \ldots, \lambda_m \in \mathbb{R}^*$, the direct sum R'' of the root systems $\lambda_i R_i$ is isomorphic to R.

18.13.3 Definition. *Let* $B = (\beta_1, \ldots, \beta_l)$ *be a base of* R. *The matrix* $[a_{\beta_i, \beta_j}]_{1 \leqslant i,j \leqslant l}$ *is called the* Cartan matrix *of* R *with respect to* B.

18.13.4 It follows from 18.8.7 that, up to a permutation, the Cartan matrix of R does not depend on the choice of the base B.

Clearly, isomorphic root systems have the same Cartan matrix, up to a permutation. The converse is given in the following proposition.

Proposition. *Let* R, R' *be root systems in* V, V', B, B' *bases of* R, R'. *Suppose that there is a bijection* $f : B \to B'$ *transforming the Cartan matrix of* R *to the Cartan matrix of* R'. *Then* R *and* R' *are isomorphic via an isomorphism* $F : V \to V'$ *which extends* f.

In particular, the Cartan matrix of a root system determines the root system up to an isomorphism.

Proof. Let $B = (\beta_1, \ldots, \beta_l)$, $B' = (\beta'_1, \ldots, \beta'_l)$ where $\beta'_i = f(\beta_i)$. Since B, B' are basis of V and V', it is clear that f extends uniquely to an isomorphism $F : V \to V'$.

Let $\alpha, \beta \in B$, $\alpha' = F(\alpha)$, $\beta' = F(\beta)$. Then $a_{\beta,\alpha} = a_{\beta',\alpha'}$ and:

$$s_{F(\alpha)}(F(\beta)) = \beta' - a_{\beta',\alpha'}\alpha' = F(\beta - a_{\beta,\alpha}\alpha) = F(s_\alpha(\beta)).$$

Since B is a basis of V, we deduce that $s_{F(\alpha)} \circ F = F \circ s_\alpha$ for $\alpha \in B$. Hence 18.7.8 implies that the map $w \mapsto F \circ w \circ F^{-1}$ is an isomorphism from $W(R)$ onto $W(R')$, sending s_α to $s_{F(\alpha)}$ if $\alpha \in B$.

Let $\beta \in R$. By 18.7.8, $\beta = w(\alpha)$ for some $\alpha \in B$ and $w \in W(R)$. It follows that $F(\beta) = (F \circ w \circ F^{-1})(F(\alpha))$. Hence $F(R) = R'$. On the other hand, since $s_{w(\alpha)} = w \circ s_\alpha \circ w^{-1}$, we have:

$$F \circ s_\beta = F \circ w \circ F^{-1} \circ F \circ s_\alpha \circ w^{-1} = (F \circ w \circ F^{-1}) \circ s_{F(\alpha)} \circ F \circ w^{-1}$$
$$= (F \circ w \circ F^{-1}) \circ s_{F(\alpha)} \circ (F \circ w \circ F^{-1})^{-1} \circ F = s_{F(w(\alpha))} \circ F = s_{F(\beta)} \circ F.$$

Finally, for $\alpha, \beta \in R$, we have:

$$s_{F(\alpha)}(F(\beta)) = F(\beta - a_{F(\beta),F(\alpha)}\alpha) , \quad F \circ s_\alpha(\beta) = F(\beta - a_{\beta,\alpha}\alpha).$$

Hence $a_{F(\beta),F(\alpha)} = a_{\beta,\alpha}$. $\square$

18.13.5 We associate to R a normed graph $\Gamma(R) = (X, f)$, called the *Dynkin diagram* as follows:

Fix a base $B = (\beta_1, \ldots, \beta_l)$ of R, and for $1 \leqslant i, j \leqslant l$, let $n_{ij} = a_{\beta_i, \beta_j}$. The vertices of the graph X is $\{1, \ldots, l\}$ (or $\{\beta_1, \ldots, \beta_l\}$) and $\{i, j\}$ is an edge if and only if $(\beta_i | \beta_j) \neq 0$ (or equivalently $n_{ij} \neq 0$ or $n_{ji} \neq 0$). If $\{i, j\}$ is an edge, we define:

$$f(i, j) = \frac{n_{ij}}{n_{ji}} = \frac{\|\beta_i\|^2}{\|\beta_j\|^2}.$$

It is clear that $f(i, j)f(j, i) = 1$.

Let θ_{ij} be the angle between β_i and β_j. Then the only possible cases, up to a permutation of i and j, are listed below:

1) i and j are not linked ; $n_{ij} = n_{ji} = 0$, $\theta_{ij} = \dfrac{\pi}{2}$.

2) $f(i, j) = f(j, i) = 1$; $n_{ij} = n_{ji} = -1$, $\|\beta_i\| = \|\beta_j\|$, $\theta_{ij} = \dfrac{2\pi}{3}$.

3) $f(i, j) = 2$, $f(j, i) = \dfrac{1}{2}$; $n_{ij} = -2$, $n_{ji} = -1$, $\|\beta_i\| = \sqrt{2}\|\beta_j\|$, $\theta_{ij} = \dfrac{3\pi}{4}$.

4) $f(i, j) = 3$, $f(j, i) = \dfrac{1}{3}$; $n_{ij} = -3$, $n_{ji} = -1$, $\|\beta_i\| = \sqrt{3}\|\beta_j\|$, $\theta_{ij} = \dfrac{5\pi}{6}$.

Thus the Dynkin diagram determines R up to an isomorphism (18.13.4).

18.13.6 In practice, we represent $\Gamma(R) = (X, f)$ by a picture with points and lines as follows: the points are the vertices of X, and we draw $n_{ij}n_{ji}$ lines from i to j. Further, if $f(i, j) > 1$ and $\theta_{ij} \neq \pi/2$ (cases 3) and 4) above), then we place the symbol $>$ on the lines joining i to j, that is from the long root to the short root:

$$f(i,j) = 2 \qquad\qquad\qquad\qquad f(i,j) = 3$$

18.13.7 Proposition. *Let R be root system in V and $\Gamma(R)$ its Dynkin diagram. Then:*

(i) R *is irreducible if and only if* $\Gamma(R)$ *is connected.*

(ii) $\Gamma(R)$ *is a forest, and if R is irreducible, then $\Gamma(R)$ is a tree.*

Proof. (i) Suppose that R is the direct sum of two non-empty root systems R_1 and R_2. Let B_1, B_2 be bases of R_1, R_2. Then $B = B_1 \cup B_2$ is a base of R and $\Gamma(R_1)$, $\Gamma(R_2)$ are full subgraphs of $\Gamma(R)$. Further there is no edge joining a vertex of $\Gamma(R_1)$ to a vertex of $\Gamma(R_2)$. So $\Gamma(R)$ is not connected.

Conversely, suppose that $\Gamma(R)$ is not connected, then R has a base of the form $B = B_1 \cup B_2$ where B_1, B_2 are non-empty orthogonal subsets. Let V_i be the subspace spanned by B_i. If $\alpha \in B_1$, then the orthogonal of α is the direct sum of V_2 and a hyperplane in V_1. It follows that $s_\alpha(V_i) = V_i$ for $i = 1, 2$. The same result applies for $\alpha \in B_2$. So V_1 and V_2 are $W(R)$-stable (18.7.8), and R is not irreducible (18.2.7).

(ii) Let B be a base of R. If $\Gamma(R)$ is not a forest, then there exist $\beta_1, \ldots, \beta_n \in B$ pairwise distinct such that $(\beta_1, \ldots, \beta_n)$ is a cycle. Let $\gamma_i = \beta_i / \|\beta_i\|$. If $\{i, j\}$ is an edge, then by 18.13.5, $(\gamma_i | \gamma_j) \leqslant -1/2$. Hence:

$$\left\| \sum_{i=1}^{n} \gamma_i \right\|^2 = n + 2 \sum_{i<j} (\gamma_i | \gamma_j) \leqslant n + 2[(\gamma_1 | \gamma_2) + \cdots + (\gamma_{n-1} | \gamma_n) + (\gamma_n | \gamma_1)]$$
$$\leqslant n - n = 0.$$

This is absurd. So $\Gamma(R)$ is a forest, and (i) implies that $\Gamma(R)$ is a tree.   □

18.13.8 Let $B = (\beta_1, \ldots, \beta_l)$ be a base of R. If $\alpha = n_1 \beta_1 + \cdots + n_l \beta_l \in R$, we set $H(\alpha)$ to be the set of $\beta_i \in B$ such that $n_i \neq 0$.

Proposition. (i) *If $\alpha \in R$, then $H(\alpha)$ is a connected subset of B (considered as the set of vertices in the Dynkin diagram of R).*
(ii) *If H is a non-empty connected subset of B, then $\sum_{\beta \in H} \beta \in R$.*

Proof. (i) We may assume that $\alpha \in R_+$ and we shall proceed by induction on the cardinality n of $H(\alpha)$. The case $n = 1$ being trivial, we shall assume that $n \geqslant 2$.

By 18.7.4, there exists $\beta \in B$ such that $\alpha - \beta \in R$. Let r be maximal such that $\gamma = \alpha - r\beta \in R$. So $r \geqslant 1$ and the β-string through γ is of the form $\{\gamma, \ldots, \gamma + q\beta\}$, with $q \geqslant r$. Thus $a_{\gamma, \beta} = -q \leqslant -r \leqslant -1$ (18.5.4), and so $(\gamma | \beta) \neq 0$. It follows that there is an edge joining β and an element of $H(\gamma)$. Since $H(\alpha) = H(\gamma) \cup \{\beta\}$, it follows from the induction hypothesis that $H(\alpha)$ is connected.

(ii) Let $\alpha = \sum_{\beta \in H} \beta$. We shall prove by induction on the cardinality n of $H(\alpha)$ that $\alpha \in R$. The case $n = 1$ being trivial, we may assume that $n \geqslant 2$.

Let $(\gamma_0, \gamma_1, \ldots, \gamma_m)$ be an injective path of maximal length in H. Since H is a tree (18.13.7) and the length of the path is maximal, γ_0 is a terminal vertex in H. Let $M = H \setminus \{\gamma_0\}$, then M is connected. By the induction hypothesis, $\alpha - \gamma_0 \in R$. Further, γ_0 and γ_1 are linked, so $(\alpha - \gamma_0 | \gamma_0) < 0$ (18.7.4). Hence $\alpha \in R$ (18.5.3).   □

18.14 Classification of root systems

18.14.1 Theorem. *Let R be an irreducible root system in V. Then the Dynkin diagram of R is isomorphic to one of the following :*

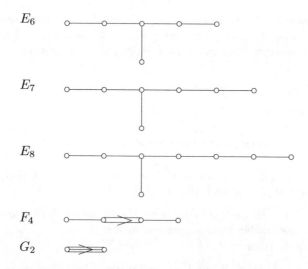

E_6

E_7

E_8

F_4

G_2

Moreover, these diagrams are pairwise non-isomorphic.

Proof. We can check easily that these diagrams are pairwise non-isomorphic. So let us prove the first part.

Let $B = (\beta_1, \ldots, \beta_l)$ be a base of R and Γ be the Dynkin diagram of R. Set $\gamma_i = \beta_i / \|\beta_i\|$. Recall from the proof of 18.13.7 that $(\gamma_i | \gamma_j) \leqslant -1/2$ if $\{i, j\}$ is an edge, and $(\gamma_i | \gamma_j) = 0$ otherwise.

Let $f(i, j)$ be as defined in 18.13.5.

1) Let us suppose that there exist vertices i, j of Γ such that $f(i, j) = 3$. Since Γ is connected (18.13.7), if Γ is not G_2, then there exists a vertex k distinct from i and j such that

$$(\gamma_i | \gamma_j) = -\frac{\sqrt{3}}{2} \ , \ (\gamma_i | \gamma_k) \leqslant 0 \ , \ (\gamma_j | \gamma_k) \leqslant -\frac{1}{2}.$$

It follows that

$$\|\sqrt{3}\gamma_i + 2\gamma_j + \gamma_k\|^2 \leqslant 3 + 4 + 1 - 2\sqrt{3} \times 2\frac{\sqrt{3}}{2} - 2 \times 2 \times \frac{1}{2} = 0.$$

This is absurd. So $\Gamma \simeq G_2$.

Let us assume from now on that $f(i, j) \neq 3$ for all i, j.

2) Let us suppose that there exist vertices i, j, k, l of Γ such that $f(i, j) = 2$ and $f(k, l) = 2$. Again, since Γ is connected, renumbering the vertices if necessary, we may assume that Γ contains the following subdiagram :

$$\underset{1}{\circ}\!\Rightarrow\!\underset{2}{\circ}\!-\!\underset{3}{\circ}\!-\!\cdots\!-\!\underset{n-2}{\circ}\!-\!\underset{n-1}{\circ}\!\Leftarrow\!\underset{n}{\circ}$$

with $n \geqslant 3$. So, $(\gamma_1 | \gamma_2) = (\gamma_{n-1} | \gamma_n) = -\frac{1}{\sqrt{2}}$. Hence :

$$\left\|\frac{1}{\sqrt{2}}\gamma_1 + \gamma_2 + \cdots + \gamma_{n-1} + \frac{1}{\sqrt{2}}\gamma_n\right\|^2 \leqslant \frac{1}{2} + \frac{1}{2} + (n-2) - 4\frac{1}{\sqrt{2}}\frac{1}{\sqrt{2}} - 2(n-3)\frac{1}{2} = 0.$$

Again, this is absurd. So we may further assume that Γ has at most one edge $\{i,j\}$ such that $f(i,j) = 2$.

3) Let us suppose that Γ has an edge $\{i,j\}$ such that $f(i,j) = 2$, and a ramification point. Then Γ contains a subdiagram of the form :

with $n \geqslant 4$. So $(\gamma_1|\gamma_2) = -\dfrac{1}{\sqrt{2}}$ and $(\gamma_i|\gamma_j) = -\dfrac{1}{2}$ for the other edges in this subdiagram. Consequently, we have

$$\left\| \frac{\gamma_1}{\sqrt{2}} + \gamma_2 + \cdots + \gamma_{n-2} + \frac{\gamma_{n-1}}{2} + \frac{\gamma_n}{2} \right\|^2 \leqslant \frac{1}{2} + (n-3) + \frac{1}{4} + \frac{1}{4} - 2\frac{1}{\sqrt{2}}\frac{1}{\sqrt{2}}$$
$$-2(n-4)\frac{1}{2} - 2\frac{1}{2}\frac{1}{2} - 2\frac{1}{2}\frac{1}{2} = 0.$$

This is absurd.

4) Suppose that Γ has an edge $\{i,j\}$ such that $f(i,j) = 2$. By the previous considerations, Γ is a chain. If Γ is not isomorphic to B_l or C_l or F_4, then it contains a subdiagram of the form :

It follows that :

$$\left\| \sqrt{2}\gamma_1 + 2\sqrt{2}\gamma_2 + 3\gamma_3 + 2\gamma_4 + \gamma_5 \right\|^2 \leqslant 2 + 8 + 9 + 4 + 1 - 2\sqrt{2}.2\sqrt{2}\frac{1}{2}$$
$$-2.2\sqrt{2}.3\frac{1}{\sqrt{2}} - 2.3.2\frac{1}{2} - 2.2\frac{1}{2} = 0.$$

Again, this is absurd. So Γ is isomorphic to either B_l, C_l or F_4.

We may therefore assume in the rest of the proof that $f(i,j) = 1$ for all edges $\{i,j\}$ of Γ.

5) If Γ has no ramification point, then it is isomorphic to A_l.

6) Let us suppose that Γ has two distinct ramification points. Then Γ contains a subdiagram of the form :

We obtain therefore that :

$$\left\| \frac{\gamma_1}{2} + \frac{\gamma_2}{2} + \gamma_3 + \cdots + \gamma_{n-2} + \frac{\gamma_{n-1}}{2} + \frac{\gamma_n}{2} \right\|^2 \leqslant 4\frac{1}{4} + (n-4) - 2(n-5)\frac{1}{2}$$
$$-2\frac{1}{2}\frac{1}{2} - 2\frac{1}{2}\frac{1}{2} - 2\frac{1}{2}\frac{1}{2} - 2\frac{1}{2}\frac{1}{2}$$
$$= 0.$$

This is absurd.

7) We are left with the case where Γ has a unique ramification point. In particular, Γ are at least 4 vertices, and it is of the form :

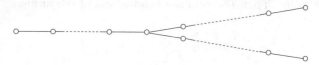

If Γ contains a subdiagram of the form :

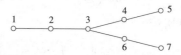

then

$$\|\gamma_1 + 2\gamma_2 + 3\gamma_3 + 2\gamma_4 + \gamma_5 + 2\gamma_6 + \gamma_7\|^2 \leqslant 24 - 2\frac{1}{2}(2 + 6 + 6 + 6 + 2 + 2) = 0.$$

This is absurd.

If Γ contains a subdiagram of the form :

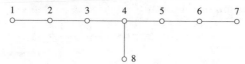

then we can check easily that

$$\|\gamma_1 + 2\gamma_2 + 3\gamma_3 + 4\gamma_4 + 3\gamma_5 + 2\gamma_6 + \gamma_7 + 2\gamma_8\|^2 \leqslant 0,$$

which is absurd.

If Γ contains a subdiagram of the form :

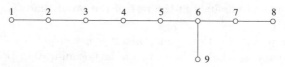

then we can check easily that

$$\|\gamma_1 + 2\gamma_2 + 3\gamma_3 + 4\gamma_4 + 5\gamma_5 + 6\gamma_6 + 4\gamma_7 + 2\gamma_8 + 3\gamma_9\|^2 \leqslant 0,$$

which is absurd.

It is now easy to see that Γ is isomorphic to either D_l, E_6, E_7 or E_8. $\quad\square$

18.14.2 In the following tables, we give explicit descriptions of irreducible root systems. The irreducible root system R is considered to be a subset of some $\mathbb{R}^n$ where n is not necessarily l, the rank of R. Let $B = (\beta_1, \ldots, \beta_l)$ be a base of R and $(\varepsilon_1, \ldots, \varepsilon_n)$ denote the canonical basis of $\mathbb{R}^n$.

The vertices of the Dynkin diagram correspond to the elements of B. We add a vertex which corresponds to $-\widetilde{\alpha}$, where $\widetilde{\alpha}$ is the highest root of R with respect to B and concerning the edges to this vertex, we use the convention of 18.13.5 and 18.13.6. This new graph is called the *extended* Dynkin diagram of R.

If $\alpha = n_1\beta_1 + \cdots + n_l\beta_l \in R$, we set:

$$|\alpha| = n_1 + \cdots + n_l.$$

We call $|\alpha|$ the *height* of α, and we denote it also by $\mathrm{ht}(\alpha)$.

We fix an order on the roots: let $\alpha, \beta \in R_+$, then $\alpha \prec \beta$ if either $|\alpha| < |\beta|$ or, $|\alpha| = |\beta|$ and α is smaller than β in the lexicographic order associated to $(\beta_l, \beta_{l-1}, \ldots, \beta_1)$.

For the graphs of type E_6, E_7, E_8, F_4, G_2, we give an explicit list of positive roots in increasing order.

References

- [13], [29], [39], [41], [80].

$$\boxed{\text{Root system of type } A_l \ (l \geqslant 1)}$$

- V is the hyperplane of $\mathbb{R}^{l+1}$ consisting of points such that the sum of its coordinates is zero.
- Roots: $\varepsilon_i - \varepsilon_j$ $(1 \leqslant i, j \leqslant l+1, \ i \neq j)$.
- Number of roots: $n = l(l+1)$.
- Base: $\beta_1 = \varepsilon_1 - \varepsilon_2$, $\beta_2 = \varepsilon_2 - \varepsilon_3, \ldots$, $\beta_l = \varepsilon_l - \varepsilon_{l+1}$.
- Highest root: $\widetilde{\alpha} = \beta_1 + \cdots + \beta_l$.
- Cartan matrix:

$$\begin{pmatrix} 2 & -1 & 0 & 0 & \cdots & 0 & 0 \\ -1 & 2 & -1 & 0 & \cdots & 0 & 0 \\ 0 & -1 & 2 & -1 & \cdots & 0 & 0 \\ 0 & 0 & -1 & 2 & \cdots & 0 & 0 \\ \cdots & \cdots & \cdots & \cdots & \cdots & \cdots & \cdots \\ 0 & 0 & 0 & 0 & \cdots & 2 & -1 \\ 0 & 0 & 0 & 0 & \cdots & -1 & 2 \end{pmatrix}$$

- Extended Dynkin diagram $(l \geqslant 2)$:

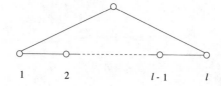

- Order of $W(R)$: $(l+1)!$.
- Positive roots:

$$\varepsilon_i - \varepsilon_j = \beta_i + \cdots + \beta_{j-1} \quad (1 \leqslant i < j \leqslant l+1).$$

- Fundamental weights:

$$\begin{aligned} \varpi_i &= (\varepsilon_1 + \cdots + \varepsilon_i) - \frac{i}{l+1}(\varepsilon_1 + \cdots + \varepsilon_{l+1}) \quad (1 \leqslant i \leqslant l) \\ &= \frac{1}{l+1}[(l-i+1)\beta_1 + 2(l-i+1)\beta_2 + \cdots + (i-1)(l-i+1)\beta_{i-1} \\ &\quad + i(l-i+1)\beta_i + i(l-i)\beta_{i+1} + \cdots + i\beta_l]. \end{aligned}$$

$$\boxed{\text{Root system of type } B_l \ (l \geqslant 2)}$$

- $V = \mathbb{R}^l$.
- Roots: $\pm \varepsilon_i \ (1 \leqslant i \leqslant l)$, $\pm \varepsilon_i \pm \varepsilon_j \ (1 \leqslant i < j \leqslant l)$.
- Number of roots: $n = 2l^2$.
- Base: $\beta_1 = \varepsilon_1 - \varepsilon_2$, $\beta_2 = \varepsilon_2 - \varepsilon_3, \dots,$ $\beta_{l-1} = \varepsilon_{l-1} - \varepsilon_l$, $\beta_l = \varepsilon_l$.
- Highest root: $\widetilde{\alpha} = \beta_1 + 2\beta_2 + 2\beta_3 + \cdots + 2\beta_l$.
- Cartan matrix:

$$
\begin{pmatrix}
2 & -1 & 0 & 0 & \cdots & 0 & 0 \\
-1 & 2 & -1 & 0 & \cdots & 0 & 0 \\
0 & -1 & 2 & -1 & \cdots & 0 & 0 \\
0 & 0 & -1 & 2 & \cdots & 0 & 0 \\
& & \cdots & & \cdots & & \\
0 & 0 & 0 & 0 & \cdots & 2 & -2 \\
0 & 0 & 0 & 0 & \cdots & -1 & 2
\end{pmatrix}
$$

- Extended Dynkin diagram:
  - ▶ $l = 2$:

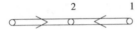

  - ▶ $l \geqslant 3$:

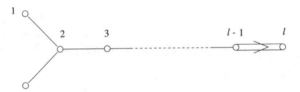

- Order of $W(R)$: $2^l . l!$.
- Positive roots:

$$
\begin{cases}
\varepsilon_i = \beta_i + \cdots + \beta_l & (1 \leqslant i \leqslant l) \\
\varepsilon_i - \varepsilon_j = \beta_i + \cdots + \beta_{j-1} & (1 \leqslant i < j \leqslant l) \\
\varepsilon_i + \varepsilon_j = \beta_i + \cdots + \beta_{j-1} + 2(\beta_j + \cdots + \beta_l) & (1 \leqslant i < j \leqslant l).
\end{cases}
$$

- Fundamental weights:

$$
\begin{aligned}
\varpi_i &= \varepsilon_1 + \cdots + \varepsilon_i \qquad (1 \leqslant i < l) \\
&= \beta_1 + 2\beta_2 + \cdots + (i-1)\beta_{i-1} + i[\beta_i + \cdots + \beta_l] \\
\varpi_l &= \frac{1}{2}(\varepsilon_1 + \cdots + \varepsilon_l) = \frac{1}{2}(\beta_1 + 2\beta_2 + \cdots + l\beta_l).
\end{aligned}
$$

$$\boxed{\text{Root system of type } C_l \ (l \geqslant 3)}$$

- $V = \mathbb{R}^l$.
- Roots: $\pm 2\varepsilon_i \quad (1 \leqslant i \leqslant l)$, $\pm\varepsilon_i \pm \varepsilon_j \quad (1 \leqslant i < j \leqslant l)$.
- Number of roots: $n = 2l^2$.
- Base: $\beta_1 = \varepsilon_1 - \varepsilon_2$, $\beta_2 = \varepsilon_2 - \varepsilon_3, \ldots$, $\beta_{l-1} = \varepsilon_{l-1} - \varepsilon_l$, $\beta_l = 2\varepsilon_l$.
- Highest root: $\widetilde{\alpha} = 2\beta_1 + 2\beta_2 + \cdots + 2\beta_{l-1} + \beta_l$.
- Cartan matrix:

$$\begin{pmatrix} 2 & -1 & 0 & 0 & \ldots & 0 & 0 \\ -1 & 2 & -1 & 0 & \ldots & 0 & 0 \\ 0 & -1 & 2 & -1 & \ldots & 0 & 0 \\ 0 & 0 & -1 & 2 & \ldots & 0 & 0 \\ \ldots & \ldots & \ldots & \ldots & \ldots & \ldots & \ldots \\ 0 & 0 & 0 & 0 & \ldots & 2 & -1 \\ 0 & 0 & 0 & 0 & \ldots & -2 & 2 \end{pmatrix}$$

- Extended Dynkin diagram:

- Order of $W(R)$: $2^l . l!$.
- Positive roots:

$$\begin{cases} 2\varepsilon_i = 2(\beta_i + \cdots + \beta_{l-1}) + \beta_l & (1 \leqslant i \leqslant l) \\ \varepsilon_i - \varepsilon_j = \beta_i + \cdots + \beta_{j-1} & (1 \leqslant i < j \leqslant l) \\ \varepsilon_i + \varepsilon_j = \beta_i + \cdots + \beta_{j-1} + 2(\beta_j + \cdots + \beta_{l-1}) + \beta_l & (1 \leqslant i < j \leqslant l). \end{cases}$$

- Fundamental weights:

$$\varpi_i = \varepsilon_1 + \cdots + \varepsilon_i \qquad (1 \leqslant i \leqslant l)$$
$$= \beta_1 + 2\beta_2 + (i-1)\beta_{i-1} + i[\beta_i + \beta_{i+1} + \cdots + \beta_{l-1} + \frac{1}{2}\beta_l].$$

$$\boxed{\text{Root system of type } D_l \ (l \geqslant 4)}$$

- $V = \mathbb{R}^l$.
- Roots: $\pm\varepsilon_i \pm \varepsilon_j$ $(1 \leqslant i < j \leqslant l)$.
- Number of roots: $n = 2l(l-1)$.
- Base: $\beta_1 = \varepsilon_1 - \varepsilon_2,\ \beta_2 = \varepsilon_2 - \varepsilon_3, \ldots, \ \beta_{l-1} = \varepsilon_{l-1} - \varepsilon_l,\ \beta_l = \varepsilon_{l-1} + \varepsilon_l$.
- Highest root: $\widetilde{\alpha} = \beta_1 + 2\beta_2 + \cdots + 2\beta_{l-2} + \beta_{l-1} + \beta_l$.
- Cartan matrix:

$$\begin{pmatrix}
2 & -1 & \ldots & 0 & 0 & 0 & 0 \\
-1 & 2 & \ldots & 0 & 0 & 0 & 0 \\
\ldots & \ldots & \ldots & \ldots & \ldots & \ldots & \ldots \\
0 & 0 & \ldots & 2 & -1 & 0 & 0 \\
0 & 0 & \ldots & -1 & 2 & -1 & -1 \\
0 & 0 & \ldots & 0 & -1 & 2 & 0 \\
0 & 0 & \ldots & 0 & -1 & 0 & 2
\end{pmatrix}$$

- Extended Dynkin diagram $(l \geqslant 4)$:

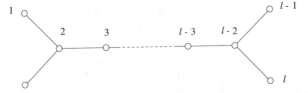

- Order of $W(R)$: $2^{l-1}.l!$.
- Positive roots:

$$\begin{cases}
\varepsilon_i - \varepsilon_j = \beta_i + \cdots + \beta_{j-1} & (1 \leqslant i < j \leqslant l) \\
\varepsilon_i + \varepsilon_l = \beta_i + \cdots + \beta_{l-2} + \beta_l & (1 \leqslant i < l) \\
\varepsilon_i + \varepsilon_j = \beta_i + \cdots + \beta_{j-1} + 2(\beta_j + \cdots + \beta_{l-2}) + \beta_{l-1} + \beta_l & (1 \leqslant i < j < l).
\end{cases}$$

- Fundamental weights:

$$\begin{aligned}
\varpi_i &= \varepsilon_1 + \cdots + \varepsilon_i \qquad (1 \leqslant i \leqslant l-2) \\
&= \beta_1 + 2\beta_2 + \cdots + (i-1)\beta_{i-1} \\
&\quad + i[\beta_i + \beta_{i+1} + \cdots + \beta_{l-2}] + \tfrac{1}{2}i[\beta_{l-1} + \beta_l] \\[4pt]
\varpi_{l-1} &= \tfrac{1}{2}(\varepsilon_1 + \varepsilon_2 + \cdots + \varepsilon_{l-2} + \varepsilon_{l-1} - \varepsilon_l) \\
&= \tfrac{1}{2}[\beta_1 + 2\beta_2 + \cdots + (l-2)\beta_{l-2}] + \tfrac{1}{4}l\beta_{l-1} + \tfrac{1}{4}(l-2)\beta_l \\[4pt]
\varpi_l &= \tfrac{1}{2}(\varepsilon_1 + \varepsilon_2 + \cdots + \varepsilon_{l-2} + \varepsilon_{l-1} + \varepsilon_l) \\
&= \tfrac{1}{2}[\beta_1 + 2\beta_2 + \cdots + (l-2)\beta_{l-2}] + \tfrac{1}{4}(l-2)\beta_{l-1} + \tfrac{1}{4}l\beta_l.
\end{aligned}$$

ROOT SYSTEM OF TYPE E_6

- V is the subspace of $\mathbb{R}^8$ consisting of points whose coordinates (ξ_i) verify $\xi_6 = \xi_7 = -\xi_8$.
- Roots:

$$\pm\varepsilon_i \pm \varepsilon_j \quad (1 \leqslant i < j \leqslant 5), \quad \pm\frac{1}{2}\left(\varepsilon_8 - \varepsilon_7 - \varepsilon_6 + \sum_{i=1}^{5}(-1)^{\nu(i)}\varepsilon_i\right),$$

where $\nu(i)$ are positive integers such that $\nu(1) + \cdots + \nu(5)$ is even.
- Number of roots: $n = 72$.
- Base:

$$\beta_1 = \frac{1}{2}(\varepsilon_1 - \varepsilon_2 - \varepsilon_3 - \varepsilon_4 - \varepsilon_5 - \varepsilon_6 - \varepsilon_7 + \varepsilon_8),$$
$$\beta_2 = \varepsilon_1 + \varepsilon_2, \quad \beta_3 = \varepsilon_2 - \varepsilon_1, \quad \beta_4 = \varepsilon_3 - \varepsilon_2, \quad \beta_5 = \varepsilon_4 - \varepsilon_3, \quad \beta_6 = \varepsilon_5 - \varepsilon_4.$$

- Highest root: $\tilde{\alpha} = \beta_1 + 2\beta_2 + 2\beta_3 + 3\beta_4 + 2\beta_5 + \beta_6$.
- Cartan matrix:

$$\begin{pmatrix} 2 & 0 & -1 & 0 & 0 & 0 \\ 0 & 2 & 0 & -1 & 0 & 0 \\ -1 & 0 & 2 & -1 & 0 & 0 \\ 0 & -1 & -1 & 2 & -1 & 0 \\ 0 & 0 & 0 & -1 & 2 & -1 \\ 0 & 0 & 0 & 0 & -1 & 2 \end{pmatrix}$$

- Extended Dynkin diagram:

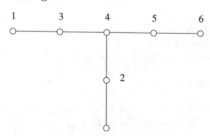

- Order of $W(R)$: $2^7.3^4.5$.
- Positive roots:

$$\begin{cases} \pm\varepsilon_i + \varepsilon_j & (1 \leqslant i < j \leqslant 5) \\ \dfrac{1}{2}\left(\varepsilon_8 - \varepsilon_7 - \varepsilon_6 + \displaystyle\sum_{i=1}^{5}(-1)^{\nu(i)}\varepsilon_i\right) & \text{with } \displaystyle\sum_{i=1}^{5}\nu(i) \text{ even.} \end{cases}$$

Let us represent the root $a\beta_1 + b\beta_2 + c\beta_3 + d\beta_4 + e\beta_5 + f\beta_6$ by $a\ c\ \overset{d}{\underset{b}{}}\ e\ f$. Then the positive roots are:

1 0 0 0 0 0	0 0 0 0 0 1	0 1 0 0 0 0	0 0 1 0 0 0	0 0 0 1 0 0	0 0 0 0 1 0	1 1 0 0 0 0	0 0 1 0 0 1
0 1 1 0 0 0	0 0 1 1 0 0	0 0 0 1 1 0	1 1 1 0 0 0	0 1 1 0 0 1	0 0 1 1 0 1	0 1 1 1 0 0	0 0 1 1 1 0
1 1 1 0 0 1	1 1 1 1 0 0	0 1 1 1 0 1	0 0 1 1 1 1	0 1 1 1 1 0	1 1 1 1 0 1	0 1 2 1 0 1	1 1 1 1 1 0
0 1 1 1 1 1	1 1 2 1 0 1	1 1 1 1 1 1	0 1 2 1 1 1	1 2 2 1 0 1	1 1 2 1 1 1	0 1 2 2 1 1	1 2 2 1 1 1
		1 1 2 2 1 1	1 2 2 2 1 1	1 2 3 2 1 1	1 2 3 2 1 2		

- Fundamental weights:

$$\varpi_1 = \frac{2}{3}(\varepsilon_8 - \varepsilon_7 - \varepsilon_6)$$

$$= \frac{1}{3}[4\beta_1 + 3\beta_2 + 5\beta_3 + 6\beta_4 + 4\beta_5 + 2\beta_6]$$

$$\varpi_2 = \frac{1}{2}(\varepsilon_1 + \varepsilon_2 + \varepsilon_3 + \varepsilon_4 + \varepsilon_5 - \varepsilon_6 - \varepsilon_7 + \varepsilon_8)$$

$$= \beta_1 + 2\beta_2 + 2\beta_3 + 3\beta_4 + 2\beta_5 + \beta_6$$

$$\varpi_3 = \frac{5}{6}(\varepsilon_8 - \varepsilon_7 - \varepsilon_6) + \frac{1}{2}(-\varepsilon_1 + \varepsilon_2 + \varepsilon_3 + \varepsilon_4 + \varepsilon_5)$$

$$= \frac{1}{3}[5\beta_1 + 6\beta_2 + 10\beta_3 + 12\beta_4 + 8\beta_5 + 4\beta_6]$$

$$\varpi_4 = \varepsilon_3 + \varepsilon_4 + \varepsilon_5 - \varepsilon_6 - \varepsilon_7 + \varepsilon_8$$

$$= 2\beta_1 + 3\beta_2 + 4\beta_3 + 6\beta_4 + 4\beta_5 + 2\beta_6$$

$$\varpi_5 = \frac{2}{3}(\varepsilon_8 - \varepsilon_7 - \varepsilon_6) + \varepsilon_4 + \varepsilon_5$$

$$= \frac{1}{3}[4\beta_1 + 6\beta_2 + 8\beta_3 + 12\beta_4 + 10\beta_5 + 5\beta_6]$$

$$\varpi_6 = \frac{1}{3}(\varepsilon_8 - \varepsilon_7 - \varepsilon_6) + \varepsilon_5$$

$$= \frac{1}{3}[2\beta_1 + 3\beta_2 + 4\beta_3 + 6\beta_4 + 5\beta_5 + 4\beta_6].$$

ROOT SYSTEM OF TYPE E_7

- V is the hyperplane of $\mathbb{R}^8$ orthogonal to $\varepsilon_7 + \varepsilon_8$.
- Roots:

$$\pm(\varepsilon_7 - \varepsilon_8), \quad \pm\varepsilon_i \pm \varepsilon_j \quad (1 \leqslant i < j \leqslant 6), \quad \pm\frac{1}{2}\left(\varepsilon_7 - \varepsilon_8 + \sum_{i=1}^{6}(-1)^{\nu(i)}\varepsilon_i\right),$$

where $\nu(i)$ are positive integers such that $\nu(1) + \cdots + \nu(6)$ is odd.
- Number of roots: $n = 126$.
- Base:

$$\beta_1 = \frac{1}{2}(\varepsilon_1 - \varepsilon_2 - \varepsilon_3 - \varepsilon_4 - \varepsilon_5 - \varepsilon_6 - \varepsilon_7 + \varepsilon_8), \quad \beta_2 = \varepsilon_1 + \varepsilon_2,$$
$$\beta_3 = \varepsilon_2 - \varepsilon_1, \quad \beta_4 = \varepsilon_3 - \varepsilon_2, \quad \beta_5 = \varepsilon_4 - \varepsilon_3, \quad \beta_6 = \varepsilon_5 - \varepsilon_4, \quad \beta_7 = \varepsilon_6 - \varepsilon_5.$$

- Highest root: $\widetilde{\alpha} = 2\beta_1 + 2\beta_2 + 3\beta_3 + 4\beta_4 + 3\beta_5 + 2\beta_6 + \beta_7$.
- Cartan matrix:

$$\begin{pmatrix}
2 & 0 & -1 & 0 & 0 & 0 & 0 \\
0 & 2 & 0 & -1 & 0 & 0 & 0 \\
-1 & 0 & 2 & -1 & 0 & 0 & 0 \\
0 & -1 & -1 & 2 & -1 & 0 & 0 \\
0 & 0 & 0 & -1 & 2 & -1 & 0 \\
0 & 0 & 0 & 0 & -1 & 2 & -1 \\
0 & 0 & 0 & 0 & 0 & -1 & 2
\end{pmatrix}$$

- Extended Dynkin diagram:

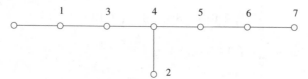

- Order of $W(R)$: $2^{10}.3^4.5.7$.
- Positive roots:

$$\begin{cases}
\varepsilon_8 - \varepsilon_7 \\
\pm\varepsilon_i + \varepsilon_j & (1 \leqslant i < j \leqslant 6) \\
\frac{1}{2}\left(\varepsilon_8 - \varepsilon_7 + \sum_{i=1}^{6}(-1)^{\nu(i)}\varepsilon_i\right) \text{ with } \sum_{i=1}^{6}\nu(i) \text{ odd.}
\end{cases}$$

Representing the root $a\beta_1 + b\beta_2 + c\beta_3 + d\beta_4 + e\beta_5 + f\beta_6 + g\beta_7$ by $\overset{a\ c\ d\ e\ f\ g}{\underset{b}{}}$,
the positive roots are:

1 0 0 0 0 0 0	0 0 0 0 0 0 1	0 1 0 0 0 0 0	0 0 1 0 0 0 0	0 0 0 1 0 0 0	0 0 0 0 1 0 0	0 0 0 0 0 1 0
1 1 0 0 0 0 0	0 0 1 0 0 0 1	0 1 1 0 0 0 0	0 0 1 1 0 0 0	0 0 0 1 1 0 0	0 0 0 0 1 1 0	1 1 1 0 0 0 0
0 1 1 0 0 0 1	0 0 1 1 0 0 1	0 1 1 1 0 0 0	0 0 1 1 1 0 0	0 0 0 1 1 1 0	1 1 1 0 0 0 1	1 1 1 1 0 0 0
0 1 1 1 0 0 1	0 0 1 1 1 0 1	0 1 1 1 1 0 0	0 0 1 1 1 1 0	1 1 1 1 0 0 1	0 1 2 1 0 0 1	1 1 1 1 1 0 0
0 1 1 1 1 0 1	0 0 1 1 1 1 1	0 1 1 1 1 1 0	1 1 2 1 0 0 1	1 1 1 1 1 0 1	0 1 2 1 1 0 1	1 1 1 1 1 1 0
0 1 1 1 1 1 1	1 2 2 1 0 0 1	1 1 2 1 1 0 1	0 1 2 2 1 0 1	1 1 1 1 1 1 1	0 1 2 1 1 1 1	1 2 2 1 1 0 1
1 1 2 2 1 0 1	1 1 2 1 1 1 1	0 1 2 2 1 1 1	1 2 2 2 1 0 1	1 2 2 1 1 1 1	1 1 2 2 1 1 1	0 1 2 2 2 1 1
1 2 3 2 1 0 1	1 2 2 2 1 1 1	1 1 2 2 2 1 1	1 2 3 2 1 0 2	1 2 3 2 1 1 1	1 2 2 2 2 1 1	1 2 3 2 1 1 2
1 2 3 2 2 1 1	1 2 3 2 2 1 2	1 2 3 3 2 1 1	1 2 3 3 2 1 2	1 2 4 3 2 1 2	1 3 4 3 2 1 2	2 3 4 3 2 1 2

- Fundamental weights:

$$\varpi_1 = \varepsilon_8 - \varepsilon_7$$
$$= 2\beta_1 + 2\beta_2 + 3\beta_3 + 4\beta_4 + 3\beta_5 + 2\beta_6 + \beta_7$$

$$\varpi_2 = \frac{1}{2}(\varepsilon_1 + \varepsilon_2 + \varepsilon_3 + \varepsilon_4 + \varepsilon_5 + \varepsilon_6 - 2\varepsilon_7 + 2\varepsilon_8)$$
$$= \frac{1}{2}[4\beta_1 + 7\beta_2 + 8\beta_3 + 12\beta_4 + 9\beta_5 + 6\beta_6 + 3\beta_7]$$

$$\varpi_3 = \frac{1}{2}(-\varepsilon_1 + \varepsilon_2 + \varepsilon_3 + \varepsilon_4 + \varepsilon_5 + \varepsilon_6 - 3\varepsilon_7 + 3\varepsilon_8)$$
$$= 3\beta_1 + 4\beta_2 + 6\beta_3 + 8\beta_4 + 6\beta_5 + 4\beta_6 + 2\beta_7$$

$$\varpi_4 = \varepsilon_3 + \varepsilon_4 + \varepsilon_5 + \varepsilon_6 - 2\varepsilon_7 + 2\varepsilon_8$$
$$= 4\beta_1 + 6\beta_2 + 8\beta_3 + 12\beta_4 + 9\beta_5 + 6\beta_6 + 3\beta_7$$

$$\varpi_5 = \frac{1}{2}(2\varepsilon_4 + 2\varepsilon_5 + 2\varepsilon_6 - 3\varepsilon_7 + 3\varepsilon_8)$$
$$= \frac{1}{2}[6\beta_1 + 9\beta_2 + 12\beta_3 + 18\beta_4 + 15\beta_5 + 10\beta_6 + 5\beta_7]$$

$$\varpi_6 = \varepsilon_5 + \varepsilon_6 - \varepsilon_7 + \varepsilon_8$$
$$= 2\beta_1 + 3\beta_2 + 4\beta_3 + 6\beta_4 + 5\beta_5 + 4\beta_6 + 2\beta_7$$

$$\varpi_7 = \frac{1}{2}(2\varepsilon_6 - \varepsilon_7 + \varepsilon_8)$$
$$= \frac{1}{2}[2\beta_1 + 3\beta_2 + 4\beta_3 + 6\beta_4 + 5\beta_5 + 4\beta_6 + 3\beta_7].$$

$$\boxed{\text{ROOT SYSTEM OF TYPE } E_8}$$

- $V = \mathbb{R}^8$.
- Roots:

$$\pm\varepsilon_i \pm \varepsilon_j \quad (1 \leqslant i < j), \quad \frac{1}{2}\sum_{i=1}^{8}(-1)^{\nu(i)}\varepsilon_i,$$

where $\nu(i)$ are positive integers such that $\nu(1) + \cdots + \nu(8)$ is even.
- Number of roots: $n = 240$.
- Base :

$$\beta_1 = \frac{1}{2}(\varepsilon_1 - \varepsilon_2 - \varepsilon_3 - \varepsilon_4 - \varepsilon_5 - \varepsilon_6 - \varepsilon_7 + \varepsilon_8), \quad \beta_2 = \varepsilon_1 + \varepsilon_2, \quad \beta_3 = \varepsilon_2 - \varepsilon_1,$$
$$\beta_4 = \varepsilon_3 - \varepsilon_2, \quad \beta_5 = \varepsilon_4 - \varepsilon_3, \quad \beta_6 = \varepsilon_5 - \varepsilon_4, \quad \beta_7 = \varepsilon_6 - \varepsilon_5, \quad \beta_8 = \varepsilon_7 - \varepsilon_6.$$

- Highest root: $\widetilde{\alpha} = 2\beta_1 + 3\beta_2 + 4\beta_3 + 6\beta_4 + 5\beta_5 + 4\beta_6 + 3\beta_7 + 2\beta_8$.
- Cartan matrix:

$$\begin{pmatrix}
2 & 0 & -1 & 0 & 0 & 0 & 0 & 0 \\
0 & 2 & 0 & -1 & 0 & 0 & 0 & 0 \\
-1 & 0 & 2 & -1 & 0 & 0 & 0 & 0 \\
0 & -1 & -1 & 2 & -1 & 0 & 0 & 0 \\
0 & 0 & 0 & -1 & 2 & -1 & 0 & 0 \\
0 & 0 & 0 & 0 & -1 & 2 & -1 & 0 \\
0 & 0 & 0 & 0 & 0 & -1 & 2 & -1 \\
0 & 0 & 0 & 0 & 0 & 0 & -1 & 2
\end{pmatrix}$$

- Extended Dynkin diagram:

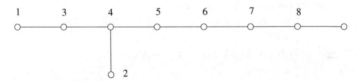

- Order of $W(R)$: $2^{14}.3^5.5^2.7$.
- Positive roots:

$$\begin{cases} \pm\varepsilon_i + \varepsilon_j & (1 \leqslant i < j \leqslant 8) \\ \dfrac{1}{2}\left(\varepsilon_8 + \sum_{i=1}^{7}(-1)^{\nu(i)}\varepsilon_i\right) & \text{with } \sum_{i=1}^{7}\nu(i) \text{ even.} \end{cases}$$

Representing the root $a\beta_1 + b\beta_2 + c\beta_3 + d\beta_4 + e\beta_5 + f\beta_6 + g\beta_7 + h\beta_8$ by $\begin{smallmatrix} a\ c\ d\ e\ f\ g\ h \\ b \end{smallmatrix}$, the positives roots are:

```
1 0 0 0 0 0    0 0 0 0 0 0    0 1 0 0 0 0    0 0 1 0 0 0    0 0 0 1 0 0 0    0 0 0 0 1 0 0
      0              1              0              0                0                  0
0 0 0 0 1 0    0 0 0 0 0 1    1 1 0 0 0 0    0 0 1 0 0 0    0 1 1 0 0 0    0 0 1 1 0 0 0
      0              0              0              1                0                  0
0 0 0 1 1 0 0    0 0 0 0 1 1 0    0 0 0 0 0 1 1    1 1 1 0 0 0    0 1 1 0 0 0    0 0 1 1 0 0 0
      0              0              0              0                1                  1
```

$$
\begin{array}{cccccc}
0\,1\,1\,1\,0\,0\,0 & 0\,0\,1\,1\,1\,0\,0 & 0\,0\,0\,1\,1\,1\,0 & 0\,0\,0\,0\,1\,1\,1 & 1\,1\,1\,0\,0\,0\,0 & 1\,1\,1\,1\,0\,0\,0 \\
0 & 0 & 0 & 0 & 1 & 0 \\[4pt]
0\,1\,1\,1\,0\,0\,0 & 0\,0\,1\,1\,1\,0\,0 & 0\,1\,1\,1\,1\,0\,0 & 0\,0\,1\,1\,1\,1\,0 & 0\,0\,0\,1\,1\,1\,1 & 1\,1\,1\,1\,0\,0\,0 \\
1 & 1 & 0 & 0 & 0 & 1 \\[4pt]
0\,1\,2\,1\,0\,0\,0 & 1\,1\,1\,1\,1\,0\,0 & 0\,1\,1\,1\,1\,0\,0 & 0\,0\,1\,1\,1\,1\,0 & 0\,1\,1\,1\,1\,1\,0 & 0\,0\,1\,1\,1\,1\,1 \\
1 & 0 & 1 & 1 & 0 & 0 \\[4pt]
1\,1\,2\,1\,0\,0\,0 & 1\,1\,1\,1\,1\,0\,0 & 0\,1\,2\,1\,1\,0\,0 & 1\,1\,1\,1\,1\,1\,0 & 0\,1\,1\,1\,1\,1\,0 & 0\,0\,1\,1\,1\,1\,1 \\
1 & 1 & 1 & 0 & 1 & 1 \\[4pt]
0\,1\,1\,1\,1\,1\,1 & 1\,2\,2\,1\,0\,0\,0 & 1\,1\,2\,1\,1\,0\,0 & 0\,1\,2\,2\,1\,0\,0 & 1\,1\,1\,1\,1\,1\,0 & 0\,1\,2\,1\,1\,1\,0 \\
0 & 1 & 1 & 1 & 1 & 1 \\[4pt]
1\,1\,1\,1\,1\,1\,1 & 0\,1\,1\,1\,1\,1\,1 & 1\,2\,2\,1\,1\,0\,0 & 1\,1\,2\,2\,1\,0\,0 & 1\,1\,2\,1\,1\,1\,0 & 0\,1\,2\,2\,1\,1\,0 \\
0 & 1 & 1 & 1 & 1 & 1 \\[4pt]
1\,1\,1\,1\,1\,1\,1 & 0\,1\,2\,1\,1\,1\,1 & 1\,2\,2\,2\,1\,0\,0 & 1\,2\,2\,1\,1\,1\,0 & 1\,1\,2\,2\,1\,1\,0 & 0\,1\,2\,2\,2\,1\,0 \\
1 & 1 & 1 & 1 & 1 & 1 \\[4pt]
1\,1\,2\,1\,1\,1\,1 & 0\,1\,2\,2\,1\,1\,1 & 1\,2\,3\,2\,1\,0\,0 & 1\,2\,2\,2\,1\,1\,0 & 1\,1\,2\,2\,2\,1\,0 & 1\,2\,2\,1\,1\,1\,1 \\
1 & 1 & 2 & 1 & 1 & 1 \\[4pt]
1\,1\,2\,2\,1\,1\,1 & 0\,1\,2\,2\,2\,1\,1 & 1\,2\,3\,2\,1\,1\,0 & 1\,2\,3\,2\,2\,1\,0 & 1\,2\,3\,2\,1\,1\,1 & 1\,2\,2\,2\,2\,1\,1 \\
1 & 1 & 2 & 1 & 1 & 1 \\[4pt]
1\,1\,2\,2\,2\,2\,1 & 1\,2\,3\,2\,2\,1\,0 & 1\,2\,3\,3\,2\,1\,0 & 1\,2\,3\,2\,1\,1\,1 & 1\,2\,3\,2\,2\,1\,1 & 1\,2\,2\,2\,2\,2\,1 \\
1 & 2 & 2 & 2 & 1 & 1 \\[4pt]
1\,2\,3\,3\,2\,1\,0 & 1\,2\,3\,2\,2\,1\,1 & 1\,2\,3\,3\,2\,1\,1 & 1\,2\,3\,2\,2\,2\,1 & 1\,2\,4\,3\,2\,1\,0 & 1\,2\,3\,3\,2\,1\,1 \\
2 & 2 & 1 & 1 & 2 & 2 \\[4pt]
1\,2\,3\,2\,2\,2\,1 & 1\,2\,3\,3\,2\,2\,1 & 1\,3\,4\,3\,2\,1\,0 & 1\,2\,4\,3\,2\,1\,1 & 1\,2\,3\,3\,2\,2\,1 & 1\,2\,3\,3\,3\,2\,1 \\
2 & 1 & 2 & 2 & 2 & 1 \\[4pt]
2\,3\,4\,3\,2\,1\,0 & 1\,3\,4\,3\,2\,1\,1 & 1\,2\,4\,3\,2\,2\,1 & 1\,2\,3\,3\,3\,2\,1 & 2\,3\,4\,3\,2\,1\,1 & 1\,3\,4\,3\,2\,2\,1 \\
2 & 2 & 2 & 2 & 2 & 2 \\[4pt]
1\,2\,4\,3\,3\,2\,1 & 2\,3\,4\,3\,2\,2\,1 & 1\,3\,4\,3\,3\,2\,1 & 1\,2\,4\,4\,3\,2\,1 & 2\,3\,4\,3\,3\,2\,1 & 1\,3\,4\,4\,3\,2\,1 \\
2 & 2 & 2 & 2 & 2 & 2 \\[4pt]
2\,3\,4\,4\,3\,2\,1 & 1\,3\,5\,4\,3\,2\,1 & 2\,3\,5\,4\,3\,2\,1 & 1\,3\,5\,4\,3\,2\,1 & 2\,3\,5\,4\,3\,2\,1 & 2\,4\,5\,4\,3\,2\,1 \\
2 & 2 & 2 & 3 & 3 & 2 \\[4pt]
2\,4\,5\,4\,3\,2\,1 & 2\,4\,6\,4\,3\,2\,1 & 2\,4\,6\,5\,3\,2\,1 & 2\,4\,6\,5\,4\,2\,1 & 2\,4\,6\,5\,4\,3\,1 & 2\,4\,6\,5\,4\,3\,2 \\
3 & 3 & 3 & 3 & 3 & 3
\end{array}
$$

- Fundamental weights:

$$
\begin{aligned}
\varpi_1 &= 2\varepsilon_8 \\
&= 4\beta_1 + 5\beta_2 + 7\beta_3 + 10\beta_4 + 8\beta_5 + 6\beta_6 + 4\beta_7 + 2\beta_8 \\
\varpi_2 &= \frac{1}{2}(\varepsilon_1 + \varepsilon_2 + \varepsilon_3 + \varepsilon_4 + \varepsilon_5 + \varepsilon_6 + \varepsilon_7 + 5\varepsilon_8) \\
&= 5\beta_1 + 8\beta_2 + 10\beta_3 + 15\beta_4 + 12\beta_5 + 9\beta_6 + 6\beta_7 + 3\beta_8 \\
\varpi_3 &= \frac{1}{2}(-\varepsilon_1 + \varepsilon_2 + \varepsilon_3 + \varepsilon_4 + \varepsilon_5 + \varepsilon_6 + \varepsilon_7 + 7\varepsilon_8) \\
&= 7\beta_1 + 10\beta_2 + 14\beta_3 + 20\beta_4 + 16\beta_5 + 12\beta_6 + 8\beta_7 + 4\beta_8 \\
\varpi_4 &= \varepsilon_3 + \varepsilon_4 + \varepsilon_5 + \varepsilon_6 + \varepsilon_7 + 5\varepsilon_8 \\
&= 10\beta_1 + 15\beta_2 + 20\beta_3 + 30\beta_4 + 24\beta_5 + 18\beta_6 + 12\beta_7 + 6\beta_8 \\
\varpi_5 &= \varepsilon_4 + \varepsilon_5 + \varepsilon_6 + \varepsilon_7 + 4\varepsilon_8 \\
&= 8\beta_1 + 12\beta_2 + 16\beta_3 + 24\beta_4 + 20\beta_5 + 15\beta_6 + 10\beta_7 + 5\beta_8 \\
\varpi_6 &= \varepsilon_5 + \varepsilon_6 + \varepsilon_7 + 3\varepsilon_8 \\
&= 6\beta_1 + 9\beta_2 + 12\beta_3 + 18\beta_4 + 15\beta_5 + 12\beta_6 + 8\beta_7 + 4\beta_8 \\
\varpi_7 &= \varepsilon_6 + \varepsilon_7 + 2\varepsilon_8 \\
&= 4\beta_1 + 6\beta_2 + 8\beta_3 + 12\beta_4 + 10\beta_5 + 8\beta_6 + 6\beta_7 + 3\beta_8 \\
\varpi_8 &= \varepsilon_7 + \varepsilon_8 \\
&= 2\beta_1 + 3\beta_2 + 4\beta_3 + 6\beta_4 + 5\beta_5 + 4\beta_6 + 3\beta_7 + 2\beta_8.
\end{aligned}
$$

$$\boxed{\text{Root system of type } F_4}$$

- $V = \mathbb{R}^4$.
- Roots:

$$\pm\varepsilon_i \quad (1 \leqslant i \leqslant 4), \quad \pm\varepsilon_i \pm \varepsilon_j \quad (1 \leqslant i < j \leqslant 4), \quad \frac{1}{2}(\pm\varepsilon_1 \pm \varepsilon_2 \pm \varepsilon_3 \pm \varepsilon_4).$$

- Number of roots: $n = 48$.
- Base:

$$\beta_1 = \varepsilon_2 - \varepsilon_3, \quad \beta_2 = \varepsilon_3 - \varepsilon_4, \quad \beta_3 = \varepsilon_4, \quad \beta_4 = \frac{1}{2}(\varepsilon_1 - \varepsilon_2 - \varepsilon_3 - \varepsilon_4).$$

- Highest root: $\widetilde{\alpha} = 2\beta_1 + 3\beta_2 + 4\beta_3 + 2\beta_4$.
- Cartan matrix:

$$\begin{pmatrix} 2 & -1 & 0 & 0 \\ -1 & 2 & -2 & 0 \\ 0 & -1 & 2 & -1 \\ 0 & 0 & -1 & 2 \end{pmatrix}$$

- Extended Dynkin diagram:

- Order of $W(R)$: $2^7.3^2$.
- Positive roots:

$$\begin{cases} \varepsilon_i & (1 \leqslant i \leqslant 4) \\ \varepsilon_i \pm \varepsilon_j & (1 \leqslant i < j \leqslant 4) \\ \dfrac{1}{2}(\varepsilon_1 \pm \varepsilon_2 \pm \varepsilon_3 \pm \varepsilon_4). \end{cases}$$

Denote by $a\,b\,c\,d$ the root $a\beta_1 + b\beta_2 + c\beta_3 + d\beta_4$, then the positive roots are:

1 0 0 0	0 1 0 0	0 0 1 0	0 0 0 1	1 1 0 0	0 1 1 0	0 0 1 1	1 1 1 0
0 1 2 0	0 1 1 1	1 1 2 0	1 1 1 1	0 1 2 1	1 2 2 0	1 1 2 1	0 1 2 2
1 2 2 1	1 1 2 2	1 2 3 1	1 2 2 2	1 2 3 2	1 2 4 2	1 3 4 2	2 3 4 2

- Fundamental weights:

$$\varpi_1 = \varepsilon_1 + \varepsilon_2 = 2\beta_1 + 3\beta_2 + 4\beta_3 + 2\beta_4$$
$$\varpi_2 = 2\varepsilon_1 + \varepsilon_2 + \varepsilon_3 = 3\beta_1 + 6\beta_2 + 8\beta_3 + 4\beta_4$$
$$\varpi_3 = \frac{1}{2}(3\varepsilon_1 + \varepsilon_2 + \varepsilon_3 + \varepsilon_3) = 2\beta_1 + 4\beta_2 + 6\beta_3 + 3\beta_4$$
$$\varpi_4 = \varepsilon_1 = \beta_1 + 2\beta_2 + 3\beta_3 + 2\beta_4.$$

$$\boxed{\text{ROOT SYSTEM OF TYPE } G_2}$$

- V is the hyperplane of $\mathbb{R}^3$ defined by $\xi_1 + \xi_2 + \xi_3 = 0$.
- Roots:

$$\pm(\varepsilon_1 - \varepsilon_2), \ \pm(\varepsilon_1 - \varepsilon_3), \ \pm(\varepsilon_2 - \varepsilon_3),$$
$$\pm(2\varepsilon_1 - \varepsilon_2 - \varepsilon_3), \ \pm(2\varepsilon_2 - \varepsilon_1 - \varepsilon_3), \ \pm(2\varepsilon_3 - \varepsilon_1 - \varepsilon_2).$$

- Number of roots: $n = 12$.
- Base: $\beta_1 = \varepsilon_1 - \varepsilon_2$, $\beta_2 = -2\varepsilon_1 + \varepsilon_2 + \varepsilon_3$.
- Highest root: $\widetilde{\alpha} = 3\beta_1 + 2\beta_2$.
- Cartan matrix:

$$\begin{pmatrix} 2 & -1 \\ -3 & 2 \end{pmatrix}$$

- Extended Dynkin diagram:

- Order of $W(R)$: 12.
- Positive roots:

$$\beta_1, \ \beta_2, \ \beta_1 + \beta_2, \ 2\beta_1 + \beta_2, \ 3\beta_1 + \beta_2, \ 3\beta_1 + 2\beta_2.$$

- Fundamental weights:

$$\varpi_1 = -\varepsilon_2 + \varepsilon_3 = 2\beta_1 + \beta_2$$
$$\varpi_2 = -\varepsilon_1 - \varepsilon_2 + 2\varepsilon_3 = 3\beta_1 + 2\beta_2.$$

Lie algebras

The aim of this chapter is to give a detailed account of the general theory of Lie algebras. The representation theory of $\mathfrak{sl}_2(\Bbbk)$ in section 19.2 will be used very often in the rest of the book. Results such as Engel's theorem, Lie's theorem and Cartan's Criterion are the backbone of the subject. The notions of regular linear forms and Cartan subalgebras are introduced in the final sections.

19.1 Generalities on Lie algebras

19.1.1 Let $\mathfrak{g}$ be a $\Bbbk$-vector space. A bilinear map $\mathfrak{g} \times \mathfrak{g} \to \mathfrak{g}$, $(x, y) \mapsto [x, y]$, is a *Lie bracket on* $\mathfrak{g}$ if for all $x, y, z \in \mathfrak{g}$, we have

- $[x, y] = -[y, x]$ (antisymmetry),
- $[x, [y, z]] = [[x, y], z] + [y, [x, z]]$ (Jacobi identity).

Endowed with a Lie bracket, $\mathfrak{g}$ is called a *Lie algebra*.

Let $\mathfrak{g}$ be a Lie algebra. A subspace $\mathfrak{h} \subset \mathfrak{g}$ is a *Lie subalgebra* (resp. an *ideal*) if $[x, y] \in \mathfrak{h}$ for all $x, y \in \mathfrak{h}$. (resp. $[x, \mathfrak{h}] \subset \mathfrak{h}$ for all $x \in \mathfrak{g}$). When $\mathfrak{h}$ is an ideal of $\mathfrak{g}$, the *quotient* $\mathfrak{g}/\mathfrak{h}$ inherits a natural structure of Lie algebra given by: $[x + \mathfrak{h}, y + \mathfrak{h}] = [x, y] + \mathfrak{h}$.

Examples. Let us give some examples of Lie algebras. Let A be an associative $\Bbbk$-algebra. Then, A has a Lie algebra structure given by: for $x, y \in A$,

$$[x, y] = xy - yx.$$

So if V is a $\Bbbk$-vector space of dimension n, then $\mathrm{End}(V)$ (resp. $\mathrm{M}_n(\Bbbk)$) endowed with the above Lie bracket is a Lie algebra that we shall denote by $\mathfrak{gl}(V)$ (resp. $\mathfrak{gl}_n(\Bbbk)$). The following subsets of $\mathfrak{gl}_n(\Bbbk)$ are clearly Lie subalgebras:

- $\mathfrak{sl}_n(\Bbbk)$ (resp. $\mathfrak{sl}(V)$): the set of matrices in $\mathfrak{gl}_n(\Bbbk)$ (resp. endomorphisms in $\mathfrak{gl}(V)$) whose trace is zero.
- $\mathfrak{t}_n(\Bbbk)$ (resp. $\mathfrak{n}_n(\Bbbk)$): the set of upper triangular matrices (resp. strictly upper triangular matrices) in $\mathfrak{gl}_n(\Bbbk)$.
- $\mathfrak{d}_n(\Bbbk)$ the set of diagonal matrices in $\mathfrak{gl}_n(\Bbbk)$.

Observe that $\mathfrak{sl}(V)$ is an ideal of $\mathfrak{gl}(V)$.

Let $\mathfrak{h}, \mathfrak{h}'$ be subsets of $\mathfrak{g}$, then we denote by $[\mathfrak{h}, \mathfrak{h}']$ the subspace spanned by the elements $[x, y]$ where $x \in \mathfrak{h}$, $y \in \mathfrak{h}'$. When $\mathfrak{h}$ and $\mathfrak{h}'$ are ideals of $\mathfrak{g}$, then clearly $[\mathfrak{h}, \mathfrak{h}']$ is also an ideal of $\mathfrak{g}$.

Let $\mathfrak{g}, \mathfrak{g}'$ be Lie algebras endowed with Lie brackets $[,]$ and $[,]'$, then a map $u : \mathfrak{g} \to \mathfrak{g}'$ is a *morphism of Lie algebras* if $u([x, y]) = [u(x), u(y)]'$ for all $x, y \in \mathfrak{g}$. So we may talk about isomorphisms and automorphisms.

A linear map $d : \mathfrak{g} \to \mathfrak{g}$ is called a *derivation* of $\mathfrak{g}$ if for all $x, y \in \mathfrak{g}$,

$$d([x, y]) = [d(x), y] + [x, d(y)].$$

We denote by $\mathrm{Der}\,\mathfrak{g}$ the set of derivations of $\mathfrak{g}$. Note that $\mathrm{Der}\,\mathfrak{g}$, endowed with the following Lie bracket,

$$[d, d'] = d \circ d' - d' \circ d,$$

for $d, d' \in \mathrm{Der}\,\mathfrak{g}$, is a Lie algebra. Let $x \in \mathfrak{g}$. The Jacobi identity implies that the linear map $y \mapsto [x, y]$, denoted $\mathrm{ad}_{\mathfrak{g}}\,x$ or $\mathrm{ad}\,x$, is a derivation of $\mathfrak{g}$, called an *inner derivation* of $\mathfrak{g}$. So it induces a linear map $\mathfrak{g} \to \mathrm{Der}(\mathfrak{g})$ which is called the *adjoint representation* of $\mathfrak{g}$.

Convention. In this chapter, we shall only consider finite-dimensional Lie algebras and $\mathfrak{g}$ will denote a finite-dimensional Lie algebra.

Set:
$$\mathfrak{g}^0(x) = \bigcup_{n \geq 0} \ker(\mathrm{ad}\,x)^n \ , \ \mathfrak{g}^{\bullet}(x) = \bigcap_{n \geq 0} (\mathrm{ad}\,x)^n(\mathfrak{g}).$$

We call $\mathfrak{g}^0(x)$ the *nilspace of* $\mathrm{ad}\,x$. We have:

$$[\mathfrak{g}^0(x), \mathfrak{g}^0(x)] \subset \mathfrak{g}^0(x) \ , \ [\mathfrak{g}^0(x), \mathfrak{g}^{\bullet}(x)] \subset \mathfrak{g}^{\bullet}(x) \ , \ \mathfrak{g} = \mathfrak{g}^0(x) \oplus \mathfrak{g}^{\bullet}(x).$$

Let $n = \dim \mathfrak{g}$ and χ_x the characteristic polynomial of $\mathrm{ad}\,x$. Then

$$\chi_x(T) = a_n(x)T^n + a_{n-1}(x)T^{n-1} + \cdots + a_1(x)T + a_0(x)$$

where a_i are polynomial functions on $\mathfrak{g}$ with $a_n = 1$ and $a_0 = 0$ (since $[x, x] = 0$). If p is the smallest integer such that $a_p(x) \neq 0$, then by decomposing $\mathfrak{g}$ into characteristic subspaces of $\mathrm{ad}\,x$, we obtain that $\dim \mathfrak{g}^0(x) = p$.

Define $\mathrm{rk}(\mathfrak{g}) = \min\{l; a_l \neq 0\}$ to be the *rank* of the Lie algebra $\mathfrak{g}$. If $x \in \mathfrak{g}$ verifies $\dim \mathfrak{g}^0(x) = \mathrm{rk}(\mathfrak{g})$, then we say that x is *generic* in $\mathfrak{g}$. Denote by $\mathfrak{g}_{\mathrm{gen}}$ the set of generic elements of $\mathfrak{g}$. It is clear from its definition that $\mathfrak{g}_{\mathrm{gen}}$ is a dense open subset of $\mathfrak{g}$.

19.1.2 Lemma. *If $\mathfrak{h}$ is a Lie subalgebra of $\mathfrak{g}$, then $\mathfrak{h} \cap \mathfrak{g}_{\mathrm{gen}} \subset \mathfrak{h}_{\mathrm{gen}}$.*

Proof. For $x \in \mathfrak{h}$, denote by $u(x)$ the endomorphism of $\mathfrak{g}/\mathfrak{h}$ induced by $\mathrm{ad}_{\mathfrak{g}}\,x$. Let $\chi_{u(x)}$ be the characteristic polynomial of $u(x)$ and $\chi_{x,\mathfrak{h}}$ the characteristic polynomial of $\mathrm{ad}_{\mathfrak{h}}\,x$. Then $\chi_x = \chi_{u(x)}\chi_{x,\mathfrak{h}}$ and the result follows easily. $\square$

19.1.3 Let $\mathfrak{p}$, $\mathfrak{q}$ be subsets of $\mathfrak{g}$. The set of elements of $\mathfrak{q}$ which commute with all the elements of $\mathfrak{p}$, denoted by $\mathfrak{c}_{\mathfrak{q}}(\mathfrak{p})$, is called the *centralizer* of $\mathfrak{p}$ in $\mathfrak{q}$. The set $\mathfrak{c}_{\mathfrak{g}}(\mathfrak{g})$, denoted by $\mathfrak{z}(\mathfrak{g})$, is called the *centre* of $\mathfrak{g}$. If $\mathfrak{g} = \mathfrak{z}(\mathfrak{g})$, then we say that $\mathfrak{g}$ is *abelian* (or *commutative*). If $x \in \mathfrak{g}$, we shall write $\mathfrak{q}^x$ for $\mathfrak{c}_{\mathfrak{q}}(\{x\})$.

The set of $x \in \mathfrak{q}$ such that $[x, \mathfrak{p}] \subset \mathfrak{p}$, denoted by $\mathfrak{n}_{\mathfrak{q}}(\mathfrak{p})$, is called the *normalizer* of $\mathfrak{p}$ in $\mathfrak{q}$. If $\mathfrak{p}$ is a Lie subalgebra, then so is $\mathfrak{n}_{\mathfrak{g}}(\mathfrak{p})$, and $\mathfrak{p}$ is an ideal of $\mathfrak{n}_{\mathfrak{g}}(\mathfrak{p})$.

19.1.4 Let us denote by $\operatorname{Aut} \mathfrak{g}$ the set of automorphisms of $\mathfrak{g}$, $\mathcal{N}$ the set of elements $x \in \mathfrak{g}$ such that $\operatorname{ad} x$ is a nilpotent endomorphism of $\mathfrak{g}$. For $x \in \mathcal{N}$, define $\gamma_x \in \operatorname{GL}(\mathfrak{g})$ as follows:

$$\gamma_x = \sum_{n=0}^{\infty} \frac{1}{n!} (\operatorname{ad} x)^n.$$

Then $\gamma_x \in \operatorname{Aut} \mathfrak{g}$. Such an automorphism is called an *elementary automorphism* of $\mathfrak{g}$. The subgroup $\operatorname{Aut}_e \mathfrak{g}$ of $\operatorname{Aut} \mathfrak{g}$ generated by elementary automorphisms is called the *group of elementary automorphisms* of $\mathfrak{g}$. Observe that if $x \in \mathcal{N}$, $\alpha \in \operatorname{Aut} \mathfrak{g}$, then $\alpha(x) \in \mathcal{N}$ and $\operatorname{ad} \alpha(x) = \alpha \circ (\operatorname{ad} x) \circ \alpha^{-1}$. So $\gamma_{\alpha(x)} = \alpha \circ \gamma_x \circ \alpha^{-1}$, and $\operatorname{Aut}_e \mathfrak{g}$ is a normal subgroup of $\operatorname{Aut} \mathfrak{g}$.

19.2 Representations

19.2.1 A *representation* of $\mathfrak{g}$, or a $\mathfrak{g}$-*module*, is a pair (V, σ), where V is $\Bbbk$-vector space and $\sigma : \mathfrak{g} \to \mathfrak{gl}(V)$ is a Lie algebra homomorphism. If V is finite-dimensional, then we say that the representation is finite-dimensional. For $x \in \mathfrak{g}$, $v \in V$, we shall denote $\sigma(x)(v)$ by $x.v$. We set $V^{\mathfrak{g}} = \{v \in V;$ $x.v = 0$ for all $x \in \mathfrak{g}\}$. We define the notions of submodule, quotient module and direct sum in the obvious way.

The $\mathfrak{g}$-module (V, σ) is said to be *simple* if $V \neq \{0\}$ and the only submodules of V are V and $\{0\}$. If V is the direct sum of simple submodules, then we say that V is *semisimple*, or *completely reducible*. The analogues of 10.7.2 and 10.7.5 are true for $\mathfrak{g}$-modules.

Let V be a $\mathfrak{g}$-module. A chain $0 = V_0 \subset V_1 \subset \cdots \subset V_n = V$ of submodules of V is called a *Jordan-Hölder series* if the modules V_i/V_{i-1} are simple for $1 \leqslant i \leqslant n$.

Let (V, σ) be a representation of $\mathfrak{g}$, then we define its *dual*, or *contragredient*, representation (V^*, π) as follows: for $x \in \mathfrak{g}$, $v \in V$ and $f \in V^*$,

$$(\pi(x)(f))(v) = -f(\sigma(x)(v)).$$

The adjoint representation is indeed a representation of $\mathfrak{g}$. Its dual representation is called the *coadjoint* representation of $\mathfrak{g}$. If $\mathfrak{h}$ is a Lie subalgebra of $\mathfrak{g}$, then the map $\mathfrak{h} \to \mathfrak{gl}(\mathfrak{g}/\mathfrak{h})$ induced by $x \mapsto \operatorname{ad} x$ is called the adjoint representation of $\mathfrak{h}$ in $\mathfrak{g}/\mathfrak{h}$.

A representation (V, σ) is said to be *faithful* if σ is injective. We shall admit the following result.

Theorem. (Ado's Theorem) *Any finite-dimensional Lie algebra $\mathfrak{g}$ has a faithful finite-dimensional representation.*

19.2.2 Let (V, σ) be a finite-dimensional representation of $\mathfrak{g}$. We define a symmetric bilinear form β_σ on $\mathfrak{g}$ associated to σ as follows: for $x, y \in \mathfrak{g}$,

$$\beta_\sigma(x, y) = \mathrm{tr}(\sigma(x)\sigma(y)).$$

The symmetric bilinear form associated to the adjoint representation is called the *Killing form* of $\mathfrak{g}$, denoted by $L_\mathfrak{g}$. If $\mathfrak{a}$ is an ideal of $\mathfrak{g}$, then the restriction of $L_\mathfrak{g}$ to $\mathfrak{a} \times \mathfrak{a}$ is the Killing form of $\mathfrak{a}$, and the orthogonal of $\mathfrak{a}$ with respect to β_σ is an ideal of $\mathfrak{g}$. Observe that if $x, y, z \in \mathfrak{g}$, then

$$\beta_\sigma([x, y], z) + \beta_\sigma(y, [x, z]) = 0.$$

If $x \in \mathfrak{g}$, denote by f_x the linear form on $\mathfrak{g}$ such that $f_x(y) = L_\mathfrak{g}(x, y)$ for all $y \in \mathfrak{g}$. It follows that the map $x \mapsto f_x$ is a homomorphism of $\mathfrak{g}$-modules, called the *Killing homomorphism* of $\mathfrak{g}$, from the adjoint representation of $\mathfrak{g}$ to the coadjoint representation of $\mathfrak{g}$.

19.2.3 Lemma. *Let V be a finite-dimensional vector space and f an endomorphism of V. Then f is nilpotent if and only if $\mathrm{tr}(f^n) = 0$ for all $n \in \mathbb{N}^*$.*

Proof. This is an elementary result from linear algebra. $\square$

19.2.4 In this rest of this section, $\mathfrak{g}$ is the Lie algebra $\mathfrak{sl}_2(\Bbbk)$. Let

$$e = \begin{pmatrix} 0 & 1 \\ 0 & 0 \end{pmatrix} , \quad h = \begin{pmatrix} 1 & 0 \\ 0 & -1 \end{pmatrix} , \quad f = \begin{pmatrix} 0 & 0 \\ 1 & 0 \end{pmatrix} .$$

Then (e, h, f) is a basis of $\mathfrak{g}$ and

$$[e, f] = h , \quad [h, e] = 2e , \quad [h, f] = -2f.$$

Let $r \in \mathbb{N}$ and for $1 \leqslant i \leqslant r$, let $\mu_i = i(r - i + 1)$ and denote by $(\Bbbk^{r+1}, \sigma_r)$ the representation of $\mathfrak{g}$ given by:

$$\sigma_r(h) = \begin{pmatrix} r & 0 & 0 & \cdots & 0 \\ 0 & r-2 & 0 & \cdots & 0 \\ 0 & 0 & r-4 & \cdots & 0 \\ \vdots & \vdots & \vdots & \ddots & \vdots \\ 0 & 0 & 0 & \cdots & -r \end{pmatrix},$$

$$\sigma_r(e) = \begin{pmatrix} 0 & \mu_1 & 0 & \cdots & 0 \\ 0 & 0 & \mu_2 & \cdots & 0 \\ \vdots & \vdots & \vdots & \ddots & \vdots \\ 0 & 0 & 0 & \cdots & \mu_r \\ 0 & 0 & 0 & \cdots & 0 \end{pmatrix} , \quad \sigma_r(f) = \begin{pmatrix} 0 & 0 & \cdots & 0 & 0 & 0 \\ 1 & 0 & \cdots & 0 & 0 & 0 \\ \vdots & \vdots & \ddots & \vdots & \vdots & \vdots \\ 0 & 0 & \cdots & 1 & 0 & 0 \\ 0 & 0 & \cdots & 0 & 1 & 0 \end{pmatrix} .$$

Lemma. *The representation σ_r is simple.*

Proof. Let $(\varepsilon_1, \ldots, \varepsilon_{r+1})$ be the canonical basis of $\Bbbk^{r+1}$. Then for any non zero vector $v \in \Bbbk^{r+1}$, there exist $n \in \mathbb{N}$ and $\lambda \in \Bbbk \setminus \{0\}$ such that $\sigma_r(f)^n(v) = \lambda \varepsilon_{r+1}$. So any non-trivial submodule contains ε_{r+1}, and the result follows since $\Bbbk \varepsilon_i = \Bbbk \sigma_r(e)^{r-i+1}(\varepsilon_{r+1})$. $\square$

19.2.5 Let (V, σ) be a finite-dimensional representation of $\mathfrak{g}$ where $\dim V \neq 0$. Set $H = \sigma(h)$, $E = \sigma(e)$ and $F = \sigma(f)$.

By a simple induction, we have the following identities for $n \in \mathbb{N}$, $m \in \mathbb{N}^*$:

(1)
$$\begin{cases} [H, E] = 2nE^n \ , \ \ [H, F^n] = -2nF^n \ , \\ [F, E^m] = -m(H - (m-1)\operatorname{id}_V) \circ E^{m-1} \end{cases}$$

Thus E and F are nilpotent (19.2.3).

Let $q = \min\{n \in \mathbb{N}; E^{q+1} = 0\}$ and $w \in V$ such that $v = E^q(w) \neq 0$. Set $v_0 = v$ and for $n \in \mathbb{N}^*$, $v_n = F^n(v)$. Then by using (1), we obtain:

$$0 = [F, E^{q+1}](w) \Rightarrow H(v_0) = qv_0.$$

Hence, again by (1), we deduce that:

(2)
$$H(v_n) = (q - 2n)v_n.$$

Further, a simple induction shows that:

(3)
$$E(v_n) = n(q - n + 1)v_{n-1}.$$

It follows that if $v_n \neq 0$, it is an eigenvector for H with eigenvalue $(q - 2n)$. Let s be such that $v_s \neq 0$ and $v_n = 0$ for $n > s$, and let W be the subspace spanned by $v_0, \ldots, v_s$. Then W is a $\mathfrak{g}$-submodule of V and $(v_0, \ldots, v_s)$ is a basis of W by (2). Denote by H_W, E_W, F_W the restrictions of H, E, F to W. Since $[E_W, F_W] = H_W$,

$$0 = \operatorname{tr}(H_W) = \sum_{n=0}^{s} (q - 2n) = (s + 1)(q - s).$$

Hence $s = q$ and the matrices of H_W, E_W, F_W in the basis $(v_0, \ldots, v_q)$ are the ones in 19.2.4 with $r = q$. We deduce that the representations $\sigma|_W$ and σ_q are equivalent. So we have obtained the following result:

Theorem. *Let $r \in \mathbb{N}$ and (V, σ) a simple representation of $\mathfrak{sl}_2(\Bbbk)$ of dimension $r + 1$. Then σ is equivalent to σ_r.*

Moreover, the eigenvalues of $\sigma(h)$ are $-r, -r + 2, \ldots, r - 2, r$, and if $v \in V \setminus \{0\}$ verifies $\sigma(e)(v) = 0$ (resp. $\sigma(f)(v) = 0$), then v is a $\sigma(h)$-eigenvector of eigenvalue r (resp. $-r$).

19.2.6 Theorem. *Any finite-dimensional representation of $\mathfrak{g} = \mathfrak{sl}_2(\Bbbk)$ is semisimple.*

Proof. Let (V, σ) be a finite-dimensional representation of $\mathfrak{g}$. If $\dim V = 0$, there is nothing to prove. So we may assume that $\dim V \geqslant 1$. By 19.2.5, there is a simple submodule W of dimension $q + 1$ in W. We shall prove that there is a submodules U of V such that $V = V \oplus W$. Then the result would follow by induction on the dimension of V.

Denote by η the dual representation of σ. Let $w \in V$ and $(v_0, \ldots, v_q)$ be the basis of W as in 19.2.5. Let $g \in V^*$ be such that $g(v_0) = 1$. Then

$$(\eta(E)^q(g))(w) = g((-E)^q(w)) = (-1)^q.$$

So $f = \eta(E)^q(g) \neq 0$ and q is the smallest integer such that $\eta(E)^{q+1} = 0$. By 19.2.5, the subspace M of V^* spanned by $f_n = \eta(F)^n(f)$, $0 \leqslant n \leqslant q$, is a simple submodule of dimension $q + 1$.

Using (3), we verify easily that:

(4)
$$\begin{aligned} f_n(v_{q-n}) &= (-1)^{q-n}(q!)^2, \\ f_n(v_p) &= 0 \ , \ \text{if } n + p > q. \end{aligned}$$

Let U be the orthogonal of M in V. Then $\dim V = \dim U + \dim W$ and U is a $\mathfrak{g}$-submodule of V. Further using (4) we deduce that $U \cap W = \{0\}$. So V is the direct sum of the submodules U and W. $\square$

19.2.7 Corollary. *Let (V, σ) be a finite-dimensional representation of* $\mathfrak{sl}_2(\mathbb{k})$.

(i) *There exist* $r_1, \ldots, r_n \in \mathbb{N}$ *such that σ is equivalent to the direct sum of the representations* σ_{r_i}, $1 \leqslant i \leqslant n$.

(ii) *We have* $V = \sigma(e)(V) \oplus \ker \sigma(f) = \sigma(f)(V) \oplus \ker \sigma(e)$.

(iii) *$\sigma(h)$ is semisimple with integer eigenvalues.*

(iv) *If V is non-trivial, then σ is irreducible if and only if the eigenvalues of $\sigma(h)$ are simple and they are either all even or all odd.*

Proof. Part (i) follows from 19.2.6 and 19.2.5. The rest is clear for σ_r. The general case follows from (i). $\square$

19.3 Nilpotent Lie algebras

19.3.1 An ideal of $\mathfrak{g}$ is said to be *characteristic* if it is invariant under all the derivations of $\mathfrak{g}$. We define by induction the following decreasing chains of characteristic ideals of $\mathfrak{g}$:

$$\begin{aligned} \mathcal{C}^1(\mathfrak{g}) &= \mathfrak{g} \ , \ \mathcal{C}^2(\mathfrak{g}) = [\mathfrak{g}, \mathfrak{g}] \ , \ldots, \ \mathcal{C}^{i+1}(\mathfrak{g}) = [\mathfrak{g}, \mathcal{C}^i(\mathfrak{g})] \ , \ldots \\ \mathcal{D}^0(\mathfrak{g}) &= \mathfrak{g} \ , \ \mathcal{D}^1(\mathfrak{g}) = [\mathfrak{g}, \mathfrak{g}] \ , \ldots, \ \mathcal{D}^{i+1}(\mathfrak{g}) = [\mathcal{D}^i(\mathfrak{g}), \mathcal{D}^i(\mathfrak{g})] \ , \ldots \end{aligned}$$

We call $(\mathcal{C}^i(\mathfrak{g}))_{i \geqslant 1}$ the *central descending series* of $\mathfrak{g}$, and $(\mathcal{D}^i(\mathfrak{g}))_{i \geqslant 0}$ the *derived series* of $\mathfrak{g}$. In particular, $\mathcal{D}(\mathfrak{g}) = \mathcal{D}^1(\mathfrak{g})$ is called the *derived ideal* of $\mathfrak{g}$.

If $u : \mathfrak{g} \to \mathfrak{h}$ is a Lie algebra homomorphism, then $u(\mathcal{C}^i(\mathfrak{g})) \subset \mathcal{C}^i(\mathfrak{h})$ and $u(\mathcal{D}^i(\mathfrak{g})) \subset \mathcal{D}^i(\mathfrak{h})$ for all i. Further, these inclusions are equalities when u is surjective.

19.3.2 Proposition. *The following conditions are equivalent:*

(i) *There exists an integer i such that $\mathcal{C}^i(\mathfrak{g}) = \{0\}$.*

(ii) *There exists an integer j such that $\operatorname{ad} x_1 \circ \operatorname{ad} x_2 \circ \cdots \circ \operatorname{ad} x_j = 0$ for all $x_1, \ldots, x_j \in \mathfrak{g}$.*

(iii) *There exists a chain $\mathfrak{g} = \mathfrak{g}_1 \supset \mathfrak{g}_2 \supset \cdots \supset \mathfrak{g}_n = \{0\}$ of ideals of $\mathfrak{g}$ such that $[\mathfrak{g}, \mathfrak{g}_i] \subset \mathfrak{g}_{i+1}$ for $i < n$.*

If these conditions are verified, we say that $\mathfrak{g}$ is nilpotent.

Proof. The equivalence (i) $\Leftrightarrow$ (ii) is straightforward by the definition of $\mathcal{C}^i(\mathfrak{g})$. Next, we have (i) $\Rightarrow$ (iii) by taking $\mathfrak{g}_i = \mathcal{C}^i(\mathfrak{g})$. Finally, to show that (iii) $\Rightarrow$ (i), it suffices to show that $\mathcal{C}^i(\mathfrak{g}) \subset \mathfrak{g}_i$, which is a simple induction. $\square$

19.3.3 Proposition. (i) *If $\mathfrak{g}$ is nilpotent, then any subalgebra (resp. quotient) of $\mathfrak{g}$ is nilpotent.*

(ii) *Let $\mathfrak{a}$ be a subalgebra of $\mathfrak{g}$ which is contained in the centre of $\mathfrak{g}$. Then $\mathfrak{g}$ is nilpotent if and only if $\mathfrak{g}/\mathfrak{a}$ is nilpotent.*

Proof. Part (i) is obvious. If $\mathfrak{g}/\mathfrak{a}$ is nilpotent, then there exists $k \in \mathbb{N}$ such that $\mathcal{C}^k(\mathfrak{g}/\mathfrak{a}) = \{0\}$, or $\mathcal{C}^k(\mathfrak{g}) \subset \mathfrak{a}$. Since $\mathfrak{a}$ is in the centre of $\mathfrak{g}$, $\mathcal{C}^{k+1}(\mathfrak{g}) = \{0\}$. The converse follows from part (i). $\square$

19.3.4 Proposition. *Let $\mathfrak{g}$ be nilpotent.*

(i) *If $\mathfrak{g} \neq \{0\}$, then $\mathfrak{z}(\mathfrak{g}) \neq \{0\}$.*

(ii) *If $\mathfrak{h}$ is a subalgebra of $\mathfrak{g}$ distinct from $\mathfrak{g}$, then $\mathfrak{n}_\mathfrak{g}(\mathfrak{h}) \neq \mathfrak{h}$.*

(iii) *The Killing form of $\mathfrak{g}$ is zero.*

Proof. (i) The last non-zero $\mathcal{C}^i(\mathfrak{g})$ is central in $\mathfrak{g}$.

(ii) Let $j = \max\{i; \mathcal{C}^i(\mathfrak{g}) + \mathfrak{h} \neq \mathfrak{h}\}$. Then $\mathcal{C}^j(\mathfrak{g}) + \mathfrak{h} \subset \mathfrak{n}_\mathfrak{g}(\mathfrak{h})$.

(iii) By 19.3.2, $\operatorname{ad} x \circ \operatorname{ad} y$ is a nilpotent endomorphism for all $x, y \in \mathfrak{g}$. $\square$

19.3.5 Lemma. *Let V be a vector space of dimension $n > 0$.*

(i) *If $x \in \operatorname{End}(V)$ is nilpotent, $\operatorname{ad} x$ is a nilpotent endomorphism of $\mathfrak{gl}(V)$. More precisely, if $x^p = 0$, then $(\operatorname{ad} x)^{2p-1} = 0$.*

(ii) *If $\mathfrak{g}$ is a Lie subalgebra of $\mathfrak{gl}(V)$ consisting of nilpotent endomorphisms, then $V^\mathfrak{g} \neq \{0\}$.*

Proof. Part (i) is clear from the fact that if $y \in \mathfrak{gl}(V)$, then $(\operatorname{ad} x)^n(y)$ is the sum of terms of the form $\pm x^i y x^j$ with $i + j = n$.

To prove (ii), we proceed by induction on $\dim \mathfrak{g} = p$. The result is obvious when $p \leqslant 1$, so let us assume that $p > 1$. Let $\mathfrak{a}$ be a non-zero Lie subalgebra of $\mathfrak{g}$ of dimension $q < p$. For $x \in \mathfrak{a}$, the endomorphism $\sigma(x)$ induced by $\operatorname{ad}_\mathfrak{g} x$ on $\mathfrak{g}/\mathfrak{a}$ is nilpotent by (i). So by induction, there exists $y \in \mathfrak{g} \setminus \mathfrak{a}$ such that $[y, \mathfrak{a}] \subset \mathfrak{a}$. Consequently, $\mathfrak{a}$ is an ideal of $\mathfrak{a} \oplus \Bbbk y$. By iterating this process, we see that $\mathfrak{g}$ contains an ideal $\mathfrak{h}$ of codimension 1.

By the induction hypothesis, $W = V^\mathfrak{h} \neq \{0\}$. Now for $y \in \mathfrak{g} \setminus \mathfrak{h}$, $z \in \mathfrak{h}$ and $w \in W$, we have,

$$z \circ y(w) = y \circ z(w) + [y, z](w) = 0,$$

since $[y, z] \in \mathfrak{h}$. Thus $y(W) \subset W$. Since y is nilpotent, $y(v) = 0$ for some $v \in W \setminus \{0\}$. Hence $V^{\mathfrak{g}} \neq \{0\}$. $\square$

19.3.6 Theorem. (Engel's Theorem) *Let* (V, σ) *be a representation of* $\mathfrak{g}$ *of dimension* $n > 0$. *Assume that* $\sigma(g)$ *is a nilpotent endomorphism for all* $g \in \mathfrak{g}$. *Then there is a flag*

$$\{0\} = V_0 \subset V_1 \subset \cdots \subset V_n = V$$

of V *such that* $\sigma(\mathfrak{g})V_i \subset V_{i-1}$ *for* $1 \leqslant i \leqslant n$.

Proof. We may assume that $\mathfrak{g} \subset \mathfrak{gl}(V)$. The result is clear when $n = 1$. So let us proceed by induction on n. By 19.3.5, there exists $v \in V^{\mathfrak{g}}$ which is non-zero. Let $\varphi : V \to V/\Bbbk v = W$ be the canonical surjection. Then by applying the induction hypothesis to W, we have a flag $(W_i)_{0 \leqslant i \leqslant n-1}$ of W with the required properties. Set $V_0 = \{0\}$ and $V_i = \varphi^{-1}(W_{i-1})$ for $1 \leqslant i \leqslant n$, then the flag $(V_i)_{0 \leqslant i \leqslant n}$ clearly satisfies $\sigma(\mathfrak{g})V_i \subset V_{i-1}$ for $1 \leqslant i \leqslant n$. $\square$

19.3.7 Corollary. *A Lie algebra* $\mathfrak{g}$ *is nilpotent if and only if for all* $x \in \mathfrak{g}$, $\operatorname{ad} x$ *is a nilpotent endomorphism of* $\mathfrak{g}$.

19.3.8 Let V be a finite-dimensional vector space and $x \in \operatorname{End}(V)$. For $\alpha \in \Bbbk$, set:

$$V_\alpha(x) = \ker(x - \alpha \operatorname{id}_V) \,, \; V^\alpha(x) = \bigcup_{n \geqslant 0} \ker(x - \alpha \operatorname{id}_V)^n.$$

The subspace $V^\alpha(x)$ is called the *generalized eigenspace* of x relative to α, and $V^0(x)$ is the *nilspace* of x. Clearly $V_\alpha(x) \subset V^\alpha(x)$ and

$$V_\alpha(x) \neq \{0\} \Leftrightarrow V^\alpha(x) \neq \{0\} \,, \; V = \bigoplus_{\alpha \in \Bbbk} V^\alpha(x).$$

Lemma. *Let* $x, y \in \operatorname{End}(V)$. *Then* $(\operatorname{ad} x)^n(y) = 0$ *for some* $n \in \mathbb{N}$ *if and only if* $y(V^\alpha(x)) \subset V^\alpha(x)$ *for all* $\alpha \in \Bbbk$.

Proof. Suppose that $(\operatorname{ad} x)^n(y) = 0$ for some $n \in \mathbb{N}$.

Since $(\operatorname{ad} x)(y) = (\operatorname{ad}(x - \alpha \operatorname{id}_V))(y)$, we only need to show that $y(V^0(x))$ is contained in $V^0(x)$. This is trivial when $n = 0$. Let us proceed by induction on n. Set $z = [x, y]$, then $(\operatorname{ad} x)^{n-1}(z) = 0$. By the induction hypothesis, $z(V^0(x)) \subset V^0(x)$. For $q \in \mathbb{N}^*$, we have the following identity:

$$x^q y = yx^q + \sum_{i=1}^{q} x^{q-i}(xy - yx)x^{-i-1} = yx^q + \sum_{i=1}^{q} x^{q-i}zx^{i-1}.$$

Since V is finite-dimensional, there exists $r \in \mathbb{N}^*$ such that $x^r(V^0(x)) = \{0\}$. It follows that $x^{2r-1}y(V^0(x)) = \{0\}$ and therefore $y(V^0(x)) \subset V^0(x)$.

Conversely, suppose that $y(V^\alpha(x)) \subset V^\alpha(x)$ for all $\alpha \in \Bbbk$. For $\alpha \in \Bbbk$, there exists $r_\alpha \in \mathbb{N}^*$ such that $(x - \alpha \operatorname{id}_V)^{r_\alpha}|_{V^\alpha(x)} = 0$. This implies that $(\operatorname{ad}(x - \alpha \operatorname{id}_V))^{2r_\alpha - 1}(y)|_{V^\alpha} = 0$ (19.3.5). But $(\operatorname{ad} x)(y) = (\operatorname{ad}(x - \alpha \operatorname{id}_V))(y)$, so $(\operatorname{ad} x)^{2r_\alpha - 1}(y)|_{V^\alpha} = 0$. Now V is finite-dimensional and V is the sum of the generalized eigenspaces $V^\alpha(x)$, so $(\operatorname{ad} x)^n(y) = 0$ for some $n \in \mathbb{N}$. $\quad\square$

19.3.9 Let $\mathfrak{g}$ be a Lie algebra, P the set of maps from $\mathfrak{g}$ to $\Bbbk$ and (V, σ) a finite-dimensional representation of $\mathfrak{g}$. For $\lambda \in P$, set:

$$V_\lambda(\mathfrak{g}) = \bigcap_{x \in \mathfrak{g}} V_{\lambda(x)}(\sigma(x)) , \quad V^\lambda(\mathfrak{g}) = \bigcap_{x \in \mathfrak{g}} V^{\lambda(x)}(\sigma(x)).$$

The sum of $V^\lambda(\mathfrak{g})$, $\lambda \in P$, is clearly direct.

Proposition. *Assume that $\mathfrak{g}$ is nilpotent.*

(i) *For all $\lambda \in P$, $V^\lambda(\mathfrak{g})$ is a submodule of V.*

(ii) *If $V^\lambda(\mathfrak{g}) \neq \{0\}$, then $V_\lambda(\mathfrak{g}) \neq \{0\}$, and λ is a linear form on $\mathfrak{g}$ which is zero on $[\mathfrak{g}, \mathfrak{g}]$.*

(iii) *We have:*

$$V = \bigoplus_{\lambda \in P} V^\lambda(\mathfrak{g}).$$

If $\lambda \in P$, there exists a basis of $V^\lambda(\mathfrak{g})$ such that for any $x \in \mathfrak{g}$, the matrix $A(x) = [a_{ij}(x)]$ of $\sigma(x)|_{V^\lambda(\mathfrak{g})}$ with respect to this basis is strictly upper triangular, with $a_{ii}(x) = \lambda(x)$ for all i.

Proof. (i) Since $\mathfrak{g}$ is nilpotent, it follows from 19.3.8 (i) that $V^{\lambda(x)}(\sigma(x))$ is a submodule.

(ii) If $W = V^\lambda(\mathfrak{g}) \neq \{0\}$, then by (i), W is a submodule. If $x \in \mathfrak{g}$, there exists $r \in \mathbb{N}^*$ such that $(\sigma(x)|_W - \lambda(x) \operatorname{id}_W)^r = 0$. It follows that $\lambda(x)$ is the unique eigenvalue of $\sigma(x)|_W$. Hence $(\dim W)\lambda(x) = \operatorname{tr} \sigma(x)|_W$, and therefore λ is a linear form on $\mathfrak{g}$ which is zero on $[\mathfrak{g}, \mathfrak{g}]$. It is then clear that the map $\mathfrak{g} \to \operatorname{End}(W)$, $x \mapsto \sigma(x)|_W - \lambda(x) \operatorname{id}_W$, is a representation of $\mathfrak{g}$. The result now follows from Engel's theorem (19.3.6).

(iii) We shall prove part (iii) by induction on $\dim V = r$. The case $r = 0$ is trivial. So let us assume that $r \geqslant 1$.

If for all $x \in \mathfrak{g}$, $\lambda(x)$ is the unique eigenvalue of $\sigma(x)$, then $V = V^\lambda(\mathfrak{g})$. From the proof of (ii), $x \mapsto \sigma(x) - \lambda(x) \operatorname{id}_V$ is a representation of $\mathfrak{g}$ in V, so by applying Engel's theorem (19.3.6) to this representation, we have the required basis.

Now suppose that there exists $x \in \mathfrak{g}$ such that $\sigma(x)$ has two distinct eigenvalues. Since $V = \bigoplus_{\alpha \in \Bbbk} V^\alpha(x)$, $\dim V^\alpha(x) < r$ and the $V^\alpha(x)$'s are submodules of V (19.3.8), the result follows by applying the induction hypothesis. $\quad\square$

19.4 Solvable Lie algebras

19.4.1 Proposition. *The following conditions are equivalent:*
(i) *There exists $k \in \mathbb{N}$ such that $\mathcal{D}^k(\mathfrak{g}) = \{0\}$.*
(ii) *There exists a chain $\mathfrak{g} = \mathfrak{g}_0 \supset \mathfrak{g}_1 \supset \cdots \supset \mathfrak{g}_n = \{0\}$ of ideals of $\mathfrak{g}$ such that $[\mathfrak{g}_i, \mathfrak{g}_i] \subset \mathfrak{g}_{i+1}$ for $0 \leqslant i \leqslant n-1$.*
If these conditions are verified, we say that $\mathfrak{g}$ is solvable.

Proof. The proof is analogue to the one for 19.3.2. □

19.4.2 Proposition. (i) *A nilpotent Lie algebra is solvable.*
(ii) *Subalgebras and quotients of a solvable Lie algebra are solvable.*
(iii) *Let $\mathfrak{a}$ be an ideal of $\mathfrak{g}$. Then $\mathfrak{g}$ is solvable if and only if $\mathfrak{a}$ and $\mathfrak{g}/\mathfrak{a}$ are solvable.*

Proof. Parts (i) and (ii) are straightforward. Now if $\mathfrak{a}$ and $\mathfrak{g}/\mathfrak{a}$ are solvable, then $\mathcal{D}^i(\mathfrak{g}) \subset \mathfrak{a}$ for some integer i and $\mathcal{D}^{i+j}(\mathfrak{g}) \subset \mathcal{D}^j(\mathfrak{a}) = \{0\}$ for some integer j. So $\mathfrak{g}$ is solvable. The converse follows from (ii). □

19.4.3 Lemma. *Let (V, σ) be a finite-dimensional representation of $\mathfrak{g}$, $\mathfrak{a}$ an ideal of $\mathfrak{g}$, $\lambda \in \mathfrak{a}^*$ and $W = \{v \in V; \sigma(x)(v) = \lambda(x)(v) \text{ for all } x \in \mathfrak{a}\}$. Then W is a submodule of V.*

Proof. We may assume that $\mathfrak{g} \subset \mathfrak{gl}(V)$ and W is non-zero. For $v_0 \in W \setminus \{0\}$ and $x \in \mathfrak{g}$, set $v_n = x^n v_0$, $n \in \mathbb{N}^*$, and p the smallest integer such that $v_0, \ldots, v_p$ are linearly independent. Denote by U the subspace spanned by $v_0, \ldots, v_p$, then it is clear that $x(U) \subset U$. Now if $y \in \mathfrak{a}$, then:

$$y(v_1) = yx(v_0) = xy(v_0) + [y,x](v_0) \in \lambda(y)v_1 + \Bbbk v_0$$
$$y(v_2) = yx(v_1) = xy(v_1) + [y,x](v_1) \in \lambda(y)v_1 + \Bbbk x(v_0) + \Bbbk v_1 + \Bbbk v_0$$
$$\subset \lambda(y)v_2 + \Bbbk v_1 + \Bbbk v_0.$$

If $0 \leqslant i \leqslant p$, then we have by induction that:

$$y(v_i) \in \lambda(y)v_i + \Bbbk v_{i-1} + \cdots + \Bbbk v_0.$$

It follows that $y(U) \subset U$ and $\operatorname{tr}(y|_U) = (p+1)\lambda(y)$. Since $[x,y] \in \mathfrak{a}$, we have that $(p+1)\lambda([x,y]) = 0$. Hence $\lambda([x,y]) = 0$. Finally,

$$yx(v_0) = xy(v_0) + [y,x](v_0) = \lambda(y)x(v_0) + \lambda([y,x])(v_0) = \lambda(y)x(v_0).$$

So $x(v_0) \in W$. □

19.4.4 Theorem. (Lie's Theorem) *Let (V, σ) be a finite-dimensional representation of a solvable Lie algebra $\mathfrak{g}$. Then there exists a flag*

$$\{0\} = V_0 \subset V_1 \subset \cdots \subset V_n = V$$

of V such that $\sigma(\mathfrak{g})(V_i) \subset V_i$ for $0 \leqslant i \leqslant n$.

Proof. We may assume that $\mathfrak{g} \subset \mathfrak{gl}(V)$ and $V \neq \{0\}$. We shall prove by induction on $\dim \mathfrak{g} = p$ that there exists $\mu \in \mathfrak{g}^*$ and $v \in V \setminus \{0\}$ such that $x(v) = \mu(x)v$ for all $x \in \mathfrak{g}$. We can then conclude as in the proof of 19.3.6.

The case $p \leqslant 1$ is obvious, so let us assume that $p \geqslant 2$. Since $\mathcal{D}(\mathfrak{g}) \neq \mathfrak{g}$, any codimension 1 subspace of $\mathfrak{g}$ containing $\mathcal{D}(\mathfrak{g})$ is an ideal of $\mathfrak{g}$. Let $\mathfrak{a}$ be such an ideal. By induction, there exist $\lambda \in \mathfrak{a}^*$ and $v \in V \setminus \{0\}$ such that $y(v) = \lambda(y)v$ for all $y \in \mathfrak{a}$. Let W be as in 19.4.3, then $W \neq \{0\}$ and $\mathfrak{g}(W) \subset W$. Let $x \in \mathfrak{g} \setminus \mathfrak{a}$, then since $\Bbbk$ is algebraically closed, $x|_W$ has an eigenvector w. It follows that $z(w) \in \Bbbk w$ for all $z \in \mathfrak{g}$. So we are done. $\square$

19.4.5 Corollary. (i) *Any finite-dimensional simple representation of a solvable Lie algebra is 1-dimensional.*

(ii) *If $\mathfrak{g}$ is solvable, then there exists a flag $\{0\} \subset \mathfrak{g}_0 \subset \cdots \subset \mathfrak{g}_n = \mathfrak{g}$ consisting of ideals of $\mathfrak{g}$.*

(iii) *If $\mathfrak{g}$ is solvable, then any element of $\mathrm{ad}[\mathfrak{g}, \mathfrak{g}]$ is nilpotent.*

(iv) *A Lie algebra $\mathfrak{g}$ is solvable if and only if $\mathcal{D}(\mathfrak{g})$ is nilpotent.*

Proof. Parts (i) and (ii) are consequences of 19.4.4, while part (iii) follows from (ii). Finally, (iv) follows from (iii) and 19.4.2 (iii). $\square$

19.4.6 Corollary. *Let (V, σ) be a representation of dimension $n > 0$ of a nilpotent Lie algebra $\mathfrak{g}$ such that $\ker \sigma(x) \neq \{0\}$ for all $x \in \mathfrak{g}$. Then there exists $v \in V \setminus \{0\}$ such that $\sigma(x)(v) = 0$ for all $x \in \mathfrak{g}$.*

Proof. We shall prove the result by induction on n. The case $n = 1$ is obvious, so we may assume that $n > 1$. By 19.4.4, there is a basis $(v_1, \dots, v_n)$ of V such that for all $x \in \mathfrak{g}$, the matrix of $\sigma(x)$ in this basis is upper triangular with diagonal entries $\lambda_1(x), \dots, \lambda_n(x)$ where $\lambda_1, \dots, \lambda_n \in \mathfrak{g}^*$. Since $\ker \sigma(x) \neq \{0\}$, $\lambda_1 \cdots \lambda_n = 0$. If $\lambda_1 \cdots \lambda_{n-1} = 0$, then by applying the induction hypothesis on the restriction of σ to $W = \Bbbk v_1 + \cdots + \Bbbk v_{n-1}$, we have our result.

So let us assume that $\lambda_1 \cdots \lambda_{n-1} \neq 0$. Then $\lambda_n = 0$ since the ring of polynomial functions on $\mathfrak{g}$ is an integral domain. Let $x \in \mathfrak{g}$ be such that $\lambda_1(x) \cdots \lambda_{n-1}(x) \neq 0$. We have $\sigma(x)(v) = 0$ for some $v \in V \setminus \{0\}$. By our choice of x, $v \notin W$, and $\sigma(\mathfrak{g})(v) \in W$ since $\lambda_n = 0$. Define $f : \mathfrak{g} \to W$ by $f(y) = \sigma(y)(v)$. Then $f([x, y]) = \sigma(x)(f(y))$ for all $y \in \mathfrak{g}$ because $\sigma(x)(v) = 0$. It follows that the restriction of $\sigma(x)$ to $f(\mathfrak{g})$ is nilpotent. If $f(\mathfrak{g}) \neq \{0\}$, then $\sigma(x)(w) = 0$ for some $w \in f(\mathfrak{g}) \setminus \{0\}$. But this is absurd since $\lambda_1(x) \cdots \lambda_{n-1}(x) \neq 0$. Thus $f(\mathfrak{g}) = \{0\}$ and $\sigma(\mathfrak{g})(v) = \{0\}$. $\square$

19.4.7 Lemma. *Let V be a finite-dimensional vector space and $\mathfrak{g} \subset \mathfrak{gl}(V)$ a Lie subalgebra such that $\mathrm{tr}(xy) = 0$ for all $x, y \in \mathfrak{g}$. Then any element of $\mathcal{D}(\mathfrak{g})$ is a nilpotent endomorphism of V.*

Proof. By 19.4.4 and 19.4.5, it suffices to prove that $\mathfrak{g}$ is solvable. We shall prove this by induction on $\dim \mathfrak{g} = r$. The case $r = 0$ is trivial. So let $r > 0$.

If $\mathcal{D}(\mathfrak{g}) \neq \mathfrak{g}$, then by the induction hypothesis, $\mathcal{D}(\mathfrak{g})$ is solvable and therefore $\mathfrak{g}$ is solvable. So let us assume that $\mathcal{D}(\mathfrak{g}) = \mathfrak{g}$ and we shall show that this case is not possible.

Let $\mathfrak{h}$ be a maximal (proper) Lie subalgebra of $\mathfrak{g}$. Then $\dim \mathfrak{h} \geqslant 1$ and $\mathfrak{h}$ is solvable by induction. Applying 19.4.4 to the adjoint representation of $\mathfrak{h}$ in $\mathfrak{g}/\mathfrak{h}$, we deduce from the maximality of $\mathfrak{h}$ that there exist $x \in \mathfrak{g} \setminus \mathfrak{h}$ and $\lambda \in \mathfrak{h}^*$ verifying $\mathfrak{g} = \mathfrak{h} \oplus \Bbbk x$ and $[y,x] - \lambda(y)x \in \mathfrak{h}$ for all $y \in \mathfrak{h}$.

1) Let (W,σ) be a simple finite-dimensional representation of $\mathfrak{g}$ and W_0 a simple $\mathfrak{h}$-submodule of W. Then by 19.4.5, $W_0 = \Bbbk w_0$ for some $w_0 \in W \setminus \{0\}$. Let $\mu \in \mathfrak{h}^*$ be such that $\sigma(y)(w_0) = \mu(y)w_0$ for all $y \in \mathfrak{h}$. Set $W_{-1} = \{0\}$ and for $n \in \mathbb{N}$, $w_n = \sigma(x)^n(w_0)$ and $W_n = \Bbbk w_0 + \cdots + \Bbbk w_n$. If $y \in \mathfrak{h}$, then:

$$\sigma(y)w_{i+1} = \sigma(y)\sigma(x)w_i = \sigma([y,x])w_i + \sigma(x)\sigma(y)w_i$$
$$= \sigma([y,x] - \lambda(y)x)w_i + \sigma(x)(\sigma(y)w_i) + \lambda(y)\sigma(x)w_i,$$

and $(\sigma(y) - \mu(y) - (i+1)\lambda(y))w_{i+1}$ is equal to

$$\sigma([y,x] - \lambda(y)x)w_i + \sigma(x)(\sigma(y) - \mu(y) - i\lambda(y))w_i.$$

It follows easily by induction that for all $i \in \mathbb{N}$ and $y \in \mathfrak{h}$, $\sigma(\mathfrak{h})(W_i) \subset W_i$ and $\sigma(y).w_i - (\mu(y) + i\lambda(y))w_i \in W_{i-1}$.

Since W is a simple $\mathfrak{g}$-module, $W = W_q$ where q is the largest integer such that $w_0, \ldots, w_q$ are linearly independent. The equality in the previous paragraph implies that

$$\text{tr}(\sigma(y)) = (q+1)\mu(y) + \frac{1}{2}q(q+1)\lambda(y).$$

Now $\mathfrak{g} = \mathcal{D}(\mathfrak{g})$, so $\text{tr}(\sigma(y)) = 0$ and we have $2\mu(y) = -q\lambda(y)$,

$$\sigma(y).w_i - (i - \frac{q}{2})\lambda(y)w_i \in W_{i-1}, \quad \text{tr}(\sigma(y)^2) = \sum_{j=0}^{q} \left(j - \frac{q}{2}\right)^2 \lambda(y)^2.$$

2) If $\{0\} = V_0 \subset \cdots \subset V_p = V$ is a Jordan-Hölder series of the $\mathfrak{g}$-module V, and $q(j) = \dim(V_{j+1}/V_j) - 1$, then, using the preceding computation, we obtain that

$$0 = \text{tr}(y^2) = \left(\sum_{j=0}^{p-1}\sum_{i=0}^{q(j)} \left(i - \frac{q(j)}{2}\right)^2\right)\lambda(y)^2.$$

Since $\mathcal{D}(\mathfrak{g}) = \mathfrak{g} \neq \mathfrak{h}$, $\lambda \neq 0$ and so $q(j) = 0$ for all j. Again, the fact that $\mathcal{D}(\mathfrak{g}) = \mathfrak{g}$ implies that $(V_{j+1}/V_j)^{\mathfrak{g}} = V_{j+1}/V_j$. Hence $\mathfrak{g}$ consists of nilpotent endomorphisms of V. By 19.3.6, $\mathfrak{g}$ is nilpotent. But this contradicts the assumption that $\mathcal{D}(\mathfrak{g}) = \mathfrak{g}$ and $\dim \mathfrak{g} > 0$. $\square$

19.4.8 Theorem. (Cartan's Criterion) *Let (V,σ) be a finite-dimensional representation of $\mathfrak{g}$ and β_σ the bilinear form associated to σ. The following conditions are equivalent:*

(i) $\sigma(\mathfrak{g})$ *is a solvable Lie algebra.*

(ii) *We have $\beta_\sigma(\mathcal{D}(\mathfrak{g}), \mathcal{D}(\mathfrak{g})) = \{0\}$.*

Proof. We may assume that $\mathfrak{g} \subset \mathfrak{gl}(V)$. Then clearly (i) $\Rightarrow$ (ii) by 19.4.5. Conversely by 19.3.6 and 19.4.7, $\mathcal{D}^2(\mathfrak{g})$ is nilpotent. So $\mathcal{D}(\mathfrak{g})$ and $\mathfrak{g}$ are solvable. Hence (ii) $\Rightarrow$ (i). $\quad\square$

19.5 Radical and the largest nilpotent ideal

19.5.1 Lemma. *Let $\mathfrak{a}$ be an ideal of $\mathfrak{g}$ and (V,σ) a finite-dimensional representation of $\mathfrak{g}$.*

(i) Assume that σ is simple and all the elements of $\sigma(\mathfrak{a})$ are nilpotent. Then $\sigma(\mathfrak{a}) = \{0\}$.

(ii) Let $\{0\} = V_0 \subset V_1 \subset \cdots \subset V_n = V$ be a Jordan-Hölder series of V. Then all the elements of $\sigma(\mathfrak{a})$ are nilpotent if and only if $\sigma(\mathfrak{a})(V_i) \subset V_{i-1}$ for $1 \leqslant i \leqslant n$.

Proof. Part (i) follows from 19.3.5 and 19.4.3, while (ii) follows from (i). $\square$

19.5.2 Lemma. *Let V be a finite-dimensional vector space, $\mathfrak{g}$ a Lie sub-algebra of $\mathfrak{gl}(V)$ and $\mathfrak{a}$ an abelian ideal of $\mathfrak{g}$. If V is a simple $\mathfrak{g}$-module, then $\mathfrak{a} \cap \mathcal{D}(\mathfrak{g}) = \{0\}$.*

Proof. 1) Let S be the associative subalgebra of $\mathrm{End}(V)$ generated by $\mathfrak{a}$. Let $\mathfrak{b} \subset \mathfrak{a}$ be an ideal of $\mathfrak{g}$ satisfying $\mathrm{tr}(bs) = 0$ for all $b \in \mathfrak{b}$ and $s \in S$. In particular, $\mathrm{tr}(b^n) = 0$ for all $b \in \mathfrak{b}$ and $n \in \mathbb{N}^*$, so b is nilpotent (19.2.3). By 19.5.1, $\mathfrak{b} = \{0\}$.

2) Since $\mathfrak{a}$ is abelian, we have for $x \in \mathfrak{g}$, $a \in \mathfrak{a}$ and $s \in S$ that

$$\mathrm{tr}([x,a]s) = \mathrm{tr}(xas - axs) = \mathrm{tr}(xas - xsa) = \mathrm{tr}(xas - xas) = 0.$$

It follows from point 1) that $[\mathfrak{g}, \mathfrak{a}] = \{0\}$, and so $\mathfrak{a} \subset \mathfrak{z}(\mathfrak{g})$. Hence $xs = sx$ for all $x \in \mathfrak{g}$ and $s \in S$.

3) Finally, we have

$$\mathrm{tr}([x,y]s) = \mathrm{tr}(xys - yxs) = \mathrm{tr}(xys - xsy) = \mathrm{tr}\left(x(ys - sy)\right) = 0$$

for $x, y \in \mathfrak{g}$ and $s \in S$. So point 1) implies that $\mathfrak{a} \cap \mathcal{D}(\mathfrak{g}) = \{0\}$. $\square$

19.5.3 Proposition. *There is a unique maximal solvable ideal in $\mathfrak{g}$. It is called the* radical *of $\mathfrak{g}$, and is denoted by* $\mathrm{rad}\,\mathfrak{g}$.

Proof. Let $\mathfrak{a}, \mathfrak{b}$ be solvable ideals of $\mathfrak{g}$. Then $\mathfrak{a} + \mathfrak{b}$ is an ideal of $\mathfrak{g}$ which is solvable (19.4.2) since $(\mathfrak{a} + \mathfrak{b})/\mathfrak{b}$ and $\mathfrak{a}/\mathfrak{a} \cap \mathfrak{b}$ are isomorphic. So there is a unique maximal solvable ideal in $\mathfrak{g}$. $\square$

19.5.4 Proposition. *The radical of $\mathfrak{g}$ is the smallest ideal $\mathfrak{a}$ of $\mathfrak{g}$ such that* $\mathrm{rad}(\mathfrak{g}/\mathfrak{a}) = \{0\}$.

Proof. Let $\mathfrak{r} = \operatorname{rad}\mathfrak{g}$ and $p : \mathfrak{g} \to \mathfrak{g}/\mathfrak{r}$ be the canonical surjection. Then $\mathfrak{o} = p^{-1}(\operatorname{rad}(\mathfrak{g}/\mathfrak{r}))$ is an ideal of $\mathfrak{g}$ such that $\mathfrak{o}/\mathfrak{r}$ is solvable. So $\mathfrak{o}$ is solvable by 19.4.2 and $\mathfrak{o} = \mathfrak{r}$.

Let $\mathfrak{a}$ be an ideal of $\mathfrak{g}$ and $\varphi : \mathfrak{g} \to \mathfrak{g}/\mathfrak{a}$ the canonical projection. If $\operatorname{rad}(\mathfrak{g}/\mathfrak{a}) = \{0\}$, then $\varphi(\mathfrak{r}) = \{0\}$. Thus $\mathfrak{a} \supset \mathfrak{r}$.   □

19.5.5 Proposition. *Let (V,σ) be a finite-dimensional representation of $\mathfrak{g}$, $\{0\} = V_0 \subset V_1 \subset \cdots \subset V_n = V$ a Jordan-Hölder series of V, β_σ the bilinear form on $\mathfrak{g}$ associated to σ and σ_i the representation of $\mathfrak{g}$ on V_i/V_{i-1} induced by σ, for $1 \leqslant i \leqslant n$.*

(i) *There is a unique maximal ideal $\mathfrak{n}$ among the ideals $\mathfrak{a}$ of $\mathfrak{g}$ such that the elements of $\sigma(\mathfrak{a})$ are nilpotent.*

(ii) *We have $\mathfrak{n} = \ker\sigma_1 \cap \cdots \cap \ker\sigma_n$ and $\beta_\sigma(\mathfrak{n},\mathfrak{g}) = \{0\}$.*

We call $\mathfrak{n}$ the largest nilpotent ideal *of σ.*

Proof. By 19.5.1, for an ideal $\mathfrak{a}$ of $\mathfrak{g}$, $\sigma(x)$ is nilpotent for all $x \in \mathfrak{a}$ if and only if $\mathfrak{a} \subset \ker\sigma_1 \cap \cdots \cap \ker\sigma_n$. So we have (i) and the first part of (ii). Let $x \in \mathfrak{g}$, $y \in \mathfrak{n}$, then $\sigma(x)\sigma(y)(V_i) \subset \sigma(x)(V_{i-1}) \subset V_{i-1}$, for $1 \leqslant i \leqslant n$. So $\sigma(x)\sigma(y)$ is nilpotent and $\beta_\sigma(\mathfrak{n},\mathfrak{g}) = \{0\}$.   □

19.5.6 Lemma. *Let (V,σ) be a finite-dimensional representation of $\mathfrak{g}$.*

(i) *If σ is simple, then $\sigma(\operatorname{rad}\mathfrak{g} \cap \mathcal{D}(\mathfrak{g})) = \{0\}$.*

(ii) *If $\mathfrak{n}$ is the largest nilpotent ideal of σ, then $\operatorname{rad}\mathfrak{g} \cap \mathcal{D}(\mathfrak{g}) \subset \mathfrak{n}$.*

Proof. (i) Let $\mathfrak{r} = \operatorname{rad}\mathfrak{g}$ and i be the smallest integer such that $\sigma(\mathcal{D}^{i+1}(\mathfrak{r})) = \{0\}$. Then the image under σ of $\mathfrak{a} = \mathcal{D}^i(\mathfrak{r})$ is an abelian ideal of $\sigma(\mathfrak{g})$. By Schur's Lemma, $\sigma(\mathfrak{a}) \subset \Bbbk\,\mathrm{id}_V$, and so $\sigma(\mathcal{D}(\mathfrak{g}) \cap \mathcal{D}^i(\mathfrak{r})) = \{0\}$.

If $i > 0$, then $\mathcal{D}^i(\mathfrak{r}) \subset \mathcal{D}(\mathfrak{g})$ and so $\sigma(\mathfrak{a}) = \{0\}$. This is impossible by the minimality of i. So $i = 0$ and $\sigma(\mathfrak{r} \cap \mathcal{D}(\mathfrak{g})) = \{0\}$.

Part (ii) follows clearly from (i) and 19.5.5.   □

19.5.7 Let us conserve the notations of 19.5.5. In general, the fact that $\sigma(x)$ is nilpotent does not imply that $x \in \mathfrak{n}$. Nevertheless, we have the following result:

Proposition. *Let (V,σ) be a finite-dimensional representation of $\mathfrak{g}$ and $\mathfrak{n}$ the largest nilpotent ideal of σ. If $x \in \operatorname{rad}\mathfrak{g}$ and $\sigma(x)$ is nilpotent, then $x \in \mathfrak{n}$.*

Proof. Let $\mathfrak{r} = \operatorname{rad}\mathfrak{g}$ and $\{0\} = V_0 \subset V_1 \subset \cdots \subset V_n = V$ be a Jordan-Hölder series of the $\mathfrak{r}$-module V. By 19.4.5, $W_i = V_i/V_{i-1}$ is 1-dimensional, and the representation of $\mathfrak{r}$ on W_i is given by a linear from $\lambda_i \in \mathfrak{r}^*$ which is zero on $\mathcal{D}(\mathfrak{r})$. The set of elements $x \in \mathfrak{r}$ such that $\sigma(x)$ is nilpotent is therefore the ideal $\mathfrak{a} = \ker\lambda_1 \cap \cdots \cap \ker\lambda_n$ of $\mathfrak{r}$. By 19.5.6, $[\mathfrak{g},\mathfrak{a}] \subset \mathfrak{r} \cap \mathcal{D}(\mathfrak{g}) \subset \mathfrak{r} \cap \mathfrak{n} \subset \mathfrak{a}$. Thus $\mathfrak{a}$ is an ideal of $\mathfrak{g}$ and the result follows from 19.5.5.   □

19.5.8 Proposition. *Let L be the Killing form of $\mathfrak{g}$ and $\mathfrak{n}$ the largest nilpotent ideal of the adjoint representation of $\mathfrak{g}$. Then $\mathfrak{n}$ is the largest nilpotent ideal of $\mathfrak{g}$, and $L(\mathfrak{g},\mathfrak{n}) = \{0\}$.*

Proof. Let $\mathfrak{a}$ be an ideal of $\mathfrak{g}$ and $x \in \mathfrak{a}$. Then $\mathrm{ad}_{\mathfrak{g}}\, x$ is nilpotent if and only if $\mathrm{ad}_{\mathfrak{a}}\, x$ is nilpotent. Thus $\mathfrak{a}$ is nilpotent if and only if $\mathrm{ad}_{\mathfrak{g}}\, x$ is nilpotent for all $x \in \mathfrak{a}$ (19.3.7). Since any nilpotent ideal is contained in the radical of $\mathfrak{g}$, the result follows from 19.5.7 and 19.5.5. $\square$

19.6 Nilpotent radical

19.6.1 It follows from 19.5.5 that the intersection of the kernels of simple finite-dimensional representations of $\mathfrak{g}$ is equal to the intersection of the largest nilpotent ideals of finite-dimensional representations of $\mathfrak{g}$. This is an ideal of $\mathfrak{g}$ that we shall call the *nilpotent radical* of $\mathfrak{g}$. In view of 19.5.8, the nilpotent radical of $\mathfrak{g}$ is contained in the radical of $\mathfrak{g}$.

19.6.2 Theorem. *Let* $\mathfrak{r} = \mathrm{rad}\,\mathfrak{g}$ *and* $\mathfrak{s}$ *the nilpotent radical of* $\mathfrak{g}$. *Then* $\mathfrak{s} = \mathfrak{r} \cap \mathcal{D}(\mathfrak{g})$. *In particular, when* $\mathfrak{g}$ *is solvable, then* $\mathfrak{s} = \mathcal{D}(\mathfrak{g})$.

Proof. Any $\lambda \in \mathfrak{g}^*$ verifying $\lambda(\mathcal{D}(\mathfrak{g})) = \{0\}$ determines a representation of $\mathfrak{g}$ of dimension 1. So $\mathfrak{s} \subset \ker \lambda$. Since $\mathfrak{s} \subset \mathfrak{r}$, we have $\mathfrak{s} \subset \mathfrak{r} \cap \mathcal{D}(\mathfrak{g})$. The reverse inclusion follows from 19.5.6. $\square$

19.6.3 Corollary. *Let* $\mathfrak{r} = \mathrm{rad}\,\mathfrak{g}$.
(i) *Let* (V, σ) *be a finite-dimensional representation of* $\mathfrak{g}$ *and* β *the bilinear form on* $\mathfrak{g}$ *associated to* σ. *Then* $\beta(\mathfrak{r}, \mathcal{D}(\mathfrak{g})) = \{0\}$.
(ii) $\mathfrak{r}$ *is the orthogonal of* $\mathcal{D}(\mathfrak{g})$ *with respect to the Killing form* L *of* $\mathfrak{g}$.
(iii) $\mathfrak{r}$ *is a characteristic ideal of* $\mathfrak{g}$.
(iv) *If* $\mathfrak{a}$ *is an ideal of* $\mathfrak{g}$, *then* $\mathrm{rad}\,\mathfrak{a} = \mathfrak{a} \cap \mathfrak{r}$.

Proof. (i) Let $\mathfrak{n}$ be the largest nilpotent ideal of $\mathfrak{g}$ and $\mathfrak{s}$ the nilpotent radical of $\mathfrak{g}$. By 19.6.2 and 19.6.1, $[\mathfrak{g}, \mathfrak{r}] \subset \mathfrak{s} \subset \mathfrak{n}$. So the result follows from 19.5.5 and 19.2.2.

(ii) Let $\mathfrak{u}$ be the orthogonal of $\mathcal{D}(\mathfrak{g})$ with respect to L. Then $\mathfrak{u}$ is an ideal of $\mathfrak{g}$ containing $\mathfrak{r}$ (19.2.2). Since the Killing form of $\mathfrak{u}$ is $L|_{\mathfrak{u} \times \mathfrak{u}}$, $\mathrm{ad}_{\mathfrak{g}}\, \mathfrak{u}$ is solvable (19.4.8). But $\mathrm{ad}_{\mathfrak{g}}\, \mathfrak{u}$ is isomorphic to $\mathfrak{u}/\mathfrak{z}(\mathfrak{g})$, so $\mathfrak{u}$ is solvable, and $\mathfrak{u} = \mathfrak{r}$.

(iii) Let $x, y \in \mathfrak{g}$ and $d \in \mathrm{Der}\,\mathfrak{g}$. Since $\mathrm{ad}\, d(x) = [d, \mathrm{ad}\, x]$, we verify easily that $L(d(x), y) + L(x, d(y)) = 0$. So the orthogonal of a characteristic ideal with respect to L is characteristic, and the result follows from (ii).

(iv) Clearly, $\mathfrak{a} \cap \mathfrak{r} \subset \mathrm{rad}\,\mathfrak{a}$. On the other hand, by applying (iii) to $\mathfrak{a}$, we obtain that $\mathrm{rad}\,\mathfrak{a}$ is a solvable ideal of $\mathfrak{g}$. So $\mathrm{rad}\,\mathfrak{a} \subset \mathfrak{r}$. $\square$

19.6.4 Proposition. *Let* $\mathfrak{r} = \mathrm{rad}\,\mathfrak{g}$ *and* $\mathfrak{n}$ *the largest nilpotent ideal of* $\mathfrak{g}$.
(i) *We have* $d(\mathfrak{r}) \subset \mathfrak{n}$ *for all derivation* d *of* $\mathfrak{g}$.
(ii) $\mathfrak{n}$ *is a characteristic ideal of* $\mathfrak{g}$.

Proof. It suffices to prove (i). We define a Lie algebra $\mathfrak{h} = \Bbbk v \oplus \mathfrak{g}$ extending the Lie bracket of $\mathfrak{g}$ by defining $[v, x] = d(x)$ for all $x \in \mathfrak{g}$. Then $\mathfrak{g}$ is an ideal of $\mathfrak{h}$ and by 19.6.3 (iii), $\mathfrak{r} \subset \mathrm{rad}\,\mathfrak{h}$, and so $d(\mathfrak{r}) \subset [v, \mathfrak{r}] \subset [\mathfrak{h}, \mathfrak{h}] \cap \mathrm{rad}\,\mathfrak{h} = \mathfrak{o}$. By 19.6.2, $\mathfrak{o}$ is a nilpotent ideal of $\mathfrak{h}$. So $d(\mathfrak{r})$ is contained in the nilpotent ideal $\mathfrak{o} \cap \mathfrak{g}$ of $\mathfrak{g}$. Hence $d(\mathfrak{r}) \subset \mathfrak{n}$ (19.5.8). $\square$

19.7 Regular linear forms

19.7.1 Let $n = \dim \mathfrak{g}$, $d \in \{0, 1, \ldots, n\}$ and $\mathrm{Gr}(\mathfrak{g}, d)$ the Grassmannian variety of d-dimensional vector subspaces of $\mathfrak{g}$. Recall from 13.6 that given $\mathcal{B} = (e_1, \ldots, e_n)$ a basis of $\mathfrak{g}$, the set $U_\mathcal{B}$ of elements $\mathfrak{o} \in \mathrm{Gr}(\mathfrak{g}, d)$ verifying $\mathfrak{o} \cap (\Bbbk e_{d+1} + \cdots + \Bbbk e_n) = \{0\}$ is open in $\mathrm{Gr}(\mathfrak{g}, d)$. Further, $\mathfrak{o} \in U_\mathcal{B}$ if and only if $\mathfrak{o}$ has a unique basis $(v_1, \ldots, v_d)$ of the form

$$v_i = e_i + \sum_{j=d+1}^{n} a_{ij} e_j.$$

The map $\mathfrak{o} \mapsto (a_{1,d+1}, \ldots, a_{1,n}, \ldots, a_{d,d+1}, \ldots, a_{d,n})$ is an isomorphism of $U_\mathcal{B}$ onto $\Bbbk^{d(n-d)}$.

19.7.2 Lemma. (i) *The set of d-dimensional Lie subalgebras of $\mathfrak{g}$ is a closed subset of $\mathrm{Gr}(\mathfrak{g}, d)$.*

(ii) *The set of d-dimensional solvable (resp. nilpotent, abelian) Lie subalgebras of $\mathfrak{g}$ is a closed subset of $\mathrm{Gr}(\mathfrak{g}, d)$.*

(iii) *The set of $(\mathfrak{o}, f) \in \mathrm{Gr}(\mathfrak{g}, d) \times \mathfrak{g}^*$ such that $f([\mathfrak{g}, \mathfrak{o}]) = \{0\}$ (resp. $f([\mathfrak{o}, \mathfrak{o}]) = \{0\}$) is a closed subset of $\mathrm{Gr}(\mathfrak{g}, d) \times \mathfrak{g}^*$.*

Proof. Let us prove (i), and (ii) for solvable Lie subalgebras. The proofs of the rest are similar. We shall use the notations of 19.7.1.

An element $\mathfrak{o} \in \mathrm{Gr}(\mathfrak{g}, d)$ is a Lie subalgebra if and only if $[v_i, v_j] \in \mathfrak{o}$ for $1 \leqslant i, j \leqslant d$. These conditions correspond to the vanishing of determinants in the a_{pq}'s. Since the open subsets $U_\mathcal{B}$, for $\mathcal{B}$ a basis of $\mathfrak{g}$, form an affine open covering of $\mathrm{Gr}(\mathfrak{g}, d)$, part (i) follows.

Further, a d-dimensional Lie subalgebra $\mathfrak{o}$ is solvable if and only if $\mathcal{D}^d(\mathfrak{o}) = \{0\}$. By expressing this condition in the basis $(v_1, \ldots, v_d)$, we see that it is equivalent to the vanishing of certain polynomials in the a_{pq}'s. □

19.7.3 For $f \in \mathfrak{g}^*$, define Φ_f to be the alternating bilinear form on $\mathfrak{g}$ given by:

$$\Phi_f(x, y) = f([x, y]).$$

For a subset $\mathfrak{a}$ of $\mathfrak{g}$, denote by $\mathfrak{a}^{(f)}$ the orthogonal of $\mathfrak{a}$ with respect to Φ_f. If $\mathfrak{a}$ is an ideal of $\mathfrak{g}$, then it is easy to check that $\mathfrak{a}^{(f)}$ is a Lie subalgebra of $\mathfrak{g}$. In particular, the kernel $\mathfrak{g}^{(f)}$ of Φ_f is a Lie subalgebra of $\mathfrak{g}$. Since Φ_f induces a non-degenerate alternating bilinear form on $\mathfrak{g}/\mathfrak{g}^{(f)}$, $\dim \mathfrak{g} - \dim \mathfrak{g}^{(f)}$ is an even integer. We define the *index* of $\mathfrak{g}$ to be the integer

$$\chi(\mathfrak{g}) = \min\{\dim \mathfrak{g}^{(f)}; f \in \mathfrak{g}^*\}.$$

We say that $f \in \mathfrak{g}^*$ is *regular* if $\dim \mathfrak{g}^{(f)} = \chi(\mathfrak{g})$, and we denote by $\mathfrak{g}^*_{\mathrm{reg}}$ the set of regular elements of $\mathfrak{g}^*$.

19.7.4 Lemma. *Let $\mathfrak{u}$ be a subspace of $\mathfrak{g}$.*

(i) *The set of elements $f \in \mathfrak{g}^*$ verifying $\mathfrak{u} \cap \mathfrak{g}^{(f)} \neq \{0\}$ is closed in $\mathfrak{g}^*$.*

(ii) *Let $f \in \mathfrak{u}^*$ and $g \in \mathfrak{g}^*$ be such that $g|_\mathfrak{u} = f$ and $\dim \mathfrak{g}^{(g)} \leqslant \dim \mathfrak{g}^{(h)}$ for all $h \in \mathfrak{g}^*$ verifying $h|_\mathfrak{u} = f$. Then $[\mathfrak{g}^{(g)}, \mathfrak{g}^{(g)}] \subset \mathfrak{u}$.*

Proof. (i) Let $\mathcal{E} = (e_1, \ldots, e_n)$ be a basis of $\mathfrak{g}$ such that $(e_1, \ldots, e_p)$ is a basis of $\mathfrak{u}$, and $(e_1^*, \ldots, e_n^*)$ the basis of $\mathfrak{g}^*$ dual to $\mathfrak{g}$. For $1 \leqslant i, j \leqslant n$, set

$$[e_i, e_j] = \sum_{k=1}^{n} \gamma_{ijk} e_k.$$

For $f = \lambda_1 e_1^* + \cdots + \lambda_n e_n^*$, we have:

$$u = \mu_1 e_1 + \cdots + \mu_p e_p \in \mathfrak{u} \cap \mathfrak{g}^{(f)} \Leftrightarrow 0 = f([u, e_j]) = \sum_{l=1}^{p} \sum_{k=1}^{n} \mu_l \gamma_{ljk} \lambda_k \text{ for all } j.$$

So $\mathfrak{u} \cap \mathfrak{g}^{(f)} \neq \{0\}$ is equivalent to the condition that the matrix $\left[\sum_{k=1}^{n} \gamma_{ljk} \lambda_k \right]$, $1 \leqslant l \leqslant p$ and $1 \leqslant j \leqslant n$, has rank strictly less than p. So we are done.

(ii) Fix a subspace $\mathfrak{w}$ such that $\mathfrak{g}$ is the direct sum of $\mathfrak{g}^{(g)}$ and $\mathfrak{w}$. Let $h \in \mathfrak{g}^*$ be such that $h|_{\mathfrak{u}} = 0$. If $t \in \Bbbk$, the assumptions on g imply that

$$\mathfrak{g} = \mathfrak{w} \oplus \mathfrak{g}^{(g+th)} \Leftrightarrow \mathfrak{w} \cap \mathfrak{g}^{(g+th)} = \{0\}.$$

By (i), this defines an open subset $\mathcal{K}$ of $\Bbbk$ containing 0. Let $(e_1, \ldots, e_n)$ be a basis of $\mathfrak{g}$ such that $(e_1, \cdots, e_p)$ is a basis of $\mathfrak{w}$. For $x = \lambda_1 e_1 + \cdots + \lambda_n e_n \in \mathfrak{g}$ and $t \in \mathcal{K}$, write $x = x_1(t) + x_2(t)$ where $x_1(t) = \mu_1 e_1 + \cdots + \mu_p e_p \in \mathfrak{w}$ and $x_2(t) \in \mathfrak{g}^{(g+th)}$. The coefficients μ_i are obtained via the following equalities:

$$\sum_{i=1}^{p} \mu_i (g + th)([e_i, e_j]) = \sum_{k=1}^{n} \lambda_k (g + th)([e_k, e_j]) , \quad 1 \leqslant j \leqslant n.$$

The set $\mathcal{K}'$ of $t \in \mathcal{K}$ such that the matrix $\left[(g + th)([e_i, e_j]) \right]$, $1 \leqslant i \leqslant p$, $1 \leqslant j \leqslant n$, has rank p, is open and it contains 0. It follows that the maps $x \mapsto x_1(t)$ and $x \mapsto x_2(t)$ are rational on $\mathcal{K}'$.

Let $x, y \in \mathfrak{g}^{(g)}$, and $x_2(t), y_2(t)$ their projections in $\mathfrak{g}^{(g+th)}$ with respect to the direct sum $\mathfrak{g}^{(g+th)} \oplus \mathfrak{w}$. Then for $t \in \mathcal{K}'$, we have:

$$(g + th)([x_2(t), y_2(t)]) = 0.$$

Taking the derivatives on both sides at $t = 0$, we obtain:

$$h([x, y]) + g([x, y_2'(0)]) + g([x_2'(0), y]) = 0.$$

Hence $h([x, y]) = 0$. This is true for all h verifying $h|_{\mathfrak{u}} = 0$, so we have $[x, y] \in \mathfrak{u}$. $\square$

19.7.5 Proposition. (i) *The set $\mathfrak{g}_{\mathrm{reg}}^*$ is open and dense in $\mathfrak{g}^*$.*

(ii) *If $f \in \mathfrak{g}_{\mathrm{reg}}^*$, then the Lie algebra $\mathfrak{g}^{(f)}$ is abelian.*

(iii) *For all $f \in \mathfrak{g}^*$, $\mathfrak{g}^{(f)}$ contains an abelian Lie subalgebra of dimension $\chi(\mathfrak{g})$.*

Proof. (i) We have $f \notin \mathfrak{g}_{\text{reg}}^*$ if and only if $\mathfrak{g}^{(f)} \cap \mathfrak{u} \neq \{0\}$ for all subspaces of dimension $\dim \mathfrak{g} - \chi(\mathfrak{g})$. So the result follows from 19.7.4 (i).

(ii) Just take $\mathfrak{u} = \{0\}$ in 19.7.4 (ii).

(iii) Let $\chi(\mathfrak{g}) = d$, $p : \mathrm{Gr}(\mathfrak{g}, d) \times \mathfrak{g}^* \to \mathfrak{g}^*$ be the canonical surjection and $\mathcal{F}$ the set of elements $(\mathfrak{u}, f) \in \mathrm{Gr}(\mathfrak{g}, d) \times \mathfrak{g}^*$ such that $f([\mathfrak{g}, \mathfrak{u}]) = \{0\}$ and $[\mathfrak{u}, \mathfrak{u}] = \{0\}$. Since p is a closed map (13.4.5 and 13.6.3), it follows from 19.7.2 (iii) that $p(\mathcal{F})$ is closed. But (ii) says that $\mathfrak{g}_{\text{reg}}^* \subset p(\mathcal{F})$. So part (i) implies that $p(\mathcal{F}) = \mathfrak{g}^*$ and we are done. $\square$

19.7.6 Fix an integer $d \geqslant 0$ and let $\mathfrak{g}_d^* = \{f \in \mathfrak{g}^*; \dim \mathfrak{g}^{(f)} = d\}$. The set $\mathfrak{g}_d^*$ is locally closed in $\mathfrak{g}^*$, so it is a subvariety of $\mathfrak{g}^*$. Let $\varphi : \mathfrak{g}_d^* \to \mathrm{Gr}(\mathfrak{g}, d)$ be the map defined by $f \mapsto \mathfrak{g}^{(f)}$.

Proposition. *The map φ is a morphism of varieties.*

Proof. Let $h \in \mathfrak{g}_d^*$, $\mathcal{E} = (e_1, \ldots, e_n)$ be a basis of $\mathfrak{g}$ such that $(e_1, \ldots, e_d)$ is a basis of $\mathfrak{g}^{(h)}$ and $(e_1^*, \ldots, e_n^*)$ the basis of $\mathfrak{g}^*$ dual to $\mathcal{E}$. Denote by V the subspace spanned by $e_{d+1}, \ldots, e_n$, $\mathfrak{u} = \{f \in \mathfrak{g}_d^*; V \cap \mathfrak{g}^{(f)} = \{0\}\}$, and $U = \{\mathfrak{p} \in \mathrm{Gr}(\mathfrak{g}, d); \mathfrak{p} \cap V = \{0\}\}$. By 19.7.1 and 19.7.4, $\mathfrak{u}$ and U are open subsets of $\mathfrak{g}_d^*$ and $\mathrm{Gr}(\mathfrak{g}, d)$. Note that $\varphi^{-1}(U) = \mathfrak{u}$.

Let $f = \lambda_1 e_1^* + \cdots + \lambda_n e_n^* \in \mathfrak{u}$. For $1 \leqslant i \leqslant n$ and $d + 1 \leqslant j \leqslant n$, set $\alpha_{i,j} = f([e_i, e_j])$. Since $\mathfrak{g}^{(f)} \cap V = \{0\}$, the condition

$$0 = f([e_i, \nu_{d+1} e_{d+1} + \cdots + \nu_n e_n]) = \nu_{d+1} \alpha_{i,d+1} + \cdots + \nu_n \alpha_{i,n}, \quad 1 \leqslant i \leqslant n$$

implies that $\nu_{d+1} = \cdots = \nu_n = 0$. So the rank of the matrix $[\alpha_{i,j}]$ is $n - d$.

Now, for $1 \leqslant k \leqslant d$, there is a unique $(\nu_{d+1,k}, \ldots, \nu_{n,k}) \in \mathbb{k}^{n-d}$ such that

$$0 = f([e_i, e_k + \nu_{d+1,k} e_{d+1} + \cdots + \nu_{n,k} e_n]), \quad 1 \leqslant i \leqslant n$$

or equivalently,

$$\nu_{d+1,k} \alpha_{i,d+1} + \cdots + \nu_{n,k} \alpha_{i,n} = (\lambda_1 e_1^* + \cdots + \lambda_n e_n^*)([e_k, e_i]), \quad 1 \leqslant i \leqslant n.$$

It follows from the formulas for solving a Cramer's system that there exists an open subset $\mathfrak{o}$ of $\mathfrak{u}$ such that the maps $f \mapsto \nu_{j,k}$, $d + 1 \leqslant j \leqslant n$, $1 \leqslant k \leqslant d$, are regular on $\mathfrak{o}$. So the result follows from the remarks in 19.7.1. $\square$

19.8 Cartan subalgebras

19.8.1 For $x \in \mathfrak{g}$, $\alpha \in \mathbb{k}$, we shall denote by $\mathfrak{g}_\alpha(x)$ and $\mathfrak{g}^\alpha(x)$ the subspaces $\mathfrak{g}_\alpha(\mathrm{ad}\, x)$ and $\mathfrak{g}^\alpha(\mathrm{ad}\, x)$ (19.3.8). Let $\mathfrak{h}$ be a Lie subalgebra of $\mathfrak{g}$. Consider $\mathfrak{g}$ as a $\mathfrak{h}$-module via the adjoint representation. If $\lambda \in \mathfrak{h}^* \setminus \{0\}$ is such that $\mathfrak{g}^\lambda(\mathfrak{h}) \neq \{0\}$ (19.3.9), then we say that λ is a *root* of $\mathfrak{g}$ relative to $\mathfrak{h}$. The set of roots, denoted by $R(\mathfrak{g}, \mathfrak{h})$, is called the *root system* of $\mathfrak{g}$ relative to $\mathfrak{h}$.

19.8.2 For $x, y, z \in \mathfrak{g}$, $n \in \mathbb{N}$ and $\lambda, \mu \in \mathbb{k}$, we have:

$$(\operatorname{ad} x - \lambda - \mu)^n([y, z]) = \sum_{i=0}^{n} \binom{n}{i} [(\operatorname{ad} x - \lambda)^i(y), (\operatorname{ad} x - \mu)^{n-i}(z)].$$

We deduce therefore that $[\mathfrak{g}^\lambda(x), \mathfrak{g}^\mu(x)] \subset \mathfrak{g}^{\lambda+\mu}(x)$. In particular, $\mathfrak{g}^0(x)$ is a Lie subalgebra of $\mathfrak{g}$. Recall also from 19.3.8 that $\mathfrak{g} = \bigoplus_{\lambda \in \Bbbk} \mathfrak{g}^\lambda(x)$ and so if $y \in \mathfrak{g}^\lambda(x)$ with $\lambda \neq 0$, then $\operatorname{ad} y$ is nilpotent.

19.8.3 Definition. *A nilpotent Lie subalgebra $\mathfrak{h}$ of $\mathfrak{g}$ verifying $\mathfrak{h} = \mathfrak{n}_\mathfrak{g}(\mathfrak{h})$ is called a Cartan subalgebra of $\mathfrak{g}$.*

19.8.4 Lemma. *A Cartan subalgebra $\mathfrak{h}$ of $\mathfrak{g}$ is a maximal nilpotent Lie subalgebra.*

Proof. If $\mathfrak{t}$ is a nilpotent Lie subalgebra containing $\mathfrak{h}$, then by applying 19.3.6 to the adjoint representation of $\mathfrak{h}$ on $\mathfrak{t}/\mathfrak{h}$, we have $\mathfrak{h} = \mathfrak{t}$ since $\mathfrak{h} = \mathfrak{n}_\mathfrak{g}(\mathfrak{h})$. $\square$

19.8.5 Lemma. *Let $x \in \mathfrak{g}$ and $\mathfrak{h}$ a Lie subalgebra of $\mathfrak{g}$.*
  (i) *If $\mathfrak{g}^0(x) \subset \mathfrak{h}$, then $\mathfrak{h} = \mathfrak{n}_\mathfrak{g}(\mathfrak{h})$.*
  (ii) *Suppose that there exists an element $x \in \mathfrak{h}$ such that $\mathfrak{h} \subset \mathfrak{g}^0(x)$ and $\dim \mathfrak{g}^0(x) \leqslant \dim \mathfrak{g}^0(y)$ for all $y \in \mathfrak{h}$. Then $\mathfrak{g}^0(x) \subset \mathfrak{g}^0(y)$ for all $y \in \mathfrak{h}$.*
  (iii) *If $x \in \mathfrak{g}_{\mathrm{gen}}$, then $\mathfrak{g}^0(x)$ is the unique Cartan subalgebra of $\mathfrak{g}$ containing x. In particular, $\mathfrak{g}$ has a Cartan subalgebra.*
  (iv) *Let $\mathfrak{t}$ be a Cartan subalgebra of $\mathfrak{g}$. There exists an element $x \in \mathfrak{g}$ such that $\mathfrak{t} = \mathfrak{g}^0(x)$.*

Proof. (i) Since $\mathfrak{g}^0(x) \subset \mathfrak{h}$, $\operatorname{ad} x$ induces a bijective endomorphism u of $\mathfrak{n}_\mathfrak{g}(\mathfrak{h})/\mathfrak{h}$. But $x \in \mathfrak{h}$ implies that $[x, \mathfrak{n}_\mathfrak{g}(\mathfrak{h})] \subset \mathfrak{h}$, so $u = 0$ and $\mathfrak{h} = \mathfrak{n}_\mathfrak{g}(\mathfrak{h})$.
  (ii) Let $\mathfrak{t} = \mathfrak{g}^0(x)$ and $y \in \mathfrak{h}$. For $\lambda \in \Bbbk$, denote by u_λ and v_λ the endomorphisms of $\mathfrak{t}$ and $\mathfrak{g}/\mathfrak{t}$ induced by $\operatorname{ad}(x + \lambda y)$. Let χ_λ, χ'_λ and χ''_λ be respectively the characteristic polynomial of $\operatorname{ad}(x + \lambda y)$, u_λ and v_λ. Then $\chi_\lambda = \chi'_\lambda \chi''_\lambda$, and if $n = \dim \mathfrak{g}$ and $r = \dim \mathfrak{t}$, then:

$$\chi'_\lambda(T) = T^r + \sum_{i=0}^{r-1} a_i(\lambda) T^i \ , \ \ \chi''_\lambda(T) = T^{n-r} + \sum_{j=0}^{n-r-1} b_j(\lambda) T^j,$$

where a_i and b_j are polynomial functions on $\Bbbk$.
  Since $\mathfrak{t} = \mathfrak{g}^0(x)$, $b_0(0) \neq 0$. So the set E of $\lambda \in \Bbbk$ such that $b_0(\lambda) \neq 0$ is open dense in $\Bbbk$; so $\Bbbk \setminus E$ is a finite set. Hence $\mathfrak{g}^0(x + \lambda y) \subset \mathfrak{t}$ if $\lambda \in E$, and by our choice of x, this implies that $\mathfrak{g}^0(x + \lambda y) = \mathfrak{t}$ if $\lambda \in E$. It follows that $\chi'_\lambda(T) = T^r$ if $\lambda \in E$, and thus for all $\lambda \in \Bbbk$. This means that $\mathfrak{t} \subset \mathfrak{g}^0(x + \lambda y)$. So we have our result by replacing y by $y - x$ and by taking $\lambda = 1$.
  (iii) If $x \in \mathfrak{g}_{\mathrm{gen}}$, then the conditions of (ii) are verified for $\mathfrak{h} = \mathfrak{g}^0(x)$. So $\mathfrak{g}^0(x) \subset \mathfrak{g}^0(y)$ for all $y \in \mathfrak{h}$. Hence $\mathfrak{g}^0(x)$ is nilpotent (19.3.7) and $\mathfrak{g}^0(x)$ is a Cartan subalgebra by (i).
  Now if $\mathfrak{t}$ is a Cartan subalgebra containing x, then $\mathfrak{t} \subset \mathfrak{g}^0(x)$ since $\mathfrak{t}$ is nilpotent. So $\mathfrak{t} = \mathfrak{g}^0(x)$ (19.8.4).
  (iv) For $y \in \mathfrak{t}$, denote by $\sigma(y)$ the endomorphism of $\mathfrak{g}/\mathfrak{t}$ induced by $\operatorname{ad} y$. Since $\mathfrak{t} = \mathfrak{n}_\mathfrak{g}(\mathfrak{t})$, 0 is the only element $v \in \mathfrak{g}/\mathfrak{t}$ such that $\sigma(y)(v) = 0$ for all

$y \in \mathfrak{t}$. But $\mathfrak{t}$ is nilpotent, so by 19.4.6, there exists $x \in \mathfrak{t}$ such that $\sigma(x)$ is bijective. Consequently $\mathfrak{g}^0(x) = \mathfrak{t}$. $\square$

19.8.6 Theorem. *Let $\mathfrak{h}$ be a nilpotent Lie subalgebra of $\mathfrak{g}$. If $\lambda \in \mathfrak{h}^*$, let us denote $\mathfrak{g}^\lambda(\mathfrak{h})$ simply by $\mathfrak{g}^\lambda$.*

(i) *We have $\mathfrak{g} = \bigoplus_{\lambda \in \mathfrak{h}^*} \mathfrak{g}^\lambda$, and $[\mathfrak{g}^\lambda, \mathfrak{g}^\mu] \subset \mathfrak{g}^{\lambda+\mu}$. In particular, $\mathfrak{g}^0$ is a Lie subalgebra of $\mathfrak{g}$ containing $\mathfrak{h}$.*

(ii) *$\mathfrak{h}$ is a Cartan subalgebra if and only if $\mathfrak{h} = \mathfrak{g}^0$.*

(iii) *If $\mathfrak{g}^0$ is nilpotent, then $\mathfrak{g}^0$ is a Cartan subalgebra of $\mathfrak{g}$.*

(iv) *Let $\lambda, \mu \in \mathfrak{h}^*$, then $L(\mathfrak{g}^\lambda, \mathfrak{g}^\mu) = \{0\}$ if $\lambda \neq -\mu$. In particular, $L(\mathfrak{h}, \mathfrak{g}^\lambda) = \{0\}$ if $\lambda \neq 0$.*

(v) *Let $x, y \in \mathfrak{h}$. Then:*

$$L(x, y) = \sum_{\lambda \in \mathfrak{h}^*} (\dim \mathfrak{g}^\lambda)\lambda(x)\lambda(y).$$

Proof. (i) This follows from 19.3.9 and 19.8.2.

(ii) Applying 19.3.6 to the adjoint representation of $\mathfrak{h}$ on $\mathfrak{g}^0/\mathfrak{h}$, we obtain that if $\mathfrak{g}^0 \neq \mathfrak{h}$, then $\mathfrak{n}_\mathfrak{g}(\mathfrak{h}) \neq \mathfrak{h}$. So $\mathfrak{h}$ is not a Cartan subalgebra.

Conversely, since $[\mathfrak{h}, \mathfrak{n}_\mathfrak{g}(\mathfrak{h})] \subset \mathfrak{h}$ and $\mathfrak{h}$ is nilpotent, $\mathfrak{h} \subset \mathfrak{n}_\mathfrak{g}(\mathfrak{h}) \subset \mathfrak{g}^0$. So if $\mathfrak{g}^0 = \mathfrak{h}$, $\mathfrak{h}$ is a Cartan subalgebra.

(iii) If $\mathfrak{t} = \mathfrak{g}^0$ is nilpotent, then $\mathfrak{t} \subset \mathfrak{g}^0(\mathfrak{t})$. Since $\mathfrak{h} \subset \mathfrak{t}$, we deduce that $\mathfrak{g}^0(\mathfrak{t}) \subset \mathfrak{t}$. Hence $\mathfrak{g}^0(\mathfrak{t}) = \mathfrak{t}$ and by (ii), $\mathfrak{t}$ is a Cartan subalgebra.

(iv) Let $\lambda, \mu, \nu \in \mathfrak{h}^*$, $x \in \mathfrak{g}^\lambda$ and $y \in \mathfrak{g}^\mu$. By (i), we have:

$$(\operatorname{ad} x \circ \operatorname{ad} y)(\mathfrak{g}^\nu) \subset \mathfrak{g}^{\lambda+\mu+\nu}.$$

If $\lambda + \mu \neq 0$, then $\operatorname{ad} x \circ \operatorname{ad} y$ is nilpotent, hence $L(x, y) = 0$. Since $\mathfrak{h} \subset \mathfrak{g}^0$, we are done.

(v) Let $x, y \in \mathfrak{h}$. There is a basis of $\mathfrak{g}^\lambda$ such that the matrix of $\operatorname{ad} x \circ \operatorname{ad} y|_{\mathfrak{g}^\lambda}$ with respect to this basis is upper triangular with a unique eigenvalue $\lambda(x)\lambda(y)$ (19.3.9). The result is now clear. $\square$

19.8.7 Corollary. *Assume that the Killing form L on $\mathfrak{g}$ is non-degenerate. Let $\mathfrak{h}$ be a Cartan subalgebra of $\mathfrak{g}$ and $R = R(\mathfrak{g}, \mathfrak{h})$.*

(i) *We have $\mathfrak{g} = \mathfrak{h} \oplus \left(\bigoplus_{\alpha \in R} \mathfrak{g}^\alpha\right)$.*

(ii) *If $\alpha \in R$, then $-\alpha \in R$ and $L|_{\mathfrak{g}^\alpha \times \mathfrak{g}^{-\alpha}}$ is non-degenerate.*

(iii) *The restriction of L to $\mathfrak{h} \times \mathfrak{h}$ is non-degenerate.*

(iv) *The elements of R span the vector space $\mathfrak{h}^*$.*

(v) *The Lie algebra $\mathfrak{h}$ is abelian.*

Proof. Parts (i), (ii) and (iii) are immediate consequences of 19.8.6.

Let $x \in \mathfrak{h}$ be such that $\alpha(x) = 0$ for all $\alpha \in R$. By 19.8.6 (v), $L(x, \mathfrak{h}) = \{0\}$. Now (iii) implies that $x = 0$. So we have proved (iv).

Finally let $x, y \in \mathfrak{h}$. If $\alpha \in R$, then $[x, \mathfrak{g}^\alpha] \subset \mathfrak{g}^\alpha$ and $[y, \mathfrak{g}^\alpha] \subset \mathfrak{g}^\alpha$. Since

$$(\operatorname{ad}[x, y])|_{\mathfrak{g}^\alpha} = [\operatorname{ad} x|_{\mathfrak{g}^\alpha}, \operatorname{ad} y|_{\mathfrak{g}^\alpha}],$$

we see that the trace of $(\mathrm{ad}[x,y])|_{\mathfrak{g}^\alpha}$ is zero. But by 19.3.9, this is equal to $(\dim \mathfrak{g}^\alpha)\alpha([x,y])$, so $\alpha([x,y]) = 0$. Thus $[x,y] = 0$ by (iv). □

References and comments

- [12], [14], [29], [38], [39], [42], [43], [80].

For a proof of Ado's theorem (19.2.1), see [12] or [29].

The notion of a Lie algebra extends to any base field k (although there is a slight ambiguity of antisymmetry when the characteristic of k is 2). Some of the results in this chapter can be extended. It is a good exercise for the reader to work out to what extent the results in this chapter remain valid. For example, Lie's theorem fails in general if k has positive characteristic.

Lie algebras over the real numbers are fundamental in the theory of Lie groups. For more details, the reader may refer to [24], [59], [84] and [87].

Semisimple and reductive Lie algebras

In this chapter, we study the structure of semisimple and reductive Lie algebras. We introduce certain important objects related to such Lie algebras such as Weyl groups, Borel subalgebras and parabolic subalgebras.

Let $\mathfrak{g}$ be a finite-dimensional Lie algebra over $\Bbbk$ and L its Killing form.

20.1 Semisimple Lie algebras

20.1.1 Definition. (i) *We say that $\mathfrak{g}$ is* simple *if* $\dim \mathfrak{g} > 1$ *and the only ideals of $\mathfrak{g}$ are $\{0\}$ and $\mathfrak{g}$.*

(ii) *A Lie algebra $\mathfrak{g}$ is* semisimple *if $\{0\}$ is the only abelian ideal of $\mathfrak{g}$.*

20.1.2 Theorem. *The following conditions are equivalent:*

(i) *$\mathfrak{g}$ is semisimple.*

(ii) *$\operatorname{rad}\mathfrak{g} = \{0\}$.*

(iii) *L is non-degenerate.*

Proof. (i) $\Rightarrow$ (ii) Since $\mathfrak{r} = \operatorname{rad}\mathfrak{g}$ is an ideal, $\mathcal{D}^r(\mathfrak{r})$ are ideals of $\mathfrak{g}$ for all $r \geqslant 0$. Moreover $\mathfrak{r}$ is solvable, so if $\mathfrak{r} \neq \{0\}$, then $\mathcal{D}^r(\mathfrak{r})$ is non-zero and commutative for some r.

(ii) $\Rightarrow$ (iii) The kernel $\mathfrak{a}$ of L is an ideal of $\mathfrak{g}$ and $L|_{\mathfrak{a} \times \mathfrak{a}}$ is the Killing form of $\mathfrak{a}$. So $L|_{\mathfrak{a} \times \mathfrak{a}} = 0$ and $\mathfrak{a}$ is solvable (19.4.8). Hence $\mathfrak{a} \subset \mathfrak{r} = \{0\}$.

(iii) $\Rightarrow$ (i) This follows from 19.5.8. $\square$

20.1.3 Corollary. (i) *$\mathfrak{g}$ is semisimple if and only if the Killing homomorphism of $\mathfrak{g}$ is an isomorphism.*

(ii) *A product of a finite number of semisimple Lie algebras is semisimple.*

(iii) *The Lie algebra $\mathfrak{g}/\operatorname{rad}\mathfrak{g}$ is semisimple.*

20.1.4 Lemma. *Let $\mathfrak{a}$ be an ideal of $\mathfrak{g}$ and $\mathfrak{b}$ the orthogonal of $\mathfrak{a}$ with respect to L. Suppose that $\{0\}$ is the only solvable ideal of $\mathfrak{g}$ contained in $\mathfrak{a}$. Then $\mathfrak{g} = \mathfrak{a} \oplus \mathfrak{b}$ and the Lie algebras $\mathfrak{g}$ and $\mathfrak{a} \times \mathfrak{b}$ are isomorphic.*

Proof. Since $\mathfrak{b}$ is an ideal, so is $\mathfrak{c} = \mathfrak{b} \cap \mathfrak{a}$. The Killing form on $\mathfrak{c}$ is zero, so it is solvable (19.4.8) and we deduce that $\mathfrak{c} = \{0\}$. So the result follows since $\dim \mathfrak{g} = \dim \mathfrak{a} + \dim \mathfrak{b}$.   □

20.1.5 Proposition. *Assume that* $\mathfrak{g}$ *is semisimple.*

(i) *The centre of* $\mathfrak{g}$ *is trivial. In particular, the adjoint representation of* $\mathfrak{g}$ *is faithful.*

(ii) *We have* $\mathfrak{g} = [\mathfrak{g}, \mathfrak{g}]$.

(iii) *All derivations of* $\mathfrak{g}$ *are inner.*

(iv) *For all finite-dimensional representation* (V, σ) *of* $\mathfrak{g}$, $\sigma(\mathfrak{g}) \subset \mathfrak{sl}(V)$. *Moreover, if* σ *is faithful, then the associated bilinear form* β_σ *on* $\mathfrak{g}$ *is non-degenerate.*

Proof. Part (i) is clear and (ii) follows from 19.6.3 and 20.1.2.

For (iii), let $\mathfrak{u} = \operatorname{ad} \mathfrak{g}$ and $\mathfrak{d} = \operatorname{Der} \mathfrak{g}$. We have $\mathfrak{u} \simeq \mathfrak{g}$, and $\mathfrak{u}$ is an ideal of $\mathfrak{d}$. It follows from 20.1.4 that $\mathfrak{d} = \mathfrak{u} \times \mathfrak{v}$ where $\mathfrak{v}$ is the orthogonal of $\mathfrak{u}$ in $\mathfrak{d}$ with respect to the Killing form. Since $[d, \operatorname{ad} x] = \operatorname{ad} d(x)$ for all $d \in \mathfrak{d}$ and $x \in \mathfrak{g}$, it is now clear that $\mathfrak{v} = \{0\}$.

The first part of (iv) is a direct consequence of (ii). Let $\mathfrak{a}$ be the orthogonal of $\mathfrak{g}$ with respect to β_σ. By Cartan's Criterion, $\sigma(\mathfrak{a})$ is solvable and so $\mathfrak{a} = \{0\}$ if σ is faithful.   □

20.1.6 Proposition. *Let* $\mathfrak{a}$ *be an ideal of a semisimple Lie algebra* $\mathfrak{g}$ *and* $\mathfrak{b}$ *the orthogonal of* $\mathfrak{a}$ *in* $\mathfrak{g}$ *with respect to the Killing form* L.

(i) $\mathfrak{a}$ *and* $\mathfrak{g}/\mathfrak{a}$ *are semisimple.*

(ii) *We have* $\mathfrak{g} = \mathfrak{a} \oplus \mathfrak{b}$ *and the Lie algebras* $\mathfrak{g}$ *and* $\mathfrak{a} \times \mathfrak{b}$ *are isomorphic.*

Proof. These are direct consequences of 19.6.3 (iv), 20.1.2 and 20.1.4.   □

20.1.7 Theorem. *The following conditions are equivalent:*

(i) $\mathfrak{g}$ *is semisimple.*

(ii) $\mathfrak{g}$ *is a direct product of simple Lie algebras.*

Proof. A simple Lie algebra is semisimple since its radical is $\{0\}$. So by 20.1.6, the implication (i) $\Rightarrow$ (ii) is a simple induction on $\dim \mathfrak{g}$.

Conversely, if $\mathfrak{g}$ is a direct product of simple Lie algebras, then its Killing form is clearly non-degenerate (since simple Lie algebras are semisimple). So $\mathfrak{g}$ is semisimple by 20.1.2.   □

20.1.8 Proposition. *Suppose that* $\mathfrak{g} = \mathfrak{a}_1 \times \cdots \times \mathfrak{a}_n$ *where the* $\mathfrak{a}_i$*'s are simple Lie algebras. Then any ideal of* $\mathfrak{g}$ *is a direct product of certain* $\mathfrak{a}_i$*'s. In particular, the* $\mathfrak{a}_i$*'s are the minimal ideals of* $\mathfrak{g}$. *We call the* $\mathfrak{a}_i$*'s the minimal components of* $\mathfrak{g}$.

Proof. Let $\mathfrak{a}$ be an ideal of $\mathfrak{g}$. By permuting the indices, we may assume that $\mathfrak{a} \cap \mathfrak{a}_i \neq \{0\}$ if and only if $1 \leqslant i \leqslant p$ for some integer p. Then by the simplicity of the $\mathfrak{a}_i$'s, $\mathfrak{a} \cap \mathfrak{a}_i = \mathfrak{a}_i$ for $1 \leqslant i \leqslant p$ and $[\mathfrak{a}, \mathfrak{a}_i] \subset \mathfrak{a} \cap \mathfrak{a}_i = \{0\}$ for $i > p$. It follows that

$$\mathfrak{a} \supset \mathfrak{a}_1 \times \cdots \times \mathfrak{a}_p \text{ and } [\mathfrak{a}, \mathfrak{a}_{p+1} \times \cdots \times \mathfrak{a}_n] = \{0\}.$$

Thus $\mathfrak{a} \cap (\mathfrak{a}_{p+1} \times \cdots \times \mathfrak{a}_n) = \{0\}$ since it is in the centre of $\mathfrak{a}_{p+1} \times \cdots \times \mathfrak{a}_n$ which is semisimple. As $\mathfrak{a} \supset \mathfrak{a}_1 \times \cdots \times \mathfrak{a}_p$, we have $\mathfrak{a} = \mathfrak{a}_1 \times \cdots \times \mathfrak{a}_p$. $\square$

20.2 Examples

20.2.1 Lemma. *Let $\mathfrak{g}$ be a Lie subalgebra of $\mathfrak{gl}(V)$ where V is a finite-dimensional vector space. Assume that $\mathfrak{z}(\mathfrak{g}) = \{0\}$ and that V is a simple $\mathfrak{g}$-module. Then $\mathfrak{g}$ is semisimple.*

Proof. By 19.5.6, an abelian ideal $\mathfrak{a}$ in $\mathfrak{g}$ verifies $[\mathfrak{a}, \mathfrak{g}] \subset \mathfrak{a} \cap \mathcal{D}(\mathfrak{g}) = \{0\}$. So $\mathfrak{a} \subset \mathfrak{z}(\mathfrak{g}) = \{0\}$. $\square$

20.2.2 Theorem. *Let V be a finite-dimensional vector space and $\mathfrak{a}$ the set of scalar endomorphisms of V.*
   (i) *We have $\mathfrak{z}(\mathfrak{gl}(V)) = \mathfrak{a}$ and $\mathfrak{z}(\mathfrak{sl}(V)) = \{0\}$.*
   (ii) *$\mathfrak{gl}(V) \simeq \mathfrak{a} \times \mathfrak{sl}(V)$.*
   (iii) *The Lie algebra $\mathfrak{sl}(V)$ is semisimple.*

Proof. Part (i) is just 10.8.4 and (ii) is clear. Part (ii) implies that any $\mathfrak{sl}(V)$-submodule of V is a $\mathfrak{gl}(V)$-submodule. So (iii) follows from (i) and 20.2.1. $\square$

20.2.3 Theorem. *Let V be a n-dimensional vector space, β a non-degenerate symmetric or alternating bilinear form on V and $\mathfrak{g}$ the Lie subalgebra of $\mathfrak{gl}(V)$ consisting of elements $f \in \mathfrak{gl}(V)$ such that*

$$\beta(f(v), w) + \beta(v, f(w)) = 0$$

for all $v, w \in V$. Then $\mathfrak{g}$ is semisimple except when β is symmetric and $n = 2$.

Proof. For a subspace W of V and we denote by $W^\perp$ its orthogonal with respect to β.
   If $n \leqslant 1$, then $\mathfrak{g} = \{0\}$ which is semisimple.
   If $n = 2$ and β is symmetric, then by choosing a suitable basis of V, we see that $\dim \mathfrak{g} = 1$. Thus $\mathfrak{g}$ is not semisimple.
   If $n = 2$ and β is alternating, then by choosing a suitable basis of V, we see that $\mathfrak{g} \simeq \mathfrak{sl}_2(\Bbbk)$ which is semisimple by 20.2.2.
   Suppose that $n \geqslant 3$. Let W be a $\mathfrak{g}$-submodule of V distinct from V and $\{0\}$ and $u, v \in V$, $w \in W$ be such that $\beta(u, w) = 0$, $\beta(v, w) \neq 0$. Define $f \in \text{End}(V)$ as follows: for $x \in V$,

$$f(x) = \beta(x, u)v - \beta(v, x)u.$$

We verify easily that $f \in \mathfrak{g}$ and since $f(w) = -\beta(v, w)u$, we have $u \in f(W) \subset W$. We conclude that $(\Bbbk w)^{\perp} \subset W$ and therefore $\dim W \geqslant n - 1$. But $V \neq W$, so $W = (\Bbbk w)^{\perp}$. This is true for all $w \in W \setminus \{0\}$. It follows that $W = W^{\perp}$ and $2(n - 1) = \dim W + \dim W^{\perp} = n$. But this is impossible for $n \geqslant 3$. We have therefore proved that V is a simple $\mathfrak{g}$-module.

In view of 20.2.1, we are left to show that $\mathfrak{z}(\mathfrak{g}) = \{0\}$. By Schur's Lemma, $\mathfrak{z}(\mathfrak{g}) \subset \Bbbk \operatorname{id}_V$ and it is clear from the hypotheses on β and $\mathfrak{g}$ that $\mathfrak{g} \cap \Bbbk \operatorname{id}_V = \{0\}$. So we are done. $\quad\square$

20.3 Semisimplicity of representations

20.3.1 Lemma. *If* $\dim \mathfrak{g} = 1$, *then* $\mathfrak{g}$ *has a 2-dimensional representation which is not semisimple.*

Proof. Let $\mathfrak{g} = \Bbbk x$ and (u, v) a basis of a 2-dimensional vector space. Define $\sigma : \mathfrak{g} \to \mathfrak{gl}(V)$ by setting $\sigma(\lambda x)(u) = \lambda v$ and $\sigma(\lambda x)(v) = 0$. Then it is clear that (V, σ) is a representation of $\mathfrak{g}$ in which $\Bbbk v$ is the unique submodule of dimension 1. In particular, V is not semisimple. $\quad\square$

20.3.2 Lemma. *Assume that* $\mathfrak{g}$ *is semisimple. Let* (V, σ) *be a simple finite-dimensional faithful representation of* $\mathfrak{g}$ *and* β *the bilinear form on* $\mathfrak{g}$ *associated to* σ.

(i) β *is non-degenerate.*

(ii) *Suppose that* $\dim \mathfrak{g} = n \neq 0$. *Let* $(x_1, \ldots, x_n)$ *be a basis of* $\mathfrak{g}$ *and* $(y_1, \ldots, y_n)$ *the basis of* $\mathfrak{g}$ *dual to* $(x_1, \ldots, x_n)$ *with respect to* β. *Then:*

$$c = \sigma(x_1)\sigma(y_1) + \cdots + \sigma(x_n)\sigma(y_n) = \frac{n}{\dim V} \operatorname{id}_V .$$

Proof. Since σ is faithful, we may assume that $\sigma(\mathfrak{g}) = \mathfrak{g} \subset \mathfrak{gl}(V)$.

(i) If $\mathfrak{a}$ is the kernel of β, then $\mathfrak{a}$ is a solvable ideal (19.4.8). So $\mathfrak{a} = \{0\}$.

(ii) For $z \in \mathfrak{g}$ and $1 \leqslant i \leqslant n$, we have:

$$[z, x_i] = \sum_{j=1}^{n} \lambda_{ij} x_j \ , \ [z, y_i] = \sum_{j=1}^{n} \mu_{ij} y_j.$$

So $\lambda_{ij} = \beta([z, x_i], y_j) = -\beta(x_i, [z, y_j]) = -\mu_{ji}$. It follows easily that $[z, c] = 0$. Since $\operatorname{tr}(c) = \beta(x_1, y_1) + \cdots + \beta(x_n, y_n) = n$, part (ii) follows from Schur's Lemma. $\quad\square$

20.3.3 Theorem. (Whitehead's First Lemma) *Let* $\mathfrak{g}$ *be a semisimple Lie algebra,* (V, σ) *a finite-dimensional representation of* $\mathfrak{g}$ *and* $f : \mathfrak{g} \to V$ *a linear map. The following conditions are equivalent:*

(i) $f([x, y]) = \sigma(x)f(y) - \sigma(y)f(x)$ *for all* $x, y \in \mathfrak{g}$.

(ii) *There exists* $v \in V$ *such that* $f(x) = \sigma(x)(v)$ *for all* $x \in \mathfrak{g}$.

Proof. The implication (ii) $\Rightarrow$ (i) is straightforward. Let us prove (i) $\Rightarrow$ (ii). Suppose that (i) is verified and we may assume that $\mathfrak{g} \neq \{0\}$.

First, suppose that σ is simple. Let $\ker \sigma = \mathfrak{a}$. By 20.1.7 and 20.1.8, $\mathfrak{g} \simeq \mathfrak{a} \times \mathfrak{b}$ where $\mathfrak{a}$, $\mathfrak{b}$ are ideals which are semisimple. So $\mathfrak{a} = [\mathfrak{a}, \mathfrak{a}]$ and by our hypothesis on f, $f(\mathfrak{a}) = \{0\}$.

Let $(x_1, \ldots, x_n)$ be a basis of $\mathfrak{b}$ and $(y_1, \ldots, y_n)$ the basis of $\mathfrak{b}$ dual to $(x_1, \ldots, x_n)$ with respect to the bilinear form on $\mathfrak{b}$ associated to the restriction of σ to $\mathfrak{b}$. By 20.3.2, $c = \sigma(x_1)\sigma(y_1) + \cdots + \sigma(x_1)\sigma(y_1) = \dfrac{n}{\dim V} \operatorname{id}_V$. Set

$$v = c^{-1}(\sigma(x_1)f(y_1) + \cdots + \sigma(x_n)f(y_n)).$$

Let $z \in \mathfrak{g}$. We have by the proof of 20.3.2 that for $1 \leqslant i \leqslant n$,

$$[z, x_i] = \sum_{j=1}^{n} \lambda_{ij} x_j \ , \ [z, y_i] = \sum_{j=1}^{n} -\lambda_{ji} y_j.$$

Since f satisfies (i), we obtain that

$$\begin{aligned}
c \circ f(z) &= \sum_{i=1}^{n} \sigma(x_i)\sigma(y_i)f(z) = \sum_{i=1}^{n} \sigma(x_i)f([y_i, z]) + \sum_{i=1}^{n} \sigma(x_i)\sigma(z)f(y_i) \\
&= \sum_{i=1}^{n} \sigma(x_i)f([y_i, z]) + \sum_{i=1}^{n} \sigma([x_i, z])f(y_i) + \sum_{i=1}^{n} \sigma(z)\sigma(x_i)f(y_i) \\
&= \sum_{i=1}^{n} \sigma(z)\sigma(x_i)f(y_i) = c \circ \sigma(z)(v).
\end{aligned}$$

So $\sigma(z)(v) = f(z)$ for all $z \in \mathfrak{g}$ as required.

In the general case, we proceed by induction on $\dim V$. From the previous paragraph, we may assume that V contains a submodule W distinct from V and $\{0\}$. Let $\pi : V \to V/W$ be the canonical surjection. Then by the induction hypothesis, there exists $v \in V$ such that $(\pi \circ f)(x) = \pi \circ \sigma(x)(v)$ for all $x \in \mathfrak{g}$. Let $\theta(x) = f(x) - \sigma(x)(v)$. Then $\theta(x) \in W$ and $\theta \in \operatorname{Hom}(\mathfrak{g}, W)$ satisfies (i). By the induction hypothesis, there exists $w \in W$ such that $\theta(x) = \sigma(x)(w)$ for all $x \in \mathfrak{g}$. So $f(x) = \sigma(x)(v + w)$ for all $x \in \mathfrak{g}$. $\square$

20.3.4 Theorem. (Weyl's Theorem) *Let $\mathfrak{g}$ be a Lie algebra. The following conditions are equivalent:*

(i) *All finite-dimensional representations of $\mathfrak{g}$ is semisimple.*

(ii) *$\mathfrak{g}$ is semisimple.*

Proof. We may assume that $\mathfrak{g} \neq \{0\}$.

(i) $\Rightarrow$ (ii) If (i) is verified, then the adjoint representation of $\mathfrak{g}$ is semisimple. If $\mathfrak{a}$ is an abelian ideal of $\mathfrak{g}$, then $\mathfrak{g} = \mathfrak{a} \oplus \mathfrak{b}$ where $\mathfrak{b}$ is an ideal of $\mathfrak{g}$. So $\mathfrak{a} \subset \mathfrak{z}(\mathfrak{g})$. If $\mathfrak{z}(\mathfrak{g}) \neq \{0\}$, then $\mathfrak{g}$ has an ideal of codimension 1. But 20.3.1 would imply that $\mathfrak{g}$ has a non-semisimple representation of dimension 2. Contradiction.

(ii) $\Rightarrow$ (i) Let (V, σ) be a finite-dimensional representation of $\mathfrak{g}$, U a nontrivial submodule of V, $\pi : V \to V/U$ the canonical surjection and $(V/U, \tau)$

the representation of $\mathfrak{g}$ induced by σ. Denote by M the set of linear maps from V/U to V and set $N = \{\phi \in M; \phi(V/U) \subset U\}$. The map $\lambda : \mathfrak{g} \to \mathrm{End}(M)$, defined by setting for $x \in \mathfrak{g}$ and $\phi \in M$:

$$\lambda(x)(\phi) = \sigma(x) \circ \phi - \phi \circ \tau(x).$$

then defines a representation of $\mathfrak{g}$ in which N is a submodule.

Let us fix $\phi_0 \in M$ such that $\pi \circ \phi_0 = \mathrm{id}_{V/U}$. The linear map $f : \mathfrak{g} \to M$, $x \mapsto \lambda(x)(\phi_0)$ verifies $f(\mathfrak{g}) \subset N$ since for any $w \in V/U$,

$$
\begin{aligned}
\pi\left(f(x)(w)\right) &= \pi \circ \sigma(x) \circ \phi_0(w) - \pi \circ \phi_0 \circ \tau(x)(w) \\
&= \tau(x) \circ \pi \circ \phi_0(w) - \tau(x)(w) = \tau(x)(w) - \tau(x)(w) = 0.
\end{aligned}
$$

So by 20.3.3, condition (i) of 20.3.3 is valid for the representation (N, μ) induced by λ. Thus there exists $\psi_0 \in N$ such that $f(x) = \mu(x)(\psi_0)$ for all $x \in \mathfrak{g}$.

Set $\theta_0 = \phi_0 - \psi_0$, then $\lambda(x)(\theta_0) = 0$ for all $x \in \mathfrak{g}$. Hence $\pi \circ \theta_0 = \mathrm{id}_{V/U}$ and $\theta_0 \circ \tau(x) = \sigma(x) \circ \theta_0$ for all $x \in \mathfrak{g}$. It follows that V is the direct sum of the submodules U and $\theta_0(V/U)$. So σ is semisimple. $\square$

20.3.5 Theorem. (Levi-Malcev Theorem) *Let $\mathfrak{g}$ be a Lie algebra and $\mathfrak{r} = \mathrm{rad}\,\mathfrak{g}$. There exists a Lie subalgebra $\mathfrak{s}$ of $\mathfrak{g}$ such that $\mathfrak{g} = \mathfrak{s} \oplus \mathfrak{r}$. Moreover, $\mathfrak{s}$ is semisimple and it is called a* Levi subalgebra *of $\mathfrak{g}$.*

Proof. Observe that such a Lie algebra is isomorphic to $\mathfrak{g}/\mathfrak{r}$, so it is semisimple by 20.1.3. We shall prove by induction on $\dim \mathfrak{r}$ that such a Lie subalgebra exists. The case $\mathfrak{r} = \{0\}$ is obvious, so we may assume that $\mathfrak{r} \neq \{0\}$.

1) Suppose that there is an ideal $\mathfrak{a}$ of $\mathfrak{g}$ contained in $\mathfrak{r}$ such that $\mathfrak{a} \neq \{0\}$ and $\mathfrak{a} \neq \mathfrak{r}$. Then $\mathrm{rad}\,\mathfrak{g}/\mathfrak{a} = \mathfrak{r}/\mathfrak{a}$ and by our induction hypothesis, there is a Lie subalgebra $\mathfrak{b}$ of $\mathfrak{g}$ such that

$$\mathfrak{b} \cap \mathfrak{r} = \mathfrak{a} \ , \ \mathfrak{g} = \mathfrak{b} + \mathfrak{r}.$$

Since $\mathfrak{b}/\mathfrak{a} \simeq \mathfrak{g}/\mathfrak{r}$ is semisimple, $\mathrm{rad}\,\mathfrak{b} \subset \mathfrak{a}$ and so $\mathrm{rad}\,\mathfrak{b} = \mathfrak{a}$. By induction, there is a Lie subalgebra $\mathfrak{s}$ of $\mathfrak{g}$ such that $\mathfrak{b} = \mathfrak{s} \oplus \mathfrak{a}$. It follows from the definition of $\mathfrak{b}$ that $\mathfrak{g} = \mathfrak{s} \oplus \mathfrak{r}$.

2) We are left with the case where $\mathfrak{r}$ is commutative. In this case, $\mathfrak{z}(\mathfrak{g}) = \{0\}$ or $\mathfrak{r}$ since $[\mathfrak{g}, \mathfrak{r}]$ is an ideal of $\mathfrak{g}$ contained in $\mathfrak{r}$.

When $\mathfrak{z}(\mathfrak{g}) = \mathfrak{r}$, the adjoint representation of $\mathfrak{g}$ is semisimple since $\mathrm{ad}\,\mathfrak{g} \simeq \mathfrak{g}/\mathfrak{r}$ is semisimple. Thus there is an ideal $\mathfrak{s}$ of $\mathfrak{g}$ such that $\mathfrak{g} = \mathfrak{s} \oplus \mathfrak{r}$.

When $\mathfrak{z}(\mathfrak{g}) = \{0\}$, $\mathfrak{g}$ is isomorphic to the Lie subalgebra $\mathrm{ad}\,\mathfrak{g}$ in $\mathrm{End}(\mathfrak{g})$. Let $(\mathrm{End}(\mathfrak{g}), \sigma)$ be the representation of $\mathfrak{g}$ defined by $\sigma(x)(u) = [\mathrm{ad}\,x, u]$ for $x \in \mathfrak{g}$ and $u \in \mathrm{End}(\mathfrak{g})$, and $M = \{u \in \mathrm{End}(\mathfrak{g}); u(\mathfrak{g}) \subset \mathfrak{r}$ and $u|_{\mathfrak{r}} = \lambda_u \,\mathrm{id}_{\mathfrak{r}}$ for some $\lambda_u \in \Bbbk\}$. Denote by N the kernel of the linear form $u \mapsto \lambda_u$ on M. Since $\mathfrak{r}$ is abelian, $R = \mathrm{ad}_{\mathfrak{g}} \mathfrak{r} \subset N$. Also, we have $\sigma(\mathfrak{g})(M) \subset N$, and $\sigma(\mathfrak{g})(R) \subset R$.

A straightforward computation shows that for $x \in \mathfrak{r}$ and $u \in M$, we have $\sigma(x)(u) = -\lambda_u \,\mathrm{ad}\,x$. So $\sigma(\mathfrak{r})(M) \subset R$. It follows that the representation

$(M/R, \theta)$ of $\mathfrak{g}$ induced by σ is semisimple since $\theta(\mathfrak{r}) = \{0\}$ (20.3.4). Moreover, $\theta(\mathfrak{g})(M/R) \subset N/R$, so there exists $v \in M$ such that $\lambda_v = -1$ and for all $x \in \mathfrak{g}$, $\sigma(x)(v) \in R$.

Since $\mathfrak{z}(\mathfrak{g}) = \{0\}$, we have a well-defined linear map $\alpha : \mathfrak{g} \to \mathfrak{r}$, given by $\operatorname{ad} \alpha(x) = \sigma(x)(v)$. We have $\alpha|_{\mathfrak{r}} = \operatorname{id}_{\mathfrak{r}}$ because $\sigma(x)(u) = -\lambda_u \operatorname{ad} x$ for $x \in \mathfrak{r}$ and $u \in M$. It follows that $\mathfrak{g} = \mathfrak{s} \oplus \mathfrak{r}$ where $\mathfrak{s} = \ker \alpha = \{x \in \mathfrak{g}; [\operatorname{ad} x, v] = 0\}$ is clearly a Lie subalgebra of $\mathfrak{g}$. $\square$

20.3.6 Corollary. *Let* $\mathfrak{r} = \operatorname{rad} \mathfrak{g}$ *and* $\mathfrak{n}$ *the nilpotent radical of* $\mathfrak{g}$. *Then*

$$\mathfrak{n} = \mathfrak{r} \cap \mathcal{D}(\mathfrak{g}) = [\mathfrak{g}, \mathfrak{r}].$$

Proof. Let $\mathfrak{s}$ be a Levi subalgebra of $\mathfrak{g}$. Then we have $\mathcal{D}(\mathfrak{g}) = [\mathfrak{s}, \mathfrak{s}] + [\mathfrak{s}, \mathfrak{r}] + [\mathfrak{r}, \mathfrak{r}] = \mathfrak{s} + [\mathfrak{s}, \mathfrak{r}] + [\mathfrak{r}, \mathfrak{r}] = \mathfrak{s} \oplus [\mathfrak{g}, \mathfrak{r}]$. So $\mathfrak{n} = \mathfrak{r} \cap \mathcal{D}(\mathfrak{g}) = [\mathfrak{g}, \mathfrak{r}]$ (19.6.2). $\square$

20.4 Semisimple and nilpotent elements

20.4.1 Lemma. *Let* V *be a finite-dimensional vector space and* x *an endomorphism of* V.

(i) *If* x *is nilpotent (resp. semisimple), then so is* $\operatorname{ad} x$.

(ii) *Let* $x = x_s + x_n$ *be the Jordan decomposition of* x. *Then* $\operatorname{ad} x_s + \operatorname{ad} x_n$ *is the Jordan decomposition of* $\operatorname{ad} x$.

Proof. Part (ii) is clear from (i). In view of 19.3.5, we only need to prove (i) when x is semisimple. Let $(v_1, \ldots, v_n)$ be a basis of eigenvectors of V for x, then it is easy to check that the endomorphisms $z_{ij} \in \operatorname{End}(V)$ defined by $z_{ij}(v_i) = v_j$, $z_{ij}(v_k) = 0$ if $k \neq i$, form a basis of eigenvectors of $\operatorname{End}(V)$ for $\operatorname{ad} x$. $\square$

20.4.2 Proposition. *Let* V *be a finite-dimensional vector space and* $\mathfrak{g}$ *a semisimple Lie subalgebra of* $\mathfrak{gl}(V)$. *Then* $\mathfrak{g}$ *contains the semisimple and nilpotent components of its elements.*

Proof. Let $\mathcal{V}$ be the set of $\mathfrak{g}$-submodules of V, and for $W \in \mathcal{V}$, we define $\mathfrak{g}_W = \{x \in \mathfrak{gl}(V); x(W) \subset W, \operatorname{tr}(x|_W) = 0\}$. Since $\mathfrak{g} = [\mathfrak{g}, \mathfrak{g}]$, $\mathfrak{g} \subset \mathfrak{g}_W$. Let

$$\mathfrak{t} = \mathfrak{n}_{\mathfrak{gl}(V)}(\mathfrak{g}) \cap \left(\bigcap_{W \in \mathcal{V}} \mathfrak{g}_W \right).$$

Let $x \in \mathfrak{t}$, $X = \operatorname{ad}_{\mathfrak{gl}(V)} x$, $S = \operatorname{ad}_{\mathfrak{gl}(V)} x_s$ and $N = \operatorname{ad}_{\mathfrak{gl}(V)} x_n$. Then the Jordan decomposition of X is $S + N$ (20.4.1). So $x_s, x_n \in \mathfrak{t}$ because S, N (resp. x_s, x_n) are polynomials in X (resp. x) with no constant term. We verify easily that $\mathfrak{t}$ is a Lie subalgebra of $\mathfrak{gl}(V)$ and $\mathfrak{g}$ is an ideal of $\mathfrak{t}$. So $\mathfrak{t} = \mathfrak{g} \times \mathfrak{a}$ for some ideal $\mathfrak{a}$ of $\mathfrak{t}$ (20.1.4). Moreover $[\mathfrak{a}, \mathfrak{g}] \subset \mathfrak{a} \cap \mathfrak{g} = \{0\}$. Take W to be a minimal element of $\mathcal{V} \setminus \{0\}$. Then W is simple. For $a \in \mathfrak{a}$, $\operatorname{tr}(a|_W) = 0$, so Schur's Lemma says that $a|_W = 0$. Since V is the sum of minimal elements of $\mathcal{V} \setminus \{0\}$, we deduce that $\mathfrak{a} = \{0\}$. Hence $\mathfrak{t} = \mathfrak{g}$ and $x_s, x_n \in \mathfrak{g}$. $\square$

20.4.3 Corollary. *With the hypotheses of 20.4.2, an element $x \in \mathfrak{g}$ is semisimple (resp. nilpotent) if and only if $\mathrm{ad}_\mathfrak{g}\, x$ is semisimple (resp. nilpotent).*

Proof. Since $(\mathfrak{g}, \mathrm{ad})$ is a $\mathfrak{g}$-submodule of $(\mathfrak{g}, \mathrm{ad}_{\mathfrak{gl}(V)})$, 20.4.1 says that if x is semisimple (resp. nilpotent), then so is $\mathrm{ad}_\mathfrak{g}\, x$. Conversely, if $\mathrm{ad}_\mathfrak{g}\, x$ is semisimple (resp. nilpotent), then since the adjoint representation is faithful, it follows by 20.4.2 that $x = x_s$ (resp. $x = x_n$). $\square$

20.4.4 Theorem. *Let $\mathfrak{g}$ be semisimple and $x \in \mathfrak{g}$. Then the following conditions are equivalent:*

   (i) $\mathrm{ad}_\mathfrak{g}\, x$ *is semisimple (resp. nilpotent).*
   (ii) *There is a finite-dimensional faithful representation (V, σ) of $\mathfrak{g}$ such that $\sigma(x)$ is semisimple (resp. nilpotent).*
   (iii) *For all finite-dimensional representation (V, σ) of $\mathfrak{g}$, $\sigma(x)$ is semisimple (resp. nilpotent).*

Proof. We have (iii) $\Rightarrow$ (ii) since the adjoint representation is faithful. Also (ii) $\Rightarrow$ (i) follows from 20.4.3. Let us prove (i) $\Rightarrow$ (iii).

Let (V, σ) be a finite-dimensional representation of $\mathfrak{g}$, $\mathfrak{h} = \mathfrak{g}/\ker \sigma$ and (V, π) the representation of $\mathfrak{h}$ obtained from σ. Let y be the image of x in $\mathfrak{h}$. If $\mathrm{ad}_\mathfrak{g}\, x$ is semisimple (resp. nilpotent), then so is $\mathrm{ad}_\mathfrak{h}\, y$. Thus $\pi(y) = \sigma(x)$ is semisimple (resp. nilpotent) by 20.4.3. $\square$

20.4.5 Let $\mathfrak{g}$ be semisimple. An element $x \in \mathfrak{g}$ is called *semisimple* (resp. *nilpotent*) if it satisfies the conditions of 20.4.4.

In view of 20.4.2, an element $x \in \mathfrak{g}$ can be written uniquely in the form $s + n$ where s is semisimple, n is nilpotent and $[s, n] = 0$. We say that s (resp. n) is the *semisimple component* (resp. *nilpotent component*) of x. By 20.4.2, we have the following result:

Corollary. *Let $\mathfrak{h}$ be a semisimple Lie subalgebra of a semisimple Lie algebra $\mathfrak{g}$, $x \in \mathfrak{h}$ and s, n the semisimple and nilpotent components of x in $\mathfrak{g}$. Then $s, n \in \mathfrak{h}$.*

20.4.6 Proposition. *Let $\mathfrak{g}$ be semisimple. Then $\mathfrak{g}$ is generated, as a Lie algebra, by its nilpotent elements.*

Proof. Let $\mathfrak{h}$ be a Cartan subalgebra of $\mathfrak{g}$ and R the root system of $\mathfrak{g}$ relative to $\mathfrak{h}$. By 19.8.6 and 19.8.7, $\mathfrak{h}$ is abelian and $[\mathfrak{h}, \mathfrak{g}^\alpha] \subset \mathfrak{g}^\alpha$ for all $\alpha \in R$. Since $\mathfrak{g} = \mathfrak{h} \oplus (\bigoplus_{\alpha \in R} \mathfrak{g}^\alpha)$, we have:

$$\mathfrak{g} = [\mathfrak{g}, \mathfrak{g}] \subset \sum_{\alpha \in R} \mathfrak{g}^\alpha + \sum_{\alpha, \beta \in R} [\mathfrak{g}^\alpha, \mathfrak{g}^\beta].$$

But all the elements of $\mathfrak{g}^\alpha$ are nilpotent (19.8.2). So the result follows. $\square$

20.5 Reductive Lie algebras

20.5.1 Definition. (i) *A Lie algebra* $\mathfrak{g}$ *is said to be* reductive *if its adjoint representation is semisimple.*

(ii) *A Lie subalgebra* $\mathfrak{h}$ *of* $\mathfrak{g}$ *is* reductive *in* $\mathfrak{g}$ *if the adjoint representation of* $\mathfrak{h}$ *in* $\mathfrak{g}$ *is semisimple.*

20.5.2 Remark. If $\mathfrak{h}$ is reductive in $\mathfrak{g}$, then $\mathfrak{h}$ is reductive since $\mathfrak{h}$ is a $\mathfrak{h}$-submodule of $\mathfrak{g}$.

20.5.3 Lemma. *An abelian Lie algebra* $\mathfrak{h}$ *has a finite-dimensional representation such that the associated bilinear form is non-degenerate.*

Proof. Let $n = \dim \mathfrak{h}$, then the representation on $\Bbbk^n$ defined by identifying $\mathfrak{h}$ with the set of diagonal matrices works. $\square$

20.5.4 Proposition. *Let* $\mathfrak{r} = \operatorname{rad} \mathfrak{g}$ *and* $\mathfrak{n}$ *the nilpotent radical of* $\mathfrak{g}$. *The following conditions are equivalent:*

(i) $\mathfrak{g}$ *is reductive.*

(ii) $\mathfrak{g} = \mathfrak{s} \times \mathfrak{a}$ *where* $\mathfrak{s}$ *is semisimple and* $\mathfrak{a}$ *is abelian.*

(iii) $\mathfrak{g}$ *has a finite-dimensional representation such that the associated bilinear form is non-degenerate.*

(iv) $\mathfrak{g}$ *has a faithful semisimple representation of finite dimension.*

(v) $\mathfrak{n} = \{0\}$.

(vi) $\mathfrak{r}$ *is the centre of* $\mathfrak{g}$.

Proof. (i) $\Rightarrow$ (ii) If $\mathfrak{g}$ is reductive, then $\mathfrak{g}$ is the direct sum of its minimal (non-trivial) ideals $\mathfrak{a}_1, \ldots, \mathfrak{a}_n$. It follows that $\mathfrak{a}_i$ is simple or $\dim \mathfrak{a}_i = 1$ for all i. So we have (ii).

(ii) $\Rightarrow$ (iii) By 20.5.3, $\mathfrak{a}$ has a finite-dimensional representation (V, σ) such that the associated bilinear form on $\mathfrak{a}$ is non-degenerate. Also the bilinear form on $\mathfrak{s}$ associated to the adjoint representation $(\mathfrak{s}, \operatorname{ad}_\mathfrak{s})$ of $\mathfrak{s}$ is non-degenerate. So the bilinear form on $\mathfrak{g}$ associated to the representation $(\mathfrak{s} \oplus V, \operatorname{ad}_\mathfrak{s} \oplus \sigma)$ of $\mathfrak{g}$ is non-degenerate.

(iii) $\Rightarrow$ (iv) In the notations of 19.5.5 (ii), take the direct sum of the σ_i's.

(iv) $\Rightarrow$ (v) This is clear by the definition of $\mathfrak{n}$ and 19.5.1.

(v) $\Rightarrow$ (vi) Since $\mathfrak{z}(\mathfrak{g}) \subset \mathfrak{r}$ and $\mathfrak{n} = [\mathfrak{g}, \mathfrak{r}]$ (20.3.6), the result is clear.

(vi) $\Rightarrow$ (i) If $\mathfrak{r} = \mathfrak{z}(\mathfrak{g})$, then $\operatorname{ad} \mathfrak{g}$ is a semisimple Lie subalgebra of $\mathfrak{gl}(\mathfrak{g})$ (20.3.5). So the adjoint representation is semisimple (20.3.4). $\square$

20.5.5 Corollary. *A Lie algebra* $\mathfrak{g}$ *is reductive if and only if* $\mathcal{D}(\mathfrak{g})$ *is semisimple. In particular, if* $\mathfrak{g}$ *is reductive, then* $\mathfrak{g} \simeq \mathcal{D}(\mathfrak{g}) \times \mathfrak{z}(\mathfrak{g})$.

Proof. If $\mathcal{D}(\mathfrak{g})$ is semisimple, then $\mathfrak{g} = \mathcal{D}(\mathfrak{g}) \times \mathfrak{a}$ for some ideal $\mathfrak{a}$ of $\mathfrak{g}$ (20.1.4). Since $[\mathfrak{a}, \mathfrak{g}] \subset \mathfrak{a} \cap \mathcal{D}(\mathfrak{g}) = \{0\}$, $\mathfrak{a} = \mathfrak{z}(\mathfrak{g})$. So the result follows from 20.5.4. $\square$

Remark. By 20.2.2, $\mathfrak{gl}(V)$ is a reductive Lie algebra.

20.5.6 Let (V, σ) and (W, π) be representations of $\mathfrak{g}$. We define the representation $(V \otimes_{\Bbbk} W, \sigma \otimes \pi)$ of $\mathfrak{g}$ by setting for $x \in \mathfrak{g}$:

$$(\sigma \otimes \pi)(x) = \sigma(x) \otimes \mathrm{id}_W + \mathrm{id}_V \otimes \pi(x).$$

Corollary. *If σ and π are semisimple of finite dimension, then $\sigma \otimes \pi$ is semisimple.*

Proof. We may assume that σ and π are simple and since the nilpotent radical $\mathfrak{n}$ of $\mathfrak{g}$ is contained in $\ker \sigma \cap \ker \pi$ (19.5.1), replacing $\mathfrak{g}$ by $\mathfrak{g}/\mathfrak{n}$, we may assume further that $\mathfrak{n} = \{0\}$, so $\mathfrak{g}$ is reductive (20.5.4). Let $\mathfrak{g} = \mathfrak{s} \times \mathfrak{a}$ be as in 20.5.4 (ii). By Schur's Lemma, $\sigma(\mathfrak{a}) \subset \Bbbk \, \mathrm{id}_V$ and $\pi(\mathfrak{a}) \subset \Bbbk \, \mathrm{id}_W$. So $\sigma \otimes \pi(\mathfrak{a}) \subset \Bbbk \, \mathrm{id}_V \otimes \mathrm{id}_W$. Since $\sigma \otimes \pi|_{\mathfrak{s}}$ is semisimple (20.3.4), the result follows. $\square$

20.5.7 Proposition. *Assume that $\mathfrak{g}$ is reductive. Let $x \in \mathcal{D}(\mathfrak{g})$, $y \in \mathfrak{z}(\mathfrak{g})$ and $z = x + y$. The following conditions are equivalent:*

(i) *x is semisimple in $\mathcal{D}(\mathfrak{g})$.*

(ii) *There exists a faithful semisimple finite-dimensional representation (V, σ) of $\mathfrak{g}$ such that $\sigma(z)$ is semisimple.*

(iii) *For all semisimple finite-dimensional representation (V, σ) of $\mathfrak{g}$, $\sigma(z)$ is semisimple.*

Proof. By Schur's Lemma, $\sigma(y)$ is a scalar multiple of id_V for any simple finite-dimensional representation (V, σ) of $\mathfrak{g}$. In particular, $\sigma(y)$ is semisimple. Since $y \in \mathfrak{z}(\mathfrak{g})$, we may assume that $y = 0$. Thus we have (i) $\Rightarrow$ (iii) by 20.4.4, (iii) $\Rightarrow$ (ii) by 20.5.4 (iv), and finally (ii) $\Rightarrow$ (i) by 20.4.4. $\square$

20.5.8 Proposition. *With the notations of 20.5.7, the following conditions are equivalent:*

(i) *x is nilpotent in $\mathcal{D}(\mathfrak{g})$ and $y = 0$.*

(ii) *There exists a faithful semisimple finite-dimensional representation (V, σ) of $\mathfrak{g}$ such that $\sigma(z)$ is nilpotent.*

(iii) *For all semisimple finite-dimensional representation (V, σ) of $\mathfrak{g}$, $\sigma(z)$ is nilpotent.*

Proof. We have (i) $\Rightarrow$ (iii) by 20.4.4 and (iii) $\Rightarrow$ (ii) by 20.5.4 (iv). So let us prove (ii) $\Rightarrow$ (i). Let (V, σ) be as in (ii).

Let s, n be the semisimple and nilpotent components of x in $\mathcal{D}(\mathfrak{g})$. By 20.4.4 and 20.5.7, $\sigma(n)$ is nilpotent and $\sigma(s + y)$ is semisimple. Furthermore, they commute and their sum $\sigma(z)$ is nilpotent. So $\sigma(s + y) = 0$. Since σ is faithful and $\mathcal{D}(\mathfrak{g}) \cap \mathfrak{z}(\mathfrak{g}) = \{0\}$, we deduce that $y = s = 0$, and hence $x = n$ is nilpotent. $\square$

20.5.9 Assume that $\mathfrak{g}$ is reductive. An element $z \in \mathfrak{g}$ is called *semisimple* (resp. *nilpotent*) if it verifies the conditions of 20.5.7 (resp. 20.5.8). We define as in 20.4.5 the semisimple and nilpotent components of z. We shall denote by $\mathcal{S}_{\mathfrak{g}}$ (resp. $\mathcal{N}_{\mathfrak{g}}$) the set of semisimple (resp. nilpotent) elements of $\mathfrak{g}$.

20.5.10 Theorem. *Let* $\mathfrak{r} = \operatorname{rad} \mathfrak{g}$, (V, σ) *a finite-dimensional representation of* $\mathfrak{g}$, $\mathfrak{g}' = \sigma(\mathfrak{g})$ *and* $\mathfrak{r}' = \sigma(\mathfrak{r})$. *The following conditions are equivalent:*
  (i) σ *is semisimple.*
  (ii) $\mathfrak{g}'$ *is reductive and the elements of its centre are semisimple.*
  (iii) *The elements of* $\mathfrak{r}'$ *are semisimple.*
  (iv) *The restriction of* σ *to* $\mathfrak{r}$ *is semisimple.*

Proof. (i) $\Rightarrow$ (ii) This is clear from 20.5.4 (iv) and 20.5.7.
  (ii) $\Rightarrow$ (iii) This is obvious.
  (iii) $\Rightarrow$ (iv) If (iii) is verified, then by 20.3.6, $[\mathfrak{g}', \mathfrak{r}'] = \{0\}$. So $\mathfrak{r}'$ is abelian, and since its elements are semisimple, (iv) follows.
  (iv) $\Rightarrow$ (i) Using the notations of 19.3.9, we have $V = \bigoplus_{\lambda \in \mathfrak{r}^*} V_\lambda(\mathfrak{r})$ by 19.4.5 (ii). Since $[\mathfrak{g}', \mathfrak{r}'] = \{0\}$ (20.3.6), each $V_\lambda(\mathfrak{r})$ is a $\mathfrak{g}$-module. So (i) follows from 20.3.4. $\square$

20.5.11 Lemma. *We have* $V = V^{\mathfrak{g}} \oplus \sigma(\mathfrak{g})(V)$ *for all semisimple representation* (V, σ) *of* $\mathfrak{g}$.

Proof. Clearly, we may assume that V is simple. Since $\sigma(\mathfrak{g})(V)$ is a submodule of V, the result follows. $\square$

20.5.12 Lemma. *Let* $\mathfrak{t}$ *be a Lie subalgebra of a semisimple Lie algebra* $\mathfrak{g}$ *verifying:*
  a) $L|_{\mathfrak{t} \times \mathfrak{t}}$ *is non-degenerate.*
  b) *If* $x \in \mathfrak{t}$, *then its semisimple and nilpotent components are in* $\mathfrak{t}$.
  Then $\mathfrak{t}$ *is reductive in* $\mathfrak{g}$.

Proof. Condition a) and 20.5.4 (iii) imply that $\mathfrak{t}$ is reductive. Let $x \in \mathfrak{z}(\mathfrak{t})$ and s, n the semisimple and nilpotent components of x. Then condition b) and 10.1.2 imply that $s, n \in \mathfrak{z}(\mathfrak{t})$. It follows that for all $y \in \mathfrak{t}$, $(\operatorname{ad} n) \circ (\operatorname{ad} y)$ is nilpotent (since $[y, n] = 0$) and hence $L(n, y) = 0$. We deduce from condition a) that $n = 0$ and so all the elements of $\mathfrak{z}(\mathfrak{t})$ are semisimple. The result now follows from 20.5.10 (ii). $\square$

20.5.13 Proposition. *Assume that* $\mathfrak{g}$ *is semisimple. Let* $\mathfrak{t}$ *be a Lie subalgebra of* $\mathfrak{g}$ *which is reductive in* $\mathfrak{g}$, *and* $\mathfrak{c} = \mathfrak{c}_{\mathfrak{g}}(\mathfrak{t})$. *Then:*
  (i) $\mathfrak{c}$ *verifies the conditions of 20.5.12, and so it is reductive in* $\mathfrak{g}$.
  (ii) $\mathfrak{g} = \mathfrak{c} \oplus [\mathfrak{t}, \mathfrak{g}]$ *and* $[\mathfrak{t}, \mathfrak{g}]$ *is the orthogonal of* $\mathfrak{c}$ *with respect to* L.

Proof. By 10.1.2 and 20.4.5, $\mathfrak{c}$ contains the semisimple and nilpotent components of its elements. For $(x, y, z) \in \mathfrak{t} \times \mathfrak{c} \times \mathfrak{g}$, we have:

$$L([x, z], y) = L(z, [y, x]) = L(z, 0) = 0.$$

Thus $\mathfrak{c}$ and $[\mathfrak{t}, \mathfrak{g}]$ are orthogonal. So by 20.5.11, we have (ii). Finally since L is non-degenerate, it is non-degenerate on $\mathfrak{c} \times \mathfrak{c}$. So (i) follows.   □

20.5.14 Proposition. *Assume that $\mathfrak{g}$ is semisimple. Let $(s, n) \in \mathcal{S}_\mathfrak{g} \times \mathcal{N}_\mathfrak{g}$ be such that $[s, n] = 0$. Then $n \in [\mathfrak{g}^s, \mathfrak{g}^s]$.*

Proof. By 20.5.13, $\mathfrak{g}^s$ is reductive in $\mathfrak{g}$. So the result follows by applying 20.5.8 to the adjoint representation of $\mathfrak{g}^s$ in $\mathfrak{g}$.   □

20.5.15 Definition. *Let $\mathfrak{g}$ be semisimple. A Lie subalgebra of $\mathfrak{g}$ is called a torus if all its elements are semisimple.*

20.5.16 Lemma. *Assume that $\mathfrak{g}$ is semisimple. Let $\mathfrak{t}$ be a torus of $\mathfrak{g}$.*
  (i) *$\mathfrak{t}$ is abelian and reductive in $\mathfrak{g}$.*
  (ii) *If $\mathfrak{t}$ is a maximal torus, then $\mathfrak{c}_\mathfrak{g}(\mathfrak{t}) = \mathfrak{t}$.*

Proof. (i) Since $\operatorname{rad} \mathfrak{t}$ consists of semisimple elements, $\mathfrak{t}$ is reductive (20.5.10) and so $\mathcal{D}(\mathfrak{t})$ is semisimple. Thus $\mathcal{D}(\mathfrak{t}) = \{0\}$ (20.4.6), and $\mathfrak{t}$ is abelian.

(ii) By 20.5.13, $\mathfrak{c} = \mathfrak{c}_\mathfrak{g}(\mathfrak{t})$ is reductive in $\mathfrak{g}$, so $\mathfrak{c}$ is the direct product of its centre $\mathfrak{z}$ and $\mathcal{D}(\mathfrak{c})$. Since $\mathfrak{t}$ is maximal, $\mathfrak{z} = \mathfrak{t}$ and $\mathcal{D}(\mathfrak{c})$ contains no non-zero semisimple elements. Hence $\mathcal{D}(\mathfrak{c}) = \{0\}$.   □

20.6 Results on the structure of semisimple Lie algebras

20.6.1 In this section, $\mathfrak{g}$ is a semisimple Lie algebra and $\mathfrak{h}$ is a Cartan subalgebra of $\mathfrak{g}$. We set $R = R(\mathfrak{g}, \mathfrak{h})$ and $\mathfrak{g}^\lambda = \mathfrak{g}^\lambda(\mathfrak{h})$. If $x, y \in \mathfrak{g}$, we shall write $\langle x, y \rangle$ for $L(x, y)$.

Let $\lambda \in \mathfrak{h}^*$. By 19.8.7 (iii), there exists a unique $h_\lambda \in \mathfrak{h}$ such that

$$\lambda(h) = \langle h_\lambda, h \rangle$$

for all $h \in \mathfrak{h}$. The map $\mathfrak{h}^* \to \mathfrak{h}$, $\lambda \mapsto h_\lambda$, is a linear isomorphism. For $\lambda, \mu \in \mathfrak{h}^*$, define $\langle \lambda, \mu \rangle = \langle h_\lambda, h_\mu \rangle$. Thus we obtain a non-degenerate symmetric bilinear form on $\mathfrak{h}^*$. Note that we have:

$$\langle \lambda, \mu \rangle = \langle h_\lambda, h_\mu \rangle = \lambda(h_\mu) = \mu(h_\lambda).$$

It follows also from 19.8.7 that h_α, $\alpha \in R$, span $\mathfrak{h}^*$.

20.6.2 Proposition. *Let $\mathfrak{t}$ be a Lie subalgebra of $\mathfrak{g}$. Then $\mathfrak{t}$ is a Cartan subalgebra of $\mathfrak{g}$ if and only if $\mathfrak{t}$ is maximal torus of $\mathfrak{g}$.*

Proof. Suppose that $\mathfrak{t}$ is a Cartan subalgebra. Let $h \in \mathfrak{t}$ and $h = s + n$ its Jordan decomposition. By 10.1.2, $s, n \in \mathfrak{n}_\mathfrak{g}(\mathfrak{t}) = \mathfrak{t}$. So 19.8.7 (iii) and 20.5.12 imply that $\mathfrak{t}$ is reductive in $\mathfrak{g}$. Hence $\mathfrak{t}$ is abelian. By 20.5.10 (ii), $\mathfrak{t}$ is a torus of $\mathfrak{g}$ and it is maximal since $\mathfrak{n}_\mathfrak{g}(\mathfrak{t}) = \mathfrak{t}$.

Conversely, suppose that $\mathfrak{t}$ is a maximal torus in $\mathfrak{g}$. By 20.5.16 (i), $\mathfrak{t}$ is abelian and there is a subspace $\mathfrak{m}$ of $\mathfrak{g}$ such that $\mathfrak{g} = \mathfrak{t} \oplus \mathfrak{m}$ and $[\mathfrak{t}, \mathfrak{m}] \subset \mathfrak{m}$. It follows that $\mathfrak{n}_\mathfrak{g}(\mathfrak{t}) = \mathfrak{c}_\mathfrak{g}(\mathfrak{t}) = \mathfrak{t}$ by 20.5.16 (ii).   □

20.6.3 Corollary. (i) *Any generic element of* $\mathfrak{g}$ *is semisimple, and any semisimple element is contained in a Cartan subalgebra of* $\mathfrak{g}$. *So* $S_{\mathfrak{g}}$ *is the union of Cartan subalgebras of* $\mathfrak{g}$.

(ii) *If* $h \in \mathfrak{h}$, $\lambda \in \mathfrak{h}^*$ *and* $x \in \mathfrak{g}^{\lambda}$, *then* $[h, x] = \lambda(h)x$.

Proof. It is immediate from 19.8.5 and 20.6.2.   □

20.6.4 Proposition. *Let* $\alpha \in R$.

(i) *If* $x \in \mathfrak{g}^{\alpha}$, $y \in \mathfrak{g}^{-\alpha}$ *and* $h \in \mathfrak{h}$, *then:*

$$\langle h, [x, y] \rangle = \alpha(h)\langle x, y \rangle = \langle h, h_{\alpha} \rangle \langle x, y \rangle \ , \ [x, y] = \langle x, y \rangle h_{\alpha}.$$

(ii) *The vector space* $[\mathfrak{g}^{\alpha}, \mathfrak{g}^{-\alpha}]$ *is* $\Bbbk h_{\alpha}$.

(iii) *We have* $\alpha(h_{\alpha}) \neq 0$.

Proof. (i) By 20.6.3, $[h, x] = \alpha(h)x$, so

$$\langle h, [x, y] \rangle = \langle [h, x], y \rangle = \alpha(h)\langle x, y \rangle = \langle h, h_{\alpha} \rangle \langle x, y \rangle = \langle h, \langle x, y \rangle h_{\alpha} \rangle.$$

Since $[x, y] \in \mathfrak{g}^0 = \mathfrak{h}$ and $L|_{\mathfrak{h} \times \mathfrak{h}}$ is non-degenerate, we have $[x, y] = \langle x, y \rangle h_{\alpha}$.

(ii) This follows from (i) since $\langle \mathfrak{g}^{\alpha}, \mathfrak{g}^{-\alpha} \rangle \neq \{0\}$ (19.8.7).

(iii) Let $(x, y) \in \mathfrak{g}^{\alpha} \times \mathfrak{g}^{-\alpha}$ be such that $\langle x, y \rangle = 1$, so $[x, y] = h_{\alpha}$. If $\alpha(h_{\alpha}) = 0$, then $\mathfrak{a} = \Bbbk h_{\alpha} + \Bbbk x + \Bbbk y$ is a nilpotent Lie algebra. Applying 19.4.4 to the adjoint representation of $\mathfrak{a}$ in $\mathfrak{g}$, we see that if $z \in [\mathfrak{a}, \mathfrak{a}]$, then $\mathrm{ad}_{\mathfrak{g}} z$ is nilpotent. Thus, h_{α} is nilpotent. By 20.6.2, we obtain that $h_{\alpha} = 0$. Contradiction.   □

20.6.5 In view of 20.6.4 (iii), for $\alpha \in R$, we may define:

$$H_{\alpha} = \frac{2h_{\alpha}}{\langle h_{\alpha}, h_{\alpha} \rangle}.$$

Then:

$$\alpha(H_{\alpha}) = 2 \ , \ \langle H_{\alpha}, H_{\alpha} \rangle = \frac{4}{\langle h_{\alpha}, h_{\alpha} \rangle} \ , \ H_{\alpha} = \frac{\langle H_{\alpha}, H_{\alpha} \rangle}{2} h_{\alpha}.$$

Let $X_{\alpha} \in \mathfrak{g}^{\alpha} \setminus \{0\}$. There exists $X_{-\alpha} \in \mathfrak{g}^{-\alpha}$ verifying $\langle X_{\alpha}, X_{-\alpha} \rangle \neq 0$. By 20.6.4, $[X_{\alpha}, X_{-\alpha}] = \langle X_{\alpha}, X_{-\alpha} \rangle h_{\alpha}$. Choose $X_{-\alpha}$ so that $[X_{\alpha}, X_{-\alpha}] = H_{\alpha}$. Then we obtain:

$$[H_{\alpha}, X_{\alpha}] = 2X_{\alpha} \ , \ [H_{\alpha}, X_{-\alpha}] = -2X_{-\alpha} \ , [X_{\alpha}, X_{-\alpha}] = H_{\alpha}.$$

So the Lie algebra $\mathfrak{s}_{\alpha} = \Bbbk H_{\alpha} + \Bbbk X_{\alpha} + \Bbbk X_{-\alpha}$ is isomorphic to $\mathfrak{sl}_2(\Bbbk)$.

20.6.6 Proposition. (i) *For* $\alpha, \beta \in R$, *we have* $\dim \mathfrak{g}^{\alpha} = 1$ *and* $\beta(H_{\alpha})$ *is an integer.*

(ii) *If* $x, y \in \mathfrak{h}$, *then* $\langle x, y \rangle = \sum_{\alpha \in R} \alpha(x)\alpha(y)$.

Proof. (i) Suppose that $\dim \mathfrak{g}^\alpha > 1$. If $y \in \mathfrak{g}^{-\alpha} \setminus \{0\}$, there exists $X_\alpha \in \mathfrak{g}^\alpha \setminus \{0\}$ such that $\langle X_\alpha, y \rangle = 0$. Let $X_{-\alpha}$ and $\mathfrak{s}_\alpha$ be as in 20.6.5. By 20.6.4, $[X_\alpha, y] = \langle X_\alpha, y \rangle h_\alpha = 0$. So it follows from 19.2.5 and 19.2.7 that y is a linear combination of eigenvectors of $\operatorname{ad} H_\alpha$ whose eigenvalues are in $\mathbb{N}$. But $[H_\alpha, y] = -\alpha(H_\alpha)y = -2y$. Contradiction. Hence $\dim \mathfrak{g}^\alpha = 1$.

By 20.6.3, $[H_\alpha, x] = \beta(H_\alpha)x$ for $\alpha, \beta \in R$, $x \in \mathfrak{g}^\beta$. So $\beta(H_\alpha) \in \mathbb{Z}$ (19.2.7 and 20.6.5).

(ii) This is clear from part (i) and 19.8.6. $\quad\square$

20.6.7 For $\alpha, \beta \in R$, set $a_{\beta\alpha} = \beta(H_\alpha)$. The integers $a_{\beta\alpha}$ are called the *Cartan integers* of $(\mathfrak{g}, \mathfrak{h})$. We have $a_{\alpha\alpha} = 2$ and:

$$a_{\beta\alpha} = \langle h_\beta, H_\alpha \rangle = 2\frac{\langle h_\beta, h_\alpha \rangle}{\langle h_\alpha, h_\alpha \rangle} = 2\frac{\langle \beta, \alpha \rangle}{\langle \alpha, \alpha \rangle} = 2\frac{\langle H_\beta, H_\alpha \rangle}{\langle H_\alpha, H_\alpha \rangle}.$$

By 20.6.6, $\langle H_\alpha, H_\beta \rangle \in \mathbb{Z}$. It follows that the ratio between h_α and H_α is rational, and that $\langle h_\alpha, h_\beta \rangle \in \mathbb{Q}$.

20.6.8 Theorem. *Let $\alpha, \beta \in R$.*

(i) *There exist $p, q \in \mathbb{N}$ such that $[-p, q]$ is the set of integers t such that $\beta + t\alpha \in R \cup \{0\}$. Moreover, we have $a_{\beta\alpha} = p - q$.*

(ii) *We have $\beta - a_{\beta\alpha}\alpha \in R$.*

(iii) *If $\beta - \alpha \notin R \cup \{0\}$, then $a_{\beta\alpha} \leqslant 0$, $p = 0$ and $q = -a_{\beta\alpha}$.*

(iv) *The only elements of R proportional to α are α and $-\alpha$.*

(v) *If $\alpha + \beta \in R$, then $[\mathfrak{g}^\alpha, \mathfrak{g}^\beta] = \mathfrak{g}^{\alpha+\beta}$.*

Proof. (i) Set $\mathfrak{u} = \sum_{t \in \mathbb{Z}} \mathfrak{g}^{\beta+t\alpha}$, $\mathfrak{s} = \mathfrak{g}^\alpha + \mathfrak{g}^{-\alpha} + \mathbb{k}H_\alpha \simeq \mathfrak{sl}_2(\mathbb{k})$. Since $[\mathfrak{s}, \mathfrak{u}] \subset \mathfrak{u}$, $\mathfrak{u}$ is a $\mathfrak{s}$-module. The set of eigenvalues for H_α in $\mathfrak{u}$ is $\{(\beta+t\alpha)(H_\alpha) = a_{\beta\alpha}+2t$, $t \in \mathbb{Z}$ and $\beta+t\alpha \in R \cup \{0\}\}$. But by 19.2.5, 19.2.6 and 19.2.7, it is also of the form $\{-r, -r+2, \ldots, r-2, r\}$ for some $r \in \mathbb{N}$. Hence the set of $t \in \mathbb{Z}$ such that $\beta+t\alpha \in R \cup \{0\}$ is $[-p, q]$ for some $p, q \in \mathbb{Z}$. Since $\beta \in R$, $0 \in [-p, q]$, so $p, q \in \mathbb{N}$. Finally, $p - q = a_{\beta\alpha}$ since $r = 2q + a_{\beta\alpha}$ and $-r = -2p + a_{\beta\alpha}$.

(ii) By (i), $-p \leqslant q - p = -a_{\beta\alpha} \leqslant q$, so $\beta - a_{\beta\alpha}\alpha \in R \cup \{0\}$. If $\beta = a_{\beta\alpha}\alpha$, then by 20.6.7, we have:

$$a_{\beta\alpha} = 2\frac{\langle \beta, \alpha \rangle}{\langle \alpha, \alpha \rangle} = 2a_{\beta\alpha}\frac{\langle \alpha, \alpha \rangle}{\langle \alpha, \alpha \rangle} = 2a_{\beta\alpha}.$$

So $a_{\beta\alpha} = 0$, which implies that $\beta = 0$. Contradiction.

(iii) This is clear from (i).

(iv) Let $\lambda \in \mathbb{k}$ be such that $\gamma = \lambda\alpha \in R$. Then $\gamma(H_\alpha) = 2\lambda \in \mathbb{Z}$. Exchanging the roles of γ and α, we deduce that $2\lambda^{-1} \in \mathbb{Z}$. Thus λ belongs to $\{-2, -1, -1/2, 1/2, 1, 2\}$. Since $R = -R$, we need only to prove that if $\alpha \in R$, then $2\alpha \notin R$.

Suppose that $2\alpha \in R$ and $y \in \mathfrak{g}^{2\alpha} \setminus \{0\}$. Then from the preceding discussion, $3\alpha \notin R$ and so $[X_\alpha, y] = 0$. Since $[X_{-\alpha}, y] \in \mathfrak{g}^\alpha = \mathbb{k}X_\alpha$, we have

$[X_\alpha, [X_{-\alpha}, y]] = 0$ and $4y = [H_\alpha, y] = [[X_\alpha, X_{-\alpha}], y] = 0$. Thus $y = 0$. Contradiction.

(v) If $\alpha + \beta \in R$, then $q \geqslant 1$. So $[X_\alpha, \mathfrak{g}^\beta] \neq 0$ (19.2.5, 19.2.6 and 19.2.7). Since $\dim \mathfrak{g}^{\alpha+\beta} = 1$ (20.6.6), we are done. $\square$

20.6.9 For $\alpha \in R$, let $s_\alpha \in \mathrm{End}(\mathfrak{h}^*)$ be defined by:

$$s_\alpha(\lambda) = \lambda - \lambda(H_\alpha)\alpha = \lambda - 2\frac{\langle \lambda, \alpha \rangle}{\langle \alpha, \alpha \rangle}\alpha.$$

Thus:

$$s_\alpha^2 = \mathrm{id}_{\mathfrak{h}^*} , \quad \langle s_\alpha(\lambda), s_\alpha(\mu) \rangle = \langle \lambda, \mu \rangle.$$

If $\beta \in R$, then $s_\alpha(\beta) = \beta - a_{\beta\alpha}\alpha \in R$. It follows that $s_\alpha(R) = R$.

The subgroup $W(\mathfrak{g}, \mathfrak{h})$ of $\mathrm{GL}(\mathfrak{h}^*)$ generated by the s_α's, $\alpha \in R$, is called the *Weyl group* relative to $(\mathfrak{g}, \mathfrak{h})$. Since $w(R) = R$ for all $w \in W(\mathfrak{g}, \mathfrak{h})$ and R spans $\mathfrak{h}^*$ (19.8.7), $W(\mathfrak{g}, \mathfrak{h})$ is a finite group.

20.6.10 From the discussion in the preceding paragraphs, we deduce that R is a reduced root system in $\mathfrak{h}^*$ (18.2.1). Moreover, the set of H_α's form a dual root system of R in $\mathfrak{h}$.

Let

$$\mathfrak{h}_\mathbb{Q} = \sum_{\alpha \in R} \mathbb{Q}H_\alpha , \quad \mathfrak{h}_\mathbb{Q}^* = \sum_{\alpha \in R} \mathbb{Q}\alpha , \quad \mathfrak{h}_\mathbb{R} = \mathbb{R} \otimes_\mathbb{Q} \mathfrak{h}_\mathbb{Q} , \quad \mathfrak{h}_\mathbb{R}^* = \mathbb{R} \otimes_\mathbb{Q} \mathfrak{h}_\mathbb{Q}^*$$

Then we may identify $\mathfrak{h}$ with $\mathbb{k} \otimes_\mathbb{Q} \mathfrak{h}_\mathbb{Q}$, and $\mathfrak{h}^*$ with $\mathbb{k} \otimes_\mathbb{Q} \mathfrak{h}_\mathbb{Q}^*$. The Weyl chambers are defined in $\mathfrak{h}_\mathbb{R}$ and $\mathfrak{h}_\mathbb{R}^*$.

The group $W(\mathfrak{g}, \mathfrak{h})$ acts on $\mathfrak{h}^*$ and $\mathfrak{h}_\mathbb{Q}^*$. It acts also on $\mathfrak{h}$ et $\mathfrak{h}_\mathbb{Q}$ (18.2.9), so on $\mathfrak{h}_\mathbb{R}$ and $\mathfrak{h}_\mathbb{R}^*$.

The form $\langle ., . \rangle$ takes on rational values on $\mathfrak{h}_\mathbb{Q}$ and $\mathfrak{h}_\mathbb{Q}^*$, and defines a non-degenerate positive-definite form on them. It can be extended to an inner product on $\mathfrak{h}_\mathbb{R}$ and $\mathfrak{h}_\mathbb{R}^*$. Finally these forms are all $W(\mathfrak{g}, \mathfrak{h})$-invariant.

20.7 Subalgebras of semisimple Lie algebras

20.7.1 In this section, $\mathfrak{g}$ is a semisimple Lie algebra, $\mathfrak{h}$ is a Cartan subalgebra of $\mathfrak{g}$. We denote $R(\mathfrak{g}, \mathfrak{h})$ by R and $W(\mathfrak{g}, \mathfrak{h})$ by W.

For $\alpha \in R$, we denote $\mathfrak{h}_\alpha = \mathbb{k}H_\alpha = [\mathfrak{g}^\alpha, \mathfrak{g}^{-\alpha}]$. For a subset P of R, we set:

$$\mathfrak{g}^P = \sum_{\alpha \in P} \mathfrak{g}^\alpha , \quad \mathfrak{h}_P = \sum_{\alpha \in P} \mathfrak{h}_\alpha.$$

If $P, Q \subset R$, then it is easy to check that:

(1) $$[\mathfrak{h}, \mathfrak{g}^P] = \mathfrak{g}^P , \quad [\mathfrak{g}^P, \mathfrak{g}^Q] = \mathfrak{h}_{P \cap (-Q)} + \mathfrak{g}^{(P+Q) \cap R}.$$

We deduce immediately the following result from these equalities.

Lemma. *Let $\mathfrak{h}'$ be a subspace of $\mathfrak{h}$ and $P \subset R$. The following conditions are equivalent:*

(i) $\mathfrak{h}' + \mathfrak{g}^P$ *is a Lie algebra of $\mathfrak{g}$.*

(ii) P *is closed subset of R and $\mathfrak{h}_{P \cap (-P)} \subset \mathfrak{h}'$.*

20.7.2 Proposition. *The following conditions are equivalent for a Lie subalgebra $\mathfrak{a}$ of $\mathfrak{g}$:*

(i) $[\mathfrak{h}, \mathfrak{a}] \subset \mathfrak{a}$.

(ii) *There exist a closed subset $P \subset R$ and a subspace $\mathfrak{h}'$ of $\mathfrak{h}$ such that $\mathfrak{h}_{P \cap (-P)} \subset \mathfrak{h}'$ and $\mathfrak{a} = \mathfrak{h}' + \mathfrak{g}^P$.*

Proof. The implication (ii) $\Rightarrow$ (i) is straightforward. Suppose that (i) is verified. Let $P = \{\alpha \in R; \mathfrak{g}^\alpha \subset \mathfrak{a}\}$ and $\mathfrak{h}' = \mathfrak{a} \cap \mathfrak{h}$. By 19.3.9, we have $\mathfrak{a} = \mathfrak{h}' + \sum_{\alpha \in P} \mathfrak{g}^\alpha$. So (ii) follows from 20.7.1. $\square$

20.7.3 Lemma. *Let $\mathfrak{a}$ be a Lie subalgebra of $\mathfrak{g}$ such that $[\mathfrak{h}, \mathfrak{a}] \subset \mathfrak{a}$, $\mathfrak{h}'$ the subspace of $\mathfrak{h}$ and $P \subset R$ such that $\mathfrak{a} = \mathfrak{h}' + \mathfrak{g}^P$. Let $\mathfrak{h}''$ be a subspace of $\mathfrak{h}$ and $Q \subset R$ verifying $\mathfrak{h}'' \subset \mathfrak{h}'$ and $Q \subset P$. Then the following conditions are equivalent:*

(i) $\mathfrak{h}'' + \mathfrak{g}^Q$ *is an ideal of $\mathfrak{a}$.*

(ii) *We have $(P + Q) \cap R \subset Q$ and $\mathfrak{h}_{P \cap (-Q)} \subset \mathfrak{h}'' \subset \bigcap_{\alpha \in P \setminus Q} \ker \alpha$.*

Proof. We have $[\mathfrak{h}' + \mathfrak{g}^P, \mathfrak{h}'' + \mathfrak{g}^Q] = \mathfrak{h}_{P \cap (-Q)} + [\mathfrak{h}', \mathfrak{g}^Q] + [\mathfrak{h}'', \mathfrak{g}^P] + \mathfrak{g}^{(P+Q) \cap R}$. Thus $\mathfrak{h}'' + \mathfrak{g}^Q$ is an ideal of $\mathfrak{a}$ if and only if $\mathfrak{h}_{P \cap (-Q)} \subset \mathfrak{h}'', [\mathfrak{h}'', \mathfrak{g}^P] \subset \mathfrak{g}^Q, \mathfrak{g}^{(P+Q) \cap R} \subset \mathfrak{g}^Q$. So the result is clear. $\square$

20.7.4 Proposition. *Let $\mathfrak{a}$ be a Lie subalgebra of $\mathfrak{g}$ such that $[\mathfrak{h}, \mathfrak{a}] \subset \mathfrak{a}$, $\mathfrak{h}'$ the subspace of $\mathfrak{h}$ and $P \subset R$ such that $\mathfrak{a} = \mathfrak{h}' + \mathfrak{g}^P$. Set:*

$$Q = \{\alpha \in P; -\alpha \notin P\}, \quad \mathfrak{t} = \{x \in \mathfrak{h}'; \alpha(x) = 0, \alpha \in P \cap (-P)\}.$$

(i) *The radical of $\mathfrak{a}$ is $\mathfrak{t} + \mathfrak{g}^Q$.*

(ii) $\mathfrak{g}^Q$ *is a nilpotent ideal of $\mathfrak{a}$.*

Proof. Let $\alpha \in P$, $\beta \in Q$ be such that $\alpha + \beta \in R$. Since P is closed, $\alpha + \beta \in P$. However, $\alpha + \beta \notin -P$, for otherwise $-\beta = -(\alpha + \beta) + \alpha \in P$, this contradicts the fact that $\beta \in Q$. So $\alpha + \beta \in Q$, and $(P + Q) \cap R \subset Q$. Since $P \cap (-Q) = \emptyset$ (so $Q \cap (-Q) = \emptyset$), it follows from 18.10.2 and 20.7.3 that $\mathfrak{g}^Q$ is a nilpotent ideal of $\mathfrak{a}$.

We have, from the definition of $\mathfrak{t}$, that $[\mathfrak{t}, \mathfrak{g}^P] \subset \mathfrak{g}^Q$. So $\mathfrak{t} + \mathfrak{g}^Q$ is an ideal of $\mathfrak{a}$ and it is solvable since $\mathfrak{g}^Q$ is nilpotent.

Let $\mathfrak{r} = \operatorname{rad} \mathfrak{a}$. Then by 19.6.3, $[\mathfrak{h}, \mathfrak{r}] \subset \mathfrak{r}$, and hence $\mathfrak{r} = \mathfrak{h}'' + \mathfrak{g}^S$ where $\mathfrak{h}'' \subset \mathfrak{h}'$ and $S \subset P$. To prove (i), it suffices to prove that $S \subset Q$ and $\mathfrak{h}'' \subset \mathfrak{t}$.

If $\alpha \in S$ is such that $-\alpha \in P$, then $\mathfrak{h}_\alpha = [\mathfrak{g}^\alpha, \mathfrak{g}^{-\alpha}] \subset \mathfrak{r}$. This implies in turn that $\mathfrak{g}^{-\alpha} = [\mathfrak{h}_\alpha, \mathfrak{g}^{-\alpha}] \subset \mathfrak{r}$. So $\mathfrak{r}$ contains a Lie subalgebra isomorphic to $\mathfrak{sl}_2(\Bbbk)$. This contradicts the fact that $\mathfrak{r}$ is solvable. Hence $S \subset Q$.

Next, if $x \in \mathfrak{h}''$ and $\alpha \in P \cap (-P)$, then $[x, \mathfrak{g}^\alpha] \subset \mathfrak{g}^\alpha \cap \mathfrak{r} = \{0\}$. So $\alpha(x) = 0$ and $\mathfrak{h}'' \subset \mathfrak{t}$. $\square$

20.7.5 Proposition. *Let* $\mathfrak{a}$ *be a Lie subalgebra of* $\mathfrak{g}$ *such that* $[\mathfrak{h}, \mathfrak{a}] \subset \mathfrak{a}$, $\mathfrak{h}'$ *the subspace of* $\mathfrak{h}$ *and* $P \subset R$ *such that* $\mathfrak{a} = \mathfrak{h}' + \mathfrak{g}^P$.

(i) *All the elements of* $\mathfrak{a}$ *are nilpotent if and only if* $\mathfrak{h}' = \{0\}$. *If this is the case, then* $P \cap (-P) = \emptyset$ *and* $\mathfrak{a}$ *is nilpotent.*

(ii) $\mathfrak{a}$ *is solvable if and only if* $P \cap (-P) = \emptyset$. *If this is the case, then* $[\mathfrak{a}, \mathfrak{a}] = \mathfrak{g}^S$ *where* $S = ((P + P) \cap R) \cup \{\alpha \in P; \alpha(\mathfrak{h}') \neq \{0\}\}$.

(iii) $\mathfrak{a}$ *is reductive in* $\mathfrak{g}$ *if and only if* $P = -P$.

(iv) $\mathfrak{a}$ *is semisimple if and only if* $P = -P$ *and* $\mathfrak{h}' = \mathfrak{h}_P$. *If this is the case,* $\mathfrak{h}_P$ *is a Cartan subalgebra of* $\mathfrak{a}$ *and the root system of* $(\mathfrak{a}, \mathfrak{h}_P)$ *is* $\{\alpha|_{\mathfrak{h}_P}; \alpha \in P\}$.

Proof. (i) If all the elements of $\mathfrak{a}$ are nilpotent, then $\mathfrak{a}$ is nilpotent (19.3.7) and $\mathfrak{h}' = \{0\}$ (20.6.2).

Conversely, if $\mathfrak{h}' = \{0\}$, then $P \cap (-P) = \emptyset$ (20.7.1). It follows from 18.10.2 that all the elements of $\mathfrak{g}^P$ are nilpotent.

(ii) If $P \cap (-P) = \emptyset$, then $\mathfrak{g}^P$ is a Lie subalgebra (20.7.1) which is nilpotent by (i). Since $[\mathfrak{a}, \mathfrak{a}] = [\mathfrak{h}', \mathfrak{g}^P] + \mathfrak{g}^{(P+P) \cap R} \subset \mathfrak{g}^P$, we see that $\mathfrak{a}$ is solvable and we have the description $[\mathfrak{a}, \mathfrak{a}] = \mathfrak{g}^S$ as required. Conversely, if $\alpha \in P \cap (-P)$, then $\mathfrak{a}$ contains the Lie subalgebra $\mathfrak{h}_\alpha + \mathfrak{g}^\alpha + \mathfrak{g}^{-\alpha}$ which is isomorphic to $\mathfrak{sl}_2(\Bbbk)$. So $\mathfrak{a}$ is not solvable.

(iii) In the notations of 20.7.4, since $\operatorname{rad} \mathfrak{a} = \mathfrak{t} + \mathfrak{g}^Q$, we have from 20.7.4 (i) that

$$\mathfrak{a} \text{ is reductive in } \mathfrak{g} \Leftrightarrow \operatorname{ad}_{\mathfrak{g}} x \text{ is semisimple for all } x \in \mathfrak{t} + \mathfrak{g}^Q$$
$$\Leftrightarrow Q = \emptyset \Leftrightarrow P = -P.$$

(iv) If $\mathfrak{a}$ is semisimple, $P = -P$ by (iii). Since $\mathfrak{a} = [\mathfrak{a}, \mathfrak{a}] = \mathfrak{h}_P + \mathfrak{g}^P$, we have $\mathfrak{h}' = \mathfrak{h}_P$.

Conversely, if $P = -P$ and $\mathfrak{h}' = \mathfrak{h}_P$, then $\mathfrak{a}$ is reductive by (iii) and its radical is $\{0\}$ by 20.7.4. So $\mathfrak{a}$ is semisimple.

Finally, suppose that $\mathfrak{a}$ is semisimple. Since $\mathfrak{h}_P$ is $\operatorname{ad} \mathfrak{h}$-stable, so is $\mathfrak{n} = \mathfrak{n}_{\mathfrak{a}}(\mathfrak{h}_P)$. It follows that $\mathfrak{n} = \mathfrak{h}_P + \mathfrak{g}^S$ where $S \subset P$. For $\alpha \in S$, we have $\mathfrak{g}^\alpha = [\mathfrak{h}_\alpha, \mathfrak{g}^\alpha] \subset [\mathfrak{h}_P, \mathfrak{g}^\alpha] \subset \mathfrak{h}_P$ which is absurd. So $S = \emptyset$ and $\mathfrak{h}_P = \mathfrak{n}$ is a Cartan subalgebra of $\mathfrak{a}$. The remaining statements now follow.   □

20.7.6 Proposition. *Let* $\mathfrak{a}$ *be a Lie subalgebra of* $\mathfrak{g}$ *containing* $\mathfrak{h}$.

(i) *We have* $\mathfrak{a} = \mathfrak{n}_{\mathfrak{g}}(\mathfrak{a})$.

(ii) $\mathfrak{a}$ *contains the semisimple and nilpotent components of its elements.*

Proof. (i) Write $\mathfrak{a} = \mathfrak{h} + \mathfrak{g}^P$ with $P \subset R$. Then $\mathfrak{n}_{\mathfrak{g}}(\mathfrak{a}) = \mathfrak{h} + \mathfrak{g}^Q$ where $Q \supset P$. Since $\mathfrak{g}^Q = [\mathfrak{h}, \mathfrak{g}^Q] \subset \mathfrak{a}$, we have $Q = P$.

(ii) Let $x \in \mathfrak{a}$, and s, n its semisimple and nilpotent components. Then $[s, \mathfrak{a}] \subset \mathfrak{a}$ and $[n, \mathfrak{a}] \subset \mathfrak{a}$ (10.1.2). Hence $s, n \in \mathfrak{a}$ by (i).   □

20.7.7 Proposition. *Let* P *be a closed subset of* R *and* $\mathfrak{a} = \mathfrak{h} + \mathfrak{g}^P$.

(i) $\mathfrak{a}$ *is solvable if and only if* $P \cap (-P) = \emptyset$. *If this is the case, then* $[\mathfrak{a}, \mathfrak{a}] = \mathfrak{g}^P$.

(ii) $\mathfrak{a}$ *is reductive if and only if* $P = -P$.

Proof. Part (i) follows from 20.7.5 (ii) and if $P = -P$, then $\mathfrak{a}$ is reductive by 20.7.5 (iii). Finally, if $\mathfrak{a}$ is reductive, then $[\mathfrak{a},\mathfrak{a}] = \mathfrak{h}' + \mathfrak{g}^P$, where $\mathfrak{h}' \subset \mathfrak{h}$, is semisimple. So $P = -P$ by 20.7.5 (iv). □

20.8 Parabolic subalgebras

20.8.1 Let us conserve the notations of 20.7.1. Let B be a base of R and denote by R_+ (resp. R_-) the corresponding set of positive (resp. negative) roots. Set:

$$\mathfrak{n}_+ = \sum_{\alpha \in R_+} \mathfrak{g}^\alpha \ , \ \mathfrak{n}_- = \sum_{\alpha \in R_-} \mathfrak{g}^\alpha \ , \ \mathfrak{b}_+ = \mathfrak{h} \oplus \mathfrak{n}_+ \ , \ \mathfrak{b}_- = \mathfrak{h} \oplus \mathfrak{n}_-.$$

By 20.7.5, $\mathfrak{n}_+$ and $\mathfrak{n}_-$ (resp. $\mathfrak{b}_+$ and $\mathfrak{b}_-$) are nilpotent (resp. solvable) Lie subalgebras of $\mathfrak{g}$. The decomposition:

$$(2) \qquad\qquad\qquad \mathfrak{g} = \mathfrak{n}_- \oplus \mathfrak{h} \oplus \mathfrak{n}_+$$

is called the *triangular decomposition* of $\mathfrak{g}$ relative to B. The Lie subalgebra $\mathfrak{b}_+$ is called the *Borel subalgebra* of $\mathfrak{g}$ relative to $\mathfrak{h}$ and B. In general, a Lie subalgebra $\mathfrak{b}$ of $\mathfrak{g}$ is called a Borel subalgebra of $(\mathfrak{g}, \mathfrak{h})$ if there exists a base B' of R such that $\mathfrak{b}$ is the Borel subalgebra of $\mathfrak{g}$ relative to $\mathfrak{h}$ and B'.

20.8.2 Proposition. (i) *The Lie algebra* $\mathfrak{n}_+$ *(resp.* $\mathfrak{n}_-$*) is generated by* $\mathfrak{g}^\alpha$, $\alpha \in B$ *(resp.* $\alpha \in -B$*).*

(ii) *The Lie algebras* $\mathfrak{b}_+$ *and* $\mathfrak{b}_-$ *are maximal solvable Lie subalgebras of* $\mathfrak{g}$*; moreover,* $\mathfrak{n}_\mathfrak{g}(\mathfrak{b}_\pm) = \mathfrak{b}_\pm$ *and* $\mathfrak{z}(\mathfrak{b}_\pm) = \{0\}$*.*

Proof. (i) Let $\alpha \in R_+$. By 18.7.7, there exist $\alpha_1, \ldots, \alpha_n \in B$ such that $\alpha = \alpha_1 + \cdots + \alpha_n$ and $\alpha_1 + \cdots + \alpha_i \in R$ for $1 \leqslant i \leqslant n$. So the result follows from 20.6.8 (v).

(ii) We have $\mathfrak{n}_\mathfrak{g}(\mathfrak{b}_\pm) = \mathfrak{b}_\pm$ by 20.7.6.

If $\mathfrak{t}$ is a Lie subalgebra of $\mathfrak{g}$ containing $\mathfrak{b}_+$, then it contains $\mathfrak{h}$. So by 20.7.2, $\mathfrak{t} = \mathfrak{h} + \mathfrak{g}^P$ for some closed subset P of R containing R_+. If P contains a negative root α, then $\mathfrak{t}$ contains the Lie subalgebra $\mathfrak{g}^\alpha + \mathfrak{g}^{-\alpha} + \mathfrak{h}_\alpha \simeq \mathfrak{sl}_2(\Bbbk)$, so $\mathfrak{t}$ is not solvable. Hence $\mathfrak{b}_+$ is a maximal solvable Lie subalgebra. The same argument applies for $\mathfrak{b}_-$.

Finally, if $X = H + \sum_{\alpha \in R_+} \lambda_\alpha X_\alpha \in \mathfrak{z}(\mathfrak{b}_+)$, then $[\mathfrak{h}, X] = \{0\}$ implies that $X = H$. On the other hand $[X, \mathfrak{n}_+] = \{0\}$, so $X = 0$. Again the same argument applies for $\mathfrak{b}_-$. □

20.8.3 Proposition. *Let* $l = \dim \mathfrak{h}$*.*

(i) *For* $(h, x) \in \mathfrak{h} \times \mathfrak{n}_+$*, the characteristic polynomial of* $\mathrm{ad}_\mathfrak{g}(h + x)$ *is:*

$$X^l \prod_{\alpha \in R} (X - \alpha(h)).$$

(ii) *The largest nilpotent ideal, the nilpotent radical and the set of nilpotent elements of* $\mathfrak{b}_+$ *are all equal to* $\mathfrak{n}_+$*. Furthermore, we have:*

$$[\mathfrak{n}_+, \mathfrak{n}_+] = \sum_{\alpha \in R_+ \setminus B} \mathfrak{g}^\alpha.$$

Proof. (i) Let us fix a lexicographic order on $\mathfrak{h}_{\mathbb{R}}^*$ defining B. For $y \in \mathfrak{g}^\alpha$, we have: $[h + x, y] = \alpha(h)y + z$ with $z \in \sum_{\beta > \alpha} \mathfrak{g}^\beta$. So in a suitable basis, the matrix of $\mathrm{ad}(h + x)$ is triangular with diagonal entries 0 (l times) and $\alpha(h)$, $\alpha \in R$.

(ii) This is clear by (i), 19.6.2, 18.7.7 and 20.6.8 (v). $\square$

20.8.4 Proposition. *Let* $\mathfrak{b} = \mathfrak{h} + \mathfrak{g}^P$ *be a Lie subalgebra of* $\mathfrak{g}$ *containing* $\mathfrak{h}$. *Then the following conditions are equivalent:*

(i) $\mathfrak{b}$ *is a Borel subalgebra of* $(\mathfrak{g}, \mathfrak{h})$.
(ii) *There exists a chamber* C *of* R *such that* $P = R_+(C)$.
(iii) R *is the disjoint union of* $-P$ *and* P.

Proof. The subset P is closed by 20.7.1. So the equivalence (ii) $\Leftrightarrow$ (iii) follows from 18.10.3. If $\mathfrak{b}$ is a Borel subalgebra of $(\mathfrak{g}, \mathfrak{h})$, then $P = R_+$, and so (i) $\Rightarrow$ (iii) $\Rightarrow$ (ii). Finally if (ii) is verified, then it is clear from the definition of a Borel subalgebra that $\mathfrak{b}$ is a Borel subalgebra of $(\mathfrak{g}, \mathfrak{h})$. Hence (ii) $\Rightarrow$ (i). $\square$

20.8.5 Proposition. *Let* $\mathfrak{p} = \mathfrak{h} + \mathfrak{g}^P$ *be a Lie subalgebra of* $\mathfrak{g}$ *containing* $\mathfrak{h}$. *Then the following conditions are equivalent:*

(i) $\mathfrak{p}$ *contains a Borel subalgebra of* $(\mathfrak{g}, \mathfrak{h})$.
(ii) *There exists a chamber* C *of* R *such that* $P \supset R_+(C)$.
(iii) P *is parabolic.*

If these conditions are verified, we say that $\mathfrak{p}$ *is a* parabolic subalgebra of $(\mathfrak{g}, \mathfrak{h})$.

Proof. We have (i) $\Leftrightarrow$ (ii) by 20.8.4 and (ii) $\Leftrightarrow$ (iii) by 18.10.5. $\square$

20.8.6 Theorem. *Let* $\mathfrak{p} = \mathfrak{h} + \mathfrak{g}^P$ *be a parabolic subalgebra of* $(\mathfrak{g}, \mathfrak{h})$, $Q = P \setminus (-P)$ *and* $\mathfrak{s} = \mathfrak{h} + \mathfrak{g}^{P \cap (-P)}$. *Then*

(i) $\mathfrak{p} = \mathfrak{s} \oplus \mathfrak{g}^Q$.
(ii) $\mathfrak{s}$ *is reductive in* $\mathfrak{g}$.
(iii) $\mathfrak{g}^Q$ *is the largest nilpotent ideal and the nilpotent radical of* $\mathfrak{p}$. *It is also the orthogonal of* $\mathfrak{p}$ *with respect to the Killing form of* $\mathfrak{g}$.
(iv) $\mathfrak{z}(\mathfrak{p}) = \{0\}$.
(v) $\mathfrak{p} = \mathfrak{n}_{\mathfrak{g}}(\mathfrak{p})$.
(vi) $\mathrm{rad}\,\mathfrak{p} = \mathfrak{t} \oplus \mathfrak{g}^Q$, *where* $\mathfrak{t} = \{x \in \mathfrak{h}; \alpha(x) = 0 \text{ for all } \alpha \in P \cap (-P)\}$.

Proof. Part (i) is clear and (ii), (v), (vi) follow from 20.7.5 (iii), 20.7.6 and 20.7.4 respectively.

(iii) By 20.7.4 (ii), $\mathfrak{g}^Q$ is a nilpotent ideal of $\mathfrak{p}$. If $\mathfrak{n}$ is the largest nilpotent ideal of $\mathfrak{p}$, then $\mathfrak{g}^Q \subset \mathfrak{n} \subset \mathfrak{h} + \mathfrak{g}^Q$ (20.7.4). If $x \in \mathfrak{h} \cap \mathfrak{n}$, then $\mathrm{ad}_{\mathfrak{p}}\, x$ is nilpotent (19.5.8), so $\alpha(x) = 0$ for all $\alpha \in P$, and $x = 0$ since P is parabolic. Thus $\mathfrak{n} = \mathfrak{g}^Q$. Since $[\mathfrak{h}, \mathfrak{g}^Q] = \mathfrak{g}^Q$, the nilpotent radical of $\mathfrak{p}$ contains $\mathfrak{g}^Q$ (20.3.6), and so it is equal to $\mathfrak{g}^Q$. Finally, part (i) and 19.8.6 imply that $\mathfrak{g}^Q$ is the orthogonal of $\mathfrak{p}$ with respect to the Killing form of $\mathfrak{g}$.

(iv) Let $z = x + y \in \mathfrak{z}(\mathfrak{p})$ where $x \in \mathfrak{h}$, $y \in \sum_{\alpha \in P} \mathfrak{g}^\alpha$. Since $[z, \mathfrak{h}] = \{0\}$, $y = 0$. Now $[z, \mathfrak{g}^\alpha] = \{0\}$ for all $\alpha \in P$. But P is parabolic, so $z = 0$. $\square$

20.8.7 Let $h \in \mathfrak{h} \setminus \{0\}$, $e, f \in \mathfrak{g}$ verifying the relations of 19.2.4. If $p \in \mathbb{Z}$, denote by $\mathfrak{g}_p$ the ad h-eigenspace of eigenvalue p. By 19.2.5, 19.2.6 and 19.2.7, we have: $\mathfrak{g} = \bigoplus_{p \in \mathbb{Z}} \mathfrak{g}_p$. Set:

$$\mathfrak{p} = \bigoplus_{p \geqslant 0} \mathfrak{g}_p \ , \ \mathfrak{s} = \mathfrak{g}_0 \ , \ \mathfrak{n} = \bigoplus_{p > 0} \mathfrak{g}_p.$$

Then $\mathfrak{h} \subset \mathfrak{s}$ and $\mathfrak{p} = \mathfrak{h} \oplus \mathfrak{g}^P$ where $P = \{\alpha \in R; \alpha(h) \geqslant 0\}$. It is easy to check that P contains a base of R. So $\mathfrak{p}$ is a parabolic subalgebra of $(\mathfrak{g}, \mathfrak{h})$ whose nilpotent radical is $\mathfrak{n}$ and $\mathfrak{s} = \mathfrak{h} \oplus \mathfrak{g}^{P \cap (-P)}$ is reductive in $\mathfrak{g}$.

20.8.8 Let B be a base of R and $\mathfrak{b}$ the Borel subalgebra of $\mathfrak{g}$ relative to B. For $J \subset B$, denote by $Q(J)$ the set of roots which can be written as a sum of elements of $-J$. Set:

$$P(J) = R_+ \cup Q(J) \ , \ \mathfrak{p}_J = \mathfrak{h} \oplus \mathfrak{g}^{P(J)}.$$

It follows from 18.10.5 that $\mathfrak{p}_J$ is a parabolic subalgebra of $\mathfrak{g}$ containing $\mathfrak{b}$, and any parabolic subalgebra of $\mathfrak{g}$ containing $\mathfrak{b}$ is obtained in this way.

It is clear that if $J, J' \subset B$, then $\mathfrak{p}_J \subset \mathfrak{p}_{J'}$ if and only if $J \subset J'$, and $\mathfrak{p}_J \neq \mathfrak{p}_{J'}$ if $J \neq J'$. Thus the parabolic subalgebras of $\mathfrak{g}$ containing $\mathfrak{b}$, distinct from $\mathfrak{g}$, and maximal with these properties are of the form $\mathfrak{p}_K$, where $K = B \setminus \{\beta\}$ for some $\beta \in B$.

References

- [14], [29], [38], [39], [42], [43], [80].

Algebraic groups

In this chapter, we present some general properties of algebraic groups. All the varieties considered in this chapter are algebraic varieties over $\Bbbk$.

21.1 Generalities

21.1.1 Let G be a group. Denote by $\mu_G : G \times G \to G$, $(\alpha, \beta) \mapsto \alpha\beta$, the group multiplication and $\iota_G : G \to G$, $\alpha \mapsto \alpha^{-1}$, the inverse. As in 10.2.1, we set, for $U, V \subset G$, $UV = \{\alpha\beta; \alpha \in U, \beta \in V\}$ and $U^{-1} = \{\alpha^{-1}; \alpha \in U\}$.

Definition. *Let G be a group endowed with a structure of algebraic variety. If the group multiplication μ_G and the inverse ι_G are morphisms of algebraic varieties, then we say that G is an* algebraic group. *Furthermore, if G is an affine algebraic variety, we say that G is an* affine algebraic group.

Remark. Note that the topology on $G \times G$ here is the Zariski topology, and not the usual product topology. So an algebraic group is not a topological group.

21.1.2 Let G, G' be algebraic groups. If $\alpha \in G$, then translations $\beta \mapsto \alpha\beta$, $\beta \mapsto \beta\alpha$, and conjugation, $\beta \mapsto \alpha\beta\alpha^{-1}$, are isomorphisms of the algebraic variety G.

A map $G \to G'$ is a *morphism of algebraic groups* if it is a homomorphism of groups and a morphism of algebraic varieties. We define in an obvious way the notion of *isomorphisms* and *automorphisms* of algebraic groups.

21.1.3 Example. 1) The additive group $\mathbf{G}_a = \Bbbk$ and the multiplicative group $\mathbf{G}_m = \Bbbk \setminus \{0\}$ are algebraic groups.

2) For $n \in \mathbb{N}^*$, $\mathrm{GL}_n(\Bbbk)$ is an algebraic group (11.7.5 and 11.8.9).

3) Any closed subgroup of an algebraic group is an algebraic group. Thus $\mathrm{SL}_n(\Bbbk)$ and the subgroup of $\mathrm{GL}_n(\Bbbk)$ consisting of diagonal (resp. upper triangular, lower triangular) matrices are algebraic groups.

4) Let $n \in \mathbb{N}^*$. Denote by $J_n = [a_{ij}] \in \mathrm{GL}_n(\Bbbk)$ the matrix where $a_{ij} = 1$ if $i + j = n + 1$ and $a_{ij} = 0$ otherwise. Set:

$$\mathrm{Sp}_{2n}(\Bbbk) = \left\{ A \in \mathrm{GL}_{2n}(\Bbbk) \; ; \; {}^t A \begin{pmatrix} 0 & J_n \\ -J_n & 0 \end{pmatrix} A = \begin{pmatrix} 0 & J_n \\ -J_n & 0 \end{pmatrix} \right\}.$$

$$\mathrm{O}_{2n+1}(\Bbbk) = \left\{ A \in \mathrm{GL}_{2n+1}(\Bbbk) \; ; \; {}^t A \begin{pmatrix} 1 & 0 & 0 \\ 0 & 0 & J_n \\ 0 & J_n & 0 \end{pmatrix} A = \begin{pmatrix} 1 & 0 & 0 \\ 0 & 0 & J_n \\ 0 & J_n & 0 \end{pmatrix} \right\}.$$

$$\mathrm{O}_{2n}(\Bbbk) = \left\{ A \in \mathrm{GL}_{2n}(\Bbbk) \; ; \; {}^t A \begin{pmatrix} 0 & J_n \\ J_n & 0 \end{pmatrix} A = \begin{pmatrix} 0 & J_n \\ J_n & 0 \end{pmatrix} \right\}.$$

Since these subgroups are closed in the corresponding $\mathrm{GL}_p(\Bbbk)$, they are algebraic groups.

5) Let $\mathfrak{A}$ be a $\Bbbk$-algebra of finite dimension (not necessarily associative). Then the group of automorphisms of $\mathfrak{A}$, denoted by $\mathrm{Aut}\,\mathfrak{A}$, is an algebraic group.

6) If G, G' are algebraic groups, then their direct product $G \times G'$ endowed with the Zariski product topology is an algebraic group.

21.1.4 Lemma. *Let G be an algebraic group. Then there is a unique irreducible component of G which contains the identity element e.*

Proof. Let X_1, X_2 be two irreducible components of G which contain e, and $X = X_1 X_2 = \mu(X_1 \times X_2)$. Since the X_i's are irreducible, so is $X_1 \times X_2$, and hence X is irreducible. Thus $X \subset X_1$ or $X \subset X_2$. On the other hand $e \in X_i$, $i = 1, 2$, it follows that $X_i \subset X$, $i = 1, 2$. Hence $X = X_1 = X_2$ and we are done. □

21.1.5 We shall denote by G° the unique irreducible component of G containing e, and we shall call G° the *identity component* of G.

21.1.6 Theorem. *Let G be an algebraic group.*

(i) *The identity component G° is a closed normal subgroup of G of finite index. The cosets of G modulo G° are at the same time the irreducible components and the connected components of G. In particular, G is a pure variety.*

(ii) *If H is a closed subgroup of finite index in G, then H contains G°.*

Proof. Since $\mu_G(G^\circ \times G^\circ) = G^\circ G^\circ$ and $\iota_G(G^\circ) = (G^\circ)^{-1}$ are irreducible subsets of G containing e, they are contained in G° (21.1.4). So G° is a subgroup of G. Similarly, for $\alpha \in G$, $\alpha G^\circ \alpha^{-1}$ is an irreducible subset of G° containing e, so $\alpha G^\circ \alpha^{-1} \subset G^\circ$ and G° is normal.

Since translations are isomorphisms of the algebraic variety G (21.1.2), the coset αG° is the unique irreducible component of G containing α. So part (i) follows from the fact that G is the disjoint union of the cosets of G modulo G°.

(ii) Let H be such a subgroup. Let $H_1, \ldots, H_n$ be the left cosets of G modulo H having non-empty intersection with G°. Then G° is the disjoint union of the closed subsets $H_1 \cap G^\circ, \ldots, H_n \cap G^\circ$. Since G^0 is irreducible, we obtain that $n = 1$ and $G^\circ \subset H$. □

21.1.7 Remarks. 1) By 21.1.6, an algebraic group is irreducible if and only if it is connected. We shall use the term *connected* algebraic groups instead of irreducible algebraic group.

2) Let G be a connected algebraic group. If $\alpha, \beta \in G$, we have $\beta = (\beta\alpha^{-1})\alpha$. Since the map $\gamma \mapsto (\beta\alpha^{-1})\gamma$ is an automorphism of the variety G, we see that the point β is smooth if and only if α is smooth. Since the set of smooth points is non-empty (16.5.5), we deduce that G is a smooth variety. We obtain in the same way that G is a normal variety (17.1.5).

21.1.8 In the rest of this chapter, G will denote an algebraic group.

21.2 Subgroups and morphisms

21.2.1 Lemma. *Let U, V be dense open subsets of G. Then $G = UV$.*

Proof. Let $\alpha \in G$. By 21.1.2, αV^{-1} is a dense open subset of G. So it follows that $\alpha V^{-1} \cap U \neq \emptyset$ and $\alpha \in UV$. $\square$

21.2.2 Proposition. *Let H be a subgroup of G.*
   (i) *$\overline{H}$ is a subgroup of G.*
   (ii) *If H contains a non-empty open subset of $\overline{H}$, then $H = \overline{H}$. In particular, if H is constructible, then H is closed.*

Proof. (i) Let $\alpha \in H$. Since $H = \alpha H \subset \alpha\overline{H}$ and $\alpha\overline{H}$ is closed by 21.1.2, $\overline{H} \subset \alpha\overline{H}$. Hence $\alpha^{-1}\overline{H} \subset \overline{H}$ and $H\overline{H} \subset \overline{H}$. If $\beta \in \overline{H}$, then $H\beta \subset \overline{H}$. It follows that $\overline{H\beta} = \overline{H}\beta \subset \overline{H}$. Hence $\overline{H}\,\overline{H} \subset \overline{H}$. Similarly, by 21.1.2, $\overline{H}^{-1} = \overline{H^{-1}} = \overline{H}$. So $\overline{H}$ is a subgroup of G.

(ii) Let U be an open subset of $\overline{H}$ contained in H. If $\alpha \in H$ and $\beta \in U$, then $\alpha \in \alpha\beta^{-1}U \subset H$ and $\alpha\beta^{-1}U$ is an open subset of $\overline{H}$. Thus H is open in $\overline{H}$. It follows that H is a constructible subset of G and therefore H contains a dense open subset of $\overline{H}$ (1.4.6). In view of 21.2.1, we have $\overline{H} = HH = H$. $\square$

21.2.3 Corollary. *If H, K are closed subgroups of G such that K normalizes H, then HK is a closed subgroup of G.*

Proof. By our hypothesis, HK is a subgroup of G. On the other hand, it is equal to $\mu_G(H \times K)$. Since μ is a morphism of varieties, HK is constructible (15.4.3). So the result follows from 21.2.2 (ii). $\square$

21.2.4 Proposition. *Let $u : G \to H$ be a morphism of algebraic groups.*
   (i) $\ker u$ *and* $\operatorname{im} u$ *are closed subgroups of G and H respectively.*
   (ii) $u(G^\circ) = u(G)^\circ$.
   (iii) $\dim G = \dim(\ker u) + \dim(\operatorname{im} u)$.

Proof. (i) Since u is a homomorphism of groups, $\ker u$ and $\operatorname{im} u$ are subgroups of G and H respectively. Since u is morphism of varieties, $\ker u$ is closed and $\operatorname{im} u$ is constructible (15.4.3). So the result follows from 21.2.2.

(ii) By (i), $u(G^\circ)$ is a closed connected subgroup of $\operatorname{im} u$. Since G° has finite index in G (21.1.6), $u(G^\circ)$ has finite index in $\operatorname{im} u$. Thus, again by 21.1.6, we obtain that $u(G^\circ) = u(G)^\circ$.

(iii) Since the fibres of the morphism $G \to u(G)$ induced by u are the cosets of G modulo $\ker u$, they all have dimension $\dim(\ker u)$. So (iii) follows from 15.5.5 (ii). $\square$

21.2.5 Let $A \in \mathrm{M}_n(\Bbbk)$ be a nilpotent matrix and $p \in \mathbb{N}$ be such that $A^{p+1} = 0$. For $t \in \Bbbk$, we set:

$$\varphi_A(t) = I_n + \frac{t}{1!}A + \frac{t^2}{2!}A^2 + \cdots + \frac{t^p}{p!}A^p.$$

Then this defines a morphism of algebraic groups $\varphi_A : \mathbf{G}_a \to \mathrm{GL}_n(\Bbbk)$. It follows from the previous results that $\varphi_A(\mathbf{G}_a)$ is a closed connected subgroup of $\mathrm{GL}_n(\Bbbk)$.

21.2.6 Proposition. *Let $u : G \to H$ be a morphism of algebraic groups. Then the following conditions are equivalent:*
(i) *u is an isomorphism of groups.*
(ii) *u is an isomorphism of algebraic groups.*

Proof. The implication (ii) $\Rightarrow$ (i) is obvious. Suppose that u is an isomorphism of groups. By 21.2.4 (ii), $u(G^\circ) = H^\circ$. Since G° and H° are normal varieties, the result is just a consequence of 17.4.8 and 21.1.6. $\square$

21.3 Connectedness

21.3.1 Theorem. *Let $(X_i)_{i \in I}$ be a family of irreducible varieties and for $i \in I$, $u_i : X_i \to G$ a morphism. Assume that $e \in Y_i = u_i(X_i)$ for all $i \in I$. Denote by H the subgroup of G generated by the Y_i's and K the smallest closed subgroup of G containing the Y_i's.*
(i) *The closed subgroup K is connected and is equal to H.*
(ii) *There exist $n \in \mathbb{N}$, $i_1, \ldots, i_n \in I$ and $\varepsilon_1, \ldots, \varepsilon_n \in \{-1, 1\}$ such that:*

$$H = Y_{i_1}^{\varepsilon_1} \cdots Y_{i_n}^{\varepsilon_n}.$$

Proof. If $i \in I$, the map $X_i \to G$, $\alpha \mapsto u_i(\alpha)^{-1}$, is again a morphism. So replacing I if necessary, we may assume that for any $i \in I$, $Y_i^{-1} = Y_j$ for some $j \in I$. Observe that we have clearly $K = \overline{H}$.

For $m \in \mathbb{N}$ and $\lambda = (i_1, \ldots, i_m) \in I^m$, set $Y_\lambda = Y_{i_1} \cdots Y_{i_m}$. Since Y_λ is the image of the morphism $X_{i_1} \times \cdots \times X_{i_m} \to G$, $(\alpha_1, \ldots, \alpha_m) \mapsto \alpha_1 \cdots \alpha_m$, it follows that Y_λ and $\overline{Y_\lambda}$ are irreducible.

Let $\mu = (j_1, \ldots, j_n) \in I^n$, $n \in \mathbb{N}$. Set $(\lambda, \mu) = (i_1, \ldots, i_m, j_1, \ldots, j_n) \in I^{m+n}$. We claim that $\overline{Y_\lambda}\,\overline{Y_\mu} \subset \overline{Y_{(\lambda,\mu)}}$. Let us prove our claim. For $\alpha \in Y_\lambda$, the translation morphism $\beta \mapsto \alpha\beta$ maps Y_μ into $Y_{(\lambda,\mu)}$. It follows that $Y_\lambda \overline{Y_\mu} \subset \overline{Y_{(\lambda,\mu)}}$. Now for $\beta \in \overline{Y_\mu}$, we have $\overline{Y_\lambda \beta} = \overline{Y_\lambda}\beta \subset \overline{Y_{(\lambda,\mu)}}$. Hence $\overline{Y_\lambda}\,\overline{Y_\mu} \subset \overline{Y_{(\lambda,\mu)}}$ as claimed.

Let $\mathcal{F} = \{\overline{Y_\lambda}; n \in \mathbb{N}, \lambda \in I^n\}$. Since the topological space G is Noetherian, $\mathcal{F}$ has a maximal element, say $\overline{Y_\mu}$. Let $n \in \mathbb{N}$ and $\lambda \in I^n$. Since $e \in \overline{Y_\lambda}$, it follows from the claim above that:

$$\overline{Y_\mu} \subset \overline{Y_\mu}\,\overline{Y_\lambda} \subset \overline{Y_{(\lambda,\mu)}}.$$

Since $\overline{Y_\mu}$ is maximal, we have

$$\overline{Y_\mu} = \overline{Y_\mu}\,\overline{Y_\lambda} = \overline{Y_{(\mu,\lambda)}}.$$

Taking $\lambda = \mu$, we deduce that $\overline{Y_\mu}$ is stable under multiplication. Similarly, taking λ such that $Y_\lambda = Y_\mu^{-1}$, we see that $\overline{Y_\mu}^{-1} \subset \overline{Y_\mu}$. So $\overline{Y_\mu}$ is a subgroup of G which is closed and connected by definition. Clearly $K = \overline{Y_\mu}$.

Furthermore, being the image of a morphism Y_μ is a constructible subset of G (15.4.3) and therefore by 1.4.6 and 21.2.1, $\overline{Y_\mu} = Y_\mu Y_\mu = Y_{(\mu,\mu)}$. Hence $H \subset \overline{H} = K = \overline{Y_\mu} = Y_{(\mu,\mu)} \subset H$. $\square$

21.3.2 Let $(G_i)_{i \in I}$ be a family of subgroups of G, and $j_i : G_i \to G$ the canonical injections. In order to apply 21.3.1 in this situation, the G_i's need to have a structure of algebraic variety and the j_i's need to be morphisms.

Corollary. *Let $(G_i)_{i \in I}$ be a family of closed connected subgroups of G and H the subgroup of G generated by the G_i's. Then H is a closed connected subgroup of G.*

21.3.3 Corollary. *With the notations of 10.8.1, the groups $\mathrm{SL}_n(\Bbbk)$, $\mathrm{D}_n(\Bbbk)$, $\mathrm{T}_n(\Bbbk)$ and $\mathrm{U}_n(\Bbbk)$ are closed connected subgroups of $\mathrm{GL}_n(\Bbbk)$.*

Proof. These are clearly closed subgroups of $\mathrm{GL}_n(\Bbbk)$. The group $\mathrm{D}_n(\Bbbk)$ is connected because it is isomorphic to $(\mathbf{G}_m)^n$. The connectedness of $\mathrm{SL}_n(\Bbbk)$ and $\mathrm{U}_n(\Bbbk)$ follows from 10.8.6 and 21.3.2. Finally, $\mathrm{T}_n(\Bbbk)$ is connected because it is isomorphic to $\mathrm{D}_n(\Bbbk) \times \mathrm{U}_n(\Bbbk)$. $\square$

21.3.4 Let $\mathfrak{g}$ be a finite-dimensional Lie algebra over $\Bbbk$. By 21.2.5 and 21.3.2, we have:

Corollary. *The group $\mathrm{Aut}_e\,\mathfrak{g}$ is a connected algebraic group which is a closed normal subgroup of $\mathrm{Aut}\,\mathfrak{g}$.*

21.3.5 Let H, K be closed subgroups of G. It is not true in general that the subgroup (H, K) is closed in G. However, we have the following result:

Proposition. *Let H, K be closed subgroups of G.*

(i) *If either H or K is connected, then (H, K) is a closed connected subgroup of G.*

(ii) *Assume that H normalizes K. Then the (H, K) is a closed subgroup of G. In particular, (G, G) is a closed subgroup of G.*

Proof. (i) Assume that H is connected. For $\beta \in K$, the map $u_\beta : H \to G$, $\alpha \mapsto \alpha\beta\alpha^{-1}\beta^{-1}$, is a morphism of varieties. Since (H, K) is the subgroup of G generated by $u_\beta(H)$, $\beta \in K$, the result follows from 21.3.1.

(ii) We may assume that $G = HK$ (21.2.3). Then K and (H, K) are normal subgroups of G. By (i), (H°, K) and (H, K°) are closed connected subgroups of G. Let L be the subgroup of G generated by $\alpha(H^\circ, K)\alpha^{-1}$ and $\beta(H, K^\circ)\beta^{-1}$, $\alpha, \beta \in G$. Since (H, K) is normal, we have $L \subset (H, K)$. Moreover, L is closed (21.3.2). So we are done if L has finite index in (H, K).

Let $\theta : G \to G/L$ be the canonical surjection. Then $\theta(H^\circ)$ centralizes $\theta(K)$, and $\theta(K^\circ)$ centralizes $\theta(H)$. Since H° (resp. K°) has finite index in H (resp. K), it follows that $S = \{(\gamma, \delta); \gamma \in \theta(H), \delta \in \theta(K)\}$ is a finite set. So $(H, K)/L$ is finite by 10.3.3. □

21.3.6 Corollary. *If G is connected, then $\mathcal{D}^n(G)$ and $\mathcal{C}^{n+1}(G)$ are closed connected subgroups of G for all $n \in \mathbb{N}$.*

21.3.7 Proposition. *Let $\alpha \in G$ and H a closed subgroup of G.*

(i) *The centralizers $C_G(\alpha)$ of α in G and $C_G(H)$ of H is G are closed subgroups of G.*

(ii) *The set $\{\beta \in G; \beta H \beta^{-1} \subset H\}$ is the closed subgroup $N_G(H)$ of G.*

Proof. (i) Let u, v denote the translations $\beta \mapsto \alpha\beta$ and $\beta \mapsto \beta\alpha$ respectively. Then $C_G(\alpha) = \{\beta \in G; u(\beta) = v(\beta)\}$. So it is closed by 12.5.8. Since $C_G(H)$ is the intersection of $C_G(\beta)$, $\beta \in H$, it is also closed.

(ii) For $\beta \in G$, the map $i_\beta : G \to G$, $\alpha \mapsto \beta\alpha\beta^{-1}$, is an automorphism of the algebraic group G. So $K_\beta = i_\beta(H)$ is a closed subgroup of G of dimension $\dim H$ and $i_\beta(H^\circ) = (K_\beta)^\circ$. On the other hand, $(H : H^\circ) = (K_\beta : (K_\beta)^\circ)$. If $\beta H \beta^{-1} \subset H$, then $K_\beta \subset H$ and K_β° is a closed connected subgroup of H containing e of dimension $\dim H^\circ$. So $(K_\beta)^\circ = H^\circ$. It follows that $H = K_\beta$, and so $N_G(H) = \{\beta \in G; \beta H \beta^{-1} \subset H\}$.

Let $\beta \in H$. The map $u_\beta : G \to G$, $\alpha \mapsto \alpha\beta\alpha^{-1}$, is a morphism because it is the composition of

$$G \to G \times G , \ \alpha \mapsto (\alpha\beta, \alpha^{-1}) \text{ and } G \times G \to G , \ (\alpha, \gamma) \mapsto \alpha\gamma.$$

So $u_\beta^{-1}(H)$ is closed in G. Since $N_G(H) = \{\beta \in G; \beta H \beta^{-1} \subset H\}$ is the intersection of $u_\beta^{-1}(H)$, $\beta \in H$, it is closed in G. □

21.4 Actions of an algebraic group

21.4.1 Definition. *Let G acts on a variety X. We say that G acts* rationally *on X, or X is a G-variety if the map*

$$G \times X \to X \ , \ (\alpha, x) \mapsto \alpha.x$$

is a morphism of algebraic varieties. So, if X is a G-variety, the map $x \mapsto \alpha.x$ is an automorphism of X for all $\alpha \in G$.

21.4.2 Proposition. *Let X be a G-variety.*

(i) *If Y and Z are subsets of X with Z closed in X, then $\mathrm{Tran}_G(Y, Z)$ is a closed subset of G.*

(ii) *For all $x \in X$, G_x is a closed subgroup of G.*

(iii) *If $\alpha \in G$, then the set of fixed points of α is closed in X. Thus X^G is closed in X.*

(iv) *If G is connected, then G stabilizes each irreducible component of X.*

Proof. (i) If $x \in X$, the map $u_x : G \to X$, $\alpha \mapsto \alpha.x$ is a morphism. So $u_x^{-1}(Z)$ is closed in G. Since $\mathrm{Tran}_G(Y, Z)$ is the intersection of $u_y^{-1}(Z)$, $y \in Y$, it is closed in G.

(ii) This is clear by (i) since $G_x = \mathrm{Tran}_G(\{x\}, \{x\})$.

(iii) The diagonal Δ of X in $X \times X$ is closed (12.5.3). For $\alpha \in G$, the map $u : X \to X \times X$, $x \mapsto (x, \alpha.x)$, is a morphism. The set of fixed points of α is $u^{-1}(\Delta)$, so it is closed in X.

(iv) Let G be connected, Y an irreducible component of X and H the stabilizer of Y in H. By (i), H is closed in G and since the elements of G permutes the irreducible components of X, we obtain that $(G : H)$ is finite. Hence $G = G^\circ \subset H$ by 21.1.6. $\square$

21.4.3 Proposition. *Let X be a G-variety, $x \in X$ and $Y = G.x$.*

(i) *The set Y is locally closed in X, so it is a subvariety of X.*

(ii) *If G is connected, then Y is irreducible and smooth.*

(iii) *We have $\dim(G.x) = \dim G - \dim G_x$.*

Proof. It is clear that $\overline{Y}$ is stable under the action of G.

(i) The map $u_x : G \to X$, $\alpha \mapsto \alpha.x$ is a morphism and $Y = \mathrm{im}\, u_x$. So Y is constructible (15.4.3) and it contains a dense open subset of $\overline{Y}$ (1.4.6). Since G acts transitively on Y and $\overline{Y}$ is stable under the action of G, we deduce that Y is open in $\overline{Y}$.

(ii) Since $Y = \mathrm{im}\, u_x$, Y is irreducible when G is connected. It is smooth since G acts transitively on Y.

(iii) Let $v_x : G \to Y$ be the morphism induced by u_x. Then the fibres of v_x are left cosets of G modulo G_x. So the results follows from 15.5.5 (ii). $\square$

21.4.4 Proposition. *Let X and Y be G-varieties. Assume that X is a G-homogeneous space.*

(i) *The G°-orbits of X are open and closed in X. They all have the same dimension and they are the pairwise disjoint irreducible components of X. In particular, X is a pure variety.*

(ii) *For $n \in \mathbb{N}$, $\{y \in Y; \dim(G.y) \leqslant n\}$ is a closed subset of Y.*

Proof. (i) Let $x \in X$. For $\alpha \in G$, the right coset $G^\circ \alpha$ defines a G°-orbit, $G^\circ \alpha.x$ of X. Since G acts transitively on X, and $(G : G^\circ)$ is finite, there exist elements $\alpha_1, \ldots, \alpha_n \in G$ such that $X = (G^\circ \alpha_1.x) \cup \cdots \cup (G^\circ \alpha_n.x)$ and $G^\circ \alpha_i.x \cap G^\circ \alpha_j.x = \emptyset$ if $i \neq j$.

Let $X_i = G^\circ \alpha_i.x$. So the X_i's are exactly the G°-orbits in X. Suppose for example that $\dim X_1 \leqslant \dim X_i$ for $2 \leqslant i \leqslant n$ and $Y = \overline{X_1} \setminus X_1$ is non-empty. Since $\overline{X_1}$ is irreducible (21.4.3) and Y is closed in $\overline{X_1}$, $\dim Y < \dim \overline{X_1} = \dim X_1$. But this is absurd because Y, being G°-stable, is the union of certain X_i's with $i \geqslant 2$. So X_1 is closed in X. Now $X_i = G^\circ \alpha_i.x = \alpha_i G^\circ.x = (\alpha_i \alpha_1^{-1}).X_1$, thus X_i is closed in X for all i. The result is now clear.

(ii) By (i), $G.y$ is a pure variety whose irreducible components are G°-orbits. So we may assume that G is connected. By 21.4.2 (iv), we may further assume that X is irreducible. Let $u : G \times X \to X \times X$ be the morphism sending (α, x) to $(x, \alpha.x)$. Then $u^{-1}(u(\alpha, x)) = (\alpha G_x, x)$ for all $\alpha \in G$ and $x \in X$. By 21.1.4 and 21.4.2 (ii), $(\alpha (G_x)^\circ, x)$ is the unique irreducible component of $(\alpha G_x, x)$ containing (α, x). Its dimension is $\dim G_x$. So the result is a consequence of 21.4.3 (iii) and 15.5.7. $\square$

21.4.5 Proposition. *Let X be a G-variety, $x \in X$ and $Y = G.x$.*

(i) *The set $\overline{Y}$ is the union of Y and orbits of dimension strictly less than $\dim Y$.*

(ii) *Orbits of minimal dimension are closed. In particular, X contains a closed orbit.*

Proof. Let $Y_1, \ldots, Y_n$ be the irreducible components of Y. They are G°-orbits by 21.4.4. By 1.1.14, $\overline{Y_1}, \ldots, \overline{Y_n}$ are the irreducible components of $\overline{Y}$. Since $\overline{Y} \setminus Y \subset (\overline{Y_1} \setminus Y_1) \cup \cdots \cup (\overline{Y_n} \setminus Y_n)$, we have $\dim(\overline{Y} \setminus Y) < \dim Y$. So we have (i) since $\overline{Y} \setminus Y$ is G-stable. Finally (ii) follows easily from (i). $\square$

21.5 Modules

21.5.1 Let V be a $\mathbb{k}$-vector space of dimension n. By identifying V with $\mathbb{k}^n$ via a basis $\mathcal{B}$ of V, we can transport the Zariski topology of $\mathbb{k}^n$ on V. It is clear that the topology obtained on V does not depend on the choice of $\mathcal{B}$. Hence we have a notion of Zariski topology on V, and in a similar way, on $\operatorname{End}(V)$ and $\operatorname{GL}(V)$.

21.5.2 Let (V, ρ) be a finite-dimensional G-module. We say that V is a *rational G-module* if the map $\rho : G \to \operatorname{GL}(V)$ is a morphism of algebraic groups.

Lemma. *The following conditions are equivalent:*

(i) V *is a rational G-module.*

(ii) *The map* $\theta : G \times V \to V$, $(\alpha, x) \mapsto \rho(\alpha)(x)$, *is a morphism of varieties.*

(iii) *For all* $x \in V$, *the map* $\theta_x : G \to V$, $\alpha \mapsto \rho(\alpha)(x)$, *is a morphism of varieties.*

Proof. Since θ is the composition of

$$G \times V \to \mathrm{End}(V) \times V \ , \ (\alpha, x) \mapsto (\rho(\alpha), x),$$
$$\mathrm{End}(V) \times V \to V \ , \ (u, x) \mapsto u(x),$$

where the second map is a morphism of varieties, we have (i) $\Rightarrow$ (ii). The implication (ii) $\Rightarrow$ (iii) is obvious.

Let $\mathcal{B} = (x_1, \ldots, x_n)$ be a basis of V and $(\varphi_1, \ldots, \varphi_n)$ the basis of V^* dual to $\mathcal{B}$. The map $V \to \Bbbk$, $v \mapsto \varphi_i(v)$, is a morphism of varieties for all i. So if (iii) is verified, then the map $G \to \mathrm{M}_n(\Bbbk)$ sending $\alpha \in G$ to the matrix $\rho(\alpha)$ with respect to $\mathcal{B}$ is a morphism of varieties. Hence (iii) $\Rightarrow$ (i). $\square$

21.5.3 Remark. Let V be a finite-dimensional rational G-module, then the map θ in 21.5.2 (ii) makes V into a G-variety. So we may apply the results of 21.4 to rational G-modules.

21.5.4 Definition. *Let* V *be a G-module. We say that* V *is* locally finite *if for all* $x \in V$, *there exists a finite-dimensional G-submodule* V_x *of* V *containing* x. *If* V *is a locally finite G-module, then* V *is a rational G-module if for all* $x \in V$, *we can choose* V_x *to be a rational G-module.*

21.5.5 Proposition. *Let* (V, ρ) *be a rational G-module and* W *a submodule of* V. *Then* W *is a rational G-module.*

Proof. For $x \in W$, let V_x be a rational G-submodule of finite dimension containing x. By 21.5.2, for $y \in V_x$, the map $G \to V$, $\alpha \mapsto \rho(\alpha)(y)$, is a morphism of varieties. If $y \in W \cap V_x$, then the map $G \to W \cap V_x$, $\alpha \mapsto \rho(\alpha)(y)$, is again a morphism of varieties. It follows from 21.5.2 that $W \cap V_x$ is a finite-dimensional rational G-submodule containing x. $\square$

21.6 Group closure

21.6.1 Let H be a subgroup of G. In this section, we study the properties of the map $H \mapsto \overline{H}$. Recall from 21.2.2 that $\overline{H}$ is a subgroup of G.

Lemma. (i) *If* H *is commutative, then so is* $\overline{H}$.

(ii) *If* H *is normal in* G, *then so is* $\overline{H}$.

Proof. (i) For $\beta \in G$, denote by u_β and v_β the morphisms from G to itself defined by $u_\beta(\alpha) = \alpha\beta$ and $v_\beta(\alpha) = \beta\alpha$. Then if $\beta \in H$, then u_β and v_β are identical on H, so they are identical on $\overline{H}$ (12.5.8). This implies in turn that for $\beta \in \overline{H}$, u_β and v_β are identical on H, hence on $\overline{H}$. Thus $\overline{H}$ is commutative.

(ii) For $\beta \in G$, the morphism of varieties $\varphi_\beta : G \to G$, $\alpha \mapsto \beta\alpha\beta^{-1}$, verifies $\varphi_\beta(H) \subset H$, and so $\varphi_\beta(\overline{H}) \subset \overline{H}$. $\square$

21.6.2 Let $P \subset G$. The intersection of all the closed subgroups of G containing P, denoted by $\mathcal{A}(P)$, is called the *group closure* of P in G. Clearly, $\mathcal{A}(P)$ is the smallest closed subgroup of G containing P. If H is a subgroup of G, then $\mathcal{A}(H) = \overline{H}$. So $\mathcal{A}(P)$ can also be identified as the closure of the subgroup of G generated by P.

21.6.3 Lemma. *Let G, H be algebraic groups, $P, Q \subset G$, $R \subset H$ and $u : G \to H$ a morphism of algebraic groups.*

(i) *If P is a dense subset of Q, then $\mathcal{A}(P) = \mathcal{A}(Q)$.*

(ii) *If P normalizes (resp. centralizes) Q, then $\mathcal{A}(P)$ normalizes (resp. centralizes) $\mathcal{A}(Q)$.*

(iii) *We have $\mathcal{A}(P \times R) = \mathcal{A}(P) \times \mathcal{A}(R)$.*

(iv) *We have $u(\mathcal{A}(P)) = \mathcal{A}(u(P))$.*

Proof. (i) Since $Q \subset \overline{P} \subset \mathcal{A}(P)$, $\mathcal{A}(Q) \subset \mathcal{A}(P)$. The reverse inclusion is obvious.

(ii) If $x \in G$ normalizes Q, then it normalizes the subgroup generated by Q, and hence $\mathcal{A}(Q)$ (same proof as in 21.6.1 (ii)). Thus $P \subset N_G(\mathcal{A}(Q))$. Since $N_G(\mathcal{A}(Q))$ is closed in G (21.3.7), $\mathcal{A}(P) \subset N_G(\mathcal{A}(Q))$.

(iii) Since $\mathcal{A}(P) \times \mathcal{A}(R)$ is a closed subgroup of $G \times H$ containing $P \times R$, it contains $\mathcal{A}(P \times R)$.

Conversely, $\mathcal{A}(P \times R)$ contains $P \times \{e_H\}$ and $\{e_G\} \times Q$, so it contains $\mathcal{A}(P) \times \{e_H\}$ and $\{e_G\} \times \mathcal{A}(Q)$. So $\mathcal{A}(P \times R) \supset \mathcal{A}(P) \times \mathcal{A}(R)$.

(iv) Since the subgroup $u(\mathcal{A}(P))$ is closed in H (21.2.4) and it contains $u(P)$, we have $\mathcal{A}(u(P)) \subset u(\mathcal{A}(P))$. Moreover,

$$P \subset u^{-1}\left(\mathcal{A}(u(P))\right) \Rightarrow \mathcal{A}(P) \subset u^{-1}\left(\mathcal{A}(u(P))\right)$$
$$\Rightarrow u(\mathcal{A}(P)) \subset \mathcal{A}(u(P)).$$

So we have proved our result. $\square$

21.6.4 Proposition. *Let H, K be subgroups of G. Then the groups (H, K) and $(\overline{H}, \overline{K})$ have the same closure in G.*

Proof. The map

$$u : G \times G \to G , \; (\alpha, \beta) \mapsto \alpha\beta\alpha^{-1}\beta^{-1},$$

is a morphism of varieties. Since $H \times K$ is dense in $\overline{H} \times \overline{K}$, $u(H \times K)$ is dense in $u(\overline{H} \times \overline{K})$. So the result follows from 21.6.3 (i) since $\mathcal{A}(u(H \times K)) = \overline{(H, K)}$ and $\mathcal{A}(u(\overline{H} \times \overline{K})) = \overline{(\overline{H}, \overline{K})}$. $\square$

21.6.5 Proposition. *Let H, K be subgroups of G. If H normalizes K, then $\overline{H}$ normalizes $\overline{K}$ and $\overline{(H, K)} = (\overline{H}, \overline{K})$.*

Proof. By 21.6.3 (ii), $\overline{H} \subset N_G(\overline{K})$, so $\overline{H}\overline{K}$ is a closed subgroup of G (21.2.3). We may therefore assume that $G = \overline{H}\overline{K}$. Then $\overline{K}$ is normal in G and $(\overline{H}, \overline{K})$ is closed in G (21.3.5). So the result follows from 21.6.4. $\square$

21.6.6 Corollary. *Let H, K be subgroups of G.*
(i) *For all $n \in \mathbb{N}$, we have:*

$$\overline{\mathcal{D}^n(H)} = \mathcal{D}^n(\overline{H}) \ , \ \overline{\mathcal{C}^{n+1}(H)} = \mathcal{C}^{n+1}(\overline{H}).$$

(ii) *Assume that K is a normal subgroup of H. If H/K is abelian (resp. nilpotent, solvable), then so is $\overline{H}/\overline{K}$.*

Proof. By 21.6.5, (i) is a simple induction on n. If H/K is solvable, then $\mathcal{D}^n(H) \subset K$ for some $n \in \mathbb{N}$. So (i) implies that $\mathcal{D}^n(\overline{H}) = \overline{\mathcal{D}^n(H)} \subset \overline{K}$. Thus $\overline{H}/\overline{K}$ is solvable. The proofs of the other cases are similar. $\square$

21.6.7 Corollary. *The following conditions are equivalent:*
(i) *G is solvable.*
(ii) *There is a sequence $G = G_0 \supset G_1 \supset \cdots \supset G_{n+1} = \{e\}$ of closed subgroups of G such that $(G_i, G_i) \subset G_{i+1}$ for $0 \leqslant i \leqslant n$.*

Proof. In view of 21.3.6 and 21.6.6 (i), we have (i) $\Rightarrow$ (ii) by taking $G_i = \mathcal{D}^i(G)$. Conversely, given (ii), a simple induction shows that $\mathcal{D}^i(G) \subset G_i$. So we have (i). $\square$

21.6.8 Corollary. *The following conditions are equivalent:*
(i) *G is nilpotent.*
(ii) *There is a sequence $G = G_0 \supset G_1 \supset \cdots \supset G_{n+1} = \{e\}$ of closed subgroups of G such that $(G, G_i) \subset G_{i+1}$ for $0 \leqslant i \leqslant n$.*

Proof. This is analogue to the proof of 21.6.7. $\square$

References

- [5], [23], [38], [40], [78].

Affine algebraic groups

In the rest of this book, all the algebraic groups that we are going to come across are affine. We shall therefore assume from now on that by an algebraic group, we mean an affine algebraic group. In fact, we shall see that under this assumption, an algebraic group can be identified with a closed subgroup of a general linear group.

By convention, G will always denote an (affine) algebraic group.

22.1 Translations of functions

22.1.1 Let G be an algebraic group, X an affine G-variety and

$$\pi : G \times X \to X \ , \ (\alpha, x) \mapsto \alpha.x$$

the morphism defining the action of G on X. The comorphism of π

$$\mathbf{A}(\pi) : \mathbf{A}(X) \to \mathbf{A}(G \times X) = \mathbf{A}(G) \otimes_{\Bbbk} \mathbf{A}(X)$$

sends $f \in \mathbf{A}(X)$ to the function F on $G \times X$ defined by $F(\alpha, x) = f(\alpha.x)$. From this, we obtain that if $\alpha \in G$ and $f \in \mathbf{A}(X)$, then the function $\tau_\alpha f$ on X defined by

$$(\tau_\alpha f)(x) = f(\alpha^{-1}.x)$$

belongs to $\mathbf{A}(X)$. We verify easily that this defines a group homomorphism $\tau : G \to \mathrm{GL}(\mathbf{A}(X))$, $\alpha \mapsto \tau_\alpha$. We shall endow $\mathbf{A}(X)$ with this G-module structure.

Proposition. *Let E be a finite-dimensional subspace of $\mathbf{A}(X)$.*
(i) There exists a finite-dimensional G-submodule of $\mathbf{A}(X)$ containing E.
(ii) The subspace E is a G-submodule of $\mathbf{A}(X)$ if and only if $\mathbf{A}(\pi)(E)$ is contained in $\mathbf{A}(G) \otimes_{\Bbbk} E$.

Proof. (i) We may assume that $\dim E = 1$. So E is spanned by an element $f \in \mathbf{A}(X)$. Let

$$\mathbf{A}(\pi)(f) = u_1 \otimes v_1 + \cdots + u_n \otimes v_n$$

where $u_1, \ldots, u_n \in \mathbf{A}(G)$ and $v_1, \ldots, v_n \in \mathbf{A}(X)$. For $\alpha \in G$ and $x \in X$,

$$(\tau_\alpha f)(x) = \mathbf{A}(\pi)(f)(\alpha^{-1}, x) = u_1(\alpha^{-1})v_1(x) + \cdots + u_n(\alpha^{-1})v_n(x).$$

Thus the G-submodule generated by f is contained in the subspace spanned by $v_1, \ldots, v_n$.

(ii) Let $(v_1, \ldots, v_n)$ be a basis of E.

Suppose that $\mathbf{A}(\pi)(E) \subset \mathbf{A}(G) \otimes_{\Bbbk} E$. If $f \in E$, then there exist $u_1, \ldots, u_n \in \mathbf{A}(G)$, such that $\mathbf{A}(\pi)(f) = u_1 \otimes v_1 + \cdots + u_n \otimes v_n$. Then for $\alpha \in G$, we have as seen in the proof of (i) that

$$\tau_\alpha f = u_1(\alpha^{-1})v_1 + \cdots + u_n(\alpha^{-1})v_n.$$

So E is a G-submodule of $\mathbf{A}(X)$.

Suppose that E is a G-submodule of $\mathbf{A}(X)$. For $f \in E$, there exist $w_1, \ldots, w_m \in \mathbf{A}(X)$, $u_1, \ldots, u_{n+m} \in \mathbf{A}(G)$ such that the elements $v_1, \ldots, v_n, w_1, \ldots, w_m$ are linearly independent and

$$\mathbf{A}(\pi)(f) = u_1 \otimes v_1 + \cdots + u_n \otimes v_n + \sum_{i=1}^{m} u_{n+i} \otimes w_i.$$

It follows that for $\alpha \in G$,

$$\tau_\alpha f = u_1(\alpha^{-1})v_1 + \cdots + u_n(\alpha^{-1})v_n + \sum_{i=1}^{m} u_{n+1}(\alpha^{-1})w_i.$$

Since E is a G-submodule, $u_{n+i}(\alpha^{-1}) = 0$ for all $\alpha \in G$ and $1 \leqslant i \leqslant m$. So $u_{n+i} = 0$ for $1 \leqslant i \leqslant m$ and we have $\mathbf{A}(\pi)(f) \in \mathbf{A}(G) \otimes_{\Bbbk} E$ as required.   □

22.1.2 Proposition. *Let G be an algebraic group and X an affine G-variety. Then the G-module $\mathbf{A}(X)$ is rational.*

Proof. Let $f \in \mathbf{A}(X)$. By 22.1.1, f is contained in a finite-dimensional G-submodule E of $\mathbf{A}(X)$. Let $(v_1, \ldots, v_n)$ be a basis of E. Then as we have seen in the proof of 22.1.1, there exist $u_1, \ldots, u_n \in \mathbf{A}(G)$, such that

$$\tau_\alpha f = u_1(\alpha^{-1})v_1 + \cdots + u_n(\alpha^{-1})v_n$$

for all $\alpha \in G$. Since the maps $G \to \Bbbk$, $\alpha \mapsto u_i(\alpha^{-1})$, are morphisms of varieties, the map $G \to E$, $\alpha \mapsto \tau_\alpha f$, is also a morphism of varieties. So E is a rational G-module (21.5.2).   □

22.1.3 Proposition. *Let G be an algebraic group and X an affine G-variety. Then there exist a finite-dimensional rational G-module V and a G-equivariant morphism $u : X \to V$ such that u is an isomorphism from X onto a G-stable closed subvariety of V.*

Proof. Let $v_1, \ldots, v_n$ be generators of the algebra $\mathbf{A}(X)$. By 22.1.2, there is a finite-dimensional rational G-submodule W of $\mathbf{A}(X)$ containing all the v_i's. The dual W^* of W is also a rational G-module. Denote by $\varphi : X \to W^*$ the map defined as follows: $\varphi(x)(w) = w(x)$ for all $w \in W$, $x \in X$. Then it is clear that φ is a G-equivariant morphism. Since the v_i's generate $\mathbf{A}(X)$, it is easy to see that $\mathbf{A}(\varphi)$ is surjective. So we can conclude by 11.6.8. $\square$

22.1.4 Let G be an affine algebraic group. The group G acts on itself by left and right translations. We obtain from the preceding discussion two G-module structures $\lambda, \rho : G \to \mathrm{GL}(\mathbf{A}(G))$ on $\mathbf{A}(G)$ defined as follows: for $\alpha, \beta \in G$, $u \in \mathbf{A}(G)$,

$$(\lambda_\alpha u)(\beta) = u(\alpha^{-1}\beta) \ , \ (\rho_\alpha u)(\beta) = u(\beta\alpha).$$

We call λ_α (resp. ρ_α) the *left translation* (resp. *right translation*) of functions.

22.1.5 Theorem. *Let G be an affine algebraic group. There exists $n \in \mathbb{N}$ such that G is isomorphic (as an algebraic group) to a closed subgroup of $\mathrm{GL}_n(\Bbbk)$.*

Proof. Applying 22.1.3 to the representation of G on itself by left translations, we obtain that there exist a rational G-module (V, ρ) and a G-equivariant morphism $u : G \to V$ such that u is an isomorphism from G onto a G-stable closed subvariety of V. Let $\alpha \in G$ be such that $\rho(\alpha) = \mathrm{id}_V$. Since u is G-equivariant, we have $u(\alpha\beta) = \rho(\alpha)u(\beta) = u(\beta)$. So $\alpha = e$ because u is injective. We conclude that ρ is injective and therefore G is isomorphic as an algebraic group to the closed subgroup $\rho(G)$ of $\mathrm{GL}(V)$ (21.2.6). $\square$

22.1.6 Lemma. *Let G be an affine algebraic group, H a closed subgroup, and $\mathfrak{a} = \mathcal{I}(H)$ the defining ideal of H in $\mathbf{A}(G)$. Then*

$$H = \{\alpha \in G; \lambda_\alpha(\mathfrak{a}) \subset \mathfrak{a}\} = \{\alpha \in G; \rho_\alpha(\mathfrak{a}) \subset \mathfrak{a}\}.$$

Proof. Let $\alpha, \beta \in H$, $u \in \mathfrak{a}$. Since $\alpha^{-1}\beta \in H$, we have $u(\alpha^{-1}\beta) = 0$, so $\lambda_\alpha u, \rho_\beta u \in \mathfrak{a}$.

Conversely, if $\lambda_\alpha(\mathfrak{a}) \subset \mathfrak{a}$, then $u(\alpha^{-1}) = (\lambda_\alpha u)(e) = 0$ for all $u \in \mathfrak{a}$. So $\alpha^{-1} \in H$ and hence $\alpha \in H$. The proof for ρ_α is analogue. $\square$

22.2 Jordan decomposition

22.2.1 Let us consider $\mathbf{A}(G)$ as a G-module by right translation. Then $\mathbf{A}(G)$ is rational (22.1.2) and by 10.1.11, ρ_α decomposes into $(\rho_\alpha)_s(\rho_\alpha)_u$ with $(\rho_\alpha)_s$ semisimple, $(\rho_\alpha)_u$ locally unipotent and $(\rho_\alpha)_s(\rho_\alpha)_u = (\rho_\alpha)_u(\rho_\alpha)_s$.

Theorem. *If $\alpha \in G$, there exists a unique pair (α_s, α_u) of elements of G such that:*

$$(\rho_\alpha)_s = \rho_{\alpha_s} \ , \ (\rho_\alpha)_u = \rho_{\alpha_u} \ , \ \alpha = \alpha_s\alpha_u = \alpha_u\alpha_s.$$

The decomposition $\alpha = \alpha_s\alpha_u$ is called the Jordan decomposition *of α and α_s (resp. α_u) is called the* semisimple *(resp.* unipotent*) component of α. If $\alpha = \alpha_s$ (resp. $\alpha = \alpha_u$), then we say that α is* semisimple *(resp.* unipotent*).*

Proof. Since ρ_α is an automorphism of the $\Bbbk$-algebra $\mathbf{A}(G)$, so are $(\rho_\alpha)_s$ and $(\rho_\alpha)_u$ (10.1.14). Thus the maps

$$f \mapsto ((\rho_\alpha)_s f)(e) \; , \; f \mapsto ((\rho_\alpha)_u f)(e)$$

from $\mathbf{A}(G) \to \Bbbk$ are $\Bbbk$-algebra homomorphisms. By 11.3.4, there exist $\alpha_s, \alpha_u \in G$ such that $((\rho_\alpha)_s f)(e) = f(\alpha_s)$ and $((\rho_\alpha)_u f)(e) = f(\alpha_u)$ for all $f \in \mathbf{A}(G)$.

Let $\beta \in G$. Since ρ_α and λ_β commute, $(\rho_\alpha)_s$ and λ_β commute also. It follows that

$$((\rho_\alpha)_s f)(\beta) = \left(\lambda_{\beta^{-1}}((\rho_\alpha)_s f)\right)(e) = ((\rho_\alpha)_s(\lambda_{\beta^{-1}} f))(e)$$
$$= (\lambda_{\beta^{-1}} f)(\alpha_s) = f(\beta\alpha_s).$$

Therefore $(\rho_\alpha)_s = \rho_{\alpha_s}$ and in the same way, we have $(\rho_\alpha)_u = \rho_{\alpha_u}$. Finally, the map $G \to \mathrm{GL}(\mathbf{A}(G))$, $\alpha \mapsto \rho_\alpha$, is injective, so $\alpha = \alpha_s\alpha_u = \alpha_u\alpha_s$ since $\rho_\alpha = (\rho_\alpha)_s(\rho_\alpha)_u = (\rho_\alpha)_u(\rho_\alpha)_s$. $\square$

22.2.2 Remark. As in 22.2.1, we have $(\lambda_\alpha)_s = \lambda_{\alpha_s}$ and $(\lambda_\alpha)_u = \lambda_{\alpha_u}$.

22.2.3 Proposition. *Let G and H be algebraic groups.*

(i) *If $\theta : G \to H$ is a morphism of algebraic groups, then $\theta(\alpha_s) = \theta(\alpha)_s$ and $\theta(\alpha_u) = \theta(\alpha)_u$ for all $\alpha \in G$.*

(ii) *Suppose that G is a closed subgroup of $\mathrm{GL}(V)$ where V is a finite-dimensional $\Bbbk$-vector space. Let $\alpha \in G$. Then its semisimple and unipotent components defined in 22.2.1 coincide with those defined in 10.1.4.*

Proof. (i) Since $K = \mathrm{im}\,\theta$ is a closed subgroup of H (21.2.4) and θ factorizes through $G \to K = \mathrm{im}\,\theta \to H$, it suffices to prove the result in the following two cases:

1) G is a closed subgroup of H and θ is the canonical injection. Since $\mathbf{A}(G) = \mathbf{A}(H)/\mathcal{I}(G)$ and $\mathcal{I}(G)$ is stable under ρ_α, $\alpha \in G$ (22.1.6), the result follows from 10.1.3 and 10.1.5.

2) θ is surjective. Then $\mathbf{A}(\theta)$ is injective and $\mathbf{A}(\theta) \circ \rho_{\theta(\alpha)} = \rho_\alpha \circ \mathbf{A}(\theta)$. Thus $\mathbf{A}(H)$ can be identified as a subspace of $\mathbf{A}(G)$ stable under ρ_α, $\alpha \in G$. So we may conclude as in the previous case.

(ii) In view of (i), it suffices to prove the result in the case where $G = \mathrm{GL}(V)$ and by fixing a basis, we may assume that $G = \mathrm{GL}_n(\Bbbk)$. Then $\mathbf{A}(G) = \Bbbk[T_{11}, T_{12}, \ldots, T_{nn}]_D$ where D is the determinant (11.7.5). Writing $\alpha = [\alpha_{ij}]$, then for $1 \leqslant i \leqslant n$,

$$\rho_\alpha T_{1i} = \alpha_{1i} T_{11} + \cdots + \alpha_{ni} T_{1n}.$$

Thus the subspace spanned by T_{1i}, $1 \leqslant i \leqslant n$, is stable under ρ_α. Again we may conclude by 10.1.3 and 10.1.5. $\square$

22.2.4 Corollary. *Let $\alpha \in G$ and θ an isomorphism of algebraic groups from G onto a closed subgroup of $\mathrm{GL}(V)$ where V is a finite-dimensional $\Bbbk$-vector space. The following conditions are equivalent:*

(i) α *is semisimple (resp. unipotent).*

(ii) $\theta(\alpha)$ *is a semisimple (resp. unipotent) automorphism of V.*

22.2.5 Proposition. *The set of unipotent elements of G is closed in G.*

Proof. By 22.1.5 and 22.2.3 (i), it suffices to prove the result for $G = \mathrm{GL}(V)$ where V is a finite-dimensional $\Bbbk$-vector space. Since $\alpha \in \mathrm{GL}(V)$ is unipotent if and only if $(\alpha - \mathrm{id}_V)^{\dim V} = 0$, the result follows. □

Remark. In general, the set of semisimple elements of G is neither open nor closed in G.

22.2.6 Proposition. *Let G be commutative and denote by G_s (resp. G_u) the set of semisimple (resp. unipotent) elements of G.*

(i) G_s *and G_u are closed subgroups of G.*

(ii) *The map $\pi : G_s \times G_u \to G$, $(\alpha, \beta) \mapsto \alpha\beta$, is an isomorphism of algebraic groups.*

(iii) *If G is connected, then G_s and G_u are also connected.*

Proof. We may assume that G is a closed subgroup of $\mathrm{GL}_n(\Bbbk)$ (22.1.5). So we may apply 22.2.4 with $V = \Bbbk^n$.

(i) Since G is commutative, we have $(\alpha\beta)_s = \alpha_s\beta_s$ and $(\alpha\beta)_u = \alpha_u\beta_u$. So G_s and G_u are subgroups of G. We saw in 22.2.5 that G_u is closed.

There exists $\alpha \in \mathrm{GL}_n(\Bbbk)$ such that $\alpha G_s \alpha^{-1} \subset \mathrm{D}_n(\Bbbk)$ because G_s is commutative. It follows that $G_s = G \cap (\alpha^{-1} \mathrm{D}_n(\Bbbk)\alpha)$ and it is closed in G since $\mathrm{D}_n(\Bbbk)$ is closed in $\mathrm{GL}_n(\Bbbk)$.

(ii) The uniqueness of the Jordan decomposition implies that π is a bijective morphism of algebraic groups. So the result follows from 21.2.6.

(iii) This follows from (ii) and 11.8.6. □

22.3 Unipotent groups

22.3.1 Definition. *Let G be an algebraic group.*

(i) *We say that G is* unipotent *if all the elements of G are unipotent.*

(ii) *A morphism $\lambda : \mathbf{G}_a \to G$ of algebraic groups is called an* additive one-parameter subgroup *of G.*

22.3.2 Proposition. *The following conditions are equivalent:*

(i) G *is unipotent.*

(ii) G *is isomorphic, as an algebraic group, to a closed subgroup of $\mathrm{U}_n(\Bbbk)$.*

Proof. (i) $\Rightarrow$ (ii) Since G is isomorphic to a closed subgroup H of $\mathrm{GL}_n(\Bbbk)$ (22.1.5) and the elements of H are unipotent automorphisms by 22.2.4, the result follows from 10.8.14.

(ii) $\Rightarrow$ (i) This is clear by 22.2.3 and 22.2.4. □

22.3.3 Let V be a finite-dimensional $\Bbbk$-vector space.

If n is a nilpotent endomorphism of V and u an unipotent automorphism of V, we may define:

$$\exp(n) = e^n = \sum_{k=0}^{\infty} \frac{1}{k!} n^k \; , \; \log(u) = \sum_{k=1}^{\infty} \frac{(-1)^{k+1}}{k} (u - \mathrm{id}_V)^k.$$

Then e^n is an unipotent automorphism of V and $\log(u)$ is a nilpotent endomorphism of V. Furthermore, it is well-known that:

$$\exp(\log(u)) = u \quad \text{and} \quad \log(\exp(n)) = n.$$

Now let

$$\varphi_n : \mathbf{G}_a \to \mathrm{GL}(V) \; , \; t \mapsto e^{tn}.$$

We saw in 21.2.5 that φ_n is a morphism of algebraic groups. It follows that $\varphi_n(\mathbf{G}_a)$ is a connected closed subgroup of $\mathrm{GL}(V)$.

Suppose that $n \neq 0$. For $t \in \Bbbk$, we have $\log(e^{tn}) = tn$. So φ_n induces a bijection between $\mathbf{G}_a$ and $\varphi_n(\mathbf{G}_a)$ whose inverse map is $\log|_{\varphi_n(\mathbf{G}_a)}$. It follows from the definition of $\log$ that φ_n is an isomorphism of algebraic groups from $\mathbf{G}_a$ onto $\varphi_n(\mathbf{G}_a)$.

22.3.4 Proposition. *Let G be a closed subgroup of $\mathrm{GL}(V)$, u an unipotent element of G and $n = \log(u)$. Then $\varphi_n(\mathbf{G}_a) \subset G$. Consequently, $\varphi_n(\mathbf{G}_a)$ is the smallest closed subgroup of G (and so of $\mathrm{GL}(V)$) containing u.*

Proof. Let $k \in \mathbb{Z}$. Since $\exp(\log(u^k)) = u^k = \exp(k\log(u))$, we deduce that $\varphi_n(\mathbb{Z}) \subset G$. Since G is closed and $\mathbb{Z}$ is dense in $\mathbf{G}_a$, we have by the continuity of φ_n that $\varphi_n(\mathbf{G}_a) = \varphi_n(\overline{\mathbb{Z}}) \subset \overline{\varphi_n(\mathbb{Z})} \subset G$. So the result follows. $\square$

22.3.5 Proposition. *Let V be a finite-dimensional $\Bbbk$-vector space and $\lambda : \mathbf{G}_a \to \mathrm{GL}(V)$ a group homomorphism. The following conditions are equivalent:*

(i) *λ is an additive one-parameter subgroup of $\mathrm{GL}(V)$.*

(ii) *There exists $n \in \mathrm{End}(V)$ nilpotent such that $\lambda(t) = e^{tn}$ for all $t \in \Bbbk$.*

Proof. The implication (ii) $\Rightarrow$ (i) is clear. So let us suppose that (i) is verified. By 22.2.4 and 22.3.3, $\mathbf{G}_a$ is unipotent. So $\lambda(\mathbf{G}_a)$ is a connected closed unipotent subgroup of $\mathrm{GL}(V)$. Let $n = \log(\lambda(1))$. Then $\varphi_n(\mathbf{G}_a) \subset \lambda(\mathbf{G}_a)$ (22.3.4) and for $k \in \mathbb{Z}$, $\lambda(k) = \lambda(1)^k = e^{kn}$. Since $\mathbb{Z}$ is dense in the irreducible variety $\mathbf{G}_a$, we obtain that $\lambda = \varphi_n$ (12.5.8). $\square$

22.3.6 Theorem. *Let G be a unipotent algebraic group.*

(i) *G is connected.*

(ii) *If V is a non-zero rational G-module, then $V^G \neq \{0\}$.*

(iii) *Let X be an affine G-variety. Then the G-orbits in X are closed.*

Proof. (i) We may assume that $G \subset \mathrm{GL}(V)$ for some finite-dimensional k-vector space V. Let N be the set of $\log(u)$, $u \in G$. By 22.3.4, G is generated by $\varphi_n(\mathbf{G}_a)$, $n \in N$. Thus the result follows from 21.2.5 and 21.3.1.

(ii) We may assume that V is finite-dimensional and the result was proved in 10.8.13.

(iii) Suppose that there exists $x \in X$ such that $Z = G.x$ is not closed in X. Set $Y = \overline{Z} \setminus Z$. Then Y is a non-empty closed subset of $\overline{Z}$ (21.4.3). Let $\mathfrak{a} = \mathcal{I}(Y)$. Then $\mathfrak{a} \neq \{0\}$ and by (ii), there exists $f \in \mathfrak{a} \setminus \{0\}$ such that $\alpha.f = f$ for all $\alpha \in G$. Hence $f(\alpha.x) = f(x)$ for all $\alpha \in G$ and f is constant on Z, and so on $\overline{Z}$. But f is zero on Y, and so $f = 0$. Contradiction. $\square$

22.3.7 Let X be an irreducible affine G-variety. We can extend the G-action on $\mathbf{A}(X)$ (22.1.1) to $\mathrm{Fract}(\mathbf{A}(X))$ as follows: for $\alpha \in G$, $u, v \in \mathbf{A}(X)$ with $v \neq 0$,

$$\alpha(uv^{-1}) = (\alpha.u)(\alpha.v)^{-1}.$$

It is clear that $\mathrm{Fract}(\mathbf{A}(X))^G$ is a subfield of $\mathrm{Fract}(\mathbf{A}(X))$. Recall also from 12.7.7 that we may identify $\mathrm{Fract}(\mathbf{A}(X))$ with $\mathbf{R}(X)$.

Corollary. *Let G be a unipotent algebraic group and X an irreducible affine G-variety. Then $\mathbf{R}(X)^G = \mathrm{Fract}\left(\mathbf{A}(X)^G\right)$.*

Proof. Clearly $\mathbf{R}(X)^G \supset \mathrm{Fract}\left(\mathbf{A}(X)^G\right)$. Let u be an element of $\mathbf{R}(X)^G$. Denote by V the set of $s \in \mathbf{A}(X)$ such that $us \in \mathbf{A}(X)$. Then V is a G-submodule of $\mathbf{A}(X)$. Since $u \in \mathrm{Fract}\,\mathbf{A}(X)$, $V \neq \{0\}$. So by 22.3.6, there exists $v \in V^G \setminus \{0\} \subset \mathbf{A}(X)^G \setminus \{0\}$. It follows that $w = uv \in \mathbf{A}(X)^G$ and $u = wv^{-1} \in \mathrm{Fract}(\mathbf{A}(X)^G)$. $\square$

22.3.8 Lemma. *Let K be a commutative field and G a group of automorphisms α of the polynomial ring $K[T]$ such that $\alpha.K \subset K$.*
(i) We have $K(T)^G = \mathrm{Fract}(K[T]^G)$.
(ii) There exists $p \in K[T]^G$ such that $K(T)^G = K^G(p)$.

Proof. Let $\alpha \in G$ and $u = \alpha.T$. If $\deg(u) = 0$, then since $\alpha.K \subset K$, we have $\alpha.K[T] \subset K$ which is absurd. Similarly, if $\deg(u) \geq 2$, then $\deg(\alpha.v) \geq 2$ for all $v \in K[T] \setminus K$, and so $\alpha.K[T] \neq K[T]$, which is again absurd. We deduce that $\deg(u) = 1$ and hence $\deg(\alpha.v) = \deg(v)$ for all $v \in K[T]$.

(i) Let $f \in K(T)^G \setminus \mathrm{Fract}\,K[T]^G$. Replacing f by f^{-1} if necessary, we may assume that $f = uv^{-1}$ where $u, v \in K[T]$ are relatively prime and $\deg(u) \geq \deg(v) > 0$.

If $\alpha \in G$, then $u(\alpha.v) - v(\alpha.u) = 0$. Since u and v are relatively prime, we deduce that there exists $\chi(\alpha) \in K \setminus \{0\}$ such that $\alpha.v = \chi(\alpha)v$ and $\alpha.u = \chi(\alpha)u$.

Now there exist $q, r \in K[T]$ such that $u = qv + r$ and $\deg(r) < \deg(v)$. Then $\chi(\alpha)u = \chi(\alpha)(\alpha.q)v + \alpha.r$. It follows that $\alpha.q = q$ and $\alpha.r = \chi(\alpha)r$. Moreover $f = q + rv^{-1}$ and $\alpha.(rv^{-1}) = rv^{-1}$. So we obtain (i) by induction on $\deg(u) + \deg(v)$.

(ii) If $K[T]^G \subset K$, then by (i), we have $K(T)^G = K^G = K^G(1)$. So let us suppose that $K[T]^G \setminus K \neq \emptyset$. Take $p \in K[T]^G \setminus K$ of minimal degree. For $u \in K[T]^G \setminus K$, there exist $q, r \in K[T]$ unique such that $u = qp + r$ and $\deg(r) < \deg(p)$. It follows from our choice of p that $r \in K^G$. By induction on $\deg(u)$, we obtain that $u \in K^G[p]$. Now the result follows from (i). $\square$

22.3.9 The following result applies in particular for unipotent groups.

Proposition. *Let $n \in \mathbb{N}$ and G a subgroup of $T_n(\Bbbk)$. Consider the natural action of G on $\Bbbk^n$, which induces an action of G on $\Bbbk(T_1, \ldots, T_n)$. Then $\Bbbk(T_1, \ldots, T_n)^G$ is a purely transcendental extension of $\Bbbk$.*

Proof. Let $K = \Bbbk(T_1, \ldots, T_{n-1})$ and $T = T_n$, then G is a group of automorphisms of $K[T]$ leaving K stable. So the result follows from 22.3.8 and induction on n. $\square$

22.4 Characters and weights

22.4.1 Lemma. *Let Γ be a group and X the set of homomorphisms from Γ to $\Bbbk \setminus \{0\}$. Then X is a family of linearly independent elements of the vector space of maps from Γ to $\Bbbk$.*

Proof. Suppose that there exist $\chi_1, \ldots, \chi_n \in X$ pairwise distinct linearly dependent with n minimal with this property. Then $\chi_i \neq 0$, $n > 1$ and there exist $\lambda_1, \ldots, \lambda_{n-1} \in \Bbbk \setminus \{0\}$ such that:

$$\chi_n = \lambda_1 \chi_1 + \cdots + \lambda_{n-1} \chi_{n-1}.$$

Since $\chi_1 \neq \chi_n$, $\chi_1(\beta) \neq \chi_n(\beta)$ for some $\beta \in \Gamma$. It follows that if $\alpha \in \Gamma$, then

$$\chi_n(\alpha)\chi_n(\beta) = \lambda_1 \chi_1(\alpha)\chi_1(\beta) + \cdots + \lambda_{n-1}\chi_{n-1}(\alpha)\chi_{n-1}(\beta)$$
$$= (\lambda_1 \chi_1(\alpha) + \cdots + \lambda_{n-1}\chi_{n-1}(\alpha))\chi_n(\beta)$$

So $\lambda_1(\chi_1(\beta) - \chi_n(\beta))\chi_1 + \cdots + \lambda_{n-1}(\chi_{n-1}(\beta) - \chi_n(\beta))\chi_{n-1} = 0$. But this is absurd by our choice of n. $\square$

22.4.2 Definition. *Let G be an algebraic group.*

(i) *A morphism $\chi : G \to \mathbf{G}_m$ of algebraic groups is called a* (multiplicative) character *of G.*

(ii) *A morphism $\lambda : \mathbf{G}_m \to G$ is called a* multiplicative one-parameter subgroup *of G.*

22.4.3 Denote by $X^*(G)$ the set of characters of G. It is a multiplicative subgroup of the set of maps from G to $\Bbbk$ and by 22.4.1, the elements of $X^*(G)$ are linearly independent in the $\Bbbk$-vector space of maps from G to $\Bbbk$.

If $\chi \in X^*(G)$ and $\alpha, \beta \in G$, then it is clear that $\chi(\alpha\beta\alpha^{-1}\beta^{-1}) = 1$, and hence $\chi|_{\mathcal{D}(G)} = 1$. In particular, $X^*(\mathrm{SL}_n(\Bbbk)) = \{1\}$ (10.8.8).

Let H be a closed normal subgroup of G. If $\chi \in X^*(H)$ and $\alpha \in G$, then the map $\chi_\alpha : H \to \mathbf{G}_m$, $\beta \mapsto \chi(\alpha\beta\alpha^{-1})$, is a character of H. Thus the map $(\alpha, \chi) \mapsto \chi_\alpha$ defines an action of G on $X^*(H)$.

22.4.4 Denote by $X_*(G)$ the set of multiplicative one-parameter subgroups of G.

If G is abelian, then $X_*(G)$ is a multiplicative subgroup of the set of maps from $\mathbf{G}_m$ to G.

Let $\lambda \in X_*(G)$ and $\alpha \in G$. Then the map $\lambda_\alpha : \mathbf{G}_m \to G$, $t \mapsto \alpha\lambda(t)\alpha^{-1}$, is again a multiplicative one-parameter subgroup of G. Thus the map $(\alpha, \lambda) \mapsto \lambda_\alpha$ defines an action of G on $X_*(G)$.

22.4.5 Let (V, π) be a rational G-module. For $\chi \in X^*(G)$, denote by $V_\chi = \{v \in V; \pi(\alpha)(v) = \chi(\alpha)v \text{ for all } \alpha \in G\}$. Then V_χ is a subspace of V. We say that χ is a *weight* of G in V if $V_\chi \neq \{0\}$. If χ is a weight of G in V, then V_χ is called the *weight space* of weight χ of G in V, and a non-zero vector of V_χ is called a *weight vector* of weight χ of G in V. A non-zero vector of V is called a weight vector if it is a weight vector for some χ.

Let $v \in V \setminus \{0\}$. Then it is easy to see that v is a weight vector if and only if the subspace spanned by $\pi(G)(v)$ is 1-dimensional.

Let H be a closed normal subgroup of G. We saw in 22.4.3 that G acts on $X^*(H)$. Let $\chi \in X^*(H)$ and denote by $V_{\chi,H}$ the weight space of weight χ of H in V. Then we see that $\pi(\alpha)(V_{\chi,H}) \subset V_{\chi_\alpha,H}$.

22.4.6 Lemma. *Let (V, π) be a rational G-module. Then the sum*

$$\sum_{\chi \in X^*(G)} V_\chi$$

is direct. In particular, if V is finite-dimensional, the set of weights of G in V is finite.

Proof. If the sum is not direct, then there exist an integer $n \geqslant 2$, pairwise distinct elements $\chi_1, \cdots, \chi_n$ of $X^*(G)$, and $v_i \in V_{\chi_i} \setminus \{0\}$, $1 \leqslant i \leqslant n$, such that $v_1 + \cdots + v_n = 0$. We may assume that n is minimal with this property. Since $\chi_1 \neq \chi_2$, there exists $\alpha \in G$ such that $\chi_1(\alpha) \neq \chi_2(\alpha)$. Then

$$0 = \pi(\alpha)(v_1 + \cdots + v_n) = \chi_1(\alpha)v_1 + \cdots + \chi_n(\alpha)v_n,$$

and hence,

$$(\chi_2(\alpha) - \chi_1(\alpha))v_2 + \cdots + (\chi_n(\alpha) - \chi_1(\alpha))v_n = 0.$$

But this contradicts the minimality of n. $\square$

22.5 Tori and diagonalizable groups

22.5.1 Definition. *Let G be an algebraic group.*

(i) *We say that G is a* torus *if it is isomorphic, as an algebraic group, to $D_n(\Bbbk)$ for some $n \in \mathbb{N}$.*

(ii) *We say that G is* diagonalizable *if it is isomorphic, as an algebraic group, to a closed subgroup of $D_n(\Bbbk)$ for some $n \in \mathbb{N}$.*

Remarks. 1) A torus is a connected diagonalizable algebraic group.

2) It is clear that a diagonalizable algebraic group is commutative, and that all the elements of a diagonalizable group are semisimple.

22.5.2 Let $n \in \mathbb{N}^*$ and $G = D_n(\Bbbk)$. We have:

$$\mathbf{A}(G) = \Bbbk[T_1, T_1^{-1}, \ldots, T_n, T_n^{-1}] = \bigoplus_{(a_1,\ldots,a_n)\in\mathbb{Z}^n} \Bbbk T_1^{a_1} \cdots T_n^{a_n}.$$

Moreover, we verify easily that invertible elements of $\mathbf{A}(G)$ are the elements $\lambda T_1^{a_1} \cdots T_n^{a_n}$ with $\lambda \in \Bbbk \setminus \{0\}$ and $(a_1, \ldots, a_n) \in \mathbb{Z}^n$.

For $(a_1, \ldots, a_n) \in \mathbb{Z}^n$, set $\chi_{(a_1,\ldots,a_n)} = T_1^{a_1} \cdots T_n^{a_n} \in \mathbf{A}(G)$. Then we verify easily that $\chi_{(a_1,\ldots,a_n)} \in X^*(G)$. Conversely, given $\chi \in X^*(G)$, then χ is invertible in $\mathbf{A}(G)$ and $\chi(1) = 1$, so it is of the form $\chi_{(a_1,\ldots,a_n)}$ for some $(a_1, \ldots, a_n) \in \mathbb{Z}^n$. It follows that the map:

$$\mathbb{Z}^n \to X^*(G) , \ (a_1, \ldots, a_n) \mapsto \chi_{(a_1,\ldots,a_n)},$$

is an isomorphism of groups, and the elements of $X^*(G)$ form a basis of $\mathbf{A}(G)$.

Similarly, for $(b_1, \ldots, b_n) \in \mathbb{Z}^n$, the map

$$\lambda_{(b_1,\ldots,b_n)} : \mathbf{G}_m \to G , \ t \mapsto \mathrm{diag}(t^{b_1}, \ldots, t^{b_n}),$$

defines an element of $X_*(G)$, and so the map

$$\mathbb{Z}^n \to X_*(G) , \ (b_1, \ldots, b_n) \mapsto \lambda_{(b_1,\ldots,b_n)},$$

is an isomorphism of groups.

Proposition. *Let $n \in \mathbb{N}^*$ and $G = D_n(\Bbbk)$.*

(i) *The groups $X^*(G)$ and $X_*(G)$ are free of rank n and the elements of $X^*(G)$ form a basis of $\mathbf{A}(G)$.*

(ii) *If $\chi \in X^*(G)$ and $\lambda \in X_*(G)$, then there exists an integer $\langle \chi, \lambda \rangle$ such that*

$$\chi(\lambda(t)) = t^{\langle \chi,\lambda \rangle}$$

for all $t \in \mathbf{G}_m$. Furthermore, the pairing

$$X^*(G) \times X_*(G) \to \mathbb{Z} , \ (\chi, \lambda) \mapsto \langle \chi, \lambda \rangle,$$

is bilinear and non-degenerate.

(iii) *If (V, π) is a rational G-module, then*

$$V = \bigoplus_{\chi\in X^*(G)} V_\chi.$$

In particular, V is the direct sum of G-submodules of dimension 1.

Proof. Part (i) is just a summary of the preceding discussion, and (ii) follows from:

$$\chi_{(a_1,\ldots,a_n)}(\lambda_{(b_1,\ldots,b_n)}(t)) = t^{a_1 b_1 + \cdots + a_n b_n}.$$

Finally for (iii), we may assume that V is finite-dimensional. Let $\mathcal{B}$ be a basis of V. For $\alpha \in G$, denote by $[\pi_{ij}(\alpha)]$ the matrix of $\pi(\alpha)$ with respect to $\mathcal{B}$. Since $\pi_{ij} \in \mathbf{A}(G)$, part (i) says that they are linear combinations of characters of G. So we have:

$$\pi(\alpha) = \sum_{\chi \in X^*(G)} \chi(\alpha) u_\chi,$$

where $u_\chi \in \operatorname{End}(V)$. If $\alpha, \beta \in G$, then:

$$\pi(\alpha\beta) = \sum_{\chi \in X^*(G)} \chi(\alpha)\chi(\beta) u_\chi = \pi(\alpha) \circ \pi(\beta) = \sum_{\chi,\chi' \in X^*(G)} \chi(\alpha)\chi'(\beta) u_\chi \circ u_{\chi'}.$$

Applying 22.4.1 to the characters of $G \times G$, we deduce that $u_\chi \circ u_\chi = u_\chi$ and $u_\chi \circ u_{\chi'} = 0$ if $\chi \neq \chi'$. On the other hand:

$$\sum_{\chi \in X^*(G)} u_\chi = \pi(e) = \operatorname{id}_V.$$

So if $M_\chi = u_\chi(V)$, then:

$$V = \bigoplus_{\chi \in X^*(G)} M_\chi.$$

Finally, if $v \in M_\chi$, then $u_\chi(v) = v$ and:

$$\pi(\alpha)(v) = \Big(\sum_{\chi' \in X^*(G)} \chi'(\alpha) u_{\chi'} \Big) \circ u_\chi(v) = \chi(\alpha) u_\chi^2(v) = \chi(\alpha) v.$$

So $M_\chi = V_\chi$ and we are done. $\square$

22.5.3 Theorem. *The following conditions are equivalent for an algebraic group G.*

(i) *G is diagonalizable.*

(ii) *$X^*(G)$ is a finitely generated abelian group whose elements form a basis of $\mathbf{A}(G)$.*

(iii) *Any rational G-module is the direct sum of G-submodules of dimension 1.*

Proof. (i) $\Rightarrow$ (ii) Suppose that G is closed subgroup of $H = \mathbf{D}_n(\Bbbk)$. Then $\mathbf{A}(G)$ is a quotient of $\mathbf{A}(H)$. If $\chi \in X^*(H)$, then its image in $\mathbf{A}(G)$ belongs to $X^*(G)$. Hence $X^*(G)$ is a basis of $\mathbf{A}(G)$ by 22.5.2 and 22.4.1.

Let $(\varepsilon_1, \ldots, \varepsilon_n)$ be the canonical basis of the $\mathbb{Z}$-module $\mathbb{Z}^n$ and ϕ_i the image of χ_{ε_i} in $\mathbf{A}(G)$. Then $\phi_1^{a_1} \cdots \phi_n^{a_n}$, $a_1, \ldots, a_n \in \mathbb{Z}$, span $\mathbf{A}(G)$, and they are characters of G. It follows that any $\phi \in X^*(G)$ is a linear combination of the $\phi_1^{a_1} \cdots \phi_n^{a_n}$'s. By 22.4.1, ϕ is one of the $\phi_1^{a_1} \cdots \phi_n^{a_n}$'s. Thus $X^*(G)$ is finitely generated.

(ii) $\Rightarrow$ (iii) The proof is the same as the one for 22.5.2.

(iii) $\Rightarrow$ (i) We may assume that G is a closed subgroup of $\mathrm{GL}(V)$ where V is a finite-dimensional $\Bbbk$-vector space (22.1.5). Then V is the direct sum of G-submodules of dimension 1, and the result is clear. $\square$

22.5.4 Corollary. *Let G be diagonalizable and H a closed subgroup of G.*

(i) *H is diagonalizable.*

(ii) *There exist $\chi_1, \ldots, \chi_n \in X^*(G)$ such that*

$$H = (\ker \chi_1) \cap \cdots \cap (\ker \chi_n).$$

(iii) *If $\theta \in X^*(H)$, then there exists $\chi \in X^*(G)$ such that $\chi|_H = \theta$.*

Proof. (i) This is obvious.

(ii) Let $X_H^*(G) = \{\chi \in X^*(G); H \subset \ker \chi\}$. If $\chi \in X_H^*(G)$, then $\chi - 1$ belongs to $\mathcal{I}(H)$.

Let $f \in \mathcal{I}(H) \setminus \{0\}$. By 22.5.3, there exist $\lambda_1, \ldots, \lambda_n \in \Bbbk \setminus \{0\}$ and $\chi_1, \ldots, \chi_n \in X^*(G)$ pairwise distinct such that $f = \lambda_1 \chi_1 + \cdots + \lambda_n \chi_n$. Reindexing if necessary, we may assume that there exist $m_0 = 0 < m_1 < \cdots < m_r = n$ such that $\chi_i|_H = \theta_j$ for $m_{j-1} + 1 \leqslant i \leqslant m_j$, where $\theta_1, \ldots, \theta_r$ are pairwise distinct elements of $X^*(H)$. Applying 22.4.1 to the θ_j's, we obtain that $\lambda_{m_{j-1}+1} + \cdots + \lambda_{m_j} = 0$ for $1 \leqslant j \leqslant r$. Set:

$$\mu_j = \sum_{i=m_{j-1}+1}^{m_j} \lambda_i (\chi_{m_j}^{-1} \chi_i - 1).$$

Then $f = \chi_{m_1} \mu_1 + \cdots + \chi_{m_r} \mu_r$, and we have $\chi_{m_j}^{-1} \chi_i \in X_H^*(G)$ if $m_{j-1} + 1 \leqslant i \leqslant m_j$. Thus the ideal $\mathcal{I}(H)$ is generated by $\chi - 1$, $\chi \in X_H^*(G)$. Since $\mathbf{A}(G)$ is Noetherian, there exist $\chi_1, \ldots, \chi_r \in X_H^*(G)$ such that

$$\mathcal{I}(H) = \mathbf{A}(G)(\chi_1 - 1) + \cdots + \mathbf{A}(G)(\chi_r - 1).$$

So the result follows.

(iii) By 22.5.3, there exist $\lambda_1, \ldots, \lambda_n \in \Bbbk \setminus \{0\}$ and $\chi_1, \ldots, \chi_n \in X^*(G)$ such that $\theta = \lambda_1(\chi_1|_H) + \cdots + \lambda_n(\chi_n|_H)$. In view of 22.4.1, $\chi_i|_H = \theta$ for all i, and the sum $\lambda_1 + \cdots + \lambda_n = 1$. $\square$

22.5.5 Lemma. (i) *The only closed connected subgroups of $\mathbf{G}_m$ are $\{1\}$ and $\mathbf{G}_m$.*

(ii) *If G is a connected algebraic group, then $X^*(G)$ is torsion free.*

Proof. (i) Let H be a closed connected subgroup of $\mathbf{G}_m$. Since $\dim H \leqslant \dim \mathbf{G}_m = 1$ (14.1.6), we have $\dim H = 0$ or 1. But H is connected, so H is either $\{1\}$ or $\mathbf{G}_m$.

(ii) Let $\chi \in X^*(G)$. Then $\chi(G)$ is a closed connected subgroup of $\mathbf{G}_m$ (21.2.4). By (i), $\chi(G) = \{1\}$ or $\mathbf{G}_m$. In the first case, $\chi = 1$. In the second case, $\chi^n \neq 1$ for all n. $\square$

22.5.6 The group algebra $\Bbbk[\mathbb{Z}^n]$ of the abelian group $\mathbb{Z}^n$ is isomorphic to $\Bbbk[T_1, T_1^{-1}, \ldots, T_n, T_n^{-1}]$. It follows from 22.5.3 that if G is diagonalizable, then the group algebra $\Bbbk[X^*(G)]$ is isomorphic to $\mathbf{A}(G)$.

22.5.7 Theorem. *Let $n \in \mathbb{N}^*$ and G an affine algebraic group. The following conditions are equivalent:*

(i) *As an algebraic group, G is isomorphic to $\mathrm{D}_n(\Bbbk)$.*

(ii) *G is a torus of dimension n.*

(iii) *G is a connected diagonalizable algebraic group of dimension n.*

(iv) *G is diagonalizable and $X^*(G)$ is isomorphic to $\mathbb{Z}^n$.*

Proof. Clearly, we have (i) $\Rightarrow$ (ii) $\Rightarrow$ (iii).

(iii) $\Rightarrow$ (iv) By 22.5.3 and 22.5.5, $X^*(G)$ is a finitely generated torsion free abelian group. So it is of the form $\mathbb{Z}^r$ for some $r \in \mathbb{N}$. In view of 22.5.6, $\mathbf{A}(G)$ is isomorphic to $\Bbbk[T_1, T_1^{-1}, \ldots, T_r, T_r^{-1}]$. Thus $r = n$ by 6.1.4, 6.2.3 and 14.1.2.

(iv) $\Rightarrow$ (i) Let $\chi_1, \ldots, \chi_n$ be a basis of the free abelian group $X^*(G)$. The map
$$u : G \to \mathrm{D}_n(\Bbbk) \ , \ \alpha \mapsto \mathrm{diag}\big(\chi_1(\alpha), \ldots, \chi_n(\alpha)\big)$$
is a morphism of algebraic groups. Since that $\mathbf{A}(G) = \Bbbk[T_1, T_1^{-1}, \ldots, T_n, T_n^{-1}]$ by 22.5.6, $\mathbf{A}(u)$ is bijective. Thus u is an isomorphism of algebraic groups (11.4.5). $\square$

22.5.8 Lemma. *Let A be a commutative ring and $f : E \to F$ a surjective homomorphism of A-modules. Suppose that F is a free A-module. Then there is a submodule E' of E such that*

(i) *$E = E' \oplus \ker f$.*

(ii) *f induces an isomorphism from E' onto F.*

Proof. Let $(y_i)_{i \in I}$ be a basis of F. For $i \in I$, fix $x_i \in E$ such that $f(x_i) = y_i$. Then $(x_i)_{i \in I}$ is a linearly independent subset. So the submodule $E' = \sum_{i \in I} A x_i$ satisfies the required properties. $\square$

22.5.9 Theorem. *Let G be diagonalizable. Then G° is a torus, and there exists a finite subgroup H (not unique in general) of G such that G is, isomorphic as an algebraic group, to the product $H \times G^\circ$.*

Proof. That G° is a torus follows from 22.5.4 and 22.5.7.

Let $\pi : G \to \mathrm{D}_n(\Bbbk)$ be a morphism of algebraic groups inducing an isomorphism from G to a closed subgroup of $\mathrm{D}_n(\Bbbk)$. By 22.5.4 (iii), restriction to H induces a surjective homomorphism from $\mathbb{Z}^n \simeq X^*(\mathrm{D}_n(\Bbbk))$ to $X^*(G^\circ)$. By 22.5.5, $X^*(G^\circ)$ is a free $\mathbb{Z}$-module. Applying 22.5.8 and using the fact that $\mathbb{Z}$ is a principal ideal domain (so any submodule of a free module is free), we obtain a basis $\chi_1, \ldots, \chi_n$ of $X^*(\mathrm{D}_n(\Bbbk))$ such that $\chi_1, \ldots, \chi_r$ generate the subgroup consisting of characters χ verifying $\chi|_{G^\circ} = 1$. The automorphism
$$\alpha \mapsto \mathrm{diag}(\chi_1(\alpha), \ldots, \chi_n(\alpha))$$

of $D_n(\Bbbk)$ then sends $\pi(G^\circ)$ to the set of matrices $\mathrm{diag}(\mu_1, \ldots, \mu_n)$ verifying $\mu_1 = \cdots = \mu_r = 1$. Thus $D_n(\Bbbk) = D_r(\Bbbk) \times \pi(G^\circ)$. So as abstract groups, $\pi(G) = H \times \pi(G^\circ)$ where $H = \pi(G) \cap D_r(\Bbbk)$. Since H is isomorphic to G/G°, it is finite (21.1.6). Finally π induces an isomorphism of algebraic groups from G onto $H \times \pi(G^\circ)$ since the restriction of π to each irreducible component of G is an isomorphism. $\square$

22.5.10 The following result is called the *rigidity of diagonalizable groups*.

Proposition. *Let G, H be diagonalizable algebraic groups, X an irreducible affine variety and $u : X \times G \to H$ a morphism of varieties. Assume that for all $x \in V$, the map*

$$v_x : G \to H \ , \ \alpha \mapsto u(x, \alpha),$$

is a morphism of algebraic groups. Then $u(x, \alpha)$ does not depend on x.

Proof. Let $(\theta, x) \in X^*(H) \times X$. By our hypotheses, the map $\alpha \mapsto \theta(u(x, \alpha))$ is a character of G and the map $(x, \alpha) \mapsto \theta(u(x, \alpha))$ is a regular function on $X \times G$. By 22.5.3, we have:

$$\theta\big(u(x, \alpha)\big) = \sum_{\chi \in X^*(G)} f_{\theta, \chi}(x) \chi(\alpha),$$

where $f_{\theta, \chi} \in \mathbf{A}(X)$ for all $\chi \in X^*(G)$. It follows from 22.4.1 that given $x \in X$, there exists $\chi_{\theta, x} \in X^*(G)$ such that $\theta(u(x, \alpha)) = \chi_{\theta, x}(\alpha)$ for all $\alpha \in G$. So $f_{\theta, \chi_{\theta, x}}(x) = 1$ and $f_{\theta, \chi}(x) = 0$ if $\chi \in X^*(G) \setminus \{\chi_{\theta, x}\}$. Thus $f_{\theta, \chi}(x) \in \{0, 1\}$. For $\varepsilon \in \{0, 1\}$, set $X_\varepsilon = \{x \in X; \ f_{\theta, \chi}(x) = \varepsilon\}$. Since X is the disjoint union of the closed subsets X_0 and X_1, we deduce that $\chi_{\theta, x}$ depends only on θ. But 11.3.3 and 22.5.3 say that the elements of $X^*(H)$ separate the points of H, so the result follows. $\square$

22.5.11 Corollary. *Let H be a closed diagonalizable subgroup of G.*
(i) *We have $\big(N_G(H)\big)^\circ = \big(C_G(H)\big)^\circ$.*
(ii) *The index of $C_G(H)$ in $N_G(H)$ is finite.*

Proof. The subgroups $N_G(H)$ and $C_G(H)$ are closed in G by 21.3.7. Applying 22.5.10 to the following map

$$u : \big(N_G(H)\big)^\circ \times H \to H \ , \ (\beta, \alpha) \to \beta \alpha \beta^{-1},$$

we obtain that $u(\beta, \alpha) = u(e, \alpha) = \alpha$ for all $\alpha \in H$ and $\beta \in \big(N_G(H)\big)^\circ$. So we have proved (i). Finally, (ii) follows from (i) and 21.1.6 (i). $\square$

22.6 Groups of dimension one

22.6.1 Lemma. *If G is connected and* $\dim G = 1$, *then G is commutative.*

Proof. Let us fix $\beta \in G$ and denote by $u : G \to G$ the morphism $\alpha \mapsto \alpha\beta\alpha^{-1}$. Since $\overline{u(G)}$ is irreducible (1.1.5 and 1.1.7) and $\dim G = 1$, either $u(G) = \{\beta\}$ or $\overline{u(G)} = G$.

Suppose that $\overline{u(G)} = G$. Then $u(G)$ contains a non-empty open subset of G (15.4.3 and 1.4.6). Thus $G \setminus u(G)$ is a finite set (14.1.3, 14.1.5 and 14.1.6). We may assume that G is a closed subgroup of $\mathrm{GL}_n(\Bbbk)$ (22.1.5). For $\alpha \in G$, denote by χ_α its characteristic polynomial. There exist $f_0, \ldots, f_{n-1} \in \mathbf{A}(G)$ such that

$$\chi_\alpha(T) = T^n + f_{n-1}(\alpha)T^{n-1} + \cdots + f_0(\alpha).$$

Since $\chi_\beta = \chi_{\alpha\beta\alpha^{-1}}$, the morphism $G \to \Bbbk^n$, $\alpha \mapsto (f_0(\alpha), \ldots, f_{n-1}(\alpha))$, is constant on $u(G)$ and hence on G. Thus $\chi_\alpha(T) = \chi_e(T) = (T-1)^n$ for all $\alpha \in G$, and G is unipotent.

Now $u(G)\beta^{-1} \subset \mathcal{D}(G)$ and so $G \setminus \mathcal{D}(G)$ is finite. It follows that $\overline{\mathcal{D}(G)} = G$, so $\mathcal{D}(G) = G$ (21.3.6). But this contradicts the fact that G is unipotent (10.8.10 and 22.3.2).

So $\alpha\beta\alpha^{-1} = \{\beta\}$ for all $\alpha, \beta \in G$ and G is commutative. $\square$

22.6.2 Theorem. *Let G be a connected algebraic group of dimension 1.*
  (i) *If G is not unipotent, then G is isomorphic to $\mathbf{G}_m$ as algebraic groups.*
  (ii) *If G is unipotent, then G is isomorphic to $\mathbf{G}_a$ as algebraic groups.*

Proof. We may assume that G is a closed subgroup of $\mathrm{GL}_n(\Bbbk)$. By 22.6.1, G is commutative. So we have $G = G_s \times G_u$. Since $\dim G = 1$ and G is connected, we have either $G = G_s$ or $G = G_u$.

If $G = G_s$, then since G is commutative, G is diagonalizable (22.2.4). Therefore G is isomorphic to $\mathbf{G}_m$ by 22.5.7.

If $G = G_u$, then we may identify G with a closed subgroup of $\mathrm{U}_n(\Bbbk)$ (22.3.2). Let $\alpha \in G \setminus \{e\}$, then the image of the additive one-parameter subgroup $\varphi_{\log(\alpha)}(\mathbf{G}_a)$ is a closed subgroup of G isomorphic to $\mathbf{G}_a$ (22.3.4). But $\dim G = 1$, so G is isomorphic to $\mathbf{G}_a$. $\square$

References and comments

- [5], [23], [38], [40], [78].

Examples of non-affine algebraic groups are abelian varieties. The reader may refer to [53] and [58] for a detailed account of the subject.

Lie algebra of an algebraic group

This chapter explains how to associate a Lie algebra to an algebraic group, and studies some basic properties of this association.

23.1 An associative algebra

23.1.1 In this chapter, G will denote an algebraic group with identity element e_G or e, μ_G or μ the group multiplication and ι_G or ι the inverse. For $\alpha \in G$, define $\chi_\alpha : \mathbf{A}(G) \to \Bbbk$, $\chi_\alpha(f) = f(\alpha)$ to be the evaluation at α. It is clear that χ_α is an element of the dual $\mathbf{A}(G)^*$ of $\mathbf{A}(G)$.

23.1.2 If $x, y \in \mathbf{A}(G)^*$, then $x \otimes y \in (\mathbf{A}(G) \otimes_\Bbbk \mathbf{A}(G))^*$ is the element defined as follows: $(x \otimes y)(f \otimes g) = x(f)y(g)$ for all $f, g \in \mathbf{A}(G)$. Denote by

$$x \cdot y = (x \otimes y) \circ \mathbf{A}(\mu).$$

So $x \cdot y \in \mathbf{A}(G)^*$.

Lemma. *Endowed with the operation $(x, y) \mapsto x \cdot y$, $\mathbf{A}(G)^*$ is an associative $\Bbbk$-algebra whose identity element is χ_e. The map $G \to \mathbf{A}(G)^*$, $\alpha \mapsto \chi_\alpha$, induces an injective homomorphism of G into the group of invertible elements of $\mathbf{A}(G)^*$.*

Proof. Let $f \in \mathbf{A}(G)$ and $\mathbf{A}(\mu)(f) = s_1 \otimes t_1 + \cdots + s_n \otimes t_n$ where $s_1, t_1, \ldots, s_n, t_n \in \mathbf{A}(G)$. For $\alpha \in G$, we have:

$$f(\alpha) = f(e\alpha) = \sum_{i=1}^n s_i(e)t_i(\alpha) = f(\alpha e) = \sum_{i=1}^n s_i(\alpha)t_i(e).$$

Thus

$$f = s_1(e)t_1 + \cdots + s_n(e)t_n = t_1(e)s_1 + \cdots + t_n(e)s_n.$$

If $x \in \mathbf{A}(G)^*$, then:

$$(\chi_e \cdot x)(f) = (\chi_e \otimes x) \circ \mathbf{A}(\mu)(f) = \sum_{i=1}^n s_i(e)x(t_i) = x\left(\sum_{i=1}^n s_i(e)t_i\right) = x(f).$$

It follows that $\chi_e \cdot x = x$. Similarly, we have $x \cdot \chi_e = x$. Consequently, we have $\chi_\alpha \cdot \chi_\beta = \chi_{\alpha\beta}$ for all $\alpha, \beta \in G$ and the map $\alpha \mapsto \chi_\alpha$ is injective.

The associativity of the group multiplication implies that:

$$\big(\mathrm{id}_{\mathbf{A}(G)} \otimes \mathbf{A}(\mu)\big) \circ \mathbf{A}(\mu) = \big(\mathbf{A}(\mu) \otimes \mathrm{id}_{\mathbf{A}(G)}\big) \circ \mathbf{A}(\mu).$$

So for all $x, y, z \in \mathbf{A}(G)^*$:

$$\begin{aligned}
(x \otimes y \otimes z) \circ \big(I \otimes \mathbf{A}(\mu)\big) \circ \mathbf{A}(\mu) &= \big(x \otimes ((y \otimes z) \circ \mathbf{A}(\mu))\big) \circ \mathbf{A}(\mu) = x \cdot (y \cdot z), \\
(x \otimes y \otimes z) \circ \big(\mathbf{A}(\mu) \otimes I\big) \circ \mathbf{A}(\mu) &= \big(((x \otimes y) \circ \mathbf{A}(\mu)) \otimes z\big) \circ \mathbf{A}(\mu) = (x \cdot y) \cdot z,
\end{aligned}$$

where $I = \mathrm{id}_{\mathbf{A}(G)}$. We have therefore proved the result. $\quad\square$

23.1.3 Lemma. *Let $u : G \to H$ be a morphism of algebraic groups and $\mathbf{A}'(u)$ the map dual to $\mathbf{A}(u)$. Then $\mathbf{A}'(u) \in \mathrm{Hom}_{\mathrm{alg}}(\mathbf{A}(G)^*, \mathbf{A}(H)^*)$, and $\mathbf{A}'(u)(\chi_\alpha) = \chi_{u(\alpha)}$ for all $\alpha \in G$.*

Proof. Since u is a morphism of algebraic groups, we have:

$$\mathbf{A}(\mu_G) \circ \mathbf{A}(u) = \big(\mathbf{A}(u) \otimes \mathbf{A}(u)\big) \circ \mathbf{A}(\mu_H).$$

So for $x, y \in \mathbf{A}(G)^*$:

$$\begin{aligned}
(x \cdot y) \circ \mathbf{A}(u) &= (x \otimes y) \circ \mathbf{A}(\mu_G) \circ \mathbf{A}(u) = (x \otimes y) \circ \big(\mathbf{A}(u) \otimes \mathbf{A}(u)\big) \circ \mathbf{A}(\mu_H) \\
&= \big(x \circ \mathbf{A}(u)\big) \cdot \big(y \circ \mathbf{A}(u)\big).
\end{aligned}$$

Thus $\mathbf{A}'(u)$ is a homomorphism of algebras and the second part follows immediately. $\quad\square$

23.2 Lie algebras

23.2.1 For $\alpha \in G$, we shall denote the left and right translation of functions by α by λ_α and ρ_α as in 22.1.4.

Since αG° is the unique irreducible component of G containing α, we have $\dim G = \dim G^\circ$. It follows from 16.3.7 and the smoothness of points in G that $T_e(G^\circ)$ can be identified with $T_e(G)$, which in turn, can be identified with $\mathrm{Der}_{\Bbbk}^e(\mathbf{A}(G), \Bbbk)$ (16.2.4).

Let $\mathfrak{Lie}(G)$ be the set of derivations X of $\mathbf{A}(G)$ verifying:

$$X \circ \lambda_\alpha = \lambda_\alpha \circ X$$

for all $\alpha \in G$. Such a derivation is said to be *left invariant* and $\mathfrak{Lie}(G)$ is a Lie subalgebra of $\mathrm{Der}_{\Bbbk}(\mathbf{A}(G))$.

Let $X \in \mathfrak{Lie}(G)$. For $f \in \mathbf{A}(G)$, set:

$$\theta(X)(f) = X(f)(e) = \chi_e\big(X(f)\big).$$

Clearly $\theta(X) \in \mathrm{Der}_{\Bbbk}^e(\mathbf{A}(G), \Bbbk)$ and so we have a linear map:

$$\theta \colon \mathfrak{Lie}(G) \to \operatorname{Der}^e_{\Bbbk}(\mathbf{A}(G),\Bbbk) = \mathrm{T}_e(G) \;,\; X \mapsto \theta(X).$$

We shall prove that θ is bijective.

23.2.2 Let $f \in \mathbf{A}(G)$ and $x \in \operatorname{Der}^e_{\Bbbk}(\mathbf{A}(G),\Bbbk)$. Define a function $f * x$ on G as follows: for $\alpha \in G$, set

$$(f * x)(\alpha) = x(\lambda_{\alpha^{-1}} f).$$

This is called the *right convolution* of f by x.

Let $\mathbf{A}(\mu)(f) = s_1 \otimes t_1 + \cdots + s_n \otimes t_n$ where $s_1, t_1, \ldots, s_n, t_n \in \mathbf{A}(G)$. For α, $\beta \in G$, we have $\mathbf{A}(\mu)(f)(\alpha,\beta) = f(\alpha\beta) = s_1(\alpha)t_1(\beta) + \cdots + s_n(\alpha)t_n(\beta)$. So

$$\lambda_{\alpha^{-1}} f = s_1(\alpha)t_1 + \cdots + s_n(\alpha)t_n.$$

Since $(f * x)(\alpha) = x(\lambda_{\alpha^{-1}} f)$, it follows that:

$$f * x = x(t_1)s_1 + \cdots + x(t_n)s_n.$$

Hence $f * x \in \mathbf{A}(G)$. So right convolution by x is an endomorphism of $\mathbf{A}(G)$. We claim that it is a derivation of $\mathbf{A}(G)$. Let $f, g \in \mathbf{A}(G)$, $x \in \operatorname{Der}^e_{\Bbbk}(\mathbf{A}(G),\Bbbk)$ and $\alpha \in G$. Then:

$$\begin{aligned}
(fg * x)(\alpha) &= x\bigl(\lambda_{\alpha^{-1}} f \lambda_{\alpha^{-1}} g\bigr) = x(\lambda_{\alpha^{-1}} f)(\lambda_{\alpha^{-1}} g)(e) + x(\lambda_{\alpha^{-1}} g)(\lambda_{\alpha^{-1}} f)(e) \\
&= x(\lambda_{\alpha^{-1}} f)g(\alpha) + x(\lambda_{\alpha^{-1}} g)f(\alpha) \\
&= (f * x)(\alpha)g(\alpha) + (g * x)(\alpha)f(\alpha).
\end{aligned}$$

Hence $(fg) * x = (f * x)g + (g * x)f$ and we have proved our claim.

For $\beta \in G$. we have:

$$\begin{aligned}
\bigl(\lambda_\beta(f * x)\bigr)(\alpha) &= (f * x)(\beta^{-1}\alpha) = x(\lambda_{\alpha^{-1}\beta} f) \\
&= x\bigl(\lambda_{\alpha^{-1}}(\lambda_\beta f)\bigr) = \bigl((\lambda_\beta f) * x\bigr)(\alpha).
\end{aligned}$$

So $\lambda_\beta(f * x) = (\lambda_\beta f) * x$ and the derivation $f \mapsto f * x$ is left invariant. It follows that $x \mapsto *x$ defines a map

$$\eta : \operatorname{Der}^e_{\Bbbk}(\mathbf{A}(G),\Bbbk) \to \mathfrak{Lie}(G).$$

We claim that η is the inverse of θ. If $X \in \mathfrak{Lie}(G)$, $f \in \mathbf{A}(G)$ and $\alpha \in G$, then:

$$\begin{aligned}
(\eta\circ\theta(X)(f))(\alpha) &= (f * \theta(X))(\alpha) = \theta(X)(\lambda_{\alpha^{-1}} f) \\
&= X(\lambda_{\alpha^{-1}} f)(e) = \bigl(\lambda_{\alpha^{-1}}(Xf)\bigr)(e) = X(f)(\alpha).
\end{aligned}$$

So $\eta\circ\theta(X) = X$. Similarly, if $x \in \operatorname{Der}^e_{\Bbbk}(\mathbf{A}(G),\Bbbk)$, then

$$\bigl(\theta\circ\eta(x)\bigr)(f) = (f * x)(e) = x(\lambda_{e^{-1}} f) = x(f).$$

Hence $\theta\circ\eta(x) = x$. So our claim follows and we have obtained the following result.

Proposition. *The linear map* $\theta : \mathfrak{Lie}(G) \to \operatorname{Der}^e_{\Bbbk}(\mathbf{A}(G), \Bbbk) = \mathrm{T}_e(G)$ *is an isomorphism. In particular,* $\mathfrak{Lie}(G)$ *is a Lie algebra over* $\Bbbk$ *of dimension* $\dim G$.

23.2.3 We can transport the Lie algebra structure on $\mathfrak{Lie}(G)$ to $\mathrm{T}_e(G)$ via the isomorphism θ, and the Lie algebra thus obtained, denoted by $\mathcal{L}(G)$, shall be called the *Lie algebra of* G. This Lie algebra structure can be recovered from the structure of associative algebra on $\mathbf{A}(G)^*$ defined in 23.1. Let $x, y \in \mathcal{L}(G) = \mathrm{T}_e(G)$, $f \in \mathbf{A}(G)$ and $\mathbf{A}(\mu)(f) = s_1 \otimes t_1 + \cdots + s_n \otimes t_n$ where $s_1, t_1, \ldots, s_n, t_n \in \mathbf{A}(G)$. Then:

$$((f * x) * y)(e) = x(t_1)(s_1 * y)(e) + \cdots + x(t_n)(s_n * y)(e)$$
$$= y(s_1)x(t_1) + \cdots + y(s_n)x(t_n) = (y \cdot x)(f).$$

Hence $\theta\big([\eta(x), \eta(y)]\big) = x \cdot y - y \cdot x$. and the Lie bracket on $\mathcal{L}(G)$ is given by:

$$[x, y] = x \cdot y - y \cdot x.$$

23.2.4 Let $u : G \to H$ be a morphism of algebraic groups and denote by $du : \mathcal{L}(G) \to \mathcal{L}(H)$ the differential du_e (or $\mathrm{T}_e(u)$) of u at e (16.3.1).

Proposition. *The map* du *is a Lie algebra homomorphism.*

Proof. Let $x, y, z \in \mathcal{L}(G)$. Since $du(z) = z \circ \mathbf{A}(u)$ (16.3.3) and we saw in the proof of 23.1.3 that:

$$(x \cdot y - y \cdot x) \circ \mathbf{A}(u) = \big(x \circ \mathbf{A}(u)\big) \cdot \big(y \circ \mathbf{A}(u)\big) - \big(y \circ \mathbf{A}(u)\big) \cdot \big(x \circ \mathbf{A}(u)\big),$$

the result follows from 23.2.3. $\square$

23.2.5 Let H be a closed subgroup of G. By 16.3.7 and 23.2.4, we may identify $\mathcal{L}(H)$ with a Lie subalgebra of $\mathcal{L}(G)$. More precisely:

Lemma. *Let* $\mathfrak{a} = \mathcal{I}(H)$. *Then:*

$$\mathcal{L}(H) = \{x \in \mathcal{L}(G); \mathfrak{a} * x \subset \mathfrak{a}\} = \{x \in \mathcal{L}(G); x(\mathfrak{a}) = \{0\}\}.$$

Proof. Let $j : H \to G$ be the canonical injection, then $\mathbf{A}(j)$ is just the restriction of an element of $\mathbf{A}(G)$ to H and if $x \in \mathcal{L}(H)$, then $dj(x) = x \circ \mathbf{A}(j)$. Hence $dj(x)(f) = 0$ for all $f \in \mathfrak{a}$. Let $f \in \mathfrak{a}$, $x \in \mathcal{L}(H)$ and $\alpha \in H$. Since $\lambda_{\alpha^{-1}} f \in \mathfrak{a}$ (22.1.6), we have $\big(f * dj(x)\big)(\alpha) = dj(x)(\lambda_{\alpha^{-1}} f) = 0$. Thus $f * dj(x) \in \mathfrak{a}$

Conversely, let $x \in \mathcal{L}(G)$ be such that $\mathfrak{a} * x \subset \mathfrak{a}$. If $f \in \mathfrak{a}$, then $(f * x)(e) = 0 = x(f)$. We deduce that x induces an element y of $\operatorname{Der}^e_{\Bbbk}(\mathbf{A}(G)/\mathfrak{a}, \Bbbk) = \operatorname{Der}^e_{\Bbbk}(\mathbf{A}(H), \Bbbk)$, because $\mathbf{A}(H) = \mathbf{A}(G)/\mathfrak{a}$. Clearly, $x = dj(y)$.

Finally if $f \in \mathfrak{a}$ and $x \in \mathcal{L}(G)$ verify $f * x \in \mathfrak{a}$, then $x(f) = (f * x)(e) = 0$. Conversely, if $x(g) = 0$ for all $g \in \mathfrak{a}$, then, since $\lambda_{\alpha^{-1}} g \in \mathfrak{a}$ for $\alpha \in H$, we have $(g * x)(\alpha) = x(\lambda_{\alpha^{-1}} g) = 0$, and so $g * x \in \mathfrak{a}$. $\square$

23.2.6 A derivation $X \in \mathrm{Der}_\Bbbk(\mathbf{A}(G), \Bbbk)$ is said to be *right invariant* if $X \circ \rho_\alpha = \rho_\alpha \circ X$ for all $\alpha \in G$. We shall denote the set of right invariant derivation of $\mathbf{A}(G)$ by $\mathfrak{Lie}'(G)$. It is a Lie subalgebra of $\mathrm{Der}_\Bbbk(\mathbf{A}(G))$. As in the case of left invariant derivations, we have a linear map $\theta' : \mathfrak{Lie}'(G) \to T_e(G)$ defined as follows: for $X \in \mathfrak{Lie}'(G)$ and $f \in \mathbf{A}(G)$, set

$$\theta'(X)(f) = X(f)(e) = \chi_e(X(f)).$$

Also for $f \in \mathbf{A}(G)$ and $x \in T_e(G)$, we have the notion of *left convolution* $x * f$ of f by x defined as follows: for $\alpha \in G$,

$$(x * f)(\alpha) = x(\rho_\alpha f).$$

Using the same arguments as in 23.2.2, right convolution defines a map η' which is the inverse of θ'. Thus we have, via θ', another Lie algebra structure on $T_e(G)$ that we shall denote by $\mathcal{L}'(G)$. Then we see that the Lie bracket on $\mathcal{L}'(G)$ is given by:

$$[x, y] = y \cdot x - x \cdot y.$$

So $\mathcal{L}'(G)$ is the *opposite* Lie algebra of $\mathcal{L}(G)$. The analogue of 23.2.5 for $\mathcal{L}'(H)$ (with left convolution) remains valid.

23.2.7 Let $\alpha \in G$, $x \in \mathcal{L}(G)$, and $f, s_1, t_1, \ldots, s_n, t_n \in \mathbf{A}(G)$ be such that

$$\mathbf{A}(\mu)(f) = s_1 \otimes t_1 + \cdots + s_n \otimes t_n.$$

Since $(\chi_\alpha \cdot x)(f) = s_1(\alpha)x(t_1) + \cdots + s_n(\alpha)x(t_n)$, it follows from 23.2.2 that:

$$(\chi_\alpha \cdot x)(f) = (f * x)(\alpha).$$

Similarly, we have:

$$(x \cdot \chi_\alpha)(f) = (x * f)(\alpha).$$

Let us evaluate $\chi_\alpha \cdot x \cdot \chi_{\alpha^{-1}}$. For $1 \leqslant i \leqslant n$, let $\mathbf{A}(\mu)(t_i) = u_{i1} \otimes v_{i1} + \cdots + u_{ip} \otimes v_{ip}$ where $u_{ij}, v_{ij} \in \mathbf{A}(G)$. Then $\rho_{\alpha^{-1}} t_i = v_{i1}(\alpha^{-1})u_{i1} + \cdots + v_{ip}(\alpha^{-1})u_{ip}$. Thus $x(\rho_{\alpha^{-1}} t_i) = x(u_{i1})v_{i1}(\alpha^{-1}) + \cdots + x(u_{ip})v_{ip}(\alpha^{-1})$ and we deduce that:

$$
\begin{aligned}
(\chi_\alpha \cdot x \cdot \chi_{\alpha^{-1}})(f) &= \big(\chi_\alpha \otimes [(x \otimes \chi_{\alpha^{-1}}) \circ \mathbf{A}(\mu)]\big) \circ \mathbf{A}(\mu)(f) \\
&= \sum_{i=1}^n s_i(\alpha)\Big(\sum_{j=1}^p x(u_{ij})v_{ij}(\alpha^{-1})\Big) = \sum_{i=1}^n s_i(\alpha)x(\rho_{\alpha^{-1}} t_i).
\end{aligned}
$$

On the other hand, let $i_\alpha : G \to G$, $\beta \mapsto \alpha\beta\alpha^{-1}$. Then:

$$(f \circ i_\alpha)(\beta) = f(\alpha\beta\alpha^{-1}) = s_1(\alpha)t_1(\beta\alpha^{-1}) + \cdots + s_n(\alpha)t_n(\beta\alpha^{-1}).$$

Hence $f \circ i_\alpha = s_1(\alpha)(\rho_{\alpha^{-1}} t_1) + \cdots + s_n(\alpha)(\rho_{\alpha^{-1}} t_n)$. So we have:

$$(\chi_\alpha \cdot x \cdot \chi_{\alpha^{-1}})(f) = x(f \circ i_\alpha).$$

23.3 Examples

23.3.1 Let $G = \mathbf{G}_a$ and $\mathbf{A}(G) = \mathbb{k}[T]$. The Lie algebras $\mathfrak{Lie}(G)$ and $\mathcal{L}(G)$ are 1-dimensional, so they are abelian. Let $X \in \mathrm{Der}_{\mathbb{k}}(\mathbf{A}(G))$ be the derivation defined by $X(P) = P'(T)$ for all $P \in \mathbb{k}[T]$. If $\alpha \in G$, then $\lambda_\alpha \circ X = X \circ \lambda_\alpha$. So X spans $\mathfrak{Lie}(G)$. It follows that $\mathcal{L}(G)$ is spanned by x, where $x(P) = P'(0)$ for $P \in \mathbb{k}[T]$.

23.3.2 Let $G = \mathbf{G}_m$ and $\mathbf{A}(G) = \mathbb{k}[T, T^{-1}]$. Again $\mathfrak{Lie}(G)$ and $\mathcal{L}(G)$ are 1-dimensional. The unique derivation X of $\mathbf{A}(G)$ verifying $X(T) = T$ is left invariant, and therefore it spans $\mathfrak{Lie}(G)$. It follows that $\mathcal{L}(G)$ is spanned by $x \in \mathrm{Der}^e_{\mathbb{k}}(\mathbf{A}(G), \mathbb{k})$ where $x(P) = P'(1)$ for $P \in \mathbf{A}(G)$.

23.3.3 Let $G = \mathrm{GL}_n(\mathbb{k})$ and denote by I the identity matrix. Since $\mathrm{GL}_n(\mathbb{k})$ is an affine open subset of $\mathrm{M}_n(\mathbb{k})$, $\mathrm{T}_I(\mathrm{GL}_n(\mathbb{k}))$ may be identified with $\mathrm{T}_I(\mathrm{M}_n(\mathbb{k}))$ (16.2.4), which is $\mathrm{M}_n(\mathbb{k})$.

Let $\mathbf{A}(G) = \mathbb{k}[T_{11}, T_{12}, \ldots, T_{nn}, D^{-1}]$ where D is the polynomial in T_{ij} such that $D(M) = \det M$ for all $M \in \mathrm{M}_n(\mathbb{k})$. It follows that an element $x \in \mathcal{L}(G)$ is completely determined by its values on T_{ij}.

Denote by $\varphi : \mathcal{L}(G) \to \mathfrak{gl}_n(\mathbb{k})$ the map given by $x \mapsto [x(T_{ij})]$. It is clear that φ is linear and injective. Since $\mathcal{L}(G)$ and $\mathfrak{gl}_n(\mathbb{k})$ are both of dimension n^2, φ is bijective.

Let $x, y \in \mathcal{L}(G)$, $X = [x_{ij}] = \varphi(x)$ and $Y = [y_{ij}] = \varphi(y)$. The formulas for matrix multiplication imply that $\mathbf{A}(\mu)(T_{ij}) = T_{i1} \otimes T_{1j} + \cdots + T_{in} \otimes T_{nj}$ for $1 \leqslant i, j \leqslant n$. Thus

$$(x \cdot y)(T_{ij}) = x_{i1}y_{1j} + \cdots + x_{in}y_{nj}.$$

It follows from 23.2.3 that

$$[x, y] = \varphi^{-1}([X, Y]).$$

So we have proved that φ is a Lie algebra isomorphism and we may identify $\mathcal{L}(G)$ with $\mathfrak{gl}_n(\mathbb{k})$ via φ.

Now let V be a finite-dimensional $\mathbb{k}$-vector space, $H = \mathrm{GL}(V)$ and $\mathfrak{h} = \mathcal{L}(H)$. Then the same applies for H. In fact, the map $\mathrm{End}(V)^* \to \mathbf{A}(H)$, $\psi \mapsto \psi|_H$, is injective since H is dense in $\mathrm{End}(V)$. So any $x \in \mathfrak{h}$ identifies with a linear form on $\mathrm{End}(V)^*$. It follows that given $x \in \mathfrak{h}$, there is a unique $X \in \mathrm{End}(V)$ such that $x(g) = g(X)$ for all $g \in \mathrm{End}(V)^*$. This defines a linear map $\varphi : \mathfrak{h} \to \mathfrak{gl}(V) = \mathrm{End}(V)$ which is injective since $\mathbf{A}(H)$ is generated by $\mathrm{End}(V)^*$ and $1/\det$. They have the same dimension, and by using the same arguments as above, it is clear that φ is a Lie algebra isomorphism. So we may identify $\mathfrak{h}$ with $\mathfrak{gl}(V)$ via φ.

23.3.4 Let H be a closed subgroup of $G = \mathrm{GL}_n(\mathbb{k})$ and $\mathfrak{a}$ (resp. $\mathfrak{b}$) the ideal of H in $\mathbf{A}(G)$ (resp. $\mathbf{A}(\mathrm{M}_n(\mathbb{k})) = \mathbb{k}[T_{11}, T_{12}, \ldots, T_{nn}]$). Then clearly $\mathfrak{a}$ is the ideal of $\mathbf{A}(G)$ generated by $\mathfrak{b}$.

Take $H = \mathrm{T}_n(\Bbbk)$ the set of upper triangular matrices in G. Then $\mathfrak{a}$ is generated by T_{ij}, $i > j$. If $x \in \mathcal{L}(G)$ and $X = \varphi(x) = [x_{ij}]$ as in 23.3.3, then by 23.2.2, we have, for $1 \leqslant i, j \leqslant n$,

$$T_{ij} * x = x_{1j}T_{i1} + x_{2j}T_{i2} + \cdots + x_{nj}T_{in}.$$

Thus if $i > j$, then $T_{ij} * x \in \mathfrak{a}$ if and only if $x_{kj} = 0$ for $k \geqslant i$. By 23.2.5, we see that $\mathcal{L}(\mathrm{T}_n(\Bbbk))$ identifies with $\mathfrak{t}_n(\Bbbk)$, the set of upper triangular matrices in $\mathrm{M}_n(\Bbbk)$.

The same arguments show that the Lie algebra of $\mathrm{D}_n(\Bbbk)$ is $\mathfrak{d}_n(\Bbbk)$, and the Lie algebra of $\mathrm{U}_n(\Bbbk)$ is $\mathfrak{n}_n(\Bbbk)$ (in the notations of 10.8.1 and 19.1.1).

23.3.5 Let $G = \mathrm{GL}_n(\Bbbk)$ and $A = [a_{ij}] \in \mathrm{M}_n(\Bbbk)$ a nilpotent matrix. Recall from 22.3.3 that the image H of the map $u : \mathbf{G}_a \to G$, $t \mapsto e^{tA}$, is a connected closed subgroup of G and u induces an isomorphism from $\mathbf{G}_a$ onto H.

We saw in 23.3.1 that $\mathcal{L}(\mathbf{G}_a)$ is spanned by x, where $x(P) = P'(0)$ for $P \in \Bbbk[T]$. We deduce that $\mathcal{L}(H)$ is spanned by $du(x)$. We have $du(x)(T_{ij}) = x(T_{ij} \circ u) = a_{ij}$, so $du(x) = \varphi^{-1}(A)$ where φ is as in 23.3.3. Identifying $\mathcal{L}(G)$ with $\mathfrak{gl}_n(\Bbbk)$, we see that $\mathcal{L}(H) = \Bbbk A$.

23.3.6 Let us use the notations of 23.3.3. We may consider the algebra $A = \mathbf{A}(\mathrm{M}_n(\Bbbk)) = \Bbbk[T_{11}, T_{12}, \ldots, T_{nn}]$ as a subalgebra of $\mathbf{A}(G)$. If $X \in \mathrm{M}_n(\Bbbk)$ and $f \in \mathbf{A}(G)$, we denote $f * \varphi^{-1}(X)$ by $f * X$ (there is no confusion since φ is a Lie algebra isomorphism). The map $f \mapsto f * X$ is a derivation of $\mathbf{A}(G)$ and if $X = [x_{ij}]$, then $T_{ij} * X = x_{1j}T_{i1} + \cdots + x_{nj}T_{in}$. We deduce that $f * X \in A$ if $f \in A$. In the same way, if $f \in A$ and $\alpha \in G$, then $\rho_\alpha f$ and $\lambda_\alpha f$ belong to A.

Proposition. *Let H be a closed subgroup of $G = \mathrm{GL}_n(\Bbbk)$ and $\mathfrak{b}$ the ideal of H in A. Then:*

$$H = \{\alpha \in G;\ \rho_\alpha(\mathfrak{b}) = \mathfrak{b}\} = \{\alpha \in G;\ \lambda_\alpha(\mathfrak{b}) = \mathfrak{b}\},$$
$$\mathcal{L}(H) = \{X \in \mathfrak{gl}_n(\Bbbk);\ \mathfrak{b} * X \subset \mathfrak{b}\}.$$

Proof. Let $S = \{D^n; n \in \mathbb{N}\}$ and $\mathfrak{a}$ the ideal of H in $\mathbf{A}(G)$. Then $\mathfrak{a} = S^{-1}\mathfrak{b}$ (2.3.7). So $\mathfrak{a} = \mathbf{A}(G)\mathfrak{b}$, $\mathfrak{b} = A \cap \mathfrak{a}$. Let $\alpha \in G$ and $X \in \mathfrak{gl}_n(\Bbbk)$.

For $f \in \mathbf{A}(G)$ and $g \in A$, we have

$$\rho_\alpha(fg) = (\rho_\alpha f)(\rho_\alpha g) ,\ (fg) * X = (f * X)g + f(g * X).$$

It follows that if $\rho_\alpha(\mathfrak{b}) \subset \mathfrak{b}$ and $\mathfrak{b} * X \subset \mathfrak{b}$, then $\rho_\alpha(\mathfrak{a}) \subset \mathfrak{a}$ and $\mathfrak{a} * X \subset \mathfrak{a}$. So 22.1.6 and 23.2.5 imply that $\alpha \in H$ and $X \in \mathcal{L}(H)$.

Conversely, given $\alpha \in H$ and $X \in \mathcal{L}(H)$. Since $\mathfrak{a} * X \subset \mathfrak{a}$ (23.2.5) and $A * X \subset A$, we have $\mathfrak{b} * X \subset \mathfrak{b}$. Similarly, we have $\rho_\alpha(\mathfrak{a}) \subset \mathfrak{a}$ and $\rho_{\alpha^{-1}}(\mathfrak{a}) \subset \mathfrak{a}$, so $\rho_\alpha(\mathfrak{a}) = \mathfrak{a}$. But $\rho_\alpha(A) = A$, we deduce therefore that $\rho_\alpha(\mathfrak{b}) = \mathfrak{b}$. $\square$

23.4 Computing differentials

23.4.1 Let G be an algebraic group and $u : G \to X$ a morphism of varieties. When there is no confusion, we shall denote du_e by du. Recall that we may identify $\mathcal{L}(\mathrm{GL}_n(\Bbbk))$ with $\mathfrak{gl}_n(\Bbbk)$ via the map $x \mapsto [x(T_{ij})]$ (23.3.3).

23.4.2 Lemma. *Let $u : G \to \mathrm{GL}_n(\Bbbk)$ be a morphism of algebraic groups and $u_{ij} \in \mathbf{A}(G)$, $1 \leqslant i, j \leqslant n$, be such that $u(\alpha) = [u_{ij}(\alpha)]$ for all $\alpha \in G$. If $x \in \mathcal{L}(G)$, then $du(x) = [x(u_{ij})]$.*

Proof. This is clear since $du(x)(T_{ij}) = x(T_{ij} \circ u) = x(u_{ij})$. $\quad\square$

23.4.3 Let V be a finite-dimensional subspace of $\mathbf{A}(G)$ such that for all $\alpha \in G$, $\rho_\alpha(V) \subset V$, and $\rho : G \to \mathrm{GL}(V)$ the morphism of algebraic groups given by $\alpha \mapsto \rho_\alpha|_V$.

Proposition. *For $x \in \mathcal{L}(G)$ and $f \in V$, we have:*

$$d\rho(x)(f) = f * x.$$

Proof. Let $\mathcal{B} = (f_1, \ldots, f_n)$ be a basis of V. If $\alpha \in G$ and $1 \leqslant j \leqslant n$, then $\rho_\alpha f_j = \rho_{1j}(\alpha)f_1 + \rho_{2j}(\alpha)f_2 + \cdots + \rho_{nj}(\alpha)f_n$, where $\rho_{ij} \in \mathbf{A}(G)$ for $1 \leqslant i, j \leqslant n$. By 23.4.2, if $x \in \mathcal{L}(G)$, then $d\rho(x)(f_j) = x(\rho_{1j})f_1 + \cdots + x(\rho_{nj})f_n$.

On the other hand, we have:

$$f_j(\beta\alpha) = (\lambda_{\beta^{-1}}f_j)(\alpha) = (\rho_\alpha f_j)(\beta) = \rho_{1j}(\alpha)f_1(\beta) + \cdots + \rho_{nj}(\alpha)f_n(\beta)$$

for $\alpha, \beta \in G$. Hence:

$$(f_j * x)(\beta) = x(\lambda_{\beta^{-1}}f_j) = x(\rho_{1j})f_1(\beta) + \cdots + x(\rho_{nj})f_n(\beta) = d\rho(x)(f_j)(\beta),$$

and we have proved the required result. $\quad\square$

23.4.4 Corollary. *Let V be a finite-dimensional subspace of $\mathbf{A}(G)$ verifying $\rho_\alpha(V) \subset V$ for all $\alpha \in G$. Then $V * x \subset V$ for all $x \in \mathcal{L}(G)$.*

23.4.5 In relation with 23.4.3, we give a result concerning the determinant. Let $n \in \mathbb{N}^*$, $\mathfrak{S}_n$ the symmetric group of $\{1, \ldots, n\}$ and if $\sigma \in \mathfrak{S}_n$, denote by $\varepsilon(\sigma)$ the signature of σ. If $\alpha \in G = \mathrm{GL}_n(\Bbbk)$, denote by $D(\alpha)$ the determinant of α. As an element of $\mathbf{A}(G)$, we have:

$$D = \sum_{\sigma \in \mathfrak{S}_n} \varepsilon(\sigma) T_{1\sigma(1)} \cdots T_{n\sigma(n)}.$$

For $\alpha, \beta \in G$ and $x = [x(T_{ij})] \in \mathcal{L}(G) = \mathfrak{gl}_n(\Bbbk)$, we have $\lambda_{\alpha^{-1}}D = D(\alpha)D$ because $(\lambda_{\alpha^{-1}}D)(\beta) = D(\alpha\beta) = D(\alpha)D(\beta)$. Since $x \in \mathcal{L}(G) = \mathrm{Der}^e_{\Bbbk}(\mathbf{A}(G), \Bbbk)$, we see that $x(\lambda_{\alpha^{-1}}D)$ is equal to

$$D(\alpha) \sum_{\sigma \in \mathfrak{S}_n} \varepsilon(\sigma) \sum_{i=1}^{n} \delta_{1\sigma(1)} \cdots \delta_{i-1,\sigma(i-1)} x(T_{i\sigma(i)}) \delta_{i+1,\sigma(i+1)} \cdots \delta_{n,\sigma(n)},$$

where δ_{ij} denotes the symbol of Kronecker. Thus $(D * x)(\alpha) = D(\alpha)\operatorname{tr}(x)$. Consequently:

$$D * x = \operatorname{tr}(x)D.$$

Now $H = \mathrm{SL}_n(\Bbbk) = \{X \in \mathrm{M}_n(\Bbbk); D(X) = 1\}$ is a connected closed subgroup of G of dimension $n^2 - 1$ (21.3.3 and 14.2.3). Denote by $\mathfrak{b}$ the ideal of H in $\mathbf{A}(\mathrm{M}_n(\Bbbk))$. Then $D - 1 \in \mathfrak{b}$. If $x \in \mathcal{L}(H)$, then $(D - 1) * x = \operatorname{tr}(x)D \in \mathfrak{b}$ (23.3.6). Since $\mathfrak{b} \neq \mathbf{A}(\mathrm{M}_n(\Bbbk))$, it follows that $\operatorname{tr}(x) = 0$. Thus $\mathcal{L}(H) \subset \mathfrak{sl}_n(\Bbbk)$. Since $\dim \mathcal{L}(H) = \dim \mathfrak{sl}_n(\Bbbk) = n^2 - 1$, we deduce that $\mathcal{L}(H) = \mathfrak{sl}_n(\Bbbk)$.

23.4.6 Let G, H be algebraic groups, $x \in \mathcal{L}(G)$ and $y \in \mathcal{L}(H)$. For $f \in \mathbf{A}(G)$ and $g \in \mathbf{A}(H)$, set:

$$\theta_{x,y}(f \otimes g) = x(f)g(e_H) + f(e_G)y(g).$$

We saw in 16.3.8 that the map $\theta : \mathcal{L}(G) \times \mathcal{L}(H) \to \mathcal{L}(G \times H)$, $(x, y) \mapsto \theta_{x,y}$, is a linear bijection. We claim that it is a Lie algebra isomorphism. Let $x, x' \in \mathcal{L}(G)$, $y, y' \in \mathcal{L}(H)$, $z = \theta_{x,y}$, $z' = \theta_{x',y'}$. Then it is easy to check that:

$$(f \otimes g) * z = (f * x) \otimes g + f \otimes (g * y).$$

It follows that $((f \otimes g) * z) * z'$ is equal to:

$$((f * x) * x') \otimes g + (f * x) \otimes (g * y') + (f * x') \otimes (g * y) + f \otimes ((g * y) * y').$$

This implies that $((f \otimes g) * z) * z' - ((f \otimes g) * z') * z$ is equal to:

$$((f * x) * x' - (f * x') * x) \otimes g + f \otimes ((g * y) * y' - (g * y') * y).$$

Hence we have proved our claim (23.2.3). We shall identify the Lie algebras $\mathcal{L}(G) \times \mathcal{L}(H)$ and $\mathcal{L}(G \times H)$ via θ.

23.4.7 Proposition. *Let $x, y \in \mathcal{L}(G)$, then:*

$$d\mu(x, y) = x + y \ , \ d\iota(x) = -x.$$

Proof. Let $f \in \mathbf{A}(G)$ and $\mathbf{A}(\mu)(f) = s_1 \otimes t_1 + \cdots + s_n \otimes t_n$ where $s_i, t_i \in \mathbf{A}(G)$ pour $1 \leqslant i \leqslant n$. Set $z = \theta_{x,y}$ (23.4.6). Then:

$$d\mu(z)(f) = z\Big(\sum_{i=1}^{n} s_i \otimes t_i\Big) = \sum_{i=1}^{n} \big(x(s_i)t_i(e) + s_i(e)y(t_i)\big).$$

On the other hand, we saw in the proof of 23.1.2 that:

$$f = s_1(e)t_1 + \cdots + s_n(e)t_n = t_1(e)s_1 + \cdots + t_n(e)s_n.$$

So it is clear that $d\mu(z)(f) = x(f) + y(f)$.

Next, let $\pi : G \to G \times G$ denote the morphism $\alpha \mapsto (\alpha, \iota(\alpha))$. Then $\mu \circ \pi(\alpha) = e$, so $d\mu \circ d\pi = 0$. But $d\pi(x) = (x, d\iota(x))$, so we obtain from the formula for $d\mu$ that $x + d\iota(x) = 0$. $\square$

23.4.8 Corollary. *Let $f, g \in \mathbf{A}(G)$ be such that $g(\alpha) = f(\alpha^{-1})$ for all $\alpha \in G$. If $x \in \mathcal{L}(G)$, then $x(g) = -x(f)$.*

Proof. Note that $g = f \circ \iota$ and f, g are morphisms from G to $\mathbf{G}_a$. So $dg = df \circ d\iota = -df$ (23.4.7). But $df(x)(h) = x(h \circ f)$ and $dg(x)(h) = x(h \circ g)$ for $x \in \mathcal{L}(G)$ and $h \in \Bbbk[T] = \mathbf{A}(\mathbf{G}_a)$. So the result follows by taking $h = T$.   □

23.4.9 Corollary. *Let (V, π) be a finite-dimensional rational representation of G, (V^*, σ) its contragredient representation (which is clearly rational), $x \in \mathcal{L}(G)$, $f \in V^*$, and $v \in V$. Then:*

$$(d\sigma(x)f)(v) = -f(d\pi(x)v).$$

Proof. Let $(v_1, \ldots, v_n)$ be a basis of V and $(\varphi_1, \ldots, \varphi_n)$ its dual basis. We have

$$\pi(\alpha)v_j = \sum_{i=1}^{n} \pi_{ij}(\alpha)v_i \; , \; \sigma(\alpha)\varphi_j = \sum_{i=1}^{n} \sigma_{ij}(\alpha)\varphi_i,$$

where $\pi_{ij}, \sigma_{ij} \in \mathbf{A}(G)$. Then we check easily that $\sigma_{ij}(\alpha) = \pi_{ji}(\alpha^{-1})$. It follows from 23.4.2 and 23.4.8 that for $x \in \mathcal{L}(G)$, we have:

$$d\sigma(x)(\varphi_j) = -x(\pi_{j1})\varphi_1 - \cdots - x(\pi_{jn})\varphi_n.$$

Hence $d\sigma(x)(\varphi_j)(v_i) = -x(\pi_{ji}) = -\varphi_j(d\pi(x)v_i)$.   □

23.4.10 Proposition. *Let $u : G \to G$ be a morphism of varieties such that $u(e) = e$, and set $v : G \to G$, $\alpha \mapsto u(\alpha)\alpha^{-1}$. Then*

$$dv = du - \mathrm{id}_{\mathcal{L}(G)}.$$

Proof. Let $\theta : G \to G \times G$ be the morphism $\alpha \mapsto (u(\alpha), \alpha^{-1})$. Then $v = \mu \circ \theta$. If $x \in \mathcal{L}(G)$, then by 23.4.7, we have $dv(x) = d\mu(du(x), -x) = du(x) - x$.   □

23.4.11 Remark. Let V be a finite-dimensional subspace of $\mathbf{A}(G)$ verifying $\lambda_\alpha(V) \subset V$ for all $\alpha \in G$. Denote by $\lambda : G \to \mathrm{GL}(V)$ the morphism of algebraic groups $\alpha \mapsto \lambda_\alpha|_V$. Using the same computations as in 23.4.3 and the result obtained in 23.4.8, we obtain that for $x \in \mathcal{L}(G)$ and $f \in V$,

$$d\lambda(x)(f) = -x * f.$$

23.4.12 Let $\pi : G \to \mathrm{GL}(U)$ and $\tau : H \to \mathrm{GL}(V)$ be finite-dimensional rational representations of the algebraic groups G and H, and $S = U \oplus V$, $T = U \otimes_{\Bbbk} V$. We define the rational representations $(S, \pi \oplus \tau)$ and $(T, \pi \otimes \tau)$ of $G \times H$ as follows: for $\alpha \in G$, $\beta \in H$, $u \in U$ and $v \in V$,

$$(\pi \oplus \tau)(\alpha, \beta)(u, v) = (\pi(\alpha)(u), \tau(\beta)(v)),$$
$$(\pi \otimes \tau)(\alpha, \beta)(u \otimes v) = \pi(\alpha)(u) \otimes \tau(\beta)(v).$$

Proposition. *Let $x \in \mathcal{L}(G)$, $y \in \mathcal{L}(H)$, $u \in U$ and $v \in V$. Then:*

$$d(\pi \oplus \tau)(x,y)(u,v) = \big(d\pi(x)(u), d\tau(y)(v)\big),$$
$$d(\pi \otimes \tau)(x,y)(u \otimes v) = \big(d\pi(x)(u)\big) \otimes v + u \otimes \big(d\tau(y)(v)\big).$$

Proof. Let $(e_1, \ldots, e_m)$ and $(f_1, \ldots, f_n)$ be bases of U and V. Then

$$\pi(\alpha)e_j = \sum_{i=1}^{m} \pi_{ij}(\alpha)e_i \ , \ \tau(\beta)f_l = \sum_{k=1}^{n} \tau_{kl}(\beta)f_k,$$

where $\alpha \in G$, $\beta \in H$, $\pi_{ij} \in \mathbf{A}(G)$ and $\tau_{kl} \in \mathbf{A}(H)$.

Let $I = \{1, \ldots, m\} \times \{1, \ldots, n\}$ and for $(j,l) \in I$, set $g_{(j,l)} = e_j \otimes f_l$. The $g_{(j,l)}$'s, for $(j,l) \in I$, form a basis of $U \otimes_{\Bbbk} V$. We have

$$(\pi \otimes \tau)(\alpha, \beta)g_{(j,l)} = \sum_{(i,k) \in I} \theta_{(i,k)}^{(j,l)}(\alpha, \beta)g_{(i,k)},$$

where $\theta_{(i,k)}^{(j,l)} \in \mathbf{A}(G \times H) = \mathbf{A}(G) \otimes_{\Bbbk} \mathbf{A}(H)$. So $\theta_{(i,k)}^{(j,l)} = \pi_{ij} \otimes \tau_{kl}$.

Let $z = (x,y) \in \mathcal{L}(G \times H)$. By 23.4.6, we have:

$$z(\theta_{(i,k)}^{(j,l)}) = \pi_{ij}(e_G)y(\tau_{kl}) + x(\pi_{ij})\tau_{kl}(e_H).$$

Hence we obtain the second equality (23.4.2). The proof for the first equality is analogue. $\square$

23.4.13 Let us conserve the notations of 23.4.12 with $H = G$. We shall also denote by $(\pi \otimes \sigma, U \otimes_{\Bbbk} V)$ the rational representation of G defined by:

$$(\pi \otimes \tau)(\alpha)(u \otimes v) = \big(\pi(\alpha)u\big) \otimes \big(\tau(\alpha)v\big).$$

It follows from the preceding discussion that for $x \in \mathcal{L}(G)$:

$$d(\pi \otimes \tau)(x) = d\pi(x) \otimes \mathrm{id}_V + \mathrm{id}_U \otimes d\tau(x).$$

23.4.14 Let (V, π) be a finite-dimensional rational representation of G. We define, as in 10.6.5, the representations $\mathrm{T}^n(\pi)$, $\mathrm{S}^n(\pi)$ and $\bigwedge^n(\pi)$ of G. They are also rational and we have for $x \in \mathcal{L}(G)$, $v_1, \ldots, v_n \in V$:

$$d\,\mathrm{T}^n(\pi)(x)(v_1 \otimes \cdots \otimes v_n) = \sum_{i=1}^{n} v_1 \otimes \cdots \otimes v_{i-1} \otimes d\pi(x)v_i \otimes v_{i+1} \otimes \cdots \otimes v_n,$$

$$d\,\mathrm{S}^n(\pi)(x)(v_1 \cdots v_n) = \sum_{i=1}^{n} v_1 \cdots v_{i-1}\big(d\pi(x)v_i\big)v_{i+1} \cdots v_n,$$

$$d{\textstyle\bigwedge}^n(\pi)(x)(v_1 \wedge \cdots \wedge v_n) = \sum_{i=1}^{n} v_1 \wedge \cdots \wedge v_{i-1} \wedge d\pi(x)v_i \wedge v_{i+1} \wedge \cdots \wedge v_n.$$

23.4.15 Let $\mathfrak{A}$ be a finite-dimensional $\Bbbk$-algebra (not necessarily associative or with unit). A $\Bbbk$-linear map $\delta : \mathfrak{A} \to \mathfrak{A}$ is called a *derivation* of $\mathfrak{A}$ if for all $x, y \in \mathfrak{A}$, $\delta(x.y) = \delta(x).y + x.\delta(y)$. Let $\operatorname{Der} \mathfrak{A}$ (resp. $\operatorname{Aut} \mathfrak{A}$) denote the set of derivations (resp automorphisms) of $\mathfrak{A}$. By 21.1.3, $\operatorname{Aut} \mathfrak{A}$ is an algebraic group.

Proposition. *Let $\pi : G \to \operatorname{Aut} \mathfrak{A}$ be a rational representation of G. Then for $x \in \mathcal{L}(G)$, $d\pi(x) \in \operatorname{Der} \mathfrak{A}$.*

Proof. Let $(e_1, \ldots, e_n)$ be a basis of $\mathfrak{A}$. There exist $\gamma_{ijk} \in \Bbbk$ and $\pi_{ij} \in \mathbf{A}(G)$ such that:

$$e_i.e_j = \sum_{k=1}^{n} \gamma_{ijk} e_k \ , \ \pi(\alpha)e_j = \sum_{i=1}^{n} \pi_{ij}(\alpha)e_i.$$

Since $\pi(\alpha)(e_i.e_j) = (\pi(\alpha)e_i).(\pi(\alpha)e_j)$, we have:

$$\sum_{k=1}^{n} \gamma_{ijk} \pi_{lk}(\alpha) = \sum_{r,s=1}^{n} \gamma_{rsl} \pi_{ri}(\alpha) \pi_{sj}(\alpha).$$

Denote by δ_{uv} the symbol of Kronecker, then:

$$\sum_{k=1}^{n} \gamma_{ijk} x(\pi_{lk}) = \sum_{r,s=1}^{n} \gamma_{rsl} \big(\delta_{ri} x(\pi_{sj}) + x(\pi_{ri}) \delta_{sj} \big)$$
$$= \sum_{s=1}^{n} \gamma_{isl} x(\pi_{sj}) + \sum_{r=1}^{n} \gamma_{rjl} x(\pi_{ri}).$$

On the other hand, 23.4.2 implies that:

$$d\pi(x)(e_i.e_j) = \sum_{l=1}^{n} \Big(\sum_{k=1}^{n} \gamma_{ijk} x(\pi_{lk}) \Big) e_l.$$

Similarly, $\big(d\pi(x)e_i\big).e_j + e_i.\big(d\pi(x)e_j\big)$ is equal to:

$$\sum_{l=1}^{n} \Big(\sum_{r=1}^{n} \gamma_{rjl} x(\pi_{ri}) \Big) e_l + \sum_{l=1}^{n} \Big(\sum_{s=1}^{n} \gamma_{isl} x(\pi_{sj}) \Big) e_l.$$

So the result follows. □

23.4.16 Proposition. *Let $\mathfrak{g}$ be a semisimple Lie algebra and G the group of elementary automorphisms of $\mathfrak{g}$. Then $\mathcal{L}(G) = \mathfrak{g}$.*

Proof. Since $G \subset \operatorname{Aut} \mathfrak{g}$, $\mathcal{L}(G) \subset \operatorname{ad} \mathfrak{g}$ (23.4.15 and 20.1.5) and $\mathcal{L}(G)$ is a Lie subalgebra of $\operatorname{ad} \mathfrak{g}$ (16.3.7 and 23.2.4). Finally, if $\operatorname{ad} x$ is nilpotent, then $\operatorname{ad} x \in \mathcal{L}(G)$ by 23.3.5. So the result is a consequence of 20.4.6. □

23.4.17 Proposition. *Let (U, π) and (V, τ) be finite-dimensional rational representations of an algebraic group G. Assume that there exists a linear map $\theta : U \to V$ such that $\theta\big(\pi(\alpha)u\big) = \tau(\alpha)\big(\theta(u)\big)$ pour $(\alpha, u) \in G \times U$. Then we have $\theta\big(d\pi(x)u\big) = d\tau(x)\big(\theta(u)\big)$ for $x \in \mathcal{L}(G)$ and $u \in U$.*

Proof. Let $\mathcal{U}$ and $\mathcal{V}$ be bases of U and V. Denote by $A(\alpha) = [\pi_{ij}(\alpha)]$ (resp. $B(\alpha) = [\tau_{kl}(\alpha)])$ the matrix of $\pi(\alpha)$ (resp. $\tau(\alpha)$) with respect to the basis $\mathcal{U}$ (resp. $\mathcal{V}$). If $x \in \mathcal{L}(G)$, 23.4.2 implies that $M(x) = [x(\pi_{ij})]$ (resp. $N(x) = [x(\tau_{kl})])$ is the matrix of $d\pi(x)$ (resp. $d\tau(x)$) with respect to the basis $\mathcal{U}$ (resp. $\mathcal{V}$). Let Λ be the matrix of θ with respect to the bases $\mathcal{U}$ and $\mathcal{V}$. Then our hypothesis says that $\Lambda A(\alpha) = B(\alpha)\Lambda$. From this, we deduce immediately that $\Lambda M(x) = N(x)\Lambda$. This proves the result. $\quad\square$

23.5 Adjoint representation

23.5.1 In the rest of this chapter, G is an algebraic group and $\mathfrak{g}$ its Lie algebra.

23.5.2 For $\alpha \in G$, denote by i_α the inner automorphism of G defined by $\beta \mapsto \alpha\beta\alpha^{-1}$. Recall that the set $\operatorname{Int} G$ of inner automorphisms of G is a subgroup of $\operatorname{Aut} G$.

Let $\operatorname{Ad}_G(\alpha)$ or $\operatorname{Ad}(\alpha)$ be the differential of i_α. By 16.3.1 and 23.2.4, it is an automorphism of the Lie algebra $\mathfrak{g}$. Since $i_{\alpha\beta} = i_\alpha \circ i_\beta$ for $\alpha, \beta \in G$, we deduce that $\operatorname{Ad}(\alpha\beta) = \operatorname{Ad}(\alpha) \circ \operatorname{Ad}(\beta)$. Thus the map

$$\operatorname{Ad} : G \to \operatorname{GL}(\mathfrak{g}) \, , \, \alpha \to \operatorname{Ad}(\alpha)$$

defines a representation of G, called the *adjoint representation* of G. By 23.2.7, if $\alpha \in G$ and $x \in \mathfrak{g}$, then:

$$\operatorname{Ad}(\alpha)(x) = \chi_\alpha \cdot x \cdot \chi_{\alpha^{-1}}.$$

Let $u : G \to H$ a morphism of algebraic groups. If $\alpha \in G$, then $u \circ i_\alpha = i_{u(\alpha)} \circ u$. It follows that:

$$du \circ \operatorname{Ad}_G(\alpha) = \operatorname{Ad}_H\big(u(\alpha)\big) \circ du.$$

Let $\alpha \in G$, $x \in \mathfrak{g}$ and $y = \operatorname{Ad}(\alpha)(x)$. In the notations of 23.2.2, let $X, Y \in \mathfrak{Lie}(G)$ be defined by $X = \eta(x)$, $Y = \eta(y)$. It is clear that $\rho_\alpha \circ X \circ \rho_{\alpha^{-1}}$ is a derivation of $\mathbf{A}(G)$. Moreover, since left and right translations commute, $\rho_\alpha \circ X \circ \rho_{\alpha^{-1}} \in \mathfrak{Lie}(G)$. Let $f \in \mathbf{A}(G)$ and $\beta \in G$. Then:

$$(\rho_\alpha \circ X \circ \rho_{\alpha^{-1}})(f)(\beta) = X(\rho_{\alpha^{-1}}f)(\beta\alpha) = x\big(\lambda_{(\beta\alpha)^{-1}}(\rho_{\alpha^{-1}}f)\big)$$
$$= x\big((\lambda_{\beta^{-1}}f)\circ i_\alpha\big) = y(\lambda_{\beta^{-1}}f) = Y(f)(\beta).$$

Hence $Y = \rho_\alpha \circ X \circ \rho_{\alpha^{-1}}$.

Let $f_1, \dots, f_r \in \mathbf{A}(G)$, be linearly independent elements which generate $\mathbf{A}(G)$ as an algebra. Complete this family to a basis $(f_n)_{n \in \mathbb{N}}$ of the $\Bbbk$-vector space $\mathbf{A}(G)$. Since the G-module $\mathbf{A}(G)$ is rational (for the representations λ and ρ), we have

$$f_k \circ i_\alpha = \sum_{n \in \mathbb{N}} a_{n,k}(\alpha) f_n$$

where, for $1 \leqslant k \leqslant r$, $(a_{n,k})_n$ is a family of regular functions on G verifying for $\alpha \in G$, $a_{n,k}(\alpha) \neq 0$ only if n belongs to a finite subset (which depends on k and α) of $\mathbb{N}$. Let $x \in \mathfrak{g}$, $y = \mathrm{Ad}(\alpha)(x)$ and $x_n = x(f_n)$. For $n \geqslant r+1$, x_n is a polynomial in $x_1, \ldots, x_r$. Moreover:

$$y(f_k) = x(f_k \circ i_\alpha) = \sum_{n \in \mathbb{N}} a_{nk}(\alpha) x_n.$$

An element $z \in \mathfrak{g}$ is completely determined by $z(f_k)$, $1 \leqslant k \leqslant r$. It follows that if $x \in \mathfrak{g}$, then the map $\alpha \mapsto \mathrm{Ad}(\alpha)(x)$ is a morphism from G to $\mathfrak{g}$. By 21.5.2, the adjoint representation of G in $\mathfrak{g}$ is rational.

23.5.3 Lemma. *Let $\alpha \in G$ and $u : G \to G$ be the morphism of varieties $\beta \to \alpha\beta\alpha^{-1}\beta^{-1}$. Then $du = \mathrm{Ad}(\alpha) - \mathrm{id}_\mathfrak{g}$.*

Proof. Apply 23.4.10 with $u = i_\alpha$. $\square$

23.5.4 Lemma. *Assume that G is a closed subgroup of $\mathrm{GL}_n(\Bbbk)$, so we may identify $\mathfrak{g}$ with a Lie subalgebra of $\mathfrak{gl}_n(\Bbbk)$ (23.2.5 and 23.3.3). If $\alpha \in G$ and $x \in \mathfrak{g}$, then $\mathrm{Ad}(\alpha)(x) = \alpha x \alpha^{-1}$.*

Proof. Let $\alpha = [\alpha_{ij}]$, $\beta = [\beta_{ij}] \in G$, $\alpha^{-1} = [\alpha'_{ij}]$ and $\gamma = \alpha\beta\alpha^{-1} = [\gamma_{ij}]$. If $x \in \mathfrak{g}$, we may identify x with the matrix $[x(T_{ij})]$. Now

$$\mathrm{Ad}(\alpha)(x)(T_{ij}) = x(T_{ij} \circ i_\alpha) = x\left(\sum_{k,l=1}^n \alpha_{ik} \alpha'_{lj} T_{kl} \right) = \sum_{k,l=1}^n \alpha_{ik} x(T_{kl}) \alpha'_{lj}.$$

On the other hand, we have

$$\gamma_{ij} = \sum_{k,l=1}^n \alpha_{ik} \beta_{kl} \alpha'_{lj}.$$

So $\mathrm{Ad}(\alpha)(x) = \alpha x \alpha^{-1}$. $\square$

23.5.5 Let us determine the differential of Ad.

Let $\alpha, \beta \in G$, $x, y \in \mathfrak{g}$, $f \in \mathbf{A}(G)$ and $\mathbf{A}(\mu)(f) = s_1 \otimes t_1 + \cdots + s_n \otimes t_n$ where $s_1, t_1, \ldots, s_n, t_n \in \mathbf{A}(G)$. We obtain $\rho_\alpha f = t_1(\alpha) s_1 + \cdots + t_n(\alpha) s_n$ as in 23.2.2. Hence

(1) $$(y * f)(\alpha) = y(\rho_\alpha f) = y(s_1) t_1(\alpha) + \cdots + y(s_n) t_n(\alpha).$$

It follows that:

(2) $$x(y * f) = y(s_1) x(t_1) + \cdots + y(s_n) x(t_n) = (y.x)(f).$$

Denote Ad by u. If $x \in \mathfrak{g}$, we saw in 23.3.3 that $du(x)$ may be identified with an element $\ell(x)$ of $\mathrm{End}(\mathfrak{g})$ and for all $\psi \in \mathrm{End}(\mathfrak{g})^* \subset \mathbf{A}(\mathrm{GL}(\mathfrak{g}))$, we have:

$$\psi(\ell(x)) = du(x)(\psi).$$

Let $f \in \mathbf{A}(\mathrm{GL}(\mathfrak{g}))$ and $y \in \mathfrak{g}$. Define $\vartheta_{f,y} \in \mathrm{End}(\mathfrak{g})^*$ as follows: for $\varphi \in \mathrm{End}(\mathfrak{g})$,

$$\vartheta_{f,y}(\varphi) = (\varphi(y))(f).$$

Then for $x, y \in \mathfrak{g}$ and $f \in \mathbf{A}(\mathrm{GL}(\mathfrak{g}))$, we have:

$$(du(x)(y))(f) = (\ell(x)(y))(f) = \vartheta_{f,y}(\ell(x)) = du(x)(\vartheta_{f,y}) = x(\vartheta_{f,y} \circ u).$$

On the other hand, if $\alpha \in G$, then:

$$(\vartheta_{f,y} \circ u)(\alpha) = (u(\alpha)(y))(f) = (di_\alpha(y))(f) = y(f \circ i_\alpha).$$

But if $\mathbf{A}(\mu)(f) = s_1 \otimes t_1 + \cdots + s_n \otimes t_n$, we saw in 23.2.7 that:

$$f \circ i_\alpha = s_1(\alpha)(\rho_{\alpha^{-1}} t_1) + \cdots + s_n(\alpha)(\rho_{\alpha^{-1}} t_n).$$

It follows from 23.2.6 that:

$$\begin{aligned} y(f \circ i_\alpha) &= s_1(\alpha)(y * t_1)(\alpha^{-1}) + \cdots + s_n(\alpha)(y * t_n)(\alpha^{-1}) \\ &= (s_1(y * t_1) \circ \iota + \cdots + s_n(y * t_n) \circ \iota)(\alpha). \end{aligned}$$

We deduce therefore that:

$$\vartheta_{f,y} \circ u = s_1(y * t_1) \circ \iota + \cdots + s_n(y * t_n) \circ \iota.$$

In view of 23.4.7, we obtain that:

$$x(\vartheta_{f,y} \circ u) = \sum_{i=1}^n x(s_i)(y * t_i)(e) - \sum_{i=1}^n s_i(e)x(y * t_i).$$

It follows from (1), (2) and the definition of $y * t_i$ that:

$$x(\vartheta_{f,y} \circ u) = \sum_{i=1}^n x(s_i)y(t_i) - \sum_{i=1}^n s_i(e)(y \cdot x)t_i.$$

Finally 23.1.2 and 23.2.3 imply that:

$$\begin{aligned} du(x)(y)(f) = x(\vartheta_{f,y} \circ u) &= (x \cdot y)(f) - (\chi_e \cdot y \cdot x)(f) \\ &= (x \cdot y)(f) - (y \cdot x)(f) = [x, y](f). \end{aligned}$$

Thus we have proved the following result:

Theorem. *The differential of the adjoint representation* Ad_G *is* $\mathrm{ad}_\mathfrak{g}$.

23.5.6 Corollary. *If H is a closed normal subgroup of G and $\mathfrak{h}$ is the Lie algebra of H. Then $\mathfrak{h}$ is an ideal of $\mathfrak{g}$.*

Proof. Let $\mathfrak{a}$ be the ideal of H in $\mathbf{A}(G)$, $\alpha \in G$, $f \in \mathfrak{a}$ and $x \in \mathfrak{h}$. If H is normal in G, then $f \circ i_\alpha \in \mathfrak{a}$, so $x(f \circ i_\alpha) = 0$ (23.2.5). Thus $\mathrm{Ad}(\alpha)(x)(f) = 0$, and this implies that $\mathrm{Ad}(\alpha)(x) \in \mathfrak{h}$ (23.2.5).

Let $(x_1, \ldots, x_n)$ be a basis of $\mathfrak{g}$ such that $(x_1, \ldots, x_p)$ is a basis of $\mathfrak{h}$. If $y \in \mathfrak{g}$, then 23.4.2 implies that for $1 \leqslant j \leqslant n$, we have $\pi_{ij} \in \mathbf{A}(G)$ such that:

$$\mathrm{Ad}(\alpha)(x_j) = \sum_{i=1}^n \pi_{ij}(\alpha)x_i \,, \quad (d\,\mathrm{Ad})(y)(x_j) = \sum_{i=1}^n y(\pi_{ij})x_i.$$

So $\mathrm{Ad}(\alpha)(\mathfrak{h}) \subset \mathfrak{h}$ implies that $\pi_{ij} = 0$ if $1 \leqslant j \leqslant p$ and $p + 1 \leqslant i \leqslant n$. Thus $(d\,\mathrm{Ad})(x)(\mathfrak{h}) \subset \mathfrak{h}$ and the result follows from 23.5.5. $\square$

23.6 Jordan decomposition

23.6.1 Let (V, π) be a representation of $\mathfrak{g}$. We say that the representation π is *locally finite* if any $v \in V$ is contained in a finite-dimensional $\mathfrak{g}$-submodule of V.

23.6.2 Let us consider $\mathbf{A}(G)$ as a rational G-module via ρ. If $x \in \mathfrak{g}$, denote by $\eta(x)$ the derivation of $\mathbf{A}(G)$ defined by $\eta(x)(f) = f * x$ for $f \in \mathbf{A}(G)$ (see 23.2.2). The map

$$\mathfrak{g} \times \mathbf{A}(G) \to \mathbf{A}(G) \ , \ (x, f) \to f * x$$

defines a representation of $\mathfrak{g}$ which is locally finite (23.4.4).

Theorem. *Let $x \in \mathfrak{g}$. There exists a unique pair (x_s, x_n) of elements of $\mathfrak{g}$ such that:*

$$x = x_s + x_n \ , \ [x_s, x_n] = 0 \ , \ \eta(x)_s = \eta(x_s) \ , \ \eta(x)_n = \eta(x_n).$$

The element x_s (resp. x_n) is called the semisimple component *(resp. nilpotent component) of x, and $x = x_s + x_n$ is called the* Jordan decomposition *of x.*

Proof. Let $X = \eta(x)$. If $\alpha \in G$, then the automorphism λ_α of $\mathbf{A}(G)$ commutes with X. By 10.1.2, λ_α commutes also with X_s and X_n. On the other hand, X_s and X_n are derivations of $\mathbf{A}(G)$ (10.1.14). So $X_s, X_n \in \mathfrak{Lie}(G)$ and we are done since $\eta : \mathfrak{g} = \mathcal{L}(G) \to \mathfrak{Lie}(G)$ is a Lie algebra isomorphism (23.2.2). $\square$

23.6.3 Proposition. *Let $u : G \to H$ be a morphism of algebraic groups. For all $x \in \mathfrak{g}$, we have:*

$$(du(x))_s = du(x_s) \ , \ (du(x))_n = du(x_n).$$

Proof. Set $X = \eta_G(x)$ and $Y = \eta_H(du(x))$. Let $f \in \mathbf{A}(H)$, then we have $du(x)(f) = x(f \circ u)$. So if $\alpha \in G$, then:

$$X(f \circ u)(\alpha) = x(\lambda_{\alpha^{-1}}(f \circ u)) = x((\lambda_{u(\alpha)^{-1}}f) \circ u) = du(x)(\lambda_{u(\alpha)^{-1}}f)$$
$$= Y(f)(u(\alpha)).$$

It follows that $X \circ \mathbf{A}(u) = \mathbf{A}(u) \circ Y$. Let $E = \mathbf{A}(u)(\mathbf{A}(H))$. Then $X(E) \subset E$. Since $\mathbf{A}(u)$ induces a surjection from $\mathbf{A}(H)$ onto E, 10.1.12 implies that $X_s \circ \mathbf{A}(u) = \mathbf{A}(u) \circ Y_s$ and $X_n \circ \mathbf{A}(u) = \mathbf{A}(u) \circ Y_n$. Thus:

$$du(x_s)(f) = x_s(f \circ u) = X_s(f \circ u)(e_G) = Y_s(f)(u(e_G))$$
$$= Y_s(f)(e_H) = (du(x))_s(f).$$

It follows that $du(x_s) = (du(x))_s$.

We obtain in a similar way $du(x_n) = (du(x))_n$. $\square$

23.6.4 Proposition. *Let V be a finite-dimensional $\Bbbk$-vector space and G a closed subgroup of $\mathrm{GL}(V)$. If $x \in \mathfrak{g}$, then the Jordan decomposition of x defined in 23.6.2 coincide with the one defined in 10.1.2.*

Proof. By 23.6.3, we may assume that $G = \mathrm{GL}(V)$ and by fixing a basis of V, we are reduced to the case where $G = \mathrm{GL}_n(\Bbbk)$. We have $\mathbf{A}(G) = \Bbbk[T_{11}, T_{12}, \ldots, T_{nn}]_D$ where D is the determinant. Let $x = [x_{ij}] \in \mathfrak{g}$. Then as in 23.3.4:

$$T_{ij} * x = x_{1j} T_{i1} + x_{2j} T_{i2} + \cdots + x_{nj} T_{in}.$$

Thus $W_i = \Bbbk T_{i1} + \cdots + \Bbbk T_{in}$ is a subspace of $\mathbf{A}(G)$ such that $\eta(x)(W_i) \subset W_i$. Moreover, the action of $\eta(x)$ on W_i may be identified as the matrix operation of x on $\Bbbk^n$. We deduce therefore that $\eta(x)|_{W_i} = \eta(x_s)|_{W_i} + \eta(x_n)|_{W_i}$ is the Jordan decomposition of $\eta(x)|_{W_i}$. It follows that if W is the sum of the W_i's, for $1 \leqslant i \leqslant n$, then $\eta(x)|_W = \eta(x_s)|_W + \eta(x_n)|_W$ is the Jordan decomposition of $\eta(x)|_W$. By 10.1.10, we obtain that $\eta(x_s)|_W = \eta(x)_s|_W$ and $\eta(x_n)|_W = \eta(x)_n|_W$. But derivations of $\mathbf{A}(G)$ which are identical on W are equal. So we are done. $\square$

References

- [5], [23], [38], [40], [78].

Correspondence between groups and Lie algebras

We describe more precisely the correspondence between algebraic groups and their Lie algebras in this chapter. We give functorial properties between them and how certain properties are preserved. We also explain how to attach to a Lie algebra an algebraic group (namely, its algebraic adjoint group).

24.1 Notations

24.1.1 Let G and H be algebraic groups. We say that G and H are isomorphic if they are isomorphic as algebraic groups. A closed subgroup of G will also be called an *algebraic subgroup* of G.

Recall that $\mathcal{L}(G)$ denote the Lie algebra of G and $\mathfrak{Lie}(G)$ the Lie algebra of left invariant derivations of $\mathbf{A}(G)$ (23.2.1).

We shall denote by μ_G or μ (resp. ι_G or ι) the group multiplication (resp. inverse) of G, and for $\alpha \in G$, i_α denotes the inner automorphism of G defined by $i_\alpha(\beta) = \alpha\beta\alpha^{-1}$.

24.1.2 Let $\mathfrak{g}$ be the Lie algebra of G. Recall that Ad_G or Ad denotes the adjoint representation of G. If $\mathfrak{a}$ is a subspace of $\mathfrak{g}$, we denote by $C_G(\mathfrak{a})$ (resp. $N_G(\mathfrak{a})$) the set of elements $\alpha \in G$ verifying $\mathrm{Ad}(\alpha)(x) = x$ (resp. $\mathrm{Ad}(\alpha)(x) \in \mathfrak{a}$) for all $x \in \mathfrak{a}$. If $x \in \mathfrak{g}$, we shall also denote $C_G(\Bbbk x)$ by $C_G(x)$.

24.2 An algebraic subgroup

24.2.1 In this section, $\mathfrak{g}$ denote the Lie algebra of an algebraic group G.

For $\alpha \in G$, let χ_α be the linear form on $\mathbf{A}(G)$ defined by $\chi_\alpha(f) = f(\alpha)$. Recall from 23.1.2 that $\mathbf{A}(G)^*$ has a structure of associative algebra whose unit is χ_e, where e denotes the identity element of G. If $x \in \mathbf{A}(G)^*$ and $n \in \mathbb{N}^*$, we shall denote by x^n the product in $\mathbf{A}(G)^*$ of n factors of x. By convention, $x^0 = \chi_e$.

24.2.2 Let us fix $x \in \mathfrak{g}$. Set $\mathfrak{j}_x$ to be the subspace of $\mathbf{A}(G)$ consisting of elements f verifying $x^n(f) = 0$ for all $n \in \mathbb{N}$.

Denote by $\eta_G(x)$ or $\eta(x)$, the left invariant derivation of $\mathbf{A}(G)$ defined by $\eta(x)(f) = f * x$ for $f \in \mathbf{A}(G)$ (see 23.2.2). Thus $x(f) = \big(\eta(x)(f)\big)(e)$.

Let $f \in \mathbf{A}(G)$, then $\mathbf{A}(\mu)(f) = s_1 \otimes t_1 + \cdots + s_p \otimes t_p$ where $s_i, t_i \in \mathbf{A}(G)$. We saw in 23.2.2 that $\eta(x)(f) = x(t_1)s_1 + \cdots + x(t_p)s_p$.

1) If $n \in \mathbb{N}$, then:

$$\eta^{n+1}(x)(f) = \eta^n(x)\big(\eta(x)(f)\big) = \eta^n(x)\big(x(t_1)s_1 + \cdots + x(t_p)s_p\big)$$
$$= \eta^n(x)(s_1)x(t_1) + \cdots + \eta^n(x)(s_p)x(t_p).$$

We have $x(f) = \big(\eta(x)(f)\big)(e)$. Now suppose that $x^n(g) = \big(\eta^n(x)(g)\big)(e)$ for all $g \in \mathbf{A}(G)$, then:

$$x^{n+1}(f) = (x^n \otimes x)\circ\mathbf{A}(\mu)(f) = x^n(s_1)x(t_1) + \cdots + x^n(s_p)x(t_p)$$
$$= \big(\eta^n(x)(s_1)\big)(e)x(t_1) + \cdots + \big(\eta^n(x)(s_p)\big)(e)x(t_p).$$

It follows that $x^{n+1}(f) = \big(\eta^{n+1}(x)(f)\big)(e)$, and so by induction, we have $x^n(f) = \big(\eta^n(x)(f)\big)(e)$ for all $n \in \mathbb{N}$ (the case $n = 0$ is clear) and $f \in \mathbf{A}(G)$.

2) Let $n \in \mathbb{N}$ and $f \in \mathbf{j}_x$. Then:

$$x^n\big(\eta(x)(f)\big) = \eta^n(x)\big(\eta(x)(f)\big)(e) = \big(\eta^{n+1}(x)(f)\big)(e) = 0.$$

Thus $\eta(x)(\mathbf{j}_x) \subset \mathbf{j}_x$.

3) Let $f \in \mathbf{j}_x$ and $g \in \mathbf{A}(G)$. Then:

$$\chi_e(fg) = f(e)g(e) = 0 \ , \ x(fg) = x(f)g(e) + f(e)x(g) = 0.$$

Set $h = \eta(x)(f)$ and $\ell = \eta(x)(g)$. Suppose that $x^n(uv) = 0$ for all $u \in \mathbf{j}_x$ and $v \in \mathbf{A}(G)$. Then:

$$x^{n+1}(fg) = \eta^n(x)\big(\eta(x)(fg)\big)(e) = \eta^n(x)\big(hg + f\ell\big)(e) = x^n(hg) + x^n(f\ell).$$

Hence $x^{n+1}(fg) = 0$ since $h \in \mathbf{j}_x$. We have therefore proved that $\mathbf{j}_x$ is an ideal of $\mathbf{A}(G)$.

4) Let $f \in \mathbf{A}(G)$ and $g = f\circ\iota$. For $\alpha, \beta \in G$, we have:

$$f(\alpha\beta) = \mathbf{A}(\mu)(f)(\alpha, \beta) = s_1(\alpha)t_1(\beta) + \cdots + s_p(\alpha)t_p(\beta).$$

It follows that:

$$\mathbf{A}(\mu)(g)(\alpha, \beta) = g(\alpha\beta) = f(\beta^{-1}\alpha^{-1}) = \mathbf{A}(\mu)(f)(\beta^{-1}, \alpha^{-1})$$
$$= s_1(\beta^{-1})t_1(\alpha^{-1}) + \cdots + s_p(\beta^{-1})t_p(\alpha^{-1}).$$

Hence:

$$\mathbf{A}(\mu)(g) = (t_1\circ\iota) \otimes (s_1\circ\iota) + \cdots + (t_p\circ\iota) \otimes (s_p\circ\iota).$$

Now $\chi_e(g) = f(e) = \chi_e(f)$ and $x(g) = -x(f)$ (23.4.8). Suppose that we have shown that $x^n(u\circ\iota) = (-1)^n x^n(u)$ for all $u \in \mathbf{A}(G)$. Then:

$$x^{n+1}(g) = (x^n \otimes x) \circ \mathbf{A}(\mu)(g) = x^n(t_1 \circ \iota)x(s_1 \circ \iota) + \cdots + x^n(t_p \circ \iota)x(s_p \circ \iota)$$
$$= (-1)^{n+1}[x^n(t_1)x(s_1) + \cdots + x^n(t_p)x(s_p)]$$
$$= (-1)^{n+1}(x \otimes x^n)\mathbf{A}(\mu)(f) = (-1)^{n+1}x^{n+1}(f).$$

Thus $x^n(g) = (-1)^n x^n(f)$ for all $n \in \mathbb{N}$. So if $f \in j_x$, then $f \circ \iota \in j_x$.

5) Let $f, g \in \mathbf{A}(G) \setminus j_x$. Denote by p and q the smallest integers such that $x^p(f)$ and $x^q(g)$ are non-zero. Since $\eta(x)$ is a derivation of $\mathbf{A}(G)$, we have:

$$x^{p+q}(fg) = \big(\eta^{p+q}(x)(fg)\big)(e) = \Big(\sum_{m+n=p+q} \frac{(p+q)!}{m!n!} \eta^m(x)(f)\,\eta^n(x)(g) \Big)(e)$$
$$= \sum_{m+n=p+q} \frac{(p+q)!}{m!n!} x^m(f)x^n(g) = \frac{(p+q)!}{p!q!} x^p(f)x^q(g) \neq 0.$$

Thus $fg \notin j_x$, and we have shown that j_x is a prime ideal of $\mathbf{A}(G)$.

6) Let $\mathcal{a}(x) = \mathcal{V}(j_x)$ be the set of $\alpha \in G$ verifying $f(\alpha) = 0$ for all $f \in j_x$. By definition, $e \in \mathcal{a}(x)$. We shall show that if $\alpha \in \mathcal{a}(x)$ and $f \in j_x$, then $\lambda_{\alpha^{-1}} f \in j_x$ where λ_α is the left translation of functions (22.1.4). Since the derivation $\eta(x)$ is left invariant (23.2.2), we have $\eta^n(x)(\lambda_{\alpha^{-1}} f) = \lambda_{\alpha^{-1}}\big(\eta^n(x)(f)\big)$. From this, we deduce that:

$$x^n(\lambda_{\alpha^{-1}} f) = \big(\eta^n(x)(\lambda_{\alpha^{-1}} f)\big)(e) = \big(\lambda_{\alpha^{-1}}\big(\eta^n(x)(f)\big)\big)(e) = \big(\eta^n(x)(f)\big)(\alpha).$$

It follows from point 2 that $\eta(x)(j_x) \subset j_x$. Hence our result.

7) Points 2, 4 and 6 prove that $\mathcal{a}(x)$ is an algebraic subgroup of G. Moreover, by 10.1.6 and point 5, the defining ideal of $\mathcal{a}(x)$ in $\mathbf{A}(G)$ is:

$$\mathcal{I}\big(\mathcal{a}(x)\big) = \sqrt{j_x} = j_x.$$

It follows that $\mathcal{a}(x)$ is connected.

8) By 11.8.1 and the preceding discussions, the ideal $\mathfrak{J}_x$ of $\mathcal{a}(x) \times \mathcal{a}(x)$ in the algebra $\mathbf{A}(G \times G) = \mathbf{A}(G) \otimes_{\Bbbk} \mathbf{A}(G)$ is equal to $\mathfrak{J}_x = j_x \otimes_{\Bbbk} \mathbf{A}(G) + \mathbf{A}(G) \otimes_{\Bbbk} j_x$.

Let $f \in j_x$. If $\alpha, \beta \in \mathcal{a}(x)$, then $\mathbf{A}(\mu)(f)(\alpha, \beta) = f(\alpha\beta) = 0$. Thus $\mathbf{A}(\mu)(j_x) \subset \mathfrak{J}_x$.

For $p, q \in \mathbb{N}$, denote by $\mathfrak{T}_{p,q,x}$ the kernel of the linear form $x^p \otimes x^q$ on $\mathbf{A}(G) \otimes_{\Bbbk} \mathbf{A}(G)$, and let $\mathfrak{T}_x$ be the intersection of the $\mathfrak{T}_{p,q,x}$'s, $p, q \in \mathbb{N}$.

We have $x^{p+q}(f) = (x^p \otimes x^q) \circ \mathbf{A}(\mu)(f)$, so $f \in j_x$ implies that $\mathbf{A}(\mu)f \in \mathfrak{T}_x$. It follows that $\mathfrak{T}_x \subset \mathfrak{J}_x$. Since we have clearly $\mathfrak{J}_x \subset \mathfrak{T}_x$, we obtained that $\mathfrak{T}_x = \mathfrak{J}_x$.

24.2.3 Theorem. *Let G be an algebraic group and $x \in \mathcal{L}(G)$.*

(i) *The group $\mathcal{a}(x)$ is a connected and commutative algebraic subgroup of G such that $x \in \mathcal{L}(\mathcal{a}(x))$.*

(ii) *Any closed subgroup H of G verifying $x \in \mathcal{L}(H)$ contains $\mathcal{a}(x)$.*

Proof. Points 2 and 7 of 24.2.2 imply that $\mathcal{I}\big(\mathcal{a}(x)\big) = j_x$ and $\eta(x)(j_x) \subset j_x$. So by 23.2.5, $x \in \mathcal{L}\big(\mathcal{a}(x)\big)$. Let us prove that $\mathcal{a}(x)$ is abelian.

Denote by $\zeta : \mathbf{A}(G) \otimes_{\Bbbk} \mathbf{A}(G) \to \mathbf{A}(G) \otimes_{\Bbbk} \mathbf{A}(G)$ the linear map defined by $\zeta(f \otimes g) = g \otimes f$. If $p, q \in \mathbb{N}$ and $f \in \mathbf{A}(G)$, we have:

$$(x^p \otimes x^q) \circ \zeta \circ \mathbf{A}(\mu)(f) = (x^q \otimes x^p) \circ \mathbf{A}(\mu)(f) = x^{q+p}(f) = (x^p \otimes x^q) \circ \mathbf{A}(\mu)(f).$$

Thus $(x^p \otimes x^q) \circ (\zeta \circ \mathbf{A}(\mu) - \mathbf{A}(\mu)) = 0$. By point 8 of 24.2.2, we deduce that $(\zeta \circ \mathbf{A}(\mu) - \mathbf{A}(\mu))(\mathbf{A}(G)) \subset \mathfrak{I}_x$. If $\alpha, \beta \in \mathcal{A}(x)$ and $f \in \mathbf{A}(G)$, then:

$$\begin{aligned}
f(\alpha\beta) - f(\beta\alpha) &= \mathbf{A}(\mu)(f)(\alpha, \beta) - \mathbf{A}(\mu)(f)(\beta, \alpha) \\
&= \sum_{i=1}^{p} s_i(\alpha) t_i(\beta) - \sum_{i=1}^{p} s_i(\beta) t_i(\alpha) \\
&= (\mathbf{A}(\mu) - \zeta \circ \mathbf{A}(\mu))(f)(\alpha, \beta) = 0.
\end{aligned}$$

So we have $\alpha\beta = \beta\alpha$.

(ii) Let H be a closed subgroup of G verifying $x \in \mathcal{L}(H)$, and $\mathfrak{a} = \mathcal{I}(H)$ the ideal of H. We have $\eta(x)(\mathfrak{a}) \subset \mathfrak{a}$ (23.2.5), so $\eta^n(x)(\mathfrak{a}) \subset \mathfrak{a}$ and $x^n(\mathfrak{a}) = \{0\}$, if $n \in \mathbb{N}$ by point 1 of 24.2.2. Thus $\mathfrak{a} \subset \mathfrak{j}_x$, and $\mathcal{A}(x) \subset H$. $\square$

24.2.4 Remark. Let V be a finite-dimensional vector space, G a closed subgroup of $\mathrm{GL}(V)$ and $x \in \mathcal{L}(G)$. If x is a nilpotent endomorphism, then by 22.3.4 and 23.3.5, the group $\mathcal{A}(x)$ is the set of e^{tx}, $t \in \Bbbk$.

24.2.5 Corollary. *Let $\mathfrak{h}$ be a Lie subalgebra of $\mathfrak{g}$, and $\mathcal{A}(\mathfrak{h})$ the intersection of all algebraic subgroups of G whose Lie algebra contains $\mathfrak{h}$. Then $\mathcal{A}(\mathfrak{h})$ is connected and $\mathfrak{h} \subset \mathcal{L}(\mathcal{A}(\mathfrak{h}))$.*

Proof. It is clear that $\mathcal{A}(\mathfrak{h})$ is an algebraic subgroup of G, and since $\mathcal{L}(\mathcal{A}(\mathfrak{h})) = \mathcal{L}(\mathcal{A}(\mathfrak{h})^\circ)$, $\mathcal{A}(\mathfrak{h})$ is connected.

By 24.2.3, we have $\mathcal{A}(x) \subset \mathcal{A}(\mathfrak{h})$ for all $x \in \mathfrak{h}$. So it follows from 16.3.7, 23.2.4 and 24.2.3 that $x \in \mathcal{L}(\mathcal{A}(x)) \subset \mathcal{L}(\mathcal{A}(\mathfrak{h}))$. Hence $\mathfrak{h} \subset \mathcal{L}(\mathcal{A}(\mathfrak{h}))$. $\square$

24.2.6 Remark. Let H be a closed connected subgroup of G, $\mathfrak{h}$ its Lie algebra. Since $\mathcal{A}(\mathfrak{h}) \subset H$, we have $\mathcal{L}(\mathcal{A}(\mathfrak{h})) \subset \mathfrak{h}$. It follows from 24.2.5 that $\mathfrak{h} = \mathcal{L}(\mathcal{A}(\mathfrak{h}))$. Thus $\dim \mathcal{A}(\mathfrak{h}) = \dim H = \dim \mathfrak{h}$. Since H is connected, we deduce that $\mathcal{A}(\mathfrak{h}) = H$.

24.3 Invariants

24.3.1 Let $\mathfrak{g}$ be the Lie algebra of an algebraic group G, (V, π) a rational representation of G of dimension n, and $\mathcal{B}$ a basis of V. If $u \in \mathrm{End}(V)$, denote by $[\varphi_{ij}(u)]$ the matrix of u with respect to $\mathcal{B}$. We define $\pi_{ij} \in \mathbf{A}(G)$ by $\pi_{ij}(\alpha) = \varphi_{ij}(\pi(\alpha))$, and if $x \in \mathfrak{g}$, set $d_{ij}(x) = \varphi_{ij}(d\pi(x))$, $1 \leqslant i, j \leqslant n$. Let $x_1, \ldots, x_r \in \mathfrak{g}$. Then:

$$\varphi_{ij}(d\pi(x_1) \circ \cdots \circ d\pi(x_r)) = \sum_{1 \leqslant k_1, \ldots, k_{r-1} \leqslant n} d_{i,k_1}(x_1) d_{k_1,k_2}(x_2) \cdots d_{k_{r-1},j}(x_r).$$

Moreover, for $x \in \mathfrak{g}$ and $\varphi \in (\mathrm{End}(V))^*$, we have by 23.3.3 that:

$$\varphi\big(d\pi(x)\big) = d\pi(x)(\varphi) = x(\varphi \circ \pi).$$

Thus:

$$\varphi_{ij}\big(d\pi(x_1) \circ \cdots \circ d\pi(x_r)\big) = \sum_{1 \leqslant k_1, \ldots, k_{r-1} \leqslant n} x_1(\pi_{i,k_1}) x_2(\pi_{k_1,k_2}) \cdots x_r(\pi_{k_{r-1},j}).$$

Hence:

$$\varphi_{ij}\big(d\pi(x_1) \circ \cdots \circ d\pi(x_r)\big) = (x_1 \cdots x_r)(\pi_{ij}) = (x_1 \cdots x_r)(\varphi_{ij} \circ \pi).$$

So by linearity, we obtain the following result:

Lemma. *Let* $r \in \mathbb{N}^*$, $x_1, \ldots, x_r \in \mathfrak{g}$ *and* $\varphi \in \big(\mathrm{End}(V)\big)^*$. *Then:*

$$\varphi\big(d\pi(x_1) \circ \cdots \circ d\pi(x_r)\big) = (x_1 \cdots x_r)(\varphi \circ \pi).$$

24.3.2 Theorem. *Let* (V, π) *be a finite-dimensional rational representation of an algebraic group* G, E *the subspace of* $\mathrm{End}(V)$ *spanned by* $\pi(\alpha) - \mathrm{id}_V$, $\alpha \in G$, *and* F *the subspace of* $\mathrm{End}(V)$ *spanned by the products* $d\pi(x_1) \circ \cdots \circ d\pi(x_r)$ *where* $r \in \mathbb{N}^*$ *and* $x_1, \ldots, x_r \in \mathfrak{g} = \mathcal{L}(G)$.
(i) *For* $x \in \mathfrak{g}$, *we have* $d\pi(x) \in E$.
(ii) *Assume that* G *is connected. If* $\alpha \in G$, *then* $\pi(\alpha) - \mathrm{id}_V \in F$.

Proof. (i) Let $\varphi \in \mathrm{End}(V)^*$ be such that $\varphi|_E = 0$. Then $\alpha \mapsto \varphi \circ \pi(\alpha)$ is a constant function on G and so $x(\varphi \circ \pi) = 0$ for all $x \in \mathfrak{g}$. So by 23.3.3, we have $0 = d\pi(x)(\varphi) = \varphi(d\pi(x))$. Hence $d\pi(x) \in E$.

(ii) Let $x \in \mathfrak{g}$, $n \in \mathbb{N}^*$ and $\varphi \in \mathrm{End}(V)^*$ verifying $\varphi|_F = 0$. We have $x^n(\varphi \circ \pi) = 0$ by 24.3.1. So $\varphi \circ \pi - \varphi(\mathrm{id}_V) \in j_x$ (notations of 24.2.2). It follows that if $\alpha \in \mathcal{A}(x)$, then $\varphi(\pi(\alpha) - \mathrm{id}_V) = 0$. Hence $\pi(\alpha) - \mathrm{id}_V \in F$ for $\alpha \in \mathcal{A}(x)$. If $\alpha, \beta \in G$, then:

$$\pi(\alpha\beta) - \mathrm{id}_V = \big(\pi(\alpha) - \mathrm{id}_V\big) \circ \big(\pi(\beta) - \mathrm{id}_V\big) + \big(\pi(\alpha) - \mathrm{id}_V\big) + \big(\pi(\beta) - \mathrm{id}_V\big).$$

So for $n \in \mathbb{N}^*$, $x_1, \ldots, x_n \in \mathfrak{g}$ and $\alpha_i \in \mathcal{A}(x_i)$, $1 \leqslant i \leqslant n$, we obtain that $\pi(\alpha_1 \cdots \alpha_n) - \mathrm{id}_V \in F$. Let H be the subgroup of G generated by $\mathcal{A}(x)$, $x \in \mathfrak{g}$. It is a connected algebraic subgroup of G (21.3.2 and 24.2.3). If $\alpha \in H$, then $\varphi(\pi(\alpha) - \mathrm{id}_V) = 0$ for all linear forms on $\mathrm{End}(V)$ which are zero on F, so $\pi(\alpha) - \mathrm{id}_V \in F$. If $x \in \mathfrak{g}$, then $x \in \mathcal{L}(\mathcal{A}(x)) \subset \mathcal{L}(H)$ (24.2.3). We have therefore $\mathfrak{g} = \mathcal{L}(H)$, hence $\dim G = \dim H$. Since G is connected, we have $G = H$, and the result follows. $\square$

24.3.3 Corollary. *Let* G *be a connected algebraic group,* $\mathfrak{g}$ *its Lie algebra and* V *a finite-dimensional rational* G-*module.*
(i) *We have* $V^G = V^{\mathfrak{g}}$.
(ii) *A subspace* W *of* V *is a* G-*submodule if and only if* W *is a* $\mathfrak{g}$-*submodule.*

24.3.4 Let (E, π) be a rational (not necessarily finite-dimensional) G-module (see 21.5.4) and V, W two finite-dimensional G-submodules of E. Denote by $\sigma : G \to \mathrm{GL}(V)$, $\tau : G \to \mathrm{GL}(W)$ and $\theta : G \to \mathrm{GL}(V + W)$ the corresponding subrepresentations of G.

If $\alpha \in G$, then $\theta(\alpha)|_V = \sigma(\alpha)$ and $\theta(\alpha)|_W = \tau(\alpha)$. We deduce easily that for $x \in \mathfrak{g}$, $d\theta(x)|_V = d\sigma(x)$ and $d\theta(x)|_W = d\tau(x)$. In particular, $d\sigma(x)|_{V \cap W} = d\tau(x)|_{V \cap W}$ and so there exists a unique $\ell(x) \in \mathrm{End}(E)$ such that $\ell(x)|_U = d(\pi|_U)(x)$ for all finite-dimensional G-stable subspace U of E. We shall also denote $\ell(x)$ by $d\pi(x)$. It is clear that the map $\mathfrak{g} \to \mathfrak{gl}(E)$, $x \mapsto d\pi(x)$, is a Lie algebra homomorphism. Thus E is a $\mathfrak{g}$-module. We deduce immediately from 24.3.3 the following result:

Proposition. *Let G be a connected algebraic group, $\mathfrak{g}$ its Lie algebra and E a rational G-module.*
   (i) *We have $E^G = E^{\mathfrak{g}}$.*
   (ii) *A subspace F of E is a G-submodule if and only if F is a $\mathfrak{g}$-submodule.*

24.3.5 Proposition. *Let H, K be algebraic subgroups of G whose Lie algebras are $\mathfrak{h}$ and $\mathfrak{k}$ respectively.*
   (i) *If $H \subset K$, then $\mathfrak{h} \subset \mathfrak{k}$.*
   (ii) *Assume that H and K and connected. If $\mathfrak{h} \subset \mathfrak{k}$, then $H \subset K$.*
   (iii) *We have $\mathcal{L}(H \cap K) = \mathfrak{h} \cap \mathfrak{k}$.*

Proof. (i) This is a consequence of 16.3.7 and 23.2.4.
   (ii) If $\mathfrak{h} \subset \mathfrak{k}$, then $\mathcal{a}(\mathfrak{h}) \subset \mathcal{a}(\mathfrak{k})$. So $H \subset K$ by 24.2.6.
   (iii) We have $\mathcal{L}(H \cap K) \subset \mathfrak{h} \cap \mathfrak{k}$ by (i). On the other hand, $\mathcal{a}(\mathfrak{h} \cap \mathfrak{k}) \subset H \cap K$ (24.2.5). So $\mathfrak{h} \cap \mathfrak{k} \subset \mathcal{L}(\mathcal{a}(\mathfrak{h} \cap \mathfrak{k})) \subset \mathcal{L}(H \cap K)$. $\square$

24.3.6 Let (V, π) be a finite-dimensional rational representation of G, W a subspace of V, $v \in V$ and:

$$N_G(W) = \{\alpha \in G;\ \pi(\alpha)(W) = W\}\ ,\quad \mathfrak{n}_{\mathfrak{g}}(W) = \{x \in \mathfrak{g};\ d\pi(x)(W) \subset W\},$$
$$C_G(v) = \{\alpha \in G;\ \pi(\alpha)(v) = v\}\ ,\quad \mathfrak{c}_{\mathfrak{g}}(v) = \{x \in \mathfrak{g};\ d\pi(x)(v) = 0\},$$
$$C_G(W) = \bigcap_{v \in W} C_G(v)\ ,\quad \mathfrak{c}_{\mathfrak{g}}(W) = \bigcap_{v \in W} \mathfrak{c}_{\mathfrak{g}}(v).$$

If H is an algebraic subgroup of G, set:

$$N_H(W) = H \cap N_G(W)\ ,\quad C_H(v) = H \cap C_G(v).$$

These are clearly algebraic subgroups of G (21.4.2).

Proposition. *With the above notations, we have:*
   (i) $\mathcal{L}(N_G(W)) = \mathfrak{n}_{\mathfrak{g}}(W)$.
   (ii) $\mathcal{L}(C_G(v)) = \mathfrak{c}_{\mathfrak{g}}(v)$.
   (iii) $\mathcal{L}(C_G(W)) = \mathfrak{c}_{\mathfrak{g}}(W)$.
   (iv) $\mathcal{L}(N_H(W)) = \mathfrak{h} \cap \mathfrak{n}_{\mathfrak{g}}(W)\ ,\quad \mathcal{L}(C_H(v)) = \mathfrak{h} \cap \mathfrak{c}_{\mathfrak{g}}(v)$.

Proof. (i) Let us fix a basis $(v_1, \ldots, v_n)$ of V such that $(v_1, \ldots, v_p)$ is a basis of W. For $\alpha \in G$, denote by $[\pi_{ij}(\alpha)]$ the matrix of $\pi(\alpha)$ with respect to this basis. Then $\alpha \in N_G(W)$ if and only if $\pi_{ij}(\alpha) = 0$ for $1 \leqslant i \leqslant p$ and $p + 1 \leqslant j \leqslant n$. So 23.2.5 and 23.4.2 imply that $\mathcal{L}(N_G(W)) \subset \mathfrak{n}_{\mathfrak{g}}(W)$.

For $f \in V^*$ and $v \in V$, define $\vartheta_{f,v} \in \mathrm{End}(V)^*$ by $\vartheta_{f,v}(\varphi) = f(\varphi(v))$ if $\varphi \in \mathrm{End}(V)$. If $n \in \mathbb{N}^*$ and $x \in \mathfrak{g}$, then by 24.3.1:

$$f\big((d\pi(x))^n(v)\big) = \vartheta_{f,v}\big((d\pi(x))^n\big) = x^n(\vartheta_{f,v} \circ \pi).$$

So $x \in \mathfrak{n}_{\mathfrak{g}}(W)$ if and only if $f\big((d\pi(x))^n(v)\big) = 0$ for all $n \in \mathbb{N}^*$, $v \in W$ and $f \in W^{\perp}$, the orthogonal of W in V^*. Let $x \in \mathfrak{n}_{\mathfrak{g}}(W)$. We have therefore $\vartheta_{f,v} \circ \pi - \vartheta_{f,v} \circ \pi(e) \in \mathfrak{j}_x$ (notations of 24.2.2). It follows that if $\alpha \in \mathcal{A}(x)$, then $f(\pi(\alpha)(v)) = \vartheta_{f,v} \circ \pi(\alpha) = \vartheta_{f,v} \circ \pi(e) = 0$ for all $v \in W$ and $f \in W^{\perp}$. Consequently, $\mathcal{A}(x) \subset N_G(W)$. By 24.2.3 and 24.3.5, we have $x \in \mathcal{L}(\mathcal{A}(x)) \subset \mathcal{L}(N_G(W))$. So we have proved (i).

(ii) As in part (i), we have $\mathcal{L}(C_G(v)) \subset \mathfrak{c}_{\mathfrak{g}}(v)$, and $x \in \mathfrak{c}_{\mathfrak{g}}(v)$ if and only if $\vartheta_{f,v}\big((d\pi(x))^n\big) = 0$ for all $n \in \mathbb{N}^*$ and $f \in V^*$. So we can finish our proof as in (i).

(iii) Since $C_G(W) = \bigcap_{i=1}^{p} C_G(v_i)$ and $\mathfrak{c}_{\mathfrak{g}}(W) = \bigcap_{i=1}^{p} \mathfrak{c}_{\mathfrak{g}}(v_i)$, the result follows from part (ii) and 24.3.5 (iii).

(iv) This follows from (i), (ii) and 24.3.5 (iii). $\square$

24.3.7 Let (V, σ) and (W, τ) be finite-dimensional rational representations of G. We have already seen that there is a natural structure of rational representation π of G on $\mathrm{Hom}(V, W)$ given by: for $f \in \mathrm{Hom}(V, W)$ and $v \in V$,

$$\big(\pi(\alpha)(f)\big)(v) = \tau(\alpha)\big(f(\sigma(\alpha^{-1})(v))\big).$$

Similarly, let (V, σ') and (W, τ') be representations of a Lie algebra $\mathfrak{g}$, then there is a natural structure of representation π' of $\mathfrak{g}$ on $\mathrm{Hom}(V, W)$ given by: for $x \in \mathfrak{g}$, $f \in \mathrm{Hom}(V, W)$ and $v \in V$,

$$\big(\pi'(x)(f)\big)(v) = \tau'(x)\big(f(v)\big) - f\big(\sigma'(x)(v)\big).$$

Suppose that $\mathfrak{g} = \mathcal{L}(G)$ and $\sigma' = d\sigma$, $\tau' = d\tau$. Then we deduce easily from 23.4 that $\pi' = d\pi$.

Now let $\mathfrak{A}$ be a finite-dimensional $\Bbbk$-algebra (not necessarily associative or with unit). The algebra structure on $\mathfrak{A}$ is defined by $\nu \in \mathrm{End}(\mathfrak{A} \otimes_{\Bbbk} \mathfrak{A}, \mathfrak{A})$ such that $\nu(a \otimes b) = ab$ for all $a, b \in \mathfrak{A}$. Denote by τ (resp. τ') the identity representation of $G = \mathrm{GL}(\mathfrak{A})$ (resp. $\mathfrak{g} = \mathfrak{gl}(\mathfrak{A})$) and let $\sigma = \tau \otimes \tau$ (see 23.4.13), $\sigma' = d\sigma$. Construct π and π' as above with $V = \mathfrak{A} \otimes_{\Bbbk} \mathfrak{A}$ and $W = \mathfrak{A}$. Then we see easily that for $\alpha \in G$ (resp. $x \in \mathfrak{g}$), $\alpha \in \mathrm{Aut}\,\mathfrak{A}$ (resp. $x \in \mathrm{Der}\,\mathfrak{A}$) if and only if $\pi(\alpha)(\nu) = \nu$ (resp. $\pi'(x)(\nu) = 0$). Since $\pi' = d\pi$, it follows from 24.3.6 (ii) that:

Proposition. *The Lie algebra of* $\mathrm{Aut}\,\mathfrak{A}$ *is* $\mathrm{Der}\,\mathfrak{A}$.

24.4 Functorial properties

24.4.1 Theorem. *Let $u : G \to H$ be a morphism of algebraic groups and K the kernel of u. Then $\mathcal{L}(K)$ is the kernel of du.*

Proof. Let $\mathfrak{a} = \mathcal{I}(K)$ be the defining ideal of K in $\mathbf{A}(G)$ and $\mathfrak{b}$ the set of elements $f \circ u - f \circ u(e_G)$ for $f \in \mathbf{A}(H)$. Clearly $K = \mathcal{V}(\mathfrak{b})$ and so $\mathfrak{a} = \sqrt{\mathfrak{b}}$.

Let $x \in \mathcal{L}(K)$ and $f \in \mathbf{A}(H)$. Then by 23.2.5, we have:

$$0 = x\big(f \circ u - f \circ u(e_G)\big) = x(f \circ u) = du(x)(f).$$

So x belongs to the kernel of du.

Let $x \in \mathcal{L}(G)$ and $f \in \mathbf{A}(H)$. Then for $\alpha \in G$,

$$\eta(x)(f \circ u)(\alpha) = \big(\lambda_{\alpha^{-1}}(\eta(x)(f \circ u))\big)(e_G) = \big(\eta(x)(\lambda_{\alpha^{-1}}(f \circ u))\big)(e_G)$$
$$= x\big(\lambda_{\alpha^{-1}}(f \circ u)\big) = x\big((\lambda_{u(\alpha)^{-1}}f)\circ u\big) = du(x)(\lambda_{u(\alpha)^{-1}}f).$$

So if $x \in \ker(du)$, then $\mathfrak{b} \subset \mathfrak{j}_x = \mathcal{I}(\mathcal{Q}(x))$ (notations as in 24.2.2). Since $\mathfrak{j}_x$ is prime, we have $\mathfrak{a} \subset \mathcal{I}(\mathcal{Q}(x))$, hence $\mathcal{Q}(x) \subset K$. So by 24.2.3 and 24.3.5, $x \in \mathcal{L}(K)$. □

24.4.2 Theorem. *Let G be an algebraic group, $Z(G)$ its centre, $\mathfrak{g}$ its Lie algebra and $\mathfrak{z}(\mathfrak{g})$ the centre of $\mathfrak{g}$.*
   (i) *We have $Z(G) \subset \ker(\mathrm{Ad}_G)$ and $\mathcal{L}\big(\ker(\mathrm{Ad}_G)\big) = \mathfrak{z}(\mathfrak{g})$.*
   (ii) *If G is connected, then $Z(G) = \ker(\mathrm{Ad}_G)$ and $\mathcal{L}\big(Z(G)\big) = \mathfrak{z}(\mathfrak{g})$.*

Proof. (i) Let $\alpha \in Z(G)$, then $i_\alpha = \mathrm{id}_G$. So for $x \in \mathfrak{g}$ and $f \in \mathbf{A}(G)$, we have:

$$\mathrm{Ad}(\alpha)(x)(f) = di_\alpha(x)(f) = x(f \circ i_\alpha) = x(f).$$

Hence $\mathrm{Ad}(\alpha) = \mathrm{id}_\mathfrak{g}$ and $\alpha \in \ker(\mathrm{Ad}_G)$. The second part of (i) follows easily from 24.4.1 and 23.5.5.

(ii) By 22.1.5, G has a faithful finite-dimensional rational representation (V, π). So 24.4.1 implies that the representation $(V, d\pi)$ of $\mathfrak{g}$ is also faithful. If $\alpha \in G$, then we have (23.5.2):

$$d\pi \circ \mathrm{Ad}_G(\alpha) = \mathrm{Ad}_{\mathrm{GL}(V)}\big(\pi(\alpha)\big) \circ d\pi.$$

So if $\alpha \in \ker(\mathrm{Ad}_G)$ and $x \in \mathfrak{g}$, then (23.5.4)

$$d\pi(x) = \pi(\alpha) \circ d\pi(x) \circ \pi(\alpha)^{-1},$$

which means that $\pi(\alpha)$ belongs to the centralizer of $d\pi(\mathfrak{g})$ in $\mathrm{End}(V)$.

If G is connected, then 24.3.2 says that the subalgebra (with unit) of $\mathrm{End}(V)$ generated by $d\pi(\mathfrak{g})$ is equal to the subalgebra (with unit) in $\mathrm{End}(V)$ generated by $\pi(G)$. So if $\alpha \in \ker \mathrm{Ad}_G$, then $\pi(\alpha\beta) = \pi(\beta\alpha)$ for all $\beta \in G$. Since π is injective, $\alpha \in Z(G)$. So by (i), we have $Z(G) = \ker(\mathrm{Ad}_G)$ and $\mathcal{L}\big(Z(G)\big) = \mathfrak{z}(\mathfrak{g})$. □

24.4.3 Remark. If G is not connected, then it may happen that $Z(G) \neq \ker(\mathrm{Ad}_G)$ and $\mathcal{L}\big(Z(G)\big) \neq \mathfrak{z}(\mathfrak{g})$. Let us give an example.

Using the notations of example 4 of 21.1.3, take $G = O_2(\Bbbk)$. We verify easily that G has two connected components G° and G_1 given by:

$$G^\circ = \left\{ \begin{pmatrix} a & 0 \\ 0 & a^{-1} \end{pmatrix} ; a \in \Bbbk \setminus \{0\} \right\} , \quad G_1 = \left\{ \begin{pmatrix} 0 & b \\ b^{-1} & 0 \end{pmatrix} ; b \in \Bbbk \setminus \{0\} \right\}.$$

Let I_2 denote the identity matrix of $M_2(\Bbbk)$, then:

$$Z(G) = \{-I_2, I_2\} , \quad \mathcal{L}\big(Z(G)\big) = \{0\}.$$

Since $G^\circ = \mathrm{SL}_2(\Bbbk) \cap D_2(\Bbbk)$, it follows from 23.3.4, 23.4.5 and 24.3.5 (iii) that $\mathfrak{g} = \mathcal{L}(G)$ is the set of matrices $\mathrm{diag}(u, -u)$ with $u \in \Bbbk$. By 23.5.4, we have $\ker(\mathrm{Ad}_G) = G^\circ$. So $Z(G) \neq \ker(\mathrm{Ad}_G)$ and $\mathcal{L}\big(Z(G)\big) \neq \mathfrak{z}(\mathfrak{g})$.

24.4.4 Theorem. *Let G be an algebraic group.*
  (i) *If G is commutative, then $\mathcal{L}(G)$ is commutative.*
  (ii) *If G is connected and $\mathcal{L}(G)$ is commutative, then G is commutative.*

Proof. (i) If G is commutative, then $G = \ker(\mathrm{Ad}_G)$. So $\mathcal{L}(G)$ is commutative (24.4.2 (i)).

(ii) By 22.1.5, G has a faithful finite-dimensional rational representation (V, π). If $\mathfrak{g} = \mathcal{L}(G)$ is commutative, then the subalgebra (with unit) of $\mathrm{End}(V)$ generated by $d\pi(\mathfrak{g})$ is also commutative. If G is connected, then 24.3.2 implies that $\pi(G)$ is commutative. Since π is injective, we conclude that G is commutative. $\square$

24.4.5 Remark. If G is not connected, then the example in 24.4.3 shows that although $\mathcal{L}(G)$ is commutative, G is not commutative.

24.4.6 Lemma. *Let $\alpha \in G$, H an algebraic subgroup of G and $\mathfrak{h}$ its Lie algebra. We have:*
$$\mathrm{Ad}(\alpha)(\mathfrak{h}) = \mathcal{L}(\alpha H \alpha^{-1}).$$

Proof. Let $\mathfrak{a} = \mathcal{I}(H)$ (resp. $\mathfrak{b} = \mathcal{I}(\alpha H \alpha^{-1})$) be the ideal of H (resp. $\alpha H \alpha^{-1}$) in $\mathbf{A}(G)$. It is clear that $\mathfrak{b} = \{f \circ i_{\alpha^{-1}}; f \in \mathfrak{a}\}$.

Let $x \in \mathcal{L}(G)$, $y = \mathrm{Ad}(\alpha)(x)$ and $f \in \mathbf{A}(G)$. Since $y(f) = x(f \circ i_\alpha)$, we have by 23.2.5 that:

$$y \in \mathcal{L}(\alpha H \alpha^{-1}) \Leftrightarrow y(\mathfrak{b}) = \{0\} \Leftrightarrow x(\mathfrak{a}) = \{0\} \Leftrightarrow x \in \mathfrak{h}.$$

So we have obtained the result. $\square$

24.4.7 Theorem. *Let G be an algebraic group, H an algebraic subgroup of G, $\mathfrak{g} = \mathcal{L}(G)$ and $\mathfrak{h} = \mathcal{L}(H)$.*
  (i) *If H is normal in G, then $\mathfrak{h}$ is an ideal of $\mathfrak{g}$.*
  (ii) *Assume that G and H are connected. If $\mathfrak{h}$ is an ideal of $\mathfrak{g}$, then H is normal in G.*

Proof. We have already seen (i) in 23.5.6. Let us prove (ii). We have $\mathrm{ad}(\mathfrak{g})(\mathfrak{h}) \subset \mathfrak{h}$, so $\mathrm{Ad}(G)(\mathfrak{h}) = \mathfrak{h}$ (24.3.3). It follows from 24.4.6 that for all $\alpha \in G$, $\mathcal{L}(\alpha H \alpha^{-1}) = \mathfrak{h}$, hence $\alpha H \alpha^{-1} = H$ (24.3.5). $\square$

24.4.8 Theorem. *Let G be an algebraic group, H an algebraic subgroup of G, $\mathfrak{g} = \mathcal{L}(G)$ and $\mathfrak{h} = \mathcal{L}(H)$.*
   (i) *We have $\mathcal{L}\big(N_G(H)\big) \subset \mathfrak{n}_\mathfrak{g}(\mathfrak{h})$ and $\mathcal{L}\big(C_G(H)\big) \subset \mathfrak{c}_\mathfrak{g}(\mathfrak{h})$.*
   (ii) *If H is connected, then:*

$$N_G(H) = N_G(\mathfrak{h}) \, , \; C_G(H) = C_G(\mathfrak{h}),$$
$$\mathcal{L}\big(N_G(H)\big) = \mathcal{L}\big(N_G(\mathfrak{h})\big) = \mathfrak{n}_\mathfrak{g}(\mathfrak{h}) \, , \; \mathcal{L}\big(C_G(H)\big) = \mathcal{L}\big(C_G(\mathfrak{h})\big) = \mathfrak{c}_\mathfrak{g}(\mathfrak{h}).$$

Proof. The groups $N_G(H)$ and $C_G(H)$ are closed in G (21.3.7). Moreover, since H is normal in $N_G(H)$, $\mathfrak{h}$ is an ideal of $\mathcal{L}\big(N_G(H)\big)$ (24.4.7 (i)), so $\mathcal{L}\big(N_G(H)\big) \subset \mathfrak{n}_\mathfrak{g}(\mathfrak{h})$.

Let (V, π) be a finite-dimensional faithful rational representation of G (22.1.5), $\mathcal{A}$ (resp. $\mathcal{B}$) the subalgebra with unit of $\mathrm{End}(V)$ generated by $\pi(H)$ (resp. $d\pi(\mathfrak{h})$). By 24.3.2 (i), we have $\mathcal{B} \subset \mathcal{A}$. It follows that $C_G(H) \subset C_G(\mathfrak{h})$. Therefore, by 24.3.5 (i) and 24.3.6 (iii), we have $\mathcal{L}\big(C_G(H)\big) \subset \mathfrak{c}_\mathfrak{g}(\mathfrak{h})$.

If H is connected, then $\mathcal{A} = \mathcal{B}$ (24.3.2). Thus $N_G(H) = N_G(\mathfrak{h})$ et $C_G(H) = C_G(\mathfrak{h})$ and the rest follows from 24.3.6. $\square$

24.4.9 Let H, K be subgroups of G. Recall that (H, K) denotes the subgroup of G generated by $\alpha\beta\alpha^{-1}\beta^{-1}$, $\alpha \in H$, $\beta \in K$. Recall also that the derived subgroup (G, G) of G is closed (21.3.5) and it is connected if G is connected (21.3.6).

Proposition. *Let H, K be closed subgroups of G, L the closure of (H, K) in G (it is a subgroup by 21.2.2), and $\mathfrak{g}, \mathfrak{h}, \mathfrak{k}, \mathfrak{l}$ the Lie algebras of G, H, K, L respectively. Let $\alpha \in H$, $\beta \in K$, $x \in \mathfrak{h}$, and $y \in \mathfrak{k}$. Then*

$$[x, y] \, , \; \mathrm{Ad}(\alpha)(y) - y \, , \; \mathrm{Ad}(\beta)(x) - x$$

are elements of $\mathfrak{l}$.

Proof. For $\alpha \in H$, denote by $u_\alpha : K \to L$ the map $\beta \mapsto \alpha\beta\alpha^{-1}\beta^{-1}$. If $y \in \mathfrak{k}$, then $du_\alpha(y) = \mathrm{Ad}(\alpha)(y) - y$ (see 23.4.10). Thus $\mathrm{Ad}(\alpha)(y) - y \in \mathfrak{l}$. Similarly, we have $\mathrm{Ad}(\beta)(x) - x \in \mathfrak{l}$.

It follows that for $x \in \mathfrak{h}$, we have a map $v_x : K \to \mathfrak{l}$ given by $v_x(\beta) = \mathrm{Ad}(\beta)(x) - x$ for $\beta \in K$. If $y \in \mathfrak{k}$, then $dv_x(y) = [x, y]$. Hence $[x, y] \in \mathfrak{l}$. $\square$

24.4.10 Corollary. *Let G be an algebraic group and $\mathfrak{g}$ its Lie algebra. Then $[\mathfrak{g}, \mathfrak{g}] \subset \mathcal{L}\big((G, G)\big)$.*

24.4.11 Remark. We shall see later that $[\mathfrak{g}, \mathfrak{g}] = \mathcal{L}\big((G, G)\big)$ if G is connected. This is however not true if G is not connected. In the example of 24.4.3, $[\mathfrak{g}, \mathfrak{g}] = \{0\}$, $(G, G) = G^\circ$ and so $\mathcal{L}\big((G, G)\big) = \mathfrak{g}$.

24.5 Algebraic Lie subalgebras

24.5.1 In this section, G will denote an algebraic group, $\mathfrak{g}$ its Lie algebra. Recall that if $x \in \mathfrak{g}$ (resp. $\mathfrak{h}$ is a subspace of $\mathfrak{g}$), then $\mathcal{A}(x)$ (resp. $\mathcal{A}(\mathfrak{h})$) is the intersection of algebraic subgroups of G whose Lie algebra contains x (resp. $\mathfrak{h}$). Moreover, $\mathcal{A}(x)$ and $\mathcal{A}(\mathfrak{h})$ are connected and $\mathcal{A}(x)$ is commutative (24.2.3 and 24.2.5).

Let $\mathfrak{m}$ be a subset of $\mathfrak{g}$ and H an algebraic subgroup of G such that $\mathfrak{m} \subset \mathcal{L}(H)$. Then $\mathcal{A}(x) \subset H$ for all $x \in \mathfrak{m}$, and so H contains the subgroup K generated by the $\mathcal{A}(x)$'s, $x \in \mathfrak{m}$. Since K is a connected algebraic subgroup of G (21.3.2) whose Lie algebra contains $\mathfrak{m}$ (24.3.5), we deduce that K is the smallest algebraic subgroup of G whose Lie algebra contains $\mathfrak{m}$. We shall denote K by $\mathcal{A}(\mathfrak{m})$.

Definition. *A Lie subalgebra $\mathfrak{h}$ of $\mathfrak{g}$ is called* algebraic *if $\mathfrak{h} = \mathcal{L}(H)$ for some algebraic subgroup H of G.*

24.5.2 Proposition. *Let $(\mathfrak{h}_i)_{i \in I}$ be a family of algebraic Lie subalgebras of $\mathfrak{g}$, and H_i an algebraic subgroup of G such that $\mathcal{L}(H_i) = \mathfrak{h}_i$. Set:*

$$\mathfrak{h} = \bigcap_{i \in I} \mathfrak{h}_i \ , \ H = \bigcap_{i \in I} H_i.$$

Then $\mathfrak{h}$ is an algebraic Lie subalgebra of $\mathfrak{g}$, and $\mathfrak{h} = \mathcal{L}(H)$.

Proof. The group H is closed in G and verifies $\mathcal{L}(H) \subset \mathfrak{h}$ by 24.3.5. Moreover, $\mathcal{A}(\mathfrak{h}) \subset H_i$, so $\mathcal{A}(\mathfrak{h}) \subset H$. Thus $\mathfrak{h} \subset \mathcal{L}\big(\mathcal{A}(\mathfrak{h})\big) \subset \mathcal{L}(H)$. $\square$

24.5.3 Proposition. *Let $u : G \to H$ be a morphism of algebraic groups, and $\mathfrak{h} = \mathcal{L}(H)$.*
  (i) *The Lie algebra of $u(G)$ is $du(\mathfrak{g})$.*
  (ii) *If $\mathfrak{k}$ is an algebraic Lie subalgebra of $\mathfrak{g}$, then $du(\mathfrak{k})$ is an algebraic Lie subalgebra of $\mathfrak{h}$.*
  (iii) *If $\mathfrak{l}$ is an algebraic Lie subalgebra of $\mathfrak{h}$, then $(du)^{-1}(\mathfrak{l})$ is an algebraic Lie subalgebra of $\mathfrak{g}$.*

Proof. (i) By 21.2.4, $u(G)$ is a closed subgroup of H, and we may assume that G is connected. The morphism $v : G \to u(G)$ induced by u is dominant. Let $\alpha, \beta \in G$. Denote by $w : G \to G$ (resp. $w' : u(G) \to u(G)$) the automorphism of varieties defined by $\gamma \mapsto (\alpha\beta^{-1})\gamma$ (resp. $\delta \mapsto v(\alpha\beta^{-1})\delta$).

We have $w(\beta) = \alpha$, $w'(v(\beta)) = v(\alpha)$ and $v \circ w = w' \circ v$. So dw_β (resp. $dw'_{v(\beta)}$) is an isomorphism from $\mathrm{T}_\beta(G)$ onto $\mathrm{T}_\alpha(G)$ (resp. from $\mathrm{T}_{v(\beta)}(u(G))$ onto $\mathrm{T}_{v(\alpha)}(u(G))$) and we have $dv_\alpha \circ dw_\beta = dw'_{v(\beta)} \circ dv_\beta$. Thus 16.5.7 (ii) implies that du induces a surjection from $\mathfrak{g}$ onto $\mathcal{L}(u(G))$.

  (ii) We may assume that $\mathfrak{k} = \mathfrak{g}$. So the result follows from (i).
  (iii) Applying (i) and 24.5.2, we may assume that u and du are surjective. Let $L = \mathcal{A}(\mathfrak{l})$ and $K = u^{-1}(L)$. Since $\ker u \subset K$, $\ker du = \mathcal{L}(\ker u) \subset \mathcal{L}(K)$ (24.3.5 and 24.4.1). Applying (i) to the surjective morphism $K \to L$ induced

by u, we obtain that $du(\mathcal{L}(K)) = \mathfrak{l}$. So we have $(du)^{-1}(\mathfrak{l}) = \mathcal{L}(K) + \ker du = \mathcal{L}(K)$. $\square$

24.5.4 Let $\mathfrak{m}$ be a non-empty subset of $\mathfrak{g}$. Denote by $\mathbf{a}(\mathfrak{m})$ the intersection of all the algebraic Lie subalgebras of $\mathfrak{g}$ containing $\mathfrak{m}$. By 24.5.2, $\mathbf{a}(\mathfrak{m})$ is the smallest algebraic Lie subalgebra of $\mathfrak{g}$ containing $\mathfrak{m}$. Clearly, we have $\mathbf{a}(\mathfrak{m}) = \mathcal{L}(\mathcal{A}(\mathfrak{m}))$. If $\mathfrak{m}$ is a Lie subalgebra of $\mathfrak{g}$, then we call $\mathbf{a}(\mathfrak{m})$ the *algebraic hull* of $\mathfrak{m}$ in $\mathfrak{g}$.

Proposition. *Let* $u : G \to H$ *be a morphism of algebraic groups and* $\mathfrak{m}$ *a non-empty subset of* $\mathfrak{g}$. *Then:*
(i) $du(\mathbf{a}(\mathfrak{m})) = \mathbf{a}(du(\mathfrak{m}))$.
(ii) $u(\mathcal{A}(\mathfrak{m})) = \mathcal{A}(du(\mathfrak{m}))$.

Proof. (i) By 24.5.3, $du(\mathbf{a}(\mathfrak{m}))$ is an algebraic Lie subalgebra of $\mathcal{L}(H)$ containing $du(\mathfrak{m})$, so it contains $\mathbf{a}(du(\mathfrak{m}))$.

Conversely, again by 24.5.3, $(du)^{-1}(\mathbf{a}(du(\mathfrak{m})))$ is an algebraic Lie subalgebra of $\mathfrak{g}$ containing $\mathfrak{m}$, so it contains $\mathbf{a}(\mathfrak{m})$. It follows that $du(\mathbf{a}(\mathfrak{m})) \subset \mathbf{a}(du(\mathfrak{m}))$.

(ii) We have $\mathcal{L}(\mathcal{A}(du(\mathfrak{m}))) = \mathbf{a}(du(\mathfrak{m}))$ and $\mathcal{L}(u(\mathcal{A}(\mathfrak{m}))) = du(\mathbf{a}(\mathfrak{m}))$ (24.5.3). Since the groups $u(\mathcal{A}(\mathfrak{m}))$ and $\mathcal{A}(du(\mathfrak{m}))$ are connected, the result follows from (i) and 24.3.5. $\square$

24.5.5 Theorem. *Let* (V, π) *be a finite-dimensional rational representation of* G, *and* E, F *vector subspaces of* V *such that* $E \subset F$. *Then* $H = \{\alpha \in G; (\pi(\alpha) - \mathrm{id}_V)(F) \subset E\}$ *is an algebraic subgroup of* G *and* $\mathfrak{h} = \{x \in \mathfrak{g}; d\pi(x)(F) \subset E\}$ *is a Lie subalgebra of* $\mathfrak{g}$ *verifying* $\mathfrak{h} = \mathcal{L}(H)$.

Proof. By definition, $H \subset N_G(F)$ and $\mathfrak{h} \subset \mathfrak{n}_\mathfrak{g}(F)$. So we may assume that $V = F$, $G = N_G(F)$ and $\mathfrak{g} = \mathfrak{n}_\mathfrak{g}(F)$ (24.3.6).

Let $\mathcal{B} = (v_1, \ldots, v_n)$ be a basis of V such that $(v_1, \ldots, v_p)$ is a basis of E. For $\alpha \in G$ and $x \in \mathfrak{g}$, denote by $[\pi_{ij}(\alpha)]$ (resp. $[x_{ij}]$) the matrix of $\pi(\alpha)$ (resp. $d\pi(x)$) with respect to $\mathcal{B}$. Then $x \in \mathfrak{h}$ (resp. $\alpha \in H$) if and only if $x_{ij} = 0$ (resp. $\pi_{ij}(\alpha) = \delta_{ij}$ where δ_{ij} denotes the Kronecker symbol) for $p + 1 \leqslant i \leqslant n$ and $1 \leqslant j \leqslant n$. We deduce therefore that H is an algebraic subgroup of G, $\mathfrak{h}$ is a Lie subalgebra of $\mathfrak{g}$, and we have $\mathcal{L}(H) \subset \mathfrak{h}$ by 23.4.2.

For $f \in V^*$ and $v \in V$, define $\vartheta_{f,v} \in (\mathrm{End}(V))^*$ as in the proof of 24.3.6, and denote by $E^\perp$ the orthogonal of E in V^*. We have:

$$x \in \mathfrak{h} \Leftrightarrow x^n(\vartheta_{f,v} \circ \pi) = 0 \text{ for all } n \in \mathbb{N}^*, (f, v) \in E^\perp \times V,$$
$$\alpha \in H \Leftrightarrow \vartheta_{f,v} \circ \pi(\alpha) = \vartheta_{f,v} \circ \pi(e) \text{ for all } (f, v) \in E^\perp \times V.$$

Proceeding as in the proof of 24.3.6, we obtain that $\mathfrak{h} \subset \mathcal{L}(H)$. $\square$

24.5.6 Corollary. *Let* $\mathfrak{m}, \mathfrak{n}$ *be subspaces of* $\mathfrak{g}$ *such that* $\mathfrak{m} \subset \mathfrak{n}$. *Then* $H = \{\alpha \in G; (\mathrm{Ad}(\alpha) - \mathrm{id}_\mathfrak{g})(\mathfrak{n}) \subset \mathfrak{m}\}$ *is an algebraic subgroup of* G, *and*

$\mathfrak{h} = \{x \in \mathfrak{g}; \mathrm{ad}(x)(\mathfrak{n}) \subset \mathfrak{m}\}$ *is an algebraic Lie subalgebra of* $\mathfrak{g}$ *verifying* $\mathfrak{h} = \mathcal{L}(H)$.

24.5.7 Theorem. *Let* $\mathfrak{h}$ *be a Lie subalgebra of* $\mathfrak{g}$.
(i) *Any ideal of* $\mathfrak{h}$ *is an ideal of* $\mathbf{a}(\mathfrak{h})$.
(ii) *We have* $[\mathbf{a}(\mathfrak{h}), \mathbf{a}(\mathfrak{h})] = [\mathfrak{h}, \mathfrak{h}]$.

Proof. (i) Let $\mathfrak{a}$ be an ideal of $\mathfrak{h}$. By 24.5.6, the set of $x \in \mathfrak{g}$ verifying $[x, \mathfrak{a}] \subset \mathfrak{a}$ is an algebraic Lie subalgebra $\mathfrak{k}$ of $\mathfrak{g}$. Since $\mathfrak{h} \subset \mathfrak{k}$, we have $\mathbf{a}(\mathfrak{h}) \subset \mathfrak{k}$. Thus $\mathfrak{a}$ is an ideal of $\mathbf{a}(\mathfrak{h})$.

(ii) As in (i), the set of $x \in \mathfrak{g}$ verifying $[x, \mathfrak{h}] \subset [\mathfrak{h}, \mathfrak{h}]$ contains $\mathbf{a}(\mathfrak{h})$. So the set of $y \in \mathfrak{g}$ such that $[y, \mathbf{a}(\mathfrak{h})] \subset [\mathfrak{h}, \mathfrak{h}]$ is an algebraic Lie subalgebra $\mathfrak{g}$ containing $\mathfrak{h}$, so it contains $\mathbf{a}(\mathfrak{h})$. We deduce therefore that $[\mathbf{a}(\mathfrak{h}), \mathbf{a}(\mathfrak{h})] \subset [\mathfrak{h}, \mathfrak{h}]$. Since $\mathfrak{h} \subset \mathbf{a}(\mathfrak{h})$, we have $[\mathbf{a}(\mathfrak{h}), \mathbf{a}(\mathfrak{h})] \supset [\mathfrak{h}, \mathfrak{h}]$. $\square$

24.5.8 Let $(X_i)_{i \in I}$ be a family of irreducible subvarieties of G. Suppose that the following conditions are satisfied:
(i) $e \in X_i$ for all $i \in I$.
(ii) If $i \in I$, there exists $j \in I$ such that $X_i^{-1} = \{\alpha^{-1}; \alpha \in X_i\} = X_j$.
Let H be the subgroup of G generated by the X_i's. By 21.3.1, H is a connected algebraic subgroup of G. Let $\mathfrak{h}$ be its Lie algebra.

Proposition. *The vector space* $\mathfrak{h}$ *is spanned by the vector subspaces* $\mathrm{Ad}(\alpha)(\mathrm{T}_e(\beta_i^{-1} X_i))$ *of* $\mathfrak{g}$, *for* $\alpha \in H$, $i \in I$ *and* $\beta_i \in X_i$.

Proof. 1) By 21.3.1, there exist $i_1, \ldots, i_n \in I$ such that the morphism

$$u : X_{i_1} \times \cdots \times X_{i_n} \to H \ , \ (\beta_1, \ldots, \beta_n) \mapsto \beta_1 \cdots \beta_n$$

is surjective. To simplify notations let us write X_k for X_{i_k} if $1 \leqslant k \leqslant n$. Let $X = X_1 \times \cdots \times X_n$.

There exists $a = (\alpha_1, \ldots, \alpha_n) \in X$ such that the map $du_a : \mathrm{T}_a(X) \to \mathrm{T}_{u(a)}(H)$ is surjective (16.5.7). For $1 \leqslant i \leqslant n$, set:

$$\gamma_i = \alpha_1 \cdots \alpha_i \ , \ Y_i = \alpha_i^{-1} X_i \ , \ Z_i = \gamma_i Y_i \gamma_i^{-1}.$$

Let $Y = Y_1 \times \cdots \times Y_n$ and $Z = Z_1 \times \cdots \times Z_n$. Define the following maps:

$$s : Y \to X \ , \ (\beta_1, \ldots, \beta_n) \mapsto (\alpha_1 \beta_1, \ldots, \alpha_n \beta_n),$$
$$t : Y \to Z \ , \ (\beta_1, \ldots, \beta_n) \mapsto (\gamma_1 \beta_1 \gamma_1^{-1}, \ldots, \gamma_n \beta_n \gamma_n^{-1}),$$
$$v : Z \to H \ , \ (\beta_1, \ldots, \beta_n) \mapsto \beta_1 \cdots \beta_n,$$
$$\theta : H \to H \ , \ \beta \mapsto \beta (u(a))^{-1}.$$

We verify easily that $\theta \circ u \circ s = v \circ t$. Moreover if $\varepsilon = (e, \ldots, e) \in Y$, then $s(\varepsilon) = a$.

The maps s and θ are isomorphisms of varieties and du_a is surjective. It follows that the differential

$$d(v \circ t)_\varepsilon \colon \mathrm{T}_\varepsilon(Y) \to \mathrm{T}_e(H) = \mathfrak{h}$$

of $v \circ t = \theta \circ u \circ s$ at the point ε is surjective.

2) Let $x = (x_1, \ldots, x_n) \in \mathrm{T}_\varepsilon(Y)$, so $x_i \in \mathrm{T}_e(\alpha_i^{-1} X_i)$ for $1 \leqslant i \leqslant n$. By 23.4.7, we have:

$$d(v \circ t)_\varepsilon(x) = \sum_{k=1}^{n} d(i_{\gamma_k})_e(x_k) = \sum_{k=1}^{n} \mathrm{Ad}(\gamma_k)(x_k).$$

Since $\gamma_k \in H$ and $d(v \circ t)_\varepsilon$ is surjective, we have proved that $\mathfrak{h}$ is contained in the sum of the vector subspaces $\mathrm{Ad}(\alpha)(\mathrm{T}_e(\beta_i^{-1} X_i))$, for $\alpha \in H$, $i \in I$ and $\beta_i \in X_i$. Now, each $\mathrm{Ad}(\alpha)(\mathrm{T}_e(\beta_i^{-1} X_i))$ is clearly contained in $\mathfrak{h}$, so the result follows. $\square$

24.5.9 Theorem. *Let $(H_i)_{i \in I}$ be a family of connected algebraic subgroups of G, H the connected algebraic subgroup of G generated by the H_i's (see 21.3.2) and $\mathfrak{h}$, $\mathfrak{h}_i$ the Lie algebras of H, H_i, $i \in I$ respectively.*

(i) *The vector space $\mathfrak{h}$ is the sum of the vector spaces $\mathrm{Ad}(\alpha)(\mathfrak{h}_i)$, $\alpha \in H$, $i \in I$.*

(ii) *As a Lie algebra, $\mathfrak{h}$ is generated by $\mathfrak{h}_i$, $i \in I$.*

Proof. (i) Since H_i is a group, we have $\beta_i^{-1} H_i = H_i$ for all $\beta_i \in H_i$ and $\mathrm{T}_e(\beta_i^{-1} H_i) = \mathfrak{h}_i$. So the result follows from 24.5.8.

(ii) Let $\mathfrak{k}$ be the Lie subalgebra of $\mathfrak{g}$ generated by $\mathfrak{h}_i$, $i \in I$. Since $H_i \subset H$, $\mathfrak{h}_i \subset \mathfrak{h}$, and as $\mathfrak{h}$ is a Lie algebra, we deduce that $\mathfrak{k} \subset \mathfrak{h}$. To show that $\mathfrak{h} \subset \mathfrak{k}$, it suffices, by (i), to establish that $\mathrm{Ad}(\alpha)(\mathfrak{k}) \subset \mathfrak{k}$ for $\alpha \in H$, that is, $H \subset N_G(\mathfrak{k})$. Since $N_G(\mathfrak{k})$ is a group, we are reduced to prove that $H_i \subset N_G(\mathfrak{k})$ for all $i \in I$. The H_i's are connected, so this is equivalent to proving that $\mathfrak{h}_i \subset \mathfrak{n}_\mathfrak{g}(\mathfrak{k})$ (24.4.8). But this is obvious since $\mathfrak{h}_i \subset \mathfrak{k}$. $\square$

24.5.10 Corollary. *The following conditions are equivalent for a Lie subalgebra $\mathfrak{h}$ of $\mathfrak{g}$:*

(i) $\mathfrak{h}$ *is an algebraic Lie subalgebra of $\mathfrak{g}$.*

(ii) *For all $x \in \mathfrak{h}$, $\mathbf{a}(x) \subset \mathfrak{h}$.*

(iii) *The vector space $\mathfrak{h}$ is a sum of algebraic Lie subalgebras of $\mathfrak{g}$.*

(iv) *As a Lie algebra, $\mathfrak{h}$ is generated by algebraic Lie subalgebras of $\mathfrak{g}$.*

Proof. The implications (i) $\Rightarrow$ (ii) $\Rightarrow$ (iii) $\Rightarrow$ (iv) are clear, while (iv) $\Rightarrow$ (i) follows from 24.5.9. $\square$

24.5.11 Let H, K be connected normal algebraic subgroups of G. By 21.3.5, $L = (H, K)$ is a connected algebraic subgroup of G. Let $\mathfrak{h}, \mathfrak{k}, \mathfrak{l}$ be the Lie algebras of H, K, L.

Theorem. *We have $\mathfrak{l} = [\mathfrak{h}, \mathfrak{k}]$.*

Proof. 1) We have $[\mathfrak{h}, \mathfrak{k}] \subset \mathfrak{l}$ by 24.4.9.

2) For $\alpha, \beta \in G$, let $c_\alpha(\beta) = \alpha\beta\alpha^{-1}\beta^{-1}$. Since

$$(\alpha\beta\alpha^{-1}\beta^{-1})^{-1} = \beta\alpha\beta^{-1}\alpha^{-1} = \alpha^{-1}(\alpha\beta\alpha^{-1})\alpha(\alpha\beta\alpha^{-1})^{-1},$$

we deduce that $c_\alpha(K)^{-1} = c_{\alpha^{-1}}(K)$.

By 21.3.1, there exist $n \in \mathbb{N}^*$ and $\alpha_1, \ldots, \alpha_n \in H$ such that the map

$$u : K^n \to L , \ (\gamma_1, \ldots, \gamma_n) \mapsto c_{\alpha_1}(\gamma_1) \cdots c_{\alpha_n}(\gamma_n)$$

is surjective. So by 16.5.7, there exists $b = (\beta_1, \ldots, \beta_n) \in K^n$ such that

$$du_b : \mathrm{T}_b(K^n) \to \mathrm{T}_{u(b)}(L)$$

is surjective.

3) Let $s : K^n \to K^n$ and $\theta : L \to L$ be defined as follows:

$$s(\gamma_1, \ldots, \gamma_n) = (\beta_1\gamma_1, \ldots, \beta_n\gamma_n) , \ \theta(\gamma) = \gamma\big(u(b)\big)^{-1}.$$

These are isomorphisms of varieties. For $1 \leqslant k \leqslant n$ and $\gamma \in K$, let:

$$C_k(\gamma) = c_{\alpha_1}(\beta_1) \cdots c_{\alpha_{k-1}}(\beta_{k-1}) c_{\alpha_k}(\beta_k\gamma)\big(c_{\alpha_k}(\beta_k)\big)^{-1} \cdots \big(c_{\alpha_1}(\beta_1)\big)^{-1}.$$

Finally, define $t : K^n \to K^n$ and $v : K^n \to H$ by:

$$t(\gamma_1, \ldots, \gamma_n) = \big(C_1(\gamma_1), \ldots, C_n(\gamma_n)\big) , \ v(\gamma_1, \ldots, \gamma_n) = \gamma_1 \cdots \gamma_n.$$

We check easily that:

$$v \circ t = \theta \circ u \circ s.$$

Denote by $\varepsilon = (e, \ldots, e) \in K^n$. Since θ and s are isomorphisms of varieties, the differential $d(v \circ t)_\varepsilon : \mathrm{T}_\varepsilon(K^n) \to \mathrm{T}_e(L) = \mathfrak{l}$ of $v \circ t$ at the point ε is surjective.

Let $\delta_k = c_{\alpha_1}(\beta_1) \cdots c_{\alpha_k}(\beta_k)$, $1 \leqslant k \leqslant n$. Then we obtain easily that:

$$C_k(\gamma) = i_{\delta_k}\big(i_{\beta_k\alpha_k}(\gamma) i_{\beta_k}(\gamma^{-1})\big).$$

If $x = (x_1, \ldots, x_n) \in \big(\mathrm{T}_e(K)\big)^n = \mathrm{T}_\varepsilon(K^n)$, then 23.4.7 implies that:

$$d(v \circ t)_\varepsilon(x) = \sum_{k=1}^n \mathrm{Ad}(\delta_k)\Big(\mathrm{Ad}(\beta_k)\big(\mathrm{Ad}(\alpha_k)(x_k) - x_k\big)\Big).$$

Since $[\mathfrak{k}, [\mathfrak{h}, \mathfrak{k}]] \subset [\mathfrak{h}, \mathfrak{k}]$ and K is connected, 24.5.5 implies that $\mathrm{Ad}(\gamma)([\mathfrak{h}, \mathfrak{k}]) \subset [\mathfrak{h}, \mathfrak{k}]$ for all $\gamma \in K$. As $\delta_k, \beta_k \in K$, to obtain the result, it suffices therefore to prove that if $x \in \mathfrak{k}$ and $\alpha \in H$, then $\mathrm{Ad}(\alpha)(x) - x \in [\mathfrak{h}, \mathfrak{k}]$. But this follows again from 24.5.5 and the fact that H is connected. $\square$

24.5.12 Theorem. *Let $\mathfrak{h}$ be a Lie subalgebra of $\mathfrak{g}$. Then $[\mathfrak{h}, \mathfrak{h}]$ is an algebraic Lie subalgebra of $\mathfrak{g}$. Furthermore, if H is a connected subgroup of G such that $\mathfrak{h} = \mathcal{L}(H)$, then $[\mathfrak{h}, \mathfrak{h}]$ is the Lie algebra of $\mathcal{D}(H)$.*

Proof. Since $[\mathfrak{h}, \mathfrak{h}] = [\mathbf{a}(\mathfrak{h}), \mathbf{a}(\mathfrak{h})]$ (24.5.7), we may assume that $\mathfrak{h}$ is algebraic. Then $H = \mathcal{A}(\mathfrak{h})$ and the result follows from 24.5.11. $\square$

24.5.13 Corollary. (i) *If G is solvable (resp. nilpotent), then $\mathfrak{g}$ is solvable (resp. nilpotent).*

(ii) *Assume that G is connected. If $\mathfrak{g}$ is solvable (resp. nilpotent), then G is solvable (resp. nilpotent).*

Proof. We have $\mathcal{L}(G) = \mathcal{L}(G^{\circ})$ and if G is solvable (resp. nilpotent), then so is G°. So we may assume that G is connected. Then 24.5.11 implies that $\mathcal{L}(\mathcal{D}^{n}(G)) = \mathcal{D}^{n}(\mathfrak{g})$ and $\mathcal{L}(\mathcal{C}^{n}(G)) = \mathcal{C}^{n}(\mathfrak{g})$ (notations of 19.3.1). So the result follows. $\square$

24.6 A particular case

24.6.1 In this section, G denotes an algebraic group and $\mathfrak{g}$ its Lie algebra. If $x \in \mathfrak{g}$, then an element of $\mathbf{a}(x)$ is called a *replica* of x.

24.6.2 Let $x \in \mathfrak{g}$ and x_s, x_n its semisimple and nilpotent components.

1) We have $x_s, x_n \in \mathbf{a}(x) = \mathcal{L}(\mathcal{A}(x))$. Thus $\mathbf{a}(x_s) \subset \mathbf{a}(x)$, $\mathbf{a}(x_n) \subset \mathbf{a}(x)$ (24.5.10), and $\mathcal{A}(x_s) \subset \mathcal{A}(x)$, $\mathcal{A}(x_n) \subset \mathcal{A}(x)$ (24.2.3). Note that the Lie algebra $\mathbf{a}(x)$ is commutative (24.4.4).

2) Let $H \subset \mathcal{A}(x)$ be the subgroup of G generated by $\mathcal{A}(x_s)$ and $\mathcal{A}(x_n)$. Then H is a connected algebraic subgroup of G (21.3.1). Moreover, $x_s \in \mathcal{L}(\mathcal{A}(x_s)) \subset \mathcal{L}(H)$ and $x_n \in \mathcal{L}(\mathcal{A}(x_n)) \subset \mathcal{L}(H)$. Hence $x \in \mathcal{L}(H)$, and $\mathcal{A}(x) \subset H$ (24.2.3). It follows that $H = \mathcal{A}(x)$. Since $\mathcal{A}(x)$ is commutative, $\mathcal{A}(x) = \mathcal{A}(x_s)\mathcal{A}(x_n)$. Applying 24.5.9 we obtain that $\mathbf{a}(x) = \mathbf{a}(x_s) + \mathbf{a}(x_n)$.

3) Let us reduce to the case where G is a closed subgroup of $\mathrm{GL}(V)$ with V a finite-dimensional vector space (22.1.5).

Suppose that x is semisimple. Using a basis of diagonalization for x, we may assume that $G \subset \mathrm{GL}_n(\Bbbk)$ and $x \in \mathfrak{d}_n(\Bbbk)$. Since $\mathfrak{d}_n(\Bbbk) = \mathcal{L}(D_n(\Bbbk))$, $\mathcal{A}(x) \subset D_n(\Bbbk)$ and $\mathbf{a}(x) \subset \mathfrak{d}_n(\Bbbk)$. Thus the elements of $\mathcal{A}(x)$ (resp. $\mathbf{a}(x)$) are semisimple.

Suppose that x is nilpotent. We may assume that $x \in \mathfrak{n}_n(\Bbbk)$. Again we see that $\mathcal{A}(x) \subset U_n(\Bbbk)$ and $\mathbf{a}(x) \subset \mathfrak{n}_n(\Bbbk)$. Thus the elements of $\mathcal{A}(x)$ are unipotent, and those of $\mathbf{a}(x)$ are nilpotent.

In the general case, the preceding discussion shows that $\mathbf{a}(x_s)$ (resp. $\mathbf{a}(x_n)$) is the set of semisimple (resp. nilpotent) elements of $\mathbf{a}(x)$. Similarly, $\mathcal{A}(x_s)$ (resp. $\mathcal{A}(x_n)$) is the set of semisimple (resp. unipotent) elements of $\mathcal{A}(x)$.

As in 22.2.6, we obtain:

Proposition. *Let us conserve the above notations.*

(i) *We have $\mathbf{a}(x) = \mathbf{a}(x_s) \oplus \mathbf{a}(x_n)$.*

(ii) *The map $\mathcal{A}(x_s) \times \mathcal{A}(x_n) \to \mathcal{A}(x)$, $(\alpha, \beta) \mapsto \alpha\beta$, is an isomorphism of algebraic groups.*

24.6.3 Suppose that G is a closed subgroup of $\mathrm{GL}(V)$. To determine $\mathcal{A}(x)$ (resp. $\mathbf{a}(x)$), it suffices to know $\mathcal{A}(x_s)$ and $\mathcal{A}(x_n)$ (resp. $\mathbf{a}(x_s)$ and $\mathbf{a}(x_n)$).

For $\mathcal{A}(x_n)$ and $\mathbf{a}(x_n)$, we may use 23.3.5 and 24.2.4. Let us consider x_s. For simplicity, let us suppose that x semi-simple.

We may assume, via a basis of diagonalization, that $G \subset \mathrm{GL}_n(\Bbbk)$ and $x = \mathrm{diag}(s_1, \ldots, s_n)$ where $s_1, \ldots, s_n \in \Bbbk$. Denote by:

$$L_x = \{(a_1, \ldots, a_n) \in \mathbb{Z}^n;\ a_1 s_1 + \cdots + a_n s_n = 0\}.$$

We saw in 24.6.2 that $\mathcal{A}(x) \subset \mathrm{D}_n(\Bbbk)$ and $\mathbf{a}(x) \subset \mathfrak{d}_n(\Bbbk)$. Moreover, by 22.5.4 and 22.5.7, $\mathcal{A}(x)$ is a torus and it is the intersection of the kernel of certain characters of $\mathrm{D}_n(\Bbbk)$.

Let $H = \mathrm{D}_n(\Bbbk)$, $\mathfrak{h} = \mathfrak{d}_n(\Bbbk)$, $\mathbf{A}(H) = \Bbbk[T_1, T_1^{-1}, \ldots, T_n, T_n^{-1}]$ and we identify $\mathcal{L}(H)$ with $\mathfrak{h}$ via the map $y \to [y(T_i)]$ (see 23.3.3 and 23.3.4). Thus if $P \in \mathbf{A}(H)$, then:

$$x(P) = \sum_{i=1}^{n} s_i \frac{\partial P}{\partial T_i}(e).$$

Let χ be a character of H. By 22.5.2, there exist integers $a_1, \ldots, a_n$ such that $\chi\big(\mathrm{diag}(t_1, \ldots, t_n)\big) = t_1^{a_1} \cdots t_n^{a_n}$ for all $t_1, \ldots, t_n \in \Bbbk \setminus \{0\}$.

If $Q \in \Bbbk[T, T^{-1}] \in \mathbf{A}(\mathbf{G}_m)$, then:

$$d\chi(x)(Q) = x(Q \circ \chi) = \sum_{i=1}^{n} s_i \frac{\partial (Q \circ \chi)}{\partial T_i}(e) = \Big(\sum_{i=1}^{n} s_i a_i \Big) Q'(1).$$

Hence (23.3.2):

$$d\chi(x) = (a_1 s_1 + \cdots + a_n s_n) \frac{d}{dT}\Big|_{T=1}.$$

By 24.4.1 and the definition of $\mathcal{A}(x)$, we deduce that $\mathcal{A}(x) \subset \ker \chi$ if and only if $(a_1, \ldots, a_n) \in L_x$. Since $\mathcal{A}(x)$ is the intersection of the kernel of certain characters of H, we have proved that $\mathcal{A}(x)$ is the set of matrices $\mathrm{diag}(t_1, \ldots, t_n)$ such that $t_1^{a_1} \cdots t_n^{a_n} = 1$ for all $(a_1, \ldots, a_n) \in L_x$. On the other hand, L_x is a subgroup of $\mathbb{Z}^n$, so it is free and finitely generated. Applying 24.3.5 (iii) and 24.4.1, we see that $\mathbf{a}(x)$ is the set of matrices $\mathrm{diag}(u_1, \ldots, u_n)$ such that $a_1 u_1 + \cdots + a_n u_n = 0$ for all $(a_1, \ldots, a_n) \in L_x$. To summarize, we have obtained the following result:

Theorem. *Let* $x = \mathrm{diag}(s_1, \ldots, s_n) \in \mathfrak{d}_n(\Bbbk)$ *and*

$$L_x = \{(a_1, \ldots, a_n) \in \mathbb{Z}^n;\ a_1 s_1 + \cdots + a_n s_n = 0\}.$$

The group $\mathcal{A}(x)$ *is the set of matrices* $\mathrm{diag}(t_1, \ldots, t_n) \in \mathrm{D}_n(\Bbbk)$ *verifying* $t_1^{a_1} \cdots t_n^{a_n} = 1$ *for all* $(a_1, \ldots, a_n) \in L_x$. *The Lie algebra* $\mathbf{a}(x)$ *is the set of matrices* $\mathrm{diag}(u_1, \ldots, u_n) \in \mathfrak{d}_n(\Bbbk)$ *such that* $a_1 u_1 + \cdots + a_n u_n = 0$ *for all* $(a_1, \ldots, a_n) \in L_x$.

24.6.4 Proposition. *Let* V *be a finite-dimensional vector space and let* $x \in \mathfrak{gl}(V)$.

(i) *Any element of* $\mathcal{A}(x)$ *can be written as* $P(x)$ *for some* $P \in \Bbbk[T]$.

(ii) *Any replica of* x *can be written as* $Q(x)$ *for some* $Q \in \Bbbk[T]$ *without constant term.*

Proof. Since $\mathcal{A}(x) = \mathcal{A}(x_s)\mathcal{A}(x_n)$, $\mathbf{a}(x) = \mathbf{a}(x_s) + \mathbf{a}(x_n)$ (24.6.2) and x_s, x_n are polynomials without constant term in x, it suffices to consider on the one hand the case where x is nilpotent, and on the other hand, the case where x is semisimple.

If x is nilpotent, then $\mathcal{A}(x) = \{e^{tx}; t \in \Bbbk\}$ and $\mathbf{a}(x) = \Bbbk x$ (23.3.5 and 24.2.4). So the result is clear.

Suppose that x is semisimple. Denote by $s_1, \ldots, s_p$ the pairwise distinct eigenvalues of x and $V_1, \ldots, V_p$ the corresponding eigenspaces. For $i \in \{1, \ldots, p\}$ and $v \in V_i$, set:

$$G_v = \{\alpha \in \mathrm{GL}(V); \alpha(\Bbbk v) \subset \Bbbk v\} \, , \; \mathfrak{g}_v = \{y \in \mathfrak{gl}(V); y(\Bbbk v) \subset \Bbbk v\}.$$

By 24.5.6, G_v is an algebraic subgroup of $\mathrm{GL}(V)$ whose Lie algebra is $\mathfrak{g}_v$. Since $x \in \mathfrak{g}_v$, we deduce easily that for all $\alpha \in \mathcal{A}(x)$ and $y \in \mathbf{a}(x)$, $V_1, \ldots, V_p$ are eigenspaces of α and y.

So let $\alpha \in \mathcal{A}(x)$ and $t_1, \ldots, t_p \in \Bbbk \setminus \{0\}$ be such that $\alpha|_{V_i} = t_i \, \mathrm{id}_{V_i}$ for $1 \leqslant i \leqslant p$. There exists $P \in \Bbbk[T]$ such that $t_i = P(s_i)$ for $1 \leqslant i \leqslant p$. Thus $\alpha = P(x)$, and we have proved (i).

We see similarly that if $y \in \mathbf{a}(x)$, then there exists $Q \in \Bbbk[T]$ such that $y = Q(x)$. If x is invertible, by using the minimal polynomial of x, we see that id_V is a polynomial without constant term in x, so we may assume that Q has no constant term. If x is not invertible, then we have, for example, $s_1 = 0$. The set of elements $y \in \mathfrak{gl}(V)$ verifying $y|_{V_1} = 0$ is an algebraic Lie subalgebra of $\mathfrak{gl}(V)$ containing x (24.5.6), hence $y|_{V_1} = 0$ for all $y \in \mathbf{a}(x)$. But $y = Q(x)$, so Q has no constant term. $\square$

24.6.5 Proposition. *Let V be a finite-dimensional vector space and $x \in \mathfrak{gl}(V)$. Then x is nilpotent if and only if $\mathrm{tr}(xy) = 0$ for all replica y of x.*

Proof. If x is nilpotent, then so is x^p for all $p \in \mathbb{N}^*$, so $\mathrm{tr}(x^p) = 0$. So 24.6.4 implies that $\mathrm{tr}(xy) = 0$ for all replica y of x.

Conversely, suppose that $\mathrm{tr}(xy) = 0$ for all replica y of x. Since $xy = yx$, $x_n y = y x_n$ and $x_n y$ is nilpotent. Consequently, $\mathrm{tr}(x_s y) = 0$ for all replica y of x. Since $\mathbf{a}(x_s) \subset \mathbf{a}(x)$, it suffices to prove that if x is semisimple and $\mathrm{tr}(xy) = 0$ for all $y \in \mathbf{a}(x)$, then $x = 0$.

Let $(v_1, \ldots, v_r)$ be a basis of V consisting of eigenvectors of x with eigenvalues $s_1, \ldots, s_r$, and E the $\mathbb{Q}$-vector subspace of $\Bbbk$ spanned by the s_i's. For a $\mathbb{Q}$-linear form φ on E, define $y \in \mathfrak{gl}(V)$ to be the element such that $y(v_i) = \varphi(s_i)v_i$ for $1 \leqslant i \leqslant r$. It is clear from 24.6.3 that $y \in \mathbf{a}(x)$. So:

$$\mathrm{tr}(xy) = s_1\varphi(s_1) + \cdots + s_r\varphi(s_r) = 0.$$

We deduce from this that:

$$\bigl(\varphi(s_1)\bigr)^2 + \cdots + \bigl(\varphi(s_r)\bigr)^2 = 0.$$

Thus $\varphi(s_1) = \cdots = \varphi(s_r) = 0$. Since the s_i's span E, we have $E = \{0\}$. $\square$

24.7 Examples

24.7.1 In this section, G denotes an algebraic group whose Lie algebra is $\mathfrak{g}$. We shall give examples of algebraic Lie subalgebras of $\mathfrak{g}$.

24.7.2 Let $\mathfrak{h}$ be a Lie subalgebra of $\mathfrak{g}$.

• If $\mathfrak{h} = [\mathfrak{h}, \mathfrak{h}]$, then $\mathfrak{h}$ is an algebraic Lie subalgebra of $\mathfrak{g}$ (24.5.12). In particular, if $\mathfrak{h}$ is semisimple, $\mathfrak{h}$ is an algebraic Lie subalgebra of $\mathfrak{g}$.

• Suppose that $\mathfrak{h} = \mathfrak{n}_\mathfrak{g}(\mathfrak{a})$, where $\mathfrak{a}$ is a Lie subalgebra of $\mathfrak{g}$. Then by 24.5.6, $\mathfrak{h}$ is algebraic in $\mathfrak{g}$. In particular, if $\mathfrak{h}$ is its own normalizer in $\mathfrak{g}$, then it is an algebraic Lie subalgebra of $\mathfrak{g}$.

• Similarly, if $\mathfrak{h} = \mathfrak{c}_\mathfrak{g}(\mathfrak{a})$, where $\mathfrak{a}$ is a subspace of $\mathfrak{g}$, then $\mathfrak{h}$ is an algebraic Lie subalgebra of $\mathfrak{g}$.

24.7.3 Lemma. *Let $\mathfrak{h}$ be an ideal of $\mathfrak{g}$. Then $[\mathfrak{g}, \mathbf{a}(\mathfrak{h})] \subset \mathfrak{h}$. In particular, $\mathbf{a}(\mathfrak{h})$ is an ideal of $\mathfrak{g}$.*

Proof. Let $\mathfrak{k} = \{x \in \mathfrak{g}; [x, \mathfrak{g}] \subset \mathfrak{h}\}$. Then 24.5.6 says that $\mathfrak{k}$ is algebraic in $\mathfrak{g}$. If $\mathfrak{h}$ is an ideal of $\mathfrak{g}$, then $\mathfrak{h} \subset \mathfrak{k}$. Hence $\mathbf{a}(\mathfrak{h}) \subset \mathfrak{k}$. □

24.7.4 Proposition. *Let $\mathfrak{h}$ be a Lie subalgebra of $\mathfrak{g}$ and $\mathfrak{r}$ the radical of $\mathfrak{h}$. The following conditions are equivalent:*

(i) $\mathfrak{h}$ *is an algebraic Lie subalgebra of $\mathfrak{g}$.*

(ii) $\mathfrak{r}$ *is an algebraic Lie subalgebra of $\mathfrak{g}$.*

Proof. (i) $\Rightarrow$ (ii) Suppose that $\mathfrak{h} = \mathbf{a}(\mathfrak{h})$. To obtain the result, we may assume that $\mathfrak{g} = \mathfrak{h}$. So $\mathfrak{r}$ is the largest solvable ideal of $\mathfrak{g}$. By 24.7.3, $\mathbf{a}(\mathfrak{r})$ is an ideal of $\mathfrak{g}$ and it is solvable since $[\mathbf{a}(\mathfrak{r}), \mathbf{a}(\mathfrak{r})] = [\mathfrak{r}, \mathfrak{r}]$ (24.5.7). Hence $\mathfrak{r} = \mathbf{a}(\mathfrak{r})$.

(ii) $\Rightarrow$ (i) By 20.3.5, $\mathfrak{h} = \mathfrak{s} \oplus \mathfrak{r}$ where $\mathfrak{s}$ is a semisimple Lie subalgebra of $\mathfrak{g}$. Since $\mathfrak{s}$ is algebraic (24.7.2), if $\mathfrak{r}$ is algebraic, then so is $\mathfrak{h}$ (24.5.10). □

24.7.5 Remark. Let $\mathfrak{r}$ be the radical of $\mathfrak{h}$ and $\mathfrak{s}$ a semisimple Lie subalgebra of $\mathfrak{g}$ verifying $\mathfrak{h} = \mathfrak{s} \oplus \mathfrak{r}$. Then we obtain easily that $\mathbf{a}(\mathfrak{h}) = \mathfrak{s} \oplus \mathbf{a}(\mathfrak{r})$. Furthermore, $\mathbf{a}(\mathfrak{r})$ is the radical of $\mathbf{a}(\mathfrak{h})$.

24.7.6 Let $\mathfrak{h}$ be a Lie subalgebra of $\mathfrak{g}$ such that $x = x_n$ for all $x \in \mathfrak{h}$. We claim that $\mathfrak{h}$ is algebraic in $\mathfrak{g}$. We may assume that $G = \mathrm{GL}(V)$. By 23.6.4, any $x \in \mathfrak{h}$ is a nilpotent endomorphism. So by 23.3.5, $\Bbbk x$ is algebraic in $\mathfrak{g}$ for all $x \in \mathfrak{h}$. Hence 24.5.10 implies that $\mathfrak{h}$ is an algebraic Lie subalgebra of $\mathfrak{g}$.

24.8 Algebraic adjoint group

24.8.1 Definition. *We define the* algebraic adjoint group *of a Lie algebra $\mathfrak{g}$ to be the smallest algebraic subgroup of $\mathrm{GL}(\mathfrak{g})$ whose Lie algebra contains $\mathrm{ad}_\mathfrak{g} \, \mathfrak{g}$.*

24.8.2 For any Lie subalgebra $\mathfrak{h}$ of $\mathfrak{gl}(\mathfrak{g})$, denote by $\mathcal{A}(\mathfrak{h})$ the smallest algebraic subgroup of $\mathrm{GL}(\mathfrak{g})$ whose Lie algebra contains $\mathfrak{h}$, and $\mathbf{a}(\mathfrak{h})$ the smallest algebraic Lie subalgebra of $\mathfrak{gl}(\mathfrak{g})$ containing $\mathfrak{h}$.

The algebraic adjoint group of $\mathfrak{g}$ is therefore $G = \mathcal{A}(\operatorname{ad}\mathfrak{g})$ and we have $\mathcal{L}(G) = \mathbf{a}(\operatorname{ad}\mathfrak{g})$. If $\operatorname{ad}\mathfrak{g}$ is an algebraic Lie subalgebra of $\mathfrak{gl}(\mathfrak{g})$, then we say that G is the *adjoint group* of $\mathfrak{g}$, and that $\mathfrak{g}$ is ad-*algebraic*.

Let $\operatorname{Aut}\mathfrak{g}$ be the group of automorphisms of $\mathfrak{g}$. We have already seen that $\operatorname{Aut}\mathfrak{g}$ is an algebraic subgroup of $\operatorname{GL}(\mathfrak{g})$ whose Lie algebra is $\operatorname{Der}\mathfrak{g}$ (24.3.7). So $G \subset \operatorname{Aut}\mathfrak{g}$. If $\alpha \in \operatorname{Aut}\mathfrak{g}$, then α normalizes $\operatorname{ad}\mathfrak{g}$, and therefore normalizes G. Thus G is a normal subgroup of $\operatorname{Aut}\mathfrak{g}$. Applying 22.3.4 and 23.3.5, we see also that G contains the group of elementary automorphisms of $\mathfrak{g}$. We have also from 23.4.16 that $G = \operatorname{Aut}_e \mathfrak{g}$ if $\mathfrak{g}$ is semisimple.

24.8.3 Let $\mathfrak{g}$ be a Lie algebra and G its algebraic adjoint group.

• Let $x \in \mathfrak{g}$. If $\alpha \in \mathcal{A}(\operatorname{ad} x)$, there exists $P \in \Bbbk[T]$ such that $\alpha = P(\operatorname{ad} x)$ (24.6.4). It follows that any ideal of $\mathfrak{g}$ is G-stable. Conversely, if a subspace of $\mathfrak{g}$ is G-stable, then it is also $\operatorname{ad}\mathfrak{g}$-stable (24.3.3). Thus G-stable subspaces of $\mathfrak{g}$ are exactly the ideals of $\mathfrak{g}$.

• The derivation $\operatorname{ad} x$ of $\mathfrak{g}$ extends to a derivation of the symmetric algebra $S(\mathfrak{g})$ of $\mathfrak{g}$, denoted by $\operatorname{ad}_{S(\mathfrak{g})} x$. An ideal of $S(\mathfrak{g})$ is said to be $\mathfrak{g}$-invariant if it is stable under $\operatorname{ad}_{S(\mathfrak{g})} x$ for all $x \in \mathfrak{g}$.

In view of 23.4.14 and 24.3.4, an ideal of $S(\mathfrak{g})$ is $\mathfrak{g}$-invariant if and only if it is G-stable.

By 24.3.4, we also have:

$$S(\mathfrak{g})^G = S(\mathfrak{g})^{\mathfrak{g}}.$$

24.8.4 Let us look at an example. Let $\alpha \in \Bbbk \setminus \{0\}$ and $\mathfrak{g}$ the Lie algebra with basis (x, y, z), and Lie bracket defined as follows:

$$[x, y] = y \ , \ [x, z] = \alpha z \ , \ [y, z] = 0,$$

all other brackets can be deduced from these.

Let $X = \operatorname{ad} x$, $Y = \operatorname{ad} y$, $Z = \operatorname{ad} z$. To determine the algebraic adjoint group G of $\mathfrak{g}$ and $\mathcal{L}(G)$, it suffices to determine $\mathcal{A}(X)$, $\mathcal{A}(Y)$, $\mathcal{A}(Z)$, and $\mathbf{a}(X)$, $\mathbf{a}(Y)$, $\mathbf{a}(Z)$ (24.5.9). We shall identify an element of $\operatorname{End}(\mathfrak{g})$ with its matrix with respect to the basis (x, y, z).

By 23.3.5 and 24.2.4, we have:

$$\mathcal{A}(Y) = \{e^{tY}; t \in \Bbbk\} \ , \ \mathcal{A}(Z) = \{e^{tZ}; t \in \Bbbk\} \ , \ \mathbf{a}(Y) = \Bbbk Y \ , \ \mathbf{a}(Z) = \Bbbk Z.$$

The eigenvalues of X are $0, 1, \alpha$. In the notations of 24.6.3, L_X is the set of elements $(a_1, a_2, a_3) \in \mathbb{Z}^3$ such that $a_2 + a_3\alpha = 0$. Let us distinguish two cases:

1) $\alpha = \dfrac{r}{s} \in \mathbb{Q}$ where $r, s \in \mathbb{Z} \setminus \{0\}$ are coprime. Then:

$$L_X = \{(n, pr, -ps); n, p \in \mathbb{Z}\}.$$

Thus $\mathbf{a}(X) = \Bbbk X$. Therefore $\mathfrak{g}$ is ad-algebraic. The group $\mathcal{A}(X)$ is the set of matrices $\operatorname{diag}(t_1, t_2, t_3)$ with $t_1^n t_2^{pr} t_3^{-ps} = 1$ for all $n, p \in \mathbb{Z}$. So it is the set of matrices $\operatorname{diag}(1, t^s, t^r)$ where $t \in \Bbbk \setminus \{0\}$. We deduce easily that:

$$G = \left\{ \begin{pmatrix} 1 & 0 & 0 \\ u & t^s & 0 \\ v & 0 & t^r \end{pmatrix} ; u, v \in \mathbb{k}, \, t \in \mathbb{k} \setminus \{0\} \right\}.$$

2) $\alpha \notin \mathbb{Q}$. Then $L_X = \mathbb{Z} \times \{0\} \times \{0\}$. Thus $\mathbf{a}(X)$ is 2-dimensional and it is spanned, as a vector space, by X and $T = \mathrm{diag}(0, 0, 1)$. We have $\mathbf{a}(\mathrm{ad}\,\mathfrak{g}) = \mathbb{k}X \oplus \mathbb{k}T \oplus \mathbb{k}Y \oplus \mathbb{k}Z$, so $\mathbf{a}(\mathrm{ad}\,\mathfrak{g})$ is of dimension 4. The group $\mathcal{A}(X)$ is the set of matrices $\mathrm{diag}(1, t, t')$ with $tt' \neq 0$. We obtain that:

$$G = \left\{ \begin{pmatrix} 1 & 0 & 0 \\ u & t & 0 \\ v & 0 & t' \end{pmatrix} ; u, v \in \mathbb{k}, \, t, t' \in \mathbb{k} \setminus \{0\} \right\}.$$

In particular, $\mathfrak{g}$ is not ad-algebraic.

24.8.5 Proposition. *Let $\mathfrak{g}$ be a Lie algebra, G its algebraic adjoint group, $\mathfrak{h}$ a Lie subalgebra of $\mathfrak{g}$, $K = \mathcal{A}(\mathrm{ad}_{\mathfrak{g}}\,\mathfrak{h})$ and $\mathfrak{k} = \mathbf{a}(\mathrm{ad}_{\mathfrak{g}}\,\mathfrak{h})$.*
  (i) For $x \in \mathfrak{h}$, $\alpha \in K$ and $X \in \mathfrak{k}$, we have $\alpha(x) \in \mathfrak{h}$ and $X(x) \in \mathfrak{h}$.
  (ii) Let $L = \{\alpha|_{\mathfrak{h}}; \alpha \in K\}$ and $\mathfrak{l} = \{X|_{\mathfrak{h}}; X \in \mathfrak{k}\}$. Then L is the algebraic adjoint group of $\mathfrak{h}$, and $\mathfrak{l}$ is the smallest algebraic Lie subalgebra of $\mathfrak{gl}(\mathfrak{h})$ containing $\mathrm{ad}_{\mathfrak{h}}\,\mathfrak{h}$.

Proof. (i) Since K is generated by the $\mathcal{A}(\mathrm{ad}_{\mathfrak{g}}\,y)$, $y \in \mathfrak{h}$, the fact that $K(\mathfrak{h}) \subset \mathfrak{h}$ follows from 24.6.4 (i). In the same way, $X(\mathfrak{h}) \subset \mathfrak{h}$ follows from 24.6.4 (ii).

  (ii) Let $u : K \to \mathrm{GL}(\mathfrak{h})$ be the map $\alpha \mapsto \alpha|_{\mathfrak{h}}$. Then $(\mathfrak{h}, u)$ is a rational representation of K. We have $L = u(K) = u\big(\mathcal{A}(\mathrm{ad}_{\mathfrak{g}}\,\mathfrak{h})\big) = \mathcal{A}(du(\mathrm{ad}_{\mathfrak{g}}\,\mathfrak{h}))$ by 24.5.4, hence $u(K) = \mathcal{A}(\mathrm{ad}_{\mathfrak{h}}\,\mathfrak{h})$. Thus L is the algebraic adjoint group of $\mathfrak{h}$. The second part of (ii) is clear since $\mathfrak{l} = \mathcal{L}(u(K))$ by 24.5.3. $\square$

References and comments

- [5], [23], [38], [40], [78].

The presentation of 24.2 given here is inspired by [38].

Most of the notions (like replica) of sections 24.5, 24.6, 24.7 and 24.8 are taken from [23].

The fact that the base field is of characteristic zero is essential in establishing the results of this chapter. When the characteristic of the base field is positive, the correspondence between algebraic groups and their Lie algebras becomes less "nice".

Homogeneous spaces and quotients

We consider in this chapter actions of an algebraic group on an algebraic variety. The notions of homogeneous space and geometric quotient are studied, and we show that the quotient of an algebraic group by a closed subgroup has a natural structure of algebraic variety.

In this chapter, G is an algebraic group.

25.1 Homogeneous spaces

25.1.1 Proposition. *Let X be an irreducible G-variety. If X is a G-homogeneous space, then X is a smooth normal variety.*

Proof. This follows from 16.5.5 and 17.1.5. $\square$

25.1.2 Theorem. *Let X, Y be G-varieties, $u : X \to Y$ a G-equivariant morphism and $r = \dim X - \dim Y$. Suppose further that X and Y are G-homogeneous spaces.*

(i) *The map u is open. For all $y \in Y$, each irreducible component of $u^{-1}(y)$ has dimension r.*

(ii) *Let W be a closed irreducible subvariety of Y and Z an irreducible component of $u^{-1}(W)$. Then Z dominates W and $r = \dim Z - \dim W$.*

(iii) *Let Z be a variety and $v : X \times Z \to Y \times Z$ be the morphism defined by $(x, z) \mapsto (u(x), z)$. Then v is open.*

(iv) *The map u is an isomorphism if and only if u is bijective.*

Proof. In view of 21.4.4, we may reduce easily to the case where G is connected and X, Y are irreducible. Note that since X and Y are G-homogeneous spaces and u is G-equivariant, u is surjective.

(i) By 15.5.4, there exists a non-empty open subset V of Y such that for all $y \in V$, all the irreducible components of $u^{-1}(y)$ have dimension r. Let $y \in Y$. Since G acts transitively on Y, we have $\alpha.y \in V$ for some $\alpha \in G$. It follows easily from the G-equivariance of u that if $Z_1, \ldots, Z_s$ are the irreducible components of $u^{-1}(y)$, then $\alpha.Z_1, \ldots, \alpha.Z_s$ are the irreducible components of

$u^{-1}(\alpha.y)$. Now the map $x \mapsto \alpha.x$ is an isomorphism of the variety X, so $\dim Z_i = r$ for all $1 \leqslant i \leqslant s$. We deduce from 17.4.11 and 25.1.1 that u is open.

(ii) By 17.1.8 and 15.5.3, there exists a non-empty open subset V of Y such that V is normal and for all $y \in V$, $x \in u^{-1}(y)$ and irreducible subvariety Y' of Y containing y, there is at least one irreducible component of $u^{-1}(Y')$ containing x which dominates Y' and for such an irreducible component X', we have $r = \dim X' - \dim Y'$.

Let Z be an irreducible component of $u^{-1}(W)$, then there exists $z \in Z$ which is contained in no other irreducible components of $u^{-1}(W)$. Moreover, there exists $\alpha \in G$ such that $u(\alpha.z) = \alpha.u(z) \in V$. It is clear that $\alpha.Z$ is an irreducible component of $u^{-1}(\alpha.W)$ containing $\alpha.z$. Applying the preceding result with $x = \alpha.z$, $y = u(\alpha.z)$, and $Y' = \alpha.W$, we obtain that $\alpha.Z$ dominates $\alpha.W$ and $\dim(\alpha.Z) - \dim(\alpha.W) = r$. So (ii) follows.

(iii) We may assume that Z is affine. Moreover, if (iii) is true for Z, then it is also true for any closed subvariety of Z. So we may further assume that $Z = \Bbbk^n$. By (i), all the fibres of v have dimension r. So the result follows from 17.4.11 since $Y \times \Bbbk^n$ is normal (17.3.2 and 25.1.1).

(iv) Since Y is normal (25.1.1), the result follows from 17.4.8. $\square$

25.1.3 Let X be a G-variety and $x \in X$. The stabilizer G_x of x in G is an algebraic subgroup of G (21.4.2) and the orbit $G.x$ of x is a locally closed subset of X (21.4.3). Further, we have by 21.4.3:

$$\dim(G.x) = \dim G - \dim G_x.$$

Recall (21.4.4) that $G.x$ is pure and its irreducible components are exactly the G°-orbits (thus pairwise disjoint) in $G.x$. Let C be an irreducible component of $G.x$. Then it is a G°-homogeneous space, so it is normal and smooth (25.1.1).

Let $\pi : G \to G.x$, $\alpha \mapsto \alpha.x$, be the orbit morphism. If we consider the G-action on G by left translation, then π is G-equivariant. By 25.1.2, π is open. Note also that if $\alpha \in G$, then $\pi^{-1}(\alpha.x) = \alpha G_x$.

Proposition. *The map $d\pi_e : \mathcal{L}(G) \to T_x(G.x)$ is surjective and its kernel is $\mathcal{L}(G_x)$.*

Proof. Let Y be an irreducible component of $G.x$ containing x. We have $T_x(G.x) = T_x(Y)$ since irreducible components of $G.x$ are pairwise disjoint, and Y is the image of the restriction of π to G°. So we may consider π to be a surjective morphism from G° onto Y. By 16.5.7, there exists a non-empty open subset U of G° such that $d\pi_\alpha : T_\alpha(G^\circ) \to T_{\alpha.x}(Y)$ is surjective for all $\alpha \in U$. Since G° and Y are both G°-homogeneous spaces, we deduce that $d\pi_e$ is surjective. Thus $\dim \ker(d\pi_e) = \dim G - \dim(G.x) = \dim \mathcal{L}(G_x)$. But it is clear that $\mathcal{L}(G_x) \subset \ker(d\pi_e)$. So we have equality. $\square$

25.2 Some remarks

25.2.1 Let $u : (X, \mathcal{O}_X) \to (Y, \mathcal{O}_Y)$ be a morphism of varieties, $U \subset X$ and $V \subset Y$ open subsets such that $u(U) \subset V$. If $g \in \mathcal{O}_Y(V)$, we define $u_V^U(g) : U \to \Bbbk$ by $u_V^U(g)(x) = g \circ u(x)$ for all $x \in U$. Clearly $u_V^U(g) \in \mathcal{O}_X(U)$.

25.2.2 In this section, X denotes a G-variety. If U is an open subset of X, we denote by $\mathcal{O}_X(U)^G$ the set of $f \in \mathcal{O}_X(U)$ such that $f(x) = f(y)$ if x and y belong to the same G-orbit.

Lemma. *Let U be an open subset of X, $f \in \mathcal{O}_X(U)^G$ and $V = \bigcup_{\alpha \in G} \alpha.U$. There exists a unique element $g \in \mathcal{O}_X(V)^G$ such that $g|_U = f$.*

Proof. The uniqueness is obvious. So let us prove the existence of g. For $\alpha \in G$, the map $u_\alpha : \alpha.U \to U$, $x \mapsto \alpha^{-1}.x$ is an isomorphism of varieties. So $g_\alpha : \alpha.U \to \Bbbk$, $x \mapsto f(\alpha^{-1}.x)$, is an element of $\mathcal{O}_X(\alpha.U)$. For $\alpha, \beta \in G$ and $x \in \alpha.U \cap \beta.U$, there exist $y, z \in U$ such that $x = \alpha.y = \beta.z$. It follows that $y, z \in G.x$, so $g_\alpha(x) = f(y) = f(z) = g_\beta(y)$. By the properties of a sheaf, there exists $g \in \mathcal{O}_X(V)$ such that $g|_{\alpha.U} = g_\alpha$ for all $\alpha \in G$. $\square$

25.2.3 Assume that U is an open G-stable subset of X. If $\alpha \in G$, the map $\alpha.f : U \to U$, $x \mapsto f(\alpha^{-1}.x)$, is an element of $\mathcal{O}_X(U)$. It is then easy to check that

$$G \times \mathcal{O}_X(U) \to \mathcal{O}_X(U) \, , \, (\alpha, f) \mapsto \alpha.f$$

is an action of G on $\mathcal{O}_X(U)$. The set of fixed points under this action is clearly $\mathcal{O}_X(U)^G$.

25.2.4 Lemma. *Assume that X is irreducible. Let U, V be non-empty open subsets of X such that $U \subset V$, and $f \in \mathcal{O}_X(U)$, $g \in \mathcal{O}_X(V)$ verifying $g|_U = f$. Then the following conditions are equivalent:*
  (i) $g \in \mathcal{O}_X(V)^G$.
  (ii) $f \in \mathcal{O}_X(U)^G$.

Proof. (i) $\Rightarrow$ (ii) This is obvious.
  (ii) $\Rightarrow$ (i) Suppose that there exist $\alpha \in G$ and $x \in V$ such that $\alpha.x \in V$ and $g(x) \neq g(\alpha.x)$. Then $x \in \alpha^{-1}.V$. Set $U' = U \cap (\alpha^{-1}.U)$ and $V' = V \cap (\alpha^{-1}.V)$.
  Let $\theta : X \to X \times X$ be the morphism $y \mapsto (y, \alpha.y)$. The map $h = g \otimes 1 - 1 \otimes g$ belongs to $\mathcal{O}_{X \times X}(V \times V)$, and the restriction of $h \circ \theta$ to V' belongs to $\mathcal{O}_X(V')$. We have $h \circ \theta(x) \neq 0$ and $h \circ \theta|_{U'} = 0$. But this is absurd since U' is dense in V'. Hence $g \in \mathcal{O}_X(V)^G$. $\square$

25.2.5 Let us assume that X is irreducible.
  Let $f \in \mathcal{O}_X(U)$ and $h \in \mathbf{R}(X)$ the rational function defined by f. Denote by V the domain of definition of h and $g \in \mathcal{O}_X(V)$ the function induced by h.
  If $f \in \mathcal{O}_X(U)^G$, then 25.2.4 says that $g \in \mathcal{O}_X(V)^G$. Hence the open subset V is G-stable (25.2.2).

Conversely, given $h \in \mathbf{R}(X)$, V its domain of definition, and $g \in \mathcal{O}_X(V)$ the function induced by h. By 25.2.2, if $g \in \mathcal{O}_X(V)^G$, then V is G-stable and $g|_U \in \mathcal{O}_X(U)^G$ for all open subsets U in V.

Let us denote by $\mathbf{R}(X)^G$ the set of rational functions h verifying the following condition: if V is the domain of definition of h, and $g \in \mathcal{O}_X(V)$ the function induced by h, then $g \in \mathcal{O}_X(V)^G$.

We can interpret $\mathbf{R}(X)^G$ in another way. Namely, let $\alpha \in G$, U an open subset of X and $u_\alpha^U : \alpha.U \to U$ the isomorphism of varieties defined by $x \mapsto \alpha^{-1}.x$. If $f \in \mathcal{O}_X(U)$, then $f \circ u_\alpha^U \in \mathcal{O}_X(\alpha.U)$, and if V is an open subset of U, then $(f|_V) \circ u_\alpha^V = (f \circ u_\alpha^U)|_{\alpha.V}$. Denote by $u_\alpha : \mathbf{R}(X) \to \mathbf{R}(X)$ the inductive limit of the family $(u_\alpha^U)_U$. Thus we obtain an action of G on $\mathbf{R}(X)$. Then $\mathbf{R}(X)^G$ is the set of invariants of G under this action.

25.2.6 Assume that X is irreducible. We deduce from the preceding discussion that for a non-empty affine open subset U of X, we have $\mathrm{Fract}(\mathcal{O}_X(U)^G) \subset \mathbf{R}(X)^G$.

25.2.7 Let $x \in X$, $Z = G.x$, $H = G_x$ and $\pi : G \to Z$ be as in 25.1.3. Endow G with the structure of H-variety via right translation.

Let V be a non-empty open subset of Z and $U = \pi^{-1}(V)$. For $\alpha \in U$ and $\beta \in H$, we have $\alpha\beta \in U$, so U is H-stable, and $\mathcal{O}_G(U)^H$ is the set of functions $h \in \mathcal{O}_G(U)$ verifying $f(\alpha\beta) = f(\alpha)$ for all $(\alpha, \beta) \in U \times H$.

Let $g \in \mathcal{O}_Z(V)$ and $f = \pi_V^U(g) \in \mathcal{O}_G(U)$. Then for $(\alpha, \beta) \in U \times H$, we have:

$$f(\alpha\beta) = g((\alpha\beta).x) = g(\alpha.x) = f(\alpha).$$

Thus $f \in \mathcal{O}_G(U)^H$ and the map $\mathcal{O}_Z(V) \to \mathcal{O}_G(U)^H$, $g \mapsto \pi_V^U(g)$, is injective. We shall show that it is bijective.

Let $C_1, \ldots, C_n$ be the irreducible components of Z. By 21.4.4 and 25.1.1, they are open and closed, smooth, normal, pairwise disjoint G°-orbits in Z. Let $V_i = V \cap C_i$ and $U_i = \pi^{-1}(V_i)$, $1 \leqslant i \leqslant n$. We may assume that $V_i \neq \emptyset$ for $1 \leqslant i \leqslant r$ and $V_i = \emptyset$ if $i > r$. Clearly, the open subsets U_i are pairwise disjoint and H-stable. So to obtain the result, it suffices to consider the case where V is irreducible.

Assume that V is irreducible. Then $\overline{V} = C_i = G^\circ.y$ for some i and $y \in Z$. Moreover, V is a normal variety by the preceding discussion. Let $f \in \mathcal{O}_G(U)^H$. Set $\Delta \subset U \times \Bbbk = \{(\alpha, f(\alpha)); \alpha \in U\}$. It is closed in $U \times \Bbbk$ (12.5.4). Let

$$u : G \times \Bbbk \to Z \times \Bbbk , \ (\alpha, \lambda) \mapsto (\pi(\alpha), \lambda).$$

By 25.1.2 (iii), u is an open morphism. Let $\Delta' = u(\Delta) \subset V \times \Bbbk$.

Let $(\pi(\beta), f(\beta)) \in \Delta'$ where $\beta \in U$, and $(\alpha, \lambda) \in (U \times \Bbbk) \setminus \Delta$. If $(\pi(\beta), f(\beta)) = u(\alpha, \lambda)$, then $\pi(\beta) = \pi(\alpha)$ and $\lambda = f(\beta)$. So $\alpha \in \beta H$ and $f(\alpha) = f(\beta) = \lambda$, hence $(\alpha, \lambda) \in \Delta$ which is absurd. It follows that $u^{-1}(\Delta') = \Delta$ and Δ' is a closed subvariety of $V \times \Bbbk$.

Let $G^\circ\beta_1, \ldots, G^\circ\beta_r$ be the irreducible components of G having non-empty intersection with U. Set $U_i = G^\circ\beta_i \cap U$. Then we have $\pi(U_i) = V$ because $\pi(G^\circ\beta_i) = G^\circ.y$.

Let $\gamma \in U \backslash U_1$, then there exists $\gamma_1 \in U_1$ such that $\gamma.x = \gamma_1.x$. So $\gamma \in \gamma_1 H$, and hence $\pi(\gamma) = \pi(\gamma_1)$ and $f(\gamma) = f(\gamma_1)$. It follows that Δ' is the image of the morphism $U_1 \to V \times \Bbbk$, $\beta \mapsto (\pi(\beta), f(\beta))$. Since U_1 is open in $G^{\circ}\beta_1$, it is irreducible. We obtain therefore that Δ' is also irreducible.

Let $p : \Delta' \to V$ (resp. $q : \Delta' \to \Bbbk$) be the restriction to Δ' of the canonical projection $Z \times \Bbbk \to Z$ (resp. $Z \times \Bbbk \to \Bbbk$). Then p is bijective, and so by 17.4.8, p is an isomorphism. Thus $q \circ p^{-1} \in \mathcal{O}_Z(V)$ and if $\alpha \in U$, then $q \circ p^{-1}(\pi(\alpha)) = f(\alpha)$. Hence $f = \pi_V^U(q \circ p^{-1})$ and we are done.

25.3 Geometric quotients

25.3.1 Definition. *Let X be a G-variety. A* geometric quotient *of X by G is a pair (Y, π) where Y is a variety and $\pi : X \to Y$ is a morphism verifying the following conditions:*

(i) *π is open, constant on G-orbits and defines a bijection from the set of G-orbits onto Y.*

(ii) *If V is an open subset of Y and $U = \pi^{-1}(V)$, π_V^U induces a bijection from $\mathcal{O}_Y(V)$ onto $\mathcal{O}_X(U)^G$.*

25.3.2 Let G and X be as in 25.3.1. If a geometric quotient (Y, π) of X by G exists, then the fibres of π are the G-orbits, so they are closed. We can further precise condition (i) of 25.3.1.

Lemma. *Let X be a G-variety and $\theta : X \to Y$ a surjective morphism of varieties whose fibres are the G-orbits in X. Then the following conditions are equivalent:*

(i) *The map θ is open.*

(ii) *For all G-stable open subset U of X, $\theta(U)$ is an open subset of Y.*

(iii) *For all G-stable closed subset F of X, $\theta(F)$ is a closed subset of Y.*

In particular, if θ is a closed map, then it is also open.

Proof. Clearly, (i) $\Rightarrow$ (ii). Conversely, if U is an open subset of X, then $U' = \bigcup_{\alpha \in G} \alpha.U$ is a G-stable open subset of X such that $\theta(U) = \theta(U')$. So (ii) $\Rightarrow$ (i).

Let F be a G-stable closed subset of X, then $U = X \setminus F$ is a G-stable open subset of X. Thus $\theta(F) = Y \setminus \theta(U)$. So (ii) $\Rightarrow$ (iii). Conversely, if U is a G-stable open subset of X, then $F = X \setminus U$ is a G-stable closed subset of X and $\theta(U) = Y \setminus \theta(F)$. Hence (iii) $\Rightarrow$ (ii). $\quad\square$

25.3.3 In the rest of this section, X is a G-variety.

Proposition. *Assume that a geometric quotient (Y, π) of X by G exists. Let Z be a variety and $u : X \to Z$ a morphism which is constant on G-orbits in X. Then there exists a unique morphism $v : Y \to Z$ such that $u = v \circ \pi$.*

Proof. The hypothesis implies that there is a unique map $v : Y \to Z$ such that $u = v \circ \pi$. Let W be an open subset of Z. Then $u^{-1}(W) = \pi^{-1}(v^{-1}(W))$

is open in X. Since π is surjective and open, $v^{-1}(W) = \pi(u^{-1}(W))$ is open in Y. Thus v is continuous.

Now let $V = v^{-1}(W)$, $U = u^{-1}(W)$ and $f \in \mathcal{O}_Z(W)$. To obtain the result, we need to prove that $f \circ v \in \mathcal{O}_Y(V)$. But $f \circ v \circ \pi = f \circ u \in \mathcal{O}_X(U)$ and since π is constant on G-orbits, $f \circ v \circ \pi \in \mathcal{O}_X(U)^G$. As π_V^U induces a bijection from $\mathcal{O}_Y(V)$ onto $\mathcal{O}_X(U)^G$, $f \circ v \in \mathcal{O}_Y(V)$. $\square$

25.3.4 Corollary. *If (Y, π) and (Y', π') are geometric quotients of X by G, then there exists a unique isomorphism $v : Y \to Y'$ such that $\pi' = v \circ \pi$.*

25.3.5 In view of 25.3.4, we can talk about, if it exists, *the* geometric quotient of X by G. The existence of a geometric quotient of X by G requires very strong conditions on the G-orbits as we shall see in the following result.

Proposition. *Assume that X is irreducible. Let $\theta : X \to Y$ be a surjective morphism of varieties whose fibres are the G-orbits in X.*

(i) *The G-orbits in X have dimension $r = \dim X - \dim Y$.*

(ii) *If Y is normal, then (Y, θ) is the geometric quotient of X by G.*

Proof. (i) By 15.5.4, $\dim G.x \geqslant r$ for all $x \in X$ and there is a non-empty open subset V of Y such that $\dim G.x = r$ for all $x \in \theta^{-1}(V)$.

Let $T = \{(\alpha, x, \alpha.x); \alpha \in G, x \in X\} \subset G \times X \times X$ be the graph of the morphism $(\alpha, x) \mapsto \alpha.x$. It is closed in $G \times X \times X$ (12.5.4). Let Δ be the diagonal of X, $Z = T \cap (G \times \Delta)$ and $p : Z \to \Delta$ the canonical projection. If $x \in X$, then $p^{-1}(x, x) = G_x \times \{(x, x)\}$, so all the irreducible components of $p^{-1}(x, x)$ are of the same dimension.

Let C be an irreducible component of Z containing $\{e\} \times \Delta$ and $q = p|_C$. Then q is surjective and if $x \in X$, then $q^{-1}(x, x)$ is the union of irreducible components of $p^{-1}(x, x)$. Applying 15.5.4, we obtain that

$$\dim G_x \geqslant \dim C - \dim \Delta$$

for all $x \in X$ and $\dim G_x = \dim C - \dim \Delta$ if x belongs to a non-empty open subset U of X. It follows that for $x \in X$,

$$r \leqslant \dim G.x = \dim G - \dim G_x \leqslant \dim G - \dim C + \dim \Delta = s.$$

Since $\dim G.x = r = s$ for $x \in U \cap \theta^{-1}(V)$, we obtain that $\dim G.x = r$ for all $x \in X$.

(ii) Assume that Y is normal. Since G-orbits are pure varieties (21.4.4), it follows from 17.4.11 and part (i) that θ is open.

Let V be a non-empty open subset of Y and $U = \theta^{-1}(V)$. The open subset U is G-stable and if $g \in \mathcal{O}_Y(V)$, then $\theta_V^U(g) \in \mathcal{O}_X(U)^G$. The map $g \mapsto \theta_V^U(g)$ is therefore an injection from $\mathcal{O}_Y(V)$ into $\mathcal{O}_X(U)^G$. We need to prove that it is surjective.

Let $f \in \mathcal{O}_X(U)^G$ and $\Delta = \{(x, f(x)); x \in U\} \subset U \times \Bbbk$. By 12.5.4, Δ is closed in $U \times \Bbbk$. Let $u : X \times \Bbbk \to Y \times \Bbbk$ be the morphism $(x, \lambda) \mapsto (\theta(x), \lambda)$.

For $x \in X$ and $\lambda \in \Bbbk$, $u^{-1}(\theta(x), \lambda) = G.x \times \{\lambda\}$. So all the irreducible components of the fibres of u have dimension r. Since $Y \times \Bbbk$ is normal (17.3.2), it follows from 17.4.11 that u is open. Since X is irreducible, so is Δ and hence $\Delta' = u(\Delta)$ is also irreducible. Proceeding as in 25.2.7, we obtain that $f \in \theta_V^U(\mathcal{O}_Y(V))$. $\square$

25.3.6 Proposition. *Assume that X is irreducible and that (Y, π) is a geometric quotient of X by G. Then the comorphism φ of π is an isomorphism from $\mathbf{R}(Y)$ onto $\mathbf{R}(X)^G$.*

Proof. Since π is dominant, the comorphism $\varphi : \mathbf{R}(Y) \to \mathbf{R}(X)$ is well-defined. Clearly, $\varphi(\mathbf{R}(Y)) \subset \mathbf{R}(X)^G$. Conversely, let $h \in \mathbf{R}(X)^G$. Its domain of definition U is G-stable and $V = \pi(U)$ is an open subset such that $\pi^{-1}(V) = U$. Since φ induces a bijection between $\mathcal{O}_Y(V)$ and $\mathcal{O}_X(U)^G$, we have $h \in \varphi(\mathbf{R}(Y))$. $\square$

25.4 Quotient by a subgroup

25.4.1 Let V be a finite-dimensional $\Bbbk$-vector space, W a subspace of dimension $d > 0$, $G = \mathrm{GL}(V)$ and $\mathfrak{g} = \mathfrak{gl}(V)$. The exterior power $\bigwedge^d V$ contains the line $L = \bigwedge^d W$. Denote by π (resp. $d\pi$) the representation of G (resp. $\mathfrak{g}$) on $\bigwedge^d V$ given in 10.6.5 (resp. 23.4.14).

Lemma. *Let $\alpha \in G$ and $x \in \mathfrak{g}$.*
(i) *We have $\alpha(W) = W$ if and only if $\pi(\alpha)(L) = L$.*
(ii) *We have $x(W) \subset W$ if and only if $d\pi(x)(L) \subset L$.*

Proof. If $\alpha(W) = W$ (resp. $x(W) \subset W$), then it is clear that $\pi(\alpha)(L) = L$ (resp. $d\pi(x)(L) \subset L$).

(i) Let $(e_1, \dots, e_n)$ be a basis of V such that $(e_1, \dots, e_d)$ is a basis of W and $(e_{l+1}, \dots, e_{l+d})$ a basis of $\alpha(W)$. Clearly $L = \Bbbk e_1 \wedge \cdots \wedge e_d$ and $\pi(\alpha)(L) = \Bbbk e_{l+1} \wedge \cdots \wedge e_{l+d}$. Since the elements $e_{i_1} \wedge \cdots \wedge e_{i_d}$, $1 \leqslant i_1 < \cdots < i_d \leqslant n$, form a basis of $\bigwedge^d V$, it is clear that if $\pi(\alpha)(L) = L$, then $l = 0$ and hence $\alpha(W) = W$.

(ii) For $1 \leqslant i \leqslant d$, let $x(e_i) = \sum\limits_{i=1}^{n} a_{ij} e_j$ and $v = e_1 \wedge \cdots \wedge v_d$. Then:

$$d\pi(x)(v) = \sum_{j=1}^{d} e_1 \wedge \cdots \wedge e_{j-1} \wedge x(e_j) \wedge e_{j+1} \wedge \cdots \wedge e_d$$
$$= \sum_{i=1}^{n} \sum_{j=1}^{d} a_{ij}\, e_1 \wedge \cdots \wedge e_{j-1} \wedge e_i \wedge e_{j+1} \wedge \cdots \wedge e_d.$$

If $x(W) \not\subset W$, then $a_{ij} \neq 0$ for some $d < i \leqslant n$, $1 \leqslant j \leqslant d$, and we obtain clearly that $d\pi(x)(L) \not\subset L$. $\square$

25.4.2 Let $\mathfrak{g}$ be the Lie algebra of G, H a closed subgroup of G and $\mathfrak{h}$ the Lie algebra of H. Let us consider the action of G on $\mathbf{A}(G)$ by right translation and conserve the notation ρ_α of 22.1.4. We shall also used the right convolution defined in 23.2.2.

Lemma. *There exists a finite-dimensional subspace V of $\mathbf{A}(G)$, stable by right translations, and a subspace W of V such that:*

$$H = \{\alpha \in G; \rho_\alpha(W) = W\} \ , \quad \mathfrak{h} = \{x \in \mathfrak{g}; W * x \subset W\}.$$

Proof. Let $(u_1, \ldots, u_r)$ be a system of generators of the defining ideal $\mathfrak{a} = \mathcal{I}(H)$ in $\mathbf{A}(G)$. There exists a finite-dimensional G-submodule V of $\mathbf{A}(G)$ containing $u_1, \ldots, u_r$ (22.1.1). Set $W = V \cap \mathfrak{a}$.

Let $\alpha \in H$. Then $\rho_\alpha(W) = W$ (22.1.6). Conversely, if $\alpha \in G$ verifies $\rho_\alpha(W) = W$, then $\rho_\alpha(u_i) \in \mathfrak{a}$ for $1 \leqslant i \leqslant r$, so $\rho_\alpha(\mathfrak{a}) \subset \mathfrak{a}$. Hence $\alpha \in H$ by 22.1.6. The second part is proved in the same manner by using 23.2.5 and 23.4.4. $\square$

25.4.3 Theorem. *Let $\mathfrak{g}$ be the Lie algebra of G, H a closed subgroup of G and $\mathfrak{h}$ its Lie algebra. There exists a rational finite-dimensional G-module V and a line L in V such that:*

$$H = \{\alpha \in G; \pi(\alpha)(L) = L\} \ , \quad \mathfrak{h} = \{x \in \mathfrak{g}; d\pi(x)(L) \subset L\}.$$

Proof. This is immediate by 25.4.1 and 25.4.2. $\square$

25.4.4 Corollary. *Let us conserve the notations of 25.4.3. There exists a quasi-projective G-variety X which is a G-homogeneous space, and an element $x \in X$ such that:*

(i) *H is the stabilizer of x in G.*

(ii) *Let $\theta : G \to X$ be the morphism $\theta(\alpha) = \alpha.x$ for $\alpha \in G$. The fibres of θ are the left cosets αH, $\alpha \in G$.*

Proof. Let V and L be as in 25.4.3, $\mathbb{P}(V)$ the projective space associated to V, $\varphi : V \setminus \{0\} \to \mathbb{P}(V)$ the canonical surjection and $x = \varphi(L)$.

Set $\alpha.\varphi(v) = \varphi\big(\pi(\alpha)(v)\big)$ for $\alpha \in G$ and $v \in V \setminus \{0\}$. This defines a rational action of G on $\mathbb{P}(V)$. Denote by X the G-orbit of x in $\mathbb{P}(V)$. By 21.4.3, X is a quasi-projective variety. So the corollary follows from 25.4.3. $\square$

25.4.5 Theorem. *Let $\mathfrak{g}$ be the Lie algebra of G, H a closed normal subgroup of G and $\mathfrak{h}$ its Lie algebra. Then there exists a finite-dimensional rational G-module (W, φ) such that $H = \ker \varphi$ and $\mathfrak{h} = \ker d\varphi$.*

Proof. Let V, π, L be as in 25.4.3 and $X^*(H)$ the group of characters of H. For $\chi \in X^*(H)$, denote by V_χ the weight space of weight χ of H in V. Recall from 22.4.5 that since H is normal, $\pi(G)$ permutes the V_χ's. Further $L \subset V_{\chi_0}$

for some $\chi_0 \in X^*(H)$, so we may assume that V is the direct sum of the weight spaces V_χ (22.4.6).

Let E be the subspace of $\mathfrak{gl}(V)$ consisting of endomorphisms ℓ such that $\ell(V_\chi) \subset V_\chi$ for all χ. Observe that E is isomorphic to the direct sum of the $\mathfrak{gl}(V_\chi)$'s. Then for $\alpha \in G$, $\ell \in E$ and $\chi \in X^*(H)$, we have

$$\pi(\alpha) \circ \ell \circ \pi(\alpha^{-1})(V_\chi) = \pi(\alpha) \circ \ell(V_{\chi_\alpha}) \subset \pi(\alpha)(V_{\chi_\alpha}) \subset V_\chi$$

where χ_α is as defined in 22.4.3. So $\pi(\alpha) \circ \ell \circ \pi(\alpha^{-1}) \in E$ and this defines a map $\varphi : G \to \mathrm{GL}(E)$ where $\varphi(\alpha)(\ell) = \pi(\alpha) \circ \ell \circ \pi(\alpha^{-1})$. It follows from 23.5.4 that $\varphi(\alpha) = \mathrm{Ad}_{\mathrm{GL}(V)}\big(\pi(\alpha)\big)$. So φ is a rational representation of G.

Since $\pi(\alpha)$, $\alpha \in H$, acts on each V_χ by scalar multiplication, we obtain that $\varphi(\alpha)(\ell) = \ell$ for all $\ell \in E$, and so $\alpha \in \ker \varphi$. Conversely, if $\alpha \in \ker \varphi$, then $\pi(\alpha) \circ \ell = \ell \circ \pi(\alpha)$ for all $\ell \in E$. It follows that $\pi(\alpha)(V_\chi) \subset V_\chi$ for all $\chi \in X^*(H)$ and $\pi(\alpha)|_{V_\chi}$ is central in $\mathfrak{gl}(V_\chi)$. Hence $\pi(\alpha)$ acts on V_χ by scalar multiplication. In particular $\pi(\alpha)(L) \subset L$ and so $\alpha \in H$ (25.4.3). The fact that $\mathfrak{h} = \ker d\varphi$ now follows from 24.4.1. $\square$

25.4.6 Let X be a G-variety, $x \in X$, $H = G_x$ and $\theta : G \to G.x$ the morphism $\alpha \mapsto \alpha.x$. If $\alpha \in G$, then $\theta^{-1}(\alpha.x) = \alpha H$. Now H acts on G in the following way: for $\alpha \in G$ and $\beta \in H$, $\beta.\alpha = \alpha\beta^{-1}$. Then the fibres of π are the H-orbits of G.

Proposition. *With the above notations, $(G.x, \theta)$ is the geometric quotient of G by H.*

Proof. By 25.1.2, θ is open. So the result is just a reformulation of 25.3.5. $\square$

25.4.7 Let H be a closed subgroup of G and $\pi : G \to G/H$ the canonical surjection. By 25.4.4 and 25.4.6, there exists a geometric quotient (X, θ) of G by H (action of H on G as in 25.4.6) and $x \in X$, verifying:
- X is a smooth quasi-projective G-variety.
- $G_x = H$.
- If $\alpha \in G$, then $\theta(\alpha) = \alpha.x$.

Define $\varphi : G/H \to X$ by $\varphi(\alpha H) = \alpha.x$. This defines a bijection from G/H onto X. Transporting the topology of X to G/H, $(G/H, \pi)$ is then a geometric quotient of G by H. Moreover, by 25.3.4, the topology thus obtained on G/H (which makes it into a quasi-projective variety) is the unique topology for which $(G/H, \pi)$ is a geometric quotient of G by H. Unless otherwise stated, we shall always endow G/H with this structure.

25.4.8 Proposition. *Let H, K be closed subgroups of G such that $H \subset K$. Then the canonical map $G/H \to G/K$ is a morphism of algebraic varieties.*

Proof. This is clear by 25.3.3 and 25.4.7. $\square$

25.4.9 For $i = 1, 2$, let G_i be an algebraic group, H_i a closed subgroup of G_i and $\pi_i : G_i \to G_i/H_i$ the canonical surjection. Let $G = G_1 \times G_2$, $H = H_1 \times H_2$ and

$$u : G/H \to (G_1/H_1) \times (G_2/H_2) , \ (\alpha_1, \alpha_2)H \mapsto (\pi_1(\alpha_1), \pi_2(\alpha_2)).$$

Proposition. (i) *The map u is an isomorphism of varieties.*

(ii) *Let $v : G_1 \to G_2$ be a morphism of algebraic groups such that $v(\alpha H_1) \subset v(\alpha)H_2$ for all $\alpha \in G$. Then the map $w : G_1/H_1 \to G_2/H_2$ such that $w \circ \pi_1 = \pi_2 \circ v$ is a morphism of varieties.*

Proof. (i) The group G acts naturally on $G_1/H_1 \times G_2/H_2$ and the stabilizer (resp. G-orbit) of (H_1, H_2) is H (resp. $(G_1/H_1) \times (G_2/H_2)$). Let $\theta : G \to (G_1/H_1) \times (G_2/H_2)$ be the morphism $(\alpha_1, \alpha_2) \mapsto (\pi_1(\alpha_1), \pi_2(\alpha_2))$. Then in view of 25.4.6, $((G_1/H_1) \times (G_2/H_2), \theta)$ is a geometric quotient of G by H. So (i) follows from 25.3.4.

(ii) Since $\pi_2 \circ v : G_1 \to G_2/H_2$ is a morphism, the result follows clearly by 25.3.3. □

25.4.10 Proposition. *Let H be a closed subgroup of G. The map*

$$u : G \times (G/H) \to G/H , \ (\alpha, \beta H) \mapsto \alpha\beta H$$

is a morphism of varieties.

Proof. Applying 25.4.9 with $G_1 = G_2 = G$, $H_1 = \{e\}$ and $H_2 = H$, we obtain that the varieties $(G \times G)/(\{e\} \times H)$ and $G \times (G/H)$ are isomorphic. Let $\pi : G \to G/H$ be the canonical surjection. Then the morphism $v = \pi \circ \mu_G : G \times G \to G/H$ is constant on the $(\{e\} \times H)$-orbits in $G \times G$ (via the right translation action). So the result follows from 25.3.3. □

25.4.11 Theorem. *Let H be a closed normal subgroup of G and $\pi : G \to G/H$ the canonical surjection.*

(i) *The group structure of G/H together with its structure of variety define a structure of affine algebraic group on G/H. The map π is a morphism of algebraic groups.*

(ii) *The map $d\pi$ induces an isomorphism from $\mathcal{L}(G)/\mathcal{L}(H)$ onto $\mathcal{L}(G/H)$.*

Proof. By 25.4.5, there is a finite-dimensional rational G-module (V, φ) such that $H = \ker \varphi$ and $\mathcal{L}(H) = \ker d\varphi$. Let $K = \varphi(G)$, then K is a closed subgroup of $\mathrm{GL}(V)$, so it is affine.

Endow $\mathrm{GL}(V)$ with the following G-variety structure: for $\alpha \in G$ and $\beta \in \mathrm{GL}(V)$, $\alpha.\beta = \varphi(\alpha) \circ \beta$. Then K is the G-orbit of id_V and $G_{\mathrm{id}_V} = H$.

Let $\pi : G \to G.\mathrm{id}_V = K$ be the morphism $\alpha \mapsto \varphi(\alpha) \circ \mathrm{id}_V$. By 25.4.6, (K, π) is a geometric quotient of G by H. Thus G/H is an affine variety. Moreover, by construction, the group structure on G/H is the one from the algebraic group K. So (i) follows. By 24.4.7, $\mathcal{L}(H)$ is an ideal of $\mathcal{L}(G)$, so (ii) follows from 25.1.3. □

25.4.12 Let H, K be closed normal subgroups of G such that $H \subset K$. Denote by $\pi : G \to G/H$ the canonical surjection.

Proposition. (i) *The set $\pi(K)$ is a closed normal subgroup of G/H, isomorphic to the algebraic group K/H.*

(ii) *The algebraic groups G/K and $(G/H)/\pi(K)$ are isomorphic.*

Proof. (i) Clearly, $\pi(K)$ is a normal subgroup of G/H. It is closed by 21.2.4 and 25.4.11 (or by 25.3.2).

Let $\theta : K \to \pi(K)$ be the morphism induced by π. By 25.3.3, there is a unique morphism of varieties $u : K/H \to \pi(K)$ such that $\theta = u \circ \pi'$ where $\pi' : K \to K/H$ is the canonical surjection. It is clear that u is a bijective group homomorphism. Hence by 21.2.6, u is an isomorphism of algebraic groups.

(ii) The kernel of the composition $G \to G/H \to (G/H)/\pi(K)$ of canonical surjections is K. By 25.3.3, this defines a morphism of varieties $v : G/K \to (G/H)/\pi(K)$. The morphism v is a bijective group homomorphism, so the result follows again from 21.2.6. $\square$

25.4.13 Let H be a closed subgroup of G. In general, if H is not normal, G/H is not affine. However, we have the following result:

Proposition. *Let G be a unipotent algebraic group and H a closed subgroup of G. Then the variety G/H is affine.*

Proof. Let V, L, π be as in 25.4.3. Since H is unipotent, $L^H \neq \{0\}$ (22.3.6). It follows that $L^H = L$ and for $x \in L \setminus \{0\}$, the map $\theta : G \to G.x$, $\alpha \mapsto \alpha.x$, induces an isomorphism from G/H to $G.x$ (25.4.6). Since $G.x$ is affine (22.3.6), we are done. $\square$

25.5 The case of finite groups

25.5.1 Proposition. *Let G be a finite group consisting of automorphisms of an affine variety X. Then the algebra of invariants $\mathbf{A}(X)^G$ is finitely generated. Moreover, if X is irreducible, then $\mathbf{R}(X)^G = \mathrm{Fract}(\mathbf{A}(X)^G)$.*

Proof. Let $f_1, \ldots, f_n$ be a system of generators for $\mathbf{A}(X)$ and $\alpha_1, \ldots, \alpha_s$ the elements of G with $\alpha_1 = e_G$. For $1 \leqslant i \leqslant n$, set:

$$P_i(T) = (T - \alpha_1.f_i) \cdots (T - \alpha_s.f_i).$$

Then G permutes the roots of P_i, so the coefficients of P_i are in $\mathbf{A}(X)^G$. Let B be the subalgebra of $\mathbf{A}(X)^G$ generated by the coefficients of the P_i's. Since $P_i(f_i) = 0$, f_i is integral over B, and so $\mathbf{A}(X)$ is integral over B (3.1.5). Moreover, $\mathbf{A}(X)$ is a finite B-algebra (3.3.5). Since B is noetherian, $\mathbf{A}(X)^G$ is a finitely generated B-module, so the algebra $\mathbf{A}(X)^G$ is finitely generated.

Suppose that X is irreducible. Let $f = uv^{-1} \in \mathbf{R}(X)^G$ where $u, v \in \mathbf{A}(X)$ and:

$$v_1 = (\alpha_1.v) \cdots (\alpha_s.v) \ , \ u_1 = u(\alpha_2.v) \cdots (\alpha_s.v).$$

Then $v_1 \in \mathbf{A}(X)^G$ and $f = u_1 v_1^{-1}$. Hence $u_1 \in \mathbf{A}(X)^G$ and so we have $f \in \mathrm{Fract}(\mathbf{A}(X)^G)$. $\square$

25.5.2 Let G be a finite group consisting of automorphisms of an irreducible affine variety X. By 25.5.1, there is an affine variety Y such that $Y = \mathrm{Spm}(\mathbf{A}(X)^G)$. Let $u : X \to Y$ be the morphism defined by the canonical injection $\mathbf{A}(X)^G \to \mathbf{A}(X)$.

Proposition. *The pair (Y, u) is the geometric quotient of X by G.*

Proof. We saw in the proof of 25.5.1 that $\mathbf{A}(X)$ is integral over $\mathbf{A}(X)^G$. If $\mathfrak{n} \in \mathrm{Spm}(\mathbf{A}(X)^G)$, then there exists $\mathfrak{m} \in \mathrm{Spm}(\mathbf{A}(X))$ such that $\mathfrak{n} = \mathfrak{m} \cap \mathbf{A}(X)^G$ (3.3.2 and 3.3.3). Thus u is surjective.

Given $x \in X$, we have $G.x \subset u^{-1}(u(x))$. Suppose that there exists $y \in u^{-1}(u(x))$ such that $y \notin G.x$. Since the finite sets $G.x$ and $G.y$ are disjoint and closed, it follows from 11.3.3 that there exists $f \in \mathbf{A}(X)$ verifying $f|_{G.x} = 1$ and $f|_{G.y} = 0$. Set $g = (\alpha_1.f) \cdots (\alpha_s.f)$ where $\alpha_1, \ldots, \alpha_s$ are the elements of G. Then $g \in \mathbf{A}(X)^G$ and $g|_{G.x} = 1$, $g|_{G.y} = 0$. This is absurd since $u(x) = u(y)$. Thus the fibres of u are the G-orbits.

By 15.3.2, u is a finite morphism, and therefore u is closed (15.3.4). Since the fibres of u are the G-orbits, 25.3.2 implies that u is open.

Let V be an open subset of Y and $U = u^{-1}(V)$. Then U is G-stable and it is clear that $u_V^U(\mathcal{O}_Y(V)) \subset \mathcal{O}_X(U)^G$. Conversely, let $f \in \mathcal{O}_X(U)^G$. Since X is irreducible, we have $\mathcal{O}_X(U)^G \subset \mathbf{R}(X)^G = \mathrm{Fract}(\mathbf{A}(X)^G)$ (12.8.2 and 25.5.1). It follows that there exist $g, h \in \mathbf{A}(X)^G = \mathbf{A}(Y)$ such that $f = g/h$ and $h|_U \neq 0$. Hence $g/h \in \mathcal{O}_Y(V)$, and u_V^U is bijective.

So we have proved that (Y, u) is the geometric quotient of X by G. $\quad\square$

25.5.3 In the case of a linear action, we have the following more precise statement:

Proposition. *Let V be a finite-dimensional vector space, G a finite subgroup of $\mathrm{GL}(V)$ and N the order of G. Then the algebra $\mathbf{A}(V)^G$ is generated by homogeneous invariants of degree at most N.*

Proof. Let $p : \mathbf{A}(V) \to \mathbf{A}(V)$ denote the map defined by:

$$p(u) = \frac{1}{N} \sum_{\alpha \in G} \alpha.u.$$

We have $up(v) = p(uv)$ if $u \in \mathbf{A}(V)^G$ and $v \in \mathbf{A}(V)$. Moreover, the image of p is $\mathbf{A}(V)^G$ and $\deg(p(u)) \leqslant \deg(u)$ for all $u \in \mathbf{A}(V)$.

Let B be the subalgebra of $\mathbf{A}(V)^G$ generated by invariants of degree at most N, and A_N the subspace of $\mathbf{A}(V)$ consisting of polynomial functions of degree at most $N-1$. We claim that $\mathbf{A}(V) = BA_N$. Since $\mathbf{A}(V)$ is generated by powers of linear forms, to prove our claim, it suffices to prove that $\ell^n \in BA_N$ for all $\ell \in V^*$ and $n \in \mathbb{N}$. This is clear if $n < N$. Now,

$$\prod_{\alpha \in G} (T - \alpha.\ell) = T^N + a_{N-1} T^{N-1} + \cdots + a_0$$

where $a_0, \ldots, a_{N-1} \in B$, so $\ell^N \in B + B\ell + \cdots + B\ell^{N-1}$. By induction, we obtain that $\ell^n \in BA_N$ if $n \geqslant N$.

Let $f \in \mathbf{A}(V)^G$. It follows from our claim that there exist $a_1, \ldots, a_r \in B$ and $f_1, \ldots, f_r \in A_N$ such that $f = a_1 f_1 + \cdots + a_r f_r$. So:

$$f = p(f) = a_1 p(f_1) + \cdots + a_r p(f_r).$$

Since $\deg(p(f_i)) \leqslant \deg(f_i) \leqslant N - 1$, we have $p(f_i) \in B$, and so $f \in B$ as required. $\square$

References and comments

- [5], [21], [38], [40], [78].

More general results concerning quotients of an algebraic variety can be found in [30].

In general, for a G-variety X, the geometric quotient of X by G does not exist. In [72], Rosenlicht proved that there exists a dense open G-stable subvariety Y of X such that the geometric quotient of Y by G exists. Other proofs of this result can be found in [35], [67], [75]. In particular, we deduce from this that

$$\operatorname{tr} \deg_{\Bbbk} \mathbf{R}(X)^G = \dim X - \rho$$

where ρ is the maximal dimension of G-orbits in X.

Solvable groups

Structure of diagonalizable groups and solvable groups are studied in this chapter. We prove in particular two important results : Lie-Kolchin Theorem and Borel's Fixed Point Theorem.

In this chapter, G denotes an algebraic group whose Lie algebra is $\mathfrak{g}$.

26.1 Conjugacy classes

26.1.1 Let H be a closed subgroup of G, $\mathfrak{h}$ its Lie algebra and $x \in \mathfrak{g}$. We set:

$$C_H(x) = \{\alpha \in H; \, (\mathrm{Ad}\,\alpha)(x) = x\} \,, \quad \mathfrak{c}_{\mathfrak{h}}(x) = \{y \in \mathfrak{h}; \, [x,y] = 0\}.$$

The set $C_H(x)$ (resp. $\mathfrak{c}_{\mathfrak{h}}(x)$) is a closed subgroup of H (resp. Lie subalgebra of $\mathfrak{h}$). We have $C_H(x) = H \cap C_G(x)$ and $\mathfrak{c}_{\mathfrak{h}}(x) = \mathfrak{h} \cap \mathfrak{c}_{\mathfrak{g}}(x)$. Moreover, by 23.5.5, 24.3.5 and 24.3.6, $\mathcal{L}\big(C_H(x)\big) = \mathfrak{c}_{\mathfrak{h}}(x)$.

For $\gamma \in G$, set:

$$C_H(\gamma) = \{\alpha \in H; \, \alpha\gamma = \gamma\alpha\} \,, \quad \mathfrak{c}_{\mathfrak{h}}(\gamma) = \{x \in \mathfrak{h}; \, (\mathrm{Ad}\,\gamma)(x) = x\}.$$

Again, $C_H(\gamma)$ (resp. $\mathfrak{c}_{\mathfrak{h}}(\gamma)$) is a closed subgroup of H (resp. Lie subalgebra of $\mathfrak{h}$). More generally, if L is a subset G, we set $C_H(L)$ (resp. $\mathfrak{c}_{\mathfrak{h}}(L)$) to be the intersection of the $C_H(\gamma)$'s, with $\gamma \in L$. (resp. $\mathfrak{c}_{\mathfrak{h}}(\gamma)$'s, $\gamma \in L$). It follows that $C_H(L)$ (resp. $\mathfrak{c}_{\mathfrak{h}}(L)$) is a closed subgroup of H (resp. Lie subalgebra of $\mathfrak{h}$).

Lemma. (i) *If $\gamma \in G$, then $\mathcal{L}\big(C_H(\gamma)\big) = \mathfrak{c}_{\mathfrak{h}}(\gamma)$.*
(ii) *If L is a subset of G, then $\mathcal{L}\big(C_H(L)\big) = \mathfrak{c}_{\mathfrak{h}}(L)$.*

Proof. By 24.5.2, it suffices to prove (i). Since $C_H(\gamma) = H \cap C_G(\gamma)$ and $\mathfrak{c}_{\mathfrak{h}}(\gamma) = \mathfrak{h} \cap \mathfrak{c}_{\mathfrak{g}}(\gamma)$, it follows from 24.3.5 (iii) that it suffices to prove the result for $H = G$.

Let $u, v, w : G \to G$ be the morphisms defined as follows: $u(\alpha) = \alpha\gamma\alpha^{-1}$, $v(\alpha) = \alpha\gamma\alpha^{-1}\gamma^{-1}$, $w(\alpha) = \alpha\gamma$, $\alpha \in G$. Then $u = w \circ v$ and $du_e = dw_e \circ dv_e = dw_e \circ (\mathrm{Ad}(\gamma) - \mathrm{id}_{\mathfrak{g}})$ (23.5.3). Since $C_G(\gamma)$ is the stabilizer of γ under the action of G given by conjugation, 25.1.3 implies that $\ker(du_e) = \mathcal{L}(C_G(\gamma))$. So the result follows because dw_e is bijective. $\quad\square$

26.1.2 For $\gamma \in G$ and $x \in \mathfrak{g}$, set:

$$\mathrm{Cl}_H(\gamma) = \{\alpha\gamma\alpha^{-1}\,;\alpha \in H\}\,,\quad \mathfrak{cl}_H(x) = \{(\mathrm{Ad}\,\alpha)(x)\,;\alpha \in H\}.$$

The set $\mathrm{Cl}_H(\gamma)$ is the H-orbit of γ under the action of H on G by conjugation, so it is locally closed in G (21.4.3). In the same way, $\mathfrak{cl}_H(x)$ is locally closed in $\mathfrak{g}$.

Lemma. *If* $(\mathrm{Ad}\,\alpha)(x) - x \in \mathfrak{h}$ *for all* $\alpha \in H$, *then* $[x, \mathfrak{h}] \subset \mathfrak{h}$.

Proof. Apply 24.5.5 to the subspaces $E = \mathfrak{h}$ and $F = \Bbbk x + \mathfrak{h}$ of $\mathfrak{g}$. $\quad\square$

26.1.3 Let us conserve the above notations and let $\pi : H \to \mathrm{Cl}_H(\gamma)$ be the morphism $\alpha \mapsto \alpha\gamma\alpha^{-1}$. By 25.1.3, $d\pi_e$ induces a surjection from $\mathfrak{h}$ onto $\mathrm{T}_\gamma(\mathrm{Cl}_H(\gamma))$ with kernel $\mathcal{L}(C_H(\gamma)) = \mathfrak{c}_{\mathfrak{h}}(\gamma)$.

Suppose further that γ is semisimple and $\gamma H\gamma^{-1} \subset H$. Then $(\mathrm{Ad}\,\gamma)(\mathfrak{h}) \subset \mathfrak{h}$ and the restriction of $\mathrm{Ad}\,\gamma$ to $\mathfrak{h}$ is semisimple. So $\mathfrak{h} = \mathfrak{c}_{\mathfrak{h}}(\gamma) \oplus \mathfrak{m}$ where $\mathfrak{m}$ is the sum of eigenspaces of $(\mathrm{Ad}\,\gamma)|_{\mathfrak{h}}$ whose eigenvalues are not equal to 1. We see, via the automorphism $\alpha \mapsto \alpha\gamma^{-1}$ of G, that $\mathrm{T}_e(\mathrm{Cl}_H(\gamma)\gamma^{-1}) = \mathfrak{m}$.

26.1.4 Theorem. *Let H be a closed subgroup of G and $\mathfrak{h}$ its Lie algebra.*

(i) Let γ be a semisimple element of G verifying $\gamma H\gamma^{-1} \subset H$. Then $\mathrm{Cl}_H(\gamma)$ is closed in G.

(ii) Let x be a semisimple element of $\mathfrak{g}$ verifying $(\mathrm{Ad}\,\alpha)(x) - x \in \mathfrak{h}$ for all $\alpha \in H$. Then $\mathfrak{cl}_{\mathfrak{h}}(\gamma)$ is closed in $\mathfrak{g}$.

Proof. We may assume that $G = \mathrm{GL}(V)$ for some finite-dimensional vector space V.

(i) For $\alpha \in G$, let μ_α be its minimal polynomial. For $\alpha \in N_G(H)$, we have $(\mathrm{Ad}\,\alpha)(\mathfrak{h}) \subset \mathfrak{h}$ and we shall denote by χ_α the characteristic polynomial of $(\mathrm{Ad}\,\alpha)|_{\mathfrak{h}}$. Let W be the set of elements $\alpha \in N_G(H)$ verifying:

$$\mu_\gamma(\alpha) = 0\,,\ \chi_\alpha = \chi_\gamma.$$

Then $\gamma \in W$, and if $\alpha \in W$, then $\beta\alpha\beta^{-1} \in W$ for all $\beta \in H$. Moreover, if $\alpha \in W$, then μ_α divides μ_γ. It follows that the elements of W are semisimple.

Let $\alpha \in W$. Then by 21.4.3 and 26.1.1, we have:

$$\dim \mathrm{Cl}_H(\alpha) = \dim H - \dim C_H(\alpha) = \dim \mathfrak{h} - \dim \mathfrak{c}_{\mathfrak{h}}(\alpha).$$

Since $\chi_\alpha = \chi_\gamma$ and $(\mathrm{Ad}\,\alpha)|_{\mathfrak{h}}$ is semisimple, $\dim \mathfrak{c}_{\mathfrak{h}}(\alpha) = \dim \mathfrak{c}_{\mathfrak{h}}(\gamma)$. Thus the H-orbits of the variety W (under conjugation) are all of the same dimension. By 21.4.5, these orbits are closed in W. Since W is clearly closed in $N_G(H)$ and $N_G(H)$ is closed in G, the result follows.

(ii) Let $\mathfrak{n}_{\mathfrak{g}}(H) = \{y \in \mathfrak{g}; (\operatorname{Ad}\alpha)(y) - y \in \mathfrak{h} \text{ for all } \alpha \in H\}$. We have $x \in \mathfrak{n}_{\mathfrak{g}}(H)$ and we saw in 26.1.2 that if $y \in \mathfrak{n}_{\mathfrak{g}}(H)$, then $[y, \mathfrak{h}] \subset \mathfrak{h}$. Let us denote by χ_y the characteristic polynomial of $(\operatorname{ad} y)|_{\mathfrak{h}}$.

Let μ_x be the minimal polynomial of x as an element of $\operatorname{End}(V)$ and set:

$$\mathfrak{w} = \{y \in \mathfrak{n}_{\mathfrak{g}}(H); \ \mu_x(y) = 0 , \ \chi_y = \chi_x\}.$$

Then $\mathfrak{w}$ contains x, and it is a closed subset of $\mathfrak{g}$ verifying $(\operatorname{Ad}\alpha)(\mathfrak{w}) \subset \mathfrak{w}$ for all $\alpha \in H$. Using the same argument as in (i), we obtain that the H-orbits of $\mathfrak{w}$ are all of the same dimension, and so they are closed. $\square$

26.1.5 Theorem. *Let U be a closed unipotent subgroup of G, γ a semisimple element of G verifying $\gamma U \gamma^{-1} \subset U$, and M the set of elements $\alpha\gamma\alpha^{-1}\gamma^{-1}$, with $\alpha \in U$.*

(i) The group $C_U(\gamma)$ is closed and connected. The set M is a closed irreducible subset of U.

(ii) The map $u : M \times C_U(\gamma) \to U$, $(\alpha, \beta) \mapsto \alpha\beta$, is an isomorphism of varieties.

(iii) The morphism $v : G \to G$, $\alpha \mapsto \alpha\gamma\alpha^{-1}\gamma^{-1}$, induces an automorphism of the variety M.

Proof. Since $\gamma U \gamma^{-1} \subset U$, we have $M \subset U$. Moreover, U is connected (22.3.6).

1) Clearly, $C_U(\gamma)$ is closed in U and it is connected (22.3.6). The fact that M is closed follows from 26.1.4 and M is irreducible because $M = v(U)$. So we have proved (i).

2) Let $\alpha \in U$, $\beta = \alpha\gamma\alpha^{-1}\gamma^{-1}$. If $\beta \in C_U(\gamma)$, then $\gamma\beta = \beta\gamma = \alpha\gamma\alpha^{-1}$. Since β is unipotent and γ is semisimple, $\beta\gamma = \gamma\beta$ is the Jordan decomposition of $\alpha\gamma\alpha^{-1}$. But this element is semisimple. So $\beta = e$ and $M \cap C_U(\gamma) = \{e\}$.

3) Let $\alpha, \beta \in U$. Then:

$$v(\alpha\beta) = \alpha\beta\gamma\beta^{-1}\alpha^{-1}\gamma^{-1} = \alpha(\beta\gamma\beta^{-1}\gamma^{-1})\alpha^{-1}(\alpha\gamma\alpha^{-1}\gamma^{-1}) = \alpha v(\beta)\alpha^{-1}v(\alpha).$$

So if $\beta \in C_U(\gamma)$, then $v(\beta) = e$ and:

$$v(\alpha\beta) = v(\alpha) = v(\alpha)v(\beta).$$

4) Using the equalities in point 3, we obtain that if α is central in U, then:

$$v(\beta\alpha) = v(\alpha\beta) = v(\beta)v(\alpha).$$

5) Let us prove (ii) in the case where U is commutative. By point 4, the map v induces a homomorphism from the group U to itself whose image is M and kernel is $C_U(\gamma)$. So $\dim U = \dim M + \dim C_U(\gamma)$ (21.2.4) and u is a homomorphism of algebraic groups. The connectedness of U, $C_U(\gamma)$ and M imply therefore that u is surjective.

Let $(\alpha, \beta) \in M \times C_U(\gamma)$ be such that $u(\alpha, \beta) = \alpha\beta = e$. Then $\alpha = \beta^{-1} \in M \cap C_U(\gamma) = \{e\}$ by point 2. It follows that $\alpha = \beta = e$. The homomorphism u

is therefore injective, and we have proved that u is an isomorphism of algebraic groups (21.2.6).

6) Let us prove (ii) in the general case. We proceed by induction on the dimension of U. The case $\dim U = 0$ is obvious, so let us assume that $\dim U > 0$. Since U is unipotent, it is nilpotent (22.3.2 and 10.8.14). Its centre Z is a non-trivial closed unipotent subgroup, so it is also connected. If $Z = U$, then the result follows from point 5. Otherwise, $Z \neq U$. Note that γ and U are contained in the normalizer $N_G(Z)$ of Z in G.

Let $\pi : N_G(Z) \to N_G(Z)/Z$ be the canonical surjection, $G' = N_G(Z)/Z$, $U' = U/Z$ and $\gamma' = \pi(\gamma)$. Set:

$$M' = \{\alpha'\gamma'(\alpha')^{-1}(\gamma')^{-1}; \alpha' \in U'\} = \pi(M) , \ N = \{\alpha\gamma\alpha^{-1}\gamma^{-1}; \alpha \in Z\}.$$

Applying the induction hypothesis to (G', U', γ') and (G, Z, γ), we obtain that the morphisms

$$u' : M' \times C_{U'}(\gamma') \to U' \ \text{ and } \ u_0 : N \times C_Z(\gamma) \to Z$$

are isomorphisms of varieties.

• Let us show that u is injective. Let $\alpha_1, \alpha_2 \in C_U(\gamma)$ and $\beta_1, \beta_2 \in M$ be such that $\beta_1\alpha_1 = \beta_2\alpha_2$. Let $\alpha = \alpha_1\alpha_2^{-1} \in C_U(\gamma)$, then $\beta_1\alpha = \beta_2$. So $\pi(\beta_1)\pi(\alpha) = \pi(\beta_2)$. Since $\pi(\alpha) \in C_{U'}(\gamma')$ and $\pi(\beta_1), \pi(\beta_2) \in M'$, we obtain that $\pi(\alpha) = e$, or equivalently $\alpha \in Z$. Let us write $\beta_1 = v(\delta_1)$, $\beta_2 = v(\delta_2)$ where $\delta_1, \delta_2 \in U$. The fact that $\alpha \in Z \cap C_U(\gamma)$ implies that

$$\delta_2\gamma\delta_2^{-1} = \beta_2\gamma = \beta_1\alpha\gamma = (\delta_1\gamma\delta_1^{-1})\alpha = \alpha(\delta_1\gamma\delta_1^{-1}).$$

The elements $\delta_1\gamma\delta_1^{-1}$ et $\delta_2\gamma\delta_2^{-1}$ are semisimple and α is unipotent. We deduce therefore that $\alpha = e$, and so $\alpha_1 = \alpha_2$ and $\beta_1 = \beta_2$.

• Next, we show that $v(Z) = Z \cap M$. If $\alpha \in Z$, then $\gamma\alpha^{-1}\gamma^{-1} \in Z$, so $v(Z) \subset Z \cap M$. Conversely, since u_0 is surjective, any $\alpha \in Z$ is of the form $\delta\beta$, where $\beta \in C_Z(\gamma)$ and $\delta \in N = v(Z)$. Furthermore, if $\alpha \in M$, then $\alpha = \delta \in v(Z)$ because u is injective.

• We claim that $C_{U'}(\gamma') = \pi\big(C_U(\gamma)\big)$. Let $\alpha \in U$ be such that $\pi(\alpha)\gamma' = \gamma'\pi(\alpha)$. Then $v(\alpha) \in M \cap \ker\pi = Z \cap M = v(Z)$. Thus $v(\alpha) = v(\beta)$ for some $\beta \in Z$. By applying point 4 to β and $\beta^{-1}\alpha$, we have $\beta^{-1}\alpha \in C_U(\gamma)$. As $\pi(\beta^{-1}\alpha) = \pi(\alpha)$, we have therefore proved our claim.

• Let us show that u is surjective. Since u' is surjective, $\pi(M) = M'$ and $\pi\big(C_U(\gamma)\big) = C_{U'}(\gamma')$, we have $U = MC_U(\gamma)Z = MZC_U(\gamma)$. The surjectivity of u_0 implies then that $U = Mv(Z)C_Z(\gamma)C_U(\gamma)$. Finally, as $Mv(Z) = M$ by point 4, we have $U = MC_U(\gamma)$.

• So u is bijective and it follows from 17.4.8 that u is an isomorphism.

7) Since u is surjective, we have $M = v(U) = v\big(MC_U(\gamma)\big) = v(M)$ by point 3. On the other hand, if $\alpha, \beta \in M$ verify $v(\alpha) = v(\beta)$, then $\beta^{-1}\alpha \in C_U(\gamma)$ and $u(\alpha, e) = u(\beta, \beta^{-1}\alpha)$. It follows from the injectivity of u that $\alpha = \beta$. Thus v induces a bijection from M onto itself. But $M = \mathrm{Cl}_U(\gamma)\gamma^{-1}$, so 25.1.1 implies that M is normal. Therefore (iii) follows from 17.4.8. $\quad\square$

26.2 Actions of diagonalizable groups

26.2.1 In this section, T denotes a diagonalizable algebraic group. Let us suppose that T acts rationally on G and that if $\alpha \in T$, the morphism

$$\theta_\alpha : G \to G , \; \beta \mapsto \alpha.\beta$$

is an automorphism of the group G.

Denote $(d\theta_\alpha)_e$ by π_α. The map π_α is an automorphism of the vector space $\mathfrak{g}$, and $\alpha \mapsto \pi_\alpha$ defines a rational representation of T on $\mathfrak{g}$.

Let G^T (resp. $\mathfrak{g}^T$) be the set of elements $\beta \in G$ (resp. $x \in \mathfrak{g}$) verifying $\theta_\alpha(\beta) = \beta$ (resp. $\pi_\alpha(x) = x$) for all $\alpha \in T$. Recall that $X^*(T)$ denotes the set of characters of T. If $\chi \in X^*(T)$, denote by $\mathfrak{g}_\chi$ the set of $x \in \mathfrak{g}$ such that $\pi_\alpha(x) = \chi(\alpha)x$ for all $\alpha \in T$. Denote by $1_T \in X^*(T)$ the trivial character of T, that is $1_T(\alpha) = 1$ for all $\alpha \in T$, and we set:

$$\Phi(T, G) = \{\chi \in X^*(T) \setminus \{1_T\}; \mathfrak{g}_\chi \neq \{0\}\}.$$

Then we have:

$$\mathfrak{g} = \mathfrak{g}^T \oplus \left(\bigoplus_{\chi \in \Phi(T,G)} \mathfrak{g}_\chi \right).$$

Let $\chi \in X^*(T)$ and T^χ the kernel of χ. Denote by $C_G(T^\chi)$ the set of $\beta \in G$ verifying $\theta_\alpha(\beta) = \beta$ for all $\alpha \in T^\chi$.

By identifying G and T as subgroups of the semidirect product $G \rtimes_\theta T$, we may always assume that G and T are closed subgroups of an algebraic group.

Proposition. (i) *We have* $\mathcal{L}(G^T) = \mathfrak{g}^T$.

(ii) *Assume that G is connected and $G \neq G^T$. Then G is generated (as a group) by the subgroups* $C_G(T^\chi)$, $\chi \in X^*(T)$.

Proof. (i) Identify G and T as closed subgroups of an algebraic group, then apply 26.1.1 (ii).

(ii) The condition $G \neq G^T$ implies that $\Phi(G, T) \neq \emptyset$. Denote by G' the subgroup generated by the $C_G(T^\chi)$, $\chi \in \Phi(G, T)$. Applying (i) with $T = T^\chi$, we obtain that $\mathcal{L}(C_G(T^\chi)) = \mathfrak{g}^{T^\chi}$. So $\mathfrak{g}^T + \mathfrak{g}_\chi \subset \mathcal{L}(G')$ if $\chi \in \Phi(T, G)$. Since $\mathfrak{g}$ is the sum of the $\mathfrak{g}^T + \mathfrak{g}_\chi$, $\chi \in \Phi(T, G)$, it follows that $\mathfrak{g} \subset \mathcal{L}(G')$. So $G = G^\circ \subset G'$ and the result follows. $\square$

26.2.2 Let us suppose that T acts rationally on the algebraic groups G and H via group automorphisms of G and H respectively, as in 26.2.1, and denote them by θ_α^G and θ_α^H for $\alpha \in T$.

Proposition. *Let $\sigma : G \to H$ be a T-equivariant surjective morphism of algebraic groups. Then σ induces a surjective morphism from $(G^T)^\circ$ onto $(H^T)^\circ$.*

Proof. Since $\sigma \circ \theta_\alpha^G = \theta_\alpha^H \circ \sigma$ for all $\alpha \in T$, we have $d\sigma \circ d\theta_\alpha^G = d\theta_\alpha^H \circ d\sigma$ for all $\alpha \in T$. The surjectivity of u implies that du is surjective. Thus du induces a surjection from $\mathcal{L}(G)^T$ onto $\mathcal{L}(H)^T$. By 26.2.1 (i), this surjection is the differential of the restriction $G^T \to H^T$. So the result follows. $\square$

26.2.3 Proposition. *Let T be a torus of G and K a closed subgroup of G such that $T \subset N_G(K)$, and $\mathfrak{k} = \mathcal{L}(K)$. Then there exists $\alpha \in T$ such that:*

$$C_K(\alpha) = C_K(T) \ , \ \mathfrak{c}_{\mathfrak{k}}(\alpha) = \mathfrak{c}_{\mathfrak{k}}(T).$$

Proof. We may assume that $G \subset \mathrm{GL}(V)$ where V is a finite-dimensional vector space. Denote by $\chi_1, \ldots, \chi_r$ the pairwise distinct weights of T in V, and let $V_i = V_{\chi_i}$, $1 \leqslant i \leqslant r$. We have $V = V_1 \oplus \cdots \oplus V_r$ (22.5.2).

If $M = \mathrm{GL}(V_1) \times \cdots \times \mathrm{GL}(V_r) \subset \mathrm{GL}(V)$ and $\mathfrak{m} = \mathcal{L}(M)$, then it is clear that $C_K(T) = M \cap K$ and $\mathfrak{c}_{\mathfrak{k}}(T) = \mathfrak{m} \cap \mathfrak{k}$.

If $1 \leqslant i, j \leqslant r$ and $i \neq j$, $\chi_i \chi_j^{-1}$ is a non-trivial character of T, its kernel T_{ij} is a subgroup of T distinct from T. Since the field $\Bbbk$ is infinite, we deduce that there exists $\alpha \in T$ which is not contained in any of the T_{ij}'s, $i \neq j$. Therefore $C_K(\alpha) = M \cap K$ and $\mathfrak{c}_{\mathfrak{k}}(\alpha) = \mathfrak{m} \cap \mathfrak{k}$. $\quad\square$

26.3 Fixed points

26.3.1 Lemma. *Let V be a $\Bbbk$-vector space of dimension $n > 0$ and G a connected solvable subgroup of $\mathrm{GL}(V)$. Then there exists $v \in V \setminus \{0\}$ such that $Gv \subset \Bbbk v$.*

Proof. In view of 21.6.6, we may assume that G is an algebraic subgroup of $\mathrm{GL}(V)$. We shall proceed by induction on $\dim G$. If $\dim G = 1$, then G is commutative (22.6.1), and the result is then clear because $G = G_s$ or $G = G_u$ (22.2.6). So let us suppose that $\dim G \geqslant 2$. Let $H = (G, G)$ be the group of commutators of G, then it is a connected solvable algebraic subgroup of G (21.3.6) verifying $\dim H < \dim G$. So by induction and 22.4.5, there exists $\chi \in X^*(H)$ such that $V_\chi \neq 0$ where V_χ denotes the weight space of weight χ of H in V.

Since H is normal in G, we saw in 22.4.5 that the (direct) sum W of the V_χ's, $\chi \in X^*(H)$, is G-stable. So we may assume that $V = W$. Note that the set of weights of H in V is finite (22.4.6).

Let us fix a basis of V consisting of weight vectors of H, and identify $\mathrm{GL}(V)$ with $\mathrm{GL}_n(\Bbbk)$ via this basis. Then the elements of H are diagonal matrices. For $\beta \in H$, the map $G \to H$, $\alpha \mapsto \alpha \beta \alpha^{-1}$, is a morphism of varieties whose image is connected since G is connected. But the image is a finite set since its elements are diagonal matrices having the same eigenvalues as for the diagonal matrix β. We deduce therefore that each V_χ is G-stable. So we may further assume that $V = V_\chi$ where $\chi \in X^*(H)$.

Now $H = (G, G)$, so $\det \beta = 1$ for all $\beta \in H$. But for $\beta \in H$, we have $\beta = \chi(\beta) \mathrm{id}_V$. Hence the connected subgroup H is finite. So $H = \{e\}$ and G is commutative. But the result is clear when G is commutative. So we are done. $\square$

26.3.2 Theorem. (Lie-Kolchin Theorem) *Let V be a finite-dimensional $\Bbbk$-vector space and G a connected solvable subgroup of $\mathrm{GL}(V)$. Then there exists a basis of V with respect to which the elements of G are upper triangular.*

Proof. Let us proceed by induction on $\dim V$. The case $\dim V \leqslant 1$ is obvious. So let us assume that $\dim V \geqslant 2$. By 26.3.1, there exists a weight vector v_1 of G in V. For $\alpha \in G$, denote by α' the automorphism of $W = V/\Bbbk v_1$ induced by α, and $v \mapsto \tilde{v}$, the canonical surjection $V \to W$. Then clearly the set G' of the α', $\alpha \in G$, is a connected solvable subgroup of $\mathrm{GL}(W)$. So by the induction hypothesis, there exists $v_2, \ldots, v_n \in V$ such that $(\tilde{v}_2, \ldots, \tilde{v}_n)$ is a basis of W with respect to which the elements of G' are upper triangular. It follows that $(v_1, \ldots, v_n)$ is a basis of V with respect to which the elements of G are upper triangular. $\square$

26.3.3 Corollary. *Let G be a connected solvable algebraic group, then its derived subgroup $\mathcal{D}(G) = (G, G)$ is a closed connected unipotent subgroup.*

Proof. In view of 21.3.6, we only need to show that $\mathcal{D}(G)$ is unipotent. We may assume that G is a closed subgroup of some $\mathrm{GL}_n(\Bbbk)$ (22.1.5). By 26.3.2, there exists $\alpha \in \mathrm{GL}_n(\Bbbk)$ such that $\alpha G \alpha^{-1} \subset \mathrm{T}_n(\Bbbk)$ (notations of 10.8.1). Hence $\alpha \mathcal{D}(G) \alpha^{-1} \subset U_n(\Bbbk)$ (10.8.9), and $\mathcal{D}(G)$ is unipotent. $\square$

26.3.4 Theorem. (Borel's Fixed Point Theorem) *Let G be a connected solvable algebraic group and X a non-empty complete G-variety. Then the set X^G is not empty.*

Proof. We shall prove the result by induction on $d = \dim G$. If $d = 0$, then $G = \{e\}$ and $X = X^G$. So let us assume that $d > 0$ and let $H = \mathcal{D}(G)$. Since $\dim H < d$, the set $Y = X^H$ is a non-empty closed subvariety of X (21.4.2). So Y is complete (13.4.3). Now H is normal in G, so if $x \in Y$, $\alpha \in G$ and $\beta \in H$, then:

$$\beta.(\alpha.x) = \alpha.\big((\alpha^{-1}\beta\alpha).x\big) = \alpha.x,$$

so Y is G-stable. There exists by 21.4.5 $x \in Y$ such that the orbit $G.x$ is closed, and therefore complete. The group G_x contains H, so G_x is normal in G. By 25.4.11 and 25.4.7, the variety G/G_x is affine and it is isomorphic to $G.x$. But $G.x$ is irreducible since G is connected, it follows therefore from 13.4.4 that $G.x = \{x\}$. $\square$

26.4 Properties of solvable groups

26.4.1 Let G_s (resp. G_u) denote the set of semisimple (resp. unipotent) elements of G. Recall from 22.2.5 that G_u is a closed subset of G.

26.4.2 Proposition. *Let G be connected and solvable.*
   (i) *The set G_u is a closed connected nilpotent subgroup of G. It contains $\mathcal{D}(G)$ and so it is normal in G.*
   (ii) *The group G/G_u is a torus.*

Proof. (i) By 22.1.5 and 26.3.2, we may assume that $G \subset \mathrm{T}_n(\Bbbk)$ for some $n \in \mathbb{N}^*$. Since the set of unipotent elements of $\mathrm{T}_n(\Bbbk)$ is $U_n(\Bbbk)$, we have

$G_u = G \cap U_n(\Bbbk)$ by 22.2.3. So part (i) follows from 10.8.9, 10.8.10, 22.2.5 and 22.3.6 (i).

(ii) Since $\mathcal{D}(G) \subset G_u$, G/G_u is abelian. It is connected because it is the image of G under the canonical surjection $G \to G/G_u$. Let $\beta \in G/G_u$ and $\alpha \in G$ such that $\pi(\alpha) = \beta$. Denote by α_s and α_u the semisimple and unipotent components of α. It follows that $\beta = \pi(\alpha_s)$. Hence β is semisimple (22.2.3) and the result follows from 22.5.7. $\square$

26.4.3 Proposition. *Let G be connected and solvable.*

(i) *There exists a chain $G = G_p \supset G_{p-1} \supset \cdots \supset G_1 \supset G_0 = G_u$ of closed connected normal subgroups of G such that $\dim(G_i/G_{i-1}) = 1$ for all $1 \leqslant i \leqslant p$.*

(ii) *There exists a chain $G = G_n \supset G_{n-1} \supset \cdots \supset G_1 \supset G_0 = \{e\}$ of closed connected normal subgroups of G such that $\dim(G_i/G_{i-1}) = 1$ for all $1 \leqslant i \leqslant n$.*

Proof. (i) Let $\pi : G \to G/G_u = T$ be the canonical surjection. By 26.4.2, T is isomorphic to $D_p(\Bbbk) \simeq (\mathbf{G}_m)^p$ for some p. It is therefore clear that there is a chain $T = T_p \supset T_{p-1} \supset \cdots \supset T_1 \supset T_0 = \{e\}$ of closed connected subgroups of T such that $T_i/T_{i-1} \simeq \mathbf{G}_m$ for $1 \leqslant i \leqslant p$.

The subgroups $H_i = \pi^{-1}(T_i)$ of G are closed and normal in G, and so are the subgroups $G_i = H_i^\circ$. Moreover, $\dim(H_i/H_{i-1}) = \dim(T_i/T_{i-1}) = 1$ (25.4.12) and $(H_{i-1} \cap G_i)/G_{i-1}$ is a finite set. So $\dim(G_i/G_{i-1}) \leqslant 1$ and the result follows.

(ii) By 26.3.2, we may assume that $G \subset T_r(\Bbbk)$ and $G_u \subset U_r(\Bbbk)$ for some $r \in \mathbb{N}^*$. Let J be the set of pairs (i,j) where $i,j \in \{1,\ldots,r\}$ and $i < j$, and endow J with the following total order: $(i,j) < (k,l)$ if $j < l$ or if $j = l$ and $i > k$. Let $\lambda_1 < \cdots < \lambda_s$ be the elements of J in ascending order, and for $(i,j) \in J$, let $U_{(i,j)}$ be the set of matrices $[x_{kl}] \in U_r(\Bbbk)$ verifying $x_{kl} = 0$ for $(k,l) \leqslant (i,j)$. Then we verify easily that $U_r(\Bbbk) = U_{\lambda_0} \supset U_{\lambda_1} \supset \cdots \supset U_{\lambda_s} = \{I_r\}$ is a chain of closed normal subgroups of $T_r(\Bbbk)$. Moreover, we have $U_{\lambda_i}/U_{\lambda_{i+1}} \simeq \mathbf{G}_a$ for $0 \leqslant i < s$. Let $G_i = (G \cap U_{\lambda_i})^\circ$, then G_i is a closed normal subgroup of G contained in G_u and we have $\dim(G_i/G_{i+1}) \leqslant 1$. So the result follows from (i). $\square$

26.4.4 Proposition. *Let G be connected and solvable. Then the following conditions are equivalent:*

(i) *The set G_s is a subgroup of G.*

(ii) *The group G is nilpotent.*

If these conditions are verified, then G_s is a connected subgroup of G contained in the centre of G, and the map $\theta : G_s \times G_u \to G$, $(\alpha, \beta) \mapsto \alpha\beta$, is an isomorphism of algebraic groups.

Proof. Let us denote by $\pi : G \to G/G_u$ the canonical surjection.

(i) $\Rightarrow$ (ii) Suppose that G_s is a subgroup of G. Since $\pi|_{G_s}$ is injective, it follows from 26.4.2 (ii) that G_s is abelian.

We may assume that $G \subset \mathrm{GL}_n(\Bbbk)$ and since G_s is abelian, we may assume further that $G_s \subset \mathrm{D}_n(\Bbbk)$. So $G_s = G \cap \mathrm{D}_n(\Bbbk)$ is closed in G. Thus the restriction of π to G_s is a bijective morphism of algebraic groups, hence an isomorphism (21.2.6). In particular, G_s is connected. Since $\alpha\beta\alpha^{-1} \in G_s$ for all $\beta \in G_s$ and $\alpha \in G$, we have $G = N_G(G_s)$. So by 22.5.11:

$$G = N_G(G_s) = N_G(G_s)^\circ = C_G(G_s)^\circ.$$

Thus G_s is contained in the centre of G. Since G/G_s is unipotent, it is nilpotent (10.8.14), and so G is nilpotent (10.5.3). Note that since G_s is contained in the centre of G, the map θ is a bijective morphism of algebraic groups, and hence it is an isomorphism (21.2.6).

(ii) $\Rightarrow$ (i) Suppose that G is nilpotent. Fix $\alpha \in G_s$ and define the following morphisms of varieties:

$$u : G \to G , \ \beta \mapsto \alpha\beta\alpha^{-1} \text{ and } v : G \to G , \ \beta \mapsto \alpha\beta\alpha^{-1}\beta^{-1}.$$

Since G is nilpotent, there exists $n \in \mathbb{N}^*$ such that $v^n(\beta) = e$ for all $\beta \in G$. So $(dv)^n = 0$, and dv is a nilpotent endomorphism of $\mathfrak{g} = \mathcal{L}(G)$. As $dv = \mathrm{Ad}(\alpha) - \mathrm{id}_\mathfrak{g}$ and $\alpha \in G_s$, it is also a semisimple endomorphism of $\mathfrak{g}$ (22.2.3). Hence $\mathrm{Ad}(\alpha) = \mathrm{id}_\mathfrak{g}$. It follows from 24.4.2 and the connectedness of G that α is central in G. It is now clear that G_s is a subgroup of G. $\square$

26.4.5 Proposition. *Let G be connected and solvable and $\mathfrak{g}$ its Lie algebra.*

(i) *$\mathcal{L}(G_u)$ is the set of nilpotent elements of $\mathfrak{g}$.*
(ii) *If G is nilpotent, then $\mathcal{L}(G_s)$ is the set of semisimple elements of $\mathfrak{g}$.*

Proof. (i) We may assume that $G \subset \mathrm{T}_n(\Bbbk)$ and $G_u \subset \mathrm{U}_n(\Bbbk)$. Then $\mathcal{L}(G_u) \subset \mathcal{L}(\mathrm{U}_n(\Bbbk)) = \mathfrak{n}_n(\Bbbk)$, so all the elements of $\mathcal{L}(G_u)$ are nilpotent.

Let $\pi : G \to G/G_u$ be the canonical surjection. If x is a nilpotent element of $\mathfrak{g}$, then since G/G_u is a torus, 23.6.3 implies that $d\pi(x) = 0$, or $x \in \ker d\pi$. So $x \in \mathcal{L}(\ker \pi) = \mathcal{L}(G_u)$ by 24.4.1.

(ii) Assume that G is nilpotent. Then π induces an isomorphism from G_s onto the torus G/G_u (26.4.2), so all the elements of $\mathcal{L}(G_s)$ are semisimple. By 26.4.4, G_s is contained in the centre of G, so $\mathcal{L}(G_s)$ is in the centre of $\mathfrak{g}$ (24.4.2). Again by 26.4.4, G and $G_s \times G_u$ are isomorphic, so $\mathfrak{g}$ is isomorphic to the direct product $\mathcal{L}(G_s) \times \mathcal{L}(G_u)$. The result follows therefore from (i) and the uniqueness of the Jordan decomposition in $\mathfrak{g}$. $\square$

26.5 Structure of solvable groups

26.5.1 In this section, G denotes a connected solvable algebraic group whose Lie algebra is $\mathfrak{g}$.

We saw in 26.4.2 that G_u is normal in G, so its Lie algebra $\mathfrak{n}$ is an ideal of $\mathfrak{g}$ (24.4.7) and $\mathfrak{n}$ is the set of nilpotent elements of $\mathfrak{g}$ (26.4.5). Moreover, since $\mathcal{D}(G) \subset G_u$ (26.4.2), we have $[\mathfrak{g}, \mathfrak{g}] \subset \mathfrak{n}$ (24.3.5 and 24.5.12).

Proposition. *There exists a Lie subalgebra* $\mathfrak{t}$ *verifying the following conditions:*

(i) $\mathfrak{t}$ *is an algebraic Lie subalgebra of* $\mathfrak{g}$.
(ii) $\mathfrak{t}$ *is abelian and all its elements are semisimple.*
(iii) *As a vector space,* $\mathfrak{g} = \mathfrak{t} \oplus \mathfrak{n}$.

Proof. We may assume that $G \subset \mathrm{GL}(V)$ and $\mathfrak{g} \subset \mathfrak{gl}(V)$ for some finite-dimensional vector space V. The elements of $\mathfrak{n}$ are then nilpotent endomorphisms of V (23.6.4).

Let $\mathcal{A}$ be the set of abelian Lie subalgebras of $\mathfrak{g}$ whose elements are semisimple endomorphisms of V. Then $\mathcal{A} \neq \emptyset$ since $\{0\} \in \mathcal{A}$. Let $\mathfrak{t}$ be an element of $\mathcal{A}$ of maximal dimension.

By 23.6.4, the representation $\mathfrak{t} \mapsto \mathrm{ad}_{\mathfrak{g}}\, \mathfrak{t}$ of $\mathfrak{t}$ in $\mathfrak{g}$ is semisimple, and $\mathfrak{t} + \mathfrak{n}$ is stable under this representation. So there exists a subspace $\mathfrak{m}$ such that $\mathfrak{g} = \mathfrak{m} \oplus (\mathfrak{t} + \mathfrak{n})$ and $[\mathfrak{t}, \mathfrak{m}] \subset \mathfrak{m}$. Since $[\mathfrak{g}, \mathfrak{g}] \subset \mathfrak{n}$, we deduce that $[\mathfrak{t}, \mathfrak{m}] = \{0\}$.

Let $x \in \mathfrak{m}$ and x_s, x_n its semisimple and nilpotent components. By 23.6.4, $x_s, x_n \in \mathfrak{g}$ and since $[x, \mathfrak{t}] = \{0\}$, we have $[x_s, \mathfrak{t}] = \{0\}$ (10.1.3 and 23.6.4). It follows that $\Bbbk x_s + \mathfrak{t}$ is an abelian subalgebra whose elements are semisimple. Our choice of $\mathfrak{t}$ implies that $x_s = 0$. So all the elements of $\mathfrak{m}$ are nilpotent. Hence $\mathfrak{m} \subset \mathfrak{n}$, which in turn implies that $\mathfrak{m} = \{0\}$. Thus $\mathfrak{g} = \mathfrak{t} + \mathfrak{n}$ and it is clear that $\mathfrak{t} \cap \mathfrak{n} = \{0\}$.

So we are left to prove that $\mathfrak{t}$ is algebraic. Let $x \in \mathfrak{t}$ and y a replica of x. Then by 24.6.4 (ii), y is semisimple and $[y, \mathfrak{t}] = \{0\}$. Again by our choice of $\mathfrak{t}$, we deduce that $y \in \mathfrak{t}$. It follows from 24.5.10 that $\mathfrak{t}$ is an algebraic Lie subalgebra of $\mathfrak{g}$. $\square$

26.5.2 Let us conserve the hypotheses and notations of 26.5.1 and let T be a connected algebraic subgroup of G verifying $\mathcal{L}(T) = \mathfrak{t}$.

Proposition. (i) *The group T is a torus of dimension* $\dim(G/G_u)$.
(ii) *The map* $u : T \times G_u \to G$, $(\alpha, \beta) \mapsto \alpha\beta$ *is an isomorphism of varieties.*

Proof. (i) We may again assume that $G \subset \mathrm{GL}(V)$ and $\mathfrak{g} \subset \mathfrak{gl}(V)$. Note that $T = \mathcal{A}(\mathfrak{t})$ (see 24.5.1). If $x \in \mathfrak{t}$, then any element of $\mathcal{A}(x)$ can written as $P(x)$ for some $P \in \Bbbk[X]$ (24.6.4) and T is generated by the $\mathcal{A}(x)$'s, $x \in \mathfrak{t}$. It follows that T is abelian and its elements are semisimple. Since T is connected, it is a torus (22.5.7) of dimension $\dim(G/G_u)$ (26.5.1 (iii)).

(ii) Since T is a torus, $T \cap G_u = \{e\}$. Let TG_u be the set of products $\alpha\beta$, $\alpha \in T$ and $\beta \in G_u$. Then TG_u is a subgroup of G because G_u is normal in G. It is also closed in G by 21.2.3 and connected since it is the image of the morphism u.

Let us consider the following actions of $T \times G_u$ on $T \times G_u$ and G:

$$(\alpha, \beta).(\alpha', \beta') = (\alpha\alpha', \beta'\beta^{-1})\ ,\ (\alpha, \beta).\gamma = \alpha\gamma\beta^{-1}.$$

where $\alpha, \alpha' \in T$, $\beta, \beta' \in G_u$ and $\gamma \in G$. Then u is injective and equivariant with respect to these actions. Moreover, the stabilizer of e in $T \times G_u$ is $\{e\}$

because $T \cap G_u = \{e\}$. As the $(T \times G_u)$-orbit of e is TG_u, it follows from 21.4.3 that:

$$\dim(TG_u) = \dim(T \times G_u) = \dim T + \dim G_u = \dim G.$$

Hence the connectedness of G implies that $G = TG_u$. So u is a bijective morphism of $(T \times G_u)$-homogeneous spaces, and 25.1.2 says that u is an isomorphism of varieties. $\square$

26.5.3 Set:

$$\mathcal{C}^\infty(G) = \bigcap_{n \geqslant 1} \mathcal{C}^n(G).$$

We have $\mathcal{C}^\infty(G) \subset \mathcal{D}(G) \subset G_u$, so $\mathcal{C}^\infty(G)$ is a normal unipotent subgroup of G.

If T is a torus of G, then $T \cap G_u = \{e\}$, so $\dim T \leqslant \dim(G/G_u)$. In particular, if $\dim T = \dim(G/G_u)$, then T is a maximal torus. Now 26.5.2 says that such a torus exists and if T is such a torus, then the map $T \times G_u \to G$ is an isomorphism of varieties.

26.5.4 Lemma. *Let $T \subset G$ be a torus of dimension $\dim(G/G_u)$. If α is an element of G_s, then there exists $\beta \in \mathcal{C}^\infty(G)$ such that $\beta\alpha\beta^{-1} \in T$.*

Proof. We have already noted that $G = TG_u$. Let us proceed by induction on $\dim G$.

Suppose that G is nilpotent. Then $T = G_s$ by 26.4.4 and so the result is obvious.

Suppose that G is not nilpotent, then $\mathcal{C}^\infty(G) \neq \{e\}$. The unipotent group $\mathcal{C}^\infty(G)$ is connected and nilpotent (10.8.10 and 21.3.6). The last non-trivial term of the central descending series of $\mathcal{C}^\infty(G)$ is connected (21.3.6), so it is contained in the identity component N of the centre of $\mathcal{C}^\infty(G)$. It follows that $N \neq \{e\}$ and N is a normal subgroup of G. In particular, $N \subset G_u$ and $N \cap T = \{e\}$.

Let $G' = G/N$ and $\pi : G \to G'$ the canonical surjection. It is clear that the map $\pi(T) \times (G')_u \to G'$, $(\gamma, \delta) \mapsto \gamma\delta$, is an isomorphism of varieties. Applying the induction hypothesis on G', we obtain that there exists $\gamma \in \mathcal{C}^\infty(G') = \pi(\mathcal{C}^\infty(G))$ such that $\gamma\pi(\alpha)\gamma^{-1} \in \pi(T)$. Let $\delta \in \mathcal{C}^\infty(G)$ be such that $\pi(\delta) = \gamma$. Then by replacing α by $\delta\alpha\delta^{-1}$, we may assume that $\alpha \in TN = NT$.

Let $\beta \in N$ and $\gamma \in T$ be such that $\alpha = \beta\gamma$. By 26.1.5, there exists $\delta \in N$ and $\varepsilon \in C_N(\gamma)$ such that $\beta = (\delta\gamma\delta^{-1}\gamma^{-1})\varepsilon$. Then $\alpha = (\delta\gamma\delta^{-1})\varepsilon = \varepsilon(\delta\gamma\delta^{-1})$ since ε commutes with γ and N is abelian. But $\delta\gamma\delta^{-1}$ is semisimple and ε is unipotent, so $\alpha = (\delta\gamma\delta^{-1})\varepsilon$ is the Jordan decomposition of α. Hence $\varepsilon = e$ and $\delta^{-1}\gamma\delta = \gamma \in T$. $\square$

26.5.5 Theorem. *Let S, T be maximal tori of a connected solvable algebraic group G.*

(i) There exists $\alpha \in \mathcal{C}^\infty(G)$ such that $S = \alpha T\alpha^{-1}$. In particular, $\dim S = \dim T = \dim(G/G_u)$.

(ii) The map $T \times G_u \to G$, $(\alpha, \beta) \mapsto \alpha\beta$, is an isomorphism of varieties.

Proof. In view of 26.5.2 and 26.5.3, it suffices to show that there exists an element $\alpha \in \mathcal{C}^{\infty}(G)$ such that $S = \alpha T \alpha^{-1}$ where T is a maximal torus of dimension $\dim(G/G_u)$.

By 26.2.3, there exists $\gamma \in S$ such that $C_G(S) = C_G(\gamma)$. By 26.5.4, we may assume that $\gamma \in T$. Then $T \subset C_G(\gamma)^{\circ} = C_G(S)^{\circ}$. Let $C = C_G(S)^{\circ}$. Any torus of C is a torus of G, and so by 26.5.2 (i) and 26.5.3, T is a torus of C of dimension $\dim(C/C_u)$. It follows by 26.5.4 that if $\beta \in S$, then $\delta\beta\delta^{-1} \in T$ for some $\delta \in C$. Since S is central in C, we deduce that $S \subset T$. But S is a maximal torus, so $S = T$. $\square$

26.5.6 Proposition. *Let G be a connected solvable algebraic group.*

(i) *If T is a maximal torus of G, then any semisimple element of G is conjugate to a unique element of T.*

(ii) *Let S be a (not necessarily closed) subgroup of G consisting of semisimple elements. Then S is contained in a torus. Moreover, the group $C_G(S)$ is connected and is equal to $N_G(S)$.*

Proof. (i) Recall that the canonical surjection $\pi : G \to G/G_u$ induces an isomorphism from T onto G/G_u. Let $\alpha \in G_s$. By 26.5.4, α is conjugate to an element of T. Now if $\beta \in G$, then $\pi(\beta\alpha\beta^{-1}) = \pi(\alpha)$ since G/G_u is abelian. So the result follows.

(ii) The restriction of the canonical surjection π to S is injective, so S is abelian. If $\alpha \in N_G(S)$ and $\beta \in S$, then $\pi(\beta\alpha\beta^{-1}) = \pi(\alpha)$ since G/G_u is abelian. So $\beta\alpha\beta^{-1} = \alpha$ from the injectivity of $\pi|_S$. Hence $C_G(S) = N_G(S)$.

Now, $\overline{S}$ is a closed diagonalizable subgroup of G verifying $C_G(S) = C_G(\overline{S})$ (21.6.3). So we may assume that S is closed in G. Let T be a maximal torus in G.

If S is central in G, then $C_G(S) = G$ is connected and since any element of S is conjugate to an element of T (26.5.4), we have $S \subset T$.

Suppose that $S \not\subset Z(G)$ and $s \in S\backslash Z(G)$. Again by 26.5.4, we may assume that $s \in T$. Then $T \subset C_G(s) = T(C_G(s))_u$, which proves that $H = C_G(s)$ is connected (22.3.6). Since $H \neq G$, we can argue by induction on $\dim G$, that S is conjugated in H to a subgroup of T and that $C_H(S) = C_G(S)$ is connected. So we have proved (ii). $\square$

References and comments

• [5], [40], [78], [79].

In this chapter, we have followed closely the presentation of solvable groups given in [5].

Reductive groups

Semisimple groups and reductive groups are very useful objects in the study of algebraic groups. In this chapter, we consider these groups and their representations. We prove Hilbert-Nagata Theorem on invariants of rational reductive group actions, and we introduce the notion of algebraic quotient.

27.1 Radical and unipotent radical

27.1.1 In this section, H is an algebraic group and G is a closed normal subgroup of H.

Denote by $\mathcal{R}_1(G)$ the set of normal solvable subgroups of G and $\mathcal{R}_2(G)$ (resp. $\mathcal{R}(G)$) the set of elements of $\mathcal{R}_1(G)$ which are closed (resp. closed and connected) in G. Let $R_1(G), R_2(G)$ and $R(G)$ be the subgroups of G generated by $\mathcal{R}_1(G)$, $\mathcal{R}_2(G)$ and $\mathcal{R}(G)$ respectively. These subgroups are normal in G and the subgroup $R(G)$ is called the *radical* of G.

Lemma. (i) *We have* $R_1(G) = R_2(G)$. *The group* $R_1(G)$ *is the largest normal solvable subgroup of* G. *It is a closed normal subgroup of* H.

(ii) *The radical* $R(G)$ *of* G *is a connected solvable algebraic subgroup of* G. *It is the largest connected normal solvable subgroup of* G. *It is normal in* H *and we have* $R(G) = R(G^\circ) = R_1(G)^\circ$.

Proof. (i) If $H \in \mathcal{R}_1(G)$, then $\overline{H} \in \mathcal{R}_2(G)$ (21.6.1 and 21.6.6). So it is clear that $R_1(G) = R_2(G)$.

If $R_1(G) = \{e\}$, then it is solvable. Otherwise, there exists $K \in \mathcal{R}_2(G)$ such that $R_1(G) \supset K \neq \{e\}$. Since the group $R_1(G)/K$ is contained in $R_1(G/K)$, we obtain by induction on $\dim G$ that $R_1(G)/K$ is solvable. Hence $R_1(G)$ is solvable (10.4.5).

Again by using 21.6.1 and 21.6.6, we see that $R_1(G)$ is closed. We deduce therefore that $R_1(G)$ is the largest normal solvable subgroup of G.

If $\alpha \in H$, then $\alpha R_1(G)\alpha^{-1}$ is again a normal solvable subgroup of G. So it is contained in $R_1(G)$. It follows that $R_1(G)$ is normal in H.

(ii) Since $R(G) \subset R_1(G)$, it is solvable. Also, it is clear that $R(G)$ is normal in G, and it is closed and connected by 21.3.2. So $R(G) \subset R_1(G)^\circ$. Now $R_1(G)$ is normal in G, so $R_1(G)^\circ$ is also normal in G. Moreover $R_1(G)^\circ$ is solvable, so $R(G) = R_1(G)^\circ$. We deduce therefore that $R(G)$ is the largest connected normal solvable subgroup of G. By (i), $R_1(G)$ is normal in H, so $R(G) = R_1(G)^\circ$ is also normal in H.

Finally, $R(G)$ is connected, so it is contained in G°. Therefore $R(G) \in \mathcal{R}(G^\circ)$. Hence $R(G) \subset R(G^\circ)$. Conversely, since G° is normal in G, $R(G^\circ)$ is a connected normal solvable algebraic subgroup of G, so $R(G^\circ) \subset R(G)$. $\square$

27.1.2 Let us conserve the notations of 27.1.1. Let $\mathcal{U}(G)$ be the set of normal unipotent algebraic subgroups of G and denote by $R_u(G)$ the subgroup of G generated by $\mathcal{U}(G)$. We call $R_u(G)$ the *unipotent radical* of G. If $K \in \mathcal{U}(G)$, then $K \in \mathcal{R}(G)$, so $R_u(G) \subset R(G)$. Moreover, by 21.3.2, $R_u(G)$ is a closed connected subgroup of G.

Lemma. *The group $R_u(G)$ is the set $R(G)_u$ of unipotent elements of $R(G)$. It is the largest normal unipotent subgroup of G. It is normal in H and we have $R_u(G) = R_u(G^\circ)$.*

Proof. If $K \in \mathcal{U}(G)$, then $K \in \mathcal{R}(G)$ and $K \subset R(G)_u$. So $R_u(G) \subset R(G)_u$. Since $R(G)$ is normal in H and $R(G)_u$ is a closed normal subgroup of $R(G)$ (26.4.2), it is clear that $R(G)_u$ is a normal algebraic subgroup of H. In particular, $R(G)_u \subset R_u(G)$, and hence $R(G)_u = R_u(G)$. Finally, $R(G) = R(G^\circ)$ by 27.1.1, so $R_u(G) = R(G)_u = R(G^\circ)_u = R_u(G^\circ)$. $\square$

27.1.3 Proposition. *Let $\mathfrak{g}$ be the Lie algebra of G, $\mathfrak{r}$ the radical of $\mathfrak{g}$ and $\mathfrak{t}$ the nilpotent radical of $\mathfrak{g}$.*
(i) *The Lie algebra of $R(G)$ is $\mathfrak{r}$.*
(ii) *We have $\mathfrak{t} \subset \mathcal{L}(R_u(G))$.*

Proof. In view of 27.1.1 and 27.1.2, we may assume that G is connected.

(i) By 24.4.7 and 24.5.13, $\mathcal{L}(R(G))$ is a solvable ideal of $\mathfrak{g}$, so $\mathcal{L}(R(G)) \subset \mathfrak{r}$. On the other hand, $\mathfrak{r}$ is an algebraic Lie subalgebra of $\mathfrak{g}$ (24.7.4). Let H be the smallest closed subgroup of G such that $\mathcal{L}(H) = \mathfrak{r}$. Then by 24.4.7 and 24.5.13, H is a connected normal solvable algebraic subgroup of G, so $H \subset R(G)$. Hence $\mathfrak{r} = \mathcal{L}(H) \subset \mathcal{L}(R(G))$. Thus $\mathfrak{r} = \mathcal{L}(R(G))$.

(ii) Let $x \in \mathfrak{t}$. Then $\rho(x)$ is nilpotent for all finite-dimensional representation ρ of $\mathfrak{g}$. It follows that $\eta(x)$ is locally nilpotent where η is the representation of $\mathfrak{g}$ in $\mathbf{A}(G)$ defined in 23.6.2. Thus x is a nilpotent element of $\mathfrak{r}$. Since $\mathcal{L}(R_u(G))$ is the set of nilpotent elements of $\mathfrak{r}$ (26.4.5 and 27.1.2), we have $x \in \mathcal{L}(R_u(G))$. $\square$

27.1.4 Remarks. 1) In general, we have $\mathfrak{t} \neq \mathcal{L}(R_u(G))$. For example, if $G = \mathbf{G}_a$, then $R_u(G) = G$, $\mathcal{L}(R_u(G)) = \mathfrak{g}$ and $\mathfrak{t} = \{0\}$.

2) The group $(G, R(G))$ is connected (21.3.5) and its Lie algebra is equal to $[\mathfrak{g}, \mathfrak{r}] = \mathfrak{t}$ (20.3.6 and 24.5.11). So 27.1.3 implies that $(G, R(G)) \subset R_u(G)$.

27.2 Semisimple and reductive groups

27.2.1 Definition. *An algebraic group G is said to be* semisimple *if $R(G) = \{e\}$. It is said to be* reductive *if $R_u(G) = \{e\}$.*

27.2.2 Proposition. *Let G be an algebraic group and $\mathfrak{g}$ its Lie algebra.*
(i) *The group G is semisimple if and only if $\mathfrak{g}$ is semisimple.*
(ii) *The group G is reductive if and only if $\mathfrak{g}$ is reductive and all the elements of the centre of $\mathfrak{g}$ are semisimple.*

Proof. Let $\mathfrak{r}$ be the radical of $\mathfrak{g}$, and $\mathfrak{t}$ the nilpotent radical of $\mathfrak{g}$.
(i) Since $\mathcal{L}\big(R(G)\big) = \mathfrak{r}$ and $R(G)$ is connected, we have $R(G) = \{e\}$ if and only if $\mathfrak{r} = \{0\}$. So the result follows from the Levi-Malcev Theorem.
(ii) Suppose that G is reductive. Then by 27.1.3, $\mathfrak{t} = \{0\}$. So $\mathfrak{g}$ is reductive and $\mathfrak{r}$ is the centre of $\mathfrak{g}$. Thus $R(G)$ is the centre of G° (24.4.2). But $R_u(G) = R(G)_u$ (27.1.2), we deduce therefore from 26.4.5 that all the elements of $\mathfrak{r}$ are semisimple.
Conversely, suppose that $\mathfrak{g}$ is reductive and all the elements of $\mathfrak{r} = \mathfrak{z}(\mathfrak{g})$ are semisimple. Since $R_u(G) = R(G)_u$ is connected and $\mathcal{L}\big(R_u(G)\big)$ is the set of nilpotent elements of $\mathfrak{r}$ (26.4.5), $R_u(G) = \{e\}$. Thus G is reductive. $\square$

27.2.3 Remarks and examples. 1) Let G is a connected algebraic group. Recall from 21.3.6 and 24.5.12 that (G,G) is a connected algebraic subgroup of G whose Lie algebra is $[\mathfrak{g}, \mathfrak{g}]$. So if G is semisimple, then $G = (G,G)$ (20.1.5). Similarly, if G is reductive, then (G,G) is semisimple (20.5.5).
2) The product of two semisimple (resp. reductive) algebraic groups is semisimple (resp. reductive).
3) Recall from 20.2.2 that $\mathfrak{sl}_n(\Bbbk)$ is semisimple. So the group $\mathrm{SL}_n(\Bbbk)$ is semisimple (23.4.5 and 27.2.2).
4) The Lie algebra $\mathfrak{g}$ can be reductive while G is not reductive. This is the case for the additive group $G = \mathbf{G}_a$ (27.1.4).

27.2.4 Proposition. *Let G be a connected algebraic group. Then the following conditions are equivalent:*
(i) *G is semisimple.*
(ii) *The only connected normal abelian algebraic subgroup of G is $\{e\}$.*

Proof. The implication (i) $\Rightarrow$ (ii) is clear since such a subgroup is contained in $R(G) = \{e\}$. Conversely if $R(G) \neq \{e\}$, then the last non-trivial term of the derived series of $R(G)$ is a connected normal abelian algebraic subgroup of G (21.3.6). $\square$

27.2.5 Proposition. *Let G be a connected reductive algebraic group.*
(i) *Its radical $R(G)$ is a torus, and it is equal to the identity component of $Z(G)$.*
(ii) *The group $(G,G) \cap R(G)$ is finite.*

Proof. (i) Since $R(G)_u = R_u(G) = \{e\}$, we have $R(G) = R(G)_s$. As $R(G)$ is connected, it is a torus (22.5.7).

By 27.1.3, $\mathcal{L}(R(G))$ is the radical of $\mathfrak{g}$, which is equal to $\mathfrak{z}(\mathfrak{g})$ (20.5.4 and 27.2.2). Since $\mathcal{L}(Z(G)) = \mathfrak{z}(\mathfrak{g})$ (24.4.2) and $R(G)$ is connected, we have $R(G) = Z(G)^\circ$ (24.3.5).

(ii) Let $H = (G, G) \cap R(G)$. Then H° is a connected normal abelian algebraic subgroup of (G, G). By 27.2.4, $H^\circ = \{e\}$. So H is finite. $\square$

27.3 Representations

27.3.1 Let us start by a remark. Let G be an algebraic group, E a rational G-module and F a simple submodule of E. If $x \in F \setminus \{0\}$, then the submodule generated by x is clearly F. It follows that F is finite-dimensional.

27.3.2 Theorem. *Let G be an algebraic group, (V, ρ) a finite-dimensional rational representation of G and σ the restriction of ρ to G°. Then the following conditions are equivalent:*

(i) *ρ is semisimple.*
(ii) *σ is semisimple.*

Proof. (i) $\Rightarrow$ (ii) By 10.7.4, we may assume that ρ is simple. So V is finite-dimensional and it contains a simple G°-submodule W. Now G° is normal in G, so $\rho(\alpha)(W)$ is a simple G°-module for all $\alpha \in G$. Set:

$$S = \sum_{\alpha \in G} \rho(\alpha)(W).$$

Then S is a non-zero G-submodule of V. Hence $S = V$ and therefore σ is semisimple.

(ii) $\Rightarrow$ (i) Let W be a d-dimensional G-submodule of V. We shall show that $V = W \oplus W'$ for some G-submodule W' of V, and the result would follow.

Let $\mathfrak{g}$ be the Lie algebra of G. By 27.2.2, $d\sigma(\mathfrak{g}) = \mathfrak{g}'$ is reductive, and the elements of its centre are semisimple. So if (E, θ) is a finite-dimensional rational representation of $\sigma(G)$, then the representation $d\theta(\mathfrak{g}')$ is semisimple.

For an integer p, denote by $(\rho_p, \bigwedge^p V)$ (resp. $(\sigma_p, \bigwedge^p V)$) the representation of G (resp. G°) associated to ρ (resp. σ) as defined in 23.4.14. Then by the above discussion and 20.5.10, $d\sigma_p$ is semisimple, and so 24.3.3 implies that σ_p is semisimple.

Let $y \in \bigwedge^d W$ be a non-zero element. Then $\bigwedge^d W = \Bbbk y$ and for $\alpha \in G$, we have $\rho_d(\alpha)(y) = \theta(\alpha)y$ where $\theta(\alpha) = \det(\rho(\alpha)|_W)$ is a character of G. Let $S = \{x \in \bigwedge^d V; \sigma_d(\alpha)(x) = \theta(\alpha)(x)$ for all $\alpha \in G^\circ\}$ and T be the subspace of $\bigwedge^d V$ spanned by $\sigma_d(\alpha)(x) - \theta(\alpha)x$, $\alpha \in G^\circ$ and $x \in \bigwedge^d V$.

Applying 10.7.8 to θ, G° and G, we obtain the decomposition $\bigwedge^d V = S \oplus T$ of $\bigwedge^d V$ into G-submodules.

We claim that the restriction λ of ρ_d to S is a semisimple representation of G. If $\alpha \in G$, set $\lambda_1(\alpha) = \theta(\alpha)^{-1}\lambda(\alpha)$. Then (λ_1, S) is a representation of G

whose kernel contains G°. Thus $\lambda_1(\alpha)$ depends only on the class of α modulo G°. This induces a representation (λ_2, S) of the finite group G/G° which is semisimple by 10.7.6. It follows that λ_1, and hence λ, is semisimple. We have therefore proved our claim.

Since $\Bbbk y$ is a G-submodule of S, it follows that $S = \Bbbk y \oplus U'$ for some G-submodule U'. This implies, in turn, that $\bigwedge^d V = \Bbbk y \oplus U$ for some G-submodule U. It follows from 10.7.7 that $V = W \oplus W'$ where W' is the set of $x \in V$ such that $x \wedge z \in U$ for all $z \in \bigwedge^{d-1} W$.

To finish the proof, it suffices therefore to prove that W' is a G-submodule. Let $x \in W'$, $\alpha \in G$ and $u \in \bigwedge^{d-1} W$. Since $\rho(\alpha)(W) = W$, we have $\rho_{d-1}(\alpha)(\bigwedge^{d-1} W) = \bigwedge^{d-1} W$. So $u = \rho_{d-1}(\alpha)(u')$ for some $u' \in \bigwedge^{d-1} W$. We deduce therefore that $\rho(\alpha)(x) \wedge u = \rho(\alpha)(x \wedge u') \in U$ since U is a G-submodule. Hence W' is a G-submodule as required. $\square$

27.3.3 Theorem. *The following conditions are equivalent for an algebraic group G:*

(i) *G is reductive.*

(ii) *Any finite-dimensional rational representation of G is semisimple.*

(iii) *Any rational representation of G is semisimple.*

Proof. (i) $\Rightarrow$ (ii) Suppose that G is reductive. Then $\mathfrak{g} = \mathcal{L}(G)$ is reductive and all the elements of $\mathfrak{r}$ are semisimple (27.2.2). Let (V, ρ) be a finite-dimensional rational G-module. By 20.5.10, the representation $(V, d\rho)$ of $\mathfrak{g}$ is semisimple. Therefore, 24.3.3 implies that the restriction of ρ to G° is semisimple. So ρ is semisimple by 27.3.2.

(ii) $\Rightarrow$ (i) We may assume that $G \subset \mathrm{GL}(V)$ for some finite-dimensional vector space V. Suppose that G is not reductive, then $H = R_u(G) \neq \{\mathrm{id}_V\}$ and the group H acts non-trivially on V. So $V^H \neq V$ and $V^H \neq \{0\}$ (22.3.6). Now since H is normal in G, V^H is a G-submodule of V. If V is a semisimple G-module, then $V = V^H \oplus W$ for some non-zero G-submodule W. But again by 22.3.6, $W^H \neq \{0\}$ which implies that $V^H \cap W \neq \{0\}$. Contradiction.

(ii) $\Rightarrow$ (iii) If V is a rational G-module, then it is the sum of finite-dimensional rational G-submodules. So if (ii) is verified, then V is the sum of simple G-modules.

(iii) $\Rightarrow$ (ii) This is obvious. $\square$

27.3.4 Let G be an algebraic group and let us consider the action ρ of G on $\mathbf{A}(G)$ via right translations (22.1.4).

Assume that G is reductive. The subspace $\mathbf{A}(G)^G$ is the set of constant functions and we may identify it with $\Bbbk$. By 10.7.9, the subspace V of $\mathbf{A}(G)$ spanned by $\rho_\alpha f - f$, $\alpha \in G$ and $f \in \mathbf{A}(G)$, is the unique submodule of $\mathbf{A}(G)$ such that $\mathbf{A}(G) = \Bbbk \oplus V$. Let us denote by I_G or I the Reynolds operator associated to $\mathbf{A}(G)$. Note that $\mathrm{I} \circ \rho_\alpha = \mathrm{I}$ for all $\alpha \in G$.

It is clear that $\lambda_\alpha(\Bbbk) = \Bbbk$ if $\alpha \in G$, where λ_α denotes the left translation of functions (22.1.4). Furthermore, we see immediately that $\lambda_\alpha(V) = V$. Thus $\mathrm{I} \circ \lambda_\alpha = \mathrm{I}$.

27.3.5 In the rest of this section, for an algebraic group G, we denote by $\widehat{G}$ the set of isomorphism classes of finite-dimensional simple rational G-modules. If $\omega \in \widehat{G}$, then M_ω denotes a representative of ω, and M_ω^* denotes the dual G-module (see 10.6.5). If E is a rational G-module, then we set $E_{(\omega)}$ to be the sum of submodules of E_ω which are isomorphic to M_ω. The G-submodule $E_{(\omega)}$ is called the *isotypic component of type* ω of E.

If E and F are finite-dimensional rational G-modules, we shall endow $\mathrm{Hom}(E, F)$ and $E \otimes_k F$ with the G-module structures as defined in 10.6.5.

27.3.6 Theorem. *Let G be a reductive algebraic group and E, F rational G-modules.*

(i) *If $\omega \in \widehat{G}$, then any simple submodule of $E_{(\omega)}$ is isomorphic to M_ω.*

(ii) *We have:*
$$E = \bigoplus_{\omega \in \widehat{G}} E_{(\omega)}.$$

(iii) *If $f \in \mathrm{Hom}_G(E, F)$ and $\omega \in \widehat{G}$, then $f(E_{(\omega)}) \subset F_{(\omega)}$.*

(iv) *For all $\omega \in \widehat{G}$, the map*
$$\theta_\omega : \mathrm{Hom}_G(M_\omega, E) \otimes_k M_\omega \to E_{(\omega)} \ , \ f \otimes v \mapsto f(v) \ ,$$

is an isomorphism of G-modules.

Proof. (i) This is clear by 10.7.5 (ii).

(ii) Since E is a semisimple G-module (27.3.3), it is equal to the sum of the $E_{(\omega)}$'s (27.3.1). Let $\theta \in \widehat{G}$, and E' the sum of the $E_{(\omega)}$'s where $\omega \in \widehat{G} \setminus \{\theta\}$. If $E' \cap E_{(\theta)}$ is non-zero, then it contains a simple submodule E''. But part (i) and 10.7.5 imply that E'' is isomorphic to M_θ and M_ω for some $\omega \neq \theta$. This is absurd (10.7.2). So the sum of the $E_{(\omega)}$'s is direct.

(iii) In view of 10.7.2, this is obvious.

(iv) By (iii), θ_ω is well-defined, and a simple verification shows that θ_ω is a homomorphism of G-modules. It is also clear that θ_ω is surjective.

Let $(V_i)_{i \in I}$ be a family of submodules of E verifying $V_i \simeq M_\omega$ for all $i \in I$, and:
$$E_{(\omega)} = \bigoplus_{i \in I} V_i.$$

Denote by $(p_i)_{i \in I}$ the family of projectors associated to this direct sum decomposition.

If $f \in \mathrm{Hom}_G(M_\omega, E)$, then the image of f is contained in $E_{(\omega)}$ and $f = \sum_{i \in I} p_i \circ f$. It follows easily that:
$$\mathrm{Hom}_G(M_\omega, E) = \bigoplus_{i \in I} \mathrm{Hom}_G(M_\omega, V_i).$$

Let us fix $u_i \in \mathrm{Hom}_G(M_\omega, V_i) \setminus \{0\}$. By 10.7.2, u_i is bijective and all the elements of $\mathrm{Hom}_G(M_\omega, V_i)$ are proportional to u_i. Let $f_1, \ldots, f_n \in \mathrm{Hom}_G(M_\omega, E)$ and $x_1, \ldots, x_n \in M_\omega$ be such that

$$\theta_\omega(f_1 \otimes x_1 + \cdots + f_n \otimes x_n) = 0.$$

For all $i \in I$, we have $p_i(f_1(x_1) + \cdots + f_n(x_n)) = 0$. So we may assume that $f_1, \ldots, f_n \in \mathrm{Hom}_G(M_\omega, W_i)$, $f_k = \lambda_k u_i$, with $\lambda_i \in \Bbbk$ for $1 \leqslant k \leqslant n$. Thus $\lambda_1 x_1 + \cdots + \lambda_n x_n = 0$ since u_i is injective. We deduce therefore that $f_1 \otimes x_1 + \cdots + f_n \otimes x_n = 0$. Hence θ_ω is injective. $\square$

27.3.7 Remarks. Let us conserve the hypotheses of 27.3.6.

1) In the proof of 27.3.6 (iv), the cardinality of the set I is called the *multiplicity* of M_ω in E, and we shall denote it by $\mathrm{mult}_\omega(E)$. The module $\mathrm{Hom}_G(M_\omega, E)$ has dimension $\mathrm{mult}_\omega(E)$, and we have the following isomorphism of G-modules:

$$E \simeq \bigoplus_{\omega \in \widehat{G}} \mathrm{mult}_\omega(E) M_\omega.$$

2) If L is a submodule of E, then $L_{(\omega)} = L \cap E_{(\omega)}$ for all $\omega \in \widehat{G}$.

27.3.8 Proposition. *Let G be a reductive algebraic group acting rationally on a commutative $\Bbbk$-algebra A by automorphisms, and let $B = A^G$.*

(i) *If $\mathfrak{a}$ is a G-stable ideal of A, then the G-modules $B/(B \cap \mathfrak{a})$ and $(A/\mathfrak{a})^G$ are isomorphic.*

(ii) *Let $(\mathfrak{a}_i)_{i \in I}$ be a family of G-stable ideals of A. Then:*

$$B \cap \left(\sum_{i \in I} \mathfrak{a}_i \right) = \sum_{i \in I} (B \cap \mathfrak{a}_i).$$

Proof. (i) By 27.3.6, $\mathfrak{a}$ is the direct sum of the $\mathfrak{a}_{(\omega)}$'s, for $\omega \in \widehat{G}$. But $\mathfrak{a}_{(\omega)} = \mathfrak{a} \cap A_{(\omega)}$ by 27.3.7. We deduce therefore that:

$$A/\mathfrak{a} = \bigoplus_{\omega \in \widehat{G}} (A_{(\omega)}/\mathfrak{a}_{(\omega)}).$$

It follows that $(A/\mathfrak{a})_{(\omega)} = A_{(\omega)}/\mathfrak{a}_{(\omega)}$, and we obtain the result by taking ω to be the isomorphism class of the trivial G-module.

(ii) We have:

$$\sum_{i \in I} \mathfrak{a}_i = \sum_{i \in I} \bigoplus_{\omega \in \widehat{G}} (\mathfrak{a}_i)_{(\omega)} = \bigoplus_{\omega \in \widehat{G}} \sum_{i \in I} (\mathfrak{a}_i)_{(\omega)}.$$

So $\left(\sum_{i \in I} \mathfrak{a}_i \right)_{(\omega)} = \sum_{i \in I} (\mathfrak{a}_i)_{(\omega)}$ for all $\omega \in \widehat{G}$, and we conclude as in (i). $\square$

27.3.9 Let E, F be G-modules and consider the action of $G \times G$ on $E \otimes_\Bbbk F$ defined in 23.4.12, and the action of $G \times G$ on $\mathbf{A}(G)$ defined as follows: for $\alpha, \beta, \gamma \in G$ and $f \in \mathbf{A}(G)$,

$$((\alpha, \beta).f)(\gamma) = f(\alpha^{-1}\gamma\beta).$$

Theorem. *Let G be a reductive algebraic group and*

$$\phi : \bigoplus_{\omega \in \widehat{G}} \left(M_\omega^* \otimes_\Bbbk M_\omega \right) \to \mathbf{A}(G)$$

the map sending $f \otimes x \in M_\omega^* \otimes_\Bbbk M_\omega$ to the function $\alpha \mapsto f(\alpha.x)$.

(i) ϕ is an isomorphism of $(G \times G)$-modules.

(ii) We have $\mathrm{mult}_\omega \big(\mathbf{A}(G) \big) = \dim M_\omega$ for all $\omega \in \widehat{G}$.

Proof. Let $x \in M_\omega$, $f \in M_\omega^*$, and $\alpha, \beta, \gamma \in G$. Then:

$$\big((\alpha, \beta).\phi(f \otimes x) \big)(\gamma) = \phi(f \otimes x)(\alpha^{-1}\gamma\beta) = f(\alpha^{-1}\gamma\beta.x)$$
$$= (\alpha.f)(\gamma\beta.x) = \phi\big((\alpha.f) \otimes (\beta.x) \big)(\gamma).$$

So ϕ is a homomorphism of $(G \times G)$-modules.

1) Let us fix $f \in M_\omega \setminus \{0\}$. Then $(e, \alpha).(f \otimes x) = f \otimes (\alpha.x)$ for $x \in M_\omega$. Thus $f \otimes_\Bbbk M_\omega$ is a simple G-module isomorphic to M_ω. It follows immediately that since

$$\rho_\alpha \big(\phi(f \otimes x) \big) = \phi \big(f \otimes (\alpha.x) \big),$$

ϕ induces a homomorphism of the simple G-module $f \otimes_\Bbbk M_\omega$ into the G-module $\mathbf{A}(G)$. It is clear that $\phi(f \otimes_\Bbbk M_\omega)$ is non-zero, so it is isomorphic to M_ω and contained in $\mathbf{A}(G)_{(\omega)}$. Hence $\mathbf{A}(G)_{(\omega)} \neq \{0\}$ for all $\omega \in \widehat{G}$.

2) Now, let V be a simple submodule of $\mathbf{A}(G)$ which is isomorphic to M_ω and $x \mapsto u_x$ a G-module isomorphism from M_ω onto V. Define $g \in M_\omega^*$ by $g(x) = u_x(e)$, $x \in M_\omega$. Then for $\alpha \in G$ and $x \in M_\omega$, we have:

$$\phi(g \otimes x)(\alpha) = g(\alpha.x) = u_{\alpha.x}(e) = (\alpha.u_x)(e) = u_x(\alpha).$$

We deduce that $\phi(g \otimes_\Bbbk M_\omega) = V$. Since $\phi(M_\omega^* \otimes_\Bbbk M_\omega) \subset \mathbf{A}(G)_{(\omega)}$ by point 1, it follows that $\phi(M_\omega^* \otimes_\Bbbk M_\omega) = \mathbf{A}(G)_{(\omega)}$. So ϕ is surjective by 27.3.6 (ii).

3) Finally, let $\omega \in \widehat{G}$ and $M = M_\omega$. If V is a submodule of M^*, its orthogonal in M is a submodule of M. So M^* is a simple G-module. We claim that $M^* \otimes_\Bbbk M$ is a simple $(G \times G)$-module.

Denote by σ and σ' the representations of G on M and M^*, and θ the associated representation of $G \times G$ on $M^* \otimes_\Bbbk M$. By 10.8.11, the subalgebra of $\mathrm{End}(M)$ (resp. $\mathrm{End}(M^*)$) generated by $\sigma(G)$ (resp. $\sigma'(G)$) is equal to $\mathrm{End}(M)$ (resp. $\mathrm{End}(M^*)$). Since $\mathrm{End}(M^* \otimes_\Bbbk M)$ is isomorphic to $\mathrm{End}(M^*) \otimes_\Bbbk \mathrm{End}(M)$, we deduce that the subalgebra of $\mathrm{End}(M^* \otimes_\Bbbk M)$ generated by $\theta(G \times G)$ is equal to $\mathrm{End}(M^* \otimes_\Bbbk M)$. So we have proved our claim.

Since $\mathbf{A}(G)_{(\omega)} = \phi(M_\omega^* \otimes_\Bbbk M_\omega) \neq \{0\}$, our claim and point 2 imply that ϕ is bijective. So we have proved (i), and (ii) follows immediately. $\square$

27.4 Finiteness properties

27.4.1 In this section, G denotes a reductive algebraic group.

Let A be a commutative $\Bbbk$-algebra on which G acts rationally by automorphisms. Denote by p the Reynolds operator associated to A.

Lemma. For $u \in A^G$ and $v \in A$, we have $p(uv) = up(v)$. Consequently, the unique complementary submodule $\ker(p)$ of A^G in A, is an A^G-module.

Proof. There exist, by 10.7.9, $\alpha_1, \ldots, \alpha_n \in G$ and $w_1, \ldots, w_n \in A$ such that:

$$v = p(v) + \sum_{i=1}^{n} (\alpha_i.w_i - w_i).$$

Since $\alpha_i.(uw_i) = u(\alpha_i.w_i)$, we have:

$$uv = up(v) + \sum_{i=1}^{n} \left(\alpha_i.(uw_i) - uw_i\right).$$

On the other hand, $up(v) \in A^G$, so the result follows from 10.7.9.   □

27.4.2 Let V be a finite-dimensional rational G-module and A the algebra of polynomial functions on V. Endow A with the G-module structure defined in 22.1.1.

Proposition. *The algebra A^G is finitely generated.*

Proof. For $n \in \mathbb{N}$, let A_n be the subspace of A consisting of homogeneous polynomial functions of degree n. Then A_n is G-stable and so:

$$A^G = \bigoplus_{n \in \mathbb{N}} A_n^G.$$

Let I be the ideal of A generated by A_n^G, $n \geqslant 1$. Since A is Noetherian, I is finitely generated. Let $f_1, \ldots, f_r \in A^G$ be a system of generators of I that we may assume to be homogeneous. Now let $f \in A^G$ be homogeneous of degree non-zero, then:

$$f = f_1 g_1 + \cdots + f_r g_r$$

where $g_1, \ldots, g_r$ are homogeneous. Let p denote the Reynolds operator of A. It follows from 27.4.1 that:

$$f = p(f) = f_1 p(g_1) + \cdots + f_r p(g_r).$$

But $h_i = p(g_i) \in A^G$ for $1 \leqslant i \leqslant r$, so if $h_i \in \Bbbk[f_1, \ldots, f_r]$ for $1 \leqslant i \leqslant r$, then $f \in \Bbbk[f_1, \ldots, f_r]$. Since $\deg(h_i) < \deg f$, we obtain by induction (on the degree of f) that $A^G = \Bbbk[f_1, \ldots, f_r]$.   □

27.4.3 Theorem. (Hilbert-Nagata Theorem) *Let G be a reductive algebraic group and A a finitely generated commutative $\Bbbk$-algebra on which G acts rationally by automorphisms.*

(i) *The algebra $B = A^G$ is finitely generated, and is a direct summand of the B-module A.*

(ii) *If $\mathfrak{b}$ is an ideal of B, then $B \cap (A\mathfrak{b}) = \mathfrak{b}$.*

Proof. The fact that B is a direct summand of the B-module A follows from 27.4.1. Let us denote by p the Reynolds operator associated to A.

Let $\mathfrak{b}$ be an ideal of B. Then $\mathfrak{b} \subset B \cap (A\mathfrak{b})$. Conversely, if $f \in B \cap (A\mathfrak{b})$, then there exist $f_1, \ldots, f_n \in \mathfrak{b}$ and $g_1, \ldots, g_n \in A$ such that $f = g_1 f_1 + \cdots + g_n f_n$. By 27.4.1, we have:

$$f = p(f) = f_1 p(g_1) + \cdots + f_n p(g_n).$$

Thus $f \in \mathfrak{b}$.

Now let $x_1, \ldots, x_n$ be a system of generators of A. There exists a finite-dimensional submodule W of A containing $x_1, \ldots, x_n$. Denote by V the dual of W. The injection $V^* = W \to A$ extends to a surjective G-module homomorphism u from $\mathbf{A}(V)$ onto A. Since $u \circ p_{\mathbf{A}(V)} = p \circ u$, u induces a surjective homomorphism $\mathbf{A}(V)^G \to A^G$. Hence the result follows from 27.4.2. $\square$

27.4.4 Let us conserve the notations and hypotheses of 27.4.3. Let V be a rational G-module. If $u \in \mathrm{Hom}_G(V, A)$ and $f \in A^G$, then clearly the map $x \mapsto fu(x)$ belongs to $\mathrm{Hom}_G(V, A)$. Thus $\mathrm{Hom}_G(V, A)$ has a natural structure of A^G-module.

Proposition. *Let V be a finite-dimensional rational G-module. Then $\mathrm{Hom}_G(V, A)$ is a finitely generated A^G-module.*

Proof. We have:

$$\begin{aligned}
\left(A \otimes \mathbf{A}(V)\right)^G &= \Big(\bigoplus_{n \geqslant 0} A \otimes \mathbf{A}(V)_n \Big)^G = \bigoplus_{n \geqslant 0} \left(A \otimes \mathbf{A}(V)_n\right)^G \\
&= A^G \oplus (A \otimes V^*)^G \oplus \bigoplus_{n \geqslant 2} \left(A \otimes \mathbf{A}(V)_n\right)^G \\
&= A^G \oplus \mathrm{Hom}_G(V, A) \oplus \bigoplus_{n \geqslant 2} \left(A \otimes \mathbf{A}(V)_n\right)^G.
\end{aligned}$$

By 27.4.3, $\left(A \otimes \mathbf{A}(V)\right)^G$ is a finitely generated commutative A^G-algebra. In particular, the ideal $\bigoplus_{n \geqslant 1} \left(A \otimes \mathbf{A}(V)_n\right)^G$ of this A^G-algebra is finitely generated. Since $\mathrm{Hom}_G(V, A) = (A \otimes \mathbf{A}(V)_1)^G$, the result follows. $\square$

27.4.5 Corollary. *Let us conserve the notations and hypotheses of 27.4.3. For all $\omega \in \widehat{G}$, the subspace $A_{(\omega)}$ is a finitely generated A^G-module.*

Proof. This is clear by 27.3.6 (iv) and 27.4.4. $\square$

27.5 Algebraic quotients

27.5.1 In this section, G is a reductive algebraic group and X an affine G-variety. Recall from 22.1.1 that G acts naturally on $\mathbf{A}(X)$.

By 27.4.3, the reduced algebra $\mathbf{A}(X)^G$ is finitely generated. So 11.6.1 implies that $\mathbf{A}(X)^G$ is the algebra of regular functions of an affine variety that we shall denote by $X /\!\!/ G$, instead of $\mathrm{Spm}(\mathbf{A}(X)^G)$. Let $\pi : X \to X /\!\!/ G$ be the morphism induced by the injection $\mathbf{A}(X)^G \to \mathbf{A}(X)$. The pair $(X /\!\!/ G, \pi)$, or $X /\!\!/ G$, is called the *algebraic quotient* of X by G.

Proposition. (i) *The map* π *is surjective and constant on* G-*orbits.*

(ii) *Let* Y *be an affine* G-*variety and* $u : X \to Y$ *be a morphism which is constant on the* G-*orbits of* X. *Then there exists a unique morphism* $v : X/\!\!/G \to Y$ *such that* $u = v \circ \pi$.

(iii) *If* X *is normal, then* $X/\!\!/G$ *is also normal.*

Proof. (i) For $\alpha \in G$, $x \in X$ et $f \in \mathbf{A}(X)^G = \mathbf{A}(X/\!\!/G)$, we have:

$$f(x) = \mathbf{A}(\pi)(f)(x) = f\big(\pi(x)\big) = f(\alpha.x) = f\big(\pi(\alpha.x)\big).$$

So $\pi(x) = \pi(\alpha.x)$.

Let $\xi \in X/\!\!/G$ and $\mathfrak{a}$ its ideal in $\mathbf{A}(X)^G$. Then $\mathfrak{a}\mathbf{A}(X) \cap \mathbf{A}(X)^G = \mathfrak{a}$ by 27.4.3. It follows that $\mathfrak{a}\mathbf{A}(X) \neq \mathbf{A}(X)$. Let $\mathfrak{b}$ be a maximal ideal of $\mathbf{A}(X)$ containing $\mathfrak{a}\mathbf{A}(X)$ and $x \in X$ be the point defined by $\mathfrak{b}$. Then we have $\xi = \pi(x)$.

(ii) The assumptions imply that $\mathbf{A}(u)\big(\mathbf{A}(Y)\big) \subset \mathbf{A}(X)^G$. Thus $\mathbf{A}(u) = \mathbf{A}(\pi) \circ \theta$ where θ is an algebra homomorphism from $\mathbf{A}(Y)$ to $\mathbf{A}(X)^G$. The morphism $v = \mathrm{Spm}(\theta)$ verifies $u = v \circ \pi$. Clearly, v is unique with this property.

(iii) If $f \in \mathrm{Fract}\big(\mathbf{A}(X)^G\big)$ is integral over $\mathbf{A}(X)^G$, then f is integral over $\mathbf{A}(X)$, so f belongs to $\mathbf{A}(X)$ since X is normal. Hence $f \in \mathbf{A}(X)^G$. $\square$

27.5.2 Proposition. *Let us conserve the hypotheses and notations of 27.5.1.*

(i) *If* Y *is a closed* G-*stable subset of* X, *then* $\pi(Y)$ *is a closed subset of* $X/\!\!/G$. *Furthermore,* $\big(\pi(Y), \pi|_Y\big)$ *is the algebraic quotient of* Y *by* G.

(ii) *Let* $(Y_i)_{i \in I}$ *be a family of closed* G-*stable subsets of* X. *Then:*

$$\pi\bigg(\bigcap_{i \in I} Y_i\bigg) = \bigcap_{i \in I} \pi(Y_i).$$

Proof. (i) Let $j : Y \to X$ be the canonical injection and $(Y/\!\!/G, \omega)$ the algebraic quotient of Y by G. The map $\mathbf{A}(j)$ is the surjection $f \mapsto f|_Y$. Let us denote by $p_{\mathbf{A}(X)}, p_{\mathbf{A}(Y)}$ the corresponding Reynolds operators. Since $\mathbf{A}(j) \circ p_{\mathbf{A}(X)} = p_{\mathbf{A}(Y)} \circ \mathbf{A}(j)$, $\mathbf{A}(j)$ induces a surjection $\theta : \mathbf{A}(X)^G \to \mathbf{A}(Y)^G$.

The map $\mathbf{A}(\omega)$ is the canonical injection $\mathbf{A}(Y)^G \to \mathbf{A}(Y)$ and we have $\mathbf{A}(\omega) \circ \theta = \mathbf{A}(j) \circ \mathbf{A}(\pi)$, so $\mathrm{Spm}(\theta) \circ \omega = \pi \circ j$. By 11.6.8, $\mathrm{Spm}(\theta) = i \circ v$ where v is an isomorphism of $Y/\!\!/G$ onto a closed subvariety Z of $X/\!\!/G$, and $i : Z \to X/\!\!/G$ is the canonical injection. So (i) follows.

(ii) If $\mathfrak{a}$ is an ideal of $\mathbf{A}(X)$ (resp. $\mathbf{A}(X)^G$), denote by $\mathcal{V}(\mathfrak{a})$ (resp. $\mathcal{V}'(\mathfrak{a})$) the variety of zeros of $\mathfrak{a}$ in X (resp. $X/\!\!/G$). We have $\bigcap_{i \in I} Y_i = \mathcal{V}(\sum_{i \in I} \mathfrak{a}_i)$. So (i) and 27.3.8 imply that:

$$\pi\bigg(\bigcap_{i \in I} Y_i\bigg) = \overline{\pi\bigg(\bigcap_{i \in I} Y_i\bigg)} = \mathcal{V}'\bigg(\mathbf{A}(X)^G \cap \big(\sum_{i \in I} \mathfrak{a}_i\big)\bigg) = \mathcal{V}'\bigg(\sum_{i \in I} (\mathbf{A}(X)^G \cap \mathfrak{a}_i)\bigg)$$
$$= \bigcap_{i \in I} \mathcal{V}'(\mathbf{A}(X)^G \cap \mathfrak{a}_i) = \bigcap_{i \in I} \pi(Y_i) = \bigcap_{i \in I} \pi(Y_i).$$

$\square$

27.5.3 Proposition. *Let us conserve the hypotheses and notations of* 27.5.1. *If* $x \in X$, *then* $\pi^{-1}(\pi(x))$ *contains a unique closed G-orbit* $\mathcal{O}$ *and we have:*

$$\pi^{-1}(\pi(x)) = \{y \in X; \mathcal{O} \subset \overline{G.y}\}.$$

Proof. The set $Y = \pi^{-1}(\pi(x))$ is closed and G-stable, so it contains a closed orbit by 21.4.5. Suppose that $\mathcal{O}_1$ and $\mathcal{O}_2$ are two distinct closed orbits contained in Y. Then 27.5.2 (ii) implies that $\pi(\mathcal{O}_1) \cap \pi(\mathcal{O}_2)$ is empty, which is absurd. So Y contains a unique closed G-orbit $\mathcal{O}$.

If $y \in Y$, then the closed G-stable subset $\overline{G.y}$ of Y contains a closed orbit (21.4.5), so $\mathcal{O} \subset \overline{G.y}$. Conversely, as π is constant on the closures of orbits, if $\mathcal{O} \subset \overline{G.y}$, then $\pi(x) = \pi(y)$. $\square$

27.5.4 Remark. In view of 27.5.3, we can interpret $X/\!/G$ as the set of closed G-orbits of X.

27.6 Characters

27.6.1 Definition. *Let G be an algebraic group.*

(i) *A function $f \in \mathbf{A}(G)$ is said to be* central *if $f(\alpha\beta\alpha^{-1}) = f(\beta)$ for all $\alpha, \beta \in G$. We shall denote by $C(G)$ the subalgebra of $\mathbf{A}(G)$ consisting of central functions.*

(ii) *Let (V, π) be a finite-dimensional rational G-module. The* character *of V is defined to be the map $\chi_V : G \to \Bbbk$, $\alpha \mapsto \operatorname{tr} \pi(\alpha)$.*

27.6.2 If V, W are finite-dimensional rational G-modules, then:

$$\chi_{V \oplus W} = \chi_V + \chi_W \ , \ \chi_{V \otimes W} = \chi_V \chi_W.$$

Moreover, it is clear that $\chi_V \in C(G)$ and that $\chi_V = \chi_W$ if V and W are isomorphic G-modules. If $\omega \in \widehat{G}$, then we denote χ_{M_ω} simply by χ_ω.

Proposition. *Let G be a reductive algebraic group. The space $C(G)$ is spanned by the χ_ω's, with $\omega \in \widehat{G}$.*

Proof. Fix $\omega \in \widehat{G}$ and set $M = M_\omega$. Let $(e_1, \ldots, e_n)$ be a basis of M, $(e_1^*, \ldots, e_n^*)$ the dual basis and ϕ as defined in 27.3.9. If $\alpha \in G$, then:

$$\chi_\omega(\alpha) = \sum_{i=1}^n e_i^*(\alpha.e_i) = \sum_{i=1}^n \phi(e_i^* \otimes e_i)(\alpha).$$

Thus $\chi_\omega \in \phi(M^* \otimes_\Bbbk M)$. By 27.3.9, to obtain the result, it suffices to prove that any central function contained in $\phi(M^* \otimes_\Bbbk M)$ is proportional to χ_ω.

Let $\theta : M^* \otimes_\Bbbk M \to \operatorname{End}_\Bbbk(M)$ be the map sending $f \otimes x$ to the endomorphism $y \mapsto f(y)x$. It is well-known that θ is a linear isomorphism and we verify easily that θ is a G-module homomorphism. Moreover, for $\alpha, \beta \in G$, $f \in M^*$ and $x \in M$, we have:

$$\phi(\alpha.(f \otimes x))(\beta) = f(\alpha^{-1}\beta\alpha x) = \phi(f \otimes x)(\alpha^{-1}\beta\alpha).$$

We deduce therefore that if $\xi \in M^* \otimes_k M$, then the following conditions are equivalent:

(i) $\phi(\xi) \in C(G)$.

(ii) $\alpha.\xi = \xi$ for all $\alpha \in G$.

(iii) $\alpha.\theta(\xi) = \theta(\xi)$ for all $\alpha \in G$.

(iv) $\alpha.\theta(\xi)(\alpha^{-1}.x) = \theta(\xi)(x)$ for all $(\alpha, x) \in G \times M$.

(v) $\theta(\xi) \in \mathrm{End}_G(M) = k\,\mathrm{id}_M$ (10.7.2).

But, we have:
$$\mathrm{id}_M = \theta(e_1^* \otimes e_1 + \cdots + e_n^* \otimes e_n).$$

Since θ is bijective, the result follows from equivalence of conditions (i) and (v). $\square$

27.6.3 Let G be a reductive algebraic group, I the Reynolds operator associated to $\mathbf{A}(G)$ and (V, σ) a finite-dimensional rational G-module.

Let $(e_1, \ldots, e_n)$ be a basis of V and $(e_1^*, \ldots, e_n^*)$ the dual basis. Set
$$\alpha.e_j = \sum_{i=1}^n \sigma_{ij}(\alpha)e_i$$

for $\alpha \in G$ and $1 \leqslant j \leqslant n$. With ϕ in the notations of 27.3.9, we have $\sigma_{ij} = \phi(e_i^* \otimes e_j)$.

If $\alpha, \beta \in G$, then:
$$
\begin{aligned}
(\alpha\beta).e_j &= \sum_{l=1}^n \sigma_{lj}(\alpha\beta)e_l = \sum_{i=1}^n \sigma_{ij}(\beta)\Big(\sum_{l=1}^n \sigma_{li}(\alpha)e_l \Big) \\
&= \sum_{l=1}^n \Big(\sum_{i=1}^n \sigma_{li}(\alpha)\sigma_{ij}(\beta) \Big)e_l.
\end{aligned}
$$

Hence:

(1) $$\rho_\beta \sigma_{lj} = \sum_{i=1}^n \sigma_{ij}(\beta)\sigma_{li} \ , \ \lambda_{\alpha^{-1}}\sigma_{lj} = \sum_{i=1}^n \sigma_{li}(\alpha)\sigma_{ij}.$$

Since $\mathrm{I} \circ \rho_\beta = \mathrm{I} \circ \lambda_{\alpha^{-1}} = \mathrm{I}$ (27.3.4), we have therefore

(2) $$\mathrm{I}(\sigma_{lj}) = \sum_{i=1}^n \sigma_{ij}(\beta)\,\mathrm{I}(\sigma_{li}) = \sum_{i=1}^n \sigma_{li}(\alpha)\,\mathrm{I}(\sigma_{ij})$$

for all $\alpha, \beta \in G$. We deduce that:

(3) $$\mathrm{I}(\sigma_{lj}) = \sum_{i=1}^n \mathrm{I}(\sigma_{li})\,\mathrm{I}(\sigma_{ij}).$$

Define $q \in \mathrm{End}(V)$ as follows: for $1 \leqslant j \leqslant n$:
$$q(e_j) = \sum_{i=1}^n \mathrm{I}(\sigma_{ij})e_i.$$

If $x \in V$ and $f \in V^*$, then:

$$(4) \qquad f\big(q(x)\big) = \mathrm{I}\big(\phi(f \otimes x)\big).$$

Lemma. *The map q is the Reynolds operator p_V associated to V.*

Proof. The equalities in (2) imply that $q \circ \sigma(\alpha) = \sigma(\alpha) \circ q = q$ for all $\alpha \in G$ and (3) implies that $q^2 = q$. Finally if $x \in V^G$, for $f \in V^*$, we have $\phi(f \otimes x)(\alpha) = f(\alpha.x) = f(x)$, so $\mathrm{I}\big(\phi(f \otimes x)\big) = f(x)$. By (4), we obtain that $f\big(q(x)\big) = f(x)$ for all $f \in V^*$, hence $q(x) = x$. Now the result follows from 10.7.9. $\square$

27.6.4 Let us define an involution $f \mapsto f^*$ in $\mathbf{A}(G)$ and a bilinear form $(f, g) \mapsto \langle f, g \rangle$ on $\mathbf{A}(G)$ as follows:

$$f^*(\alpha) = f(\alpha^{-1}) \, , \; \langle f, g \rangle = \mathrm{I}(fg^*).$$

If $h \in \mathbf{A}(G)$ and $\alpha \in G$, then $(\rho_\alpha h - h)^* = \lambda_\alpha h^* - h^*$. By 10.7.9 and 27.3.4, we deduce that $\mathrm{I}(f) = \mathrm{I}(f^*)$ for $f \in \mathbf{A}(G)$. The form $\langle . \, , . \rangle$ is therefore symmetric. Let us endow $\mathbf{A}(G)$ with this bilinear form in the rest of this section.

27.6.5 Let (V, σ) and (W, τ) be non-isomorphic simple rational G-modules of dimension m and n, $\mathcal{E}$ (resp. $\mathcal{F}$) a basis of V (resp. W). For $\alpha \in G$, denote by $S(\alpha) = [\sigma_{ij}(\alpha)]$ and $T(\alpha) = [\tau_{ij}(\alpha)]$ the matrices of $\sigma(\alpha)$ and $\tau(\alpha)$ with respect to the basis $\mathcal{E}$ et $\mathcal{F}$.

Lemma. (i) *For $1 \leqslant i, j \leqslant m$ and $1 \leqslant k, l \leqslant n$, we have $\langle \sigma_{ij}, \tau_{kl} \rangle = 0$.*
(ii) *For $1 \leqslant i, j, k, l \leqslant m$, we have $m \langle \sigma_{ij}, \sigma_{kl} \rangle = \delta_{il} \delta_{jk}$, where δ denotes the Kronecker symbol.*

Proof. (i) For $A \in \mathrm{M}_{m,n}(\Bbbk)$, set $A(\alpha) = S(\alpha) A T(\alpha^{-1}) = [\theta_{ij}(\alpha)]$ and let $B = [\mathrm{I}(\theta_{ij})] \in \mathrm{M}_{m,n}(\Bbbk)$. If $\alpha, \beta \in G$, then $A(\alpha\beta) = S(\alpha) A(\beta) T(\alpha^{-1})$. It follows as in 27.6.3 that $S(\alpha) B = BT(\alpha)$. So 10.7.2 implies that $B = 0$. But if $A = [a_{ij}]$, then:

$$\theta_{ij}(\alpha) = \sum_{k,l} \sigma_{ik}(\alpha) a_{kl} \tau_{lj}^*(\alpha).$$

Hence:

$$\mathrm{I}(\theta_{ij}) = \sum_{k,l} a_{kl} \langle \sigma_{ik}, \tau_{lj} \rangle.$$

So (i) follows by taking A to be the matrices of the canonical basis of $\mathrm{M}_{m,n}(\Bbbk)$.
(ii) If $\sigma = \tau$, then 10.7.2 implies that $B = \mu_A I_m$, with $\mu_A \in \Bbbk$. So $\mathrm{I}(\theta_{ij}) = 0$ if $i \neq j$, and the same argument as in (i) proves the result for $i \neq j$. On the other hand, we have $\mathrm{tr}\big(A(\alpha)\big) = \mathrm{tr}(A)$ for all $\alpha \in G$, so $\mathrm{tr}(B) = \mathrm{tr}(A)$. It follows that $\mu_A = m^{-1} \mathrm{tr}(A)$. So (ii) follows as in (i) by taking A to be the matrices of the canonical basis of $\mathrm{M}_m(\Bbbk)$. $\square$

27.6.6 Theorem. *Let G be a reductive algebraic group.*
(i) *The decomposition*

$$\mathbf{A}(G) = \bigoplus_{\omega \in \widehat{G}} (M_\omega^* \otimes M_\omega)$$

in 27.3.9 is orthogonal. Moreover, the restriction of $\langle\,.\,,\,.\,\rangle$ to each $M_\omega^* \otimes M_\omega$ is non-degenerate.
(ii) *The characters χ_ω, $\omega \in \widehat{G}$, form an orthonormal basis of $C(G)$.*
(iii) *If V is a finite-dimensional rational G-module and $\omega \in \widehat{G}$, then:*

$$\mathrm{mult}_\omega(V) = \langle \chi_\omega, \chi_V \rangle.$$

(iv) *Any finite-dimensional rational G-module is determined up to an isomorphism by its character.*

Proof. Let V and W be two simple rational G-modules of finite dimension, $(e_1, \ldots, e_m)$ (resp. $(\varepsilon_1, \ldots, \varepsilon_n)$) a basis of V (resp. W), and $(e_1^*, \ldots, e_m^*)$ (resp. $(\varepsilon_1^*, \ldots, \varepsilon_n^*)$) the dual basis. Let ϕ be as in 27.3.9.

Suppose that V and W are not isomorphic. Then $\langle \phi(e_i^* \otimes e_j), \phi(\varepsilon_k^* \otimes \varepsilon_l) \rangle = 0$ by 27.6.3 and 27.6.5. Hence we have (i).

Let $\omega, \theta \in \widehat{G}$, where $\omega \neq \theta$. We saw in the proof of 27.6.2 that $\chi_\omega \in \phi(M_\omega^* \otimes M_\omega)$. So (i) implies that $\langle \chi_\omega, \chi_\theta \rangle = 0$. Moreover:

$$\langle \chi_\omega, \chi_\omega \rangle = \sum_{i,j} \langle \phi(e_i^* \otimes e_i), \phi(e_j^* \otimes e_j) \rangle.$$

So 27.6.5 says that $\langle \chi_\omega, \chi_\omega \rangle = 1$, and (ii) follows from 27.6.2.

Part (iii) follows from (ii) and 27.3.7, and (iv) is an immediate consequence of (iii). $\square$

References

- [5], [38], [40], [51], [77], [78].

Borel subgroups, parabolic subgroups, Cartan subgroups

In this chapter, our interest turns to certain fundamental notions of the theory of algebraic groups : Borel subgroups, parabolic subgroups, Cartan subgroups. One of the main result we prove in this chapter is that Borel subgroups (resp. Cartan subgroups) of an algebraic group are conjugate.

Let G be a connected algebraic group and $\mathfrak{g}$ its Lie algebra. Two elements $x, y \in \mathfrak{g}$ are *conjugate* if there exists $\alpha \in G$ such that $y = \mathrm{Ad}(\alpha)(x)$.

28.1 Borel subgroups

28.1.1 Definition. *A maximal connected solvable subgroup of G is called a* Borel subgroup *of G.*

28.1.2 Remarks. Let H be a subgroup of G. If H is solvable and connected, then so is $\overline{H}$. Thus a Borel subgroup is a maximal connected solvable algebraic subgroup of G. In particular, for reasons of dimension, G has a Borel subgroup.

If H is unipotent, then $\overline{H}$ is also unipotent (10.8.14), so it is nilpotent and connected (10.8.10 and 22.3.6). Thus H is contained in a Borel subgroup of G. In the same way, a torus of G is contained in a Borel subgroup of G.

28.1.3 Theorem. (i) *If B is a Borel subgroup, then G/B is projective.*
(ii) *All the Borel subgroups of G are conjugate.*

Proof. Let P be a Borel subgroup of G of maximal dimension. By 25.4.3, there exists a finite-dimensional rational representation (V, π) of G and a line L of V such that:

$$P = \{\alpha \in G; \pi(\alpha)(L) = L\} \ , \ \mathfrak{p} = \mathcal{L}(P) = \{x \in \mathfrak{g}; d\pi(x)(L) \subset L\}.$$

Applying 26.3.2 to the action of P on V/L, we obtain that P fixes a flag $\mathfrak{H}_0 : L = V_1 \subset V_2 \subset \cdots \subset V_n = V$ of V. Since G acts rationally on the flag variety $\mathfrak{F}$ of V, our choice of L implies that P is the stabilizer of $\mathfrak{H}_0$ in G.

Now let $\mathfrak{H} \in \mathfrak{F}$ and H the stabilizer of $\mathfrak{H}$ in G. Then $\pi(H)$ is solvable. On the other hand, $\ker \pi \subset P$, so $\ker \pi$ is solvable. It follows that H is solvable. Hence $\dim H \leqslant \dim P$. We deduce therefore that the G-orbit of $\mathfrak{H}_0$ is of minimal dimension. So it is closed (21.4.5) and projective (13.6.7). Since the morphism $G/P \to G.\mathfrak{H}_0$, $\alpha P \mapsto \alpha.\mathfrak{H}_0$, is G-equivariant and bijective, it follows from 25.1.2 that it is an isomorphism. Thus G/P is projective.

Finally, let B be a Borel subgroup of G. Then B acts on G/P by $(\beta, \alpha P) \mapsto \beta \alpha P$. By 13.4.5 and 26.3.4, there exists $\alpha \in G$ such that $B\alpha P = \alpha P$. So $\alpha^{-1}B\alpha \subset P$, or $B \subset \alpha P \alpha^{-1}$. It follows from the definition of a Borel subgroup that $\alpha^{-1}B\alpha = P$. $\square$

28.1.4 Corollary. *Let P be a closed subgroup of G. Then the following conditions are equivalent:*
  (i) *The variety G/P is projective.*
  (ii) *The variety G/P is complete.*
  (iii) *The group P contains a Borel subgroup of G.*
If these conditions are verified, then P is called a parabolic subgroup *of G.*

Proof. (i) $\Rightarrow$ (ii) This is clear by 13.4.5.

(ii) $\Rightarrow$ (iii) Let B be a Borel subgroup of G. Then by 26.3.4, the action of B on G/P, given by $(\beta, \alpha P) \mapsto \beta \alpha P$, has a fixed point, say αP. So $B\alpha P = \alpha P$ and P contains the Borel subgroup $\alpha^{-1}B\alpha$ of G.

(iii) $\Rightarrow$ (i) Let B be a Borel subgroup contained in P. Since G/B is projective (28.1.3) and the canonical morphism $G/B \to G/P$ is surjective, 13.4.3 implies that G/P is complete. But G/P is quasi-projective (25.4.7), so it is a projective variety. $\square$

28.1.5 Remark. Let H be a closed connected subgroup of G. By 28.1.4, the following conditions are equivalent:
  (i) H is a Borel subgroup of G
  (ii) H is solvable and G/H is projective.

28.1.6 Corollary. *Let B be a Borel subgroup of G.*
  (i) *Let $u : G \to G$ be an automorphism of the algebraic group G. If $u(\alpha) = \alpha$ for all $\alpha \in B$, then $u = \mathrm{id}_G$.*
  (ii) *We have $Z(G)^\circ \subset Z(B) \subset C_G(B) = Z(G)$.*

Proof. (i) Let $\pi : G \to G/B$ be the canonical surjection and $v : G \to G$ the morphism $\alpha \mapsto u(\alpha)\alpha^{-1}$. If $u(\alpha) = \alpha$ for all $\alpha \in B$, then v is constant on αB for all $\alpha \in G$. By 25.3.3 and 25.4.7, there exists a morphism $w : G/B \to G$ such that $w \circ \pi = v$. It follows therefore from 13.4.3, 13.4.5 and 28.1.3 that $v(G)$ is closed in G. So $v(G)$ is affine and complete. Moreover, $v(G)$ is connected. Thus $v(G) = v(B) = \{e\}$ by 13.4.4, and $u = \mathrm{id}_G$.

(ii) The subgroup $Z(G)^\circ$ is connected and solvable. So $Z(G)^\circ$ is contained in a Borel subgroup P of G. By 28.1.3, there exists $\alpha \in G$ such that $\alpha P \alpha^{-1} = B$. So $Z(G)^\circ \subset Z(B)$ since $\alpha Z(G)^\circ \alpha^{-1} = Z(G)^\circ$. Clearly, $Z(G) \cup Z(B) \subset$

$C_G(B)$. Finally, let $\alpha \in C_G(B)$ and i_α the inner automorphism of G defined by α. Then $i_\alpha|_B = \mathrm{id}_B$, so $i_\alpha \in \mathrm{id}_G$ by part (i). Hence $\alpha \in Z(G)$. $\square$

Remark. We shall see in 28.2.3 that $Z(B) = Z(G)$.

28.1.7 Corollary. (i) *The set of maximal tori of G is equal to the set of maximal tori of the Borel subgroups of G. Two maximal tori of G are conjugate.*

(ii) *The set of maximal unipotent subgroups of G is equal to the set of maximal unipotent subgroups of the Borel subgroups of G. Two maximal unipotent subgroups of G are conjugate.*

Proof. (i) Let T be a maximal torus. We observed in 28.1.2 that T is contained in a Borel subgroup B and T is a maximal torus in B. By 26.5.5, any maximal torus of B is conjugate to T. So the result follows from the fact that Borel subgroups are conjugate (28.1.3).

(ii) Again, we saw in 28.1.2 that any maximal unipotent subgroup U is contained in a Borel subgroup B and U is a maximal unipotent subgroup in B. So $U = B_u$ by 26.4.2. The result follows again from 28.1.3. $\square$

28.1.8 Corollary. *Let B be a Borel subgroup of G.*
(i) *If $B = B_s$, then G is a torus.*
(ii) *If $\{e\}$ is the only torus of B, then G is unipotent.*

Proof. (i) By 26.5.5, $B = TB_u$ where T is a maximal torus of B. If $B = B_s$, then $B = T$, and $T = Z(B) \subset Z(G)$ by 28.1.6. So T is normal in G, and the variety G/T is irreducible, affine and complete (25.4.11 and 28.1.3). So it is reduced to a single point (13.4.4).

(ii) If $B = B_u$, then $H = Z(B)^\circ$ is not trivial unless $G = \{e\}$, and $H \subset Z(G)$ (28.1.6). The group B/H is a unipotent Borel subgroup of G/H. By induction on $\dim G$, we deduce that $G/H = B/H$. Hence $G = B$ is unipotent. $\square$

28.1.9 Corollary. *The following conditions are equivalent:*
(i) *G is nilpotent.*
(ii) *G has a nilpotent Borel subgroup.*
(iii) *G has a unique maximal torus.*
(iv) *There exists a maximal torus of G contained in $Z(G)$.*

Proof. (i) $\Rightarrow$ (ii) This is obvious.

(ii) $\Rightarrow$ (iii) Let B be a nilpotent Borel subgroup. By 26.4.4, $B = T \times B_u$ where $T = B_s$ is a maximal torus of G. By 28.1.6, $T \subset Z(B) \subset Z(G)$. Since any maximal torus of G is conjugate to T (28.1.7), T is the unique maximal torus of G.

(iii) $\Rightarrow$ (iv) If G has a unique maximal torus T, then T is normal in G (28.1.7). For $\alpha \in G$, the map $T \to T$, $\beta \mapsto \alpha\beta\alpha^{-1}$, is a morphism of the algebraic group T. So by 22.5.10, we have $T \subset Z(G)$.

(iv) $\Rightarrow$ (i) Let T be a maximal torus of G contained in $Z(G)$. By 26.5.5 and 28.1.7, we have $B = T \times G_u$ and $T = B_s$. So B is nilpotent (26.4.4). The group B/T is then a nilpotent Borel subgroup of G/T. By induction on $\dim G$, we deduce that G/T is nilpotent, and since $T \subset Z(G)$, G is nilpotent (10.5.3).   □

28.1.10 Corollary. (i) *If T is a maximal torus of G, then $C = C_G(T)^{\circ}$ is nilpotent and $C = N_G(C)^{\circ}$.*
(ii) *If $\dim G \leqslant 2$, then G is solvable.*

Proof. (i) By 28.1.7 and 28.1.9, T is the unique maximal torus of C and C is nilpotent. Moreover if $\alpha \in N_G(C)$, then $\alpha T \alpha^{-1}$ is a maximal torus of C, so $\alpha T \alpha^{-1} = T$. Thus T is normal in $N_G(C)$, and so T commutes with $N_G(C)^{\circ}$ by 22.5.10. Hence $N_G(C)^{\circ} \subset C$ and the result follows.

(ii) Let B be a Borel subgroup of G. Then $B = TB_u$ where T is a maximal torus of G. If $B \neq G$, then $\dim B \leqslant 1$, so $B = T$ or $B = B_u$. But 28.1.8 would imply that $G = B$. Contradiction.   □

28.1.11 Theorem. *Let $\mathfrak{B}$ be the set of Borel subgroups of G. Then:*

$$R(G) = \left(\bigcap_{B \in \mathfrak{B}} B \right)^{\circ} , \quad R_u(G) = \left(\bigcap_{B \in \mathfrak{B}} B_u \right)^{\circ}.$$

Proof. Let H (resp. K) be the identity component of the intersection of B (resp. B_u), $B \in \mathfrak{B}$. Since $R(G)$ is connected, normal and solvable, it is contained in H (28.1.3). On the other hand, H is normal (28.1.3), connected and solvable. So $H \subset R(G)$. The equality $K = R_u(G)$ is proved in the same way.
□

28.2 Theorems of density

28.2.1 Proposition. *Let H be a closed subgroup of G and*

$$X = \bigcup_{\alpha \in H} \alpha H \alpha^{-1}.$$

(i) *X contains a dense open subset of $\overline{X}$.*
(ii) *If the variety G/H is complete, then X is closed in G.*
(iii) *Suppose that there exists $\gamma \in G$ which is contained only in a finite number of conjugates of H. Then $\overline{H} = G$.*

Proof. Denote by $\mathrm{pr}_1 : (G/H) \times G \to G/H$ and $\mathrm{pr}_2 : (G/H) \times G \to G$ the canonical projections, $\pi : G \to G/H$ the canonical surjection and u, v the morphisms:

$$u : G \times G \to G \times G , \quad (\alpha, \beta) \mapsto (\alpha, \alpha\beta\alpha^{-1}),$$
$$v : G \times G \to (G \times G)/(H \times \{e\}) = (G/H) \times G , \quad (\alpha, \beta) \mapsto (\pi(\alpha), \beta).$$

Finally, let $M = \{(\alpha, \beta) \in G \times G; \beta \in \alpha H \alpha^{-1}\}$, $N = v(M)$.

(i) Since $X = \mathrm{pr}_2 \circ v \circ u(G \times H)$, it is a constructible subset of G (15.4.3). So the result follows from 1.4.6.

(ii) We have $M = u(G \times H)$ and u is an isomorphism of varieties. Thus M is closed in $G \times G$. It follows from 25.3.2 and 25.4.7 that N is closed in $(G/H) \times G$. So if G/H is complete, then $X = \mathrm{pr}_2(N)$ is closed.

(iii) Let $p = \mathrm{pr}_1|_N$. If $\alpha \in G$, then $p^{-1}(\pi(\alpha)) = \{\pi(\alpha)\} \times (\alpha H \alpha^{-1})$. We obtain therefore from 15.5.5 that $\dim N = \dim(G/H) + \dim H = \dim G$.

Consider the action of G on G/H given by $\alpha.\pi(\beta) = \pi(\alpha\beta)$ and denote $q = \mathrm{pr}_2|_N$. If $\alpha \in G$, then $q^{-1}(\alpha) = E_\alpha \times \{\alpha\}$ where

$$E_\alpha = \{\pi(\beta); \beta^{-1}\alpha\beta \in H\} = \{\pi(\beta); \alpha.\pi(\beta) = \pi(\beta)\}.$$

The set $q^{-1}(\gamma)$ is finite by assumption. Let us consider the morphism $N \to \overline{X} = \mathrm{pr}_2(N)$ induced by q. Then 15.5.4 (i) says that $\dim X = \dim N$. Thus $\dim \overline{X} = \dim G$ and, since G is connected, we have $\overline{X} = G$ by 1.2.3. $\square$

28.2.2 Theorem. *Let B be a Borel subgroup of G, T a maximal torus in G and $C = C_G(T)^\circ$. Then:*

$$G = \bigcup_{\alpha \in G} \alpha B \alpha^{-1} \; , \; G_u = \bigcup_{\alpha \in G} \alpha B_u \alpha^{-1} \; , \; G_s = \bigcup_{\alpha \in G} \alpha T \alpha^{-1}.$$

Furthermore, the set $\bigcup_{\alpha \in G} \alpha C \alpha^{-1}$ contains a dense open subset of G.

Proof. By 28.1.10, C is nilpotent. So 26.4.4 implies that $C = T \times C_u$. Fix $\alpha \in T$ such that $C_G(\alpha) = C_G(T)$ (26.2.3) and let $\beta \in G$ be such that $\beta\alpha\beta^{-1} \in C$. Then $\beta\alpha\beta^{-1} \in T$ and $C_G(T) \subset C_G(\beta\alpha\beta^{-1})$. But $\dim C_G(\beta\alpha\beta^{-1}) = \dim C$, so we deduce that $C_G(\beta\alpha\beta^{-1}) = C_G(\beta\alpha\beta^{-1})^\circ = C$. Thus $\beta C \beta^{-1} = C$, that is $\beta \in N_G(C)$. Since $C = N_G(C)^\circ$ (28.1.10), we have proved that the set of conjugates of α which are contained in C is finite. We deduce therefore from 28.2.1 that the union of the conjugates of C contains a dense open subset of G.

The group C is connected and nilpotent, so up to conjugation, we may assume that $C \subset B$ (28.1.3). So the union D of the set of Borel subgroups contains a dense open subset of G. But G/B is projective (28.1.3), so D is closed in G by 28.2.1. Hence $D = G$. The other equalities follow now from 26.5.6 and 28.1.7. $\square$

28.2.3 Corollary. (i) *We have $Z(B) = Z(G)$ for any Borel subgroup B of the group G.*

(ii) *The set $Z(G)_s$ is the intersection of all the maximal tori of G.*

Proof. (i) Let $\alpha \in Z(G)$. By 28.2.2, a conjugate of α is in B, so $\alpha \in B$ and $Z(G) \subset Z(B)$. The result follows therefore from 28.1.6.

(ii) Let $\alpha \in Z(G)_s$. As in the proof of (i), we have $\alpha \in B$. So by 26.5.6, α is contained in a maximal torus of B. Hence α belongs to the intersection H of all the maximal tori of G (28.1.7). Conversely, H is a diagonalizable subgroup of G and clearly H is normal in G. So 22.5.10 implies that H is central in G. Hence $H = Z(G)_s$. $\square$

28.3 Centralizers and tori

28.3.1 Recall that for $\alpha \in G$, α_s (resp. α_u) denotes the semisimple (resp. unipotent) component of α.

Proposition. (i) *Let H be a connected solvable subgroup of G. For any $\alpha \in C_G(H)$, there exists a Borel subgroup of G containing H and α.*

(ii) *Let S be a torus of G and $\alpha \in C_G(S)$. There exists a torus of G containing S and α_s. The group $C_G(S)$ is connected.*

(iii) *If $\alpha \in G$, then $\alpha \in C_G(\alpha_s)^\circ$.*

Proof. (i) The group $\overline{H}$ is solvable, connected and $\alpha \in C_G(\overline{H})$. We may therefore assume that H is closed. Let B be a Borel subgroup containing α (28.2.2). Consider the action of G on G/B given by: $(\beta, \gamma B) \mapsto (\beta\gamma)B$. The variety X of fixed points of α is non-empty since it contains eB, and it is H-stable. It follows from 26.3.4 that there exists $\beta \in G$ such that $\beta B \in X$ and $\gamma.\beta B = \beta B$ for all $\gamma \in H$. So $\beta B \beta^{-1}$ is a Borel subgroup of G containing H and α.

(ii) By (i), S and α are contained in a Borel subgroup B of G. We have $\alpha_s \in B$ and the first part follows from 26.5.6. Again by 26.5.6, $C_B(S)$ is connected, so $\alpha \in C_G(S)^\circ$. Hence $C_G(S) = C_G(S)^\circ$.

(iii) Let B be a Borel subgroup containing α. Then it contains also α_s, and $\alpha \in C_B(\alpha_s) = C_B(\alpha_s)^\circ$ (26.5.6). Hence $\alpha \in C_G(\alpha_s)^\circ$. $\square$

28.3.2 Proposition. *Let S be a torus of G and $C = C_G(S)$.*

(i) *If B is a Borel subgroup of G containing S, then $C \cap B$ is a Borel subgroup of C.*

(ii) *Any Borel subgroup of C is of the form $C \cap B$, where B is a Borel subgroup of G containing S.*

Similar results hold for maximal tori and maximal unipotent subgroups.

Proof. If D is a Borel subgroup of C, then $S \subset Z(C) \subset Z(D)$ (28.2.3). Since all the Borel subgroups of C are conjugate (28.1.3), it suffices to prove (i).

The group $C \cap B = C_B(S)$ is solvable and connected (26.5.6), so the result would follows from 28.1.4 if the variety $C/(C \cap B)$ is complete.

Observe that SB_u is a connected algebraic subgroup of G (21.2.3).

Consider the action of C on G/B given by $(\alpha, \beta B) \mapsto (\alpha\beta)B$. Then $C \cap B$ is the stabilizer of $e_B = eB$ in C. So the orbit map $C/(C \cap B) \to Ce_B$ is a bijective C-equivariant morphism, hence, it is an isomorphism by 25.1.2. It suffices therefore to show that Ce_B is closed in G/B. Let $\pi : G \to G/B$ be the canonical surjection. Then $X = CB = \pi^{-1}(Ce_B)$ is B-stable (on the right), it follows that $\overline{X}$ is also B-stable. Since π is open (25.4.7), π sends B-stable closed subsets to closed subsets of G/B (25.3.2). So we are reduced to show that X is closed in G.

If $\alpha \in X$, then $\alpha^{-1}S\alpha \subset B$, and this is still true for $\alpha \in \overline{X}$ by continuity. Let us consider the morphism

$$\overline{X} \times S \to B/B_u \ , \ (\alpha, \beta) \mapsto \alpha^{-1}\beta\alpha B_u.$$

It verifies the hypotheses of 22.5.10. It follows that $\alpha^{-1}\beta\alpha \in \beta B_u$ for $\alpha \in \overline{X}$ and $\beta \in S$. Thus $\alpha^{-1}S\alpha$ is a maximal torus of the group SB_u. So 28.1.7 implies that there exists $\gamma \in SB_u$ such that $\alpha^{-1}S\alpha = \gamma^{-1}S\gamma$. Hence $\alpha\gamma^{-1} \in N_G(S)$, and so $\alpha \in N_G(S)B$. Let $Y = N_G(S)B$. We obtain therefore that $CB = X \subset \overline{X} \subset Y$.

Since $Y = \pi^{-1}(N_G(S)e_B)$ is the inverse image of a locally closed subset (21.4.3), it is locally closed, and hence it is a subvariety of G.

Let us consider the action of $C \times B$ on Y given by $(\alpha, \beta) \cdot \gamma = \alpha\gamma\beta^{-1}$. The orbits under this action are of the form $C\gamma B$, $\gamma \in N_G(S)$.

By 22.5.11 and 28.3.1, $C = N_G(S)^\circ$ and so C is normal in $N_G(S)$ (21.1.6). It follows that for $\alpha \in N_G(S)$, $\alpha CB = C\alpha B$, and for $\alpha, \beta \in N_G(S)$, the morphism $\alpha CB \to \beta CB$, $\gamma \mapsto \beta\alpha^{-1}\gamma$, is an isomorphism of irreducible subvarieties of G.

Consequently, since C has finite index in $N_G(S)$ (21.1.6), it follows from the preceding discussion that Y has a finite number of $C \times B$-orbits all of the same dimension. In particular, these orbits are closed (21.4.5). Since $\overline{X}$ is also the closure of X in Y, we have $X = \overline{X}$ (21.4.5).

The proofs for the cases of maximal tori and maximal unipotent subgroups are analogue. □

28.3.3 Corollary. *Let T be a maximal torus of G. If B is a Borel subgroup of G containing T, then B contains $C_G(T)$.*

Proof. The group $C_G(T)$ is connected (28.3.1) and nilpotent (28.1.9). So the result follows from 28.3.2. □

28.4 Properties of parabolic subgroups

28.4.1 Lemma. *Let $u : G \to H$ be a surjective morphism of algebraic groups and $B = TB_u$ a Borel subgroup of G where T is a maximal torus of G. We have $u(B) = u(T)u(B_u)$ and $u(B)$ is a Borel subgroup of H. Moreover, $u(T)$ is a maximal torus of H and $u(B_u) = u(B)_u$. In particular, any Borel subgroup of H is the image of a Borel subgroup of G.*

Proof. The composition $G \to H \to H/u(B)$ of morphisms induces a surjective morphism $G/B \to H/u(B)$. So $H/u(B)$ is a complete variety (28.1.3 and 13.4.3). Thus $u(B)$ is a parabolic subgroup of H (28.1.4). Since $u(B)$ is solvable and connected, it is a Borel subgroup of H. Finally, it is clear that $u(B) = u(T)u(B_u)$, and we have $u(B_u) = u(B)_u$ by 22.2.3. So $u(T)$ is a maximal torus of $u(B)$, and hence of H. □

28.4.2 Theorem. *Let G be a connected algebraic group.*

(i) *If B is a Borel subgroup of G, then $N_G(B) = B$.*

(ii) *For any parabolic subgroup P of G, we have $N_G(P) = P = P^\circ$.*

Proof. (i) Let $N = N_G(B)$ and proceed by induction on $\dim G$. The result is clear if $\dim G \leqslant 2$ by 28.1.10.

Let $\alpha \in N$ and T a maximal torus of B. Since $\alpha T \alpha^{-1}$ is a maximal torus of B, there exists $\beta \in B$ such that $\alpha T \alpha^{-1} = \beta T \beta^{-1}$ (28.1.7). So $\alpha \beta^{-1} \in N_G(T)$ and by replacing α by $\alpha \beta^{-1}$, we may assume that $\alpha \in N_G(T)$. Consider the morphism

$$u : T \to T , \ \gamma \mapsto \alpha \gamma \alpha^{-1} \gamma^{-1}.$$

Since T is commutative and α normalizes T, we see that u is an endomorphism of the group T. We have two cases:

1) u is not surjective. Then $S = (\ker u)^\circ$ is a non-trivial torus of T and $\alpha \in C_G(S)$. Moreover, α normalizes $C_G(S) \cap B$ which is a Borel subgroup of $C_G(S)$ (28.3.2).

If $C_G(S) \neq G$, then $\alpha \in B$ by the induction hypothesis. Otherwise if $C_G(S) = G$, then $S \subset Z(G) = Z(B)$. So by passing to the quotient G/S, we obtain the result by 28.4.1 and the induction hypothesis.

2) u is surjective. By 25.4.3, there exists a finite-dimensional rational G-module V and $v \in V \setminus \{0\}$ such that $N = \{\alpha \in G; \alpha.v \in \Bbbk v\}$.

If $\beta \in B_u$, then $\beta.v = v$ by 22.3.6. If $\beta \in T$, then the surjectivity of u implies that β is the commutator of two elements of N, so $\beta.v = v$. It follows that $\beta.v = v$ for all $\beta \in B$. The map $\gamma \mapsto \gamma.v$ induces therefore a morphism $G/B \to V$. By 13.4.3, 13.4.4 and 13.4.5, this morphism is constant. So $\gamma.v = v$ for all $\gamma \in G$. Hence $G = N$, and therefore $G = N = B$ by 28.2.2.

(ii) Let P be a parabolic subgroup of G and B a Borel subgroup of G contained in P. In particular, $B \subset P^\circ$. Let $\alpha \in N_G(P)$. Then $\alpha B \alpha^{-1}$ is a Borel subgroup of P°, so there exists $\beta \in P^\circ$ such that $\alpha B \alpha^{-1} = \beta B \beta^{-1}$. Now part (i) says that $\beta^{-1}\alpha \in N_G(B) = B$. So $\alpha \in \beta B \subset P^\circ$. We have therefore proved that $N_G(P) = P^\circ = P$. $\square$

28.4.3 Corollary. *Let B be a Borel subgroup of G.*

(i) *We have $N_G(B_u) = B$.*

(ii) *Any parabolic subgroup of G is conjugate to a unique parabolic subgroup containing B.*

(iii) *If H is a solvable subgroup (not necessarily connected or closed) of G containing B, then $H = B$.*

Proof. (i) Let $U = B_u$ and $N = N_G(U)$. Since B normalizes U, it is a Borel subgroup of N°. By 28.2.2, any unipotent element of N° is conjugate in N° to an element of U. As U is normal in N°, it follows that $U = N_u$. We deduce therefore that the group N°/U is a torus, and so, it is solvable. Thus N° is solvable and $B = N^\circ$. Now N normalizes N°, so $N \subset N_G(B) = B$.

(ii) Let P, Q be parabolic subgroups containing B verifying $Q = \alpha P \alpha^{-1}$ with $\alpha \in G$. Then B and $\alpha B \alpha^{-1}$ are Borel subgroups of Q. So there exists $\beta \in Q$ such that $\alpha B \alpha^{-1} = \beta B \beta^{-1}$ (28.1.3). Thus $\beta^{-1} \alpha \in N_G(B) = B$, and so $\alpha \in \beta B \subset Q$. Hence $P = Q$.

(iii) It suffices to show that $\overline{H} = B$. But $\overline{H}$ is a parabolic subgroup. So $\overline{H}$ is connected 28.4.2, and since it is solvable, we have $\overline{H} = B$ (28.1.5). $\square$

Remark. Condition (iii) of 28.4.3 does not say that a maximal solvable subgroup of G is a Borel subgroup of G.

28.5 Cartan subgroups

28.5.1 Definition. *A subgroup of G is called a* Cartan subgroup *of G if it is equal to the centralizer of a maximal torus of G.*

28.5.2 Cartan subgroups of G are connected and nilpotent (28.1.10 and 28.3.1). Two Cartan subgroups are conjugate (28.1.7) and their common dimension is called the *rank* of G, denoted by $\mathrm{rk}(G)$. By 28.2.2, if C is a Cartan subgroup of G, then the union of the conjugates of C contains a dense open subset of G.

By 28.1.7 and 28.3.2, the map $T \mapsto C_G(T)$ is a bijection from the set of maximal tori of G onto the set of Cartan subgroups of G. We have also, for a maximal torus T, $C_G(T) = T \times C_G(T)_u$.

28.5.3 Lemma. (i) *Let H be a connected nilpotent algebraic group and K a closed subgroup of H distinct from H. Then $\dim K < \dim N_H(K)$.*

(ii) *If C is a Cartan subgroup of G, then it is a maximal connected nilpotent subgroup of G.*

Proof. (i) Since $H \neq K$, $H \neq \{e\}$. The last non-trivial term in the central descending series of H is connected (21.3.6) and central in H. So $Z = Z(H)^\circ \neq \{e\}$.

If $Z \not\subset K$, then the connected algebraic subgroup ZK contains strictly K and normalizes K. So $\dim N_H(K) \geqslant \dim ZK > \dim K$.

Suppose that $Z \subset K$. Let $\pi : H \to H/Z$ be the canonical surjection. We have:

$$N_{H/Z}(K/Z) = \pi(N_H(K)) \ , \ \dim N_H(K) = \dim Z + \dim N_{H/Z}(K/Z).$$

So the result follows by induction on $\dim H$.

(ii) Suppose that H is a connected nilpotent subgroup of G containing strictly C. We may assume that H is closed. By (i), $C \neq N_H(C)^\circ$. Let $\alpha \in N_H(C)$ and write $C = C_G(T)$ where T is a maximal torus of G. Then $\alpha T \alpha^{-1}$ is a maximal torus of C, and so $\alpha T \alpha^{-1} = T$ since T is central in C. Thus $N_H(C)^\circ \subset N_H(T)^\circ = C_H(T) = C$ (28.1.10 and 28.3.1). We have therefore obtained a contradiction. $\square$

28.5.4 Let $\alpha \in G$. By 28.2.2, α_s belongs to a maximal torus of G. We deduce therefore that $\dim C_G(\alpha_s) \geqslant \operatorname{rk}(G)$. We say that α is *generic* in G if $\dim C_G(\alpha_s) = \operatorname{rk}(G)$. Denote by G_{gen} the set of generic elements of G.

Lemma. *Let T be a maximal torus of G and $\alpha \in T$. Then the following conditions are equivalent:*

(i) $\alpha \in G_{\mathrm{gen}}$.

(ii) $C_G(\alpha)^\circ = C_G(T)$.

(iii) $\chi(\alpha) \neq 1$ for all $\chi \in \Phi(T, G)$ *(in the notations of 26.2.1).*

Proof. The equivalence (i) $\Leftrightarrow$ (ii) is clear since $C_G(T)$ is connected (28.3.1), while (ii) $\Leftrightarrow$ (iii) follows from 26.1.1 and 26.2.1. $\square$

28.5.5 Proposition. *Let α be a semisimple element of G. The following conditions are equivalent:*

(i) $\alpha \in G_{\mathrm{gen}}$.

(ii) $C_G(\alpha)^\circ$ *is a Cartan subgroup of G.*

(iii) $C_G(\alpha)^\circ$ *is a nilpotent subgroup of G.*

(iv) α *belongs to a unique maximal torus of G.*

(v) α *belongs to a finite number of maximal torus of G.*

Proof. The equivalence (i) $\Leftrightarrow$ (ii) (resp. (ii) $\Leftrightarrow$ (iii)) follows from 28.5.4 (resp. 26.4.4 and 28.5.3). Let T be a maximal torus of G containing α (28.2.2). Then $T \subset C_G(\alpha)^\circ$. If $C_G(\alpha)^\circ$ is nilpotent, then T is its unique maximal torus (26.4.4). So (iii) $\Rightarrow$ (iv) $\Rightarrow$ (v). To finish the proof, we shall show that (v) $\Rightarrow$ (ii).

Let $H = C_G(\alpha)^\circ$ and T a maximal torus of G containing α. Since maximal tori of H are conjugate in H, the connected variety $H/N_H(T)$ is finite, and hence it is reduced to a point. Thus T is normal in H. It follows from 22.5.10 that T is central in H. So $H \subset C_G(T)$. But $C_G(T) \subset C_G(\alpha)^\circ = H$, hence $H = C_G(T)$. $\square$

28.5.6 Lemma. *Let H be a connected algebraic group which contains a unipotent element α belonging to a unique Cartan subgroup C of H. Then H is nilpotent.*

Proof. Let T be a maximal torus such that $C_G(T) = C$, and B a Borel subgroup of H containing T. By 28.3.3, $C \subset B$.

Let $B_u = N_n \supset N_{n-1} \supset \cdots \supset N_0 = \{e\}$ be the central descending series of B_u. We have $N_0 \subset C$. Suppose that $N_i \subset C$. Since $\alpha \in B_u$, we have $\alpha^{-1}\beta\alpha\beta^{-1} \in N_i$ if $\beta \in N_{i+1}$, so $\beta\alpha\beta^{-1} \in \alpha N_i \subset C$, and $\alpha \in \beta^{-1}C\beta$. Since $\beta^{-1}C\beta$ is a Cartan subgroup of H containing α, our hypothesis on α implies that $\beta^{-1}C\beta = C$. We deduce therefore that $N_{i+1} \subset N_H(C)$. Since N_{i+1} is connected (21.3.6), we have $N_{i+1} \subset N_H(C)^\circ = C$.

We have therefore proved that $B_u \subset C$. Now $T \subset C$ and $B = TB_u$, so we have $B = C$. Consequently B is nilpotent (28.1.10 and 28.3.1), and so H is nilpotent (28.1.9). $\square$

28.5.7 Theorem. *Let G be a connected algebraic group.*

(i) *An element $\alpha \in G$ is generic if and only if it belongs to a unique Cartan subgroup of G.*

(ii) *The set G_{gen} contains a dense open subset of G.*

Proof. (i) If α is generic, then $C = C_G(\alpha_s)^{\circ}$ is the unique Cartan subgroup containing α_s (28.5.5), and $\alpha \in C$ by 28.3.1. Now let D be a Cartan subgroup containing α. Then $\alpha_s \in D_s$ and $D = D_s \times D_u$. Thus $D \subset C_G(D_s)^{\circ} \subset C_G(\alpha_s)^{\circ}$, and hence $D = C$.

Conversely, suppose that α belongs to a unique Cartan subgroup C. Since $\alpha_s \in C_s \subset Z(C)$, we deduce that $H = C_G(\alpha_s)^{\circ}$ contains C. So any Cartan subgroup of H contains $\alpha_s \in Z(H)$. Since α is contained in a unique Cartan subgroup, it follows that $\alpha_u = \alpha_s^{-1}\alpha$ is contained in a unique Cartan subgroup of H. Hence H is nilpotent by 28.5.6, and so H is a Cartan subgroup of G (28.5.5) which implies that α_s, and hence α, is generic.

(ii) Observe that by 28.5.4 (iii), the set G_{gen} is non-empty. Let $C = C_G(T) = T \times C_u$ be a Cartan subgroup of G and T_0 the set of $\alpha \in T$ verifying $\chi(\alpha) \neq 1$ for all $\chi \in \Phi(T,G)$ (notations of 26.2.1). The set $C_0 = T_0 \times C_u$ is dense and open in C and by 28.5.4, $C_0 = C \cap G_{\mathrm{gen}}$. Since any generic element belong to a Cartan subgroup, we deduce that G_{gen} is the image of the morphism

$$u : G \times C_0 \to G , \ (\alpha, \gamma) \mapsto \alpha\gamma\alpha^{-1}.$$

It follows from the inclusion $C_0 \subset u(G \times C_0) = G_{\mathrm{gen}}$ that $C = \overline{C_0} \subset \overline{G_{\mathrm{gen}}}$. As $\overline{G_{\mathrm{gen}}}$ is stable by conjugation, it contains the union of the conjugates of C, which contains a dense open subset of G (28.5.2). □

28.5.8 Let H be a nilpotent algebraic group. By 26.4.4, $T = (H^{\circ})_s$ is the unique maximal torus of H° and $H^{\circ} = T \times (H^{\circ})_u$. So $T \subset Z(H^{\circ})$ and T is normal in H.

Lemma. *We have $T \subset Z(H)$.*

Proof. Let $\mathfrak{t} = \mathcal{L}(T)$ and for $\alpha \in H$, let u_{α} and v_{α} be the morphisms of T into itself defined by:

$$u_{\alpha}(\gamma) = \alpha\gamma\alpha^{-1} , \ v_{\alpha}(\gamma) = \alpha\gamma\alpha^{-1}\gamma^{-1}.$$

Then $u_{\alpha}(e) = v_{\alpha}(e) = e$, so du_{α} and dv_{α} are endomorphisms of $\mathfrak{t}$. Since H/H° is finite, u_{α} and du_{α} are automorphisms of finite order. So du_{α} is a semisimple automorphism of $\mathfrak{t}$. Similarly, since H is nilpotent, dv_{α} is nilpotent. But $dv_{\alpha} = du_{\alpha} - \mathrm{id}_{\mathfrak{t}}$, so $u_{\alpha} = \mathrm{id}_T$. Hence $T \subset Z(H)$. □

28.5.9 Theorem. *Let G be a connected algebraic group and C a subgroup (not necessarily closed) of G. The following conditions are equivalent:*

(i) *C is a Cartan subgroup of G.*

(ii) *C is a maximal nilpotent subgroup of G, and any subgroup of C of finite index in C, has finite index in $N_G(C)$.*

(iii) *C is a closed connected nilpotent subgroup of G and $C = N_G(C)^{\circ}$.*

Proof. (i) $\Rightarrow$ (ii) Let $C = C_G(T)$ be a Cartan subgroup of G where T is a maximal torus of G, and H a nilpotent subgroup containing C. We may assume that H is closed. Then T is a maximal torus of H°. By 28.5.8, we have $T \subset Z(H)$, and so $H \subset C_G(T) = C$.

Now C is connected, so if K is a subgroup of C of finite index in C, then $\overline{K} = C$. Thus $N_G(K) \subset N_G(C)$. Since $N_G(C)^\circ = C$ by 28.1.10, C has finite index in $N_G(C)$. The inclusions $K \subset C \subset N_G(C)$ and 10.2.3 show that K has finite index in $N_G(C)$.

(i) $\Rightarrow$ (iii) This follows from 28.1.10, 28.3.1 and 28.5.3.

(iii) $\Rightarrow$ (i) We have $C = S \times C_u$ where $S = C_s$ (26.4.4). Let B be a Borel subgroup containing C, T a maximal torus of B containing S, and D the centralizer of S in B. The group D is connected (28.3.1) and $B = TB_u$, $D = TD_u$.

Since $S \subset Z(D)$, the group SD_u is connected, nilpotent and it contains C. Now $C = N_G(C)^\circ$, it follows that $SD_u = C$ (28.5.3). The group D is connected and solvable. So $(D, D) \subset D_u \subset C$ (26.4.2), and hence C is normal in D. It follows from the equality $C = N_G(C)^\circ$ that $D = C$, so $T \subset C$, and $S = T$. Thus $C = C_B(T)$ is a Cartan subgroup of B, and hence of G.

(ii) $\Rightarrow$ (i) If C verifies (ii), then C is closed since $\overline{C}$ is nilpotent. Next, C° has finite index in C and therefore C° has finite index in $N_G(C)$. Hence $C^\circ = N_G(C)^\circ$. It follows from (iii) $\Rightarrow$ (i) that C° is a Cartan subgroup of G, and (i) $\Rightarrow$ (ii) shows that C° is maximal nilpotent. So $C = C^\circ$. $\square$

References

- [5], [38], [40], [78].

Cartan subalgebras, Borel subalgebras and parabolic subalgebras

In this chapter, we apply the results in Chapter 28 on Cartan subgroups, Borel subgroups and parabolic subgroups to their Lie algebras. We generalize the definitions of Borel subalgebras and parabolic subalgebras to arbitrary Lie algebras, and we establish the relations between the group objects and Lie algebra objects. In particular, we prove analogues of results on conjugacy of such subalgebras.

29.1 Generalities

29.1.1 In this chapter, $\mathfrak{g}$ is a finite-dimensional $\Bbbk$-Lie algebra and G its algebraic adjoint group (24.8.1). By 24.5.7 and 24.5.12, $\mathrm{ad}[\mathfrak{g},\mathfrak{g}] = [\mathrm{ad}\,\mathfrak{g}, \mathrm{ad}\,\mathfrak{g}]$ is an algebraic Lie subalgebra of $\mathfrak{gl}(\mathfrak{g})$. Let us denote by G_1 the connected algebraic subgroup of G such that $\mathcal{L}(G_1) = \mathrm{ad}[\mathfrak{g},\mathfrak{g}]$.

29.1.2 As in 19.8.1, we let $\mathfrak{g}^\lambda(x)$ to be the generalized eigenspace of $\mathrm{ad}\,x$ relative to $\lambda \in \Bbbk$, and denote by $\mathfrak{N}(\mathfrak{g})$ the set of $x \in \mathfrak{g}$ verifying $x \in \mathfrak{g}^\lambda(y)$ for some $y \in \mathfrak{g}$ and $\lambda \in \Bbbk \setminus \{0\}$. We saw in 19.8.2 that if $x \in \mathfrak{N}(\mathfrak{g})$, then $\mathrm{ad}\,x$ is nilpotent. Let $\mathrm{Aut}_s\,\mathfrak{g}$ be the subgroup of $\mathrm{Aut}_e\,\mathfrak{g}$ generated by $e^{\mathrm{ad}\,x}$, $x \in \mathfrak{N}(\mathfrak{g})$. If $\mathfrak{h}$ is a Lie subalgebra of $\mathfrak{g}$, then $\mathfrak{N}(\mathfrak{h}) \subset \mathfrak{N}(\mathfrak{g})$. So if we denote by $\mathrm{Aut}_s(\mathfrak{g},\mathfrak{h})$ the group generated by $e^{\mathrm{ad}_\mathfrak{g}\,x}$, $x \in \mathfrak{N}(\mathfrak{h})$, then we may identify $\{\gamma|_\mathfrak{h}; \gamma \in \mathrm{Aut}_s(\mathfrak{g},\mathfrak{h})\}$ with $\mathrm{Aut}_s\,\mathfrak{h}$.

The group $\mathrm{Aut}_s\,\mathfrak{g}$ is a connected algebraic subgroup of $\mathrm{Aut}_e\,\mathfrak{g}$ (21.2.5 and 21.3.2).

Lemma. *Let $f : \mathfrak{g} \to \mathfrak{h}$ be a surjective morphism of Lie algebras and $\beta \in \mathrm{Aut}_s\,\mathfrak{h}$. Then there exists $\alpha \in \mathrm{Aut}_s\,\mathfrak{g}$ such that $\beta \circ f = f \circ \alpha$.*

Proof. Let $y \in \mathfrak{h}$ and $x \in \mathfrak{g}$ be such that $y = f(x)$. Then $f(\mathfrak{g}^\lambda(x)) = \mathfrak{h}^\lambda(y)$, so $f(\mathfrak{N}(\mathfrak{g})) = \mathfrak{N}(\mathfrak{h})$.

To obtain the result, we may assume that $\beta = e^{\mathrm{ad}\,y}$ where $y \in \mathfrak{N}(\mathfrak{h})$. Then $y = f(x)$ for some $x \in \mathfrak{N}(\mathfrak{g})$. It is then easy to see that $\alpha = e^{\mathrm{ad}\,x}$ verifies the conclusion of the lemma. $\square$

29.1.3 Assume that $\mathfrak{g}$ is semisimple. Let $\mathfrak{h}$ be a Cartan subalgebra of $\mathfrak{g}$, $R = R(\mathfrak{g}, \mathfrak{h})$ the root system of $\mathfrak{g}$ relative to $\mathfrak{h}$, and $W = W(\mathfrak{g}, \mathfrak{h})$ its Weyl group. Recall from 18.2.9 that W acts on $\mathfrak{h}$.

Lemma. *Let $w \in W$ be considered as an automorphism of $\mathfrak{h}$. There exists $\theta \in \mathrm{Aut}_s\, \mathfrak{g}$ such that $\theta|_{\mathfrak{h}} = w$.*

Proof. We may assume that $w = s_\alpha$, where $\alpha \in R$. Let $X_\alpha \in \mathfrak{g}^\alpha$, $X_{-\alpha} \in \mathfrak{g}^{-\alpha}$ be such that $[X_\alpha, X_{-\alpha}] = H_\alpha$. Let:

$$\theta = e^{\mathrm{ad}\, X_\alpha} \circ e^{-\,\mathrm{ad}\, X_{-\alpha}} \circ e^{\mathrm{ad}\, X_\alpha}.$$

Then $\theta \in \mathrm{Aut}_s\, \mathfrak{g}$. To obtain the result, it suffices to prove that $\theta(h) = h - \alpha(h)H_\alpha$ for all $h \in \mathfrak{h}$.

This is clear if $\alpha(h) = 0$. On the other hand, we have:

$$\theta(H_\alpha) = e^{\mathrm{ad}\, X_\alpha} \circ e^{-\,\mathrm{ad}\, X_{-\alpha}}(H_\alpha - 2X_\alpha) = -H_\alpha + 2X_\alpha - 2X_\alpha = -H_\alpha = s_\alpha(H_\alpha).$$

So we are done.   □

29.1.4 Proposition. *Let H be an algebraic group, $\mathfrak{h}$ its Lie algebra and (V, π) a finite-dimensional rational H-module. Define θ to be the morphism of varieties:*

$$\theta : H \times V \to V \ , \ (\beta, w) \mapsto \pi(\beta)(w).$$

(i) *If $(\alpha, v) \in H \times V$ and $(x, w) \in T_\alpha(H) \times V$, then:*

$$d\theta_{(\alpha,v)}(x, w) = d\pi_\alpha(x)(v) + \pi(\alpha)(w).$$

(ii) *For all $(\alpha, v) \in H \times V$, the linear maps $d\theta_{(\alpha,v)}$ and $d\theta_{(e_H,v)}$ have the same rank.*

Proof. (i) Define the morphisms:

$$p : H \to H \times V \ , \ \beta \mapsto (\beta, v) \text{ and } q : V \to H \times V \ , \ w \mapsto (\alpha, w).$$

Then $\theta \circ p(\beta) = \pi(\beta)(v)$ and $\theta \circ q(w) = \pi(\alpha)(w)$ and so:

$$d(\theta \circ p)_\alpha(x) = d\pi_\alpha(x)(v) \ , \ d(\theta \circ q)_v(w) = \pi(\alpha)(w).$$

On the other hand, we have $dp_\alpha(x) = (x, 0)$ and $dq_v(w) = (0, w)$. Thus:

$$d\pi_\alpha(x)(v) = d(\theta \circ p)_\alpha(x) = d\theta_{(\alpha,v)} \circ dp_\alpha(x) = d\theta_{(\alpha,v)}(x, 0),$$
$$\pi(\alpha)(w) = d(\theta \circ q)_v(w) = d\theta_{(\alpha,v)} \circ dq_v(w) = d\theta_{(\alpha,v)}(0, w).$$

Since $d\theta_{(\alpha,v)}(x, w) = d\theta_{(\alpha,v)}(x, 0) + d\theta_{(\alpha,v)}(0, w)$, the result follows.

(ii) Fix $\alpha \in H$ and consider the following isomorphisms of varieties:

$$\sigma : H \times V \to H \times V \ , \ (\beta, w) \mapsto (\alpha\beta, w) \ ,$$
$$\rho : V \to V, \ w \mapsto \pi(\alpha)(w).$$

Then $\theta \circ \sigma = \rho \circ \theta$. If $v \in V$, then $d(\theta \circ \sigma)_{(e_H,v)} = d(\rho \circ \theta)_{(e_H,v)}$, that is, $d\theta_{(\alpha,v)} \circ d\sigma_{(e_H,v)} = \pi(\alpha) \circ d\theta_{(e_H,v)}$. So the result follows.   □

29.2 Cartan subalgebras

29.2.1 We shall use the notations of chapter 19.

Lemma. *Let G_1 be as in 29.1.1, $\mathfrak{h}$ a Lie subalgebra of $\mathfrak{g}$ and $x \in \mathfrak{h}$ be such that $\operatorname{ad} x$ induces an automorphism of $\mathfrak{g}/\mathfrak{h}$. The image of the morphism*

$$\theta : G_1 \times \mathfrak{h} \to \mathfrak{g} , \ (\beta, y) \mapsto \beta(y)$$

contains a dense open subset of $\mathfrak{g}$.

Proof. The hypothesis on x implies that $\mathfrak{g} = \mathfrak{h} + [x, \mathfrak{g}]$. In particular, we have $\mathfrak{g} = \mathfrak{h} + [\mathfrak{g}, \mathfrak{g}]$ and so $\mathfrak{g} = \mathfrak{h} + [x, \mathfrak{g}] = \mathfrak{h} + [x, \mathfrak{h} + [\mathfrak{g}, \mathfrak{g}]] = \mathfrak{h} + [x, [\mathfrak{g}, \mathfrak{g}]]$. Now 29.1.4 says that $d\theta_{(e,x)}(u, y) = u(x) + y$ for $u \in \mathcal{L}(G_1)$ and $y \in \mathfrak{h}$. So the image $\mathfrak{t}$ of $d\theta_{(e,x)}$ is $\mathfrak{h} + \mathcal{L}(G_1)(x)$. But $\mathcal{L}(G_1) = \operatorname{ad}[\mathfrak{g}, \mathfrak{g}]$, and so $\mathfrak{t} = \mathfrak{g}$. It follows from 16.5.7 that θ is dominant, and so the result follows from 15.4.2. $\square$

29.2.2 Lemma. *Any Cartan subalgebra $\mathfrak{h}$ of $\mathfrak{g}$ contains a generic element of $\mathfrak{g}$.*

Proof. For $x \in \mathfrak{h}$, let $\pi(x)$ denote the endomorphism of $\mathfrak{g}/\mathfrak{h}$ induced by $\operatorname{ad} x$. Since $\mathfrak{h} = \mathfrak{n}_\mathfrak{g}(\mathfrak{h})$, 0 is the only element v of $\mathfrak{g}/\mathfrak{h}$ verifying $\pi(x)(v) = 0$ for all $x \in \mathfrak{h}$. As $\mathfrak{h}$ is nilpotent, it follows from 19.4.6 that there exists $x_0 \in \mathfrak{h}$ such that $\pi(x_0)$ is bijective. By applying 29.2.1 and using the fact that the set of generic elements $\mathfrak{g}_{\mathrm{gen}}$ of $\mathfrak{g}$ contains a dense open subset of $\mathfrak{g}$ (19.1.1), we deduce that $G_1(\mathfrak{h}) \cap \mathfrak{g}_{\mathrm{gen}} \neq \emptyset$. Hence we obtain $\mathfrak{h} \cap \mathfrak{g}_{\mathrm{gen}} \neq \emptyset$ because $G_1(\mathfrak{g}_{\mathrm{gen}}) = \mathfrak{g}_{\mathrm{gen}}$. $\square$

29.2.3 Theorem. (i) *A Lie subalgebra of $\mathfrak{g}$ is a Cartan subalgebra of $\mathfrak{g}$ if and only if $\mathfrak{h} = \mathfrak{g}^0(x)$ for some $x \in \mathfrak{g}_{\mathrm{gen}}$.*

(ii) *Let $\mathfrak{h}$ and $\mathfrak{t}$ be Cartan subalgebras of $\mathfrak{g}$. There exists $\alpha \in \operatorname{Aut}_s \mathfrak{g}$ such that $\mathfrak{t} = \alpha(\mathfrak{h})$. In particular, all the Cartan subalgebras of $\mathfrak{g}$ have dimension the rank of $\mathfrak{g}$.*

Proof. (i) This is clear by 19.8.5 (iii) and 29.2.2.

(ii) By (i), $\mathfrak{h} = \mathfrak{g}^0(x)$ for some $x \in \mathfrak{g}_{\mathrm{gen}}$. Let $\lambda_1, \ldots, \lambda_n$ be the pairwise distinct non-zero eigenvalues of $\operatorname{ad} x$. Then:

$$\mathfrak{g} = \mathfrak{g}^0(x) \oplus \mathfrak{g}^{\lambda_1}(x) \oplus \cdots \oplus \mathfrak{g}^{\lambda_n}(x).$$

If $y_i \in \mathfrak{g}^{\lambda_i}$, then $y_i \in \mathfrak{N}(\mathfrak{g})$, and so $\operatorname{ad} y_i$ belongs to the Lie algebra of $\operatorname{Aut}_s \mathfrak{g}$. Consider the morphism

$$\theta : (\operatorname{Aut}_s \mathfrak{g}) \times \mathfrak{h} \to \mathfrak{g} , \ (\beta, z) \mapsto \beta(z).$$

If $z \in \mathfrak{h}$ et $y_i \in \mathfrak{g}^{\lambda_i}$, then 29.1.4 implies that:

$$d\theta_{(e,x)}(0, y) = y , \ d\theta_{(e,x)}(\operatorname{ad} y_i, 0) = [y_i, x].$$

Since $\operatorname{ad} x$ induces a bijective endomorphism of $\mathfrak{g}^{\lambda_i}$, the image of $d\theta_{(e,x)}$ is $\mathfrak{g}$. The same argument as in 29.2.1 shows that $(\operatorname{Aut}_s \mathfrak{g})(\mathfrak{h})$ contains a dense open subset of $\mathfrak{g}$. So the same is true for $(\operatorname{Aut}_s \mathfrak{g})(\mathfrak{k})$. We deduce therefore from 26.2.2 that $\mathfrak{h} \cap \mathfrak{g}_{\mathrm{gen}} \cap (\operatorname{Aut}_s \mathfrak{g})(\mathfrak{k}) \neq \emptyset$. So the result follows easily by 19.8.5 (iii). $\square$

29.2.4 Proposition. *Let $f : \mathfrak{g} \to \mathfrak{g}'$ be a surjective morphism of Lie algebras.*

(i) *The image under f of a Cartan subalgebra $\mathfrak{h}$ of $\mathfrak{g}$ is a Cartan subalgebra of $\mathfrak{g}'$.*

(ii) *Let $\mathfrak{h}'$ be a Cartan subalgebra of $\mathfrak{g}'$. Then any Cartan subalgebra of $f^{-1}(\mathfrak{h}')$ is a Cartan subalgebra of $\mathfrak{g}$.*

(iii) *Any Cartan subalgebra of $\mathfrak{g}'$ is the image under f of a Cartan subalgebra of $\mathfrak{g}$.*

Proof. (i) Let $\mathfrak{a} = \ker f$ and $\pi : \mathfrak{g} \to \mathfrak{g}/\mathfrak{a} \simeq \mathfrak{g}'$ be the canonical surjection. If $\pi(x) \in \mathfrak{n}_{\mathfrak{g}/\mathfrak{a}}(\pi(\mathfrak{h}))$, then $x \in \mathfrak{n}_{\mathfrak{g}}(\mathfrak{h} + \mathfrak{a}) = \mathfrak{h} + \mathfrak{a}$ (19.8.5 (i) and 29.2.3 (i)). Thus $\mathfrak{n}_{\mathfrak{g}/\mathfrak{a}}(\pi(\mathfrak{h})) = \pi(\mathfrak{h})$, and $\mathfrak{n}_{\mathfrak{g}'}(f(\mathfrak{h})) = f(\mathfrak{h})$. Since $f(\mathfrak{h})$ is nilpotent, it is a Cartan subalgebra of $\mathfrak{g}'$.

(ii) Let $\mathfrak{h}$ be a Cartan subalgebra of $\mathfrak{t} = f^{-1}(\mathfrak{h}')$. By (i), $f(\mathfrak{h})$ is a Cartan subalgebra of $\mathfrak{h}'$, so $f(\mathfrak{h}) = \mathfrak{h}'$ (19.8.4). If $x \in \mathfrak{n}_{\mathfrak{g}}(\mathfrak{h})$, then $f(x) \in \mathfrak{n}_{\mathfrak{g}'}(\mathfrak{h}') = \mathfrak{h}'$. Thus $x \in \mathfrak{t}$, and therefore $\mathfrak{n}_{\mathfrak{g}}(\mathfrak{h}) = \mathfrak{n}_{\mathfrak{t}}(\mathfrak{h}) = \mathfrak{h}$. As $\mathfrak{h}$ is nilpotent, it is a Cartan subalgebra of $\mathfrak{g}$.

(iii) This is direct consequence of (i) and (ii). $\square$

29.2.5 Proposition. *Let L be a connected algebraic group and $\mathfrak{l}$ its Lie algebra. A Lie subalgebra of $\mathfrak{l}$ is a Cartan subalgebra of $\mathfrak{l}$ if and only if it is the Lie algebra of a Cartan subgroup of L.*

Proof. Let H be a Cartan subgroup of L and $\mathfrak{h}$ its Lie algebra. Then H is a connected nilpotent closed subgroup verifying $H = N_L(H)^\circ$ (28.5.9). So 24.4.8 and 24.5.13 imply that $\mathfrak{h}$ is a Cartan subalgebra of $\mathfrak{l}$.

Conversely, let $\mathfrak{h}$ be a Cartan subalgebra of $\mathfrak{l}$. Since $\mathfrak{h} = \mathfrak{n}_{\mathfrak{l}}(\mathfrak{h})$, $\mathfrak{h}$ is an algebraic Lie algebra of $\mathfrak{l}$ (24.7.2). Let H be a connected algebraic subgroup of L such that $\mathcal{L}(H) = \mathfrak{h}$. Then H is nilpotent (24.5.13) and $H = N_L(H)^\circ$ by 24.3.5 since $\mathcal{L}(N_L(H)^\circ) = \mathcal{L}(N_L(H)) = \mathfrak{n}_{\mathfrak{l}}(\mathfrak{h}) = \mathfrak{h}$ (24.4.8). So H is a Cartan subgroup of L by 28.5.9. $\square$

29.2.6 Proposition. *Let $\mathfrak{l}$ be the Lie algebra of a connected algebraic group L.*

(i) *An element $x \in \mathfrak{l}$ is semisimple if and only if x belongs to the Lie algebra of a torus of L.*

(ii) *If $x \in \mathfrak{l}$ is semisimple, then x belongs to a Cartan subalgebra of $\mathfrak{l}$.*

Proof. (i) If T is a torus of L, then all the elements of $\mathcal{L}(T)$ are semisimple (26.4.5). Conversely, suppose that $x \in \mathfrak{l}$ is semisimple. Let $H = C_L(x)$. Then

$\mathfrak{c}_{\mathfrak{l}}(x) = \mathcal{L}(H)$ (24.3.6), so $x \in \mathcal{L}(H)$. Let T be a maximal torus of H and $C = C_H(T)^\circ$. Then $\mathcal{L}(C) = \mathfrak{c}_{\mathfrak{h}}(T)$ (26.1.1), so $x \in \mathcal{L}(C)$. Now C is nilpotent (28.1.10) and so $C = T \times C_u$ (26.4.4). It follows therefore from 26.4.5 that $x \in \mathcal{L}(T)$.

(ii) Let T be a maximal torus of L such that $x \in \mathcal{L}(T)$. Then $C = C_L(T)$ is a Cartan subgroup of L containing T. Hence 29.2.5 implies that $\mathcal{L}(C)$ is a Cartan subalgebra of $\mathfrak{l}$ containing x. $\square$

29.2.7 Proposition. *Let* $\mathfrak{m}$ *be the Lie algebra of a connected algebraic group* M, $\mathfrak{a}$ *a Lie subalgebra of* $\mathfrak{m}$ *and* $\mathfrak{s} = \mathbf{a}(\mathfrak{a})$.

(i) *We have* $\mathrm{rk}(\mathfrak{s}) = \mathrm{rk}(\mathfrak{a}) + \dim(\mathfrak{s}/\mathfrak{a})$.

(ii) *If* $\mathfrak{k}$ *is a Cartan subalgebra of* $\mathfrak{s}$, *then* $\mathfrak{a} \cap \mathfrak{k}$ *is a Cartan subalgebra of* $\mathfrak{a}$.

(iii) *If* $\mathfrak{h}$ *is a Cartan subalgebra of* $\mathfrak{a}$, *then* $\mathfrak{k} = \mathbf{a}(\mathfrak{h})$ *is a Cartan subalgebra of* $\mathfrak{s}$ *such that* $\mathfrak{h} = \mathfrak{k} \cap \mathfrak{a}$.

Proof. 1) Let $\mathfrak{k}$ be a Cartan subalgebra of $\mathfrak{s}$ and $\mathfrak{h} = \mathfrak{a} \cap \mathfrak{k}$. Since $\mathfrak{k}$ is algebraic (29.2.5), we have $\mathbf{a}(\mathfrak{h}) \subset \mathfrak{k}$.

Let $x \in \mathfrak{k} \cap \mathfrak{s}_{\mathrm{gen}}$ (29.2.2). We have $\mathfrak{s} = \mathfrak{k} \oplus \mathfrak{l}$ where $\mathfrak{l}$ is a subspace of $\mathfrak{s}$ which is $(\mathrm{ad}\,x)$-stable, and $\mathrm{ad}\,x$ induces an automorphism of $\mathfrak{l}$. Thus $\mathfrak{l} = [x, \mathfrak{l}] \subset [\mathfrak{s}, \mathfrak{s}]$, and 24.5.7 implies that $\mathfrak{s} = \mathfrak{k} + [\mathfrak{a}, \mathfrak{a}]$. We deduce therefore that $\mathfrak{a} = \mathfrak{h} + [\mathfrak{a}, \mathfrak{a}]$.

Moreover, $\mathbf{a}(\mathfrak{h})$ and $[\mathfrak{a}, \mathfrak{a}]$ are algebraic, so $\mathbf{a}(\mathfrak{h}) + [\mathfrak{a}, \mathfrak{a}]$ is an algebraic Lie subalgebra of $\mathfrak{m}$ containing $\mathfrak{a}$ (24.5.10). It follows that $\mathfrak{s} = \mathbf{a}(\mathfrak{h}) + [\mathfrak{a}, \mathfrak{a}]$. Consequently, if $x \in \mathfrak{k}$, then $x = y + z$ where $y \in \mathbf{a}(\mathfrak{h})$ and $z \in [\mathfrak{a}, \mathfrak{a}]$. Hence $z = x - y \in \mathfrak{k}$, and so $z \in \mathfrak{k} \cap \mathfrak{a} = \mathfrak{h}$ which in turn implies that $x \in \mathbf{a}(\mathfrak{h})$. We have therefore proved that $\mathbf{a}(\mathfrak{h}) = \mathfrak{k}$.

2) Denote by S the connected algebraic subgroup of M such that $\mathcal{L}(S) = \mathfrak{s}$. Let $N = \{\alpha \in S; \mathrm{Ad}(\alpha)(\mathfrak{h}) = \mathfrak{h}\}$ and $\mathfrak{n} = \mathcal{L}(N)$. Then $\mathfrak{n} = \mathfrak{n}_{\mathfrak{s}}(\mathfrak{h})$ (24.3.6). But if $\alpha \in N$ and $\mathfrak{p}$ is an algebraic Lie subalgebra of $\mathfrak{s}$ containing $\mathfrak{h}$. Then $\mathrm{Ad}(\alpha)(\mathfrak{p})$ also contains $\mathfrak{h}$. It follows from the definition of $\mathbf{a}(\mathfrak{h}) = \mathfrak{k}$ that $\mathrm{Ad}(\alpha)(\mathfrak{k}) = \mathfrak{k}$. We deduce therefore that $\mathfrak{n} \subset \mathfrak{n}_{\mathfrak{s}}(\mathfrak{k}) = \mathfrak{k}$ (24.3.5). Hence $\mathfrak{n} \cap \mathfrak{a} \subset \mathfrak{h}$, and so $\mathfrak{h} = \mathfrak{n}_{\mathfrak{a}}(\mathfrak{h})$. As $\mathfrak{h}$ is nilpotent, it is a Cartan subalgebra of $\mathfrak{a}$. We have therefore proved (ii).

3) Now $\mathfrak{s} = \mathfrak{k} + \mathfrak{a}$, so $\dim(\mathfrak{s}/\mathfrak{a}) = \dim(\mathfrak{k}/\mathfrak{h})$. So (i) follows.

4) Finally, let $\mathfrak{h}$ be a Cartan subalgebra of $\mathfrak{a}$, $x \in \mathfrak{h} \cap \mathfrak{a}_{\mathrm{gen}}$, and u the endomorphism of $\mathfrak{s}/\mathfrak{a}$ induced by $\mathrm{ad}_{\mathfrak{s}}\,x$. Since $\mathfrak{a}$ is an ideal of $\mathfrak{s}$ containing $[\mathfrak{s}, \mathfrak{s}] = [\mathfrak{a}, \mathfrak{a}]$ (24.5.7), we have $u = 0$. Thus

$$\dim \mathfrak{s}^0(x) = \dim \mathfrak{a}^0(x) + \dim(\mathfrak{s}/\mathfrak{a}),$$

and by (i), we deduce that $x \in \mathfrak{s}_{\mathrm{gen}}$. It follows that $\mathfrak{k} = \mathfrak{s}^0(x)$ is a Cartan subalgebra of $\mathfrak{s}$ (19.8.5). Since $\mathfrak{h} = \mathfrak{a}^0(x)$, we have $\mathfrak{h} = \mathfrak{k} \cap \mathfrak{a}$. So part (iii) follows from point 1. $\square$

29.3 Applications to semisimple Lie algebras

29.3.1 Assume that $\mathfrak{g}$ is semisimple, and denote by L its Killing form. Since L is non-degenerate, the Killing homomorphism $x \mapsto f_x$ defined in 19.2.2 is an isomorphism. In the notations of 19.7.3, we have for $x, y, z \in \mathfrak{g}$:

$$\Phi_{f_x}(y, z) = f_x([y, z]) = L(x, [y, z]).$$

We deduce immediately that:

$$\mathfrak{g}^{(f_x)} = \mathfrak{g}^x.$$

Thus we say that $x \in \mathfrak{g}$ is *regular* if $f_x \in \mathfrak{g}^*_{\mathrm{reg}}$. Let us denote by $\mathfrak{g}_{\mathrm{reg}}$ the set of regular elements of $\mathfrak{g}$; it is a dense open subset of $\mathfrak{g}$.

29.3.2 Proposition. *If $\mathfrak{g}$ is semisimple, then its rank is equal to its index.*

Proof. Since $\mathfrak{g}_{\mathrm{gen}}$ and $\mathfrak{g}_{\mathrm{reg}}$ are both dense open subsets of $\mathfrak{g}$, there exists $x \in \mathfrak{g}$ which is regular and generic. So the result is clear since $\mathfrak{g}^x = \mathfrak{g}^0(x)$ (19.8.7 (v)). $\square$

29.3.3 Theorem. *Let $\mathfrak{g}$ be a semisimple Lie algebra, $\mathfrak{h}_1$ and $\mathfrak{h}_2$ two Cartan subalgebras of $\mathfrak{g}$. Then the root systems $R_1 = R(\mathfrak{g}, \mathfrak{h}_1)$ and $R_2 = R(\mathfrak{g}, \mathfrak{h}_2)$ are isomorphic.*

Proof. By 29.2.3, there exists $\theta \in \mathrm{Aut}_e \, \mathfrak{g}$ such that $\theta(\mathfrak{h}_1) = \mathfrak{h}_2$. Let us define $\tau : \mathfrak{h}_1^* \to \mathfrak{h}_2^*$ by $\lambda \mapsto \lambda \circ \theta^{-1}$.

Let $x \in \mathfrak{g}$ and $\alpha \in R_1$ be such that $[h, x] = \alpha(h)x$ for all $h \in \mathfrak{h}_1$. If $t \in \mathfrak{h}_2$, then:

$$[t, \theta(x)] = \theta\big([\theta^{-1}(t), x]\big) = \alpha\big(\theta^{-1}(t)\big)\theta(x) = \big(\tau(\alpha)(t)\big)\theta(x).$$

Thus $\tau(R_1) = R_2$, and τ induces an isomorphism from $(\mathfrak{h}_1^*)_{\mathbb{R}}$ onto $(\mathfrak{h}_2^*)_{\mathbb{R}}$. If $x, y \in \mathfrak{h}_1$, then:

$$L(x, y) = \sum_{\alpha \in R_1} \alpha(x)\alpha(y) = \sum_{\alpha \in R_1} \tau(\alpha)\big(\theta(x)\big)\,\tau(\alpha)\big(\theta(y)\big)$$
$$= \sum_{\beta \in R_2} \beta\big(\theta(x)\big)\beta\big(\theta(y)\big) = L\big(\theta(x), \theta(y)\big).$$

For $i = 1, 2$, $\lambda \in \mathfrak{h}_i^*$, let h_λ^i be the unique element of $\mathfrak{h}_i$ verifying $L(h, h_\lambda^i) = \lambda(h)$ for all $h \in \mathfrak{h}_i$. It follows that $\theta(h_\lambda^1) = h_{\tau(\lambda)}^2$. Hence in the notations of 20.6.1,

$$\langle \lambda, \mu \rangle = L(h_\lambda^1, h_\mu^1) = L(h_{\tau(\lambda)}^2, h_{\tau(\mu)}^2) = \langle \tau(\lambda), \tau(\mu) \rangle$$

for all $\lambda, \mu \in \mathfrak{h}_1^*$. So 18.13.1 implies that R_1 and R_2 are isomorphic. $\square$

29.3.4 Let $\mathfrak{h}$ be a Cartan subalgebra of a semisimple Lie algebra $\mathfrak{g}$. Then 29.3.3 says that up to an isomorphism, $R = R(\mathfrak{g}, \mathfrak{h})$ does not depend on the choice of $\mathfrak{h}$. We shall say that R is the root system of $\mathfrak{g}$ and similarly, for all the objects relative to R such as Weyl group, weights and so on.

29.3.5 Proposition. *Assume that $\mathfrak{g}$ is semisimple and denote by R its root system.*

(i) *If $\mathfrak{g} = \mathfrak{g}_1 \times \mathfrak{g}_2$ and R_1, R_2 are the root systems of $\mathfrak{g}_1$ and $\mathfrak{g}_2$, then R is the direct sum of R_1 and R_2.*

(ii) *If R is the direct sum of root systems R_1 and R_2, then $\mathfrak{g} = \mathfrak{g}_1 \times \mathfrak{g}_2$ where*

$$\mathfrak{g}_1 = \sum_{\alpha \in R_1} \Bbbk H_\alpha + \sum_{\alpha \in R_1} \mathfrak{g}^\alpha \; , \; \mathfrak{g}_2 = \sum_{\alpha \in R_2} \Bbbk H_\alpha + \sum_{\alpha \in R_2} \mathfrak{g}^\alpha,$$

where H_α is as defined in 20.6.5.

Proof. (i) The ideal $\mathfrak{g}_1$ and $\mathfrak{g}_2$ are orthogonal with respect to L. So if $\mathfrak{h}_i$ is a Cartan subalgebra of $\mathfrak{g}_i$, then $\mathfrak{h} = \mathfrak{h}_1 \times \mathfrak{h}_2$ is a Cartan subalgebra of $\mathfrak{g}$. We have $\mathfrak{h}^* = \mathfrak{h}_1^* \times \mathfrak{h}_2^*$, and by taking $R_i = R(\mathfrak{g}_i, \mathfrak{h}_i)$ and $R = R(\mathfrak{g}, \mathfrak{h})$, it is clear that $R = R_1 \cup R_2$ and that R_1 and R_2 are orthogonal.

(ii) If $\alpha, \beta \in R_i$ verify $\alpha + \beta \in R$, then $\alpha + \beta \in R_i$. So $\mathfrak{g}_i$ is a subalgebra of $\mathfrak{g}$.

Let $\alpha \in R_1$ and $\beta \in R_2$. Then $\alpha + \beta \notin R$, so $[\mathfrak{g}^\alpha, \mathfrak{g}^\beta] = \{0\}$. Moreover, $(\alpha|\beta) = 0$, so $\beta(H_\alpha) = 0$ and $[H_\alpha, \mathfrak{g}^\beta] = \{0\}$. Hence $[\mathfrak{g}_1, \mathfrak{g}_2] = \{0\}$. $\square$

29.3.6 Corollary. *Let $\mathfrak{g}$ be a semisimple Lie algebra and R its root system. Then $\mathfrak{g}$ is simple if and only if R is irreducible.*

29.3.7 We shall admit the following result:

Theorem. (i) *Two semisimple Lie algebras are isomorphic if and only if their root systems are isomorphic.*

(ii) *Each of the diagrams A_l, B_l, C_l, D_l, E_6, E_7, E_8, F_4, G_2 of 18.14 corresponds to an isomorphism class of simple Lie algebras.*

29.3.8 Let $\mathfrak{g}$ be a simple Lie algebra. We shall say, for example, that $\mathfrak{g}$ is of type A_l if its root system is of type A_l.

29.4 Borel subalgebras

29.4.1 Definition. (i) *A maximal solvable Lie subalgebra of a Lie algebra $\mathfrak{g}$ is called a* Borel subalgebra *of $\mathfrak{g}$.*

(ii) *A Lie subalgebra of a Lie algebra $\mathfrak{g}$ is a* parabolic subalgebra *of $\mathfrak{g}$ if it contains a Borel subalgebra of $\mathfrak{g}$.*

Remark. By 20.8.2, these definitions are compatible with the ones for semisimple Lie algebras in 20.8.1 and 20.8.5.

29.4.2 Lemma. *Let $\mathfrak{b}$ be a Borel subalgebra of a Lie algebra $\mathfrak{g}$. Then $\mathfrak{n}_\mathfrak{g}(\mathfrak{b}) = \mathfrak{b}$.*

Proof. If $x \in \mathfrak{n}_\mathfrak{g}(\mathfrak{b})$, then $\mathfrak{b} + \Bbbk x$ is a solvable Lie subalgebra of $\mathfrak{h}$. So $x \in \mathfrak{b}$.
$\square$

29.4.3 Proposition. *Let $\mathfrak{h}$ be the Lie algebra of a connected algebraic group H.*

(i) *A Borel subalgebra of $\mathfrak{h}$ is algebraic in $\mathfrak{h}$.*

(ii) *A Lie subalgebra of $\mathfrak{h}$ is a Borel subalgebra if and only if it is the Lie algebra of a Borel subgroup of H. In particular, all the Borel subalgebras of $\mathfrak{h}$ are conjugate.*

(iii) *If $\mathfrak{p}$ is a parabolic subalgebra of $\mathfrak{h}$, then it is algebraic in $\mathfrak{h}$, it contains a Cartan subalgebra of $\mathfrak{h}$, and $\mathfrak{p} = \mathfrak{n}_{\mathfrak{h}}(\mathfrak{p})$.*

(iv) *A Lie subalgebra of $\mathfrak{h}$ is a parabolic subalgebra if and only if it is the Lie algebra of a parabolic subgroup of H.*

Proof. (i) This follows from 24.7.2 and 29.4.2.

(ii) Let $\mathfrak{b}$ be a Borel subalgebra of $\mathfrak{h}$. By (i), there exists a connected algebraic subgroup B of H such that $\mathcal{L}(B) = \mathfrak{b}$. So B is solvable by 24.5.13. If R is a connected solvable algebraic subgroup of H containing B, then $\mathcal{L}(R)$ is a solvable Lie algebra containing $\mathfrak{b}$. It follows that $R = B$. Hence B is a Borel subgroup of H.

Conversely, let B be a Borel subgroup of H and $\mathfrak{b} = \mathcal{L}(B)$. Then $\mathfrak{b}$ is solvable. If $\mathfrak{r}$ is a solvable Lie subalgebra of $\mathfrak{h}$ containing $\mathfrak{b}$, then $\mathcal{Q}(\mathfrak{r})$ is a connected solvable algebraic subgroup of H containing B. Thus $\mathfrak{r} = \mathfrak{b}$ and $\mathfrak{b}$ is a Borel subalgebra of $\mathfrak{h}$.

(iii) The group $\mathcal{Q}(\mathfrak{p})$ contains a Borel subgroup of H (24.3.5), and so a Cartan subgroup of H (28.1.7 and 28.3.3). By 29.2.5, $\mathfrak{p}$ contains a Cartan subalgebra of $\mathfrak{h}$. The fact that $\mathfrak{p} = \mathfrak{n}_{\mathfrak{h}}(\mathfrak{p})$ follows from 19.8.5 and 29.2.2, hence $\mathfrak{p}$ is algebraic (24.7.2).

(iv) This is an immediate consequence of (iii). $\square$

29.4.4 Proposition. *Let $\mathfrak{m}$ be the Lie algebra of a connected algebraic group M, $\mathfrak{a}$ a Lie subalgebra of $\mathfrak{m}$ and $\mathfrak{s} = \mathbf{a}(\mathfrak{a})$.*

(i) *Let $\mathfrak{k}$ be a Borel subalgebra of $\mathfrak{s}$. Then $\mathfrak{b} = \mathfrak{k} \cap \mathfrak{a}$ is a Borel subalgebra of $\mathfrak{a}$ such that $\mathfrak{k} = \mathbf{a}(\mathfrak{b})$. Moreover, we have:*

$$\dim(\mathfrak{k}/\mathfrak{b}) = \dim(\mathfrak{s}/\mathfrak{a}).$$

(ii) *If $\mathfrak{b}$ is a Borel subalgebra of $\mathfrak{a}$, then $\mathbf{a}(\mathfrak{b})$ is a Borel subalgebra of $\mathfrak{s}$ such that $\mathbf{a}(\mathfrak{b}) \cap \mathfrak{a} = \mathfrak{b}$.*

Proof. (i) Since $\mathfrak{k}$ is an algebraic Lie subalgebra of $\mathfrak{s}$, $\mathbf{a}(\mathfrak{b}) \subset \mathfrak{k}$ and $\mathfrak{k}$ contains a Cartan subalgebra $\mathfrak{l}$ of $\mathfrak{s}$ (29.4.3). By 29.2.7 and its proof, we have $\mathbf{a}(\mathfrak{h}) = \mathfrak{l}$ and $\mathfrak{s} = \mathfrak{l} + [\mathfrak{a}, \mathfrak{a}]$ where $\mathfrak{h} = \mathfrak{l} \cap \mathfrak{a}$. So an element $x \in \mathfrak{k}$ is of the form $y + z$ where $y \in \mathfrak{l}$ and $z \in [\mathfrak{a}, \mathfrak{a}]$. So $z = x - y \in \mathfrak{k} \cap \mathfrak{a} = \mathfrak{b}$. Hence $x \in \mathbf{a}(\mathfrak{b})$ and we obtain that $\mathbf{a}(\mathfrak{b}) = \mathfrak{k}$.

Let $\mathfrak{b}'$ be a solvable Lie subalgebra of $\mathfrak{a}$ containing $\mathfrak{b}$. Then $\mathbf{a}(\mathfrak{b}')$ is a solvable Lie subalgebra of $\mathfrak{s}$ containing $\mathbf{a}(\mathfrak{b}) = \mathfrak{k}$. Thus $\mathbf{a}(\mathfrak{b}') = \mathfrak{k}$ and $\mathfrak{b}' \subset \mathfrak{k} \cap \mathfrak{a} = \mathfrak{b}$. We have therefore proved that $\mathfrak{b}$ is a Borel subalgebra of $\mathfrak{a}$.

Observe that the Lie subalgebras $\mathfrak{l}$ and $\mathfrak{h}$ above are Cartan subalgebras of $\mathfrak{s}$ and $\mathfrak{a}$ (29.2.7). We deduce therefore that $\mathrm{rk}(\mathfrak{a}) = \mathrm{rk}(\mathfrak{b})$ and $\mathrm{rk}(\mathfrak{s}) = \mathrm{rk}(\mathfrak{k})$. Since $\mathfrak{k} = \mathfrak{a}(\mathfrak{b})$, the equality of dimensions follows immediately from 29.2.7 (i).

(ii) The Lie algebra $\mathfrak{k} = \mathfrak{a}(\mathfrak{b})$ is solvable (24.5.7). If $\mathfrak{k}'$ is a Borel subalgebra of $\mathfrak{s}$ containing $\mathfrak{k}$, then $\mathfrak{b} = \mathfrak{a} \cap \mathfrak{k}'$. Hence $\mathfrak{k}' = \mathfrak{a}(\mathfrak{b}) = \mathfrak{k}$ by (i). $\square$

29.4.5 Corollary. (i) *Any Borel subalgebra of $\mathfrak{g}$ contains a Cartan subalgebra of $\mathfrak{g}$.*

(ii) *All the Borel subalgebras of $\mathfrak{g}$ have the same dimension.*

Proof. By 19.2.1, we may assume that $\mathfrak{g}$ is a subalgebra of $\mathfrak{gl}(V) = \mathcal{L}(\mathrm{GL}(V))$ for some finite-dimensional vector space V. Then the results are consequences of 29.2.7, 29.4.3 and 29.4.4. $\square$

29.4.6 Assume that $\mathfrak{g}$ is semisimple. Let $\mathfrak{b}$ be a Borel subalgebra of $\mathfrak{g}$. Then 29.4.5 says that $\mathfrak{b}$ contains a Cartan subalgebra $\mathfrak{h}$ of $\mathfrak{g}$. Let $R = R(\mathfrak{g}, \mathfrak{h})$. Then 20.7.1 and 20.7.2 imply that there is a closed subset P of R such that $\mathfrak{b} = \mathfrak{h} + \mathfrak{g}^P$ (notations of 20.7.1). Since $\mathfrak{b}$ is solvable, $P \cap (-P) = \emptyset$ (20.7.5). By 18.10.2, there exists a chamber C of R such that $P \subset R_+(C)$. Furthermore, $P = R_+(C)$ since $\mathfrak{b}$ is maximal solvable.

We have therefore proved that a Lie subalgebra of $\mathfrak{g}$ is a Borel subalgebra if and only if it is a Borel subalgebra of $(\mathfrak{g}, \mathfrak{h})$ (20.8.1) for some Cartan subalgebra $\mathfrak{h}$ of $\mathfrak{g}$.

29.4.7 Theorem. *Let $\mathfrak{b}$ and $\mathfrak{b}'$ be Borel subalgebras of $\mathfrak{g}$. Then there exists $\theta \in \mathrm{Aut}_s\, \mathfrak{g}$ such that $\mathfrak{b}' = \theta(\mathfrak{b})$.*

Proof. Let us proceed by induction on $\dim \mathfrak{g}$. The result is clear if $\mathfrak{g}$ is solvable. So let us assume that $\mathfrak{g}$ is not solvable, and let $\mathfrak{r} = \mathrm{rad}\,\mathfrak{g}$.

Suppose that $\mathfrak{r} \neq \{0\}$. Let $\mathfrak{a} = \mathfrak{g}/\mathfrak{r}$ and $f : \mathfrak{g} \to \mathfrak{a}$ be the canonical surjection. The Lie algebra $\mathfrak{b} + \mathfrak{r}$ is solvable, so $\mathfrak{r} \subset \mathfrak{b}$. Similarly $\mathfrak{r} \subset \mathfrak{b}'$. Since $f(\mathfrak{b})$ and $f(\mathfrak{b}')$ are Borel subalgebras of $\mathfrak{a}$, by the induction hypothesis, there exists $\sigma \in \mathrm{Aut}_s\, \mathfrak{a}$ such that $f(\mathfrak{b}') = \sigma(f(\mathfrak{b}))$. As $\mathfrak{r} \subset \mathfrak{b} \cap \mathfrak{b}'$, it follows from 29.1.2 that there exists $\theta \in \mathrm{Aut}_s\, \mathfrak{g}$ such that $\mathfrak{b}' = \theta(\mathfrak{b})$.

We may therefore assume that $\mathfrak{g}$ is semisimple. Then by 29.4.6, there exists a Cartan subalgebra $\mathfrak{h}$ (resp. $\mathfrak{h}'$) of $\mathfrak{g}$ such that $\mathfrak{b}$ (resp. $\mathfrak{b}'$) is a Borel subalgebra of $(\mathfrak{g}, \mathfrak{h})$ (resp. $(\mathfrak{g}, \mathfrak{h}')$). Let B (resp. B') be a base of $R(\mathfrak{g}, \mathfrak{h})$ (resp. $R(\mathfrak{g}, \mathfrak{h}')$) which defines $\mathfrak{b}$ (resp. $\mathfrak{b}'$). By 29.2.3 (ii), we may assume that $\mathfrak{h} = \mathfrak{h}'$. Now 18.8.7 says that there exists $w \in W(\mathfrak{g}, \mathfrak{h})$ such that $w(B) = B'$. So the result follows from 29.1.3. $\square$

29.4.8 Corollary. *Assume that $\mathfrak{g}$ is semisimple. Let L denote its Killing form.*

(i) *If $\mathfrak{n}$ is a Lie subalgebra of $\mathfrak{g}$ whose elements are nilpotent, then $\mathfrak{n}$ is contained in the nilpotent radical of a Borel subalgebra of $\mathfrak{g}$.*

(ii) *If $x \in \mathfrak{g}$, then x is contained in a Borel subalgebra of $\mathfrak{g}$. In particular, there exists a Borel subalgebra of $\mathfrak{g}$ verifying the following conditions:*

$$x \in \mathfrak{b} \ , \ 2 \dim \mathfrak{b} = \dim \mathfrak{g} + \mathrm{rk}(\mathfrak{g}) \ , \ L(x, [\mathfrak{b}, \mathfrak{b}]) = \{0\}.$$

Proof. (i) By 19.3.7, $\mathfrak{n}$ is nilpotent, so it is contained in a Borel subalgebra $\mathfrak{b}$ of $\mathfrak{g}$. So by 20.8.3 and 29.4.6, $\mathfrak{n}$ is contained in the nilpotent radical of $\mathfrak{b}$.

(ii) The first part is clear and the second follows from 29.4.6.   □

29.4.9 Proposition. *Let $\mathfrak{b}$ and $\mathfrak{b}'$ be Borel subalgebras of a semisimple Lie algebra $\mathfrak{g}$. Then $\mathfrak{b} \cap \mathfrak{b}'$ contains a Cartan subalgebra of $\mathfrak{g}$.*

Proof. Let $\mathfrak{n} = [\mathfrak{b}, \mathfrak{b}]$, $\mathfrak{n}' = [\mathfrak{b}', \mathfrak{b}']$, $\mathfrak{p} = \mathfrak{b} \cap \mathfrak{b}'$, $\mathfrak{s}$ a complementary subspace of $\mathfrak{b} + \mathfrak{b}'$ in $\mathfrak{g}$, and $\mathfrak{h}$ a Cartan subalgebra of $\mathfrak{b}$ (and so of $\mathfrak{g}$ by 29.4.5). Also let l, n, p be the dimensions of $\mathfrak{h}, \mathfrak{n}$ and $\mathfrak{p}$. We have $\dim \mathfrak{b} = \dim \mathfrak{b}' = l + n$, and:

$$\dim \mathfrak{s}^{\perp} = \dim(\mathfrak{b} + \mathfrak{b}') = 2(l + n) - p,$$

where $\mathfrak{s}^{\perp}$ is the orthogonal of $\mathfrak{s}$ with respect to the Killing form of $\mathfrak{g}$. Since $\dim(\mathfrak{s}^{\perp} \cap \mathfrak{p}) \geqslant \dim \mathfrak{s}^{\perp} + \dim \mathfrak{p} - \dim \mathfrak{g}$, we have:

$$(1) \qquad \dim(\mathfrak{s}^{\perp} \cap \mathfrak{p}) \geqslant 2(l + n) - p + p - (l + 2n) = l.$$

By 20.8.3 and 20.8.6, we have $\mathfrak{n} = \mathfrak{b}^{\perp}$, $\mathfrak{n}' = (\mathfrak{b}')^{\perp}$ and $\mathfrak{p} \cap \mathfrak{n} \subset \mathfrak{n}'$. It follows that $\mathfrak{p} \cap \mathfrak{n} \subset \mathfrak{n} \cap \mathfrak{n}' = \mathfrak{b}^{\perp} \cap (\mathfrak{b}')^{\perp}$. Consequently, $\mathfrak{s}^{\perp} \cap \mathfrak{p} \cap \mathfrak{n} = \{0\}$. We deduce therefore from (1) that $\mathfrak{b} = \mathfrak{n} \oplus (\mathfrak{s}^{\perp} \cap \mathfrak{p})$ and $\mathfrak{b}' = \mathfrak{n}' \oplus (\mathfrak{s}^{\perp} \cap \mathfrak{p})$.

Let $z \in \mathfrak{h} \cap \mathfrak{g}_{\mathrm{gen}}$ (29.2.2). There exists $y \in \mathfrak{n}$ such that $x = y + z \in \mathfrak{s}^{\perp} \cap \mathfrak{p}$. By 20.8.3 (i), $\mathrm{ad}_{\mathfrak{g}} x$ and $\mathrm{ad}_{\mathfrak{g}} z$ have the same characteristic polynomial, so $x \in \mathfrak{g}_{\mathrm{gen}}$. It follows therefore from 19.1.2 that x is generic simultaneously in $\mathfrak{g}$, $\mathfrak{b}$ and $\mathfrak{b}'$. As $\mathfrak{g}$, $\mathfrak{b}$ and $\mathfrak{b}'$ have the same rank (29.4.5), we deduce that $\mathfrak{g}^{0}(x) = \mathfrak{b}^{0}(x) = (\mathfrak{b}')^{0}(x)$ is a Cartan subalgebra of $\mathfrak{g}$, $\mathfrak{b}$ and $\mathfrak{b}'$.   □

29.4.10 Corollary. *Assume that $\mathfrak{g}$ is semisimple. The group $\mathrm{Aut}_{e} \mathfrak{g}$ acts transitively on the set of pairs $(\mathfrak{h}, \mathfrak{b})$ where $\mathfrak{b}$ is a Borel subalgebra of $\mathfrak{g}$ and $\mathfrak{h}$ is a Cartan subalgebra of $\mathfrak{g}$ contained in $\mathfrak{b}$.*

Proof. Given $(\mathfrak{h}_1, \mathfrak{b}_1)$ and $(\mathfrak{h}_2, \mathfrak{b}_2)$ two such pairs. By 29.4.9, there exists a Cartan subalgebra $\mathfrak{h}$ of $\mathfrak{g}$ contained in $\mathfrak{b}_1 \cap \mathfrak{b}_2$. By 29.2.3, we may therefore assume that $\mathfrak{h} = \mathfrak{h}_1 = \mathfrak{h}_2$.

Let $R = R(\mathfrak{g}, \mathfrak{h})$ and B_i a base of R associated to $\mathfrak{b}_i$, $i = 1, 2$. Then there exists $w \in W(\mathfrak{g}, \mathfrak{h})$ such that $w(B_1) = B_2$ (18.8.7). By 29.1.3, there exists $\theta \in \mathrm{Aut}_{e} \mathfrak{g}$ such that $\theta|_{\mathfrak{h}} = w$. Thus $\theta(\mathfrak{h}) = \mathfrak{h}$ and $\theta(\mathfrak{b}_1) = \mathfrak{b}_2$.   □

29.5 Properties of parabolic subalgebras

29.5.1 Definition. *A special automorphism of a Lie algebra $\mathfrak{a}$ is an automorphism of the form $e^{\mathrm{ad} x}$ where x belongs to the nilpotent radical of $\mathfrak{a}$.*

29.5.2 The following result is a refinement of 20.3.5.

Theorem. *Let $\mathfrak{a}$ be a Lie algebra and $\mathfrak{s}$, $\mathfrak{t}$ be Levi subalgebras of $\mathfrak{a}$. Then there exists a special automorphism of $\mathfrak{a}$ which sends $\mathfrak{s}$ to $\mathfrak{t}$.*

Proof. Let $\mathfrak{r} = \mathrm{rad}\,\mathfrak{a}$ and $\mathfrak{n}$ the nilpotent radical of $\mathfrak{a}$. By 20.3.6, we have $\mathfrak{n} = [\mathfrak{a}, \mathfrak{r}] = [\mathfrak{a}, \mathfrak{a}] \cap \mathfrak{r}$.

1) If $[\mathfrak{a}, \mathfrak{r}] = \mathfrak{n} = \{0\}$, then $\mathfrak{s} = [\mathfrak{a}, \mathfrak{a}] = \mathfrak{t}$ (20.5.4), and the result is clear.

2) Suppose that $[\mathfrak{a}, \mathfrak{r}] \neq \{0\}$ and the only ideals of $\mathfrak{a}$ contained in $\mathfrak{r}$ are $\mathfrak{r}$ and $\{0\}$. Then $[\mathfrak{a}, \mathfrak{r}] = \mathfrak{r}$, $[\mathfrak{r}, \mathfrak{r}] = \{0\}$, and the centre of $\mathfrak{a}$ is $\{0\}$.

For $x \in \mathfrak{t}$, let $\varphi(x)$ be the unique element of $\mathfrak{r}$ such that $x + \varphi(x) \in \mathfrak{s}$. If $x, y \in \mathfrak{t}$, then:

$$[x + \varphi(x), y + \varphi(y)] = [x, y] + [x, \varphi(y)] + [\varphi(x), y] \in \mathfrak{s}.$$

Hence:

$$\varphi([x, y]) = (\mathrm{ad}\,x)(\varphi(y)) - (\mathrm{ad}\,y)(\varphi(x)).$$

By 20.3.3, there exists $r \in \mathfrak{r}$ such that $\varphi(x) = -[x, r]$ for all $x \in \mathfrak{t}$. Since $\mathfrak{r}$ is commutative, we obtain, for $x \in \mathfrak{t}$, that:

$$x + \varphi(x) = x + [r, x] = e^{\mathrm{ad}\,r}(x).$$

Hence our result since we have $\mathfrak{r} = \mathfrak{n}$ here.

3) Let us consider the general case. We proceed by induction on the dimension n of $\mathfrak{r}$. We may suppose that $n > 0$ and $\mathfrak{n} = [\mathfrak{a}, \mathfrak{r}] \neq \{0\}$. The centre $\mathfrak{c}$ of $\mathfrak{n}$ is non-trivial (19.3.4). Let $\mathfrak{m}$ be a minimal non-zero ideal of $\mathfrak{a}$ contained in $\mathfrak{c}$. If $\mathfrak{m} = \mathfrak{r}$, then the result follows from point 2. So let us assume that $\mathfrak{m} \neq \mathfrak{r}$.

Let $\mathfrak{a}' = \mathfrak{a}/\mathfrak{m}$, $\mathfrak{r}' = \mathfrak{r}/\mathfrak{m} = \mathrm{rad}(\mathfrak{a}')$ and $f : \mathfrak{a} \to \mathfrak{a}'$ the canonical surjection. Since the Lie algebras $f(\mathfrak{s})$ and $f(\mathfrak{t})$ are Levi subalgebras of $\mathfrak{a}'$, it follows from the induction hypothesis that there exists $a' \in [\mathfrak{a}', \mathfrak{r}']$ such that $f(\mathfrak{t}) = e^{\mathrm{ad}\,a'}(f(\mathfrak{s}))$.

Fix $a \in [\mathfrak{a}, \mathfrak{r}]$ such that $a' = f(a)$. Then:

$$\mathfrak{s}_1 = e^{\mathrm{ad}\,a}(\mathfrak{s}) \subset \mathfrak{m} + \mathfrak{t} = \mathfrak{h}.$$

Since $\mathfrak{s}_1$ and $\mathfrak{t}$ are Levi subalgebras of $\mathfrak{h}$, it follows again by the induction hypothesis that $e^{\mathrm{ad}\,b}(\mathfrak{s}_1) = \mathfrak{t}$ for some $b \in \mathfrak{m}$. The element b belongs to the centre of $[\mathfrak{a}, \mathfrak{r}]$, so:

$$\mathfrak{t} = e^{\mathrm{ad}\,b}e^{\mathrm{ad}\,a}(\mathfrak{s}) = e^{\mathrm{ad}(a+b)}(\mathfrak{s}).$$

Hence the result since $a + b \in [\mathfrak{a}, \mathfrak{r}]$.   □

29.5.3 Corollary. *Let $\mathfrak{s}$ be a Levi subalgebra of $\mathfrak{a}$ and $\mathfrak{h}$ a semisimple subalgebra of $\mathfrak{a}$.*

(i) There is a special automorphism of $\mathfrak{a}$ which sends $\mathfrak{h}$ into a subalgebra of $\mathfrak{s}$.

(ii) $\mathfrak{h}$ is contained in a Levi subalgebra of $\mathfrak{a}$.

(iii) The Levi subalgebras of $\mathfrak{a}$ are the maximal semisimple subalgebras of the Lie algebra $\mathfrak{a}$.

Proof. It suffices to prove (i). Let $\mathfrak{b} = \mathfrak{h} + \operatorname{rad}\mathfrak{a}$. It is clear that $\mathfrak{h}$ is a Levi subalgebra of $\mathfrak{b}$ and $\mathfrak{b} \cap \mathfrak{s}$ is a complementary Lie subalgebra of $\operatorname{rad}\mathfrak{a}$ in $\mathfrak{b}$. So $\mathfrak{b} \cap \mathfrak{s}$ is a Levi subalgebra of $\mathfrak{b}$. It follows from 29.5.2 that $e^{\operatorname{ad}a}(\mathfrak{h}) = \mathfrak{b} \cap \mathfrak{s}$ for some $a \in [\mathfrak{b}, \operatorname{rad}\mathfrak{a}]$. $\square$

29.5.4 Proposition. *Let $\mathfrak{g}$ be a semisimple Lie algebra, $\mathfrak{p}$ a parabolic subalgebra of $\mathfrak{g}$, $\mathcal{H}$ the set of Cartan subalgebras of $\mathfrak{p}$, $\mathcal{S}$ the set of commutative Lie subalgebras of $\mathfrak{p}$ consisting of semisimple elements and $\mathcal{S}_0$ the set of maximal elements of $\mathcal{S}$. Then $\mathcal{H} = \mathcal{S}_0$.*

Proof. Since Cartan subalgebras of $\mathfrak{p}$ are exactly the Cartan subalgebras of $\mathfrak{g}$ contained in $\mathfrak{p}$ (29.4.3 and 29.2.3), it follows from 20.6.2 and 20.5.16 that $\mathcal{H} \subset \mathcal{S}_0$.

Now let $\mathfrak{t} \in \mathcal{S}_0$ and $\mathfrak{c} = \mathfrak{c}_{\mathfrak{p}}(\mathfrak{t})$, and $\mathfrak{h}$ a Cartan subalgebra of $\mathfrak{c}$. Then $\mathfrak{t} \subset \mathfrak{h}$.

Since $\mathfrak{t}$ consists of semisimple elements which commute pairwise, we have a decomposition $\mathfrak{q} = \mathfrak{n}_{\mathfrak{p}}(\mathfrak{h}) = \mathfrak{m} \oplus \mathfrak{h}$ of $(\operatorname{ad}_{\mathfrak{p}}\mathfrak{t})$-stable subspaces. So:

$$[\mathfrak{t}, \mathfrak{m}] \subset [\mathfrak{t}, \mathfrak{q}] \subset [\mathfrak{h}, \mathfrak{q}] \subset \mathfrak{h}.$$

Thus $[\mathfrak{t}, \mathfrak{m}] \subset \mathfrak{h} \cap \mathfrak{m} = \{0\}$, and $\mathfrak{m} \subset \mathfrak{c}$. Consequently, $\mathfrak{q} = \mathfrak{n}_{\mathfrak{c}}(\mathfrak{h})$, hence $\mathfrak{q} = \mathfrak{h}$. It follows that $\mathfrak{h}$ is a Cartan subalgebra of $\mathfrak{p}$, hence of $\mathfrak{g}$. Thus $\mathfrak{h}$ is a commutative Lie subalgebra of $\mathfrak{p}$ consisting of semisimple elements. So $\mathfrak{t} = \mathfrak{h}$ and $\mathcal{S}_0 \subset \mathcal{H}$. $\square$

29.5.5 Corollary. *Let $\mathfrak{g}$ be a semisimple Lie algebra, $\mathfrak{p}$ a parabolic subalgebra of $\mathfrak{g}$ and $\mathfrak{t}$ a Lie subalgebra of $\mathfrak{p}$ consisting of semisimple elements. Then $\mathfrak{t}$ is contained in a Cartan subalgebra of $\mathfrak{p}$.*

Proof. This follows from 29.5.4 since $\mathfrak{t}$ is commutative (20.5.16). $\square$

29.5.6 Definition. *Let $\mathfrak{g}$ be a semisimple Lie algebra, $\mathfrak{p}$ a parabolic subalgebra of $\mathfrak{g}$ and $\mathfrak{n}$ the nilpotent radical of $\mathfrak{p}$. A Lie subalgebra $\mathfrak{l}$ of $\mathfrak{p}$ is called a Levi factor of $\mathfrak{p}$ if it is reductive in $\mathfrak{g}$ and $\mathfrak{p} = \mathfrak{l} \oplus \mathfrak{n}$.*

29.5.7 Theorem. *Let $\mathfrak{g}$ be a semisimple Lie algebra and $\mathfrak{p}$ a parabolic subalgebra of $\mathfrak{g}$.*

  (i) *$\mathfrak{p}$ has a Levi factor.*

  (ii) *If $\mathfrak{l}$ and $\mathfrak{s}$ are Levi factors of $\mathfrak{p}$, then there is an elementary automorphism of $\mathfrak{p}$ which sends $\mathfrak{l}$ into $\mathfrak{s}$.*

  (iii) *Let $\mathfrak{c}$ be the centre of $\mathfrak{l}$. Then $\mathfrak{c} = \mathfrak{c}_{\mathfrak{g}}(\mathfrak{l})$ and $\mathfrak{l} = \mathfrak{c}_{\mathfrak{g}}(\mathfrak{c})$.*

Proof. (i) This follows from 20.8.6.

(ii) Let $\mathfrak{r} = \operatorname{rad}\mathfrak{p}$ and $\mathfrak{n}$ the nilpotent radical of $\mathfrak{p}$. We have $\mathfrak{l} = \mathfrak{l}' \oplus \mathfrak{c}$, $\mathfrak{s} = \mathfrak{s}' \oplus \mathfrak{d}$, where $\mathfrak{l}' = [\mathfrak{l}, \mathfrak{l}]$ and $\mathfrak{s}' = [\mathfrak{s}, \mathfrak{s}]$ are semisimple, and $\mathfrak{c} = \mathfrak{z}(\mathfrak{l})$ (resp. $\mathfrak{d} = \mathfrak{z}(\mathfrak{s})$) consists of semisimple elements (20.5.5 and 20.5.10).

Take $\mathfrak{l}$ to be the Levi factor of $\mathfrak{p}$ defined in 20.8.6, we see that if $\mathfrak{h}$ and $\mathfrak{k}$ are Cartan subalgebras of $\mathfrak{l}'$ and $\mathfrak{s}'$, then $\mathfrak{h} \oplus \mathfrak{c}$ and $\mathfrak{k} \oplus \mathfrak{d}$ are Cartan subalgebras of $\mathfrak{p}$. By 29.2.3, there exists $\alpha \in \operatorname{Aut}_e \mathfrak{p}$ such that $\alpha(\mathfrak{h} \oplus \mathfrak{c}) = \alpha(\mathfrak{k} \oplus \mathfrak{d})$. But:

$$\alpha(\mathfrak{r}) = \mathfrak{r} \, , \; (\mathfrak{h} \oplus \mathfrak{c}) \cap \mathfrak{r} = \mathfrak{c} \, , \; (\mathfrak{k} \oplus \mathfrak{d}) \cap \mathfrak{r} = \mathfrak{d}.$$

We deduce therefore that $\alpha(\mathfrak{c}) = \mathfrak{d}$. So we may assume that $\mathfrak{d} = \mathfrak{c}$.

Let $\mathfrak{z} = \mathfrak{c}_\mathfrak{p}(\mathfrak{c})$. Then $\mathfrak{l}'$ and $\mathfrak{c}'$ are Levi subalgebras of $\mathfrak{z}$. So by 29.5.2, there exists an element x in the nilpotent radical of $\mathfrak{z}$ such that $\mathfrak{s}' = e^{\operatorname{ad} x}(\mathfrak{l}')$. Since $[x, \mathfrak{c}] = \{0\}$, we also have $e^{\operatorname{ad} x}(\mathfrak{c}) = \mathfrak{c}$. Hence $e^{\operatorname{ad} x}(\mathfrak{l}) = \mathfrak{s}$.

(iii) This follows easily from (ii) and 20.8.6. □

Remark. Note that Levi *factors* and Levi *subalgebras* are not the same.

29.6 More on reductive Lie algebras

29.6.1 Let (V, σ) and (W, τ) be finite-dimensional $\mathfrak{g}$-modules and (V^*, σ^*) the contragredient representation of σ (19.2.1). We define the the representation π of $\mathfrak{g}$ on $\operatorname{Hom}(V, W)$ as follows:

$$\pi(x)(f) = \tau(x) \circ f - f \circ \sigma(x).$$

The map $V^* \otimes_{\mathbb{k}} W \to \operatorname{Hom}(V, W)$ sending $\varphi \otimes y$, where $\varphi \in V^*$ and $y \in W$, to the map $v \mapsto \varphi(v)y$, is a vector space isomorphism. We verify easily that it is a homomorphism of $\mathfrak{g}$-modules if we endow $V^* \otimes_{\mathbb{k}} W$ with the representation $\sigma^* \otimes \tau$ (20.5.6). It follows therefore from 20.5.6 that we have the following result:

Lemma. *If σ and τ are semisimple, then so is π.*

29.6.2 Proposition. *Let (V, σ) be a $\mathfrak{g}$-module, $\mathfrak{h}$ a reductive Lie subalgebra of $\mathfrak{g}$ and W the sum of simple $\mathfrak{h}$-submodules in V. Then $\sigma(\mathfrak{g})(W) \subset W$.*

Proof. Let S be a simple $\mathfrak{h}$-submodule of V and consider $\mathfrak{g}$ as a $\mathfrak{h}$-module via the adjoint representation. Then 20.5.6 implies that $\mathfrak{g} \otimes_{\mathbb{k}} S$ is a semisimple $\mathfrak{h}$-module. Let $\theta : \mathfrak{g} \otimes_{\mathbb{k}} S \to V$ be the linear map defined by $\theta(x \otimes v) = \sigma(x)(v)$. Then we verify easily that θ is homomorphism of $\mathfrak{h}$-modules. It follows that $\theta(\mathfrak{g} \otimes_{\mathbb{k}} S)$ is a semisimple $\mathfrak{h}$-submodule of V, and so it is contained in W. Hence $\sigma(\mathfrak{g})(S) \subset W$ and the result follows. □

29.6.3 Corollary. *Let $\mathfrak{h}$ be a reductive Lie subalgebra of $\mathfrak{g}$ and (V, σ) a finite-dimensional semisimple representation of $\mathfrak{g}$. The restriction of $\mathfrak{g}$ to $\mathfrak{h}$ is semisimple.*

Proof. We may assume that σ is simple. Let W be as in 29.6.2 and S a non-zero $\mathfrak{h}$-submodule of V of minimal dimension. Then S is a simple $\mathfrak{h}$-module, so $S \subset W$ and $W \neq \{0\}$. It follows again by 29.6.2 and the simplicity of V that $W = V$. □

29.6.4 Proposition. *Let V be a finite-dimensional vector space and $\mathfrak{g}$ a Lie subalgebra of $\mathfrak{gl}(V)$. The following conditions are equivalent:*

(i) *The natural representation of $\mathfrak{g}$ on V is semisimple.*

(ii) *$\mathfrak{g}$ is reductive and all the elements of the centre of $\mathfrak{g}$ are semisimple endomorphisms of V.*

(iii) *$\mathfrak{g}$ is reductive in $\mathfrak{gl}(V)$.*

Proof. We have (i) $\Rightarrow$ (iii) by 29.6.1, (iii) $\Rightarrow$ (i) by 29.6.3, while the equivalence (i) $\Leftrightarrow$ (ii) was proved in 20.5.10. $\square$

29.7 Other applications

29.7.1 Let $\mathfrak{a}, \mathfrak{b}$ be Lie algebras and $y \mapsto \delta_y$ a homomorphism from $\mathfrak{b}$ to the Lie algebra of derivations of $\mathfrak{a}$. Then we verify easily that for $x, x' \in \mathfrak{a}$ and $y, y' \in \mathfrak{b}$,

$$[(y, x), (y', x')] = \big([y, y'], [x, x'] + \delta_y(x') - \delta_{y'}(x)\big)$$

defines a Lie bracket on $\mathfrak{b} \times \mathfrak{a}$. We obtain therefore a Lie algebra $\mathfrak{g}$ that we call the *semi-direct product* of $\mathfrak{b}$ by $\mathfrak{a}$ (with respect to the homomorphism $y \mapsto \delta_y$). The underlying vector space of $\mathfrak{g}$ is $\mathfrak{b} \times \mathfrak{a}$ in which we may identify canonically $\mathfrak{a}$ and $\mathfrak{b}$ as subspaces, and $\mathfrak{a}$ (resp. $\mathfrak{b}$) is an ideal (resp. Lie subalgebra) of $\mathfrak{g}$.

29.7.2 Lemma. *Let $\mathfrak{a}$ be a commutative Lie subalgebra of $\mathfrak{g}$ and $\mathfrak{c} = \mathfrak{c}_{\mathfrak{g}}(\mathfrak{a})$. Suppose that $\mathrm{ad}_{\mathfrak{g}}\, x$ is semisimple for all $x \in \mathfrak{a}$. Then Cartan subalgebras of $\mathfrak{c}$ are exactly the Cartan subalgebras of $\mathfrak{g}$ which contains $\mathfrak{a}$.*

Proof. Let $\mathfrak{h}$ be a Cartan subalgebra of $\mathfrak{c}$. Then $\mathfrak{a} \subset \mathfrak{z}(\mathfrak{c}) \subset \mathfrak{h}$. Let $\mathfrak{n} = \mathfrak{n}_{\mathfrak{g}}(\mathfrak{h})$. We have $[\mathfrak{a}, \mathfrak{n}] \subset [\mathfrak{h}, \mathfrak{n}] \subset \mathfrak{h}$. Since $\mathrm{ad}_{\mathfrak{g}}\, x$ is semisimple for all $x \in \mathfrak{a}$ and $\mathfrak{a}$ is commutative, we have the decomposition $\mathfrak{n} = \mathfrak{h} \oplus \mathfrak{m}$ of $\mathfrak{a}$-stable subspaces. It follows that $[\mathfrak{a}, \mathfrak{m}] \subset \mathfrak{h} \cap \mathfrak{m} = \{0\}$, so $\mathfrak{m} \subset \mathfrak{c}$. Hence $\mathfrak{n} = \mathfrak{n}_{\mathfrak{c}}(\mathfrak{h})$ and $\mathfrak{n} = \mathfrak{h}$. Thus $\mathfrak{h}$ is a Cartan subalgebra of $\mathfrak{g}$.

Conversely, let $\mathfrak{h}$ be a Cartan subalgebra of $\mathfrak{g}$ which contains $\mathfrak{a}$. Then $\mathfrak{h} = \mathfrak{g}^0(\mathfrak{h}) \subset \mathfrak{g}^0(\mathfrak{a}) = \mathfrak{c}$. Since $\mathfrak{h} = \mathfrak{n}_{\mathfrak{g}}(\mathfrak{h})$, we have $\mathfrak{h} = \mathfrak{n}_{\mathfrak{c}}(\mathfrak{h})$, and $\mathfrak{h}$ is a Cartan subalgebra of $\mathfrak{c}$. $\square$

29.7.3 In the rest of this section, V is a finite-dimensional $\Bbbk$-vector space, $\mathfrak{g}$ is a Lie subalgebra of $\mathfrak{gl}(V)$, $\mathfrak{r} = \mathrm{rad}\,\mathfrak{g}$, $\mathfrak{u}$ the set of nilpotent endomorphisms of V in $\mathfrak{r}$ and $\mathfrak{n}$ the largest nilpotent ideal of the natural representation of $\mathfrak{g}$ in V. By 19.3.5, 19.3.7 and 19.5.7, we have $\mathfrak{n} = \mathfrak{u}$. Also $[\mathfrak{g}, \mathfrak{g}] \cap \mathfrak{r} \subset \mathfrak{u}$ by 19.6.2.

Finally, if $\mathfrak{h}$ is a Lie subalgebra of $\mathfrak{gl}(V)$, we denote by $\mathfrak{u}(\mathfrak{h})$ the set of nilpotent endomorphisms of V in $\mathrm{rad}\,\mathfrak{h}$. Thus $\mathfrak{u} = \mathfrak{u}(\mathfrak{g}) = \mathfrak{u}(\mathfrak{r})$.

29.7.4 Lemma. *Let $\mathfrak{g}$ be nilpotent and algebraic in $\mathfrak{gl}(V)$ and $\mathfrak{t}$ the set of semisimple endomorphisms of V in $\mathfrak{g}$. Then $\mathfrak{t} \subset \mathfrak{z}(\mathfrak{g})$ and $\mathfrak{g}$ is the direct product of $\mathfrak{t}$ and $\mathfrak{u}$.*

Proof. This follows easily from 24.5.13, 26.4.4 and 26.4.5. $\square$

29.7.5 Proposition. *Let $\mathfrak{g}$ be algebraic in $\mathfrak{gl}(V)$ and denote by $\mathcal{S}$ the set of commutative Lie subalgebras of $\mathfrak{g}$ consisting of semisimple elements, $\mathcal{S}_0$ the set of maximal elements of $\mathcal{S}$ and $\mathcal{H}$ the set of Cartan subalgebras of $\mathfrak{g}$.*

(i) For $\mathfrak{h} \in \mathcal{H}$, the set $\varphi(\mathfrak{h})$ of semisimple elements of $\mathfrak{h}$ belongs to $\mathcal{S}_0$.

(ii) If $\mathfrak{t} \in \mathcal{S}_0$, then $\psi(\mathfrak{t}) = \mathfrak{c}_{\mathfrak{g}}(\mathfrak{t}) \in \mathcal{H}$.

(iii) The maps $\varphi : \mathcal{H} \to \mathcal{S}_0$, $\psi : \mathcal{S}_0 \to \mathcal{H}$ are bijections and they are mutually inverse.

(iv) The group $\mathrm{Aut}_e\,\mathfrak{g}$ acts transitively on $\mathcal{S}_0$.

Proof. (i) Since $\mathfrak{h} = \mathfrak{n}_\mathfrak{g}(\mathfrak{h})$, $\mathfrak{h}$ is algebraic in $\mathfrak{g}$. By 29.7.4, $\varphi(\mathfrak{h}) \in \mathcal{S}$ and $\mathfrak{h} = \varphi(\mathfrak{h}) \times \mathfrak{u}$. It follows that $\mathfrak{h} \subset \mathfrak{c}_\mathfrak{g}(\varphi(\mathfrak{h})) \subset \mathfrak{g}^0(\mathfrak{h}) = \mathfrak{h}$ (19.8.6), because the elements of $\mathfrak{u}$ are nilpotent. Thus $\mathfrak{h} = \mathfrak{c}_\mathfrak{g}(\varphi(\mathfrak{h}))$.

Let $\mathfrak{t} \in \mathcal{S}$ be such that $\varphi(\mathfrak{h}) \subset \mathfrak{t}$. Then $\mathfrak{t} \subset \mathfrak{c}_\mathfrak{g}(\varphi(\mathfrak{h})) \subset \mathfrak{h}$. So $\mathfrak{t} = \varphi(\mathfrak{h})$, and we obtain that $\varphi(\mathfrak{h}) \in \mathcal{S}_0$.

(ii) Let $\mathfrak{t} \in \mathcal{S}_0$ and $\mathfrak{h}$ a Cartan subalgebra of $\psi(\mathfrak{t})$. By 29.7.2, we have $\mathfrak{h} \in \mathcal{H}$ and $\mathfrak{t} \subset \mathfrak{h}$. So $\varphi(\mathfrak{h}) \in \mathcal{S}$ and $\mathfrak{t} \subset \varphi(\mathfrak{h})$. Hence $\mathfrak{t} = \varphi(\mathfrak{h})$, and $\mathfrak{h} = \psi(\varphi(\mathfrak{h})) = \psi(\mathfrak{t})$. Thus $\psi(\mathfrak{t}) \in \mathcal{H}$ and $\varphi(\psi(\mathfrak{t})) = \mathfrak{t}$.

(iii) This is clear from the proof of (ii).

(iv) This follows from (iii) and 29.2.3.   □

29.7.6 Corollary. *Let us conserve the hypotheses of* 29.7.5 *and suppose further that* $\mathfrak{g}$ *is solvable. If* $\mathfrak{t} \in \mathcal{S}_0$, *then* $\mathfrak{g}$ *is the semi-direct product of* $\mathfrak{t}$ *by* $\mathfrak{u}$.

Proof. Since $\mathfrak{t} \cap \mathfrak{u} = \{0\}$, it suffices to prove that $\mathfrak{g} = \mathfrak{t} + \mathfrak{u}$.

Let $\mathfrak{h} = \psi(\mathfrak{t}) \in \mathcal{H}$. Since $\mathfrak{g}/[\mathfrak{g}, \mathfrak{g}]$ is commutative, by applying 29.2.4 to the canonical surjection $\mathfrak{g} \to \mathfrak{g}/[\mathfrak{g}, \mathfrak{g}]$, we obtain that $\mathfrak{g} = \mathfrak{h} + [\mathfrak{g}, \mathfrak{g}]$. But $\mathfrak{g}$ is solvable, so $[\mathfrak{g}, \mathfrak{g}] \subset \mathfrak{u}$. So $\mathfrak{g} = \mathfrak{h} + \mathfrak{u}$.

Using again the fact that $\mathfrak{g}$ is solvable, it is clear that $\mathfrak{u}(\mathfrak{h}) \subset \mathfrak{u}$. Since $\mathfrak{h} = \mathfrak{t} + \mathfrak{u}(\mathfrak{h})$ by 29.7.4, we have $\mathfrak{g} = \mathfrak{t} + \mathfrak{u}$ as required.   □

29.7.7 Let $\mathfrak{g}$ be algebraic in $\mathfrak{gl}(V)$. Then for $x \in \mathfrak{g}$, its semisimple component s and nilpotent component n are also in $\mathfrak{g}$ (23.6.2 and 23.6.4). Moreover, 10.1.3 and 20.4.1 say that $\mathrm{ad}_\mathfrak{g}\, s$ and $\mathrm{ad}_\mathfrak{g}\, n$ are the semisimple and nilpotent components of $\mathrm{ad}_\mathfrak{g}\, x$.

Let $\mathfrak{h}$ be a Lie subalgebra of $\mathfrak{gl}(V)$ which contains $\mathfrak{g}$. If $\mathrm{ad}_\mathfrak{g}\, x$ is nilpotent, then $\mathrm{ad}_\mathfrak{g}\, x = \mathrm{ad}_\mathfrak{g}\, n$, and $\mathrm{ad}_\mathfrak{h}\, n$ extends $\mathrm{ad}_\mathfrak{g}\, x$. Since n is nilpotent and $n \in \mathfrak{h}$, it follows again from 10.1.3 and 20.4.1 that $\mathrm{ad}_\mathfrak{h}\, n$ is nilpotent.

Thus we have obtained the following result:

Lemma. *Any element of* $\mathrm{Aut}_e\, \mathfrak{g}$ *extends to an element of* $\mathrm{Aut}_e\, \mathfrak{h}$.

29.7.8 Proposition. *Let* $\mathfrak{g}$ *be algebraic in* $\mathfrak{gl}(V)$.

(i) *There exists a Lie subalgebra* $\mathfrak{m}$ *of* $\mathfrak{g}$, *reductive in* $\mathfrak{gl}(V)$, *such that* $\mathfrak{g}$ *is the semi-direct product of* $\mathfrak{m}$ *by* $\mathfrak{u}$.

(ii) *If* $\mathfrak{m}$ *and* $\mathfrak{m}'$ *are two Lie subalgebras verifying the conditions of* (i), *then there exists* $\theta \in \mathrm{Aut}_e\, \mathfrak{g}$ *such that* $\mathfrak{m}' = \theta(\mathfrak{m})$. *Moreover,* $\mathfrak{m}$ *and* $\mathfrak{m}'$ *are algebraic Lie subalgebras of* $\mathfrak{g}$.

Proof. (i) We have $\mathfrak{r} = \mathfrak{t} \oplus \mathfrak{u}$ by 29.7.6, where $\mathfrak{t}$ is a commutative Lie subalgebra consisting of semisimple elements. Also, by 20.5.11, we have $\mathfrak{g} = \mathfrak{z} \oplus [\mathfrak{t}, \mathfrak{g}]$ where $\mathfrak{z} = \mathfrak{c}_\mathfrak{g}(\mathfrak{t})$. Since $[\mathfrak{t}, \mathfrak{g}] \subset \mathfrak{r}$, we have obtained that $\mathfrak{g} = \mathfrak{z} + \mathfrak{r}$.

Let $\mathfrak{s}$ be a Levi subalgebra of $\mathfrak{z}$ (20.3.5) and $\mathfrak{r}_\mathfrak{z}$ the radical of $\mathfrak{z}$. Then $\mathfrak{z} = \mathfrak{s} \oplus \mathfrak{r}_\mathfrak{z}$, and $\mathfrak{g} = \mathfrak{s} + \mathfrak{r}_\mathfrak{z} + \mathfrak{r}$. As $\mathfrak{r}_\mathfrak{z} + \mathfrak{r}$ is clearly a solvable ideal of $\mathfrak{g}$, we deduce that $\mathfrak{r}_\mathfrak{z} + \mathfrak{r} = \mathfrak{r}$. Thus $\mathfrak{g} = \mathfrak{s} \oplus \mathfrak{r}$, and $\mathfrak{s}$ is a Levi subalgebra of $\mathfrak{g}$. Now let $\mathfrak{m} = \mathfrak{s} \oplus \mathfrak{t}$. Then $\mathfrak{g} = \mathfrak{m} \oplus \mathfrak{u}$, and since the centre of $\mathfrak{m}$ is $\mathfrak{t}$, 29.6.4 implies that $\mathfrak{m}$ is reductive in $\mathfrak{gl}(V)$.

(ii) Let $\mathfrak{m}'$ be another such Lie subalgebra of $\mathfrak{g}$. We have $\mathfrak{m}' = \mathfrak{s}' \oplus \mathfrak{t}'$ where $\mathfrak{s}' = [\mathfrak{m}', \mathfrak{m}']$ is semisimple and $\mathfrak{t}' = \mathfrak{z}(\mathfrak{m}')$ consists of semisimple elements.

Then $\mathfrak{r} = \mathfrak{t} \oplus \mathfrak{u} = \mathfrak{t}' \oplus \mathfrak{u}$. So by 29.7.5, there exists $\sigma \in \mathrm{Aut}_e\, \mathfrak{r}$ such that $\mathfrak{t}' = \sigma(\mathfrak{t})$ and $\sigma(\mathfrak{u}) = \mathfrak{u}$. So 29.7.7 says that σ extends to an element of $\mathrm{Aut}_e\, \mathfrak{g}$. Thus we may assume that $\mathfrak{t} = \mathfrak{t}'$.

So $\mathfrak{s}' \subset \mathfrak{c}_\mathfrak{g}(\mathfrak{t})$ and since $\dim \mathfrak{s} = \dim \mathfrak{s}'$, $\mathfrak{s}'$ is a Levi subalgebra of $\mathfrak{c}_\mathfrak{g}(\mathfrak{t})$. By 29.5.2, there exists an elementary automorphism θ of $\mathfrak{c}_\mathfrak{g}(\mathfrak{t})$ such that $\theta(\mathfrak{s}) = \mathfrak{s}'$. Since $\mathfrak{t}$ is central in $\mathfrak{c}_\mathfrak{g}(\mathfrak{t})$, we have $\theta(\mathfrak{t}) = \mathfrak{t}$, and so $\theta(\mathfrak{m}) = \mathfrak{m}'$. Finally, $\mathfrak{c}_\mathfrak{g}(\mathfrak{t})$ is an algebraic Lie subalgebra of $\mathfrak{g}$ (24.7.2), so we may extend θ to an element of $\mathrm{Aut}_e\, \mathfrak{g}$ (29.7.7).

To show that $\mathfrak{m} = \mathfrak{s} \oplus \mathfrak{t}$ is algebraic, it suffices to show that $\mathfrak{t}$ is algebraic (24.7.4). Since $\mathfrak{r}$ is algebraic (24.7.4), $\mathbf{a}(\mathfrak{t})$ is a commutative Lie subalgebra of $\mathfrak{r}$ consisting of semisimple elements (24.5.7, 24.5.10 and 24.6.4). Hence $\mathbf{a}(\mathfrak{t}) \cap \mathfrak{u} = \{0\}$, and so $\mathfrak{t} = \mathbf{a}(\mathfrak{t})$ is algebraic. $\square$

29.7.9 As in 29.5.6, a Lie subalgebra $\mathfrak{m}$ of $\mathfrak{g}$ verifying condition (i) of 29.7.8 is called a *Levi factor* of $\mathfrak{g}$.

29.7.10 Corollary. *Let $\mathfrak{g}$ be algebraic in $\mathfrak{gl}(V)$. There exist a Borel subalgebra $\mathfrak{b}$ of $\mathfrak{g}$ and $\theta \in \mathrm{Aut}_e\, \mathfrak{g}$ such that $\mathfrak{u} = \mathfrak{u}\big(\mathfrak{b} \cap \theta(\mathfrak{b})\big)$.*

Proof. Let $\mathfrak{m}$ be a Levi factor of $\mathfrak{g}$, $\mathfrak{s} = [\mathfrak{m}, \mathfrak{m}]$ and $\mathfrak{t} = \mathfrak{z}(\mathfrak{m})$.

Let $\mathfrak{h}$ be a Cartan subalgebra of $\mathfrak{s}$ and R_+ a positive root system associated to the root system of $(\mathfrak{s}, \mathfrak{h})$. Set:

$$\mathfrak{b} = \mathfrak{t} \oplus \mathfrak{u} \oplus \sum_{\alpha \in R_+} \mathfrak{g}^\alpha\ , \quad \mathfrak{b}' = \mathfrak{t} \oplus \mathfrak{u} \oplus \sum_{\alpha \in R_+} \mathfrak{g}^{-\alpha}.$$

Then it is clear that $\mathfrak{b}$ and $\mathfrak{b}'$ are Borel subalgebras of $\mathfrak{g}$, and $\mathfrak{b} \cap \mathfrak{b}' = \mathfrak{t} \oplus \mathfrak{u}$. So $\mathfrak{u} = \mathfrak{u}(\mathfrak{b} \cap \mathfrak{b}')$, and the result follows from 29.4.7. $\square$

29.8 Maximal subalgebras

29.8.1 Theorem. *Let $\mathfrak{g}$ be a semisimple Lie algebra, $\mathfrak{n}$ a Lie subalgebra of $\mathfrak{g}$ consisting of nilpotent elements and $\mathfrak{p} = \mathfrak{n}_\mathfrak{g}(\mathfrak{n})$. If $\mathfrak{n}$ is the set of nilpotent elements of $\mathrm{rad}\, \mathfrak{p}$, then $\mathfrak{p}$ is a parabolic subalgebra of $\mathfrak{g}$.*

Proof. Since the adjoint representation is faithful, we may assume that $\mathfrak{g}$ is an algebraic Lie subalgebra of $\mathfrak{gl}(V)$ for some finite-dimensional vector space V (24.7.2), and we shall use the notations of 29.7.3. By 20.4.4, $x \in \mathfrak{g}$ is semisimple (resp. nilpotent) if and only if it is a semisimple (resp. nilpotent) endomorphism of V. So $\mathfrak{n} = \mathfrak{u}(\mathfrak{p})$. Note also that $\mathfrak{p}$ is algebraic in $\mathfrak{g}$ (24.7.2).

Let $\mathfrak{b}$ be a Borel subalgebra of $\mathfrak{g}$ whose nilpotent radical contains $\mathfrak{n}$ (29.4.8), and $\mathfrak{s} = \mathfrak{b} \cap \mathfrak{p}$. By 29.7.10, there exist a Borel subalgebra $\mathfrak{b}_1$ of $\mathfrak{p}$ and $\theta \in \mathrm{Aut}_e\, \mathfrak{p}$ such that $\mathfrak{n} = \mathfrak{u}\big(\mathfrak{b}_1 \cap \theta(\mathfrak{b}_1)\big)$. It is clear that $\mathfrak{n}$ is stable under elementary automorphisms of $\mathfrak{p}$, and as $\mathfrak{s}$ is solvable, we may assume that $\mathfrak{s} \subset \mathfrak{b}_1$ (29.4.7). Further, θ extends to an element $\sigma \in \mathrm{Aut}_e\, \mathfrak{g}$ (29.7.7). Let $\mathfrak{b}' = \sigma(\mathfrak{b})$.

Since $\theta(\mathfrak{n}) = \mathfrak{n}$, we deduce that $\mathfrak{n} \subset \mathfrak{p} \cap \mathfrak{b} \cap \mathfrak{b}'$. As $\mathfrak{p} \cap \mathfrak{b} \cap \mathfrak{b}'$ is solvable, $\mathfrak{n} \subset \mathfrak{u}(\mathfrak{p} \cap \mathfrak{b} \cap \mathfrak{b}')$.

On the other hand, $\mathfrak{b} \cap \mathfrak{p} = \mathfrak{s} \subset \mathfrak{b}_1$, so $\mathfrak{p} \cap \mathfrak{b} \cap \mathfrak{b}' \subset \mathfrak{b}_1 \cap \theta(\mathfrak{b}_1)$. It follows that $\mathfrak{u}(\mathfrak{p} \cap \mathfrak{b} \cap \mathfrak{b}') \subset \mathfrak{n}$.

Thus we have proved that $\mathfrak{n} = \mathfrak{u}(\mathfrak{p} \cap \mathfrak{b} \cap \mathfrak{b}')$ and hence $\mathfrak{n} \subset \mathfrak{u}(\mathfrak{b} \cap \mathfrak{b}')$. Suppose that $\mathfrak{n} \ne \mathfrak{u}(\mathfrak{b} \cap \mathfrak{b}')$, then by applying Lie's Theorem (19.4.4) to the adjoint representation of $\mathfrak{n}$ in $\mathfrak{u}(\mathfrak{b} \cap \mathfrak{b}')$, we obtain an element $x \in \mathfrak{u}(\mathfrak{b} \cap \mathfrak{b}') \backslash \mathfrak{u}(\mathfrak{p} \cap \mathfrak{b} \cap \mathfrak{b}')$ such that $[x, \mathfrak{n}] \subset \mathfrak{n}$. This is absurd since $\mathfrak{p} = \mathfrak{n}_{\mathfrak{g}}(\mathfrak{n})$ and $\mathfrak{n} = \mathfrak{u}(\mathfrak{p} \cap \mathfrak{b} \cap \mathfrak{b}')$. Consequently, we have $\mathfrak{n} = \mathfrak{u}(\mathfrak{b} \cap \mathfrak{b}')$.

Since $\mathfrak{b} \cap \mathfrak{b}'$ normalizes $\mathfrak{u}(\mathfrak{b} \cap \mathfrak{b}') = \mathfrak{n}$, we deduce that $\mathfrak{b} \cap \mathfrak{b}' \subset \mathfrak{p}$. It follows from 29.4.9 that $\mathfrak{b} \cap \mathfrak{b}'$, and hence $\mathfrak{p}$, contains a Cartan subalgebra $\mathfrak{h}$ of $\mathfrak{g}$. Let $R = R(\mathfrak{g}, \mathfrak{h})$, B the base of R defined by $\mathfrak{b}$ and R_+ the set of corresponding positive roots. Then there exists a closed subset P of R such that $\mathfrak{p} = \mathfrak{h} + \mathfrak{g}^P$ (in the notations of 20.7.1). Let $\alpha \in B$ and $Q = R_+ \backslash \{\alpha\}$. Then 18.7.4 and 18.8.7 say that $Q \cup \{-\alpha\}$ is a system of positive roots, and $\mathfrak{b}'' = \mathfrak{h} \oplus \mathfrak{g}^{Q \cup \{-\alpha\}}$ is a Borel subalgebra of $\mathfrak{g}$.

Let $S \subset R$ be the closed subset such that $\mathfrak{b}' = \mathfrak{h} \oplus \mathfrak{g}^S$. Clearly, we have $\mathfrak{u}(\mathfrak{b}) = \mathfrak{g}^{R_+}$, $\mathfrak{u}(\mathfrak{b}') = \mathfrak{g}^S$, $\mathfrak{n} = \mathfrak{u}(\mathfrak{b} \cap \mathfrak{b}') = \mathfrak{g}^{R_+ \cap S}$ and $\mathfrak{u}(\mathfrak{b} \cap \mathfrak{b}'') = \mathfrak{g}^Q$.

If $\alpha \notin R_+ \cap S$, then $\alpha \notin S$, and we have $-\alpha \in S$ and $R_+ \cap S \cap Q = R_+ \cap S$. Thus $\mathfrak{u}(\mathfrak{b} \cap \mathfrak{b}' \cap \mathfrak{b}'') = \mathfrak{u}(\mathfrak{b} \cap \mathfrak{b}') = \mathfrak{n}$. It follows that $[\mathfrak{g}^{-\alpha}, \mathfrak{n}] \subset \mathfrak{n}$ because S and Q are closed subsets not containing α. Thus $-\alpha \in P$. As $\pm\alpha \notin R_+ \cap S$, 20.7.4 implies that $\alpha \in P$.

We have therefore proved that $B \subset P$. So $\mathfrak{p}$ is a parabolic subalgebra of $\mathfrak{g}$ containing $\mathfrak{b}$. $\square$

29.8.2 Theorem. *Let $\mathfrak{g}$ be a semisimple Lie algebra and $\mathfrak{a}$ a Lie subalgebra of $\mathfrak{g}$. If $\mathfrak{a}$ is maximal in the set of Lie subalgebras of $\mathfrak{g}$ distinct from $\mathfrak{g}$, then $\mathfrak{a}$ is either parabolic or reductive in $\mathfrak{g}$.*

Proof. We may assume that $\mathfrak{g}$ is an algebraic Lie subalgebra of $\mathfrak{gl}(V)$ for some finite-dimensional vector space V. Recall that $\mathfrak{a}(\mathfrak{a})$ is the smallest algebraic Lie subalgebra of $\mathfrak{g}$ which contains $\mathfrak{a}$.

If $\mathfrak{a}(\mathfrak{a}) = \mathfrak{g}$, then by 24.5.7, $\mathfrak{a}$ is an ideal of $\mathfrak{g}$, so it is semisimple (20.1.6), and therefore reductive in $\mathfrak{g}$ (20.3.4).

If $\mathfrak{a}(\mathfrak{a}) \ne \mathfrak{g}$, then $\mathfrak{a}(\mathfrak{a}) = \mathfrak{a}$. Let $\mathfrak{m}$ be a Levi factor of $\mathfrak{a}$. Then $\mathfrak{a} = \mathfrak{m} \oplus \mathfrak{u}(\mathfrak{a})$ (29.7.8). If $\mathfrak{u}(\mathfrak{a}) = \{0\}$, then $\mathfrak{a} = \mathfrak{m}$ is reductive in $\mathfrak{g}$. Otherwise, the normalizer $\mathfrak{p}$ of $\mathfrak{u}(\mathfrak{a})$ in $\mathfrak{g}$ contains $\mathfrak{a}$ and $\mathfrak{p} \ne \mathfrak{g}$ because $\mathfrak{g}$ is semisimple. So $\mathfrak{p} = \mathfrak{a}$, and 29.8.1 says that $\mathfrak{a}$ is parabolic in $\mathfrak{g}$. $\square$

29.8.3 Theorem. *Let $\mathfrak{g}$ be a semisimple Lie algebra and $\mathfrak{n}$ a Lie subalgebra of $\mathfrak{g}$ consisting of nilpotent elements. There exists a parabolic subalgebra $\mathfrak{p}$ of $\mathfrak{g}$ verifying the following conditions:*

   (i) *$\mathfrak{n}$ is contained in the nilpotent radical of $\mathfrak{p}$.*

   (ii) *The normalizer of $\mathfrak{n}$ in $\mathfrak{g}$ is contained in $\mathfrak{p}$.*

   (iii) *$\mathfrak{p}$ is stable under any automorphism of $\mathfrak{g}$ which leaves $\mathfrak{n}$ invariant.*

Proof. We may assume that $\mathfrak{g}$ is an algebraic Lie subalgebra of $\mathfrak{gl}(V)$ for some finite-dimensional vector space V. Let $\mathfrak{n}_0 = \mathfrak{n}$ and $\mathfrak{p}_1 = \mathfrak{n}_{\mathfrak{g}}(\mathfrak{n})$. Define by induction the two sequences $(\mathfrak{n}_i)_{i \geqslant 0}$ and $(\mathfrak{p}_i)_{i \geqslant 1}$ as follows:

$$\mathfrak{p}_i = \mathfrak{n}_{\mathfrak{g}}(\mathfrak{n}_{i-1}) \;, \; \mathfrak{n}_i = \mathfrak{u}(\mathfrak{p}_i).$$

They are both increasing. So there exists an integer j such that $\mathfrak{p}_j = \mathfrak{p}_{j+1}$, that is $\mathfrak{p}_j = \mathfrak{n}_{\mathfrak{g}}(\mathfrak{u}(\mathfrak{p}_j))$. It follows from 29.8.1 that $\mathfrak{p}_j$ is a parabolic subalgebra of $\mathfrak{g}$. We have therefore $\mathfrak{n} \subset \mathfrak{n}_j = \mathfrak{u}(\mathfrak{p}_j)$ and $\mathfrak{p}_1 \subset \mathfrak{p}$. Finally, an automorphism of $\mathfrak{g}$, which leaves $\mathfrak{n}$ invariant, also leaves invariant the sequences $(\mathfrak{n}_i)_{i \geqslant 0}$ and $(\mathfrak{p}_i)_{i \geqslant 1}$. $\quad\square$

29.8.4 Corollary. *Let $\mathfrak{g}$ be a semisimple Lie algebra, $\mathfrak{n}$ a Lie subalgebra of $\mathfrak{g}$ consisting of nilpotent elements and $\mathfrak{s}$ a semisimple Lie subalgebra of $\mathfrak{g}$ which normalizes $\mathfrak{n}$. Then there exists a parabolic subalgebra $\mathfrak{p}$ of $\mathfrak{g}$ verifying the following conditions:*

(i) *$\mathfrak{n}$ is contained in the nilpotent radical of $\mathfrak{p}$.*
(ii) *There is a Levi factor of $\mathfrak{p}$ which contains $\mathfrak{s}$.*

Proof. By 29.8.3, there is a parabolic subalgebra $\mathfrak{p}$ of $\mathfrak{g}$ such that $\mathfrak{s} \subset \mathfrak{p}$ and $\mathfrak{n}$ is contained in the nilpotent radical of $\mathfrak{p}$. Let $\mathfrak{p} = \mathfrak{l} \oplus \mathfrak{m}$ where $\mathfrak{m}$ is the nilpotent radical of $\mathfrak{p}$ and $\mathfrak{l}$ a Levi factor of $\mathfrak{p}$ (29.5.7). The Lie algebra $\mathfrak{t} = [\mathfrak{l}, \mathfrak{l}]$ is a Levi subalgebra of $\mathfrak{p}$. So by 29.5.3, there exists $\theta \in \mathrm{Aut}_e\,\mathfrak{p}$ such that $\theta(\mathfrak{s}) \subset \mathfrak{t}$. Thus $\theta^{-1}(\mathfrak{l})$ is a Levi factor of $\mathfrak{p}$ containing $\mathfrak{s}$. $\quad\square$

References and comments

- [5], [14], [29], [32], [38], [39], [40], [42], [43], [78], [80].

For a proof of 29.3.7, see [14], [59].

Representations of semisimple Lie algebras

We define the enveloping algebra of a Lie algebra in this chapter. This infinite-dimensional associative algebra is a very useful tool to studying representations of the Lie algebra. We obtain in the case of semisimple Lie algebras a complete description, up to isomorphisms, of finite-dimensional irreducible representations.

30.1 Enveloping algebra

30.1.1 Let $\mathfrak{g}$ be a Lie algebra. Denote by $T(\mathfrak{g})$ the tensor algebra of $\mathfrak{g}$ and I the bilateral ideal of $T(\mathfrak{g})$ generated by the elements

$$x \otimes y - y \otimes x - [x, y]$$

for $x, y \in \mathfrak{g}$. The quotient $T(\mathfrak{g})/I$, denote by $U(\mathfrak{g})$, is called the *universal enveloping algebra* of $\mathfrak{g}$.

We shall state without proof in this section some properties of $U(\mathfrak{g})$.

Theorem. (Poincaré-Birkhoff-Witt Theorem) *The map* $\mathfrak{g} \to U(\mathfrak{g})$, $x \mapsto x + I$, *is injective. We shall identify* $\mathfrak{g}$ *as a subspace of* $U(\mathfrak{g})$ *via this map. If* $(x_1, \ldots, x_n)$ *is a basis of* $\mathfrak{g}$, *then the elements* $x_1^{\nu_1} \cdots x_n^{\nu_n}$, *with* $\nu_1, \ldots, \nu_n \in \mathbb{N}$, *form a basis of* $U(\mathfrak{g})$.

30.1.2 We shall call the injection of $\mathfrak{g}$ in $U(\mathfrak{g})$ the canonical injection. It is clear that this injection is a Lie algebra homomorphism if we endow $U(\mathfrak{g})$ with the Lie bracket given by the commutator (19.1.1).

30.1.3 Proposition. *Any representation of* $\mathfrak{g}$ *extends uniquely to a representation of the associative algebra* $U(\mathfrak{g})$. *In particular, this induces a bijection from the set of representations of* $\mathfrak{g}$ *onto the set of representations of* $U(\mathfrak{g})$.

30.1.4 Let $\varphi : \mathfrak{g} \to \mathfrak{h}$ be a Lie algebra homomorphism. Then φ extends uniquely to a homomorphism $U(\varphi) : U(\mathfrak{g}) \to U(\mathfrak{h})$ of associative algebras with unit.

Proposition. *Let $\varphi : \mathfrak{g} \to \mathfrak{h}$ and $\psi : \mathfrak{h} \to \mathfrak{k}$ be homomorphisms of Lie algebras.*

(i) *We have* $\mathrm{U}(\psi \circ \varphi) = \mathrm{U}(\psi) \circ \mathrm{U}(\varphi)$.

(ii) *φ is injective (resp. surjective) if and only if $\mathrm{U}(\varphi)$ is injective (resp. surjective).*

30.1.5 Let $\mathfrak{h}$ be a Lie subalgebra of $\mathfrak{g}$ and $j : \mathfrak{h} \to \mathfrak{g}$ the canonical injection. Then 30.1.4 says that $\mathrm{U}(j) : \mathrm{U}(\mathfrak{h}) \to \mathrm{U}(\mathfrak{g})$ is injective. We may therefore identify $\mathrm{U}(\mathfrak{h})$ as a subalgebra of $\mathrm{U}(\mathfrak{g})$ via $\mathrm{U}(j)$.

Proposition. *The algebra $\mathrm{U}(\mathfrak{g})$ is a free (left or right) $\mathrm{U}(\mathfrak{h})$-module. If $(x_1, \ldots, x_p)$ is a basis of a complementary subspace of $\mathfrak{h}$ in $\mathfrak{g}$, then the elements $x_1^{\nu_1} \cdots x_p^{\nu_p}$, with $\nu_1, \ldots, \nu_p \in \mathbb{N}$, form a basis of $\mathrm{U}(\mathfrak{g})$ as a (left or right) $\mathrm{U}(\mathfrak{h})$-module.*

30.1.6 The adjoint representation of $\mathfrak{g}$ extends naturally to a representation on $\mathrm{T}^n(\mathfrak{g})$ as follows:

$$x.(x_1 \otimes \cdots \otimes x_n) = \sum_{i=1}^n x_1 \otimes \cdots \otimes x_{i-1} \otimes [x, x_i] \otimes x_{i+1} \otimes \cdots \otimes x_n.$$

So $\mathfrak{g}$ acts on $\mathrm{T}(\mathfrak{g})$, and it is easy to check that the bilateral ideal I is stable under this action. It follows that $\mathrm{U}(\mathfrak{g})$ is a $\mathfrak{g}$-module. Moreover, the natural grading of $\mathrm{T}(\mathfrak{g})$ induces a filtration $\left(\mathrm{U}_n(\mathfrak{g})\right)_{n \geqslant 0}$ of $\mathfrak{g}$-submodules of $\mathrm{U}(\mathfrak{g})$. We can describe this filtration explicitly: if $n \in \mathbb{N}$, then $\mathrm{U}_n(\mathfrak{g})$ is the subspace of $\mathrm{U}(\mathfrak{g})$ spanned by the products $x_1 \cdots x_p$, where $x_1, \ldots, x_p \in \mathfrak{g}$ and $p \leqslant n$. In particular, $\mathrm{U}_0(\mathfrak{g}) = \Bbbk$. We shall call this filtration the canonical filtration of $\mathrm{U}(\mathfrak{g})$.

It follows from 30.1.1 that if $(x_1, \ldots, x_r)$ is a basis of $\mathfrak{g}$, then the elements $x_1^{\nu_1} \cdots x_r^{\nu_r}$, with $\nu_1, \ldots, \nu_r \in \mathbb{N}$ and $\nu_1 + \cdots + \nu_r \leqslant n$, form a basis of $\mathrm{U}_n(\mathfrak{g})$.

30.1.7 Let $\mathrm{U}_{-1}(\mathfrak{g}) = \{0\}$. For $n \geqslant 0$, let $G^n = \mathrm{U}_n(\mathfrak{g}) / \mathrm{U}_{n-1}(\mathfrak{g})$ and

$$G = \bigoplus_{n \geqslant 0} G^n$$

is the graded algebra associated to the canonical filtration of $\mathrm{U}(\mathfrak{g})$. It follows from the explicit description of $\mathrm{U}_n(\mathfrak{g})$ in 30.1.6 that we may identify G^0 with $\Bbbk$, G^1 with $\mathfrak{g}$ and G^n with the space spanned by the product of exactly n elements of $G^1 = \mathfrak{g}$. Moreover, it is not difficult to see from the defining relations of I that G is commutative.

Thus the canonical injection $\mathfrak{g} \to G$ extends uniquely to a homomorphism, that we shall call canonical, of associative algebras $\mathrm{S}(\mathfrak{g}) \to G$, where $\mathrm{S}(\mathfrak{g})$ denotes the symmetric algebra of $\mathfrak{g}$.

Theorem. *The canonical homomorphism $\mathrm{S}(\mathfrak{g}) \to G$ is an isomorphism.*

30.2 Weights and primitive elements

30.2.1 In the rest of this chapter, $\mathfrak{g}$ is a semisimple Lie algebra, $\mathfrak{h}$ is a Cartan subalgebra of $\mathfrak{g}$, $R = R(\mathfrak{g}, \mathfrak{h})$ and $W = W(\mathfrak{g}, \mathfrak{h})$ (see 19.8.1 and 20.6.9).

Let us fix a base B of R, and denote by R_+ (resp. R_-) the corresponding set of positive (resp. negative) roots of R. Let P, P_{++}, Q be respectively the weight lattice of R, the set of dominant weights of R relative to B and the root lattice of R (18.12). Denote by $Q_+ \subset Q$ the set of linear combination with coefficients in $\mathbb{N}$ of the elements of B.

We define the following partial order on $\mathfrak{h}^*$:

$$\lambda \leqslant \mu \Leftrightarrow \mu - \lambda \in Q_+.$$

Let L be the Killing form on $\mathfrak{g}$ and we shall use the notations h_λ, H_α, X_α in 20.6.1 and 20.6.5. Finally, we let:

$$\mathfrak{n}_+ = \sum_{\alpha \in R_+} \mathfrak{g}^\alpha \ , \ \mathfrak{n}_- = \sum_{\alpha \in R_-} \mathfrak{g}^\alpha \ , \ \mathfrak{b}_+ = \mathfrak{h} + \mathfrak{n}_+ \ , \ \mathfrak{b}_- = \mathfrak{h} + \mathfrak{n}_-.$$

30.2.2 Let (V, σ) be a representation of $\mathfrak{g}$. For simplicity, we shall denote $\sigma(x)(v)$ by $x.v$. If $\mu \in \mathfrak{h}^*$, let V_μ be the set of $v \in V$ verifying $x.v = \mu(x)v$ for all $x \in \mathfrak{h}$. Then V_μ is a subspace of V and the sum of the V_μ's is direct (19.3.9). A non-zero element of V is called a *weight vector of weight* μ. The dimension of V_μ, denote by $m_V(\mu)$, is called the *multiplicity* of μ in V (or σ). If $V_\mu \neq \{0\}$, then μ is said to be a *weight* of V (or σ), and we denote by $\mathcal{P}(V)$ the set of weights of V.

Lemma. *Let $\alpha \in R$ and $\mu \in \mathfrak{h}^*$. Then $\sigma(\mathfrak{g}^\alpha)(V_\mu) \subset V_{\mu+\alpha}$.*

Proof. Let $x \in \mathfrak{h}$, $y \in \mathfrak{g}^\alpha$ and $v \in V_\mu$. Then:

$$x.(y.v) = y.(x.v) + [x, y].v = \mu(x)y.v + \alpha(x)y.v = (\mu(x) + \alpha(x))y.v.$$

So $y.v \in V_{\mu+\alpha}$. $\square$

30.2.3 Lemma. *Let (V, σ) be a representation of $\mathfrak{g}$ and $v \in V$. The following conditions are equivalent:*
   (i) $\sigma(\mathfrak{b}_+)(v) \subset \mathbb{k}v$.
   (ii) $\sigma(\mathfrak{h})(v) \subset \mathbb{k}v$ *and* $\sigma(\mathfrak{n}_+)(v) = \{0\}$.
   (iii) $\sigma(\mathfrak{h})(v) \subset \mathbb{k}v$ *and* $\sigma(\mathfrak{g}^\alpha)(v) = \{0\}$ *for all* $\alpha \in B$.
   If these conditions are verified and $v \neq 0$, then v is called a primitive vector *of V.*

Proof. Suppose that (i) is verified, then $v \in V_\mu$ for some $\mu \in \mathfrak{h}^*$. So if $\alpha \in R_+$, then $\sigma(\mathfrak{g}^\alpha)(v) \in V_\mu \cap V_{\mu+\alpha} = \{0\}$ (30.2.2). Hence we have (ii).

The implication (ii) $\Rightarrow$ (iii) is clear and (iii) $\Rightarrow$ (i) follows from the fact that $\mathfrak{n}_+$ is generated by the $\mathfrak{g}^\alpha$, $\alpha \in B$ (20.8.2). $\square$

30.2.4 Remark. It follows from the proof of 30.2.3 that if v is a primitive vector of V, then $v \in V_\mu$ for some $\mu \in \mathfrak{h}^*$.

30.2.5 Proposition. *Let (V, σ) be a $\mathfrak{g}$-module, v a primitive vector of V of weight λ and U the $\mathfrak{n}_-$-submodule of V generated by v. Suppose further that V is generated by v as a $\mathfrak{g}$-module.*
(i) *We have $U = V$.*
(ii) *For all $\mu \in \mathfrak{h}^*$, V_μ is finite-dimensional and $\dim V_\lambda = 1$.*
(iii) *We have $V = \bigoplus_{\mu \in \mathfrak{h}^*} V_\mu$.*
(iv) *For all $\mu \in \mathcal{P}(V)$, we have $\lambda \geqslant \mu$.*
(v) *The centralizer of $\sigma(\mathfrak{g})$ in $\mathrm{End}(V)$ is $\Bbbk \, \mathrm{id}_V$.*

Proof. For $\alpha \in R$, let $Y_\alpha = \sigma(X_\alpha)$. The space U is spanned by the vectors $u(\alpha_1, \ldots, \alpha_n) = Y_{\alpha_1} \circ \cdots \circ Y_{\alpha_n}(v)$, with $n \in \mathbb{N}$, $\alpha_1, \ldots, \alpha_n \in R_-$.

If $x \in \mathfrak{g}$, then $\sigma(x)(u(\alpha_1, \ldots, \alpha_n))$ is equal to:

$$\sum_{j=1}^n Y_{\alpha_1} \circ \cdots \circ Y_{\alpha_{j-1}} \circ [\sigma(x), Y_{\alpha_j}] \circ Y_{\alpha_{j+1}} \circ \cdots \circ Y_{\alpha_n}(v) + Y_{\alpha_1} \circ \cdots \circ Y_{\alpha_n} \circ \sigma(x)(v).$$

Hence for $h \in \mathfrak{h}$:

$$(1) \qquad \sigma(h)(u(\alpha_1, \ldots, \alpha_n)) = \big(\alpha_1(h) + \cdots + \alpha_n(h) + \lambda(h)\big) u(\alpha_1, \ldots, \alpha_n).$$

Similarly, if $x \in \mathfrak{n}_+$, then:
$$(2)$$
$$\sigma(x)(u(\alpha_1, \ldots, \alpha_n)) = \sum_{j=1}^n Y_{\alpha_1} \circ \cdots \circ Y_{\alpha_{j-1}} \circ [\sigma(x), Y_{\alpha_j}] \circ Y_{\alpha_{j+1}} \circ \cdots \circ Y_{\alpha_n}(v).$$

The equality (1) shows that U is $\sigma(\mathfrak{h})$-stable. On the other hand, in (2), we have:
$$[\sigma(x), Y_{\alpha_i}] \in \sigma(\mathfrak{h}) + \sum_{\alpha > \alpha_i} \Bbbk Y_\alpha.$$

A simple induction then proves that $\sigma(x)(u(\alpha_1, \ldots, \alpha_n)) \in U$. Thus U is a $\mathfrak{g}$-submodule of V, so $U = V$ by the hypothesis on v. In view of (1), we have also obtained (iii) and (iv).

Let $\alpha_1, \ldots, \alpha_r$ be pairwise distinct elements of R_+. By (1), if $\mu \in \mathfrak{h}^*$, $\dim V_\mu$ is at most the cardinality of the set of elements $(p_1, \ldots, p_r) \in \mathbb{N}^r$ verifying $p_1 \alpha_1 + \cdots + p_r \alpha_r = \lambda - \mu$. So (ii) follows.

Finally suppose that $c \in \mathrm{End}(V)$ commutes with $\sigma(\mathfrak{g})$. Then for $h \in \mathfrak{h}$, we have:
$$\sigma(h) \circ c(v) = c \circ \sigma(h)(v) = \lambda(h) c(v).$$

Thus $c(v) \in V_\lambda$. By (ii), there exists $t \in \Bbbk$ such that $c(v) = tv$. So we obtain that $c(u(\alpha_1, \ldots, \alpha_n)) = tu(\alpha_1, \ldots, \alpha_n)$. Hence $c = t \, \mathrm{id}_V$ by (i). $\square$

30.2.6 Lemma. *Let (V, σ) be a simple representation of $\mathfrak{g}$ and $\lambda \in \mathcal{P}(V)$. The following conditions are equivalent:*

(i) *For all $\mu \in \mathcal{P}(V)$, we have $\mu \leqslant \lambda$.*
(ii) *λ is the largest weight in V.*
(iii) *If $\alpha \in B$, then $\alpha + \lambda \notin \mathcal{P}(V)$.*
(iv) *There exists a primitive vector of V of weight λ.*
If these conditions are verified, then we say that λ is the highest weight *of V.*

Proof. The implications (i) $\Rightarrow$ (ii) $\Rightarrow$ (iii) are clear, while (iii) $\Rightarrow$ (iv) follows from 30.2.2 and 30.2.3. Finally, if v is a primitive vector of V of weight λ, then since V is simple, V is generated by v, and (i) follows from 30.2.5. $\square$

30.2.7 Consequently, we have obtained the following result:

Proposition. *Let (V, σ) be a simple representation of $\mathfrak{g}$ such that $\mathcal{P}(V)$ has a maximal element λ.*
    (i) *For all $\mu \in \mathfrak{h}^*$, V_μ is finite-dimensional and $\dim V_\lambda = 1$. Primitive vectors of V are the non-zero vectors of V_λ. We have $\mu \leqslant \lambda$ for all $\mu \in \mathcal{P}(V)$, or equivalently, λ is the highest weight of V.*
    (ii) *The $\mathfrak{h}$-module V is semisimple and $V = \bigoplus_{\mu \in \mathfrak{h}^*} V_\mu$.*
    (iii) *The centralizer of $\sigma(\mathfrak{g})$ in $\mathrm{End}(V)$ is $\Bbbk \, \mathrm{id}_V$.*

30.3 Finite-dimensional modules

30.3.1 Theorem. *Let (V, σ) be a finite-dimensional representation of $\mathfrak{g}$.*
    (i) *We have $V = \bigoplus_{\mu \in \mathfrak{h}^*} V_\mu$.*
    (ii) *Any weight of V belongs to the weight lattice P.*
    (iii) *If $\mu \in \mathfrak{h}^*$ and $w \in W$, then $\dim V_\mu = \dim V_{w(\mu)}$. In particular, $\mathcal{P}(V)$ is W-stable.*
    (iv) *Suppose that V is simple. Then the centralizer of $\sigma(\mathfrak{g})$ in $\mathrm{End}(V)$ is $\Bbbk \, \mathrm{id}_V$. The $\mathfrak{g}$-module V has a highest weight λ and we have $\lambda \in P_{++}$, $\dim V_\lambda = 1$.*

Proof. (i) Let $\alpha \in B$. By 19.2.7 and 20.6.5, $\sigma(H_\alpha)$ is diagonalizable with integer eigenvalues. Since $\mathfrak{h}$ is commutative, the elements $\sigma(h)$, $h \in \mathfrak{h}$, can be diagonalized simultaneously. So $V^\mu(\mathfrak{h}) = V_\mu$ (notations of 19.3.9). Hence we have (i).

(ii) We have already noted that the eigenvalues of $\sigma(H_\alpha)$ are integers. So $\mu(H_\alpha) \in \mathbb{Z}$ for all $\alpha \in B$. Hence $\mu \in P$ (18.11.5).

(iii) Let $\alpha \in B$, $n = \mu(H_\alpha)$ and $v \in V_\mu \setminus \{0\}$.
If $n \geqslant 0$, the vector $\sigma(X_{-\alpha})^n(v)$ is non-zero (19.2) and it belongs to $V_{\mu - n\alpha}$ (30.2.2). Thus $\sigma(X_{-\alpha})^n|_{V_\mu}$ is injective, hence $m_V(\mu - n\alpha) \geqslant m_V(\mu)$.
If $n \leqslant 0$, then $\sigma(X_\alpha)^{-n}|_{V(\mu)}$ is injective, so $m_V(\mu - n\alpha) \geqslant m_V(\mu)$.
But $\mu - n\alpha = \mu - \mu(H_\alpha)\alpha = s_\alpha(\mu)$. Consequently, if $\mu \in \mathcal{P}(V)$, then $s_\alpha(\mu) \in \mathcal{P}(V)$ and $m_V(s_\alpha(\mu)) \geqslant m_V(\mu)$. By 18.7.8, for $w \in W$, we have $w(\mu) \in \mathcal{P}(V)$ and $m_V(w(\mu)) \geqslant m_V(\mu)$. By exchanging the roles of μ et $w(\mu)$, we obtain (iii).

(iv) If V is simple, then (i) implies that $\mathcal{P}(V)$ is a finite non-empty set. So by 30.2.7, V has a highest weight λ and $\dim V_\lambda = 1$.

Let $v \in V_\lambda \setminus \{0\}$ and $\alpha \in B$. Then $X_\alpha.v = 0$, so $\lambda(H_\alpha) \geqslant 0$ (19.2). Hence $\lambda \in P_{++}$ by (ii) and 18.11.7. $\square$

30.3.2 Let $(.\,|\,.)$ be a W-invariant inner product on $\mathfrak{h}_\mathbb{R}^*$ and $\|.\|$ the norm associated to it.

Proposition. *Let (V,σ) be a finite-dimensional simple representation of $\mathfrak{g}$ and λ its highest weight. Then for all $\mu \in \mathcal{P}(V)$, we have $\|\mu\| \leqslant \|\lambda\|$.*

Proof. Let $\mu \in \mathcal{P}(V)$. There exists $w \in W$ such that $\nu = w(\mu) \in P_{++}$ (18.8.9). By 30.3.1 (iii), we have $\nu \in \mathcal{P}(V)$, so $\lambda - \nu \in Q_+$, $\lambda + \nu \in P_{++}$. Thus:

$$0 \leqslant (\lambda + \nu | \lambda - \nu) = \|\lambda\|^2 - \|\nu\|^2.$$

Since $\|\mu\| = \|\nu\|$, the result follows. $\square$

30.3.3 Remark. Recall from 18.8.7 (iii) that there exists a unique element $w_0 \in W$ such that $w_0(B) = -B$. It follows that if λ is the highest weight of the simple $\mathfrak{g}$-module V, then $w_0(\lambda)$ is the lowest weight of V. By 30.3.1, $m_V(w_0(\lambda)) = 1$.

30.3.4 Proposition. *Let (V,σ) be a finite-dimensional representation of $\mathfrak{g}$ and $w_0 \in W$ the unique element such that $w_0(B) = -B$. Endow V^* with the contragredient representation θ of σ.*

(i) *If $\mu \in \mathfrak{h}^*$, the orthogonal of V_μ in V^* is $\sum_{\nu \neq -\mu} V_\nu^*$, so we may identify $V_{-\mu}^*$ as the dual of V_μ.*

(ii) *Assume that V is simple with highest weight λ. Then V^* is simple with highest weight $-w_0(\lambda)$.*

Proof. (i) Since $V = \bigoplus_{\nu \in \mathfrak{h}^*} V_\nu$ (30.3.1), we may identify V^* with the direct sum $\bigoplus_{\nu \in \mathfrak{h}^*} (V_\nu)^*$. If $h \in \mathfrak{h}$, then $\theta(h) = -{}^t\sigma(h)$, so $\theta(h)|_{(V_\nu)^*} = -\nu(h)\,\mathrm{id}_{(V_\nu)^*}$. So the result follows.

(ii) Assume that V is simple. If M^* is a $\mathfrak{g}$-submodule of V^*, then its orthogonal M in V is a $\mathfrak{g}$-submodule. So V^* is simple. By (i), the highest weight of V^* is $-\xi$, where ξ is the lowest weight of V. By 30.3.3, we have $\xi = w_0(\lambda)$. $\square$

30.4 Verma modules

30.4.1 Recall from 30.1.3 that there is a canonical bijection between the set of representations of a Lie algebra $\mathfrak{a}$ and the set of representations of its universal enveloping algebra. By abuse of notation, if σ is a representation of $\mathfrak{a}$, then we shall also denote by σ the representation of $\mathrm{U}(\mathfrak{a})$. The representation θ of $\mathrm{U}(\mathfrak{a})$ on itself by left multiplication is called the *left regular representation* of $\mathrm{U}(\mathfrak{a})$.

Let $\mathfrak{k}$ be a Lie subalgebra of $\mathfrak{g}$ and $(x_1,\ldots,x_n)$ a basis of a complementary subspace of $\mathfrak{k}$ in $\mathfrak{g}$. For $\nu = (\nu_1,\ldots,\nu_n) \in \mathbb{N}^n$, we set $x^\nu = x_1^{\nu_1}\cdots x_n^{\nu_n} \in \mathrm{U}(\mathfrak{g})$. Recall from 30.1.5 that the elements x^ν, $\nu \in \mathbb{N}^n$, form a basis of the right $\mathrm{U}(\mathfrak{k})$-module $\mathrm{U}(\mathfrak{g})$.

Let (M,π) be a representation of $\mathfrak{k}$, so M is a left $\mathrm{U}(\mathfrak{k})$-module. By considering $\mathrm{U}(\mathfrak{g})$ as a left $\mathrm{U}(\mathfrak{g})$-module via the left regular representation, and as a right $\mathrm{U}(\mathfrak{k})$-module, we obtain a left $\mathrm{U}(\mathfrak{g})$-module:

$$V = \mathrm{U}(\mathfrak{g}) \otimes_{\mathrm{U}(\mathfrak{k})} M.$$

Then $V = \bigoplus_{\nu\in\mathbb{N}^n} \Bbbk x^\nu \otimes M$.

Let (V,σ) denote the corresponding $\mathrm{U}(\mathfrak{g})$-module. If $m \in M$, then we may identify m with $1\otimes m$, and we have $x^\nu \otimes m = \sigma(x^\nu)(m)$. Thus $\sigma(x^\nu)|_M$ induces an isomorphism from M onto $\sigma(x^\nu)(M)$. Therefore, we have:

$$V = \bigoplus_{\nu\in\mathbb{N}^n} \sigma(x^\nu)(M).$$

Let $u \in \mathrm{U}(\mathfrak{g})$. Then there exist elements $a_{\mu\nu}$ of $\mathrm{U}(\mathfrak{k})$ such that:

$$ux^\nu = \sum_{\mu\in\mathbb{N}^n} x^\mu a_{\mu\nu}.$$

So if $m \in M$, then:

$$\sigma(u)(x^\nu \otimes m) = \sigma\Big(\sum_\mu x^\mu a_{\mu\nu}\Big)(1\otimes m) = \sum_\mu x^\mu \otimes \pi(a_{\mu\nu})(m)$$
$$= \sum_\mu \sigma(x^\mu)\big(\pi(a_{\mu\nu})(m)\big).$$

30.4.2 Let $\lambda \in \mathfrak{h}^*$. We define a representation $(\tau_\lambda, \Bbbk_\lambda)$ of $\mathfrak{b}_+$ as follows: as a vector space $\Bbbk_\lambda = \Bbbk$, and if $h \in \mathfrak{h}$ and $n \in \mathfrak{n}_+$,

$$\tau_\lambda(h+n)(1) = \lambda(h)1.$$

As in 30.4.1, we can define a left $\mathrm{U}(\mathfrak{g})$-module:

$$M(\lambda) = \mathrm{U}(\mathfrak{g}) \otimes_{\mathrm{U}(\mathfrak{b}_+)} \Bbbk_\lambda.$$

We call $M(\lambda)$ the *Verma module* associated to $\mathfrak{g}$, $\mathfrak{h}$, B (or $\mathfrak{b}_+$) and λ.

30.4.3 Remark. If we consider $\mathrm{U}(\mathfrak{n}_-)$ as a left $\mathrm{U}(\mathfrak{n}_-)$-module via its left regular representation, then 30.4.1 implies that the map $\mathrm{U}(\mathfrak{n}_-) \to M(\lambda)$, $u \mapsto u \otimes 1$, is an isomorphism of $\mathrm{U}(\mathfrak{n}_-)$-modules.

30.4.4 If $\mu \in \mathfrak{h}^*$, denote by $\mathfrak{B}(\mu)$ the number of families $(n_\alpha)_{\alpha\in R_+}$ of elements of $\mathbb{N}$ verifying $\mu = \sum_{\alpha\in R_+} n_\alpha \alpha$. We have $\mathfrak{B}(\mu) > 0$ if and only if $\mu \in Q_+$.

Proposition. *Let $\lambda \in \mathfrak{h}^*$ and $\alpha_1,\ldots,\alpha_n$ the pairwise distinct elements of R_+.*

(i) *The module $M(\lambda)$ has highest weight λ, and $1 \otimes 1$ is a primitive vector of V of weight λ. In particular, $M(\lambda)_\lambda = 1 \otimes \Bbbk$ and $M(\lambda) = U(\mathfrak{n}_-)M(\lambda)_\lambda$.*

(ii) *The weights of $M(\lambda)$ are the elements of $\mathfrak{h}^*$ of the form $\lambda - \nu$, with $\nu \in Q_+$.*

(iii) *If $\mu \in \mathfrak{h}^*$, then $\dim M(\lambda)_\mu = \mathfrak{B}(\lambda - \mu)$.*

(iv) *For $\mu \in \mathfrak{h}^*$, let $S(\mu)$ be the set of elements $(\nu_1, \ldots, \nu_n) \in \mathbb{N}^n$ such that $\lambda - \nu_1 \alpha_1 - \cdots - \nu_n \alpha_n = \mu$. Then:*

$$M(\lambda)_\mu = \sum_{(\nu_1, \ldots, \nu_n) \in S(\mu)} X_{-\alpha_1}^{\nu_1} \cdots X_{-\alpha_n}^{\nu_n} \otimes \Bbbk.$$

Proof. If $(\nu_1, \ldots, \nu_n) \in \mathbb{N}^n$, denote $X^\nu = X_{-\alpha_1}^{\nu_1} \cdots X_{-\alpha_n}^{\nu_n}$. For $h \in \mathfrak{h}$, we have:

$$\begin{aligned} h.(X^\nu \otimes 1) &= [h, X^\nu] \otimes 1 + X^\nu h \otimes 1 \\ &= (-\nu_1 \alpha_1 - \cdots - \nu_n \alpha_n)(h) X^\nu \otimes 1 + X^\nu \otimes \lambda(h) \\ &= (\lambda - \nu_1 \alpha_1 - \cdots - \nu_n \alpha_n)(h)(X^\nu \otimes 1). \end{aligned}$$

Hence (iv). The other parts follows immediately. □

30.4.5 We shall denote the element $1 \otimes 1$ of $M(\lambda)$ by v_λ, and we shall call v_λ the *canonical generator* of $M(\lambda)$.

30.4.6 The module $M(\lambda)$ is universal in the following sense:

Proposition. *Let $\lambda \in \mathfrak{h}^*$ and (V, σ) a representation of $\mathfrak{g}$ which has a primitive vector v of weight λ.*

(i) *There exists a unique $\mathfrak{g}$-module homomorphism $\varphi : M(\lambda) \to V$ such that $\varphi(v_\lambda) = v$.*

(ii) *Suppose further that V is generated by v as a $\mathfrak{g}$-module. Then φ is surjective. A necessary and sufficient condition for φ to be bijective is that $\sigma(u)$ is injective for all $u \in U(\mathfrak{n}_-) \setminus \{0\}$.*

Proof. (i) Since v_λ generates $M(\lambda)$, it is clear that if φ exists, it is unique.

Now let $\psi \in \mathrm{Hom}(\Bbbk_\lambda, V)$ be such that $\psi(1) = v$. Then if $h \in \mathfrak{h}$ and $n \in \mathfrak{n}_+$, we have:

$$\psi \circ \tau_\lambda(h + n)(1) = \lambda(h)v = \sigma(h + n)v.$$

So ψ is a homomorphism of $\mathfrak{b}_+$-modules. Now by the universal property of the tensor product, ψ extends uniquely to a $\mathfrak{g}$-module homomorphism $\varphi : M(\lambda) \to V$ verifying $\varphi(v_\lambda) = v$.

(ii) Suppose that V is generated by v as a $\mathfrak{g}$-module. Then clearly φ is surjective. If φ is bijective, then 30.4.3 implies that $\sigma(u)$ is injective for all $u \in U(\mathfrak{n}_-) \setminus \{0\}$.

If φ is not bijective, then there exists $u \in U(\mathfrak{n}_-)$ such that $\varphi(u \otimes 1) = 0$. Hence:

$$\sigma(u)(v) = \sigma(u)\varphi(v_\lambda) = \varphi(u.v_\lambda) = \varphi(u \otimes 1) = 0.$$

Thus $\sigma(u)$ is not injective. □

30.4.7 Proposition. *Let $\lambda \in \mathfrak{h}^*$ and:*

$$M(\lambda)_+ = \sum_{\mu \neq \lambda} M(\lambda)_\mu.$$

(i) *Any $\mathfrak{g}$-submodule of $M(\lambda)$ distinct from $M(\lambda)$ is contained in $M(\lambda)_+$.*

(ii) *There exists a largest $\mathfrak{g}$-submodule $K(\lambda)$ of $M(\lambda)$ distinct from $M(\lambda)$. The quotient $M(\lambda)/K(\lambda)$ is a simple $\mathfrak{g}$-module.*

Proof. Let E be a $\mathfrak{g}$-submodule of $M(\lambda)$. By 30.2.5 and 30.4.4, we have:

$$E = \sum_{\mu \in \mathfrak{h}^*} (E \cap M(\lambda)_\mu).$$

Since $M(\lambda)_\lambda = \Bbbk v_\lambda$ and $M(\lambda)$ is generated by v_λ, it follows that if $E \neq M(\lambda)$, then $E \cap M(\lambda)_\lambda = \{0\}$. Thus $E \subset M(\lambda)_+$.

Now let $K(\lambda)$ be the sum of all the $\mathfrak{g}$-submodules of $M(\lambda)$ which are distinct from $M(\lambda)$. Then we have just proved that $K(\lambda) \subset M(\lambda)_+$. Hence (ii) follows easily. $\square$

30.4.8 In the notations of 30.4.7, we shall denote:

$$L(\lambda) = M(\lambda)/K(\lambda),$$

and $\overline{v_\lambda}$ the image of v_λ in $L(\lambda)$. It is clear that $\overline{v_\lambda}$ is a primitive vector of $L(\lambda)$ of weight λ, and it generates the $\mathfrak{g}$-module $L(\lambda)$. We shall call $\overline{v_\lambda}$ the *canonical generator* of $L(\lambda)$.

30.4.9 Proposition. *Let $\lambda \in \mathfrak{h}^*$ and V a simple $\mathfrak{g}$-module with highest weight λ. Then V is isomorphic to $L(\lambda)$.*

Proof. This follows from 30.4.6 and 30.4.7. $\square$

30.5 Results on existence and uniqueness

30.5.1 Lemma. *Let $\beta \in B$, $\mathfrak{s}_\beta = \Bbbk H_\beta + \Bbbk X_\beta + \Bbbk X_{-\beta}$ and (V, σ) a representation of $\mathfrak{g}$.*

(i) *The sum V_β of $\mathfrak{s}_\beta$-submodules of finite dimension of V is a $\mathfrak{g}$-submodule.*

(ii) *Suppose that there exist $m \in \mathbb{N}$ and $v \in V$ verifying $\sigma(X_\beta)(v) = 0$ and $\sigma(H_\beta)(v) = mv$. For $j \geqslant 0$, let $v_j = \sigma(X_{-\beta})^j(v)$. Then $\sigma(X_\beta)(v_{m+1}) = 0$. Moreover, if $v_{m+1} = 0$, then $\Bbbk v_0 + \cdots + \Bbbk v_m$ is a $\mathfrak{s}_\beta$-submodule of V.*

Proof. (i) This follows from 19.2.6 and 29.6.2.

(ii) Set $v_{-1} = 0$. We verify easily that $\sigma(H_\beta)(v_j) = (m - 2j)v_j$ for $l \geqslant 0$. Suppose that $\sigma(X_\beta)(v_j) = j(m - j + 1)v_{j-1}$ (This is true if $j = 0$). Then:

$$\begin{aligned}
\sigma(X_\beta)(v_{j+1}) &= \sigma(X_{-\beta})\sigma(X_\beta)(v_j) + \sigma(H_\beta)(v_j) \\
&= j(m - j + 1)\sigma(X_{-\beta})(v_{j-1}) + (m - 2j)v_j \\
&= (j + 1)(m - j)v_j.
\end{aligned}$$

In particular, $\sigma(X_\beta)(v_{m+1}) = 0$. The last part follows immediately. $\square$

30.5.2 Theorem. *Let* $\mathfrak{g}$ *be a semisimple Lie algebra and* (V, σ) *a simple* $\mathfrak{g}$-*module with highest weight* λ. *The following conditions are equivalent:*

(i) V *is finite-dimensional.*

(ii) $\lambda \in P_{++}$.

Proof. We have already seen that (i) implies (ii) in 30.3.1. So let us assume that $\lambda \in P_{++}$ and let $v \in V$ be a primitive vector of weight λ.

Let $\beta \in B$, $\mathfrak{s}_\beta = \Bbbk H_\beta + \Bbbk X_\beta + \Bbbk X_{-\beta}$, $m = \lambda(H_\beta) \in \mathbb{N}$. For $j \geqslant 0$, let $v_j = \sigma(X_{-\beta})^j(v)$. By 30.5.1 (ii), $\sigma(X_\beta)(v_{m+1}) = 0$. Let $\alpha \in B \setminus \{\beta\}$. Since $[X_\alpha, X_{-\beta}] = 0$ (18.7.4), we have:

$$\sigma(X_\alpha)(v_{m+1}) = \sigma(X_\alpha)\sigma(X_{-\beta})^{m+1}(v) = \sigma(X_{-\beta})^{m+1}\sigma(X_\alpha)(v) = 0.$$

If $v_{m+1} \neq 0$, then it is a primitive vector of V of weight distinct from λ (because $\sigma(H_\beta)(v_{m+1}) = -(m-2)v_{m+1}$). This is absurd by 30.2.7, hence $v_{m+1} = 0$, and $\Bbbk v_0 + \cdots + \Bbbk v_m$ is a $\mathfrak{s}_\beta$-submodule of V. The sum of all the finite-dimensional simple $\mathfrak{s}_\beta$-submodules of V is therefore non-zero. and by 30.5.1 (i), it is a $\mathfrak{g}$-submodule, so it is equal to V. It follows that there is family $(M_i)_{i \in I}$ of finite-dimensional simple $\mathfrak{s}_\beta$-modules such that $V = \bigoplus_{i \in I} M_i$.

Let $u \in V_\mu \setminus \{0\}$ and write $u = \sum_i u_i$, where $u_i \in M_i$ for all i. Then $\sigma(H_\beta)(u_i) = \mu(H_\beta)u_i$ for all i. It follows from 19.2.7 that $n = \mu(H_\beta) \in \mathbb{Z}$ because the u_i's are not all equal to zero. This being true for all $\beta \in B$, we deduce that $\mathcal{P}(V) \subset P$.

If $n \geqslant 0$ and $u_i \neq 0$, then $\sigma(X_{-\beta})^n(u_i) \neq 0$ (19.2), so $\sigma(X_{-\beta})^n(u) \neq 0$. If $n \leqslant 0$, then $\sigma(X_\beta)^n(u) \neq 0$. So in both cases, $s_\beta(\mu) = \mu - n\beta \in \mathcal{P}(V)$. Since the s_β's generate W (18.7.8), we have $W(\mathcal{P}(V)) \subset \mathcal{P}(V)$.

Now any W-orbit of an element of P meets P_{++} (18.8.9 and 18.11.7). Moreover, by 30.2.7, if $\mu \in \mathcal{P}(V) \cap P_{++}$, then $\lambda - \mu \in Q_+$ and $\lambda + \mu \in P_{++}$, so $\|\lambda\| \geqslant \|\mu\|$ as in the proof of 30.3.2. It follows that $\mathcal{P}(V)$ is contained in the intersection of the discrete set P and the compact set $\{\mu \in \mathfrak{h}_\mathbb{R}^*; \|\mu\| \leqslant \|\lambda\|\}$. So $\mathcal{P}(V)$ is finite.

Finally, for $\mu \in \mathcal{P}(V)$, V_μ is finite-dimensional (30.2.7), so V is finite-dimensional. $\square$

30.5.3 If V is a $\mathfrak{g}$-module, denote by $[V]$ its isomorphism class.

Theorem. *The map* $\lambda \mapsto [L(\lambda)]$ *is a bijection from* P_{++} *to the set of isomorphism classes of finite-dimensional simple* $\mathfrak{g}$-*modules.*

Proof. This follows from 30.3.1, 30.4.9 and 30.5.2. $\square$

30.5.4 Let $B = (\beta_1, \ldots, \beta_l)$ and $(\varpi_1, \ldots, \varpi_l)$ the set of fundamental weights of R with respect to B (18.11.6). By 30.5.3, for $1 \leqslant i \leqslant l$, there is, up to an isomorphism, a unique finite-dimensional simple representation σ_i of $\mathfrak{g}$ with highest weight ϖ_i. The representations $\sigma_1, \ldots, \sigma_l$ are called the *fundamental simple representations* of $\mathfrak{g}$.

30.6 A property of the Weyl group

30.6.1 Denote by $\mathrm{Aut}_e(\mathfrak{g}, \mathfrak{h})$ (resp. $\mathrm{Aut}_e^0(\mathfrak{g}, \mathfrak{h})$) the subgroup of $\mathrm{Aut}_e\,\mathfrak{g}$ consisting of elements $\theta \in \mathrm{Aut}_e\,\mathfrak{g}$ verifying $\theta(\mathfrak{h}) = \mathfrak{h}$ (resp. $\theta(h) = h$ for all $h \in \mathfrak{h}$). It is clear that $\mathrm{Aut}_e^0(\mathfrak{g}, \mathfrak{h})$ is a normal subgroup of $\mathrm{Aut}_e(\mathfrak{g}, \mathfrak{h})$.

The group $\mathrm{Aut}_e\,\mathfrak{g}$ acts on $\mathfrak{g}^*$ as follows:

$$\theta(\mu) = \mu \circ \theta^{-1}$$

for $\theta \in \mathrm{Aut}_e\,\mathfrak{g}$ and $\mu \in \mathfrak{g}^*$.

30.6.2 Let (V, σ) be a finite-dimensional representation of $\mathfrak{g}$. If $x \in \mathfrak{g}$ is nilpotent, then $\sigma(x)$ is a nilpotent endomorphism of V (20.4.4). We may therefore define $e^{\sigma(x)} \in \mathrm{End}(V)$.

Lemma. *Let $x, y \in \mathfrak{g}$ be such that x is nilpotent. Then:*

$$\sigma\!\left(e^{\mathrm{ad}\,x}(y)\right) = e^{\sigma(x)} \circ \sigma(y) \circ e^{-\sigma(x)}.$$

Proof. We obtain easily that $\sigma(e^{\mathrm{ad}\,x}(y)) = e^{\mathrm{ad}\,\sigma(x)}(\sigma(y))$. So it suffices to prove that if $f, g \in \mathrm{End}(V)$ with f nilpotent, then $e^{\mathrm{ad}\,f}(g) = e^f \circ g \circ e^{-f}$.

Define the endomorphisms $\mathbf{L}, \mathbf{R}$ of $\mathrm{End}(V)$ as follows: for $g \in \mathrm{End}(V)$:

$$\mathbf{L}(g) = f \circ g \ , \ \ \mathbf{R}(g) = -g \circ f.$$

We have $\mathbf{L} \circ \mathbf{R} = \mathbf{R} \circ \mathbf{L}$, $\mathrm{ad}\,f = \mathbf{L} + \mathbf{R}$, and $\mathbf{L}, \mathbf{R}$ are nilpotent. We deduce therefore that:

$$e^{\mathrm{ad}\,f}(g) = e^{\mathbf{L}+\mathbf{R}}(g) = e^{\mathbf{L}}e^{\mathbf{R}}(g) = \sum_{i,j \geqslant 0} \frac{\mathbf{L}^i}{i!}\frac{\mathbf{R}^j}{j!}(g) = e^f \circ g \circ e^{-f}.$$

Hence the result. □

30.6.3 Let (V, σ) be a finite-dimensional representation of $\mathfrak{g}$. If $\theta \in \mathrm{Aut}_e\,\mathfrak{g}$, then 30.6.2 says that there exists $S_\theta \in \mathrm{GL}(V)$ such that for all $x \in \mathfrak{g}$:

$$\sigma\!\left(\theta(x)\right) = S_\theta \circ \sigma(x) \circ S_\theta^{-1}.$$

Suppose that $\theta \in \mathrm{Aut}_e(\mathfrak{g}, \mathfrak{h})$. Let $\mu \in \mathfrak{h}^*$, $v \in V_\mu$ and $h \in \mathfrak{h}$. Then:

$$\sigma(h)\big(S_\theta(v)\big) = (\mu \circ \theta^{-1})(h)S_\theta(v) = \theta(\mu)(h)S_\theta(v).$$

It follows that $S_\theta(V_\mu) = V_{\theta(\mu)}$, so $\mathcal{P}(V)$ is stable under θ. Applying this to the adjoint representation of $\mathfrak{g}$, we see that R and Q are stable under θ.

30.6.4 Lemma. *Let $\theta \in \mathrm{Aut}_e(\mathfrak{g}, \mathfrak{h})$ be such that $\theta(B) = B$. Then $\theta(\beta) = \beta$ for all $\beta \in B$ and $\theta \in \mathrm{Aut}_e^0(\mathfrak{g}, \mathfrak{h})$.*

Proof. We saw in 30.6.3 that $\theta(Q) = Q$, and since $\theta(B) = B$, we have $\theta(Q_+) = Q_+$.

Let $\lambda \in P_{++}$. There exists a finite-dimensional representation (V, σ) of $\mathfrak{g}$ with highest weight λ (30.5.3). By 30.6.3, $\theta(\lambda)$ is a weight of V, so $\theta(\lambda) \leqslant \lambda$ (30.2.7). In the same way, $\theta^{-1}(\lambda) \leqslant \lambda$. From this, we deduce that $\theta(\lambda) = \lambda$ since $\theta(Q_+) = Q_+$. Using, for example, the fundamental weights of R, we see that $\theta|_{\mathfrak{h}^*} = \mathrm{id}_{\mathfrak{h}^*}$, and so $\theta|_{\mathfrak{h}} = \mathrm{id}_{\mathfrak{h}}$.   □

30.6.5 Theorem. *Let us conserve the hypotheses and notations of 30.2.1 and 30.6.1. We define a homomorphism of groups*

$$\varphi : \mathrm{Aut}_e(\mathfrak{g}, \mathfrak{h}) \to \mathrm{GL}(\mathfrak{h}^*)$$

by setting $\varphi(\theta)(\mu) = \mu \circ \theta^{-1}$ for $\theta \in \mathrm{Aut}_e(\mathfrak{g}, \mathfrak{h})$ and $\mu \in \mathfrak{h}^*$. Then φ induces an isomorphism of groups from $\mathrm{Aut}_e(\mathfrak{g}, \mathfrak{h}) / \mathrm{Aut}_e^0(\mathfrak{g}, \mathfrak{h})$ onto W.

Proof. We have $\ker \varphi = \mathrm{Aut}_e^0(\mathfrak{g}, \mathfrak{h})$, so φ induces an injective homomorphism

$$\overline{\varphi} : \mathrm{Aut}_e(\mathfrak{g}, \mathfrak{h}) / \mathrm{Aut}_e^0(\mathfrak{g}, \mathfrak{h}) \to \mathrm{GL}(\mathfrak{h}^*).$$

It follow easily from 29.1.3 that $W \subset \mathrm{im}\, \varphi$.

Let $\theta \in \mathrm{Aut}_e(\mathfrak{g}, \mathfrak{h})$. Then $\varphi(\theta)(R) = R$ (30.6.3). Also, it is clear that $B' = \varphi(\theta)(B)$ is a base of R. Let $w \in W$ be such that $w(B') = B$ (18.8.7) and $\theta' \in \mathrm{Aut}_e(\mathfrak{g}, \mathfrak{h})$ verifying $\varphi(\theta') = w$. Then $\varphi(\theta' \circ \theta)(B) = B$. So we have $\theta' \circ \theta \in \mathrm{Aut}_e^0(\mathfrak{g}, \mathfrak{h})$ (30.6.4). Hence $\mathrm{im}\, \varphi = W$, and we are done.   □

References and comments

- [12], [14], [29], [38], [39], [42], [43], [80].

We have only given a few properties of enveloping algebras in this chapter. A complete treatment of these algebras, in particular for the properties admitted in section 30.1, can be found in [29].

Symmetric invariants

In this chapter, we prove two fundamental theorems of Chevalley and Kostant on invariants.

31.1 Invariants of finite groups

31.1.1 Lemma. *Let $l \in \mathbb{N}$ and $P, Q \in A = \Bbbk[T_1, \ldots, T_l]$. Suppose that P is homogeneous of degree 1 and $Q(x) = 0$ for all $x \in \Bbbk^l$ such that $P(x) = 0$. Then there exists $R \in A$ such that $Q = PR$.*

Proof. We may assume that P is of the form $T_1 + a_2 T_2 + \cdots + a_l T_l$. Then there exist $U \in A$ and $V \in \Bbbk[T_2, \ldots, T_l]$ such that $Q = PU + V$. If $V \neq 0$, then since $\Bbbk$ is infinite, there exist $\alpha_2, \ldots, \alpha_l \in \Bbbk$ such that $V(\alpha_2, \ldots, \alpha_l) \neq 0$. Let $\alpha_1 \in \Bbbk$ be such that $P(\alpha_1, \ldots, \alpha_l) = 0$ (recall that $\Bbbk$ is algebraically closed). Then $Q(\alpha_1, \ldots, \alpha_l) = V(\alpha_2, \ldots, \alpha_l) \neq 0$. Contradiction. $\square$

31.1.2 In this section, V is a $\Bbbk$-vector space of dimension l, $S = S(V^*)$ the symmetric algebra of V^* and $F = \text{Fract}(S)$. The algebra S is isomorphic to $\Bbbk[T_1, \ldots, T_l]$, and for $n \in \mathbb{N}$, we denote by S^n the set of homogeneous elements of degree n in S.

Let G be a finite subgroup of $\text{GL}(V)$ of order m. Then G acts on S as follows: for $\alpha \in G$, $P \in S$ and $x \in V$,

$$(\alpha.P)(x) = P(\alpha^{-1}(x)).$$

This action extends to an action of G on F by setting $\alpha.(u^{-1}v) = (\alpha.u)^{-1}(\alpha.v)$ for $\alpha \in G$, $u, v \in S$ and $u \neq 0$. Let us denote by $R = S^G$ and $K = F^G$ the set of fixed points under these actions. Observe that R is a graded subalgebra of S and K is a subfield of F. Denote by R_+ the set of elements of R without constant term and $I = SR_+$. Since S is Noetherian, there exist homogeneous elements $u_1, \ldots, u_r$ of R_+ such that $I = Su_1 + \cdots + Su_r$.

Since the characteristic of $\Bbbk$ is zero, we may define for $u \in S$:

$$u^\bullet = \frac{1}{m} \sum_{\alpha \in G} \alpha.u.$$

Then $u^\bullet \in R$. Also if $u \in S$ and $v \in R$, then $(uv)^\bullet = u^\bullet v$.

31.1.3 Lemma. (i) *The algebra S is integral over R.*

(ii) *The field K is the field of fractions of R.*

(iii) *The extension $K \subset F$ is finite of degree m, and the transcendence degree of K over $\Bbbk$ is equal to l.*

Proof. For $x \in F$, let

$$P_x(T) = \prod_{\alpha \in G} (T - \alpha.x).$$

(i) If $u \in S$, then P_u is G-invariant, so its coefficients are in R. Since $P_u(u) = 0$, we deduce that u is integral over R.

(ii) This follows from 25.5.1.

(iii) By 5.4.10, it suffices to prove that $[F : K] = m$. Let $x \in F$. As in (i), $P_x(x) = 0$ and $P_x \in K[T]$, so x is algebraic over K, and $[K(x) : K] \leqslant m$. Let $x \in F$ be such that $[K(x) : K]$ is maximal. For any $y \in F$, there exists, by 5.6.7, $z \in F$ such that $K(x,y) = K(z)$. It follows from the choice of x that $F = K(x)$ and $[F : K] \leqslant m$.

Let $Q \in K[T]$ be an irreducible polynomial which has a root $x \in F$. Then Q divides P_x, and so Q splits over F. Thus the extension $K \subset F$ is normal. Let $H = \mathrm{Gal}(F/K)$ and Ω an algebraic closure of K. We have $K = F^H$ (5.7.6). It follows from 5.6.6 and the normality of the extension $K \subset F$ that:

$$[F : K] \leqslant m = \mathrm{card}\, G \leqslant \mathrm{card}\, H \leqslant \mathrm{card}\, \mathrm{Hom}_K(F, \Omega) = [F : K].$$

So we have proved the lemma. $\square$

31.1.4 Lemma. *Let $u_1, \ldots, u_r$ be homogeneous elements of R_+ which generate the ideal $I = SR_+$. Then the $\Bbbk$-algebra R is generated by $u_1, \ldots, u_r$.*

Proof. We shall show by induction on n that if $u \in R$ is homogeneous of degree n, then it is a polynomial in the u_i's. This is obvious if $n = 0$. If $n > 0$, then $u \in I$, so $u = s_1 u_1 + \cdots + s_r u_r$ with $s_1, \ldots, s_r \in S$.

Since u and the u_i's are homogeneous, we may also assume that the s_i's are homogeneous of degree $\deg u - \deg u_i < n$. Now as observed earlier, we have:

$$u = u^\bullet = s_1^\bullet u_1 + \cdots + s_r^\bullet u_r.$$

Since $s_i^\bullet \in R$, the induction hypothesis implies that u is a polynomial in the u_i's. $\square$

31.1.5 Lemma. *Let us suppose that G is generated by reflections. Let $u_1, \ldots, u_r \in R$ be such that $u_1 \notin Ru_2 + \cdots + Ru_r$, and $v_1 \ldots, v_r \in S$ homogeneous such that:*

(1) $u_1 v_1 + \cdots + u_r v_r = 0.$

Then $v_1 \in I$.

Proof. Let $J = Su_2 + \cdots + Su_r$. If $u_1 \in J$, then $u_1 = w_2u_2 + \cdots + w_ru_r$ with $w_2, \ldots, w_r \in S$. But then, we would have $u_1 = u_1^\bullet = w_2^\bullet u_2 + \cdots + w_r^\bullet u_r \in Ru_2 + \cdots + Ru_r$ which contradicts the hypothesis. Thus $u_1 \notin J$.

We shall prove $v_1 \in I$ by induction on $\deg v_1$. If $\deg v_1 \leqslant 0$, then $v_1 = 0$ because $u_1 \notin J$. So let us suppose that $\deg v_1 > 0$.

Let $s \in G$ be a reflection with respect to the hyperplane H and $u \in V^*$ verifying $\ker u = H$. Since $s.v_i - v_i$ is zero on H, it follows from 31.1.1 that there exists $w_i \in S$ such that $sv_i - v_i = uw_i$. As v_i and $s.v_i$ are homogeneous of the same degree, w_i is homogeneous of degree $< \deg v_i$. On the other hand, we have:

$$(2) \qquad u_1(s.v_1) + \cdots + u_r(s.v_r) = 0.$$

Equations 1 and 2 imply therefore that $u(u_1w_1 + \cdots + u_rw_r) = 0$. It follows by the induction hypothesis that $w_1 \in I$, so $s.v_1 \equiv v_1 \pmod{I}$.

Since R_+ is stable under the action of G, so is I. Thus G acts on the quotient S/I. Let $\overline{v_1}$ be the image of v_1 in S/I, then we have just seen that $s.\overline{v_1} = \overline{v_1}$. But G is generated by reflections, so $\alpha.\overline{v_1} = \overline{v_1}$ for all $\alpha \in G$. Hence $v_1^\bullet \equiv v_1 \pmod{I}$. As $v_1^\bullet \in R_+ \subset I$, we have $v_1 \in I$. $\qquad \square$

31.1.6 Theorem. (Chevalley's Theorem) *Let V be a vector space of dimension l, and G a finite subgroup of $\mathrm{GL}(V)$ generated by reflections. Then the $\Bbbk$-algebra of G-invariants of $\mathrm{S}(V^*)$ is generated by l algebraically independent homogeneous elements.*

Proof. By fixing a basis of V^*, we shall identify $S = \mathrm{S}(V^*)$ with $\Bbbk[T_1, \ldots, T_l]$. Let $u_1, \ldots, u_r \in R_+$ be homogeneous elements which generate the ideal I, and let r be minimal with this property. By 31.1.4, R is generated by the u_i's, so it suffices to prove that the u_i's are algebraically independent over $\Bbbk$, and 31.1.3 would imply that $r = l$.

Let us suppose that $P(u_1, \ldots, u_r) = 0$ for some $P \in \Bbbk[Y_1, \ldots, Y_r] \setminus \{0\}$.

Let $\lambda Y_1^{m_1} \cdots Y_r^{m_r}$ be a monomial which appears in P. Then the degree of $\lambda u_1^{m_1} \cdots u_r^{m_r}$ is $d = m_1d_1 + \cdots + m_rd_r$ where $d_i = \deg u_i$. We may therefore assume that P is the sum of monomials $\lambda Y_1^{m_1} \cdots Y_r^{m_r}$ such that the sum $m_1d_1 + \cdots + m_rd_r$ are all equal to d. For $1 \leqslant k \leqslant l$, we have:

$$\sum_{i=1}^{r} P_i \frac{\partial u_i}{\partial T_k} = 0, \quad \text{where } P_i = \frac{\partial P}{\partial Y_i}(u_1, \ldots, u_r).$$

Note that P_i is homogeneous of degree $d - d_i$ in R and that the $\partial u_i/\partial T_k$'s are homogeneous elements of S.

Without loss of generality, we may assume that $(P_1, \ldots, P_m)$ is a minimal set of generators of the ideal of R generated by the P_i's. So for $m < i \leqslant r$, there exist $Q_{ij} \in R$, $1 \leqslant j \leqslant m$, such that

$$P_i = Q_{i1}P_1 + \cdots + Q_{im}P_m.$$

We may further assume that the Q_{ij}'s are homogeneous of degree $d_j - d_i$. For $1 \leqslant k \leqslant l$, we have:

$$\sum_{i=1}^{m} P_i \left(\frac{\partial u_i}{\partial T_k} + \sum_{j=m+1}^{r} Q_{ji} \frac{\partial u_j}{\partial T_k} \right) = 0.$$

Applying 31.1.5, we see that there exist $t_{1k}, \ldots, t_{rk} \in S$ such that:

$$(3) \qquad \frac{\partial u_1}{\partial T_k} + \sum_{j=m+1}^{r} Q_{j1} \frac{\partial u_j}{\partial T_k} = \sum_{i=1}^{r} t_{ik} u_i.$$

Multiplying both sides of equation (3) by T_k, and sum over k, we obtain by Euler's formula that

$$(4) \qquad d_1 u_1 + \sum_{j=m+1}^{r} d_j Q_{j1} u_j = \sum_{i=1}^{r} v_i u_i,$$

where $v_1, \ldots, v_s \in S$. But $v_1 = T_1 t_{11} + \cdots + T_l t_{1l}$ has no constant term, and the left hand side of (4) is homogeneous of degree d_1. So by comparing the homogeneous components of degree d_1 in equation (4), we see that $d_1 u_1$, and hence u_1, belongs to the ideal of S generated by $u_2, \ldots, u_r$. This contradicts our choice of r. $\square$

31.1.7 Proposition. *Let us conserve the hypotheses and notations of 31.1.6. Let $u_1, \ldots, u_l$ and $v_1, \ldots, v_l$ be two sets of algebraically independent homogeneous generators of R. Then there exists a permutation σ of $\{1, \ldots, l\}$ such that $\deg u_i = \deg v_{\sigma(i)}$ for $1 \leqslant i \leqslant l$.*

Proof. Each u_i is a polynomial in the v_i's and vice versa. So we have for $1 \leqslant i, j \leqslant l$:

$$\sum_{k=1}^{l} \frac{\partial u_i}{\partial v_k} \frac{\partial v_k}{\partial u_j} = \delta_{ij}.$$

This proves that the matrices $[\partial u_i / \partial v_j]$ and $[\partial v_i / \partial u_j]$ are invertible, so their determinant are non-zero. It follows from the definition of the determinant that there is a permutation σ of $\{1, \ldots, l\}$ such that

$$\prod_{i=1}^{l} \frac{\partial u_i}{\partial v_{\sigma(i)}} \neq 0.$$

Up to a reindexation, we may assume that σ is the identity permutation. So if we write $u_i = P_i(v_1, \ldots, v_l)$, then v_i appears in P_i. Removing if necessary the terms which are redundant in P_i, we may assume that for each monomial $v_1^{m_1} \cdots v_l^{m_l}$ in P_i, we have:

$$\deg u_i = m_1 \deg v_1 + \cdots + m_l \deg v_l.$$

Hence $\deg u_i \geqslant \deg v_i$. Now by exchanging the roles of u_i and v_i, we obtain that $\deg u_i = \deg v_i$. $\square$

31.1.8 Proposition. *With the hypotheses of* 31.1.6, S *is a free R-module of rank $m = \operatorname{card} G$.*

Proof. Let $\theta : S \to S/I$ denote the canonical surjection.

1) Let $(v_\lambda)_{\lambda \in \Lambda}$ be a set of homogeneous elements of S such that the $\theta(v_\lambda)$'s span the $\Bbbk$-vector space S/I. Let T be the R-submodule of S generated by the v_λ's. It is a graded submodule of S. We claim that $S = T$. We shall prove our claim by showing, by induction, that $S^n = T^n$ for all $n \in \mathbb{N}$ where S^n and T^n denote the homogeneous component of degree n of S and T.

Since $S^0 \cap I = \{0\}$, one of the v_λ's is of degree 0, so $S^0 = T^0$.

Let $u \in S^n$, $n > 0$. Then there exist $\alpha_\lambda \in \Bbbk$, $w_1, \ldots, w_r \in R_+$ and $t_1, \ldots, t_r \in S$ homogeneous of degree $< n$ such that

$$u = \sum_{\lambda \in \Lambda} \alpha_\lambda v_\lambda + \sum_{i=1}^{r} t_i w_i.$$

By the induction hypothesis, $t_i \in T$ for all i. Hence $u \in T^n$. So we have proved our claim.

2) Let $v_1, \ldots, v_r \in S$ be homogeneous elements such that the $\theta(v_i)$'s are linearly independent. We shall show by induction on r that the v_i's are independent over R. The result is obvious if $r = 1$. So let us suppose that $r > 1$ and that $u_1 v_1 + \cdots + u_r v_r = 0$ with $u_i \in R$ homogeneous. Since $v_1 \notin I$, there exist homogeneous elements $w_2, \ldots, w_r \in R$ such that $u_1 = w_2 u_2 + \cdots + w_r u_r$ (31.1.5). Thus

$$u_2(v_2 + w_2 v_1) + \cdots + u_r(w_r + w_r v_1) = 0.$$

But the elements $v_i + w_i v_1$, $2 \leqslant i \leqslant r$, are homogeneous and their image under θ are linearly independent. By the induction hypothesis, $u_2 = \cdots = u_r = 0$. Hence $u_1 = 0$.

3) We deduce from points 1 and 2 that S is a free R-module. Moreover, 31.1.3 says that $K = \operatorname{Fract}(R)$ and $[F : K] = m$. It is now clear that the rank of S over R is m. $\quad\square$

31.2 Invariant polynomial functions

31.2.1 In this section, V is a vector space of dimension l. For $n \in \mathbb{N}$, we denote by $\mathrm{S}^n(V)$ be the set of homogeneous elements of degree n in the symmetric algebra $\mathrm{S}(V)$ of V. The algebra $\mathrm{S}(V^*)$ can be identified as the algebra of polynomial functions on V with values in $\Bbbk$.

31.2.2 Let $\mathfrak{a}$ be a Lie algebra and (V, σ) a representation of $\mathfrak{a}$. This induces a representation θ of $\mathfrak{a}$ in $\mathrm{S}(V)$ given by: for $x \in \mathfrak{a}$ and $u = u_1 \cdots u_n \in \mathrm{S}^n(V)$:

$$x.u = \theta(x)u = \sum_{i=1}^{n} u_1 \cdots u_{i-1}(\sigma(x)u_i)u_{i+1} \cdots u_n.$$

Thus for $x \in \mathfrak{a}$ and $u, v \in \mathrm{S}(V)$, we have:

(5) $\theta(x)(uv) = (\theta(x)u)v + u(\theta(x)v).$

Recall that $S(V)^{\mathfrak{a}}$ (resp. $S^n(V)^{\mathfrak{a}}$) is the set of elements $u \in S(V)$ (resp. $S^n(V)$) verifying $\theta(x)(u) = 0$ for all $x \in \mathfrak{a}$. By (5), $S(V)^{\mathfrak{a}}$ is a graded subalgebra of $S(V)$.

In this section, we are interested in the case where $V = \mathfrak{a}$ the adjoint representation or $V = \mathfrak{a}^*$ the coadjoint representation. An element of $S(\mathfrak{a}^*)^{\mathfrak{a}}$ is called an *invariant polynomial function on* $\mathfrak{a}$.

31.2.3 Lemma. *Let* $\mathfrak{a}$ *be a Lie algebra,* (V, σ) *a finite-dimensional* $\mathfrak{a}$-*module and* $n \in \mathbb{N}$. *The map* $f : \mathfrak{a} \to \Bbbk$, $x \mapsto \mathrm{tr}(\sigma(x)^n)$, *is an invariant polynomial function on* $\mathfrak{a}$.

Proof. It is clear that $f \in S(\mathfrak{a}^*)$. Now if $x, y \in \mathfrak{a}$, then:

$$-(x.f)(y) = \sum_{i=1}^{n} \mathrm{tr}\left(\sigma(y)^{i-1}\sigma([x, y])\sigma(y)^{n-i}\right)$$
$$= \sum_{i=1}^{n} \mathrm{tr}\left(\sigma(y)^{i-1}\sigma(x)\sigma(y)^{n-i+1}\right) - \sum_{i=1}^{n} \mathrm{tr}\left(\sigma(y)^i\sigma(x)\sigma(y)^{n-i}\right)$$
$$= \mathrm{tr}\left(\sigma(x)\sigma(y)^n - \sigma(y)^n\sigma(x)\right) = 0.$$

So the result follows. $\square$

31.2.4 In the rest of this section, $\mathfrak{g}$ is a semisimple Lie algebra and $\mathfrak{h}$ a Cartan subalgebra of $\mathfrak{g}$.

Proposition. *For* $f \in S(\mathfrak{g}^*)$, *we have* $f \in S(\mathfrak{g}^*)^{\mathfrak{g}}$ *if and only if* $f \circ \alpha = f$ *for all* $\alpha \in \mathrm{Aut}_e\, \mathfrak{g}$.

Proof. This follows from 23.4.14, 23.4.16 and 24.3.4. $\square$

31.2.5 We shall use the notations of 30.2.1. The Weyl group $W = W(\mathfrak{g}, \mathfrak{h})$ can be considered as a group of automorphisms of $\mathfrak{h}$ (18.2.9). Let $S(\mathfrak{h}^*)^W$ (resp. $S^n(\mathfrak{h}^*)^W$) be the set of $f \in S(\mathfrak{h}^*)$ (resp. $f \in S^n(\mathfrak{h}^*)$) verifying $f \circ w = f$ for all $w \in W$.

Lemma. *If* $n \in \mathbb{N}$, *the vector space* $S^n(\mathfrak{h}^*)^W$ *is spanned by polynomial functions on* $\mathfrak{h}$ *of the form* $x \mapsto \mathrm{tr}(\rho(x)^n)$ *where* ρ *is a finite-dimensional representation of* $\mathfrak{g}$.

Proof. If $f \in S(\mathfrak{h}^*)$, we set $\theta(f) \in S(\mathfrak{h}^*)^W$ to be:

$$\theta(f) = \sum_{w \in W} f \circ w.$$

By 18.11.5, P is a sublattice of $\mathfrak{h}^*$ spanning $\mathfrak{h}^*$. It follows that the λ^n's, for $\lambda \in P$, span the vector space $S^n(\mathfrak{h}^*)$. Moreover, by 18.8.9 and 18.11.7, $\{w(\lambda)\,;\, w \in W\} \cap P_{++} \neq \emptyset$ for all $\lambda \in P$. We deduce therefore that $\theta(\lambda^n)$, with $\lambda \in P_{++}$, span the vector space $S^n(\mathfrak{h}^*)^W$.

For $\lambda \in P_{++}$, let $E_\lambda = \{\mu \in P_{++} ; \mu < \lambda\}$ (with respect to the partial order defined in 30.2.1). Recall from the proof of 30.5.2 that E_λ is a finite set.

Let (V, σ) be a finite-dimensional representation of $\mathfrak{g}$ with highest weight λ. Define $g : \mathfrak{h} \to \Bbbk$, $x \mapsto \mathrm{tr}(\sigma(x)^n)$.

Let $\mathcal{P}(V)$ be the set of weights of V. By 30.3.1, $g \in S(\mathfrak{h}^*)^W$. Hence $g = \sum_{\mu \in \mathcal{P}(V) \cap P_{++}} a_\mu \theta(\mu^n)$ with $a_\mu \in \Bbbk$ are non-zero for all $\mu \in \mathcal{P}(V) \cap P_{++}$. Thus $\theta(\lambda^n)$ is a linear combination of g and the $\theta(\mu^n)$'s, with $\mu \in E_\lambda$ (30.2.7). So the result follows by induction on the cardinality of E_λ. $\quad\square$

31.2.6 Theorem. *Let $\mathfrak{g}$ be a semisimple Lie algebra, $\mathfrak{h}$ a Cartan subalgebra of $\mathfrak{g}$, $W = W(\mathfrak{g}, \mathfrak{h})$ and $i : S(\mathfrak{g}^*) \to S(\mathfrak{h}^*)$ the homomorphism of restriction.*

   (i) *The map i induces an isomorphism from $S(\mathfrak{g}^*)^{\mathfrak{g}}$ onto $S(\mathfrak{h}^*)^W$.*

   (ii) *If $n \in \mathbb{N}$, the vector space $S^n(\mathfrak{g}^*)^{\mathfrak{g}}$ is spanned by functions of the form $x \mapsto \mathrm{tr}(\sigma(x)^n)$ where σ is a finite-dimensional representation of $\mathfrak{g}$.*

Proof. Let $w \in W$. By 29.1.3, there exists $\alpha \in \mathrm{Aut}_e\, \mathfrak{g}$ such that $\alpha|_{\mathfrak{h}} = w$. Thus 31.2.4 implies that:

$$f(x) = f \circ \alpha(x) = f \circ w(x)$$

for $f \in S(\mathfrak{g}^*)^{\mathfrak{g}}$ and $x \in \mathfrak{h}$. Hence $i(S(\mathfrak{g}^*)^{\mathfrak{g}}) \subset S(\mathfrak{h}^*)^W$.

Now let $f \in S(\mathfrak{g}^*)^{\mathfrak{g}}$ be such that $i(f) = 0$. If $x \in \mathfrak{g}$ is semisimple, then $\alpha(x) \in \mathfrak{h}$ for some $\alpha \in \mathrm{Aut}_e\, \mathfrak{g}$ (29.2.3 and 29.2.6). So $f(x) = f \circ \alpha(x) = 0$ and hence by 29.2.2 and the fact that $\mathfrak{g}_{\mathrm{gen}}$ is dense in $\mathfrak{g}$, we have $f = 0$. So i is injective on $S(\mathfrak{g}^*)^{\mathfrak{g}}$.

For $n \in \mathbb{N}$, let T_n be the linear span of the functions $x \mapsto \mathrm{tr}(\sigma(x)^n)$, where σ is a finite-dimensional representation of $\mathfrak{g}$. By 31.2.3 and 31.2.5, we have:

$$T_n \subset S^n(\mathfrak{g}^*)^{\mathfrak{g}} \ , \ S^n(\mathfrak{h}^*)^W \subset i(T_n).$$

So the theorem follows. $\quad\square$

31.2.7 Let $\varphi : \mathfrak{g} \to \mathfrak{g}^*$ be the Killing isomorphism of $\mathfrak{g}$. Then φ extends to an isomorphism, also denoted by φ, from $S(\mathfrak{g})$ onto $S(\mathfrak{g}^*)$, that we shall call the *Killing isomorphism* from $S(\mathfrak{g})$ onto $S(\mathfrak{g}^*)$.

Denote by $\psi : S(\mathfrak{h}) \to S(\mathfrak{h}^*)$ the isomorphism induced by restriction to $\mathfrak{h}$ of the Killing form L of $\mathfrak{g}$. Since $L|_{\mathfrak{h} \times \mathfrak{h}}$ is W-invariant, ψ commutes with the actions of W on $S(\mathfrak{h})$ and $S(\mathfrak{h}^*)$.

Let $\mathfrak{g} = \mathfrak{h} \oplus \mathfrak{n}_+ \oplus \mathfrak{n}_-$ be as in 30.2.1.

Lemma. *Let J be the ideal of $S(\mathfrak{g})$ generated by $\mathfrak{n}_+ \cup \mathfrak{n}_-$.*

   (i) *We have $S(\mathfrak{g}) = S(\mathfrak{h}) \oplus J$. Let $j : S(\mathfrak{g}) \to S(\mathfrak{h})$ be the homomorphism of algebras defined by the projection with respect to this decomposition.*

   (ii) *Let $i : S(\mathfrak{g}^*) \to S(\mathfrak{h}^*)$ be the restriction map. Then $i \circ \varphi = \psi \circ j$.*

Proof. Part (i) is obvious. Let us prove part (ii). Let $x \in \mathfrak{g}$ and $h \in \mathfrak{h}$. We have $L(\mathfrak{h}, \mathfrak{n}_+ + \mathfrak{n}_-) = \{0\}$, so:

$$i \circ \varphi(x)(h) = \varphi(x)(h) = L(x, h) = L(j(x), h) = \psi \circ j(x)(h).$$

Thus (ii) follows since i, j, φ, ψ are homomorphisms of algebras.   □

31.2.8 Theorem. *Let $\mathfrak{g}$ be a semisimple Lie algebra of rank l and $\mathfrak{h}$ a Cartan subalgebra of $\mathfrak{g}$. Let us conserve the notations of 31.2.7.*

(i) *The homomorphism j induces an isomorphism from the algebra $S(\mathfrak{g})^{\mathfrak{g}}$ onto the algebra $S(\mathfrak{h})^W$ of W-invariants in $S(\mathfrak{h})$.*

(ii) *There exist l algebraically independent elements of $S(\mathfrak{g})^{\mathfrak{g}}$ which generate $S(\mathfrak{g})^{\mathfrak{g}}$.*

Proof. Part (i) follows from 31.2.6 and 31.2.7, while (ii) follows from (i) and 31.1.6.   □

31.3 A free module

31.3.1 Let E be a finite-dimensional $\Bbbk$-vector space and $\mathbf{A}(E) = S(E^*)$. For $n \in \mathbb{N}$, let $\mathbf{A}^n(E)$ be the set of homogeneous elements of degree n of $\mathbf{A}(E)$. For $n < 0$, set $\mathcal{F}_n(E) = \{0\}$, and for $n \geqslant 0$:

$$\mathcal{F}_n(E) = \mathbf{A}^0(E) + \mathbf{A}^1(E) + \cdots + \mathbf{A}^n(E).$$

Thus $(\mathcal{F}_n(E))_n$ defines a filtration on $\mathbf{A}(E)$.

31.3.2 Let V be a vector space of dimension l and U a subspace of V. Denote by $i : \mathbf{A}(V) \to \mathbf{A}(U)$ the restriction map. Since $(V/U)^*$ identifies with linear forms on V which are zero on U, we have an injective homomorphism of algebras $s : \mathbf{A}(V/U) \to \mathbf{A}(V)$. We define a filtration on $\mathbf{A}(V)$ as follows: for $n \in \mathbb{Z}$,

$$\mathcal{G}_n(V) = s\big(\mathbf{A}(V/U)\big).\mathcal{F}_n(V) = s\big(\mathbf{A}(V/U)\big).\mathcal{F}_n(U).$$

Let $\mathrm{gr}_\mathcal{G}(V)$ be the graded algebra associated to this filtration. If $u \in \mathbf{A}(V)$, denote by $\sigma_\mathcal{G}(u)$ its principal symbol.

Define a grading $(\mathcal{H}_n)_{n \in \mathbb{Z}}$ on the algebra $\mathbf{A}(V/U) \otimes_\Bbbk \mathbf{A}(U)$ as follows:

$$\mathcal{H}_n = \mathbf{A}(V/U) \otimes_\Bbbk \mathbf{A}^n(U).$$

Lemma. (i) *The graded algebras $\mathrm{gr}_\mathcal{G}(V)$ and $\mathbf{A}(V/U) \otimes_\Bbbk \mathbf{A}(U)$ are isomorphic.*

(ii) *If f is a homogeneous element of $\mathbf{A}(V)$ such that $i(f)$ is non-zero. Then $\sigma_\mathcal{G}(f) = i(f) \in \mathbf{A}(U) \subset \mathbf{A}(V/U) \otimes_\Bbbk \mathbf{A}(U)$.*

Proof. Let $\mathcal{E}$ be a basis of V which contains a basis of U. Then $\mathbf{A}(V)$ is generated by the dual basis of $\mathcal{E}$. The proof of the lemma is now straightforward.   □

31.3.3 Proposition. *Let A be a graded subalgebra of $\mathbf{A}(V)$, $A' = i(A)$ and $R = \mathbf{A}(V/U) \otimes_{\Bbbk} A$. Suppose that the following conditions are verified:*

(i) *The restriction of i to A is injective.*

(ii) *The A'-module $\mathbf{A}(U)$ is free.*

Then $\mathbf{A}(V)$ is a free R-module. In particular, it is a free A-module.

Let $(v_j)_{j \in J}$ be a basis of the A'-module $\mathbf{A}(U)$ consisting of homogeneous elements. For $j \in J$, let $u_j \in \mathbf{A}(V)$ be homogeneous such that $i(u_j) = v_j$. then $(u_j)_{j \in J}$ is a basis of the R-module $\mathbf{A}(V)$.

Proof. Using the notations of 31.3.2, let us consider the filtration induced on A by the $\mathcal{G}_n(V)$'s. Since $i|_A$ is injective, the subalgebra $\sigma_{\mathcal{G}}(A)$ of $\mathrm{gr}_{\mathcal{G}}(V)$ can be identified with $A' \subset \mathbf{A}(U)$, and we have $\sigma_{\mathcal{G}}(u_j) = i(u_j) = v_j$ for all $j \in J$ (31.3.2).

Now $(v_j)_{j \in J}$ is also a basis of the $\mathbf{A}(V/U) \otimes_{\Bbbk} A'$-module $\mathbf{A}(V/U) \otimes_{\Bbbk} \mathbf{A}(U)$. Since $\mathbf{A}(V/U) \otimes_{\Bbbk} \mathbf{A}(U)$ is isomorphic to $\mathrm{gr}_{\mathcal{G}}(V)$ (31.3.2), we deduce by 7.5.10 that the R-module $\mathbf{A}(V)$ is free, and $(u_j)_{j \in J}$ is a basis of this module. $\square$

31.3.4 Theorem. *Let $\mathfrak{g}$ be a semisimple Lie algebra, $\mathfrak{h}$ a Cartan subalgebra of $\mathfrak{g}$ and $W = W(\mathfrak{g}, \mathfrak{h})$.*

(i) *The algebra $\mathrm{S}(\mathfrak{g}^*)$ is a free $\mathrm{S}((\mathfrak{g}/\mathfrak{h})^*) \otimes_{\Bbbk} \mathrm{S}(\mathfrak{g}^*)^{\mathfrak{g}}$-module of rank* card W.

(ii) *The algebra $\mathrm{S}(\mathfrak{g}^*)$ is a free $\mathrm{S}(\mathfrak{g}^*)^{\mathfrak{g}}$-module, and the algebra $\mathrm{S}(\mathfrak{g})$ is a free $\mathrm{S}(\mathfrak{g})^{\mathfrak{g}}$-module.*

Proof. Part (i) is a consequence of 31.1.8, 31.2.8 and 31.3.3, while (ii) follows from (i) and by using the Killing isomorphism. $\square$

References and comments

- [4], [29], [41], [48], [49], [80].

Theorem 31.2.8 is due to Chevalley.

Theorem 31.3.4 is due to Kostant, and the proof given here is taken from [4].

S-triples

In the first section, we prove the Jacobson-Morosov Theorem on Lie subalgebras of a semisimple Lie algebra which are isomorphic to $\mathfrak{sl}_2(\Bbbk)$. We then show how such Lie subalgebras can be used to obtain information on nilpotent elements in a semisimple Lie algebra. The results in this chapter will be used to study certain orbits in semisimple Lie algebras.

Let $\mathfrak{g}$ be a finite-dimensional $\Bbbk$-Lie algebra and L its Killing form. If $\mathfrak{a}$ is a subset of $\mathfrak{g}$, $\mathfrak{a}^\perp$ denotes the orthogonal of $\mathfrak{a}$ with respect to L.

32.1 Jacobson-Morosov Theorem

32.1.1 Definition. *A triple (x,t,y) of elements of $\mathfrak{g}$ is called a S-triple if $(x,t,y) \neq (0,0,0)$ and*

$$[t,x] = 2x \ , \ [t,y] = -2y \ , \ [x,y] = t.$$

If (x,t,y) is a S-triple of $\mathfrak{g}$, then t is called its simple *element, x its* positive *element and y its* negative *element. An element of $\mathfrak{g}$ is simple if it is the simple element of a S-triple.*

32.1.2 Lemma. *Let u,v be endomorphisms of a finite-dimensional vector space V. Suppose that $[u,[u,v]] = 0$. Then:*
(i) *$[u,v]$ is nilpotent.*
(ii) *If u is nilpotent, then so is $u \circ v$.*

Proof. (i) Let $w = [u,v]$ and $p \in \mathbb{N}$. Since $[u,w] = 0$, we have:

$$[u, v \circ w^p] = [u,v] \circ w^p = w^{p+1}.$$

Hence $\operatorname{tr}(w^{p+1}) = 0$ for all $p \in \mathbb{N}$. So w is nilpotent (19.2.3).
(ii) Let $p \in \mathbb{N}^*$. Again since $[u,w] = 0$, we have:

$$[v, u^p] = \sum_{i=0}^{p-1} u^i \circ [v,u] \circ u^{p-i-1} = p[v,u] \circ u^{p-1}.$$

Let $\lambda \in \Bbbk$ and $x \in V \setminus \{0\}$ be such that $u \circ v(x) = \lambda x$. If $r \in \mathbb{N}^*$ is the smallest integer such that $u^r(x) = 0$, then:

$$\lambda u^{r-1}(x) = u^r \circ v(x) = v \circ u^r(x) - [v, u^r](x) = -r[v, u] \circ u^{r-1}(x).$$

As $[v, u]$ is nilpotent and $u^{r-1}(x) \neq 0$, we have $\lambda = 0$. Thus 0 is the only eigenvalue of $u \circ v$, so $u \circ v$ is nilpotent. $\square$

32.1.3 Lemma. *Let $t, x \in \mathfrak{g}$ be such that $[t, x] = 2x$ and $t \in [x, \mathfrak{g}]$. Then there exists $y \in \mathfrak{g}$ such that $[t, y] = -2y$ and $[x, y] = t$.*

Proof. The Lie algebra $\mathfrak{h} = \Bbbk t + \Bbbk x$ is solvable and $x \in [\mathfrak{h}, \mathfrak{h}]$. So it follows from 19.4.4 that $\operatorname{ad} x$ is nilpotent. Let $z \in \mathfrak{g}$ be such that $[x, z] = t$, and set:

$$I = \operatorname{id}_{\mathfrak{g}} \ , \ X = \operatorname{ad} x \ , \ T = \operatorname{ad} t \ , \ Z = \operatorname{ad} z.$$

Recall from 19.2.5 that

$$[T, X^p] = 2p X^p \ , \ [Z, X^p] = -p(T - (p-1)I)X^{p-1},$$

so $\mathfrak{n} = \ker X$ and $\mathfrak{m}_p = \operatorname{im} X^p$ are T-stable. Furthermore, if $u \in \mathfrak{m}_{p-1}$, then:

$$-p(T - (p-1)I)(u) \in ZX(u) + \mathfrak{m}_p.$$

Moreover if $u \in \mathfrak{n}$, then $X(u) = 0$ and we obtain for $p \geqslant 1$,

$$(T - (p-1)I)(\mathfrak{n} \cap \mathfrak{m}_{p-1}) \subset \mathfrak{n} \cap \mathfrak{m}_p.$$

As X is nilpotent, there exists $k \in \mathbb{N}$ such that $\mathfrak{m}_{k+1} = \{0\}$. So

$$(T - kI) \cdots (T - I)T(\mathfrak{n}) = \{0\}.$$

This shows that the eigenvalues of $T|_{\mathfrak{n}}$ are greater than or equal to zero, and so $(T + 2I)|_{\mathfrak{n}}$ is invertible. But $[x, [t, z] + 2z] = -2[x, z] + 2t = 0$. Thus $[t, z] + 2z \in \mathfrak{n}$. So there exists $v \in \mathfrak{n}$ such that $[t, z] + 2z = (T + 2I)(v) = [t, v] + 2v$. Let $y = z - v$, then we have $[x, y] = t$, $[t, y] = -2y$. $\square$

32.1.4 Lemma. *Let $\mathfrak{g}$ be a semisimple Lie algebra and x a nilpotent element of $\mathfrak{g}$. Then $x \in [x, [x, \mathfrak{g}]]$.*

Proof. Let $\mathfrak{n} = \ker(\operatorname{ad} x)^2$, $\mathfrak{m} = (\operatorname{ad} x)^2(\mathfrak{g})$. If $y \in \mathfrak{n}$, then $0 = \operatorname{ad}[x, [x, y]] = [\operatorname{ad} x, [\operatorname{ad} x, \operatorname{ad} y]]$, so by 32.1.2, $(\operatorname{ad} x) \circ (\operatorname{ad} y)$ is nilpotent. Hence $L(x, y) = 0$. Thus $x \in \mathfrak{n}^{\perp}$. Now for $z \in \mathfrak{g}$, we have:

$$0 = L([x, [x, y]], z) = -L([x, y], [x, z]) = L(y, [x, [x, z]]).$$

So $\mathfrak{m} \subset \mathfrak{n}^{\perp}$ and since L is non-degenerate, we have $\mathfrak{m} = \mathfrak{n}^{\perp}$, and $x \in \mathfrak{m}$. $\square$

32.1.5 Theorem. (Jacobson-Morosov Theorem) *Let* $\mathfrak{g}$ *be a semisimple Lie algebra and* x *a non-zero nilpotent element of* $\mathfrak{g}$. *Then there exist* $t, y \in \mathfrak{g}$ *such that* (x, t, y) *is a S-triple of* $\mathfrak{g}$.

Proof. By 32.1.4, there exists $z \in \mathfrak{g}$ such that $[x, [x, z]] = x$. Let $t = -2[x, z]$. Then $[t, x] = 2x$ and $t \in [x, \mathfrak{g}]$. So the result follows from 32.1.3.   □

32.1.6 Corollary. *If* $\mathfrak{g}$ *is semisimple, then* $\mathrm{Aut}_s\,\mathfrak{g} = \mathrm{Aut}_e\,\mathfrak{g}$.

Proof. Let x be a non-zero nilpotent element of $\mathfrak{g}$. By 32.1.5, there exists $t \in \mathfrak{g}$ such that $[t, x] = 2x$. So the result follows.   □

32.1.7 Proposition. *Let* $\mathfrak{g}$ *be a semisimple Lie algebra and* $h, x \in \mathfrak{g}$. *Suppose that* h *is semisimple and* x *is a non-zero nilpotent element such that* $[h, x] = \lambda x$, *with* $\lambda \in \Bbbk$. *Then there is a S-triple* (x, t, y) *such that:*

$$[h, t] = 0 \ , \ [h, y] = -\lambda y.$$

Proof. For $\mu \in \Bbbk$, denote by $\mathfrak{g}_\mu$ the $(\mathrm{ad}\,h)$-eigenspace of eigenvalue μ. By 32.1.5, there is a S-triple (x, t, y). Let us write:

$$t = \sum_{\nu \in \Bbbk} t_\nu \ , \ y = \sum_{\nu \in \Bbbk} y_\nu,$$

with $t_\nu, y_\nu \in \mathfrak{g}_\nu$. Since $[t_\nu, x] \in \mathfrak{g}_{\lambda + \nu}$, we deduce from $[t, x] = 2x$ that $[t_0, x] = 2x$ and $[t_\nu, x] = 0$ if $\nu \neq 0$. Moreover:

$$[x, y] = t = \sum_{\nu \in \Bbbk} [x, y_\nu].$$

So $t_0 = [x, y_{-\lambda}]$.

Thus $[t_0, x] = 2x$ and $t_0 \in [x, \mathfrak{g}]$. By 32.1.3, there is a S-triple (x, t_0, z). If we write $z = \sum_{\nu \in \Bbbk} z_\nu$ where $z_\nu \in \mathfrak{g}_\nu$, then it follows from the relations of a S-triple that $[t_0, z_{-\lambda}] = -2z_{-\lambda}$ and $[x, z_{-\lambda}] = t_0$. Hence $(x, t_0, z_{-\lambda})$ is a S-triple with the required properties.   □

32.1.8 Proposition. *Let* $\mathfrak{g}$ *be a semisimple Lie algebra,* $\lambda \in \Bbbk$ *and* $x, h, t \in \mathfrak{g} \setminus \{0\}$ *verifying:*

$$[h, x] = 2x \ , \ [h, t] = 0 \ , \ [t, x] = \lambda x.$$

Let us suppose that t *is semisimple and* $h \in (\mathfrak{g}^t \cap \mathfrak{g}^x)^\perp$. *Then there exists* $y \in \mathfrak{g}$ *such that* (x, h, y) *is a S-triple.*

Proof. Let $\mathfrak{g}_\nu$ be the $(\mathrm{ad}\,t)$-eigenspace of eigenvalue ν. Since $[t, x] = \lambda x$, $\mathfrak{g}^x$ is $(\mathrm{ad}\,t)$-stable. Hence we have:

$$\mathfrak{g}^x = \bigoplus_{\nu \in \Bbbk} (\mathfrak{g}^x \cap \mathfrak{g}_\nu).$$

Note that $\mathfrak{g}^t = \mathfrak{g}_0$ and $L(\mathfrak{g}_\mu, \mathfrak{g}_\nu) = \{0\}$ if $\mu + \nu \neq 0$. So $h \in (\mathfrak{g}^x)^\perp$ if $h \in (\mathfrak{g}^t \cap \mathfrak{g}^x)^\perp$. Since $(\mathfrak{g}^x)^\perp = [x, \mathfrak{g}]$, the result follows from 32.1.3.   □

32.1.9 Proposition. *Let t be a simple element of a semisimple Lie algebra $\mathfrak{g}$. Let $q \in \mathbb{Z}$, then the dimension of the $(\operatorname{ad} t)$-eigenspace of eigenvalue $2q+1$ is even.*

Proof. For $q \in \mathbb{Z}$, denote by $\mathfrak{g}_q$ the $(\operatorname{ad} t)$-eigenspace of eigenvalue q. Let $p \in \mathbb{N}$ and (x,t,y) be a S-triple of $\mathfrak{g}$. We define an alternating bilinear form on $\mathfrak{g}_{2p+1}$ as follows:

$$B_p(u,v) = L\big(u, (\operatorname{ad} y)^{2p+1}(v)\big)$$

where $u,v \in \mathfrak{g}_{2p+1}$. Since $(\operatorname{ad} y)^{2p+1}(\mathfrak{g}_{2p+1}) = \mathfrak{g}_{-2p-1}$ and the restriction of L to $\mathfrak{g}_{2p+1} \times \mathfrak{g}_{-2p-1}$ is non-degenerate, B_p is non-degenerate. It follows that $\dim \mathfrak{g}_{2p+1}$ is even, and since $\dim \mathfrak{g}_{2p+1} = \dim \mathfrak{g}_{-2p-1}$, the result follows. $\square$

32.2 Some lemmas

32.2.1 Lemma. *Let $\mathfrak{h}_1$, $\mathfrak{h}_2$ be Cartan subalgebras of a semisimple Lie algebra $\mathfrak{g}$. There exists $\theta \in \operatorname{Aut}_e \mathfrak{g}$ such that $\mathfrak{h}_1 = \theta(\mathfrak{h}_2)$ and $\theta(x) = x$ for all $x \in \mathfrak{h}_1 \cap \mathfrak{h}_2$.*

Proof. The algebra $\mathfrak{c} = \mathfrak{c}_\mathfrak{g}(\mathfrak{h}_1 \cap \mathfrak{h}_2)$ is reductive in $\mathfrak{g}$ (20.5.13 and 20.5.16). So $\mathfrak{c} = [\mathfrak{c},\mathfrak{c}] \oplus \mathfrak{z}(\mathfrak{c})$. Let $\mathfrak{k}_1, \mathfrak{k}_2$ be Cartan subalgebras of $[\mathfrak{c},\mathfrak{c}]$ such that $\mathfrak{h}_i = \mathfrak{k}_i \oplus \mathfrak{z}(\mathfrak{c})$ for $i = 1,2$. By 29.2.3, there exists $\alpha \in \operatorname{Aut}_s[\mathfrak{c},\mathfrak{c}]$ such that $\alpha(\mathfrak{k}_1) = \mathfrak{k}_2$. Let $x_1,\dots,x_n$ be nilpotent elements of $[\mathfrak{c},\mathfrak{c}]$ such that $\alpha = \exp(\operatorname{ad}_{[\mathfrak{c},\mathfrak{c}]} x_1) \cdots \exp(\operatorname{ad}_{[\mathfrak{c},\mathfrak{c}]} x_n)$. It follows from 20.5.8 that $\operatorname{ad}_\mathfrak{g} x_i$ is a nilpotent endomorphism of $\mathfrak{g}$ for $1 \leqslant i \leqslant n$. Thus $\theta = \exp(\operatorname{ad}_\mathfrak{g} x_1) \cdots \exp(\operatorname{ad}_\mathfrak{g} x_n)$ satisfies the required properties. $\square$

32.2.2 Lemma. *Let $\mathfrak{s} = \mathfrak{sl}_2(\Bbbk)$ and e,f,h be as in 19.2.4. Let x be a non-zero nilpotent element of $\mathfrak{s}$. Then there exists $\alpha \in \operatorname{Aut}_e \mathfrak{s}$ such that $\alpha(x) = e$.*

Proof. Let $x = ae + bh + cf$ with $a,b,c \in \Bbbk$. Since x is a nilpotent endomorphism of $\Bbbk^2$ (20.4.4), we have $b^2 + ac = 0$. For $\lambda \in \Bbbk \setminus \{0\}$, let $\theta_\lambda = e^{\lambda \operatorname{ad} e} \circ e^{-\lambda^{-1} \operatorname{ad} f} \circ e^{\lambda \operatorname{ad} e}$. Then:

(1) $$\theta_\lambda(e) = -\lambda^{-2} f \; , \; \theta_\lambda(f) = -\lambda^2 e \; , \; \theta_\lambda(h) = -h.$$

If $a = 0$, then $b = 0$, $c \neq 0$ and $\theta_\lambda(x) = -c\lambda^2 e$. So we are reduced to the case where $a \neq 0$. Since $\Bbbk$ is algebraically closed, the formulas (1) imply that we may assume $a = 1$. Then $e^{b \operatorname{ad} f}(x) = e + (c + b^2)f = e$. $\square$

32.2.3 Lemma. *If (x,t,y) and (x,t,z) are S-triples in $\mathfrak{g}$, then $y = z$.*

Proof. We have $[x, y - z] = 0$, $[t, y - z] = -2(y - z)$, and the restriction of $\operatorname{ad} x$ to $\ker(\operatorname{ad} t + 2\operatorname{id}_\mathfrak{g})$ is injective (19.2), so $y = z$. $\square$

32.2.4 Lemma. *Let $x \in \mathfrak{g}$, $\mathfrak{p} = \ker(\operatorname{ad} x)$, $\mathfrak{q} = [x,\mathfrak{g}]$ and $\mathfrak{n} = \mathfrak{p} \cap \mathfrak{q}$.*
  (i) *We have $[\mathfrak{p},\mathfrak{q}] \subset \mathfrak{q}$ and $\mathfrak{n}$ is a Lie subalgebra of $\mathfrak{g}$.*
  (ii) *If x is a positive element of a S-triple (x,t,y), then $\mathfrak{n}$ is $(\operatorname{ad} t)$-stable, and for all $u \in \mathfrak{n}$, $\operatorname{ad} u$ is nilpotent.*

Proof. (i) Let $u \in \mathfrak{p}$, $v \in \mathfrak{q}$ and $w \in \mathfrak{g}$ be such that $v = [x, w]$. Then:

$$[u, v] = [u, [x, w]] = [[u, x], w] + [x, [u, w]] = [x, [u, w]] \in \mathfrak{q}.$$

As $\mathfrak{p}$ is a Lie subalgebra of $\mathfrak{g}$, $[\mathfrak{p} \cap \mathfrak{q}, \mathfrak{p} \cap \mathfrak{q}] \subset \mathfrak{p} \cap \mathfrak{q}$.

(ii) Since $[t, x] = 2x$, $\mathfrak{p}$ and $\mathfrak{q}$ are $(\operatorname{ad} t)$-stable, so $\mathfrak{n}$ is also $(\operatorname{ad} t)$-stable. For $t \in \mathbb{Z}$, let $\mathfrak{g}_p = \ker(\operatorname{ad} t - p \operatorname{id}_{\mathfrak{g}})$. Then:

$$\mathfrak{g} = \sum_{p \in \mathbb{Z}} \mathfrak{g}_p \ , \ [\mathfrak{g}_p, \mathfrak{g}_q] \subset \mathfrak{g}_{p+q}.$$

We deduce therefore that if $u \in \mathfrak{r} = \sum_{p>0} \mathfrak{g}_p$, then $\operatorname{ad} u$ is nilpotent. So the result follows since $\mathfrak{n} \subset \mathfrak{r}$ by 19.2. □

32.2.5 Lemma. *Let $\mathfrak{n}$ be a Lie subalgebra of $\mathfrak{g}$ such that $\operatorname{ad}_{\mathfrak{g}} x$ is nilpotent for all $x \in \mathfrak{n}$. Let $t \in \mathfrak{g}$ be such that $[t, \mathfrak{n}] = \mathfrak{n}$. Then $\exp(\operatorname{ad}_{\mathfrak{g}} \mathfrak{n})(t) = t + \mathfrak{n}$.*

Proof. We have clearly $\exp(\operatorname{ad} \mathfrak{n})(t) \subset t + \mathfrak{n}$. Let $v \in \mathfrak{n}$. Since $\mathfrak{n}$ is nilpotent, it suffices to prove that for all $p \geqslant 1$:

$$t + v \in e^{\operatorname{ad} \mathfrak{n}}(t) + \mathcal{C}^p(\mathfrak{n}).$$

This is clear if $p = 1$. Suppose that there exist $x \in \mathfrak{n}$ and $y \in \mathcal{C}^p(\mathfrak{n})$ such that $t + v = e^{\operatorname{ad} x}(t) + y$. Since $[t, \mathfrak{n}] = \mathfrak{n}$, $\operatorname{ad}_{\mathfrak{n}} t$ is bijective, and as the subspaces $\mathcal{C}^p(\mathfrak{n})$ are $(\operatorname{ad} t)$-stable, the restriction of $\operatorname{ad} t$ to $\mathcal{C}^p(\mathfrak{n})$ is also bijective. So there exists $z \in \mathcal{C}^p(\mathfrak{n})$ such that $y = [z, t]$, and so:

$$e^{\operatorname{ad}(x+z)}(t) - e^{\operatorname{ad} x}(t) \in [z, t] + \mathcal{C}^{p+1}(\mathfrak{n}).$$

Hence:

$$e^{\operatorname{ad}(x+z)}(t) \in t + v - y + [z, t] + \mathcal{C}^{p+1}(\mathfrak{n}) = t + v + \mathcal{C}^{p+1}(\mathfrak{n}).$$

We have therefore proved the lemma. □

32.2.6 Lemma. *Let (x, t, y) and (x, t', y') be S-triples of $\mathfrak{g}$. Denote by $\mathfrak{n} = \ker(\operatorname{ad} x) \cap \operatorname{im}(\operatorname{ad} x)$. Then there exists $z \in \mathfrak{n}$ such that:*

$$e^{\operatorname{ad} z}(x) = x \ , \ e^{\operatorname{ad} z}(t) = t' \ , \ e^{\operatorname{ad} z}(y) = y'.$$

Proof. By 32.2.4, $\operatorname{ad} z$ is nilpotent for any $z \in \mathfrak{n}$, and we have $[t, \mathfrak{n}] = \mathfrak{n}$ by 19.2. Also, $[x, t - t'] = 0$ and $[x, y - y'] = t - t'$, so $t' \in t + \mathfrak{n}$. Thus 32.2.5 says that $e^{\operatorname{ad} z}(t) = t'$ for some $z \in \mathfrak{n}$. Since $z \in \mathfrak{n}$, we have $e^{\operatorname{ad} z}(x) = x$. Finally by 32.2.3, we obtain that $e^{\operatorname{ad} z}(y) = y'$. □

32.2.7 Lemma. *Let (x, t, y) be a S-triple of $\mathfrak{g}$. For $p \in \mathbb{Z}$, set $\mathfrak{g}_p = \ker(\operatorname{ad} t - p \operatorname{id}_{\mathfrak{g}})$, and for $\lambda \in \mathbb{k} \setminus \{0\}$, let τ_λ be the automorphism of $\mathfrak{g}$ verifying $\tau_\lambda|_{\mathfrak{g}_p} = \lambda^p \operatorname{id}_{\mathfrak{g}_p}$. Then $\tau_\lambda \in \operatorname{Aut}_e \mathfrak{g}$.*

Proof. Let $\mathfrak{s} = \Bbbk x + \Bbbk t + \Bbbk y$ and $V \subset \mathfrak{g}$ a simple $\mathfrak{s}$-module of dimension $r + 1$. By 19.2.5 and 19.2.7, there is a basis $(u_0, \ldots, u_r)$ of V such that

$$[t, u_i] = (r - 2i)u_i \ , \ [x, u_i] = i(r - i + 1)u_{i-1} \ , \ [y, u_i] = u_{i+1},$$

where by convention, we set $u_{-1} = u_{r+1} = 0$. The endomorphisms $\operatorname{ad} x$, $\operatorname{ad} y$ of $\mathfrak{g}$ are nilpotent (19.2.5). So if $\lambda \in \Bbbk \setminus \{0\}$, we may define $\theta_\lambda = e^{\lambda \operatorname{ad} x} \circ e^{-\lambda^{-1} \operatorname{ad} y} \circ e^{\lambda \operatorname{ad} x} \in \operatorname{Aut}_e \mathfrak{g}$.

By the relations (1) of 32.2.2, there exist $\nu_i \in \Bbbk \setminus \{0\}$ such that $\theta_\lambda(u_i) = \nu_i u_{r-i}$ for $0 \leqslant i \leqslant r$. Since $u_i = (\operatorname{ad} y)^i(u_0)$, we obtain (by using again (1)):

$$\theta_\lambda(u_i) = (-1)^i \lambda^{2i} (\operatorname{ad} x)^i(\theta_\lambda(u_0)) = (-1)^i \lambda^{2i} \nu_0 (\operatorname{ad} x)^i(u_r)$$
$$= (-1)^i \lambda^{2i} \nu_0 r(r - 1) \cdots (r - i + 1) i! u_{r-i}.$$

Let us compute ν_0. We have $e^{\lambda \operatorname{ad} x}(u_0) = u_0$ and:

$$e^{-\lambda^{-1} \operatorname{ad} y}(u_0) = u_0 + \frac{(-1)}{\lambda} u_1 + \cdots + \frac{(-1)^r}{r! \lambda^r} u_r.$$

Since $\theta_\lambda(u_0)$ is proportional to u_r, we have:

$$\theta_\lambda(u_0) = \frac{(-1)^r}{r! \lambda^r} u_r.$$

Thus we have obtained that:

$$\theta_\lambda(u_i) = \frac{(-1)^{r-i} i!}{(r - i)!} \lambda^{2i-r} u_{r-i}.$$

It follows that $\theta_\lambda \circ \theta_1^3(u_i) = \lambda^{r-2i} u_i = \tau_\lambda(u_i)$. So by 19.2, we are done. $\square$

32.2.8 Let $\mathfrak{g}$ be semisimple, $G = \operatorname{Aut}_e \mathfrak{g}$ (it is the adjoint group of $\mathfrak{g}$ by 21.3.4 and 23.4.16), t a semisimple element of $\mathfrak{g}$, G_t the stabilizer of t in G and:

$$\mathfrak{g}_0 = \ker(\operatorname{ad} t) \ , \ \mathfrak{g}_2 = \ker(\operatorname{ad} t - 2 \operatorname{id}_\mathfrak{g}).$$

Then by 24.3.6 (ii), $\mathcal{L}(G_t) = \mathcal{L}(G_t^\circ) = \mathfrak{g}_0$. It is also clear that $G_t^\circ(\mathfrak{g}_2) \subset \mathfrak{g}_2$.

Lemma. *Assume that there exists* $x \in \mathfrak{g}_2$ *such that* $[x, \mathfrak{g}_0] = \mathfrak{g}_2$. *Then* $G_t(x)$ *contains a dense open subset of* $\mathfrak{g}_2$.

Proof. The morphism $u : G_t^\circ \to \mathfrak{g}_2$, $\alpha \mapsto \alpha(x)$, is dominant by 16.5.7. So the result follows from 15.4.2. $\square$

32.2.9 Lemma. *Let* $\mathfrak{g}$ *be semisimple,* (x, t, y) *a S-triple of* $\mathfrak{g}$ *and* G_t, $\mathfrak{g}_0$, $\mathfrak{g}_2$ *be as in 32.2.8. Then* $\widehat{\mathfrak{g}}_2 = \{u \in \mathfrak{g}_2; [u, \mathfrak{g}_0] = \mathfrak{g}_2\}$ *is a dense open subset of* $\mathfrak{g}_2$. *Moreover, it is a* G_t-*orbit.*

Proof. The set $\widehat{\mathfrak{g}}_2$ is non-empty since it contains x (19.2). The condition $u \in \mathfrak{g}_2 \setminus \widehat{\mathfrak{g}}_2$ is equivalent to the condition $\dim[u, \mathfrak{g}_0] < \dim \mathfrak{g}_2$. But it is equivalent to the vanishing of certain polynomials. So $\widehat{\mathfrak{g}}_2$ is a dense open subset of $\mathfrak{g}_2$.

The fact that $\widehat{\mathfrak{g}}_2$ is a G_t-orbit follows easily from 32.2.8. $\square$

32.3 Conjugation of S-triples

32.3.1 In this section, $\mathfrak{g}$ is a semisimple Lie algebra and $G = \mathrm{Aut}_e \, \mathfrak{g}$.

If (x, t, y) and (x', t', y') are S-triples of $\mathfrak{g}$, then we say that they are G-conjugate if there exists $\theta \in G$ such that $x' = \theta(x)$, $t' = \theta(t)$ and $y' = \theta(y)$.

Theorem. *Let (x, t, y) and (x', t', y') be S-triples of $\mathfrak{g}$, $t = \Bbbk x + \Bbbk t + \Bbbk y$ and $t' = \Bbbk x' + \Bbbk t' + \Bbbk y'$. Then the following conditions are equivalent:*

(i) *There exists $\theta \in G$ such that $x' = \theta(x)$.*

(ii) *There exists $\theta \in G$ such that $t' = \theta(t)$.*

(iii) *The S-triples (x, t, y) and (x', t', y') are G-conjugate.*

(iv) *There exists $\theta \in G$ such that $t' = \theta(t)$.*

Proof. The implications (iii) $\Rightarrow$ (i), (iii) $\Rightarrow$ (ii) and (iii) $\Rightarrow$ (iv) are clear, while (i) $\Rightarrow$ (iii) follows from 32.2.6.

(ii) $\Rightarrow$ (i) If (ii) is verified, we may assume that $t = t'$. Then $[x, \mathfrak{g}_0] = [x', \mathfrak{g}_0] = \mathfrak{g}_2$, and 32.2.9 implies that $x' = \theta(x)$ for some $\theta \in G_t$.

(iv) $\Rightarrow$ (i) If (iv) is verified, we may assume that $t = t'$. Since t is isomorphic to $\mathfrak{sl}_2(\Bbbk)$ and x, x' are nilpotent, it follows from 32.2.2 that $x' = \theta(x)$ for some $\theta \in \mathrm{Aut}_e \, t$. But if $u \in t$ is nilpotent, then $\mathrm{ad}_{\mathfrak{g}} \, u$ is also nilpotent (20.4.4). So θ extends to an element of G. $\square$

32.3.2 Corollary. (i) *The map sending a S-triple of $\mathfrak{g}$ to its positive element induces a bijection from the set of G-conjugacy classes of S-triples onto the set of G-orbits of non-zero nilpotent elements of $\mathfrak{g}$.*

(ii) *The map sending a S-triple of $\mathfrak{g}$ to its simple element induces a bijection from the set of G-conjugacy classes of S-triples onto the set of G-orbits of non-zero simple elements of $\mathfrak{g}$.*

Proof. This is clear by 32.1.5 and 32.3.1. $\square$

32.3.3 Proposition. *Let (x, t, y) be a S-triple of $\mathfrak{g}$.*

(i) *Let $\mathfrak{n} = \ker(\mathrm{ad} \, x) \cap \mathrm{im}(\mathrm{ad} \, x)$. Then $t + \mathfrak{n}$ is the set of elements t' of $\mathfrak{g}$ such that there exists a S-triple of the form (x, t', y'). The map $\mathfrak{n} \to \mathfrak{g}$, $u \mapsto t + u$, induces a bijection from $\mathfrak{n}$ onto the set of S-triples of $\mathfrak{g}$ whose positive element is x.*

(ii) *Let $\widehat{\mathfrak{g}}_2$ be defined as in 32.2.9. There exists a S-triple of the form (x', t, y') if and only if $x' \in \widehat{\mathfrak{g}}_2$. The map $(x', t, y') \mapsto x'$ is a bijection from the set of S-triples of $\mathfrak{g}$ whose simple element is t onto $\widehat{\mathfrak{g}}_2$.*

Proof. Part (ii) follows immediately from 32.2.9. Let us prove (i). If $u \in \mathfrak{n}$, then $t + u \in \mathrm{im}(\mathrm{ad} \, x)$ and $[t + u, x] = 2x$. So by 32.1.3, there is a S-triple of the form $(x, t + u, y')$. Conversely, if (x, t', y') is a S-triple, then $[t - t', x] = 0$ and $[x, y - y'] = t - t'$, so $t' - t \in \mathfrak{n}$. $\square$

32.3.4 Remark. In the above notations, we have (32.2.5 and 32.2.9):

$$t + \mathfrak{n} = \{\exp(\mathrm{ad} \, z)(t); z \in \mathfrak{n}\} \ , \ \widehat{\mathfrak{g}}_2 = G_t(x).$$

32.4 Characteristic

32.4.1 In this section, $\mathfrak{g}$ is assumed to be semisimple. Let $G = \mathrm{Aut}_e\,\mathfrak{g}$. If $x \in \mathfrak{g}$, denote by $\mathcal{O}_x$ the G-orbit of x. Let us also fix a Cartan subalgebra $\mathfrak{h}$ of $\mathfrak{g}$ and a base B of the root system $R = R(\mathfrak{g}, \mathfrak{h})$.

32.4.2 Let $t \in \mathfrak{h}$. If t is a simple element of $\mathfrak{g}$, then its eigenvalues are integers. It follows that $\alpha(t) \in \mathbb{Z}$ for all $\alpha \in R$.

Lemma. *Let $t \in \mathfrak{h}$ be a simple element of $\mathfrak{g}$ such that $\beta(t) \in \mathbb{N}$ for all $\beta \in B$. Then $\beta(t) \in \{0, 1, 2\}$ for all $\beta \in B$.*

Proof. Let $\mathfrak{g}_p = \ker(\mathrm{ad}\,t - p\,\mathrm{id}_\mathfrak{g})$, $p \in \mathbb{Z}$, $S = \{\alpha \in B; \alpha(t) = 2\}$ and for $\alpha \in R$, $X_\alpha \in \mathfrak{g}^\alpha \setminus \{0\}$. Denote by $\mathfrak{b}$ the Borel subalgebra associated to $\mathfrak{h}$ and B, and $\mathfrak{n}$ the nilpotent radical of $\mathfrak{b}$. Recall from 18.7.4 that $[\mathfrak{n}, X_{-\beta}] \subset \mathfrak{b}$ for all $\beta \in B$.

Let (x, t, y) be a S-triple of $\mathfrak{g}$. Since $x \in \mathfrak{g}_2$, we have $x \in \mathfrak{n}$ and we can write $x = \sum_{\alpha \in S} \lambda_\alpha X_\alpha$ where $\lambda_\alpha \in \Bbbk$. If $\beta \in B$, then:

$$[x, X_{-\beta}] = \sum_{\alpha \in S} \lambda_\alpha [X_\alpha, X_{-\beta}] \in \mathfrak{b} \cap \mathfrak{g}_{2-\beta(t)}.$$

If $\lambda_\alpha \neq 0$ and $[X_\alpha, X_{-\beta}] \neq 0$ for some $\alpha \in S$, then $\beta(t) \leqslant 2$. Otherwise, we have $[x, X_{-\beta}] = 0$. But $(\mathrm{ad}\,x)|_{\mathfrak{g}^\gamma}$ is injective for all $\gamma \in R$ verifying $\gamma(t) < 0$. Hence $\beta(t) = 0$. $\square$

32.4.3 Proposition. *If t is a simple element of $\mathfrak{g}$, then there exists $\theta \in G$ such that $\theta(t) \in \mathfrak{h}$ and $\beta(\theta(t)) \in \{0, 1, 2\}$ for all $\beta \in B$.*

Proof. By 29.2.3 and 29.2.6, we may assume that $t \in \mathfrak{h}$ and so $t \in \mathfrak{h}_\mathbb{Q}$ because t is simple. There exists $w \in W(\mathfrak{g}, \mathfrak{h})$ such that $\beta(w(t)) \geqslant 0$ for all $\beta \in B$ (18.8.7). So the result follows from 32.4.2 and 29.1.3. $\square$

32.4.4 Corollary. *Let l be the rank of $\mathfrak{g}$.*

(i) *The number of G-orbits of simple elements of $\mathfrak{g}$ is at most 3^l.*

(ii) *The number of G-orbits of nilpotent elements of $\mathfrak{g}$ is at most 3^l.*

(iii) *The number of G-conjugacy classes of Lie subalgebras of $\mathfrak{g}$ which are isomorphic to $\mathfrak{sl}_2(\Bbbk)$ is at most 3^l.*

Proof. Part (i) follows from 32.4.3 and the others from 32.3.1. $\square$

32.4.5 Lemma. *Let t be a simple element of $\mathfrak{g}$. There is a unique $t_1 \in \mathcal{O}_t$ such that $t_1 \in \mathfrak{h}$ and $\beta(t_1) \geqslant 0$ for all $\beta \in B$.*

Proof. The existence of t_1 follows from 32.4.3. Now suppose that t_1, t_2 verify the required properties. Let $\sigma_1, \sigma_2 \in G$ be such that $\sigma_1(t) = t_1$, $\sigma_2(t) = t_2$, and let $\mathfrak{h}_i = \sigma_i^{-1}(\mathfrak{h})$ for $i = 1, 2$. We have $t \in \mathfrak{h}_1 \cap \mathfrak{h}_2$. By 32.2.1, there exists $\sigma \in G$ such that $\sigma(\mathfrak{h}_1) = \mathfrak{h}_2$ and $\sigma(t) = t$. So $\theta = \sigma_2 \circ \sigma \circ \sigma_1$ verifies $\theta(t_1) = t_2$ and $\theta(\mathfrak{h}) = \mathfrak{h}$. Thus $\theta \in \mathrm{Aut}_e(\mathfrak{g}, \mathfrak{h})$ (in the notations of 30.6.1). By 30.6.5, there exists $w \in W(\mathfrak{g}, \mathfrak{h})$ such that $\theta|_\mathfrak{h} = w|_\mathfrak{h}$ (18.2.9). In particular, $w(t_1) = t_2$, hence $t_1 = t_2$ (18.8.9). $\square$

32.4.6 Let $B = (\beta_1, \ldots, \beta_l)$ and x a non-zero nilpotent element of $\mathfrak{g}$. Then 32.1.5 says that there is a S-triple of the form (x, t, y). Let t_1 be the unique element of $\mathcal{O}_t$ such that $t_1 \in \mathfrak{h}$ and $\beta_i(t_1) \geqslant 0$ for all $1 \leqslant i \leqslant l$ (32.4.5). Let $n_i = \beta_i(t_1)$, $1 \leqslant i \leqslant l$. We have $n_i \in \{0, 1, 2\}$ by 32.4.2. By 32.2.6, we see that the sequence $(n_1, \ldots, n_l)$ depends only on x (or $\mathcal{O}_x$). We shall denote this sequence by $C(x)$, and we call $C(x)$ the *characteristic* of x. Note that if x' is a nilpotent element of $\mathfrak{g}$ such that $C(x') = C(x)$, then $x' \in \mathcal{O}_x$ (32.3.1). By convention, we set $C(0) = (0, \ldots, 0)$. The map

$$\mathcal{O}_x \mapsto C(x)$$

defines an injection of the set of G-orbits of nilpotent elements of $\mathfrak{g}$ into the set of sequence of l elements of $\{0, 1, 2\}$. This map is not bijective in general.

We often represent a nilpotent orbit of $\mathfrak{g}$ by attaching to each vertex β_i of the Dynkin diagram of $\mathfrak{g}$, the number n_i.

32.5 Regular and principal elements

32.5.1 Let us conserve the hypotheses and notations of 32.4.1.

Let V be a finite-dimensional $\mathfrak{g}$-module. Recall that if

$$V = V_1 \oplus \cdots \oplus V_n = W_1 \oplus \cdots \oplus W_p$$

are two decompositions of V into simple $\mathfrak{g}$-submodules, then $n = p$ and it is equal to the *length* of V.

Recall from 29.3.1 and 29.3.2 that an element $x \in \mathfrak{g}$ is regular if and only if $\dim \mathfrak{g}^x = \operatorname{rk} \mathfrak{g}$.

32.5.2 Definition. (i) *A simple element t of $\mathfrak{g}$ is called* principal *if it is regular and the eigenvalues of* $\operatorname{ad} t$ *are even integers.*

(ii) *A S-triple (x, t, y) is called* principal *if the length of $\mathfrak{g}$ as a module over $\Bbbk x + \Bbbk t + \Bbbk y$ is equal to the rank of $\mathfrak{g}$.*

32.5.3 Proposition. *Let (x, t, y) be a S-triple of $\mathfrak{g}$. The following conditions are equivalent:*

(i) *x is regular.*

(ii) *t is principal.*

(iii) *(x, t, y) is principal.*

Proof. Let $\mathfrak{s} = \Bbbk x + \Bbbk t + \Bbbk y$, $\mathfrak{g}_p = \ker(\operatorname{ad} t - \operatorname{id}_{\mathfrak{g}})$, $p \in \mathbb{Z}$, and $\mathfrak{a}$ the sum of the subspaces $\mathfrak{g}_{2p}$ for $p \in \mathbb{Z}$. The subspace $\mathfrak{a}$ is a $\mathfrak{s}$-submodule of $\mathfrak{g}$. Denote by $l_{\mathfrak{g}}$ and $l_{\mathfrak{a}}$ the lengths of the $\mathfrak{s}$-modules $\mathfrak{g}$ and $\mathfrak{a}$. From the results of 19.2, 19.7, 29.3.1 and 29.3.2, we have:

$$(2) \qquad \dim \mathfrak{g}^x = l_{\mathfrak{g}} \geqslant l_{\mathfrak{a}} = \dim \mathfrak{g}^t \geqslant \operatorname{rk}(\mathfrak{g}).$$

So it follows from the definitions that all three conditions are equivalent to the condition that the inequalities in (2) are equalities. □

32.5.4 Let $B = (\beta_1, \ldots, \beta_l)$ and R_+ the set of positive roots with respect to B. Let us use the notations of 20.6.5 and for $\alpha \in R$, let $X_\alpha \in \mathfrak{g}^\alpha$ be elements such that $[X_\alpha, X_{-\alpha}] = H_\alpha$. For $1 \leqslant i \leqslant l$, denote $H_i = H_{\beta_i}$. Let

$$t^0 = \sum_{\alpha \in R_+} H_\alpha.$$

Lemma. (i) *We have $\beta_i(t^0) = 2$ for $1 \leqslant i \leqslant l$, and there exist non-zero positive integers $n_1, \ldots, n_l$ such that:*

$$t^0 = n_1 H_1 + \cdots + n_l H_l.$$

(ii) *Let $(a_i)_{1 \leqslant i \leqslant l}$ and $(b_i)_{1 \leqslant i \leqslant l}$ be families of scalars such that $a_i b_i = n_i$ for $1 \leqslant i \leqslant l$, and:*

$$x = a_1 X_{\beta_1} + \cdots + a_l X_{\beta_l} \ , \ y = b_1 X_{-\beta_1} + \cdots + b_l X_{-\beta_l}.$$

Then (x, t^0, y) is a S-triple of $\mathfrak{g}$.

Proof. (i) By 18.11.5 and 20.6.10, $(H_i)_{1 \leqslant i \leqslant l}$ is a base of the root system $(H_\alpha)_{\alpha \in R}$. Hence the existence of the n_i's, and $\beta_i(t^0) = 2$ by 18.7.6.

(ii) It follows from 18.7.4 (ii) that:

$$[x, y] = \sum_{i,j=1}^{l} a_i b_j [X_{\beta_i}, X_{-\beta_j}] = \sum_{i=1}^{l} a_i b_i [X_{\beta_i}, X_{-\beta_i}] = \sum_{i=1}^{l} n_i H_i = t^0.$$

Hence (x, t^0, y) is a S-triple of $\mathfrak{g}$. □

32.5.5 Proposition. *Suppose that $\mathfrak{g} \neq \{0\}$ and let us conserve the notations of 32.5.4.*

(i) *The element t^0 is principal.*

(ii) *The elements u of $\mathfrak{g}$ contained in a S-triple (u, t^0, v) are the elements of the form $\sum_{\beta \in B} x_\beta$ where $x_\beta \in \mathfrak{g}^\beta \setminus \{0\}$ for $\beta \in B$.*

(iii) *The Lie algebra $\mathfrak{g}$ contains regular nilpotent elements.*

Proof. (i) By 32.5.4, $\beta(t^0) = 2$ for $\beta \in B$. It follows that the eigenvalues of $\mathrm{ad}\, t^0$ are even, and the centralizer of t^0 in $\mathfrak{g}$ is $\mathfrak{h}$. So t^0 is principal.

(ii) Let $u = \sum_{\beta \in B} x_\beta$ where $x_\beta \in \mathfrak{g}^\beta \setminus \{0\}$ for $\beta \in B$. Then it is contained in a S-triple (u, t^0, v) by 32.5.4.

Conversely, let (u, t^0, v) be a S-triple. Since $[t^0, u] = 2u$ and $[t^0, v] = -2v$, we deduce that:

$$u \in \sum_{\beta \in B} \mathfrak{g}^\beta \ , \ v \in \sum_{\beta \in B} \mathfrak{g}^{-\beta}.$$

Let

$$u = \sum_{\beta \in B} x_\beta \ , \ v = \sum_{\beta \in B} x_{-\beta} \ , \ t^0 = \sum_{\beta \in B} n_\beta H_\beta,$$

with $x_\beta \in \mathfrak{g}^\beta$, $x_{-\beta} \in \mathfrak{g}^{-\beta}$ and $n_\beta \in \mathbb{N}^*$ for all $\beta \in B$ (32.5.4). It follows from 18.7.4 (ii) that:

$$\sum_{\beta \in B} n_\beta H_\beta = t^0 = [u,v] = \sum_{\alpha,\beta \in B} [x_\alpha, x_{-\beta}] = \sum_{\beta \in B} [x_\beta, x_{-\beta}].$$

Hence $[x_\beta, x_{-\beta}] \neq 0$, and so $x_\beta \neq 0$ for all $\beta \in B$.

(iii) This is clear by (i), (ii) and 32.5.3. $\square$

32.5.6 Theorem. *Let $\mathfrak{g}$ be a semisimple Lie algebra, G its adjoint group.*

(i) *The set of simple principal elements of $\mathfrak{g}$ is a G-orbit.*

(ii) *The set of regular nilpotent elements of $\mathfrak{g}$ is a G-orbit.*

Proof. (i) Let t be a simple principal element of $\mathfrak{g}$. Up to a conjugation by an element of G, we may assume that $t \in \mathfrak{h}$ and $\beta(t) \in \{0,1,2\}$ for all $\beta \in B$ (32.4.3). Since t is regular, $\mathfrak{g}^t = \mathfrak{h}$, so $\beta(t) \neq 0$ if $\beta \in B$. Also $\beta(t) \in 2\mathbb{Z}$ (t being principal). Thus, $\beta(t) = 2$ for all $\beta \in B$, so $t = t^0$.

(ii) Let x, x' be regular nilpotent elements and (x,t,y), (x',t',y') S-triples of $\mathfrak{g}$. By 32.5.3, t and t' are principal. So the result follows from part (i) and 32.3.1. $\square$

32.5.7 Remark. If $\mathfrak{g} \neq \{0\}$, then 32.5.6 implies that the characteristic of a regular nilpotent element of $\mathfrak{g}$ is $(2,2,\ldots,2)$ with $\mathrm{rk}(\mathfrak{g})$ factors.

References and comments

- [14], [20], [24], [32], [48], [49], [80].

Many results on S-triples presented in this chapter are due to Kostant ([48] and [49]).

We can find explicit tables for the characteristic of nilpotent elements of simple Lie algebras in [20], [24] and [32].

33

Polarizations

In this chapter, we introduce certain subalgebras of a Lie algebra which generalize Cartan subalgebras. One of the main results proved in this chapter is Richardson's Theorem (33.5.6).

Throughout this chapter, $\mathfrak{g}$ will denote a finite-dimensional semisimple Lie algebra over $\Bbbk$ and L its Killing form. If $\mathfrak{a} \subset \mathfrak{g}$, $\mathfrak{a}^{\perp}$ denotes the orthogonal of $\mathfrak{a}$ with respect to L.

33.1 Definition of polarizations

33.1.1 Let $\mathfrak{a}$ be a Lie algebra. Recall from 19.7.3 that if $f \in \mathfrak{a}^*$, Φ_f denotes the alternating bilinear form on $\mathfrak{a}$ defined by:

$$\Phi_f(x,y) = f([x,y]).$$

If $\mathfrak{b}$ is an ideal of $\mathfrak{a}$, then $\mathfrak{b}^{(f)}$ denotes the orthogonal of $\mathfrak{b}$ with respect to Φ_f.

33.1.2 Let $\mathfrak{a}$ be ad-algebraic (24.8.2) and A its adjoint group. Then A acts naturally on $\mathfrak{a}$ and $\mathfrak{a}^*$ which correspond to the adjoint and coadjoint representations of $\mathfrak{a}$ (23.4.9 and 23.5.5). Let $f \in \mathfrak{a}^*$ and $A.f$ its orbit. By 24.3.6, $\mathrm{ad}_{\mathfrak{a}}\,\mathfrak{a}^{(f)}$ is the Lie algebra of the stabilizer of f in A. We deduce from 21.4.3 that:

$$\dim A.f = \dim(\mathrm{ad}_{\mathfrak{a}}\,\mathfrak{a}) - \dim(\mathrm{ad}_{\mathfrak{a}}\,\mathfrak{a}^{(f)}).$$

Since $\mathfrak{z}(\mathfrak{a})$ is in the kernel of the adjoint representation, it is clear that $\mathfrak{z}(\mathfrak{a})$ is contained in $\mathfrak{a}^{(f)}$. So:

$$\dim A.f = \dim \mathfrak{a} - \dim \mathfrak{a}^{(f)}.$$

In particular, $\dim A.f$ is even (19.7.3).

33.1.3 Definition. *Let $f \in \mathfrak{a}^*$.*

(i) A Lie subalgebra $\mathfrak{b}$ of $\mathfrak{a}$ is said to be subordinate *to f if it is a totally isotropic subspace with respect to Φ_f, that is $f([\mathfrak{b}, \mathfrak{b}]) = \{0\}$.*

(ii) *A polarization of $\mathfrak{a}$ at f is a Lie subalgebra which is a maximal totally isotropic subspace with respect to Φ_f of $\mathfrak{a}$. In particular, it contains $\mathfrak{a}^{(f)}$, and a subordinate Lie subalgebra $\mathfrak{b}$ for f is a polarization for f if and only if:*

$$2 \dim \mathfrak{b} = \dim \mathfrak{a} + \dim \mathfrak{a}^{(f)}.$$

33.1.4 Recall from 19.7.3 that $\mathfrak{a}^*_{\text{reg}}$ denotes the set of regular linear forms. We shall state without proof the following result which will not used in the sequel.

Theorem. (Duflo's Theorem) *Let $\mathfrak{b}$ be a solvable ideal of a Lie algebra $\mathfrak{a}$ and $f \in \mathfrak{a}^*$ verifying $\dim \mathfrak{a}^{(f)} \leqslant \dim \mathfrak{a}^{(g)}$ for all $g \in \mathfrak{a}^*$ such that $g|_{\mathfrak{b}} = f|_{\mathfrak{b}}$. Then there exists a solvable polarization of $\mathfrak{a}$ at f. In particular, if $f \in \mathfrak{a}^*_{\text{reg}}$, then there is a solvable polarization of $\mathfrak{a}$ at f.*

33.2 Polarizations in the semisimple case

33.2.1 Recall from 29.3.1 that when $\mathfrak{g}$ is semisimple, the Killing homomorphism $x \mapsto f_x$ of $\mathfrak{g}$ is an isomorphism, and $\mathfrak{g}^{(f_x)} = \mathfrak{g}^x$. We shall say that a Lie subalgebra $\mathfrak{a}$ of $\mathfrak{g}$ is *subordinate* to $x \in \mathfrak{g}$ if $L(x, [\mathfrak{a}, \mathfrak{a}]) = \{0\}$. Similarly, a *polarization* of $\mathfrak{g}$ at x is a Lie subalgebra $\mathfrak{p}$ of $\mathfrak{g}$ verifying:

$$f_x([\mathfrak{p}, \mathfrak{p}]) = L(x, [\mathfrak{p}, \mathfrak{p}]) = \{0\} \ , \ 2 \dim \mathfrak{p} = \dim \mathfrak{g} + \dim \mathfrak{g}^x.$$

We shall denote by $\text{Pol}(x, \mathfrak{g})$ or $\text{Pol}(x)$ the set of polarizations of $\mathfrak{g}$ at x, and by $\text{Pol}_r(x, \mathfrak{g})$ or $\text{Pol}_r(x)$ the set of solvable polarizations of $\mathfrak{g}$ at x. An element $x \in \mathfrak{g}$ is *polarizable* if $\text{Pol}(x) \neq \emptyset$.

Clearly, if $\mathfrak{p}$ is subordinate to x, then $\mathfrak{p} + \mathfrak{g}^x$ is a totally isotropic subspace with respect to Φ_{f_x}. So by definition, if $\mathfrak{p} \in \text{Pol}(x)$, then $\mathfrak{p} \supset \mathfrak{g}^x$.

By 29.3.1, 29.4.5 and 29.4.8, we have the following result:

Proposition. *If $x \in \mathfrak{g}$, then $\text{Pol}_r(x) \neq \emptyset$ if and only if $x \in \mathfrak{g}_{\text{reg}}$. Moreover, if $x \in \mathfrak{g}_{\text{reg}}$, then there is a Borel subalgebra of $\mathfrak{g}$ which is a polarization of $\mathfrak{g}$ at x.*

33.2.2 Lemma. *Let $n \in \mathbb{N}^*$ and $(\mu_1, \ldots, \mu_n), (\nu_1, \ldots, \nu_n) \in \mathbb{k}^n$ be such that the μ_i's are pairwise distinct and if $\mu_i = 0$, then $\nu_i = 0$. There exists $P \in \mathbb{k}[T]$ without constant term such that $P(\mu_i) = \nu_i$ for $1 \leqslant i \leqslant n$.*

Proof. Replacing $(\mu_1, \ldots, \mu_n), (\nu_1, \ldots, \nu_n)$ by $(\mu_1, \ldots, \mu_n, 0), (\nu_1, \ldots, \nu_n, 0)$ if necessary, we may assume that one of the μ_i's is equal to 0. Then:

$$P(T) = \sum_{i=1}^{n} \Big[\prod_{j \neq i} \frac{T - \mu_j}{\mu_i - \mu_j} \nu_i \Big]$$

is a polynomial which verifies $P(\mu_i) = \nu_i$ and $P(0) = 0$. $\square$

33.2.3 Lemma. *Let V be a finite-dimensional vector space, X, Y subspaces of $\mathfrak{gl}(V)$ such that $Y \subset X$, and $T = \{t \in \mathfrak{gl}(V); [t, X] \subset Y\}$. If $z \in T$ verifies $\text{tr}(zt) = 0$ for all $t \in T$, then z is nilpotent.*

Proof. Let $d = \dim V$ and $z = s + n$ the Jordan decomposition of x with s semisimple and n nilpotent. Let $(e_1, \ldots, e_d)$ be a basis of V such that $s(e_i) = \lambda_i e_i$ for $1 \leqslant i \leqslant d$, and denote $(g_{ij})_{1 \leqslant i,j \leqslant d}$ the basis of $\mathfrak{gl}(V)$ such that $g_{ij}(e_k) = \delta_{jk} e_i$ for $1 \leqslant i, j, k \leqslant d$.

Let U be the $\mathbb{Q}$-subspace of $\Bbbk$ spanned by the λ_i's and f a $\mathbb{Q}$-linear form on U. We shall prove that $f = 0$, for it would imply that $U = \{0\}$ and therefore $z = n$ is nilpotent.

Let $t \in \mathfrak{gl}(V)$ be defined by $t(e_i) = f(\lambda_i) e_i$, $1 \leqslant i \leqslant d$. If $1 \leqslant i, j \leqslant d$, then:

$$(\operatorname{ad} s)(g_{ij}) = (\lambda_i - \lambda_j) g_{ij} \ , \ (\operatorname{ad} t)(g_{ij}) = (f(\lambda_i) - f(\lambda_j)) g_{ij}.$$

Note also that since f is $\mathbb{Q}$-linear, we have:

$$\begin{cases} \lambda_i - \lambda_j = 0 \Rightarrow f(\lambda_i) - f(\lambda_j) = 0 \\ \lambda_i - \lambda_j = \lambda_h - \lambda_l \Rightarrow f(\lambda_i) - f(\lambda_j) = f(\lambda_h) - f(\lambda_l). \end{cases}$$

So by 33.2.2, there exists $P \in \Bbbk[T]$ without constant term such that $\operatorname{ad} t = P(\operatorname{ad} s)$. But $\operatorname{ad} s$ is a polynomial in $\operatorname{ad} z$ without constant term. Since $(\operatorname{ad} z)(X) \subset Y$ and $Y \subset X$, we have $t \in T$. As n is also a polynomial in z, we deduce that $[t, n] = 0$, so nt is nilpotent. It follows that:

$$0 = \operatorname{tr}(zt) = \operatorname{tr}(st) = \lambda_1 f(\lambda_1) + \cdots + \lambda_d f(\lambda_d).$$

Hence:

$$0 = f(\operatorname{tr}(zt)) = (f(\lambda_1))^2 + \cdots + (f(\lambda_d))^2.$$

Since $f(\lambda_i) \in \mathbb{Q}$ and the λ_i's span U, we have $f = 0$. $\quad\square$

33.2.4 Lemma. *Let $x \in \mathfrak{g}$ and $\mathfrak{p}$ a Lie subalgebra of $\mathfrak{g}$. The following conditions are equivalent:*

(i) $\mathfrak{p} \in \operatorname{Pol}(x)$.

(ii) $[x, \mathfrak{p}]$ *is the orthogonal of $\mathfrak{p}$ with respect to L.*

If these conditions are verified, then $[x, \mathfrak{p}]$ is an ideal of $\mathfrak{p}$.

Proof. Let $\mathfrak{q} = \mathfrak{p}^\perp$. Then $\dim \mathfrak{q} = \dim \mathfrak{g} - \dim \mathfrak{p}$.

(i) $\Rightarrow$ (ii) Suppose that $\mathfrak{p} \in \operatorname{Pol}(x)$. Then:

$$\{0\} = L(x, [\mathfrak{p}, \mathfrak{p}]) = L([x, \mathfrak{p}], \mathfrak{p}) \ , \ \dim \mathfrak{q} = \dim \mathfrak{p} - \dim \mathfrak{g}^x.$$

So $[x, \mathfrak{p}] \subset \mathfrak{q}$. Since $\mathfrak{p} \in \operatorname{Pol}(x)$, we have $\mathfrak{g}^x \subset \mathfrak{p}$. Thus:

$$\dim[x, \mathfrak{p}] = \dim \mathfrak{p} - \dim \mathfrak{g}^x = \dim \mathfrak{q}.$$

Hence $[x, \mathfrak{p}] = \mathfrak{q}$.

(ii) $\Rightarrow$ (i) Suppose that $[x, \mathfrak{p}] = \mathfrak{q}$. Since $L([x, \mathfrak{p}], \mathfrak{g}^x) = L(\mathfrak{p}, [x, \mathfrak{g}^x]) = \{0\}$, we deduce that $\mathfrak{g}^x \subset \mathfrak{q}^\perp = \mathfrak{p}$. So:

$$\dim \mathfrak{q} = \dim \mathfrak{g} - \dim \mathfrak{p} = \dim[x, \mathfrak{p}] = \dim \mathfrak{p} - \dim \mathfrak{g}^x.$$

Hence $2 \dim \mathfrak{p} = \dim \mathfrak{g} + \dim \mathfrak{g}^x$. Moreover:

$$\{0\} = L(\mathfrak{q}, \mathfrak{p}) = L([x, \mathfrak{p}], \mathfrak{p}) = L(x, [\mathfrak{p}, \mathfrak{p}]).$$

So $\mathfrak{p} \in \mathrm{Pol}(x)$. Finally if conditions (i) and (ii) are verified, then:

$$L([[x, \mathfrak{p}], \mathfrak{p}], \mathfrak{p}) = L([x, \mathfrak{p}], [\mathfrak{p}, \mathfrak{p}]) \subset L([x, \mathfrak{p}], \mathfrak{p}) = \{0\}.$$

Thus $[[x, \mathfrak{p}], \mathfrak{p}] \subset \mathfrak{p}^\perp = [x, \mathfrak{p}]$, and $[x, \mathfrak{p}]$ is an ideal of $\mathfrak{p}$. $\square$

33.2.5 Lemma. *Let x be a polarizable element of $\mathfrak{g}$ and $\mathfrak{p} \in \mathrm{Pol}(x)$. Then all the elements of $[x, \mathfrak{p}]$ are nilpotent.*

Proof. For $x \in \mathfrak{g}$, let $\sigma(x) = \mathrm{ad}_{\mathfrak{g}}\, x$, and T the set of elements $t \in \mathfrak{gl}(\mathfrak{g})$ verifying $[t, \sigma(\mathfrak{p})] \subset \sigma([x, \mathfrak{p}])$. Let $z \in \mathfrak{p}$ and $u \in T$. Then by 33.2.4:

$$\mathrm{tr}\left(u\sigma([x, z])\right) = \mathrm{tr}\left([u, \sigma(x)]\sigma(z)\right) \in \mathrm{tr}\left(\sigma([x, \mathfrak{p}])\sigma(z)\right) \subset L([x, \mathfrak{p}], \mathfrak{p}) = \{0\}.$$

So the result follows by 33.2.3. $\square$

33.2.6 Theorem. *Let x be a polarizable element of $\mathfrak{g}$ and $\mathfrak{p} \in \mathrm{Pol}(x)$. Then:*
  (i) *$\mathfrak{p}$ is a parabolic subalgebra of $\mathfrak{g}$ whose nilpotent radical is $[x, \mathfrak{p}]$.*
  (ii) *If x is nilpotent, then $L(x, \mathfrak{p}) = \{0\}$.*

Proof. (i) By 33.2.4 and 33.2.5, $[x, \mathfrak{p}]$ is an ideal of $\mathfrak{p}$ consisting of nilpotent elements. So there exists a Borel subalgebra $\mathfrak{b}$ of $\mathfrak{g}$ such that $[x, \mathfrak{p}] \subset \mathfrak{n}$ where $\mathfrak{n}$ denotes the nilpotent radical of $\mathfrak{b}$ (29.4.8). Since $\mathfrak{n}$ is the orthogonal of $\mathfrak{b}$ with respect to L, we have $\mathfrak{b} \subset [x, \mathfrak{p}]^\perp = \mathfrak{p}$ (33.2.4), and so $\mathfrak{p}$ is a parabolic subalgebra. The second part of (i) follows from 33.2.4 and 20.8.6.

(ii) In view of (i) and the results of 20.8, there exists a Cartan subalgebra $\mathfrak{h}$ of $\mathfrak{g}$, a positive root system R_+ of $R(\mathfrak{g}, \mathfrak{h})$, and $R_1, R_2 \subset R_+$ such that:

$$R_+ = R_1 \cup R_2 \, , \; R_1 \cap R_2 = \emptyset \, , \; \mathfrak{p} = \mathfrak{h} \oplus \sum_{\alpha \in S} \mathfrak{g}^\alpha$$

where $S = R_+ \cup (-R_1)$. Let us write $x = h + \sum_{\alpha \in S} X_\alpha$ with $h \in \mathfrak{h}$ and $X_\alpha \in \mathfrak{g}^\alpha$. Since $L(x, [\mathfrak{p}, \mathfrak{p}]) = \{0\}$, we have $X_\alpha = 0$ for $\alpha \in \pm R_1$. If x is nilpotent, then $h = 0$ (20.8.3). Thus x belongs to the nilpotent radical of $\mathfrak{p}$, hence $L(x, \mathfrak{p}) = \{0\}$ (20.8.6). $\square$

33.2.7 Corollary. *Let $x \in \mathfrak{g}_{\mathrm{reg}}$. Then $\mathrm{Pol}_r(x) = \mathrm{Pol}(x)$ is the set of Borel subalgebras of $\mathfrak{g}$ containing x.*

33.2.8 Proposition. (i) *Any semisimple element of $\mathfrak{g}$ is polarizable.*
  (ii) *If $\mathfrak{p}$ is a parabolic subalgebra of $\mathfrak{g}$, then there is a semisimple element $x \in \mathfrak{g}$ such that $\mathfrak{p} \in \mathrm{Pol}(x)$.*

Proof. (i) If x is semisimple, then $\mathfrak{g}^x$ contains a Cartan subalgebra $\mathfrak{h}$ of $\mathfrak{g}$ (20.6.3). Let $R = R(\mathfrak{g}, \mathfrak{h})$ and $R' = \{\alpha \in R; \alpha(x) = 0\}$. Then

$$\mathfrak{g}^x = \mathfrak{h} \oplus \sum_{\alpha \in R'} \mathfrak{g}^\alpha.$$

Since $R' = R \cap V'$ where $V' = \{\lambda \in \mathfrak{h}^*; \lambda(x) = 0\}$, it is a root system in V' (18.2.5). Let B' be a base of R', B a base of R containing B' (18.7.9) and R_+ the corresponding positive root system. Set $S = R_+ \cup R'$ and:

$$\mathfrak{p} = \mathfrak{h} \oplus \sum_{\alpha \in S} \mathfrak{g}^\alpha.$$

We obtain therefore that $2 \dim \mathfrak{p} = \dim \mathfrak{g} + \dim \mathfrak{g}^x$, and $\mathfrak{p}$ is a parabolic subalgebra of $\mathfrak{g}$. Moreover, in the notations of 20.7.1, we have:

$$[\mathfrak{p}, \mathfrak{p}] = \mathfrak{h}_{R'} \oplus \sum_{\alpha \in S} \mathfrak{g}^\alpha.$$

We can now verify easily that $L(x, [\mathfrak{p}, \mathfrak{p}]) = \{0\}$, and hence $\mathfrak{p} \in \mathrm{Pol}(x)$.

(ii) There is a Cartan subalgebra $\mathfrak{h}$ of $\mathfrak{g}$ and a base B of $R(\mathfrak{g}, \mathfrak{h})$ such that $\mathfrak{p} = \mathfrak{h} \oplus \mathfrak{g}^{P(J)}$ where $J \subset B$ (notations of 20.8.8). Let $x \in \mathfrak{h}$ be such that $\alpha(x) = 0$ if $\alpha \in J$, and $\alpha(x) = 1$ if $\alpha \in B \setminus J$. Then it is clear that $\mathfrak{p} \in \mathrm{Pol}(x)$. $\square$

33.3 A non-polarizable element

33.3.1 In this section, $\mathfrak{g}$ is assumed to be *simple* of rank l. Let $\mathfrak{h}$ be a Cartan subalgebra of $\mathfrak{g}$, $R = R(\mathfrak{g}, \mathfrak{h})$, B a base of R and R_+, R_- the corresponding system of positive and negative roots. Let us conserve the notation H_α of 20.6.5.

Let θ be the highest root of R with respect to B and $H = H_\theta$. Fix $E \in \mathfrak{g}^\theta$ and $F \in \mathfrak{g}^{-\theta}$ such that $[E, F] = H$. Then (E, H, F) is a S-triple. Set:

$$\mathfrak{p} = \sum_{i \in \mathbb{N}} \mathfrak{g}(i) \text{ where for } i \in \mathbb{Z} , \ \mathfrak{g}(i) = \{x \in \mathfrak{g}; [H, x] = ix\}.$$

By 20.8.7, $\mathfrak{p}$ is a parabolic subalgebra of $\mathfrak{g}$ and it is a polarization of $\mathfrak{g}$ at H. In view of the results of 19.2, we have:

$$\dim \mathfrak{g}^E = \dim \mathfrak{g}(0) + \dim \mathfrak{g}(1).$$

If $\alpha \in R_+$, then $\alpha(H) \in \{0, 1, 2\}$ (18.9.2). Thus $\mathfrak{g}(i) = \{0\}$ if $i \geqslant 3$. Let $\alpha \in R_+ \setminus \{\theta\}$ be such that $\alpha(H) > 0$. Then $\alpha(H) = 1$ (18.9.2). Hence:

$$\mathfrak{g}(2) = \Bbbk E , \ \mathfrak{p} = \mathfrak{g}_0 + \mathfrak{g}(1) + \Bbbk E.$$

Since $[H, E] = 2E$, we deduce that $\mathfrak{p} = \mathfrak{g}^E \oplus \Bbbk H$.

33.3.2 Let

$$\pi = \{\alpha \in B; \alpha(H) > 0\} \ , \ \delta = \sum_{\alpha \in B} \alpha.$$

Since $\mathfrak{g}$ is simple, we have $\delta \in R_+$ (18.13.8). We have, in the notation of 20.8.8, $\mathfrak{p} = \mathfrak{p}_J$ where $J = B \setminus \pi$.

Lemma. *Suppose that $l \geqslant 2$.*
(i) *If $\alpha \in B$, then $\alpha(H) \in \{0, 1\}$.*
(ii) *We have $0 < \mathrm{card}(\pi) \leqslant 2$ and $\mathrm{card}(\pi) = 2$ if and only if $\delta = \theta$, which is equivalent to the condition that $\mathfrak{g}$ is of type A_l.*

Proof. (i) This is clear since $\mathfrak{g}(2) = \Bbbk E$, $l \neq 1$ and $\theta \notin B$.
(ii) Since $\mathfrak{g}(2) \neq \{0\}$, $\pi \neq \emptyset$. Now if $\mathrm{card}(\pi) \geqslant 3$, then $\delta(H) \geqslant 3$ which is absurd.
Suppose that $\pi = \{\alpha, \beta\}$, with $\alpha \neq \beta$. Then $\alpha(H) = \beta(H) = 1$ by (i), so $\delta(H) = 2$ and $\delta = \theta$. The converse is clear by (i). Finally by using the classification of irreducible root systems in 18.14, we deduce that $\delta = \theta$ if and only if $\mathfrak{g}$ is of type A_l. $\square$

33.3.3 Lemma. *With the above notations.*
(i) *If E is polarizable, then $\mathfrak{p} \subset \mathfrak{q}$ for all $\mathfrak{q} \in \mathrm{Pol}(E)$.*
(ii) *If $l \geqslant 2$ and $\mathrm{card}(\pi) = 1$, then E is not polarizable.*

Proof. (i) For $x, y \in \mathfrak{g}$, let $\Phi(x, y) = L(E, [x, y])$. We have $L(E, H) = 0$ and $L(E, \mathfrak{q}) = \{0\}$ (33.2.6). So it follows easily that $\mathfrak{q} + \Bbbk H$ is a totally isotropic subspace with respect to Φ. But $\mathfrak{q} \in \mathrm{Pol}(E)$, so $H \in \mathfrak{q}$, and as $\mathfrak{g}^E \subset \mathfrak{q}$, we deduce from 33.3.1 that $\mathfrak{p} \subset \mathfrak{q}$.
(ii) Since $\mathfrak{p} = \mathfrak{p}_{B \setminus \pi}$, if $\mathrm{card}(\pi) = 1$, then $\mathfrak{p}$ is a maximal (proper) parabolic subalgebra of $\mathfrak{g}$. If E is polarizable, then (i) implies that $\mathrm{Pol}(E) = \{\mathfrak{p}\}$. But $\mathfrak{p} \in \mathrm{Pol}(H)$, so

$$\dim \mathfrak{g}(0) = \dim \mathfrak{g}^H = \dim \mathfrak{g}^E = \dim \mathfrak{g}(0) + \dim \mathfrak{g}(1).$$

Thus $\dim \mathfrak{g}(1) = 0$ which is absurd because $\mathrm{card}(\pi) \leqslant \dim \mathfrak{g}(1)$ by 33.3.2. $\square$

33.3.4 The preceding result shows that if $\mathfrak{g}$ is not of type A_l, then it contains a non-polarizable nilpotent element.

33.3.5 When $\mathrm{card}(\pi) = 2$, or equivalently, $\mathfrak{g}$ is of type A_l with $l \geqslant 2$, it is easy to check that the elements of π are exactly the two terminal vertices α_1 and α_l. The parabolic subalgebras containing $\mathfrak{p}_{B \setminus \pi}$ are exactly $\mathfrak{p}_{B \setminus \{\alpha_1\}}$ and $\mathfrak{p}_{B \setminus \{\alpha_l\}}$. They are both subordinate subalgebras of $\mathfrak{g}$ at E, and they are both of codimension l in $\mathfrak{g}$. By 33.3.3 (i), they are the only polarizations of $\mathfrak{g}$ at E.

33.4 Polarizable elements

33.4.1 Lemma. *Let $\mathfrak{p}$ be a parabolic subalgebra of $\mathfrak{g}$, $\mathfrak{n}$ its nilpotent radical, h a semisimple element of $\mathfrak{p}$ and $\mathfrak{s} = [\mathfrak{g}^h, \mathfrak{g}^h]$. Then $\mathfrak{p} \cap \mathfrak{s}$ is a parabolic subalgebra of $\mathfrak{s}$ with nilpotent radical $\mathfrak{n} \cap \mathfrak{s}$.*

Proof. Let us fix a Cartan subalgebra $\mathfrak{h}$ of $\mathfrak{p}$ which contains h (29.5.5). Let $R = R(\mathfrak{g}, \mathfrak{h})$ and $R' = \{\alpha \in R; \alpha(h) = 0\}$. Then in the notation of 20.7.1,

$$\mathfrak{g}^h = \mathfrak{h} \oplus \sum_{\alpha \in R'} \mathfrak{g}^\alpha \ , \ \mathfrak{s} = \mathfrak{h}_{R'} \oplus \sum_{\alpha \in R'} \mathfrak{g}^\alpha.$$

There is a positive root system R_+ of R and a partition $R_+ = R_1 \cup R_2$ such that

$$\mathfrak{p} = \mathfrak{h} \oplus \sum_{\alpha \in S} \mathfrak{g}^\alpha \ , \ \mathfrak{n} = \sum_{\alpha \in R_2} \mathfrak{g}^\alpha$$

where $S = R_+ \cup (-R_1)$. Let $R'_+ = R' \cap R_+$, $R'_i = R' \cap R_i$, $i = 1, 2$, and $S' = R' \cap S$. Then $S' = R'_+ \cup (-R'_1)$ is a parabolic subset of R', and:

$$\mathfrak{p} \cap \mathfrak{s} = \mathfrak{h}_{R'} \oplus \sum_{\alpha \in S'} \mathfrak{g}^\alpha \ , \ \mathfrak{n} \cap \mathfrak{s} = \sum_{\alpha \in R'_2} \mathfrak{g}^\alpha.$$

Hence the result follows.   □

33.4.2 Let $x \in \mathfrak{g}$, and h, e be the semisimple and nilpotent components of x. The Lie algebra $\mathfrak{g}^h$ is reductive in $\mathfrak{g}$ (20.5.13), and $\mathfrak{g}^h = \mathfrak{g}' \oplus \mathfrak{c}$ where $\mathfrak{c} = \mathfrak{z}(\mathfrak{g}^h)$ and $\mathfrak{g}' = [\mathfrak{g}^h, \mathfrak{g}^h]$. Also, we have $e \in \mathfrak{g}'$ (20.5.14).

Theorem. *With the above notations, the following conditions are equivalent:*

(i) *x is polarizable in $\mathfrak{g}$.*
(ii) *e is polarizable in $\mathfrak{g}'$.*

Proof. (i) $\Rightarrow$ (ii) Let $\mathfrak{p} \in \mathrm{Pol}(x)$. Then $\mathfrak{p}$ is a parabolic subalgebra of $\mathfrak{g}$ with nilpotent radical $\mathfrak{n} = [x, \mathfrak{p}]$ (33.2.6).

Since $h \in \mathfrak{c} \subset \mathfrak{g}^h \cap \mathfrak{g}^e = \mathfrak{g}^x \subset \mathfrak{p}$, we have $\mathfrak{p} \cap \mathfrak{g}^h = \mathfrak{c} + \mathfrak{p} \cap \mathfrak{g}'$. So by 33.4.1, $\mathfrak{p} \cap \mathfrak{g}'$ is a parabolic subalgebra of $\mathfrak{g}'$ with nilpotent radical $\mathfrak{n} \cap \mathfrak{g}'$, and $e \in \mathfrak{p} \cap \mathfrak{g}'$ since $e \in \mathfrak{g}^x$.

Let $\varphi = (\mathrm{ad}\, x)|_\mathfrak{p}$ and $\psi = (\mathrm{ad}\, h)|_\mathfrak{p}$. Then ψ is semisimple and $\varphi \circ \psi = \psi \circ \varphi$. Since $\mathfrak{p} = \ker \psi \oplus \psi(\mathfrak{p})$, we have:

$$\varphi(\mathfrak{p}) = \varphi(\ker \psi) \oplus \varphi(\psi(\mathfrak{p})) = \varphi(\ker \psi) \oplus \psi(\varphi(\mathfrak{p})).$$

So $\varphi(\ker \psi) = \varphi(\mathfrak{p}) \cap \ker \psi$, that is $[x, \mathfrak{p} \cap \mathfrak{g}^h] = \mathfrak{g}^h \cap [x, \mathfrak{p}] = \mathfrak{g}^h \cap \mathfrak{n}$. Thus we obtain:

$$[e, \mathfrak{p} \cap \mathfrak{g}'] = [e, \mathfrak{c} + (\mathfrak{p} \cap \mathfrak{g}')] = [e, \mathfrak{p} \cap \mathfrak{g}^h] = [x, \mathfrak{p} \cap \mathfrak{g}^h] = \mathfrak{g}^h \cap \mathfrak{n}.$$

Since $[e, \mathfrak{p} \cap \mathfrak{g}'] \subset \mathfrak{g}'$, we deduce therefore that:

$$[e, \mathfrak{p} \cap \mathfrak{g}'] = \mathfrak{g}^h \cap \mathfrak{n} = \mathfrak{g}' \cap \mathfrak{n}.$$

It follows from 33.2.4 that $\mathfrak{p} \cap \mathfrak{g}'$ is a polarization of $\mathfrak{g}'$ at e.

(ii) $\Rightarrow$ (i) Let $\mathfrak{p}' \in \mathrm{Pol}(e, \mathfrak{g}')$. Then $\mathfrak{p}'$ is a parabolic subalgebra of $\mathfrak{g}'$ with nilpotent radical $\mathfrak{n}' = [e, \mathfrak{p}']$, and $\mathfrak{g}^e \cap \mathfrak{g}' \subset \mathfrak{p}'$.

Let $\mathfrak{q} \in \mathrm{Pol}(h, \mathfrak{g})$ (33.2.8) with nilpotent radical $\mathfrak{m}$. We have:

$$\mathfrak{q} = \mathfrak{g}^h \oplus \mathfrak{m} \ , \ [h, \mathfrak{q}] = \mathfrak{m} \ , \ \mathfrak{g} = \mathfrak{g}^h \oplus [h, \mathfrak{g}] \ , \ \mathfrak{g}^h = \mathfrak{g}' \oplus \mathfrak{c},$$

so we may define:

$$\mathfrak{p} = \mathfrak{p}' \oplus \mathfrak{c} \oplus \mathfrak{m} \ , \ \mathfrak{n} = \mathfrak{n}' \oplus \mathfrak{m}.$$

We shall prove that $\mathfrak{p}$ is a polarization of $\mathfrak{g}$ at x.

1) We have $\mathfrak{p}' + \mathfrak{c} \subset \mathfrak{g}^h \subset \mathfrak{q}$, so $[\mathfrak{p}' + \mathfrak{c}, \mathfrak{m}] \subset \mathfrak{m}$ and $\mathfrak{p}$ is a Lie subalgebra of $\mathfrak{g}$.

2) Let $\mathfrak{h}'$ be a Cartan subalgebra of $\mathfrak{p}'$. Then $\mathfrak{h} = \mathfrak{h}' \oplus \mathfrak{c}$ is a Cartan subalgebra of $\mathfrak{g}^h$, and hence of $\mathfrak{g}$. Let $R = R(\mathfrak{g}, \mathfrak{h})$ and $R' = \{\alpha \in R; \alpha(h) = 0\}$. In the notations of 20.7.1, we have:

$$\mathfrak{p}' = \mathfrak{h}' \oplus (\mathfrak{g}')^P \ , \ \mathfrak{q} = \mathfrak{h} \oplus \mathfrak{g}^{R' \cup S},$$

with $P \cup (-P) = R'$, $R' \cup S \cup (-S) = R$. We deduce therefore that $\mathfrak{p} = \mathfrak{h} \oplus \mathfrak{g}^T$ where $T \cup (-T) = R$. Thus $\mathfrak{p}$ is a parabolic subalgebra of $\mathfrak{g}$ and it is easy to see that $\mathfrak{n}$ is the nilpotent radical of $\mathfrak{p}$.

3) We obtain

$$\begin{aligned}
2 \dim \mathfrak{p} &= 2(\dim \mathfrak{p}' + \dim \mathfrak{c} + \dim \mathfrak{m}) \\
&= \dim \mathfrak{g}' + \dim(\mathfrak{g}^e \cap \mathfrak{g}') + 2 \dim \mathfrak{c} + \dim \mathfrak{g} - \dim \mathfrak{g}^h \\
&= \dim \mathfrak{g} + \dim \mathfrak{g}^x,
\end{aligned}$$

because $\mathfrak{g}^h = \mathfrak{g}' \oplus \mathfrak{c}$ and $\mathfrak{g}^x = \mathfrak{g}^h \cap \mathfrak{g}^e = (\mathfrak{g}' \cap \mathfrak{g}^e) \oplus \mathfrak{c}$.

4) Since $e \in \mathfrak{g}^h$ and $\mathfrak{g}^h \subset \mathfrak{q}$, we have:

$$\begin{aligned}
[x, \mathfrak{p}] = [x, \mathfrak{c} + \mathfrak{p}' + \mathfrak{m}] &= [x, \mathfrak{p}'] + [h + e, \mathfrak{m}] \\
&\subset [x, \mathfrak{p}'] + \mathfrak{m} = [e, \mathfrak{p}'] + \mathfrak{m} = \mathfrak{n}' + \mathfrak{m} = \mathfrak{n}.
\end{aligned}$$

Hence:

$$L(x, [\mathfrak{p}, \mathfrak{p}]) = L([x, \mathfrak{p}], \mathfrak{p}) \subset L(\mathfrak{n}, \mathfrak{p}) = \{0\}.$$

We have therefore obtained our result. □

33.4.3 Theorem 33.4.2 shows that in the search of polarizable elements of $\mathfrak{g}$, it is important to determine polarizable nilpotent elements. Let us give some sufficient conditions for a nilpotent element to be polarizable.

Let (e, h, f) be a S-triple of $\mathfrak{g}$ and set:

$$\mathfrak{p} = \sum_{i \in \mathbb{N}} \mathfrak{g}(i) \text{ where for } i \in \mathbb{Z} \ , \ \mathfrak{g}(i) = \{x \in \mathfrak{g}; [h, x] = ix\}.$$

Then:

$$\mathfrak{g}(0) = \mathfrak{g}^h \ , \ L(e, \mathfrak{g}(i)) = \{0\} \text{ if } i \neq -2,$$
$$2 \dim \mathfrak{p} = \dim \mathfrak{g} + \dim \mathfrak{g}(0) \ , \ \dim \mathfrak{g}^e = \dim \mathfrak{g}(0) + \dim \mathfrak{g}(1).$$

Lemma. (i) *If* $\mathfrak{g}(2i+1) = \{0\}$ *for all* $i \in \mathbb{Z}$, *then* e *is polarizable.*
(ii) *If* $\mathfrak{g}$ *is simple of type* A_l, *then* e *is polarizable.*

Proof. (i) If $\mathfrak{g}(2i+1) = \{0\}$ for all $i \in \mathbb{Z}$, then $\dim \mathfrak{g}^e = \dim \mathfrak{g}(0)$, and so $\mathfrak{p} \in \text{Pol}(e)$.

(ii) Let $\mathfrak{g}$ be simple of type A_l, $\mathfrak{h}$ a Cartan subalgebra of $\mathfrak{g}$ containing h, $R = R(\mathfrak{g}, \mathfrak{h})$ and B a base of R such that $\beta(h) \in \{0, 1, 2\}$ for all $\beta \in B$ (32.4.6). Let $\beta_1, \ldots, \beta_l$ be the pairwise distinct elements of B such that $\beta_i + \cdots + \beta_j \in R$ for $1 \leqslant i \leqslant j \leqslant l$ (see 18.14), and $(h_1, \ldots, h_l)$ the basis of $\mathfrak{h}$ dual to $(\beta_1, \ldots, \beta_l)$. We have $h = \lambda_1 h_1 + \cdots + \lambda_l h_l$ with $\lambda_1, \ldots, \lambda_l \in \{0, 1, 2\}$.

Let $r_1 < r_2 < \cdots < r_m$ be the indices i such that $\lambda_i = 2$. Set $r_0 = 0$, $r_{m+1} = l+1$ and for $0 \leqslant j \leqslant l$, $\mathcal{E}_j = \{k \in \mathbb{N}; r_j < k < r_{j+1}\}$. Thus $\lambda_k \in \{0, 1\}$ for all $k \in \mathcal{E}_j$.

For $0 \leqslant j \leqslant m$, let $\mathcal{E}_j^1 = \{s \in \mathcal{E}_j; \lambda_s = 1\}$. If $s_1 < s_2 < \cdots < s_{p_j}$ are the elements of $\mathcal{E}_j^1$ in increasing order, we define:

$$H_j^+ = \sum_{i \text{ even}} h_{s_i} \ , \ H_j^- = \sum_{i \text{ odd}} h_{s_i}.$$

Now let:

$$\mathcal{R}_j^+ = \{\alpha \in R; \, \alpha(h) = \alpha(H_j^+) = 1\} \ , \ \mathcal{R}_j^- = \{\alpha \in R; \, \alpha(h) = \alpha(H_j^-) = 1\},$$
$$\mathcal{R}_+ = \bigcup_{j=0}^m \mathcal{R}_j^+ \ , \ \mathfrak{s}_+ = \sum_{\alpha \in \mathcal{R}_+} \mathfrak{g}^{-\alpha} \ , \ \mathcal{R}_- = \bigcup_{j=0}^m \mathcal{R}_j^- \ , \ \mathfrak{s}_- = \sum_{\alpha \in \mathcal{R}_-} \mathfrak{g}^{-\alpha}.$$

Since $\mathfrak{g}$ is of type A_l, any positive root is of the form $\beta_i + \cdots + \beta_j$ with $1 \leqslant i \leqslant j \leqslant l$. It follows that if $(\beta_i + \cdots + \beta_j)(h) = 1$, then there exist a unique k and a unique $s \in \mathcal{E}_k^1$ such that $i, j \in \mathcal{E}_k$ and $i \leqslant s \leqslant j$. We deduce easily from this that:

$$\mathfrak{g}(-1) = \mathfrak{s}_+ \oplus \mathfrak{s}_-, [\mathfrak{s}_+, \mathfrak{s}_+] = [\mathfrak{s}_-, \mathfrak{s}_-] = \{0\}, [\mathfrak{g}(0), \mathfrak{s}_+] \subset \mathfrak{s}_+, [\mathfrak{g}(0), \mathfrak{s}_-] \subset \mathfrak{s}_-.$$

From this, we deduce that $\mathfrak{p} \oplus \mathfrak{s}_+$ and $\mathfrak{p} \oplus \mathfrak{s}_-$ are Lie subalgebras of $\mathfrak{g}$ such that $L(e, \mathfrak{p} \oplus \mathfrak{s}_+) = L(e, \mathfrak{p} \oplus \mathfrak{s}_-) = \{0\}$. Thus:

$$2 \dim \mathfrak{s}_+ \leqslant \dim \mathfrak{g}(-1) \ , \ 2 \dim \mathfrak{s}_- \leqslant \dim \mathfrak{g}(-1).$$

Consequently, $2 \dim \mathfrak{s}_+ = 2 \dim \mathfrak{s}_- = \dim \mathfrak{g}(-1)$, and $\mathfrak{p} \oplus \mathfrak{s}_+$, $\mathfrak{p} \oplus \mathfrak{s}_-$ are polarizations of $\mathfrak{g}$ at e. $\square$

33.4.4 Remark. Let $\mathfrak{g}$ be simple of type A_l. Then the proof of 33.4.3 says that:
1) $\dim \mathfrak{g}(-1)$ is even (we have a more general result in 32.1.9).

2) If 1 appears in the characteristic (32.4.6) of e, then there exists at least two distinct polarizations of $\mathfrak{g}$ at e.

33.4.5 Lemma. *Let* $\mathfrak{g} = \mathfrak{g}_1 \times \cdots \times \mathfrak{g}_n$ *and* $x = x_1 + \cdots + x_n$ *with* $x_i \in \mathfrak{g}_i$, *for* $1 \leqslant i \leqslant n$. *The following conditions are equivalent:*
   (i) *For* $1 \leqslant i \leqslant n$, x_i *is polarizable in* $\mathfrak{g}$.
   (ii) x *is polarizable in* $\mathfrak{g}_i$.
If these conditions are verified, then any polarization of $\mathfrak{g}$ *at* x *is of the form* $\mathfrak{p}_1 \times \cdots \times \mathfrak{p}_n$ *where* $\mathfrak{p}_i \in \mathrm{Pol}(x_i, \mathfrak{g}_i)$.

Proof. Note that the restriction of L to $\mathfrak{g}_i \times \mathfrak{g}_i$ is the Killing form of $\mathfrak{g}_i$. Moreover, it is clear that $\mathfrak{g}^x = \mathfrak{g}_1^{x_1} \times \cdots \times \mathfrak{g}_n^{x_n}$.
   (i) $\Rightarrow$ (ii) If $\mathfrak{p}_i \in \mathrm{Pol}(x_i, \mathfrak{g}_i)$, then clearly $\mathfrak{p}_1 \times \cdots \times \mathfrak{p}_n \in \mathrm{Pol}(x)$.
   (ii) $\Rightarrow$ (i) Let $\mathfrak{p} \in \mathrm{Pol}(x)$. By 33.2.6, $\mathfrak{p}$ is a parabolic subalgebra of $\mathfrak{g}$. It follows from 20.8.8 that $\mathfrak{p} = \mathfrak{p}_1 \times \cdots \times \mathfrak{p}_n$ where $\mathfrak{p}_i$ is a parabolic subalgebra of $\mathfrak{g}_i$.
   Moreover, $L(x, [\mathfrak{p}, \mathfrak{p}]) = \{0\}$, so $L(x_i, [\mathfrak{p}_i, \mathfrak{p}_i]) = \{0\}$ and $\mathfrak{p}_i$ is a subordinate Lie subalgebra of $\mathfrak{g}_i$ at x_i, and $2 \dim \mathfrak{p}_i \leqslant \dim \mathfrak{g}_i + \dim \mathfrak{g}_i^{x_i}$. But:

$$2 \dim \mathfrak{p} = 2 \sum_{i=1}^{n} \dim \mathfrak{p}_i = \dim \mathfrak{g} + \dim \mathfrak{g}^x = \sum_{i=1}^{n} (\dim \mathfrak{g}_i + \dim \mathfrak{g}_i^{x_i}).$$

Hence $2 \dim \mathfrak{p}_i = \dim \mathfrak{g}_i + \dim \mathfrak{g}_i^{x_i}$, and $\mathfrak{p}_i \in \mathrm{Pol}(x_i, \mathfrak{g}_i)$. $\square$

33.4.6 Let $\mathfrak{g}_1, \ldots, \mathfrak{g}_r$ be the minimal components of the semisimple Lie algebra $\mathfrak{g}$. We say that $\mathfrak{g}$ is of type A if there exist $l_1, \ldots, l_r \in \mathbb{N}^*$ such that $\mathfrak{g}_i$ is simple of type A_{l_i} for $1 \leqslant i \leqslant r$.

Proposition. *Let* $\mathfrak{g}$ *be of type* A, *then all the elements of* $\mathfrak{g}$ *are polarizable.*

Proof. Let $x \in \mathfrak{g}$. If x is nilpotent, then it is polarizable by 33.4.3 and 33.4.5. Otherwise, let h, e be the semisimple and nilpotent components of x. We have $h \neq 0$ and $e \in \mathfrak{g}' = [\mathfrak{g}^h, \mathfrak{g}^h]$. Since $\mathfrak{g}'$ is also of type A, the result follows from 33.4.2. $\square$

33.5 Richardson's Theorem

33.5.1 Let $G = \mathrm{Aut}_e\, \mathfrak{g}$ and $\mathcal{N}_\mathfrak{g}$ the set of nilpotent elements of $\mathfrak{g}$. One of the goals of this section is to prove that if $\mathfrak{p}$ is a parabolic subalgebra of $\mathfrak{g}$, then $\mathfrak{p} \in \mathrm{Pol}(x)$ for some $x \in \mathcal{N}_\mathfrak{g}$. This is much harder to prove than part (ii) of 33.2.8.

33.5.2 Lemma. *Let* $x \in \mathfrak{g}$.
   (i) *We have* $x \in [\mathfrak{g}, x]$ *if and only if* $x \in \mathcal{N}_\mathfrak{g}$.
   (ii) *If* $x \notin \mathcal{N}_\mathfrak{g}$, *then* $\dim G.(\Bbbk x) = 1 + \dim G(x)$.

Proof. (i) We may assume that $x \neq 0$. Then the result follows from 32.1.5 and 19.8.2.

(ii) Let $\Bbbk^* = \Bbbk \setminus \{0\}$. The map $G \times \Bbbk^* \times \mathfrak{g} \to \mathfrak{g}$, $(\alpha, \lambda, y) \mapsto \lambda\alpha(y)$, defines an action of the algebraic group $H = G \times \Bbbk^*$ on $\mathfrak{g}$. The H-orbit of x is $G(\Bbbk^*x)$ and we have $\overline{G(\Bbbk^*x)} = \overline{G(\Bbbk x)}$.

Denote by $\pi : H \to \mathfrak{g}$, the orbit map $(\alpha, \lambda) \mapsto \lambda\alpha(x)$. By 25.1.3, $d\pi_{(e,1)}$ is a surjection from $\mathcal{L}(H) = (\operatorname{ad}\mathfrak{g}) \times \Bbbk$ onto $T_e\left(\overline{G(\Bbbk^*x)}\right)$. But π is the composition of

$$H \to \operatorname{End}(\mathfrak{g}) \,,\, (\alpha, \lambda) \mapsto \lambda\alpha \,,\, \text{and } \operatorname{End}(\mathfrak{g}) \to \mathfrak{g} \,,\, u \mapsto u(x).$$

We deduce therefore that for $\mu \in \Bbbk$ and $y \in \mathfrak{g}$:

$$d\pi_{(e,1)}(\operatorname{ad} y, \mu) = \mu x + [y, x].$$

The image of $d\pi_{(e,1)}$ is therefore $\Bbbk x + [\mathfrak{g}, x]$. So the result follows from (i). $\quad\square$

33.5.3 Proposition. *Let $z \in \mathfrak{g}$. There exists $x \in \overline{G(\Bbbk z)}$ such that:*

$$x \in \mathcal{N}_{\mathfrak{g}} \text{ and } \dim G(x) = \dim G(z).$$

Proof. We may assume that z is not nilpotent. For $x \in \mathfrak{g}$, let χ_x denote the characteristic polynomial of $\operatorname{ad} x$. We have

$$\chi_x(T) = T^n + f_{n-1}(x)T^{n-1} + \cdots + f_1(x)T + f_0(x),$$

where $n = \dim \mathfrak{g}$ and $f_0, \ldots, f_{n-1}$ are G-invariant polynomial functions on $\mathfrak{g}$. Since z is not nilpotent, there exist $r \in \{0, 1, \ldots, n-1\}$ and $a_r, \ldots, a_{n-1} \in \Bbbk$ such that $a_r \neq 0$ and:

$$\chi_z(T) = T^n + a_{n-1}T^{n-1} + \cdots + a_r T^r.$$

If $\lambda \in \Bbbk$ and $\alpha \in G$, then:

$$\chi_{\alpha(\lambda z)}(T) = T^n + a_{n-1}\lambda T^{n-1} + \cdots + a_r \lambda^{n-r} T^r.$$

We deduce therefore that if $x \in \overline{G(\Bbbk z)}$, then:

(1) $$\begin{cases} f_0(x) = \cdots = f_{r-1}(x) = 0, \\ a_r^{n-k}[f_k(x)]^{n-r} = a_k^{n-r}[f_r(x)]^{n-k} \,,\, r \leqslant k \leqslant n-1. \end{cases}$$

The relations in (1) are clearly verified if $x \in \overline{G(\Bbbk z)}$. But x is nilpotent if and only if $\chi_x(T) = T^n$. So:

$$\overline{G(\Bbbk z)} \cap \mathcal{N}_{\mathfrak{g}} = \{x \in \overline{G(\Bbbk z)}\,;\, f_r(x) = 0\}.$$

Since $z \notin \overline{G(\Bbbk z)} \cap \mathcal{N}_{\mathfrak{g}}$ and $0 \in \overline{G(\Bbbk z)} \cap \mathcal{N}_{\mathfrak{g}}$, it follows from 14.2.3 and 33.5.2 that any irreducible component Z of $\overline{G(\Bbbk z)} \cap \mathcal{N}_{\mathfrak{g}}$ verifies:

$$\dim Z = \dim G(\Bbbk z) - 1 = \dim G(z).$$

The set Z is closed and G-stable. We have $Z = \overline{G(x_1)} \cup \cdots \cup \overline{G(x_s)}$, with $x_1, \ldots, x_s \in \overline{G(\Bbbk z)} \cap \mathcal{N}_{\mathfrak{g}}$, because the nilpotent orbits are irreducible and the number of nilpotent orbits is finite (32.4.4). As Z is irreducible, $Z = \overline{G(x_i)}$ for some i and $\dim G(x_i) = \dim G(z)$. $\quad\square$

33.5.4 Let $\mathfrak{p}$ be a parabolic subalgebra of $\mathfrak{g}$. By 29.4.3 (iv), $\operatorname{ad}_{\mathfrak{g}} \mathfrak{p}$ is the Lie algebra of a parabolic subgroup P of G. In view of 28.4.2, 29.4.3 (iii) and 24.3.6, P is the identity component of the normalizer of $\mathfrak{p}$ in G.

Let $\mathfrak{m}$ be the nilpotent radical of $\mathfrak{p}$, $d = \dim \mathfrak{p}$ and $\operatorname{Gr}(\mathfrak{g}, d)$ the Grassmannian variety of d-dimensional subspaces of $\mathfrak{g}$. Then G acts naturally on $\operatorname{Gr}(\mathfrak{g}, d)$. Let $\mathcal{P}$ be the G-orbit of $\mathfrak{p}$. Then by 21.4.3, we obtain that:

$$\dim \mathcal{P} = \dim G - \dim P = \dim \mathfrak{g} - \dim \mathfrak{p} = \dim \mathfrak{m}.$$

Proposition. *Let $z \in \mathfrak{g}$ be a polarizable element and $\mathfrak{p} \in \operatorname{Pol}(z)$. There exists $x \in \overline{G(\Bbbk z)}$ verifying the following conditions:*
  (i) x *is nilpotent and* $\dim G(x) = \dim G(z)$.
  (ii) x *is polarizable and* $\mathfrak{p} \in \operatorname{Pol}(x)$.

Proof. We may assume that z is not nilpotent. Let $d = \dim \mathfrak{p}$, $\mathcal{P}$ the G-orbit of $\mathfrak{p}$ in $\operatorname{Gr}(\mathfrak{g}, d)$ and $p : \operatorname{Gr}(\mathfrak{g}, d) \times \mathfrak{g} \to \mathfrak{g}$ the canonical projection on $\mathfrak{g}$.

Recall from 21.4.3 and 21.4.5 that $\mathcal{P}$ is open in $\overline{\mathcal{P}}$ and $\overline{\mathcal{P}} \setminus \mathcal{P}$ is a union of G-orbits of dimension strictly less than $\dim \mathcal{P}$.

Since $\mathfrak{p}$ is a Lie subalgebra, the elements of $\overline{\mathcal{P}}$ are Lie subalgebras of $\mathfrak{g}$ (19.7.2). Denote by $\mathcal{L}$ the set of elements $(\mathfrak{q}, y) \in \overline{\mathcal{P}} \times \overline{G(\Bbbk z)}$ verifying:

$$L(y, [\mathfrak{q}, \mathfrak{q}]) = \{0\}.$$

Then $\mathcal{L}$ is a closed subset of $\overline{\mathcal{P}} \times \overline{G(\Bbbk z)}$ containing $(\mathfrak{p}, z)$, and $p(\mathcal{L})$ contains $G(\Bbbk z)$. As $\operatorname{Gr}(\mathfrak{g}, d)$ is a complete variety (13.6.1 and 13.4.5), $p(\mathcal{L})$ contains $\overline{G.(\Bbbk z)}$. Thus if $y \in \overline{G.(\Bbbk z)}$, then there exists a d-dimensional Lie subalgebra $\mathfrak{q}$ of $\mathfrak{g}$ such that $\mathfrak{q} \in \overline{\mathcal{P}}$ and $L(y, [\mathfrak{q}, \mathfrak{q}]) = \{0\}$.

By 33.5.3, there exists $x \in \overline{G.(\Bbbk z)}$ verifying (i). Let $\mathfrak{q} \in \overline{\mathcal{P}}$ be such that $L(x, [\mathfrak{q}, \mathfrak{q}]) = \{0\}$. Then:

$$\dim \mathfrak{q} = \dim \mathfrak{p} = \frac{1}{2}(\dim \mathfrak{g} + \dim \mathfrak{g}^z) = \frac{1}{2}(\dim \mathfrak{g} + \dim \mathfrak{g}^x).$$

So $\mathfrak{q} \in \operatorname{Pol}(x)$. Let $\mathfrak{n}$ be the nilpotent radical of $\mathfrak{q}$. We have $[x, \mathfrak{q}] = \mathfrak{n}$ by 33.2.6 and $\mathfrak{g}^x \subset \mathfrak{q}$. Hence:

$$\dim \mathfrak{n} = \dim \mathfrak{q} - \dim \mathfrak{g}^x = \dim \mathfrak{p} - \dim \mathfrak{g}^z = \dim \mathfrak{m}.$$

Thus the G-orbits of $\mathfrak{p}$ and $\mathfrak{q}$ in $\operatorname{Gr}(\mathfrak{g}, d)$ have the same dimension. It follows that $\mathfrak{q} \in \mathcal{P}$, and so there exists $\alpha \in G$ such that $\mathfrak{q} = \alpha(\mathfrak{p})$. The element $\alpha^{-1}(x)$ verifies therefore the required conditions. □

33.5.5 Corollary. *If $z \in \mathfrak{g}$ is polarizable, then there exists a polarizable nilpotent element $x \in \mathfrak{g}$ such that $\dim G(x) = \dim G(z)$.*

33.5.6 Theorem. (Richardson's Theorem) *Let $\mathfrak{g}$ be a semisimple Lie algebra, G its adjoint group, $\mathfrak{p}$ a parabolic subalgebra of $\mathfrak{g}$ with nilpotent radical $\mathfrak{m}$, and P a parabolic subgroup of G such that $\mathcal{L}(P) = \operatorname{ad}_{\mathfrak{g}} \mathfrak{p}$.*

(i) *There exists a unique nilpotent G-orbit $\mathcal{O}$ such that $\mathcal{O} \cap \mathfrak{m}$ is a dense open subset of $\mathfrak{m}$.*

(ii) *The set $\mathcal{O} \cap \mathfrak{m}$ is a P-orbit.*

(iii) *If $x \in \mathcal{O} \cap \mathfrak{m}$, then $[x, \mathfrak{p}] = \mathfrak{m}$ and $\mathfrak{p} \in \mathrm{Pol}(x)$.*

Proof. By 32.4.4, there exist nilpotent orbits $\mathcal{O}_1, \ldots, \mathcal{O}_r$ such that

$$\mathfrak{m} = (\mathcal{O}_1 \cap \mathfrak{m}) \cup \cdots \cup (\mathcal{O}_r \cap \mathfrak{m}).$$

Hence:

$$\mathfrak{m} = \overline{\mathcal{O}_1 \cap \mathfrak{m}} \cup \cdots \cup \overline{\mathcal{O}_r \cap \mathfrak{m}}.$$

Since $\mathfrak{m}$ is irreducible, there is a nilpotent orbit $\mathcal{O}$ such that $\mathfrak{m} = \overline{\mathcal{O} \cap \mathfrak{m}}$. So $\mathfrak{m} \subset \overline{\mathcal{O}}$, and as an orbit is open in its closure, $\mathcal{O} \cap \mathfrak{m}$ is a dense open subset of $\mathfrak{m}$. It is now clear that $\mathcal{O}$ is the unique G-orbit $\mathcal{U}$ such that $\mathcal{U} \cap \mathfrak{m}$ is a dense open subset of $\mathfrak{m}$.

Let

$$s = \inf\{\dim \mathfrak{g}^x \, ; \, x \in \mathfrak{m}\} \, , \; \mathcal{M} = \{x \in \mathfrak{m} \, ; \, \dim \mathfrak{g}^x = s\}.$$

The set $\mathcal{M}$ is a non-empty open subset of $\mathfrak{m}$. So it has a non-empty intersection with $\mathcal{O}$, hence $\dim \mathcal{O} = \dim \mathfrak{g} - s$, and $\mathcal{O} \cap \mathfrak{m} \subset \mathcal{M}$.

Conversely, if $y \in \mathcal{M}$, then $\dim \mathcal{O}_y = \dim \mathcal{O}$, $y \in \overline{\mathcal{O}}$, so $y \in \mathcal{O}$ (21.4.5). Thus $\mathcal{M} = \mathcal{O} \cap \mathfrak{m}$.

There is a Cartan subalgebra $\mathfrak{h}$ of $\mathfrak{g}$ such that $\mathfrak{h} \subset \mathfrak{p}$. Let $R = R(\mathfrak{g}, \mathfrak{h})$, B a base of R and R_+ the corresponding set of positive roots. Since $\mathfrak{p}$ is parabolic, for a suitable choice of B, there is a partition $R_+ = R_1 \cup R_2$ of R_+ such that:

$$\mathfrak{p} = \mathfrak{h} \oplus \sum_{\alpha \in R_+ \cup (-R_1)} \mathfrak{g}^\alpha \, , \; \mathfrak{m} = \sum_{\alpha \in R_2} \mathfrak{g}^\alpha.$$

If $\mathfrak{h}' = \{h \in \mathfrak{h} ; \alpha(h) = 0 \text{ for all } \alpha \in R_1\}$, then the radical $\mathfrak{r}$ of $\mathfrak{p}$ is $\mathfrak{r} = \mathfrak{h}' \oplus \mathfrak{m}$ and

$$\mathfrak{l} = \mathfrak{h} \oplus \sum_{\alpha \in R_1 \cup (-R_1)} \mathfrak{g}^\alpha$$

is a Levi factor of $\mathfrak{p}$.

Let $z \in \mathfrak{h}$ be such that $\alpha(z) = 0$ if $\alpha \in R_1$ and $\alpha(z) \neq 0$ if $\alpha \in R_2$. We have $\mathfrak{g}^z = \mathfrak{l}$, and since $z \in \mathfrak{r}$, $\mathfrak{p}$ is subordinate to z (19.6.3). We deduce therefore that $\mathfrak{p} \in \mathrm{Pol}(z)$. It follows from 33.5.4 that there exists $x \in \mathcal{N}_\mathfrak{g}$ such that $\dim \mathfrak{g}^x = \dim \mathfrak{g}^z$ and $\mathfrak{p} \in \mathrm{Pol}(x)$. But $L(x, \mathfrak{p}) = \{0\}$ (33.2.6), so $x \in \mathfrak{m}$ (20.8.6).

Now, as $\mathfrak{p} \in \mathrm{Pol}(x)$, we have $\mathfrak{g}^x \subset \mathfrak{p}$. So $\dim \mathfrak{g}^x = \dim \mathfrak{g}^z = \dim \mathfrak{l}$ implies that:

$$\dim P(x) = \dim[x, \mathfrak{p}] = \dim \mathfrak{p} - \dim \mathfrak{g}^x = \dim \mathfrak{p} - \dim \mathfrak{l} = \dim \mathfrak{m}.$$

Since $(\mathrm{ad}_\mathfrak{g} \, \mathfrak{p})(\mathfrak{m}) \subset \mathfrak{m}$, we have $P(\mathfrak{m}) = \mathfrak{m}$. So the orbit $P(x)$ is a dense open subset of $\mathfrak{m}$, and hence $P(x) \cap \mathcal{O} \neq \emptyset$. Consequently, $s = \dim \mathfrak{l}$ and $\dim \mathcal{O} = 2 \dim \mathfrak{m}$.

If $y \in \mathcal{O} \cap \mathfrak{m}$, the same arguments show that $\mathfrak{p} \in \mathrm{Pol}(y)$, and again we obtain that $P(y)$ is a dense open subset of $\mathfrak{m}$. It follows that $P(y) = P(x)$. Thus $P(x) = \mathcal{O} \cap \mathfrak{m}$. □

33.5.7 Corollary. *Let $\mathfrak{p}$ be a parabolic subalgebra of $\mathfrak{g}$ and $\mathfrak{m}$ its nilpotent radical.*
   (i) *There exists $x \in \mathfrak{m}$ such that $\mathfrak{p} \in \mathrm{Pol}(x)$.*
   (ii) *The set $G(\mathfrak{m})$ is irreducible and its dimension is $2 \dim \mathfrak{m}$.*

33.5.8 Remarks. 1) The proof of 33.5.6 shows that $\mathcal{O}$ is the unique orbit $\mathcal{U}$ verifying $\dim \mathcal{U} = 2 \dim \mathfrak{m}$ and $\mathcal{U} \cap \mathfrak{m} \neq \emptyset$.
   2) An element x verifying the conditions of 33.5.6 is called a *Richardson element* of $\mathfrak{p}$, and the corresponding orbit $\mathcal{O}$ is called the *Richardson orbit* associated to $\mathfrak{p}$.
   3) Let $x \in \mathcal{O} \cap \mathfrak{m}$ and:
$$\mathfrak{m}_- = \sum_{\alpha \in R_2} \mathfrak{g}^{-\alpha}.$$

Then:
$$[x, \mathfrak{g}] = [x, \mathfrak{p}] + [x, \mathfrak{m}_-] = \mathfrak{m} + [x, \mathfrak{m}_-].$$

Since $\dim[x, \mathfrak{g}] = 2 \dim \mathfrak{m}$, it follows that the restriction of $\mathrm{ad}\, x$ to $\mathfrak{m}_-$ is injective and $\mathfrak{m} \cap [x, \mathfrak{m}_-] = \{0\}$ because $[x, \mathfrak{m}_-] \subset \mathfrak{l}$.

References and comments

   • [27], [28], [29], [31], [70], [80].

Polarizations are very useful for the determination of primitive ideals of the enveloping algebra. In the case of a solvable Lie algebra, they provide a way to determine all primitive ideals of the enveloping algebra via the orbit method of Kirillov-Dixmier ([29]).

Duflo's Theorem stated in 33.1.4 is proved in [29] and [31].

Richardson's Theorem (33.5.6) is fundamental in the study of semisimple Lie algebras. The result formulated in this chapter is in the Lie algebra setting. For other formulations, see [70].

Results on orbits

This chapter is devoted to the description of properties of adjoint orbits in a semisimple Lie algebra.

34.1 Notations

34.1.1 In this chapter, $\mathfrak{g}$ is a semisimple Lie algebra of rank l, L is the Killing form of $\mathfrak{g}$ and G is the adjoint group of $\mathfrak{g}$. If $\mathfrak{a}$ is a subspace of $\mathfrak{g}$, then $\mathfrak{a}^{\perp}$ denotes the orthogonal of $\mathfrak{a}$ with respect to L.

Denote by $\mathcal{S}_{\mathfrak{g}}$ and $\mathcal{N}_{\mathfrak{g}}$ respectively the set of semisimple and nilpotent elements of $\mathfrak{g}$. Recall also that $\mathfrak{g}_{\text{gen}}$ and $\mathfrak{g}_{\text{reg}}$ denote the set of generic and regular elements of $\mathfrak{g}$.

Let us fix a Cartan subalgebra $\mathfrak{h}$ of $\mathfrak{g}$. Let $R = R(\mathfrak{g}, \mathfrak{h})$, $W = W(\mathfrak{g}, \mathfrak{h})$, B a base of R and R_{+} (resp. R_{-}) the corresponding set of positive (resp. negative) roots. Let

$$\mathfrak{n} = \sum_{\alpha \in R_{+}} \mathfrak{g}^{\alpha} \ , \ \mathfrak{b} = \mathfrak{h} \oplus \mathfrak{n}.$$

A S-triple (e, h, f) is called *standard* (with respect to $\mathfrak{h}$ and B) if $h \in \mathfrak{h}$ and $\beta(h) \in \{0, 1, 2\}$ for all $\beta \in B$.

By 32.3.2 and 32.4.3, we obtain that if u is a non-zero nilpotent element of $\mathfrak{g}$, then there exist $\theta \in G$ and $t, v \in \mathfrak{g}$ such that $(\theta(u), t, v)$ is a standard S-triple.

34.1.2 We shall denote the G-orbit of an element $x \in \mathfrak{g}$ by $\mathcal{O}_{x}$, and we say that $\mathcal{O}_{x}$ is *nilpotent* (resp. *semisimple, polarizable*) if x is nilpotent (resp. semisimple, polarizable).

Note that $\dim \mathcal{O}_{x}$ is even since $\dim \mathcal{O}_{x} = \dim \mathfrak{g} - \dim \mathfrak{g}^{x}$ (29.3.1 and 19.7.3). Also, by 25.1.3, the tangent space of $\mathcal{O}_{x}$ at x is $[x, \mathfrak{g}]$.

34.2 Some lemmas

34.2.1 Lemma. *Let* $x, y \in \mathfrak{h}$ *be such that* $y \notin W(x)$. *Then there exists* $f \in S(\mathfrak{h}^*)^W$ *such that* $f(x) \neq f(y)$.

Proof. If $w \in W$, then there exists $g_w \in S(\mathfrak{h}^*)$ such that $g_w(y) = 1$ and $g_w(w(x)) = 0$. Set:

$$g = 1 - \prod_{w \in W} g_w \;,\; f = \prod_{w \in W} g \circ w.$$

Then $g(y) = 0$ and $g(z) = 1$ if $z \in W(x)$. Hence $f(y) = 0$ and $f(x) = 1$. Finally, it is clear that $f \in S(\mathfrak{h}^*)^W$.   □

34.2.2 Lemma. *Let* $x, y \in S_{\mathfrak{g}}$ *be such that* $f(x) = f(y)$ *for all* $f \in S(\mathfrak{g}^*)^{\mathfrak{g}}$. *Then there exists* $\theta \in G$ *such that* $y = \theta(x)$.

Proof. By 29.2.3 and 31.2.4, we may assume that $x, y \in \mathfrak{h}$. Then $f(x) = f(y)$ for all $f \in S(\mathfrak{h}^*)^W$ (31.2.6). So the result follows by 34.2.1 and 29.1.3.   □

34.2.3 Lemma. *Let* $x \in \mathfrak{g}$ *and* s, n *its semisimple and nilpotent components. Then* $f(x) = f(s)$ *for all* $f \in S(\mathfrak{g}^*)^{\mathfrak{g}}$.

Proof. Let (V, σ) be a finite-dimensional representation of $\mathfrak{g}$. Then

$$\sigma(x)^p = \sum_{k=0}^{p} \binom{p}{k} \sigma(s)^k \sigma(n)^{p-k}.$$

for $p \in \mathbb{N}^*$. But $\sigma(s)^k \sigma(n)^{p-k}$ is nilpotent for $0 \leqslant k < p$. Thus $\operatorname{tr}[\sigma(x)^p] = \operatorname{tr}[\sigma(s)^p]$, and the result follows by 31.2.6.   □

34.2.4 Lemma. *Let* $x, y \in \mathfrak{g}$. *The following conditions are equivalent:*
  (i) *The semisimple components of* x *and* y *are* G-*conjugate.*
  (ii) *We have* $f(x) = f(y)$ *for all* $f \in S(\mathfrak{g}^*)^{\mathfrak{g}}$.

Proof. This is clear by 34.2.2 and 34.2.3.   □

34.2.5 Lemma. *Let us conserve the notations* x, s, n *of 34.2.3.*
  (i) *If* $n \neq 0$, *then* $\dim \mathfrak{g}^x < \dim \mathfrak{g}^s$.
  (ii) *If* $\lambda \in \Bbbk \setminus \{0\}$, *then there exists* $\theta_\lambda \in G$ *such that* $\theta_\lambda(x) = s + \lambda^2 n$.
  (iii) *We have* $s \in \overline{\mathcal{O}_x}$.
  (iv) *If* $n \neq 0$, *then* $x \notin \overline{\mathcal{O}_s}$.

Proof. (i) We have $\mathfrak{g}^x = \mathfrak{g}^s \cap \mathfrak{g}^n$, $\mathfrak{g}^s$ is reductive in $\mathfrak{g}$ and $n \in \mathfrak{a} = [\mathfrak{g}^s, \mathfrak{g}^s]$. Thus $\mathfrak{g}^x = \mathfrak{z}(\mathfrak{g}^s) \oplus \mathfrak{a}^n$. If $n \neq 0$, then $\mathfrak{a}^n \neq \mathfrak{a}$. Hence $\mathfrak{g}^x \neq \mathfrak{g}^s$.

(ii) We may assume that $n \neq 0$. Let $\mathfrak{a}$ be as in the proof of (i). Then by 32.1.5, there exist $h, m \in \mathfrak{a}$ such that (n, h, m) is a S-triple. For $\lambda \in \Bbbk \setminus \{0\}$, let:

$$\gamma_\lambda = \exp(\operatorname{ad} \lambda^{-1} n) \exp(-\operatorname{ad} \lambda m) \exp(\operatorname{ad} \lambda^{-1} n) \text{ and } \theta_\lambda = \gamma_1 \circ \gamma_\lambda.$$

Then $\theta_\lambda(n) = \lambda^2 n$. Since $m, n \in \mathfrak{a}$, we have $\theta_\lambda(s) = s$. Hence $\theta_\lambda(x) = s + \lambda^2 n$.

(iii) By (ii), we have $s + \mu n \subset \mathcal{O}_x$ if $\mu \in \Bbbk \setminus \{0\}$, so $s + \Bbbk n \subset \overline{\mathcal{O}_x}$.

(iv) This follows from (i) and 34.1.2. □

34.2.6 Lemma. *Let $x, y \in \mathfrak{g}$ be such that $y \in \overline{\mathcal{O}_x}$. Then:*

(i) $\dim \mathfrak{g}^x \leqslant \dim \mathfrak{g}^y$, *or equivalently,* $\dim[y, \mathfrak{g}] \leqslant \dim[x, \mathfrak{g}]$.

(ii) $\dim(\operatorname{ad} y)^2(\mathfrak{g}) \leqslant \dim(\operatorname{ad} x)^2(\mathfrak{g})$.

Proof. (i) Since $\dim \mathcal{O}_x = \dim \mathfrak{g} - \dim \mathfrak{g}^x$, this is clear by 21.4.5.

(ii) For $z \in \mathfrak{g}$, denote by $M(z)$ the matrix of $(\operatorname{ad} z)^2$ with respect to some basis of $\mathfrak{g}$ and $\rho(z) = \dim(\operatorname{ad} z)^2(\mathfrak{g})$ its rank. Then (ii) follows from the fact that the map $z \mapsto \rho(z)$ is lower semi-continuous, and constant on $\mathcal{O}_x$. □

34.3 Generalities on orbits

34.3.1 Proposition. *If $\mathfrak{g} \neq \{0\}$, then it contains an infinite number of semisimple G-orbits.*

Proof. For $x \in \mathfrak{g}$, let $V(x)$ be the set of eigenvalues of $\operatorname{ad} x$. If $u, v \in \mathfrak{g}$ verify $\mathcal{O}_u = \mathcal{O}_v$, then $V(u) = V(v)$. So if x is a non-zero semisimple element of $\mathfrak{g}$, then there is a strictly increasing sequence of integers $(n_p)_{p \geqslant 0}$ such that $V(n_p x) \neq V(n_q x)$ if $p \neq q$. □

34.3.2 Proposition. *The following conditions are equivalent for $x \in \mathfrak{g}$:*

(i) x *is semisimple.*

(ii) $\mathcal{O}_x$ *is closed.*

Proof. (i) $\Rightarrow$ (ii) Let $y \in \overline{\mathcal{O}_x}$ and $y = y_s + y_n$ its Jordan decomposition. Then 34.2.5 (iii) implies that $y_s \in \overline{\mathcal{O}_y} \subset \overline{\mathcal{O}_x}$.

Now 34.2.4 says that if $f \in S(\mathfrak{g}^*)^{\mathfrak{g}}$, then f is constant on $\mathcal{O}_x$, and hence on $\overline{\mathcal{O}_x}$. We deduce therefore that $f(x) = f(y_s)$. Thus 34.2.4 implies that $y_s \in \mathcal{O}_x$, and so $y \in \overline{\mathcal{O}_x} = \overline{\mathcal{O}_{y_s}}$. It follows from 34.2.5 (iv) that $y = y_s$. Hence $y \in \mathcal{O}_x$, and $\mathcal{O}_x$ is closed.

(ii) $\Rightarrow$ (i) This is clear by 34.2.5 (iii). □

34.3.3 Let us make a few remarks on nilpotent elements.

1) Let $x \in \mathcal{N}_{\mathfrak{g}}$ and $\lambda \in \Bbbk \setminus \{0\}$. Then by 34.2.5 (ii), $\lambda x \in \mathcal{O}_x$.

2) If (e, h, f) is a S-triple in $\mathfrak{g}$, then $\mathcal{O}_e = \mathcal{O}_f$ by point 1 and the computation in the proof of 32.2.7.

3) It follows from the first remark that $\mathcal{N}_{\mathfrak{g}}$ is a cone. It is closed in $\mathfrak{g}$ since:

$$\mathcal{N}_{\mathfrak{g}} = \{x \in \mathfrak{g}; (\operatorname{ad} x)^{\dim \mathfrak{g}} = 0\}.$$

Let J_+ be the set of elements of $S(\mathfrak{g}^*)^{\mathfrak{g}}$ having no constant term. By 19.2.3 and 20.4.4, $x \in \mathcal{N}_{\mathfrak{g}}$ is equivalent to the condition $\operatorname{tr}(\sigma(x)^n) = 0$ for all $n \in \mathbb{N}^*$

and all finite-dimensional representation σ of $\mathfrak{g}$. In view of 31.2.6, we have that $\mathcal{N}_\mathfrak{g} = \mathcal{V}(S(\mathfrak{g}^*)J_+)$ (notations of 11.1.3).

4) By 29.4.7 and 29.4.8, any nilpotent element is conjugate to an element of $\mathfrak{n}$, thus $\mathcal{N}_\mathfrak{g} = G.\mathfrak{n}$.

5) The set of regular nilpotent elements of $\mathfrak{g}$ is a G-orbit (32.5.6) whose characteristic is $(2, \ldots, 2)$. We shall denote this orbit by $\mathcal{O}_{\text{reg}}$, and it is called the *regular nilpotent orbit* of $\mathfrak{g}$.

34.3.4 Proposition. *We have $\mathcal{N}_\mathfrak{g} = \overline{\mathcal{O}_{\text{reg}}}$. In particular, $\mathcal{N}_\mathfrak{g}$ is irreducible.*

Proof. Since $\mathfrak{n} \cap \mathfrak{g}_{\text{reg}}$ is a dense open subset of $\mathfrak{n}$, we have $\mathfrak{n} \subset \overline{\mathcal{O}_{\text{reg}}}$. So the result follows from remark 4 of 34.3.3. $\square$

34.3.5 Proposition. *The following conditions are equivalent for $x \in \mathfrak{g}$:*
   (i) *x is nilpotent.*
   (ii) *$0 \in \overline{\mathcal{O}_x}$.*

Proof. The implication (i) $\Rightarrow$ (ii) follows from 34.2.5 (ii). If (ii) is verified, then $f(x) = f(0)$ for all $f \in S(\mathfrak{g}^*)^\mathfrak{g}$. In particular, $\text{tr}(\text{ad}\, x)^n = 0$ for all $n \in \mathbb{N}^*$ (31.2.3). So x is nilpotent (19.2.3). $\square$

34.3.6 Proposition. *Let $x \in \mathfrak{g}$. Then:*
   (i) *$\overline{\mathcal{O}_x}$ is a union of a finite number of G-orbits.*
   (ii) *$\overline{\mathcal{O}_x}$ contains a unique closed G-orbit. It is the G-orbit of the semisimple component of x.*

Proof. (i) By 21.4.5, $\overline{\mathcal{O}_x}$ is a union of G-orbits. If x is semisimple, then $\overline{\mathcal{O}_x} = \mathcal{O}_x$ is a single G-orbit (34.3.2). If x is nilpotent, then $\overline{\mathcal{O}_x} \subset \mathcal{N}_\mathfrak{g}$ (34.3.4) is a union of a finite number of nilpotent G-orbits (32.4.4).

So let us assume that $x = s+n$ where s, n are the semisimple and nilpotent components of x which are both non-zero. Let $F = \{y \in \mathfrak{g}; f(x) = f(y)$ for all $f \in S(\mathfrak{g}^*)^\mathfrak{g}\}$. It is a closed subset of $\mathfrak{g}$ which contains $\mathcal{O}_x$ (31.2.4), so $\overline{\mathcal{O}_x} \subset F$.

Let $y \in \mathfrak{g}$, and u, v its semisimple and nilpotent components. By 34.2.2 and 34.2.3, $y \in F$ if and only if $u \in \mathcal{O}_s$. We deduce therefore that if $y \in F$, then there exists $\theta \in G$ such that $y = \theta(s + w)$ where w is nilpotent and $[s, w] = 0$. Moreover, w belongs to the semisimple Lie algebra $\mathfrak{s} = [\mathfrak{g}^s, \mathfrak{g}^s]$.

Let $v_1, \ldots, v_r$ be representatives of the set of nilpotent $(\text{Aut}_e\, \mathfrak{s})$-orbits in $\mathfrak{s}$. Then there exist $i \in \{1, 2, \ldots, r\}$ and $\gamma \in \text{Aut}_e\, \mathfrak{s}$ such that $y = \theta(s + \gamma(v_i)) = \theta \circ \gamma(s + v_i)$. Hence F contains at most 3^m G-orbits where m is the rank of $\mathfrak{s}$ (32.4.4).

(ii) The result is clear if x is semisimple (34.3.2), and if x is nilpotent, $\{0\}$ is the unique closed orbit in $\overline{\mathcal{O}_x}$ (34.3.5).

So let us assume that $x = s+n$ where s, n are the semisimple and nilpotent components of x which are both non-zero. By 34.2.5 and 34.3.2, $\mathcal{O}_s$ is a closed orbit in $\overline{\mathcal{O}_x}$. Moreover, we saw in the proof of (i) that the orbits contained in $\overline{\mathcal{O}_x}$ are of the form $\mathcal{O}_y$ where $y = s + v$ with v nilpotent and $[v, s] = 0$. By 34.3.2, such an orbit is closed if and only if $v = 0$. So we are done. $\square$

34.4 Minimal nilpotent orbit

34.4.1 Lemma. *Let (e, h, f) be a S-triple of $\mathfrak{g}$ and $\mathfrak{s} = \Bbbk e + \Bbbk h + \Bbbk f$. The following conditions are equivalent:*

(i) *The image of $(\operatorname{ad} e)^2$ is $\Bbbk e$.*

(ii) *The $\mathfrak{s}$-module $\mathfrak{g}/\mathfrak{s}$ is the sum of simple modules of dimension 1 or 2.*

(iii) *The only eigenvalues of $\operatorname{ad} h$ distinct from -1, 0 and 1 are -2 and 2, and they are both of multiplicity 1.*

Proof. By 19.2.6, we have a decomposition $\mathfrak{g} = \mathfrak{s} \oplus \mathfrak{a}_1 \oplus \cdots \oplus \mathfrak{a}_p$ of $\mathfrak{g}$ into simple $\mathfrak{s}$-submodules. So the lemma follows from the classification of simple finite-dimensional representations of $\mathrm{sl}_2(\Bbbk)$ in 19.2. $\square$

34.4.2 In the rest of this section, $\mathfrak{g}$ is assumed to be simple. Denote by θ the highest root of R and H the unique element h of $[\mathfrak{g}^\theta, \mathfrak{g}^{-\theta}]$ such that $\theta(h) = 2$. Fix also $E \in \mathfrak{g}^\theta$ and $F \in \mathfrak{g}^{-\theta}$ verifying $[E, F] = H$.

Lemma. *Let (e, h, f) be a standard S-triple of $\mathfrak{g}$. The following conditions are equivalent:*

(i) *(e, h, f) verifies the conditions of 34.4.1.*

(ii) *$h = H$.*

If these conditions are verified, then $e \in \mathfrak{g}^\theta$ and $f \in \mathfrak{g}^{-\theta}$.

Proof. Since (E, H, F) is a standard S-triple verifying the conditions of 34.4.1, we have (ii) $\Rightarrow$ (i) by 32.3.1 (ii). Conversely, suppose that (i) is verified. If $l = 1$, then the result is clear. Suppose that $l \geqslant 2$. As (e, h, f) is standard, we have $\alpha(h) \in \{0, 1, 2\}$ for all $\alpha \in B$. So $\alpha(h) \leqslant \theta(h)$ for all $\alpha \in R_+$ and we deduce from condition (iii) of 34.4.1 that $\theta(h) = 2$ and $\alpha(h) \in \{0, 1\}$ if $\alpha \in R_+ \setminus \{\theta\}$. Hence $e \in \mathfrak{g}^\theta$, $f \in \mathfrak{g}^{-\theta}$ and since $[e, f] = h \in [\mathfrak{g}^\theta, \mathfrak{g}^{-\theta}]$, we have $h = H$. $\square$

34.4.3 Lemma. (i) *The S-triples verifying the conditions of 34.4.2 are G-conjugate.*

(ii) *The set of non-zero nilpotent elements of $\mathfrak{g}$ verifying the conditions of 34.4.1 is the G-orbit $\mathcal{O}_E$. We have $\overline{\mathcal{O}_E} = \{0\} \cup \mathcal{O}_E$. We shall denote $\mathcal{O}_E$ also by $\mathcal{O}_{\min}$, and $\mathcal{O}_{\min}$ is called the* (non-zero) minimal nilpotent orbit *of $\mathfrak{g}$.*

Proof. (i) Any S-triple is G-conjugate to a standard S-triple (32.4.3). So the result follows by 34.4.2 and 32.3.1.

(ii) The first part is clear by (i) and 32.3.1. Now if $y \in \overline{\mathcal{O}_E}$, then $\dim \operatorname{im}(\operatorname{ad} y)^2 \leqslant 1$ (34.2.6). So it follows that $\overline{\mathcal{O}_E} \subset \{0\} \cup \mathcal{O}_E$. The reverse inclusion is clear by 34.3.5. $\square$

34.4.4 Let us define

$$n(\mathfrak{g}) = \min\{\dim \mathcal{O}_x \, ; \, x \in \mathfrak{g} \setminus \{0\}\} \, , \quad \mathcal{E} = \{x \in \mathfrak{g} \, ; \, \dim \mathcal{O}_x = n(\mathfrak{g})\},$$
$$n_1(\mathfrak{g}) = \min\{\dim \mathcal{O}_x \, ; \, x \in \mathcal{N}_\mathfrak{g} \setminus \{0\}\} \, , \quad n_2(\mathfrak{g}) = \min\{\dim \mathcal{O}_x \, ; \, x \in \mathcal{S}_\mathfrak{g} \setminus \{0\}\},$$

and $l(\mathfrak{g})$ the minimal codimension of parabolic subalgebras of $\mathfrak{g}$.

Lemma. (i) *An element of $\mathcal{E}$ is either nilpotent or semisimple.*

(ii) *We have $n(\mathfrak{g}) = n_1(\mathfrak{g}) \leqslant 2l(\mathfrak{g}) = n_2(\mathfrak{g})$.*

Proof. (i) This is clear by 34.2.5 (i).

(ii) By 33.2.8, we have $2l(\mathfrak{g}) = n_2(\mathfrak{g})$.

Let $\mathfrak{p}$ be a parabolic subalgebra of $\mathfrak{g}$ and x a non-zero element of its nilpotent radical. Then $L(x, \mathfrak{p}) = \{0\}$ (20.8.6), so $\mathfrak{p}$ is a Lie subalgebra of $\mathfrak{g}$ subordinate to x. Thus $2 \dim \mathfrak{p} \leqslant \dim \mathfrak{g} + \dim \mathfrak{g}^x$. Hence $n_1(\mathfrak{g}) \leqslant 2l(\mathfrak{g})$. Finally (i) implies that $n(\mathfrak{g}) = n_1(\mathfrak{g})$. $\square$

34.4.5 Let $n \in \mathcal{N}_{\mathfrak{g}} \setminus \{0\}$ and (n, s, m) a S-triple of $\mathfrak{g}$. Up to a conjugation by G, we may assume that it is standard, that is, $n \in \mathfrak{n}$, $s \in \mathfrak{h}$ and for all $\alpha \in B$, $\alpha(s) \in \{0, 1, 2\}$ (32.4.3). Then $\theta(s) \geqslant \alpha(s)$ for all $\alpha \in R_+$. For $p \in \mathbb{Z}$, let $\mathfrak{g}_s(p) = \ker(\operatorname{ad} s - p\operatorname{id}_{\mathfrak{g}})$ and

$$\mathfrak{p}_s = \sum_{p \geqslant 0} \mathfrak{g}_s(p) \, , \ \mathfrak{g}_s^2 = \sum_{p \geqslant 2} \mathfrak{g}_s(p).$$

Then $\mathfrak{g}^\theta \subset \mathfrak{g}_s^2$ and $\mathfrak{p}_s \in \operatorname{Pol}(s)$, so $\mathfrak{p}_s$ is a parabolic subalgebra of $\mathfrak{g}$ (33.2.6). Recall that in view of 19.2, we have

$$\dim \mathfrak{g}^n = \dim \mathfrak{g}_s(0) + \dim \mathfrak{g}_s(1),$$

and the map $\mathfrak{p}_s \to \mathfrak{g}_s^2$, $x \mapsto [x, n]$, is surjective.

34.4.6 Proposition. *The orbit $\mathcal{O}_{\min}$ is the unique nilpotent G-orbit of dimension $n(\mathfrak{g})$. For any $n \in \mathcal{N}_{\mathfrak{g}} \setminus \{0\}$, we have $\mathcal{O}_{\min} \subset \overline{\mathcal{O}_n}$.*

Proof. Let $n \in \mathcal{N}_{\mathfrak{g}} \setminus \{0\}$. Up to a conjugation by G, we may assume that we are in the situation of 34.4.5. Let P_s be the parabolic subgroup of G whose Lie algebra is $\operatorname{ad}_{\mathfrak{g}} \mathfrak{p}_s$ (29.4.3). Then by 24.3.3 (ii), we have $P_s(n) \subset \mathfrak{g}_s^2$. We deduce therefore a morphism $P_s \to \mathfrak{g}_s^2$, $\sigma \mapsto \sigma(n)$, whose differential is the surjective map $\operatorname{ad}_{\mathfrak{g}} \mathfrak{p}_s \to \mathfrak{g}_s^2$, $x \mapsto [x, n]$. Hence $\mathfrak{g}_s^2 \subset \overline{P_s(n)} \subset \overline{\mathcal{O}_n}$ (16.5.7). Since $E \in \mathfrak{g}_s^2$, we deduce that $\mathcal{O}_{\min} = \mathcal{O}_E \subset \overline{\mathcal{O}_n}$. Finally, as $n(\mathfrak{g}) = n_1(\mathfrak{g})$ (34.4.4), the result follows. $\square$

34.4.7 Let us consider the case where $(n, s, m) = (E, H, F)$ and let us use the notations π, δ of 33.3.2.

Suppose that $\operatorname{card}(\pi) = 1$ and $l \geqslant 2$. We have $n(\mathfrak{g}) \leqslant 2l(\mathfrak{g})$ (34.4.4). If $n(\mathfrak{g}) = 2l(\mathfrak{g})$, then any parabolic subalgebra of codimension $l(\mathfrak{g})$ in $\mathfrak{g}$ which contains $\mathfrak{b}$ belongs to $\operatorname{Pol}(E)$. But this is absurd by 33.3.3. Hence $n(\mathfrak{g}) < 2l(\mathfrak{g})$.

We deduce from this and from the results of 33.3 the following result:

Proposition. *If $\mathfrak{g}$ is a simple Lie algebra which is not of type A_l, then there is a unique non-zero G-orbit of minimal dimension. This orbit is nilpotent and it is contained in the closure of any non-zero nilpotent orbit of $\mathfrak{g}$.*

34.5 Subregular nilpotent orbit

34.5.1 In this section, $\mathfrak{g}$ is assumed to be simple of dimension n.

Definition. (i) *An element $x \in \mathfrak{g}$ is* subregular *if* $\dim \mathfrak{g}^x = l + 2$, *or equivalently,* $\dim \mathcal{O}_x = n - l - 2$.

(ii) *A G-orbit $\mathcal{O}_x$ of $\mathfrak{g}$ is* subregular *if x is subregular.*

34.5.2 Let $\alpha \in B$. Define $h \in \mathfrak{h}$ by $\alpha(h) = 0$ and $\beta(h) = 1$ if $\beta \in B \setminus \{\alpha\}$. Then it is clear that h is subregular. Thus, subregular semisimple orbits exist. The goal of this section is to prove that there is a unique subregular nilpotent orbit in $\mathfrak{g}$.

34.5.3 For $\alpha \in B$, we set:

$$\mathfrak{p}_\alpha = \mathfrak{b} + \mathbb{k}\mathfrak{g}^{-\alpha} \ , \ \mathfrak{n}_\alpha = \sum_{\beta \in R_+ \setminus \{\alpha\}} \mathfrak{g}^\beta.$$

Thus $\mathfrak{p}_\alpha$ is a parabolic subalgebra of $\mathfrak{g}$ whose nilpotent radical is $\mathfrak{n}_\alpha$.

Similarly, let α, β be distinct elements of B, $S = (\mathbb{Z}\alpha + \mathbb{Z}\beta) \cap R$ and:

$$\mathfrak{p}_{\alpha\beta} = \mathfrak{b} + \sum_{\gamma \in S} \mathfrak{g}^\gamma \ , \ \mathfrak{n}_{\alpha\beta} = \sum_{\gamma \in R_+ \setminus S} \mathfrak{g}^\gamma.$$

Then $\mathfrak{p}_{\alpha\beta}$ is a parabolic subalgebra of $\mathfrak{g}$ whose nilpotent radical is $\mathfrak{n}_{\alpha\beta}$.

Let x be a Richardson element of $\mathfrak{p}_\alpha$ (33.5.8). We have $x \in \mathfrak{n}_\alpha$, $[x, \mathfrak{p}_\alpha] = \mathfrak{n}_\alpha$, and $\mathfrak{p}_\alpha \in \mathrm{Pol}(x)$ (33.5.6). Thus:

$$\dim \mathfrak{g}^x = 2 \dim \mathfrak{p}_\alpha - \dim \mathfrak{g} = l + 2.$$

It follows that x is subregular, so there is a subregular nilpotent orbit in $\mathfrak{g}$.

34.5.4 Let (x, h, y) be a standard S-triple such that x is subregular. If $p \in \mathbb{Z}$, let $\mathfrak{g}_h(p) = \ker(\mathrm{ad}\, h - p\,\mathrm{id}_\mathfrak{g})$. For $\alpha \in R$, let X_α be a non-zero element of $\mathfrak{g}^\alpha$. We have:

$$\dim \mathfrak{g}^h \geqslant l \ , \ \dim \mathfrak{g}^x = \dim \mathfrak{g}^h + \dim \mathfrak{g}_h(1) = l + 2.$$

So $\dim \mathfrak{g}_h(1) \in \{0, 2\}$ since $\dim \mathfrak{g} - \dim \mathfrak{g}^u$ is even for all $u \in \mathfrak{g}$ (34.1.2).

If $\dim \mathfrak{g}_h(1) = 0$, then $\dim \mathfrak{g}^x = \dim \mathfrak{g}^h$, and $\mathfrak{g}_h(i) = \{0\}$ if i is odd. As $\dim \mathfrak{g}^h = l + 2$, $\mathfrak{g}^h = \mathfrak{h} + \mathbb{k}X_\alpha + \mathbb{k}X_{-\alpha}$ for some $\alpha \in B$. Thus

$$\mathfrak{p}_\alpha = \sum_{p \geqslant 0} \mathfrak{g}_h(p)$$

and $\mathfrak{p}_\alpha$ is a polarization of $\mathfrak{g}$ at h.

If $\dim \mathfrak{g}_h(1) = 2$, then h is regular, so $\alpha(h) \in \{1, 2\}$ for all $\alpha \in B$. So there exist $\alpha, \beta \in B$ distinct such that $\alpha(h) = \beta(h) = 1$ and $\gamma(h) = 0$ for $\gamma \in B \setminus \{\alpha, \beta\}$. We have $x \in \mathfrak{n}_\alpha$, and $\mathfrak{p}_\alpha \in \mathrm{Pol}(x)$.

If $\alpha + \beta \notin R$, then $x \in \mathfrak{n}_{\alpha\beta}$ and $\mathfrak{p}_{\alpha\beta}$ is a Lie subalgebra of $\mathfrak{g}$ subordinate to x. We deduce then that:

$$2 + \dim \mathfrak{b} \leqslant \dim \mathfrak{p}_{\alpha\beta} \leqslant \frac{1}{2}(\dim \mathfrak{g} + \dim \mathfrak{g}^x) = 1 + \dim \mathfrak{b}.$$

Contradiction. So $\alpha + \beta \in R$.

34.5.5 It follows from 34.5.4 that any subregular nilpotent element x is polarizable. Since a polarization is a parabolic subalgebra of $\mathfrak{g}$ (33.2.6), by a dimension argument, we obtain that if $\mathfrak{p}$ is a polarization of $\mathfrak{g}$ at x, then $\mathfrak{p}$ is G-conjugate to $\mathfrak{p}_{\alpha}$ for some $\alpha \in B$. Note that x is a Richardson element in $\mathfrak{p}$.

34.5.6 Let Γ denote the Dynkin diagram of R and we shall identify B with the set of vertices of Γ.

Lemma. *Let $\alpha, \beta \in B$ be linked in Γ (see 18.12.2). Then $\mathfrak{p}_{\alpha}$ and $\mathfrak{p}_{\beta}$ have a common Richardson element.*

Proof. Let $S = (\mathbb{Z}\alpha + \mathbb{Z}\beta) \cap R$. Then S is a root system of rank 2 in the subspace of $\mathfrak{h}^*$ spanned by α and β. Let $T = R_+ \setminus S$ and :

$$\mathfrak{a} = \sum_{\gamma \in T} \mathfrak{g}^{\gamma}.$$

For $\lambda \in \Bbbk$, $e^{\lambda \operatorname{ad} X_{\alpha}}$ and $e^{\lambda \operatorname{ad} X_{-\alpha}}$ leave $\mathfrak{a}$ stable.

1) Suppose that S is of type G_2. Then by the classification given in 18.14, we have $R = S$. So $B = \{\alpha, \beta\}$, and we may assume that α is a short root. Let

$$x = X_{\alpha+\beta} + X_{2\alpha+\beta},$$

and $h \in \mathfrak{h}$ be defined by $\alpha(h) = 0$, $\beta(h) = 2$. Then:

$$[h, x] = 2x \ , \ \mathfrak{g}^h = \mathfrak{h} + \Bbbk X_{\alpha} + \Bbbk X_{-\alpha}.$$

Let $u \in \mathfrak{h}$, $\lambda, \mu \in \Bbbk$ be such that $t = u + \lambda X_{\alpha} + \mu X_{-\alpha}$ verifies $[t, x] = 0$. Since the coefficient of $X_{3\alpha+\beta}$ (resp. X_{β}) in $[t, x]$ is zero, we obtain $\lambda = \mu = 0$. Hence $u = 0$. Thus $\mathfrak{g}^h \cap \mathfrak{g}^x = \{0\}$. By 32.1.8, there is a S-triple of the form (x, h, y). As $\mathfrak{g}_h(1) = \{0\}$, we have $\dim \mathfrak{g}^x = \dim \mathfrak{g}^h = l + 2$, and x is subregular. Note that $x \in \mathfrak{n}_{\alpha} \cap \mathfrak{n}_{\beta}$, and:

$$[x, \mathfrak{p}_{\alpha}] = \mathfrak{n}_{\alpha} \ , \ [x, \mathfrak{p}_{\beta}] = \mathfrak{n}_{\beta}.$$

We have therefore proved that x is a Richardson element for both $\mathfrak{p}_{\alpha}$ and $\mathfrak{p}_{\beta}$.

In the rest of the proof, x is a Richardson element of $\mathfrak{p}_{\alpha}$.

2) Suppose that S is of type A_2, and $x \notin \mathfrak{n}_{\beta}$. Then

$$x = aX_{\beta} + bX_{\alpha+\beta} + u,$$

with $u \in \mathfrak{a}$, $a, b \in \Bbbk$ and $a \neq 0$. By replacing x by $e^{\operatorname{ad} X_{\alpha}}(x)$, we may assume that $b \neq 0$. Then, for a suitable $\lambda \in \Bbbk$,

$$y = e^{\lambda \operatorname{ad} X_{-\alpha}}(x) = bX_{\alpha+\beta} + v,$$

where $v \in \mathfrak{a}$. Thus $y \in \mathfrak{n}_\alpha \cap \mathfrak{n}_\beta$. Since y is a Richardson element for $\mathfrak{p}_\alpha$ and $\dim \mathfrak{p}_\alpha = \dim \mathfrak{p}_\beta$, y is also a Richardson element for $\mathfrak{p}_\beta$.

3) Suppose that S is of type B_2, α is long and $x \notin \mathfrak{n}_\beta$. Then

$$x = aX_\beta + bX_{\alpha+\beta} + cX_{\alpha+2\beta} + u,$$

with $u \in \mathfrak{a}$, $a, b, c \in \mathbb{k}$ and $a \neq 0$. Again we may assume that $b \neq 0$, and for a suitable $\lambda \in \mathbb{k}$,

$$y = e^{\lambda \operatorname{ad} X_{-\alpha}}(x) = bX_{\alpha+\beta} + cX_{\alpha+2\beta} + v,$$

where $v \in \mathfrak{a}$. So y is the element with the required properties.

4) Suppose that S is of type B_2, α is short and $x \notin \mathfrak{n}_\beta$. Then

$$x = aX_\beta + bX_{\alpha+\beta} + cX_{2\alpha+\beta} + u,$$

where $u \in \mathfrak{a}$, $a, b, c \in \mathbb{k}$ and $a \neq 0$. Again we may assume $b \neq 0$. Let $\lambda \in \mathbb{k}$. Then $e^{\lambda \operatorname{ad} X_{-\alpha}}(aX_\beta + bX_{\alpha+\beta} + cX_{2\alpha+\beta})$ is of the form

$$(a + b\lambda\mu + \lambda^2 c\nu)X_\beta + dX_{\alpha+\beta} + cX_{2\alpha+\beta},$$

where $d \in \mathbb{k}$ and $\mu, \nu \in \mathbb{k} \setminus \{0\}$ do not depend on λ. So there is a λ such that $y = e^{\lambda \operatorname{ad} X_{-\alpha}}(x) \in \mathfrak{n}_\beta$. Thus y is a Richardson element for $\mathfrak{p}_\alpha$ and $\mathfrak{p}_\beta$. $\square$

34.5.7 Proposition. (i) *There is a unique subregular nilpotent orbit $\mathcal{O}$ in $\mathfrak{g}$.*

(ii) *We have $\overline{\mathcal{O}} = \mathcal{N}'_\mathfrak{g}$, the set of non-regular nilpotent elements of $\mathfrak{g}$.*

Proof. (i) We saw in 34.5.3 that there are subregular nilpotent elements. Let x, y be subregular nilpotent elements. Up to a conjugation by G, we may assume that there exist $\alpha, \beta \in B$ such that $\mathfrak{p}_\alpha$ and $\mathfrak{p}_\beta$ are polarizations of $\mathfrak{g}$ at x and at y (34.5.5). The Dynkin diagram of $\mathfrak{g}$ is connected and Richardson elements in a parabolic subalgebra are conjugate (33.5.6), so by repeated applications of 34.5.6, we obtain that $\mathcal{O}_x = \mathcal{O}_y$.

(ii) Let (x, h, y) a standard S-triple with x non-regular. Then there is an $\alpha \in B$ such that $\alpha(h) \neq 2$ (32.5.7). We have $x \in \mathfrak{n}_\alpha$. If $z \in \mathfrak{n}_\alpha$, then $L(z, \mathfrak{p}_\alpha) = \{0\}$ (20.8.6). So $\mathfrak{p}_\alpha$ is subordinate to z, and hence z is not regular. So $x \in \mathfrak{n}_\alpha \subset \mathcal{N}'_\mathfrak{g}$.

By (i), there is a subregular element in $\mathfrak{n}_\alpha$. So it follows that the set of subregular elements of $\mathfrak{g}$ in $\mathfrak{n}_\alpha$ is the set of elements $z \in \mathfrak{n}_\alpha$ verifying $\mathfrak{g}^z \leqslant \mathfrak{g}^t$ for all $t \in \mathfrak{n}_\alpha$, which is a non-empty open subset of $\mathfrak{n}_\alpha$. We deduce again from (i) that $\mathfrak{n}_\alpha \subset \overline{\mathcal{O}}$. In particular $x \in \mathfrak{n}_\alpha \subset \overline{\mathcal{O}}$, and hence $\mathcal{N}'_\mathfrak{g} \subset \overline{\mathcal{O}}$.

Finally, if $z \in \overline{\mathcal{O}}$, then $\dim \mathcal{O}_z \leqslant \dim \mathcal{O}$, so z is not a regular element and $z \in \mathcal{N}'_\mathfrak{g}$. $\square$

34.5.8 We list below, for the different types of simple Lie algebra, the characteristic of the unique subregular nilpotent orbit.

- Type A_{2r+1}, $r \geqslant 0$:

- Type A_{2r}, $r \geqslant 1$:

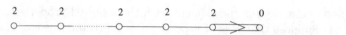

- Type B_l, $l \geqslant 2$:

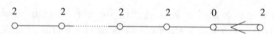

- Type C_l, $l \geqslant 3$:

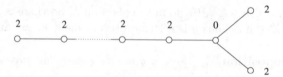

- Type D_l, $l \geqslant 4$:

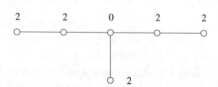

- Type E_6 :

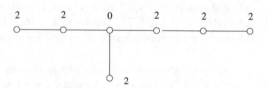

- Type E_7 :

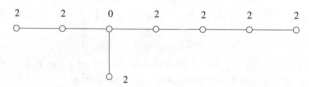

- Type E_8 :

- Type F_4 :

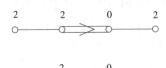

- Type G_2 :

34.6 Dimension of nilpotent orbits

34.6.1 In this section, $\mathfrak{g}$ is assumed to be simple of dimension n. Let I be the set of integers $p \in \mathbb{N}$ such that $n - l - 2p$ is the dimension of a nilpotent orbit. Note that, in general, I is not an interval of $\mathbb{N}$. We shall determine an interval J inside I.

34.6.2 Lemma. *Let* $m \in \mathbb{N}$ *and* $P \subset B$.
(i) *If* card $P = 2m$, *then there is a subset* Q *of* P *of cardinal* m *such that* $\alpha + \beta \notin R$ *for all* $\alpha, \beta \in Q$.
(ii) *If* card $P = 2m + 1$, *then there is a subset* Q *of* P *of cardinal* $m + 1$ *such that* $\alpha + \beta \notin R$ *for all* $\alpha, \beta \in Q$.

Proof. Let Γ be the Dynkin diagram of R. The results are clear if $m = 0$ or if $m = 1$ since Γ has no cycles. Let us proceed by induction on m.

Let Λ be the full subgraph of Γ whose vertices are the elements in P. Then Λ has a terminal vertex. Let $\alpha \in P$ be a terminal vertex of Λ. If $\beta \in P$ is linked to α, then set $P' = P \setminus \{\alpha, \beta\}$. If α is not linked to any vertex of Λ, then let $\beta \in P \setminus \{\alpha\}$ and set $P' = P \setminus \{\alpha, \beta\}$.

By the induction hypothesis, there is a subset Q' of P' verifying the conclusion of the lemma for P'. We obtain therefore the result by letting $Q = Q' \cup \{\alpha\}$. □

34.6.3 Proposition. *Let* r *be the largest integer such that* $2r \leqslant l + 1$. *For* $0 \leqslant p \leqslant r$, *there is a polarizable nilpotent orbit of dimension* $n - l - 2p$.

Proof. By 34.6.2, there exist $\alpha_1, \ldots, \alpha_r \in B$ such that $\alpha_i + \alpha_j \notin R$ for $1 \leqslant i, j \leqslant r$. For $0 \leqslant p \leqslant r$, set:

$$\mathfrak{p}_p = \mathfrak{b} + \mathfrak{g}^{-\alpha_1} + \cdots + \mathfrak{g}^{-\alpha_p}.$$

Then $\mathfrak{p}_p$ is a parabolic subalgebra of $\mathfrak{g}$. Let x_p be a Richardson element of $\mathfrak{p}_p$ (33.5.8), then $\mathfrak{p}_p \in \mathrm{Pol}(x_p)$ (33.5.6), and its orbit is a polarizable nilpotent orbit of dimension $n - l - 2p$. □

34.7 Prehomogeneous spaces of parabolic type

34.7.1 In this section, $\mathfrak{g}$ is a semisimple Lie algebra and we assume that there is a decomposition of $\mathfrak{g}$ into subspaces

$$\mathfrak{g} = \bigoplus_{i \in \mathbb{Z}} \mathfrak{g}_i \text{ such that } [\mathfrak{g}_i, \mathfrak{g}_j] \subset \mathfrak{g}_{i+j} \text{ for } i, j \in \mathbb{Z}.$$

Let $\partial \in \mathrm{End}(\mathfrak{g})$ be defined by $\partial(x) = ix$ if $x \in \mathfrak{g}_i$. Then ∂ is a semisimple derivation of $\mathfrak{g}$. By 20.1.5, there is a unique semisimple element t_0 of $\mathfrak{g}$ such that $\partial = \mathrm{ad}\, t_0$. Thus $\mathfrak{g}_i$ is the $(\mathrm{ad}\, t_0)$-eigenspace with eigenvalue i. Also, we have $L(\mathfrak{g}_i, \mathfrak{g}_j) = \{0\}$ if $i + j \neq 0$ and $L|_{\mathfrak{g}_i \times \mathfrak{g}_{-i}}$ is non-degenerate.

The Lie subalgebra $\mathfrak{g}_0 = \mathfrak{g}^{t_0}$ is a reductive in $\mathfrak{g}$ (20.5.13) and it contains a Cartan subalgebra of $\mathfrak{g}$ (20.6.3). Denote by K the identity component of the stabilizer of t_0 in G, and if $x \in \mathfrak{g}$, we denote by K_x the stabilizer of x in K.

34.7.2 Definition. *Let* $i \in \mathbb{Z}$.

(i) *A S-triple* (e, h, f) *of* $\mathfrak{g}$ *is called a* S_i-triple *if* $e \in \mathfrak{g}_i$ *and* $f \in \mathfrak{g}_{-i}$ *(so* $h \in \mathfrak{g}_0$*)*.

(ii) *We say that* $h \in \mathfrak{g}$ *is* i-simple *if there is a* S_i-triple of the form (e, h, f).

34.7.3 Remark. Let e be a non-zero nilpotent element of $\mathfrak{g}_i$. Then 32.1.7 implies that there is a S_i-triple of the form (e, h, f).

34.7.4 Let $h \in \mathcal{S}_{\mathfrak{g}} \cap \mathfrak{g}_0$, and for $\lambda \in \Bbbk$, let $\mathfrak{g}^\lambda = \ker(\mathrm{ad}\, h - \lambda \, \mathrm{id}_{\mathfrak{g}})$. Let us assume that there exist $i \in \mathbb{Z}$ and $e \in \mathfrak{g}_i$ such that

$$e \in \mathfrak{g}^2 \ , \ [e, \mathfrak{g}^0] = \mathfrak{g}^2.$$

Lemma. *The set* $K_h(e)$ *contains a dense open subset of* $\mathfrak{g}_i \cap \mathfrak{g}^2$.

Proof. First of all, note that from the definition of K, we have $\alpha(\mathfrak{g}_n) = \mathfrak{g}_n$ for all $\alpha \in K$ and $n \in \mathbb{Z}$. Similarly, $H = K_h^\circ$ leaves $\mathfrak{g}^\lambda$ stable for all $\lambda \in \Bbbk$. We may therefore define the morphism

$$\varphi : H \to \mathfrak{g}_i \cap \mathfrak{g}^2 \ , \ \alpha \mapsto \alpha(e).$$

By 24.3.6, $\mathcal{L}(K) = \mathrm{ad}\, \mathfrak{g}_0$ and $\mathcal{L}(H) = \mathrm{ad}(\mathfrak{g}_0 \cap \mathfrak{g}^0)$ (24.3.5). So the image of the differential of φ at the identity of H is $[e, \mathfrak{g}_0 \cap \mathfrak{g}^0]$. So by 15.4.2 and 16.5.7, it suffices to prove that $[e, \mathfrak{g}_0 \cap \mathfrak{g}^0] = \mathfrak{g}_i \cap \mathfrak{g}^2$.

Let $x \in \mathfrak{g}_i \cap \mathfrak{g}^2$. Then by our assumption on e, we have $x = [e, y]$ for some $y \in \mathfrak{g}^0$. Since $[t_0, h] = 0$, we can write

$$y = \sum_j y_j$$

with $y_j \in \mathfrak{g}_j \cap \mathfrak{g}^0$ for all $j \in \mathbb{Z}$. So we deduce from $x = [e, y]$ that $x = [e, y_0]$. Thus we have $[e, \mathfrak{g}_0 \cap \mathfrak{g}^0] = \mathfrak{g}_i \cap \mathfrak{g}^2$ as required. $\square$

34.7.5 Two S-triples (e, h, f) and (e', h', f') are K-conjugate if there exists $\alpha \in K$ such that $e' = \alpha(e)$, $h' = \alpha(h)$ and $f' = \alpha(f)$.

Proposition. *Let* $i \in \mathbb{Z} \setminus \{0\}$ *and* (e, h, f), (e', h', f') *two* S_i-*triples. The following conditions are equivalent:*

(i) *There exists* $\alpha \in K$ *such that* $\alpha(e) = e'$.
(ii) *There exists* $\alpha \in K$ *such that* $\alpha(h) = h'$.
(iii) *The* S-*triples* (e, h, f) *and* (e', h', f') *are* K-*conjugate.*

Proof. (i) $\Rightarrow$ (ii) Since $K(\mathfrak{g}_i) = \mathfrak{g}_i$ for all $i \in \mathbb{Z}$, we may assume that $e = e'$. Let $\mathfrak{n} = \ker(\mathrm{ad}\, e) \cap \mathrm{im}(\mathrm{ad}\, e) \cap \mathfrak{g}_0$. By 32.2.4, $\mathfrak{n}$ is a Lie subalgebra of $\mathfrak{g}$ whose elements are nilpotent. Since $[e, h - h'] = 0$ and $[e, f - f'] = h - h'$, we have $h - h' \in \mathfrak{n}$. But $[h, \mathfrak{n}] = \mathfrak{n}$ (19.2.7). So there exists $z \in \mathfrak{n}$ such that $e^{\mathrm{ad}\, z}(h) = h'$ (32.2.5). Since $z \in \mathfrak{g}_0$, we have $e^{\mathrm{ad}\, z} \in K$.

(ii) $\Rightarrow$ (iii) Again, we may assume that $h = h'$. Since e and e' satisfy the hypotheses of 34.7.4, we have $K_h(e) \cap K_h(e') \neq \emptyset$. Thus if $\alpha \in K_h$ verifies $\alpha(e) = e'$, then $\alpha(h) = h$ and $\alpha(f) = f'$ (32.2.3).

(iii) $\Rightarrow$ (i) This is obvious. $\square$

34.7.6 Lemma. *Let* $\mathfrak{h}$ *be a Cartan subalgebra of* $\mathfrak{g}$, $\mathfrak{c}$ *a subspace of* $\mathfrak{h}$ *and* θ *an automorphism of* $\mathfrak{g}$ *verifying* $\theta^{-1}(\mathfrak{c}) \subset \mathfrak{h}$. *Then there exists* $\alpha \in G$ *such that* $\alpha \circ \theta(\mathfrak{h}) = \mathfrak{h}$ *and* $\alpha(x) = x$ *for all* $x \in \mathfrak{c}$.

Proof. Since $\theta(\mathfrak{h})$ is a Cartan subalgebra of $\mathfrak{g}$ containing $\mathfrak{c}$, there exists, by 32.2.1, $\alpha \in G$ such that $\alpha \circ \theta(\mathfrak{h}) = \mathfrak{h}$ and $\alpha(x) = x$ for all $x \in \mathfrak{h} \cap \theta(\mathfrak{h})$. $\square$

34.7.7 Lemma. *Let* $\mathfrak{h}$ *be a Cartan subalgebra of* $\mathfrak{g}$ *containing* t_0 *and* h *a semisimple element of* $\mathfrak{g}_0$. *There exists* $\theta \in \mathrm{Aut}_e([\mathfrak{g}_0, \mathfrak{g}_0])$ *such that* $\theta(h) \in \mathfrak{h}$.

Proof. We have $\mathfrak{g}_0 = [\mathfrak{g}_0, \mathfrak{g}_0] \oplus \mathfrak{z}(\mathfrak{g})$ and $\mathfrak{z}(\mathfrak{g}) \subset \mathfrak{h}$ (20.5.5). Thus $h = h_1 + h_2$ with $h_1 \in [\mathfrak{g}_0, \mathfrak{g}_0]$ and $h_2 \in \mathfrak{z}(\mathfrak{g})$. Since $\mathfrak{h}' = \mathfrak{h} \cap [\mathfrak{g}_0, \mathfrak{g}_0]$ is a Cartan subalgebra of $[\mathfrak{g}_0, \mathfrak{g}_0]$, it follows from 29.2.3 and 29.2.6 that there exists $\theta \in \mathrm{Aut}_e([\mathfrak{g}_0, \mathfrak{g}_0])$ such that $h_1' = \theta(h_1) \in \mathfrak{h}'$. Hence $\theta(h) = h_1' + h_2 \in \mathfrak{h}$. $\square$

34.7.8 Let $\mathfrak{h}$ be a Cartan subalgebra of $\mathfrak{g}$ containing t_0, H the set of $\alpha \in G$ verifying $\alpha(\mathfrak{h}) = \mathfrak{h}$ and W_0 the set of $w \in W$ such that $w(t_0) = t_0$ (see 18.2.9).

Denote by φ the natural map from H to W as defined in 30.6.5. Thus $W_0 = \varphi(H_{t_0})$, where H_{t_0} is the stabilizer of t_0 in H.

Proposition. *If* $i \in \mathbb{Z} \setminus \{0\}$, *then there exists a natural injection from the set of* K-*orbits of* i-*simple elements into the set of double cosets* $W_0 \backslash W / W_0$.

Proof. Let $h \in \mathfrak{g}$ be i-simple. By 34.7.7, $K(h) \cap \mathfrak{h} \neq \emptyset$, so we may assume that $h \in \mathfrak{h}$. Let (e, h, f) be a S_i-triple and $\sigma = e^{-\mathrm{ad}\, e} \circ e^{\mathrm{ad}\, f} \circ e^{-\mathrm{ad}\, e} \in G$. Then:

$$\sigma(h) = -h \ , \ \sigma(t_0) = t_0 - ih.$$

Let $\mathfrak{c} = \Bbbk t_0 \oplus \Bbbk h$. By 34.7.6, there exists $\rho \in G$ such that $\rho(h) = h$, $\rho(t_0) = t_0$ and $\rho \circ \sigma(\mathfrak{h}) = \mathfrak{h}$. Set $\theta = \rho \circ \sigma$, then:

(1) $\theta \in H$, $\theta(h) = -h$, $\theta(t_0) = t_0 - ih$.

Let $h, h' \in \mathfrak{h}$ be i-simple elements verifying $h' \in K(h)$. In view of 34.7.6, we may assume that $h' = \alpha(h)$ with $\alpha \in H_{t_0}$. Let $\theta_h \in H$ (resp. $\theta_{h'} \in H$) be the element of G associated to h (resp. h') verifying the relations in (1). Then:

$$(\theta_h)^{-1} \circ \alpha^{-1} \circ \theta_{h'}(t_0) = (\theta_h)^{-1} \circ \alpha^{-1}(t_0 - ih') = (\theta_h)^{-1}(t_0 - ih) = t_0.$$

Thus

$$\varphi[(\theta_h)^{-1} \circ \alpha^{-1} \circ \theta_{h'}] = [\varphi(\theta_h)]^{-1} [\varphi(\alpha)]^{-1} \varphi(\theta_{h'}) \in W_0.$$

Hence:

$$W_0 \varphi(\theta_h) W_0 = W_0 \varphi(\theta_{h'}) W_0.$$

So this induces a map from the set of K-orbits of i-simple elements to set of double cosets $W_0 \backslash W / W_0$ given by $K(h) \mapsto W_0 \varphi(\theta_h) W_0$. Let us show that this map is injective.

If h, h' are i-simple elements such that $W_0 \varphi(\theta_h) W_0 = W_0 \varphi(\theta_{h'}) W_0$, then there exist $\alpha_1, \alpha_2 \in H_{t_0}$ such that:

$$\varphi(\alpha_1 \circ \theta_h \circ \alpha_2) = \varphi(\theta_{h'}).$$

Hence:

$$t_0 - i\alpha_1(h) = \alpha_1 \circ \theta_h \circ \alpha_2(t_0) = \theta_{h'}(t_0) = t_0 - ih'.$$

Thus $\alpha_1(h) = h'$ (since $i \neq 0$), and so $K(h) = K(h')$. $\square$

34.7.9 Corollary. *Let $i \in \mathbb{Z} \setminus \{0\}$. The number of K-orbits in $\mathfrak{g}_i$ is finite, and $\mathfrak{g}_i$ contains an open K-orbit.*

Proof. If $i \neq 0$, then all the elements of $\mathfrak{g}_i$ are nilpotent. So 34.7.3, 34.7.5 and 34.7.8 imply that the number of K-orbits in $\mathfrak{g}_i$ is finite. If $\mathcal{O}_1, \dots, \mathcal{O}_s$ are the K-orbits in $\mathfrak{g}_i$, then $\mathfrak{g}_i = \overline{\mathcal{O}_1} \cup \dots \cup \overline{\mathcal{O}_s}$. Since $\mathfrak{g}_i$ is irreducible, the result follows from 21.4.3. $\square$

References and comments

- [24], [29], [44], [48], [49], [57], [79], [80], [85], [86].

The proofs given in 34.5 on subregular orbits are inspired by the methods of [79].

The reader can find in [44] results relating the minimal nilpotent orbit to primitive ideals of the enveloping algebra of $\mathfrak{g}$.

Centralizers

We study the centralizer of an element in a semisimple Lie algebra in this chapter. We characterize in particular certain nilpotent elements which are called distinguished.

Let $\mathfrak{g}$ be a semisimple Lie algebra. We shall use the notations and conventions of 34.1.

35.1 Distinguished elements

35.1.1 Definition. (i) *An element $x \in \mathfrak{g}$ is called* distinguished *(in $\mathfrak{g}$) if $\mathfrak{g}^x$ contains only nilpotent elements.*

(ii) *Let $x \in \mathcal{N}_{\mathfrak{g}}$. We say that x is* even *if its characteristic contains only the integers 0 and 2.*

35.1.2 Lemma. *If $x \in \mathfrak{g}$ is nilpotent, then all the elements of $\mathfrak{z}(\mathfrak{g}^x)$ are nilpotent.*

Proof. Let $u \in \mathfrak{z}(\mathfrak{g}^x)$ and $u = y + z$ be its Jordan decomposition, with y semisimple and z nilpotent. Then $y, z \in \mathfrak{z}(\mathfrak{g}^x)$. If $y \neq 0$, then there exist $\lambda \in \Bbbk$ and $v \in \mathfrak{g}$, both non-zero, such that $[y, v] = \lambda v$. Thus:

$$[y, [x, v]] = [[y, x], v] + [x, [y, v]] = [x, [y, v]] = \lambda[x, v].$$

Let $\mathfrak{a} = \{w \in \mathfrak{g} \, ; \, [y, w] = \lambda w\}$, then we have shown that $[x, \mathfrak{a}] \subset \mathfrak{a}$. Since x is nilpotent, there exists $w \in \mathfrak{a} \setminus \{0\}$ such that $[x, w] = 0$, or equivalently, $w \in \mathfrak{g}^x$. So $[y, w] = \lambda w \neq 0$ and $y \in \mathfrak{z}(\mathfrak{g}^x)$. Contradiction. $\square$

35.1.3 Proposition. *The following conditions are equivalent for $x \in \mathfrak{g}$:*

(i) *x is distinguished in $\mathfrak{g}$.*

(ii) *0 is the only semisimple element of $\mathfrak{g}$ which commutes with x.*

(iii) *$[x, \mathfrak{g}]$ contains a parabolic subalgebra of $\mathfrak{g}$.*

(iv) *$\mathfrak{g}^x \subset [x, \mathfrak{g}]$.*

(v) *x is nilpotent and $\mathfrak{g}^x$ is a nilpotent Lie algebra.*

Proof. (i) ⇔ (ii) This is clear since $\mathfrak{g}^x$ contains the semisimple and nilpotent components of its elements.

(i) ⇒ (iii) Let x be distinguished in $\mathfrak{g}$, then 29.4.8 implies that $\mathfrak{g}^x$ is contained in the nilpotent radical $\mathfrak{n}$ of a Borel subalgebra $\mathfrak{q}$ of $\mathfrak{g}$. Taking the orthogonals with respect to the Killing form on $\mathfrak{g}$, we obtain that $\mathfrak{q} \subset [\mathfrak{g}, x]$.

(iii) ⇒ (iv) Let $\mathfrak{p}$ be a parabolic subalgebra of $\mathfrak{g}$, and $\mathfrak{m}$ its nilpotent radical. If $\mathfrak{p} \subset [\mathfrak{g}, x]$, then, as above, we have $\mathfrak{g}^x \subset \mathfrak{m}$, so $\mathfrak{g}^x \subset [\mathfrak{g}, x]$ as required.

(iv) ⇒ (i) If $y \in \mathfrak{g}$ verifies $[x, [x, y]] = 0$, then $[x, y]$ is nilpotent (32.1.2). Hence we have (i).

(i) ⇒ (v) This is obvious.

(v) ⇒ (ii) Let us suppose that $\mathfrak{g}^x$ contains a non-zero semisimple element h. Then ad h leaves $\mathfrak{g}^x$ stable, and since $\mathfrak{g}^x$ is nilpotent, we have clearly $h \in \mathfrak{z}(\mathfrak{g}^x)$. But this contradicts 35.1.2. □

35.1.4 Remarks. 1) It follows from 35.1.3 (v) and 19.7.5 that a regular nilpotent element of $\mathfrak{g}$ is distinguished.

2) Let $x \in \mathcal{N}_{\mathfrak{g}} \setminus \{0\}$. Assume that x is not distinguished in $\mathfrak{g}$, and $h \in \mathfrak{g}^x$ is a non-zero semisimple element. Then $x \in \mathfrak{s} = [\mathfrak{g}^h, \mathfrak{g}^h]$, and $\mathfrak{s}$ is semisimple. If x is not distinguished in $\mathfrak{s}$, then we repeat the above procedure by replacing $\mathfrak{g}$ by $\mathfrak{s}$. By repeating this process, we see that there exists a semisimple Lie subalgebra $\mathfrak{t}$ of $\mathfrak{g}$ containing x such that x is distinguished in $\mathfrak{t}$.

35.1.5 Let (x, h, y) be a S-triple of $\mathfrak{g}$. For $p \in \mathbb{Z}$, we set $\mathfrak{g}_p$ to be the eigenspace of ad h associated to the eigenvalue p, and:

$$\mathfrak{l} = \mathfrak{g}_0 \ , \ \mathfrak{p} = \bigoplus_{p \geqslant 0} \mathfrak{g}_p \ , \ \mathfrak{m} = \bigoplus_{p > 0} \mathfrak{g}_p.$$

Then $\mathfrak{p}$ is a parabolic subalgebra of $\mathfrak{g}$ with nilpotent radical $\mathfrak{m}$, and $\mathfrak{l}$ is a Levi factor of $\mathfrak{p}$.

Recall from 19.2 that the map $\mathfrak{g}_0 \to \mathfrak{g}_2$, $z \mapsto [x, z]$, is surjective.

Proposition. *The following conditions are equivalent:*

(i) x *is distinguished in* $\mathfrak{g}$.

(ii) ad x *induces a bijection from* $\mathfrak{g}_0$ *onto* $\mathfrak{g}_2$.

(iii) dim $\mathfrak{g}_0 =$ dim $\mathfrak{g}_2$.

Proof. Let $\mathfrak{s} = \Bbbk x + \Bbbk h + \Bbbk y$. By the theory of representations of $\mathfrak{s}$ described in 19.2, we have:

$$\mathfrak{g}^x = \bigoplus_{p \geqslant 0} (\mathfrak{g}^x \cap \mathfrak{g}_p) = (\mathfrak{l} \cap \mathfrak{g}^x) \oplus \mathfrak{m}^x.$$

Since $[\mathfrak{g}_i, \mathfrak{g}_j] \subset \mathfrak{g}_{i+j}$ if $i, j \in \mathbb{Z}$, the elements of $\mathfrak{m}^x$ are nilpotent. Moreover, $\mathfrak{l} \cap \mathfrak{g}^x = \mathfrak{c}_{\mathfrak{g}}(\mathfrak{s})$, so it is reductive in $\mathfrak{g}$ (20.5.13). Hence all the elements of $\mathfrak{g}^x$ are nilpotent if and only if $\mathfrak{l} \cap \mathfrak{g}^x = \{0\}$. The proposition is now clear. □

35.1.6 Theorem. (Bala-Carter Theorem) *Any distinguished element in a semisimple Lie algebra* $\mathfrak{g}$ *is even.*

Proof. Let us use the notations in 35.1.5 and assume that x is distinguished in $\mathfrak{g}$. By 33.5.6, there exists $u \in \mathfrak{m}$ such that $[\mathfrak{p}, u] = \mathfrak{m}$. Let us write

$$u = \sum_{i>0} u_i,$$

with $u_i \in \mathfrak{g}_i$ for $i \in \mathbb{N}^*$. From

$$\left[\bigoplus_{i \geqslant 0} \mathfrak{g}_i , \sum_{i>0} u_i \right] = \bigoplus_{i>0} \mathfrak{g}_i,$$

we deduce that:

$$[\mathfrak{g}_0, u_1 + u_2] + [\mathfrak{g}_1, u_1] = \mathfrak{g}_1 \oplus \mathfrak{g}_2.$$

Suppose that $\mathfrak{g}_1 \neq \{0\}$. Since $u_1 \in \mathfrak{g}_1$, we have $\dim[\mathfrak{g}_1, u_1] < \dim \mathfrak{g}_1$. We obtain therefore from 35.1.5 that:

$$\dim[\mathfrak{g}_0, u_1 + u_2] + \dim[\mathfrak{g}_1, u_1] < \dim \mathfrak{g}_0 + \dim \mathfrak{g}_1 = \dim \mathfrak{g}_2 + \dim \mathfrak{g}_1.$$

Contradiction. Thus $\mathfrak{g}_1 = \{0\}$, and hence $\mathfrak{g}_i = \{0\}$ if i is odd. So x is even. $\square$

35.1.7 Corollary. *Any distinguished element in $\mathfrak{g}$ is polarizable.*

Proof. In the notations of 35.1.5, if x is even, then $\dim \mathfrak{g}^h = \dim \mathfrak{g}^x$. So $2 \dim \mathfrak{p} = \dim \mathfrak{g} + \dim \mathfrak{g}^x$, and $\mathfrak{p} \in \mathrm{Pol}(x)$ since $x \in \mathfrak{m}$. $\square$

35.2 Distinguished parabolic subalgebras

35.2.1 Let J be a subset of B, denote by R_J the set of roots which are linear combinations of elements of J. and:

$$\mathfrak{l}_J = \mathfrak{h} \oplus \sum_{\alpha \in R_J} \mathfrak{g}^\alpha , \quad \mathfrak{p}_J = \mathfrak{b} + \mathfrak{l}_J , \quad \mathfrak{m}_J = \sum_{\alpha \in R_+ \setminus R_J} \mathfrak{g}^\alpha.$$

Then $\mathfrak{p}_J$ is a parabolic subalgebra of $\mathfrak{g}$ with nilpotent radical $\mathfrak{m}_J$, and $\mathfrak{l}_J$ is a Levi factor of $\mathfrak{p}_J$.

35.2.2 For $\beta \in B$, set:

$$\eta_J(\beta) = \begin{cases} 2 & \text{if } \beta \in B \setminus J, \\ 0 & \text{if } \beta \in J. \end{cases}$$

We define a map $\eta_J : R \to 2\mathbb{Z}$ by setting for $\alpha \in R$:

$$\alpha = \sum_{\beta \in B} n_\beta \beta \Rightarrow \eta_J(\alpha) = \sum_{\beta \in B} n_\beta \eta_J(\beta).$$

For $i \in \mathbb{Z}$, let:

$$\mathfrak{g}_J(i) = \begin{cases} \displaystyle\sum_{\eta_J(\alpha)=i} \Bbbk\mathfrak{g}^\alpha & \text{if } i \neq 0, \\ \displaystyle\mathfrak{h} + \sum_{\eta_J(\alpha)=0} \Bbbk\mathfrak{g}^\alpha & \text{si } i = 0. \end{cases}$$

We have $[\mathfrak{g}_J(i), \mathfrak{g}_J(j)] \subset \mathfrak{g}_J(i+j)$ for all $i, j \in \mathbb{Z}$. Moreover, $\mathfrak{l}_J = \mathfrak{g}_J(0)$, and $\mathfrak{g}$ is the direct sum of the $\mathfrak{g}_J(i)$ for $i \in \mathbb{Z}$.

Recall that if $\mathfrak{a}$ is a Lie algebra and $k \in \mathbb{N}^*$, then $\mathcal{C}^k(\mathfrak{a})$ denotes the k-th term of the central descending series of $\mathfrak{a}$. The following result is a generalization of 20.8.3.

Proposition. *For $k \in \mathbb{N}^*$, we have:*

$$\mathcal{C}^k(\mathfrak{m}_J) = \sum_{p \geqslant k} \mathfrak{g}_J(2p).$$

In particular, $\dim \mathfrak{g}_J(2) = \dim(\mathfrak{m}_J/[\mathfrak{m}_J, \mathfrak{m}_J])$.

Proof. For $k \in \mathbb{N}^*$, let

$$\mathcal{G}_k = \sum_{p \geqslant k} \mathfrak{g}_J(2p).$$

Clearly $\mathfrak{m}_J = \mathcal{C}^1(\mathfrak{m}_J) = \mathcal{G}_1$ and $\mathcal{C}^k(\mathfrak{m}_J) \subset \mathcal{G}_k$. To establish the reverse inclusion, we proceed by induction on k.

Let $\alpha \in R_+$ be such that $\eta_J(\alpha) \geqslant 2k$. By 18.7.7, $\alpha = \beta_1 + \cdots + \beta_r$, with $\beta_1, \ldots, \beta_r \in B$ and $\beta_1 + \cdots + \beta_i \in R$ for $1 \leqslant i \leqslant r$. Let $s = \max\{i; \beta_i \in B \setminus J\}$. We have $s \geqslant 2$ since $\eta_J(\alpha) \geqslant 2k$. Set $\gamma = \beta_1 + \cdots + \beta_{s-1}$. Then $\gamma \in R$ and $\alpha = \gamma + \beta_s + \cdots + \beta_r$. It follows that we may choose non-zero vectors X_δ in $\mathfrak{g}^\delta$, for $\delta \in R$, such that

$$X_\alpha = [[\cdots [[X_\gamma, X_{\beta_s}], X_{\beta_{s+1}}] \cdots], X_{\beta_r}].$$

From this, we deduce that $X_\alpha \in [[\cdots [[\mathcal{G}_k, \mathcal{G}_1], \mathfrak{g}_J(0)] \cdots], \mathfrak{g}_J(0)] \subset [\mathcal{G}_k, \mathcal{G}_1]$. By the induction hypothesis, we obtain that $\mathcal{G}_{k+1} \subset \mathcal{C}^{k+1}(\mathfrak{m}_J)$. $\square$

35.2.3 Definition. *A parabolic subalgebra $\mathfrak{p}$ of $\mathfrak{g}$ is called* distinguished *if* $\dim(\mathfrak{p}/\mathfrak{m}) = \dim(\mathfrak{m}/[\mathfrak{m}, \mathfrak{m}])$, *where $\mathfrak{m}$ is the nilpotent radical of $\mathfrak{p}$.*

35.2.4 Proposition. (i) *We have $\dim \mathfrak{l}_J \geqslant \dim(\mathfrak{m}_J/[\mathfrak{m}_J, \mathfrak{m}_J])$.*

(ii) *The parabolic subalgebra $\mathfrak{p}_J$ of $\mathfrak{g}$ is distinguished if and only if we have* $\dim \mathfrak{g}_J(0) = \dim \mathfrak{g}_J(2)$.

Proof. By 35.2.2, it suffices to prove (i). Let x be a Richardson element of $\mathfrak{p}_J$ (33.5.6). We have $[\mathfrak{p}_J, x] = \mathfrak{m}_J$.

The adjoint representation of $\mathfrak{p}_J$ on $\mathfrak{m}_J$ induces a representation σ of $\mathfrak{p}_J/\mathfrak{m}_J$ on $\mathfrak{m}_J/[\mathfrak{m}_J, \mathfrak{m}_J]$. It follows that $\sigma(\mathfrak{p}_J/\mathfrak{m}_J)(\widetilde{x}) = \mathfrak{m}_J/[\mathfrak{m}_J, \mathfrak{m}_J]$ where $\widetilde{x}$ denotes the image of x in $\mathfrak{m}_J/[\mathfrak{m}_J, \mathfrak{m}_J]$. So we have the result. $\square$

35.2.5 We shall now give the relation between distinguished elements and distinguished parabolic subalgebras.

Proposition. *Let x be a distinguished element in $\mathfrak{g}$, (x, h, y) a S-triple and $\mathfrak{p}$ the parabolic subalgebra of $\mathfrak{g}$ associated to this S-triple as defined in 35.1.5.*

(i) *$\mathfrak{p}$ is a distinguished parabolic subalgebra of $\mathfrak{g}$.*

(ii) *x belongs to the Richardson orbit of $\mathfrak{p}$.*

Proof. Using the notations of 35.1.5, we have:

$$[\mathfrak{p}, x] = \bigoplus_{i \geqslant 2} \mathfrak{g}_i.$$

Since x is even (35.1.6), we have $[\mathfrak{p}, x] = \mathfrak{m}$ and $\mathfrak{p} \in \mathrm{Pol}(x)$. We deduce therefore that $\dim \mathcal{O}_x = 2 \dim \mathfrak{m}$ and x is a Richardson element of $\mathfrak{p}$ (33.5.8).

Now there exists a Cartan subalgebra $\mathfrak{h}$ such that $h \in \mathfrak{h} \subset \mathfrak{p}$. For a suitable choice of base B of the root system $(\mathfrak{g}, \mathfrak{h})$, we have $\mathfrak{p} = \mathfrak{p}_J$, where $J \subset B$. It is then obvious that $\mathfrak{g}_J(i) = \mathfrak{g}_i$ for all $i \in \mathbb{Z}$. So (i) follows from 35.1.5 (iii) and 35.2.4 (ii). $\square$

35.2.6 Proposition. *Suppose that $\mathfrak{p}_J$ is a distinguished parabolic subalgebra of $\mathfrak{g}$. If x is a Richardson element of $\mathfrak{p}_J$, then it is distinguished in $\mathfrak{g}$ and $\mathfrak{g}^x = \mathfrak{p}_J^x = \mathfrak{m}_J^x$.*

Proof. We have $[\mathfrak{p}_J, x] = \mathfrak{m}_J$, $\mathfrak{p}_J \in \mathrm{Pol}(x)$, and so $\mathfrak{g}^x \subset \mathfrak{p}_J$, and $\mathfrak{g}^x = \mathfrak{p}_J^x$.

Let $y \in \mathfrak{g}_J(-2) \setminus \{0\}$. If $[x, y] \in \mathfrak{m}_J$, then $y \in [\mathfrak{p}_J, x]^\perp = \mathfrak{m}_J^\perp = \mathfrak{p}_J$ which is absurd. So $[x, y] \notin \mathfrak{m}_J$. It follows easily from (35.2.4) that $[\mathfrak{p}_J + \mathfrak{g}_J(-2), x] = \mathfrak{p}_J$. Hence by 35.1.3 (iii), x is a distinguished element in $\mathfrak{g}$. Finally since $\mathfrak{p}_J \subset [\mathfrak{g}, x]$, we have $\mathfrak{g}^x \subset \mathfrak{m}_J$ by taking orthogonals with respect to the Killing form. We have therefore $\mathfrak{g}^x = \mathfrak{p}_J^x = \mathfrak{m}_J^x$. $\square$

35.2.7 Corollary. *The following conditions are equivalent for $x \in \mathfrak{g}$:*
   (i) *x is distinguished in $\mathfrak{g}$.*
   (ii) *We have $\mathfrak{g}^x \subset \overline{\mathcal{O}_x}$.*

Proof. (i) $\Rightarrow$ (ii) Let (x, h, y) be a S-triple containing x and $\mathfrak{p}$ the associated parabolic subalgebra as defined in 35.1.5. Denote by $\mathfrak{m}$ the nilpotent radical of $\mathfrak{p}$. By 35.2.5 and 35.2.6, we have $\mathfrak{g}^x \subset \mathfrak{m}$ and $\mathfrak{m} \subset \overline{\mathcal{O}_x}$.

(ii) $\Rightarrow$ (i) Suppose that (ii) is verified, and let f be a polynomial function which vanishes on $\overline{\mathcal{O}_x}$. If $t \in \Bbbk$ and $y \in \mathfrak{g}^x$, then $f(x+ty) = 0$, so $D_x(f)(y) = 0$. Thus $\mathfrak{g}^x \subset \mathrm{T}_x(\overline{\mathcal{O}_x})$ (see 16.1). Now $\mathrm{T}_x(\overline{\mathcal{O}_x}) = \mathrm{T}_x(\mathcal{O}_x) = [x, \mathfrak{g}]$ (16.2.4, 21.4.3 and 34.1.2). So we may conclude by 35.1.3 (iv). $\square$

35.3 Double centralizers

35.3.1 Proposition. *Let $x \in \mathfrak{g}$. Then:*
   (i) $[x, \mathfrak{g}] = [\mathfrak{z}(\mathfrak{g}^x), \mathfrak{g}] = (\mathfrak{g}^x)^\perp$.
   (ii) $\mathfrak{z}(\mathfrak{g}^x) = [\mathfrak{g}^x, \mathfrak{g}]^\perp$.

Proof. If $y \in \mathfrak{g}$, then:

$$L([\mathfrak{g}^x, \mathfrak{g}], y) = \{0\} \Leftrightarrow L(\mathfrak{g}, [y, \mathfrak{g}^x]) = \{0\} \Leftrightarrow [y, \mathfrak{g}^x] = \{0\} \Leftrightarrow y \in \mathfrak{z}(\mathfrak{g}^x).$$

Hence $\mathfrak{z}(\mathfrak{g}^x) = [\mathfrak{g}^x, \mathfrak{g}]^\perp$. Similarly:

$$L([x, \mathfrak{g}], y) = \{0\} \Leftrightarrow L([x, y], \mathfrak{g}) = \{0\} \Leftrightarrow y \in \mathfrak{g}^x,$$
$$L([\mathfrak{z}(\mathfrak{g}^x), \mathfrak{g}], y) = \{0\} \Leftrightarrow L([\mathfrak{z}(\mathfrak{g}^x), y], \mathfrak{g}) = \{0\} \Leftrightarrow y \in \mathfrak{g}^x.$$

So we are done. $\square$

35.3.2 Corollary. *The following conditions are equivalent for $x, y \in \mathfrak{g}$:*

(i) $y \in \mathfrak{z}(\mathfrak{g}^x)$.

(ii) $\mathfrak{g}^x \subset \mathfrak{g}^y$.

(iii) $[\mathfrak{g}, \mathfrak{g}^x] \subset [\mathfrak{g}, \mathfrak{g}^y]$.

(iv) $[\mathfrak{g}, y] \subset [\mathfrak{g}, x]$.

(v) $\mathfrak{z}(\mathfrak{g}^y) \subset \mathfrak{z}(\mathfrak{g}^x)$.

(vi) $[\mathfrak{z}(\mathfrak{g}^y), \mathfrak{g}] \subset [\mathfrak{z}(\mathfrak{g}^x), \mathfrak{g}]$.

Proof. The equivalences (ii) $\Leftrightarrow$ (iv), (iv) $\Leftrightarrow$ (vi) and (iii) $\Leftrightarrow$ (v) follow from 35.3.1, and it is clear that (ii) $\Rightarrow$ (iii).

(i) $\Rightarrow$ (ii) If $[y, \mathfrak{g}^x] = \{0\}$, then $\mathfrak{g}^x \subset \mathfrak{g}^y$.

(v) $\Rightarrow$ (i) If $\mathfrak{z}(\mathfrak{g}^y) \subset \mathfrak{z}(\mathfrak{g}^x)$, then $y \in \mathfrak{z}(\mathfrak{g}^x)$ since $y \in \mathfrak{z}(\mathfrak{g}^y)$. $\square$

35.3.3 Let $x \in \mathfrak{g}$. If $y \in \mathfrak{g}$ verifies $\mathfrak{g}^y = \mathfrak{g}^x$, then $y \in \mathfrak{z}(\mathfrak{g}^x)$ (35.3.2). Set:

$$\mathfrak{z}(\mathfrak{g}^x)^{\bullet} = \{y \in \mathfrak{g} \,;\, \mathfrak{g}^x = \mathfrak{g}^y\}.$$

Proposition. *The set $\mathfrak{z}(\mathfrak{g}^x)^{\bullet}$ is a non-empty open subset of $\mathfrak{z}(\mathfrak{g}^x)$, and:*

$$\mathfrak{z}(\mathfrak{g}^x)^{\bullet} = \{y \in \mathfrak{g} \,;\, \mathfrak{z}(\mathfrak{g}^y) = \mathfrak{z}(\mathfrak{g}^x)\} = \{y \in \mathfrak{z}(\mathfrak{g}^x) \,;\, \mathrm{rk}(\mathrm{ad}\, y) = \mathrm{rk}(\mathrm{ad}\, x)\}.$$

Proof. Let $y \in \mathfrak{z}(\mathfrak{g}^x)$. By 35.3.2, we have $\mathrm{rk}(\mathrm{ad}\, y) \leqslant \mathrm{rk}(\mathrm{ad}\, x)$ and:

$$\mathrm{rk}(\mathrm{ad}\, y) = \mathrm{rk}(\mathrm{ad}\, x) \Leftrightarrow \mathfrak{z}(\mathfrak{g}^y) = \mathfrak{z}(\mathfrak{g}^x) \Leftrightarrow \mathfrak{g}^y = \mathfrak{g}^x.$$

The proposition is now clear. $\square$

35.3.4 Proposition. *Let $x \in \mathfrak{g}$ and $\mathcal{O}$ its G-orbit.*

(i) *If x is semisimple, then the set $\mathcal{O} \cap \mathfrak{z}(\mathfrak{g}^x)$ is finite.*

(ii) *If x is nilpotent, then $\mathcal{O} \cap \mathfrak{z}(\mathfrak{g}^x) = \mathfrak{z}(\mathfrak{g}^x)^{\bullet}$.*

Proof. (i) If x is semi-simple, then there exist a Cartan subalgebra $\mathfrak{h}$ of $\mathfrak{g}$, a base B of the root system of $(\mathfrak{g}, \mathfrak{h})$ and a subset J of B such that $\mathfrak{z}(\mathfrak{g}^x)$ is the set of $h \in \mathfrak{h}$ such that $\alpha(h) = 0$ for all $\alpha \in J$. By 34.2.4, 31.2.6 and 34.2.1, the set $\mathcal{O} \cap \mathfrak{h}$ is finite.

(ii) Let us suppose that x is nilpotent. Since the number of nilpotent orbits is finite, there is a nilpotent orbit $\mathcal{U}$ such that $\mathcal{U} \cap \mathfrak{z}(\mathfrak{g}^x) = \mathcal{U}'$ is an open dense subset of $\mathfrak{z}(\mathfrak{g}^x)$ (35.1.2). By 35.3.3, we have $\mathcal{U}' \cap \mathfrak{z}(\mathfrak{g}^x) \neq \emptyset$, $\dim \mathcal{U} = \dim \mathcal{O}$, and $\mathcal{U}' \subset \mathfrak{z}(\mathfrak{g}^x)^{\bullet}$. Conversely, if $y \in \mathfrak{z}(\mathfrak{g}^x)^{\bullet}$, then $y \in \overline{\mathcal{U}}$ and $\dim \mathcal{O}_y = \dim \mathcal{U}$, so $y \in \mathcal{U}$ (21.4.5). Thus $\mathfrak{z}(\mathfrak{g}^x)^{\bullet} = \mathcal{U}'$ and $\mathcal{U} = \mathcal{O}$. $\square$

35.3.5 Proposition. *Let $x \in \mathfrak{g}$. Then:*

(i) $\mathfrak{n}_{\mathfrak{g}}(\mathfrak{g}^x) = \{y \in \mathfrak{g} \,;\, [x, y] \in \mathfrak{z}(\mathfrak{g}^x)\}$.

(ii) $\mathfrak{n}_{\mathfrak{g}}(\mathfrak{g}^x) = \mathfrak{n}_{\mathfrak{g}}\big(\mathfrak{z}(\mathfrak{g}^x)\big)$.

Proof. (i) Let $y \in \mathfrak{g}$ be such that $[x, y] \in \mathfrak{z}(\mathfrak{g}^x)$. If $u \in \mathfrak{g}^x$, then $[[y, x], u] = 0$. So:

$$0 = [[y, u], x] + [y, [x, u]] = [[y, u], x].$$

Thus $[y, u] \in \mathfrak{g}^x$, and $y \in \mathfrak{n}_{\mathfrak{g}}(\mathfrak{g}^x)$.

Conversely, let $y \in \mathfrak{n}(\mathfrak{g}^x)$. If $u \in \mathfrak{g}^x$ and $v \in \mathfrak{z}(\mathfrak{g}^x)$, then:

$$0 = [y, [u, v]] = [[y, u], v] + [u, [y, v]] = [u, [y, v]].$$

We deduce therefore that $[y, v] \in \mathfrak{c}_{\mathfrak{g}}(\mathfrak{g}^x) = \mathfrak{z}(\mathfrak{g}^x)$. In particular, $[y, x] \in \mathfrak{z}(\mathfrak{g}^x)$.

(ii) From the preceding argument, we obtain that $\mathfrak{n}_{\mathfrak{g}}(\mathfrak{g}^x) \subset \mathfrak{n}_{\mathfrak{g}}(\mathfrak{z}(\mathfrak{g}^x))$. Conversely, if $y \in \mathfrak{n}_{\mathfrak{g}}(\mathfrak{z}(\mathfrak{g}^x))$, then $[y, x] \in \mathfrak{z}(\mathfrak{g}^x)$, and so $y \in \mathfrak{n}_{\mathfrak{g}}(\mathfrak{g}^x)$ by (i). $\square$

35.3.6 Corollary. *If x is semisimple, then* $\mathfrak{n}_{\mathfrak{g}}(\mathfrak{g}^x) = \mathfrak{n}_{\mathfrak{g}}(\mathfrak{z}(\mathfrak{g}^x)) = \mathfrak{g}^x$.

Proof. Here we have $\mathfrak{g} = [x, \mathfrak{g}] \oplus \mathfrak{g}^x$. So the result is clear (35.3.5 (i)). $\square$

35.3.7 We shall study in detail the structure of $\mathfrak{z}(\mathfrak{g}^x)$ and $\mathfrak{n}(\mathfrak{g}^x)$ when x is nilpotent.

Let (e, h, f) be a S-triple of $\mathfrak{g}$ and $\mathfrak{s} = \Bbbk e + \Bbbk h + \Bbbk f$.

If t is a semisimple element of $\mathfrak{g}$, $\mathfrak{a}$ an $(\operatorname{ad} t)$-stable subspace of $\mathfrak{g}$ and $\lambda \in \Bbbk$, we set:

$$\mathfrak{a}_{t,\lambda} = \{x \in \mathfrak{a} \ ; \ [t, x] = \lambda x\}.$$

When $t = h$, we shall denote $\mathfrak{a}_{h,\lambda}$ by $\mathfrak{a}_{\lambda}$.

35.3.8 It is clear that $\mathfrak{g}^e$ and $\mathfrak{z}(\mathfrak{g}^e)$ are $(\operatorname{ad} h)$-stable.

Lemma. *We have* $\mathfrak{z}(\mathfrak{g}^e)_n = \{0\}$ *if* $n \leqslant 1$.

Proof. By the theory of representations of $\mathfrak{s}$ (see 19.2), we have $(\mathfrak{g}^e)_n = \{0\}$ if $n < 0$. So $\mathfrak{z}(\mathfrak{g}^e)_n = \{0\}$ if $n < 0$.

Since $(\mathfrak{g}^e)_0 = \mathfrak{c}_{\mathfrak{g}}(\mathfrak{s})$, we have $\mathfrak{z}(\mathfrak{g}^e)_0 \subset \mathfrak{c}_{\mathfrak{g}}(\mathfrak{s})$. If $y \in \mathfrak{z}(\mathfrak{g}^e)_0$ and $z \in \mathfrak{c}_{\mathfrak{g}}(\mathfrak{s})$, then $[z, y] = 0$ because $z \in \mathfrak{g}^e$. We deduce therefore that $y \in \mathfrak{z}(\mathfrak{c}_{\mathfrak{g}}(\mathfrak{s}))$. But $\mathfrak{c}_{\mathfrak{g}}(\mathfrak{s})$ is reductive in $\mathfrak{g}$, so the elements of $\mathfrak{z}(\mathfrak{c}_{\mathfrak{g}}(\mathfrak{s}))$ are semisimple. In view of 35.1.2, we have $y = 0$. Hence $\mathfrak{z}(\mathfrak{g}^e)_0 = \{0\}$.

Suppose that $\mathfrak{z}(\mathfrak{g}^e)_1 \neq \{0\}$. Then the sum $\mathfrak{r}$ of 2-dimensional simple $\mathfrak{s}$-submodules of $\mathfrak{g}$ is non-zero. We see easily that $L|_{\mathfrak{r} \times \mathfrak{r}}$ is non-degenerate.

Let $u \in \mathfrak{z}(\mathfrak{g}^e)_1 \setminus \{0\}$. We have $[f, u] \in \mathfrak{r}_{-1} \setminus \{0\}$. So there exists $v \in \mathfrak{r}_1$ such that $L(v, [f, u]) = 1$. Thus $L([u, v], f) = 1$. But it is clear that $v \in \mathfrak{g}^e$. So we have a contradiction. Hence $\mathfrak{z}(\mathfrak{g}^e)_1 = \{0\}$. $\square$

35.3.9 Proposition. *Let* $e \in \mathcal{N}_{\mathfrak{g}} \setminus \{0\}$ *and* (e, h, f) *a S-triple. Then:*

$$\mathfrak{z}(\mathfrak{g}^e) = \sum_{n \geqslant 1} \mathfrak{z}(\mathfrak{g}^e)_{2n}.$$

Proof. If e is distinguished, then the result follows from 35.1.6 and 35.3.8. Otherwise, there is a non-zero semisimple element $x \in \mathfrak{g}^e$. Let $\mathfrak{a} = [\mathfrak{g}^x, \mathfrak{g}^x]$, then:

$$\mathfrak{g}^x = \mathfrak{a} \oplus \mathfrak{z}(\mathfrak{g}^x).$$

The Lie algebra $\mathfrak{a}$ is semisimple and all the elements of $\mathfrak{z}(\mathfrak{g}^x)$ are semisimple. Now $\mathfrak{z}(\mathfrak{g}^e) \subset \mathfrak{g}^x$, so by 35.1.2, we obtain that $\mathfrak{z}(\mathfrak{g}^e) \subset \mathfrak{a}$. Hence:

$$\mathfrak{z}(\mathfrak{g}^e) \subset \mathfrak{z}(\mathfrak{a}^e) \subset \mathfrak{a}^e \subset \mathfrak{g}^e.$$

If e is not distinguished in $\mathfrak{a}$, then we repeat the same argument by replacing $\mathfrak{g}$ by $\mathfrak{a}$. So we are reduced to the following situation: there exists a semisimple subalgebra $\mathfrak{t}$ of $\mathfrak{g}$ containing e such that:

(i) $\mathfrak{z}(\mathfrak{g}^e) \subset \mathfrak{z}(\mathfrak{t}^e)$.

(ii) e is distinguished in $\mathfrak{t}$.

Let (e, h_1, f_1) be a S-triple of $\mathfrak{t}$. The eigenvalues of $\operatorname{ad} h_1|_{\mathfrak{t}}$ are even (35.1.6), so the eigenvalues of $\operatorname{ad} h_1|_{\mathfrak{z}(\mathfrak{g}^e)}$ are also even. Now there exists $\theta \in G$ such that $\theta(e) = e$ and $\theta(h) = h_1$ (32.3.1). Clearly, $\theta(\mathfrak{g}^e) = \mathfrak{g}^e$ and $\theta(\mathfrak{z}(\mathfrak{g}^e)) = \mathfrak{z}(\mathfrak{g}^e)$. Consequently, the eigenvalues of $\operatorname{ad} h|_{\mathfrak{z}(\mathfrak{g}^e)}$ are even. Since $\mathfrak{z}(\mathfrak{g}^e)_0 = \{0\}$ (35.3.8), we have obtained the result. $\square$

35.3.10 Corollary. *Let $e \in \mathcal{N}_{\mathfrak{g}} \setminus \{0\}$, $\lambda \in \Bbbk$ and $t \in \mathcal{S}_{\mathfrak{g}}$ be such that $[t, e] = \lambda e$. Then $\mathfrak{z}(\mathfrak{g}^e)$ is $(\operatorname{ad} t)$-stable and:*

$$\mathfrak{z}(\mathfrak{g}^e) = \sum_{n \geqslant 1} \mathfrak{z}(\mathfrak{g}^e)_{t, n\lambda}.$$

Proof. It is obvious that $\mathfrak{g}^e$ and $\mathfrak{z}(\mathfrak{g}^e)$ are $(\operatorname{ad} t)$-stable. We may assume that $[t, h] = 0$ and $[t, f] = -\lambda f$ (32.1.7). So $2t - \lambda h \in \mathfrak{g}^e$. By 35.3.9, the eigenvalues of $\operatorname{ad} h|_{\mathfrak{z}(\mathfrak{g}^e)}$ are of the form $2n$, with $n \in \mathbb{N}^*$. Since t and h are semisimple, $[t, h] = 0$, and $\operatorname{ad}(2t - \lambda h)|_{\mathfrak{z}(\mathfrak{g}^e)} = 0$, the eigenvalues of $\operatorname{ad} t|_{\mathfrak{z}(\mathfrak{g}^e)}$ are of the form $n\lambda$ where $n \in \mathbb{N}^*$. $\square$

35.4 Normalizers

35.4.1 Let us conserve the hypotheses and notations of 35.3.7 and set:

$$\mathfrak{d} = [f, \mathfrak{z}(\mathfrak{g}^e)].$$

It is clear that $\mathfrak{d}$ is $(\operatorname{ad} h)$-stable, and by 35.3.9, we have:

$$\mathfrak{d} = \sum_{n \geqslant 0} \mathfrak{d}_n \quad \text{and} \quad \operatorname{ad} h|_{\mathfrak{z}(\mathfrak{g}^e)} \text{ is injective.}$$

Hence:

$$[e, \mathfrak{d}] = [e, [f, \mathfrak{z}(\mathfrak{g}^e)]] = [h, \mathfrak{z}(\mathfrak{g}^e)] = \mathfrak{z}(\mathfrak{g}^e).$$

It follows that $\dim \mathfrak{d} = \dim \mathfrak{z}(\mathfrak{g}^e)$, $\mathfrak{d} \cap \mathfrak{g}^e = \{0\}$ and $\mathfrak{d} \subset \mathfrak{n}_{\mathfrak{g}}(\mathfrak{g}^e)$ (35.3.5). In particular, the sum $\mathfrak{d} + \mathfrak{g}^e$ is direct.

Proposition. *We have:*
(i) $[\mathfrak{n}_{\mathfrak{g}}(\mathfrak{g}^e), e] = \mathfrak{z}(\mathfrak{g}^e)$.
(ii) $\mathfrak{n}_{\mathfrak{g}}(\mathfrak{g}^e) = \mathfrak{g}^e \oplus \mathfrak{d}$, *so* $\dim \mathfrak{n}_{\mathfrak{g}}(\mathfrak{g}^e) = \dim \mathfrak{g}^e + \dim \mathfrak{z}(\mathfrak{g}^e)$. *Furthermore:*

$$\mathfrak{n}_{\mathfrak{g}}(\mathfrak{g}^e) = \sum_{n \geqslant 0} \mathfrak{n}(\mathfrak{g}^e)_n.$$

Proof. (i) We have $\mathfrak{z}(\mathfrak{g}^e) = [e, \mathfrak{d}] \subset [e, \mathfrak{n}_{\mathfrak{g}}(\mathfrak{g}^e)] \subset \mathfrak{z}(\mathfrak{g}^e)$ by 35.3.5 (i). So part (i) follows.

(ii) Since $\mathfrak{g}^e \subset \mathfrak{n}_{\mathfrak{g}}(\mathfrak{g}^e)$, part (i) implies that:

$$\dim \mathfrak{n}_{\mathfrak{g}}(\mathfrak{g}^e) = \dim \mathfrak{z}(\mathfrak{g}^e) + \dim \left(\mathfrak{n}_{\mathfrak{g}}(\mathfrak{g}^e) \cap \mathfrak{g}^e\right) = \dim \mathfrak{z}(\mathfrak{g}^e) + \dim \mathfrak{g}^e.$$

So $\mathfrak{n}_{\mathfrak{g}}(\mathfrak{g}^e) = \mathfrak{g}^e \oplus \mathfrak{d}$ since $\dim \mathfrak{d} = \dim \mathfrak{z}(\mathfrak{g}^e)$ and $\mathfrak{d} \cap \mathfrak{g}^e = \{0\}$. The last part follows from 35.3.9. □

35.4.2 We have $\mathfrak{c}_{\mathfrak{g}}(\mathfrak{s}) = \mathfrak{g}^e \cap \mathfrak{g}^h = \mathfrak{g}^f \cap \mathfrak{g}^h = \mathfrak{g}^e \cap \mathfrak{g}^f$, or $\mathfrak{c}_{\mathfrak{g}}(\mathfrak{s}) = (\mathfrak{g}^e)_0 = (\mathfrak{g}^f)_0$. So we have by 35.3.9 and 35.4.1 that:

$$\mathfrak{n}_{\mathfrak{g}}(\mathfrak{g}^e)_0 = (\mathfrak{g}^e)_0 \oplus \mathfrak{d}_0 = \mathfrak{c}_{\mathfrak{g}}(\mathfrak{s}) \oplus \mathfrak{d}_0 = \mathfrak{c}_{\mathfrak{g}}(\mathfrak{s}) \oplus [f, \mathfrak{z}(\mathfrak{g}^e)_2].$$

Note also that $\mathfrak{n}_{\mathfrak{g}}(\mathfrak{g}^e)_0$ is a Lie subalgebra of $\mathfrak{g}$.

Proposition. *The vector space $\mathfrak{d}_0$ is the orthogonal of $c_{\mathfrak{g}}(\mathfrak{s})$ in $\mathfrak{n}_{\mathfrak{g}}(\mathfrak{g}^e)_0$ with respect to L, and it is an abelian Lie algebra. We have $[\mathfrak{d}_0, c_{\mathfrak{g}}(\mathfrak{s})] = \{0\}$, and so $\mathfrak{n}_{\mathfrak{g}}(\mathfrak{g}^e)_0$ is the product of the Lie algebras $\mathfrak{d}_0$ and $c_{\mathfrak{g}}(\mathfrak{s})$.*

Proof. 1) We deduce from $\mathfrak{c}_{\mathfrak{g}}(\mathfrak{s}) \subset \mathfrak{g}^f$ that:

$$L(\mathfrak{d}, \mathfrak{c}_{\mathfrak{g}}(\mathfrak{s})) = L([f, \mathfrak{z}(\mathfrak{g}^e)], \mathfrak{c}_{\mathfrak{g}}(\mathfrak{s})) = L(\mathfrak{z}(\mathfrak{g}^e), [f, \mathfrak{c}_{\mathfrak{g}}(\mathfrak{s})]) = \{0\}.$$

Thus $\mathfrak{d}_0$ is contained in the orthogonal of $\mathfrak{c}_{\mathfrak{g}}(\mathfrak{s})$. Since $L|_{\mathfrak{c}_{\mathfrak{g}}(\mathfrak{s}) \times \mathfrak{c}_{\mathfrak{g}}(\mathfrak{s})}$ is non-degenerate (20.5.13), the first part follows.

2) Let $u \in \mathfrak{z}(\mathfrak{g}^e)$, $v \in \mathfrak{c}_{\mathfrak{g}}(\mathfrak{s}) = \mathfrak{g}^e \cap \mathfrak{g}^f$. We have $[u, v] = [f, v] = 0$, and so:

$$[[f, u], v] = [[f, v], u] + [f, [u, v]] = 0.$$

It follows that $[\mathfrak{d}, \mathfrak{c}_{\mathfrak{g}}(\mathfrak{s})] = \{0\}$. In particular, $[\mathfrak{d}_0, \mathfrak{c}_{\mathfrak{g}}(\mathfrak{s})] = \{0\}$.

3) If $u, v \in \mathfrak{d}$, $w \in \mathfrak{c}_{\mathfrak{g}}(\mathfrak{s})$, then point 2 implies that:

$$L([u, v], w) = L(u, [v, w]) = 0.$$

Since $\mathfrak{n}_{\mathfrak{g}}(\mathfrak{g}^e)_0$ is a Lie subalgebra of $\mathfrak{g}$, it follows from point 1 that $[u, v] \in \mathfrak{d}_0$. Thus $\mathfrak{d}_0$ is a Lie subalgebra of $\mathfrak{g}$.

4) Let $w_1, w_2 \in \mathfrak{d}_0$. There exist $z_1, z_2 \in \mathfrak{z}(\mathfrak{g}^e)_2$ such that:

$$w_1 = [f, z_1] \ , \ w_2 = [f, z_2].$$

For $i = 1, 2$, we have:

$$[e, w_i] = [e, [f, z_i]] = [[e, f], z_i] + [f, [e, z_i]] = [[e, f], z_i] = [h, z_i] = 2z_i.$$

So:

$$[e, [w_1, w_2]] = [[e, w_1], w_2] + [w_1, [e, w_2]]$$
$$= 2[z_1, [f, z_2]] + 2[[f, z_1], z_2] = 2[f, [z_1, z_2]] = 0.$$

Thus $[w_1, w_2] \in \mathfrak{g}^e$, and point 3 implies that $[w_1, w_2] \in \mathfrak{d}_0 \cap \mathfrak{g}^e$. As the sum $\mathfrak{d} + \mathfrak{g}^e$ is direct, we have proved that $\mathfrak{d}_0$ is commutative. $\square$

35.5 A semisimple Lie subalgebra

35.5.1 Lemma. *Let $\mathfrak{a}$ be a reductive Lie subalgebra of $\mathfrak{g}$, $\mathfrak{c} = \mathfrak{c}_{\mathfrak{g}}(\mathfrak{a})$, $\mathfrak{c}^{\perp}$ the orthogonal of $\mathfrak{c}$ with respect to the Killing form of $\mathfrak{g}$ and $\mathfrak{t}$ the Lie subalgebra of $\mathfrak{g}$ generated by $\mathfrak{c}^{\perp}$. Then $\mathfrak{t}$ is an ideal of $\mathfrak{g}$.*

Proof. We have $[\mathfrak{c}^{\perp}, \mathfrak{t}] \subset \mathfrak{t}$. Also we verify easily that $[\mathfrak{c}, \mathfrak{c}^{\perp}] \subset \mathfrak{c}^{\perp}$, so $[\mathfrak{c}, \mathfrak{t}] \subset \mathfrak{t}$. Since $\mathfrak{g} = \mathfrak{c} \oplus \mathfrak{c}^{\perp}$ by 20.5.13, $\mathfrak{t}$ is an ideal of $\mathfrak{g}$. $\square$

35.5.2 Let us conserve the hypotheses of 35.3.7, and set $d = \dim \mathfrak{z}(\mathfrak{g}^e)_2$. Let $\theta = e^{\operatorname{ad} e} o e^{-\operatorname{ad} f} o e^{\operatorname{ad} e} \in G$. We saw in 32.2.2 that:

$$\theta(e) = -f \ , \ \theta(h) = -h \ , \ \theta(f) = -e.$$

Hence for $n \in \mathbb{Z}$:

$$\theta\big(\mathfrak{z}(\mathfrak{g}^e)_n\big) = \mathfrak{z}(\mathfrak{g}^f)_{-n}.$$

Let $\mathfrak{v}$ be the $\mathfrak{s}$-submodule of $\mathfrak{g}$ generated by $\mathfrak{z}(\mathfrak{g}^e)_2$. By the definition of θ, we see easily that $\mathfrak{v}$ is θ-stable, so it contains $\mathfrak{z}(\mathfrak{g}^f)_{-2}$. It contains also $\mathfrak{d}_0 = [f, \mathfrak{z}(\mathfrak{g}^e)_2]$, which is of dimension d, because $\operatorname{ad} f|_{\mathfrak{g}_i}$ is injective if $i > 0$.

Let $v_0 \in \mathfrak{z}(\mathfrak{g}^e)_2 \setminus \{0\}$. For $j \geqslant 0$, set $v_j = (\operatorname{ad} f)^j(v_0)$. By 30.5.1, $v_3 = 0$, and since $[f, v_0] \neq 0$, $\Bbbk v_0 + \Bbbk v_1 + \Bbbk v_2$ is a 3-dimensional simple $\mathfrak{s}$-submodule of $\mathfrak{g}$ (19.2). Thus $\mathfrak{v} = \mathfrak{v}_{-2} \oplus \mathfrak{v}_0 \oplus \mathfrak{v}_2$ where:

$$\mathfrak{v}_2 = \mathfrak{z}(\mathfrak{g}^e)_2 \ , \ \mathfrak{v}_0 = \mathfrak{d}_0 \ , \ \mathfrak{v}_{-2} = \mathfrak{z}(\mathfrak{g}^f)_{-2}.$$

Proposition. *The subspace $\mathfrak{v}$ is a Lie subalgebra of $\mathfrak{g}$. It is isomorphic to the product of d copies of $\mathfrak{sl}_2(\Bbbk)$, and $\mathfrak{d}_0$ is a Cartan subalgebra of $\mathfrak{v}$.*

Proof. First, let us prove that $\mathfrak{v}$ is a Lie subalgebra of $\mathfrak{g}$ and $\mathfrak{d}_0$ is a Cartan subalgebra of $\mathfrak{v}$.

We have $[\mathfrak{z}(\mathfrak{g}^e)_2, \mathfrak{z}(\mathfrak{g}^e)_2] = [\mathfrak{z}(\mathfrak{g}^f)_{-2}, \mathfrak{z}(\mathfrak{g}^f)_{-2}] = \{0\}$, and by 35.4.2, $\mathfrak{d}_0$ is an abelian Lie subalgebra of $\mathfrak{g}$. Further, 35.4.1 and 35.3.5 (ii) say that $[\mathfrak{d}_0, \mathfrak{z}(\mathfrak{g}^e)_2] \subset \mathfrak{z}(\mathfrak{g}^e)_2$.

Taking $r = 2$ in the computations of 32.2.7, we obtain that $\theta(x) = -x$ for all $x \in \mathfrak{d}_0$. It follows that $[\mathfrak{d}_0, \mathfrak{z}(\mathfrak{g}^f)_{-2}] \subset \mathfrak{z}(\mathfrak{g}^f)_{-2}$.

Since $\mathfrak{d}_0 \cap \mathfrak{c}_{\mathfrak{g}}(\mathfrak{s}) = \{0\}$, we have by the theory of representations of $\mathfrak{sl}_2(\Bbbk)$ (see 19.2) that:

(1) $$[e, \mathfrak{d}_0] = \mathfrak{z}(\mathfrak{g}^e)_2 \ , \ [e, \mathfrak{z}(\mathfrak{g}^f)_{-2}] = \mathfrak{d}_0.$$

We deduce therefore that:

$$[\mathfrak{z}(\mathfrak{g}^e)_2, \mathfrak{z}(\mathfrak{g}^f)_{-2}] = [[e, \mathfrak{d}_0], \mathfrak{z}(\mathfrak{g}^f)_{-2}]$$
$$= [e, [\mathfrak{d}_0, \mathfrak{z}(\mathfrak{g}^f)_{-2}]] + [\mathfrak{d}_0, [e, \mathfrak{z}(\mathfrak{g}^f)_{-2}]]$$
$$\subset [e, \mathfrak{z}(\mathfrak{g}^f)_{-2}] + [\mathfrak{d}_0, \mathfrak{d}_0] \subset \mathfrak{d}_0.$$

So we have proved that $\mathfrak{v}$ is a Lie subalgebra of $\mathfrak{g}$, and $\mathfrak{d}_0$ is a Cartan subalgebra of $\mathfrak{g}$ since it is abelian (35.4.2) and $\mathfrak{d}_0 = \mathfrak{n}_{\mathfrak{v}}(\mathfrak{d}_0)$ because $h \in \mathfrak{d}_0$, $[h, \mathfrak{z}(\mathfrak{g}^e)_2] = \mathfrak{z}(\mathfrak{g}^e)_2$, $[h, \mathfrak{z}(\mathfrak{g}^f)_{-2}] = \mathfrak{z}(\mathfrak{g}^f)_{-2}$.

Next, let us prove that $\mathfrak{v}$ is semisimple.

Applying θ to the first equality in (1), we obtain that $[f, \mathfrak{d}_0] = \mathfrak{z}(\mathfrak{g}^f)_2$. It follows therefore again by (1) that $[\mathfrak{v}, \mathfrak{v}] = \mathfrak{v}$. Thus the radical $\mathfrak{r}$ of $\mathfrak{v}$ is exactly the nilpotent radical of $\mathfrak{v}$, and the elements of $\mathfrak{r}$ are all nilpotent in $\mathfrak{g}$.

Let us write $\mathfrak{v} = \mathfrak{t} \oplus \mathfrak{r}$ where $\mathfrak{t}$ is a Levi subalgebra of $\mathfrak{v}$ containing $\mathfrak{s}$ (29.5.3). So by 29.8.4, there exists a parabolic subalgebra $\mathfrak{p}$ of $\mathfrak{g}$ such that its nilpotent radical contains $\mathfrak{r}$, and it has a Levi factor $\mathfrak{l}$ which contains $\mathfrak{t}$.

If $\mathfrak{r}$ is non-zero, then since $\mathfrak{v}$ is the sum of simple $\mathfrak{s}$-submodules of dimension 3, $\mathfrak{r}$ is also the sum of simple $\mathfrak{s}$-submodules of dimension 3. Thus $\mathfrak{r} \cap \mathfrak{z}(\mathfrak{g}^e)_2$ contains a non-zero element w. But if $\mathfrak{c}$ denotes the centre of $\mathfrak{l}$, then $\mathfrak{c} = \mathfrak{c}_{\mathfrak{g}}(\mathfrak{l})$ and $\mathfrak{l} = \mathfrak{c}_{\mathfrak{g}}(\mathfrak{c})$ (29.5.7). Thus there exists $x \in \mathfrak{g}$ such that $[x, \mathfrak{t}] = \{0\}$, hence $x \in \mathfrak{g}^e$, and $[x, w] \neq 0$. This is absurd since $w \in \mathfrak{z}(\mathfrak{g}^e)$. So $\mathfrak{r} = \{0\}$ and $\mathfrak{v}$ is semisimple.

Finally, as $\mathfrak{d}_0$ is a Cartan subalgebra of $\mathfrak{v}$ and $d = \dim \mathfrak{d}_0$, a base of a root system of $\mathfrak{v}$ has d elements. Since $\dim \mathfrak{v} = 3d$, it is obvious that $\mathfrak{v}$ is a product of d copies of $\mathfrak{sl}_2(\Bbbk)$. $\square$

35.5.3 Theorem. *Let e be a non-zero nilpotent element of a semisimple Lie algebra $\mathfrak{g}$ and $d = \dim \mathfrak{z}(\mathfrak{g}^e)_2$ be as defined in 35.5.2. Then d is the number of simple ideals of $\mathfrak{g}$ on which the projection of e is non-zero.*

Proof. We shall often use the results on the representations of semisimple Lie algebras (see 19.2 and chapter 30). Also, it is easy to see that to prove the theorem, it suffices to verify that if $\mathfrak{g}$ is simple, then $d = 1$. We shall use the notations of 35.5.2.

Let $\mathfrak{a}_1, \ldots, \mathfrak{a}_d$ be the simple components of $\mathfrak{v}$; they are isomorphic to $\mathfrak{sl}_2(\Bbbk)$ (35.5.2). For $1 \leqslant i \leqslant d$, let $p_i : \mathfrak{v} \to \mathfrak{a}_i$ be the projector associated to the decomposition $\mathfrak{v} = \mathfrak{a}_1 \oplus \cdots \oplus \mathfrak{a}_d$; the p_i's are Lie algebra homomorphisms. Set:

$$e_i = p_i(e) \ , \ h_i = p_i(h) \ , \ f_i = p_i(f).$$

As the Lie algebra $\mathfrak{s}$ is simple, $p_i(\mathfrak{s})$ is either simple or is equal to $\{0\}$. In the latter, we would have $[\mathfrak{s}, \mathfrak{a}_i] = \{0\}$, hence $\mathfrak{a}_i \subset \mathfrak{v}_0 = \mathfrak{d}_0$ which is absurd since $\mathfrak{d}_0$ is abelian (35.5.2). So $p_i(\mathfrak{s})$ is simple.

Note that $[h_i, e_i] = 2e_i$, so $[h, e_i] = 2e_i$. Thus $e_i \in \mathfrak{z}(\mathfrak{g}^e)_2$. Similarly, $f_i \in \mathfrak{z}(\mathfrak{g}^f)_{-2}$, and $h_i \in \mathfrak{d}_0$.

Let us consider the bases of the root systems of $(\mathfrak{s}, \Bbbk h)$ and $(\mathfrak{v}, \mathfrak{d}_0)$ associated to the Borel subalgebras $\Bbbk h + \Bbbk e$ and $\mathfrak{d}_0 + \mathfrak{z}(\mathfrak{g}^e)_2$.

Let V be a simple $\mathfrak{v}$-submodule of $\mathfrak{g}$ and u a primitive vector of V (u is unique up to multiplication by a non-zero scalar). For $1 \leqslant i \leqslant d$, we have $[e_i, u] = 0$ and $[h_i, u] = k_i u$, with $k_i \in \mathbb{N}$. Set $F_i = \operatorname{ad} f_i|_V$. We verify easily that the set of vectors $F_1^{r_1} \circ \cdots \circ F_d^{r_d}(u)$, with $0 \leqslant r_i \leqslant k_i$ for all i, is a basis of V. We deduce immediately that $\mathbb{k}u$ is the set of $v \in V$ such that $[e_i, v] = 0$ for all i, and $\dim V = (k_1 + 1) \cdots (k_d + 1)$.

Now consider V as a $\mathfrak{s}$-module, and let $w \in V$ be a primitive vector for $\mathfrak{s}$. Then $[e, w] = 0$, that is $w \in \mathfrak{g}^e$. Since $e_i \in \mathfrak{z}(\mathfrak{g}^e)$, we have $[e_i, w] = 0$. Thus w is a scalar multiple of u. It follows that V is a simple $\mathfrak{s}$-module. Now $[h, u] = (k_1 + \cdots + k_d)u$, so the dimension of V is $k_1 + \cdots + k_d + 1$. Hence:

$$k_1 + \cdots + k_d + 1 = (k_1 + 1) \cdots (k_d + 1).$$

Consequently, if $\dim V > 1$, all except one of the k_i's are equal to zero, and if $k_i \neq 0$, then V is a simple $\mathfrak{a}_i$-module such that $[\mathfrak{a}_j, V] = \{0\}$ for $j \neq i$.

The $\mathfrak{v}$-module $\mathfrak{g}$ being semisimple, the discussion above implies that

$$(2) \qquad \mathfrak{g} = \mathfrak{c}_\mathfrak{g}(\mathfrak{v}) \oplus \mathfrak{r}_1 \oplus \cdots \oplus \mathfrak{r}_d$$

where, for $1 \leqslant i \leqslant d$, $\mathfrak{r}_i$ is a $\mathfrak{a}_i$-module, which is the sum of simple $\mathfrak{a}_i$-modules of dimensions not equal to 1 (so $\mathfrak{c}_{\mathfrak{r}_i}(\mathfrak{a}_i) = \{0\}$), and $[\mathfrak{a}_j, \mathfrak{r}_i] = \{0\}$ for $j \neq i$. We have $\mathfrak{r}_i \neq \{0\}$ since $\mathfrak{a}_i = p_i(\mathfrak{s}) \subset \mathfrak{r}_i$. Set:

$$\mathfrak{b}_i = \bigoplus_{j \neq i} \mathfrak{a}_j , \quad \mathfrak{m}_i = \bigoplus_{j \neq i} \mathfrak{r}_j.$$

Thus $\mathfrak{c}_\mathfrak{g}(\mathfrak{a}_i) = \mathfrak{c}_\mathfrak{g}(\mathfrak{v}) \oplus \mathfrak{m}_i$.

Let $i, j \in \{1, \ldots, d\}$ be distinct. Since $[\mathfrak{a}_i, \mathfrak{r}_i] = \mathfrak{r}_i$ and $[\mathfrak{a}_i, \mathfrak{r}_j] = \{0\}$, we have:

$$L(\mathfrak{r}_i, \mathfrak{r}_j) = L([\mathfrak{a}_i, \mathfrak{r}_i], \mathfrak{r}_j) = L(\mathfrak{r}_i, [\mathfrak{a}_i, \mathfrak{r}_j]) = \{0\},$$
$$L(\mathfrak{r}_i, \mathfrak{c}_\mathfrak{g}(\mathfrak{v})) = L(\mathfrak{r}_i, [\mathfrak{a}_i, \mathfrak{c}_\mathfrak{g}(\mathfrak{v})]) = \{0\}.$$

The factors in the decomposition (2) are therefore pairwise orthogonal. As $\mathfrak{r}_i = [\mathfrak{a}_i, \mathfrak{g}]$ by (2), we obtain that:

$$[\mathfrak{r}_i, \mathfrak{r}_j] \subset [[\mathfrak{a}_i, [\mathfrak{a}_j, \mathfrak{g}]], \mathfrak{g}] + [\mathfrak{a}_i, [\mathfrak{g}, [\mathfrak{a}_j, \mathfrak{g}]]]$$
$$\subset [[\mathfrak{a}_i, \mathfrak{r}_j], \mathfrak{g}] + [\mathfrak{a}_i, [\mathfrak{g}, [\mathfrak{a}_j, \mathfrak{g}]]] \subset \mathfrak{r}_i.$$

Similarly, $[\mathfrak{r}_i, \mathfrak{r}_j] \subset \mathfrak{r}_j$. Hence $[\mathfrak{r}_i, \mathfrak{r}_j] = \{0\}$. We deduce therefore that:

$$L([\mathfrak{r}_i, \mathfrak{c}_\mathfrak{g}(\mathfrak{v}) + \mathfrak{r}_i], \mathfrak{m}_i) = L\left(\mathfrak{c}_\mathfrak{g}(\mathfrak{v}) + \mathfrak{r}_i, \sum_{j \neq i}[\mathfrak{r}_i, \mathfrak{r}_j]\right) = \{0\}.$$

Thus $[\mathfrak{r}_i, \mathfrak{c}_\mathfrak{g}(\mathfrak{v}) + \mathfrak{r}_i] \subset \mathfrak{m}_i^\perp = \mathfrak{c}_\mathfrak{g}(\mathfrak{v}) + \mathfrak{r}_i$.

The Lie algebra $\mathfrak{b}_i$ is reductive in $\mathfrak{g}$, and:

$$\mathfrak{c}_\mathfrak{g}(\mathfrak{b}_i) = \mathfrak{c}_\mathfrak{g}(\mathfrak{v}) \oplus \mathfrak{r}_i , \quad \mathfrak{c}_\mathfrak{g}(\mathfrak{b}_i)^\perp = \mathfrak{m}_i.$$

Let $\mathfrak{t}$ be the Lie subalgebra of $\mathfrak{g}$ generated by $\mathfrak{c}_\mathfrak{g}(\mathfrak{b}_i)^\perp$. Then the inclusion $\mathfrak{r}_i \subset \mathfrak{c}_\mathfrak{g}(\mathfrak{v}) + \mathfrak{r}_i$ implies that:

$$\mathfrak{m}_i \subset \mathfrak{t} \subset \mathfrak{r}_i^{\perp} = \mathfrak{c}_\mathfrak{g}(\mathfrak{v}) \oplus \mathfrak{m}_i = \mathfrak{c}_\mathfrak{g}(\mathfrak{a}_i).$$

If $d > 1$, then $\mathfrak{t} \neq \{0\}$ and $\mathfrak{t} \neq \mathfrak{g}$. By 35.5.1, $\mathfrak{t}$ is an ideal of $\mathfrak{g}$. But this is impossible if $\mathfrak{g}$ is simple. So we are done. $\square$

35.6 Centralizers and regular elements

35.6.1 Lemma. *Let τ be an automorphism of finite order of a non-zero semisimple Lie algebra $\mathfrak{a}$. Then there exists $x \in \mathfrak{a} \setminus \{0\}$ such that $\tau(x) = x$.*

Proof. Let $r > 1$ be the order of τ. We may assume that $\mathfrak{a}$ does not contain a non-zero semisimple τ-stable Lie subalgebra other than $\mathfrak{a}$ itself. Denote $S = \{\lambda \in \Bbbk; \lambda^r = 1\}$ and for $\lambda \in S$, let $\mathfrak{a}(\lambda)$ be the eigenspace of τ associated to the eigenvalue λ. Then $\mathfrak{a}$ is the direct sum of the vector subspaces $\mathfrak{a}(\lambda)$, for $\lambda \in P = \{\lambda \in S; \mathfrak{a}(\lambda) \neq \{0\}\}$.

We verify easily that $[\mathfrak{a}(\lambda), \mathfrak{a}(\mu)] \subset \mathfrak{a}(\lambda\mu)$, and the restriction of the Killing form L of $\mathfrak{a}$ on $\mathfrak{a}(\lambda) \times \mathfrak{a}(\mu)$ is zero if $\lambda\mu \neq 1$, while $L|_{\mathfrak{a}(\lambda) \times \mathfrak{a}(\lambda^{-1})}$ is non-degenerate.

Let us suppose that $\mathfrak{a}(1) = \{0\}$. Then $[\mathfrak{a}(\lambda), \mathfrak{a}(\lambda^{-1})] = \{0\}$ for all $\lambda \in S$.

1) Let $\lambda \in P$. It is clear that $\mathfrak{a}(\lambda)$ contains the semisimple and nilpotent components of its elements. Let x be a nilpotent element of $\mathfrak{a}(\lambda)$. Since $[\mathfrak{a}(\lambda), \mathfrak{a}(\lambda^{-1})] = \{0\}$, $(\mathrm{ad}\, x) \circ (\mathrm{ad}\, y)$ is nilpotent for all $y \in \mathfrak{a}(\lambda^{-1})$, so $L(x, \mathfrak{a}(\lambda^{-1})) = \{0\}$. Since $L|_{\mathfrak{a}(\lambda) \times \mathfrak{a}(\lambda^{-1})}$ is non-degenerate, $x = 0$.

Consequently any element $x \in \mathfrak{a}(\lambda)$ is semisimple, and so $\mathfrak{a}^x$ is a τ-stable reductive Lie subalgebra and it has the same rank as $\mathfrak{a}$. Our assumption on $\mathfrak{a}$ implies that the τ-stable semisimple Lie algebra is equal to $\{0\}$. Thus $\mathfrak{a}^x$ is a Cartan subalgebra of $\mathfrak{a}$, and so, x is regular.

We have therefore shown that any non-zero element of $\mathfrak{a}(\lambda)$ is generic in $\mathfrak{a}$.

2) Let $\lambda \in P$ and:

$$\mathfrak{u}(\lambda) = \sum_{m \in \mathbb{Z}} \mathfrak{a}(\lambda^m).$$

Let $x \in \mathfrak{a}(\lambda) \setminus \{0\}$. If $m \in \mathbb{Z}$, then there exists $p \in \mathbb{N}$ such that $\lambda^{m+p} = 1$. Thus $\mathrm{ad}\, x|_{\mathfrak{u}(\lambda)}$ is nilpotent, so it is zero because x is generic. Thus $\mathfrak{u}(\lambda)$ is contained in the Cartan subalgebra $\mathfrak{a}^x$ of $\mathfrak{a}$.

3) Let λ be a primitive r-th root of unity. We have:

$$\mathfrak{a} = \bigoplus_{i=0}^{r-1} \mathfrak{a}(\lambda^i) = \bigoplus_{i=1}^{r-1} \mathfrak{a}(\lambda^i).$$

If $\lambda \in P$, then point 2 implies that $\mathfrak{a}$ is abelian, which is absurd. So $\lambda \notin P$.

Let p be the smallest positive integer such that $\mathfrak{a}(\lambda^p) \neq \{0\}$. By using the earlier remarks on the Killing form, we have:

$$\mathfrak{a}(1) = \cdots = \mathfrak{a}(\lambda^{p-1}) = \mathfrak{a}(\lambda^{r-p+1}) = \cdots = \mathfrak{a}(\lambda^r) = \{0\} \ , \ \mathfrak{a}(\lambda^{r-p}) \neq \{0\}.$$

Let us fix a non-zero element x in $\mathfrak{a}(\lambda^{r-p}) = \mathfrak{a}(\lambda^{-p})$, and q an integer such that $p \leqslant q \leqslant r-1$. If q is a multiple of p, then $\mathrm{ad}\, x|_{\mathfrak{a}(\lambda^q)} = 0$ by point 2. Otherwise,

$q = np + s$, with $n, s \in \mathbb{N}^*$ and $1 \leqslant s \leqslant p - 1$. Then $q + n(r - p) = s + nr$, which proves that $(\operatorname{ad} x)^n|_{\mathfrak{a}(\lambda^q)} = 0$. We have therefore shown that $\operatorname{ad} x$ is nilpotent. It follows from point 1 that $x = 0$. Contradiction. $\square$

35.6.2 Lemma. *Let $\mathfrak{a}$ be a non-solvable Lie algebra and τ an automorphism of $\mathfrak{a}$ of finite order. There exists $x \in \mathfrak{a} \setminus \{0\}$ such that $\tau(x) = x$.*

Proof. By 35.6.1, we may assume that $\mathfrak{a}$ is not semisimple. Let $\mathfrak{r}$ be the radical of $\mathfrak{a}$. It is τ-stable and there exists a τ-stable subspace $\mathfrak{f}$ of $\mathfrak{a}$ such that $\mathfrak{a} = \mathfrak{f} \oplus \mathfrak{r}$. Let $\bar{\tau}$ be the automorphism of $\mathfrak{a}/\mathfrak{r}$ induced by τ. By 35.6.1, there exists $x \in \mathfrak{f}$ such that $\bar{\tau}(x + \mathfrak{r}) = x + \mathfrak{r}$. This implies that $\tau(x) \in x + \mathfrak{r}$, and hence $\tau(x) = x$ since $\mathfrak{f}$ is τ-stable. $\square$

35.6.3 Let us use the notations of 35.3.7 and let us assume that e is a distinguished element in $\mathfrak{g}$. So $\mathfrak{g}_{2i+1} = \{0\}$ for all $i \in \mathbb{Z}$ (35.1.6). Let m be the largest integer such that $\mathfrak{g}_{2m} \neq \{0\}$; it is also the largest integer such that $\mathfrak{g}_{-2m} \neq \{0\}$;

Lemma. *Let $y \in \mathfrak{g}_{-2m}$ and $x = e + y$.*
(i) *The element x is nilpotent if and only if $y = 0$.*
(ii) *We have $\mathfrak{g}^x \cap \mathfrak{g}_0 = \{0\}$.*

Proof. (i) Note that x is non-zero. If x is nilpotent, then there exists $t \in \mathfrak{g}$ such that $[t, x] = 2x$ (32.1.5). Let us write $t = \sum t_n$ with $t_n \in \mathfrak{g}_n$ for all $n \in \mathbb{Z}$. Then $[t_0, e] = 2e$ and $[t_0, y] = 2y$. So $t_0 - h \in \mathfrak{g}^e \cap \mathfrak{g}_0$. Hence $t_0 = h$ (35.1.5). Consequently $2y = -2my$, and so $y = 0$. The converse is clear.
    (ii) If $z \in \mathfrak{g}^x \cap \mathfrak{g}_0$, then $[z, e] = [z, y] = 0$. Therefore $z \in \mathfrak{g}^e \cap \mathfrak{g}_0 = \{0\}$ by 35.1.5. $\square$

35.6.4 Let us conserve the hypotheses of 35.6.3. Let λ be a primitive $(2m + 2)$-th root of unity. Denote by τ the automorphism of $\mathfrak{g}$ such that $\tau(x) = \lambda^{2n} x$ for $x \in \mathfrak{g}_{2n}$, $-m \leqslant n \leqslant m$. We have $\tau \in \operatorname{Aut}_e \mathfrak{g}$ (32.2.7), and the eigenspaces of τ are $\mathfrak{a}_0, \ldots, \mathfrak{a}_m$ where

$$\mathfrak{a}_i = \mathfrak{g}_{2i} + \mathfrak{g}_{2i-2m-2}$$

for $0 \leqslant i \leqslant m$. In particular, $\mathfrak{g}_0$ is the space of fixed points of τ.
    Let $y \in \mathfrak{g}_{-2m}$ and $x = e + y$. Then $\tau(x) = \lambda^2 x$, so $\mathfrak{g}^x$ is τ-stable.

Lemma. *The Lie algebra $\mathfrak{g}^x$ is solvable.*

Proof. By 35.6.3 (ii), 0 is the only fixed point of τ in $\mathfrak{g}^x$. So the result follows from 35.6.2. $\square$

35.6.5 Lemma. *Let us conserve the hypotheses of 35.6.3. The following conditions are equivalent:*
(i) *$\mathfrak{g}^x$ is a Cartan subalgebra of $\mathfrak{g}$.*

(ii) x *is a generic element of* $\mathfrak{g}$.

(iii) x *is semisimple.*

There exists $y \in \mathfrak{g}_{-2m}$ *such that* x *verifies these conditions if and only if* $\mathfrak{a}_1$ *contains a generic element of* $\mathfrak{g}$.

Proof. The implications (i) $\Rightarrow$ (ii) $\Rightarrow$ (iii) are obvious. So let x be semisimple. Then $\mathfrak{g}^x$ is a reductive subalgebra whose rank is the rank of $\mathfrak{g}$. As $\mathfrak{g}^x$ is solvable by 35.6.4, it is a Cartan subalgebra of $\mathfrak{g}$.

Finally, if x verifies these conditions, it is a generic element of $\mathfrak{g}$ contained in $\mathfrak{a}_1$. Conversely, let $\mathfrak{u}$ be the set of generic elements of $\mathfrak{g}$ in $\mathfrak{a}_1$ and let us suppose that $\mathfrak{u} \neq \emptyset$. Then $\mathfrak{u}$ is a dense open subset of $\mathfrak{a}_1$. Let G_h be the stabilizer of h in G, then by 32.2.9, $G_h(e) + \mathfrak{g}_{-2m}$ is a dense open subset of $\mathfrak{a}_1$. So there exist $\alpha \in G_h$ and $z \in \mathfrak{g}_{-2m}$ such that $\alpha(e) + z$ is generic. Hence $e + \alpha^{-1}(z)$ is generic and $\alpha^{-1}(z) \in \mathfrak{g}_{-2m}$. $\square$

35.6.6 Lemma. *Let us conserve the hypotheses of 35.6.3. Suppose that*

$$\dim \mathfrak{g}_2 - \dim \mathfrak{g}_4 = 1.$$

For all $y \in \mathfrak{g}_{-2m} \setminus \{0\}$, $x = e + y$ *is a semisimple element of* $\mathfrak{g}$.

Proof. Let x_s and x_n be the semisimple and nilpotent component of x. It is clear that $x_s, x_n \in \mathfrak{a}_1$. Let us write $x_n = u + v$ with $u \in \mathfrak{g}_2$ and $v \in \mathfrak{g}_{-2m}$. Since $[x_n, x] = 0$, we have $[e, u] = 0$. Since the map $\operatorname{ad} e : \mathfrak{g}_2 \to \mathfrak{g}_4$ is surjective with kernel $\Bbbk e$, we have $u \in \Bbbk e$. If $u = 0$, then $[x_n, x] = 0$ implies that $v \in \mathfrak{g}^e \cap \mathfrak{g}_{-2m} = \{0\}$. Thus $x_n = 0$ and x is semisimple. If $u = \lambda e$ with $\lambda \neq 0$, then $v = 0$ (35.6.3 (i)). The condition $[x_n, x] = 0$ implies in this case that $y \in \mathfrak{g}^e \cap \mathfrak{g}^{-2m} = \{0\}$. $\square$

35.6.7 Lemma. *Let us assume that* $\mathfrak{g}$ *is simple and suppose that* e *is a non-zero nilpotent element in* $\mathfrak{g}$ *such that* $\mathfrak{g}^e$ *is a commutative Lie algebra. Then* e *is regular.*

Proof. By 35.1.3 (v), e is distinguished. For $i \in \mathbb{N}$, we have:

$$\dim \mathfrak{g}_{2i} - \dim \mathfrak{g}_{2(i+1)} = \dim(\mathfrak{g}_{2i} \cap \mathfrak{g}^e) = \dim \mathfrak{z}(\mathfrak{g}^e)_{2i}.$$

Since $\mathfrak{g}$ is simple, it follows from 35.5.3 that:

$$\dim \mathfrak{g}_2 - \dim \mathfrak{g}_4 = 1.$$

Let m be as in 35.6.3. Then 35.6.5 and 35.6.6 imply that there exists $y \in \mathfrak{g}_{-2m}$ such that $x = e + y$ is a generic element of $\mathfrak{g}$.

1) Let us first show that for all $u \in \mathfrak{g}^e$, there is a unique $v \in \mathfrak{g}$ verifying:

$$v \in \bigoplus_{i<0} \mathfrak{g}_i \ , \ u + v \in \mathfrak{g}^x.$$

If v_1, v_2 verify these conditions, then $v_1 - v_2$ is a nilpotent element of the Cartan subalgebra $\mathfrak{g}^x$ of $\mathfrak{g}$. So $v_1 = v_2$.

Now consider the following equation in v:

$$0 = [u + v, e + y] = [u, y] + [v, e].$$

Note that $[y, v] = 0$ by the hypotheses on y and v. Since $\mathfrak{g}^e$ is abelian, $[u, y] \in (\mathfrak{g}^e)^\perp = [\mathfrak{g}, e]$ (35.3.1). So there exists $v \in \mathfrak{g}$ such that $[v, e] + [u, y] = 0$. We have $[u, y] \in \bigoplus_{i \leqslant 0} \mathfrak{g}_i$ and $e \in \mathfrak{g}_2$. So we may assume that $v \in \bigoplus_{i < 0} \mathfrak{g}_i$.

2) It follows from point 1 that the map $\mathfrak{g}^e \to \mathfrak{g}^x$, $u \mapsto u + v$ is linear and injective, so $\dim \mathfrak{g}^e \leqslant \dim \mathfrak{g}^x$. Since x is regular, so is e. $\square$

35.6.8 Let $\mathfrak{g}$ be a reductive Lie algebra of rank r, $\mathfrak{s} = [\mathfrak{g}, \mathfrak{g}]$ and $\mathfrak{c} = \mathfrak{z}(\mathfrak{g})$. If $x \in \mathfrak{s}$ and $y \in \mathfrak{c}$, then $\mathfrak{g}^{x+y} = \mathfrak{c} + \mathfrak{s}^x$. This implies that $\dim \mathfrak{g}^{x+y} \geqslant r$, and that $\dim \mathfrak{g}^{x+y} = r$ if and only if x is regular in $\mathfrak{s}$. If this is the case, we shall say that $x + y$ is a *regular* element of $\mathfrak{g}$.

Theorem. *Let $\mathfrak{g}$ be a reductive Lie algebra and $x \in \mathfrak{g}$. The following conditions are equivalent:*

(i) *x is a regular element of $\mathfrak{g}$.*

(ii) *The Lie algebra $\mathfrak{g}^x$ is abelian.*

Proof. (i) $\Rightarrow$ (ii) This follows from 19.7.5 and 29.3.1.

(ii) $\Rightarrow$ (i) We shall prove this by induction on $\dim \mathfrak{g}$. The case $\mathfrak{g} = \{0\}$ is obvious. We may assume that $\mathfrak{g}$ is semisimple.

If x is semisimple, then $\mathfrak{g}^x$ is a reductive subalgebra whose rank is equal to the rank of $\mathfrak{g}$. So if $\mathfrak{g}^x$ is abelian, then it is a Cartan subalgebra of $\mathfrak{g}$, and x is regular.

If x is nilpotent, then the result follows easily from 35.6.7.

Now suppose that x is neither semisimple nor nilpotent. Let x_s and x_n be the semisimple and nilpotent component of x. Then $\mathfrak{g}^{x_s}$ is a reductive subalgebra whose rank is equal to the rank of $\mathfrak{g}$ and it contains x_n. Now $(\mathfrak{g}^{x_s})^{x_n} = \mathfrak{g}^x$ is abelian. By the induction hypothesis, $\dim(\mathfrak{g}^{x_s})^{x_n}$ is equal to the rank of $\mathfrak{g}^{x_s}$, and so it is equal to the rank of $\mathfrak{g}$. Consequently, the element x is regular. $\square$

References and comments

- [2], [3], [17], [18], [20], [46], [63], [76], [80].

The proof of the Bala-Carter Theorem (35.1.6) given here is due to Jantzen, and it is taken from [46].

A great number of results presented in sections 35.3 to 35.6 are proved in [17] and [18]. Certain proofs given here are due to Panyushev ([63]), for example 35.5.3.

The proof of 35.6.8 is also due to Panyushev ([63]), and the proof of 35.6.6 is due to Springer ([76]).

σ-root systems

In this chapter, we study properties of certain root systems related to involutive automorphisms. We shall use these properties in Chapters 37 and 38.

Let k be a commutative field of characteristic zero, V be a k-vector space of dimension l with $l \neq 0$ and R be a *reduced* root system in V. If $v \in V$ and $f \in V^*$, we denote $f(v)$ by $\langle v, f \rangle$. We shall use the notations of chapter 18.

36.1 Definition

36.1.1 Let us fix an involutive automorphism σ of V such that $\sigma \neq \pm \mathrm{id}_V$. Set:

$$V' = \{x \in V; \sigma(x) = x\} \ , \ V'' = \{x \in V; \sigma(x) = -x\},$$
$$R' = R \cap V' = \{\alpha \in R; \sigma(\alpha) = \alpha\} \ , \ R'' = \{\alpha \in R; \sigma(\alpha) \neq \alpha\}.$$

By 18.2.5, R' is a root system in the subspace of V' spanned by R'. Moreover, if B' is a base of R', then there exists a base B of R containing B', and R' is the set of elements of R which are linear combinations of elements of B' (18.7.9).

Let us identify V'^* (resp. V''^*) with the set of elements of V^* which vanishes on V'' (resp. V'). Thus $V^* = V'^* \oplus V''^*$. If θ is the transpose of σ, then:

$$V'^* = \{\varphi \in V^*; \theta(\varphi) = \varphi\} \ , \ V''^* = \{\varphi \in V^*; \theta(\varphi) = -\varphi\}.$$

For $x \in V$ and $\varphi \in V^*$, we shall write

$$x = x' + x'' \ , \ \varphi = \varphi' + \varphi'',$$

with $x' \in V'$, $x'' \in V''$, $\varphi' \in V'^*$, $\varphi'' \in V''^*$.

36.1.2 Definition. (i) *We say that R is a σ-root system if $\sigma(R) = R$.*
(ii) *A σ-root system R is called* normal *if for all $\alpha \in R$, we have:*

$$\alpha + \sigma(\alpha) \notin R.$$

36.1.3 In the rest of this section, we assume that R is a σ-root system in V.

Let $\alpha, \beta \in R$, $x \in V$. Then:

$$s_\alpha\big(\sigma(x)\big) = \sigma(x) - \langle \sigma(x), \alpha^\vee \rangle \alpha = \sigma\big(x - \langle x, \theta(\alpha^\vee) \rangle \sigma(\alpha)\big),$$
$$\langle \sigma(\alpha), \theta(\alpha^\vee) \rangle = \langle \alpha, \alpha^\vee \rangle = 2 \,, \ \langle \beta, \theta(\alpha^\vee) \rangle = \langle \sigma(\beta), \alpha^\vee \rangle \in \mathbb{Z},$$
$$\beta - \langle \beta, \theta(\alpha^\vee) \rangle \sigma(\alpha) = \sigma\big(\sigma(\beta) - \langle \sigma(\beta), \alpha^\vee \rangle \alpha\big) \in R.$$

Thus:

$$\sigma \circ s_\alpha \circ \sigma = s_{\sigma(\alpha)} \,, \ \sigma(\alpha)^\vee = \theta(\alpha^\vee).$$

36.1.4 Let B be a base of R and R_+ (resp. R_-) the corresponding set of positive (resp. negative) roots. Set:

$$R''_+ = R'' \cap R_+ \,, \ R''_- = R'' \cap R_-.$$

Lemma. *In the notations above, we can choose B so that the following conditions are verified:*
   (i) $\sigma(R''_+) = R''_-$.
   (ii) *If $\alpha \in R''_+$ and $\beta \in R$ verify $\alpha - \beta \in V'$, then $\beta \in R''_+$.*
   (iii) $(R''_+ + R''_+) \cap R \subset R''_+$.

Proof. Denote by $V_{\mathbb{Q}}$ the $\mathbb{Q}$-vector subspace of V spanned by R; we know that $V = V_{\mathbb{Q}} \otimes_{\mathbb{Q}} k$, and we have $\sigma(V_{\mathbb{Q}}) = V_{\mathbb{Q}}$ (because R is a σ-root system). Set $V'_{\mathbb{Q}} = V' \cap V_{\mathbb{Q}}$ et $V''_{\mathbb{Q}} = V'' \cap V_{\mathbb{Q}}$.
   1) We claim that $V' = V'_{\mathbb{Q}} \otimes_{\mathbb{Q}} k$ et $V'' = V''_{\mathbb{Q}} \otimes_{\mathbb{Q}} k$.
   Let $(\alpha_1, \ldots, \alpha_l)$ be a base of R. If $x \in V$, then there exist scalars $\lambda_1, \ldots, \lambda_l$ such that $x = \lambda_1 \alpha_1 + \cdots + \lambda_l \alpha_l$. Thus

$$x = \sum_{i=1}^{l} \frac{\lambda_i}{2} \big(\alpha_i + \sigma(\alpha_i)\big) + \sum_{i=1}^{l} \frac{\lambda_i}{2} \big(\alpha_i - \sigma(\alpha_i)\big).$$

So the claim follows easily.
   2) Let $(e_1, \ldots, e_p)$ be a basis of $V''_{\mathbb{Q}}$ and $(e_{p+1}, \ldots, e_l)$ a basis of $V'_{\mathbb{Q}}$. Endow $V_{\mathbb{Q}}$ with the lexicographic order defined by the basis $(e_1, \ldots, e_l)$ of $V_{\mathbb{Q}}$. Let R_+ (resp. R_-) be the corresponding set of positive (resp. negative) roots, and $R''_+ = V'' \cap R_+$, $R''_- = V'' \cap R_-$.
   Let $\alpha = \lambda_r e_r + \cdots + \lambda_l e_l \in R''_+$, with $\lambda_r \neq 0$. So we have $\lambda_r > 0$ and $r \leqslant p$. It follows that:

$$\sigma(\alpha) = -\lambda_r e_r - \cdots - \lambda_p e_p + \lambda_{p+1} e_{p+1} + \cdots + \lambda_l e_l.$$

We deduce from $\lambda_r > 0$ that $\sigma(\alpha) \in R''_-$. So we have (i).
   If $\beta = \mu_s e_s + \cdots + \mu_l e_l \in R$, with $s \leqslant l$ and $\mu_s \neq 0$, then $\alpha - \beta \in V'$ implies that $s = r$ and $\mu_s = \lambda_r$. Thus $\beta \in R''_+$, and we have (ii). Finally, (iii) follows from our choice of basis. $\square$

36.1.5 Let us conserve the notations of 36.1.4. Let $x \in V_{\mathbb{Q}}$. There exist $\lambda_1, \ldots, \lambda_l \in \mathbb{Q}$ such that $x = \lambda_1 \alpha_1 + \cdots + \lambda_l \alpha_l$. So:

$$x' = \frac{x + \sigma(x)}{2} = \sum_{i=1}^{l} \frac{\lambda_i}{2}(\alpha_i + \sigma(\alpha_i)) \ , \ x'' = \frac{x - \sigma(x)}{2} = \sum_{i=1}^{l} \frac{\lambda_i}{2}(\alpha_i - \sigma(\alpha_i)).$$

Let $S = \{\alpha''; \alpha \in R''\}$. Then it follows that $V_{\mathbb{Q}}''$ is the $\mathbb{Q}$-vector subspace of V spanned by S.

36.2 Restricted root systems

36.2.1 Theorem. *Let R be a normal σ-root system in V and*

$$S = \{\alpha''; \alpha \in R''\}.$$

Then S is a (not necessarily reduced) root system in V''.

Proof. It is clear that S is finite, does not contain 0 and spans V''.

Let $\lambda, \mu \in S$. Fix $\alpha, \beta \in R$ verifying $\alpha'' = \lambda$, $\beta'' = \mu$. Let us consider the following cases.

1) We have $\sigma(\alpha) = -\alpha$, or $\alpha = \lambda$. Then $\sigma(\alpha)^{\vee} = -\alpha^{\vee}$, so $\alpha^{\vee} = \alpha^{\vee''}$ belongs to V''^{*}. For $x \in V''$, set:

$$s(x) = x - \langle x, \alpha^{\vee''} \rangle \lambda.$$

Then the fact that $\langle \lambda, \alpha^{\vee''} \rangle = \langle \alpha, \alpha^{\vee} \rangle = 2$, implies that s is a reflection; in particular, $s(\mu) \neq 0$. Moreover:

$$\langle \mu, \alpha^{\vee''} \rangle = \langle \beta' + \mu, \alpha^{\vee''} \rangle = \langle \beta, \alpha^{\vee} \rangle \in \mathbb{Z},$$
$$s_{\alpha}(\beta) = \beta' + s(\mu) \Leftrightarrow s(\mu) = [s_{\alpha}(\beta)]''.$$

2) We have $\sigma(\alpha) \neq -\alpha$ and $\langle \alpha, \sigma(\alpha)^{\vee} \rangle = 0$. We have:

$$2\lambda = \alpha - \sigma(\alpha) \ , \ 2\alpha^{\vee''} = \alpha^{\vee} - \sigma(\alpha)^{\vee}.$$

For $x \in V''$, set:

$$s(x) = x - \langle x, 2\alpha^{\vee''} \rangle \lambda.$$

Since $\langle \alpha, \sigma(\alpha)^{\vee} \rangle = \langle \sigma(\alpha), \alpha^{\vee} \rangle = 0$, we obtain that:

$$\langle \lambda, 2\alpha^{\vee''} \rangle = \frac{1}{2}\langle \alpha, \alpha^{\vee} \rangle + \frac{1}{2}\langle \sigma(\alpha), \sigma(\alpha)^{\vee} \rangle = 2.$$

Thus s is a reflection, so $s(\mu) \neq 0$. Moreover:

$$\langle \mu, 2\alpha^{\vee''} \rangle = \langle \beta, \alpha^{\vee} - \sigma(\alpha)^{\vee} \rangle - \langle \beta', \alpha^{\vee} - \sigma(\alpha)^{\vee} \rangle$$
$$= \langle \beta, \alpha^{\vee} \rangle - \langle \beta, \sigma(\alpha)^{\vee} \rangle \in \mathbb{Z}.$$

Set $\gamma = s_{\sigma(\alpha)} \circ s_{\alpha}(\beta)$. As $\langle \alpha, \sigma(\alpha)^{\vee} \rangle = \langle \sigma(\alpha), \alpha^{\vee} \rangle = 0$, we obtain that:

$$\gamma = \beta - \langle \beta, \alpha^\vee \rangle \alpha - \langle \beta, \sigma(\alpha)^\vee \rangle \sigma(\alpha).$$

So $\gamma = \nu' + s(\mu)$ where $\nu' \in V'$. Hence $s(\mu) = \gamma''$.

3) We have $\sigma(\alpha) \neq -\alpha$ and $\langle \sigma(\alpha), \alpha^\vee \rangle \neq 0$. Since $\alpha + \sigma(\alpha) \notin R$, we have $\langle \sigma(\alpha), \alpha^\vee \rangle > 0$ (18.5.3). Moreover:

$$\langle \sigma(\alpha), \alpha^\vee \rangle = \langle \alpha, \theta(\alpha^\vee) \rangle = \langle \alpha, \sigma(\alpha)^\vee \rangle.$$

So, by 18.5.2:

$$\begin{cases} \text{either} & \langle \alpha, \sigma(\alpha)^\vee \rangle = \langle \sigma(\alpha), \alpha^\vee \rangle = 1 \\ \text{or} & \langle \alpha, \sigma(\alpha)^\vee \rangle = \langle \sigma(\alpha), \alpha^\vee \rangle = 2 \ \text{and} \ \alpha = \sigma(\alpha) \end{cases}$$

Since $\alpha \neq \sigma(\alpha)$ (because $\alpha \in R''$), it follows that:

$$\langle \alpha, \sigma(\alpha)^\vee \rangle = \langle \sigma(\alpha), \alpha^\vee \rangle = 1.$$

For $x \in V''$, set:

$$s(x) = x - \langle x, 4\alpha^{\vee ''} \rangle \lambda.$$

We have:

$$\begin{aligned} \langle \lambda, 4\alpha^{\vee ''} \rangle &= \langle \alpha - \sigma(\alpha), \alpha^\vee - \sigma(\alpha)^\vee \rangle \\ &= \langle \alpha, \alpha^\vee \rangle + \langle \sigma(\alpha), \sigma(\alpha)^\vee \rangle - \langle \alpha, \sigma(\alpha)^\vee \rangle - \langle \sigma(\alpha), \alpha^\vee \rangle \\ &= 2 + 2 - 1 - 1 = 2. \end{aligned}$$

Thus s is a reflection, so $s(\mu) \neq 0$. Moreover:

$$\begin{aligned} \langle \mu, 4\alpha^{\vee ''} \rangle &= 2\langle \beta, \alpha^\vee - \sigma(\alpha)^\vee \rangle - 2\langle \beta', \alpha^\vee - \sigma(\alpha)^\vee \rangle \\ &= 2\langle \beta, \alpha^\vee - \sigma(\alpha)^\vee \rangle \in \mathbb{Z}. \end{aligned}$$

Set $\gamma = s_\alpha \circ s_{\sigma(\alpha)} \circ s_\alpha(\beta)$. As $\langle \alpha, \sigma(\alpha)^\vee \rangle = \langle \sigma(\alpha), \alpha^\vee \rangle$, we obtain that:

$$\begin{aligned} \gamma &= \beta - \langle \beta, \alpha^\vee - \sigma(\alpha)^\vee \rangle (\alpha - \sigma(\alpha)) \\ &= \beta - \langle \beta, 2\alpha^{\vee ''} \rangle (2\lambda) = \beta' + s(\mu). \end{aligned}$$

Hence $s(\mu) = \gamma''$.

4) It follows finally from the preceding discussion that S is a root system in V''. $\square$

36.2.2 Let us assume again that R is a normal σ-root system in V. Denote by W the Weyl group of R and set:

$$W_\sigma = \{w \in W; \ w \circ \sigma = \sigma \circ w\}.$$

Let W' be the subgroup of W generated by the s_α's, $\alpha \in R'$. The group W' can be identified as the Weyl group of R', and we verify easily that W' is a normal subgroup of W_σ.

Lemma. Let $\lambda \in S$. There exists $w \in W_\sigma$ such that $s_\lambda = w|_{V''}$.

Proof. Let us fix $\alpha \in R''$ such that $\alpha'' = \lambda$. We shall consider the same cases as in 36.2.1.

1) We have $\sigma(\alpha) = -\alpha$. We saw in 36.2.1 that $s_\lambda = s_\alpha|_{V''}$, and it follows easily that $s_\alpha \in W_\sigma$.

2) We have $\sigma(\alpha) \neq -\alpha$ and $\langle \alpha, \sigma(\alpha)^\vee \rangle = \langle \sigma(\alpha), \alpha^\vee \rangle = 0$. By 36.2.1, we have $\lambda^\vee = 2\alpha^{\vee''}$.

Let $w = s_{\sigma(\alpha)} \circ s_\alpha$. Since $\langle \alpha, \sigma(\alpha)^\vee \rangle = 0$, it follows that for $x \in V''$:

$$\begin{aligned} w(x) &= x - \langle x, \alpha^\vee \rangle \alpha - \langle \sigma(x), \alpha^\vee \rangle \sigma(\alpha) \\ &= x - \langle x, \alpha^\vee \rangle \big(\alpha - \sigma(\alpha) \big) = x - \langle x, \alpha^\vee \rangle (2\alpha'') \\ &= x - \langle x, \alpha^{\vee''} \rangle (2\alpha'') = s_\lambda(x). \end{aligned}$$

We have therefore obtained that $w(V'') \subset V''$ and $w|_{V''} = s_\lambda$. On the other hand, if $x \in V'$, then:

$$\begin{aligned} w(x) &= x - \langle x, \alpha^\vee \rangle \alpha - \langle x, \sigma(\alpha)^\vee \rangle \sigma(\alpha) \\ &= x - \langle x, \alpha^\vee \rangle (2\alpha') \in V'. \end{aligned}$$

So we have $w \in W_\sigma$.

3) We have $\sigma(\alpha) \neq -\alpha$ and $\langle \sigma(\alpha), \alpha^\vee \rangle \neq 0$. Again we saw in the proof of 36.2.1 that we have $\langle \alpha, \sigma(\alpha)^\vee \rangle = \langle \sigma(\alpha), \alpha^\vee \rangle = 1$ and $\lambda^\vee = 4\alpha^{\vee''}$.

Let $w = s_\alpha \circ s_{\sigma(\alpha)} \circ s_\alpha$. If $x \in V$, then:

$$w(x) = x - \langle x, 4\alpha^{\vee''} \rangle \alpha''.$$

This shows that w stabilizes V' and V'', and so $w \in W_\sigma$ and $w|_{V''} = s_\lambda$. $\quad\square$

36.2.3 Let us conserve the notations of 36.2.2.

Let l' be the rank of R' and $(\alpha_{l-l'+1}, \ldots, \alpha_l)$ a base of R'. In the construction of 36.1.4, take $e_i = \alpha_i$, for $l - l' + 1 \leqslant i \leqslant l$. We see easily that $\alpha_{l-l'+1}, \ldots, \alpha_l$ are simple roots with respect to the lexicographic order on $V_{\mathbb{Q}}$ defined by the basis $(e_1, \ldots, e_l)$. By 36.1.4, there exists a base $B = (\alpha_1, \ldots, \alpha_l)$ of R verifying the following conditions:

(i) $(\alpha_{l-l'+1}, \ldots, \alpha_l)$ is a base of R' (so $\alpha_1, \ldots, \alpha_{l-l'} \in R''_+$).
(ii) $\sigma(R''_+) = R''_-$.

Lemma. *There exists a permutation of order 2, $i \mapsto i^\bullet$, of $\{1, \ldots, l-l'\}$ such that, for $1 \leqslant i \leqslant l - l'$, we have*

$$\sigma(\alpha_i) = -\alpha_{i^\bullet} + \sum_{j=l-l'+1}^{l} n_{ij}\alpha_j,$$

where $-n_{ij} \in \mathbb{N}$ for $l - l' + 1 \leqslant j \leqslant l$.

Proof. Let $i \in \{1, \ldots, l - l'\}$. We have

$$\sigma(\alpha_i) = \sum_{j=1}^{l} n_{ij}\alpha_j,$$

with $-n_{ij} \in \mathbb{N}$. Moreover, since $\sigma(\alpha_i) \in R''_-$, there exists $i^\bullet \in \{1, \ldots, l - l'\}$ such that $n_{ii^\bullet} < 0$. By applying σ, we obtain:

$$\alpha_i = \sum_{j=l-l'+1}^{l} n_{ij}\alpha_j + \sum_{j=1}^{l-l'} n_{ij}\left(\sum_{k=1}^{l} n_{jk}\alpha_k\right).$$

So:

$$\alpha_i = \sum_{j=1}^{l-l'} n_{ij}\left(\sum_{k=1}^{l-l'} n_{jk}\alpha_k\right) = \sum_{k=1}^{l-l'}\left(\sum_{j=1}^{l-l'} n_{ij}n_{jk}\right)\alpha_k.$$

If $i, k \in \{1, \ldots, l - l'\}$ are distinct, then:

$$1 = \sum_{j=1}^{l-l'} n_{ij}n_{ji} \ , \ \sum_{j=1}^{l-l'} n_{ij}n_{jk} = 0.$$

Since $n_{ij} \in -\mathbb{N}$ and $n_{ii^\bullet} \neq 0$, we have $n_{i^\bullet k} = 0$ for $k \neq i$. But:

$$1 = \sum_{k=1}^{l-l'} n_{i^\bullet k}n_{ki^\bullet}.$$

We deduce therefore that $n_{ii^\bullet} = n_{i^\bullet i} = -1$.

As $n_{i^\bullet i} \neq 0$, we have shown that $n_{ik} = 0$ for $k \neq i^\bullet$. So the result follows immediately. $\square$

36.2.4 Let us conserve the notations of 36.2.3. From the preceding discussions, we may assume that there exist integers l_1, l_2, l'' such that

$$l - l' = l_1 + 2l_2 \ , \ l'' = l_1 + l_2$$

and:

$$i^\bullet = \begin{cases} i & \text{if } 1 \leqslant i \leqslant l_1 \\ i + l_2 & \text{if } l_1 + 1 \leqslant i \leqslant l_1 + l_2 \\ i - l_2 & \text{if } l_1 + l_2 + 1 \leqslant i \leqslant l_1 + 2l_2 \end{cases}$$

If $x \in V$, then $2x'' = x - \sigma(x)$. Consequently:

$$(1) \qquad \alpha_i'' = \begin{cases} \alpha_i & \text{if } 1 \leqslant i \leqslant l_1 \\ \dfrac{1}{2}(\alpha_i + \alpha_{i+l_2}) & \text{if } l_1 + 1 \leqslant i \leqslant l_1 + l_2 \\ \dfrac{1}{2}(\alpha_i + \alpha_{i-l_2}) & \text{if } l_1 + l_2 + 1 \leqslant i \leqslant l_1 = 2l_2. \end{cases}$$

From this, we deduce that $(\alpha_1'', \ldots, \alpha_{l''}'')$ is a basis of V''. Hence:

$$\dim V'' = l'' = l_1 + l_2.$$

On the other hand, it is clear that any element of S is a linear combination (with integer coefficients of the same sign) of $\alpha_1'', \ldots, \alpha_{l''}''$. So $(\alpha_1'', \ldots, \alpha_{l''}'')$ is a base of S (18.8.11).

Lemma. *Let $w \in W_\sigma$. Then $w \in W'$ if and only if $w|_{V''} = \mathrm{id}_{V''}$, and for w to be an element of W', it suffices that:*

$$\{w(\alpha_1''), \ldots, w(\alpha_{l''}'')\} = \{\alpha_1'', \ldots, \alpha_{l''}''\}.$$

Proof. If $\alpha \in V'$, then $\alpha^\vee \in V'^*$, so $s_\alpha|_{V''} = \mathrm{id}_{V''}$. It follows that if $w \in W'$, then $w|_{V''} = \mathrm{id}_{V''}$.

Suppose that $w|_{V''} = \mathrm{id}_{V''}$. Since $w \in W_\sigma$, we have $w(R') = R'$, and it is clear that $\{w(\alpha_{l-l'+1}), \ldots, w(\alpha_l)\}$ is a base of R'. It follows from 18.8.7 that there exists $w' \in W'$ such that:

$$\{w(\alpha_{l-l'+1}), \ldots, w(\alpha_l)\} = \{w'(\alpha_{l-l'+1}), \ldots, w(\alpha_l)\}.$$

Replace w by $w'^{-1} \circ w$, we are reduced to the case where:

$$\{w(\alpha_{l-l'+1}), \ldots, w(\alpha_l)\} = \{\alpha_{l-l'+1}, \ldots, \alpha_l\}.$$

Let $i \in \{1, \ldots, l-l'\}$. We have $\alpha_i - \sigma(\alpha_i) \in V''$, and so:

$$w(\alpha_i) - \sigma(w(\alpha_i)) = w(\alpha_i - \sigma(\alpha_i)) = \alpha_i - \sigma(\alpha_i).$$

As $w(R') = R'$, we have $w(R'') = R''$. Also, $\sigma(\alpha_i) \in R''_-$ (36.2.3). We deduce therefore from the above equality and 36.2.3 (i) that $w(\alpha_i) \in R_+$. Consequently, the positive roots with respect to the base $\{w(\alpha_1), \ldots, w(\alpha_l)\}$ are the positive roots with respect to the base $\{\alpha_1, \ldots, \alpha_l\}$. It follows from 18.8.7 that $w = \mathrm{id}_V$.

Now suppose that $\{w(\alpha_1''), \ldots, w(\alpha_{l''}'')\} = \{\alpha_1'', \ldots, \alpha_{l''}''\}$. Replace w by $w'^{-1} \circ w$, with $w' \in W'$, we may assume that $\{\alpha_{l-l'+1}, \ldots, \alpha_l\}$ is invariant by w.

Let $i \in \{1, \ldots, l-l'\}$. In view of our assumption, there exists $k \in \{1, \ldots, l-l'\}$ such that $w(\alpha_i'') = \alpha_k''$. So $w(\alpha_i) = \alpha_k - \alpha_k' + w(\alpha_i')$. We deduce therefore from $w(V') \subset V'$ that $\alpha_k - w(\alpha_i) = \alpha_k' - w(\alpha_i') \in V'$. By 36.1.4 (ii), $w(\alpha_i) \in R''_+$. So we can conclude as in the previous case. $\square$

36.2.5 Theorem. *Let R be a normal σ-root system in V, S be defined as in 36.2.1 and W_S the Weyl group of S. The map $w \mapsto w|_{V''}$ induces a surjective homomorphism $W_\sigma \to W_S$, with kernel W'.*

Proof. Let us use the notations of 36.2.3 and 36.2.4. Let $w \in W_\sigma$. We have $w(R') = R'$ and $w(R'') = R''$, so $w(S) = S$. The set $\{w(\alpha_1''), \ldots, w(\alpha_{l''}'')\}$ is a base of S (18.8.11). So there exists $w' \in W_S$ such that

$$\{w(\alpha_1''), \ldots, w(\alpha_{l''}'')\} = \{w'(\alpha_1''), \ldots, w'(\alpha_l'')\}.$$

Let $w_1 \in W_\sigma$ be such that $w' = w_1|_{V''}$ (36.2.2). We have $w_1^{-1} \circ w \in W_\sigma$, and $w_1^{-1} \circ w$ stabilizes $\{\alpha_1'', \ldots, \alpha_l''\}$. By 36.2.4, we have $w = w_1 \circ w_2$, with $w_2 \in W'$. So $w|_{V''} = w_1|_{V''} = w'$. The map $w \mapsto w|_{V''}$ induces therefore a surjective map from W onto W_S (36.2.2), with kernel W' (36.2.4). $\square$

36.2.6 We shall use the notation $V_{\mathbb{Q}}''$ of 36.1.4 and 36.1.5.

Proposition. *Let* $\lambda, \mu \in V_{\mathbb{Q}}''$. *The following conditions are equivalent:*
(i) $\mu \in W_S(\lambda)$.
(ii) $\mu \in W(\lambda)$.
Thus if $x \in V_{\mathbb{Q}}''$, *then* $W_S(x) = W(x) \cap V_{\mathbb{Q}}''(x)$.

Proof. (i) $\Rightarrow$ (ii) This is clear by 36.2.5.

(ii) $\Rightarrow$ (i) In the notations of 36.2.3 and 36.2.4, there exists a base $\{\alpha_1, \ldots, \alpha_l\}$ of R verifying the following conditions:
a) $\{\alpha_{l-l'+1}, \ldots, \alpha_l\}$ is a base of R'.
b) $\alpha_1, \ldots, \alpha_{l-l'} \in R_+''$.
c) $\{\alpha_1'', \ldots, \alpha_{l''}''\}$ is a base of S.
d) The relations (1) of 36.2.4 are verified.
Set:
$$D' = \{x \in V_{\mathbb{Q}};\ \langle x, \alpha_i^\vee \rangle \geqslant 0,\ 1 \leqslant i \leqslant l\},$$
$$D_S' = \{x \in V_{\mathbb{Q}}'';\ \langle x, (\alpha_i'')^\vee \rangle \geqslant 0,\ 1 \leqslant i \leqslant l''\}.$$

For $l_1 + 1 \leqslant i \leqslant l_1 + l_2$, we have $\alpha_i'' = \alpha_{i+l_2}''$. Thus:
$$D_S' = \{x \in V_{\mathbb{Q}}'';\ \langle x, (\alpha_i'')^\vee \rangle \geqslant 0,\ 1 \leqslant i \leqslant l - l'\}.$$

On the other hand, if $\alpha \in R''$, then $(\alpha'')^\vee \in \{\alpha^{\vee''}, 2\alpha^{\vee''}, 4\alpha^{\vee''}\}$ (see the proof of 36.2.1). Hence $D_S' \subset D_S$.

Let $w \in W$ be such that $\mu = w(\lambda)$. By 18.8.11 (where we take only the intersections of the chambers with $V_{\mathbb{Q}}$), there exist $w_1, w_2 \in W_S$ and $\nu \in D_S'$ such that $w_2(\mu) \in D_S'$ and $\lambda = w_1(\nu)$. Using 36.2.5, we see that there exist $w_1', w_2' \in W_\sigma$ verifying $w_1 = w_1'|_{V''}, w_2 = w_2'|_{V''}$. So $w_2'(\mu) = w_2' \circ w \circ w_1'(\nu)$. Since $w_2'(\mu), \nu \in D_S' \subset D'$, it follows from 18.8.9 that $w_2'(\mu) = \nu$. Hence $\mu = w_2'^{-1} \circ w_1'^{-1}(\lambda)$, and so $\mu \in W_S(\lambda)$ (36.2.5). $\square$

36.3 Restriction of a root

36.3.1 Let R be a normal σ-root system in V, and let us conserve the notations of the preceding sections.

If $\lambda \in S$, denote by $S(\lambda)$ the set of $\alpha \in R''$ such that $\alpha'' = \lambda$. Set

$$m(\lambda) = \operatorname{card} S(\lambda).$$

We say that $\lambda \in S$ is *special* if $\lambda \in R$. So $\lambda \in S$ is special if and only if there exists $\alpha \in S(\lambda)$ verifying $\sigma(\alpha) = -\alpha$.

36.3.2 Lemma. *The following conditions are equivalent for* $\lambda \in S$:
(i) λ *is special.*
(ii) $m(\lambda)$ *is odd.*

Proof. If $\alpha = \alpha' + \alpha'' \in R$, then $\sigma(\alpha) = \alpha' - \alpha''$. Thus:
$$\alpha \in S(\lambda) \Leftrightarrow -\sigma(\alpha) \in S(\lambda).$$

It follows that if λ is special, then $S(\lambda) \setminus \{\lambda\}$ is the disjoint union of sets of the form $\{\alpha, -\sigma(\alpha)\}$. So $m(\lambda)$ is odd. The converse is analogue. $\square$

36.3.3 Lemma. *Let* $\lambda \in S$. *Then:*
  (i) *If* $m(\lambda)$ *is odd, then* $2\lambda \notin S$.
  (ii) *If* $m(\lambda)$ *is even and* $2\lambda \in S$, *then* $m(2\lambda)$ *is odd.*

Proof. (i) Suppose that $m(\lambda)$ is odd and $2\lambda \in S$. Then $\lambda \in R$ (36.3.2). Since $\sigma(\lambda) = -\lambda$, we have $(-\lambda)^\vee = -\lambda^\vee = \sigma(\lambda)^\vee = \theta(\lambda^\vee)$, so $\lambda^\vee \in V''^*$. Let $\alpha \in R$ verify $\alpha'' = 2\lambda$. Then:
$$\langle \alpha, \lambda^\vee \rangle = \langle \alpha'', \lambda^\vee \rangle = 4.$$
This implies, by 18.5.2, that $\alpha = 2\lambda$. This is absurd since R is reduced.
  (ii) Suppose that $m(\lambda)$ and $m(2\lambda)$ are even. Let us fix $\alpha \in S(\lambda)$ and $\beta \in S(2\lambda)$. It follows from 36.3.2 that $\alpha \neq -\sigma(\alpha)$ and $\beta \neq -\sigma(\beta)$.
  • Since R is normal, we have $\alpha + \sigma(\alpha) \notin R$, hence $\langle \sigma(\alpha), \alpha^\vee \rangle \geqslant 0$ (18.5.3). On the other hand, $\alpha - \sigma(\alpha) = 2\lambda \notin R$, because $m(2\lambda)$ is even (36.3.2). We deduce therefore that $\langle \sigma(\alpha), \alpha^\vee \rangle \leqslant 0$. Thus $\langle \sigma(\alpha), \alpha^\vee \rangle = 0$. Hence:
$$2 = \langle \alpha' + \lambda, \alpha^{\vee'} + \alpha^{\vee''} \rangle = \langle \alpha', \alpha^{\vee'} \rangle + \langle \lambda, \alpha^{\vee''} \rangle$$
$$0 = \langle \alpha' - \lambda, \alpha^{\vee'} + \alpha^{\vee''} \rangle = \langle \alpha', \alpha^{\vee'} \rangle - \langle \lambda, \alpha^{\vee''} \rangle$$

Consequently:
$$\langle \alpha', \alpha^{\vee'} \rangle = \langle \lambda, \alpha^{\vee''} \rangle = 1.$$

  • Similarly, $\beta + \sigma(\beta) \notin R$, where $\langle \sigma(\beta), \beta^\vee \rangle \geqslant 0$. If $\langle \sigma(\beta), \beta^\vee \rangle > 0$, then $\beta - \sigma(\beta) = 4\lambda \in R$. So $4\lambda \in S$ which is absurd because $\lambda \in S$ (18.5.3). We deduce therefore that $\langle \sigma(\beta), \beta^\vee \rangle = 0$. As in the preceding point, we obtain that:
$$\langle \beta', \beta^{\vee'} \rangle = \langle 2\lambda, \beta^{\vee''} \rangle = 1.$$

  • Suppose that $\langle \alpha, \beta^\vee \rangle < 0$. Then $\alpha + \beta \in R$, so $3\lambda \in S$ which is absurd. Hence $\langle \alpha, \beta^\vee \rangle \geqslant 0$. Similarly, we have:
$$\langle \alpha, \beta^\vee \rangle \geqslant 0 \ , \ \langle \beta, \alpha^\vee \rangle \geqslant 0 \ , \ \langle \sigma(\alpha), \beta^\vee \rangle \leqslant 0 \ , \ \langle \sigma(\beta), \alpha^\vee \rangle \leqslant 0.$$

Now:
$$\langle \alpha, \beta^\vee \rangle = \langle \alpha', \beta^{\vee'} \rangle + \langle \lambda, \beta^{\vee''} \rangle = \frac{1}{2} + \langle \alpha', \beta^{\vee'} \rangle \geqslant 0$$
$$\langle \sigma(\alpha), \beta^\vee \rangle = \langle \alpha', \beta^{\vee'} \rangle - \langle \lambda, \beta^{\vee''} \rangle = -\frac{1}{2} + \langle \alpha', \beta^{\vee'} \rangle \leqslant 0$$

Since $\langle \alpha, \beta^\vee \rangle$ and $\langle \sigma(\alpha), \beta^\vee \rangle$ are integers, we have $\langle \alpha', \beta^{\vee'} \rangle = \pm\frac{1}{2}$. Also, we have:
$$\langle \beta, \alpha^\vee \rangle = \langle \beta', \alpha^{\vee'} \rangle + \langle 2\lambda, \alpha^{\vee''} \rangle = \langle \beta', \alpha^{\vee'} \rangle + 2 \geqslant 0$$
$$\langle \sigma(\beta), \alpha^\vee \rangle = \langle \beta', \alpha^{\vee'} \rangle - \langle 2\lambda, \alpha^{\vee''} \rangle = \langle \beta', \alpha^{\vee'} \rangle - 2 \leqslant 0.$$

  If $\langle \alpha', \beta^{\vee'} \rangle = -\frac{1}{2}$, then $\langle \alpha, \beta^\vee \rangle = 0$, so $\langle \beta, \alpha^\vee \rangle = 0$. We deduce that $\langle \beta', \alpha^{\vee'} \rangle = -2$, and so $\langle \sigma(\beta), \alpha^\vee \rangle = -4$. By 18.5.2, we obtain that $\sigma(\beta) = 2\alpha$. This is absurd since R is reduced.
  If $\langle \alpha', \beta^{\vee'} \rangle = \frac{1}{2}$, then we obtain $\beta = 2\alpha$. Contradiction.
  • We have therefore proved that $m(2\lambda)$ is odd. □

36.3.4 Lemma. *Let* $\lambda \in S$. *If* $m(\lambda)$ *is odd, then* $\langle \sigma(\alpha), \alpha^\vee \rangle = 0$ *for all* $\alpha \in S(\lambda) \setminus \{\lambda\}$.

Proof. If $\langle \sigma(\alpha), \alpha^\vee \rangle > 0$, then $\alpha - \sigma(\alpha) \in R$, so $2\lambda \in S$. But this contradicts 36.3.3 (i).

If $\langle \sigma(\alpha), \alpha^\vee \rangle < 0$, then since $\alpha \neq -\sigma(\alpha)$ (because $\alpha \neq \lambda$), we have $\alpha + \sigma(\alpha) \in R$. This is absurd because R is normal. $\square$

36.3.5 Lemma. *Let* $\lambda \in S$ *be such that* $m(\lambda)$ *is even.*

(i) *We have* $2\lambda \in S$ *if and only if there exists* $\alpha \in S(\lambda)$ *such that* $\langle \sigma(\alpha), \alpha^\vee \rangle > 0$. *If these conditions are verified,* $\langle \sigma(\beta), \beta^\vee \rangle > 0$ *for all* $\beta \in S(\lambda)$.

(ii) *We have* $2\lambda \notin S$ *if and only if there exists* $\alpha \in S(\lambda)$ *such that* $\langle \sigma(\alpha), \alpha^\vee \rangle = 0$. *If these conditions are verified,* $\langle \sigma(\beta), \beta^\vee \rangle = 0$ *for all* $\beta \in S(\lambda)$.

Proof. (i) If $\langle \sigma(\alpha), \alpha \rangle > 0$, then $\alpha - \sigma(\alpha) = 2\lambda \in R$, so $2\lambda \in S$.

Conversely, suppose that $2\lambda \in S$. Since $m(\lambda)$ is even, we have $\lambda \notin R$ (36.3.2), so $\sigma(\alpha) \neq -\alpha$ for $\alpha \in S(\lambda)$. The system R being normal, $\alpha + \sigma(\alpha) \notin R$, hence $\langle \alpha, \sigma(\alpha)^\vee \rangle \geqslant 0$. On the other hand, $m(2\lambda)$ is odd (36.3.3), hence $2\lambda = \alpha - \sigma(\alpha) \in R$ (36.3.2). So the $\sigma(\alpha)$-string of α is of the form

$$\alpha - q\sigma(\alpha), \ldots, \alpha - \sigma(\alpha), \alpha,$$

with $q \geqslant 1$. But $-q = -\langle \alpha, \sigma(\alpha)^\vee \rangle$ (18.5.4). So $\langle \alpha, \sigma(\alpha)^\vee \rangle > 0$.

(ii) Suppose that $\langle \sigma(\alpha), \alpha^\vee \rangle = 0$. By (i), $2\lambda \notin S$.

Conversely, if $2\lambda \notin S$, then (i) implies that $\langle \sigma(\alpha), \alpha^\vee \rangle \leqslant 0$. If $\langle \sigma(\alpha), \alpha^\vee \rangle < 0$, then $\alpha + \sigma(\alpha) \in R$, unless $\sigma(\alpha) = -\alpha$. But if $\sigma(\alpha) = -\alpha$, then $\alpha \in S$, so $m(\lambda)$ is odd (36.3.2). Contradiction. Since the system is normal, the result follows. $\square$

36.3.6 Lemma. *Let* $\lambda \in S$ *and* $\alpha, \beta \in S(\lambda)$ *be such that* $\alpha \neq \beta$ *and* $\alpha \neq -\sigma(\beta)$.

(i) *If* $2\lambda \notin S$, *then* $\langle \alpha, \beta^\vee \rangle > 0$.

(ii) *If* $2\lambda \in S$, *then* $\langle \alpha, \beta^\vee \rangle \geqslant 0$. *Furthermore,* $\langle \alpha, \beta^\vee \rangle > 0$ *if and only if* $\langle \alpha, \sigma(\beta)^\vee \rangle = 0$.

Proof. (i) Suppose that $2\lambda \notin S$.

a) If $\lambda = \alpha$, then $\langle \alpha, \beta^\vee \rangle = \langle \lambda, \beta^{\vee\prime\prime} \rangle$. Moreover, $m(\lambda)$ is odd (36.3.2) and $\langle \sigma(\beta), \beta^\vee \rangle = 0$ (36.3.5). We deduce therefore that $\langle \beta', \beta^{\vee\prime} \rangle = \langle \lambda, \beta^{\vee\prime\prime} \rangle$. So:

$$\langle \alpha, \beta^\vee \rangle = \langle \lambda, \beta^{\vee\prime\prime} \rangle = \frac{1}{2}(\langle \beta', \beta^{\vee\prime} \rangle + \langle \lambda, \beta^{\vee\prime\prime} \rangle) = \frac{1}{2}\langle \beta, \beta^\vee \rangle = 1.$$

b) Suppose that $\alpha \neq \lambda$ and $\beta \neq \lambda$. By 36.3.4 and 36.3.5:

$$\langle \sigma(\alpha), \alpha^\vee \rangle = 0 \Rightarrow \langle \alpha', \alpha^{\vee\prime} \rangle = \langle \lambda, \alpha^{\vee\prime\prime} \rangle$$
$$\langle \sigma(\beta), \beta^\vee \rangle = 0 \Rightarrow \langle \beta', \beta^{\vee\prime} \rangle = \langle \lambda, \beta^{\vee\prime\prime} \rangle$$

Since

$$2 = \langle \alpha, \alpha^{\vee} \rangle = \langle \alpha', \alpha^{\vee'} \rangle + \langle \lambda, \alpha^{\vee''} \rangle$$
$$= \langle \beta, \beta^{\vee} \rangle = \langle \beta', \beta^{\vee'} \rangle + \langle \lambda, \beta^{\vee''} \rangle,$$

we obtain that:

$$\langle \alpha, \beta^{\vee} \rangle = \langle \alpha', \beta^{\vee'} \rangle + \langle \lambda, \beta^{\vee''} \rangle = \langle \alpha', \beta^{\vee'} \rangle + 1$$
$$\langle \sigma(\alpha), \beta^{\vee} \rangle = \langle \alpha', \beta^{\vee'} \rangle - \langle \lambda, \beta^{\vee''} \rangle = \langle \alpha', \beta^{\vee'} \rangle - 1.$$

Hence:

$$\langle \alpha, \beta^{\vee} \rangle - \langle \sigma(\alpha), \beta^{\vee} \rangle = 2.$$

Exchanging the roles of α and β, we have also:

$$\langle \beta, \alpha^{\vee} \rangle - \langle \sigma(\beta), \alpha^{\vee} \rangle = 2.$$

Now, since $2\lambda \notin S$, we have $\alpha + \beta \notin R$, so $\langle \alpha, \beta^{\vee} \rangle \geqslant 0$. Similarly, from $\sigma(\alpha) - \beta \notin R$, we deduce that $\langle \sigma(\alpha), \beta^{\vee} \rangle \leqslant 0$.

If $\langle \alpha, \beta^{\vee} \rangle = 0$, then $\langle \beta, \alpha^{\vee} \rangle = 0$ and $\langle \sigma(\beta), \alpha^{\vee} \rangle = -2$. By 18.5.2, we have either $\alpha = -\sigma(\beta)$, which is not possible by our hypothesis, or $\langle \alpha, \sigma(\beta)^{\vee} \rangle = -1$, so $\langle \sigma(\alpha), \beta^{\vee} \rangle = -1$. But this is absurd since $\langle \sigma(\alpha), \beta^{\vee} \rangle = -2$. So we have proved that $\langle \alpha, \beta^{\vee} \rangle > 0$.

(ii) Suppose that $2\lambda \in S$. Then $m(\lambda)$ is even, and by 36.3.5:

$$\langle \sigma(\alpha), \alpha^{\vee} \rangle > 0 \; , \; \langle \sigma(\beta), \beta^{\vee} \rangle > 0.$$

Also, we have $2\lambda \in R$ (36.3.2 and 36.3.3).

Since $2\lambda + \alpha \notin R$ (otherwise $3\lambda \in S$), $\langle 2\lambda, \alpha^{\vee} \rangle \geqslant 0$. Moreover:

$$0 \leqslant \langle 2\lambda, \alpha^{\vee} \rangle = \langle \alpha, \alpha^{\vee} \rangle - \langle \sigma(\alpha), \alpha^{\vee} \rangle = 2 - \langle \sigma(\alpha), \alpha^{\vee} \rangle < 2.$$

So $\langle 2\lambda, \alpha^{\vee} \rangle \in \{0, 1\}$.

If $\langle 2\lambda, \alpha^{\vee} \rangle = 0$, then $\langle \sigma(\alpha), \alpha^{\vee} \rangle = 2$. By 18.5.2, either $\sigma(\alpha) = \alpha$, which is absurd, or $\langle \alpha, \sigma(\alpha)^{\vee} \rangle = 1$, that is $\langle \sigma(\alpha), \alpha^{\vee} \rangle = 1$. Contradiction.

Thus $\langle 2\lambda, \alpha^{\vee} \rangle = 1$. Similarly $\langle 2\lambda, \beta^{\vee} \rangle = 1$. It follows that:

$$\langle \beta, \alpha^{\vee} \rangle - \langle \sigma(\beta), \alpha^{\vee} \rangle = 1 = \langle \alpha, \beta^{\vee} \rangle - \langle \sigma(\alpha), \beta^{\vee} \rangle.$$

Since

$$\langle 2\lambda, \alpha^{\vee} \rangle = 1 = 2 - \langle \sigma(\alpha), \alpha^{\vee} \rangle = 2 - \langle \sigma(\beta), \beta^{\vee} \rangle,$$

we have:

$$\langle \sigma(\alpha), \alpha^{\vee} \rangle = \langle \sigma(\beta), \beta^{\vee} \rangle = 1.$$

Suppose that $\langle \beta, \alpha^{\vee} \rangle < 0$. Then $\alpha + \beta \in R$. Moreover $2\alpha + \beta \notin R$ because $3\lambda \notin S$. So the α-string of β is of the form

$$\beta - q\alpha, \dots, \beta, \beta + \alpha.$$

We have $1 - q = -\langle \beta, \alpha^{\vee} \rangle > 0$ (18.5.4), hence $q = 0$. Thus $\beta - \alpha \notin R$ and $\langle \beta, \alpha^{\vee} \rangle = -1$. We deduce therefore that $\langle \sigma(\beta), \alpha^{\vee} \rangle = -2$. As $\alpha \neq -\sigma(\beta)$, we have $\langle \alpha, \sigma(\beta)^{\vee} \rangle = -1$ (18.5.2), hence $\langle \sigma(\alpha), \beta^{\vee} \rangle = -1$.

Since $1 = \langle \alpha, \beta^{\vee} \rangle - \langle \sigma(\alpha), \beta^{\vee} \rangle$, we deduce that $\langle \alpha, \beta^{\vee} \rangle = 0$, and so $\langle \beta, \alpha^{\vee} \rangle = 0$. Contradiction.

We have therefore proved that $\langle \beta, \alpha^{\vee} \rangle \geqslant 0$. Replacing β by $-\sigma(\beta)$, we obtain that $\langle \sigma(\beta), \alpha^{\vee} \rangle \leqslant 0$.

As $\langle \beta, \alpha^{\vee} \rangle$ and $\langle \sigma(\beta), \alpha^{\vee} \rangle$ are integers and

$$\langle \beta, \alpha^{\vee} \rangle - \langle \sigma(\beta), \alpha^{\vee} \rangle = 1 = \langle \alpha, \beta^{\vee} \rangle - \langle \sigma(\alpha), \beta^{\vee} \rangle,$$

it follows that:

$$\langle \beta, \alpha^{\vee} \rangle > 0 \Leftrightarrow \langle \beta, \alpha^{\vee} \rangle = 1 \Leftrightarrow \langle \sigma(\beta), \alpha^{\vee} \rangle = 0.$$

So we have obtained (ii). $\square$

References and comments

- [1], [29], [87].

Most of the results presented in this chapter were first obtained in [1].

Symmetric Lie algebras

37.1 Primary subspaces

37.1.1 In this section, we shall generalize certain notions in chapter 19. Let $\mathfrak{g}$ be a Lie algebra and (V, σ) a finite-dimensional representation of $\mathfrak{g}$. When $V = \mathfrak{g}$, we shall assume that σ is the adjoint representation of $\mathfrak{g}$.

Let A be a subset of $\mathfrak{g}$ and denote by P the set of maps from A to the field $\Bbbk$. For $\lambda \in P$, set $V_\lambda(A)$ (resp. $V^\lambda(A)$) to be the subspace of V consisting of vectors v such that for all $x \in A$, we have $\sigma(x)(v) = \lambda(x)v$ (resp. $(\sigma(x) - \lambda(x)\,\mathrm{id}_V)^n(v) = 0$ for some integer n). We shall call $V_\lambda(A)$ the *eigenspace* of V relative to λ (and σ), and $V^\lambda(A)$ the *primary subspace* of V relative to λ (and σ).

If $V^\lambda(A)$ is non-zero, we say that λ is a *weight* of A in V, and any non-zero vector in $V^\lambda(A)$ shall be called a *weight vector*. The set of weights of A in V will be denoted by $\Delta_A(V)$, and the subspace $V^0(A)$, corresponding to the map $x \mapsto 0$ for all $x \in A$, is called the *nilspace* of V (with respect to σ).

When A is reduced to a singleton x, we can identify P with $\Bbbk$. In this case, we shall denote $V_\lambda(A)$ and $V^\lambda(A)$ simply by $V_{\lambda(x)}(x)$ and $V^{\lambda(x)}(x)$. It follows in the general case that:

$$V_\lambda(A) = \bigcap_{x \in A} V_{\lambda(x)}(x) \; , \; V^\lambda(A) = \bigcap_{x \in A} V^{\lambda(x)}(x).$$

Let $u \in \mathrm{End}(V)$, then by considering the abelian Lie algebra $\Bbbk u$ and the representation induced by the inclusion $\Bbbk u \subset \mathrm{End}(V)$, we may define for $\alpha \in \Bbbk$ the subspaces $V_\alpha(u)$ and $V^\alpha(u)$ as above. Thus we recover the definitions of 19.3.8.

37.1.2 Proposition. *The sums*

$$\sum_{\lambda \in P} V_\lambda(A) \quad and \quad \sum_{\lambda \in P} V^\lambda(A)$$

are direct, and the set $\Delta_A(V)$ is finite.

Proof. Since V is finite-dimensional, it is clear that $\Delta_A(V)$ is finite. Also, it is clear that we only need to prove that the second sum is direct.

1) If A is empty, then there is nothing to prove.

2) If $A = \{x\}$, then this is just a consequence of the decomposition of V into generalized eigenspaces of $\sigma(x)$.

3) Let us assume that card $A \geqslant 2$ and suppose that there exists a non-empty subset I of $\Delta_A(V)$ minimal with the property that we can find a non-zero vector v_λ in $V^\lambda(A)$ verifying $\sum_{\lambda \in I} v_\lambda = 0$. Clearly card $I \geqslant 2$. Let $\mu, \nu \in I$ be distinct. Then there exists $x \in A$ such that $\mu(x) \neq \nu(x)$. Let $Q = \{\lambda(x); \lambda \in I\}$ and $J = \{\lambda \in I; \lambda(x) = \mu(x)\}$. Since $\mu(x) \neq \nu(x)$, card $Q \geqslant 2$. Moreover, for $\alpha \in Q$, we verify easily that $v_\lambda \in V^\alpha(x)$ if and only if $\lambda(x) = \alpha$.

We saw in point 2 that the sum $\sum_{\alpha \in Q} V^\alpha(x)$ is direct. This implies for $\alpha = \mu(x)$ that $\sum_{\lambda \in J} v_\lambda = 0$. Since $\nu \notin J$, this contradicts the minimality of the set I. $\square$

37.1.3 Proposition. *The following conditions are equivalent:*
  (i) $A \subset \mathfrak{g}^0(A)$.
  (ii) *For all $\lambda \in P$, $V^\lambda(A)$ is $\sigma(A)$-stable, and we have:*

$$V = \sum_{\lambda \in \Delta_A(V)} V^\lambda(A).$$

Proof. (i) $\Rightarrow$ (ii) The subspace $V^\lambda(A)$ is the intersection of the $V^{\lambda(x)}(x)$, for $x \in A$. By the hypothesis and 19.3.8, each $V^\lambda(A)$ is $\sigma(A)$-stable. We shall prove by induction on dim V that V is the sum of the $V^\lambda(A)$'s.

If $\sigma(x)$ has a unique eigenvalue $\lambda(x)$ for all $x \in A$, then clearly $V = V^\lambda(A)$.

Otherwise, there exists $x \in A$ such that $\sigma(x)$ has at least two distinct eigenvalues. Now V is the direct sum of the subspaces $V^\alpha(x)$, $\alpha \in \Bbbk$, and dim $V^\alpha(x) < $ dim V for all α. Since each $V^\alpha(x)$ is $\sigma(A)$-stable (19.3.8), the result follows by the induction hypothesis.

(ii) $\Rightarrow$ (i) This follows from 19.3.8. $\square$

37.1.4 Proposition. (i) *Let $\lambda, \mu \in P$. Then:*

$$\sigma(\mathfrak{g}^\lambda(A))(V^\mu(A)) \subset V^{\lambda+\mu}(A) \ , \ [\mathfrak{g}^\lambda(A), \mathfrak{g}^\mu(A)] \subset \mathfrak{g}^{\lambda+\mu}(A).$$

In particular, $\mathfrak{g}^0(A)$ is a Lie subalgebra of $\mathfrak{g}$.

  (ii) *Suppose that A is a subspace of $\mathfrak{g}$ and $A \subset \mathfrak{g}^0(A)$. Then all the elements of $\Delta_A(V)$ are linear forms on A.*

Proof. (i) Let $x \in A$, $\alpha = \lambda(x)$ and $\beta = \mu(x)$. Set $S = \operatorname{ad} x - \alpha \operatorname{id}_\mathfrak{g}$ and $T = \sigma(x) - \beta \operatorname{id}_V$. If $y \in \mathfrak{g}^\lambda(A)$ and $v \in V^\mu(A)$, then we obtain by induction on n that:

$$\left(\sigma(x) - (\alpha + \beta)\operatorname{id}_V\right)^n \left(\sigma(y)(v)\right) = \sum_{i=0}^n \binom{n}{i} \sigma\left(S^i(y)\right)\left(T^{n-i}(v)\right).$$

Hence the first inclusion. The second inclusion is just a special case of the first.

(ii) Let $\lambda \in \Delta_A(V)$. By the hypotheses and 19.3.8, $V^\lambda(A)$ is $\sigma(A)$-stable. So we are reduced to the case where $V = V^\lambda(A)$. If $x \in A$, then the characteristic polynomial χ_x of $\sigma(x)$ verifies $\chi_x(X) = (X - \lambda(x))^n$, where $n = \dim V$. On the other hand,

$$\chi_x(X) = X^n + a_1(x)X^{n-1} + \cdots + a_n(x),$$

where a_i is a polynomial function on A, of degree i, $0 \leqslant i \leqslant n$. Thus $n\lambda(x) = -a_1(x)$ for all $x \in A$. Since the characteristic of $\Bbbk$ is zero, we are done. $\square$

37.1.5 Proposition. *Let A be a subset of $\mathfrak{g}$ verifying $A \subset \mathfrak{g}^0(A)$, and $\mathfrak{a}$ the subspace of $\mathfrak{g}$ spanned by A. Then there exists $x \in \mathfrak{a}$ such that $\sigma(x)$ induces an automorphism of $V/V^0(A)$.*

Proof. Let $\lambda \in \Delta_A(V) \setminus \{0\}$. If we set $d_\lambda(y) = \det(\sigma(y)|_{V^\lambda(A)})$, then d_λ is a non-zero polynomial function on $\mathfrak{a}$. Since $\Bbbk$ is infinite, the product d of the d_λ's, for $\lambda \in \Delta_A(V) \setminus \{0\}$, is non-zero on $\mathfrak{a}$. So any $x \in \mathfrak{a}$ verifying $d(x) \neq 0$ works. $\square$

37.1.6 Set:
$$V^\bullet(A) = \sum_{x \in A} \left(\bigcap_{i \geqslant 1} (\sigma(x))^i(V) \right).$$

Proposition. *Suppose that $A \subset \mathfrak{g}^0(A)$. Then:*
(i) *$V^0(A)$ and $V^\bullet(A)$ are $\sigma(A)$-stable.*
(ii) *$V = V^0(A) \oplus V^\bullet(A)$.*
(iii) *Any $\sigma(A)$-stable subspace W of V verifying $W^0(A) = \{0\}$, is contained in $V^\bullet(A)$.*
(iv) *We have:*
$$\sum_{x \in A} \sigma(x)(V^\bullet(A)) = V^\bullet(A).$$

Moreover, $V^\bullet(A)$ is the only $\sigma(A)$-stable subspace of V which is also a complementary subspace of $V^0(A)$ in V.

Proof. Each $V^\lambda(A)$ is $\sigma(A)$-stable and V is the sum of the $V^\lambda(A)$'s, for $\lambda \in P$ (37.1.3).

If $x \in A$, the characteristic polynomial of $\sigma(x)|_{V^\lambda(A)}$ is $(X - \lambda(x))^{n(\lambda)}$, where $n(\lambda) = \dim V^\lambda(A)$. Thus the intersection $\bigcap_{i \geqslant 1}(\sigma(x)^i)(V^\lambda(A))$ is zero if $\lambda(x) = 0$, and is equal to $V^\lambda(A)$ if $\lambda(x) \neq 0$. It follows that:

$$V^\bullet(A) = \sum_{\lambda \in P \setminus \{0\}} V^\lambda(A).$$

Hence (i), (ii) and (iv).

Let W be a subspace of V verifying the conditions of (iii). Then:

$$W = \sum_{\lambda \in P} W^\lambda(A) \, , \; W^\lambda(A) = W \cap V^\lambda(A).$$

Since $W^0(A) = \{0\}$, we have (iii).

Finally, let U be a $\sigma(A)$-stable subspace of V verifying $U \cap V^0(A) = \{0\}$. Then $U^0(A) = \{0\}$, and so $U \subset V^\bullet(A)$ by (iii). The last assertion is now clear. $\square$

37.1.7 In the notations of 37.1.6, the decomposition $V = V^0(A) \oplus V^\bullet(A)$ is called the *Fitting decomposition* of V (with respect to σ). When $A = \{x\}$, we shall denote $V^0(A)$ and $V^\bullet(A)$ simply by $V^0(x)$ and $V^\bullet(x)$.

As in 37.1.1, if $u \in \mathrm{End}(V)$, we shall also use the notations $V^0(u)$ and $V^\bullet(u)$.

Proposition. *Let $u \in \mathrm{End}(V)$, s and n the semisimple and nilpotent components of u. For all $\lambda \in \Bbbk$, we have $V^\lambda(u) = V^\lambda(s) = V_\lambda(s)$.*

Proof. Let $\mathfrak{g} = \Bbbk u$. Then $u \in \mathfrak{g}^0(u)$ and V is the sum of the $V^\lambda(u)$, $\lambda \in \Bbbk$. Since s is a polynomial without constant term in u, $V^\lambda(u)$ is s-stable. But $s|_{V^\lambda(u)} = \lambda \, \mathrm{id}_{V^\lambda(u)}$, so the result follows. $\square$

37.1.8 A bilinear form b on V is called *A-invariant* if

$$b(\sigma(x)v, w) + b(v, \sigma(x)w) = 0$$

for all $v, w \in V$ and $x \in A$.

Proposition. *Let b be an A-invariant bilinear form on V and $\lambda, \mu \in P$. If $\lambda + \mu \neq 0$, then $V^\lambda(A)$ and $V^\mu(A)$ are orthogonal with respect to b.*

Proof. Let $x \in A$, $f = \sigma(x) - \lambda(x) \, \mathrm{id}_V$, $g = \sigma(x) - \mu \, \mathrm{id}_V$, $v \in V^\lambda(A)$ and $w \in V^\mu(A)$. Since b is A-invariant, we have:

$$-\big(\lambda(x) + \mu(x)\big)b(v, w) = b\big(f(v), w\big) + b\big(v, g(w)\big).$$

By induction on n, we obtain that:

$$(-1)^n \big(\lambda(x) + \mu(x)\big)^n b(v, w) = \sum_{i=0}^{n} \binom{n}{i} b\big(f^i(v), g^{n-i}(w)\big).$$

The result follows immediately. $\square$

37.2 Definition of symmetric Lie algebras

37.2.1 Definition. *A* symmetric Lie algebra *is a triple* $(\mathfrak{g}; \mathfrak{k}, \mathfrak{p})$ *where* $\mathfrak{g}$ *is a Lie algebra and* $\mathfrak{k}, \mathfrak{p}$ *are subspaces of* $\mathfrak{g}$ *verifying:*

$$\mathfrak{g} = \mathfrak{k} \oplus \mathfrak{p} \ , \ [\mathfrak{k}, \mathfrak{k}] \subset \mathfrak{k} \ , \ [\mathfrak{k}, \mathfrak{p}] \subset \mathfrak{p} \ , \ [\mathfrak{p}, \mathfrak{p}] \subset \mathfrak{k}.$$

The direct sum decomposition $\mathfrak{g} = \mathfrak{k} \oplus \mathfrak{p}$ *will be called the* symmetric decomposition *of* $\mathfrak{g}$.

37.2.2 Let $(\mathfrak{g}; \mathfrak{k}, \mathfrak{p})$ be a symmetric Lie algebra. By definition, $\mathfrak{k}$ is a Lie subalgebra of $\mathfrak{g}$.

Let $\theta \in \operatorname{End}(\mathfrak{g})$ be defined by $\theta(x) = x$ if $x \in \mathfrak{k}$, and $\theta(x) = -x$ if $x \in \mathfrak{p}$. It is easy to check that θ is an involution of $\mathfrak{g}$, and $\mathfrak{k}$, $\mathfrak{p}$ are its eigenspaces with eigenvalues 1 and -1. It is obvious that we have the following converse:

Proposition. *Let* θ *be an involutive automorphism of a Lie algebra* $\mathfrak{g}$, *and:*

$$\mathfrak{k} = \{x \in \mathfrak{g}; \theta(x) = x\} \ , \ \mathfrak{p} = \{x \in \mathfrak{g}; \theta(x) = -x\}.$$

Then $(\mathfrak{g}; \mathfrak{k}, \mathfrak{p})$ *be a symmetric Lie algebra.*

37.2.3 In other words, 37.2.2 says that a symmetric Lie algebra is a pair $(\mathfrak{g}, \theta)$ where $\mathfrak{g}$ is a Lie algebra and θ an involutive automorphism of $\mathfrak{g}$. The involution θ determines $\mathfrak{k}$ and $\mathfrak{p}$, and vice versa. We shall also use the notation $(\mathfrak{g}, \theta)$ for a symmetric Lie algebra.

Let L be the Killing form on $\mathfrak{g}$. Since, for $x \in \mathfrak{k}$ and $y \in \mathfrak{p}$, we have:

$$(\operatorname{ad} x) \circ (\operatorname{ad} y)(\mathfrak{k}) \subset \mathfrak{p} \ \text{ and } \ (\operatorname{ad} x) \circ (\operatorname{ad} y)(\mathfrak{p}) \subset \mathfrak{k},$$

it follows that $L(x, y) = \{0\}$. Thus $\mathfrak{k}$ and $\mathfrak{p}$ are orthogonal with respect to L. In particular, if L is non-degenerate, then $\mathfrak{k}$ is the orthogonal of $\mathfrak{p}$, and vice versa.

A Lie subalgebra of a Lie algebra $\mathfrak{g}$ will be called *symmetrizing* if it is the set of fixed points of an involutive automorphism of $\mathfrak{g}$.

37.2.4 Examples. 1) Let n be an integer greater than or equal to 2, and $\mathfrak{g} = \mathfrak{sl}_n(\Bbbk)$. For $x \in \mathfrak{g}$, set $\theta(x) = -{}^t x$ where ${}^t x$ is the transpose of the matrix x. Then θ is an involutive automorphism of $\mathfrak{g}$, and $\mathfrak{k}$ (resp. $\mathfrak{p}$) is the set of skew-symmetric (resp. symmetric) matrices in $\mathfrak{g}$.

2) Let $\mathfrak{s}$ be a Lie algebra, $\mathfrak{g} = \mathfrak{s} \times \mathfrak{s}$ and θ the involutive automorphism of $\mathfrak{g}$ defined by $\theta(x, y) = (y, x)$. The symmetric Lie algebra $(\mathfrak{g}, \theta)$ is called the *diagonal* defined by $\mathfrak{s}$. Here $\mathfrak{k} = \{(x, x) \, ; \, x \in \mathfrak{s}\}$ and $\mathfrak{p} = \{(x, -x) \, ; \, x \in \mathfrak{s}\}$.

3) A Lie algebra $\mathfrak{g}$ is said to be *graded* if $\mathfrak{g} = \bigoplus_{p \in \mathbb{Z}} \mathfrak{g}_p$ is $\mathbb{Z}$-graded as a vector space, and for all $p, q \in \mathbb{Z}$, we have $[\mathfrak{g}_p, \mathfrak{g}_q] \subset \mathfrak{g}_{p+q}$.

Let $\mathfrak{g}$ be such a Lie algebra. Take $\mathfrak{k}$ (resp. $\mathfrak{p}$) to be the direct sum of the $\mathfrak{g}_p$'s for p even (resp. odd). Then $(\mathfrak{g}; \mathfrak{k}, \mathfrak{p})$ is a symmetric Lie algebra.

37.3 Natural subalgebras

37.3.1 Let $(\mathfrak{g}; \mathfrak{k}, \mathfrak{p})$ or $(\mathfrak{g}, \theta)$ be a symmetric Lie algebra.
For $A \subset \mathfrak{g}$, using the notations of 37.1, let:

$$\mathfrak{k}^0(A) = \mathfrak{k} \cap \mathfrak{g}^0(A) , \ \mathfrak{k}^\bullet(A) = \mathfrak{k} \cap \mathfrak{g}^\bullet(A),$$
$$\mathfrak{p}^0(A) = \mathfrak{p} \cap \mathfrak{g}^0(A) , \ \mathfrak{p}^\bullet(A) = \mathfrak{p} \cap \mathfrak{g}^\bullet(A).$$

If $A = \{x\}$, then we shall denote $\mathfrak{p}^0(A)$ by $\mathfrak{p}^0(x)$.

37.3.2 Proposition. *Let $A \subset \mathfrak{k}$ or $A \subset \mathfrak{p}$ be such that $A \subset \mathfrak{g}^0(A)$. Then:*

$$\mathfrak{g}^0(A) = \mathfrak{k}^0(A) \oplus \mathfrak{p}^0(A) , \ \mathfrak{g}^\bullet(A) = \mathfrak{k}^\bullet(A) \oplus \mathfrak{p}^\bullet(A),$$
$$\mathfrak{k} = \mathfrak{k}^0(A) \oplus \mathfrak{k}^\bullet(A) , \ \mathfrak{p} = \mathfrak{p}^0(A) \oplus \mathfrak{p}^\bullet(A),$$
$$\mathfrak{g} = \mathfrak{k}^0(A) \oplus \mathfrak{k}^\bullet(A) \oplus \mathfrak{p}^0(A) \oplus \mathfrak{p}^\bullet(A).$$

Proof. Since $\mathfrak{g} = \mathfrak{g}^0(A) \oplus \mathfrak{g}^\bullet(A)$ (37.1.6), we only need to prove the equalities on the first line. Let $y = k + p$, with $k \in \mathfrak{k}$, $p \in \mathfrak{p}$. Now $y \in \mathfrak{g}^0(A)$ if and only if for all $x \in A$, there exists $n \in \mathbb{N}$ such that:

$$0 = (\operatorname{ad} x)^{2n}(y) = (\operatorname{ad} x)^{2n}(k) + (\operatorname{ad} x)^{2n}(p).$$

As $(\operatorname{ad} x)^{2n}(\mathfrak{k}) \subset \mathfrak{k}$ and $(\operatorname{ad} x)^{2n}(\mathfrak{p}) \subset \mathfrak{p}$ (because $A \subset \mathfrak{k}$ or $A \subset \mathfrak{p}$), we deduce that $\mathfrak{g}^0(A) = \mathfrak{k}^0(A) \oplus \mathfrak{p}^0(A)$. Similarly:

$$\mathfrak{g}^\bullet(A) = \sum_{x \in A} \left(\bigcap_{i \geqslant 0} (\operatorname{ad} x)^{2i}(\mathfrak{g}) \right).$$

So $\mathfrak{g}^\bullet(A) = \mathfrak{k}^\bullet(A) \oplus \mathfrak{p}^\bullet(A)$ since $(\operatorname{ad} x)^{2i}(\mathfrak{g}) = (\operatorname{ad} x)^{2i}(\mathfrak{k}) \oplus (\operatorname{ad} x)^{2i}(\mathfrak{p})$. $\square$

37.3.3 Definition. *Let $(\mathfrak{g}; \mathfrak{k}, \mathfrak{p})$ be a symmetric Lie algebra.*
(i) A subspace $\mathfrak{q}$ of $\mathfrak{p}$ is called a $\mathfrak{p}$-subalgebra of $\mathfrak{g}$ (or of $\mathfrak{p}$) if $(\operatorname{ad} x)^2(\mathfrak{q}) \subset \mathfrak{q}$ for all $x \in \mathfrak{q}$.
(ii) A $\mathfrak{p}$-subalgebra $\mathfrak{q}$ of $\mathfrak{g}$ is said to be natural *if there exists $x_0 \in \mathfrak{q}$ such that $(\operatorname{ad} x_0)^2$ induces an automorphism of the vector space $\mathfrak{p}/\mathfrak{q}$.*

37.3.4 Theorem. *Let $(\mathfrak{g}; \mathfrak{k}, \mathfrak{p})$ be a symmetric Lie algebra, $A \subset \mathfrak{p}$ verifying $A \subset \mathfrak{g}^0(A)$, and $\mathfrak{a}$ the subspace of $\mathfrak{p}$ spanned by A.*
(i) The subspace $\mathfrak{p}^0(A)$ is a natural $\mathfrak{p}$-subalgebra of $\mathfrak{g}$.
(ii) There exists $x \in \mathfrak{a}$ such that $(\operatorname{ad} x)^2$ induces an automorphism of the vector space $\mathfrak{p}/\mathfrak{p}^0(A)$.

Proof. By 37.1.4, $\mathfrak{g}^0(A)$ is a Lie subalgebra of $\mathfrak{g}$. If $x \in \mathfrak{p}^0(A)$, then $(\operatorname{ad} x)^2(\mathfrak{p}) \subset \mathfrak{p}$ and $(\operatorname{ad} x)^2(\mathfrak{g}^0(A)) \subset \mathfrak{g}^0(A)$. So $\mathfrak{p}^0(A)$ is a $\mathfrak{p}$-subalgebra of $\mathfrak{g}$.
By 37.1.5, there exists $x \in \mathfrak{a} \subset \mathfrak{p}^0(A)$ such that $\operatorname{ad} x$ induces an automorphism of $\mathfrak{g}/\mathfrak{g}^0(A) = (\mathfrak{k}/\mathfrak{k}^0(A)) \oplus (\mathfrak{p}/\mathfrak{p}^0(A))$. So $(\operatorname{ad} x)^2$ induces an automorphism of $\mathfrak{p}/\mathfrak{p}^0(A)$. So we have part (ii) and $\mathfrak{p}^0(A)$ is a natural $\mathfrak{p}$-subalgebra of $\mathfrak{g}$. $\square$

37.4 Cartan subspaces

37.4.1 Definition. *Let* $(\mathfrak{g}; \mathfrak{k}, \mathfrak{p})$ *be a symmetric Lie algebra. An element* $x \in \mathfrak{p}$ *is called* $\mathfrak{p}$-*generic in* $\mathfrak{g}$ *if for all* $y \in \mathfrak{p}$:

$$\dim \mathfrak{p}^0(x) \leqslant \dim \mathfrak{p}^0(y).$$

37.4.2 For $x \in \mathfrak{p}$, denote by χ_x the characteristic polynomial of $(\operatorname{ad} x)^2|_{\mathfrak{p}}$:

$$\chi_x(X) = X^r + a_{r-1}(x)X^{r-1} + \cdots + a_s(x)X^s$$

where $\dim \mathfrak{p} = r$ and $a_{r-1}, \ldots, a_s$ are polynomial functions on $\mathfrak{p}$, with $a_s \neq 0$. The subspace $\mathfrak{p}^0(x)$ is the generalized eigenspace of $(\operatorname{ad} x)^2$ associated to the eigenvalue 0. So we have $\dim \mathfrak{p}^0(x) = s$ if x is $\mathfrak{p}$-generic, and the set of $\mathfrak{p}$-generic elements in $\mathfrak{g}$ is a non-empty Zariski-open subset of $\mathfrak{p}$.

37.4.3 Definition. *Let* $(\mathfrak{g}; \mathfrak{k}, \mathfrak{p})$ *be a symmetric Lie algebra. A subset* $\mathfrak{a}$ *of* $\mathfrak{p}$ *is called a* Cartan subspace *of* $\mathfrak{g}$ *if* $\mathfrak{a} = \mathfrak{p}^0(\mathfrak{a})$.

Remark. It is clear that a Cartan subspace of $\mathfrak{g}$ is indeed a subspace of the vector space $\mathfrak{p}$.

37.4.4 Theorem. *Let* $(\mathfrak{g}; \mathfrak{k}, \mathfrak{p})$ *be a symmetric Lie algebra and* $\mathfrak{a}$ *a subset of* $\mathfrak{p}$. *The following conditions are equivalent:*
   (i) $\mathfrak{a}$ *is a Cartan subspace of* $\mathfrak{g}$.
   (ii) $\mathfrak{a}$ *is a minimal natural* $\mathfrak{p}$-*subalgebra of* $\mathfrak{g}$.
   (iii) $\mathfrak{a} \subset \mathfrak{g}^0(\mathfrak{a})$ *and there exists* $x \in \mathfrak{p}$ *such that* $\mathfrak{a} = \mathfrak{p}^0(x)$.
   (iv) $\mathfrak{a}$ *is a minimal natural* $\mathfrak{p}$-*subalgebra of* $\mathfrak{g}$ *verifying* $\mathfrak{a} \subset \mathfrak{g}^0(\mathfrak{a})$.

Proof. (i) $\Rightarrow$ (ii) Let $\mathfrak{a}$ be a Cartan subspace of $\mathfrak{g}$. By 37.3.4, $\mathfrak{a}$ is a natural $\mathfrak{p}$-subalgebra of $\mathfrak{g}$.

Let $\mathfrak{q} \subset \mathfrak{a}$ be a natural $\mathfrak{p}$-subalgebra of $\mathfrak{g}$ and $x \in \mathfrak{q}$ be such that $(\operatorname{ad} x)^2$ induces an automorphism of $\mathfrak{p}/\mathfrak{q}$. Then $(\operatorname{ad} x)^2$ induces an automorphism of $\mathfrak{a}/\mathfrak{q}$. Since $(\operatorname{ad} x)^n(\mathfrak{a}) = \{0\}$ for some large n, we deduce that $\mathfrak{q} = \mathfrak{a}$.

(ii) $\Rightarrow$ (iii) For $x \in \mathfrak{a}$, let $\delta(x)$ be the determinant of the endomorphism of $\mathfrak{p}/\mathfrak{a}$ induced by $(\operatorname{ad} x)^2$. If (ii) is verified, then there exists $x_0 \in \mathfrak{a}$ such that $\delta(x_0) \neq 0$. It follows that $\mathfrak{p}^0(x_0) \subset \mathfrak{a}$. But $\mathfrak{p}^0(x_0)$ is a natural $\mathfrak{p}$-subalgebra by 37.3.4. Thus $\mathfrak{a} = \mathfrak{p}^0(x_0)$.

Let $n = \dim \mathfrak{a}$. For $x \in \mathfrak{a}$, let $[a_{ij}(x)]$ be the matrix of $(\operatorname{ad} x)^{2n}|_{\mathfrak{a}}$ with respect to a basis of $\mathfrak{a}$. If $\mathfrak{a} \not\subset \mathfrak{g}^0(\mathfrak{a})$, then there exists $x \in \mathfrak{a}$ such that $(\operatorname{ad} x)^{2n}|_{\mathfrak{a}} \neq 0$. So we can find integers p, q such that $a_{pq}(x) \neq 0$. As $\Bbbk$ is infinite, the polynomial function $x \mapsto a_{pq}(x)\delta(x)$ is non-zero on $\mathfrak{a}$. Therefore if $y \in \mathfrak{a}$ verifies $a_{pq}(y)\delta(y) \neq 0$, then $\mathfrak{p}^0(y) \subset \mathfrak{a}$ and $(\operatorname{ad} y)^{2n}|_{\mathfrak{a}} \neq 0$, so $(\operatorname{ad} y)^2|_{\mathfrak{a}}$ is not nilpotent. Thus $\mathfrak{p}^0(y) \neq \mathfrak{a}$. This is absurd since $\mathfrak{p}^0(y)$ is natural (37.3.4).

(iii) $\Rightarrow$ (iv) By 37.3.4, $\mathfrak{a}$ is a natural $\mathfrak{p}$-subalgebra of $\mathfrak{g}$, and we may conclude as in the proof of (i) $\Rightarrow$ (ii).

(iv) $\Rightarrow$ (i) Since $\mathfrak{a} \subset \mathfrak{g}^0(\mathfrak{a})$, we have $\mathfrak{a} \subset \mathfrak{p}^0(\mathfrak{a})$. If $\mathfrak{a}$ is a natural $\mathfrak{p}$-subalgebra of $\mathfrak{g}$, then $\mathfrak{a} = \mathfrak{p}^0(\mathfrak{a})$ as in the proof of (i) $\Rightarrow$ (ii). $\square$

37.4.5 Corollary. (i) *Any symmetric Lie algebra has a Cartan subspace.*
(ii) *If $x \in \mathfrak{g}$ is $\mathfrak{p}$-generic, then $\mathfrak{p}^0(x)$ is a Cartan subspace of $\mathfrak{g}$.*

Proof. (i) Since $\mathfrak{g}$ has natural $\mathfrak{p}$-subalgebras, (i) follows from 37.4.4.

(ii) Let x be a $\mathfrak{p}$-generic element of $\mathfrak{g}$. If $\mathfrak{p}^0(x)$ is not a Cartan subspace, then there exists a natural $\mathfrak{p}$-subalgebra $\mathfrak{q}$ which is contained strictly in $\mathfrak{p}^0(x)$. Let $x_0 \in \mathfrak{q}$ be such that $(\operatorname{ad} x_0)^2$ induces an automorphism of $\mathfrak{p}/\mathfrak{q}$. Then $\mathfrak{p}^0(x_0) \subset \mathfrak{q} \subsetneqq \mathfrak{p}^0(x)$. This is absurd since x is $\mathfrak{p}$-generic. $\square$

37.4.6 Proposition. *Let $\mathfrak{q}$ be a $\mathfrak{p}$-subalgebra of $\mathfrak{g}$. The following conditions are equivalent:*

(i) $\mathfrak{q}$ *is natural.*
(ii) $\mathfrak{q}$ *contains a natural $\mathfrak{p}$-subalgebra of $\mathfrak{g}$.*
(iii) $\mathfrak{q}$ *contains a Cartan subspace of $\mathfrak{g}$.*
(iv) *There exists $x \in \mathfrak{p}$ such that $\mathfrak{p}^0(x) \subset \mathfrak{q}$.*

Proof. (i) $\Rightarrow$ (ii) This is clear.

(ii) $\Rightarrow$ (iii) If (ii) is verified, then $\mathfrak{q}$ contains a minimal natural $\mathfrak{p}$-subalgebra of $\mathfrak{g}$. So $\mathfrak{q}$ contains a Cartan subspace (37.4.4).

(iii) $\Rightarrow$ (iv) This follows clearly from the equivalence (i) $\Leftrightarrow$ (iii) of 37.4.4.

(iv) $\Rightarrow$ (i) If $\mathfrak{p}^0(x) \subset \mathfrak{q}$, then as $(\operatorname{ad} x)^2$ induces an automorphism of $\mathfrak{p}/\mathfrak{p}^0(x)$ (37.3.4), it induces an automorphism of $\mathfrak{p}/\mathfrak{q}$. $\square$

37.4.7 Theorem. *Let A be a subset of $\mathfrak{p}$ verifying the following conditions:*
a) $[x, y] = 0$ *for all $x, y \in A$.*
b) $\operatorname{ad}_{\mathfrak{g}} x$ *is semisimple for all $x \in A$.*
Then A is contained in a Cartan subspace $\mathfrak{a}$ of $\mathfrak{g}$. In particular, if $x \in \mathfrak{p}$ and $\operatorname{ad}_{\mathfrak{g}} x$ is semisimple, then x belongs to a Cartan subspace of $\mathfrak{g}$.

Proof. Condition a) implies that $A \subset \mathfrak{g}^0(A)$, so $\mathfrak{p}^0(A)$ is a natural $\mathfrak{p}$-subalgebra of $\mathfrak{g}$ (37.3.4). Now condition b) implies that $\mathfrak{p}^0(A)$ is the centralizer of A in $\mathfrak{p}$. Let $\mathfrak{a}$ be a Cartan subspace of $\mathfrak{g}$ contained in $\mathfrak{p}^0(A)$ (37.4.6), then $[\mathfrak{a}, x] = \{0\}$ for all $x \in A$, hence $A \subset \mathfrak{p}^0(\mathfrak{a}) = \mathfrak{a}$. $\square$

37.4.8 Proposition. *Let $\mathfrak{a}$ be a Cartan subspace of $\mathfrak{g}$ and $\mathfrak{h}$ the Lie subalgebra of $\mathfrak{g}$ generated by $\mathfrak{a}$. Then $\mathfrak{h} = \mathfrak{a} \oplus [\mathfrak{a}, \mathfrak{a}]$, and $\mathfrak{h}$ is nilpotent.*

Proof. We have $\mathfrak{a} = \mathfrak{p}^0(\mathfrak{a}) \subset \mathfrak{g}^0(\mathfrak{a})$, and $\mathfrak{g}^0(\mathfrak{a})$ is a Lie subalgebra of $\mathfrak{g}$. It follows that:

$$[\mathfrak{a}, [\mathfrak{a}, \mathfrak{a}]] \subset \mathfrak{g}^0(\mathfrak{a}) \cap \mathfrak{p} = \mathfrak{a} \; , \; [[\mathfrak{a}, \mathfrak{a}], [\mathfrak{a}, \mathfrak{a}]] \subset [\mathfrak{a}, \mathfrak{a}].$$

Thus $\mathfrak{h} = \mathfrak{a} \oplus [\mathfrak{a}, \mathfrak{a}]$. Let L be the Killing form of $\mathfrak{h}$. Since $(\mathfrak{h}; [\mathfrak{a}, \mathfrak{a}], \mathfrak{a})$ is a symmetric Lie algebra, $\mathfrak{a}$ and $[\mathfrak{a}, \mathfrak{a}]$ are orthogonal with respect to L (37.2.3). As $(\operatorname{ad} x)^2|_{\mathfrak{h}}$ is nilpotent for $x \in \mathfrak{a}$, $L(x, x) = 0$. It follows that $L(\mathfrak{a}, \mathfrak{a}) = \{0\}$, and hence $L(\mathfrak{h}, \mathfrak{a}) = \{0\}$. This implies that $\mathfrak{a}$ is contained in the radical of $\mathfrak{h}$. Since $\operatorname{ad} x|_{\mathfrak{h}}$ is nilpotent for all $x \in \mathfrak{a}$, it follows from 19.5.7 that $\mathfrak{a}$ is contained in the largest nilpotent ideal of $\mathfrak{h}$. But $\mathfrak{h} = \mathfrak{a} \oplus [\mathfrak{a}, \mathfrak{a}]$, so $\mathfrak{h}$ is nilpotent. $\square$

37.4.9 Corollary. *Let $\mathfrak{a}$ be a Cartan subspace of $\mathfrak{g}$ and (V,σ) a finite-dimensional representation of $\mathfrak{g}$. There exists a basis $\mathcal{E}$ of V such that with respect to $\mathcal{E}$, the matrix of the $\sigma(x)$'s, $x \in \mathfrak{a}$, are upper triangular.*

Proof. This is clear by 19.4.4 and 37.4.8. $\square$

37.4.10 Let K be the smallest algebraic subgroup of $\mathrm{Aut}\,\mathfrak{g}$ such that its Lie algebra contains $\mathrm{ad}_{\mathfrak{g}}\,\mathfrak{k}$. Then the set $\{\alpha|_{\mathfrak{k}}; \alpha \in K\}$ is the algebraic adjoint group of $\mathfrak{k}$ (24.8.5), and we have $\alpha(\mathfrak{k}) = \mathfrak{k}$, $\alpha(\mathfrak{p}) = \mathfrak{p}$ for all $\alpha \in K$.

Theorem. *Let $(\mathfrak{g}; \mathfrak{k}, \mathfrak{p})$ be a symmetric Lie algebra and K as above. If $\mathfrak{a}_1, \mathfrak{a}_2$ are Cartan subspaces of $\mathfrak{g}$, then there exists $\alpha \in K$ such that $\mathfrak{a}_2 = \alpha(\mathfrak{a}_1)$.*

Proof. For $i = 1, 2$, let U_i be the set of $x \in \mathfrak{a}_i$ such that $(\mathrm{ad}\,x)^2$ induces an automorphism of $\mathfrak{p}/\mathfrak{a}_i$. Then U_i is a non-empty Zariski-open subset of $\mathfrak{a}_i$, and if $x \in U_i$, then $\mathfrak{a}_i = \mathfrak{p}^0(x)$. Let

$$\varphi_i : K \times U_i \to \mathfrak{p} , \ (\alpha, x) \mapsto \alpha(x).$$

If $x \in U_i$, denote by $f_{i,x}$ the differential of φ_i at the point (e, x). Since U_i is open, if $x \in U_i$, the image of $f_{i,x}$ contains $\mathfrak{a}_i$ and $[\mathfrak{k}, x]$ (see 29.1.4), so it contains $(\mathrm{ad}\,x)^2(\mathfrak{p})$. As $\mathfrak{p} = \mathfrak{a}_i + (\mathrm{ad}\,x)^2(\mathfrak{p})$, the image of $f_{i,x}$ contains also $\mathfrak{p}$, and the interior of the image of φ_i is non-empty (15.4.2 and 16.5.7). From this, we deduce that there exist $x_0 \in \mathfrak{p}$, $\alpha_i \in K$ and $x_i \in U_i$ such that $x_0 = \alpha_1(x_1) = \alpha_2(x_2)$. If $\alpha = \alpha_2^{-1}\alpha_1$, then $x_2 = \alpha(x_1)$. So we have $\mathfrak{a}_2 = \alpha(\mathfrak{a}_1)$ because $\mathfrak{a}_i = \mathfrak{p}^0(x_i)$. $\square$

37.4.11 Corollary. *Let $(\mathfrak{g}; \mathfrak{k}, \mathfrak{p})$ be a symmetric Lie algebra. Then:*
  (i) *All the Cartan subspaces of $\mathfrak{g}$ have the same dimension.*
  (ii) *Any Cartan subspace of $\mathfrak{g}$ is of the form $\mathfrak{p}^0(x)$ where x is a $\mathfrak{p}$-generic element of $\mathfrak{g}$.*

Proof. (i) This is clear by 37.4.10.
  (ii) If x is $\mathfrak{p}$-generic, then $\mathfrak{p}^0(x)$ is a Cartan subspace (37.4.5). Now if $\mathfrak{a}$ is a Cartan subspace of $\mathfrak{g}$, then there exists $y \in \mathfrak{p}$ such that $\mathfrak{a} = \mathfrak{p}^0(y)$ (37.4.4). By (i), y is $\mathfrak{p}$-generic. $\square$

37.5 The case of reductive Lie algebras

37.5.1 Proposition. *Let $(\mathfrak{g}; \mathfrak{k}, \mathfrak{p})$ or $(\mathfrak{g}, \theta)$ be a semisimple symmetric Lie algebra, that is, $\mathfrak{g}$ is semisimple.*
  (i) *Let $x \in \mathfrak{g}$ and s, n its semisimple and nilpotent components. If $x \in \mathfrak{k}$ (resp. $x \in \mathfrak{p}$), then $s, n \in \mathfrak{k}$ (resp. $s, n \in \mathfrak{p}$).*
  (ii) *The Lie algebra $\mathfrak{k}$ is reductive in $\mathfrak{g}$ and $\mathfrak{k} = \mathfrak{n}_{\mathfrak{g}}(\mathfrak{k})$.*

Proof. (i) This is obvious since $\theta(x) = \theta(s) + \theta(n)$ is the Jordan decomposition of $\theta(x)$.

(ii) Let L be the Killing form on $\mathfrak{g}$. By 37.2.3, the restriction of L to $\mathfrak{k}$ and $\mathfrak{p}$ are non-degenerate. So $\mathfrak{k}$ is reductive in $\mathfrak{g}$ by (i) and 20.5.12.

Now the normalizer of $\mathfrak{k}$ in $\mathfrak{g}$ is of the form $\mathfrak{k} \oplus \mathfrak{u}$ where $\mathfrak{u}$ is a subspace of $\mathfrak{p}$. We have $[\mathfrak{k}, \mathfrak{u}] \subset \mathfrak{k} \cap \mathfrak{p} = \{0\}$, and as $\mathfrak{k}$ is reductive in $\mathfrak{g}$, there exists a subspace $\mathfrak{o}$ of $\mathfrak{p}$ complementary to $\mathfrak{u}$ such that $[\mathfrak{k}, \mathfrak{o}] \subset \mathfrak{o}$. From:

$$[\mathfrak{g}, \mathfrak{k} + \mathfrak{v}] = [\mathfrak{k} + \mathfrak{u} + \mathfrak{v}, \mathfrak{k} + \mathfrak{v}] \subset \mathfrak{k} + [\mathfrak{k}, \mathfrak{u} + \mathfrak{v}] \subset \mathfrak{k} + \mathfrak{v},$$

we deduce that $\mathfrak{k} \oplus \mathfrak{o}$ is an ideal of $\mathfrak{g}$. But $[\mathfrak{u}, \mathfrak{u}] \subset \mathfrak{k}$, so $\mathfrak{g}/(\mathfrak{k} \oplus \mathfrak{u})$ is abelian. Hence $\mathfrak{g} = \mathfrak{k} \oplus \mathfrak{o}$ (because $\mathfrak{g}$ is semisimple), and $\mathfrak{u} = \{0\}$. □

37.5.2 Theorem. *Let* $(\mathfrak{g}; \mathfrak{k}, \mathfrak{p})$ *be a semisimple symmetric Lie algebra,* L *its Killing form,* $\mathfrak{a}$ *a Cartan subspace of* $\mathfrak{g}$ *and* $\mathfrak{m}$ *the centralizer of* $\mathfrak{a}$ *in* $\mathfrak{k}$.

(i) $\mathfrak{a}$ *is an abelian Lie subalgebra of* $\mathfrak{g}$ *consisting of semisimple elements.*

(ii) *The centralizer of* $\mathfrak{a}$ *in* $\mathfrak{g}$ *is* $\mathfrak{m} \oplus \mathfrak{a}$.

(iii) *The restrictions of* L *to* $\mathfrak{a} \times \mathfrak{a}$ *and* $\mathfrak{m} \times \mathfrak{m}$ *are non-degenerate.*

(iv) *The Lie algebras* $\mathfrak{a}$ *and* $\mathfrak{m}$ *are reductive in* $\mathfrak{g}$.

(v) $\mathfrak{a}$ *is contained in the centre of* $\mathfrak{g}^0(\mathfrak{a})$ *and* $(\mathfrak{g}^0(\mathfrak{a}); \mathfrak{m}, \mathfrak{a})$ *is a reductive symmetric Lie algebra, that is,* $\mathfrak{g}^0(\mathfrak{a})$ *is reductive.*

(vi) *Let* $\mathfrak{l}$ *be a Cartan subalgebra of* $\mathfrak{m}$. *Then* $\mathfrak{l} \oplus \mathfrak{a}$ *is a Cartan subalgebra of* $\mathfrak{g}$.

Proof. By 37.1.3 and 37.1.4, we have:

$$\mathfrak{g} = \bigoplus_{\lambda \in \mathfrak{a}^*} \mathfrak{g}^\lambda(\mathfrak{a}).$$

Also $L(\mathfrak{g}^\lambda(\mathfrak{a}), \mathfrak{g}^\mu(\mathfrak{a})) = \{0\}$ if $\lambda + \mu \neq 0$ (37.1.8). It follows that $L|_{\mathfrak{g}^0(\mathfrak{a}) \times \mathfrak{g}^0(\mathfrak{a})}$ is non-degenerate, and so $\mathfrak{g}^0(\mathfrak{a})$ is reductive (20.5.4 (iii)).

Let L_0 be the Killing form of $\mathfrak{g}^0(\mathfrak{a})$. If $x \in \mathfrak{a}$, $(\operatorname{ad} x)^2|_{\mathfrak{g}^0(\mathfrak{a})}$ is nilpotent, hence $L_0(x, x) = 0$. Thus $L_0|_{\mathfrak{a} \times \mathfrak{a}}$ is both symmetric and alternating, so it is identical zero. As $L_0(\mathfrak{k}^0(\mathfrak{a}), \mathfrak{a}) = \{0\}$ (37.2.3) and $\mathfrak{g}^0(\mathfrak{a}) = \mathfrak{k}^0(\mathfrak{a}) \oplus \mathfrak{a}$, we see that $\mathfrak{a}$ is contained in the kernel of L_0. Since $\mathfrak{g}^0(\mathfrak{a})$ is reductive, $\mathfrak{a}$ is contained in the centre of $\mathfrak{g}^0(\mathfrak{a})$. In particular, $\mathfrak{a}$ is abelian.

It follows that the centralizer of $\mathfrak{a}$ in $\mathfrak{g}$ (resp. $\mathfrak{k}, \mathfrak{p}$) is $\mathfrak{g}^0(\mathfrak{a})$ (resp. $\mathfrak{m}, \mathfrak{a}$). So we have $\mathfrak{g}^0(\mathfrak{a}) = \mathfrak{m} \oplus \mathfrak{a}$, and part (v) is clear.

Recall from above that $L|_{\mathfrak{g}^0(\mathfrak{a}) \times \mathfrak{g}^0(\mathfrak{a})}$ is non-degenerate. Since $L(\mathfrak{m}, \mathfrak{a}) = \{0\}$, the restrictions of L to $\mathfrak{a} \times \mathfrak{a}$ and $\mathfrak{m} \times \mathfrak{m}$ are therefore non-degenerate. So by 20.5.4 (iii), $\mathfrak{a}$ and $\mathfrak{m}$ are reductive.

Let $\mathfrak{l}$ be as in (vi) and $\mathfrak{h} = \mathfrak{l} \oplus \mathfrak{a}$. The Lie algebras $\mathfrak{l}$ and $\mathfrak{a}$ are nilpotent and $[\mathfrak{l}, \mathfrak{a}] = \{0\}$. So $\mathfrak{h}$ is nilpotent. Now $\mathfrak{g}^0(\mathfrak{h}) \subset \mathfrak{g}^0(\mathfrak{a}) = \mathfrak{a} \oplus \mathfrak{m}$. Thus if $x \in \mathfrak{g}^0(\mathfrak{h})$, we can write $x = y + z$ with $y \in \mathfrak{a}$ and $z \in \mathfrak{m}$. But $z \in \mathfrak{m}^0(\mathfrak{l})$, so $z \in \mathfrak{l}$ because $\mathfrak{l}$ is a Cartan subalgebra of $\mathfrak{m}$. It follows that $\mathfrak{h}$ is a Cartan subalgebra of $\mathfrak{g}$.

As $\mathfrak{a} \subset \mathfrak{h}$, the elements of $\mathfrak{a}$ are semi-simple. So $\mathfrak{a}$ is a torus of $\mathfrak{g}$, hence it is reductive in $\mathfrak{g}$ (20.5.16). By 20.5.13, $\mathfrak{m} \oplus \mathfrak{a}$ is also reductive in $\mathfrak{g}$. We obtain therefore by 20.5.12 that $\mathfrak{m}$ is reductive in $\mathfrak{g}$. □

37.5.3 Theorem. *Let* $(\mathfrak{g}; \mathfrak{k}, \mathfrak{p})$ *be a reductive symmetric Lie algebra,* $\mathfrak{a}$ *a Cartan subspace of* $\mathfrak{g}$ *and* $\mathfrak{m}$ *the centralizer of* $\mathfrak{a}$ *in* $\mathfrak{k}$.

(i) $\mathfrak{a}$ *is an abelian Lie subalgebra of* $\mathfrak{g}$ *consisting of semisimple elements.*

(ii) *The centralizer of* $\mathfrak{a}$ *in* $\mathfrak{g}$ *is* $\mathfrak{m} \oplus \mathfrak{a}$.

(iii) *The Lie algebras* $\mathfrak{a}$ *and* $\mathfrak{m}$ *are reductive in* $\mathfrak{g}$.

(iv) $\mathfrak{a}$ *is contained in the centre of* $\mathfrak{g}^0(\mathfrak{a})$ *and* $(\mathfrak{g}^0(\mathfrak{a}); \mathfrak{m}, \mathfrak{a})$ *is a reductive symmetric Lie algebra.*

(v) *Let* $\mathfrak{l}$ *be a Cartan subalgebra of* $\mathfrak{m}$. *Then* $\mathfrak{l} \oplus \mathfrak{a}$ *is a Cartan subalgebra of* $\mathfrak{g}$.

Proof. All the subspaces $\mathfrak{u}$ considered here verify:

$$\mathfrak{u} = (\mathfrak{u} \cap [\mathfrak{g}, \mathfrak{g}]) \oplus (\mathfrak{u} \cap \mathfrak{z}(\mathfrak{g})).$$

So the assertions are direct consequences of 37.5.2. □

37.5.4 Corollary. *The following conditions are equivalent for a subset* $\mathfrak{a}$ *of* $\mathfrak{p}$:

(i) $\mathfrak{a}$ *is a Cartan subspace of* $\mathfrak{g}$.

(ii) $\mathfrak{a}$ *is maximal among the set of subspaces of* $\mathfrak{p}$ *consisting of semisimple elements which commute pairwise.*

(iii) $\mathfrak{a}$ *is a maximal subset of pairwise commuting semisimple elements in* $\mathfrak{p}$.

Proof. This is obvious by 37.4.7 and 37.5.3. □

37.5.5 Remarks. 1) Let $\mathfrak{g} = \Bbbk x \oplus \Bbbk y$ be the Lie algebra defined by $[x, y] = y$. Set $\mathfrak{k} = \Bbbk x$ and $\mathfrak{p} = \Bbbk y$, then $(\mathfrak{g}; \mathfrak{k}, \mathfrak{p})$ is a symmetric Lie algebra. Clearly $\mathfrak{a} = \mathfrak{p}$ is a Cartan subspace of $\mathfrak{g}$ and $c_{\mathfrak{k}}(\mathfrak{p}) = \{0\}$. But $\mathfrak{a}$ is not a Cartan subalgebra of $\mathfrak{g}$.

2) Let $(E_{ij})_{1 \leqslant i,j \leqslant 3}$ be the canonical basis of $\mathfrak{gl}_3(\Bbbk)$. Set:

$$\mathfrak{g}_0 = \Bbbk E_{11} + \Bbbk E_{22} + \Bbbk E_{33} \ , \ \mathfrak{g}_1 = \Bbbk E_{12} + \Bbbk E_{23} \ , \ \mathfrak{g}_2 = \Bbbk E_{13},$$
$$\mathfrak{g}_i = \{0\} \ \text{if} \ i \in \mathbb{Z} \setminus \{0, 1, 2\}.$$

The direct sum $\mathfrak{g}$ of the $\mathfrak{g}_i$'s, $i \in \mathbb{Z}$, is a graded Lie algebra. As we have seen in example 3 of 37.2.4, $(\mathfrak{g}; \mathfrak{g}_0 \oplus \mathfrak{g}_2, \mathfrak{g}_1)$ is a symmetric Lie algebra. It is easy to see that $\mathfrak{p} = \mathfrak{g}_1$ is a Cartan subspace of $\mathfrak{g}$ which is not abelian.

37.6 Linear forms

37.6.1 Let us identify $\mathfrak{k}^*$ (resp. $\mathfrak{p}^*$) with the set of linear forms on $\mathfrak{g}$ which is zero on $\mathfrak{p}$ (resp. $\mathfrak{k}$). Then $\mathfrak{g}^* = \mathfrak{k}^* \oplus \mathfrak{p}^*$.

Let us consider the coadjoint representation of $\mathfrak{g}$ on $\mathfrak{g}^*$ and denote the action by $(x, f) \mapsto x.f$. If $x \in \mathfrak{k}$, $y \in \mathfrak{p}$, $f \in \mathfrak{k}^*$ and $g \in \mathfrak{p}^*$, then:

$$x.f \in \mathfrak{k}^* \ , \ y.f \in \mathfrak{p}^* \ , \ x.g \in \mathfrak{p}^* \ , \ y.g \in \mathfrak{k}^*.$$

Using the notations of 19.7.3, we set, for $f \in \mathfrak{g}^*$:

$$\mathfrak{k}^{(f)} = \mathfrak{k} \cap \mathfrak{g}^{(f)} \; , \; \mathfrak{p}^{(f)} = \mathfrak{p} \cap \mathfrak{g}^{(f)}.$$

If $f \in \mathfrak{k}^* \cup \mathfrak{p}^*$, then we obtain easily that $\mathfrak{g}^{(f)} = \mathfrak{k}^{(f)} + \mathfrak{p}^{(f)}$.

Proposition. *If $f \in \mathfrak{p}^*$, then:*

$$\dim \mathfrak{k} - \dim \mathfrak{k}^{(f)} = \dim \mathfrak{p} - \dim \mathfrak{p}^{(f)}.$$

Proof. The bilinear form Φ_f of 19.7.3 induces a non-degenerate alternating bilinear form $\widetilde{\Phi}_f$ on $\mathfrak{g}/\mathfrak{g}^{(f)} = (\mathfrak{k}/\mathfrak{k}^{(f)}) \times (\mathfrak{p}/\mathfrak{p}^{(f)})$. Since $\mathfrak{k}/\mathfrak{k}^{(f)}$ et $\mathfrak{p}/\mathfrak{p}^{(f)}$ are totally isotropic with respect to $\widetilde{\Phi}_f$, the result is clear. $\square$

37.6.2 In view of 37.6.1, we have the following result:

Proposition. *The following conditions are equivalent for $f \in \mathfrak{p}^*$:*
   (i) $\dim \mathfrak{k}^{(f)} \leqslant \dim \mathfrak{k}^{(g)}$ *for all $g \in \mathfrak{p}^*$.*
   (ii) $\dim \mathfrak{p}^{(f)} \leqslant \dim \mathfrak{p}^{(g)}$ *for all $g \in \mathfrak{p}^*$.*
   (iii) $\dim \mathfrak{g}^{(f)} \leqslant \dim \mathfrak{g}^{(g)}$ *for all $g \in \mathfrak{p}^*$.*

37.6.3 We say that $f \in \mathfrak{p}^*$ is $\mathfrak{p}$-*regular* if it verifies the conditions of 37.6.2. Let us denote by $\mathfrak{p}^*_{\mathrm{reg}}$ the set of $\mathfrak{p}$-regular elements of $\mathfrak{p}^*$. It is an open dense subset of $\mathfrak{p}^*$.

37.6.4 Proposition. *Let $f \in \mathfrak{p}^*_{\mathrm{reg}}$. Then:*
   (i) $[\mathfrak{k}^{(f)}, \mathfrak{p}^{(f)}] = \{0\}$.
   (ii) $\mathfrak{k}^{(f)}$ *is an ideal of $\mathfrak{g}^{(f)}$, and $[\mathfrak{p}^{(f)}, \mathfrak{p}^{(f)}]$ is an abelian ideal of $\mathfrak{g}^{(f)}$.*

Proof. Since $f \in \mathfrak{p}^*_{\mathrm{reg}}$, we have $\dim \mathfrak{g}^{(f)} \leqslant \dim \mathfrak{g}^{(g)}$ for all $g \in \mathfrak{p}^*$, or equivalently, for all $g \in \mathfrak{g}^*$ such that $f|_\mathfrak{k} = g|_\mathfrak{k} = 0$. It follows from 19.7.4 that $[\mathfrak{g}^{(f)}, \mathfrak{g}^{(f)}] \subset \mathfrak{k}$. Thus $[\mathfrak{p}^{(f)}, \mathfrak{k}^{(f)}] = \{0\}$, and $\mathfrak{k}^{(f)}$ is an ideal of $\mathfrak{g}^{(f)}$. Now, if $x \in \mathfrak{k}^{(f)}$, $y, z \in \mathfrak{p}^{(f)}$, then:

$$[x, [y, z]] = [[x, y], z] + [y, [x, z]] = 0.$$

So $[\mathfrak{p}^{(f)}, \mathfrak{p}^{(f)}]$ is an abelian ideal of $\mathfrak{g}^{(f)}$. $\square$

37.6.5 Let $\mathfrak{g} = \mathbb{k}x \oplus \mathbb{k}y \oplus \mathbb{k}z$ be the Lie algebra defined as follows:

$$[x, y] = z, [x, z] = [y, z] = 0.$$

Then $(\mathfrak{g}; \mathfrak{k} = \mathbb{k}z, \mathfrak{p} = \mathbb{k}x \oplus \mathbb{k}y)$ is a symmetric Lie algebra. For any $f \in \mathfrak{p}^*$, we have $\mathfrak{g}^{(f)} = \mathfrak{g}$, $\mathfrak{p}^{(f)} = \mathfrak{p}$. Thus f is $\mathfrak{p}$-regular, but $\mathfrak{p}^{(f)}$ is not abelian.
   Note that:

$$\mathfrak{k}^x = \mathfrak{k} \; , \; \mathfrak{p}^x = \mathbb{k}x \; , \; \dim \mathfrak{k} - \dim \mathfrak{k}^x = 0 \; , \; \dim \mathfrak{p} - \dim \mathfrak{p}^x = 1.$$

So the result of 37.6.1 does not extend to the adjoint representation.

References and comments

   • [29], [50], [54], [55].

The notion of natural $\mathfrak{p}$-subalgebra comes from [55].

38

Semisimple symmetric Lie algebras

In this chapter, we examine to what extent the results for in Chapters 31, 32, 33, 34 and 35 for semisimple Lie algebras have analogues for semisimple symmetric Lie algebras.

38.1 Notations

38.1.1 In this chapter, $(\mathfrak{g};\mathfrak{k},\mathfrak{p})$ or $(\mathfrak{g},\theta)$ (see 37.2.1 and 37.2.3) will be a semisimple symmetric Lie algebra whose Killing form will be denoted by L.

We shall denote by $\mathfrak{a}$ a Cartan subspace of $\mathfrak{g}$, $\mathfrak{m}$ the centralizer of $\mathfrak{a}$ in $\mathfrak{k}$ and $\mathfrak{l}$ a Cartan subalgebra of $\mathfrak{m}$. We saw in 37.5.2 that $\mathfrak{h} = \mathfrak{l} \oplus \mathfrak{a}$ is a Cartan subalgebra of $\mathfrak{g}$.

38.1.2 Let us denote by R the root system of $(\mathfrak{g}, \mathfrak{h})$ and W its Weyl group. Then $\theta(\mathfrak{h}) = \mathfrak{h}$, and if σ is the transpose of $\theta|_{\mathfrak{h}}$, then $\sigma(R) = R$.

For $\alpha \in R$, $\mathfrak{g}^\alpha$ denotes the generalized eigenspace of $\mathfrak{g}$ relative to α. Set:

$$R' = \{\alpha \in R;\ \alpha|_{\mathfrak{a}} = 0\}\ ,\ R'' = \{\alpha \in R;\ \alpha|_{\mathfrak{a}} \neq 0\}.$$

We have:

$$R' = \{\alpha \in R;\ \sigma(\alpha) = \alpha\}.$$

38.1.3 Let B be a base of R and R_+ the corresponding set of positive roots. Set:

$$R''_+ = R'' \cap R_+\ ,\ R''_- = R'' \cap R_-.$$

From now on, we shall assume that B is chosen in such a way so that the conditions of 36.1.4 are verified.

38.2 Iwasawa decomposition

38.2.1 Proposition. (i) *The set* $\{\alpha|_{\mathfrak{l}}; \alpha \in R'\}$ *is the root system of* $(\mathfrak{m}, \mathfrak{l})$.
(ii) *If* $\alpha \in R'$, *then* $\mathfrak{m}^{\alpha|_{\mathfrak{l}}} = \mathfrak{g}^{\alpha}$, *and:*

$$\mathfrak{m} = \mathfrak{l} \oplus \Big(\bigoplus_{\alpha \in R'} \mathfrak{g}^{\alpha} \Big).$$

Proof. Let $\alpha \in R'$ and $x \in \mathfrak{g}^{\alpha}$. Then $[x, \mathfrak{a}] = \{0\}$, so $x = y + z$ where $y \in \mathfrak{m}$ and $z \in \mathfrak{a}$ (37.5.2). On the other hand, if $l \in \mathfrak{l}$, then:

$$\alpha(l)(y + z) = [l, y + z] = [l, y] \in \mathfrak{m}.$$

Thus $z = 0$. So $x \in \mathfrak{m}$ and $\alpha|_{\mathfrak{l}}$ is a root of $(\mathfrak{m}, \mathfrak{l})$ and $\mathfrak{g}^{\alpha} = \mathfrak{m}^{\alpha|_{\mathfrak{l}}}$.

Conversely, let β be a root of $(\mathfrak{m}, \mathfrak{l})$ and $x \in \mathfrak{m}^{\beta}$. Denote by α the linear form on $\mathfrak{h}$ extending β which is zero on $\mathfrak{a}$. For $a \in \mathfrak{a}$ and $l \in \mathfrak{l}$, we have:

$$[a + l, x] = [l, x] = \beta(l)x = \alpha(a + l)x.$$

Thus $\alpha \in R'$ and the result follows. $\square$

38.2.2 Lemma. *If* $\alpha \in R$, *then* $\alpha + \sigma(\alpha) \notin R$.

Proof. Let $x \in \mathfrak{g}^{\alpha} \setminus \{0\}$. We have $\theta(x) \in \mathfrak{g}^{\sigma(\alpha)} \setminus \{0\}$. If $\beta = \alpha + \sigma(\alpha) \in R$, then $[x, \theta(x)] \in \mathfrak{g}^{\beta} \setminus \{0\}$. But $\beta \in R'$, so $\mathfrak{g}^{\beta} \subset \mathfrak{m} \subset \mathfrak{k}$ (38.2.1). Hence:

$$[x, \theta(x)] = \theta([\theta(x), x]) = [\theta(x), x] = -[x, \theta(x)].$$

Thus $[x, \theta(x)] = 0$. Contradiction. $\square$

38.2.3 Let S be the set of restrictions to $\mathfrak{a}$ of the elements of R''. By 36.2.1 and 38.2.2, S is a (not necessarily reduced) root system in $\mathfrak{a}^*$. We call S the *root system of* $(\mathfrak{g}, \mathfrak{a})$ or the *restricted root system of* $\mathfrak{h}$ *to* $\mathfrak{a}$. We shall denote the Weyl group of S by W_S.

38.2.4 Lemma. *The following conditions are equivalent for* $\lambda \in \mathfrak{a}^* \setminus \{0\}$*:*
   (i) $\lambda \in S$.
   (ii) *There exists* $x \in \mathfrak{g} \setminus \{0\}$ *such that* $[a, x] = \lambda(a)x$ *for all* $a \in \mathfrak{a}$.
   (iii) *There exists* $p \in \mathfrak{p} \setminus \{0\}$ *such that* $[a, [a, p]] = \lambda(a)^2 p$ *for all* $a \in \mathfrak{a}$.
   (iv) *There exists* $k \in \mathfrak{k} \setminus \{0\}$ *such that* $[a, [a, k]] = \lambda(a)^2 k$ *for all* $a \in \mathfrak{a}$.

Proof. (i) $\Rightarrow$ (ii) This is obvious.
   (ii) $\Rightarrow$ (i) Let $\mathfrak{r}$ be the set of $x \in \mathfrak{g}$ verifying $[a, x] = \lambda(a)x$ for all $a \in \mathfrak{a}$. For $(a, x, m) \in \mathfrak{a} \times \mathfrak{r} \times \mathfrak{m}$, we have:

$$\lambda(a)[m, x] = [m, [a, x]] = [a, [m, x]].$$

Thus $\mathfrak{r}$ is ad $\mathfrak{m}$-stable, so it is ad $\mathfrak{h}$-stable. So there exist $\mu \in \mathfrak{h}^*$ and $x \in \mathfrak{r} \setminus \{0\}$ such that $[h, x] = \mu(h)x$ for all $h \in \mathfrak{h}$. As $\mu|_{\mathfrak{a}} = \lambda$, we have $\lambda \in S$.

(ii) $\Rightarrow$ (iii) Let us write $x = k + p$, with $(k, p) \in \mathfrak{k} \times \mathfrak{p}$. If $a \in \mathfrak{a}$, then $[a, \mathfrak{k}] \subset \mathfrak{p}$ and $[a, \mathfrak{p}] \subset \mathfrak{k}$. If $[a, x] = \lambda(a)x$ for all $a \in \mathfrak{a}$, then:

$$[a, k] = \lambda(a)p \ , \ [a, p] = \lambda(a)k.$$

This proves that k et p are non-zero and:

$$[a, [a, k]] = \lambda^2(a)k \ , \ [a, [a, p]] = \lambda^2(a)p.$$

So we have also proved (ii) $\Rightarrow$ (iv).

(iii) $\Rightarrow$ (i) Let $p \in \mathfrak{p} \setminus \{0\}$ be such that $[a, [a, p]] = \lambda^2(a)p$ for all $a \in \mathfrak{a}$. Let us write $p = \sum_i x_i$, where the x_i's are linear independent elements of $\mathfrak{g}$ verifying $[a, x_i] = \mu_i(a)x_i$ for all $a \in \mathfrak{a}$ and the μ_i's are pairwise distinct linear forms on $\mathfrak{a}$. Then:

$$\lambda^2(a) \sum_i x_i = \sum_i \mu_i^2(a)x_i.$$

So $\lambda^2(a) = \mu_i^2(a)$, which implies that the μ_i's have the same kernel as λ. Thus there exists $\varepsilon_i \in \{-1, 1\}$ such that $\mu_i = \varepsilon_i \lambda$. In view of the implication (ii) $\Rightarrow$ (i), we deduce that $\lambda \in S$ or $-\lambda \in S$, hence $\lambda \in S$.

(iv) $\Rightarrow$ (i) The proof is analogue to the proof of (iii) $\Rightarrow$ (i). $\square$

38.2.5 From now on, if $\lambda \in \mathfrak{a}^*$, we shall denote by $\mathfrak{g}_\mathfrak{a}^\lambda$ the set of $x \in \mathfrak{g}$ such that $[a, x] = \lambda(a)x$ for all $a \in \mathfrak{a}$. Thus:

$$\mathfrak{g} = \mathfrak{h} \oplus \left(\bigoplus_{\lambda \in S} \mathfrak{g}_\mathfrak{a}^\lambda \right).$$

If $\lambda \in S$, then $\mathfrak{g}_\mathfrak{a}^\lambda$ is the sum of the $\mathfrak{g}^\alpha$'s for $\alpha \in R$ verifying $\alpha|_\mathfrak{a} = \lambda$.

38.2.6 Lemma. *Let $\lambda \in S$ and $w \in W_S$. Then $\dim \mathfrak{g}_\mathfrak{a}^\lambda = \dim \mathfrak{g}_\mathfrak{a}^{w(\lambda)}$.*

Proof. Let A (resp. B) be the set of roots of R whose restriction to $\mathfrak{a}$ is λ (resp. $w(\lambda)$). By 36.2.5, there exists $w' \in W$ such that $w'(\mathfrak{l}^*) = \mathfrak{l}^*$, $w'(\mathfrak{a}^*) = \mathfrak{a}^*$ and $w'|_{\mathfrak{a}^*} = w$. Thus $w'(A) = B$. Since:

$$\mathfrak{g}_\mathfrak{a}^\lambda = \bigoplus_{\nu \in A} \mathfrak{g}^\nu \ , \ \mathfrak{g}_\mathfrak{a}^{w(\lambda)} = \bigoplus_{\nu \in B} \mathfrak{g}^\nu$$

and $\dim \mathfrak{g}^\nu = 1$ for all $\nu \in R$, the result follows. $\square$

38.2.7 Let S_+ be a system of positive roots of S and:

$$\mathfrak{n} = \sum_{\lambda \in S_+} \mathfrak{g}_\mathfrak{a}^\lambda.$$

Proposition. (i) *The subspace $\mathfrak{n}$ of $\mathfrak{g}$ is a nilpotent Lie subalgebra and:*

(1) $$\mathfrak{g} = \mathfrak{k} \oplus \mathfrak{a} \oplus \mathfrak{n}.$$

This is called the Iwasawa decomposition *of $\mathfrak{g}$ defined by $\mathfrak{k}$, $\mathfrak{a}$ and S_+.*
(ii) *The orthogonal of $\mathfrak{a}$ with respect to L is $\mathfrak{k} \oplus \mathfrak{n}$.*

Proof. (i) We have:

$$(2) \qquad \mathfrak{g} = \mathfrak{m} \oplus \mathfrak{a} \oplus \left(\sum_{\lambda \in S_+} \mathfrak{g}_\mathfrak{a}^\lambda \right) \oplus \left(\sum_{\lambda \in S_+} \mathfrak{g}_\mathfrak{a}^{-\lambda} \right).$$

To establish $\mathfrak{g} = \mathfrak{k} + \mathfrak{a} + \mathfrak{n}$, it suffices to prove that $\mathfrak{g}_\mathfrak{a}^{-\lambda} \subset \mathfrak{k} + \mathfrak{n}$ for all $\lambda \in S_+$. Let $x \in \mathfrak{g}_\mathfrak{a}^{-\lambda}$, then $\theta(x) \in \theta(\mathfrak{g}_\mathfrak{a}^{-\lambda}) = \mathfrak{g}_\mathfrak{a}^\lambda$. It follows that:

$$x = (x + \theta(x)) - \theta(x) \in \mathfrak{k} + \mathfrak{n}.$$

Now if $(x, y, z) \in \mathfrak{k} \times \mathfrak{a} \times \mathfrak{n}$ verifies $x + y + z = 0$, then:

$$0 = \theta(x + y + z) = x - y + \theta(z),$$

and $2y + (z - \theta(z)) = 0$. So (2) implies that $y = z = 0$, hence $x = 0$. Thus $\mathfrak{g} = \mathfrak{k} \oplus \mathfrak{a} \oplus \mathfrak{n}$. Finally, it is clear that $\mathfrak{n}$ is a nilpotent Lie subalgebra.

(ii) We have $L(\mathfrak{a}, \mathfrak{k}) = \{0\}$ (37.2.3) and $L(\mathfrak{a}, \mathfrak{n}) = \{0\}$ (19.8.6). So (ii) follows from (i). □

38.2.8 Let B be a base of R verifying the conditions of 36.1.4. The restrictions of the elements of R''_+ to $\mathfrak{a}$ form a system of positive roots S_+ of S. The corresponding decomposition $\mathfrak{g} = \mathfrak{k} \oplus \mathfrak{a} \oplus \mathfrak{n}$ will be called the *Iwasawa decomposition of $\mathfrak{g}$ defined by* $\mathfrak{k}, \mathfrak{a}, \mathfrak{h}, B$.

Proposition. *The subspace* $\mathfrak{q} = \mathfrak{m} \oplus \mathfrak{a} \oplus \mathfrak{n}$ *is a Lie subalgebra of $\mathfrak{g}$ and $\mathfrak{n}$ is an ideal in $\mathfrak{q}$.*

Proof. We have $[\mathfrak{a}, \mathfrak{n}] \subset \mathfrak{n}$. Let $\alpha \in R''_+$ and $\beta \in R'$. If $\alpha + \beta \in R$, then $\alpha + \beta \in R''_+$ (36.1.4), so $[\mathfrak{g}^\alpha, \mathfrak{g}^\beta] \subset \mathfrak{g}^{\alpha+\beta} \subset \mathfrak{n}$. We deduce from 38.2.1 that $[\mathfrak{m}, \mathfrak{n}] \subset \mathfrak{n}$. So the result follows. □

38.2.9 In the notations of example 2 of 37.2.4 with $\mathfrak{s}$ semisimple, let $\mathfrak{t}$ be a Cartan subalgebra of $\mathfrak{s}$ and $\mathfrak{a} = \{(x, -x); x \in \mathfrak{t}\}$. Then $\mathfrak{m} = \{(x, x); x \in \mathfrak{t}\}$, and $\mathfrak{h} = \mathfrak{a} \oplus \mathfrak{m} \simeq \mathfrak{t} \times \mathfrak{t}$. We shall identify $\mathfrak{h}^*$ with $\mathfrak{t}^* \times \mathfrak{t}^*$.

Let $T = R(\mathfrak{s}, \mathfrak{t})$, $R = R(\mathfrak{g}, \mathfrak{h})$. Then $R = (T \times \{0\}) \times (\{0\} \times T)$ and $R' = \emptyset$.

If C is a base of T, then $B = (C \times \{0\}) \times (\{0\} \times -C)$ is a base of R verifying the conditions of 36.1.4. Moreover, if T_+, T_-, R_+, R_- are the corresponding sets of positive and negative roots, then $R_+ = (T_+ \times \{0\}) \times (\{0\} \times T_-)$.

Let $\mathfrak{s} = \mathfrak{t} \oplus \mathfrak{n}_+ \oplus \mathfrak{n}_-$ be the triangular decomposition of $\mathfrak{s}$ defined by C, $\mathfrak{n}_+ = \mathfrak{n}_+ \times \mathfrak{n}_-$ and $\mathfrak{n}_- = \mathfrak{n}_- \times \mathfrak{n}_+$. Then $\mathfrak{g} = \mathfrak{h} \oplus \mathfrak{n}_+ \oplus \mathfrak{n}_-$ is the triangular decomposition of $\mathfrak{g}$ defined by B and $\mathfrak{g} = \mathfrak{k} \oplus \mathfrak{a} \oplus \mathfrak{n}_+$ is the Iwasawa decomposition of $\mathfrak{g}$ defined by $\mathfrak{k}, \mathfrak{a}, \mathfrak{h}, B$.

38.2.10 Now let us consider example 1 of 37.2.4. Let $\mathfrak{a}$ be the set of diagonal matrices of $\mathfrak{g}$. Then $\mathfrak{a}$ is a Cartan subalgebra of $\mathfrak{g}$ which is also a Cartan subspace of $\mathfrak{g}$. Here, we have $R' = \emptyset$. Let $(E_{ij})_{1 \leqslant i,j \leqslant n}$ be the canonical basis of $\mathfrak{gl}_n(\Bbbk)$ and for $1 \leqslant i \leqslant n$, let ε_i be the linear form on $\mathfrak{a}$ given by $\lambda_1 E_{11} + \cdots + \lambda_n E_{nn} \mapsto \lambda_i$. Set $\beta_i = \varepsilon_i - \varepsilon_{i+1}$, $1 \leqslant i \leqslant n - 1$. Then $B = \{\beta_1, \ldots, \beta_{n-1}\}$ is a base of R. The Lie subalgebra $\mathfrak{n}$ associated to B is therefore the set of strictly upper triangular matrices of $\mathfrak{g}$.

38.3 Coroots

38.3.1 We define a non-degenerate symmetric bilinear form L_θ on $\mathfrak{g}$ by setting for $x, y \in \mathfrak{g}$:

$$L_\theta(x, y) = L(x, \theta(y)).$$

If $x \in \mathfrak{g}_\mathfrak{a}^\lambda$ and $y \in \mathfrak{g}_\mathfrak{a}^\mu$, then $L(x, y) = 0$ when $\lambda + \mu \neq 0$. Thus the decomposition

$$\mathfrak{g} = \mathfrak{m} \oplus \mathfrak{a} \oplus \left(\sum_{\lambda \in S} \mathfrak{g}_\mathfrak{a}^\lambda \right)$$

is L_θ-orthogonal. Note that the restrictions of L_θ to $\mathfrak{a} \times \mathfrak{a}$ and to $\mathfrak{g}_\mathfrak{a}^\lambda \times \mathfrak{g}_\mathfrak{a}^\lambda$ for $\lambda \in S$, are non-degenerate.

38.3.2 Since $L|_{\mathfrak{a} \times \mathfrak{a}}$ is non-degenerate, if $\phi \in \mathfrak{a}^*$, there exists a unique $v_\phi \in \mathfrak{a}$ such that for all $a \in \mathfrak{a}$:

$$L(v_\phi, a) = \phi(a).$$

Lemma. *Let* $\lambda \in S$.
(i) *If* $e \in \mathfrak{g}_\mathfrak{a}^\lambda$, *then* $[e, \theta(e)] = L(e, \theta(e)) v_\lambda = L_\theta(e, e) v_\lambda$.
(ii) *We have* $L(v_\lambda, v_\lambda) = \lambda(v_\lambda) \neq 0$.

Proof. As $\theta(e) \in \theta(\mathfrak{g}_\mathfrak{a}^\lambda) = \mathfrak{g}_\mathfrak{a}^{-\lambda}$, we have $[e, \theta(e)] \in \mathfrak{m} \oplus \mathfrak{a}$. Also:

$$\theta([e, \theta(e)]) = [\theta(e), e] = -[e, \theta(e)].$$

Hence $[e, \theta(e)] \in \mathfrak{a}$. So for $a \in \mathfrak{a}$:

$$L([e, \theta(e)], a) = L([a, e], \theta(e)) = \lambda(a) L(e, \theta(e)) = \lambda(a) L_\theta(e, e).$$

The result follows since $L|_{\mathfrak{a} \times \mathfrak{a}}$ is non-degenerate.

(ii) There exists $e \in \mathfrak{g}_\mathfrak{a}^\lambda$ such that $L_\theta(e, e) = L(e, \theta(e)) \neq 0$ (because $L_\theta|_{\mathfrak{g}_\mathfrak{a}^\lambda \times \mathfrak{g}_\mathfrak{a}^\lambda}$ is non-degenerate). We deduce therefore from (i) that there exists $(x, y) \in \mathfrak{g}_\mathfrak{a}^\lambda \times \mathfrak{g}_\mathfrak{a}^{-\lambda}$ such that $[x, y] = v_\lambda$.

Suppose that $\lambda(v_\lambda) = 0$. Then $[v_\lambda, x] = [v_\lambda, y] = 0$. Thus $\mathfrak{u} = \Bbbk v_\lambda \oplus \Bbbk x \oplus \Bbbk y$ is a nilpotent Lie algebra. If $z \in [\mathfrak{u}, \mathfrak{u}]$, then the eigenvalues of $\mathrm{ad}_\mathfrak{g} z$ are equal to zero by Lie's Theorem. Since $v_\lambda \in [\mathfrak{u}, \mathfrak{u}]$, we have $\alpha(v_\lambda) = 0$ for all $\alpha \in R$. Contradiction. $\square$

38.3.3 We set, for $\lambda \in S$,

$$t_\lambda = \frac{2 v_\lambda}{\lambda(v_\lambda)}.$$

Thus $\lambda(t_\lambda) = 2$. It follows from the proof of 38.3.2 that there exist $e_\lambda \in \mathfrak{g}_\mathfrak{a}^\lambda$ and $f_\lambda \in \mathfrak{g}_\mathfrak{a}^{-\lambda}$ such that $(e_\lambda, t_\lambda, f_\lambda)$ is a S-triple of $\mathfrak{g}$.

Lemma. *Let* $\lambda, \mu \in S$.
(i) *We have* $\mu(t_\lambda) \in \mathbb{Z}$.
(ii) *The set of* $n \in \mathbb{Z}$ *such that* $\mu + n\lambda \in S \cup \{0\}$ *is an interval* $[-n', n'']$ *with* $n', n'' \in \mathbb{N}$. *Moreover,* $n'' - n' = \mu(t_\lambda)$.
(iii) *We have* $\mu - \mu(t_\lambda)\lambda \in S$.

Proof. Let:

$$\mathfrak{g}_\lambda = \Bbbk t_\lambda \oplus \Bbbk e_\lambda \oplus \Bbbk f_\lambda \simeq \mathfrak{sl}(2,\Bbbk) \ , \quad \mathfrak{b} = \sum_{n \in \mathbb{Z}} \mathfrak{g}_{\mathfrak{a}}^{\mu + n\lambda}.$$

The Lie algebra $\mathfrak{g}_\lambda$ acts on $\mathfrak{b}$ via the adjoint representation, that we shall denote by ρ.

The eigenvalues of $\rho(t_\lambda)$ are of the form $\mu(t_\lambda) + 2n$. They are consecutive even (or odd) integers from $-r$ to r. So $\mu(t_\lambda) \in \mathbb{Z}$, and the set of $n \in \mathbb{Z}$ such that $\mu + n\lambda \in S \cup \{0\}$ is an interval $[-n', n'']$. Since $0 \in [-n', n'']$, $n', n'' \in \mathbb{N}$. Furthermore:

$$\mu(t_\lambda) + 2n'' = r \ , \quad \mu(t_\lambda) + 2n' = -r.$$

We deduce therefore that $\mu(t_\lambda) = n'' - n'$.

As $-n' \leqslant n'' - n' \leqslant n''$, we have $-n' \leqslant \mu(t_\lambda) \leqslant n''$; hence $\mu - \mu(t_\lambda)\lambda \in S \cup \{0\}$. If $\mu(t_\lambda)\lambda = \mu$, then:

$$\mu(t_\lambda) = \mu(t_\lambda)\lambda(t_\lambda) = 2\mu(t_\lambda).$$

It follows that $\mu(t_\lambda) = 0$, hence $\mu = \mu(t_\lambda)\lambda = 0$ which is absurd. $\square$

38.3.4 Let $\lambda \in S$ and s_λ be the reflection of $\mathfrak{a}^*$ associated to λ. By 38.3.3, we have, for all $\phi \in \mathfrak{a}^*$,

$$s_\lambda(\phi) = \phi - \phi(t_\lambda)\lambda.$$

38.3.5 Let us use the notations h_α and H_α of 20.6.1 and 20.6.5 for the pair $(\mathfrak{g}, \mathfrak{h})$.

For $\alpha \in R$, we have $\theta(H_\alpha) = H_{\sigma(\alpha)}$. In particular, if $\sigma(\alpha) = \alpha$, then $H_\alpha \in \mathfrak{l}$, and if $\sigma(\alpha) = -\alpha$, then $\theta(H_\alpha) = H_{-\alpha} = -H_\alpha$, so $H_\alpha \in \mathfrak{a}$.

If $h \in \mathfrak{h}$, we can write $h = h' + h''$, with $h' \in \mathfrak{l}$ and $h'' \in \mathfrak{a}$. Let us identify $\mathfrak{l}^*$ (resp. $\mathfrak{a}^*$) with the set of elements of $\mathfrak{h}^*$ which are zero on $\mathfrak{a}$ (resp. $\mathfrak{l}$). If $\mu \in \mathfrak{h}^*$, then we may write $\mu = \mu' + \mu''$, with $\mu' \in \mathfrak{l}^*$ and $\mu'' \in \mathfrak{a}^*$.

Let $\lambda \in S$ and $\alpha \in R$ be such that $\alpha'' = \lambda$. We shall express t_λ in terms of H_α. There are three cases.

1) If $\sigma(\alpha) = -\alpha$, then $\lambda = \alpha$ and $H_\alpha \in \mathfrak{a}$. So it is clear that $h_\alpha = v_\lambda$, and we have $t_\lambda = H_\alpha$.

2) If $\sigma(\alpha) \neq -\alpha$ and $\langle \alpha, \sigma(\alpha)^\vee \rangle = 0$, then:

$$0 = \alpha(H_{\sigma(\alpha)}) = \alpha(\theta(H_\alpha)) = \sigma(\alpha)(H_\alpha).$$

In other words:

$$0 = L(\theta(h_\alpha), h_\alpha) = L(h'_\alpha, h'_\alpha) - L(h''_\alpha, h''_\alpha).$$

It follows that:

$$L(h_\alpha, h_\alpha) = L(h'_\alpha, h'_\alpha) + L(h''_\alpha, h''_\alpha) = 2L(h''_\alpha, h''_\alpha).$$

If $a \in \mathfrak{a}$, then:

$$\lambda(a) = L(v_\lambda, a) = \alpha(a) = L(h'_\alpha + h''_\alpha, a) = L(h''_\alpha, a).$$

We deduce therefore that:

$$2v_\lambda = 2h''_\alpha = h_\alpha - \theta(h_\alpha).$$

Hence:

$$H_\alpha = \frac{2h_\alpha}{L(h_\alpha, h_\alpha)} = \frac{2h_\alpha}{2L(v_\lambda, v_\lambda)} = \frac{h_\alpha}{L(v_\lambda, v_\lambda)}.$$

Now:

$$t_\lambda = \frac{2v_\lambda}{L(v_\lambda, v_\lambda)} = \frac{2h''_\alpha}{L(v_\lambda, v_\lambda)}.$$

Consequently, we obtain that:

$$t_\lambda = H_\alpha - \theta(H_\alpha) = H_\alpha - H_{\sigma(\alpha)}.$$

3) If $\sigma(\alpha) \neq -\alpha$ and $\langle \alpha, \sigma(\alpha)^\vee \rangle \neq 0$, then we saw in the proof of 36.2.1 that:

$$\langle \alpha, \sigma(\alpha)^\vee \rangle = \langle \sigma(\alpha), \alpha^\vee \rangle = 1,$$

that is:

$$\alpha(H_{\sigma(\alpha)}) = \sigma(\alpha)(H_\alpha) = \alpha(\theta(H_\alpha)) = 1.$$

This implies that:

$$L(h_\alpha, 2h'_\alpha - 2h''_\alpha) = L(h_\alpha, 2\theta(h_\alpha)) = L(h_\alpha, h_\alpha).$$

Hence:

$$2L(h'_\alpha, h'_\alpha) - 2L(h''_\alpha, h''_\alpha) = 2L(h_\alpha, \theta(h_\alpha)) = L(h'_\alpha + h''_\alpha, h'_\alpha + h''_\alpha)$$
$$= L(h'_\alpha, h'_\alpha) + L(h''_\alpha, h''_\alpha).$$

Consequently:

$$L(h'_\alpha, h'_\alpha) = 3L(h''_\alpha, h''_\alpha) \ , \ L(h_\alpha, h_\alpha) = 4L(h''_\alpha, h''_\alpha).$$

As in case 2, we have $2v_\lambda = 2h''_\alpha = h_\alpha - \theta(h_\alpha)$. Hence:

$$H_\alpha = \frac{2h_\alpha}{L(h_\alpha, h_\alpha)} = \frac{h_\alpha}{2L(v_\lambda, v_\lambda)}.$$

It follows that:

$$H_\alpha - \theta(H_\alpha) = \frac{2h''_\alpha}{2\,L(v_\lambda, v_\lambda)} = \frac{v_\lambda}{L(v_\lambda, v_\lambda)}.$$

We have therefore obtained that:

$$t_\lambda = 2H_\alpha - 2\theta(H_\alpha) = 2H_\alpha - 2H_{\sigma(\alpha)}.$$

38.3.6 Let $\lambda \in S$. Since $L_\theta|_{\mathfrak{g}_a^\lambda \times \mathfrak{g}_a^\lambda}$ is non-degenerate, there exists $e \in \mathfrak{g}_a^\lambda$ such that $L(e,e) \neq 0$. As $\Bbbk$ is algebraically closed, there exists $r \in \Bbbk$ such that:

$$r^2 = \frac{2}{\lambda(v_\lambda)L_\theta(e,e)}.$$

Set:

$$x_\lambda = re , \quad y_\lambda = \theta(x_\lambda).$$

We verify easily that:

(3) $[t_\lambda, x_\lambda] = 2x_\lambda , \quad [t_\lambda, y_\lambda] = -2y_\lambda , \quad [x_\lambda, y_\lambda] = t_\lambda.$

Similarly, we obtain:

(4)
$$\begin{cases} \left[\frac{1}{2}(x_\lambda + y_\lambda), t_\lambda\right] = -x_\lambda + y_\lambda \\[2mm] \left[\frac{1}{2}(x_\lambda + y_\lambda), x_\lambda - y_\lambda\right] = -t_\lambda \\[2mm] \left[\frac{1}{2}(x_\lambda + y_\lambda), z\right] = 0 \text{ if } z \in \ker \lambda \subset \mathfrak{a} \end{cases}$$

38.4 Centralizers

38.4.1 We shall denote by $\mathcal{S}$ (resp. $\mathcal{N}$) the set of semisimple (resp. nilpotent) elements of $\mathfrak{g}$ in $\mathfrak{p}$, and $\mathcal{G}$ the set of $\mathfrak{p}$-generic elements of $\mathfrak{g}$ (37.4.1). By 37.4.5 and 37.5.2, we have $\mathcal{G} \subset \mathcal{S}$ and $\mathcal{G}$ is a dense open subset of $\mathfrak{p}$ (37.4.2).

Let $G = \mathrm{Aut}_e\, \mathfrak{g}$. Then $\mathcal{L}(G) = \mathrm{ad}_\mathfrak{g}\, \mathfrak{g}$ and G is the algebraic adjoint group of $\mathfrak{g}$ (23.4.16 and 24.8.2). We saw in 37.5.1 that $\mathfrak{k} = \mathfrak{n}_\mathfrak{g}(\mathfrak{k})$. Since the adjoint representation of $\mathfrak{g}$ is injective, it follows from 24.7.2 that there exists a connected algebraic subgroup K of G such that $\mathcal{L}(K) = \mathrm{ad}_\mathfrak{g}\, \mathfrak{k}$. As $[\mathfrak{k}, \mathfrak{k}] \subset \mathfrak{k}$ and $[\mathfrak{k}, \mathfrak{p}] \subset \mathfrak{p}$, we have $K(\mathfrak{k}) = \mathfrak{k}$ and $K(\mathfrak{p}) = \mathfrak{p}$ (24.3.4). Moreover, by 24.3.6, K is the identity component of the normalizer of $\mathfrak{k}$ in G. Finally, it follows from 24.8.5 that we may identify $\{\alpha|_\mathfrak{k}; \alpha \in K\}$ with the algebraic adjoint group of $\mathfrak{k}$.

38.4.2 For $x \in \mathfrak{g}$, we set:

$$\mathfrak{k}^x = \mathfrak{k} \cap \mathfrak{g}^x , \quad \mathfrak{p}^x = \mathfrak{p} \cap \mathfrak{g}^x.$$

If $x \in \mathfrak{k} \cup \mathfrak{p}$, then clearly, $\mathfrak{g}^x = \mathfrak{k}^x \oplus \mathfrak{p}^x$.

Proposition. *For all* $x \in \mathfrak{p}$, *we have:*

$$\dim \mathfrak{k}^x - \dim \mathfrak{p}^x = \dim \mathfrak{k} - \dim \mathfrak{p}.$$

In particular,

$$\dim \mathfrak{g} - \dim \mathfrak{g}^x = 2(\dim \mathfrak{k} - \dim \mathfrak{k}^x) = 2(\dim \mathfrak{p} - \dim \mathfrak{p}^x).$$

Proof. Since $\mathfrak{g}$ is semisimple, the Killing homomorphism of $\mathfrak{g}$ is an isomorphism of $\mathfrak{g}$ onto $\mathfrak{g}^*$. The result is therefore a reformulation of 37.6.1. □

38.4.3 Corollary. *The following conditions are equivalent for $x \in \mathfrak{p}$:*
(i) $\dim \mathfrak{g}^x \leqslant \dim \mathfrak{g}^y$ *for all $y \in \mathfrak{p}$.*
(ii) $\dim \mathfrak{k}^x \leqslant \dim \mathfrak{k}^y$ *for all $y \in \mathfrak{p}$.*
(iii) $\dim \mathfrak{p}^x \leqslant \dim \mathfrak{p}^y$ *for all $y \in \mathfrak{p}$.*

38.4.4 An element $x \in \mathfrak{p}$ is said to be $\mathfrak{p}$-*regular* if x satisfies the conditions of 38.4.3. We shall denote by $\mathcal{R}$ the set of $\mathfrak{p}$-regular elements of $\mathfrak{p}$. It is a dense open subset of $\mathfrak{p}$. So $\mathcal{R} \cap \mathcal{G} \neq \emptyset$, and $\mathcal{G} = \mathcal{R} \cap \mathcal{S}$.

Corollary. *The following conditions are equivalent for $x \in \mathfrak{p}$:*
(i) $x \in \mathcal{R}$.
(ii) $\dim G(x) \geqslant \dim G(y)$ *for all $y \in \mathfrak{p}$.*
(iii) $\dim K(x) \geqslant \dim K(y)$ *for all $y \in \mathfrak{p}$.*

Proof. Let G_x (resp. K_x) be the stabilizer of x in G (resp. K). Then $\mathcal{L}(G_x) = \mathrm{ad}_\mathfrak{g} \, \mathfrak{g}^x$ and $\mathcal{L}(K_x) = \mathrm{ad}_\mathfrak{g} \, \mathfrak{k}^x$ (24.3.6). So the result follows from 21.4.3 and 38.4.3. □

38.4.5 Let $\mathfrak{a}$ be a Cartan subspace of $\mathfrak{g}$ and $\mathfrak{m} = \mathfrak{c}_\mathfrak{k}(\mathfrak{a})$. Recall from 37.5.2 that $\mathfrak{c}_\mathfrak{g}(\mathfrak{a}) = \mathfrak{m} \oplus \mathfrak{a}$.

Proposition. *We have:*
$$\dim \mathfrak{a} - \dim \mathfrak{m} = \dim \mathfrak{p} - \dim \mathfrak{k}.$$

Proof. Since $\mathfrak{a}$ is an abelian Lie subalgebra of $\mathfrak{g}$ consisting of semisimple elements (37.5.2), there exist pairwise distinct linear forms $\lambda_1, \ldots, \lambda_n$ on $\mathfrak{a}$ such that:
$$\mathfrak{g} = \mathfrak{a} \oplus \mathfrak{m} \oplus \mathfrak{g}_\mathfrak{a}^{\lambda_1} \oplus \cdots \oplus \mathfrak{g}_\mathfrak{a}^{\lambda_n}.$$
If $x \in \mathfrak{a} \setminus (\ker \lambda_1 \cup \cdots \cup \ker \lambda_n)$, then $\mathfrak{k}^x = \mathfrak{m}$ and $\mathfrak{p}^x = \mathfrak{a}$. So by 38.4.2, we are done. □

38.4.6 Corollary. *The following conditions are equivalent for $x \in \mathfrak{p}$:*
(i) $x \in \mathcal{R}$.
(ii) $\dim \mathfrak{k}^x = \dim \mathfrak{m}$.
(iii) $\dim \mathfrak{p}^x = \dim \mathfrak{a}$.
(iv) $\dim \mathfrak{g}^x = \dim \mathfrak{a} + \dim \mathfrak{m}$.

Proof. The equivalence (i) $\Leftrightarrow$ (iii) follows from the fact that $\mathcal{G} \cap \mathcal{R} \neq \emptyset$, while (ii) $\Leftrightarrow$ (iii) is clear by 38.4.2 and 38.4.5. Finally (iv) $\Leftrightarrow$ (i) is obvious by the previous equivalences and 38.4.3. □

38.4.7 Proposition. *Let* $(\mathfrak{g}; \mathfrak{k}, \mathfrak{p})$ *be a semisimple symmetric Lie algebra and* r *the dimension of Cartan subspaces of* $\mathfrak{g}$.

(i) *If* $x \in \mathfrak{p}$, *then* $\mathfrak{p}^x$ *contains an abelian Lie subalgebra of dimension* r.

(ii) *If* $x \in \mathcal{R}$, *then* $[\mathfrak{k}^x, \mathfrak{p}^x] = [\mathfrak{p}^x, \mathfrak{p}^x] = \{0\}$, *so* $\mathfrak{p}^x \subset \mathfrak{z}(\mathfrak{g}^x)$, *and we may identify* $\mathfrak{g}^x$ *with the direct product of the Lie algebras* $\mathfrak{k}^x$ *and* $\mathfrak{p}^x$.

(iii) *If* $x \in \mathcal{R} \cap \mathcal{N}$, *then any element of* $\mathfrak{p}^x$ *is nilpotent.*

(iv) *Any element of* $\mathcal{R}$ *is polarizable.*

Proof. For $n \in \mathbb{N}$, we shall denote by $\mathrm{Gr}(\mathfrak{p}, n)$ the Grassmannian variety of n-dimensional subspaces of $\mathfrak{p}$, and let $p : \mathrm{Gr}(\mathfrak{p}, r) \times \mathfrak{p} \to \mathfrak{p}$ be the canonical projection.

(i) The set $\mathcal{L}$ of $(\mathfrak{q}, x) \in \mathrm{Gr}(\mathfrak{p}, r) \times \mathfrak{p}$ verifying $[\mathfrak{q}, \mathfrak{q}] = [\mathfrak{q}, x] = \{0\}$ is a closed subset of $\mathrm{Gr}(\mathfrak{p}, r) \times \mathfrak{p}$. As $\mathrm{Gr}(\mathfrak{p}, r)$ is a complete variety, it follows that $p(\mathcal{L})$ is closed in $\mathfrak{p}$. But by 37.4.11 and 37.5.4, $p(\mathcal{L})$ contains the dense subset $\mathcal{G}$ of $\mathfrak{p}$. So $p(\mathcal{L}) = \mathfrak{p}$, and we have proved (i).

(ii) This is obvious since for $x \in \mathcal{R}$, we have $[\mathfrak{p}^x, \mathfrak{p}^x] = \{0\}$ by (i), while $[\mathfrak{k}^x, \mathfrak{p}^x] = \{0\}$ by 37.6.4.

(iii) This is clear by (ii) and 35.1.2.

(iv) For $x \in \mathcal{R}$, 38.4.6 says that:

$$\dim \mathfrak{g} + \dim \mathfrak{g}^x = \dim \mathfrak{g} + \dim \mathfrak{m} + \dim \mathfrak{a}.$$

Thus $\dim \mathfrak{g} + \dim \mathfrak{m} + \dim \mathfrak{a}$ is an even integer, say $2s$.

The set $\mathcal{M}$ of $(\mathfrak{q}, x) \in \mathrm{Gr}(\mathfrak{p}, s) \times \mathfrak{p}$ such that $\mathfrak{q}$ is a Lie subalgebra of $\mathfrak{g}$ verifying $L(x, [\mathfrak{q}, \mathfrak{q}]) = \{0\}$, is a closed subset of $\mathrm{Gr}(\mathfrak{p}, s)$. The elements of $\mathcal{G}$, being semisimple, are polarizable (33.2.8), so $\mathcal{G} \subset p(\mathcal{M})$. Since $\mathcal{G}$ is dense in $\mathfrak{p}$ and $p(\mathcal{M})$ is closed in $\mathfrak{p}$, we have $p(\mathcal{M}) = \mathfrak{p}$. The result is now clear. $\square$

38.4.8 In general, if $x \in \mathfrak{p}$ is polarizable, there is no polarization $\mathfrak{q}$ of $\mathfrak{g}$ in x which verifies $\mathfrak{q} = (\mathfrak{q} \cap \mathfrak{k}) \oplus (\mathfrak{q} \cap \mathfrak{p})$. For example, let $(E_{ij})_{1 \leqslant i, j \leqslant 2}$ be the canonical bases of $\mathfrak{gl}_2(\Bbbk)$,

$$\mathfrak{g} = \mathfrak{sl}_2(\Bbbk) \, , \ \mathfrak{k} = \Bbbk(E_{12} - E_{21}) \, , \ \mathfrak{p} = \Bbbk(E_{11} - E_{22}) \oplus \Bbbk(E_{12} + E_{21}).$$

The only polarizations of $\mathfrak{g}$ in $x = E_{11} - E_{22}$ is $\Bbbk x \oplus E_{12}$ and $\Bbbk x \oplus E_{21}$. Neither of them is the sum of their intersections with $\mathfrak{k}$ and $\mathfrak{p}$.

38.5 S-triples

38.5.1 Let us denote by $\mathfrak{U}$ the set of Lie subalgebras of $\mathfrak{g}$ which are isomorphic to $\mathfrak{sl}_2(\Bbbk)$.

Definition. (i) *An element* $\mathfrak{s} \in \mathfrak{U}$ *is called normal if* $\mathfrak{s} \not\subset \mathfrak{k}$ *and* $\theta(\mathfrak{s}) = \mathfrak{s}$.

(ii) *A S-triple* (e, h, f) *is called normal if* $h \in \mathfrak{k}$ *and* $e, f \in \mathfrak{p}$.

(iii) *Let* H *be a subgroup of* $\mathrm{Aut}\,\mathfrak{g}$. *Two S-triples* (e, h, f) *and* (e', h', f') *are* H-*conjugate if there exists* $\alpha \in H$ *such that* $e' = \alpha(e)$, $h' = \alpha(h)$ *and* $f' = \alpha(f)$.

(iv) *An element $\mathfrak{s} \in \mathfrak{U}$ is said to be* principal *if it is normal and it contains a nilpotent $\mathfrak{p}$-regular element.*

(v) *A normal S-triple (e, h, f) is called* principal *if e is $\mathfrak{p}$-regular.*

38.5.2 Remarks. Let $\mathfrak{s} \in \mathfrak{U}$. Then it is easy to prove the following results:

1) If $\mathfrak{s}$ is normal, then:

$$\mathfrak{s} = (\mathfrak{s} \cap \mathfrak{k}) \oplus (\mathfrak{s} \cap \mathfrak{p}) \ , \ \dim(\mathfrak{s} \cap \mathfrak{k}) = 1 \ , \ \dim(\mathfrak{s} \cap \mathfrak{p}) = 2.$$

2) Since $\Bbbk$ is algebraically closed, $\mathfrak{s}$ is normal if and only if there exists a normal S-triple (e, h, f) such that $\mathfrak{s} = \Bbbk e + \Bbbk h + \Bbbk f$.

38.5.3 Proposition. *Let e be a non-zero nilpotent element of $\mathfrak{p}$. Then there exists a normal S-triple (e, h, f) containing e.*

Proof. By 32.1.5, there exists a S-triple (e, h, f) containing e. Let us write $h = h_1 + h_2$, $f = f_1 + f_2$, with $h_1, f_1 \in \mathfrak{k}$ and $h_2, f_2 \in \mathfrak{p}$. Then:

$$2e = [h, e] = [h_1, e] + [h_2, e] \ , \ h_1 + h_2 = h = [e, f] = [e, f_1] + [e, f_2].$$

Hence $[h_1, e] = 2e$, $[e, f_2] = h_1$. It follows from 32.1.3 that there exists $f_3 \in \mathfrak{g}$ such that (e, h_1, f_3) is a S-triple. Let $f_3 = f_4 + f_5$, with $f_4 \in \mathfrak{k}$, $f_5 \in \mathfrak{p}$. Now

$$-2f_3 = [h_1, f_3] = [h_1, f_4] + [h_1, f_5] \ , \ h_1 = [e, f_3] = [e, f_4] + [e, f_5],$$

so $[h_1, f_5] = -2f_5$, $[e, f_5] = h_1$. Thus (e, h_1, f_5) is a normal S-triple. $\quad\square$

38.5.4 Lemma. *Let (e, h, f) and (e, h', f') be normal S-triples having the same positive element. Then there exists a nilpotent element $x \in \mathfrak{k}^e$ such that:*

$$e = e^{\mathrm{ad}\,x}(e) \ , \ h' = e^{\mathrm{ad}\,x}(h) \ , \ f' = e^{\mathrm{ad}\,x}(f).$$

Proof. The subspace $\mathfrak{k}^e$ is $\mathrm{ad}\,h$-stable. Let $\mathfrak{t}^e_+$ be the subspace of $\mathfrak{t}^e$ spanned by the elements $x \in \mathfrak{t}^e$ verifying $[h, x] = \lambda x$, with $\lambda \in \mathbb{N}^*$. Then $\mathfrak{t}^e_+ = \mathfrak{t}^e \cap [e, \mathfrak{g}]$ and by 32.2.4, $\mathfrak{t}^e_+$ is a Lie subalgebra of $\mathfrak{g}$ whose elements are nilpotent. As $[h, \mathfrak{t}^e_+] = \mathfrak{t}^e_+$, 32.2.5 implies that $\exp(\mathfrak{t}^e_+)(h) = h + \mathfrak{t}^e_+$.

Now $[h' - h, e] = 0$ and $h' - h = [e, f - f']$, so $h' - h \in \mathfrak{k}^e_+$. Consequently, there exists $x \in \mathfrak{t}^e_+$ such that $h' = e^{\mathrm{ad}\,x}(h)$. Hence $f' = e^{\mathrm{ad}\,x}(f)$ (32.2.3). $\quad\square$

38.5.5 Let (e, h, f) be a normal S-triple. For $n \in \mathbb{Z}$, set:

$$\mathfrak{g}_n = \{x \in \mathfrak{g}; \ [h, x] = nx\} \ , \ \mathfrak{k}_n = \mathfrak{k} \cap \mathfrak{g}_n \ , \ \mathfrak{p}_n = \mathfrak{p} \cap \mathfrak{g}_n.$$

We have $\mathfrak{g}_n = \mathfrak{k}_n \oplus \mathfrak{p}_n$. The following result is analogue to the result of 32.1.9.

Proposition. *Let n be an integer.*

(i) *We have $\dim \mathfrak{k}_n = \dim \mathfrak{k}_{-n}$ and $\dim \mathfrak{p}_n = \dim \mathfrak{p}_{-n}$.*

(ii) *If n is odd, then $\dim \mathfrak{t}_n = \dim \mathfrak{p}_n$.*

Proof. Recall that $L|_{\mathfrak{g}_n \times \mathfrak{g}_{-n}}$ is non-degenerate. Since $\mathfrak{k}$ and $\mathfrak{p}$ are orthogonal with respect to L, we deduce that the restrictions of L to $\mathfrak{k}_n \times \mathfrak{k}_{-n}$ and $\mathfrak{p}_n \times \mathfrak{p}_{-n}$ are non-degenerate. This proves (i).

Let $j \in \mathbb{N}$ and $\varphi_j : \mathfrak{g}_{-2j-1} \to \mathfrak{g}_{2j+1}$, $x \mapsto (\operatorname{ad} e)^j(x)$. The map φ_j is bijective, and since $2j + 1$ is odd, we obtain:

$$\varphi_j(\mathfrak{k}_{-2j-1}) = \mathfrak{p}_{2j+1} \, , \ \varphi_j(\mathfrak{p}_{-2j-1}) = \mathfrak{k}_{2j+1}.$$

So (ii) follows from (i). $\square$

38.5.6 Lemma. *Let* (e, h, f) *be a S-triple of* $\mathfrak{g}$ *and* $\mathfrak{s}$ *be the element of* $\mathfrak{U}$ *spanned by* e, h, f. *Set* $h_c = e + f$.

(i) *There exists* $\alpha \in G$ *such that* $h_c = \alpha(h)$.

(ii) *There exists a S-triple* (x, h_c, y), *G-conjugate to* (e, h, f), *such that* x, h_c, y *span* $\mathfrak{s}$.

(iii) *Assume that* (e, h, f) *is normal. Then there exists a S-triple* (e_c, h_c, f_c) *verifying* $f_c = \theta(e_c)$ *and* e_c, h_c, f_c *span* $\mathfrak{s}$. *Furthermore,* e_c *is uniquely determined up to a scalar multiplication by* ± 1.

Proof. Let $\alpha \in G$ be defined by:

$$\alpha = \exp(\operatorname{ad} f) \circ \exp\left(-\frac{1}{2} \operatorname{ad} e\right).$$

Then:

$$\alpha(h) = e + f \, , \ \alpha(e) = e - h - f \, , \ \alpha(f) = -\frac{1}{4}(e + h - f).$$

From this, we deduce immediately (i) and (ii).

Set:

$$e_c = \frac{1}{2}(e - h - f) \, , \ f_c = -\frac{1}{2}(e + h - f).$$

Then (e_c, h_c, f_c) is a S-triple. By part (i) and 32.3.1, it is G-conjugate to (e, h, f). Moreover, if (e, h, f) is normal, then $f_c = \theta(e_c)$.

Let us assume that (e, h, f) is normal, and let (e', h_c, f') be another S-triple verifying the conclusion of (iii). From $\theta(h_c) = -h_c$, we deduce that $[h_c, \theta(e')] = -2\theta(e')$. So $\theta(e') = \lambda f_c$, with $\lambda \in \mathbb{k} \setminus \{0\}$. Therefore $e' = \theta^2(e') = \lambda e_c$. But:

$$h_c = [e', f'] = [\lambda e_c, \theta(e')] = \lambda[e_c, \theta(\lambda e_c)] = \lambda^2[e_c, f_c] = \lambda^2 h_c.$$

So $\lambda = \pm 1$. $\square$

38.5.7 Lemma. *Let* $\mathfrak{s} \in \mathfrak{U}$ *and* (e, h, f) *a S-triple spanning* $\mathfrak{s}$.

(i) *Let* x *be a non-zero nilpotent element of* $\mathfrak{s}$. *Then there exists* $\alpha \in G$ *such that* $\alpha(x) = e$.

(ii) *Assume that* $\mathfrak{s}$ *is principal. Any element* $x \in \mathfrak{s} \cap \mathcal{N} \setminus \{0\}$ *is* $\mathfrak{p}$*-regular.*

Proof. (i) There exists $\alpha \in \mathrm{Aut}_e \, \mathfrak{s}$ such that $\alpha(x) = e$ (32.2.2). If $z \in \mathfrak{s}$ is such that $\mathrm{ad}_\mathfrak{s} \, z$ is nilpotent, then $\mathrm{ad}_\mathfrak{g} \, z$ is also nilpotent (20.4.4). Thus α extends to an element of G.

(ii) Let $x \in \mathfrak{s} \cap \mathcal{N} \setminus \{0\}$, then (i) implies that $x = \alpha(e)$ for some $\alpha \in G$. So $\dim \mathfrak{g}^e = \dim \mathfrak{g}^x$ and (ii) follows from 38.4.6. $\square$

38.5.8 Proposition. *Let (e, h, f) be a normal S-triple and $\mathfrak{s}$ the element of $\mathfrak{U}$ spanned by e, h, f.*

(i) *Assume that $\mathfrak{s}$ is principal. Then $h_c = e + f \in \mathcal{G}$, and if $x \in \mathfrak{s} \cap \mathcal{S} \setminus \{0\}$, then $x \in \mathcal{G}$. Moreover, the dimension of any irreducible $\mathfrak{s}$-submodule of $\mathfrak{g}$ is odd.*

(ii) *Conversely, let us assume that the following conditions are verified:*

    a) *$\mathfrak{s} \cap \mathcal{G} \neq \emptyset$.*

    b) *The dimension of any irreducible $\mathfrak{s}$-submodule of $\mathfrak{g}$ is odd.*

Then $\mathfrak{s}$ is principal in $\mathfrak{U}$.

Proof. (i) By the theory of representations of $\mathfrak{sl}_2(\Bbbk)$, we have $\dim \mathfrak{g}^h \leqslant \dim \mathfrak{g}^e$ with equality if and only if the dimensions of all the irreducible $\mathfrak{s}$-submodules of $\mathfrak{g}$ are odd. On the other hand, $\dim \mathfrak{g}^h = \dim \mathfrak{g}^{e+f}$ (38.5.6). As $\mathfrak{s}$ is normal, we have $e + f \in \mathfrak{p}$. Finally, $e \in \mathcal{R}$ (38.5.7). Hence:

$$\dim \mathfrak{g}^e \leqslant \dim \mathfrak{g}^h = \dim \mathfrak{g}^{e+f} \leqslant \dim \mathfrak{g}^e.$$

This shows that $h_c \in \mathcal{G}$, and the dimension of any irreducible $\mathfrak{s}$-submodule of $\mathfrak{g}$ is odd.

Let $x = ae + bh + cf$ be a non-zero semisimple element of $\mathfrak{s} \simeq \mathfrak{sl}_2(\Bbbk)$. As an endomorphism of $\Bbbk^2$, x is semisimple (20.4.4). So $b^2 + ac \neq 0$. We deduce therefore that there exist $\lambda, \mu \in \Bbbk$ such that:

$$c\lambda^2 + 2b\lambda - a = 0 \,, \ \mu^2(c\lambda^2 + 2b\lambda - a) + 2\mu(b + c\lambda) + c = 0.$$

A simple computation shows that $\exp(\mu \, \mathrm{ad} \, f) \circ \exp(\lambda \, \mathrm{ad} \, e)(x) = \nu h$, where $\nu \in \Bbbk \setminus \{0\}$. So $\dim \mathfrak{g}^x = \dim \mathfrak{g}^h$, and $x \in \mathcal{G}$.

(ii) Condition b) implies that $\dim \mathfrak{g}^e = \dim \mathfrak{g}^h$, while condition a) implies that $x \in \mathcal{G}$ for $x \in \mathfrak{s} \cap \mathcal{S}$ non-zero. Hence $e \in \mathcal{N} \cap \mathcal{R}$, and $\mathfrak{s}$ is normal. $\square$

38.6 Orbits

38.6.1 Throughout this section, $\mathcal{S}, \mathcal{N}, \mathcal{G}, G$ and K are as in 38.4.1.

Theorem. (i) *Let $\mathcal{O}$ be the G-orbit of an element of $\mathfrak{p}$. Any irreducible component Z of $\mathcal{O} \cap \mathfrak{p}$ is a K-orbit and verifies:*

$$2 \dim Z = \dim \mathcal{O}.$$

(ii) *The number of K-orbits of the elements of $\mathcal{N}$ is finite.*

Proof. By 32.4.4, it suffices to prove (i). Let Z be an irreducible component of $\mathcal{O} \cap \mathfrak{p}$. Since $K.Z$ is an irreducible subset of $\mathfrak{p} \cap \mathcal{O}$, Z is K-stable.

Let $x \in Z$ and $\mathcal{K} = K(x)$. Then we have by 38.4.2:

$$\dim \mathcal{O} = \dim \mathfrak{g} - \dim \mathfrak{g}^x$$
$$= 2(\dim \mathfrak{k} - \dim \mathfrak{k}^x) = 2 \dim \mathcal{K}.$$

It follows that all the K-orbits in Z have the same dimension.

Now:

$$\mathrm{T}_x(\mathcal{O}) = [x, \mathfrak{g}] = [x, \mathfrak{k}] + [x, \mathfrak{p}] \ , \ \mathrm{T}_x(\mathcal{K}) = [x, \mathfrak{k}].$$

Thus $\mathrm{T}_x(\mathcal{O}) \cap \mathfrak{p} = [x, \mathfrak{k}]$. On the other hand:

$$\mathrm{T}_x(Z) \subset \mathrm{T}_x(\mathcal{O}) \cap \mathfrak{p} = [x, \mathfrak{k}] = \mathrm{T}_x(\mathcal{K}) \subset \mathrm{T}_x(Z).$$

We deduce therefore that $\mathrm{T}_x(Z) = \mathrm{T}_x(\mathcal{K})$.

Let y be a smooth point in Z. It follows from the previous paragraph that $\dim Z = \dim K(y)$, and so $K(y)$ is dense in Z (14.1.6). Since all the K-orbits in Z have the same dimension, we conclude by 21.4.5 that $Z = K(y)$, and the result follows from the equality of dimensions above. $\qquad\square$

38.6.2 Let (e, h, f) be a normal S-triple. For $n \in \mathbb{Z}$, set:

$$\mathfrak{g}_n = \{x \in \mathfrak{g}; [h, x] = nx\} \ , \ \mathfrak{k}_n = \mathfrak{k} \cap \mathfrak{g}_n \ , \ \mathfrak{p}_n = \mathfrak{p} \cap \mathfrak{g}_n.$$

Then $\mathfrak{g}_n = \mathfrak{k}_n \oplus \mathfrak{p}_n$.

Let $\xi \in \Bbbk \setminus \{0\}$. The automorphism τ_ξ of $\mathfrak{g}$, given by $\tau_\xi(x) = \xi^n x$ for $x \in \mathfrak{g}_n$, is an element of G (32.2.7). Let $H = \{\tau_\xi; \xi \in \Bbbk \setminus \{0\}\}$; it is an irreducible subgroup of G.

Let $s_1, \ldots, s_m$ be the eigenvalues of $\mathrm{ad}_{\mathfrak{k}} h$ and L the smallest algebraic subgroup of G whose Lie algebra contains $\Bbbk \, \mathrm{ad}_{\mathfrak{g}} h$. If $\xi \in \Bbbk \setminus \{0\}$ and $(a_1, \ldots, a_m) \in \mathbb{Z}^m$ verifying $a_1 s_1 + \cdots + a_m s_m = 0$, then:

$$(\xi^{s_1})^{a_1} \cdots (\xi^{s_m})^{a_m} = \xi^{a_1 s_1 + \cdots + a_m s_m} = 1.$$

It follows from 24.6.3 that $\tau_\xi \in L$. Since $\tau_\xi(\mathfrak{k}) = \mathfrak{k}$, by 38.4.1, we have $H \subset K$ (because H is irreducible).

38.6.3 Let us conserve the notations of 38.6.2 and denote by K_h the stabilizer of h in K.

If $x \in \mathfrak{p}_2$ and $\alpha \in K_h$, then $2\alpha(x) = \alpha([h, x]) = [\alpha(h), \alpha(x)] = [h, \alpha(x)]$. Thus $\alpha(x) \in \mathfrak{g}_2$, and so $\alpha(x) \in \mathfrak{p}_2$ because $\alpha(\mathfrak{p}) = \mathfrak{p}$. Now $[e, \mathfrak{g}_0] = \mathfrak{g}_2$, so $[e, \mathfrak{k}_0] = \mathfrak{p}_2$.

Lemma. *The set $K_h(e)$ contains a dense open subset of $\mathfrak{p}_2$.*

Proof. The proof is analogue to the proof of 32.2.8. $\qquad\square$

38.6.4 Lemma. *Two normal S-triples (e, h, f), (e', h, f') having the same simple element are K-conjugate.*

Proof. By 38.6.3, we have $K_h(e) \cap K_h(e') \neq \emptyset$. So there exists $\alpha \in K$ such that $\alpha(e) = e'$ and $\alpha(h) = h$. So by 32.2.3, $\alpha(f) = f'$. $\square$

38.6.5 Theorem. *Let* $(\mathfrak{g}; \mathfrak{k}, \mathfrak{p})$ *be a semisimple symmetric Lie algebra, and* K *be as in 38.4.1.*

(i) *Let* (e, h, f) *and* (e', h', f') *be normal S-triples of* $\mathfrak{g}$. *The following conditions are equivalent:*
   a) *The S-triples* (e, h, f) *and* (e', h', f') *are K-conjugate.*
   b) *There exists* $\alpha \in K$ *such that* $\alpha(e) = e'$.
   c) *There exists* $\alpha \in K$ *such that* $\alpha(h) = h'$.

(ii) *The map* $(e, h, f) \mapsto e$ *induces a bijection between the set of K-conjugacy classes of normal S-triples of* $\mathfrak{g}$ *and the set of K-orbits of non-zero nilpotent elements of* $\mathfrak{p}$.

Proof. This follows immediately from 38.5.3, 38.5.4 and 38.6.4. $\square$

38.6.6 Remarks. 1) If (e, h, f) is a normal S-triple, then e and f are not in general K-conjugate.

2) Let (e, h, f) and (e', h', f') be normal S-triples and $\mathfrak{s}, \mathfrak{s}'$ the corresponding elements in $\mathfrak{U}$. If there exists $\alpha \in K$ such that $\alpha(\mathfrak{s}) = \mathfrak{s}'$, then the two S-triples are not necessarily K-conjugate. So there is no natural bijection between the set of K-orbits of non-zero nilpotent elements of $\mathfrak{p}$ and the set of normal elements of $\mathfrak{U}$.

38.6.7 Lemma. *Let* $x \in \mathfrak{p}$ *and* s, n *its semisimple and nilpotent components.*
   (i) *If* $\xi \in \Bbbk \setminus \{0\}$, *then* $s + \xi^2 n \in \overline{K(x)}$.
   (ii) *We have* $s \in \overline{K(x)}$.

Proof. Using 38.5.3 and the fact that $\tau_\xi \in K$ (38.6.2), the proof is the same as in 34.2.5. $\square$

38.6.8 Proposition. *Let* $x \in \mathfrak{p}$.
   (i) *We have* $x \in \mathcal{N}$ *if and only if* $0 \in \overline{K(x)}$. *If these conditions are satisfied, then* $\lambda x \in K(x)$ *for all* $\lambda \in \Bbbk \setminus \{0\}$.
   (ii) *We have* $x \in \mathcal{S}$ *if and only if* $K(x)$ *is closed in* $\mathfrak{p}$.

Proof. Since $K(x) \subset G(x)$, part (i) follows from 34.3.5 and 38.6.7. If $x \in \mathcal{S}$, then $G(x)$ is closed by 34.3.2, and 38.6.1 implies that $K(x)$ is closed. If x is not semisimple, then 38.6.7 says that $K(x)$ is not closed. $\square$

38.6.9 Theorem. *Let* $x \in \mathfrak{p}$.
   (i) *There exists* $z \in \mathcal{N}$ *such that:*

$$z \in \overline{K(\Bbbk x)} \ , \ \dim K(z) = \dim K(x).$$

(ii) *If x is polarizable, then we may choose z to be polarizable.*

(iii) *Assume that x is polarizable and there exists a polarization $\mathfrak{r}$ of $\mathfrak{g}$ in x such that $\mathfrak{r} = (\mathfrak{r} \cap \mathfrak{k}) \oplus (\mathfrak{r} \cap \mathfrak{p})$. Then we may choose z such that $\mathfrak{r}$ is a polarization of $\mathfrak{g}$ in z.*

Proof. (i) The proof is the same as in 33.5.3 by replacing G by K.

(ii) Assume that x is polarizable. Let d be the dimension of a polarization of $\mathfrak{g}$ in x and $p : \mathrm{Gr}(\mathfrak{g}, d) \times \mathfrak{p} \to \mathfrak{p}$ the canonical surjection. The set:

$$\mathcal{F} = \{(\mathfrak{q}, y) \in \mathrm{Gr}(\mathfrak{g}, d) \times \mathfrak{p}; [\mathfrak{q}, \mathfrak{q}] \subset \mathfrak{q} \,, \; L(y, [\mathfrak{q}, \mathfrak{q}]) = \{0\}\}$$

is closed in $\mathrm{Gr}(\mathfrak{g}, d) \times \mathfrak{p}$ and the closed subset $p(\mathcal{F})$ contains $K(\Bbbk x)$. It follows that $\overline{K(\Bbbk x)} \subset p(\mathcal{F})$. Let $z \in \overline{K(\Bbbk x)} \cap \mathcal{N}$ be such that $\dim K(z) = \dim K(x)$. Then $\dim G(z) = \dim G(x)$ by 38.6.1, so $\dim \mathfrak{g}^z = \dim \mathfrak{g}^x$. But there exists a d-dimensional Lie subalgebra $\mathfrak{q}$ of $\mathfrak{g}$ such that $L(z, [\mathfrak{q}, \mathfrak{q}]) = \{0\}$. Hence $\mathfrak{q}$ is a polarization of $\mathfrak{g}$ in z.

(iii) Let $\mathfrak{r}$ be a parabolic subalgebra of $\mathfrak{g}$ such that $\mathfrak{r} = (\mathfrak{r} \cap \mathfrak{k}) \oplus (\mathfrak{r} \cap \mathfrak{p})$. Set $r = \dim \mathfrak{r}$, $k = \dim(\mathfrak{r} \cap \mathfrak{k})$, $p = \dim(\mathfrak{r} \cap \mathfrak{p})$, and let $q : \mathrm{Gr}(\mathfrak{g}, r) \times \mathfrak{p} \to \mathfrak{p}$ be the canonical surjection. Let $\mathcal{H}$ be the K-orbit of $\mathfrak{r}$ in $\mathrm{Gr}(\mathfrak{g}, r)$. The normalizer of $\mathfrak{r}$ in $\mathfrak{k}$ is $\mathfrak{k} \cap \mathfrak{r}$. So:

$$\dim \mathcal{H} = \dim \mathfrak{k} - k.$$

The K-orbit $\mathcal{H}$ is open in its closure $\overline{\mathcal{H}}$ and $\overline{\mathcal{H}} \setminus \mathcal{H}$ is the union of K-orbits of dimension strictly smaller than $\dim \mathcal{H}$ (21.4.3 and 21.4.5). For $\mathfrak{r}' \in \mathcal{H}$, we have:

$$\mathfrak{r}' = (\mathfrak{k} \cap \mathfrak{r}') \oplus (\mathfrak{p} \cap \mathfrak{r}') \,, \;\; \dim(\mathfrak{k} \cap \mathfrak{r}') = k \,, \;\; \dim(\mathfrak{p} \cap \mathfrak{r}') = p.$$

Let $\mathfrak{q} \in \overline{\mathcal{H}}$. Then $\mathfrak{q}$ is a Lie subalgebra of $\mathfrak{g}$ (19.7.2) and:

$$\mathfrak{q} = (\mathfrak{k} \cap \mathfrak{q}) \oplus (\mathfrak{p} \cap \mathfrak{q}) \,, \;\; \dim(\mathfrak{k} \cap \mathfrak{q}) = k \,, \;\; \dim(\mathfrak{p} \cap \mathfrak{q}) = p.$$

Now, let $\mathfrak{r}$ be a polarization of $\mathfrak{g}$ in x. The set $\mathcal{L}$ of $(\mathfrak{q}, y) \in \overline{\mathcal{H}} \times \overline{K(\Bbbk x)}$ verifying $L(y, [\mathfrak{q}, \mathfrak{q}]) = \{0\}$ is a closed subset of $\mathrm{Gr}(\mathfrak{g}, r) \times \mathfrak{p}$, and the closed subset $q(\mathcal{L})$ contains $K(\Bbbk x)$, hence $\overline{K(\Bbbk x)}$.

Let $z \in \overline{K(\Bbbk x)} \cap \mathcal{N}$ as in part (i). There exists $\mathfrak{q} \in \overline{\mathcal{H}}$ such that $L(z, [\mathfrak{q}, \mathfrak{q}]) = \{0\}$. We have $\dim K(z) = \dim K(x)$, so $\dim \mathfrak{k}^z = \dim \mathfrak{k}^x$, hence $\dim \mathfrak{g}^z = \dim \mathfrak{g}^x$. It follows that $\mathfrak{q}$ is a polarization of $\mathfrak{g}$ in x, so it is a parabolic subalgebra of $\mathfrak{g}$ (33.2.6). Since $\dim(\mathfrak{q} \cap \mathfrak{k}) = \dim(\mathfrak{r} \cap \mathfrak{k})$, the discussion above says that $\mathfrak{q}$ and $\mathfrak{r}$ are in the same K-orbit of $\mathrm{Gr}(\mathfrak{g}, r)$. Thus there exists $\alpha \in K$ such that $\mathfrak{q} = \alpha(\mathfrak{r})$. Taking $z' = \alpha^{-1}(z)$, we have that $z' \in \overline{K(\Bbbk x)} \cap \mathcal{N}$, $\dim K(z') = \dim K(x)$ and $\mathfrak{r}$ is a polarization of $\mathfrak{g}$ in z'.   □

38.6.10 Remarks. 1) In the notation $x = E_{11} - E_{22}$ of 38.4.8, we can verify easily that if $\mathfrak{q}$ is a polarization of $\mathfrak{g}$ in x, then there does not exist a nilpotent element $z \in \mathfrak{p}$ such that $\mathfrak{q}$ is a polarization of $\mathfrak{g}$ in z.

2) Let $\mathfrak{g}$ be a simple Lie algebra of type B_2, with basis, in the notations 18.6.1 and 20.6.5:

$$H_\alpha, H_\beta, X_\alpha, X_\beta, X_{\alpha+\beta}, X_{2\alpha+\beta}, X_{-\alpha}, X_{-\beta}, X_{-\alpha-\beta}, X_{-2\alpha-\beta}.$$

Denote by $\mathfrak{k}$ (resp. $\mathfrak{p}$) the subspace of $\mathfrak{g}$ spanned by $H_\alpha, H_\beta, X_{\pm\beta}, X_{\pm(2\alpha+\beta)}$ (resp. $X_{\pm\alpha}, X_{\pm(\alpha+\beta)}$). Then $(\mathfrak{g}; \mathfrak{k}, \mathfrak{p})$ is a symmetric Lie algebra. Set:

$$\mathfrak{b} = \Bbbk H_\alpha + \Bbbk H_\beta + \Bbbk X_\alpha + \Bbbk X_\beta + \Bbbk X_{\alpha+\beta} + \Bbbk X_{2\alpha+\beta}.$$

The Lie algebra $\mathfrak{b}$ is a Borel subalgebra of $\mathfrak{g}$ verifying $\mathfrak{b} = (\mathfrak{b} \cap \mathfrak{k}) \oplus (\mathfrak{b} \cap \mathfrak{p})$. We verify easily that $\Bbbk X_\alpha + \Bbbk X_{\alpha+\beta}$ does not contain any regular element of $\mathfrak{g}$. So if $x \in \mathfrak{p}$, then $\mathfrak{b}$ is not a polarization of $\mathfrak{g}$ in x (33.2.1).

38.6.11 The following result is a version of Richardson's Theorem for semisimple symmetric Lie algebras.

Theorem. *Let $\mathfrak{q}$ be a parabolic subalgebra of $\mathfrak{g}$, $\mathfrak{m}$ its radical nilpotent. Assume that the following conditions are verified:*
   a) $\mathfrak{q} = (\mathfrak{q} \cap \mathfrak{k}) \oplus (\mathfrak{q} \cap \mathfrak{p})$.
   b) *There exists $z \in \mathfrak{p}$ such that $\mathfrak{q}$ is a polarization of $\mathfrak{g}$ in z.*
   Let us denote by K_0 the identity component of the stabilizer of $\mathfrak{q}$ in K.
   (i) *There exists a unique nilpotent K-orbit $\mathcal{K}$ in $\mathfrak{p}$ such that $\mathfrak{m} \cap \mathfrak{p} \cap \mathcal{K}$ is a dense open subset of $\mathfrak{m} \cap \mathfrak{p}$.*
   (ii) *The set $\mathfrak{m} \cap \mathfrak{p} \cap \mathcal{K}$ is a K_0-orbit.*
   (iii) *If $x \in \mathfrak{m} \cap \mathfrak{p} \cap \mathcal{K}$, then*

$$[x, \mathfrak{q} \cap \mathfrak{k}] = \mathfrak{m} \cap \mathfrak{p} \; , \; [x, \mathfrak{q} \cap \mathfrak{p}] = \mathfrak{m} \cap \mathfrak{k},$$

and $\mathfrak{q}$ is a polarization of $\mathfrak{g}$ in x.

Proof. Part (i) is clear because the number of nilpotent K-orbits of $\mathfrak{p}$ is finite. Set:

$$s = \min\{\dim \mathfrak{k}^x; \, x \in \mathfrak{m} \cap \mathfrak{p}\} \; , \; \mathcal{M} = \{x \in \mathfrak{m} \cap \mathfrak{p}; \dim \mathfrak{k}^x = s\}.$$

The set $\mathcal{M}$ is a non-empty open subset of $\mathfrak{m} \cap \mathfrak{p}$, so $\mathcal{K} \cap \mathcal{M} \neq \emptyset$. Hence $\dim \mathcal{K} = \dim \mathfrak{k} - s$, and $\mathcal{K} \cap \mathfrak{m} \cap \mathfrak{p} \subset \mathcal{M}$.
   Conversely, if $u \in \mathcal{M}$, then $\dim K(u) = \dim \mathcal{K}$ and $u \in \overline{\mathcal{K}}$. So $u \in \mathcal{K}$. Thus $\mathcal{M} = \mathfrak{m} \cap \mathfrak{p} \cap \mathcal{K}$.
   By 38.6.9 (iii), there exists $y \in \mathfrak{p} \cap \mathcal{N}$ such that $\mathfrak{q}$ is a polarization of $\mathfrak{g}$ in y. As y is nilpotent, we have $L(y, \mathfrak{q}) = \{0\}$ (33.2.6), hence $y \in \mathfrak{m} \cap \mathfrak{p}$.
   Now, for $u \in \mathcal{M}$, we have $L(u, \mathfrak{q}) = \{0\}$. Hence (38.4.2):

$$2 \dim \mathfrak{q} \leqslant \dim \mathfrak{g} + \dim \mathfrak{g}^u = 2 \dim \mathfrak{p} + 2s.$$

Since $\mathfrak{q}$ is a polarization of $\mathfrak{g}$ in y, we have:

$$2 \dim \mathfrak{q} = \dim \mathfrak{g} + \dim \mathfrak{g}^y = 2 \dim \mathfrak{p} + 2 \dim \mathfrak{k}^y.$$

Thus $s = \dim \mathfrak{k}^y$. This proves that if $u \in \mathfrak{m} \cap \mathfrak{p} \cap \mathcal{K}$, then $\mathfrak{q}$ is a polarization of $\mathfrak{g}$ in u. So $[u, \mathfrak{q}] = \mathfrak{m}$ (33.2.6). Part (iii) now follows.

Let $u \in \mathfrak{m} \cap \mathfrak{p} \cap \mathcal{K}$. Since $\mathfrak{q}$ is a polarization of $\mathfrak{g}$ in u, we have $\mathfrak{g}^u \subset \mathfrak{q}$. From $[u, \mathfrak{q} \cap \mathfrak{k}] = \mathfrak{m} \cap \mathfrak{p}$, we deduce that $\dim(\mathfrak{m} \cap \mathfrak{p}) = \dim(\mathfrak{q} \cap \mathfrak{k}) - \dim \mathfrak{k}^u$. The Lie algebra of K_0 is $\mathrm{ad}_\mathfrak{q}(\mathfrak{k} \cap \mathfrak{q})$. So we have:

$$\dim K_0(u) = \dim(\mathfrak{k} \cap \mathfrak{q}) - \dim(\mathfrak{k} \cap \mathfrak{q} \cap \mathfrak{k}^u)$$
$$= \dim(\mathfrak{k} \cap \mathfrak{q}) - \dim \mathfrak{k}^u = \dim(\mathfrak{m} \cap \mathfrak{p}).$$

Thus $K_0(u)$ is a dense open subset of $\mathfrak{m} \cap \mathfrak{p}$ contained in $\mathfrak{m} \cap \mathfrak{p} \cap \mathcal{K}$. This being true for all $u \in \mathfrak{m} \cap \mathfrak{p} \cap \mathcal{K}$, we have (ii). $\square$

38.6.12 Let $\mathfrak{q}$ be a parabolic subalgebra of $\mathfrak{g}$ verifying $\mathfrak{q} = (\mathfrak{q} \cap \mathfrak{k}) \oplus (\mathfrak{q} \cap \mathfrak{p})$, and $\mathfrak{m}$ its nilpotent radical. It is natural to ask whether there exists $x \in \mathfrak{p} \cap \mathfrak{m}$ such that $[x, \mathfrak{k} \cap \mathfrak{q}] = \mathfrak{p} \cap \mathfrak{m}$. This would give a more interesting version of Richardson's Theorem than 38.6.11. We shall give an example to show that the answer to the above question is in general negative.

Let $\mathfrak{g}$ be a simple Lie algebra of type D_4 and $B = \{\alpha_1, \alpha_2, \alpha_3, \alpha_4\}$ a base of the root system of $\mathfrak{g}$. In the notations of 18.14.2 with $\beta_i = \alpha_i$ and 20.6.5, let $\mathfrak{k}$ be the subspace of $\mathfrak{g}$ spanned by:

$$H_1, H_2, H_3, H_4, X_{\pm \alpha_1}, X_{\pm \alpha_3}, X_{\pm \alpha_4}, X_{\pm(\alpha_1 + 2\alpha_2 + \alpha_3 + \alpha_4)}.$$

Let $\mathfrak{p}$ be the subspace of $\mathfrak{g}$ spanned by:

$$X_{\pm \alpha_2}, X_{\pm(\alpha_1 + \alpha_2)}, X_{\pm(\alpha_2 + \alpha_3)}, X_{\pm(\alpha_2 + \alpha_4)},$$
$$X_{\pm(\alpha_1 + \alpha_2 + \alpha_3)}, X_{\pm(\alpha_1 + \alpha_2 + \alpha_4)}, X_{\pm(\alpha_2 + \alpha_3 + \alpha_4)}, X_{\pm(\alpha_1 + \alpha_2 + \alpha_3 + \alpha_4)}.$$

The triple $(\mathfrak{g}; \mathfrak{k}, \mathfrak{p})$ is a symmetric Lie algebra. Let $\mathfrak{q}$ be the Borel subalgebra of $\mathfrak{g}$ with respect to B. Then:

$$\mathfrak{q} = (\mathfrak{k} \cap \mathfrak{q}) \oplus (\mathfrak{k} \cap \mathfrak{p}) \ , \ \dim(\mathfrak{k} \cap \mathfrak{q}) = \dim(\mathfrak{p} \cap \mathfrak{q}) = 8.$$

The subspace $\mathfrak{p} \cap \mathfrak{q} = \mathfrak{p} \cap \mathfrak{m}$ of $\mathfrak{q}$ is spanned by:

$$X_{\alpha_2}, X_{\alpha_1 + \alpha_2}, X_{\alpha_2 + \alpha_3}, X_{\alpha_2 + \alpha_4},$$
$$X_{\alpha_1 + \alpha_2 + \alpha_3}, X_{\alpha_1 + \alpha_2 + \alpha_4}, X_{\alpha_2 + \alpha_3 + \alpha_4}, X_{\alpha_1 + \alpha_2 + \alpha_3 + \alpha_4}.$$

If $x \in \mathfrak{p} \cap \mathfrak{m}$, then $[x, X_{\alpha_1 + 2\alpha_2 + \alpha_3 + \alpha_4}] = 0$. Hence:

$$\dim([x, \mathfrak{k} \cap \mathfrak{q}]) \leqslant 7 < \dim(\mathfrak{p} \cap \mathfrak{m}).$$

Thus $[x, \mathfrak{k} \cap \mathfrak{q}] \neq \mathfrak{p} \cap \mathfrak{m}$.

38.7 Symmetric invariants

38.7.1 In this section, we shall use the notations $\mathfrak{a}, \mathfrak{m}, \mathfrak{l}, \mathfrak{h}, R, S, W, W_S$ of 38.1.1 and 38.2.3.

Lemma. *The set $K(\mathfrak{a})$ is dense in $\mathfrak{p}$.*

Proof. Since $\mathcal{G}$ is dense in $\mathfrak{p}$, the result follows from 37.4.10 and 37.4.11. □

38.7.2 Let us consider the action of W_S on $\mathfrak{a}$ given by:

$$\langle w(a), \lambda \rangle = \langle a, w^{-1}(\lambda) \rangle$$

for $(a, \lambda, w) \in \mathfrak{a} \times \mathfrak{a}^* \times W_S$. Thus $w(a) = {}^t w^{-1}(a)$. In particular, if $\lambda \in S$, then:

$$s_\lambda(a) = a - \lambda(a) t_\lambda,$$

where t_λ is as in 38.3.3. Let $\mathfrak{u}_\lambda = \ker \lambda \subset \mathfrak{a}$, then:

$$s_\lambda(t_\lambda) = -t_\lambda \ , \ s_\lambda|_{\mathfrak{u}_\lambda} = \mathrm{id}_{\mathfrak{u}_\lambda} .$$

Lemma. *Let $w \in W_S$. There exists $\alpha \in K$ such that $\alpha(a) = w(a)$ for all $a \in \mathfrak{a}$.*

Proof. It suffices to consider the case where $w = s_\lambda$, with $\lambda \in S$. In the notations of 38.3.6, set:

$$h = x_\lambda + y_\lambda \ , e = \frac{1}{2}(x_\lambda - y_\lambda - t_\lambda) \ , \ f = -\frac{1}{2}(x_\lambda - y_\lambda + t_\lambda).$$

We see easily that (e, h, f) is a S-triple of $\mathfrak{g}$. Moreover, $t_\lambda \in \mathfrak{a}$ and $\theta(x_\lambda) = y_\lambda$ (see 38.3.6). It follows that (e, h, f) is normal. Let $\tau = \tau_\xi \in K$ be defined with respect to (e, h, f) as in 38.6.2, with $\xi^2 = -1$. Then $\tau(e) = -e$, $\tau(f) = -f$, $t_\lambda = -(e + f)$, so $\tau(t_\lambda) = -t_\lambda$. Now if $z \in \mathfrak{u}_\lambda$, then $[h, z] = 0$, because $[z, x_\lambda] = [z, y_\lambda] = 0$. Hence $\tau(z) = z$. So we have proved that $\tau|_{\mathfrak{a}} = s_\lambda$. □

38.7.3 The Lie algebra $\mathfrak{k}$ acts on $\mathfrak{p}^*$ via the coadjoint representation. This action extends to a representation of $\mathfrak{k}$ in $S(\mathfrak{p}^*)$ via derivations of $S(\mathfrak{p}^*)$. Denote by $S(\mathfrak{p}^*)^\mathfrak{k}$ the algebra of $\mathfrak{k}$-invariants, that is, the set of $f \in S(\mathfrak{p}^*)$ such that $x.f = 0$ for all $x \in \mathfrak{k}$. Since the group K acts on $\mathfrak{p}^*$, it acts on $S(\mathfrak{p}^*)$ and by 24.3.4, we have $S(\mathfrak{p}^*)^\mathfrak{k} = S(\mathfrak{p}^*)^K$.

In the same way, W_S acts on $\mathfrak{a}^*$, and so on $S(\mathfrak{a}^*)$. Denote by $S(\mathfrak{a}^*)^{W_S}$ the set of W_S-invariants.

38.7.4 Lemma. *Let $x, y \in \mathfrak{a}$ be such that $y \notin W_S(x)$. Then there exists $f \in S(\mathfrak{a}^*)^{W_S}$ such that $f(x) \neq f(y)$.*

Proof. This is the same as in 34.2.1. □

38.7.5 Lemma. *Let $x \in \mathfrak{p}$ and s, n its semisimple and nilpotent components. Then $f(x) = f(s)$ for all $f \in S(\mathfrak{p}^*)^{\mathfrak{k}}$.*

Proof. Since $S(\mathfrak{p}^*)^{\mathfrak{k}} = S(\mathfrak{p}^*)^K$, if $f \in S(\mathfrak{p}^*)^{\mathfrak{k}}$ and $\alpha \in K$, then $f(x) = f(\alpha(x))$. So if $y \in \overline{K(x)}$, then $f(y) = f(x)$. The result is therefore a consequence of 38.6.7. □

38.7.6 Let $j : S(\mathfrak{p}^*) \to S(\mathfrak{a}^*)$ be the restriction map.

Lemma. (i) *The restriction of j to $S(\mathfrak{p}^*)^{\mathfrak{k}}$ is injective.*
(ii) *We have $j(S(\mathfrak{p}^*)^{\mathfrak{k}}) \subset S(\mathfrak{a}^*)^{W_S}$.*

Proof. Let $f \in S(\mathfrak{p}^*)^{\mathfrak{k}} = S(\mathfrak{p}^*)^K$ be such that $f|_{\mathfrak{a}} = 0$. If $a \in \mathfrak{a}$ and $\alpha \in K$, then $f(\alpha(a)) = f(a) = 0$. Since $K(\mathfrak{a})$ is dense in $\mathfrak{p}$ (38.7.1), we have $f = 0$. Finally part (ii) follows from 38.7.2. □

38.7.7 Let us denote by $\pi : S(\mathfrak{p}^*)^{\mathfrak{k}} \to S(\mathfrak{a}^*)^{W_S}$ the injective map induced by j.
Let $r = \dim \mathfrak{p}$. For $x \in \mathfrak{p}$, set:

$$T_x = (\operatorname{ad} x)^2|_{\mathfrak{p}}.$$

The characteristic polynomial χ_x of T_x is of the form:

$$\chi_x(\lambda) = \det(\lambda \operatorname{id}_{\mathfrak{p}} - T_x) = \lambda^r + p_{r-1}(x)\lambda^{r-1} + \cdots + p_s(x)\lambda^s,$$

where $p_{r-1}, \ldots, p_s \in S(\mathfrak{p}^*)$. It is clear that $p_{r-1}, \ldots, p_s \in S(\mathfrak{p}^*)^K$.
For $x \in \mathfrak{a}$ and $\lambda \in S$, we have $\chi_x(\lambda^2(x)) = 0$. Thus λ^2, and hence λ is integral over $\operatorname{im} \pi$. Since S spans $\mathfrak{a}^*$, we have proved the following result:

Lemma. *The ring $S(\mathfrak{a}^*)$ is integral over $\operatorname{im} \pi$.*

38.7.8 Lemma. *Let $f, g \in S(\mathfrak{p}^*)^{\mathfrak{k}} \setminus \{0\}$ and $p \in S(\mathfrak{p}^*)$ be such that $f = gp$. Then $p \in S(\mathfrak{p}^*)^{\mathfrak{k}}$.*

Proof. For $x \in \mathfrak{k}$, we have:

$$0 = x.f = (x.g)p + g(x.p) = g(x.p).$$

So the result is clear. □

38.7.9 Lemma. *The ring $S(\mathfrak{p}^*)^{\mathfrak{k}}$ is integrally closed.*

Proof. Let Q be the field of fractions of $S(\mathfrak{p}^*)^{\mathfrak{k}}$. If $x \in Q$ is integral over $S(\mathfrak{p}^*)^{\mathfrak{k}}$, then it is integral over $S(\mathfrak{p}^*)$. Since $S(\mathfrak{p}^*)$ is factorial, we have $x \in S(\mathfrak{p}^*)$. Hence $x \in S(\mathfrak{p}^*)^{\mathfrak{k}}$ by 38.7.8. □

38.7.10 Since π is injective, we have therefore obtained the following result:

Lemma. *The ring* $\operatorname{im}\pi$ *is integrally closed.*

38.7.11 Let us denote by $\mathfrak{a}_\mathbb{Q}$ the $\mathbb{Q}$-vector subspace of $\mathfrak{a}$ spanned by the t_λ's, $\lambda \in S$ (38.3.3).

Lemma. *Let* $x \in \mathfrak{a}_\mathbb{Q}$ *and* $y \in \mathfrak{a}$ *be such that* $f(x) = f(y)$ *for all* $f \in \operatorname{im}\pi$. *Then* $y \in \mathfrak{a}_\mathbb{Q}$ *and* $y \in W_S(x)$.

Proof. The restriction maps $S(\mathfrak{g}^*) \to S(\mathfrak{p}^*) \to S(\mathfrak{a}^*)$ induce the homomorphisms of algebras $S(\mathfrak{g}^*)^\mathfrak{g} \to S(\mathfrak{p}^*)^{\mathfrak{k}} \to S(\mathfrak{a}^*)^{W_S}$.

By assumption, $g(x) = g(y)$ for all $g \in S(\mathfrak{g}^*)^\mathfrak{g}$. It follows therefore from 31.2.6 and 34.2.1 that $y \in W(x)$. In particular, $y \in \mathfrak{a}_\mathbb{Q}$. Thus by 36.2.6, we have $y \in W_S(x)$. $\square$

38.7.12 Theorem. *Let* $(\mathfrak{g}; \mathfrak{k}, \mathfrak{p})$ *be a semisimple symmetric Lie algebra, $\mathfrak{a}$ a Cartan subspace of $\mathfrak{g}$, S the root system of $(\mathfrak{g}, \mathfrak{a})$, W_S the Weyl group of $(\mathfrak{g}, \mathfrak{a})$. The restriction map $S(\mathfrak{p}^*) \to S(\mathfrak{a}^*)$ induces an isomorphism of algebras from $S(\mathfrak{p}^*)^{\mathfrak{k}}$ onto $S(\mathfrak{a}^*)^{W_S}$.*

Proof. We shall use the notations of 38.7.7. Let $\mathbb{E}$ (resp. $\mathbb{F}$) be the field of fractions of $S(\mathfrak{a}^*)$ (resp. $\operatorname{im}\pi$) and set $q_i = \pi(p_i)$, $s \leqslant i < r$. We define a polynomial F with coefficients in $\operatorname{im}\pi$ by:

$$F(X) = X^{2r} + q_{r-1}X^{2r-2} + \cdots + q_s X^{2s}.$$

In a suitable algebraic closure of $\mathbb{F}$, the roots of F are 0 and the λ's, where $\lambda \in S$. Since S spans $\mathfrak{a}^*$, $\mathbb{E}$ is the splitting field of F over $\mathbb{F}$. So the extension $\mathbb{F} \subset \mathbb{E}$ is normal and finite. Let $H = \operatorname{Gal}(\mathbb{E}/\mathbb{F})$.

Let $\sigma \in H$. Then σ fixes each element of $\mathbb{F}$ and permutes the non-zero roots of F. Thus $S(\mathfrak{a}^*)$ is σ-invariant. If $f \in S(\mathfrak{a}^*)$, let us denote by f^σ its image under σ.

Let $x \in \mathfrak{a}_\mathbb{Q}$ and $\sigma \in H$. We define a homomorphism of $\mathbb{k}$-algebras:

$$\theta : S(\mathfrak{a}^*) \to \mathbb{k} , \ f \mapsto f^\sigma(x).$$

So there exists $y \in \mathfrak{a}$ such that

$$\theta(f) = f^\sigma(x) = f(y)$$

for all $f \in S(\mathfrak{a}^*)$. In particular, if $g \in \operatorname{im}\pi$, then:

$$g(y) = g^\sigma(x) = g(x).$$

It follows from 38.7.11 that $y \in \mathfrak{a}_\mathbb{Q}$ and $y \in W_S(x)$. If $h \in S(\mathfrak{a}^*)^{W_S}$, then:

$$h^\sigma(x) = h(y) = h(x).$$

This being true for all $x \in \mathfrak{a}_\mathbb{Q}$, we have obtained that $h^\sigma = h$, and so $h \in \mathbb{F}$.

Taking into account the results of 38.7.7 and 38.7.10, we have the following results:

a) $S(\mathfrak{a}^*)^{W_S} \subset \mathbb{F}$.

b) $\operatorname{im} \pi$ is integrally closed.

c) The ring $S(\mathfrak{a}^*)^{W_S}$ is integral over $\operatorname{im} \pi$.

Consequently $S(\mathfrak{a}^*)^{W_S} \subset \operatorname{im} \pi$, and the result follows from 38.7.6.   □

38.7.13 Let us denote by I the set of elements of $S(\mathfrak{p}^*)^\mathfrak{k}$ without constant term.

Proposition. *Let $x \in \mathfrak{p}$. Then $x \in \mathcal{N}$ if and only if $f(x) = 0$ for all $f \in S(\mathfrak{p}^*)I$.*

Proof. Let J_+ be as in 34.3.3. The restriction map $S(\mathfrak{g}^*) \to S(\mathfrak{p}^*)$ sends J_+ into I. If $f(x) = 0$ for all $f \in S(\mathfrak{p}^*)I$, then $x \in \mathcal{N}$ (34.3.3).

Conversely, let $x \in \mathcal{N}$ and $f \in I$. Then for $\alpha \in K$, we have $f(\alpha(x)) = f(x)$. Thus f is constant on $K(x)$, and hence on $\overline{K(x)}$. Since $0 \in \overline{K(x)}$ (38.6.8), we have $f(x) = f(0) = 0$.   □

38.7.14 Lemma. *If $x, y \in \mathcal{S}$ verify $f(x) = f(y)$ for all $f \in S(\mathfrak{p}^*)^\mathfrak{k}$, then $y \in K(x)$.*

Proof. We may assume that x and y belong to a Cartan subspace of $\mathfrak{g}$ (37.4.10). By 38.7.12, we have $f(x) = f(y)$ for all $f \in S(\mathfrak{a}^*)^{W_S}$. So $y \in W_S(x)$ (38.7.4), and hence by 38.7.2, $y \in K(x)$.   □

38.7.15 Lemma. *The following conditions are equivalent for $x, y \in \mathfrak{p}$:*

(i) *The semisimple components of x and y belong to the same K-orbit in $\mathfrak{p}$.*

(ii) *We have $f(x) = f(y)$ for all $f \in S(\mathfrak{p}^*)^\mathfrak{k}$.*

Proof. This is a direct consequence of 38.7.5 and 38.7.14.   □

38.7.16 Proposition. *Let $x \in \mathfrak{p}$ and s, n its semisimple and nilpotent components.*

(i) *If n is non-zero, then $\dim \mathfrak{p}^x < \dim \mathfrak{p}^s$.*

(ii) *$\overline{K(x)}$ is the union of a finite number of K-orbits.*

(iii) *$K(s)$ is the unique closed K-orbit contained in $\overline{K(x)}$.*

Proof. (i) This follows from 34.2.5 and 38.4.2.

(ii) The set $\overline{K(x)}$ is a union of K-orbits in $\mathfrak{p}$.

If x is nilpotent or semisimple, then the result follows from 38.6.1 or 38.6.8.

Suppose now that s and n are non-zero. The set F of $y \in \mathfrak{p}$ verifying $f(x) = f(y)$ for all $f \in S(\mathfrak{p}^*)^\mathfrak{k} = S(\mathfrak{p}^*)^K$ is a closed subset of $\mathfrak{p}$ containing $\overline{K(x)}$.

By 38.7.15, $y \in \mathfrak{p}$ belongs to F if and only if its semisimple component u belongs to $K(s)$. From this, we deduce that if $y \in F$, then there exists $\alpha \in K$ such that $y = \alpha(s+w)$ for some w nilpotent verifying $[s, w] = 0$. In particular, w belongs to the semisimple Lie algebra $\mathfrak{s} = [\mathfrak{g}^s, \mathfrak{g}^s]$.

The triple $(\mathfrak{s}; \mathfrak{s} \cap \mathfrak{k}, \mathfrak{s} \cap \mathfrak{p})$ is a semisimple symmetric Lie algebra and $w \in \mathfrak{s} \cap \mathfrak{p}$. Let K' be the smallest algebraic subgroup of G whose Lie algebra contains $\operatorname{ad}_{\mathfrak{g}} \mathfrak{s} \cap \mathfrak{k}$. Clearly $K' \subset K$ and $\gamma(s) = s$ for all $\gamma \in K'$. By 38.6.1, the number of nilpotent K'-orbits in $\mathfrak{s} \cap \mathfrak{p}$ is finite. Let $w_1, \ldots, w_n$ be representatives of these nilpotent orbits. Then there exist $\gamma \in K'$ and an index i such that $y = \alpha(s + \gamma(w_i)) = \alpha \circ \gamma(s + w_i)$. Thus $y \in K(s + w_i)$, and the result follows.

(iii) Any closed K-orbit in $\overline{K(x)}$ is of the form $K(y)$ with y semisimple (38.6.8). So $f(x) = f(s) = f(y)$ for all $f \in S(\mathfrak{p}^*)^{\mathfrak{k}}$. Hence we have $y \in K(s)$ by 38.7.15. $\square$

38.7.17 Let $r = \dim \mathfrak{a}$. By 38.7.12 and 31.1.6, there exist homogeneous and algebraically independent elements $u_1, \ldots, u_r$ in $S(\mathfrak{p}^*)^{\mathfrak{k}}$ which generate the algebra $S(\mathfrak{p}^*)^{\mathfrak{k}}$. Thus:

$$S(\mathfrak{p}^*)I = \sum_{i=1}^{r} S(\mathfrak{p}^*)u_i.$$

Theorem. *The set $\mathcal{N}$ of nilpotent elements in $\mathfrak{p}$ is a closed subset of codimension r in $\mathfrak{p}$. Let $\mathcal{N}_1, \ldots, \mathcal{N}_k$ be its irreducible components. For $1 \leqslant i \leqslant k$, we have that:*

(i) $\mathcal{N}_i$ is a K-invariant subset of codimension r in $\mathfrak{p}$. In particular, it is the union of a finite number of K-orbits.

(ii) $\mathcal{R} \cap \mathcal{N}_i$ is non-empty and there exists $e_i \in \mathcal{R} \cap \mathcal{N}_i$ such that:

$$K(e_i) = \mathcal{R} \cap \mathcal{N}_i.$$

The orbit $K(e_i)$ is a dense open subset of $\mathcal{N}_i$. The set

$$\mathcal{R} \cap \mathcal{N} = \bigcup_{i=1}^{k} (\mathcal{R} \cap \mathcal{N}_i)$$

is a dense open subset of $\mathcal{N}$.

Proof. It is clear that $\mathcal{N}$ is closed in $\mathfrak{p}$ and that each $\mathcal{N}_i$ is the union of a finite number of nilpotent K-orbits (38.6.1). It follows that $\mathcal{N}_i = \overline{K(e_i)}$ for some $e_i \in \mathcal{N}_i$, and $K(e_i)$ is the unique dense open K-orbit in $\mathcal{N}_i$.

Now by 38.4.2, 38.4.3 and 38.4.6, we have:

$$\operatorname{codim}_{\mathfrak{p}} K(e_i) = \dim \mathfrak{p} - \dim K(e_i) = \dim \mathfrak{p} - \dim \mathfrak{k} + \dim \mathfrak{k}^{e_i}$$
$$= \dim \mathfrak{p}^{e_i} \geqslant \dim \mathfrak{a} = r.$$

Moreover, since $\mathcal{N}$ is the closed subset defined by $u_1, \ldots, u_r$ (38.7.13), it follows that $\operatorname{codim}_{\mathfrak{p}} \mathcal{N}_i \leqslant r$ (14.2.6). Thus:

$$\text{codim}_{\mathfrak{p}}\, \mathcal{N}_i = \text{codim}_{\mathfrak{p}}\, K(e_i) = r.$$

We deduce therefore that $e_i \in \mathcal{R} \cap \mathcal{N}_i$. The other parts are now clear. $\square$

38.7.18 Let us consider example 1 of 37.2.4 with $n = 2$. Set:

$$e = \frac{1}{2}\begin{pmatrix} i & 1 \\ 1 & -i \end{pmatrix}, \quad h = i\begin{pmatrix} 0 & 1 \\ -1 & 0 \end{pmatrix}, \quad f = \frac{1}{2}\begin{pmatrix} -i & 1 \\ 1 & i \end{pmatrix},$$

where $i^2 = -1$. Then (e, h, f) is a normal S-triple of $\mathfrak{g}$. We see readily that $\mathcal{N}$ is the set of matrices of the form:

$$\begin{pmatrix} a & b \\ b & -a \end{pmatrix},$$

where $a, b \in \Bbbk$ verify $a^2 + b^2 = 0$.

From this, we deduce the following results:

a) We have $\mathcal{N} = (\Bbbk e) \cup (\Bbbk f)$, so $\mathcal{N}$ is not irreducible.

b) The K-orbits of e and f are distinct.

38.8 Double centralizers

38.8.1 We shall conserve the notations of 38.4.1. Recall that if $\mathfrak{a}$ and $\mathfrak{b}$ are subspaces of $\mathfrak{g}$, then:

$$\mathfrak{n}_{\mathfrak{a}}(\mathfrak{b}) = \{x \in \mathfrak{a};\ [x, \mathfrak{b}] \subset \mathfrak{b}\}, \quad \mathfrak{c}_{\mathfrak{a}}(\mathfrak{b}) = \{x \in \mathfrak{a};\ [x, \mathfrak{b}] = \{0\}\}.$$

The proofs of the following two lemmas are analogue to the ones for 35.3.1 and 35.3.2.

Lemma. *Let $x \in \mathfrak{p}$. Then:*
(i) $[x, \mathfrak{k}] + \mathfrak{k} = [\mathfrak{c}_{\mathfrak{p}}(\mathfrak{p}^x), \mathfrak{k}] + \mathfrak{k} = (\mathfrak{p}^x)^{\perp}$.
(ii) $\mathfrak{c}_{\mathfrak{p}}(\mathfrak{p}^x) = ([\mathfrak{p}^x, \mathfrak{k}] + \mathfrak{k})^{\perp}$.
(iii) $[x, \mathfrak{p}] + \mathfrak{p} = [\mathfrak{c}_{\mathfrak{p}}(\mathfrak{k}^x), \mathfrak{p}] + \mathfrak{p} = (\mathfrak{k}^x)^{\perp}$.
(iv) $\mathfrak{c}_{\mathfrak{k}}(\mathfrak{p}^x) = ([\mathfrak{p}^x, \mathfrak{p}] + \mathfrak{p})^{\perp}$.

38.8.2 Lemma. *The following conditions are equivalent for $x, y \in \mathfrak{p}$:*
(i) $y \in \mathfrak{c}_{\mathfrak{p}}(\mathfrak{p}^x)$.
(ii) $\mathfrak{p}^x \subset \mathfrak{p}^y$.
(iii) $[\mathfrak{p}^x, \mathfrak{k}] \subset [\mathfrak{p}^y, \mathfrak{k}]$.
(iv) $[y, \mathfrak{k}] \subset [x, \mathfrak{k}]$.
(v) $\mathfrak{c}_{\mathfrak{p}}(\mathfrak{p}^y) \subset \mathfrak{c}_{\mathfrak{p}}(\mathfrak{p}^x)$.
(vi) $[\mathfrak{c}_{\mathfrak{p}}(\mathfrak{p}^y), \mathfrak{k}] \subset [\mathfrak{c}_{\mathfrak{p}}(\mathfrak{p}^x), \mathfrak{k}]$.

38.8.3 Proposition. *Let $x \in \mathfrak{p}$.*

(i) *The element x is semisimple if and only if $\mathfrak{c}_{\mathfrak{p}}(\mathfrak{p}^x)$ is an abelian Lie subalgebra consisting of semisimple elements. If this is the case, then $\mathfrak{c}_{\mathfrak{p}}(\mathfrak{p}^x)$ is contained in $\mathfrak{z}(\mathfrak{g}^x)$, and so $\mathfrak{c}_{\mathfrak{p}}(\mathfrak{p}^x) = \mathfrak{p} \cap \mathfrak{z}(\mathfrak{g}^x)$.*

(ii) *The element x is nilpotent if and only if $\mathfrak{c}_{\mathfrak{p}}(\mathfrak{p}^x)$ is an abelian Lie subalgebra consisting of nilpotent elements.*

(iii) *Let s, n be the semisimple and nilpotent components of x. Let us write $[\mathfrak{g}^s, \mathfrak{g}^s] = \mathfrak{k}_1 \oplus \mathfrak{p}_1$ where $\mathfrak{k}_1 \subset \mathfrak{k}$ and $\mathfrak{p}_1 \subset \mathfrak{p}$. Then:*

$$\mathfrak{c}_{\mathfrak{p}}(\mathfrak{p}^x) = \mathfrak{c}_{\mathfrak{p}}(\mathfrak{p}^s) \oplus \mathfrak{c}_{\mathfrak{p}_1}(\mathfrak{p}_1^n).$$

Proof. Note that $x \in \mathfrak{p}^x$, so $\mathfrak{c}_{\mathfrak{p}}(\mathfrak{p}^x) \subset \mathfrak{p}^x$ and the Jacobi identity says that $\mathfrak{c}_{\mathfrak{p}}(\mathfrak{p}^x)$ is an abelian ideal of the Lie algebra $\mathfrak{g}^x$.

(i) If x is semisimple, then $\mathfrak{g}^x$ is reductive in $\mathfrak{g}$. So $\mathfrak{c}_{\mathfrak{p}}(\mathfrak{p}^x) \subset \mathfrak{z}(\mathfrak{g}^x)$, and the result is clear since $x \in \mathfrak{c}_{\mathfrak{p}}(\mathfrak{p}^x)$.

(ii) Let x be nilpotent, $u \in \mathfrak{c}_{\mathfrak{p}}(\mathfrak{p}^x)$ and y, z the semisimple and nilpotent components of u. We have $y, z \in \mathfrak{c}_{\mathfrak{p}}(\mathfrak{p}^x)$. If $y \neq 0$, then there exist $\lambda \in \Bbbk \setminus \{0\}$ and $v \in \mathfrak{g} \setminus \{0\}$ such that $[y, v] = \lambda v$. Then:

$$[y, [x, v]] = [[y, x], v] + [x, [y, v]] = [x, [y, v]] = \lambda [x, v].$$

Thus $\mathfrak{a} = \{w \in \mathfrak{g}; [y, w] = \lambda w\}$ is ad x-stable. Since x is nilpotent, there exists $w \in \mathfrak{a} \setminus \{0\}$ such that $[x, w] = 0$, hence $w \in \mathfrak{g}^x$. Let us write $w = w_1 + w_2$ with $w_1 \in \mathfrak{k}^x$ et $w_2 \in \mathfrak{p}^x$. As $y \in \mathfrak{p}$ and $w \in \mathfrak{a}$, we obtain that:

$$[y, w_1] = \lambda w_2 \ , \ [y, w_2] = \lambda w_1.$$

But $[y, w_2] = 0$ since $w_2 \in \mathfrak{p}^x$. So $w_1 = w = 0$. Contradiction.

(iii) We have $\mathfrak{p}^x \subset \mathfrak{p}^s$ and $\mathfrak{p}^x \subset \mathfrak{p}^n$, hence $\mathfrak{c}_{\mathfrak{p}}(\mathfrak{p}^s) \subset \mathfrak{c}_{\mathfrak{p}}(\mathfrak{p}^x)$ and $\mathfrak{c}_{\mathfrak{p}}(\mathfrak{p}^n) \subset \mathfrak{c}_{\mathfrak{p}}(\mathfrak{p}^x)$. By parts (i) and (ii), we see that $\mathfrak{c}_{\mathfrak{p}}(\mathfrak{p}^s) \oplus \mathfrak{c}_{\mathfrak{p}}(\mathfrak{p}^n) \subset \mathfrak{c}_{\mathfrak{p}}(\mathfrak{p}^x)$. On the other hand:

$$\mathfrak{g}^s = \mathfrak{k}^s \oplus \mathfrak{p}^s = \mathfrak{z}(\mathfrak{g}^s) \oplus [\mathfrak{g}^s, \mathfrak{g}^s] = \mathfrak{z}(\mathfrak{g}^s) \oplus \oplus \mathfrak{k}_1 \oplus \mathfrak{p}_1.$$

We know also that $n \in [\mathfrak{g}^s, \mathfrak{g}^s]$, hence $n \in \mathfrak{p}_1$. Finally, since $\mathfrak{g}^x = \mathfrak{g}^s \cap \mathfrak{g}^n$, we deduce that $\mathfrak{p}^x = \mathfrak{p}^s \cap \mathfrak{p}^n = \mathfrak{p}_1^n \oplus (\mathfrak{z}(\mathfrak{g}^s) \cap \mathfrak{p})$. So by (i), we obtain that $\mathfrak{c}_{\mathfrak{p}}(\mathfrak{p}^x) = (\mathfrak{z}(\mathfrak{g}^s) \cap \mathfrak{p}) \oplus \mathfrak{c}_{\mathfrak{p}_1}(\mathfrak{p}_1^n) = \mathfrak{c}_{\mathfrak{p}}(\mathfrak{p}^s) \oplus \mathfrak{c}_{\mathfrak{p}_1}(\mathfrak{p}_1^n)$. □

38.8.4 Let $x, y \in \mathfrak{p}$. If $\mathfrak{p}^y = \mathfrak{p}^x$, then $\mathfrak{c}_{\mathfrak{p}}(\mathfrak{p}^y) = \mathfrak{c}_{\mathfrak{p}}(\mathfrak{p}^x)$ (38.8.2). In particular, $y \in \mathfrak{c}_{\mathfrak{p}}(\mathfrak{p}^x)$. Set:
$$\mathfrak{c}_{\mathfrak{p}}(\mathfrak{p}^x)^{\bullet} = \{y \in \mathfrak{p}; \mathfrak{p}^y = \mathfrak{p}^x\} \subset \mathfrak{c}_{\mathfrak{p}}(\mathfrak{p}^x).$$

Lemma. *The set $\mathfrak{c}_{\mathfrak{p}}(\mathfrak{p}^x)^{\bullet}$ is a non-empty open subset of $\mathfrak{c}_{\mathfrak{p}}(\mathfrak{p}^x)$ and:*

$$\mathfrak{c}_{\mathfrak{p}}(\mathfrak{p}^x)^{\bullet} = \{y \in \mathfrak{p}; \mathfrak{c}_{\mathfrak{p}}(\mathfrak{p}^y) = \mathfrak{c}_{\mathfrak{p}}(\mathfrak{p}^x)\} = \{y \in \mathfrak{c}_{\mathfrak{p}}(\mathfrak{p}^x); \mathrm{rk}(\mathrm{ad}\, y) = \mathrm{rk}(\mathrm{ad}\, x)\}.$$

Proof. The first equality follows from 38.8.2.

Let $y \in \mathfrak{c}_{\mathfrak{p}}(\mathfrak{p}^x)$. Again by 38.8.2, we have that $\dim \mathfrak{p}^x \leqslant \dim \mathfrak{p}^y$, hence $\dim \mathfrak{g}^x \leqslant \dim \mathfrak{g}^y$ (38.4.2) and $\mathrm{rk}(\mathrm{ad}\, x) \geqslant \mathrm{rk}(\mathrm{ad}\, y)$. Similarly, $\mathrm{rk}(\mathrm{ad}\, x) = \mathrm{rk}(\mathrm{ad}\, y)$ if and only if $\dim \mathfrak{p}^x = \dim \mathfrak{p}^y$. But since $y \in \mathfrak{c}_{\mathfrak{p}}(\mathfrak{p}^x)$, we have $\mathfrak{p}^x \subset \mathfrak{p}^y$. Thus $\mathrm{rk}(\mathrm{ad}\, x) = \mathrm{rk}(\mathrm{ad}\, y)$ if and only if $\mathfrak{p}^x = \mathfrak{p}^y$. □

38.8.5 Proposition. *Let $x \in \mathfrak{p}$ and $\mathcal{K}$ its K-orbit.*

(i) *If x is nilpotent, then $\mathcal{K} \cap \mathfrak{c}_{\mathfrak{p}}(\mathfrak{p}^x) = \mathfrak{c}_{\mathfrak{p}}(\mathfrak{p}^x)^{\bullet}$.*

(ii) *If x is semisimple, then $\mathcal{K} \cap \mathfrak{c}_{\mathfrak{p}}(\mathfrak{p}^x)$ is a finite set.*

Proof. (i) The number of nilpotent K-orbits in $\mathfrak{p}$ being finite, it follows from 38.8.3 (ii) that there exists such an orbit $\mathcal{U}$ such that $\mathcal{U}' = \mathcal{U} \cap \mathfrak{c}_{\mathfrak{p}}(\mathfrak{p}^x)$ is a dense open subset of $\mathfrak{c}_{\mathfrak{p}}(\mathfrak{p}^x)$. Now 38.8.4 implies that $\mathcal{U}' \cap \mathcal{K} \neq \emptyset$. Hence $\dim \mathcal{U} = \dim \mathcal{K}$, and $\mathcal{U}' \subset \mathfrak{c}_{\mathfrak{p}}(\mathfrak{p}^x)^{\bullet}$.

Conversely, if $y \in \mathfrak{c}_{\mathfrak{p}}(\mathfrak{p}^x)^{\bullet}$, then $y \in \overline{\mathcal{U}}$ and $\dim \mathcal{K}(y) = \dim \mathcal{U}$. It follows that $y \in \mathcal{U}$. Thus $\mathfrak{c}_{\mathfrak{p}}(\mathfrak{p}^x)^{\bullet} = \mathcal{U}'$ and $\mathcal{U} = \mathcal{K}$.

(ii) If x is semisimple, then 35.3.4 says that the set $\mathfrak{z}(\mathfrak{g}^x) \cap G(x)$ is finite, and the result follows since $\mathfrak{c}_{\mathfrak{p}}(\mathfrak{p}^x) \subset \mathfrak{z}(\mathfrak{g}^x)$ (38.8.3 (i)). $\square$

38.8.6 Let (e, h, f) be a normal S-triple in $\mathfrak{g}$ and $\mathfrak{s} = \Bbbk e + \Bbbk h + \Bbbk f$. Denote by $\mathfrak{g}^{\mathfrak{s}}$ the centralizer of $\mathfrak{s}$ in $\mathfrak{g}$ and $\mathfrak{k}^{\mathfrak{s}} = \mathfrak{k} \cap \mathfrak{g}^{\mathfrak{s}}$, $\mathfrak{p}^{\mathfrak{s}} = \mathfrak{p} \cap \mathfrak{g}^{\mathfrak{s}}$. In the notations of 35.3.7, we have $\mathfrak{g}^{\mathfrak{s}} = (\mathfrak{g}^e)_0$, $\mathfrak{k}^{\mathfrak{s}} = (\mathfrak{k}^e)_0$ and $\mathfrak{p}^{\mathfrak{s}} = (\mathfrak{p}^e)_0$.

38.8.7 Proposition. *In the above notations, we have:*

$$\mathfrak{c}_{\mathfrak{p}}(\mathfrak{p}^e)_0 = \mathfrak{c}_{\mathfrak{k}}(\mathfrak{p}^e)_1 = \{0\}.$$

Proof. 1) Let $x \in \mathfrak{c}_{\mathfrak{p}}(\mathfrak{p}^e)_0$. Then $x \in \mathfrak{p}^{\mathfrak{s}}$ and $[x, (\mathfrak{p}^e)_0] = \{0\} = [x, \mathfrak{p}^{\mathfrak{s}}]$.

If $y \in \mathfrak{k}^{\mathfrak{s}}$ and $z \in \mathfrak{p}^e$, then:

$$[[y, x], z] = [[y, z], x] + [y, [x, z]].$$

We have $[x, z] = 0$ and $[y, z] \in \mathfrak{p}^e$, so $[[y, z], x] = 0$, hence $[y, x] \in \mathfrak{c}_{\mathfrak{p}}(\mathfrak{p}^e)$. Thus $\mathfrak{c}_{\mathfrak{p}}(\mathfrak{p}^e)_0$ is an ideal of $\mathfrak{g}^{\mathfrak{s}}$ which is reductive in $\mathfrak{g}$. It follows from 38.8.3 that any element of $\mathfrak{c}_{\mathfrak{p}}(\mathfrak{p}^e)_0$ is nilpotent. Hence $\mathfrak{c}_{\mathfrak{p}}(\mathfrak{p}^e)_0 \subset \mathfrak{a} = [\mathfrak{g}^{\mathfrak{s}}, \mathfrak{g}^{\mathfrak{s}}]$. Thus $\mathfrak{c}_{\mathfrak{p}}(\mathfrak{p}^e)_0$ is a nilpotent ideal of the semisimple Lie algebra $\mathfrak{a}$. Consequently, $\mathfrak{c}_{\mathfrak{p}}(\mathfrak{p}^e)_0 = \{0\}$.

2) Let $\mathfrak{r}$ be the sum of 2-dimensional simple $\mathfrak{s}$-submodules of $\mathfrak{g}$. The restriction of L to $\mathfrak{r} \times \mathfrak{r}$ is non-degenerate and $\mathfrak{r} = \mathfrak{r}_1 \oplus \mathfrak{r}_{-1}$. Moreover, it is clear that $L(\mathfrak{r}_1, \mathfrak{r}_1) = L(\mathfrak{r}_{-1}, \mathfrak{r}_{-1}) = \{0\}$.

If $x \in \mathfrak{c}_{\mathfrak{k}}(\mathfrak{p}^e)_1 \setminus \{0\}$, then $x \in \mathfrak{r}_1 \cap \mathfrak{k} \setminus \{0\}$. Since $L(\mathfrak{k}, \mathfrak{p}) = \{0\}$, there exists $y \in \mathfrak{r}_{-1} \cap \mathfrak{k}$ such that $L(x, y) \neq 0$. We may write $y = [f, z]$ where $z \in \mathfrak{r}_1 \cap \mathfrak{p}$. In particular, $z \in \mathfrak{p}^e$. So:

$$L([z, x], f) = L(x, [f, z]) = L(x, y) \neq 0.$$

This is absurd because $[z, x] = 0$. $\square$

38.8.8 Examples and remarks. 1) Let $\mathfrak{g} = \mathfrak{sl}_3(\Bbbk)$ and $(E_{ij})_{1 \leqslant i,j \leqslant 3}$ the canonical basis of $\mathfrak{gl}_3(\Bbbk)$. Set:

$$\mathfrak{k} = \Bbbk(E_{11} - E_{22}) + \Bbbk(E_{22} - E_{33}) + \Bbbk E_{13} + \Bbbk E_{31},$$
$$\mathfrak{p} = \Bbbk E_{12} + \Bbbk E_{21} + \Bbbk E_{23} + \Bbbk E_{32}.$$

Then $(\mathfrak{g}; \mathfrak{k}, \mathfrak{p})$ is a symmetric Lie algebra. Let (e, h, f) be the normal S-triple defined by $e = E_{12}$, $h = E_{11} - E_{22}$, $f = E_{21}$. Then:

$$\mathfrak{p}^e = \mathfrak{c}_\mathfrak{p}(\mathfrak{p}^e) = \Bbbk E_{12} + \Bbbk E_{32} \ , \ \mathfrak{c}_\mathfrak{p}(\mathfrak{p}^e)_2 = \Bbbk E_{12} \ , \ \mathfrak{c}_\mathfrak{p}(\mathfrak{p}^e)_1 = \Bbbk E_{32}.$$

In particular, $\mathfrak{c}_\mathfrak{p}(\mathfrak{p}^e)_1 \neq \{0\}$.

We verify easily that the dimension of a Cartan subspace of $\mathfrak{g}$ is 1. So this example shows also that $\mathfrak{p}^x$ can be abelian without the condition that $x \in \mathcal{R}$ (compare with 35.6.8).

2) Let $\mathfrak{g} = \mathfrak{sl}_{2n}(\Bbbk)$. For

$$X = \begin{pmatrix} A & B \\ C & D \end{pmatrix} \in \mathfrak{sl}_{2n}(\Bbbk),$$

with $A, B, C, D \in \mathfrak{gl}_n(\Bbbk)$, set:

$$\theta(X) = \begin{pmatrix} A & -B \\ -C & D \end{pmatrix}.$$

This defines an involutive automorphism of the Lie algebra $\mathfrak{g}$. Let $\mathfrak{k}$ be the set of fixed points of θ and $\mathfrak{p}$ the set of $X \in \mathfrak{g}$ such that $\theta(X) = -X$. Then $(\mathfrak{g}; \mathfrak{k}, \mathfrak{p})$ is a symmetric Lie algebra. More precisely, we have:

$$\mathfrak{k} = \left\{ \begin{pmatrix} A & 0 \\ 0 & D \end{pmatrix} ; \ A, D \in \mathfrak{gl}_n(\Bbbk), \mathrm{tr}(A + D) = 0 \right\},$$

$$\mathfrak{p} = \left\{ \begin{pmatrix} 0 & B \\ C & 0 \end{pmatrix} ; \ B, C \in \mathfrak{gl}_n(\Bbbk) \right\}.$$

Suppose that $n = 3$. We define a normal S-triple (e, h, f) by:

$$e = E_{15} + E_{26} + E_{63} \ , \ h = E_{11} + 2E_{22} - 2E_{33} - E_{55} \ , \ f = E_{51} + 2E_{62} + 2E_{36}.$$

Then $\mathfrak{k}^e$ is spanned by:

$$E_{13}, E_{21} + E_{65}, E_{23}, E_{45}, E_{11} - 2E_{44} + E_{55}, E_{22} + E_{33} - 3E_{44} + E_{66},$$

and $\mathfrak{p}^e$ is spanned by:

$$E_{14}, E_{15}, E_{16} + E_{53}, E_{24}, E_{25}, E_{26} + E_{63}, E_{43}.$$

It follows that:

$$\mathfrak{c}_\mathfrak{k}(\mathfrak{p}^e) = \Bbbk E_{13} + \Bbbk E_{23} \ , \ \mathfrak{c}_\mathfrak{p}(\mathfrak{p}^e) = \Bbbk e.$$

Thus $\mathfrak{c}_\mathfrak{k}(\mathfrak{p}^e)_3 = \Bbbk E_{13}$, $\mathfrak{c}_\mathfrak{k}(\mathfrak{p}^e)_4 = \Bbbk E_{23}$. In particular, $\mathrm{ad}\, h|_{\mathfrak{c}_\mathfrak{k}(\mathfrak{p}^e)}$ has even and odd eigenvalues (compare with 35.3.9).

3) Let us consider example 2 of 38.6.10 with:

$$e = X_{\alpha+\beta} \ , \ h = H_\alpha + 2H_\beta \ , \ f = X_{-\alpha-\beta}.$$

Then $\mathfrak{p}^e = \Bbbk e$, $\mathfrak{c}_\mathfrak{k}(\mathfrak{p}^e) = \Bbbk H_\alpha + \Bbbk X_\beta + \Bbbk X_{2\alpha+\beta}$. So $\mathfrak{c}_\mathfrak{k}(\mathfrak{p}^e)_0 = \Bbbk H_\alpha$, $\mathfrak{c}_\mathfrak{k}(\mathfrak{p}^e)_2 = \Bbbk X_\beta + \Bbbk X_{2\alpha+\beta}$.

38.9 Normalizers

38.9.1 Proposition. *Let $x \in \mathfrak{p}$. Then:*
(i) $\mathfrak{n}_{\mathfrak{k}}(\mathfrak{p}^x) = \{y \in \mathfrak{k};\ [x, y] \in \mathfrak{c}_{\mathfrak{p}}(\mathfrak{p}^x)\}$.
(ii) $\mathfrak{n}_{\mathfrak{k}}(\mathfrak{p}^x) = \mathfrak{n}_{\mathfrak{k}}(\mathfrak{c}_{\mathfrak{p}}(\mathfrak{p}^x))$.
(iii) *If x is semisimple, then* $\mathfrak{n}_{\mathfrak{k}}(\mathfrak{p}^x) = \mathfrak{k}^x$.

Proof. (i) Let $y \in \mathfrak{k}$, $u \in \mathfrak{p}^x$ and suppose that $[x, y] \in \mathfrak{c}_{\mathfrak{p}}(\mathfrak{p}^x)$. Then $[[y, x], u] = 0$, so:
$$0 = [[y, u], x] + [y, [x, u]] = [[y, u], x].$$
Thus $y \in \mathfrak{n}_{\mathfrak{k}}(\mathfrak{p}^x)$.

Conversely, if $y \in \mathfrak{n}_{\mathfrak{k}}(\mathfrak{p}^x)$, then for $u \in \mathfrak{p}^x$ and $v \in \mathfrak{c}_{\mathfrak{p}}(\mathfrak{p}^x)$, we have:
$$0 = [[y, u], v] = [[y, v], u] + [y, [u, v]] = [[y, v], u].$$
Thus $[y, v] \in \mathfrak{c}_{\mathfrak{p}}(\mathfrak{p}^x)$. In particular, $[y, x] \in \mathfrak{c}_{\mathfrak{p}}(\mathfrak{p}^x)$.

(ii) We have just seen that $\mathfrak{n}_{\mathfrak{k}}(\mathfrak{p}^x) \subset \mathfrak{n}_{\mathfrak{k}}(\mathfrak{c}_{\mathfrak{p}}(\mathfrak{p}^x))$. Conversely if $y \in \mathfrak{n}(\mathfrak{c}_{\mathfrak{p}}(\mathfrak{p}^x))$, then $[y, x] \in \mathfrak{c}_{\mathfrak{p}}(\mathfrak{p}^x)$. Hence $y \in \mathfrak{n}_{\mathfrak{k}}(\mathfrak{p}^x)$ by (i).

(iii) If x is semisimple, then $\mathfrak{g} = \mathfrak{g}^x \oplus [x, \mathfrak{g}]$. We deduce therefore that $\mathfrak{k} = \mathfrak{k}^x \oplus [x, \mathfrak{p}]$ and $\mathfrak{p} = \mathfrak{p}^x \oplus [x, \mathfrak{k}]$. Let $u \in \mathfrak{k}^x$, $v \in [x, \mathfrak{p}]$ and $y = u + v$. We have $[y, x] = [v, x] \in [x, \mathfrak{k}]$. So $[y, x] \in \mathfrak{c}_{\mathfrak{p}}(\mathfrak{p}^x) \subset \mathfrak{p}^x$ if and only if $[v, x] = 0$, that is $v = 0$. So the result follows from (i). $\square$

38.9.2 Let (e, h, f) be a normal S-triple of $\mathfrak{g}$. Since $\mathfrak{c}_{\mathfrak{p}}(\mathfrak{p}^e)_0 = \{0\}$ by 38.8.7, we have:
$$[e, [f, \mathfrak{c}_{\mathfrak{p}}(\mathfrak{p}^e)]] = [h, \mathfrak{c}_{\mathfrak{p}}(\mathfrak{p}^e)] = \mathfrak{c}_{\mathfrak{p}}(\mathfrak{p}^e).$$
This shows that $\dim[f, \mathfrak{c}_{\mathfrak{p}}(\mathfrak{p}^e)] = \dim \mathfrak{c}_{\mathfrak{p}}(\mathfrak{p}^e)$, and $\mathfrak{k}^e \cap [f, \mathfrak{c}_{\mathfrak{p}}(\mathfrak{p}^e)] = \{0\}$. So we deduce that the sum $\mathfrak{k}^e + [f, \mathfrak{c}_{\mathfrak{p}}(\mathfrak{p}^e)]$ is direct.

Moreover, it follows from 38.9.1 (i) that:
$$[f, \mathfrak{c}_{\mathfrak{p}}(\mathfrak{p}^e)] \subset \mathfrak{n}_{\mathfrak{k}}(\mathfrak{p}^e).$$

Proposition. *Let (e, h, f) be a normal S-triple of $\mathfrak{g}$. Then:*
(i) $[\mathfrak{n}_{\mathfrak{k}}(\mathfrak{p}^e), e] = \mathfrak{c}_{\mathfrak{p}}(\mathfrak{p}^e)$.
(ii) $\mathfrak{n}_{\mathfrak{k}}(\mathfrak{p}_e) = \mathfrak{k}^e \oplus [f, \mathfrak{c}_{\mathfrak{p}}(\mathfrak{p}^e)]$. *In particular:*
$$\mathfrak{n}_{\mathfrak{k}}(\mathfrak{p}^e) = \sum_{n \geqslant -1} \mathfrak{n}_{\mathfrak{k}}(\mathfrak{p}^e)_n.$$

Proof. (i) Since $[f, \mathfrak{c}_{\mathfrak{p}}(\mathfrak{p}^e)] \subset \mathfrak{n}_{\mathfrak{k}}(\mathfrak{p}^e)$ and $[e, [f, \mathfrak{c}_{\mathfrak{p}}(\mathfrak{p}^e)]] = \mathfrak{c}_{\mathfrak{p}}(\mathfrak{p}^e)$, the result is a consequence of 38.9.1 (i).

(ii) Since $\mathfrak{k}^e \subset \mathfrak{n}_{\mathfrak{k}}(\mathfrak{p}^e)$, it follows from (i) that:
$$\dim \mathfrak{n}_{\mathfrak{k}}(\mathfrak{p}^e) = \dim \mathfrak{k}^e + \dim \mathfrak{c}_{\mathfrak{p}}(\mathfrak{p}^e) = \dim \mathfrak{k}^e + \dim[f, \mathfrak{c}_{\mathfrak{p}}(\mathfrak{p}^e)].$$

So we have proved the first part. The second part is then a consequence of 38.8.7. $\square$

38.9.3 In view of 38.9.2, we have:

$$\mathfrak{n}_\mathfrak{g}(\mathfrak{p}^e) = \mathfrak{k}^e \oplus [f, \mathfrak{c}_\mathfrak{p}(\mathfrak{p}^e)] \oplus \mathfrak{c}_\mathfrak{p}(\mathfrak{p}^e) \ , \ [e, \mathfrak{n}_\mathfrak{g}(\mathfrak{p}^e)] = \mathfrak{c}_\mathfrak{p}(\mathfrak{p}^e).$$

38.10 Distinguished elements

38.10.1 Definition. *An element* $x \in \mathfrak{p}$ *is* $\mathfrak{k}$-*distinguished (resp.* $\mathfrak{p}$-*distinguished) if all the elements of* $\mathfrak{k}^x$ *(resp.* $\mathfrak{p}^x$*) are nilpotent.*

38.10.2 In the rest of this section, (e, h, f) is a normal S-triple of $\mathfrak{g}$, and $\mathfrak{s} = \Bbbk e + \Bbbk h + \Bbbk f$. We shall denote by $\mathfrak{g}^\mathfrak{s}$ (resp. $\mathfrak{k}^\mathfrak{s}, \mathfrak{p}^\mathfrak{s}$) the centralizer of $\mathfrak{s}$ in $\mathfrak{g}$ (resp. $\mathfrak{k}, \mathfrak{p}$). We have:

$$\mathfrak{g}^\mathfrak{s} = \mathfrak{g}^e \cap \mathfrak{g}^f = \mathfrak{g}^e \cap \mathfrak{g}^h = \mathfrak{g}^f \cap \mathfrak{g}^h.$$

Since the S-triplet is normal, we have analogous identities for $\mathfrak{k}$ and $\mathfrak{p}$.

38.10.3 Let $x \in \mathfrak{p}$. It is clear that $\mathfrak{p}^x$ contains the semisimple and nilpotent components of its elements. Thus x is $\mathfrak{p}$-distinguished if and only if 0 is the only semisimple element of $\mathfrak{p}^x$. Note that, by 38.4.7 (iii), a $\mathfrak{p}$-regular element is $\mathfrak{p}$-distinguished.

Similarly, x is $\mathfrak{k}$-distinguished if and only if 0 is the only semisimple element of $\mathfrak{k}^x$.

Proposition. *Let* $x \in \mathfrak{p}$. *Then* x *is* $\mathfrak{p}$-*distinguished if and only if it is nilpotent and the Lie subalgebra of* $\mathfrak{g}$ *generated by* $\mathfrak{p}^x$ *is nilpotent.*

Proof. We can check easily that $\mathfrak{b} = \mathfrak{p}^x \oplus [\mathfrak{p}^x, \mathfrak{p}^x]$ is the Lie subalgebra generated by $\mathfrak{p}^x$.

If x is nilpotent and $\mathfrak{b}$ is nilpotent, then any semisimple element $y \in \mathfrak{p}^x$ is central in $\mathfrak{b}$. Hence $y \in \mathfrak{c}_\mathfrak{p}(\mathfrak{p}^x)$. By 38.8.3, $y = 0$. Thus x is $\mathfrak{p}$-distinguished.

Conversely, suppose that x is $\mathfrak{p}$-distinguished. Let L' denote the Killing form of $\mathfrak{b}$. Since $(\mathfrak{b}; [\mathfrak{p}^x, \mathfrak{p}^x], \mathfrak{p}^x)$ is a symmetric Lie subalgebra of $(\mathfrak{g}, \theta)$, $[\mathfrak{p}^x, \mathfrak{p}^x]$ and $\mathfrak{p}^x$ are orthogonal with respect to L'. Let $y \in \mathfrak{p}^x$. Then $\operatorname{ad} y|_\mathfrak{b}$ is nilpotent, and therefore $L'(y, y) = 0$. Hence $L'(\mathfrak{p}^x, \mathfrak{p}^x) = \{0\}$. It follows that $L'(\mathfrak{p}^x, \mathfrak{b}) = \{0\}$, and $\mathfrak{p}^x$ is contained in the radical of $\mathfrak{b}$. We deduce then from 19.5.7 that $\mathfrak{p}^x$ is contained in the largest nilpotent ideal of $\mathfrak{b}$. We conclude therefore from the definition of $\mathfrak{b}$ that it is nilpotent. $\square$

38.10.4 Proposition. *In the notations of* 38.10.2, *the following conditions are equivalent:*

   (i) e *is* $\mathfrak{p}$-*distinguished.*
   (ii) f *is* $\mathfrak{p}$-*distinguished.*
   (iii) *The restriction of* L *to* $\mathfrak{p}^e \times \mathfrak{p}^e$ *is identically zero.*
   (iv) *The restriction of* L *to* $\mathfrak{p}^f \times \mathfrak{p}^f$ *is identically zero.*
   (v) *The restriction of* L *to* $\mathfrak{p}^\mathfrak{s} \times \mathfrak{p}^\mathfrak{s}$ *is identically zero.*
   (vi) *The vector space* $\mathfrak{p}^\mathfrak{s}$ *is* $\{0\}$.

Proof. (v) $\Leftrightarrow$ (vi) The restriction of L to $\mathfrak{g}^{\mathfrak{s}} \times \mathfrak{g}^{\mathfrak{s}}$ is non-degenerate (20.5.13) and $\mathfrak{g}^{\mathfrak{s}} = \mathfrak{k}^{\mathfrak{s}} \oplus \mathfrak{p}^{\mathfrak{s}}$. Since $\mathfrak{k}$ and $\mathfrak{p}$ are orthogonal with respect to L, the equivalence of the two conditions is clear.

(iii) $\Leftrightarrow$ (v) The subspace $\mathfrak{p}^{\mathfrak{s}}$ is ad h-stable, and the eigenvalues of ad $h|_{\mathfrak{p}^e}$ are in $\mathbb{N}$. Since $L(\mathfrak{g}_i, \mathfrak{g}_j) = \{0\}$ if $i + j \neq 0$ and $\mathfrak{p}^{\mathfrak{s}} = \mathfrak{p}^e \cap \mathfrak{g}_0$, we see that $L|_{\mathfrak{p}^e \times \mathfrak{p}^e}$ is identically zero if and only if $L|_{\mathfrak{p}^{\mathfrak{s}} \times \mathfrak{p}^{\mathfrak{s}}}$ is identically zero.

(i) $\Rightarrow$ (vi) Suppose that e is $\mathfrak{p}$-distinguished and $\mathfrak{p}^{\mathfrak{s}} \neq \{0\}$. Then from the equivalence of (iii), (v) and (vi), we deduce that $L|_{\mathfrak{p}^e \times \mathfrak{p}^e}$ is not identically zero. Since

$$2L(x, y) = L(x + y, x + y) - L(x, x) - L(y, y),$$

for $x, y \in \mathfrak{p}^e$, it follows that there exists $x \in \mathfrak{p}^e$ such that $L(x, x) = \operatorname{tr}(\operatorname{ad} x)^2 \neq 0$. This is absurd since x is nilpotent.

(vi) $\Rightarrow$ (i) Let $x \in \mathfrak{p}^e$ be semisimple and non-zero. Then $\mathfrak{g}^x = \mathfrak{k}^x \oplus \mathfrak{p}^x$ is reductive in $\mathfrak{g}$. It follows from 32.1.7 that there is a S-triple (e, h', f') such that $h', f' \in \mathfrak{g}^x$.

Let us write $h' = h_1 + h_2$, $f' = f_1 + f_2$ with $h_1, f_1 \in \mathfrak{k}^x$ and $h_2, f_2 \in \mathfrak{p}^x$. Then:

$$2e = [h_1, e] + [h_2, e] \ , \ h_1 + h_2 = [e, f_1] + [e, f_2].$$

Thus $[h_1, e] = 2e$, $[e, f_2] = h_1$. By 32.1.3, there exists $f_3 \in \mathfrak{g}$ such that (e, h_1, f_3) is a S-triple. As in 38.5.3, we may assume that $f_3 \in \mathfrak{p}$. Let $\mathfrak{t} = \Bbbk e + \Bbbk h_1 + \Bbbk f_3$. Then $x \in \mathfrak{p}^{\mathfrak{t}}$. But the S-triples (e, h, f) and (e, h_1, f_3) are K-conjugate (38.6.5). Consequently $\mathfrak{p}^{\mathfrak{s}} \neq \{0\}$.

The proofs of the equivalences (vi) $\Leftrightarrow$ (ii) and (iv) $\Leftrightarrow$ (v) are similar. $\square$

38.10.5 Corollary. *The following conditions are equivalent:*

(i) *e is $\mathfrak{p}$-distinguished.*

(ii) $\dim \mathfrak{p}_0 = \dim \mathfrak{k}_2$.

(iii) $\mathfrak{p}^e \subset [e, \mathfrak{k}]$.

(iv) $[e, \mathfrak{k}] + \mathfrak{k}$ *contains a parabolic subalgebra of* $\mathfrak{g}$.

Proof. (i) $\Leftrightarrow$ (ii) The map ad e induces a surjection from $\mathfrak{g}_0$ onto $\mathfrak{g}_2$, so it induces a surjective map φ from $\mathfrak{p}_0$ onto $\mathfrak{k}_2$. The kernel of φ is $\mathfrak{p}^{\mathfrak{s}}$. The equivalence is therefore clear by 38.10.4.

(i) $\Rightarrow$ (iii) If $\mathfrak{p}^{\mathfrak{s}} = \{0\}$, then by the theory of representations of $\mathfrak{sl}_2(\Bbbk)$, we have $\mathfrak{p}^e \subset [e, \mathfrak{g}] = [e, \mathfrak{k}] + [e, \mathfrak{p}]$. Hence $\mathfrak{p}^e \subset [e, \mathfrak{k}]$.

(iii) $\Rightarrow$ (i) Suppose that (iii) is verified, and $x \in \mathfrak{p}^e$. There exists $y \in \mathfrak{k}$ such that $x = [e, y]$. We have $[e, [e, y]] = 0$, so $x = [e, y]$ is nilpotent (32.1.2).

(i) $\Rightarrow$ (iv) Set

$$\mathfrak{q} = \sum_{n \geqslant 0} \mathfrak{g}_n \ , \ \mathfrak{m} = \sum_{n \geqslant 1} \mathfrak{g}_n.$$

Then $\mathfrak{q}$ is a parabolic subalgebra of $\mathfrak{g}$ and $\mathfrak{m}$ is its nilpotent radical.

If e is $\mathfrak{p}$-distinguished, then $\mathfrak{p}^{\mathfrak{s}} = \{0\}$, so $\mathfrak{p}^e \subset \mathfrak{m}$. Taking the orthogonals with respect to L, we obtain by 38.8.1 that $\mathfrak{q} \subset [e, \mathfrak{k}] + \mathfrak{k}$.

(iv) $\Rightarrow$ (i) Let $\mathfrak{q}'$ be a parabolic subalgebra of $\mathfrak{g}$ contained in $[e, \mathfrak{k}] + \mathfrak{k}$ and $\mathfrak{m}'$ its nilpotent radical. By taking the orthogonals with respect to L, we have $\mathfrak{p}^e \subset \mathfrak{m}'$. So the elements of $\mathfrak{p}^e$ are nilpotent. $\square$

38.10.6 We have also the following analogue of 35.2.7:

Proposition. *The following conditions are equivalent:*
(i) *e is $\mathfrak{p}$-distinguished.*
(ii) *$\mathfrak{p}^e \subset \overline{K(e)}$.*

Proof. (i) $\Rightarrow$ (ii) Recall that $\mathcal{N}$ denotes the set of nilpotent elements in $\mathfrak{p}$. Let $\mathcal{N}_2 = \{(x, y) \in \mathcal{N} \times \mathcal{N}; [x, y] = 0\}$ and $\pi : \mathcal{N}_2 \to \mathcal{N}$, $(x, y) \mapsto x$, the canonical projection. Note that K acts diagonally on $\mathcal{N}_2$ and π is K-equivariant.

1) By 38.6.1, the number of K-orbits of the elements of $\mathcal{N}$ is finite. Let $\mathcal{O}_1, \ldots, \mathcal{O}_r$ be the distinct nilpotent K-orbits of $\mathfrak{p}$, and for $1 \leqslant i \leqslant r$, fix $x_i \in \mathcal{O}_i$. For $x \in \mathcal{N}$ and $\alpha \in K$, we have

$$\pi^{-1}(\alpha.x) = \{\alpha.x\} \times (\mathfrak{p}^{\alpha.x} \cap \mathcal{N}) = \alpha\big(\{x\} \times (\mathfrak{p}^x \cap \mathcal{N})\big) = \alpha\big(\pi^{-1}(x)\big).$$

So $\dim \pi^{-1}(x) = \dim \pi^{-1}(\alpha.x)$ and $\pi^{-1}(\mathcal{O}_i) = K.\big(\{x_i\} \times (\mathfrak{p}^{x_i} \cap \mathcal{N})\big)$. Consequently,

$$\dim \pi^{-1}(\mathcal{O}_i) = \dim \mathcal{O}_i + \dim(\mathfrak{p}^{x_i} \cap \mathcal{N}) \leqslant \dim \mathcal{O}_i + \dim \mathfrak{p}^{x_i} = \dim \mathfrak{p}.$$

Since $\mathfrak{p}^{x_i} \cap \mathcal{N}$ is closed in $\mathfrak{p}^{x_i}$, we deduce further that $\dim \pi^{-1}(\mathcal{O}_i) = \dim \mathfrak{p}$ if and only if x_i is $\mathfrak{p}$-distinguished. In particular, we have $\dim \mathcal{N}_2 = \dim \mathfrak{p}$.

2) The map

$$\begin{pmatrix} a & b \\ c & d \end{pmatrix} \cdot (x, y) \mapsto (ax + by, cx + dy),$$

defines an action of $\mathrm{GL}_2(\Bbbk)$ on $\mathcal{N}_2$. Since $\mathrm{GL}_2(\Bbbk)$ is irreducible, if $\mathcal{C}$ is an irreducible component of $\mathcal{N}_2$, then $\mathrm{GL}_2(\Bbbk) \cdot \mathcal{C}$ is an irreducible subset of $\mathcal{N}_2$ containing $\mathcal{C}$. So $\mathrm{GL}_2(\Bbbk) \cdot \mathcal{C} = \mathcal{C}$.

3) If e is $\mathfrak{p}$-distinguished and $\mathcal{O} = K(e)$, then $\pi^{-1}(\mathcal{O}) = K.(\{e\} \times \mathfrak{p}^e)$ is irreducible of dimension $\dim \mathfrak{p}$. It follows from point 1 that $\overline{\pi^{-1}(\mathcal{O})}$ is an irreducible component of $\mathcal{N}_2$. Now, by point 2, we have:

$$\mathfrak{p}^e \times \{e\} = \begin{pmatrix} 0 & 1 \\ 1 & 0 \end{pmatrix} \cdot (\{e\} \times \mathfrak{p}^e) \subset \begin{pmatrix} 0 & 1 \\ 1 & 0 \end{pmatrix} \cdot \overline{\pi^{-1}(\mathcal{O})} = \overline{\pi^{-1}(\mathcal{O})}.$$

Hence by applying π, we obtain that:

$$\mathfrak{p}^e \subset \pi\big(\overline{\pi^{-1}(\mathcal{O})}\big) \subset \overline{\mathcal{O}}.$$

(ii) $\Rightarrow$ (i) If $\mathfrak{p}^e \subset \overline{K(e)}$, then $0 \in \overline{K(e)}$ and e is nilpotent (38.6.8). Hence $\mathfrak{p}^e \subset \overline{K(e)} \subset \mathcal{N}$. So e is $\mathfrak{p}$-distinguished. $\square$

38.10.7 As in 38.10.4 and 38.10.5, we have the following result:

Proposition. *The following conditions are equivalent:*
(i) *e is $\mathfrak{k}$-distinguished.*
(ii) *f is $\mathfrak{k}$-distinguished.*
(iii) *The restriction of L to $\mathfrak{k}^e \times \mathfrak{k}^e$ is identically zero.*
(iv) *The restriction of L to $\mathfrak{k}^f \times \mathfrak{k}^f$ is identically zero.*
(v) *The restriction of L to $\mathfrak{k}^s \times \mathfrak{k}^s$ is identically zero.*
(vi) *The vector space $\mathfrak{k}^s$ is $\{0\}$.*
(vii) *$\mathfrak{k}^e \subset [e, \mathfrak{p}]$.*
(viii) *$[e, \mathfrak{p}] + \mathfrak{p}$ contains a parabolic subalgebra of $\mathfrak{g}$.*

38.10.8 Let us consider the example given in the second remark of 38.8.8 with $n = 2$. The elements:

$$e = E_{14} + E_{42} \ , \ h = 2E_{11} - 2E_{22} \ , \ f = 2E_{41} + 2E_{24}.$$

form a normal S-triple, and:

$$\mathfrak{k}^e = \Bbbk E_{12} + \Bbbk(E_{11} + E_{22} - 3E_{33} + E_{44}).$$

So $\mathfrak{k}^e$ is an abelian Lie algebra which contains non-zero semisimple elements Thus in contrast 38.10.3, the following conditions are not equivalent:
(i) *e is $\mathfrak{k}$-distinguished.*
(ii) *e is nilpotent and the Lie algebra $\mathfrak{k}^e$ is nilpotent.*

References and comments

* [29], [50], [54], [55].

Many results presented in sections 38.4 and 38.5 are taken from [50].

The proof of 38.7.12 is due to Lepowsky ([54]).

Semisimple symmetric Lie algebras appear essentially in the theory of symmetric spaces. The reader may refer to [36] for a detailed account of the theory of symmetric spaces.

The notion of $\mathfrak{p}$-distinguished elements was also considered in [64], [65], [66]. In particular, a classification of $\mathfrak{p}$-distinguished elements was given in [66].

Sheets of Lie algebras

In this chapter, we study the geometry of orbits via the notions of Jordan classes and sheets. We treat both the semisimple and semisimple symmetric cases.

Throughout this chapter, $\mathfrak{g}$ is a semisimple Lie algebra with Killing form L and adjoint group G. We shall denote by $\mathcal{S}_{\mathfrak{g}}$ (resp. $\mathcal{N}_{\mathfrak{g}}$) the set of semisimple (resp. nilpotent) elements of $\mathfrak{g}$.

39.1 Jordan classes

39.1.1 Proposition. *Let $x \in \mathfrak{g}$, and s, n its semisimple and nilpotent components.*

(i) $x \in \mathcal{N}_{\mathfrak{g}}$ (resp. $x \in \mathcal{S}_{\mathfrak{g}}$) if and only if $\mathfrak{z}(\mathfrak{g}^x)$ is an abelian Lie subalgebra contained in $\mathcal{N}_{\mathfrak{g}}$ (resp. in $\mathcal{S}_{\mathfrak{g}}$).

(ii) If $\mathfrak{s} = [\mathfrak{g}^s, \mathfrak{g}^s]$, then we have:

$$\mathfrak{z}(\mathfrak{g}^x) = \mathfrak{z}(\mathfrak{g}^s) \oplus \mathfrak{z}(\mathfrak{s}^n).$$

Moreover, $\mathfrak{z}(\mathfrak{g}^s)$ (resp. $\mathfrak{z}(\mathfrak{s}^n)$) is the set of semisimple (resp. nilpotent) elements of $\mathfrak{z}(\mathfrak{g}^x)$.

Proof. (i) If $x \in \mathcal{S}_{\mathfrak{g}}$, then $\mathfrak{g}^x$ is reductive in $\mathfrak{g}$ (20.5.13). So the elements of $\mathfrak{z}(\mathfrak{g}^x)$ are semisimple. The converse is clear.

The case where x is nilpotent follows from 35.1.2.

(ii) We have $n \in \mathfrak{s}$ (20.5.14), $\mathfrak{g}^x = \mathfrak{g}^s \cap \mathfrak{g}^n$ and $\mathfrak{g}^s = \mathfrak{s} \oplus \mathfrak{z}(\mathfrak{g}^s)$. Thus $\mathfrak{g}^x = \mathfrak{z}(\mathfrak{g}^s) \oplus \mathfrak{s}^n$. Hence $\mathfrak{z}(\mathfrak{g}^x) = \mathfrak{z}(\mathfrak{g}^s) \oplus \mathfrak{z}(\mathfrak{s}^n)$. The last statement follows from part (i). □

39.1.2 For $x \in \mathfrak{g}$, denote by $\mathfrak{z}(\mathfrak{g}^x)^\bullet$ the set of elements $y \in \mathfrak{g}$ verifying $\mathfrak{g}^y = \mathfrak{g}^x$. By 35.3.3, $\mathfrak{z}(\mathfrak{g}^x)^\bullet$ is a dense open subset of $\mathfrak{z}(\mathfrak{g}^x)$, and:

$$\mathfrak{z}(\mathfrak{g}^x)^\bullet = \{y \in \mathfrak{g}; \mathfrak{z}(\mathfrak{g}^x) = \mathfrak{z}(\mathfrak{g}^y)\} = \{y \in \mathfrak{z}(\mathfrak{g}^x); \mathrm{rk}(\mathrm{ad}_{\mathfrak{g}}\, y) = \mathrm{rk}(\mathrm{ad}_{\mathfrak{g}}\, x)\}.$$

More generally, let $\mathfrak{s}$ be a semisimple Lie subalgebra of $\mathfrak{g}$ and $x \in \mathfrak{s}$. Then we denote by $\mathfrak{z}(\mathfrak{s}^x)^\bullet$ the set of elements $y \in \mathfrak{s}$ verifying $\mathfrak{s}^y = \mathfrak{s}^x$.

Lemma. *Let* $x \in \mathfrak{g}$, s, n *its semisimple and nilpotent components and* $\mathfrak{s} = [\mathfrak{g}^s, \mathfrak{g}^s]$. *We have:*

$$\mathfrak{z}(\mathfrak{g}^x)^\bullet = \{y + z; \, y \in \mathfrak{z}(\mathfrak{g}^s)^\bullet \, , \, z \in \mathfrak{z}(\mathfrak{s}^n)^\bullet\}.$$

Proof. Let $y \in \mathfrak{z}(\mathfrak{g}^s)$, $z \in \mathfrak{z}(\mathfrak{s}^n)$ and $u = y + z$. By 39.1.1, y and z are respectively the semisimple and nilpotent components of u.

Suppose that $y \in \mathfrak{z}(\mathfrak{g}^s)^\bullet$ and $z \in \mathfrak{z}(\mathfrak{s}^n)^\bullet$. Then we have $\mathfrak{g}^y = \mathfrak{g}^s$, $\mathfrak{s}^z = \mathfrak{s}^n$. Thus $\mathfrak{g}^u = \mathfrak{z}(\mathfrak{g}^s) + \mathfrak{s}^n = \mathfrak{g}^x$, and $u \in \mathfrak{z}(\mathfrak{g}^x)^\bullet$.

Let $\mathfrak{t} = [\mathfrak{g}^y, \mathfrak{g}^y]$. By the first paragraph of the proof and 39.1.1, $\mathfrak{z}(\mathfrak{g}^y)$ (resp. $\mathfrak{z}(\mathfrak{t}^z)$) is the set of semisimple (resp. nilpotent) elements of $\mathfrak{z}(\mathfrak{g}^u)$. If $u \in \mathfrak{z}(\mathfrak{g}^x)^\bullet$, then $\mathfrak{z}(\mathfrak{g}^y) = \mathfrak{z}(\mathfrak{g}^s)$ and $\mathfrak{z}(\mathfrak{t}^z) = \mathfrak{z}(\mathfrak{s}^n)$. It follows from 35.3.2 that $\mathfrak{g}^y = \mathfrak{g}^s$, hence $\mathfrak{s} = \mathfrak{t}$. This proves that $y \in \mathfrak{z}(\mathfrak{g}^s)^\bullet$ and $z \in \mathfrak{z}(\mathfrak{s}^n)^\bullet$. $\square$

39.1.3 Let $x, y \in \mathfrak{g}$. Denote respectively by x_s, y_s (resp. x_n, y_n) the semisimple (resp. nilpotent) components of x and y.

We say that x and y are *G-Jordan equivalent,* in which case we write $x \overset{G}{\sim} y$, if there exists $\alpha \in G$ such that:

$$\mathfrak{g}^{y_s} = \mathfrak{g}^{\alpha(x_s)} = \alpha(\mathfrak{g}^{x_s}) \, , \, y_n = \alpha(x_n).$$

This defines an equivalence relation on $\mathfrak{g}$. The equivalence class of x, that we shall denote by $J_G(x)$, is called the *G-Jordan class* or the *Jordan class* of x in $\mathfrak{g}$. A Jordan class is clearly a G-stable set.

39.1.4 Examples. 1) If $x \in \mathcal{N}_\mathfrak{g}$, then $J_G(x) = G(x)$.

2) Let $\mathfrak{g}_{\text{gen}}$ be the set of generic elements of $\mathfrak{g}$. Then $x \in \mathfrak{g}_{\text{gen}}$ if and only if $\mathfrak{g}^x$ is a Cartan subalgebra of $\mathfrak{g}$. By 29.2.2 and 29.2.3, $\mathfrak{g}_{\text{gen}}$ is a G-Jordan class in $\mathfrak{g}$.

39.1.5 Proposition. *Let* $x \in \mathfrak{g}$.

(i) *If* s, n *are the semisimple and nilpotent components of* x, *then:*

$$J_G(x) = G\big(\mathfrak{z}(\mathfrak{g}^s)^\bullet + n\big).$$

(ii) *The set* $J_G(x)$ *is irreducible in* $\mathfrak{g}$.

Proof. (i) Let $z \in \mathfrak{z}(\mathfrak{g}^s)^\bullet$ and $y = z + n$. By 39.1.1, z and n are the semisimple and nilpotent components of y, and we have $\mathfrak{g}^z = \mathfrak{g}^s$ as in the proof of 39.1.2. Hence $y \in J_G(x)$, and so $G(y) \subset J_G(x)$.

Conversely, let $y \in J_G(x)$, t and m its semisimple and nilpotent components, and $\alpha \in G$ verifying $\mathfrak{g}^t = \mathfrak{g}^{\alpha(s)}$, $m = \alpha(n)$. We have $t \in \mathfrak{z}(\mathfrak{g}^{\alpha(s)})^\bullet$, so $t \in \alpha\big(\mathfrak{z}(\mathfrak{g}^s)^\bullet\big)$. We have therefore obtained (i).

(ii) Let us consider the morphism $\varphi : G \times \mathfrak{g} \to \mathfrak{g}$, $(\alpha, z) \mapsto \alpha(z + n)$. By (i), we have $J_G(x) = \varphi\big(\mathfrak{z}(\mathfrak{g}^s)^\bullet\big)$. Since $\mathfrak{z}(\mathfrak{g}^s)^\bullet$ is irreducible, part (ii) follows. $\square$

39.1.6 Proposition. *Let* $x, y \in \mathfrak{g}$. *The following conditions are equivalent:*

(i) $x \overset{G}{\sim} y$.

(ii) *There exists* $\alpha \in G$ *such that* $\mathfrak{g}^y = \alpha(\mathfrak{g}^x)$.

(iii) *There exists* $\alpha \in G$ *such that* $\mathfrak{z}(\mathfrak{g}^y) = \alpha\big(\mathfrak{z}(\mathfrak{g}^x)\big)$.

Proof. The equivalence (ii) $\Leftrightarrow$ (iii) is immediate by 35.3.2.

(i) $\Rightarrow$ (ii) We shall use the notations of 39.1.3. If there exists $\alpha \in G$ such that $\mathfrak{g}^{y_s} = \mathfrak{g}^{\alpha(x_s)}$ and $y_n = \alpha(x_n)$, then:

$$\mathfrak{g}^y = \mathfrak{g}^{y_s} \cap \mathfrak{g}^{y_n} = \alpha(\mathfrak{g}^{x_s}) \cap \alpha(\mathfrak{g}^{x_n}) = \alpha(\mathfrak{g}^{x_s} \cap \mathfrak{g}^{x_n}) = \alpha(\mathfrak{g}^x).$$

(iii) $\Rightarrow$ (i) It suffices to prove that if $\mathfrak{z}(\mathfrak{g}^y) = \mathfrak{z}(\mathfrak{g}^x)$, then $x \overset{G}{\sim} y$.

If x is nilpotent, then it follows from 35.3.4 and 39.1.4 that $y \in G(x) = J_G(x)$.

Let us consider the general case. Let s, n be the semisimple and nilpotent components of x, $\mathfrak{s} = [\mathfrak{g}^s, \mathfrak{g}^s]$, and S the smallest algebraic subgroup of G whose Lie algebra is $\mathrm{ad}_{\mathfrak{g}}\, \mathfrak{s}$ (24.7.2). The set $\{\alpha|_{\mathfrak{s}}; \alpha \in S\}$ is the adjoint group of $\mathfrak{s}$ (24.8.5). We have $S\big(\mathfrak{z}(\mathfrak{g}^s)\big) = \mathfrak{z}(\mathfrak{g}^s)$, so $S\big(\mathfrak{z}(\mathfrak{g}^s)^\bullet\big) = \mathfrak{z}(\mathfrak{g}^s)^\bullet$. In view of 39.1.2, 39.1.4, 39.1.5, and the case where x is nilpotent, we obtain that:

$$y \in \mathfrak{z}(\mathfrak{g}^s)^\bullet + \mathfrak{z}(\mathfrak{s}^n)^\bullet \subset \mathfrak{z}(\mathfrak{g}^s)^\bullet + S(n) = S\big(\mathfrak{z}(\mathfrak{g}^s)^\bullet + n\big)$$
$$\subset G\big(\mathfrak{z}(\mathfrak{g}^s)^\bullet + n\big) = J_G(x).$$

Hence the result. $\quad\square$

39.1.7 Corollary. (i) *We have*

$$J_G(x) = G\big(\mathfrak{z}(\mathfrak{g}^x)^\bullet\big).$$

(ii) *The set* $J_G(x)$ *is locally closed in* $\mathfrak{g}$, *so it is a subvariety of* $\mathfrak{g}$.

Proof. Part (i) is clear by 39.1.6. Let us prove (ii).

Let $d = \dim \mathfrak{g}^x$ and $\mathfrak{u} = \{y \in \mathfrak{g}; \dim \mathfrak{g}^y = d\}$. By 19.7.6 and 29.3.1, the map $\varphi : \mathfrak{u} \to \mathrm{Gr}(\mathfrak{g}, d)$, $y \mapsto \mathfrak{g}^y$ is a morphism of varieties.

Now 39.1.6 says that $J_G(x) = \varphi^{-1}(\Omega)$ where Ω is the G-orbit of $\mathfrak{g}^x$ in $\mathrm{Gr}(\mathfrak{g}, d)$. Since Ω is locally closed in $\mathrm{Gr}(\mathfrak{g}, d)$ (21.4.3), so is $J_G(x)$. $\quad\square$

39.1.8 Let us fix a Cartan subalgebra $\mathfrak{h}$ of $\mathfrak{g}$, and denote by R the root system of $\mathfrak{g}$ relative to $\mathfrak{h}$.

1) Let $\mathcal{P}$ be a subset of R which is the intersection of R with a vector subspace of $\mathfrak{h}^*$. Denote by $\mathfrak{h}_{\mathcal{P}}$ the set of elements x of $\mathfrak{h}$ such that $\alpha(x) = 0$ for all $\alpha \in \mathcal{P}$, and $\mathfrak{h}_{\mathcal{P}}^\bullet$ the set of elements x of $\mathfrak{h}_{\mathcal{P}}$ verifying $\alpha(x) \neq 0$ for all $\alpha \in R \setminus \mathcal{P}$. The set of the subsets $\mathfrak{h}_{\mathcal{P}}$ of $\mathfrak{h}$ thus constructed is finite.

Let $x \in \mathfrak{h}$ and R' the set of elements of R which are zero at x. Then R' is the intersection of R with a vector subspace of $\mathfrak{h}^*$. We have $\mathfrak{z}(\mathfrak{g}^x) = \mathfrak{h}_{R'}$ and $\mathfrak{z}(\mathfrak{g}^x)^\bullet = \mathfrak{h}_{R'}^\bullet$.

2) Let $\mathfrak{s}_x = [\mathfrak{g}^x, \mathfrak{g}^x]$ and S_x the smallest algebraic subgroup of G whose Lie algebra is $\mathrm{ad}_{\mathfrak{g}}\,\mathfrak{s}_x$. We saw in 39.1.6 that the sets $\mathfrak{z}(\mathfrak{g}^x)$ and $\mathfrak{z}(\mathfrak{g}^x)^\bullet$ are stable under the action of S_x.

By 32.4.4, there exist $n_1, \ldots, n_r \in \mathfrak{s}_x$ such that $S_x(n_i)$, $1 \leqslant i \leqslant r$, are the distinct nilpotent S_x-orbits of S_x in $\mathfrak{s}_x$.

Let $n \in \mathfrak{g}^x \cap \mathcal{N}_{\mathfrak{g}}$. We have $n \in \mathfrak{s}_x$, so $n = \theta(n_i)$, for some $\theta \in S_x$ and $1 \leqslant i \leqslant r$. So:

$$\mathfrak{z}(\mathfrak{g}^x)^\bullet + n = \mathfrak{z}(\mathfrak{g}^x)^\bullet + \theta(n_i) = \theta\big(\mathfrak{z}(\mathfrak{g}^x)^\bullet + n_i\big).$$

By 39.1.5, we obtain that $J_G(x + n_i) = J_G(x + n)$.

3) Finally let $y \in \mathfrak{g}$, and t, u its semisimple and nilpotent components. There exists $\theta \in G$ such that $\theta(t) \in \mathfrak{h}$. Consequently, $J_G(y) = J_G(x + v)$, with $x \in \mathfrak{h}$ and $v \in \mathfrak{g}^x$ nilpotent. It follows therefore from points 1, 2 and 3 that:

Proposition. *The set of Jordan classes in $\mathfrak{g}$ is finite.*

39.2 Topology of Jordan classes

39.2.1 Lemma. *Let H be an algebraic group, K a parabolic subgroup of H, X an H-variety and Y a K-stable closed subset of X. Then $H(Y)$ is a closed subset of X.*

Proof. Endow H/K with the structure of variety as defined in 25.4.7, and denote by $\pi : H \to H/K$ the canonical surjection. The group H acts rationally on H and H/K by left multiplication (25.4.10). Thus the H-varieties H and H/K are H-homogeneous and π is an H-equivariant morphism. It follows from 25.1.2 that the morphism

$$u : H \times X \to (H/K) \times X, \ (\alpha, x) \mapsto (\pi(\alpha), x)$$

is open.

Let $S = \{(\alpha, x) \in H \times X; \alpha^{-1}(x) \in Y\}$. Since S is the inverse image of Y of the morphism $H \times X \to X$, $(\alpha, x) \mapsto \alpha^{-1}(x)$, it is closed in $H \times X$. Set $S' = u(S)$.

Let us consider the action of K on $H \times X$ given by $\beta(\alpha, x) = (\alpha\beta^{-1}, x)$, $(\alpha, \beta, x) \in H \times K \times X$. The fibres of u are the K-orbits in $H \times X$. It follows that S is K-stable because Y is K-stable. Hence by 25.3.2, S' is a closed subset of $(H/K) \times X$.

Let $p : H \times X \to X$ and $q : (H/K) \times X \to X$ denote the canonical surjections. Then clearly, we have $q(S') = p(S) = H(Y)$. Now, H/K is complete (28.1.4), so $q(S')$ is closed in X, and we have obtained the result. $\square$

39.2.2 Let $x \in \mathfrak{g}$, and s, n its semisimple and nilpotent components.

Let us fix a Cartan subalgebra $\mathfrak{h}$ of $\mathfrak{g}$ contained in $\mathfrak{g}^s$, and let R be the root system of $\mathfrak{g}$ relative to $\mathfrak{h}$. Set $R' = \{\alpha \in R; \alpha(s) = 0\}$. Then in the notations of 39.1.8, we have:

$$\mathfrak{g}^s = \mathfrak{h} \oplus \sum_{\alpha \in R'} \mathfrak{g}^\alpha \ , \ \mathfrak{z}(\mathfrak{g}^s) = \mathfrak{h}_{R'} \ , \ \mathfrak{z}(\mathfrak{g}^s)^\bullet = \mathfrak{h}_{R'}^\bullet.$$

The set R' is a root system in the subspace that it spans. Let B' be a base of R', B a base of R containing B' (18.7.9), R_+ the set of positive roots with respect to B and $R'' = R' \cup R_+$. Then

$$\mathfrak{p}_1 = \mathfrak{h} \oplus \sum_{\alpha \in R''} \mathfrak{g}^\alpha$$

is a parabolic subalgebra of $\mathfrak{g}$ having $\mathfrak{g}^s$ as a Levi factor, and whose nilpotent radical is:

$$\mathfrak{n}_1 = \sum_{\alpha \in R_+ \setminus R'} \mathfrak{g}^\alpha.$$

Let P_1 be a parabolic subgroup of G such that $\mathcal{L}(P_1) = \mathrm{ad}_\mathfrak{g}\,\mathfrak{p}_1$ (29.4.3), and N_1 the subgroup of G generated by $\exp(\mathrm{ad}_\mathfrak{g}\,u)$, $u \in \mathfrak{n}_1$. By 21.3.2, N_1 is a closed connected subgroup of G whose Lie algebra is $\mathrm{ad}_\mathfrak{g}\,\mathfrak{n}_1$ (23.3.5 and 24.5.9). Thus N_1 is a normal subgroup of P_1 (24.4.7).

On the other hand, by 24.7.2, $\mathrm{ad}_\mathfrak{g}\,\mathfrak{g}^s$ is an algebraic Lie subalgebra of $\mathrm{ad}_\mathfrak{g}\,\mathfrak{p}_1$. So there is a closed connected subgroup M of P_1 such that $\mathcal{L}(M) = \mathrm{ad}_\mathfrak{g}\,\mathfrak{g}^s$.

Since N_1 is normal in P_1, it follows from 24.5.9 that $MN_1 = N_1 M$ is a closed connected subgroup of P_1 whose Lie algebra is $\mathrm{ad}_\mathfrak{g}\,\mathfrak{g}^s + \mathrm{ad}_\mathfrak{g}\,\mathfrak{n}_1 = \mathrm{ad}_\mathfrak{g}\,\mathfrak{p}_1$. We have therefore obtained that $P_1 = N_1 M = M N_1$.

Now let $u \in \mathfrak{z}(\mathfrak{g}^s)^\bullet$ and $y = u + n$. Then u and n are the semisimple and nilpotent components of y, so $\mathfrak{g}^y = \mathfrak{g}^u \cap \mathfrak{g}^n = \mathfrak{g}^s \cap \mathfrak{g}^n \subset \mathfrak{g}^s$. Hence $\mathfrak{g}^y \cap \mathfrak{n}_1 = \{0\}$. We deduce therefore that $[y, \mathfrak{n}_1] = \mathfrak{n}_1$, and by 32.2.5, $\exp(\mathrm{ad}_\mathfrak{g}\,\mathfrak{n}_1)(y) = y + \mathfrak{n}_1$. Thus $y + \mathfrak{n}_1 \subset N_1(y)$. On the other hand, it is clear from the definition of N_1 that $N_1(y) \subset y + \mathfrak{n}_1$, so we have $N_1(y) = y + \mathfrak{n}_1$. It follows that:

$$N_1\big(\mathfrak{z}(\mathfrak{g}^s)^\bullet + n\big) = \mathfrak{z}(\mathfrak{g}^s)^\bullet + n + \mathfrak{n}_1 \ , \ \overline{N_1\big(\mathfrak{z}(\mathfrak{g}^s)^\bullet + n\big)} = \mathfrak{z}(\mathfrak{g}^s) + n + \mathfrak{n}_1.$$

On the other hand:

$$M\big(\mathfrak{z}(\mathfrak{g}^s) + n + \mathfrak{n}_1\big) = \mathfrak{z}(\mathfrak{g}^s) + M(n) + \mathfrak{n}_1,$$

and,

$$\overline{M\big(\mathfrak{z}(\mathfrak{g}^s) + n + \mathfrak{n}_1\big)} = \mathfrak{z}(\mathfrak{g}^s) + \overline{M(n)} + \mathfrak{n}_1.$$

Consequently:

$$\overline{P_1\big(\mathfrak{z}(\mathfrak{g}^s)^\bullet + n\big)} = \mathfrak{z}(\mathfrak{g}^s) + \overline{M(n)} + \mathfrak{n}_1.$$

In view of 39.1.5 and 39.2.1, we have obtained the following result:

Proposition. *We have:*

$$\overline{J_G(x)} = G\big(\mathfrak{z}(\mathfrak{g}^s) + \overline{M(n)} + \mathfrak{n}_1\big).$$

39.2.3 Remark. Let $x \in \mathfrak{g}$. Since $J_G(x) = G\big(\mathfrak{z}(\mathfrak{g})^\bullet\big)$ (39.1.7), it follows from 39.1.2 that $\dim G(y) = \dim G(x)$ for all $y \in J_G(x)$. Hence by 21.4.4, $\dim G(y) \leqslant \dim G(x)$ if $y \in \overline{J_G(x)}$.

It may happen that $\dim G(y) = \dim G(x)$ and $y \in \overline{J_G(x)} \setminus J_G(x)$. For example, if $\mathfrak{g} \neq \{0\}$ and $x \in \mathfrak{g}_{\mathrm{gen}}$, then $J_G(x) = \mathfrak{g}_{\mathrm{gen}}$ and $\overline{J_G(x)} = \mathfrak{g}$. Now, for any $y \in \mathcal{N}_{\mathfrak{g}} \cap \mathfrak{g}_{\mathrm{reg}}$, we have $\dim G(y) = \dim G(x)$.

39.2.4 Proposition. *The following conditions are equivalent for a subset $\mathfrak{J}$ of $\mathfrak{g}$:*

(i) *There exists $s \in S_{\mathfrak{g}}$ such that $\mathfrak{J} = \overline{J_G(s)}$.*

(ii) *There exists a parabolic subalgebra $\mathfrak{p}$ of $\mathfrak{g}$ such that $\mathfrak{J} = G(\mathfrak{r})$, where $\mathfrak{r}$ is the radical of $\mathfrak{p}$.*

Proof. (i) $\Rightarrow$ (ii) Using 39.2.2 with $n = 0$, we have $\overline{J_G(s)} = G\big(\mathfrak{z}(\mathfrak{g}^s) + \mathfrak{n}_1\big)$, and the result follows from the fact that $\mathfrak{z}(\mathfrak{g}^s) + \mathfrak{n}_1$ is the radical of $\mathfrak{p}_1$.

(ii) $\Rightarrow$ (i) Let $\mathfrak{l}$ be a Levi factor of $\mathfrak{p}$, $\mathfrak{k}$ the centre of $\mathfrak{l}$ and $\mathfrak{n}$ the nilpotent radical of $\mathfrak{p}$. Then $\mathfrak{p} = \mathfrak{k} \oplus [\mathfrak{l}, \mathfrak{l}] \oplus \mathfrak{n}$, and $\mathfrak{r} = \mathfrak{k} \oplus \mathfrak{n}$ is the radical of $\mathfrak{p}$. There exists $s \in \mathfrak{k}$ such that $\mathfrak{g}^s = \mathfrak{l}$ and $\mathfrak{z}(\mathfrak{g}^s) = \mathfrak{k}$ (see for example the proof of 33.2.8 (ii)). By 39.2.2, we have $G(\mathfrak{r}) = \overline{J_G(s)}$. $\square$

39.2.5 Theorem. *Let $\mathfrak{J}$ be a Jordan class in $\mathfrak{g}$. There exist a parabolic subalgebra $\mathfrak{p}$ of $\mathfrak{g}$ and a solvable ideal $\mathfrak{w}$ of $\mathfrak{p}$ such that $\overline{\mathfrak{J}} = G(\mathfrak{w})$.*

Proof. Let $x \in \mathfrak{g}$ be such that $J_G(x) = \mathfrak{J}$, and s, n its semisimple and nilpotent components. We have already treated the case $n = 0$ in 39.2.4. So we may assume that $n \neq 0$. Set $\mathfrak{s} = [\mathfrak{g}^s, \mathfrak{g}^s]$, $\mathfrak{k} = \mathfrak{z}(\mathfrak{g}^s)$. Then $n \in \mathfrak{s}$. The Lie algebras $\mathrm{ad}_{\mathfrak{g}} \, \mathfrak{s}$ and $\mathrm{ad}_{\mathfrak{g}} \, \mathfrak{k}$ are algebraic Lie subalgebras of $\mathrm{ad}_{\mathfrak{g}} \, \mathfrak{g}$. Let us denote by S and K the closed connected subgroups of G such that $\mathcal{L}(S) = \mathrm{ad}_{\mathfrak{g}} \, \mathfrak{s}$ and $\mathcal{L}(K) = \mathrm{ad}_{\mathfrak{g}} \, \mathfrak{k}$.

1) Let us use the notations $\mathfrak{p}_1, \mathfrak{n}_1, P_1, N_1, M$ of 39.2.2. As in 39.2.2, we see that $M = KS = SK$, and K is central in M.

Let (n, h, m) be a S-triple in $\mathfrak{s}$ containing n as positive element. For $i \in \mathbb{Z}$, let $\mathfrak{s}_i$ be the set of elements $y \in \mathfrak{s}$ verifying $[h, y] = iy$. If

$$\mathfrak{p}_2' = \bigoplus_{i \geqslant 0} \mathfrak{s}_i \, , \quad \mathfrak{p}_2 = \mathfrak{k} \oplus \mathfrak{p}_2' \, , \quad \mathfrak{n}_2 = \bigoplus_{i \geqslant 2} \mathfrak{s}_i,$$

then $\mathfrak{p}_2'$ is a parabolic subalgebra of $\mathfrak{s}$, and $\mathfrak{n}_2$ is an ideal of $\mathfrak{p}_2'$ contained in the nilpotent radical of $\mathfrak{p}_2'$. Set P_2', P_2 and N_2 to be the connected algebraic subgroups of G whose Lie algebras are respectively $\mathrm{ad}_{\mathfrak{g}} \, \mathfrak{p}_2', \mathrm{ad}_{\mathfrak{g}} \, \mathfrak{p}_2$ and $\mathrm{ad}_{\mathfrak{g}} \, \mathfrak{n}_2$. We have $P_2 = KP_2' = P_2'K$, and the set $P_2'(n) = P_2(n)$ is open and dense in $\mathfrak{n}_2$ (see the proof of 34.4.6).

2) Set:

$$\mathfrak{p} = \mathfrak{k} + \mathfrak{p}_2' + \mathfrak{n}_1 \, , \quad \mathfrak{n} = \mathfrak{n}_1 + \mathfrak{n}_2 \, , \quad \mathfrak{w} = \mathfrak{k} + \mathfrak{n}.$$

We have $[\mathfrak{k} + \mathfrak{p}_2', \mathfrak{n}_1] \subset [\mathfrak{g}^s, \mathfrak{n}_1] \subset \mathfrak{n}_1$. From this, we deduce that $\mathfrak{p}$ is a Lie subalgebra of $\mathfrak{g}$ which is clearly parabolic. Moreover:

$$[\mathfrak{p}, \mathfrak{w}] \subset [\mathfrak{p}, \mathfrak{k} + \mathfrak{n}_1] + [\mathfrak{p}, \mathfrak{n}_2] \subset \mathfrak{n}_1 + [\mathfrak{n}_1 + \mathfrak{p}'_2, \mathfrak{n}_2] \subset \mathfrak{n}_1 + \mathfrak{n}_2 \subset \mathfrak{w}.$$

So $\mathfrak{w}$ is an ideal of $\mathfrak{p}$ which is clearly solvable. If P is the parabolic subgroup of G whose Lie algebra is $\mathrm{ad}_{\mathfrak{g}}\,\mathfrak{p}$, then $P(\mathfrak{w}) = \mathfrak{w}$. It follows from 39.2.1 that $G(\mathfrak{w})$ is a closed subset of $\mathfrak{g}$. Hence 39.1.5 and 39.2.2 imply that $\overline{J_G(x)} = \overline{G(\mathfrak{k} + n)}$. Thus $\overline{J_G(x)} \subset \overline{G(\mathfrak{w})} = G(\mathfrak{w})$.

3) Now $P = P_2 N_1 = N_1 P_2$, and N_1 is a normal subgroup of P. Let $\alpha \in P_2$ and $u \in \mathfrak{z}(\mathfrak{g}^s)^\bullet$. We saw in 39.2.2 that $N_1(u + n) = u + n + \mathfrak{n}_1$. But $\mathfrak{p}_2 \subset \mathfrak{c}_{\mathfrak{g}}(\mathfrak{k}) = \mathfrak{g}^u$, so $\alpha(u) = u$. Consequently:

$$\begin{aligned} N_1\big(u + \alpha(n)\big) = N_1\big(\alpha(u + n)\big) &= \alpha\big(N_1(u + n)\big) \\ &= \alpha(u + n + \mathfrak{n}_1) = u + \alpha(n) + \mathfrak{n}_1. \end{aligned}$$

Thus:

$$P(u + n) = N_1 P_2(u + n) = N_1\big(u + P_2(n)\big) = u + P_2(n) + \mathfrak{n}_1.$$

Hence:

$$P\big(\mathfrak{z}(\mathfrak{g}^s)^\bullet + n\big) = \mathfrak{z}(\mathfrak{g}^s)^\bullet + P_2(n) + \mathfrak{n}_1.$$

Recall from point 1 that $\overline{P_2(n)} = \mathfrak{n}_2$. So:

$$\overline{P\big(\mathfrak{z}(\mathfrak{g}^s)^\bullet + n\big)} = \mathfrak{k} + \overline{P_2(n)} + \mathfrak{n}_1 = \mathfrak{k} + \mathfrak{n}_2 + \mathfrak{n}_1 = \mathfrak{w}.$$

We have therefore $\overline{J_G(x)} = G(\mathfrak{w})$ as required. $\square$

39.2.6 Corollary. *Let $\mathfrak{J}$ be a Jordan class in $\mathfrak{g}$. There exists a unique nilpotent G-orbit $\mathcal{O}$ in $\mathfrak{g}$ such that $\overline{\mathfrak{J}} \cap \mathcal{N}_{\mathfrak{g}} = \overline{\mathcal{O}}$.*

Proof. Let us conserve the notations of 39.2.5 so that $\overline{\mathfrak{J}} = G(\mathfrak{w})$. By the construction of $\mathfrak{w}$, and by 20.8.3 (i), $\mathfrak{n}$ is the set of nilpotent elements of $\mathfrak{w}$. This implies that $\overline{\mathfrak{J}} \cap \mathcal{N}_{\mathfrak{g}} = G(\mathfrak{n})$. Thus $\overline{\mathfrak{J}} \cap \mathcal{N}_{\mathfrak{g}}$ is the image of the morphism $G \times \mathfrak{n} \to \mathfrak{g}$, $(\alpha, x) \mapsto \alpha(x)$. It is therefore an irreducible subset of $\mathfrak{g}$.

Since $\mathfrak{J}$ is G-stable, it follows from 32.4.4 that

$$\overline{\mathfrak{J}} \cap \mathcal{N}_{\mathfrak{g}} = \mathcal{O}_1 \cup \cdots \cup \mathcal{O}_r = \overline{\mathcal{O}_1} \cup \cdots \cup \overline{\mathcal{O}_r},$$

where $\mathcal{O}_1, \dots, \mathcal{O}_r$ are nilpotent G-orbits. Hence $\overline{\mathfrak{J}} \cap \mathcal{N}_{\mathfrak{g}} = \overline{\mathcal{O}_j}$ for some j. Finally, uniqueness is a consequence of the fact that any G-orbit is open in its closure. $\square$

39.2.7 Proposition. *Let $x \in \mathfrak{g}$, s, n its semisimple and nilpotent components, and $\mathfrak{J} = J_G(x)$.*

(i) $\overline{\mathfrak{J}}$ is a union of Jordan classes in $\mathfrak{g}$.

(ii) $\overline{\mathfrak{J}}$ contains the semisimple and nilpotent components of all its elements, and the set of semisimple elements of $\overline{\mathfrak{J}}$ is $G\big(\mathfrak{z}(\mathfrak{g}^s)\big)$.

Proof. We shall use the notations of 39.2.5. Let $u \in \mathfrak{w}$ be a semisimple element which commutes with the elements of $\mathfrak{k}$. Let us write $u = t + n_1 + n_2$, with $t \in \mathfrak{k}$, $n_1 \in \mathfrak{n}_1$ and $n_2 \in \mathfrak{n}_2$. Since $[u, \mathfrak{k}] = [n_1, \mathfrak{k}]$, $n_1 = 0$. So t and n_2 are the semisimple and nilpotent components of u. Hence $u = t$. Thus $\mathfrak{k}$ is maximal among the set of commutative Lie subalgebras of $\mathfrak{w}$ consisting of semisimple elements.

Let $y \in \overline{\mathfrak{J}} = G(\mathfrak{w})$. To prove (i), it suffices to show that $J_G(y) \subset \overline{\mathfrak{J}}$. Since Jordan classes are G-stable (39.1.3), so we may assume that $y \in \mathfrak{w}$.

Let t, m be the semisimple and nilpotent components of y. Then $t, m \in \mathfrak{w}$ because, by construction, $\mathrm{ad}_{\mathfrak{g}}\, \mathfrak{w}$ is the Lie algebra of an algebraic subgroup of G. It follows from the preceding discussion, 29.7.5 (iv) and 29.7.7 that we may further assume that $t \in \mathfrak{k}$. So $\mathfrak{g}^t \supset \mathfrak{g}^s$, and hence $\mathfrak{z}(\mathfrak{g}^t) \subset \mathfrak{z}(\mathfrak{g}^s) = \mathfrak{k}$. Consequently, $\mathfrak{z}(\mathfrak{g}^t)^{\bullet} + m \subset \mathfrak{k} + \mathfrak{n}_1 + \mathfrak{n}_2 = \mathfrak{w}$. Hence $J_G(y) \subset G(\mathfrak{w}) = \overline{\mathfrak{J}}$. Finally, part (ii) follows easily from the above statements.   □

39.2.8 Lemma. *Let $x \in \mathfrak{g}$, and s, n its semisimple and nilpotent components. Then $[x, \mathfrak{g}] + \mathfrak{z}(\mathfrak{g}^x) = [x, \mathfrak{g}] + \mathfrak{z}(\mathfrak{g}^s)$, and the sum $[x, \mathfrak{g}] + \mathfrak{z}(\mathfrak{g}^s)$ is direct.*

Proof. Let $\mathfrak{s} = [\mathfrak{g}^s, \mathfrak{g}^s]$. We saw in 39.1.1 that $\mathfrak{z}(\mathfrak{g}^x) = \mathfrak{z}(\mathfrak{g}^s) + \mathfrak{z}(\mathfrak{s}^n)$. But by 35.3.9, $\mathfrak{z}(\mathfrak{s}^n) \subset [n, \mathfrak{s}] = [s + n, \mathfrak{s}] = [x, \mathfrak{s}] \subset [x, \mathfrak{g}]$. So we have the required equality.

The Lie algebra $\mathfrak{g}^s$ is reductive in $\mathfrak{g}$, so the restriction of L to $\mathfrak{z}(\mathfrak{g}^s)$ is non-degenerate (20.5.13). Since $L(\mathfrak{z}(\mathfrak{g}^s), [s + n, \mathfrak{g}]) = L([\mathfrak{z}(\mathfrak{g}^s), s + n], \mathfrak{g}) = \{0\}$, $[x, \mathfrak{g}] \cap \mathfrak{z}(\mathfrak{g}^s) = \{0\}$.   □

39.2.9 Proposition. *Let $x \in \mathfrak{g}$, s, n its semisimple and nilpotent components and N the normalizer of $\mathfrak{g}^x$ in G. Then:*

$$\dim J_G(x) = \dim G(x) + \dim \mathfrak{z}(\mathfrak{g}^s) = \dim[x, \mathfrak{g}] + \dim \mathfrak{z}(\mathfrak{g}^s)$$
$$= \dim \mathfrak{g} - \dim N + \dim \mathfrak{z}(\mathfrak{g}^x).$$

Proof. 1) The map $\varphi \colon G \times \mathfrak{z}(\mathfrak{g}^x)^{\bullet} \to J_G(x)$, $(\alpha, y) \mapsto \alpha(y)$ is a dominant morphism (39.1.7). So there exists a smooth point (α, y) of $G \times \mathfrak{z}(\mathfrak{g}^x)^{\bullet}$ such that $\alpha(y)$ is a smooth point of $J_G(x)$, and the tangent space $T_{\alpha(y)}\big(J_G(x)\big)$ is the image of $d\varphi_{(\alpha, y)}$ (16.5.7).

For $\beta \in G$, the maps $(\gamma, z) \mapsto (\beta\gamma, z)$ and $\gamma(z) \mapsto \beta\gamma(z)$ are isomorphisms of varieties. We may therefore assume that α is the identity element e of G. As in 29.1.4, the image of $d\varphi_{(e,y)}$ is $[y, \mathfrak{g}] + \mathfrak{z}(\mathfrak{g}^y)$. If t is the semisimple component of y, then $\dim J_G(x) = \dim[y, \mathfrak{g}] + \dim \mathfrak{z}(\mathfrak{g}^t)$ (39.2.8). Hence we have the first two equalities because $y \overset{G}{\sim} x$.

2) Set $d = \dim \mathfrak{g}^x$, and let $\mathcal{O}$ be the G-orbit of $\mathfrak{g}^x$ in $\mathrm{Gr}(\mathfrak{g}, d)$. As we have seen in 39.1.7, the map $\psi : J_G(x) \to \mathcal{O}$, $y \mapsto \mathfrak{g}^y$ is a surjective morphism of varieties. If $y \in J_G(x)$, then $\psi^{-1}(\mathfrak{g}^y) = \mathfrak{z}(\mathfrak{g}^y)^{\bullet}$. So the fibres of ψ have dimension $\dim \mathfrak{z}(\mathfrak{g}^x)^{\bullet} = \dim \mathfrak{z}(\mathfrak{g}^x)$. But $\dim \mathcal{O} = \dim G - \dim N$ (21.4.3). So the last equality follows by 15.5.4.   □

39.3 Sheets

39.3.1 Recall that a cone $\mathcal{C}$ (of $\mathfrak{g}$) is a subset of $\mathfrak{g}$ such that $\lambda x \in \mathcal{C}$ for all $x \in \mathcal{C}$, $\lambda \in \mathbb{k} \setminus \{0\}$.

Let $n \in \mathbb{N}$ and $\mathcal{U}$ a subset of $\mathfrak{g}$. We shall denote by $\mathcal{U}^{(n)}$ the set of elements $x \in \mathcal{U}$ verifying $\dim G(x) = n$. If p is the largest integer such that $\mathcal{U}^{(p)} \neq \emptyset$, then we shall write $\mathcal{U}^{\bullet}$ for $\mathcal{U}^{(p)}$ (this generalizes the notation of 39.1.2).

Definition. *An irreducible component of $\mathfrak{g}^{(n)}$, $n \in \mathbb{N}$, is called a* sheet *of* $\mathfrak{g}$.

39.3.2 Remarks. Let $\mathfrak{N}$ be a sheet of $\mathfrak{g}$.

1) By 21.4.4, $\mathfrak{N}$ is a locally closed subset of $\mathfrak{g}$.

2) Since $G.\mathfrak{N}$ is an irreducible subset in the corresponding $\mathfrak{g}^{(n)}$, $\mathfrak{N}$ is G-stable.

Let $x \in \mathfrak{N}$ and $F = (G(\mathbb{k} \setminus \{0\})).x$. Since $\mathfrak{N}$ is a G-stable cone, $F \subset \mathfrak{N}$. On the other hand, the set of $y \in \overline{F} = \overline{G(\mathbb{k}x)}$ verifying $\dim G(y) = \dim G(x)$ is irreducible, it is therefore contained in $\mathfrak{N}$. Consequently, $\mathfrak{N}$ contains a nilpotent G-orbit (33.5.3).

3) The set $\mathfrak{g}_{\mathrm{reg}} = \mathfrak{g}^{\bullet}$ is a sheet of $\mathfrak{g}$, called the *regular sheet* of $\mathfrak{g}$.

39.3.3 Proposition. *Let $\mathfrak{N}$ be a sheet of $\mathfrak{g}$. There exists a unique Jordan class $\mathfrak{J}$ in $\mathfrak{g}$ such that $\mathfrak{J} \subset \mathfrak{N}$ and $\overline{\mathfrak{J}} = \overline{\mathfrak{N}}$. Moreover $\mathfrak{N} = (\overline{\mathfrak{J}})^{\bullet}$.*

Proof. Let $n \in \mathbb{N}$. If x, y belong to the same Jordan class, then the G-orbits of x and y have the same dimension (39.1.6 (ii)). We deduce therefore from 39.1.8 that there exist Jordan classes $\mathfrak{J}_1, \ldots, \mathfrak{J}_r$ in $\mathfrak{g}$ such that $\mathfrak{g}^{(n)} = \mathfrak{J}_1 \cup \cdots \cup \mathfrak{J}_r$.

Let us denote by $\mathfrak{N}_1, \ldots, \mathfrak{N}_s$ the irreducible components of $\mathfrak{g}^{(n)}$. For $1 \leqslant i \leqslant s$, we have $\mathfrak{N}_i \subset \overline{\mathfrak{J}_1} \cup \cdots \cup \overline{\mathfrak{J}_r}$. Since $\mathfrak{N}_i$ is irreducible, $\mathfrak{N}_i \subset \overline{\mathfrak{J}_t}$, for some t. Hence $\overline{\mathfrak{N}_i} \subset \overline{\mathfrak{J}_t}$. Conversely, since $\mathfrak{J}_t$ is irreducible, $\mathfrak{J}_t \subset \mathfrak{N}_j$ for some j. So $\overline{\mathfrak{N}_i} \subset \overline{\mathfrak{J}_t} \subset \overline{\mathfrak{N}_j}$. Thus $i = j$ (1.1.14). Uniqueness follows from 39.2.7 (i) and 39.1.8.

Finally, let $\mathfrak{J}$ be a Jordan class verifying $\mathfrak{J} \subset \mathfrak{N}$ and $\overline{\mathfrak{J}} = \overline{\mathfrak{N}}$. For $x \in \mathfrak{J}$ and $y \in \overline{\mathfrak{J}}$, we have $\dim G(y) \leqslant \dim G(x)$ (21.4.4). Consequently, $\mathfrak{N} \subset (\overline{\mathfrak{J}})^{\bullet}$. The reverse inclusion follows from the fact that $(\overline{\mathfrak{J}})^{\bullet}$ is an open subset of $\overline{\mathfrak{J}}$, so it is irreducible. $\square$

Remark. With the above notations, we have $\mathfrak{N} \neq \mathfrak{J}$ in general. This is the case if $\mathfrak{g} \neq \{0\}$, and $\mathfrak{N} = \mathfrak{g}_{\mathrm{reg}}$, $\mathfrak{J} = \mathfrak{g}_{\mathrm{gen}}$.

39.3.4 Corollary. *Let $\mathfrak{N}$ be a sheet of $\mathfrak{g}$.*

(i) *There exists a parabolic subalgebra $\mathfrak{p}$ of $\mathfrak{g}$ and a solvable ideal $\mathfrak{w}$ of $\mathfrak{p}$ verifying:*

$$\overline{\mathfrak{N}} = G(\mathfrak{w}) \ , \ \ \mathfrak{N} = G(\mathfrak{w}^{\bullet}) = \big(G(\mathfrak{w})\big)^{\bullet}.$$

(ii) *$\mathfrak{N}$ is a union of Jordan classes in $\mathfrak{g}$.*

Proof. (i) This is clear by 39.2.5 and 39.3.3.

(ii) Let $\mathfrak{J}$ be the unique Jordan class verifying the conditions of 39.3.3 for $\mathfrak{N}$. By 39.2.7, $\overline{\mathfrak{J}}$ is a union of Jordan classes. Since $\mathfrak{N} = (\overline{\mathfrak{J}})^{\bullet}$ (39.3.3) and the elements of a Jordan class belong to a single $\mathfrak{g}^{(n)}$ (39.1.6 (ii)), the result is clear. $\square$

39.3.5 Proposition. *Let $\mathfrak{N}$ be a sheet of $\mathfrak{g}$. Then $\mathfrak{N}$ contains a unique nilpotent G-orbit.*

Proof. Let d be the dimension of the G-orbits contained in $\mathfrak{N}$ and $\mathcal{U} \subset \mathfrak{N}$ a nilpotent G-orbit. Such an orbit exists by 39.3.2. Denote by $\mathcal{O}$ the unique nilpotent G-orbit such that $\overline{\mathfrak{N}} \cap \mathcal{N}_{\mathfrak{g}} = \overline{\mathcal{O}}$ (39.2.6 and 39.3.3). Then $\dim \overline{\mathcal{O}} = \dim \mathcal{O} \leqslant d$ (21.4.4), and $\mathcal{U} \subset \overline{\mathcal{O}}$. So $\overline{\mathcal{U}} = \overline{\mathcal{O}}$, and hence $\mathcal{U} = \mathcal{O}$ (21.4.5). $\square$

39.3.6 Let $\mathfrak{N}$ be a sheet of $\mathfrak{g}$. Denote by $\mathfrak{J} = J_G(x)$ the unique Jordan class verifying $\mathfrak{N} = (\overline{\mathfrak{J}})^{\bullet}$ (39.3.3). Let s, n be the semisimple and nilpotent components of x, and we shall use the notations of 39.2.5.

We saw in 39.2.6 that $G(\mathfrak{n}) = \overline{\mathfrak{J}} \cap \mathcal{N}_{\mathfrak{g}}$. By 39.3.4 and 39.3.5, we deduce that $G(\mathfrak{n}^{\bullet}) = \big(G(\mathfrak{n})\big)^{\bullet}$ is the unique nilpotent G-orbit contained in $\mathfrak{N}$.

39.3.7 Proposition. *Let $y \in \mathfrak{w}$ and $z \in \mathfrak{w}^{\bullet} \subset \mathfrak{N}$. Then:*
(i) $\underline{\dim} G(z) = \dim \mathfrak{g} + \dim \mathfrak{n} - \dim \mathfrak{p}$.
(ii) $\overline{P(z)} = z + \mathfrak{n}$.
(iii) $\mathfrak{w} \cap G(z)$ *is a finite union of P-orbits.*
(iv) $\mathfrak{N} \cap (y + \mathfrak{n}) \neq \emptyset$.

Proof. (i) We may assume that $z = x$ because x, z belong to the same sheet, so $G(x)$ and $G(z)$ have the same dimension. We have $\mathfrak{g}^{x} = \mathfrak{g}^{s} \cap \mathfrak{g}^{n} = \mathfrak{k} \oplus \mathfrak{s}^{n}$. Also:
$$\dim \mathfrak{s}^{n} = \dim \mathfrak{s}_0 + \dim \mathfrak{s}_1 = \dim \mathfrak{p}_2' - \dim \mathfrak{n}_2.$$
Since $\mathfrak{p} = \mathfrak{k} \oplus \mathfrak{p}_2' \oplus \mathfrak{n}_1$ and $\mathfrak{n} = \mathfrak{n}_1 \oplus \mathfrak{n}_2$, we obtain that $\dim \mathfrak{g}^{x} = \dim(\mathfrak{p}/\mathfrak{n})$.

(ii) We verify easily that $[\mathfrak{p}, \mathfrak{w}] \subset \mathfrak{n}$. Consequently, $P(u) \subset u + \mathfrak{n}$ for $u \in \mathfrak{w}$ (24.5.5). It follows from part (i) that:
$$\dim(\mathfrak{p}/\mathfrak{n}) = \dim \mathfrak{g}^{z} \geqslant \dim \mathfrak{p}^{z} = \dim \mathfrak{p} - \dim P(z)$$
$$\geqslant \dim \mathfrak{p} - \dim(z + \mathfrak{n}) = \dim(\mathfrak{p}/\mathfrak{n}).$$
Thus $\dim P(z) = \dim \mathfrak{n}$, and hence $\overline{P(z)} = z + \mathfrak{n}$.

(iii) Let $\mathfrak{h}$ be a Cartan subalgebra of $\mathfrak{g}$ containing $\mathfrak{k}$ and W the Weyl group relative to $(\mathfrak{g}, \mathfrak{h})$. Recall that $\mathfrak{w} = \mathfrak{k} \oplus \mathfrak{n}$. Also, we saw in 39.2.6 that $\mathfrak{n}$ is the set of nilpotent elements of $\mathfrak{w}$, and in 39.2.7, we saw that for any semisimple element u of $\mathfrak{w}$, there exists $\alpha \in \mathrm{Aut}_e\,\mathfrak{w} \subset P$ such that $\alpha(u) \in \mathfrak{k}$.

Let $r \in \mathfrak{w}$, and t, v its semisimple and nilpotent components. We have $v \in \mathfrak{n}$, and there exists $\alpha \in P$ such that $\alpha(t) \in \mathfrak{k}$. But $P(t) \in t + \mathfrak{n}$ by the proof of (ii). From this, we deduced that $r + \mathfrak{n} = \tau_r + \mathfrak{n}$, where $\tau_r \in \mathfrak{k}$ is P-conjugate to the semisimple component of r.

If $r, z \in \mathfrak{w}$ are G-conjugate, then $\tau_r, \tau_z \in \mathfrak{k}$ are G-conjugate. It follows from 31.2.6, 34.2.1 and 34.2.4 that τ_r and τ_z belong to the same W-orbit in $\mathfrak{h}$. We deduce that $G(z)$ meets only a finite number of classes in $\mathfrak{w}$ modulo $\mathfrak{n}$. Let us denote these classes by $z_1 + \mathfrak{n}, \ldots, z_s + \mathfrak{n}$, with $z_1, \ldots, z_s \in \mathfrak{w} \cap G(z)$. Then:

$$P(z_1) \cup \cdots \cup P(z_s) \subset \mathfrak{w} \cap G(z) \subset (z_1 + \mathfrak{n}) \cup \cdots \cup (z_s + \mathfrak{n}).$$

For $1 \leqslant i \leqslant s$, $z_i \in G(z)$, so $z_i \in \mathfrak{w}^\bullet$, and $\overline{P(z_i)} = z_i + \mathfrak{n}$ by (ii). For $u \in \overline{P(z_i)} \setminus P(z_i)$, we have $\dim P(u) < \dim P(z_i) = \dim P(z) = \dim \mathfrak{n}$ (see (ii) and 21.4.5), hence $u \notin \mathfrak{w}^\bullet$, and $u \notin G(z)$. Consequently:

$$\mathfrak{w} \cap G(z) = P(z_1) \cup \cdots \cup P(z_s).$$

(iv) Since $G(\mathfrak{n}^\bullet) \subset \mathfrak{N}$, the result is clear if $y \in \mathfrak{n}$ (39.3.6). Let us suppose that $y \notin \mathfrak{n}$. We have $\Bbbk y + \mathfrak{n} \subset \mathfrak{w} \subset G(\mathfrak{w}) = \overline{\mathfrak{N}}$. Since $\mathfrak{N}$ is locally closed and $\mathfrak{n}^\bullet \subset \mathfrak{N}$, we deduce that $\mathfrak{N} \cap (\Bbbk y + \mathfrak{n})$ is a non-empty open subset of $\Bbbk y + \mathfrak{n}$. There exists therefore $t \in \Bbbk \setminus \{0\}$ such that $(ty + \mathfrak{n}) \cap \mathfrak{N} \neq \emptyset$. But $\mathfrak{N}$ is a cone, so we have also $(y + \mathfrak{n}) \cap \mathfrak{N} \neq \emptyset$. $\square$

39.4 Dixmier sheets

39.4.1 Lemma. *Let X, Y be irreducible smooth varieties such that Y is normal, and $\varphi : X \to Y$ a dominant morphism. Assume that for all $x \in X$, the differential of φ at the point x is surjective.*

(i) *Let $x \in X$ and Z an irreducible component of $\varphi^{-1}(\varphi(x))$. Then:*

$$\dim Z = \dim X - \dim Y.$$

(ii) *The map φ is open.*

Proof. By 17.4.11, it suffices to prove (i). Our assumptions imply that $\dim X \geqslant \dim Y$ and $\dim Z \geqslant \dim X - \dim Y$ for all irreducible component Z of $\varphi^{-1}(\varphi(x))$ (15.5.4). Let $z \in Z$. Then:

$$\dim \ker d\varphi_z = \dim \mathrm{T}_z(X) - \dim \mathrm{T}_{\varphi(z)}(Y) = \dim X - \dim Y.$$

Since the restriction ψ of φ to Z is constant, $d\psi_z = 0$. So $\mathrm{T}_z(Z) \subset \ker d\varphi_z$, and $\dim Z \leqslant \dim \mathrm{T}_z(Z) \leqslant \dim X - \dim Y$. Hence (i). $\square$

39.4.2 Lemma. *Let $x \in \mathcal{S}_\mathfrak{g}$ and $y \in \mathfrak{g}^x$. The following conditions are equivalent:*

(i) $\mathfrak{g}^y \subset \mathfrak{g}^x$.

(ii) $\mathfrak{g} = [y, \mathfrak{g}] + \mathfrak{g}^x$.

Proof. (i) $\Rightarrow$ (ii) Since $x \in \mathcal{S}_\mathfrak{g}$, $\mathfrak{g} = \mathfrak{g}^x + [x, \mathfrak{g}]$. So if $\mathfrak{g}^y \subset \mathfrak{g}^x$, then $[x, \mathfrak{g}] \subset [y, \mathfrak{g}]$ (35.3.1), and hence (i) $\Rightarrow$ (ii).

(ii) $\Rightarrow$ (i) For $\lambda \in \Bbbk$, set $\mathfrak{g}_\lambda$ to be the eigenspace of $\mathrm{ad}\, x$ corresponding to the eigenvalue λ. We have:

$$\mathfrak{g} = \bigoplus_{\lambda \in \Bbbk} \mathfrak{g}_\lambda \ , \ [x, \mathfrak{g}] = \bigoplus_{\lambda \in \Bbbk \setminus \{0\}} \mathfrak{g}_\lambda.$$

Let $y \in \mathfrak{g}^x$ be such that $\mathfrak{g} = [y, \mathfrak{g}] + \mathfrak{g}^x$. Taking the orthogonals with respect to L, we obtain that $\mathfrak{g}^y \cap [x, \mathfrak{g}] = \{0\}$. But since $y \in \mathfrak{g}^x$, $\mathfrak{g}^y$ is $(\operatorname{ad} x)$-stable, so $\mathfrak{g}^y$ is the sum of its intersections with the $\mathfrak{g}_\lambda$'s. Thus $\mathfrak{g}^y \subset \mathfrak{g}_0 = \mathfrak{g}^x$. $\quad\square$

39.4.3 Lemma. *Let $s \in \mathcal{S}_\mathfrak{g}$ and $\mathfrak{u}_x$ the set of $y \in \mathfrak{g}^x$ verifying $\mathfrak{g}^y \subset \mathfrak{g}^x$. Then $\mathfrak{u}_x$ is an open subset of $\mathfrak{g}^x$.*

Proof. Let us consider the map $\varphi : G \times \mathfrak{g}^x \to \mathfrak{g}$, $(\alpha, y) \mapsto \alpha(y)$. For $y \in \mathfrak{g}^x$, the image of $d\varphi_{(e,y)}$ is $[y, \mathfrak{g}] + \mathfrak{g}^x$. By 39.4.2, $y \in \mathfrak{u}_x$ if and only if the rank of $d\varphi_{(e,y)}$ is equal to $\dim \mathfrak{g}$. So the result follows. $\quad\square$

39.4.4 Lemma. *Let $x \in \mathcal{S}_\mathfrak{g}$ and $\mathfrak{v}_x$ the set of $y \in \mathfrak{g}$ such that $\mathfrak{g}^y$ is G-conjugate to a subspace of $\mathfrak{g}^x$. Then $\mathfrak{v}_x$ is an open subset of $\mathfrak{g}$.*

Proof. Let $\mathfrak{u}_x$ be as in 39.4.3. It is open in $\mathfrak{g}^x$, so it is a smooth variety. Let us consider the map $\varphi : G \times \mathfrak{u}_x \to \mathfrak{g}$, $(\alpha, y) \mapsto \alpha(y)$. For $(\alpha, y) \in G \times \mathfrak{u}_x$, the image of $d\varphi_{(\alpha,y)}$ is $\alpha([y, \mathfrak{g}] + \mathfrak{g}^x)$. So 39.4.2 implies that $d\varphi_{(\alpha,y)}$ is surjective, hence φ is dominant (16.5.7). The result follows therefore from 39.4.1 because $\mathfrak{v}_x$ is the image of φ. $\quad\square$

39.4.5 Theorem. *Let $x \in \mathcal{S}_\mathfrak{g}$. Then $\left(\overline{G(\mathfrak{z}(\mathfrak{g}^x))}\right)^{\bullet}$ is a sheet of $\mathfrak{g}$. It is the unique sheet of $\mathfrak{g}$ containing x and its dimension is $\dim G(x) + \dim \mathfrak{z}(\mathfrak{g}^x)$.*

Proof. Let $\mathfrak{N}$ be a sheet containing x, and $\mathfrak{v}_x$ be as in 39.4.4. Then $\mathfrak{N} \cap \mathfrak{v}_x$ is contained in the set of $y \in \mathfrak{g}$ such that $\mathfrak{g}^y$ is G-conjugate to $\mathfrak{g}^x$, that is in $G\left(\mathfrak{z}(\mathfrak{g}^x)^{\bullet}\right)$ (39.1.6 and 39.1.7). By 39.4.4, $\mathfrak{N} \cap \mathfrak{v}_x$ is a non-empty open subset of $\mathfrak{N}$. Hence:

$$\overline{\mathfrak{N}} = \overline{\mathfrak{N} \cap \mathfrak{v}_x} \subset \overline{G(\mathfrak{z}(\mathfrak{g}^x)^{\bullet})} \subset \overline{G(\mathfrak{z}(\mathfrak{g}^x))}.$$

Let $\mathcal{U} = \left(\overline{G(\mathfrak{z}(\mathfrak{g}^x))}\right)^{\bullet}$. Then $\mathfrak{N} \subset \mathcal{U}$. Since $\mathcal{U}$ is open in $\overline{G(\mathfrak{z}(\mathfrak{g}^x))}$, it meets $G(\mathfrak{z}(\mathfrak{g}^x))$. Now if $y \in G(\mathfrak{z}(\mathfrak{g}^x))$, then $\dim \mathfrak{g}^y \geqslant \dim \mathfrak{g}^x$. We deduce therefore that $\mathcal{U}$ is contained in the $\mathfrak{g}^{(n)}$ which contains $\mathfrak{N}$. So $\mathfrak{N} = \mathcal{U}$ because $\mathcal{U}$ is irreducible. Finally, 39.2.9 says that its dimension is $\dim G(x) + \dim \mathfrak{z}(\mathfrak{g}^x)$. $\quad\square$

39.4.6 Definition. *A sheet of $\mathfrak{g}$ which contains a semisimple element is called a* Dixmier sheet *of $\mathfrak{g}$.*

Remarks. 1) By 39.4.5, a semisimple element of $\mathfrak{g}$ is contained in a unique (Dixmier) sheet of $\mathfrak{g}$.

2) Let $\mathfrak{h}$ be a Cartan subalgebra of $\mathfrak{g}$ and R the root system of $\mathfrak{g}$ relative to $\mathfrak{h}$. In the notations of 39.1.8, 39.4.5 says that Dixmier sheets of $\mathfrak{g}$ are subsets of $\mathfrak{g}$ of the form $\left(\overline{G(\mathfrak{h}_\mathcal{P})}\right)^{\bullet}$, where $\mathcal{P}$ is the intersection of R with a vector subspace of $\mathfrak{h}^*$.

39.4.7 Theorem. *Dixmier sheets of $\mathfrak{g}$ are precisely the subsets of $\mathfrak{g}$ of the form $G(\mathfrak{r})^{\bullet}$, where $\mathfrak{r}$ is the radical of a parabolic subalgebra of $\mathfrak{g}$.*

Proof. Let $\mathfrak{p}$ be a parabolic subalgebra of $\mathfrak{g}$, $\mathfrak{r}$ its radical, $\mathfrak{l}$ a Levi factor of $\mathfrak{p}$, and $\mathfrak{c}$ the centre of $\mathfrak{l}$. If $x \in \mathfrak{c}^\bullet$, then $\mathfrak{z}(\mathfrak{g}^x) = \mathfrak{c}$. It follows from 39.4.5 that $\big(G(\mathfrak{c})\big)^\bullet$ is a Dixmier sheet of $\mathfrak{g}$. But $G(\mathfrak{r}) = J_G(x) = \overline{G(\mathfrak{c})}$ (39.2.2, 39.2.4 and 39.1.7). So $G(\mathfrak{r})^\bullet$ is a Dixmier sheet of $\mathfrak{g}$. The proof of the converse is analogue by using again 39.4.5. $\square$

39.4.8 Theorem. *Let $\mathfrak{g}$ be a semisimple Lie algebra. The set of polarizable elements of $\mathfrak{g}$ is the union of the Dixmier sheets of $\mathfrak{g}$.*

Proof. Let us fix a Cartan subalgebra $\mathfrak{h}$ of $\mathfrak{g}$ and denote by R the root system of $\mathfrak{g}$ relative to $\mathfrak{h}$.

1) Let $\mathfrak{D} = \big(\overline{G(\mathfrak{h}_\mathcal{P})}\big)^\bullet$ be a Dixmier sheet of $\mathfrak{g}$ (see Remark 2 of 39.4.6), B' a base of $\mathcal{P}$, B a base of R containing B', R_+ the set of positive roots with respect to B and

$$\mathfrak{p} = \mathfrak{h} \oplus \sum_{\alpha \in \mathcal{P} \cup R_+} \mathfrak{g}^\alpha.$$

Let $q = \dim \mathfrak{p}$, $\mathrm{Gr}(\mathfrak{g}, q)$ the Grassmannian variety of q-dimensional subspaces of $\mathfrak{g}$, and $p : \mathrm{Gr}(\mathfrak{g}, q) \times \mathfrak{g} \to \mathfrak{g}$ the canonical surjection. Denote by $\mathfrak{F}$ the set of elements of $\mathrm{Gr}(\mathfrak{g}, q)$ of the form $\theta(\mathfrak{p})$, with $\theta \in G$, and:

$$\mathcal{L} = \{(\mathfrak{q}, x) \in \overline{\mathfrak{F}} \times \overline{\mathfrak{D}}; \ L(x, [\mathfrak{q}, \mathfrak{q}]) = \{0\}\}.$$

The set $\mathcal{L}$ is closed in $\overline{\mathfrak{F}} \times \overline{\mathfrak{D}}$ and we check easily that the closed subset $p(\mathcal{L})$ contains $G(\mathfrak{h}_\mathcal{P})$. Thus $\overline{\mathfrak{D}} \subset p(\mathcal{L})$.

Let $x \in \mathfrak{D}$. Then there exists $\mathfrak{q} \in \overline{\mathfrak{F}}$ such that $L(x, [\mathfrak{q}, \mathfrak{q}]) = \{0\}$. Since $\overline{\mathfrak{F}}$ consists of Lie subalgebras of $\mathfrak{g}$ (19.7.2), we see that $\mathfrak{q}$ is a polarization of $\mathfrak{g}$ at x.

2) Let x be a polarizable element of $\mathfrak{g}$ and $\mathfrak{p}$ a polarization of $\mathfrak{g}$ at x. We saw in 33.2.6 that $\mathfrak{p}$ is a parabolic subalgebra of $\mathfrak{g}$ whose nilpotent radical is $\mathfrak{n} = [x, \mathfrak{p}]$. It follows that x belong to the radical $\mathfrak{r}$ of $\mathfrak{p}$. Up to a conjugation by an element of G, we may assume that $\mathfrak{h} \subset \mathfrak{p}$. For a suitable choice of positive roots R_+, there exists a subset $\mathcal{P}$ of R which is the intersection of R with a vector subspace of $\mathfrak{h}^*$ such that:

$$\mathfrak{p} = \mathfrak{h} \oplus \sum_{\alpha \in \mathcal{P} \cup R_+} \mathfrak{g}^\alpha, \quad \mathfrak{n} = \sum_{\alpha \in R_+ \setminus \mathcal{P}} \mathfrak{g}^\alpha.$$

As in 33.2.8, if $h \in \mathfrak{h}_\mathcal{P}^\bullet$, then $\mathfrak{p}$ is a polarization of $\mathfrak{g}$ at h. Consequently, $\dim \mathfrak{g}^h = \dim \mathfrak{g}^x$. Since $\exp(\mathrm{ad}\,\mathfrak{n})(h) = h + \mathfrak{n}$ (32.2.5), it is clear that $x \in \big(\overline{G(\mathfrak{h}_\mathcal{P})}\big)^\bullet$. $\square$

39.5 Jordan classes in the symmetric case

39.5.1 In the rest of this chapter, $(\mathfrak{g}; \mathfrak{k}, \mathfrak{p})$ or $(\mathfrak{g}, \theta)$ is a semisimple symmetric Lie algebra (see Chapter 38). We shall denote by K the connected algebraic subgroup of G such that $\mathcal{L}(K) = \mathrm{ad}_\mathfrak{g}\,\mathfrak{k}$. Finally, set $\mathcal{S}_\mathfrak{p} = \mathcal{S}_\mathfrak{g} \cap \mathfrak{p}$, and $\mathcal{N}_\mathfrak{p} = \mathcal{N}_\mathfrak{g} \cap \mathfrak{p}$.

Let $x \in \mathfrak{p}$. By 38.8.3, $\mathfrak{c}_{\mathfrak{p}}(\mathfrak{p}^x)$ is an abelian ideal of $\mathfrak{g}^x$. If $x \in \mathcal{N}_{\mathfrak{p}}$, then $\mathfrak{c}_{\mathfrak{p}}(\mathfrak{p}^x)$ contains only nilpotent elements. If $x \in \mathcal{S}_{\mathfrak{p}}$, then $\mathfrak{c}_{\mathfrak{p}}(\mathfrak{p}^x)$ contains only semisimple elements and $\mathfrak{c}_{\mathfrak{p}}(\mathfrak{p}^x) = \mathfrak{p} \cap \mathfrak{z}(\mathfrak{g}^x)$.

Recall also that $\mathfrak{c}_{\mathfrak{p}}(\mathfrak{p}^x)^\bullet$ is the set of elements $y \in \mathfrak{p}$ verifying $\mathfrak{p}^y = \mathfrak{p}^x$. It is an open subset of $\mathfrak{c}_{\mathfrak{p}}(\mathfrak{p}^x)$, and we have (38.8.4):

$$\mathfrak{c}_{\mathfrak{p}}(\mathfrak{p}^x)^\bullet = \{y \in \mathfrak{c}_{\mathfrak{p}}(\mathfrak{p}^x); \operatorname{rk}(\operatorname{ad} y) = \operatorname{rk}(\operatorname{ad} x)\}.$$

By using 38.8.3, the proof of the following result is analogue to the one for 39.1.2.

Lemma. *Let* $x \in \mathfrak{p}$, s, n *its semisimple and nilpotent components and* $[\mathfrak{g}^s, \mathfrak{g}^s] = \mathfrak{k}_1 \oplus \mathfrak{p}_1$, *with* $\mathfrak{k}_1 \subset \mathfrak{k}$ *et* $\mathfrak{p}_1 \subset \mathfrak{p}$. *Then:*

$$\mathfrak{c}_{\mathfrak{p}}(\mathfrak{p}^x)^\bullet = \{y + z; \; y \in \mathfrak{c}_{\mathfrak{p}}(\mathfrak{p}^s)^\bullet, \; z \in \mathfrak{c}_{\mathfrak{p}_1}(\mathfrak{p}_1^n)^\bullet\}.$$

39.5.2 Let $x, y \in \mathfrak{p}$. We have $x = x_s + x_n$ and $y = y_s + y_n$, where x_s, y_s are semisimple, x_n, y_n nilpotent, and $[x_s, x_n] = [y_s, y_n] = 0$.

We say that x and y are *K-Jordan equivalent* if there exists $\alpha \in K$ such that:

$$\mathfrak{p}^{y_s} = \mathfrak{p}^{\alpha(x_s)} = \alpha(\mathfrak{p}^{x_s}) \;, \; y_n = \alpha(x_n).$$

This defines an equivalence relation on $\mathfrak{p}$. The class of x, denoted by $J_K(x)$, is called the *K-Jordan class* of x in $\mathfrak{p}$. We shall write $x \overset{K}{\sim} y$ if x and y are K-Jordan equivalent.

The proof of the following result is analogue to the one for 39.1.5.

Proposition. *Let* $x \in \mathfrak{p}$ *and* s, n *its semisimple and nilpotent components. The set* $J_K(x)$ *is an irreducible subset of* $\mathfrak{p}$ *and:*

$$J_K(x) = K\big(\mathfrak{c}_{\mathfrak{p}}(\mathfrak{p}^s)^\bullet + n\big).$$

39.5.3 Remark. Let $x \in \mathfrak{p}$ and s, n its semisimple and nilpotent components. If $t \in \mathfrak{c}_{\mathfrak{p}}(\mathfrak{p}^s)^\bullet$, then $\mathfrak{p}^t = \mathfrak{p}^s$, and $\operatorname{rk}(\operatorname{ad} s) = \operatorname{rk}(\operatorname{ad} t)$ (39.5.1). Since $\mathfrak{c}_{\mathfrak{p}}(\mathfrak{p}^s) \subset \mathfrak{z}(\mathfrak{g}^s)$ (39.5.1), $\mathfrak{c}_{\mathfrak{p}}(\mathfrak{p}^s)^\bullet \subset \mathfrak{z}(\mathfrak{g}^s)^\bullet$ (notation of 39.1.2). It follows from 39.1.5 and 39.5.2 that $J_K(x) \subset J_G(x)$. Thus, for $x, y \in \mathfrak{p}$, the condition $x \overset{K}{\sim} y$ implies that $x \overset{G}{\sim} y$.

However, it may happen that $x \overset{G}{\sim} y$, but x and y are not K-Jordan equivalent. This is the case for the elements e and f of 38.7.18.

39.5.4 Proposition. *Let* $x, y \in \mathfrak{p}$. *The following conditions are equivalent:*

(i) $x \overset{K}{\sim} y$.

(ii) *There exists* $\alpha \in K$ *such that* $\mathfrak{p}^y = \alpha(\mathfrak{p}^x)$.

(iii) *There exists* $\alpha \in K$ *such that* $\mathfrak{c}_{\mathfrak{p}}(\mathfrak{p}^y) = \alpha\big(\mathfrak{c}_{\mathfrak{p}}(\mathfrak{p}^x)\big)$.

Proof. The equivalence (ii) $\Leftrightarrow$ (iii) is straightforward by 38.8.2, and the implication (i) $\Rightarrow$ (ii) is proved as in 39.1.6 by replacing G by K and $\mathfrak{g}$ by $\mathfrak{p}$.

Let us suppose that there exists $\alpha \in K$ such that $\mathfrak{p}^y = \alpha(\mathfrak{p}^x) = \mathfrak{p}^{\alpha(x)}$.

If x is nilpotent, then 38.8.4 and 38.8.5 imply that $y \in K(\alpha(x)) = K(x)$, so $x \overset{K}{\sim} y$.

Let us consider the general case. Let s, n be the semisimple and nilpotent components of $\alpha(x)$. We have $y \in \mathfrak{c}_{\mathfrak{p}}(\mathfrak{p}^{\alpha(x)})$. Let us write $[\mathfrak{g}^s, \mathfrak{g}^s] = \mathfrak{k}_1 \oplus \mathfrak{p}_1$, with $\mathfrak{k}_1 \subset \mathfrak{k}$ and $\mathfrak{p}_1 \subset \mathfrak{p}$, and denote by K_1 the smallest algebraic subgroup of K with Lie algebra $\mathrm{ad}_{\mathfrak{g}} \mathfrak{k}_1$. Then $K_1(\mathfrak{z}(\mathfrak{g}^s)) = \mathfrak{z}(\mathfrak{g}^s)$, so $K_1(\mathfrak{c}_{\mathfrak{p}}(\mathfrak{p}^s)) = \mathfrak{c}_{\mathfrak{p}}(\mathfrak{p}^s)$ by 38.8.3 (i). Hence $K_1(\mathfrak{c}_{\mathfrak{p}}(\mathfrak{p}^s)^\bullet) = \mathfrak{c}_{\mathfrak{p}}(\mathfrak{p}^s)^\bullet$. It follows from the preceding paragraph and 39.5.1 that:

$$y \in \mathfrak{c}_{\mathfrak{p}}(\mathfrak{p}^s)^\bullet + \mathfrak{c}_{\mathfrak{p}_1}(\mathfrak{p}_1^n)^\bullet \subset \mathfrak{c}_{\mathfrak{p}}(\mathfrak{p}^s)^\bullet + K_1(n) = K_1(\mathfrak{c}_{\mathfrak{p}}(\mathfrak{p}^s)^\bullet + n)$$
$$\subset K(\mathfrak{c}_{\mathfrak{p}}(\mathfrak{p}^s)^\bullet + n) = J_K(\alpha(x)) = J_K(x).$$

Thus $x \overset{K}{\sim} y$. $\square$

39.5.5 Corollary. *Let $x \in \mathfrak{p}$.*

(i) *We have:*
$$J_K(x) = K(\mathfrak{c}_{\mathfrak{p}}(\mathfrak{p}^x)^\bullet).$$

(ii) *The set $J_K(x)$ is locally closed in $\mathfrak{p}$, so it is a subvariety of $\mathfrak{p}$.*

Proof. (i) This is clear by 39.5.4.

(ii) Let $d \in \mathbb{N}$ and $\mathfrak{u}$ the set of $y \in \mathfrak{p}$ such that $\dim \mathfrak{p}^y = d$. The set $\mathfrak{u}$ is a locally closed subset of $\mathfrak{p}$, so it is a subvariety of $\mathfrak{p}$. As in 19.7.6 and identifying $\mathfrak{p}$ with $\mathfrak{p}^*$ via the Killing form, the map $\varphi : \mathfrak{u} \to \mathrm{Gr}(\mathfrak{p}, d)$, $y \mapsto \mathfrak{p}^y$, is a morphism of varieties. In view of 39.5.4, we may finish the proof as in 39.1.7 (ii). $\square$

39.5.6 Let us fix a Cartan subspace $\mathfrak{a}$ of $\mathfrak{p}$, and denote by S the root system of $(\mathfrak{g}, \mathfrak{a})$. If $\mathcal{P}$ is a subset of S which is the intersection of S with a vector subspace of $\mathfrak{a}^*$, denote by $\mathfrak{a}_{\mathcal{P}}$ the set of $x \in \mathfrak{a}$ such that $\alpha(x) = 0$ for all $\alpha \in \mathcal{P}$. Let $\mathfrak{a}_{\mathcal{P}}^\bullet$ be the set of $x \in \mathfrak{a}_{\mathcal{P}}$ verifying $\alpha(x) \neq 0$ for all $\alpha \in S \setminus \mathcal{P}$. The set of the subsets $\mathfrak{a}_{\mathcal{P}}$ and $\mathfrak{a}_{\mathcal{P}}^\bullet$ of $\mathfrak{a}$ thus constructed is finite.

Let $x \in \mathfrak{a}$ and $\mathcal{P}$ the set of elements of S which are zero at x. We obtain easily that $\mathfrak{c}_{\mathfrak{p}}(\mathfrak{p}^x) = \mathfrak{a}_{\mathcal{P}}$.

Let us write $[\mathfrak{g}^x, \mathfrak{g}^x] = \mathfrak{k}_x \oplus \mathfrak{p}_x$, with $\mathfrak{k}_x \subset \mathfrak{k}$ and $\mathfrak{p}_x \subset \mathfrak{p}$. Let K_x be the smallest algebraic subgroup of K with Lie algebra $\mathrm{ad}_{\mathfrak{g}} \mathfrak{k}_x$. We have $K_x(\mathfrak{z}(\mathfrak{g}^x)) = \mathfrak{z}(\mathfrak{g}^x)$, $K_x(\mathfrak{c}_{\mathfrak{p}}(\mathfrak{p}^x)) = \mathfrak{c}_{\mathfrak{p}}(\mathfrak{p}^x)$ and $K_x(\mathfrak{c}_{\mathfrak{p}}(\mathfrak{p}^x)^\bullet) = \mathfrak{c}_{\mathfrak{p}}(\mathfrak{p}^x)^\bullet$.

By 38.6.1, there exist $n_1, \ldots, n_r \in \mathfrak{p}_x$ such that $K_x(n_1), \ldots, K_x(n_r)$ are the nilpotent K_x-orbits in $\mathfrak{p}_x$.

Let $n \in \mathfrak{p}$ be a nilpotent element which commutes with x. Then $n \in \mathfrak{p}_x$, so $n = \alpha(n_i)$, for some $1 \leqslant i \leqslant r$ and $\alpha \in K_x$. Hence:

$$\mathfrak{c}_{\mathfrak{p}}(\mathfrak{p}^x)^\bullet + n = \mathfrak{c}_{\mathfrak{p}}(\mathfrak{p}^x)^\bullet + \alpha(n_i) = \alpha(\mathfrak{c}_{\mathfrak{p}}(\mathfrak{p}^x)^\bullet + n_i).$$

Consequently, $J_K(x + n) = J_K(x + n_i)$.

Finally, let $y \in \mathfrak{p}$ and s, n its semisimple and nilpotent components. There exists $\alpha \in K$ such that $\alpha(s) \in \mathfrak{a}$ (37.4.10 and 37.5.4). We deduce therefore that $J_K(y) = J_K(x + u)$, where $x \in \mathfrak{a}$ and $u \in \mathfrak{p}$ is a nilpotent element which commutes with x. So we have obtained the following result:

Proposition. *The number of K-Jordan classes in $\mathfrak{p}$ is finite.*

39.5.7 Lemma. *Let $x \in \mathfrak{p}$, and s, n its semisimple and nilpotent components. We have $[x, \mathfrak{k}] + \mathfrak{c}_\mathfrak{p}(\mathfrak{p}^x) = [x, \mathfrak{k}] + \mathfrak{c}_\mathfrak{p}(\mathfrak{p}^s)$, and the sum $[x, \mathfrak{k}] + \mathfrak{c}_\mathfrak{p}(\mathfrak{p}^s)$ is direct.*

Proof. Since $\mathfrak{c}_\mathfrak{p}(\mathfrak{p}^s) \subset \mathfrak{z}(\mathfrak{g}^s)$ (39.5.1), 39.2.8 implies that the sum $[x, \mathfrak{k}] + \mathfrak{c}_\mathfrak{p}(\mathfrak{p}^s)$ is direct.

Let us write $\mathfrak{s} = [\mathfrak{g}^s, \mathfrak{g}^s] = \mathfrak{k}_1 + \mathfrak{p}_1$, with $\mathfrak{k}_1 \subset \mathfrak{k}$ and $\mathfrak{p}_1 \subset \mathfrak{p}$. By 38.8.3, we have $\mathfrak{c}_\mathfrak{p}(\mathfrak{p}^x) = \mathfrak{c}_\mathfrak{p}(\mathfrak{p}^s) + \mathfrak{c}_{\mathfrak{p}_1}(\mathfrak{p}_1^n)$. Denote by $S \subset K$ the connected algebraic subgroup with Lie algebra $\mathrm{ad}_\mathfrak{g}\, \mathfrak{s}$, and let $\mathcal{O}$ be the S-orbit of n. It follows from 38.8.5 that $\mathfrak{c}_{\mathfrak{p}_1}(\mathfrak{p}_1^n) \subset \overline{\mathcal{O}}$. Hence:

$$\mathfrak{c}_{\mathfrak{p}_1}(\mathfrak{p}_1^n) \subset \mathrm{T}_n\left(\mathfrak{c}_{\mathfrak{p}_1}(\mathfrak{p}_1^n)\right) \subset \mathrm{T}_n(\overline{\mathcal{O}}) = \mathrm{T}_n(\mathcal{O})$$
$$= [n, \mathfrak{s}] = [s + n, \mathfrak{s}] = [x, \mathfrak{s}] \subset [x, \mathfrak{g}].$$

Since $\mathfrak{c}_{\mathfrak{p}_1}(\mathfrak{p}_1^n) \subset \mathfrak{p}$, $\mathfrak{c}_{\mathfrak{p}_1}(\mathfrak{p}_1^n) \subset [x, \mathfrak{k}]$. Hence the result. $\square$

39.5.8 In view of 39.5.7, the proof of the following result is analogue to the one for 39.2.9.

Proposition. *Let $x \in \mathfrak{p}$, s, n its semisimple and nilpotent components and N the normalizer of $\mathfrak{p}^x$ in K. Then:*

$$\dim J_K(x) = \dim K(x) + \dim \mathfrak{c}_\mathfrak{p}(\mathfrak{p}^s) = \dim[x, \mathfrak{k}] + \dim \mathfrak{c}_\mathfrak{p}(\mathfrak{p}^s)$$
$$= \dim \mathfrak{k} - \dim N + \dim \mathfrak{c}_\mathfrak{p}(\mathfrak{p}^x).$$

39.6 Sheets in the symmetric case

39.6.1 We shall conserve the hypotheses and notations of 39.5.1. For $n \in \mathbb{N}$, we denote by $\mathfrak{p}^{(n)}$ the set of elements $x \in \mathfrak{p}$ such that $\dim K(x) = n$. It is a locally closed subset of $\mathfrak{p}$. We shall call an irreducible component of $\mathfrak{p}^{(n)}$ a *sheet* of $\mathfrak{p}$. As in 39.3.2, we observe that a sheet of $\mathfrak{p}$ is a K-stable cone. The set $\mathcal{R}$ of $\mathfrak{p}$-regular elements of $\mathfrak{p}$ (see 38.4.4) is a sheet of $\mathfrak{p}$ that we shall call the *regular sheet* of $\mathfrak{p}$.

39.6.2 Proposition. *If $\mathcal{Q}$ is a sheet of $\mathfrak{p}$, then $\mathcal{Q}$ contains a nilpotent K-orbit.*

Proof. We may proceed as in remark 2 of 39.3.2 by using 38.6.9 (i) instead of 33.5.3. $\square$

39.6.3 Remark. Contrary to the Lie algebra case (39.3.5), a sheet of $\mathfrak{p}$ can contain several nilpotent K-orbits. For example, the regular sheet of example 38.7.18 contains two distinct nilpotent K-orbits.

39.6.4 Lemma. *Let $x \in \mathcal{S}_\mathfrak{p}$ and $y \in \mathfrak{p}^x$. The following conditions are equivalent:*

(i) $\mathfrak{p}^y \subset \mathfrak{p}^x$.

(ii) $\mathfrak{p} = [y, \mathfrak{k}] + \mathfrak{p}^x$.

Proof. Since $x \in \mathcal{S}_\mathfrak{p}$, we have $\mathfrak{g} = [x, \mathfrak{g}] \oplus \mathfrak{g}^x$. Hence:

$$\mathfrak{k} = [x, \mathfrak{p}] \oplus \mathfrak{k}^x \ , \ \mathfrak{p} = [x, \mathfrak{k}] \oplus \mathfrak{p}^x \ , \ [x, \mathfrak{k}] = [x, [x, \mathfrak{p}]].$$

If $\mathfrak{p}^y \subset \mathfrak{p}^x$, then $[x, \mathfrak{k}] \subset [y, \mathfrak{k}]$ (38.8.2). So (i) $\Rightarrow$ (ii).

Let us prove (ii) $\Rightarrow$ (i). Consider the restriction u of $(\operatorname{ad} x)^2$ to $\mathfrak{p}$. It is a semisimple endomorphism of $\mathfrak{p}$. For $\lambda \in \Bbbk$, denote by $\mathfrak{p}_\lambda$ the eigenspace of u corresponding to the eigenvalue λ. Then:

$$\mathfrak{p} = \bigoplus_{\lambda \in \Bbbk} \mathfrak{p}_\lambda \ , \ u(\mathfrak{p}) = \bigoplus_{\lambda \in \Bbbk \setminus \{0\}} \mathfrak{p}_\lambda.$$

Let $y \in \mathfrak{p}^x$ be such that $\mathfrak{p} = [y, \mathfrak{k}] + \mathfrak{p}^x$. Then by 38.8.1, we obtain:

$$\mathfrak{k} = \mathfrak{p}^\perp = ([y, \mathfrak{k}] + \mathfrak{p}^x)^\perp = (\mathfrak{p}^y + \mathfrak{k}) \cap ([x, \mathfrak{k}] + \mathfrak{k}).$$

From this, we deduce that $\mathfrak{p}^y \cap [x, \mathfrak{k}] = \{0\}$, that is $\mathfrak{p}^y \cap u(\mathfrak{p}) = \{0\}$. But since $y \in \mathfrak{p}^x$, $\mathfrak{p}^y$ is u-stable, so it is the sum of the $\mathfrak{p}^y \cap \mathfrak{p}_\lambda$ for $\lambda \in \Bbbk$. Hence $\mathfrak{p}^y \subset \mathfrak{p}_0 = \mathfrak{p}^x$. $\square$

39.6.5 Lemma. *Let $x \in \mathcal{S}_\mathfrak{p}$ and $\mathfrak{u}_x$ the set of $y \in \mathfrak{p}^x$ verifying $\mathfrak{p}^y \subset \mathfrak{p}^x$. Then $\mathfrak{u}_x$ is an open subset of $\mathfrak{p}^x$.*

Proof. Let $\varphi : K \times \mathfrak{p}^x \to \mathfrak{p}$ be the map $(\alpha, y) \mapsto \alpha(y)$. The image of $d\varphi_{(e,y)}$ is $[y, \mathfrak{k}] + \mathfrak{p}^x$. By 39.6.4, $y \in \mathfrak{u}_x$ if and only if the rank of $d\varphi_{(e,y)}$ is $\dim \mathfrak{p}$. So the result follows. $\square$

39.6.6 Lemma. *Let $x \in \mathcal{S}_\mathfrak{p}$ and $\mathfrak{v}_x$ the set of $y \in \mathfrak{p}^x$ such that $\mathfrak{p}^y$ is K-conjugate to a subspace of $\mathfrak{p}^x$. Then $\mathfrak{v}_x$ is an open subset of $\mathfrak{p}$.*

Proof. In view of 39.6.5, the proof is analogue to the one for 39.4.4. $\square$

39.6.7 Theorem. *Let $x \in \mathcal{S}_\mathfrak{p}$. Then $(\overline{K(\mathfrak{c}_\mathfrak{p}(\mathfrak{p}^x))})^\bullet$ is a sheet of $\mathfrak{p}$. It is the unique sheet of $\mathfrak{p}$ containing x, and its dimension is $\dim \mathfrak{c}_\mathfrak{p}(\mathfrak{p}^x) + \dim K(x)$.*

Proof. In view of the preceding results, the proof is analogue to the one for 39.4.5. $\square$

39.6.8 Proposition. *Let $\mathcal{Q}_1, \ldots, \mathcal{Q}_r$ be the sheets of $\mathfrak{p}$ which contain a semisimple element. Then any element $\mathcal{Q}_1 \cup \cdots \cup \mathcal{Q}_r$ is polarizable.*

Proof. If $x \in \mathcal{S}_{\mathfrak{p}}$, then $\mathfrak{c}_{\mathfrak{p}}(\mathfrak{p}^x) \subset \mathfrak{z}(\mathfrak{g}^x)$ (39.5.1). Hence:

$$(\overline{K(\mathfrak{c}_{\mathfrak{p}}(\mathfrak{p}^x))})^\bullet \subset (\overline{G(\mathfrak{z}(\mathfrak{g}^x))})^\bullet.$$

Since $(\overline{G(\mathfrak{z}(\mathfrak{g}^x))})^\bullet$ is a sheet of G in $\mathfrak{g}$ containing x (39.4.5), the result is a consequence of 39.4.8.   □

39.6.9 Let us consider example 1 of 38.8.8. By 33.4.6, any element of $\mathfrak{g}$ is polarizable.

Let $x \in \mathfrak{p}$ be a non-zero semisimple element. Then it is contained in a Cartan subspace $\mathfrak{a}$ of $\mathfrak{p}$ (37.4.7). By 37.4.11, $\mathfrak{a} = \mathfrak{p}^s$ where s is $\mathfrak{p}$-generic. But $\dim \mathfrak{a} = 1$ (38.8.8 and 37.4.11). So x is $\mathfrak{p}$-generic.

We have therefore shown that any non-zero semisimple element of $\mathfrak{p}$ is $\mathfrak{p}$-generic, and so, the regular sheet $\mathcal{R}$ is the unique sheet of $\mathfrak{p}$ which contain a non-zero semisimple element of $\mathfrak{p}$.

Let $y = E_{12} \in \mathfrak{p}$. Then $\mathfrak{p}^y = \Bbbk E_{12} + \Bbbk E_{32}$, so $y \notin \mathcal{R}$ but y is polarizable.

Thus, contrary to the Lie algebra case (39.4.8), the union of the sheets of $\mathfrak{p}$ which contain a semisimple element is not the set of polarizable elements of $\mathfrak{p}$.

References and comments

- [6], [7], [16], [28], [81]

The notion of sheets was first introduced in [28] for determining polarizable elements. The definition of a sheet was generalized in [6] and [7].

For the presentation of Jordan classes in this chapter, we have followed [16].

The fact that the set of polarizable elements is the union of Dixmier sheets (39.4.8) was used in [71] to prove the irreducibility of the commuting variety $\{(x, y) \in \mathfrak{g} \times \mathfrak{g}; [x, y] = 0\}$ of a reductive Lie algebra $\mathfrak{g}$.

Generalization of the problem of the irreducibility of the commuting variety to the symmetric case was studied in [60], [73], [61], [74]. In particular, the notion of sheets and $\mathfrak{p}$-distinguished elements were used in [74].

Index and linear forms

In this chapter, we study certain properties related to the coadjoint representation of a Lie algebra. In particular, we determine the index of certain classes of Lie algebras.

Throughout this chapter, $\mathfrak{g}$ will denote a finite-dimensional Lie algebra.

40.1 Stable linear forms

40.1.1 Let us consider the coadjoint representation of $\mathfrak{g}$ on $\mathfrak{g}^*$ and let us conserve the notations of section 19.7. For a subspace $\mathfrak{a}$ of $\mathfrak{g}$, we denote by $\mathfrak{a}^\perp$ the orthogonal of $\mathfrak{a}$ in $\mathfrak{g}^*$.

Recall that for $f \in \mathfrak{g}^*$, we denote by Φ_f the alternating bilinear form on $\mathfrak{g}$ defined by

$$\Phi_f(x,y) = f([x,y])$$

for $x, y \in \mathfrak{g}$, and $\mathfrak{a}^{(f)}$ denote the orthogonal of $\mathfrak{a}$ with respect to Φ_f. In particular, $\mathfrak{g}^{(f)}$ is the kernel of Φ_f, and we have:

$$\left(\mathfrak{g}^{(f)}\right)^\perp = \mathfrak{g}.f \ , \ [\mathfrak{g}, \mathfrak{g}^{(f)}]^\perp = \{g \in \mathfrak{g}^*; \mathfrak{g}^{(f)} \subset \mathfrak{g}^{(g)}\}.$$

In the rest of this section, K will be an algebraic subgroup of $\mathrm{Aut}\,\mathfrak{g}$ with Lie algebra $\mathfrak{k}$. The group K acts rationally on $\mathfrak{g}$ and $\mathfrak{g}^*$. For $\alpha \in K$, $f \in \mathfrak{g}^*$ and $x \in \mathfrak{g}$, we have $(\alpha.f)(x) = f(\alpha^{-1}(x))$. The differential of this action defines an action of $\mathfrak{k}$ on $\mathfrak{g}^*$ given by $(X.f)(x) = f(-X(x))$ for $X \in \mathfrak{k}$. We set:

$$\mathfrak{k}_f = \{X \in \mathfrak{k}; X.f = 0\} \ , \ \mathfrak{k}^f = \{x \in \mathfrak{g}; (\mathfrak{k}.f)(x) = \{0\}\}.$$

Thus we have $(\mathfrak{k}^f)^\perp = \mathfrak{k}.f$. If K contains the algebraic adjoint group of $\mathfrak{g}$, then $\mathrm{ad}_\mathfrak{g}\,\mathfrak{g} \subset \mathfrak{k}$, and so $\mathfrak{k}^f \subset \mathfrak{g}^{(f)}$.

The bilinear form $(X, x) \mapsto f(X(x))$ on $\mathfrak{k} \times \mathfrak{g}$ induces a non-degenerate bilinear form on $(\mathfrak{k}/\mathfrak{k}_f) \times (\mathfrak{g}/\mathfrak{k}^f)$. It follows that:

$$\dim \mathfrak{k} - \dim \mathfrak{k}_f = \dim \mathfrak{g} - \dim \mathfrak{k}^f.$$

Definition. *An element $f \in \mathfrak{g}^*$ is said to be K-stable if there exists a neighbourhood V of f in $\mathfrak{g}^*$ such that for all $g \in V$, $\mathfrak{g}^{(f)}$ and $\mathfrak{g}^{(g)}$ are K-conjugate. If K is the algebraic adjoint group of G (resp. if $K = \operatorname{Aut}\mathfrak{g}$), a K-stable linear form is called* stable *(resp.* weakly stable*).*

40.1.2 Lemma. *Let $f \in \mathfrak{g}^*$. Suppose that K is connected and that the morphism*

$$\psi : K \times [\mathfrak{g}, \mathfrak{g}^{(f)}]^{\perp} \longrightarrow \mathfrak{g}^* , \ (\alpha, h) \mapsto \alpha(h)$$

is dominant. Then:

(i) $f \in \mathfrak{g}^*_{\mathrm{reg}}$.

(ii) *Let $W = [\mathfrak{g}, \mathfrak{g}^{(f)}]^{\perp} \cap \mathfrak{g}^*_{\mathrm{reg}}$. The restriction of ψ to $K \times W$ is open.*

Proof. (i) The image of ψ contains an open dense subset of $\mathfrak{g}^*$ by 15.4.2, so it contains a regular element g of $\mathfrak{g}^*$ because $\mathfrak{g}^*_{\mathrm{reg}}$ is a non-empty open subset of $\mathfrak{g}^*$. If $\alpha \in K$ and $h \in [\mathfrak{g}, \mathfrak{g}^{(f)}]^{\perp}$ verify $g = \alpha.h$, then $\alpha(\mathfrak{g}^{(f)}) \subset \alpha(\mathfrak{g}^{(h)}) = \mathfrak{g}^{(g)}$. Hence $f \in \mathfrak{g}^*_{\mathrm{reg}}$.

(ii) Let $\varphi : K \times W \to \mathfrak{g}^*$ denote the restriction of ψ to $K \times W$. By (i), W is a non-empty open subset of $[\mathfrak{g}, \mathfrak{g}^{(f)}]^{\perp}$. So φ is dominant.

Let N be the normalizer of $\mathfrak{g}^{(f)}$ in K. If $\alpha, \beta \in K$ and $g, h \in W$ verify $\beta.h = \alpha.g$, or equivalently $h = (\beta^{-1}\alpha).g$, then since $\mathfrak{g}^{(h)} = \mathfrak{g}^{(g)} = \mathfrak{g}^{(f)}$, we have $\beta^{-1}\alpha \in N$, or $\beta \in \alpha N$. It follows that the fibre $S = \varphi^{-1}\big(\varphi(\alpha, g)\big)$ is equal to $\{(\alpha\beta, \beta^{-1}.g); \beta \in N\}$.

Now S is also the graph of the morphism $\alpha N \to \mathfrak{g}^*$, $\gamma \mapsto (\gamma^{-1}\alpha).g$, so it is isomorphic to N (12.5.4). As N is a pure variety (21.1.6), it follows from 15.5.4 and 17.4.11 that φ is an open map. $\square$

40.1.3 Lemma. *Let K° be the identity component of K and $f \in \mathfrak{g}^*$. The following conditions are equivalent:*

(i) f *is K°-stable.*

(ii) f *is K-stable.*

If these conditions are verified, then $f \in \mathfrak{g}^*_{\mathrm{reg}}$.

Proof. (i) $\Rightarrow$ (ii) This is clear.

(ii) $\Rightarrow$ (i) Let us fix an open neighbourhood U of f such that $\mathfrak{g}^{(h)}$ and $\mathfrak{g}^{(f)}$ are K-conjugate for all $h \in U$. Since $U \cap \mathfrak{g}^*_{\mathrm{reg}} \neq \emptyset$, $f \in \mathfrak{g}^*_{\mathrm{reg}}$, and therefore $W = [\mathfrak{g}, \mathfrak{g}^{(f)}]^{\perp} \cap \mathfrak{g}^*_{\mathrm{reg}}$ is a non-empty open subset of $[\mathfrak{g}, \mathfrak{g}^{(f)}]^{\perp}$; W is in fact the set of $h \in \mathfrak{g}^*$ verifying $\mathfrak{g}^{(h)} = \mathfrak{g}^{(f)}$.

If $g \in U$, then $\mathfrak{g}^{(g)} = \alpha(\mathfrak{g}^{(f)})$ for some $\alpha \in K$. So $\mathfrak{g}^{(\alpha^{-1}.g)} = \mathfrak{g}^{(f)}$, and $g \in \alpha(W)$. Thus the image of the morphism $\psi : K \times W \to \mathfrak{g}^*$, $(\alpha, h) \mapsto \alpha.h$, contains U.

Now let $K_0 = K^{\circ}$, $K_1 = \alpha_1 K^{\circ}$, ...,$K_r = \alpha_r K^{\circ}$ be the (distinct) irreducible components of K, where $\alpha_1, \ldots, \alpha_r \in K$. Set $T_i = \psi(K_i \times W)$. Since $U \subset T_0 \cup \cdots \cup T_r$, we obtain that $\mathfrak{g}^* = \overline{T_0} \cup \cdots \cup \overline{T_r}$. It follows from the irreducibility of $\mathfrak{g}^*$ that $\mathfrak{g}^* = \overline{T_i}$ for some i. But $\overline{T_i} = \overline{\alpha_i(T_0)} = \alpha_i \overline{T_0}$, so $\mathfrak{g}^* = \overline{T_0}$, and the restriction φ of ψ to $K^{\circ} \times W$ is dominant. Now 40.1.2 implies that φ

is an open map, so T_0 is an open neighbourhood of f in $\mathfrak{g}^*$ with the property that for any $g \in T_0$, $\mathfrak{g}^{(g)}$ and $\mathfrak{g}^{(f)}$ are K°-conjugate. $\square$

40.1.4 Theorem. *Let* $\mathfrak{g}$ *be a Lie algebra,* $f \in \mathfrak{g}^*$, K *an algebraic subgroup of* $\operatorname{Aut} \mathfrak{g}$ *and* $\mathfrak{k}$ *the Lie algebra of* K. *Then the following conditions are equivalent:*

(i) $[\mathfrak{g}, \mathfrak{g}^{(f)}] \cap \mathfrak{k}^f = \{0\}$.

(ii) *The linear form* f *is* K*-stable.*

Proof. By 40.1.3, we may assume that K is connected. Let e denote the identity element of K.

(i) $\Rightarrow$ (ii) Let us consider the morphism $\psi : K \times [\mathfrak{g}, \mathfrak{g}^{(f)}]^\perp \to \mathfrak{g}^*$ given by $(\alpha, h) \mapsto \alpha.h$. By 29.1.4, we have, for $X \in \mathfrak{k}$ and $h \in [\mathfrak{g}, \mathfrak{g}^{(f)}]^\perp$:

$$d\psi_{(e,f)}(X, h) = h + X.f.$$

The image of $d\psi_{(e,f)}$ is therefore $[\mathfrak{g}, \mathfrak{g}^{(f)}]^\perp + \mathfrak{k}.f$, which is $\mathfrak{g}^*$ by our hypothesis. Thus ψ is dominant (16.5.7). Let $W = [\mathfrak{g}, \mathfrak{g}^{(f)}]^\perp \cap \mathfrak{g}^*_{\mathrm{reg}}$. Then 40.1.2 says that $\psi(K \times W)$ is an open subset of $\mathfrak{g}^*$ containing f, and for any $g \in \psi(K \times W)$, $\mathfrak{g}^{(g)}$ and $\mathfrak{g}^{(f)}$ are K-conjugate.

(ii) $\Rightarrow$ (i) We have $f \in \mathfrak{g}^*_{\mathrm{reg}}$ (40.1.3). So $W = [\mathfrak{g}, \mathfrak{g}^{(f)}]^\perp \cap \mathfrak{g}^*_{\mathrm{reg}}$ is a nonempty open subset of $[\mathfrak{g}, \mathfrak{g}^{(f)}]^\perp$; recall that W is the set of $h \in \mathfrak{g}^*$ such that $\mathfrak{g}^{(h)} = \mathfrak{g}^{(f)}$. The image of the morphism $\varphi : K \times W \to \mathfrak{g}^*$, $(\alpha, h) \mapsto \alpha.h$ is therefore the set of $h \in \mathfrak{g}^*$ such that $\mathfrak{g}^{(h)}$ is K-conjugate to $\mathfrak{g}^{(f)}$.

For $g \in W$ and $(X, h) \in \mathfrak{k} \times [\mathfrak{g}, \mathfrak{g}^{(f)}]^\perp$, we have:

$$d\varphi_{(e,g)}(X, h) = h + X.g.$$

Thus the image T_g of $d\varphi_{(e,g)}$ is $[\mathfrak{g}, \mathfrak{g}^{(f)}]^\perp + \mathfrak{k}.g$.

Let us denote by $\mathfrak{n}$ the normalizer of $\mathfrak{g}^{(f)}$ in $\mathfrak{k}$. As:

$$(X.g)([\mathfrak{g}, \mathfrak{g}^{(f)}]) = g([X(\mathfrak{g}), \mathfrak{g}^{(f)}]) + g([\mathfrak{g}, X(\mathfrak{g}^{(f)})]) = g([\mathfrak{g}, X(\mathfrak{g}^{(f)})]),$$

and $\mathfrak{g}^{(g)} = \mathfrak{g}^{(f)}$, we deduce that $X.g \in [\mathfrak{g}, \mathfrak{g}^{(f)}]^\perp$ if and only if $X \in \mathfrak{n}$. Hence:

$$\dim(\mathfrak{k}.g \cap [\mathfrak{g}, \mathfrak{g}^{(f)}]^\perp) = \dim \mathfrak{n}.g.$$

On the other hand, if $X \in \mathfrak{k}_g$, then $X \in \mathfrak{n}$ because $X.g = 0$. So:

$$\dim \mathfrak{n}.g = \dim \mathfrak{n} - \dim \mathfrak{k}_g = \dim \mathfrak{n} - \dim \mathfrak{k} + \dim \mathfrak{k}.g.$$

We deduce therefore that:

$$\dim T_g = \dim \mathfrak{k} - \dim \mathfrak{n} + \dim([\mathfrak{g}, \mathfrak{g}^{(f)}]^\perp).$$

In particular, $\dim T_g$ does not depend on g. It follows from 29.1.4 (ii) that the dimension of the image $T_{(\alpha,g)}$ of $d\varphi_{(\alpha,g)}$ does not depend on α or on g. Now, our hypothesis implies that φ is dominant, so there exists $(\alpha, g) \in K \times W$ such that $T_{(\alpha,g)} = \mathfrak{g}^*$ (16.5.7). Hence $T_{(e,f)} = \mathfrak{g}^*$, and $[\mathfrak{g}, \mathfrak{g}^{(f)}]^\perp + \mathfrak{k}.f = \mathfrak{g}^*$. Consequently, $[\mathfrak{g}, \mathfrak{g}^{(f)}] \cap \mathfrak{k}^f = \{0\}$. $\square$

40.1.5 Corollary. *Let $f \in \mathfrak{g}^*$.*

(i) *If $\mathfrak{g}$ is ad-algebraic, then f is stable if and only if $[\mathfrak{g}, \mathfrak{g}^{(f)}] \cap \mathfrak{g}^{(f)} = \{0\}$.*

(ii) *If $[\mathfrak{g}, \mathfrak{g}^{(f)}] \cap \mathfrak{g}^{(f)} = \{0\}$, then f is K-stable for any algebraic subgroup K of $\operatorname{Aut} \mathfrak{g}$ whose Lie algebra contains $\operatorname{ad}_\mathfrak{g} \mathfrak{g}$, in particular, this is the case if K contains the algebraic adjoint group of $\mathfrak{g}$.*

40.1.6 The following proposition gives a sufficient condition for applying part (ii) of 40.1.5:

Proposition. *Let $\mathfrak{g}$ be a Lie subalgebra of a semisimple Lie algebra and $f \in \mathfrak{g}^*$. If $\mathfrak{g}^{(f)}$ is a commutative Lie algebra consisting of semisimple elements, then $[\mathfrak{g}, \mathfrak{g}^{(f)}] \cap \mathfrak{g}^{(f)} = \{0\}$.*

Proof. If $\mathfrak{g}^{(f)}$ is a commutative Lie algebra consisting of semisimple elements, then there exists a vector subspace $\mathfrak{r}$ of $\mathfrak{g}$ such that $\mathfrak{g} = \mathfrak{g}^{(f)} \oplus \mathfrak{r}$ and $[\mathfrak{g}^{(f)}, \mathfrak{r}] \subset \mathfrak{r}$. So $[\mathfrak{g}, \mathfrak{g}^{(f)}] \subset \mathfrak{r}$, and the result follows. $\square$

Remark. More generally, if $\mathfrak{a}$ is a commutative Lie subalgebra of $\mathfrak{g}$ consisting of semisimple elements, then $[\mathfrak{g}, \mathfrak{a}] \cap \mathfrak{a} = \{0\}$.

40.1.7 We shall now give an example of a Lie algebra which does not admit any weakly stable linear form.

Let $\mathfrak{s} = \mathfrak{sl}_2(\Bbbk)$ and V be a simple $\mathfrak{s}$-module of dimension 5. The Lie bracket of the semi-direct product $\mathfrak{s} \ltimes V$ is given by: for $x_1, x_2 \in \mathfrak{s}$ and $v_1, v_2 \in V$,

$$[(x_1, v_1), (x_2, v_2)] = ([x_1, x_2], x_1.v_2 - x_2.v_1).$$

The endomorphism t of $\mathfrak{s} \ltimes V$, $(x, v) \mapsto v$, is a derivation of the Lie algebra $\mathfrak{s} \ltimes V$. We may therefore define the semi-direct product $\mathfrak{g} = \Bbbk t \ltimes (\mathfrak{s} \ltimes V)$ where the Lie bracket is given by: for $\lambda_1, \lambda_2 \in \Bbbk$ and $z_1, z_2 \in \mathfrak{s} \ltimes V$,

$$[(\lambda_1 t, z_1), (\lambda_2 t, z_2)] = (0, [z_1, z_2] + \lambda_1 t(z_2) - \lambda_2 t(z_1)).$$

It is easy to check that $\Bbbk t \ltimes V$ and V are respectively the radical and the nilpotent radical of $\mathfrak{g}$. We shall show that $\mathfrak{g}$ does not admit any weakly stable linear form.

The Lie algebra of $\operatorname{Aut} \mathfrak{g}$ is $\mathfrak{d} = \operatorname{Der} \mathfrak{g}$. We claim that $\mathfrak{d} = \operatorname{ad}_\mathfrak{g} \mathfrak{g}$. Let us proof our claim. Let $\delta \in \mathfrak{d}$. By 20.3.3, there exists $w \in \mathfrak{g}$ such that $\delta(x) = [x, w]$ for all $x \in \mathfrak{s}$. Replacing δ by $\delta - \operatorname{ad}_\mathfrak{g} w$, we may assume that $\delta|_\mathfrak{s} = 0$.

Since the ideals $\Bbbk t \ltimes V$ and V are stable under δ, we have, for $v \in V$:

$$\delta(v) = \delta([t, v]) = [\delta(t), v] + [t, \delta(v)] = [\delta(t), v] + \delta(v).$$

So $[\delta(t), V] = \{0\}$, and hence $\delta(t) \in V$. Now, if $x \in \mathfrak{s}$, then:

$$0 = \delta([x, t]) = [\delta(x), t] + [x, \delta(t)] = [x, \delta(t)].$$

Since V is simple and $\delta(t) \in V$, we deduce that $\delta(t) = 0$.

For $x \in \mathfrak{s}$ and $v \in V$, we have $\delta([x,v]) = [x, \delta(v)]$. So $\delta|_V$ is an endomorphism of V which commutes with the action of $\mathfrak{s}$. It follows by Schur's lemma that $\delta = \lambda \operatorname{ad}_\mathfrak{g} t$ for some $\lambda \in \mathbb{k}$. So we have proved our claim.

Let $f \in \mathfrak{g}^*$. The subspace $\mathfrak{g}^{(f)}$ is the kernel of the linear map $u_f : \mathfrak{g} \to \mathfrak{g}^*$, $x \mapsto x.f$. Since $[V,V] = \{0\}$, the restriction of u_f to V induce a linear map from V to $(\mathbb{k}t \oplus \mathfrak{s})^*$. But $\dim V = 5$ and $\dim(\mathbb{k}t \oplus \mathfrak{s}) = 4$, so $\mathfrak{g}^{(f)} \cap V \neq \{0\}$, and $\mathfrak{g}^{(f)} \cap V \subset [\mathfrak{g}, \mathfrak{g}^{(f)}]$ because $[t,v] = v$ for all $v \in V$.

Finally, our claim implies that $\mathfrak{g}^{(f)} = \mathfrak{d}^f$, so condition (i) of 40.1.4 is not satisfied, and f is not weakly stable.

40.2 Index of a representation

40.2.1 Let V be a finite-dimensional $\mathfrak{g}$-module. We set:

$$\mathfrak{g}_v = \{x \in \mathfrak{g}; x.v = 0\} \ , \ \mathfrak{g}.v = \{x.v; x \in \mathfrak{g}\}.$$

We say that v is $\mathfrak{g}$-*regular* or simply *regular* if:

$$\dim \mathfrak{g}_v = \min\{\dim \mathfrak{g}_w; w \in V\}.$$

Since $\dim \mathfrak{g}_v + \dim \mathfrak{g}.v = \dim \mathfrak{g}$, v is regular if and only if:

$$\dim \mathfrak{g}.v = \max\{\dim \mathfrak{g}.w; w \in V\}.$$

The set of regular elements of V is a non-empty Zariski open subset of V.

40.2.2 In the notations of 40.2.1, we consider the contragredient representation of V on V^*.

Definition. *The integer*

$$\dim V - \max\{\dim \mathfrak{g}.f; f \in V^*\}$$

is called the index *of the $\mathfrak{g}$-module V, and we shall denote it by $\chi(\mathfrak{g}, V)$.*

40.2.3 Remarks. 1) The index of the adjoint representation of $\mathfrak{g}$ is just the index $\chi(\mathfrak{g})$ of $\mathfrak{g}$ as defined in 19.7.3.

2) Let G be a connected algebraic group, $\mathfrak{g}$ its Lie algebra, and V a rational G-module. Then V has a natural structure of $\mathfrak{g}$-module, and we have

$$\chi(\mathfrak{g}, V) = \dim V - \max\{\dim G.f; f \in V^*\} = \min\{\dim G_f; f \in V^*\}.$$

40.2.4 Let $f \in V^*$. We define a linear map $\mathcal{L}_f : \mathfrak{g} \to V^*$ as follows: for $x \in \mathfrak{g}$ and $v \in V$,
$$\mathcal{L}_f(x)(v) = f(x.v).$$
The kernel (resp. image) of $\mathcal{L}_f$ is $\mathfrak{g}_f$ (resp. $\mathfrak{g}.f$). Thus:

$$\chi(\mathfrak{g}, V) = \dim V - \max\{\operatorname{rk}(\mathcal{L}_f); f \in V^*\}.$$

40.2.5 Let $f \in V^*$. We define a bilinear form Ψ_f on $\mathfrak{g} \times V$ as follows: for $x \in \mathfrak{g}$ and $v \in V$,

$$\Psi_f(x, v) = \mathcal{L}_f(x)(v) = f(x.v).$$

Let us fix bases $\mathcal{X} = (x_1, \ldots, x_n)$ of $\mathfrak{g}$ and $\mathcal{V} = (v_1, \ldots, v_m)$ of V. Let $M(\mathcal{X}, \mathcal{V})$ be the matrix $[x_i.v_j]_{1 \leqslant i \leqslant n, 1 \leqslant j \leqslant m}$, considered as a matrix with coefficients in the field of fractions of the symmetric algebra $S(V)$ of V. Then $M_f(\mathcal{X}, \mathcal{V}) = [f(x_i.v_j)] \in \mathrm{M}_{n,m}(\Bbbk)$ is the matrix of Ψ_f with respect to the bases $\mathcal{X}$ and $\mathcal{V}$. Thus $\mathrm{rk}(\mathcal{L}_f) = \mathrm{rk}(M_f(\mathcal{X}, \mathcal{V}))$. Consequently, we have:

$$\chi(\mathfrak{g}, V) = \dim V - \mathrm{rk}(M(\mathcal{X}, \mathcal{V})).$$

40.3 Some useful inequalities

40.3.1 Let V be a finite-dimensional $\mathfrak{g}$-module and $v \in V$. For $x \in \mathfrak{g}_v$ and $y \in \mathfrak{g}$, we have:

$$x.(y.v) = [x, y]v + y.(x.v) = [x, y].v.$$

So $\mathfrak{g}.v$ is a $\mathfrak{g}_v$-module. Let us denote by $\overline{V}$ the quotient $\mathfrak{g}_v$-module $V/\mathfrak{g}.v$.

Proposition. (Vinberg's Lemma) *In the above notations, we have:*

$$\max\{\dim \mathfrak{g}.u; u \in V\} \geqslant \dim \mathfrak{g}.v + \max\{\dim \mathfrak{g}_v.\xi; \xi \in \overline{V}\}.$$

Proof. 1) Let $(x_1, \ldots, x_n)$ be a basis of $\mathfrak{g}$ such that $(x_1, \ldots, x_p)$ is a basis of $\mathfrak{g}_v$. For $t \in \Bbbk$, the dimension of $W_t = \mathfrak{g}.(v + tu) + \mathfrak{g}_v.u$ is the rank of the family of vectors $(x_1.(v+tu), \ldots, x_n.(v+tu), x_1.u, \ldots, x_p.u)$. So there exists an open subset $\mathcal{O}$ of $\Bbbk$ containing 0 such that for all $t \in \mathcal{O}$, we have:

$$\dim W_0 \leqslant \dim W_t.$$

2) Let $\mathcal{O}^\bullet = \mathcal{O} \setminus \{0\}$. For $t \in \Bbbk \setminus \{0\}$, we have:

$$\mathfrak{g}_v.u = t\mathfrak{g}_v.u = \mathfrak{g}_v(v + tu) \subset \mathfrak{g}.(v + tu).$$

Thus for $t \in \mathcal{O}^\bullet$, we have:

$$\dim W_0 \leqslant \dim((\mathfrak{g}.(v + tu)).$$

3) Let $\pi : V \to \overline{V}$ be the canonical surjection. Since $\pi(\mathfrak{g}_v.u) = \mathfrak{g}_v.\pi(u)$, we have, for $\xi = \pi(u)$, that:

$$\dim(\mathfrak{g}_v.\xi) = \dim(\mathfrak{g}_v.u) - \dim((\mathfrak{g}.v) \cap (\mathfrak{g}_v.u)).$$

Hence

$$\dim W_0 = \dim \mathfrak{g}.v + \dim(\mathfrak{g}_v.\xi).$$

4) It follows from point 2 that if $t \in \mathcal{O}^\bullet$, then:

$$\dim \mathfrak{g}.v + \dim(\mathfrak{g}_v.\xi) \leqslant \dim((\mathfrak{g}.(v + tu)).$$

So the result follows. $\square$

40.3.2 Corollary. *Let* $\mathfrak{g}$ *be a Lie algebra and* $f \in \mathfrak{g}^*$. *Then:*

$$\chi(\mathfrak{g}) \leqslant \chi(\mathfrak{g}^{(f)}).$$

Proof. Let $R : \mathfrak{g}^* \to (\mathfrak{g}^{(f)})^*$ be the restriction map. The kernel of R is $\mathfrak{g}.f$. So R induces an isomorphism

$$\widetilde{R} : \mathfrak{g}^*/\mathfrak{g}.f \longrightarrow (\mathfrak{g}^{(f)})^*.$$

Let $\pi : \mathfrak{g}^* \to \mathfrak{g}^*/\mathfrak{g}.f$ denote the canonical surjection. Then for any $\lambda \in \mathfrak{g}^*$, we have $R(\lambda) = \widetilde{R}(\pi(\lambda))$.

For $x \in \mathfrak{g}^{(f)}$, we have $x.\pi(\lambda) = \pi(x.\lambda)$, and for $y \in \mathfrak{g}^{(f)}$, we have:

$$R(x.\lambda)(y) = (x.\lambda)(y) = \lambda([y,x]) = R(\lambda)([y,x]) = (x.R(\lambda))(y).$$

Thus

$$\widetilde{R}(x.\pi(\lambda)) = \widetilde{R}(\pi(x.\lambda)) = R(x.\lambda) = x.R(\lambda) = x.\widetilde{R}(\pi(\lambda)).$$

This shows that $\widetilde{R}$ is an isomorphism of $\mathfrak{g}^{(f)}$-modules, and the action of $\mathfrak{g}^{(f)}$ on $\mathfrak{g}^*/\mathfrak{g}.f$ can be identified with the coadjoint representation of $\mathfrak{g}^{(f)}$. So the result follows from 40.3.1. $\square$

40.3.3 Remark. If f is regular in $\mathfrak{g}^*$, then 40.3.2 implies that:

$$\dim \mathfrak{g}^{(f)} = \chi(\mathfrak{g}) \leqslant \chi(\mathfrak{g}^{(f)}) \leqslant \dim \mathfrak{g}^{(f)}.$$

Thus $\chi(\mathfrak{g}^{(f)}) = \dim \mathfrak{g}^{(f)}$, which proves that $\mathfrak{g}^{(f)}$ is a commutative Lie algebra. So we recover 19.7.5 (ii).

40.3.4 Theorem. *Let* $\mathfrak{g}$ *be a Lie algebra and* $\mathfrak{a}$ *an ideal of* $\mathfrak{g}$. *Then:*

$$\chi(\mathfrak{g}) + \chi(\mathfrak{a}) \leqslant \dim(\mathfrak{g}/\mathfrak{a}) + 2\chi(\mathfrak{g}, \mathfrak{a}).$$

Proof. Let $f \in \mathfrak{g}^*$, and $h \in \mathfrak{a}^*$ the restriction of f to $\mathfrak{a}$. Then it is clear that $\mathfrak{a}^{(h)} + \mathfrak{g}^{(f)} \subset \mathfrak{a}^{(f)}$. It follows therefore that $\dim(\mathfrak{a}^{(h)} + \mathfrak{g}^{(f)}) \leqslant \dim \mathfrak{a}^{(f)}$. On the other hand, we have also:

$$\dim \mathfrak{a}^{(f)} = \dim \mathfrak{g} - \dim \mathfrak{a} + \dim \mathfrak{a} \cap \mathfrak{g}^{(f)} = \dim(\mathfrak{g}/\mathfrak{a}) + \dim \mathfrak{a} \cap \mathfrak{g}^{(f)}.$$

Now, $\mathfrak{a}^{(h)} \cap \mathfrak{g}^{(f)} = \mathfrak{a} \cap \mathfrak{g}^{(f)}$ is the orthogonal in $\mathfrak{a}$ of the subspace $\mathfrak{g}.h$ of $\mathfrak{a}^*$. Consequently, $\dim(\mathfrak{a}^{(h)} \cap \mathfrak{g}^{(f)}) = \dim \mathfrak{a} - \dim \mathfrak{g}.h$. Hence:

$$\dim \mathfrak{a}^{(h)} + \dim \mathfrak{g}^{(f)} \leqslant \dim(\mathfrak{g}/\mathfrak{a}) + 2 \dim \mathfrak{a} - 2 \dim \mathfrak{g}.h.$$

From this, we deduce that:

$$\chi(\mathfrak{g}) + \chi(\mathfrak{a}) \leqslant \dim(\mathfrak{g}/\mathfrak{a}) + 2(\dim \mathfrak{a} - \dim \mathfrak{g}.h).$$

This inequality holds for all $h \in \mathfrak{a}^*$, so the result follows. $\square$

40.4 Index and semi-direct products

40.4.1 Let $\mathfrak{q}$ be a Lie algebra, $\mathfrak{a}$ an ideal of $\mathfrak{q}$, $f \in \mathfrak{q}^*$, and f_0 the restriction of f to $\mathfrak{a}$. Since $\mathfrak{a}$ is an ideal, the Lie subalgebra $\mathfrak{h} = \mathfrak{a}^{(f)}$ depends only on f_0. Let us denote by h the restriction of f to $\mathfrak{h}$. As $\mathfrak{q}^{(f)} \subset \mathfrak{a}^{(f)}$, we have:

$$\mathfrak{h}^{(h)} = \mathfrak{h} \cap \mathfrak{h}^{(f)} = \mathfrak{a}^{(f)} \cap (\mathfrak{a} + \mathfrak{q}^{(f)}) = (\mathfrak{a} \cap \mathfrak{a}^{(f)}) + \mathfrak{q}^{(f)}.$$

Consequently:

$$(1) \qquad\qquad \mathfrak{h}^{(h)} = \mathfrak{a}^{(f_0)} + \mathfrak{q}^{(f)}.$$

Suppose that $\mathfrak{a}$ is commutative. Then $\mathfrak{a} = \mathfrak{a}^{(f_0)}$ and $\mathfrak{a} \subset \mathfrak{h}^{(h)}$. Moreover, $\mathfrak{a} \cap \mathfrak{q}^{(f)}$ is the orthogonal of $\mathfrak{q}.f_0$ in $\mathfrak{a}$. It follows therefore from (1) that:

$$(2) \qquad\qquad \dim \mathfrak{q}^{(f)} = \dim \mathfrak{h}^{(h)} - \dim(\mathfrak{q}.f_0).$$

40.4.2 Let V be a finite-dimensional $\mathfrak{g}$-module. Endow $V \times \mathfrak{g}$ with the Lie algebra structure given by: for $x_1, x_2 \in \mathfrak{g}$ and $v_1, v_2 \in V$,

$$[(v_1, x_1), (v_2, x_2)] = (x_1.v_2 - x_2.v_1, [x_1, x_2]).$$

Let $\mathfrak{q} = V \times \mathfrak{g}$ and we may identify $\mathfrak{q}^*$ with $V^* \times \mathfrak{g}^*$.

Let $f \in \mathfrak{q}^*$, $f_0 = f|_V$ and $f_1 = f|_{\mathfrak{g}}$. Then in the notations of 40.4.1, $\mathfrak{h} = \mathfrak{g}_{f_0} \times V$ and $\mathfrak{q}.f_0 = \mathfrak{g}.f_0$.

Let $\ell = (\ell_0, \ell_1) \in V^* \times \mathfrak{g}^* = \mathfrak{q}^*$ be such that $\ell_0 = f_0$. Set $h = \ell|_{\mathfrak{h}}$ and $g = \ell|_{\mathfrak{g}_{f_0}} = \ell_1|_{\mathfrak{g}_{f_0}}$.

For $(v, x), (w, y) \in \mathfrak{h}$, we have:

$$\ell\big([(v, x), (w, y)]\big) = \ell_1([x, y]) = g([x, y])$$

It follows that:

$$\mathfrak{h}^{(h)} = V \times (\mathfrak{g}_{f_0})^{(g)}.$$

In view of (2), we obtain that:

$$\dim \mathfrak{q}^{(\ell)} = \dim(\mathfrak{g}_{f_0})^{(g)} + \dim V - \dim(\mathfrak{g}.f_0).$$

From this, we deduce immediately the following result:

Proposition. *With the hypotheses and notations above, let $f = (f_0, f_1)$ be a regular element of $\mathfrak{q}^*$.*
(i) *We have:*

$$\chi(\mathfrak{q}) = \dim \mathfrak{q}^{(f)} = \chi(\mathfrak{g}_{f_0}) + \dim V - \dim(\mathfrak{g}.f_0).$$

(ii) *Let us assume further that f_0 is $\mathfrak{g}$-regular in V^*. Then:*

$$\chi(\mathfrak{q}) = \chi(\mathfrak{g}_{f_0}) + \chi(\mathfrak{g}, V).$$

40.4.3 Let G be a connected algebraic group, $\mathfrak{g}$ its Lie algebra and V a finite-dimensional rational G-module. Then V has a natural structure of $\mathfrak{g}$-module, and we construct the Lie algebra $\mathfrak{q}$ as in 40.4.2.

Proposition. *Assume that $\chi(\mathfrak{g}, V) = 0$. Then for any $\mathfrak{g}$-regular element f_0 in V^*, we have:*

$$\chi(\mathfrak{q}) = \chi(\mathfrak{g}_{f_0}).$$

Proof. The assumption $\chi(\mathfrak{g}, V) = 0$ implies that the G-orbit of any $\mathfrak{g}$-regular element is an open subset of V^*. Thus $\mathfrak{g}$-regular elements are G-conjugate. So the result follows by 40.4.2. □

40.4.4 Let us now illustrate how the above considerations give estimations on the index of certain Lie algebras.

Let us suppose that there exist a Lie subalgebra $\mathfrak{a}$ of $\mathfrak{g}$ and an ideal $\mathfrak{m}$ of $\mathfrak{g}$ such that $\mathfrak{g} = \mathfrak{a} \oplus \mathfrak{m}$.

For $\lambda \in \Bbbk$, we define a Lie algebra $\mathfrak{g}_\lambda$ whose underlying vector space is $\mathfrak{g}$, and whose Lie bracket $[.,.]_\lambda$ is given as follows:

$$[x, y]_\lambda = [x, y] \ \text{ if } \ (x, y) \in (\mathfrak{a} \times \mathfrak{a}) \cup (\mathfrak{a} \times \mathfrak{m}) \cup (\mathfrak{m} \times \mathfrak{a}) ,$$
$$[x, y]_\lambda = \lambda[x, y] \ \text{ if } \ (x, y) \in \mathfrak{m} \times \mathfrak{m}.$$

Thus $\mathfrak{g}_1 = \mathfrak{g}$, while $\mathfrak{g}_0$ verifies the hypotheses of 40.4.2.

For $\lambda \neq 0$, the linear map $f_\lambda : \mathfrak{g} \to \mathfrak{g}_\lambda$ defined by $f_\lambda(x) = x$ if $x \in \mathfrak{a}$ and $f_\lambda(x) = \lambda^{-1}x$ if $x \in \mathfrak{m}$, defines an isomorphism of Lie algebras. Thus $\chi(\mathfrak{g}_\lambda) = \chi(\mathfrak{g})$ for all $\lambda \neq 0$.

Let $\mathcal{X} = (e_1, \ldots, e_n)$ be a basis of $\mathfrak{g}$ such that $(e_1, \ldots, e_p)$ is a basis of $\mathfrak{a}$ and $(e_{p+1}, \ldots, e_n)$ is a basis of $\mathfrak{m}$. In the notations of 40.2.5, we denote by M_λ the matrix $M(\mathcal{X}, \mathcal{X})$ relative to the Lie algebra $\mathfrak{g}_\lambda$. We have:

$$M_1 = \begin{pmatrix} A & B \\ -{}^tB & C \end{pmatrix} , \ M_\lambda = \begin{pmatrix} A & B \\ -{}^tB & \lambda C \end{pmatrix}$$

where $A \in \mathrm{M}_p(\mathrm{S}(\mathfrak{g}))$ and $C \in \mathrm{M}_{n-p}(\mathrm{S}(\mathfrak{g}))$.

Since $|\operatorname{rk}(M_0) - \operatorname{rk}(M_1)| \leqslant \operatorname{rk}(M_0 - M_1) = \operatorname{rk} C$, we obtain that:

$$|\chi(\mathfrak{g}_0) - \chi(\mathfrak{g})| \leqslant \dim \mathfrak{m} - \chi(\mathfrak{m}).$$

As there is an open subset $\mathcal{O}$ of $\Bbbk$ containing 0 verifying $\operatorname{rk}(M_0) \leqslant \operatorname{rk}(M_\lambda)$ for all $\lambda \in \mathcal{O}$, we deduce that $\operatorname{rk}(M_0) \leqslant \operatorname{rk}(M_1)$. Hence $\chi(\mathfrak{g}) \leqslant \chi(\mathfrak{g}_0)$, and we have proved:

Proposition. *In the above notations, we have:*

$$\chi(\mathfrak{g}_0) + \chi(\mathfrak{m}) - \dim \mathfrak{m} \leqslant \chi(\mathfrak{g}) \leqslant \chi(\mathfrak{g}_0).$$

40.4.5 We shall now give another application of 40.4.2. Suppose that $\mathfrak{g}$ is of the form:

$$\mathfrak{g} = \mathfrak{g}_0 \oplus \mathfrak{g}_1 \oplus \mathfrak{g}_2,$$

where $[\mathfrak{g}_i, \mathfrak{g}_j] \subset \mathfrak{g}_{i+j}$ where we set $\mathfrak{g}_k = \{0\}$ if $k \geqslant 3$.

The subspaces $\mathfrak{g}_0$ and $\mathfrak{h} = \mathfrak{g}_0 \oplus \mathfrak{g}_2$ are Lie subalgebras of $\mathfrak{g}$, while $\mathfrak{g}_2$ is an abelian ideal of $\mathfrak{g}$.

For $i = 0, 1, 2$, we can identify $(\mathfrak{g}_i)^*$ with the subspace $\mathfrak{g}^*_{-i}$ of $\mathfrak{g}^*$ consisting of elements $f \in \mathfrak{g}^*$ which are identically zero on $\mathfrak{g}_j$ for all $j \neq i$. If $x \in \mathfrak{g}_i$ and $f \in \mathfrak{g}^*_{-j}$, then $x.f \in \mathfrak{g}^*_{i-j}$.

Proposition. *Suppose that there exists a $\mathfrak{g}_0$-regular element f_2 in $\mathfrak{g}^*_{-2}$ verifying $\mathfrak{g}^{(f_2)} \cap \mathfrak{g}_1 = \{0\}$.*

(i) *There exists a $\mathfrak{g}_0$-regular element $g \in \mathfrak{g}^*_{-2}$ such that:*

$$\chi(\mathfrak{g}) = \chi(\mathfrak{h}) = \chi\big((\mathfrak{g}_0)_g\big) + \chi(\mathfrak{g}_0, \mathfrak{g}_2).$$

(ii) *Let us suppose that $\mathfrak{g}_0$ is the Lie algebra of an algebraic group G_0, $\mathfrak{g}_2$ is a rational G_0-module and that the $\mathfrak{g}_0$-action on $\mathfrak{g}_2$ is the one given by the differential of the G_0-action on $\mathfrak{g}_2$. If $\chi(\mathfrak{g}_0, \mathfrak{g}_2) = 0$, then*

$$\chi(\mathfrak{g}) = \chi(\mathfrak{h}) = \chi\big((\mathfrak{g}_0)_g\big)$$

for any $\mathfrak{g}_0$-regular element $g \in \mathfrak{g}^*_{-2}$.

Proof. (i) We need only to prove $\chi(\mathfrak{g}) = \chi(\mathfrak{h})$ since the second equality follows from 40.4.2.

Let $g = g_0 + g_1 + g_2 \in \mathfrak{g}^*_{\text{reg}}$ where $g_i \in \mathfrak{g}^*_{-i}$ for $i = 0, 1, 2$. Then in view of 19.7.4 (i) and our hypothesis, we may assume that g_2 is $\mathfrak{g}_0$-regular and $\mathfrak{g}^{(g_2)} \cap \mathfrak{g}_1 = \{0\}$. We have therefore:

$$\mathfrak{g}_1.\mathfrak{g}^*_0 = \{0\} \ , \ \mathfrak{g}_1.\mathfrak{g}^*_{-1} \subset \mathfrak{g}^*_0 \ , \ \mathfrak{g}_1.g_2 = \mathfrak{g}^*_{-1}.$$

Thus there exists $x \in \mathfrak{g}_1$ such that $h = e^{\text{ad}\,x}.g$ verifies $h = h_0 + g_2$ where $h_0 \in \mathfrak{g}^*_0$. So $h \in \mathfrak{g}^*_{\text{reg}}$, and $\mathfrak{h}^{(h|_\mathfrak{h})} \subset \mathfrak{g}^{(h)}$. Hence $\chi(\mathfrak{h}) \leqslant \chi(\mathfrak{g})$.

Conversely, let S be the set of $\mathfrak{g}_0$-regular elements $f_2 \in \mathfrak{g}^*_{-2}$ verifying $\mathfrak{g}^{(f_2)} \cap \mathfrak{g}_1 = \{0\}$. Our hypothesis implies that S is a non-empty open subset of $\mathfrak{g}^*_{-2}$. So there exists $\ell \in \mathfrak{h}^*_{\text{reg}}$ such that $\ell = \ell_0 + \ell_2$ with $\ell_0 \in \mathfrak{g}^*_0$ and $\ell_2 \in S$. Denote by $\ell' \in \mathfrak{g}^*$ the linear form such that $\ell'|_\mathfrak{h} = \ell$ and $\ell'|_{\mathfrak{g}_1} = 0$. Since $\mathfrak{g}_1.\ell_2 = \mathfrak{g}^*_{-1}$, we deduce that $\mathfrak{g}^{(\ell')} \subset \mathfrak{h}$, and hence $\mathfrak{g}^{(\ell')} \subset \mathfrak{h}^{(\ell)}$. Therefore $\chi(\mathfrak{g}) \leqslant \chi(\mathfrak{h})$.

(ii) In view of (i), the proof is the same as the one for 40.4.3. $\quad\square$

40.5 Heisenberg algebras in semisimple Lie algebras

40.5.1 Let $\mathfrak{g}$ be a semisimple Lie algebra, G its adjoint group, L its Killing form, $\mathfrak{h}$ a Cartan subalgebra of $\mathfrak{g}$ and R the root system of $\mathfrak{g}$ relative to $\mathfrak{h}$. Let us fix a base Π of R, and denote by R_+ (resp. R_-) the corresponding set of positive (resp. negative) roots. Set:

$$\mathfrak{n} = \sum_{\alpha \in R_+} \mathfrak{g}^\alpha \ , \ \mathfrak{n}_- = \sum_{\alpha \in R_+} \mathfrak{g}^{-\alpha} \ , \ \mathfrak{b} = \mathfrak{h} \oplus \mathfrak{n} \ , \ \mathfrak{b}_- = \mathfrak{h} \oplus \mathfrak{n}_-.$$

Let $\alpha, \beta \in R$. Denote by H_α the unique element of $[\mathfrak{g}^\alpha, \mathfrak{g}^{-\alpha}]$ such that $\alpha(H_\alpha) = 2$. Recall that $\beta(H_\alpha) \in \mathbb{Z}$. For $\lambda \in \mathfrak{h}^*$, we shall write $\langle \lambda, \alpha^\vee \rangle$ for $\lambda(H_\alpha)$. So in the notations of 18.3.3 and 20.6.7, we have $\langle \beta, \alpha^\vee \rangle = a_{\beta\alpha}$. If P is a subset of R, we set:

$$\mathfrak{h}_P = \sum_{\alpha \in P} \Bbbk H_\alpha.$$

For any subset S of Π, denote by $\mathbb{Z}S$ (resp. $\mathbb{N}S$) the subgroup (resp. subsemigroup) of $\mathfrak{h}^*$ generated by S. We set:

$$R^S = R \cap \mathbb{Z}S \ , \ R_+^S = R_+ \cap R^S = R \cap \mathbb{N}S.$$

The set R^S is clearly closed and symmetric, so by 18.10.6, it is a root system in the vector subspace that it spans in $\mathfrak{h}^*$. Moreover, S is a base of R^S and R_+^S is the corresponding set of positive roots. When S is a connected subset of Π, the root system R^S is irreducible, and we shall denote the highest root of R^S by ε_S.

Let us suppose that S is connected. For any root $\alpha \in R_+^S \setminus \{\varepsilon_S\}$, we have $\langle \alpha, \varepsilon_S^\vee \rangle \in \{0, 1\}$ (18.9.2). If $T = \{\alpha \in R^S; \langle \alpha, \varepsilon_S^\vee \rangle = 0\}$, then T is a root system in the subspace that it spans in $\mathfrak{h}^*$, and $S \cap T$ is a base of T.

Let $\alpha \in T \cap R_+^S$. Then $\alpha + \varepsilon_S \notin R$. It follows by 18.5.4 that $\alpha - \varepsilon_S \notin R$. Hence α and ε_S are strongly orthogonal (see 18.5.3).

40.5.2 Let us suppose that $\mathfrak{g}$ is simple, or equivalently, R is irreducible. Denote by P the set of roots that are not orthogonal to ε_Π, and $P_+ = P \cap R_+$. Set:

$$\mathfrak{g}' = \mathfrak{h}_{R\setminus P} \oplus \sum_{\alpha \in R\setminus P} \mathfrak{g}^\alpha \ , \ \mathfrak{m} = \sum_{\alpha \in R_+ \setminus P_+} \mathfrak{g}^\alpha \ , \ \mathfrak{m}_- = \sum_{\alpha \in R_+ \setminus P_+} \mathfrak{g}^{-\alpha} \ ,$$

$$\mathfrak{p} = \sum_{\alpha \in P_+} \mathfrak{g}^\alpha \ , \ \mathfrak{p}_0 = \sum_{\alpha \in P_+ \setminus \{\varepsilon_\Pi\}} \mathfrak{g}^\alpha .$$

Proposition. (i) *The Lie algebra $\mathfrak{g}'$ is semisimple and we have the direct sum decomposition $\mathfrak{g}' = \mathfrak{h}_{R\setminus P} \oplus \mathfrak{m} \oplus \mathfrak{m}_-$.*

(ii) *We have:*

$$\mathfrak{n} = \mathfrak{m} \oplus \mathfrak{p} \ , \ [\mathfrak{m}, \mathfrak{p}_0] \subset \mathfrak{p}_0 \ , \ [\mathfrak{n}, \mathfrak{g}^{\varepsilon_\Pi}] = \{0\}.$$

In particular, $\mathfrak{p}$ is an ideal of $\mathfrak{n}$.

(iii) *If $\alpha \in P_+ \setminus \{\varepsilon_\Pi\}$, then there exists a unique $\alpha' \in P_+ \setminus \{\varepsilon_\Pi\}$ such that $\alpha + \alpha' = \varepsilon_\Pi$. For any $\alpha \in P_+ \setminus \{\varepsilon_\Pi\}$, we may choose $X_\alpha \in \mathfrak{g}^\alpha \setminus \{0\}$ so that for $\alpha, \beta \in P_+ \setminus \{\varepsilon_\Pi\}$:*

$$[X_\alpha, X_\beta] = \begin{cases} \pm X_{\varepsilon_\Pi} & \text{if } \beta = \alpha', \\ 0 & \text{if } \beta \neq \alpha'. \end{cases}$$

Proof. (i) This is obvious.

(ii) The two equalities are clear. Let $\alpha \in P_+ \setminus \{\varepsilon_\Pi\}$ and $\beta \in R_+ \setminus P_+$. Then $\langle \alpha, \varepsilon_\Pi^\vee \rangle = 1$ and $\langle \beta, \varepsilon_\Pi^\vee \rangle = 0$. So $\langle \alpha + \beta, \varepsilon_\Pi^\vee \rangle = 1$, and hence $[\mathfrak{m}, \mathfrak{p}_0] \subset \mathfrak{p}_0$.

(iii) Let $\alpha \in P_+ \setminus \{\varepsilon_\Pi\}$. Since $\langle \alpha, \varepsilon_\Pi^\vee \rangle = 1$, the ε_Π-string through α is $\{\alpha - \varepsilon_\Pi, \alpha\}$. Set $\alpha' = \varepsilon_\Pi - \alpha$. Then $\alpha' \in R_+$ (18.9.2) and:

$$\langle \alpha', \varepsilon_\Pi^\vee \rangle = \langle \varepsilon_\Pi, \varepsilon_\Pi^\vee \rangle - \langle \alpha, \varepsilon_\Pi^\vee \rangle = 2 - 1 = 1.$$

So $\alpha' \in P_+ \setminus \{\varepsilon_\Pi\}$. It is obvious that α' is unique.

Finally, let $\alpha, \beta \in P_+ \setminus \{\varepsilon_\Pi\}$ be such that $\alpha + \beta \neq \varepsilon_\Pi$. Then $\alpha + \beta \notin R$ because $\langle \alpha + \beta, \varepsilon_\Pi^\vee \rangle = 2$. So the last part follows. □

40.5.3 Remark. A *Heisenberg algebra* is a Lie algebra admitting a basis $(x_1, y_1, \ldots, x_n, y_n, z)$, where $n \in \mathbb{N}$, such that: for $1 \leqslant i \neq j \leqslant n$,

$$[x_i, y_i] = z \ , \ [x_i, y_j] = [x_i, z] = [y_j, z] = 0.$$

So 40.5.2 says that $\mathfrak{p}$ is a Heisenberg algebra.

40.5.4 Let S be a subset of Π. We define a set $\mathcal{K}(S)$ by induction on the the cardinality of S as follows:

a) $\mathcal{K}(\emptyset) = \emptyset$.

b) If $S_1, \ldots, S_r$ are the connected components of S, then:

$$\mathcal{K}(S) = \mathcal{K}(S_1) \cup \cdots \cup \mathcal{K}(S_r).$$

c) If S is connected, then:

$$\mathcal{K}(S) = \{S\} \cup \mathcal{K}(\{\alpha \in S; \langle \alpha, \varepsilon_S^\vee \rangle = 0\}).$$

An element of $\mathcal{K}(S)$ is therefore a subset of S. The following results are immediate consequences of the definition.

Lemma. (i) *Each $M \in \mathcal{K}(S)$ is a connected subset of Π.*

(ii) *If $M, M' \in \mathcal{K}(S)$, then either $M' \subset M$, or $M \subset M'$, or M and M' are disjoint subsets of S verifying $\alpha + \beta \notin R$ for all $\alpha \in R^M$ and $\beta \in R^{M'}$.*

(iii) *If $M, M' \in \mathcal{K}(S)$ and $M \neq M'$, then ε_M and $\varepsilon_{M'}$ are strongly orthogonal.*

40.5.5 Using the classification of irreducible root systems in 18.14, we give, in the table below, the cardinality k of $\mathcal{K}(\Pi)$ for the different types of simple Lie algebras. Here, for $r \in \mathbb{Q}$, $[r]$ denotes the largest integer $\leqslant r$.

	$A_\ell, \ell \geqslant 1$	$B_\ell, \ell \geqslant 2$	$C_\ell, \ell \geqslant 3$	$D_\ell, \ell \geqslant 4$	E_6	E_7	E_8	F_4	G_2
k	$\left[\dfrac{\ell+1}{2}\right]$	ℓ	ℓ	$2\left[\dfrac{\ell}{2}\right]$	4	7	8	4	2

40.5.6 Using the numbering of simple roots in 18.14, we describe below the elements of the set $\{\varepsilon_M \; ; \; M \in \mathcal{K}(\Pi)\}$ for the different types of simple Lie algebras.

For a strictly positive integer n, we denote by $\mathbb{N}_n^{\mathrm{odd}}$ the set of odd positive integers less than or equal to n.

• Type A_ℓ, $\ell \geqslant 1$:

$$\left\{\beta_i + \cdots + \beta_{i+(\ell-2i+1)} \; ; \; 1 \leqslant i \leqslant \left[\frac{\ell+1}{2}\right]\right\}.$$

• Type B_ℓ, $\ell \geqslant 2$:

$$\{\beta_i + 2\beta_{i+1} + \cdots + 2\beta_\ell \; ; \; i \in \mathbb{N}_{\ell-1}^{\mathrm{odd}}\} \cup \{\beta_i \; ; \; i \in \mathbb{N}_\ell^{\mathrm{odd}}\}.$$

• Type C_ℓ, $\ell \geqslant 3$:

$$\{2\beta_i + \cdots + 2\beta_{\ell-1} + \beta_\ell \; ; \; 1 \leqslant i \leqslant \ell - 1\} \cup \{\beta_\ell\}.$$

• Type D_ℓ, $\ell \geqslant 4$ and ℓ even:

$$\{\beta_i + 2\beta_{i+1} + \cdots + 2\beta_{\ell-2} + \beta_{\ell-1} + \beta_\ell \; ; \; i \in \mathbb{N}_{\ell-3}^{\mathrm{odd}}\} \cup \{\beta_i \; ; \; i \in \mathbb{N}_\ell^{\mathrm{odd}}\} \cup \{\beta_\ell\}.$$

• Type D_ℓ, $\ell \geqslant 5$ and ℓ odd:

$$\{\beta_i + 2\beta_{i+1} + \cdots + 2\beta_{\ell-2} + \beta_{\ell-1} + \beta_\ell \; ; \; i \in \mathbb{N}_{\ell-3}^{\mathrm{odd}}\}$$
$$\cup \{\beta_i \; ; \; i \in \mathbb{N}_{\ell-2}^{\mathrm{odd}}\} \cup \{\beta_{\ell-2} + \beta_{\ell-1} + \beta_\ell\}.$$

• Type E_6: (in the notations of chapter 18)

$$\begin{matrix} 1\;2\;3\;2\;1 \\ 2 \end{matrix} \, , \, \begin{matrix} 1\;1\;1\;1\;1 \\ 0 \end{matrix} \, , \, \begin{matrix} 0\;1\;1\;1\;0 \\ 0 \end{matrix} \, , \, \begin{matrix} 0\;0\;1\;0\;0 \\ 0 \end{matrix} \, .$$

• Type E_7: (in the notations of chapter 18)

$$\begin{matrix} 2\;3\;4\;3\;2\;1 \\ 2 \end{matrix} \, , \, \begin{matrix} 0\;1\;2\;2\;2\;1 \\ 1 \end{matrix} \, , \, \begin{matrix} 0\;1\;2\;1\;0\;0 \\ 1 \end{matrix} \, , \, \begin{matrix} 0\;0\;0\;0\;0\;1 \\ 0 \end{matrix} \, ,$$
$$\begin{matrix} 0\;0\;0\;0\;0\;0 \\ 1 \end{matrix} \, , \, \begin{matrix} 0\;1\;0\;0\;0\;0 \\ 0 \end{matrix} \, , \, \begin{matrix} 0\;0\;0\;1\;0\;0 \\ 0 \end{matrix} \, .$$

• Type E_8: (in the notations of chapter 18)

$$\begin{matrix} 2\;4\;6\;5\;4\;3\;2 \\ 3 \end{matrix} \, , \, \begin{matrix} 2\;3\;4\;3\;2\;1\;0 \\ 2 \end{matrix} \, , \, \begin{matrix} 0\;1\;2\;2\;2\;1\;0 \\ 1 \end{matrix} \, , \, \begin{matrix} 0\;1\;2\;1\;0\;0\;0 \\ 1 \end{matrix} \, ,$$
$$\begin{matrix} 0\;0\;0\;0\;0\;1\;0 \\ 0 \end{matrix} \, , \, \begin{matrix} 0\;0\;0\;0\;0\;0\;0 \\ 1 \end{matrix} \, , \, \begin{matrix} 0\;1\;0\;0\;0\;0\;0 \\ 0 \end{matrix} \, , \, \begin{matrix} 0\;0\;0\;1\;0\;0\;0 \\ 0 \end{matrix} \, .$$

• Type F_4: (in the notations of chapter 18)

$$2\;3\;4\;2 \; , \; 0\;1\;2\;2 \; , \; 0\;1\;2\;0 \; , \; 0\;1\;0\;0 \, .$$

- Type G_2:
$$3\beta_1 + 2\beta_2 \ , \ \beta_1.$$

40.5.7 In the rest of this section, we shall denote $\mathcal{K}(\Pi)$ by $\mathcal{K}$. For $M \in \mathcal{K}$, set:
$$\Gamma^M = \{\alpha \in R^M; \langle \alpha, \varepsilon_M^\vee \rangle > 0\} \ , \ \Gamma_0^M = \Gamma^M \setminus \{\varepsilon_M\}.$$

Lemma. *Let M, M' be distinct elements of $\mathcal{K}$, $\alpha, \beta \in \Gamma^M$ and $\gamma \in \Gamma^{M'}$.*
(i) *We have $\Gamma^M = R_+^M \setminus \{\delta \in R_+^M; \langle \delta, \varepsilon_M^\vee \rangle = 0\}$.*
(ii) *R_+ is the disjoint union of Γ^M, for $M \in \mathcal{K}$.*
(iii) *If $\alpha + \beta \in R$, then $\alpha + \beta = \varepsilon_M$.*
(iv) *If $\alpha + \gamma \in R$, then either we have $M \subset M'$ and $\alpha + \gamma \in \Gamma_0^{M'}$ or $M' \subset M$ and $\alpha + \gamma \in \Gamma_0^M$.*

Proof. Part (i) is clear by 18.9.2. Part (ii) is obvious and (iii) follows from 40.5.2. Let us prove (iv).

By 40.5.4, we have either $M' \subset M$ or $M \subset M'$. Let us suppose, for example, that $M' \subset M$. Then $\langle \gamma, \varepsilon_M^\vee \rangle = 0$. Hence $\langle \alpha + \gamma, \varepsilon_M^\vee \rangle > 0$, so $\alpha + \gamma \in \Gamma^M$. Finally, if $\alpha + \gamma = \varepsilon_M$, then $\gamma = \varepsilon_M - \alpha$. So $\alpha \in \Gamma_0^M$, and
$$1 = \langle \alpha, \varepsilon_M^\vee \rangle = \langle \alpha + \gamma, \varepsilon_M^\vee \rangle = \langle \varepsilon_M, \varepsilon_M^\vee \rangle = 2,$$
which is absurd. So $\alpha + \gamma \in \Gamma_0^M$. $\square$

40.5.8 For $M \in \mathcal{K}$, set:
$$\mathfrak{a}_M = \sum_{\alpha \in \Gamma^M} \mathfrak{g}^\alpha \ , \ \mathfrak{n}_M = \sum_{\alpha \in R_+^M} \mathfrak{g}^\alpha.$$

By 40.5.2, $\mathfrak{a}_M$ is a Heisenberg algebra. The following result is an immediate consequence of 40.5.4:

Lemma. *Let $M \in \mathcal{K}$.*
(i) *The set R_+^M is the disjoint union of $\Gamma^{M'}$, for $M' \in \mathcal{K}$ verifying $M' \subset M$.*
(ii) *The subspace $\mathfrak{n}_M$ is the direct sum of $\mathfrak{a}_{M'}$, for $M' \in \mathcal{K}$ verifying $M' \subset M$.*

40.5.9 In the rest of this section, we shall fix a vector $X_\alpha \in \mathfrak{g}^\alpha \setminus \{0\}$ for each $\alpha \in R$. If $\mathfrak{m}$ is an ideal of $\mathfrak{b}$ contained in $\mathfrak{n}$, we set:
$$\mathcal{L}(\mathfrak{m}) = \{M \in \mathcal{K}; \mathfrak{m} \cap \mathfrak{n}_M \neq \{0\}\}.$$

Lemma. (i) *Let $M' \in \mathcal{K}$ and $M \in \mathcal{L}(\mathfrak{m})$ be such that $M \subset M'$. Then $M' \subset \mathcal{L}(\mathfrak{m})$.*
(ii) *We have:*
$$\mathcal{L}(\mathfrak{m}) = \{M \in \mathcal{K}; X_{\varepsilon_M} \in \mathfrak{m}\}.$$

(iii) *Suppose that $\mathfrak{m}$ is the nilpotent radical of a parabolic subalgebra of $\mathfrak{g}$ containing $\mathfrak{b}$. Let $M \in \mathcal{L}(\mathfrak{m})$ and $\alpha \in \Gamma_0^M$. Then we have either $X_\alpha \in \mathfrak{m}$ or $X_{\varepsilon_M - \alpha} \in \mathfrak{m}$.*

Proof. (i) This is clear because if $M \subset M'$, then $\mathfrak{n}_M \subset \mathfrak{n}_{M'}$.

(ii) Since $X_{\varepsilon_M} \in \mathfrak{n}_M$, we have $\{M \in \mathcal{K}; X_{\varepsilon_M} \in \mathfrak{m}\} \subset \mathcal{L}(\mathfrak{m})$. Conversely, let $M \in \mathcal{L}(\mathfrak{m})$. The subspace $\mathfrak{m} \cap \mathfrak{n}_M$ is an $(\operatorname{ad}\mathfrak{h})$-stable Lie subalgebra of $\mathfrak{g}$. It follows by 20.7.2 that there exists $\alpha \in R_+^M$ such that $\mathfrak{g}^\alpha \subset \mathfrak{m}$. But, in view of 18.7.7 and 18.9.2 (i), X_{ε_M} belongs to the $\mathfrak{n}_M$-module generated by X_α. Since $\mathfrak{m}$ is an ideal of $\mathfrak{b}$, we obtain that $X_{\varepsilon_M} \in \mathfrak{m}$. So we have proved (ii).

(iii) Let $M \in \mathcal{L}(\mathfrak{m})$ and $\alpha \in \Gamma_0^M$. We have $\varepsilon_M - \alpha \in \Gamma_0^M$ (40.5.2). Now, if $X_\alpha \notin \mathfrak{m}$, then $X_{-\alpha} \in \mathfrak{p}$. Since $X_{\varepsilon_M} \in \mathfrak{m}$ by (ii), we deduce that $[X_{-\alpha}, X_{\varepsilon_M}] \in \mathfrak{m}$. So we are done. $\square$

40.5.10 Let $\mathfrak{m}$ be an ideal of $\mathfrak{b}$ contained in $\mathfrak{n}$. We set:

$$\mathfrak{l}(\mathfrak{m}) = \sum_{M \in \mathcal{L}(\mathfrak{m})} \Bbbk H_{\varepsilon_M} \; , \; \mathfrak{d}(\mathfrak{m}) = \sum_{M \in \mathcal{L}(\mathfrak{m})} \mathfrak{a}_M \; , \; \mathfrak{c}(\mathfrak{m}) = \mathfrak{l}(\mathfrak{m}) \oplus \mathfrak{d}(\mathfrak{m}).$$

Note that the ε_M's, being pairwise strongly orthogonal, form a linearly independent family. In particular, we have $\dim \mathfrak{l}(\mathfrak{m}) = \operatorname{card} \mathcal{L}(\mathfrak{m})$.

Lemma. *The subspaces $\mathfrak{d}(\mathfrak{m})$ and $\mathfrak{c}(\mathfrak{m})$ are ideals of $\mathfrak{b}$ containing $\mathfrak{m}$.*

Proof. Let $\alpha \in R_+$, $M \in \mathcal{L}(\mathfrak{m})$ and $\beta \in \Gamma^M$.

We have $[H_{\varepsilon_M}, X_\alpha] = \alpha(H_{\varepsilon_M})X_\alpha = \langle \alpha, \varepsilon_M^\vee \rangle X_\alpha$. It follows from the definitions that $[\mathfrak{b}, \mathfrak{l}(\mathfrak{m})] \subset \mathfrak{d}(\mathfrak{m})$.

We have $\alpha \in \Gamma^{M'}$ for some $M' \in \mathcal{K}$. If $[X_\alpha, X_\beta] \neq 0$, then 40.5.7 implies that either $M \subset M'$ or $M' \subset M$. If $M \subset M'$, then $M' \in \mathcal{L}(\mathfrak{m})$ (40.5.9), and $\alpha + \beta \in \Gamma^{M'}$ (40.5.7). So $[X_\alpha, X_\beta] \in \mathfrak{d}(\mathfrak{m})$. If $M' \subset M$, then $\alpha + \beta \in \Gamma^M$ and again $[X_\alpha, X_\beta] \in \mathfrak{d}(\mathfrak{m})$. Hence we have proved that $[\mathfrak{b}, \mathfrak{d}(\mathfrak{m})] \subset \mathfrak{d}(\mathfrak{m})$. $\square$

40.6 Index of Lie subalgebras of Borel subalgebras

40.6.1 In this section, we shall conserve the notations and assumptions of section 40.5, and we shall denote by $\mathcal{K}$ the set $\mathcal{K}(\Pi)$.

Let $\mathfrak{a}$ be a Lie subalgebra of $\mathfrak{g}$. If $Y \in \mathfrak{g}$, then we define a linear form $\varphi_{\mathfrak{a}}^Y$ on $\mathfrak{a}$ by setting for $X \in \mathfrak{a}$:

$$\varphi_{\mathfrak{a}}^Y (X) = L(X, Y).$$

If $\mathfrak{c}$ is a Lie subalgebra of $\mathfrak{g}$ such that $[\mathfrak{c}, \mathfrak{a}] \subset \mathfrak{a}$, then $\mathfrak{c}$ acts on $\mathfrak{a}^*$ via the coadjoint representation, and if $Z \in \mathfrak{c}$, then:

$$Z.\varphi_{\mathfrak{a}}^Y = \varphi_{\mathfrak{a}}^{[Z,Y]}.$$

Thus:

$$\mathfrak{g}^Y \cap \mathfrak{a} \subset \mathfrak{a}^{(\varphi_{\mathfrak{a}}^Y)}.$$

40.6.2 For $\alpha \in R$, we fix a non-zero element X_α of $\mathfrak{g}^\alpha$, and we set:

$$U = \sum_{K \in \mathcal{K}} X_{-\varepsilon_K}.$$

Lemma. *Let $X \in \mathfrak{b}$. The following conditions are equivalent:*
(i) $X \in \mathfrak{h}$ *and* $\varepsilon_K(X) = 0$ *for all* $K \in \mathcal{K}$.
(ii) $[X, U] \in \mathfrak{n}$.

Proof. (i) $\Rightarrow$ (ii) If (i) is verified, then $[X, U] = 0$.
(ii) $\Rightarrow$ (i) There exist $H \in \mathfrak{h}$, $a_\alpha \in \Bbbk$ for $\alpha \in R_+$, such that

$$X = H + \sum_{\alpha \in R_+} a_\alpha X_\alpha.$$

So

$$[X, U] \in \sum_{K \in \mathcal{K}} a_{\varepsilon_K} \lambda_K H_{\varepsilon_K} + \mathfrak{n} + \mathfrak{n}^-$$

where $\lambda_K \in \Bbbk \setminus \{0\}$ for $K \in \mathcal{K}$. As the ε_K's are pairwise strongly orthogonal, the H_{ε_K}'s are linearly independent. It follows that $a_{\varepsilon_K} = 0$ for all $K \in \mathcal{K}$.
Let $K \in \mathcal{K}$, $\alpha \in \Gamma_0^K$ and $\beta = \varepsilon_K - \alpha$. Then

$$\begin{aligned}
[X, U] = \lambda a_\alpha X_{-\beta} + a_\alpha & \sum_{K' \in \mathcal{K} \setminus \{K\}} [X_\alpha, X_{-\varepsilon_{K'}}] \\
+ \sum_{K' \in \mathcal{K}, \gamma \in R_+ \setminus \{\alpha\}} & a_\gamma [X_\gamma, X_{-\varepsilon_{K'}}] - \sum_{K' \in \mathcal{K}} \varepsilon_{K'}(H) X_{-\varepsilon_{K'}},
\end{aligned}$$

where $\lambda \in \Bbbk \setminus \{0\}$.
Observe that since $\beta \in \Gamma_0^K$, $\langle \beta, \varepsilon_K^\vee \rangle = 1$, so $\beta \neq \varepsilon_{K'}$ for any $K' \in \mathcal{K}$. So if $[X, U] \in \mathfrak{n}$, then either $a_\alpha = 0$ or $\gamma - \varepsilon_{K'} = -\beta$ for some $K' \in \mathcal{K}$ and $\gamma \in R_+ \setminus \{\alpha\}$.
Suppose that $a_\alpha \neq 0$, then we would have $\beta + \gamma = \varepsilon_{K'}$ for some $K' \in \mathcal{K}$ and $\gamma \in R_+ \setminus \{\alpha\}$. By 40.5.7, we deduce that $K = K'$ and $\gamma = \alpha$ which is absurd.
We have therefore proved that $a_\alpha = 0$ for all $\alpha \in R_+$. Hence $X \in \mathfrak{h}$, and since

$$[X, U] = -\sum_{K \in \mathcal{K}} \varepsilon_K(X) X_{-\varepsilon_K},$$

we deduce that $\varepsilon_K(X) = 0$ for all $K \in \mathcal{K}$. $\square$

40.6.3 Proposition. *Let U be as in 40.6.2, $\mathfrak{m}$ an ideal of $\mathfrak{b}$ contained in $\mathfrak{n}$ and $f = \varphi_\mathfrak{m}^U$.*
(i) *We have $\mathfrak{b}.f = \mathfrak{m}^*$, so f is a $\mathfrak{b}$-regular element of $\mathfrak{m}^*$. In other words, $\chi(\mathfrak{b}, \mathfrak{m}) = 0$.*
(ii) *Let B be the Borel subgroup of G whose Lie algebra is $\mathrm{ad}_\mathfrak{g} \mathfrak{b}$. Then the B-orbit of f is open and dense in $\mathfrak{m}^*$.*
(iii) *Any $\mathfrak{b}$-regular element of $\mathfrak{m}^*$ is a regular element of $\mathfrak{m}^*$.*

Proof. (i) It is clear that we may assume $\mathfrak{m} = \mathfrak{n}$. Then it suffices to show that $\mathfrak{N} = \{X \in \mathfrak{n}; f([\mathfrak{b}, X]) = \{0\}\} = \{0\}$. But $\mathfrak{N}$ is the set of elements $X \in \mathfrak{n}$ verifying $L([U, X], \mathfrak{b}) = \{0\}$. Since the orthogonal of $\mathfrak{b}$ with respect to L is $\mathfrak{n}$, it is the set of elements $X \in \mathfrak{n}$ such that $[X, U] \in \mathfrak{n}$. So the result follows from 40.6.2.

(ii) This is clear by (i), 16.5.7 and 21.4.3.

(iii) Since $\mathfrak{m}^{(\alpha \cdot f)} = \alpha(\mathfrak{m}^{(f)})$ for $\alpha \in B$, this is an immediate consequence of (i), (ii) and 19.7.5. $\square$

40.6.4 Proposition. *Let $\mathfrak{m}$ be an ideal of $\mathfrak{b}$ contained in $\mathfrak{n}$, and $\mathcal{L}(\mathfrak{m})$, $\mathfrak{d}(\mathfrak{m})$, $\mathfrak{l}(\mathfrak{m})$ and $\mathfrak{c}(\mathfrak{m})$ be as in 40.5.9 and 40.5.10.*

(i) *We have $\chi(\mathfrak{d}(\mathfrak{m})) = \mathrm{card}(\mathcal{L}(\mathfrak{m}))$. In particular, $\chi(\mathfrak{n}) = \mathrm{card}(\mathcal{K})$.*

(ii) *Suppose that $\mathfrak{m}$ is the nilpotent radical of a parabolic subalgebra $\mathfrak{p}$ of $\mathfrak{g}$ containing $\mathfrak{b}$. Then:*

$$\chi(\mathfrak{m}) = \dim \mathfrak{c}(\mathfrak{m}) - \dim \mathfrak{m}.$$

Proof. Let $K_1, \ldots, K_r$ be the elements of $\mathcal{L}(\mathfrak{m})$. For $1 \leqslant i \leqslant r$, let p_i be the integer such that $2p_i + 1 = \dim \mathfrak{a}_{K_i}$. Let $\alpha_{i,1}, \ldots, \alpha_{i,p_i} \in \Gamma_0^{K_i}$ be such that $\{\alpha_{i,1}, \varepsilon_{K_i} - \alpha_{i,1}, \ldots, \alpha_{i,p_i}, \varepsilon_{K_i} - \alpha_{i,p_i}\} = \Gamma_0^{K_i}$. Choose $X_{i,j} \in \mathfrak{g}^{\alpha_{i,j}}$ and $Y_{i,j} \in \mathfrak{g}^{\varepsilon_{K_i} - \alpha_{i,j}}$ such that $[X_{i,j}, Y_{i,j}] = X_{\varepsilon_{K_i}}$ (This is possible by 40.5.2). Let $\mathcal{B}_i = (X_{i,1}, Y_{i,1}, \ldots, X_{i,p_i}, Y_{i,p_i})$. Then $\mathcal{B}_i \cup \{X_{\varepsilon_{K_i}}\}$ is a basis of $\mathfrak{a}_{K_i}$.

(i) Let $f = \varphi_{\mathfrak{d}(\mathfrak{m})}^U$, and denote by Φ_f the alternating bilinear form $(X, Y) \mapsto f([X, Y])$ on $\mathfrak{d}(\mathfrak{m}) \times \mathfrak{d}(\mathfrak{m})$. By 40.5.10 and 40.6.3, f is a regular element of $\mathfrak{d}(\mathfrak{m})^*$.

Let $1 \leqslant i < j \leqslant r$. If $K_i \cap K_j = \emptyset$, then $[\mathfrak{a}_{K_i}, \mathfrak{a}_{K_j}] = \{0\}$ (40.5.4). Otherwise, we have either $K_i \subset K_j$ or $K_j \subset K_i$ (40.5.4).

Let us suppose for example that $K_j \subset K_i$. Let $\alpha \in \Gamma^{K_i}$ and $\beta \in \Gamma^{K_j}$ be such that $\alpha + \beta \in R$. Then by 40.5.7, $\alpha + \beta \in \Gamma_0^{K_i}$. So $f([X_\alpha, X_\beta]) = 0$.

Now let $\mathcal{B}' = (X_{\varepsilon_{K_1}}, \ldots, X_{\varepsilon_{K_r}})$ and $\mathcal{B}'' = \mathcal{B}_1 \cup \cdots \cup \mathcal{B}_r$. Then $\mathcal{B} = \mathcal{B}' \cup \mathcal{B}''$ is a basis of $\mathfrak{d}(\mathfrak{m})$. We deduce from the arguments above that the matrix of Φ_f with respect to the basis $\mathcal{B}$ is of the form:

$$\begin{pmatrix} 0_{r,r} & 0_{r,2} & \cdots & \cdots & 0_{r,2} \\ 0_{2,r} & J & 0_{2,2} & \cdots & 0_{2,2} \\ \vdots & 0_{2,2} & J & \ddots & \vdots \\ \vdots & \vdots & \ddots & \ddots & 0_{2,2} \\ 0_{2,r} & 0_{2,2} & \cdots & 0_{2,2} & J \end{pmatrix}$$

where $0_{r,s}$ denotes the zero r by s matrix, and $J = \begin{pmatrix} 0 & 1 \\ -1 & 0 \end{pmatrix}$.

Hence

$$\mathrm{rk}(\Phi_f) = \dim \mathfrak{d}(\mathfrak{m}) - \mathrm{card}(\mathcal{L}(\mathfrak{m})).$$

We have therefore proved the first part of (i). If $\mathfrak{m} = \mathfrak{n}$, then $\mathfrak{d}(\mathfrak{m}) = \mathfrak{n}$ and $\mathcal{L}(\mathfrak{m}) = \mathcal{K}$. So the second part follows.

(ii) Let $g = \varphi_{\mathfrak{m}}^U$ and Φ_g be the bilinear form $(X, Y) \mapsto g([X, Y])$ on $\mathfrak{m}$. For $1 \leqslant i \leqslant r$, $\mathfrak{m}_{K_i} = \mathfrak{m} \cap \mathfrak{a}_{K_i}$ is $\mathfrak{h}$-stable because $\mathfrak{m}$ is an ideal of $\mathfrak{b}$. So there exists a subset $\mathcal{B}'_i$ of $\mathcal{B}_i$ such that $\mathcal{B}'_i \cup \{X_{-\varepsilon_{K_i}}\}$ is a basis of $\mathfrak{m}_{K_i}$.

Let E_1 (resp. E_2) be the set of indices k such that $\{X_{i,k}, Y_{i,k}\} \notin \mathcal{B}_i'$ (resp. $\{X_k, Y_k\} \in \mathcal{B}_i'$). By 40.5.9 (iii), we have $\operatorname{card}(\mathcal{B}_i' \cap \{X_{i,k}, Y_{i,k}\}) = 1$ if $k \in E_1$. Since $X_{\varepsilon K_i} \in \mathfrak{m}$ (40.5.9 (ii)), it follows that:

$$\dim \mathfrak{a}_{K_i} = 1 + 2\operatorname{card} E_1 + 2\operatorname{card} E_2,$$
$$\dim \mathfrak{m}_{K_i} = 1 + \operatorname{card} E_1 + 2\operatorname{card} E_2.$$

On the other hand, by considering the matrix of $\Phi_g|_{\mathfrak{m}_{K_i} \times \mathfrak{m}_{K_i}}$ with respect to $\mathcal{B}_i' \cup \{X_{-\varepsilon K_i}\}$, we verify easily that:

$$r_i = \operatorname{rk}(\Phi_g|_{\mathfrak{m}_{K_i} \times \mathfrak{m}_{K_i}}) = 2\operatorname{card} E_2.$$

Consequently, we deduce that:

$$1 + \dim \mathfrak{a}_{K_i} - \dim \mathfrak{m}_{K_i} = \dim \mathfrak{m}_{K_i} - r_i.$$

In view of the proof of part (i), we obtain that:

$$\operatorname{card}(\mathcal{L}(\mathfrak{m})) + \dim \mathfrak{d}(\mathfrak{m}) - \dim \mathfrak{m} = \dim \mathfrak{m} - \operatorname{rk}(\Phi_g).$$

So we are done because g is a regular element of $\mathfrak{m}^*$ (40.6.3). $\quad\square$

40.6.5 Let $\mathfrak{m}$ be an ideal of $\mathfrak{b}$ contained in $\mathfrak{n}$, and $\mathfrak{c}(\mathfrak{m})$ be as in 40.5.10. By 40.5.10, $\mathfrak{c}(\mathfrak{m})$ is an ideal of $\mathfrak{b}$. Set $\mathfrak{t}(\mathfrak{m}) = \mathfrak{h} + \mathfrak{c}(\mathfrak{m}) = \mathfrak{h} \oplus \mathfrak{d}(\mathfrak{m})$.

Proposition. *We have* $\chi(\mathfrak{c}(\mathfrak{m})) = 0$ *and* $\chi(\mathfrak{t}(\mathfrak{m})) = \operatorname{rk}(\mathfrak{g}) - \operatorname{card}(\mathcal{L}(\mathfrak{m}))$. *Moreover,* $\mathfrak{c}(\mathfrak{m})^*$ *and* $\mathfrak{t}(\mathfrak{m})^*$ *contain a stable element.*

Proof. Let $\mathcal{B}$ be the basis of $\mathfrak{d}(\mathfrak{m})$ as in the proof of 40.6.4 (i), and $\mathcal{C} = (H_{\varepsilon K_1}, \ldots, H_{\varepsilon K_r})$ where $K_1, \ldots, K_r$ are the elements of $\mathcal{L}(\mathfrak{m})$. Then $\mathcal{D} = \mathcal{C} \cup \mathcal{B}$ is a basis of $\mathfrak{c}(\mathfrak{m})$. Let $f = \varphi_{\mathfrak{c}(\mathfrak{m})}^U$. Then the matrix of Φ_f with respect to $\mathcal{D}$ is of the form:

$$M = \begin{pmatrix} 0_{r,r} & A & 0_{r,2} & \cdots & \cdots & 0_{r,2} \\ -^t A\, 0_{r,r} & 0_{r,2} & \cdots & \cdots & 0_{r,2} \\ 0_{2,r} & 0_{2,r} & J & 0_{2,2} & \cdots & 0_{2,2} \\ \vdots & \vdots & 0_{2,2} & J & \ddots & \vdots \\ \vdots & \vdots & \vdots & \ddots & \ddots & 0_{2,2} \\ 0_{2,r} & 0_{2,r} & 0_{2,2} & \cdots & 0_{2,2} & J \end{pmatrix}$$

where $0_{r,s}$ and J are as in the proof of 40.6.4, and $A \in M_r(\Bbbk)$ is a diagonal matrix of rank r. Hence $\operatorname{rk}(\Phi_f) = \dim \mathfrak{c}(\mathfrak{m})$, and so $\chi(\mathfrak{c}(\mathfrak{m})) = 0$.

Finally, let $g = \varphi_{\mathfrak{t}(\mathfrak{m})}^U$. Let $\mathfrak{z} = \{H \in \mathfrak{h}; \varepsilon_K(H) = 0\}$ for all $K \in \mathcal{L}(\mathfrak{m})$. Then $\dim \mathfrak{z} = \operatorname{rk}(\mathfrak{g}) - \operatorname{card}(\mathcal{L}(\mathfrak{m})) = s$. Let $\mathcal{E}$ be a basis of $\mathfrak{z}$. Then $\mathcal{E} \cup \mathcal{D}$ is a basis of $\mathfrak{t}(\mathfrak{m})$, and the matrix of Φ_g with respect to this basis is of the form:

$$\begin{pmatrix} 0_{s,s} & 0_{s,m} \\ 0_{m,s} & M \end{pmatrix}$$

where $m = \dim \mathfrak{c}(\mathfrak{m})$.

It follows that $\mathfrak{t}(\mathfrak{m})^{(g)} = \mathfrak{z}$. Since $\mathfrak{z}$ is a commutative Lie algebra consisting of semisimple elements, g is stable by 40.1.6 and 40.1.5. So we are done since a stable linear form is regular (40.1.3). $\square$

40.6.6 Proposition. *Let $f = \varphi_\mathfrak{b}^U$ be as in 40.6.2.*

(i) *f is a stable element of $\mathfrak{b}^*$. In particular, it is regular in $\mathfrak{b}^*$.*

(ii) *We have $\chi(\mathfrak{b}) = \mathrm{rk}(\mathfrak{g}) - \mathrm{card}(\mathcal{K})$. In particular, $\chi(\mathfrak{b}) + \chi(\mathfrak{n}) = \mathrm{rk}(\mathfrak{g})$.*

Proof. By applying 40.6.5 with $\mathfrak{m} = \mathfrak{n}$, we have (i) and the first part of (ii). The second part of (ii) follows from 40.6.4. $\square$

40.7 Seaweed Lie algebras

40.7.1 In the rest of this chapter, $\mathfrak{g}$ will be a semisimple Lie algebra, and G its adjoint group.

If P is a subset of R, set:

$$\mathfrak{g}^P = \sum_{\alpha \in P} \mathfrak{g}^\alpha.$$

Definition. (i) *Two parabolic subalgebras $\mathfrak{p}$ and $\mathfrak{p}'$ are said to be* weakly opposite *if $\mathfrak{p} + \mathfrak{p}' = \mathfrak{g}$.*

(ii) *A* seaweed Lie algebra *of $\mathfrak{g}$ is a Lie subalgebra $\mathfrak{q}$ of $\mathfrak{g}$ of the form $\mathfrak{q} = \mathfrak{p} \cap \mathfrak{p}'$ where $\mathfrak{p}$ and $\mathfrak{p}'$ are weakly opposite parabolic subalgebras of $\mathfrak{g}$.*

40.7.2 Let S, T be subsets of Π. We check easily that $R_+^S \cup R_-^T$ is a closed subset of R. Set:

$$\mathfrak{p} = \mathfrak{g}^{R_+^S} \oplus \mathfrak{b}_- \ , \ \mathfrak{p}' = \mathfrak{g}^{R_-^T} \oplus \mathfrak{b}.$$

Then $\mathfrak{p}$ and $\mathfrak{p}'$ are weakly opposite parabolic subalgebras of $\mathfrak{g}$, and

$$\mathfrak{g}_{S,T} = \mathfrak{p} \cap \mathfrak{p}' = \mathfrak{h} \oplus \mathfrak{g}^{R_+^S} \oplus \mathfrak{g}^{R_-^T}$$

is a seaweed Lie algebra of $\mathfrak{g}$.

Conversely, let $\mathfrak{p}$ and $\mathfrak{p}'$ be parabolic subalgebras verifying $\mathfrak{b} \subset \mathfrak{p}$ and $\mathfrak{b}_- \subset \mathfrak{p}'$. Then $\mathfrak{p} + \mathfrak{p}' = \mathfrak{g}$, and there exist subsets S and T of Π such that $\mathfrak{p} \cap \mathfrak{p}' = \mathfrak{h} \oplus \mathfrak{g}^{R_+^S} \oplus \mathfrak{g}^{R_-^T} = \mathfrak{g}_{S,T}$.

We shall say that $\mathfrak{q}$ is a *standard* seaweed Lie algebra of $\mathfrak{g}$ (relative to $\mathfrak{h}$ and Π) if there exist $S, T \subset \Pi$ such that $\mathfrak{q} = \mathfrak{h} \oplus \mathfrak{g}^{R_+^S} \oplus \mathfrak{g}^{R_-^T} = \mathfrak{g}_{S,T}$.

40.7.3 Proposition. *Let $\mathfrak{q}$ be a Lie subalgebra of $\mathfrak{g}$. The following conditions are equivalent:*

(i) *$\mathfrak{q}$ is a seaweed Lie algebra of $\mathfrak{g}$.*

(ii) *$\mathfrak{q}$ is G-conjugate to a standard seaweed Lie algebra of $\mathfrak{g}$.*

Proof. It is clear that (ii) implies (i). Let us suppose that (i) is verified. Let $\mathfrak{p}$ and $\mathfrak{p}'$ be weakly opposite parabolic subalgebras of $\mathfrak{g}$ such that $\mathfrak{q} = \mathfrak{p} \cap \mathfrak{p}'$. By 29.4.9, $\mathfrak{q}$ contains a Cartan subalgebra $\mathfrak{t}$ of $\mathfrak{g}$. Denote by R' the root system of $\mathfrak{g}$ relative to $\mathfrak{t}$. Then there exist parabolic subsets P, Q of R' such that $\mathfrak{p} = \mathfrak{t} \oplus \mathfrak{g}^P$ and $\mathfrak{p}' = \mathfrak{t} \oplus \mathfrak{g}^Q$. Since $\mathfrak{q}$ is a seaweed Lie algebra, we have:

$$P \cup (-P) = Q \cup (-Q) = P \cup Q = R'.$$

So we have also $(-P) \cup (-Q) = R'$. Let $T = P \cap (-Q)$. It is a closed subset of R'. We shall show that T is in fact parabolic, that is $T \cup (-T) = R'$.

Let $\alpha \in R' \setminus T$. We have three cases:
a) If $\alpha \notin P$ and $\alpha \notin -Q$, then $\alpha \in -P$ and $\alpha \in Q$.
b) If $\alpha \in P$ and $\alpha \notin -Q$, then $\alpha \in -P$ and $\alpha \in Q$ since $(-P) \cup (-Q) = R'$.
c) If $\alpha \notin P$ and $\alpha \in -Q$, then $\alpha \in -P$ and $\alpha \in Q$ because $P \cup Q = R'$.

So in all the cases, $\alpha \in -T$. Thus T is a parabolic subset of R', and the implication (i) $\Rightarrow$ (ii) follows immediately. $\square$

40.8 An upper bound for the index

40.8.1 The notations we introduce in this section will be used in the rest of this chapter.

Let us conserve the assumptions of 40.5. Denote by ℓ the rank of $\mathfrak{g}$. We shall fix for each $\alpha \in R$, a non-zero element X_α of $\mathfrak{g}^\alpha$, and we shall also fix a total order $>$ on R_+.

40.8.2 Let $(H_1, \ldots, H_\ell)$ be a basis of $\mathfrak{h}$, $\{H_i; 1 \leqslant i \leqslant \ell\} \cup \{X_\alpha; \alpha \in R\}$, is then a basis of $\mathfrak{g}$, and we shall denote by $\{H_i^*; 1 \leqslant i \leqslant \ell\} \cup \{X_\alpha^*; \alpha \in R\}$ the corresponding dual basis.

40.8.3 Let S, T be subsets of Π, $S = \mathcal{K}(S)$, $T = \mathcal{K}(T)$ and E_S (resp. $E_T, E_{S,T}$) the subspace of $\mathfrak{h}^*$ spanned by ε_M, $M \in S$ (resp. $T, S \cup T$).

For $M \in S$ (resp. $N \in T$), we denote by a_M (resp. b_N) a non-zero element of $\Bbbk$. Set:

$$f = \sum_{M \in S} a_M X_{\varepsilon_M}^* + \sum_{N \in T} b_N X_{-\varepsilon_N}^* \in \mathfrak{q}^*,$$

where $\mathfrak{q} = \mathfrak{g}_{S,T}$ is the standard seaweed Lie algebra of $\mathfrak{g}$ associated to S and T as defined in 40.7.

We have also $f = \varphi_{\mathfrak{q}}^{X_{S,T}}$, where

$$X_{S,T} = \sum_{M \in S} a_M' X_{-\varepsilon_M} + \sum_{N \in T} b_N' X_{\varepsilon_N},$$

where

$$a_M' = \frac{a_M}{L(X_{\varepsilon_M}, X_{-\varepsilon_M})} , \quad b_N' = \frac{b_N}{L(X_{\varepsilon_N}, X_{-\varepsilon_N})}.$$

Let m_S (resp. m_T) be the cardinality of S (resp. T) and $m = m_S + m_T$. We define

$$\Omega_f = (a_{M_1}, \ldots, a_{M_{m_S}}, b_{N_1}, \ldots, b_{N_{m_T}}) \in (\Bbbk \setminus \{0\})^m,$$

where $\varepsilon_{M_1} < \cdots < \varepsilon_{M_{m_S}}$ and $\varepsilon_{N_1} < \cdots < \varepsilon_{N_{m_T}}$.

We shall say that Ω_f *defines* f.

40.8.4 For $M \in \mathcal{S}$ and $N \in \mathcal{T}$, we define the following sets:

- $\mathcal{H}_1(M)$ is the set of pairs $(\alpha, \varepsilon_M - \alpha)$ with $\alpha \in \Gamma_0^M$ and $\alpha < \varepsilon_M - \alpha$.
- $\mathcal{H}_2(N)$ is the set of pairs $(-\beta, -\varepsilon_N + \beta)$ with $\beta \in \Gamma_0^N$ and $\beta < \varepsilon_N - \beta$.
- $\mathcal{I}_1(M, N)$ is the set of pairs $(-\beta, \varepsilon_M)$ with $\beta \in \Gamma_0^N$, and $\varepsilon_M - \beta = -\varepsilon_{N'}$ for some $N' \in \mathcal{T}$.
- $\mathcal{I}_2(M, N)$ is the set of pairs $(\alpha, -\varepsilon_N)$ with $\alpha \in \Gamma_0^M$, and $\alpha - \varepsilon_N = \varepsilon_{M'}$ for some $M' \in \mathcal{S}$.
- $\mathcal{J}_1(M, N)$ is the set of pairs $(\alpha, -\beta)$ with $\alpha \in \Gamma_0^M$ and $\beta \in \Gamma_0^N$, such that $\alpha - \beta = \varepsilon_{M'}$ for some $M' \in \mathcal{S}$.
- $\mathcal{J}_2(M, N)$ is the set of pairs $(\alpha, -\beta)$ with $\alpha \in \Gamma_0^M$ and $\beta \in \Gamma_0^N$, such that $\alpha - \beta = -\varepsilon_{N'}$ for some $N' \in \mathcal{T}$.

Let us write:

$$\mathcal{H}_1 = \bigcup_{M \in \mathcal{S}} \mathcal{H}_1(M) , \ \mathcal{H}_2 = \bigcup_{N \in \mathcal{T}} \mathcal{H}_2(N) , \ \mathcal{H} = \mathcal{H}_1 \cup \mathcal{H}_2,$$
$$\mathcal{I}_1 = \bigcup_{(M,N) \in \mathcal{S} \times \mathcal{T}} \mathcal{I}_1(M, N) , \ \mathcal{I}_2 = \bigcup_{(M,N) \in \mathcal{S} \times \mathcal{T}} \mathcal{I}_2(M, N) ,$$
$$\mathcal{J} = \Big(\bigcup_{(M,N) \in \mathcal{S} \times \mathcal{T}} \mathcal{J}_1(M, N) \Big) \cup \Big(\bigcup_{(M,N) \in \mathcal{S} \times \mathcal{T}} \mathcal{J}_2(M, N) \Big) ,$$
$$\mathcal{Z} = \mathcal{H} \cup \mathcal{I}_1 \cup \mathcal{I}_2 \cup \mathcal{J} ,$$
$$r = \operatorname{card} \mathcal{H} , \ s = \dim E_{\mathcal{S},\mathcal{T}}.$$

If $z = (\alpha, \beta) \in \mathcal{Z}$, then we shall denote by $\widetilde{z} = \{\alpha, \beta\}$ the underlying set of z.

40.8.5 Lemma. *Let* $M \in \mathcal{S}$ *and* $N \in \mathcal{T}$. *Then:*

(i) *If* $(-\beta, \varepsilon_M) \in \mathcal{I}_1(M, N)$ *verifies* $\varepsilon_M - \beta = -\varepsilon_{N'}$, *then* $N' \subsetneq N$ *and* $\varepsilon_M \in \Gamma_0^N$.

(ii) *If* $(\alpha, -\varepsilon_N) \in \mathcal{I}_2(M, N)$ *verifies* $\alpha - \varepsilon_N = \varepsilon_{M'}$, *then* $M' \subsetneq M$ *and* $\varepsilon_N \in \Gamma_0^M$.

(iii) *If* $(\alpha, -\beta) \in \mathcal{J}_1(M, N)$ *verifies* $\alpha - \beta = \varepsilon_{M'}$, *then* $M' \subsetneq M$ *and* $\beta \in \Gamma_0^M$.

(iv) *If* $(\alpha, -\beta) \in \mathcal{J}_2(M, N)$ *verifies* $\alpha - \beta = -\varepsilon_{N'}$, *then* $N' \subsetneq N$ *and* $\alpha \in \Gamma_0^N$.

Proof. This is clear by lemmas 40.5.4 and 40.5.7. $\square$

40.8.6 For $M \in \mathcal{S}$ and $N \in \mathcal{T}$, we set:

$$\mathfrak{a}_M^\bullet = \sum_{\alpha \in \Gamma_0^M} \mathfrak{g}^\alpha , \ \mathfrak{b}_N^\bullet = \sum_{\alpha \in \Gamma_0^N} \mathfrak{g}^{-\alpha} , \mathfrak{a}_M = \Bbbk X_{\varepsilon_M} + \mathfrak{a}_M^\bullet , \ \mathfrak{b}_N = \Bbbk X_{-\varepsilon_N} + \mathfrak{b}_N^\bullet ,$$
$$\mathfrak{u} = \sum_{M \in \mathcal{S}} \Bbbk X_{\varepsilon_M} + \sum_{N \in \mathcal{T}} \Bbbk X_{-\varepsilon_N} , \ \mathfrak{v} = \sum_{M \in \mathcal{S}} \mathfrak{a}_M^\bullet + \sum_{N \in \mathcal{T}} \mathfrak{b}_N^\bullet.$$

Recall from 40.5.8 that $\mathfrak{a}_M$ and $\mathfrak{b}_N$ are Heisenberg algebras whose centres are $\Bbbk X_{\varepsilon_M}$ and $\Bbbk X_{-\varepsilon_N}$ respectively.

The dimension of $\mathfrak{a}_M^{\bullet}$ (resp. $\mathfrak{b}_N^{\bullet}$) is an even integer $2n_M$ (resp. $2n_N$), and we have:

$$r = \sum_{M \in \mathcal{S}} n_M + \sum_{N \in \mathcal{T}} n_N.$$

We shall consider $\mathfrak{h}^*$, $\mathfrak{u}^*$ and $\mathfrak{v}^*$ as vector subspaces of $\mathfrak{q}^*$ via the direct sum decomposition $\mathfrak{q} = \mathfrak{h} \oplus \mathfrak{u} \oplus \mathfrak{v}$.

40.8.7 Using the notations of 40.8.2, for $z \in (\alpha, \beta) \in \mathcal{Z}$, we set:

$$v_z = X_\alpha^* \wedge X_\beta^* \in \bigwedge{}^2 \mathfrak{q}^*.$$

Let $M, M' \in \mathcal{S}$, $N, N' \in \mathcal{T}$, $\alpha \in \Gamma_0^M$ and $\beta \in \Gamma_0^N$. Recall from lemmas 40.5.4 and 40.5.7 that:

- $\varepsilon_M - \varepsilon_N \notin \{\varepsilon_{M'}, -\varepsilon_N\}$.
- $\alpha - \varepsilon_N \neq -\varepsilon_{N'}$.
- $\varepsilon_M - \beta \neq \varepsilon_{M'}$.

It follows that if we identify the alternating bilinear form Φ_f on $\mathfrak{q}$ with an element of $\bigwedge^2 \mathfrak{q}^*$, then

$$\Phi_f = \Psi_f + \Theta_f,$$

where $\Theta_f \in E_{\mathcal{S},\mathcal{T}} \bigwedge \mathfrak{u}^*$ and

$$\Psi_f = \sum_{z \in \mathcal{Z}} \lambda_z v_z$$

with $\lambda_z \in \Bbbk$ for all $z \in \mathcal{Z}$.

If $z = (\alpha, \beta) \in \mathcal{Z}$, then $\Theta_f(X_\alpha, X_\beta) = 0$. Moreover, we have either $[X_\alpha, X_\beta] = \mu_z X_{\varepsilon_M}$ for some $M \in \mathcal{S}$ and $\mu_z \in \Bbbk \setminus \{0\}$, or $[X_\alpha, X_\beta] = \mu_z X_{-\varepsilon_N}$ for some $N \in \mathcal{T}$ and non zero scalar μ_z. Consequently,

$$\lambda_z = \Phi_f(X_\alpha, X_\beta) = f([X_\alpha, X_\beta]) = \begin{cases} \mu_z a_M & \text{if } [X_\alpha, X_\beta] = \mu X_{\varepsilon_M}, \\ \mu_z b_N & \text{if } [X_\alpha, X_\beta] = \mu X_{-\varepsilon_N}. \end{cases}$$

Thus λ_z is non-zero.

40.8.8 Lemma. *In the above notations, we have:*

(i) $\mathfrak{q}^{(f)}$ *contains a commutative Lie subalgebra of* $\mathfrak{g}$ *consisting of semi-simple elements of dimension:*

$$\dim \mathfrak{h} - \dim E_{\mathcal{S},\mathcal{T}} + \operatorname{card}(\mathcal{S} \cap \mathcal{T}).$$

(ii) We have $\wedge^s \Theta_f \neq 0$ *and* $\wedge^{s+1} \Theta_f = 0$.

(iii) There exists a non-empty open subset U *of* $(\Bbbk \setminus \{0\})^m$ *such that* $\wedge^r \Psi_f \neq 0$ *whenever* $\Omega_f \in U$.

(iv) If $\Omega_f \in U$, *then* $\wedge^{r+s} \Phi_f \neq 0$.

Proof. (i) Let $\mathfrak{t}$ be the orthogonal of $E_{S,T}$ in $\mathfrak{h}$. Then:

$$\dim \mathfrak{t} = \dim \mathfrak{h} - \dim E_{S,T}.$$

Let $X_{S,T}$ be as in 40.8.3. If $H \in \mathfrak{t}$, then $[H, X_{S,T}] = 0$, and so $H.f = 0$.
Next let $M \in \mathcal{S} \cap \mathcal{T}$, and

$$Y_M = a'_M X_{\varepsilon M} + b'_M X_{-\varepsilon M}.$$

Set $e = a'_M X_{\varepsilon M}$, $f = b'_M X_{-\varepsilon M}$ and $h = [e, f]$. Then 38.5.6 says that Y_M is G-conjugate to a non-zero multiple of h. Thus Y_M is semisimple. Now by the definition of $X_{S,T}$ and part (iii) of lemma 40.5.4, we obtain that $Y_M.f = 0$.

In view of the previous paragraphs, we have proved (i).

(ii) Let $(\alpha_1, \ldots, \alpha_s)$ be a basis of $E_{S,T}$ such that $\alpha_i = \varepsilon_{M_i}$ if $1 \leqslant i \leqslant p$, and $\alpha_i = -\varepsilon_{N_i}$ if $p + 1 \leqslant i + 1 \leqslant s$, where the M_i's and the N_i's are elements of $\mathcal{S}$ and $\mathcal{T}$ respectively. Complete this basis to a basis $\mathcal{B}' = (\alpha_1, \ldots, \alpha_\ell)$ of $\mathfrak{h}^*$, and let $\mathcal{B} = (h_1, \ldots, h_\ell)$ be the basis of $\mathfrak{h}$ dual to $\mathcal{B}'$. In the same manner, we complete the system $(X_{\varepsilon M_1}, \ldots, X_{\varepsilon M_p}, X_{-\varepsilon N_{p+1}}, \ldots, X_{-\varepsilon N_s})$ of linearly independent vectors to a basis $\mathcal{C}$ of $\mathfrak{u}$. Then $\mathcal{D} = \mathcal{B} \cup \mathcal{C}$ is a basis of $\mathfrak{h} + \mathfrak{u}$. Observe that $[h_k, \mathfrak{u}] = \{0\}$ if $s + 1 \leqslant k \leqslant \ell$.

Since the ε_M's (resp. ε_N's) are pairwise strongly orthogonal, it follows from 40.8.7 that in the basis $\mathcal{D}$, the matrix of the restriction of Φ_f, which is also the restriction of Θ_f, to $\mathfrak{h} + \mathfrak{u}$ is of the form:

$$M = \begin{pmatrix} 0_{s,s} & 0_{s,\ell-s} & D & A \\ 0_{\ell-s,s} & 0_{\ell-s,\ell-s} & 0_{\ell-s,s} & 0_{\ell-s,m-s} \\ -{}^t D & 0_{s,\ell-s} & 0_{s,s} & 0_{s,m-s} \\ -{}^t A & 0_{m-s,\ell-s} & 0_{m-s,s} & 0_{m-s,m-s} \end{pmatrix},$$

where $A \in M_{s,m-s}(\Bbbk)$, (m is as defined in 40.8.3), and $D \in M_{s,s}(\Bbbk)$ is diagonal with entries:

$$(a_{M_1}, \ldots, a_{M_p}, b_{N_{p+1}}, \ldots, b_{N_s}).$$

Thus the rank of M is $2s$, and therefore $\wedge^s \Theta_f \neq 0$ and $\wedge^{s+1} \Theta_f = 0$.

(iii) Let $z_1, \ldots, z_n$ be the elements of $\mathcal{Z}$ such that $z_1, \ldots, z_r$ are the elements of $\mathcal{H}$. For simplicity, let us write $\lambda_i v_i$ and μ_i, instead of $\lambda_{z_i} v_{z_i}$ and μ_{z_i}.

Observe that for $1 \leqslant i, j \leqslant n$, $v_i \wedge v_j = v_j \wedge v_i$ and $v_i \wedge v_i = 0$. Consequently:

$$(3) \qquad \wedge^r \Psi_f = r! \sum_{1 \leqslant i_1 < \cdots < i_r \leqslant n} \lambda_{i_1} \cdots \lambda_{i_r} v_{i_1} \wedge \cdots \wedge v_{i_r}.$$

So the coefficient $\lambda_1 \cdots \lambda_r$ of $v_1 \wedge \cdots \wedge v_r$ in the sum of (3) is:

$$\prod_{M \in \mathcal{S}} a_M^{n_M} \Big(\prod_{z \in \mathcal{H}_1(M)} \mu_z \Big) \prod_{N \in \mathcal{T}} b_N^{n_N} \Big(\prod_{z \in \mathcal{H}_2(N)} \mu_z \Big).$$

Now suppose that $v_{i_1} \wedge \cdots \wedge v_{i_r} = \lambda v_1 \wedge \cdots \wedge v_r$ with $\lambda \in \Bbbk \setminus \{0\}$, $1 \leqslant i_1 < \cdots < i_r \leqslant n$ and $(i_1, \ldots, i_r) \neq (1, \ldots, r)$. Then we have:

$$\mathcal{E} = \widetilde{z_1} \cup \cdots \cup \widetilde{z_r} = \widetilde{z_{i_1}} \cup \cdots \cup \widetilde{z_{i_r}}.$$

It follows that if $z_{i_t} \notin \mathcal{H}$, then $z_{i_t} \in \mathcal{J}$ and there exist $M \in \mathcal{S}$ and $N \in \mathcal{T}$ such that $\Gamma_0^M \cap \widetilde{z_{i_t}} \neq \emptyset$ and $(-\Gamma_0^N) \cap \widetilde{z_{i_t}} \neq \emptyset$. Let $z_{j_1}, \ldots, z_{j_k}$ be the elements z_{i_t} which do not belong to $\mathcal{H}$.

Let $M_0 \in \mathcal{S}$ be maximal by inclusion among the elements M of $\mathcal{S}$ verifying:

$$\Gamma_0^M \cap (\widetilde{z_{j_1}} \cup \cdots \cup \widetilde{z_{j_k}}) \neq \emptyset.$$

There exists $t \in \{1, \ldots, k\}$ such that if $z = z_{j_t}$, then $\Gamma_0^{M_0} \cap \widetilde{z} \neq \emptyset$. Let $\alpha \in \Gamma_0^{M_0} \cap \widetilde{z}$. Then $x \in \mathcal{J}$ and $(\alpha, \varepsilon_{M_0} - \alpha) \neq z_{i_l}$ for $1 \leqslant l \leqslant r$. It follows therefore from 40.8.7 and lemma 40.8.5 that either $\lambda_z = \mu_z a_M$ for some $M \in \mathcal{S}$, $M \neq M_0$, or $\lambda_z = \mu_z b_N$ for some $N \in \mathcal{T}$.

In both cases, we deduce that the coefficient $\lambda_{i_1} \cdots \lambda_{i_r}$ in the sum of (3) is of the form:

$$\mu_{i_1} \cdots \mu_{i_r} \prod_{M \in \mathcal{S}} a_M^{m_M} \prod_{N \in \mathcal{T}} b_N^{m_N}$$

where $m_{M_0} < n_{M_0}$.

It is now clear that there exists a non-empty open subset U of $(\Bbbk \setminus \{0\})^m$ verifying $\wedge^r \Psi_f \neq 0$ if $\Omega_f \in U$.

(iv) We have:

$$\wedge^{r+s} \Phi_f = \sum_{k=0}^{r+s} \binom{r+s}{k} (\wedge^k \Psi_f) \wedge (\wedge^{r+s-k} \Theta_f).$$

Since $\wedge^j \Theta_f \in (\bigwedge^j E_{S,T}) \bigwedge (\bigwedge^j \mathfrak{u}^*)$, to show that $\wedge^{r+s} \Phi_f \neq 0$, it suffices to prove that $(\wedge^r \Psi_f) \wedge (\wedge^s \Theta_f) \neq 0$.

We saw in the proof of (iii) that if $\Omega_f \in U$, then

$$\wedge^r \Psi_f = \lambda v_1 \wedge \cdots \wedge v_r + w,$$

where $\lambda \in \Bbbk \setminus \{0\}$ and w is a linear combination of terms of the form $v_{z_{i_1}} \wedge \cdots \wedge v_{z_{i_r}}$ with $\widetilde{z_{i_1}} \cup \cdots \cup \widetilde{z_{i_r}} \neq \mathcal{E}$. It is therefore clear that $(\wedge^r \Psi_f) \wedge (\wedge^s \Theta_f) \neq 0$ if $\Omega_f \in U$. $\square$

Remarks. 1) Observe that the preceding proof shows that if $\Omega_f \in U$, then the restriction of Ω_f to $\mathfrak{v} \times \mathfrak{v}$ is non-degenerate.

2) If S or T is empty, then 40.6.6 says that $\wedge^{r+s} \Phi_f \neq 0$ for all $\Omega_f \in (\Bbbk \setminus \{0\})^m$. However, this is not true in general.

40.8.9 Theorem. *In the notations above, we have:*

$$\chi(\mathfrak{g}_{S,T}) \leqslant \mathrm{rk}(\mathfrak{g}) + \mathrm{card}\,\mathcal{S} + \mathrm{card}\,\mathcal{T} - 2\dim E_{S,T}.$$

Proof. Recall from 40.8.6 that $\mathfrak{g}_{S,T} = \mathfrak{h} \oplus \mathfrak{u} \oplus \mathfrak{v}$. So:

$$\dim \mathfrak{g} = \mathfrak{h} + \mathrm{card}\,\mathcal{S} + \mathrm{card}\,\mathcal{T} + 2r.$$

If $\Omega_f \in U$, then the fact that $\wedge^{r+s}\Phi_f \neq 0$ implies that $\mathrm{rk}(\Phi_f) \geqslant 2(r+s)$. Thus:

$$\dim \mathfrak{g}_{S,T}^{(f)} \leqslant \dim \mathfrak{g} - 2(r+s).$$

Hence:

$$\dim \mathfrak{g}_{S,T}^{(f)} \leqslant \dim \mathfrak{h} + \mathrm{card}\,\mathcal{S} + \mathrm{card}\,\mathcal{T} - 2s.$$

So we are done.   □

40.8.10 Remark. Since $\mathrm{card}\,\mathcal{S} = \dim E_S$, $\mathrm{card}\,\mathcal{T} = \dim E_T$ and $E_{S,T} = E_S + E_T$, the preceding inequality can be rewritten as follows:

$$\chi(\mathfrak{g}_{S,T}) \leqslant \mathrm{rk}(\mathfrak{g}) + \dim E_S + \dim E_T - 2\dim(E_S + E_T).$$

40.8.11 Corollary. *Let* $\mathfrak{q}$ *be a seaweed Lie algebra of* $\mathfrak{g}$. *Then:*
  (i) $\chi(\mathfrak{q}) \leqslant \mathrm{rk}(\mathfrak{g})$.
  (ii) $\chi(\mathfrak{q}) = \mathrm{rk}(\mathfrak{g})$ *if and only if* $\mathfrak{q}$ *is a Levi factor of* $\mathfrak{g}$.

Proof. By proposition 40.7.3, we may assume that $\mathfrak{q}$ is standard. Let S, T be subsets of Π such that $\mathfrak{q} = \mathfrak{h} \oplus \mathfrak{g}^{R_+^S} \oplus \mathfrak{g}^{R_-^T}$.

(i) This is a direct consequence of 40.8.10.

(ii) If $\mathfrak{q}$ is a Levi factor of $\mathfrak{g}$, then $\mathfrak{q}$ is reductive, and it is clear by 29.3.2 that $\chi(\mathfrak{q}) = \mathrm{rk}(\mathfrak{g})$.

Conversely, suppose that $\chi(\mathfrak{q}) = \mathrm{rk}(\mathfrak{g})$. Then 40.8.10 says that $E_S = E_T$. Let N be a connected component of T. We have $\varepsilon_N \in E_T = E_S \subset \Bbbk S$. Since ε_N is a linear combination of all the roots of N with strictly positive integral coefficients and Π is a basis of $\mathfrak{h}^*$, it follows that $N \subset S$, and hence $T \subset S$. By exchanging the roles of S and T, we obtain that $S \subset T$. Thus $S = T$, and $\mathfrak{q}$ is a Levi factor of $\mathfrak{g}$.   □

40.8.12 Corollary. *If the elements* ε_M, $M \in \mathcal{S} \cup \mathcal{T}$, *form a basis of* $\mathfrak{h}^*$, *then* $\chi(\mathfrak{g}_{S,T}) = 0$.

Proof. By 40.8.10, $\chi(\mathfrak{g}_{S,T}) \leqslant 0$. So the result follows.   □

40.9 Cases where the bound is exact

40.9.1 Lemma. *Let* V *be a finite-dimensional vector space,* V' *a hyperplane of* V, Φ *an alternating bilinear form on* V *and* Φ' *its restriction to* V'. *Denote by* N *and* N' *the kernel of* Φ *and* Φ'.
  (i) *If* $N \subset V'$, *then* N *is a hyperplane of* N'.
  (ii) *If* $N \not\subset V'$, *then* $N' = N \cap V'$ *and* N' *is a hyperplane of* N.

Proof. (i) If $N \subset V'$, then $N \subset N'$, and if $x \in V \setminus V'$, then:

$$N = \{v \in N'; \Phi(v, x) = 0\}.$$

It follows that $\dim(N'/N) \leqslant 1$. Being the rank of Φ and Φ', the integers $\dim V - \dim N$ and $\dim V' - \dim N'$ are even. So $\dim(N'/N) = 1$.

(ii) Suppose that $N \not\subset V'$, and $x \in N \setminus V'$. If $v \in N'$, then $\Phi(x,v) = 0$ and $\Phi(v, V') = \{0\}$. So $v \in N$, and hence $N' \subset N \cap V'$. Clearly, $N \cap V' \subset N'$. Thus $N' = N \cap V'$, and N' is a hyperplane of N. $\quad\square$

40.9.2 In the notations of 40.8, we set:

$$d_{S,T} = \mathrm{rk}(\mathfrak{g}) + \mathrm{card}\,\mathcal{K}(S) + \mathrm{card}\,\mathcal{K}(T) - 2\dim E_{S,T}$$
$$= \mathrm{rk}(\mathfrak{g}) + \dim E_S + \dim E_T - 2\dim(E_S + E_T).$$

By 40.8.9, we have $\chi(\mathfrak{g}_{S,T}) \leqslant d_{S,T}$. We shall show that equality holds in certain cases.

40.9.3 Proposition. *If $d_{S,T} \in \{0,1\}$, then $\chi(\mathfrak{g}_{S,T}) = d_{S,T}$.*

Proof. The case $d_{S,T} = 0$ is clear by 40.8.9. So let us suppose that $d_{S,T} = 1$.

We have $\dim \mathfrak{g}_{S,T} - d_{S,T} = 2(r+s)$. Since $\dim \mathfrak{g}_{S,T} - \chi(\mathfrak{g}_{S,T})$ is an even integer (it is the rank of an alternating bilinear form on $\mathfrak{g}_{S,T}$), we deduce that $d_{S,T}$ and $\chi(\mathfrak{g}_{S,T})$ are of the same parity. So $\chi(\mathfrak{g}_{S,T}) = 1$. $\quad\square$

40.9.4 Proposition. *Let S and T be subsets of Π. Suppose that the set $\{\varepsilon_M; M \in \mathcal{K}(S) \cup \mathcal{K}(T)\}$ consists of linear independent elements. Then, if $\Omega_f \in U$, f is a stable element of $\mathfrak{g}_{S,T}^*$ and $\chi(\mathfrak{g}_{S,T}) = d_{S,T}$.*

Proof. Our hypothesis implies that

$$\dim E_{S,T} = \mathrm{card}(\mathcal{K}(S) \cup \mathcal{K}(T)).$$

So we deduce from the definition of $d_{S,T}$ that:

$$d_{S,T} = \mathrm{rk}(\mathfrak{g}) - \dim E_{S,T} + \mathrm{card}(\mathcal{K}(S) \cap \mathcal{K}(T)).$$

It follows from 40.8.8 and 40.8.9 that if $\Omega_f \in U$, then $\dim \mathfrak{g}_{S,T}^{(f)} = d_{S,T}$ and $\mathfrak{g}_{S,T}^{(f)}$ is a commutative Lie algebra consisting of semisimple elements. So the result follows by 40.1.6 and 40.1.5. $\quad\square$

40.9.5 Proposition. *If $\mathrm{card}(S) = 1$ or $\mathrm{card}(T) = 1$, then $\chi(\mathfrak{g}_{S,T}) = d_{S,T}$.*

Proof. Without loss of generality, we may assume that $\mathrm{card}(T) = 1$ and $T = \{\alpha\}$. For simplicity, we shall denote $\mathfrak{g}_{S,T}$ by $\mathfrak{q}$.

In view of 40.9.4, we may further assume that $\alpha \in S$, $\{\alpha\} \notin \mathcal{K}(S)$, and $\alpha \in E_S$. Set:

$$\mathfrak{a} = \mathfrak{h} \oplus \mathfrak{g}^{R_+^S}.$$

Let

$$U = \sum_{M \in \mathcal{S}} a_M X_{-\varepsilon_M} \ , \quad V = X_\alpha + U \ , \quad f = \varphi_{\mathfrak{a}}^U \ , \quad g = \varphi_{\mathfrak{q}}^V \ ,$$

where the coefficients a_M are chosen so that f is a stable element of $\mathfrak{a}^*$ (this is possible by 40.9.4). The restriction of Φ_g to $\mathfrak{a}$ is therefore Φ_f. The proof of 40.8.8 (i) shows that:

$$\mathfrak{a}^{(f)} = \bigcap_{M \in \mathcal{S}} \ker \varepsilon_M.$$

As $\alpha \in E_S$, we deduce that $[X_\alpha, \mathfrak{a}^{(f)}] = \{0\}$, so $\mathfrak{a}^{(f)} \subset \mathfrak{q}^{(g)}$. By lemma 40.9.1, we obtain that:

$$\mathfrak{q}^{(g)} = \mathbb{k}X \oplus \mathfrak{a}^{(f)}$$

where $X = X_{-\alpha} + Y$ for some $Y \in \mathfrak{a}$.

Since $\mathfrak{a}^{(f)} \subset \mathfrak{h}$, we have $[\mathfrak{a}, \mathfrak{a}^{(f)}] \cap \mathfrak{a}^{(f)} = \{0\}$, and therefore $[X, \mathfrak{a}^{(f)}] = \{0\}$. Thus $\mathfrak{q}^{(g)}$ is a commutative Lie algebra.

Let A be the smallest algebraic subgroup of the adjoint group G of $\mathfrak{g}$ whose Lie algebra is $\mathrm{ad}_{\mathfrak{g}}\, \mathfrak{a}$. Then the set of restrictions of the elements of A to $\mathfrak{a}$ is the adjoint group of $\mathfrak{a}$.

Since the set of stable elements of $\mathfrak{a}^*$ is a non-empty A-invariant open subset of $\mathfrak{a}^*$, we deduce that there exists $h \in \mathfrak{q}^*$ regular such that $\lambda = h|_{\mathfrak{a}}$ is stable. If $\theta \in A$, then $\theta(h)|_{\mathfrak{a}} = \theta(\lambda)$. Consequently, we may assume that:

$$\mathfrak{a}^{(\lambda)} = \mathfrak{a}^{(f)} = \bigcap_{M \in \mathcal{S}} \ker \varepsilon_M \subset \mathfrak{h}.$$

Now $h([\mathfrak{a}, \mathfrak{a}^{(f)}]) = \{0\}$, and $X_{-\alpha}$ commutes with $\mathfrak{a}^{(f)} = \mathfrak{a}^{(\lambda)}$. It follows therefore that $h([X_{-\alpha}, \mathfrak{a}^{(\lambda)}]) = \{0\}$. Hence $h([\mathfrak{q}, \mathfrak{a}^{(\lambda)}]) = \{0\}$ and $\mathfrak{a}^{(\lambda)} \subset \mathfrak{q}^{(h)}$.

It follows by 40.9.1 that $\dim \mathfrak{q}^{(h)} = 1 + \dim \mathfrak{a}^{(\lambda)}$, hence $\chi(\mathfrak{q}) = 1 + \chi(\mathfrak{a})$.

Since $E_{S,T} = E_{S,\emptyset}$, $\mathrm{card}\, \mathcal{K}(T) = 1$ and $\chi(\mathfrak{a}) = d_{S,\emptyset}$, we deduce that $\chi(\mathfrak{q}) = d_{S,T}$. $\square$

40.9.6 A Lie algebra verifying the hypothesis of 40.9.5 does not possess necessarily a stable linear form. Let us give an example of such a Lie algebra.

Suppose that $\mathfrak{g}$ is of type D_4, and using the notations of 18.14, let $S = \{\beta_1, \beta_2, \beta_3, \beta_4\}$, $T = \{\beta_2\}$. Then $\mathfrak{q} = \mathbb{k}X_{-\beta_2} + \mathfrak{b}$. Let B (resp. Q) be the smallest algebraic subgroup of G whose Lie algebra is $\mathrm{ad}_{\mathfrak{g}}\, \mathfrak{b}$ (resp. $\mathrm{ad}_{\mathfrak{g}}\, \mathfrak{q}$). Then the set of restrictions of the elements of B (resp. Q) to $\mathfrak{b}$ (resp. $\mathfrak{q}$) is the adjoint group of $\mathfrak{b}$ (resp. $\mathfrak{q}$).

Set:

$$U = X_{-\beta_1} + X_{-\beta_3} + X_{-\beta_4} + X_{-\beta_1 - 2\beta_2 - \beta_3 - \beta_4} \ , \ g = \varphi_{\mathfrak{b}}^U.$$

By 40.5.6, 40.6.2 and 40.6.6, g is a stable element of $\mathfrak{b}^*$. Moreover, $\chi(\mathfrak{b}) = 0$ (40.6.6). It follows that the B-orbit of g is the open set $\mathfrak{b}_{\mathrm{reg}}^*$ of $\mathfrak{b}^*$.

Now suppose that the open subset $\mathfrak{E}$ of stable elements of $\mathfrak{q}^*$ is non-empty. Then there exists $f \in \mathfrak{E}$ such that $f|_{\mathfrak{b}} \in \mathfrak{b}_{\mathrm{reg}}^*$.

If $\theta \in B$, then $\theta(f) \in \mathfrak{E}$ and $\theta(f)|_{\mathfrak{b}} = \theta(f|_{\mathfrak{b}})$. So we may assume that f is of the form $f = \varphi_{\mathfrak{q}}^V$ where $V = \lambda X_{\beta_2} + U$ where $\lambda \in \mathbb{k}$.

If $\lambda \neq 0$, then we can check easily that $\mathfrak{q}^{(f)} = \mathbb{k}X$, where

$$X = X_{-\beta_2} + \mu_1 X_{\beta_1} + \mu_2 X_{\beta_3} + \mu_3 X_{\beta_4}$$
$$+ \mu_4 X_{\beta_1+\beta_2+\beta_3} + \mu_5 X_{\beta_1+\beta_2+\beta_4} + \mu_6 X_{\beta_2+\beta_3+\beta_4} + \mu_7 X_{\beta_1+2\beta_2+\beta_3+\beta_4},$$

and $\mu_1, \ldots, \mu_7$ are non-zero elements of $\Bbbk$.

If $\lambda = 0$, then again, we can check easily that $\mathfrak{q}^{(f)} = \Bbbk X$, where

$$X = X_{-\beta_2} + \mu_1 X_{\beta_1+\beta_2+\beta_3} + \mu_2 X_{\beta_1+\beta_2+\beta_4} + \mu_3 X_{\beta_2+\beta_3+\beta_4},$$

and μ_1, μ_2, μ_3 are non-zero elements of $\Bbbk$.

Let $H \in \mathfrak{h}$ be the element such that $\beta_1(H) = \beta_3(H) = \beta_4(H) = -\beta_2(H) = 1$. Then in both cases, $[H, X] = X$. So $[\mathfrak{q}, \mathfrak{q}^{(f)}] \cap \mathfrak{q}^{(f)} \neq \{0\}$. But this contradicts 40.1.5. Thus, $\mathfrak{q}^*$ does not possess any stable linear form.

40.10 On the index of parabolic subalgebras

40.10.1 Lemma. *Let S be a subset of Π and $n = \operatorname{card} \mathcal{K}(S)$. Then there exist subsets $S_1, \ldots, S_n$ of Π verifying the following conditions:*

(i) $S_1 \subset S_2 \subset \cdots \subset S_n = S$.

(ii) $\mathcal{K}(S_1) \subset \mathcal{K}(S_2) \subset \cdots \subset \mathcal{K}(S_n)$.

(iii) $\operatorname{card} \mathcal{K}(S_i) = i$ for $1 \leqslant i \leqslant n$.

Proof. Let M be a connected component of S and M' the set of elements $\alpha \in M$ such that $\langle \alpha, \varepsilon_M^\vee \rangle = 0$. Set $S_{n-1} = S \setminus (M \setminus M')$. Then it is clear by 40.5.4 that $\operatorname{card} \mathcal{K}(S_{n-1}) = n - 1$ and $\mathcal{K}(S_{n-1}) \subset \mathcal{K}(S)$. So the result follows. $\square$

40.10.2 Theorem. *Let $\mathfrak{g}$ be a semisimple Lie algebra of rank ℓ. For any integer $i \in \{0, 1, \ldots, \ell\}$, there exists a parabolic subalgebra $\mathfrak{p}$ of $\mathfrak{g}$ such that $\chi(\mathfrak{p}) = i$.*

Proof. It is clear that we may assume $\mathfrak{g}$ to be simple. Let $n = \operatorname{card} \mathcal{K}(\Pi)$. Using the numbering of root systems in 18.14, we set $\Pi = \{\beta_1, \ldots, \beta_\ell\}$.

There exist subsets $S_0, S_1, \ldots, S_n$ of Π such that $S_0 = \emptyset$, $S_n = \Pi$ and $S_1, \ldots, S_n$ verify the conditions (i), (ii) and (iii) of lemma 40.10.1.

By 40.9.4, for $0 \leqslant i \leqslant n$, we have:

$$\chi(\mathfrak{g}_{\Pi, S_i}) = \ell + i - n.$$

Recall from table 40.5.5 that if $\mathfrak{g}$ is of type $B_k, C_k, D_{2k}, E_7, E_8, F_4$ or G_2, then $n = \ell$. So the theorem follows in these cases.

1) Let us suppose that $\mathfrak{g}$ is of type D_{2k+1}. Then $\operatorname{card} \mathcal{K}(\Pi) = 2k$ and for $1 \leqslant i \leqslant \ell$, $\mathfrak{g}_{\Pi, S_{i-1}}$ is a parabolic subalgebra of $\mathfrak{g}$ of index i.

As $\beta_{2k} \notin E_\Pi$ (40.5.6), by 40.9.4 (iii) or 40.9.5, we have $\chi(\mathfrak{g}_{\Pi, \{\beta_{2k}\}}) = 0$.

2) Let us suppose that $\mathfrak{g}$ is of type E_6. Then $\operatorname{card} \mathcal{K}(\Pi) = 4$ and again by using the S_j's, we obtain parabolic subalgebras of $\mathfrak{g}$ of index i for $2 \leqslant i \leqslant 6$.

By 40.5.6, $\{\beta_1, \beta_5\} \cup \{\varepsilon_M; M \in \mathcal{K}(M)\}$ is basis of $\mathfrak{h}^*$. Again by using 40.9.4 (iii), we obtain that:

$$\chi(\mathfrak{g}_{\Pi,\{\beta_1\}}) = 1 \ , \ \chi(\mathfrak{g}_{\Pi,\{\beta_1,\beta_5\}}) = 0.$$

3) Finally, let us suppose that $\mathfrak{g}$ is of type A_ℓ. Let ρ be the largest integer $\leqslant (\ell+1)/2$. By 40.5.5, $\rho = \operatorname{card}\mathcal{K}(\Pi)$. So by using the S_j's, we obtain parabolic subalgebras of $\mathfrak{g}$ of index i for $\ell - \rho \leqslant i \leqslant \ell$. We are therefore left to construct parabolic subalgebras of $\mathfrak{g}$ of index $0, 1, \ldots, \ell - \rho - 1$.

Set $T_0 = \emptyset$, and define T_k, for $1 \leqslant k \leqslant \ell - \rho = \ell'$, by:

$$T_k = \begin{cases} T_{k-1} \cup \{\beta_k\} & \text{if } k \text{ is odd,} \\ T_{k-1} \cup \{\beta_{\ell+1-k}\} & \text{if } k \text{ is even.} \end{cases}$$

By 40.5.6, the ε_M's, $M \in \mathcal{K}(\Pi) \cup \mathcal{K}(T_k)$, are linearly independent. So 40.9.4 (iii) implies that $\chi(\mathfrak{g}_{\Pi,T_k}) = \ell' - k$ for $0 \leqslant k \leqslant \ell'$. So we are done because $\mathfrak{g}_{\Pi,T_k}$ is a parabolic subalgebra of $\mathfrak{g}$. $\quad\square$

References and comments

- [25], [33], [43], [45], [47], [63], [62], [68], [82], [83].

The notion of stable linear form was introduced in [47], and was studied in an algebraic point of view in [82].

The results 40.3.4 and 40.4.5 are taken from [63], and 40.4.2 is taken from [68]. The proof of 40.3.4 given here is due to Raïs.

The construction given in 40.5.4 can be found in [45] and [43], and it is often called the Kostant's cascade construction.

The results of 40.6.3 was established in [45], and 40.6.5 in [82]

The notion of seaweed Lie algebra was first introduced in [25] for type A, and in [62] in the general case.

If $\mathfrak{g}$ is of type A, the index of a seaweed Lie algebra was computed in [25] via a combinatorial formula. For types A, B, C, D, inductive formulas relating the index of seaweed Lie algebras were established in [62].

The results of sections 40.8 to 40.10 are taken from [83]. Note also that 40.10.2 was announced without proof in [33].

The index of a certain class of Lie algebras related to semisimple Lie algebras is computed in [69].

Let $\mathfrak{g}$ be an ad-algebraic Lie algebra. Then by Rosenlicht's Theorem stated in the comments of Chapter 25, we have the equality $\chi(\mathfrak{g}) = \operatorname{tr\,deg}_{\Bbbk} \mathbf{R}(\mathfrak{g}^*)$.

References

1. ARAKI S., On root systems and an infinitesimal classification of irreducible symmetric spaces, *J. of Math., Osaka City Univ.*, 13, 1962, p. 1-34.
2. BALA P., CARTER R. W., Classes of unipotent elements in simple algebraic groups I, *Math. Proc. Cambridge Phi. Soc.*, 79, 1976, p. 401-425.
3. BALA P., CARTER R. W., Classes of unipotent elements in simple algebraic groups II, *Math. Proc. Cambridge Phi. Soc.*, 80, 1976, p. 1-17.
4. BERNSTEIN J., LUNTS V., A simple proof of Kostant's theorem that $U(\mathfrak{g})$ is free over its center, *American J. of Math.*, 118, 1996, p. 979-987.
5. BOREL A., *Linear algebraic groups*, Graduate texts in math., 126, Springer-Verlag, 1991.
6. BORHO W., KRAFT H., Über bahen und deren Deformationen bei linearen Aktionen reduktiver Gruppen, *Comment. math. Helvetici*, 54, 1979, p. 61-104
7. BORHO W., Über Schichten halbeinfacher Lie-Algebren, *Invent. Math.*, 65, 1981, p. 283-317.
8. BOURBAKI N., *Algèbre*, chap. 1,2,3, Hermann, 1970.
9. BOURBAKI N., *Algèbre*, chap. 4,5,6,7, Masson, 1981.
10. BOURBAKI N., *Algèbre commutative*, chap. 1,2,3,4, Masson, 1985.
11. BOURBAKI N., *Algèbre commutative*, chap. 5,6,7, Masson, 1985.
12. BOURBAKI N., *Groupes et algèbres de Lie*, chap. 1, Hermann, 1971.
13. BOURBAKI N., *Groupes et algèbres de Lie*, chap. 4,5,6, Masson 1981.
14. BOURBAKI N., *Groupes et algèbres de Lie*, chap. 7,8, CCLS Diffusion, 1975.
15. BREDON G., *Sheaf theory 2nd Edition*, Graduate texts in math., 170, Springer-Verlag, 1997.
16. BROER A., Lectures in decomposition classes, *NATO ASI Series C*, Kluwer Academic Publishers, Dordrecht, 1998, p. 39-83.
17. BRYLINSKI R., KOSTANT B., The variety of all invariant symplectic structures on an homogeneous space and normalizers of isotropy groups, *Symplectic geometry and mathematical physics*, Birkhäuser, 1991, p. 80-113.
18. BRYLINSKI R., KOSTANT B., Nilpotent orbits, normality, and hamiltonian group actions, *J. American Math. Soc.*, 1994, p. 269-298.
19. CARTAN H., CHEVALLEY C., *Géométrie algébrique*, mimeographed lecture notes, Secrétariat mathématique, Paris, 1956.
20. CARTER R. W., *Finite groups of Lie type, Conjugacy classes and complex characters*, John Wiley, 1993.

21. CHEVALLEY C., *Classification des groupes de Lie algébriques*, mimeographed lecture notes, Secrétariat mathématique, Paris, 1958.

22. CHEVALLEY C., *Fondements de la géométrie algébrique*, Cours multigraphié, Secrétariat mathématique, Paris, 1958.

23. CHEVALLEY C., *Théorie des groupes de Lie*, Hermann, 1968.

24. COLLINGWOOD D. H., MCGOVERN W. M., *Nilpotent orbits in semisimple Lie algebras*, Van Nostrand Reinhold, 1993.

25. DERGACHEV V., KIRILLOV A., Index of Lie algebras of seaweed type, *J. of Lie Theory*, 10, 2000, p. 331-343

26. DIEUDONNÉ J., *Cours de géométrie algébrique*, tome 2, PUF, 1974.

27. DIXMIER J., Polarisations dans les algèbres de Lie semi-simples complexes, *Bull. Sc. math.*, 99, 1975, p. 45-63.

28. DIXMIER J., Polarisations dans les algèbres de Lie II, *Bull. Soc. Math. France*, 104, 1976, p. 145-164.

29. DIXMIER J., *Enveloping algebras*, Graduate Studies in Math., 11, AMS, 1996.

30. DIXMIER J., RAYNAUD M., Sur le quotient d'une variété algébrique par un groupe algébrique, *Mathematical analysis and applications, Part A*, p. 327-344, Adv. in Math. Suppl. Stud., 7a, Academic Press, 1981.

31. DUFLO M., Construction of primitive ideals in an enveloping algebra, *Publi. of 1971 Summer School in Math.*, Janos Bolyai Math. Soc. Budapest, p. 77-93.

32. DYNKIN E. B., Semisimple Lie algebras of semisimple Lie algebras, *American Math. Soc. Trans.*, 6, 1957, p. 111-243.

33. ELASHVILI A. G., On the index of parabolic subalgebras of semisimple Lie algebras, *Preprint*, 1990.

34. GODEMENT R., *Théorie des faisceaux*, Hermann, 1973.

35. GROSSHANS F., *Algebraic homogeneous spaces and invariant theory*, Lecture Notes in Mathematics, 1673, Springer-Verlag, 1997.

36. HELGASON S., *Differential geometry, Lie groups, and symmetric spaces*, Graduate Studies in Mathematics, 34. American Mathematical Society, 2001.

37. HARTSHORNE R., *Algebraic geometry*, Graduate texts in math., 52, Springer-Verlag, 1997.

38. HOCHSCHILD G. P., *Basic theory of algebraic groups and Lie algebras*, Graduate texts in math., 75, Springer-Verlag, 1981.

39. HUMPHREYS J. E., *Introduction to Lie algebras and representation theory*, Graduate texts in math., 9, Springer-Verlag, 1972.

40. HUMPHREYS J. E., *Linear algebraic groups*, Graduate texts in math., 21, Springer-Verlag, 1975.

41. HUMPHREYS J. E., *Reflection groups and Coxeter groups*, Cambridge studies in advanced mathematics, 1990.

42. JACOBSON N., *Lie algebras*, John Wiley and Sons, 1962.

43. JANTZEN J. C., *Einhüllenden Algebren halbeinfacher Lie-Algebren*, Ergebnisse der Mathematik und iher Grenzgebiete, 3, Springer-Verlag, 1983.

44. JOSEPH A., The minimal orbit in a simple Lie algebra and its associated maximal ideal, *Ann. Scient. Ec. Norm. Sup.*, 9, 1976, p. 1-30.

45. JOSEPH A., A preparation theorem for the prime spectrum of a semisimple Lie algebra, *J. of Algebra*, 48, 1977, p. 241-289.

46. KAC V. G., Some remarks on nilpotent orbits, *J. of Algebra*, 64, 1980, p. 190-213.

47. KOSMANN Y., STERNBERG S., Conjugaison des sous-algèbres d'isotropie, *C. R. Acad. Sci. Paris*, A, t. 279, 1974, p. 777-779.

48. KOSTANT B., The principal three-dimensional subgroup and the Betti numbers of a complex simple Lie group, *American J. of Math.*, 81, 1959, p. 973-1032.

49. KOSTANT B., Lie group representations on polynomial rings, *American J. of Math.*, 85, 1963, p. 327-402.

50. KOSTANT B., RALLIS S., Orbits and representations associated with symmetric spaces, *American J. of Math.*, 93, 1971, p. 753-809.

51. KRAFT H., *Geometrische Methoden in der Invariantentheorie*, Aspects of Mathematics, Band D1, Vieweg, Braunscheweig, 1985.

52. LANG S., *Algebra*, Addison-Wesley, 1993.

53. LANG S., *Abelian varieties*, Springer-Verlag, 1983.

54. LEPOWSKY J., Generalized Verma modules, the Cartan-Helgason theorem, and the Harish-Chandra homomorphism, *J. of Algebra*, 49, 1977, p. 470-495.

55. LEPOWSKY J., MCCOLLUM G. W., Cartan subspaces of symmetric Lie algebras, *Trans. of the American Math. Soc.*, 216, 1976, p. 217-228.

56. MALLIAVIN M.-P., *Algèbre commutative*, Masson, 1985.

57. MULLER I., *Systèmes de racines orthogonales et orbites d'espaces préhomogènes*, Thesis, Université de Strasbourg, 1996.

58. MUMFORD D., *Abelian varieties*, Tata Institute of Fundamental Research Studies in Mathematics, No. 5, Oxford University Press, 1970.

59. ONISHCHIK A. L., VINBERG E. B., *Lie groups and algebraic groups*, translated from the Russian and with a preface by D. A. Leites, Springer Series in Soviet Mathematics, Springer-Verlag, 1990.

60. PANYUSHEV D., The Jacobian modules of a representation of a Lie algebra and geometry of commuting varieties, *Compositio Math.*, 94, 1994, p. 181-199.

61. PANYUSHEV D., On the irreducibility of Commuting varieties associated with involutions of simple Lie algebras, *Functional Analysis and its applications*, 38, 2004, p. 38-44.

62. PANYUSHEV D., Inductive formulas for the index of seaweed Lie algebras, *Moscow Math. Journal*, 1, 2001, p. 221-241.

63. PANYUSHEV D. , The index of a Lie algebra, the centralizer of a nilpotent element, and the normalizer of the centralizer, *Math. Proc. Cambridge Philos. Soc.*, 134, 2003, p. 41-59.

64. POPOV V. L., Self-dual algebraic varieties and nilpotent orbits, *Proc. of the Intern. Colloquium on Algebra*, Arithmetic and Geometry, Mumbai 2000, Ed. by R.Parimala, Tata Inst. Fund. Research, Narosa Publ. House, 2002, p. 509-533.

65. POPOV V. L., Projective duality and principal nilpotent elements of symmetric pairs, *Amer. Math. Soc. Transl., Ser.* (2), Lie Groups and Invariant Theory, Ed. by E.B. Vinberg, 2004 (to appear).

66. POPOV V. L., TEVELEV E.A., Self-dual projective algebraic varieties associated with symmetric spaces, *Algebraic Transformation Groups and Algebraic Varieties*, Ed. V. Popov, Encycl. Math. Sci., Vol. 132, Springer-Verlag, 2004, p. 131-167.

67. POPOV V. L., VINBERG E. B., Invariant theory, *Algebraic geometry IV*, Eds. A. Parshin, I. Shafarevich, Encycl. Math. Sci., Vol. 55, Springer-Verlag, 1994, p. 123-278.

68. RAÏS M., L'indice des produits semi-directs $E \times_\rho \mathfrak{g}$, *C. R. Acad. Sci. Paris*, A, 1978, p. 195-197.

69. RAÏS M., TAUVEL P., L'indice d'une certaine classe d'algèbres de Lie, *J. reine angew. math.*, 425, 1992, p. 123-140.

70. RICHARDSON R. W., Conjugacy classes in parabolic subgroups of semi-simple algebraic groups, *Bull. London Math. Soc.*, 6, t. 287, 1974, p. 21-24.

71. RICHARDSON R. W., Commuting varieties of semisimple Lie algebras and algebraic groups, *Compositio Math.*, 38, 1979, p. 311-327.

72. ROSENLICHT M., A remark on quotient spaces, *An. Acad. Brasil Cienc.*, 35, 1963, p. 487-489.

73. SABOURIN H., YU R. W. T., Sur l'irréductibilité de la variété commutante d'une paire symétrique réductive de rang 1, *Bulletin des Sciences Math.*, 126, 2002, p. 143-150.

74. SABOURIN H., YU R. W. T., On the irreducibility of the commuting variety of a symmetric pair associated to a parabolic subalgebra with abelian unipotent radical, *preprint math.RT/0407354*, 2004.

75. SESHADRI C. S., Some results on the quotient space by an algebraic group of automorphisms, *Math. Ann.*, 149, 1963, p. 286-301.

76. SPRINGER T. A., Regular elements of finite reflection groups, *Inv. Math.*, 25, 1974, p. 159-198.

77. SPRINGER T. A., Aktionen reduktiven Gruppen auf Varietäten, *Algebraische Transformationsgruppen und Invariantentheorie*, DMV Seminar, Band 13, Birkhäuser, 1989.

78. SPRINGER T. A., *Linear algebraic groups*, Progress in Mathematics 9, Birkhäuser, 1998.

79. STEINBERG R., *Conjugacy classes in algebraic groups*, Lectures notes in math., 366, Springer-Verlag, 1974.

80. TAUVEL P., *Introduction à la théorie des algèbres de Lie*, Diderot, 1998.

81. TAUVEL P., Quelques résultats sur les algèbres de Lie symétriques, *Bull. Sci. Math.*, 125, 2001, p. 641-665.

82. TAUVEL P., YU R. W. T., Indice et formes linéaires stables dans les algèbres de Lie, *J. of Algebra*, 273, 2004, p. 507-516.

83. TAUVEL P., YU R. W. T., Sur l'indice de certaines algèbres de Lie, *to appear in Annales de l'Institut Fourier (preprint math.RT/0311104)*.

84. VARADARAJAN V. S., *Lie groups, Lie algebras, and their representations*, Graduate texts in math., 102, Springer-Verlag, 1984.

85. VINBERG E. B., On the classification of homogeneous nilpotent elements of a graded Lie algebra, *Soviet. Math. Dokl*, 16, 1975, p. 1517-1520.

86. VINBERG E. B., Classification of homogeneous nilpotent elements of a simple graded Lie algebra, *Selecta Math. Sovietica*, 6, 1987, p. 15-35.

87. WARNER G., *Harmonic Analysis on Semi-Simple Lie groups* I, Springer-Verlag, 1972.

List of notations

Index

Printing: Strauss GmbH, Mörlenbach
Binding: Schäffer, Grünstadt

Springer Monographs in Mathematics

This series publishes advanced monographs giving well-written presentations of the "state-of-the-art" in fields of mathematical research that have acquired the maturity needed for such a treatment. They are sufficiently self-contained to be accessible to more than just the intimate specialists of the subject, and sufficiently comprehensive to remain valuable references for many years. Besides the current state of knowledge in its field, an SMM volume should also describe its relevance to and interaction with neighbouring fields of mathematics, and give pointers to future directions of research.

Abhyankar, S.S. **Resolution of Singularities of Embedded Algebraic Surfaces** 2nd enlarged ed. 1998
Alexandrov, A.D. **Convex Polyhedra** 2005
Andrievskii, V.V.; Blatt, H.-P. **Discrepancy of Signed Measures and Polynomial Approximation** 2002
Angell, T. S.; Kirsch, A. **Optimization Methods in Electromagnetic Radiation** 2004
Ara, P.; Mathieu, M. **Local Multipliers of C*-Algebras** 2003
Armitage, D.H.; Gardiner, S.J. **Classical Potential Theory** 2001
Arnold, L. **Random Dynamical Systems** corr. 2nd printing 2003 (1st ed. 1998)
Arveson, W. **Noncommutative Dynamics and E-Semigroups** 2003
Aubin, T. **Some Nonlinear Problems in Riemannian Geometry** 1998
Auslender, A.; Teboulle M. **Asymptotic Cones and Functions in Optimization and Variational Inequalities** 2003
Bang-Jensen, J.; Gutin, G. **Digraphs** 2001
Baues, H.-J. **Combinatorial Foundation of Homology and Homotopy** 1999
Brown, K.S. **Buildings** 3rd printing 2000 (1st ed. 1998)
Cherry, W.; Ye, Z. **Nevanlinna's Theory of Value Distribution** 2001
Ching, W.K. **Iterative Methods for Queuing and Manufacturing Systems** 2001
Crabb, M.C.; James, I.M. **Fibrewise Homotopy Theory** 1998
Chudinovich, I. **Variational and Potential Methods for a Class of Linear Hyperbolic Evolutionary Processes** 2005
Dineen, S. **Complex Analysis on Infinite Dimensional Spaces** 1999
Dugundji, J.; Granas, A. **Fixed Point Theory** 2003
Elstrodt, J.; Grunewald, F. Mennicke, J. **Groups Acting on Hyperbolic Space** 1998
Edmunds, D.E.; Evans, W.D. **Hardy Operators, Function Spaces and Embeddings** 2004
Fadell, E.R.; Husseini, S.Y. **Geometry and Topology of Configuration Spaces** 2001
Fedorov, Y.N.; Kozlov, V.V. **A Memoir on Integrable Systems** 2001
Flenner, H.; O'Carroll, L. Vogel, W. **Joins and Intersections** 1999
Gelfand, S.I.; Manin, Y.I. **Methods of Homological Algebra** 2nd ed. 2003
Griess, R.L. Jr. **Twelve Sporadic Groups** 1998
Gras, G. **Class Field Theory** corr. 2nd printing 2005
Hida, H. **p-Adic Automorphic Forms on Shimura Varieties** 2004
Ischebeck, F.; Rao, R.A. **Ideals and Reality** 2005
Ivrii, V. **Microlocal Analysis and Precise Spectral Asymptotics** 1998
Jech, T. **Set Theory** (3rd revised edition 2002)
Jorgenson, J.; Lang, S. **Spherical Inversion on SLn (R)** 2001
Kanamori, A. **The Higher Infinite** corr. 2nd printing 2005 (2nd ed. 2003)
Kanovei, V. **Nonstandard Analysis, Axiomatically** 2005
Khoshnevisan, D. **Multiparameter Processes** 2002
Koch, H. **Galois Theory of p-Extensions** 2002
Komornik, V. **Fourier Series in Control Theory** 2005
Kozlov, V.; Maz'ya, V. **Differential Equations with Operator Coefficients** 1999
Landsman, N.P. **Mathematical Topics between Classical & Quantum Mechanics** 1998
Leach, J.A.; Needham, D.J. **Matched Asymptotic Expansions in Reaction-Diffusion Theory** 2004
Lebedev, L.P.; Vorovich, I.I. **Functional Analysis in Mechanics** 2002
Lemmermeyer, F. **Reciprocity Laws: From Euler to Eisenstein** 2000
Malle, G.; Matzat, B.H. **Inverse Galois Theory** 1999

Mardesic, S. **Strong Shape and Homology** 2000
Margulis, G.A. **On Some Aspects of the Theory of Anosov Systems** 2004
Murdock, J. **Normal Forms and Unfoldings for Local Dynamical Systems** 2002
Narkiewicz, W. **Elementary and Analytic Theory of Algebraic Numbers** 3rd ed. 2004
Narkiewicz, W. **The Development of Prime Number Theory** 2000
Parker, C.; Rowley, P. **Symplectic Amalgams** 2002
Peller, V. (Ed.) **Hankel Operators and Their Applications** 2003
Prestel, A.; Delzell, C.N. **Positive Polynomials** 2001
Puig, L. **Blocks of Finite Groups** 2002
Ranicki, A. **High-dimensional Knot Theory** 1998
Ribenboim, P. **The Theory of Classical Valuations** 1999
Rowe, E.G.P. **Geometrical Physics in Minkowski Spacetime** 2001
Rudyak, Y.B. **On Thom Spectra, Orientability and Cobordism** 1998
Ryan, R.A. **Introduction to Tensor Products of Banach Spaces** 2002
Saranen, J.; Vainikko, G. **Periodic Integral and Pseudodifferential Equations with Numerical Approximation** 2002
Schneider, P. **Nonarchimedean Functional Analysis** 2002
Serre, J-P. **Complex Semisimple Lie Algebras** 2001 (reprint of first ed. 1987)
Serre, J-P. **Galois Cohomology** corr. 2nd printing 2002 (1st ed. 1997)
Serre, J-P. **Local Algebra** 2000
Serre, J-P. **Trees** corr. 2nd printing 2003 (1st ed. 1980)
Smirnov, E. **Hausdorff Spectra in Functional Analysis** 2002
Springer, T.A. Veldkamp, F.D. **Octonions, Jordan Algebras, and Exceptional Groups** 2000
Sznitman, A.-S. **Brownian Motion, Obstacles and Random Media** 1998
Taira, K. **Semigroups, Boundary Value Problems and Markov Processes** 2003
Talagrand, M. **The Generic Chaining** 2005
Tauvel, P.; Yu, R.W.T. **Lie Algebras and Algebraic Groups** 2005
Tits, J.; Weiss, R.M. **Moufang Polygons** 2002
Uchiyama, A. **Hardy Spaces on the Euclidean Space** 2001
Üstünel, A.-S.; Zakai, M. **Transformation of Measure on Wiener Space** 2000
Vasconcelos, W. **Integral Closure. Rees Algebras, Multiplicities, Algorithms** 2005
Yang, Y. **Solitons in Field Theory and Nonlinear Analysis** 2001